Calculus with Applications *is a solid, application-oriented text for students majoring in business, management, economics, or the life or social sciences. The problems and examples included in the text should help students learn how extensively mathematics is applied, as well as motivate those who wonder how the mathematics they study relates to the outside world and where they might use it. A few featured applications include:*

Accident Rate According to data from the National Highway Traffic Safety Administration, the accident rate as a function of the age of the driver in years x can be approximated by the function

$$f(x) = 60.0 - 2.28x + .0232x^2$$

for $16 \le x \le 85$. Find the age at which the accident rate is a minimum and the minimum rate.

Murder Rate The number of murders in Chicago from 1990–2002 is approximated by $f(x) = .817x^3 - 15.4x^2 + 49.4x + 874$, where x corresponds to the number of years after 1990—that is, 1991 corresponds to $x = 1$, 1992 corresponds to $x = 2$, and so on.

a. In what years did a relative maximum or a relative minimum number of murders occur? How many murders occurred in those years?

b. In what year did the rate of decrease in murders start to slow down?

Fungal Growth Because of the time that many people spend indoors, there is a concern about the health risk of being exposed to harmful fungi that thrive in buildings. The risk appears to increase in damp environments. Researchers have discovered that by controlling both the temperature and the relative humidity in a building, the growth of the fungus *A. versicolor* can be limited. The relationship between temperature and relative humidity, which limits growth, can be described by

$$R(T) = -.00007T^3 + .0401T^2 - 1.6572T + 97.086, \quad 15 \le T \le 46,$$

where $R(T)$ is the relative humidity (in percent) and T is the temperature (in degrees Celsius).Find the temperature at which the relative humidity is minimized.

Spherical Radius A large weather balloon is being inflated with air at the rate of 1.2 ft^3 per minute. Find the rate of change of the radius when the radius is 1.2 ft.

Foot-and-Mouth Epidemic In 2001, the United Kingdom suffered an epidemic of foot-and-mouth disease. The graph shows the number of reported cases each day since February 18, as well as the number of cases epidemiologists project would have occurred had they culled all livestock on infected farms within 24 hours, and all livestock on neighboring farms within 48 hours of the infection.*

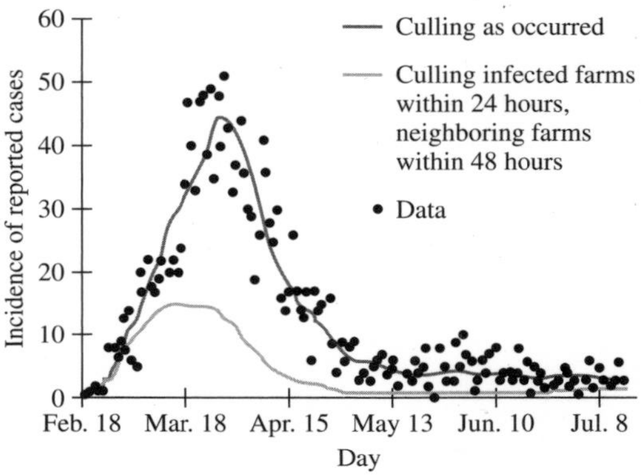

Oil Consumption The graph at the bottom of the page shows U.S. oil production and consumption rates (in millions of barrels per day) for the years 1950–2001.†

a. Estimate the amount of oil produced in the United States (domestic supply) between 1990 and 2000. Use rectangles with widths of 5 years.

b. Estimate the amount of oil imported in the United States between 1990 and 2000. Use rectangles with widths of 5 years.

c. Estimate the amount of oil consumed in the United States between 1990 and 2000 and then compare it to the total of parts a and b. Comment on the accuracy of these calculations.

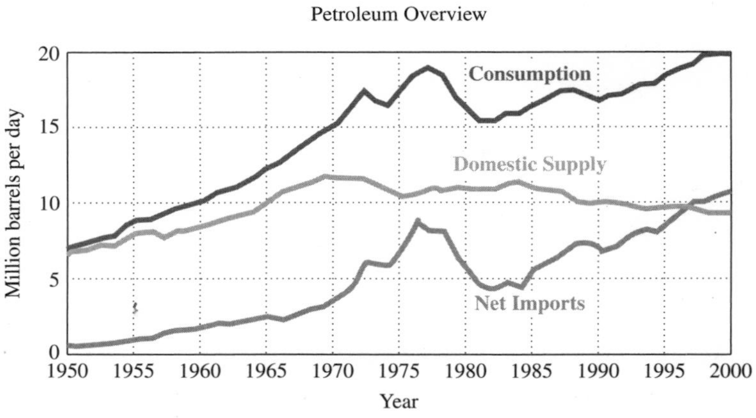

*U.S. Department of Energy, www.eia.doe.gov.

†*Science,* Vol. 294, Oct. 5, 2001, p. 26. Peter Morrison/AP. Reprinted by permission of the Associated Press.

KEY DEFINITIONS, THEOREMS, AND FORMULAS

3.1 Rules for Limits

Let a, A, and B be real numbers, and let f and g be functions such that

$$\lim_{x \to a} f(x) = A \qquad \text{and} \qquad \lim_{x \to a} g(x) = B.$$

1. If k is a constant, then $\lim_{x \to a} k = k$ and $\lim_{x \to a} [k \cdot f(x)] = k \cdot \lim_{x \to a} f(x) = k \cdot A$.
2. $\lim_{x \to a} [f(x) \pm g(x)] = \lim_{x \to a} f(x) \pm \lim_{x \to a} g(x) = A \pm B$
   (The limit of a sum or difference is the sum or difference of the limits.)
3. $\lim_{x \to a} [f(x) \cdot g(x)] = [\lim_{x \to a} f(x)] \cdot [\lim_{x \to a} g(x)] = A \cdot B$
   (The limit of a product is the product of the limits.)
4. $\lim_{x \to a} \dfrac{f(x)}{g(x)} = \dfrac{\lim_{x \to a} f(x)}{\lim_{x \to a} g(x)} = \dfrac{A}{B}$ if $B \neq 0$
   (The limit of a quotient is the quotient of the limits, provided the limit of the denominator is not zero.)
5. If $p(x)$ is a polynomial, then $\lim_{x \to a} p(x) = p(a)$.
6. For any real number k, $\lim_{x \to a} [f(x)]^k = [\lim_{x \to a} f(x)]^k = A^k$, provided this limit exists.
7. $\lim_{x \to a} f(x) = \lim_{x \to a} g(x)$ if $f(x) = g(x)$ for all $x \neq a$.
8. For any real number $b > 0$, $\lim_{x \to a} b^{f(x)} = b^{[\lim f(x)]} b^A$.
9. For any real number b such that $0 < b < 1$ or $1 < b$,
   $\lim_{x \to a} [\log_b f(x)] = \log_b [\lim_{x \to a} f(x)] = \log_b A$ if $A > 0$.

3.1 Limits at Infinity

For any positive real number n,

$$\lim_{x \to \infty} \frac{1}{x^n} = 0 \qquad \text{and} \qquad \lim_{x \to -\infty} \frac{1}{x^n} = 0.$$

3.3 Instantaneous Rate of Change

The instantaneous rate of change for a function f when $x = a$ is

$$\lim_{h \to 0} \frac{f(a+h) - f(a)}{h}, \qquad \text{provided this limit exists.}$$

3.4 Derivative

The derivative of the function f at x, written $f'(x)$, is defined as

$$f'(x) = \lim_{h \to 0} \frac{f(x+h) - f(x)}{h}, \qquad \text{provided this limit exists.}$$

Rules for Derivatives

The following rules for derivatives are valid when all the indicated derivatives exist.

4.1 *Constant Rule* If $f(x) = k$, where k is any real number, then $f'(x) = 0$.

4.1 *Power Rule* If $f(x) = x^n$ for any real number n, then $f'(x) = nx^{n-1}$.

4.1 *Constant Times a Function* Let k be a real number. Then the derivative of $y = k \cdot f(x)$ is

$$dy/dx = k \cdot f'(x).$$

More key definitions, theorems, and formulas appear at the back of the book.

Calculus with Applications

EIGHTH EDITION

Calculus with Applications

EIGHTH EDITION

MARGARET L. LIAL
American River College

RAYMOND N. GREENWELL
Hofstra University

NATHAN P. RITCHEY
Youngstown State University

PEARSON
Addison Wesley

Boston San Francisco New York
London Toronto Sydney Tokyo Singapore Madrid
Mexico City Munich Paris Cape Town Hong Kong Montreal

Publisher:	Greg Tobin
Acquisitions Editor:	Carter Fenton
Executive Project Manager:	Christine O'Brien
Editorial Project Assistant:	Sarah Santoro
Managing Editor:	Karen Wernholm
Production Supervisor:	Cindy Cody
Senior Marketing Manager:	Becky Anderson
Marketing Coordinator:	Carolyn Buddeke
Senior Prepress Supervisor:	Caroline Fell
Associate Media Producer:	Lynne Blaszak
Software Development:	Bob Carroll/Marty Wright
Production Supervisor, Supplements:	Sheila Spinney
Senior Technical Art Specialist:	Joe Vetere
Senior Manufacturing Buyer:	Evelyn Beaton
Cover Designer:	Barbara T. Atkinson
Rights and Permissions Advisor:	Dana Weightman
Production Coordination and Text Design:	Elm Street Publishing Services, Inc.
Compositor:	WestWords, Inc.
Illustrations:	Techsetters, Inc.
Cover Photograph:	ImageBank/Getty

For permission to use copyrighted material, grateful acknowledgment is made to the copyright holders on page A51 which is hereby made part of this copyright page.

Library of Congress Cataloging-in-Publication Data

Lial, Margaret L.
Calculus with applications / Margaret L. Lial, Raymond N. Greenwell, Nathan P. Ritchey.—8th ed.
p. cm.
Includes bibliographical references and index.
ISBN 0-321-22814-6
1. Calculus. I. Greenwell, Raymond N. II. Ritchey, Nathan P. III. Title.

QA303.2.L53 2005
515—dc22

2004045097

3 4 5 6 7 8 9 10—DOW—070605

Contents

CHAPTER 12 SEQUENCES AND SERIES 653

CHAPTER 13 THE TRIGONOMETRIC FUNCTIONS 709

Tables

Preface

Calculus with Applications is a solid, application-oriented text for students majoring in business, management, economics, or the life or social sciences. A prerequisite of two years of high school algebra is assumed. New exercises, new applications, and the addition of new technology supplements make this latest edition a richer, stronger learning resource for students.

NEW AND ENHANCED FEATURES

Updating of Real-Data Applications This edition has many updated application exercises and examples using real data, as well as many new applications, with references to articles appearing in newspapers, books, and journals. For examples, see pages 75, 168, 294, 323, 334, 367, 378, 413–414, and 427. Examples with recent data help students learn how extensively mathematics is applied, how it relates to the world around them and where they might use it. We believe the quantity and quality of real-data applications set this book apart from others available for this course.

Group Work Exercises for the Extended Applications As in previous editions, we have included in-depth applied exercises, called Extended Applications, at the end of most chapters to stimulate student interest. In this edition, we have added a feature called **Directions for Group Projects** for instructors who wish to use the Extended Applications in that way. A larger collection of Extended Applications is available in the MyMathLab online course for this book.

CONTINUING FEATURES

This edition continues to offer the many popular features of the previous edition.

Pedagogical Features

- careful explanation of the mathematics
- fully developed examples with explanatory annotations in color on the side
- an algebra reference (Chapter R), designed to be used either in class or by students on their own
- thought-provoking questions open most sections and are answered in an application within the section or in the section exercises
- "just in time" margin reviews give short explanations or comments reminding students of skills or techniques learned earlier that are needed at this point
- common student difficulties and errors highlighted under the heading "Caution"

- important treatments and asides highlighted with the heading "Notes"
- summaries of rules or formulas for chapters where students may have trouble deciding which of several techniques to use
- an index of applications showing the abundant variety of real-data applications used in the text and allowing direct reference to particular topics
- multiple representations of a topic whenever possible, where each topic is examined symbolically, numerically, graphically, and verbally
- multiple methods of solutions for some topics
- use of graphing calculators and spreadsheets wherever appropriate

Exercises

- exercises carefully arranged according to the material in the section, with the more challenging exercises placed near the end
- applied exercises (labeled "Applications") grouped by subject, with subheadings indicating the specific topic
- writing exercises, labeled with the icon, to provide students with an opportunity to explain important mathematical ideas
- connections exercises, labeled with the icon, to integrate topics presented in different sections or chapters
- problems from entrance examinations for Japanese universities to give students a glimpse of international standards in math education

STUDENT SUPPLEMENTS

Student's Solutions Manual

- Provides detailed solutions to all odd-numbered text exercises and sample chapter tests with answers

ISBN: 0-321-22815-4

Graphing Calculator and Excel Spreadsheet Manual

- Provides instructions and keystroke operations for the TI-83/84 Plus, TI-85, TI-86, and TI-89 as well as for the Excel spreadsheet program

ISBN: 0-321-26843-1

Digital Video Tutor

- Complete set of digitized videos on CD for student use at home or on campus
- Ideal for distance learning or supplemental instruction

ISBN: 0-321-26775-3

Addison-Wesley Math Tutor Center

- Provides free tutoring through a registration number that can be packaged with a new textbook or purchased separately
- Staffed by qualified college mathematics instructors
- Accessible via toll-free telephone, toll-free fax, e-mail, and the Internet
- www.aw-bc.com/tutorcenter

INSTRUCTOR SUPPLEMENTS

Instructor's Resource Guide and Solutions Manual

- Provides complete solutions to even-numbered exercises, two versions of a pre-test and final exam as well as teaching tips

ISBN: 0-321-22816-2

Videotape Series

- Features an engaging team of lecturers
- Provides coverage of most topics in the text

ISBN: 0-321-26805-9

TestGen-EQ with Quizmaster-EQ

- Enables instructors to build, edit, print, and administer tests
- Features a computerized bank of questions developed to cover all text objectives
- Available on a dual-platform Windows/Macintosh CD-ROM

ISBN: 0-321-23840-0

PowerPoint Lecture Presentation

- Classroom presentation software correlates to the topic sequence of this textbook
- Available within MyMathLab or at www.aw-bc.com/suppscentral

TECHNOLOGY SUPPLEMENTS

MathXL

MathXL is an online homework, tutorial, and assessment system that uses algorithmically generated exercises correlated to the objectives in the textbook. Instructors can assign tests and homework provided by Addison-Wesley or create and customize their own tests and homework assignments. Instructors can also track their students' results and tutorial work in an online gradebook. Students can take chapter tests and receive personalized study plans that diagnose weaknesses and link students to areas they need to study and retest. Students can also work unlimited practice exercises that provide tutorial instruction and can access animations and video clips directly from selected exercises. MathXL can be packaged with new copies of the textbook. Please go to www.mathxl.com for details.

MyMathLab

MyMathLab is a complete online course available with this text and is perfect for a lecture-based, self-paced, or online course. This site offers instructors and students a wide variety of resources from dynamic multimedia—video clips, animations, and more—to course management tools. With MyMathLab, instructors can customize their course and help increase student comprehension and success!

- MyMathLab provides a powerful system for creating and assigning tests and homework, as well as a gradebook for tracking all student performance.
- The entire textbook is available online and is supplemented by multimedia content, such as videos and animations, which is used to explain concepts. With MyMathLab, students can work with tutorial exercises tied directly to those in their textbook.
- MyMathLab allows students to do practice work and to complete instructor-assigned tests and homework assignments online. Based on their results, MyMathLab automatically builds individual study plans that students can use to improve their skills.
- The MyMathLab site for this book contains additional Extended Applications and downloadable TI graphing calculator programs.

MyMathLab requires a student access code. For more information about MyMathLab, go to www.mymathlab.com.

InterActMath® Tutorial Web site www.interactmath.com

Get practice and tutorial help online! This interactive tutorial Web site provides algorithmically generated practice exercises that correlate directly to the exercises in the text. A detailed worked-out example and guided solution accompany each practice exercise. The Web site recognizes student errors and provides feedback.

Acknowledgments

We wish to thank the following professors for their contributions in reviewing portions of this text.

Kristin Barker, *Clark College*
Lynn Brooks Cade, *Pensacola Junior College*
Valeree Falduto, *Lynn University*
Homi Fatemi, *Santa Clara University*
Gary McCracken, *Shelton State College*
Sam Northshield, *Plattsburgh State University*
Jackie Robertson, *Brigham Young University*
Andreas H. Soemadi, *Kirkwood Community College*
Dr. H. Edward Stone, *University of Texas, Dallas*

We are grateful to LaurelTech Integrated Publishing Services, especially Teri Lovelace, for doing an excellent job coordinating the *Student's Solutions Manual* and *Instructor's Resource Guide and Solutions Manual,* an enormous and time-consuming task. We also thank Sheri Minkner and Judy Martinez for typesetting these manuals. Tim Mogill and Rebecca Cointin conducted a careful accuracy check of the answer section. Bernice Eisen has created an accurate and complete index for us, and Becky Troutman has compiled the Index of Applications. We thank the following Hofstra University professors for their numerous suggestions: Seymour Berg, Barry Brenin, Gillian Elston, Peter Grassi, and Dan Seabold. We are indebted to J. Laurie Snell of Dartmouth College, whose electronic newsletter *Chance* alerted us to many applications in probability and statistics. We also want to thank Karla Harby and Mary Ann Ritchey for their editorial assistance. We especially appreciate the staff at Pearson/Addison-Wesley, whose contributions have been very important in bringing this project to a successful conclusion: Greg Tobin, Christine O'Brien, Bill Hoffman, and Carter Fenton. Finally, we wish to thank Phyllis Crittenden of Elm Street Publishing Services for a great job as project editor.

Margaret L. Lial
Raymond N. Greenwell
Nathan P. Ritchey

Index of Applications

CONSTRUCTION

CONSUMER INFORMATION

ECONOMICS

EDUCATION

ENGINEERING

ENVIRONMENT

FINANCE

GENERAL INTEREST

GEOLOGY

GEOMETRY

GOVERNMENT

HEALTH AND LIFE SCIENCES

LABOR

PHYSICS

SOCIAL SCIENCE

SPORTS AND ENTERTAINMENT

STATISTICS AND DEMOGRAPHICS

TECHNOLOGY

TRANSPORTATION

WILDLIFE

CHAPTER

R Algebra Reference

In this chapter, we will review the most important topics in algebra. Knowing algebra is a fundamental prerequisite to success in higher mathematics. This algebra reference is designed for self-study; study it all at once or refer to it when needed throughout the course. Since this is a review, answers to all exercises are given in the answer section at the back of the book.

R.1 POLYNOMIALS

An expression such as $9p^4$ is a **term;** the number 9 is the **coefficient,** p is the **variable,** and 4 is the **exponent.** The expression p^4 means $p \cdot p \cdot p \cdot p$, while p^2 means $p \cdot p$, and so on. Terms having the same variable and the same exponent, such as $9x^4$ and $-3x^4$, are **like terms.** Terms that do not have both the same variable and the same exponent, such as m^2 and m^4, are **unlike terms.**

A **polynomial** is a term or a finite sum of terms in which all variables have whole number exponents, and no variables appear in denominators. Examples of polynomials include

$$5x^4 + 2x^3 + 6x, \qquad 8m^3 + 9m^2n - 6mn^2 + 3n^3, \qquad 10p, \qquad \text{and} \qquad -9.$$

Adding and Subtracting Polynomials The following properties of real numbers are useful for performing operations on polynomials.

PROPERTIES OF REAL NUMBERS

For all real numbers a, b, and c:

1. $a + b = b + a$; $ab = ba$;	**Commutative properties**
2. $(a + b) + c = a + (b + c)$; $(ab)c = a(bc)$;	**Associative properties**
3. $a(b + c) = ab + ac$.	**Distributive property**

EXAMPLE 1 *Properties of Real Numbers*

(a) $2 + x = x + 2$ — Commutative property of addition

(b) $x \cdot 3 = 3x$ — Commutative property of multiplication

(c) $(7x)x = 7(x \cdot x) = 7x^2$ — Associative property of multiplication

(d) $3(x + 4) = 3x + 12$ — Distributive property

One use of the distributive property is to add or subtract polynomials. Only like terms may be added or subtracted. For example,

$$12y^4 + 6y^4 = (12 + 6)y^4 = 18y^4,$$

and

$$-2m^2 + 8m^2 = (-2 + 8)m^2 = 6m^2,$$

but the polynomial $8y^4 + 2y^5$ cannot be further simplified. To subtract polynomials, we use the facts that $-(a + b) = -a - b$ and $-(a - b) = -a + b$. In the next example, we show how to add and subtract polynomials.

EXAMPLE 2 *Adding and Subtracting Polynomials*

Add or subtract as indicated.

(a) $(8x^3 - 4x^2 + 6x) + (3x^3 + 5x^2 - 9x + 8)$

Solution Combine like terms.

$$\begin{aligned}(8x^3 - 4x^2 + 6x) &+ (3x^3 + 5x^2 - 9x + 8)\\ &= (8x^3 + 3x^3) + (-4x^2 + 5x^2) + (6x - 9x) + 8\\ &= 11x^3 + x^2 - 3x + 8\end{aligned}$$

(b) $(-4x^4 + 6x^3 - 9x^2 - 12) + (-3x^3 + 8x^2 - 11x + 7)$

Solution Combining like terms as before yields

$$-4x^4 + 3x^3 - x^2 - 11x - 5.$$

(c) $(2x^2 - 11x + 8) - (7x^2 - 6x + 2)$

Solution Distributing the minus sign yields

$$\begin{aligned}(2x^2 - 11x + 8) &+ (-7x^2 + 6x - 2)\\ &= -5x^2 - 5x + 6.\end{aligned}$$

Multiplying Polynomials The distributive property is also used to multiply polynomials, along with the fact that $a^m \cdot a^n = a^{m+n}$. For example,

$$x \cdot x = x^1 \cdot x^1 = x^2 \qquad \text{and} \qquad x^2 \cdot x^5 = x^7.$$

EXAMPLE 3 *Multiplying Polynomials*

Multiply.

(a) $8x(6x - 4)$

Solution

$$\begin{aligned}8x(6x - 4) &= 8x(6x) - 8x(4)\\ &= 48x^2 - 32x\end{aligned}$$

(b) $(3p - 2)(p^2 + 5p - 1)$

Solution

$$\begin{aligned}(3p - 2)&(p^2 + 5p - 1)\\ &= 3p(p^2 + 5p - 1) - 2(p^2 + 5p - 1)\\ &= 3p(p^2) + 3p(5p) + 3p(-1) - 2(p^2) - 2(5p) - 2(-1)\\ &= 3p^3 + 15p^2 - 3p - 2p^2 - 10p + 2\\ &= 3p^3 + 13p^2 - 13p + 2\end{aligned}$$

(c) $(x + 2)(x + 3)(x - 4)$

Solution

$$\begin{aligned}(x + 2)&(x + 3)(x - 4)\\ &= [(x + 2)(x + 3)](x - 4)\\ &= (x^2 + 2x + 3x + 6)(x - 4)\\ &= (x^2 + 5x + 6)(x - 4)\\ &= x^3 + 5x^2 + 6x - 4x^2 - 20x - 24\\ &= x^3 + x^2 - 14x - 24\end{aligned}$$

A **binomial** is a polynomial with exactly two terms, such as $2x + 1$ or $m + n$. When two binomials are multiplied, the FOIL method (First, Outer, Inner, Last) is used as a memory aid.

EXAMPLE 4 *Multiplying Polynomials*

Find $(2m - 5)(m + 4)$ using the FOIL method.

Solution

$$\begin{aligned}(2m - 5)(m + 4) &= \overset{\text{F}}{(2m)(m)} + \overset{\text{O}}{(2m)(4)} + \overset{\text{I}}{(-5)(m)} + \overset{\text{L}}{(-5)(4)} \\ &= 2m^2 + 8m - 5m - 20 \\ &= 2m^2 + 3m - 20\end{aligned}$$

EXAMPLE 5 *Multiplying Polynomials*

Find $(2k - 5)^2$.

Solution Use FOIL.

$$\begin{aligned}(2k - 5)^2 &= (2k - 5)(2k - 5) \\ &= 4k^2 - 10k - 10k + 25 \\ &= 4k^2 - 20k + 25\end{aligned}$$

Notice that the product of the square of a binomial is the square of the first term, $(2k)^2$, plus twice the product of the two terms, $(2)(2k)(-5)$, plus the square of the last term, $(-5)^2$.

CAUTION Avoid the common error of writing $(x + y)^2 = x^2 + y^2$. As Example 5 shows, the square of a binomial has three terms, so

$$(x + y)^2 = x^2 + 2xy + y^2.$$

Furthermore, higher powers of a binomial also result in more than two terms. For example, verify by multiplication that

$$(x + y)^3 = x^3 + 3x^2y + 3xy^2 + y^3.$$

Remember, for any value of $n \neq 1$,

$$(x + y)^n \neq x^n + y^n.$$

R.1 EXERCISES

Perform the indicated operations.

1. $(2x^2 - 6x + 11) + (-3x^2 + 7x - 2)$

2. $(-4y^2 - 3y + 8) - (2y^2 - 6y - 2)$

3. $-3(4q^2 - 3q + 2) + 2(-q^2 + q - 4)$

4. $2(3r^2 + 4r + 2) - 3(-r^2 + 4r - 5)$

5. $(.613x^2 - 4.215x + .892) - .47(2x^2 - 3x + 5)$

6. $.83(5r^2 - 2r + 7) - (7.12r^2 + 6.423r - 2)$

7. $-9m(2m^2 + 3m - 1)$

8. $(6k - 1)(2k - 3)$

9. $(5r - 3s)(5r + 4s)$

10. $(9k + q)(2k - q)$

11. $\left(\frac{2}{5}y + \frac{1}{8}z\right)\left(\frac{3}{5}y + \frac{1}{2}z\right)$

12. $\left(\frac{3}{4}r - \frac{2}{3}s\right)\left(\frac{5}{4}r + \frac{1}{3}s\right)$

13. $(12x - 1)(12x + 1)$

14. $(6m + 5)(6m - 5)$

15. $(3p - 1)(9p^2 + 3p + 1)$

16. $(2p - 1)(3p^2 - 4p + 5)$

17. $(2m + 1)(4m^2 - 2m + 1)$

18. $(k + 2)(12k^3 - 3k^2 + k + 1)$

19. $(m - n + k)(m + 2n - 3k)$

20. $(r - 3s + t)(2r - s + t)$

21. $(x + 1)(x + 2)(x + 3)$

22. $(x - 1)(x + 2)(x - 3)$

23. $(3a + b)^2$

24. $(x - 2y)^3$

R.2 FACTORING

Multiplication of polynomials relies on the distributive property. The reverse process, where a polynomial is written as a product of other polynomials, is called **factoring.** For example, one way to factor the number 18 is to write it as the product $9 \cdot 2$; both 9 and 2 are **factors** of 18. Usually, only integers are used as factors of integers. The number 18 can also be written with three integer factors as $2 \cdot 3 \cdot 3$.

The Greatest Common Factor To factor the algebraic expression $15m + 45$, first note that both $15m$ and 45 are divisible by 15; $15m = 15 \cdot m$ and $45 = 15 \cdot 3$. By the distributive property,

$$15m + 45 = 15 \cdot m + 15 \cdot 3 = 15(m + 3).$$

Both 15 and $m + 3$ are factors of $15m + 45$. Since 15 divides into both terms of $15m + 45$ (and is the largest number that will do so), 15 is the **greatest common factor** for the polynomial $15m + 45$. The process of writing $15m + 45$ as $15(m + 3)$ is often called **factoring out** the greatest common factor.

EXAMPLE 1 *Factoring*

Factor out the greatest common factor.

(a) $12p - 18q$

Solution Both $12p$ and $18q$ are divisible by 6. Therefore,

$$12p - 18q = 6 \cdot 2p - 6 \cdot 3q = 6(2p - 3q).$$

(b) $8x^3 - 9x^2 + 15x$

Solution Each of these terms is divisible by x.

$$
\begin{aligned}
8x^3 - 9x^2 + 15x &= (8x^2) \cdot x - (9x) \cdot x + 15 \cdot x \\
&= x(8x^2 - 9x + 15) \quad \text{or} \quad (8x^2 - 9x + 15)x
\end{aligned}
$$

One can always check factorization by finding the product of the factors and comparing it to the original expression.

CAUTION When factoring out the greatest common factor in an expression like $2x^2 + x$, be careful to remember the 1 in the second term.

$$2x^2 + x = 2x^2 + 1x = x(2x + 1), \quad \text{not } x(2x). \quad ■$$

Factoring Trinomials A polynomial that has no greatest common factor (other than 1) may still be factorable. For example, the polynomial $x^2 + 5x + 6$ can be factored as $(x + 2)(x + 3)$. To see that this is correct, find the product $(x + 2)(x + 3)$; you should get $x^2 + 5x + 6$. A polynomial such as this with three terms is called a **trinomial.** To factor the trinomial $x^2 + 5x + 6$, where the coefficient of x^2 is 1, we use FOIL backwards.

EXAMPLE 2 *Factoring a Trinomial*

Factor $y^2 + 8y + 15$.

Solution Since the coefficient of y^2 is 1, factor by finding two numbers whose *product* is 15 and whose *sum* is 8. Since the constant and the middle term are positive, the numbers must both be positive. Begin by listing all pairs of positive integers having a product of 15. As you do this, also form the sum of each pair of numbers.

Products	Sums
$15 \cdot 1 = 15$	$15 + 1 = 16$
$5 \cdot 3 = 15$	$5 + 3 = 8$

The numbers 5 and 3 have a product of 15 and a sum of 8. Thus, $y^2 + 8y + 15$ factors as

$$y^2 + 8y + 15 = (y + 5)(y + 3).$$

The answer also can be written as $(y + 3)(y + 5)$.

If the coefficient of the squared term is *not* 1, work as shown below.

EXAMPLE 3 *Factoring a Trinomial*

Factor $4x^2 + 8xy - 5y^2$.

Solution The possible factors of $4x^2$ are $4x$ and x or $2x$ and $2x$; the possible factors of $-5y^2$ are $-5y$ and y or $5y$ and $-y$. Try various combinations of these factors until one works (if, indeed, any work). For example, try the product $(x + 5y)(4x - y)$.

$$\begin{aligned}(x + 5y)(4x - y) &= 4x^2 - xy + 20xy - 5y^2 \\ &= 4x^2 + 19xy - 5y^2\end{aligned}$$

This product is not correct, so try another combination.

$$\begin{aligned}(2x - y)(2x + 5y) &= 4x^2 + 10xy - 2xy - 5y^2 \\ &= 4x^2 + 8xy - 5y^2\end{aligned}$$

Since this combination gives the correct polynomial,

$$4x^2 + 8xy - 5y^2 = (2x - y)(2x + 5y).$$

Special Factorizations Four special factorizations occur so often that they are listed here for future reference.

SPECIAL FACTORIZATIONS

$x^2 - y^2 = (x + y)(x - y)$	**Difference of two squares**
$x^2 + 2xy + y^2 = (x + y)^2$	**Perfect square**
$x^3 - y^3 = (x - y)(x^2 + xy + y^2)$	**Difference of two cubes**
$x^3 + y^3 = (x + y)(x^2 - xy + y^2)$	**Sum of two cubes**

A polynomial that cannot be factored is called a **prime polynomial.**

EXAMPLE 4 *Factoring Polynomials*

Factor each polynomial, if possible.

(a) $64p^2 - 49q^2 = (8p)^2 - (7q)^2 = (8p + 7q)(8p - 7q)$

(b) $x^2 + 36$ is a prime polynomial.

(c) $x^2 + 12x + 36 = (x + 6)^2$

(d) $9y^2 - 24yz + 16z^2 = (3y - 4z)^2$

(e) $y^3 - 8 = y^3 - 2^3 = (y - 2)(y^2 + 2y + 4)$

(f) $m^3 + 125 = m^3 + 5^3 = (m + 5)(m^2 - 5m + 25)$

(g) $8k^3 - 27z^3 = (2k)^3 - (3z)^3 = (2k - 3z)(4k^2 + 6kz + 9z^2)$

CAUTION In factoring, always look for a common factor first. Since $36x^2 - 4y^2$ has a common factor of 4,

$$36x^2 - 4y^2 = 4(9x^2 - y^2) = 4(3x + y)(3x - y).$$

It would be incomplete to factor it as

$$36x^2 - 4y^2 = (6x + 2y)(6x - 2y)$$

since each factor can be factored still further. To *factor* means to factor completely, so that each polynomial factor is prime. ■

R.2 EXERCISES

Factor each polynomial. If a polynomial cannot be factored, write prime. *Factor out the greatest common factor as necessary.*

1. $8a^3 - 16a^2 + 24a$

2. $3y^3 + 24y^2 + 9y$

3. $25p^4 - 20p^3q + 100p^2q^2$

4. $60m^4 - 120m^3n + 50m^2n^2$

5. $m^2 + 9m + 14$

6. $x^2 + 4x - 5$

7. $z^2 + 9z + 20$

8. $b^2 - 8b + 7$

9. $a^2 - 6ab + 5b^2$

10. $s^2 + 2st - 35t^2$

11. $y^2 - 4yz - 21z^2$

12. $6a^2 - 48a - 120$

13. $3m^3 + 12m^2 + 9m$

14. $2x^2 - 5x - 3$

15. $3a^2 + 10a + 7$

16. $2a^2 - 17a + 30$

17. $15y^2 + y - 2$

18. $21m^2 + 13mn + 2n^2$

19. $24a^4 + 10a^3b - 4a^2b^2$

20. $32z^5 - 20z^4a - 12z^3a^2$

21. $x^2 - 64$

22. $9m^2 - 25$

23. $121a^2 - 100$

24. $9x^2 + 64$

25. $z^2 + 14zy + 49y^2$

26. $m^2 - 6mn + 9n^2$

27. $9p^2 - 24p + 16$

28. $a^3 - 216$

29. $8r^3 - 27s^3$

30. $64m^3 + 125$

31. $x^4 - y^4$

32. $16a^4 - 81b^4$

R.3 RATIONAL EXPRESSIONS

Many algebraic fractions are **rational expressions,** which are quotients of polynomials with nonzero denominators. Examples include

$$\frac{8}{x - 1}, \quad \frac{3x^2 + 4x}{5x - 6}, \quad \text{and} \quad \frac{2y + 1}{y^2}.$$

Properties for working with rational expressions are summarized next.

PROPERTIES OF RATIONAL EXPRESSIONS

For all mathematical expressions P, Q, R, and S, with Q and $S \neq 0$:

$$\frac{P}{Q} = \frac{PS}{QS}$$ **Fundamental property**

$$\frac{P}{Q} + \frac{R}{Q} = \frac{P + R}{Q}$$ **Addition**

$$\frac{P}{Q} - \frac{R}{Q} = \frac{P - R}{Q}$$ **Subtraction**

$$\frac{P}{Q} \cdot \frac{R}{S} = \frac{PR}{QS}$$ **Multiplication**

$$\frac{P}{Q} \div \frac{R}{S} = \frac{P}{Q} \cdot \frac{S}{R} \quad (R \neq 0)$$ **Division**

When writing a rational expression in lowest terms, we may need to use the fact that $\frac{a^m}{a^n} = a^{m-n}$. For example,

$$\frac{x^2}{3x} = \frac{1x^2}{3x} = \frac{1}{3} \cdot \frac{x^2}{x} = \frac{1}{3}x.$$

EXAMPLE 1 *Reducing Rational Expressions*

Write each rational expression in lowest terms, that is, reduce the expression as much as possible.

(a) $$\frac{8x + 16}{4} = \frac{8(x + 2)}{4} = \frac{4 \cdot 2(x + 2)}{4} = 2(x + 2)$$

Factor both the numerator and denominator in order to identify any common factors, which have a quotient of 1. The answer could also be written as $2x + 4$.

(b) $$\frac{k^2 + 7k + 12}{k^2 + 2k - 3} = \frac{(k + 4)(k + 3)}{(k - 1)(k + 3)} = \frac{k + 4}{k - 1}$$

The answer cannot be further reduced.

CAUTION One of the most common errors in algebra involves incorrect use of the fundamental property of rational expressions. Only common *factors* may be divided or "canceled." It is essential to factor rational expressions before writing them in lowest terms. In Example 1(b), for instance, it is not correct to "cancel" k^2 (or cancel k, or divide 12 by -3) because the additions and subtraction must be performed first. Here they cannot be performed, so it is not possible to divide. After factoring, however, the fundamental property can be used to write the expression in lowest terms. ■

EXAMPLE 2 *Combining Rational Expressions*

Perform each operation.

(a) $\dfrac{3y+9}{6} \cdot \dfrac{18}{5y+15}$

Solution Factor where possible, then multiply numerators and denominators and reduce to lowest terms.

$$\begin{aligned}\frac{3y+9}{6} \cdot \frac{18}{5y+15} &= \frac{3(y+3)}{6} \cdot \frac{18}{5(y+3)} \\ &= \frac{3 \cdot 18(y+3)}{6 \cdot 5(y+3)} \\ &= \frac{3 \cdot 6 \cdot 3(y+3)}{6 \cdot 5(y+3)} = \frac{3 \cdot 3}{5} = \frac{9}{5}\end{aligned}$$

(b) $\dfrac{m^2+5m+6}{m+3} \cdot \dfrac{m}{m^2+3m+2}$

Solution Factor where possible.

$$\begin{aligned}&\frac{(m+2)(m+3)}{m+3} \cdot \frac{m}{(m+2)(m+1)} \\ &\quad = \frac{m(m+2)(m+3)}{(m+3)(m+2)(m+1)} = \frac{m}{m+1}\end{aligned}$$

(c) $\dfrac{9p-36}{12} \div \dfrac{5(p-4)}{18}$

Solution Use the division property of rational expressions.

$$\begin{aligned}&\frac{9p-36}{12} \cdot \frac{18}{5(p-4)} \qquad \text{Invert and multiply.} \\ &\quad = \frac{9(p-4)}{6 \cdot 2} \cdot \frac{6 \cdot 3}{5(p-4)} = \frac{27}{10}\end{aligned}$$

(d) $\dfrac{4}{5k} - \dfrac{11}{5k}$

Solution As shown in the list of properties, to subtract two rational expressions that have the same denominators, subtract the numerators while keeping the same denominator.

$$\frac{4}{5k} - \frac{11}{5k} = \frac{4-11}{5k} = -\frac{7}{5k}$$

(e) $\dfrac{7}{p} + \dfrac{9}{2p} + \dfrac{1}{3p}$

Solution These three fractions cannot be added until their denominators are the same. A **common denominator** into which p, $2p$, and $3p$ all divide is $6p$. Note that $12p$ is also a common denominator, but $6p$ is the **least common denominator.** Use the fundamental property to rewrite each rational expression with a denominator of $6p$.

$$\frac{7}{p}+\frac{9}{2p}+\frac{1}{3p}=\frac{6\cdot 7}{6\cdot p}+\frac{3\cdot 9}{3\cdot 2p}+\frac{2\cdot 1}{2\cdot 3p}$$
$$=\frac{42}{6p}+\frac{27}{6p}+\frac{2}{6p}$$
$$=\frac{42+27+2}{6p}$$
$$=\frac{71}{6p}$$

(f) $\dfrac{x+1}{x^2+5x+6}-\dfrac{5x-1}{x^2-x-12}$

Solution To find the least common denominator, we first factor each denominator. Then we change each fraction so they all have the same denominator, being careful to multiply only by quotients that equal 1.

$$\frac{x+1}{x^2+5x+6}-\frac{5x-1}{x^2-x-12}$$
$$=\frac{x+1}{(x+2)(x+3)}-\frac{5x-1}{(x+3)(x-4)}$$
$$=\frac{x+1}{(x+2)(x+3)}\cdot\frac{(x-4)}{(x-4)}-\frac{5x-1}{(x+3)(x-4)}\cdot\frac{(x+2)}{(x+2)}$$
$$=\frac{(x^2-3x-4)-(5x^2+9x-2)}{(x+2)(x+3)(x-4)}$$
$$=\frac{-4x^2-12x-2}{(x+2)(x+3)(x-4)}$$
$$=\frac{-2(2x^2+6x+1)}{(x+2)(x+3)(x-4)}$$

Because the numerator cannot be factored further, we leave our answer in this form. We could also multiply out the denominator, but factored form is usually more useful.

R.3 EXERCISES

Write each rational expression in lowest terms.

1. $\dfrac{7z^2}{14z}$

2. $\dfrac{25p^3}{10p^2}$

3. $\dfrac{8k+16}{9k+18}$

4. $\dfrac{3(t+5)}{(t+5)(t-3)}$

5. $\dfrac{8x^2+16x}{4x^2}$

6. $\dfrac{36y^2+72y}{9y}$

7. $\dfrac{m^2-4m+4}{m^2+m-6}$

8. $\dfrac{r^2-r-6}{r^2+r-12}$

9. $\dfrac{x^2+3x-4}{x^2-1}$

10. $\dfrac{z^2-5z+6}{z^2-4}$

11. $\dfrac{8m^2+6m-9}{16m^2-9}$

12. $\dfrac{6y^2+11y+4}{3y^2+7y+4}$

Perform the indicated operations.

13. $\dfrac{9k^2}{25}\cdot\dfrac{5}{3k}$

14. $\dfrac{15p^3}{9p^2}\div\dfrac{6p}{10p^2}$

15. $\dfrac{a+b}{2p}\cdot\dfrac{12}{5(a+b)}$

16. $\frac{a-3}{16} \div \frac{a-3}{32}$

17. $\frac{2k+8}{6} \div \frac{3k+12}{2}$

18. $\frac{9y-18}{6y+12} \cdot \frac{3y+6}{15y-30}$

19. $\frac{4a+12}{2a-10} \div \frac{a^2-9}{a^2-a-20}$

20. $\frac{6r-18}{9r^2+6r-24} \cdot \frac{12r-16}{4r-12}$

21. $\frac{k^2-k-6}{k^2+k-12} \cdot \frac{k^2+3k-4}{k^2+2k-3}$

22. $\frac{m^2+3m+2}{m^2+5m+4} \div \frac{m^2+5m+6}{m^2+10m+24}$

23. $\frac{2m^2-5m-12}{m^2-10m+24} \div \frac{4m^2-9}{m^2-9m+18}$

24. $\frac{6n^2-5n-6}{6n^2+5n-6} \cdot \frac{12n^2-17n+6}{12n^2-n-6}$

25. $\frac{a+1}{2} - \frac{a-1}{2}$

26. $\frac{3}{p} + \frac{1}{2}$

27. $\frac{2}{y} - \frac{1}{4}$

28. $\frac{1}{6m} + \frac{2}{5m} + \frac{4}{m}$

29. $\frac{1}{m-1} + \frac{2}{m}$

30. $\frac{6}{r} - \frac{5}{r-2}$

31. $\frac{8}{3(a-1)} + \frac{2}{a-1}$

32. $\frac{2}{5(k-2)} + \frac{3}{4(k-2)}$

33. $\frac{2}{x^2-2x-3} + \frac{5}{x^2-x-6}$

34. $\frac{2y}{y^2+7y+12} - \frac{y}{y^2+5y+6}$

35. $\frac{3k}{2k^2+3k-2} - \frac{2k}{2k^2-7k+3}$

36. $\frac{4m}{3m^2+7m-6} - \frac{m}{3m^2-14m+8}$

37. $\frac{2}{a+2} + \frac{1}{a} + \frac{a-1}{a^2+2a}$

38. $\frac{5x+2}{x^2-1} + \frac{3}{x^2+x} - \frac{1}{x^2-x}$

R.4 EQUATIONS

Linear Equations Equations that can be written in the form $ax + b = 0$, where a and b are real numbers, with $a \neq 0$, are **linear equations.** Examples of linear equations include $5y + 9 = 16$, $8x = 4$, and $-3p + 5 = -8$. Equations that are *not* linear include absolute value equations such as $|x| = 4$. The following properties are used to solve linear equations.

PROPERTIES OF EQUALITY

For all real numbers a, b, and c:

1. **If $a = b$, then $a + c = b + c$.** — **Addition property of equality**
   (The same number may be added to both sides of an equation.)
2. **If $a = b$, then $ac = bc$.** — **Multiplication property of equality**
   (Both sides of an equation may be multiplied by the same number.)

EXAMPLE 1 *Solving Linear Equations*

(a) If $x - 2 = 3$, then $x = 2 + 3 = 5$ — Addition property of equality

(b) If $x/2 = 3$, then $x = 2 \cdot 3 = 6$ — Multiplication property of equality

The following example shows how these properties are used to solve linear equations. Of course, the solutions should always be checked by substitution in the original equation.

EXAMPLE 2 *Solving a Linear Equation*
Solve $2x - 5 + 8 = 3x + 2(2 - 3x)$.

Solution

$$\begin{aligned} 2x - 5 + 8 &= 3x + 4 - 6x && \text{Distributive property} \\ 2x + 3 &= -3x + 4 && \text{Combine like terms.} \\ 5x + 3 &= 4 && \text{Add } 3x \text{ to both sides.} \\ 5x &= 1 && \text{Add } -3 \text{ to both sides.} \\ x &= \frac{1}{5} && \text{Multiply both sides by } \tfrac{1}{5}. \end{aligned}$$

Check by substituting in the original equation. The left side becomes $2(1/5) - 5 + 8$ and the right side becomes $3(1/5) + 2[2 - 3(1/5)]$. Verify that both of these expressions simplify to $17/5$.

Quadratic Equations An equation with 2 as the highest exponent of the variable is a *quadratic equation.* A **quadratic equation** has the form $ax^2 + bx + c = 0$, where a, b, and c are real numbers and $a \neq 0$. A quadratic equation written in the form $ax^2 + bx + c = 0$ is said to be in **standard form.**

The simplest way to solve a quadratic equation, but one that is not always applicable, is by factoring. This method depends on the **zero-factor property.**

ZERO-FACTOR PROPERTY

If a and b are real numbers, with $ab = 0$, then

$$a = 0, \quad b = 0, \quad \text{or both.}$$

EXAMPLE 3 *Solving a Quadratic Equation*
Solve $6r^2 + 7r = 3$.

Solution First write the equation in standard form.

$$6r^2 + 7r - 3 = 0$$

Now factor $6r^2 + 7r - 3$ to get

$$(3r - 1)(2r + 3) = 0.$$

By the zero-factor property, the product $(3r - 1)(2r + 3)$ can equal 0 if and only if

$$3r - 1 = 0 \quad \text{or} \quad 2r + 3 = 0.$$

Solve each of these equations separately to find that the solutions are $1/3$ and $-3/2$. Check these solutions by substituting them in the original equation.

CAUTION Remember, the zero-factor property requires that the product of two (or more) factors be equal to *zero,* not some other quantity. It would be incorrect to use the zero-factor property with an equation in the form $(x + 3)(x - 1) = 4$, for example. ■

If a quadratic equation cannot be solved easily by factoring, use the *quadratic formula.* (The derivation of the quadratic formula is given in most algebra books.)

QUADRATIC FORMULA

The solutions of the quadratic equation $ax^2 + bx + c = 0$, where $a \neq 0$, are given by

$$x = \frac{-b \pm \sqrt{b^2 - 4ac}}{2a}.$$

EXAMPLE 4 *Quadratic Formula*

Solve $x^2 - 4x - 5 = 0$ by the quadratic formula.

Solution The equation is already in standard form (it has 0 alone on one side of the equals sign), so the values of a, b, and c from the quadratic formula are easily identified. The coefficient of the squared term gives the value of a; here, $a = 1$. Also, $b = -4$ and $c = -5$. (Be careful to use the correct signs.) Substitute these values into the quadratic formula.

$$x = \frac{-(-4) \pm \sqrt{(-4)^2 - 4(1)(-5)}}{2(1)} \qquad a = 1, b = -4, c = -5$$

$$x = \frac{4 \pm \sqrt{16 + 20}}{2} \qquad (-4)^2 = (-4)(-4) = 16$$

$$x = \frac{4 \pm 6}{2} \qquad \sqrt{16 + 20} = \sqrt{36} = 6$$

The $\pm$ sign represents the two solutions of the equation. To find both of the solutions, first use $+$ and then use $-$.

$$x = \frac{4 + 6}{2} = \frac{10}{2} = 5 \quad \text{or} \quad x = \frac{4 - 6}{2} = \frac{-2}{2} = -1$$

The two solutions are 5 and -1.

CAUTION Notice in the quadratic formula that the square root is added to or subtracted from the value of $-b$ *before* dividing by $2a$. ■

EXAMPLE 5 *Quadratic Formula*

Solve $x^2 + 1 = 4x$.

Solution First, add $-4x$ on both sides of the equals sign in order to get the equation in standard form.

$$x^2 - 4x + 1 = 0$$

Now identify the letters a, b, and c. Here $a = 1$, $b = -4$, and $c = 1$. Substitute these numbers into the quadratic formula.

$$\begin{aligned} x &= \frac{-(-4) \pm \sqrt{(-4)^2 - 4(1)(1)}}{2(1)} \\ &= \frac{4 \pm \sqrt{16 - 4}}{2} \\ &= \frac{4 \pm \sqrt{12}}{2} \end{aligned}$$

Simplify the solutions by writing $\sqrt{12}$ as $\sqrt{4 \cdot 3} = \sqrt{4} \cdot \sqrt{3} = 2\sqrt{3}$. Substituting $2\sqrt{3}$ for $\sqrt{12}$ gives

$$\begin{aligned} x &= \frac{4 \pm 2\sqrt{3}}{2} \\ &= \frac{2(2 \pm \sqrt{3})}{2} && \text{Factor } 4 \pm 2\sqrt{3}. \\ &= 2 \pm \sqrt{3}. && \text{Reduce to lowest terms.} \end{aligned}$$

The two solutions are $2 + \sqrt{3}$ and $2 - \sqrt{3}$.

The exact values of the solutions are $2 + \sqrt{3}$ and $2 - \sqrt{3}$. The $\sqrt{\ }$ key on a calculator gives decimal approximations of these solutions (to the nearest thousandth):

$$\begin{aligned} 2 + \sqrt{3} &\approx 2 + 1.732 = 3.732^* \\ 2 - \sqrt{3} &\approx 2 - 1.732 = .268 \end{aligned}$$

NOTE Sometimes the quadratic formula will give a result with a negative number under the radical sign, such as $3 \pm \sqrt{-5}$. A solution of this type is a complex number. Since this text deals only with real numbers, such solutions cannot be used. ■

Equations with Fractions When an equation includes fractions, first eliminate all denominators by multiplying both sides of the equation by a common denominator, a number that can be divided (with no remainder) by each denominator in the equation. When an equation involves fractions with variable denominators, it is *necessary* to check all solutions in the original equation to be sure that no solution will lead to a zero denominator.

EXAMPLE 6 *Solving Rational Equations*

Solve each equation.

(a) $\dfrac{r}{10} - \dfrac{2}{15} = \dfrac{3r}{20} - \dfrac{1}{5}$

*The symbol $\approx$ means "is approximately equal to."

Solution The denominators are 10, 15, 20, and 5. Each of these numbers can be divided into 60, so 60 is a common denominator. Multiply both sides of the equation by 60 and use the distributive property. (If a common denominator cannot be found easily, all the denominators in the problem can be multiplied together to produce one.)

$$\frac{r}{10} - \frac{2}{15} = \frac{3r}{20} - \frac{1}{5}$$

$$60\left(\frac{r}{10} - \frac{2}{15}\right) = 60\left(\frac{3r}{20} - \frac{1}{5}\right) \qquad \text{Multiply by the common denominator.}$$

$$60\left(\frac{r}{10}\right) - 60\left(\frac{2}{15}\right) = 60\left(\frac{3r}{20}\right) - 60\left(\frac{1}{5}\right) \qquad \text{Distributive property}$$

$$6r - 8 = 9r - 12$$

Add $-9r$ and 8 to both sides.

$$6r - 8 + (-9r) + 8 = 9r - 12 + (-9r) + 8$$

$$-3r = -4$$

$$r = \frac{4}{3} \qquad \text{Multiply each side by } -\tfrac{1}{3}.$$

Check by substituting into the original equation.

(b) $\dfrac{3}{x^2} - 12 = 0$

Solution Begin by multiplying both sides of the equation by x^2 to get $3 - 12x^2 = 0$. This equation could be solved by using the quadratic formula with $a = -12$, $b = 0$, and $c = 3$. Another method, which works well for the type of quadratic equation in which $b = 0$, is shown below.

$$3 - 12x^2 = 0$$

$$3 = 12x^2 \qquad \text{Add } 12x^2.$$

$$\frac{1}{4} = x^2 \qquad \text{Multiply by } \tfrac{1}{12}.$$

$$\pm\frac{1}{2} = x \qquad \text{Take square roots.}$$

Verify that there are two solutions, $-1/2$ and $1/2$.

(c) $\dfrac{2}{k} - \dfrac{3k}{k + 2} = \dfrac{k}{k^2 + 2k}$

Solution Factor $k^2 + 2k$ as $k(k + 2)$. The least common denominator for all the fractions is $k(k + 2)$. Multiplying both sides by $k(k + 2)$ gives the following.

$$2(k + 2) - 3k(k) = k$$
$$2k + 4 - 3k^2 = k \quad \text{Distributive property}$$
$$-3k^2 + k + 4 = 0 \quad \text{Add } -k; \text{ rearrange terms.}$$
$$3k^2 - k - 4 = 0 \quad \text{Multiply by } -1.$$
$$(3k - 4)(k + 1) = 0 \quad \text{Factor.}$$
$$3k - 4 = 0 \qquad \text{or} \qquad k + 1 = 0$$
$$k = \frac{4}{3} \qquad\qquad\qquad k = -1$$

Verify that the solutions are $4/3$ and -1.

CAUTION It is possible to get, as a solution of a rational equation, a number that makes one or more of the denominators in the original equation equal to zero. That number is not a solution, so it is *necessary* to check all potential solutions of rational equations. These introduced solutions are called **extraneous solutions.** ■

EXAMPLE 7 *Solving a Rational Equation*

Solve $\dfrac{2}{x - 3} + \dfrac{1}{x} = \dfrac{6}{x(x - 3)}$.

Solution The common denominator is $x(x - 3)$. Multiply both sides by $x(x - 3)$ and solve the resulting equation.

$$\frac{2}{x - 3} + \frac{1}{x} = \frac{6}{x(x - 3)}$$
$$2x + x - 3 = 6$$
$$3x = 9$$
$$x = 3$$

Checking this potential solution by substitution in the original equation shows that 3 makes two denominators 0. Thus 3 cannot be a solution, so there is no solution for this equation.

R.4 EXERCISES

Solve each equation.

1. $.2m - .5 = .1m + .7$

2. $\dfrac{5}{6}k - 2k + \dfrac{1}{3} = \dfrac{2}{3}$

3. $2x + 8 = x - 4$

4. $5x + 2 = 8 - 3x$

5. $3r + 2 - 5(r + 1) = 6r + 4$

6. $5(a + 3) + 4a - 5 = -(2a - 4)$

7. $2[m - (4 + 2m) + 3] = 2m + 2$

8. $4[2p - (3 - p) + 5] = -7p - 2$

Solve each equation by factoring or by using the quadratic formula. If the solutions involve square roots, give both the exact solutions and the approximate solutions to three decimal places.

9. $x^2 + 5x + 6 = 0$

10. $x^2 = 3 + 2x$

11. $m^2 + 16 = 8m$

12. $2k^2 - k = 10$

13. $6x^2 - 5x = 4$

14. $m(m - 7) = -10$

15. $9x^2 - 16 = 0$

16. $z(2z + 7) = 4$

17. $12y^2 - 48y = 0$

18. $3x^2 - 5x + 1 = 0$

19. $2m^2 = m + 4$

20. $p^2 + p - 1 = 0$

21. $k^2 - 10k = -20$

22. $2x^2 + 12x + 5 = 0$

23. $2r^2 - 7r + 5 = 0$

24. $2x^2 - 7x + 30 = 0$

25. $3k^2 + k = 6$

26. $5m^2 + 5m = 0$

Solve each equation.

27. $\frac{3x - 2}{7} = \frac{x + 2}{5}$

28. $\frac{x}{3} - 7 = 6 - \frac{3x}{4}$

29. $\frac{4}{x - 3} - \frac{8}{2x + 5} + \frac{3}{x - 3} = 0$

30. $\frac{5}{2p + 3} - \frac{3}{p - 2} = \frac{4}{2p + 3}$

31. $\frac{2}{m} + \frac{m}{m + 3} = \frac{3m}{m^2 + 3m}$

32. $\frac{2y}{y - 1} = \frac{5}{y} + \frac{10 - 8y}{y^2 - y}$

33. $\frac{1}{x - 2} - \frac{3x}{x - 1} = \frac{2x + 1}{x^2 - 3x + 2}$

34. $\frac{5}{a} + \frac{-7}{a + 1} = \frac{a^2 - 2a + 4}{a^2 + a}$

35. $\frac{2b^2 + 5b - 8}{b^2 + 2b} + \frac{5}{b + 2} = -\frac{3}{b}$

36. $\frac{2}{x^2 - 2x - 3} + \frac{5}{x^2 - x - 6} = \frac{1}{x^2 + 3x + 2}$

37. $\frac{2}{y^2 + 7y + 12} - \frac{1}{y^2 + 5y + 6} = \frac{5}{y^2 + 6y + 8}$

R.5 INEQUALITIES

To write that one number is greater than or less than another number, we use the following symbols.

INEQUALITY SYMBOLS

$<$ means *is less than* $\leq$ means *is less than or equal to*

$>$ means *is greater than* $\geq$ means *is greater than or equal to*

Linear Inequalities An equation states that two expressions are equal; an **inequality** states that they are unequal. A **linear inequality** is an inequality that can be simplified to the form $ax < b$. (Properties introduced in this section are given only for $<$, but they are equally valid for $>$, $\leq$, or $\geq$.) Linear inequalities are solved with the following properties.

PROPERTIES OF INEQUALITY

For all real numbers a, b, and c:

1. If $a < b$, then $a + c < b + c$.
2. If $a < b$ and if $c > 0$, then $ac < bc$.
3. If $a < b$ and if $c < 0$, then $ac > bc$.

Pay careful attention to property 3; it says that if both sides of an inequality are multiplied by a negative number, the direction of the inequality symbol must be reversed.

EXAMPLE 1 *Solving a Linear Inequality*
Solve $4 - 3y \le 7 + 2y$.

Solution Use the properties of inequality.

$$4 - 3y + (-4) \le 7 + 2y + (-4) \quad \text{Add } -4 \text{ to both sides.}$$
$$-3y \le 3 + 2y$$

Remember that *adding* the same number to both sides never changes the direction of the inequality symbol.

$$-3y + (-2y) \le 3 + 2y + (-2y) \quad \text{Add } -2y \text{ to both sides.}$$
$$-5y \le 3$$

Multiply both sides by $-1/5$. Since $-1/5$ is negative, change the direction of the inequality symbol.

$$-\frac{1}{5}(-5y) \ge -\frac{1}{5}(3)$$
$$y \ge -\frac{3}{5}$$

CAUTION It is a common error to forget to reverse the direction of the inequality sign when multiplying or dividing by a negative number. For example, to solve $-4x \le 12$, we must multiply by $-1/4$ on both sides *and* reverse the inequality symbol to get $x \ge -3$. ■

The solution $y \ge -3/5$ in Example 1 represents an interval on the number line. **Interval notation** often is used for writing intervals. With interval notation, $y \ge -3/5$ is written as $[-3/5, \infty)$. This is an example of a **half-open interval,** since one endpoint, $-3/5$, is included. The **open interval** $(2, 5)$ corresponds to $2 < x < 5$, with neither endpoint included. The **closed interval** $[2, 5]$ includes both endpoints and corresponds to $2 \le x \le 5$.

The **graph** of an interval shows all points on a number line that correspond to the numbers in the interval. To graph the interval $[-3/5, \infty)$, for example, use a solid circle at $-3/5$, since $-3/5$ is part of the solution. To show that the solution includes all real numbers greater than or equal to $-3/5$, draw a heavy arrow pointing to the right (the positive direction). See Figure 1.

FIGURE 1

EXAMPLE 2 *Graphing a Linear Inequality*
Solve $-2 < 5 + 3m < 20$. Graph the solution.

Solution The inequality $-2 < 5 + 3m < 20$ says that $5 + 3m$ is *between* -2 and 20. Solve this inequality with an extension of the properties given above. Work as follows, first adding -5 to each part.

$$-2 + (-5) < 5 + 3m + (-5) < 20 + (-5)$$
$$-7 < 3m < 15$$

Now multiply each part by $1/3$.

$$-\frac{7}{3} < m < 5$$

A graph of the solution is given in Figure 2; here open circles are used to show that $-7/3$ and 5 are *not* part of the graph.*

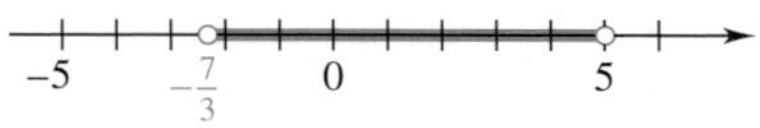

FIGURE 2

Quadratic Inequalities A **quadratic inequality** has the form $ax^2 + bx + c > 0$ (or $<$, or $\le$, or $\ge$). The highest exponent is 2. The next few examples show how to solve quadratic inequalities.

EXAMPLE 3 *Solving a Quadratic Inequality*

Solve the quadratic inequality $x^2 - x < 12$.

Solution Write the inequality with 0 on one side, as $x^2 - x - 12 < 0$. This inequality is solved with values of x that make $x^2 - x - 12$ negative (<0). The quantity $x^2 - x - 12$ changes from positive to negative or from negative to positive at the points where it equals 0. For this reason, first solve the *equation* $x^2 - x - 12 = 0$.

$$x^2 - x - 12 = 0$$
$$(x - 4)(x + 3) = 0$$
$$x = 4 \quad \text{or} \quad x = -3$$

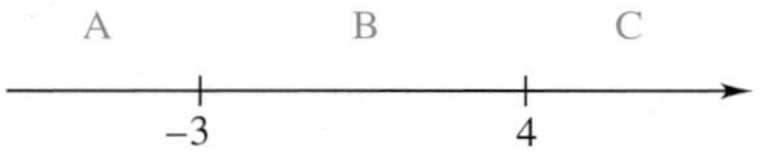

FIGURE 3

Locating -3 and 4 on a number line, as shown in Figure 3, determines three intervals A, B, and C. Decide which intervals include numbers that make $x^2 - x - 12$ negative by substituting any number from each interval in the polynomial. For example,

choose -4 from interval A: $(-4)^2 - (-4) - 12 = 8 > 0$;

choose 0 from interval B: $0^2 - 0 - 12 = -12 < 0$;

choose 5 from interval C: $5^2 - 5 - 12 = 8 > 0$.

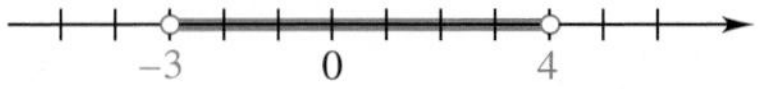

FIGURE 4

Only numbers in interval B satisfy the given inequality, so the solution is $(-3, 4)$. A graph of this solution is shown in Figure 4.

EXAMPLE 4 *Solving a Polynomial Inequality*

Solve the inequality $x(x - 1)(x + 3) \ge 0$.

Solution This is not a quadratic inequality. If the three factors are multiplied, the highest-degree term is x^3. However, it can be solved in the same way as a quadratic inequality because it is in factored form. First solve the corresponding equation.

$$x(x - 1)(x + 3) = 0$$
$$x = 0 \quad \text{or} \quad x - 1 = 0 \quad \text{or} \quad x + 3 = 0$$
$$x = 1 \qquad\qquad x = -3$$

These three solutions determine four intervals on the number line: $(-\infty, -3)$, $(-3, 0)$, $(0, 1)$, and $(1, \infty)$. Substitute a number from each interval into the original inequality to determine that the solution consists of the numbers between -3

*Some textbooks use brackets in place of solid circles for the graph of a closed interval, and parentheses in place of open circles for the graph of an open interval.

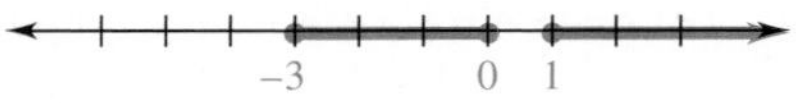

FIGURE 5

and 0 (including the endpoints) and all numbers that are greater than or equal to 1. See Figure 5. In interval notation, the solution is

$$[-3, 0] \cup [1, \infty).*$$

Inequalities with Fractions   Inequalities with fractions are solved in a similar manner as quadratic inequalities.

EXAMPLE 5   *Solving a Rational Inequality*

Solve $\dfrac{2x - 3}{x} \geq 1$.

Solution   First solve the corresponding equation.

$$\frac{2x - 3}{x} = 1$$
$$2x - 3 = x$$
$$x = 3$$

The solution, $x = 3$, determines the intervals on the number line where the fraction may change from greater than 1 to less than 1. This change also may occur on either side of a number that makes the denominator equal 0. Here, the x-value that makes the denominator 0 is $x = 0$. Test each of the three intervals determined by the numbers 0 and 3.

$$\text{For } (-\infty, 0), \text{ choose } -1\text{:} \frac{2(-1) - 3}{-1} = 5 \geq 1.$$

$$\text{For } (0, 3), \quad \text{choose } \quad 1\text{:} \frac{2(1) - 3}{1} = -1 \ngeq 1.$$

$$\text{For } (3, \infty), \quad \text{choose } \quad 4\text{:} \frac{2(4) - 3}{4} = \frac{5}{4} \geq 1.$$

The symbol $\ngeq$ means "is *not* greater than or equal to." Testing the endpoints 0 and 3 shows that the solution is $(-\infty, 0) \cup [3, \infty)$.

CAUTION A common error is to try to solve the inequality in Example 5 by multiplying both sides by x. The reason this is wrong is that we don't know in the beginning whether x is positive or negative. If x is negative, the $\geq$ would change to $\leq$ according to the third property of inequality listed at the beginning of this section. ■

EXAMPLE 6   *Solving a Rational Inequality*

Solve $\dfrac{(x - 1)(x + 1)}{x} \leq 0$.

*The symbol $\cup$ indicates the *union* of two sets, which includes all elements in either set.

Solution We first solve the corresponding equation.

$$\frac{(x - 1)(x + 1)}{x} = 0$$

$$(x - 1)(x + 1) = 0 \qquad \text{Multiply both sides by } x.$$

$$x = 1 \quad \text{or} \quad x = -1 \qquad \text{Use the zero-factor property.}$$

Setting the denominator equal to 0 gives $x = 0$, so the intervals of interest are $(-\infty, -1)$, $(-1, 0)$, and $(0, \infty)$. Testing a number from each region in the original inequality and checking the endpoints, we find the solution is

$$(-\infty, -1] \cup (0, 1].$$

CAUTION Remember to solve the equation formed by setting the *denominator* equal to zero. Any number that makes the denominator zero always creates two intervals on the number line. For instance, in Example 6, substituting $x = 0$ makes the denominator of the rational inequality equal to 0, so we know that there may be a sign change from one side of 0 to the other (as was indeed the cse). ■

EXAMPLE 7 *Solving a Rational Inequality*

Solve $\dfrac{x^2 - 3x}{x^2 - 9} < 4$.

Solution Solve the corresponding equation.

$$\frac{x^2 - 3x}{x^2 - 9} = 4$$

$$x^2 - 3x = 4x^2 - 36 \qquad \text{Multiply by } x^2 - 9.$$

$$0 = 3x^2 + 3x - 36 \qquad \text{Get 0 on one side.}$$

$$0 = x^2 + x - 12 \qquad \text{Multiply by } \tfrac{1}{3}.$$

$$0 = (x + 4)(x - 3) \qquad \text{Factor.}$$

$$x = -4 \quad \text{or} \quad x = 3$$

Now set the denominator equal to 0 and solve that equation.

$$x^2 - 9 = 0$$

$$(x - 3)(x + 3) = 0$$

$$x = 3 \quad \text{or} \quad x = -3$$

The intervals determined by the three (different) solutions are $(-\infty, -4)$, $(-4, -3)$, $(-3, 3)$, and $(3, \infty)$. Testing a number from each interval in the given inequality shows that the solution is

$$(-\infty, -4) \cup (-3, 3) \cup (3, \infty).$$

For this example, none of the endpoints are part of the solution because $x = 3$ and $x = -3$ make the denominator zero and $x = -4$ produces an equality.

R.5 EXERCISES

Write each expression in interval notation. Graph each interval.

1. $x < 0$ **2.** $x \geq -3$ **3.** $1 \leq x < 2$

4. $-5 < x \leq -4$ **5.** $-9 > x$ **6.** $6 \leq x$

Using the variable x, write each interval as an inequality.

7. $(-4, 3)$ **8.** $[2, 7)$ **9.** $(-\infty, -1]$ **10.** $(3, \infty)$

11.

12.

13.
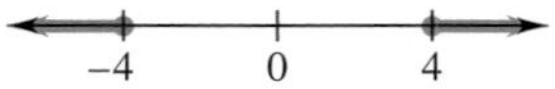

14.

Solve each inequality and graph the solution.

15. $-3p - 2 \geq 1$ **16.** $6k - 4 < 3k - 1$

17. $m - (4 + 2m) + 3 < 2m + 2$ **18.** $-2(3y - 8) \geq 5(4y - 2)$

19. $3p - 1 < 6p + 2(p - 1)$ **20.** $x + 5(x + 1) > 4(2 - x) + x$

21. $-7 < y - 2 < 4$ **22.** $8 \leq 3r + 1 \leq 13$

23. $-4 \leq \dfrac{2k - 1}{3} \leq 2$ **24.** $-1 \leq \dfrac{5y + 2}{3} \leq 4$

25. $\dfrac{3}{5}(2p + 3) \geq \dfrac{1}{10}(5p + 1)$ **26.** $\dfrac{8}{3}(z - 4) \leq \dfrac{2}{9}(3z + 2)$

Solve each quadratic inequality. Graph each solution.

27. $(m + 2)(m - 4) < 0$ **28.** $(t + 6)(t - 1) \geq 0$ **29.** $y^2 - 3y + 2 < 0$

30. $2k^2 + 7k - 4 > 0$ **31.** $q^2 - 7q + 6 \leq 0$ **32.** $2k^2 - 7k - 15 \leq 0$

33. $6m^2 + m > 1$ **34.** $10r^2 + r \leq 2$ **35.** $2y^2 + 5y \leq 3$

36. $3a^2 + a > 10$ **37.** $x^2 \leq 25$ **38.** $p^2 - 16p > 0$

Solve each inequality.

39. $\dfrac{m - 3}{m + 5} \leq 0$ **40.** $\dfrac{r + 1}{r - 1} > 0$ **41.** $\dfrac{k - 1}{k + 2} > 1$

42. $\dfrac{a - 5}{a + 2} < -1$ **43.** $\dfrac{2y + 3}{y - 5} \leq 1$ **44.** $\dfrac{a + 2}{3 + 2a} \leq 5$

45. $\dfrac{7}{k + 2} \geq \dfrac{1}{k + 2}$ **46.** $\dfrac{5}{p + 1} > \dfrac{12}{p + 1}$ **47.** $\dfrac{3x}{x^2 - 1} < 2$

48. $\dfrac{8}{p^2 + 2p} > 1$ **49.** $\dfrac{z^2 + z}{z^2 - 1} \geq 3$ **50.** $\dfrac{a^2 + 2a}{a^2 - 4} \leq 2$

R.6 EXPONENTS

Integer Exponents Recall that $a^2 = a \cdot a$, while $a^3 = a \cdot a \cdot a$, and so on. In this section a more general meaning is given to the symbol a^n.

DEFINITION OF EXPONENT

If n is a natural number, then

$$a^n = a \cdot a \cdot a \cdot \cdots \cdot a,$$

where a appears as a factor n times.

In the expression a^n, the power n is the **exponent** and a is the **base.** This definition can be extended by defining a^n for zero and negative integer values of n.

ZERO AND NEGATIVE EXPONENTS

If a is any nonzero real number, and if n is a positive integer, then

$$a^0 = 1 \quad \text{and} \quad a^{-n} = \frac{1}{a^n}.$$

(The symbol 0^0 is meaningless.)

EXAMPLE 1 *Exponents*

(a) $6^0 = 1$

(b) $(-9)^0 = 1$

(c) $3^{-2} = \frac{1}{3^2} = \frac{1}{9}$

(d) $9^{-1} = \frac{1}{9^1} = \frac{1}{9}$

(e) $\left(\frac{3}{4}\right)^{-1} = \frac{1}{(3/4)^1} = \frac{1}{3/4} = \frac{4}{3}$

The following properties follow from the definitions of exponents given above.

PROPERTIES OF EXPONENTS

For any integers m and n, and any real numbers a and b for which the following exist:

1. $a^m \cdot a^n = a^{m+n}$

2. $\frac{a^m}{a^n} = a^{m-n}$

3. $(a^m)^n = a^{mn}$

4. $(ab)^m = a^m \cdot b^m$

5. $\left(\frac{a}{b}\right)^m = \frac{a^m}{b^m}$.

EXAMPLE 2 *Simplifying Exponential Expressions*

Use the properties of exponents to simplify each expression. Leave answers with positive exponents. Assume that all variables represent positive real numbers.

(a) $7^4 \cdot 7^6 = 7^{4+6} = 7^{10}$ (or 282,475,249) — Property 1

(b) $\dfrac{9^{14}}{9^6} = 9^{14-6} = 9^8$ (or 43,046,721) — Property 2

(c) $\dfrac{r^9}{r^{17}} = r^{9-17} = r^{-8} = \dfrac{1}{r^8}$ — Property 2

(d) $(2m^3)^4 = 2^4 \cdot (m^3)^4 = 16m^{12}$ — Properties 3 and 4

(e) $(3x)^4 = 3^4 \cdot x^4 = 81x^4$ — Property 4

(f) $\left(\dfrac{x^2}{y^3}\right)^6 = \dfrac{(x^2)^6}{(y^3)^6} = \dfrac{x^{2\cdot 6}}{y^{3\cdot 6}} = \dfrac{x^{12}}{y^{18}}$ — Properties 4 and 5

(g) $\dfrac{a^{-3}b^5}{a^4b^{-7}} = \dfrac{b^{5-(-7)}}{a^{4-(-3)}} = \dfrac{b^{5+7}}{a^{4+3}} = \dfrac{b^{12}}{a^7}$ — Property 2

(h) $p^{-1} + q^{-1} = \dfrac{1}{p} + \dfrac{1}{q} = \dfrac{1}{p} \cdot \dfrac{q}{q} + \dfrac{1}{q} \cdot \dfrac{p}{p} = \dfrac{q}{pq} + \dfrac{p}{pq} = \dfrac{p+q}{pq}$

(i)

$$\frac{x^{-2} - y^{-2}}{x^{-1} - y^{-1}} = \frac{\dfrac{1}{x^2} - \dfrac{1}{y^2}}{\dfrac{1}{x} - \dfrac{1}{y}}$$ Definition of a^{-n}

$$= \frac{\dfrac{y^2 - x^2}{x^2y^2}}{\dfrac{y - x}{xy}}$$ Get common denominators and combine terms.

$$= \frac{y^2 - x^2}{x^2y^2} \cdot \frac{xy}{y - x}$$ Invert and multiply.

$$= \frac{(y - x)(y + x)}{x^2y^2} \cdot \frac{xy}{y - x}$$ Factor.

$$= \frac{x + y}{xy}$$ Simplify.

CAUTION If Example 2(e) were written $3x^4$, the properties of exponents would not apply. When no parentheses are used, the exponent refers only to the factor closest to it. Also notice in Examples 2(c), 2(g), 2(h), and 2(i) that a negative exponent does *not* indicate a negative number. ■

Roots For *even* values of n, the expression $a^{1/n}$ is defined to be the **positive *n*th root** of a or the **principal *n*th root** of a. For example, $a^{1/2}$ denotes the positive second root, or **square root,** of a, while $a^{1/4}$ is the positive fourth root of a. When n is *odd,* there is only one nth root, which has the same sign as a. For example, $a^{1/3}$, the **cube root** of a, has the same sign as a. By definition, if $b = a^{1/n}$, then $b^n = a$. On a calculator, a number is raised to a power using a key labeled x^y, y^x, or $\wedge$. For example, to take the fourth root of 6 on a TI-83/84 Plus calculator, enter $6 \wedge (1/4)$, to get the result 1.56508458.

EXAMPLE 3 *Calculations with Exponents*

(a) $121^{1/2} = 11$, since 11 is positive and $11^2 = 121$.

(b) $625^{1/4} = 5$, since $5^4 = 625$.

(c) $256^{1/4} = 4$

(d) $64^{1/6} = 2$

(e) $27^{1/3} = 3$

(f) $(-32)^{1/5} = -2$

(g) $128^{1/7} = 2$

(h) $(-49)^{1/2}$ is not a real number.

Rational Exponents In the following definition, the domain of an exponent is extended to include all rational numbers.

DEFINITION OF $a^{m/n}$

For all real numbers a for which the indicated roots exist, and for any rational number m/n,

$$a^{m/n} = (a^{1/n})^m.$$

EXAMPLE 4 *Calculations with Exponents*

(a) $27^{2/3} = (27^{1/3})^2 = 3^2 = 9$

(b) $32^{2/5} = (32^{1/5})^2 = 2^2 = 4$

(c) $64^{4/3} = (64^{1/3})^4 = 4^4 = 256$

(d) $25^{3/2} = (25^{1/2})^3 = 5^3 = 125$

NOTE $27^{2/3}$ could also be evaluated as $(27^2)^{1/3}$, but this is more difficult to perform without a calculator because it involves squaring 27 and then taking the cube root of this large number. On the other hand, when we evaluate it as $(27^{1/3})^2$, we know that the cube root of 27 is 3 without using a calculator, and squaring 3 is easy. ■

All the properties for integer exponents given in this section also apply to any rational exponent on a nonnegative real-number base.

EXAMPLE 5 *Simplifying Exponential Expressions*

(a) $\dfrac{y^{1/3}y^{5/3}}{y^3} = \dfrac{y^{1/3+5/3}}{y^3} = \dfrac{y^2}{y^3} = y^{2-3} = y^{-1} = \dfrac{1}{y}$

(b) $m^{2/3}(m^{7/3} + 2m^{1/3}) = m^{2/3+7/3} + 2m^{2/3+1/3} = m^3 + 2m$

(c) $\left(\dfrac{m^7n^{-2}}{m^{-5}n^2}\right)^{1/4} = \left(\dfrac{m^{7-(-5)}}{n^{2-(-2)}}\right)^{1/4} = \left(\dfrac{m^{12}}{n^4}\right)^{1/4} = \dfrac{(m^{12})^{1/4}}{(n^4)^{1/4}} = \dfrac{m^{12/4}}{n^{4/4}} = \dfrac{m^3}{n}$

In calculus, it is often necessary to factor expressions involving fractional exponents.

EXAMPLE 6 *Simplifying Exponential Expressions*

Factor out the smallest power of the variable, assuming all variables represent positive real numbers.

(a) $4m^{1/2} + 3m^{3/2} = m^{1/2}(4 + 3m)$

Solution To check this result, multiply $m^{1/2}$ by $4 + 3m$.

(b) $9x^{-2} - 6x^{-3}$

Solution The smallest exponent here is -3. Since 3 is a common numerical factor, factor out $3x^{-3}$.

$$9x^{-2} - 6x^{-3} = 3x^{-3}(3x^{-2-(-3)} - 2x^{-3-(-3)}) = 3x^{-3}(3x - 2)$$

Check by multiplying. The factored form can be written without negative exponents as

$$\frac{3(3x - 2)}{x^3}.$$

(c) $(x^2 + 5)(3x - 1)^{-1/2}(2) + (3x - 1)^{1/2}(2x)$.

Solution There is a common factor of 2. Also, $(3x - 1)^{-1/2}$ and $(3x - 1)^{1/2}$ have a common factor. Always factor out the quantity to the *smallest* exponent. Here $-1/2 < 1/2$, so the common factor is $2(3x - 1)^{-1/2}$ and the factored form is

$$2(3x - 1)^{-1/2}[(x^2 + 5) + (3x - 1)x] = 2(3x - 1)^{-1/2}(4x^2 - x + 5).$$

R.6 EXERCISES

Evaluate each expression. Write all answers without exponents.

1. 8^{-2} **2.** 3^{-4} **3.** 5^0 **4.** $(-12)^0$

5. $-(-3)^{-2}$ **6.** $-(-3^{-2})$ **7.** $\left(\dfrac{2}{7}\right)^{-2}$ **8.** $\left(\dfrac{4}{3}\right)^{-3}$

Simplify each expression. Assume that all variables represent positive real numbers. Write answers with only positive exponents.

9. $\dfrac{3^{-4}}{3^2}$ **10.** $\dfrac{8^9 \cdot 8^{-7}}{8^{-3}}$ **11.** $\dfrac{10^8 \cdot 10^{-10}}{10^4 \cdot 10^2}$ **12.** $\left(\dfrac{5^{-6} \cdot 5^3}{5^{-2}}\right)^{-1}$

13. $\dfrac{x^4 \cdot x^3}{x^5}$ **14.** $\dfrac{y^9 \cdot y^7}{y^{13}}$ **15.** $\dfrac{(4k^{-1})^2}{2k^{-5}}$ **16.** $\dfrac{(3z^2)^{-1}}{z^5}$

17. $\dfrac{2^{-1}x^3y^{-3}}{xy^{-2}}$

18. $\dfrac{5^{-2}m^2y^{-2}}{5^2m^{-1}y^{-2}}$

19. $\left(\dfrac{a^{-1}}{b^2}\right)^{-3}$

20. $\left(\dfrac{2c^2}{d^3}\right)^{-2}$

21. $\left(\dfrac{x^6y^{-3}}{x^{-2}y^5}\right)^{1/2}$

22. $\left(\dfrac{a^{-7}b^{-1}}{b^{-4}a^2}\right)^{1/3}$

Simplify each expression, writing the answer as a single term without negative exponents.

23. $a^{-1} + b^{-1}$

24. $b^{-2} - a$

25. $\dfrac{2n^{-1} - 2m^{-1}}{m + n^2}$

26. $\left(\dfrac{m}{3}\right)^{-1} + \left(\dfrac{n}{2}\right)^{-2}$

27. $(x^{-1} - y^{-1})^{-1}$

28. $(x^{-2} + y^{-2})^{-2}$

Write each number without exponents.

29. $81^{1/2}$

30. $27^{1/3}$

31. $32^{2/5}$

32. $-125^{2/3}$

33. $\left(\dfrac{4}{9}\right)^{1/2}$

34. $\left(\dfrac{64}{27}\right)^{1/3}$

35. $16^{-5/4}$

36. $625^{-1/4}$

37. $\left(\dfrac{27}{64}\right)^{-1/3}$

38. $\left(\dfrac{121}{100}\right)^{-3/2}$

Simplify each expression. Write all answers with only positive exponents. Assume that all variables represent positive real numbers.

39. $2^{1/2} \cdot 2^{3/2}$

40. $27^{2/3} \cdot 27^{-1/3}$

41. $\dfrac{4^{2/3} \cdot 4^{5/3}}{4^{1/3}}$

42. $\dfrac{3^{-5/2} \cdot 3^{3/2}}{3^{7/2} \cdot 3^{-9/2}}$

43. $\dfrac{7^{-1/3} \cdot 7r^{-3}}{7^{2/3} \cdot (r^{-2})^2}$

44. $\dfrac{12^{3/4} \cdot 12^{5/4} \cdot y^{-2}}{12^{-1} \cdot (y^{-3})^{-2}}$

45. $\dfrac{6k^{-4} \cdot (3k^{-1})^{-2}}{2^3 \cdot k^{1/2}}$

46. $\dfrac{8p^{-3} \cdot (4p^2)^{-2}}{p^{-5}}$

47. $\dfrac{a^{4/3} \cdot b^{1/2}}{a^{2/3} \cdot b^{-3/2}}$

48. $\dfrac{x^{1/3} \cdot y^{2/3} \cdot z^{1/4}}{x^{5/3} \cdot y^{-1/3} \cdot z^{3/4}}$

49. $\dfrac{k^{-3/5} \cdot h^{-1/3} \cdot t^{2/5}}{k^{-1/5} \cdot h^{-2/3} \cdot t^{1/5}}$

50. $\dfrac{m^{7/3} \cdot n^{-2/5} \cdot p^{3/8}}{m^{-2/3} \cdot n^{3/5} \cdot p^{-5/8}}$

Factor each expression.

51. $12x^2(x^2 + 2)^2 - 4x(4x^3 + 1)(x^2 + 2)$

52. $6x(x^3 + 7)^2 - 6x^2(3x^2 + 5)(x^3 + 7)$

53. $(x^2 + 2)(x^2 - 1)^{-1/2}(x) + (x^2 - 1)^{1/2}(2x)$

54. $9(6x + 2)^{1/2} + 3(9x - 1)(6x + 2)^{-1/2}$

55. $x(2x + 5)^2(x^2 - 4)^{-1/2} + 2(x^2 - 4)^{1/2}(2x + 5)$

56. $(4x^2 + 1)^2(2x - 1)^{-1/2} + 16x(4x^2 + 1)(2x - 1)^{1/2}$

R.7 RADICALS

We have defined $a^{1/n}$ as the positive or principal nth root of a for appropriate values of a and n. An alternative notation for $a^{1/n}$ uses radicals.

RADICALS

If n is an even natural number and $a > 0$, or n is an odd natural number, then

$$a^{1/n} = \sqrt[n]{a}.$$

The symbol $\sqrt[n]{}$ is a **radical sign,** the number a is the **radicand,** and n is the **index** of the radical. The familiar symbol $\sqrt{a}$ is used instead of $\sqrt[2]{a}$.

EXAMPLE 1 *Radical Calculations*

(a) $\sqrt[4]{16} = 16^{1/4} = 2$

(b) $\sqrt[5]{-32} = -2$

(c) $\sqrt[3]{1000} = 10$

(d) $\sqrt[6]{\dfrac{64}{729}} = \dfrac{2}{3}$

With $a^{1/n}$ written as $\sqrt[n]{a}$, the expression $a^{m/n}$ also can be written using radicals.

$$a^{m/n} = (\sqrt[n]{a})^m \qquad \text{or} \qquad a^{m/n} = \sqrt[n]{a^m}$$

The following properties of radicals depend on the definitions and properties of exponents.

PROPERTIES OF RADICALS

For all real numbers a and b and natural numbers m and n such that $\sqrt[n]{a}$ and $\sqrt[n]{b}$ are real numbers:

1. $(\sqrt[n]{a})^n = a$
2. $\sqrt[n]{a^n} = \begin{cases} |a| & \text{if } n \text{ is even} \\ a & \text{if } n \text{ is odd} \end{cases}$
3. $\sqrt[n]{a} \cdot \sqrt[n]{b} = \sqrt[n]{ab}$
4. $\dfrac{\sqrt[n]{a}}{\sqrt[n]{b}} = \sqrt[n]{\dfrac{a}{b}} \qquad (b \neq 0)$
5. $\sqrt[m]{\sqrt[n]{a}} = \sqrt[mn]{a}$

Property 3 can be used to simplify certain radicals. For example, since $48 = 16 \cdot 3$,

$$\sqrt{48} = \sqrt{16 \cdot 3} = \sqrt{16} \cdot \sqrt{3} = 4\sqrt{3}.$$

To some extent, simplification is in the eye of the beholder, and $\sqrt{48}$ might be considered as simple as $4\sqrt{3}$. In this textbook, we will consider an expression to be simpler when we have removed as many factors as possible from under the radical.

EXAMPLE 2 *Radical Calculations*

(a) $\sqrt{1000} = \sqrt{100 \cdot 10} = \sqrt{100} \cdot \sqrt{10} = 10\sqrt{10}$

(b) $\sqrt{128} = \sqrt{64 \cdot 2} = 8\sqrt{2}$

(c) $\sqrt{108} = \sqrt{36 \cdot 3} = 6\sqrt{3}$

(d) $\sqrt[3]{54} = \sqrt[3]{27 \cdot 2} = \sqrt[3]{27} \cdot \sqrt[3]{2} = 3\sqrt[3]{2}$

(e) $\sqrt{288m^5} = \sqrt{144 \cdot m^4 \cdot 2m} = 12m^2\sqrt{2m}$

(f)
$$\begin{aligned} 2\sqrt{18} - 5\sqrt{32} &= 2\sqrt{9 \cdot 2} - 5\sqrt{16 \cdot 2} \\ &= 2\sqrt{9} \cdot \sqrt{2} - 5\sqrt{16} \cdot \sqrt{2} \\ &= 2(3)\sqrt{2} - 5(4)\sqrt{2} = -14\sqrt{2} \end{aligned}$$

Rationalizing Denominators The next example shows how to *rationalize* (remove all radicals from) the denominator in an expression containing radicals.

EXAMPLE 3 *Rationalizing Denominators*

Simplify each expression by rationalizing the denominator.

(a) $\dfrac{4}{\sqrt{3}}$

Solution To rationalize the denominator, multiply by $\sqrt{3}/\sqrt{3}$ (or 1) so that the denominator of the product is a rational number.

$$\frac{4}{\sqrt{3}} \cdot \frac{\sqrt{3}}{\sqrt{3}} = \frac{4\sqrt{3}}{3} \qquad \sqrt{3} \cdot \sqrt{3} = \sqrt{9} = 3$$

(b) $\dfrac{2}{\sqrt[3]{x}}$

Solution Here, we need a perfect cube under the radical sign to rationalize the denominator. Multiplying by $\sqrt[3]{x^2}/\sqrt[3]{x^2}$ gives

$$\frac{2}{\sqrt[3]{x}} \cdot \frac{\sqrt[3]{x^2}}{\sqrt[3]{x^2}} = \frac{2\sqrt[3]{x^2}}{\sqrt[3]{x^3}} = \frac{2\sqrt[3]{x^2}}{x}.$$

(c) $\dfrac{1}{1 - \sqrt{2}}$

Solution The best approach here is to multiply both numerator and denominator by the number $1 + \sqrt{2}$. The expressions $1 + \sqrt{2}$ and $1 - \sqrt{2}$ are conjugates,* and their product is $1^2 - (\sqrt{2})^2 = 1 - 2 = -1$. Thus,

$$\frac{1}{1 - \sqrt{2}} = \frac{1(1 + \sqrt{2})}{(1 - \sqrt{2})(1 + \sqrt{2})} = \frac{1 + \sqrt{2}}{1 - 2} = -1 - \sqrt{2}.$$

Sometimes it is advantageous to rationalize the *numerator* of a rational expression. The following example arises in calculus when evaluating a *limit.*

*If a and b are real numbers, the *conjugate* of $a + b$ is $a - b$.

xlviii ■ Chapter R Algebra Reference

EXAMPLE 4 *Rationalizing Numerators*
Rationalize each numerator.

(a) $\dfrac{\sqrt{x} - 3}{x - 9}$.

Solution Multiply the numerator and denominator by the conjugate of the numerator, $\sqrt{x} + 3$.

$$\frac{\sqrt{x} - 3}{x - 9} \cdot \frac{\sqrt{x} + 3}{\sqrt{x} + 3} = \frac{(\sqrt{x})^2 - 3^2}{(x - 9)(\sqrt{x} + 3)} \qquad (a - b)(a + b) = a^2 - b^2$$

$$= \frac{x - 9}{(x - 9)(\sqrt{x} + 3)}$$

$$= \frac{1}{\sqrt{x} + 3}$$

(b) $\dfrac{\sqrt{3} + \sqrt{x + 3}}{\sqrt{3} - \sqrt{x + 3}}$

Solution Multiply the numerator and denominator by the conjugate of the numerator, $\sqrt{3} - \sqrt{x + 3}$.

$$\frac{\sqrt{3} + \sqrt{x + 3}}{\sqrt{3} - \sqrt{x + 3}} \cdot \frac{\sqrt{3} - \sqrt{x + 3}}{\sqrt{3} - \sqrt{x + 3}} = \frac{3 - (x + 3)}{3 - 2\sqrt{3}\sqrt{x + 3} + (x + 3)}$$

$$= \frac{-x}{6 + x - 2\sqrt{3(x + 3)}}$$

When simplifying a square root, keep in mind that $\sqrt{x}$ is positive by definition. Also, $\sqrt{x^2}$ is not x, but $|x|$, the **absolute value of x,** defined as

$$|x| = \begin{cases} x & \text{if } x \geq 0 \\ -x & \text{if } x < 0. \end{cases}$$

For example, $\sqrt{(-5)^2} = |-5| = 5$.

EXAMPLE 5 *Simplifying by Factoring*
Simplify $\sqrt{m^2 - 4m + 4}$.

Solution Factor the polynomial as $m^2 - 4m + 4 = (m - 2)^2$. Then by property 2 of radicals and the definition of absolute value,

$$\sqrt{(m - 2)^2} = |m - 2| = \begin{cases} m - 2 & \text{if } m - 2 \geq 0 \\ -(m - 2) = 2 - m & \text{if } m - 2 < 0. \end{cases}$$

CAUTION Avoid the common error of writing $\sqrt{a^2 + b^2}$ as $\sqrt{a^2} + \sqrt{b^2}$. We must add a^2 and b^2 *before* taking the square root. For example, $\sqrt{16 + 9} =$

$\sqrt{25} = 5$, *not* $\sqrt{16} + \sqrt{9} = 4 + 3 = 7$. This idea applies as well to higher roots. For example, in general,

$$\sqrt[3]{a^3 + b^3} \neq \sqrt[3]{a^3} + \sqrt[3]{b^3},$$
$$\sqrt[4]{a^4 + b^4} \neq \sqrt[4]{a^4} + \sqrt[4]{b^4}.$$

Also, $$\sqrt{a + b} \neq \sqrt{a} + \sqrt{b}.$$ ■

R.7 EXERCISES

Simplify each expression by removing as many factors as possible from under the radical. Assume that all variables represent positive real numbers.

1. $\sqrt[3]{125}$
2. $\sqrt[4]{1296}$
3. $\sqrt[5]{-3125}$
4. $\sqrt{50}$
5. $\sqrt{2000}$
6. $\sqrt{32y^5}$
7. $7\sqrt{2} - 8\sqrt{18} + 4\sqrt{72}$
8. $4\sqrt{3} - 5\sqrt{12} + 3\sqrt{75}$
9. $2\sqrt{5} - 3\sqrt{20} + 2\sqrt{45}$
10. $3\sqrt{28} - 4\sqrt{63} + \sqrt{112}$
11. $\sqrt[3]{2} - \sqrt[3]{16} + 2\sqrt[3]{54}$
12. $2\sqrt[3]{3} + 4\sqrt[3]{24} - \sqrt[3]{81}$
13. $\sqrt[3]{32} - 5\sqrt[3]{4} + 2\sqrt[3]{108}$
14. $\sqrt{2x^3y^2z^4}$
15. $\sqrt{98r^3s^4t^{10}}$
16. $\sqrt[3]{16x^8y^4z^5}$
17. $\sqrt[4]{x^8y^7z^{11}}$
18. $\sqrt{a^3b^5} - 2\sqrt{a^7b^3} + \sqrt{a^3b^9}$
19. $\sqrt{p^7q^3} - \sqrt{p^5q^9} + \sqrt{p^9q}$

Rationalize each denominator. Assume that all radicands represent positive real numbers.

20. $\dfrac{5}{\sqrt{7}}$
21. $\dfrac{-2}{\sqrt{3}}$
22. $\dfrac{-3}{\sqrt{12}}$
23. $\dfrac{4}{\sqrt{8}}$
24. $\dfrac{3}{1 - \sqrt{5}}$
25. $\dfrac{5}{2 - \sqrt{6}}$
26. $\dfrac{-2}{\sqrt{3} - \sqrt{2}}$
27. $\dfrac{1}{\sqrt{10} + \sqrt{3}}$
28. $\dfrac{1}{\sqrt{r} - \sqrt{3}}$
29. $\dfrac{5}{\sqrt{m} - \sqrt{5}}$
30. $\dfrac{y - 5}{\sqrt{y} - \sqrt{5}}$
31. $\dfrac{z - 11}{\sqrt{z} - \sqrt{11}}$
32. $\dfrac{\sqrt{x} + \sqrt{x + 1}}{\sqrt{x} - \sqrt{x + 1}}$
33. $\dfrac{\sqrt{p} + \sqrt{p^2 - 1}}{\sqrt{p} - \sqrt{p^2 - 1}}$

Rationalize each numerator. Assume that all radicands represent positive real numbers.

34. $\dfrac{1 + \sqrt{2}}{2}$
35. $\dfrac{1 - \sqrt{3}}{3}$
36. $\dfrac{\sqrt{x} + \sqrt{x + 1}}{\sqrt{x} - \sqrt{x + 1}}$
37. $\dfrac{\sqrt{p} + \sqrt{p^2 - 1}}{\sqrt{p} - \sqrt{p^2 - 1}}$

Simplify each root, if possible.

38. $\sqrt{16 - 8x + x^2}$
39. $\sqrt{4y^2 + 4y + 1}$
40. $\sqrt{4 - 25z^2}$
41. $\sqrt{9k^2 + h^2}$

CHAPTER

1 Linear Functions

Over short time intervals, many changes in the economy are well modeled by linear functions. In an exercise in the first section of this chapter we will examine a linear model that predicts airline passenger traffic in the year 2005 at some of the fastest-growing airports in the United States. Such predictions are important tools for airline executives and airport planners.

Before using mathematics to solve a real-world problem, we must usually set up a **mathematical model,** a mathematical description of the situation. Constructing such a model requires a solid understanding of the situation to be modeled, as well as familiarity with relevant mathematical ideas and techniques.

Much mathematical theory is available for building models, but the very richness and diversity of contemporary mathematics often prevents people in other fields from finding the mathematical tools they need. There are so many useful parts of mathematics that it can be hard to know which to choose.

To avoid this problem, it is helpful to have a thorough understanding of the most basic and useful mathematical tools that are available for constructing mathematical models. In this chapter we look at some mathematics of *linear* models, which are used for data whose graphs can be approximated by straight lines.

1.1 SLOPES AND EQUATIONS OF LINES

THINK ABOUT IT *How fast has tuition at public colleges been increasing in recent years, and how well can we predict tuition in the future?*

In Example 15 of this section, we will answer these questions using the equation of a line.

There are many everyday situations in which two quantities are related. For example, if a bank account pays 6% simple interest per year, then the interest I that a deposit of P dollars would earn in one year is given by

$$I = .06 \cdot P, \qquad \text{or} \qquad I = .06P.$$

The formula $I = .06P$ describes the relationship between interest and the amount of money deposited.

Using this formula, we see, for example, that if $P = \$100$, then $I = \$6$, and if $I = \$12$, then $P = \$200$. These corresponding pairs of numbers can be written as **ordered pairs,** $(100, 6)$ and $(200, 12)$, pairs of numbers whose order is important. The first number denotes the value of P and the second number the value of I.

Ordered pairs are **graphed** with the perpendicular number lines of a **Cartesian coordinate system,** shown in Figure 1. The horizontal number line, or ***x*-axis,** represents the first components of the ordered pairs, while the vertical or ***y*-axis** represents the second components. The point where the number lines cross is the zero point on both lines; this point is called the **origin.**

The name "Cartesian" honors René Descartes (1596–1650), one of the greatest mathematicians of the seventeenth century. According to legend, Descartes was lying in bed when he noticed an insect crawling on the ceiling and realized that if he could determine the distance from the bug to each of two perpendicular walls, he could describe its position at any given moment. The same idea can be used to locate a point in a plane.

Each point on the xy-plane corresponds to an ordered pair of numbers, where the x-value is written first. From now on, we will refer to the point corresponding to the ordered pair (a, b) as "the point (a, b)."

Locate the point $(-2, 4)$ on the coordinate system by starting at the origin and counting 2 units to the left on the horizontal axis and 4 units upward, parallel to the vertical axis. This point is shown in Figure 1, along with several other sample points. The number -2 is the ***x*-coordinate** and the number 4 is the ***y*-coordinate** of the point $(-2, 4)$.

The x-axis and y-axis divide the plane into four parts, or **quadrants.** For example, quadrant I includes all those points whose x- and y-coordinates are both positive. The quadrants are numbered as shown in Figure 1. The points on the axes themselves belong to no quadrant. The set of points corresponding to the ordered pairs of an equation is the **graph** of the equation.

The x- and y-values of the points where the graph of an equation crosses the axes are called the ***x*-intercept** and ***y*-intercept,** respectively.* See Figure 2.

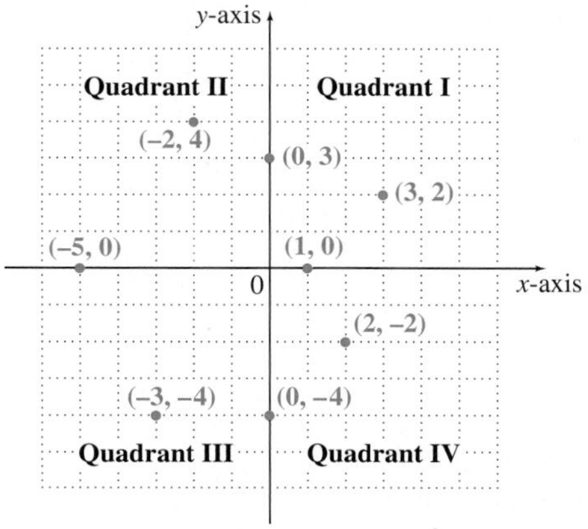

FIGURE 1

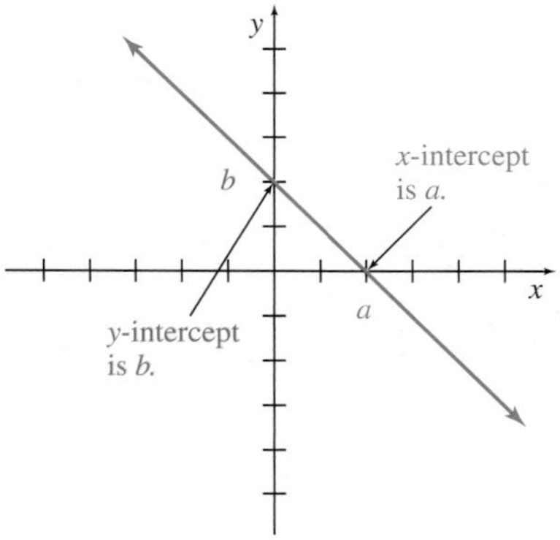

FIGURE 2

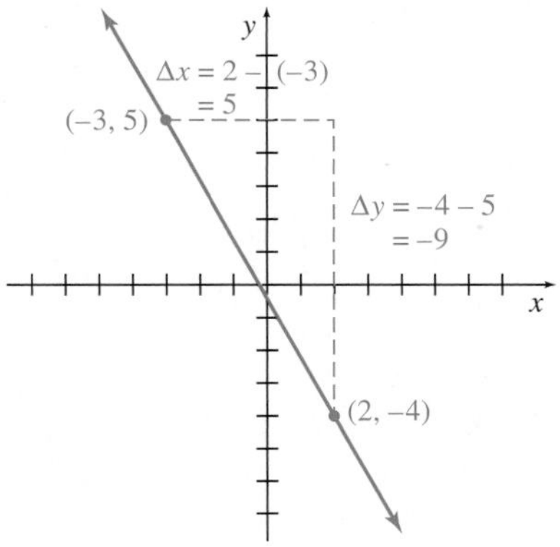

FIGURE 3

Slope of a Line

An important characteristic of a straight line is its *slope,* a number that represents the "steepness" of the line. To see how slope is defined, look at the line in Figure 3. The line goes through the points $(x_1, y_1) = (-3, 5)$ and $(x_2, y_2) = (2, -4)$. The difference in the two x-values,

$$x_2 - x_1 = 2 - (-3) = 5$$

in this example, is called the **change in *x*.** The symbol Δx (read "delta x") is used to represent the change in x. In the same way, Δy represents the **change in *y*.** In our example,

$$\begin{aligned}\Delta y &= y_2 - y_1 \\ &= -4 - 5 \\ &= -9.\end{aligned}$$

These symbols, Δx and Δy, are used in the following definition of slope.

SLOPE OF A LINE

The **slope** of a line is defined as the vertical change (the "rise") over the horizontal change (the "run") as one travels along the line. In symbols, taking two different points (x_1, y_1) and (x_2, y_2) on the line, the slope is

$$m = \frac{\textbf{Change in } y}{\textbf{Change in } x} = \frac{\Delta y}{\Delta x} = \frac{y_2 - y_1}{x_2 - x_1},$$

where $x_1 \neq x_2$.

*Some people prefer to define the intercepts as ordered pairs, rather than as numbers.

By this definition, the slope of the line in Figure 3 is

$$m = \frac{\Delta y}{\Delta x} = \frac{-4 - 5}{2 - (-3)} = -\frac{9}{5}.$$

The slope of a line tells how fast y changes for each unit of change in x.

NOTE Using similar triangles, it can be shown that the slope of a line is independent of the choice of points on the line. That is, the same slope will be obtained for *any* choice of two different points on the line.   ■

EXAMPLE 1   *Slope*

Find the slope of the line through each pair of points.

(a) $(-7, 6)$ and $(4, 5)$

Solution Let $(x_1, y_1) = (-7, 6)$ and $(x_2, y_2) = (4, 5)$. Use the definition of slope.

$$m = \frac{\Delta y}{\Delta x} = \frac{5 - 6}{4 - (-7)} = -\frac{1}{11}$$

(b) $(5, -3)$ and $(-2, -3)$

Solution Let $(x_1, y_1) = (5, -3)$ and $(x_2, y_2) = (-2, -3)$. Then

$$m = \frac{-3 - (-3)}{-2 - 5} = \frac{0}{-7} = 0.$$

Lines with zero slope are horizontal (parallel to the x-axis).

(c) $(2, -4)$ and $(2, 3)$

Solution Let $(x_1, y_1) = (2, -4)$ and $(x_2, y_2) = (2, 3)$. Then

$$m = \frac{3 - (-4)}{2 - 2} = \frac{7}{0},$$

which is undefined. This happens when the line is vertical (parallel to the y-axis).   ■

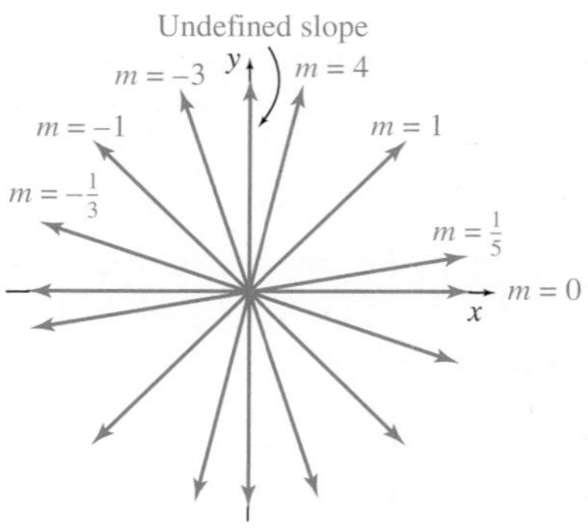

CAUTION The phrase "no slope" should be avoided; specify instead whether the slope is zero or undefined.   ■

In finding the slope of the line in Example 1(a), we could have let $(x_1, y_1) = (4, 5)$ and $(x_2, y_2) = (-7, 6)$. In that case,

$$m = \frac{6 - 5}{-7 - 4} = \frac{1}{-11} = -\frac{1}{11},$$

the same answer as before. The order in which coordinates are subtracted does not matter, as long as it is done consistently.

Figure 4 shows examples of lines with different slopes. Lines with positive slopes go up from left to right, while lines with negative slopes go down from left to right.

FIGURE 4

It might help you to compare slope with the percent grade of a hill. If a sign says a hill has a 10% grade uphill, this means the slope is .10, or 1/10, so the hill rises 1 foot for every 10 feet horizontally. A 15% grade downhill means the slope is $-.15$.

FOR REVIEW

For review on solving a linear equation, see Section R.4.

Equations of a Line An equation in two first-degree variables, such as $4x + 7y = 20$, has a line as its graph, so it is called a **linear equation.** In the rest of this section, we consider various forms of the equation of a line.

EXAMPLE 2 *Equation of a Line*

Find the equation of the line through $(0, -3)$ with slope $3/4$.

Solution We can use the definition of slope, letting $(x_1, y_1) = (0, -3)$ and (x, y) represent another point on the line.

$$m = \frac{y_2 - y_1}{x_2 - x_1}$$

$$\frac{3}{4} = \frac{y - (-3)}{x - 0} = \frac{y + 3}{x} \quad \text{Substitute.}$$

$$3x = 4(y + 3) \quad \text{Cross multiply.}$$

$$3x = 4y + 12$$

A generalization of the method of Example 2 can be used to find the equation of any line, given its y-intercept and slope. Assume that a line has y-intercept b, so that it goes through the point $(0, b)$. Let the slope of the line be represented by m. If (x, y) is any point on the line *other* than $(0, b)$, then the definition of slope can be used with the points $(0, b)$ and (x, y) to get

$$m = \frac{y - b}{x - 0}$$

$$m = \frac{y - b}{x}$$

$$mx = y - b$$

$$y = mx + b.$$

This result is called the *slope-intercept form* of the equation of a line, because b is the y-intercept of the graph of the line.

SLOPE-INTERCEPT FORM

If a line has slope m and y-intercept b, then the equation of the line in **slope-intercept form** is

$$y = mx + b.$$

When $b = 0$, we say that y is **proportional** to x.

EXAMPLE 3 *Slope-Intercept Form*

Find the equation of the line in slope-intercept form having y-intercept $7/2$ and slope $-5/2$.

Solution Use the slope-intercept form with $b = 7/2$ and $m = -5/2$.

$$y = mx + b$$
$$y = -\frac{5}{2}x + \frac{7}{2}$$

The slope-intercept form shows that we can find the slope of a line by solving its equation for y. In that form the coefficient of x is the slope and the constant term is the y-intercept. For instance, in Example 2 the slope of the line $3x = 4y + 12$ was given as $3/4$. This slope also could be found by solving the equation for y.

$$4y + 12 = 3x$$
$$4y = 3x - 12$$
$$y = \frac{3}{4}x - 3$$

The coefficient of x, $3/4$, is the slope of the line. The y-intercept is -3.

The slope-intercept form of the equation of a line involves the slope and the y-intercept. Sometimes, however, the slope of a line is known, together with one point (perhaps *not* the y-intercept) that the line goes through. The *point-slope form* of the equation of a line is used to find the equation in this case. Let (x_1, y_1) be any fixed point on the line and let (x, y) represent any other point on the line. If m is the slope of the line, then by the definition of slope,

$$\frac{y - y_1}{x - x_1} = m,$$

or

$$y - y_1 = m(x - x_1). \quad \text{Multiply both sides by } x - x_1.$$

POINT-SLOPE FORM

If a line has slope m and passes through the point (x_1, y_1), then an equation of the line is given by

$$\boldsymbol{y - y_1 = m(x - x_1),}$$

the **point-slope form** of the equation of a line.

EXAMPLE 4 *Point-Slope Form*

Find an equation of the line that passes through the point $(3, -7)$ and has slope $m = 5/4$.

Solution Use the point-slope form.

$$y - y_1 = m(x - x_1)$$
$$y - (-7) = \frac{5}{4}(x - 3) \quad y_1 = -7, m = \tfrac{5}{4}, x_1 = 3$$
$$y + 7 = \frac{5}{4}(x - 3)$$
$$4y + 28 = 5(x - 3) \quad \text{Multiply both sides by 4.}$$
$$4y + 28 = 5x - 15$$
$$4y = 5x - 43 \quad \text{Combine constants.}$$

FOR REVIEW

See Section R.4 for details on eliminating denominators in an equation.

The equation of the same line can be given in many forms. To avoid confusion, the linear equations used in the rest of this section will be written in slope-intercept form, $y = mx + b$, which is often the most useful form.

The point-slope form also can be useful to find an equation of a line if we know two different points that the line goes through. The procedure for doing this is shown in the next example.

EXAMPLE 5 *Using Point-Slope Form to Find an Equation*

Find an equation of the line through $(5, 4)$ and $(-10, -2)$.

Solution Begin by using the definition of slope to find the slope of the line that passes through the given points.

$$\text{Slope} = m = \frac{-2 - 4}{-10 - 5} = \frac{-6}{-15} = \frac{2}{5}$$

Either $(5, 4)$ or $(-10, -2)$ can be used in the point-slope form with $m = 2/5$. If $(x_1, y_1) = (5, 4)$, then

$$y - y_1 = m(x - x_1)$$

$$y - 4 = \frac{2}{5}(x - 5) \qquad y_1 = 4, m = \tfrac{2}{5}, x_1 = 5.$$

$$5y - 20 = 2(x - 5) \qquad \text{Multiply both sides by 5.}$$

$$5y - 20 = 2x - 10 \qquad \text{Distributive property}$$

$$5y = 2x + 10 \qquad \text{Add 20 to both sides.}$$

$$y = \frac{2}{5}x + 2 \qquad \text{Divide by 5 to put in slope-intercept form.}$$

Check that the same result is found if $(x_1, y_1) = (-10, -2)$.

EXAMPLE 6 *Horizontal Line*

Find an equation of the line through $(8, -4)$ and $(-2, -4)$.

Solution Find the slope.

$$m = \frac{-4 - (-4)}{-2 - 8} = \frac{0}{-10} = 0$$

Choose, say, $(8, -4)$ as (x_1, y_1).

$$y - y_1 = m(x - x_1)$$

$$y - (-4) = 0(x - 8) \qquad y_1 = -4, m = 0, x_1 = 8$$

$$y + 4 = 0 \qquad 0(x - 8) = 0$$

$$y = -4$$

Plotting the given ordered pairs and drawing a line through the points, show that the equation $y = -4$ represents a horizontal line. See Figure 5a on the next page. Every horizontal line has a slope of zero and an equation of the form $y = k$, where k is the y-value of all ordered pairs on the line.

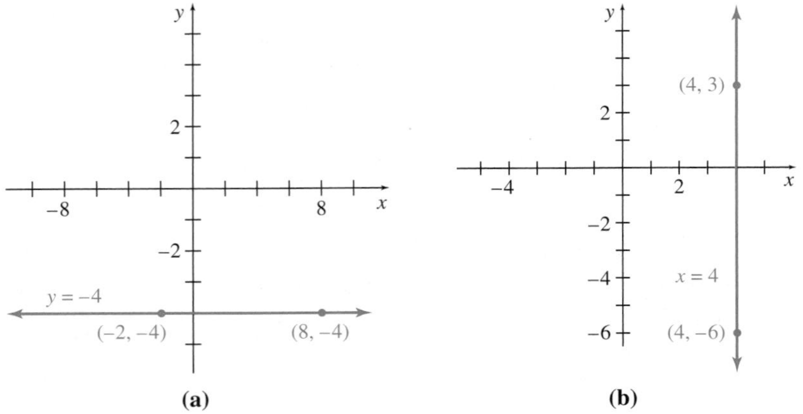

FIGURE 5

EXAMPLE 7 *Vertical Line*

Find an equation of the line through $(4, 3)$ and $(4, -6)$.

Solution The slope of the line is

$$m = \frac{-6 - 3}{4 - 4} = \frac{-9}{0},$$

which is undefined. Since both ordered pairs have x-coordinate 4, the equation is $x = 4$. Because the slope is undefined, the equation of this line cannot be written in the slope-intercept form.

Again, plotting the given ordered pairs and drawing a line through them show that the graph of $x = 4$ is a vertical line. See Figure 5(b).

> The slope of a horizontal line is 0.
> The slope of a vertical line is undefined.

The different forms of linear equations discussed in this section are summarized below. The slope-intercept and point-slope forms are equivalent ways to express the equation of a nonvertical line. The slope-intercept form is simpler for a final answer, but you may find the point-slope form easier to use when you know the slope of a line and a point through which the line passes.

EQUATIONS OF LINES

Equation	Description
$y = mx + b$	**Slope-intercept form:** slope m, y-intercept b
$y - y_1 = m(x - x_1)$	**Point-slope form:** slope m, line passes through (x_1, y_1)
$x = k$	**Vertical line:** x-intercept k, no y-intercept (except when $k = 0$), undefined slope
$y = k$	**Horizontal line:** y-intercept k, no x-intercept (except when $k = 0$), slope 0

Parallel and Perpendicular Lines One application of slope involves deciding whether two lines are parallel. Since two parallel lines are equally "steep," they should have the same slope. Also, two lines with the same "steepness" are parallel.

PARALLEL LINES

Two lines are **parallel** if and only if they have the same slope, or if they are both vertical.

EXAMPLE 8 *Parallel Line*

Find the equation of the line that passes through the point $(3, 5)$ and is parallel to the line $2x + 5y = 4$.

Solution The slope of $2x + 5y = 4$ can be found by writing the equation in slope-intercept form.

$$2x + 5y = 4$$
$$y = -\frac{2}{5}x + \frac{4}{5}$$

This result shows that the slope is $-2/5$. Since the lines are parallel, $-2/5$ is also the slope of the line whose equation we want. This line passes through $(3, 5)$. Substituting $m = -2/5$, $x_1 = 3$, and $y_1 = 5$ into the point-slope form gives

$$y - y_1 = m(x - x_1)$$
$$y - 5 = -\frac{2}{5}x + \frac{6}{5}$$
$$y = -\frac{2}{5}x + \frac{6}{5} + 5$$
$$y = -\frac{2}{5}x + \frac{31}{5}$$

As already mentioned, two nonvertical lines are parallel if and only if they have the same slope. Two lines having slopes with a product of -1 are perpendicular. A proof of this fact, which depends on similar triangles from geometry, is given as Exercise 43 in this section.

PERPENDICULAR LINES

Two lines are **perpendicular** if and only if the product of their slopes is -1, or if one is vertical and the other horizontal.

EXAMPLE 9 *Perpendicular Line*

Find the slope of the line L perpendicular to the line having the equation $5x - y = 4$.

Solution To find the slope, write $5x - y = 4$ in slope-intercept form:

$$y = 5x - 4.$$

The slope is 5. Since the lines are perpendicular, if line L has slope m, then

$$5m = -1$$
$$m = -\frac{1}{5}.$$

The next two examples use different forms of the equation of a line to analyze real-world data.

EXAMPLE 10 *Workforce*

In recent decades, the percentage of the U.S. civilian population age 16 and over that is in the workforce has risen at a roughly constant rate, from 59.4% in 1960 to 66.6% in 2002.* Find the equation describing this linear relationship.

Solution For this example, let x represent time in years, with $x = 0$ for 1960. Such rescaling of a variable is often used to simplify the arithmetic, although computers and calculators have made rescaling less important than in the past. Here it allows us to work with smaller numbers, and, as you will see, find the y-intercept of the line more easily. We will use such rescaling on many examples throughout this book. When we do, it is important to be consistent. In this example, if we want to refer to the year 1975, we must let $x = 15$, and not $x = 1975$. Let y represent the percent of the population in the workforce.

With 1960 corresponding to $x = 0$, the year 2002 corresponds to $x = 2002 - 1960 = 42$. The two ordered pairs representing the given information are $(0, 59.4)$ and $(42, 66.6)$. The slope of the line through these points is

$$m = \frac{66.6 - 59.4}{42 - 0} = \frac{7.2}{42} = .1714286 \approx .171.^{\dagger}$$

This means that, on average, the percent of the population in the workforce has gone up by about .17% per year.

Using $m = .171$ and $(x_1, y_1) = (0, 59.4)$ in the point-slope form gives the required equation,

$$y - 59.4 = .171(x - 0)$$
$$y = .171x + 59.4.$$

This result could also have been obtained by observing that $(0, 59.4)$ is the y-intercept.

Notice that if this formula is valid for all nonnegative x, then eventually y becomes 100:

$$.171x + 59.4 = 100$$
$$.171x = 40.6 \qquad \text{Subtract 59.4 from both sides.}$$
$$x = 40.6/.171 \approx 237, \qquad \text{Divide both sides by .171.}$$

which indicates that 237 years from 1960 (in the year 2197), 100% of the population will be in the workforce.

*U.S. Department of Labor, *Bureau of Labor Statistics Data.*
†The symbol ≈ means "is approximately equal to."

In Example 10, of course, it is still possible that in 2197 there will be people who are not in the workforce; the trend of recent decades may not continue. Most equations are valid for some specific set of numbers. It is highly speculative to extrapolate beyond those values. On the other hand, people in business and government often need to make some prediction about what will happen in the future, so a tentative conclusion based on past trends may be better than no conclusion at all. There are also circumstances, particularly in the physical sciences, in which theoretical reasons imply that the trend will continue.

EXAMPLE 11 *Antibiotic Resistance*

The linear equation $y = 4.9x - 9783.9$, where x represents the year, can be used to estimate the percent of the gonorrhea cases diagnosed in Hawaii from 1997 to 2001 that are resistant to the commonly prescribed antibiotic ciprofloxacin.*

(a) Determine this percent in 2003.

Solution Substitute 2003 for x in the equation.

$$\begin{aligned} y &= 4.9x - 9783.9 \\ &= 4.9(2003) - 9783.9 \\ &\approx 30.8 \end{aligned}$$

This means that about 30.8% of gonorrhea cases diagnosed in Hawaii in 2003 were resistant to the antibiotic ciprofloxacin.

(b) Find and interpret the slope of the line.

Solution The equation is given in slope-intercept form, so the slope is the coefficient of x, which is 4.9. Since

$$m = 4.9 = \frac{\text{change in } y}{\text{change in } x} = \frac{4.9}{1},$$

the slope indicates the change in the percent of ciprofloxacin-resistant cases in Hawaii per year from 1997 to 2001. Because the slope is positive, the percent of ciprofloxacin-resistant cases in Hawaii increased by 4.9% per year.

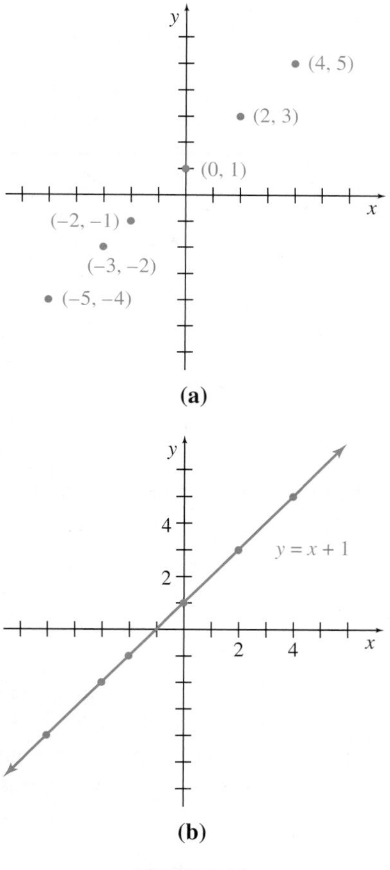

FIGURE 6

Graph of a Line We can graph the linear equation defined by $y = x + 1$ by finding several ordered pairs. For example, if $x = 2$, then $y = 2 + 1 = 3$, giving the ordered pair $(2, 3)$. Also, $(0, 1)$, $(4, 5)$, $(-2, -1)$, $(-5, -4)$, $(-3, -2)$, among many others, are ordered pairs that satisfy the equation.

To graph $y = x + 1$, we begin by locating the ordered pairs obtained above, as shown in Figure 6(a). All the points of this graph appear to lie on a straight line, as in Figure 6(b). This straight line is the graph of $y = x + 1$.

It can be shown that every equation of the form $ax + by = c$ has a straight line as its graph. Although just two points are needed to determine a line, it is a good idea to plot a third point as a check. It is often convenient to use the x- and y-intercepts as the two points, as in the following example.

*U.S. Department of Defense, Gonococcal Isolate Surveillance Project, April 2002.

EXAMPLE 12 *Graph of a Line*

Graph $3x + 2y = 12$.

Solution To find the y-intercept, let $x = 0$.

$$3(0) + 2y = 12$$
$$2y = 12 \quad \text{Divide both sides by 2.}$$
$$y = 6$$

Similarly, find the x-intercept by letting $y = 0$, which gives $x = 4$. Verify that when $x = 2$, the result is $y = 3$. These three points are plotted in Figure 7(a). A line is drawn through them in Figure 7(b).

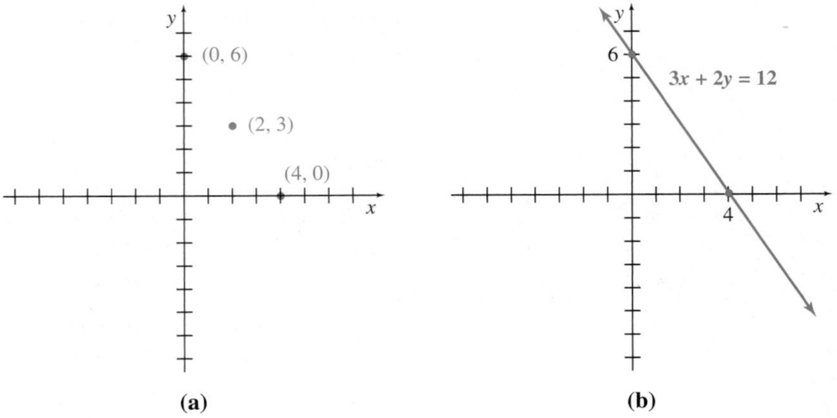

FIGURE 7

Not every line has two distinct intercepts; the graph in the next example does not cross the x-axis, and so it has no x-intercept.

EXAMPLE 13 *Graph of a Horizontal Line*

Graph $y = -3$.

Solution The equation $y = -3$, or equivalently, $y = 0x - 3$, always gives the same y-value, -3, for any value of x. Therefore, no value of x will make $y = 0$, so the graph has no x-intercept. As we saw in Example 6, the graph of such an equation is a horizontal line parallel to the x-axis. In this case the y-intercept is -3, as shown in Figure 8.

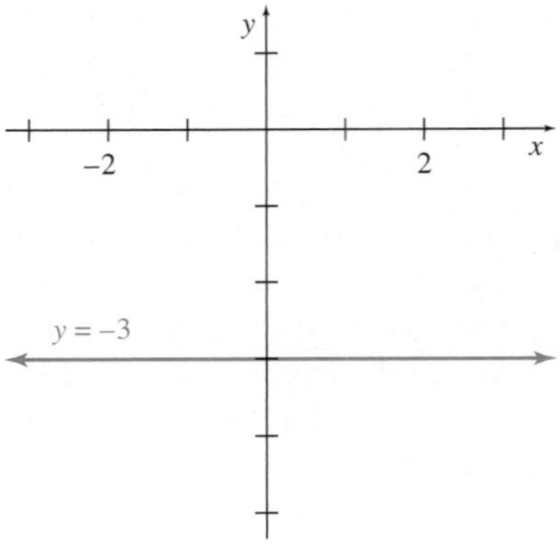

FIGURE 8

In general, the graph of $y = k$, where k is a real number, is the horizontal line having y-intercept k.

The graph in Example 13 had only one intercept. Another type of linear equation with coinciding intercepts is graphed in Example 14.

EXAMPLE 14 *Graph of a Line Through the Origin*

Graph $y = -3x$.

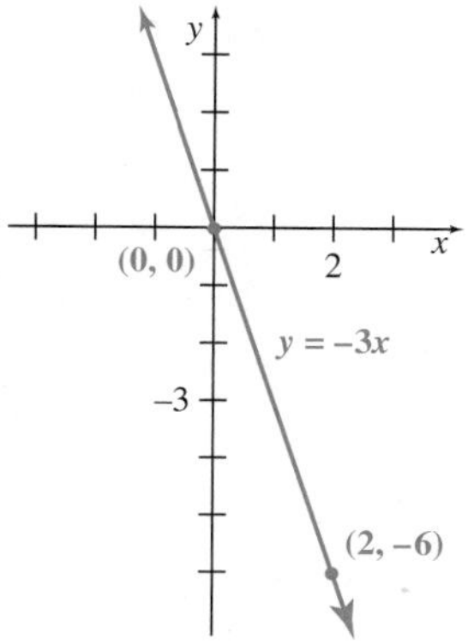

FIGURE 9

Solution Begin by looking for the x-intercept. If $y = 0$, then

$$y = -3x$$
$$0 = -3x \quad \text{Let } y = 0.$$
$$0 = x. \quad \text{Divide both sides by } -3.$$

We have the ordered pair $(0, 0)$. Starting with $x = 0$ gives exactly the same ordered pair, $(0, 0)$. Two points are needed to determine a straight line, and the intercepts have led to only one point. To get a second point, we choose some other value of x (or y). For example, if $x = 2$, then

$$y = -3x = -3(2) = -6, \quad \text{Let } x = 2.$$

giving the ordered pair $(2, -6)$. These two ordered pairs, $(0, 0)$ and $(2, -6)$, were used to get the graph shown in Figure 9.

Linear equations allow us to set up simple mathematical models for real-life situations. In almost every case, linear (or any other reasonably simple) equations provide only approximations to real-world situations. Nevertheless, these are often remarkably useful approximations.

EXAMPLE 15 *Tuition*

The table lists the average annual cost (in dollars) of tuition and fees at public four-year colleges for selected years.*

Year	Tuition and Fees
1990	1908
1992	2334
1994	2705
1996	2975
1998	3243
2000	3510
2002	4081

(a) Plot the cost of public colleges by letting $x = 90$ correspond to 1990. Is the data *exactly* linear? Could the data be *approximated* by a linear equation?

Solution Use a scale from 1500 to 4500 on the y-axis. The graph is shown in Figure 10(a) in a figure known as a **scatterplot.** Although it is not exactly linear, it is approximately linear and could be approximated by a linear equation.

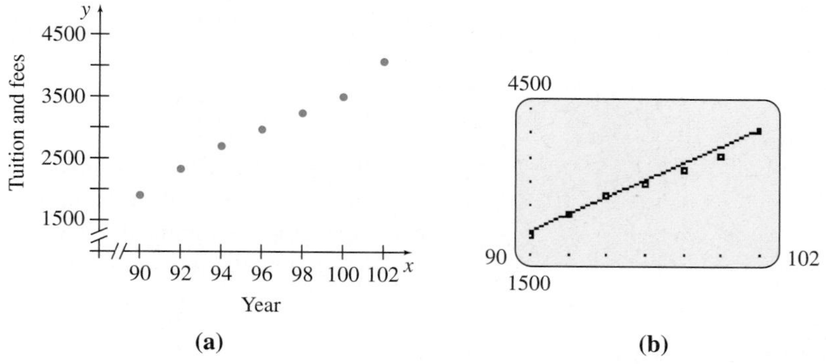

FIGURE 10

*The College Board.

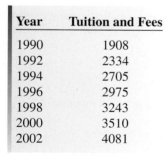

(b) Use the points $(92, 2334)$ and $(102, 4081)$ to determine an equation that models the data.

Solution After finding the slope, we substitute one of the given ordered pairs into a form of the equation of a line. Using $y = mx + b$, we proceed as follows.

The slope is

$$m = \frac{4081 - 2334}{102 - 92} = \frac{1747}{10} = 174.7.$$

From $y = mx + b$, with $x = 92$ and $y = 2334$,

$$\begin{aligned} 2334 &= 174.7(92) + b = 16{,}072.4 + b \\ b &= -13{,}738.4, \end{aligned}$$

so the equation is $y = 174.7x - 13{,}738.4$. (We could have used the other ordered pair for the values of x and y. Verify that we get the same equation.)

(c) Discuss the accuracy of using this equation to estimate the cost of public colleges in the year 2020.

Solution The year 2020 corresponds to $x = 120$, for which the equation predicts a cost of $174.7(120) - 13{,}738.4 = 7225.6$, or about \$7226. The year 2020 is many years in the future, however. Many factors could affect the tuition, and the actual figure for 2020 could be very different from our prediction.

You can plot data with a TI-83/84 Plus graphing calculator using the following steps.

1. Store the data in lists.
2. Define the stat plot.
3. Turn off $Y =$ functions (unless you also want to graph a function).
4. Turn on the plot you want to display.
5. Define the viewing window.
6. Display the graph.

Consult the calculator's instruction booklet or *The Graphing Calculator Manual* that is available with this book for specific instructions. See the calculator-generated graph in Figure 10(b), which includes the points and line from Example 15. Notice how the line closely approximates the data.

1.1 EXERCISES

Find the slope of each line that has a slope.

1. Through $(4, 5)$ and $(-1, 2)$

2. Through $(5, -4)$ and $(1, 3)$

3. Through $(8, 4)$ and $(8, -7)$

4. Through $(1, 5)$ and $(-2, 5)$

5. $y = 2x$

6. $y = 3x - 2$

7. $5x - 9y = 11$

8. $4x + 7y = 1$

9. $x = -6$

10. The x-axis

11. $y = 8$

12. $y = -4$

13. A line parallel to $2y - 4x = 7$

14. A line perpendicular to $6x = y - 3$

Find an equation in slope-intercept form (where possible) for each line in Exercises 15–34.

15. Through $(1, 3)$, $m = -2$

16. Through $(2, 4)$, $m = -1$

17. Through $(6, 1)$, $m = 0$

18. Through $(-8, 1)$, with undefined slope

19. Through $(4, 2)$ and $(1, 3)$

20. Through $(8, -1)$ and $(4, 3)$

21. Through $(1/2, 5/3)$ and $(3, 1/6)$

22. Through $(-2, 3/4)$ and $(2/3, 5/2)$

23. Through $(-8, 4)$ and $(-8, 6)$

24. Through $(-1, 3)$ and $(0, 3)$

25. x-intercept 3, y-intercept -2

26. x-intercept -2, y-intercept 4

27. Vertical, through $(-6, 5)$

28. Horizontal, through $(8, 7)$

29. Through $(-1, 4)$, parallel to $x + 3y = 5$

30. Through $(2, -5)$, parallel to $y - 4 = 2x$

31. Through $(3, -4)$, perpendicular to $x + y = 4$

32. Through $(-2, 6)$, perpendicular to $2x - 3y = 5$

33. The line with y-intercept 2 and perpendicular to $3x + 2y = 6$

34. The line with x-intercept $-2/3$ and perpendicular to $2x - y = 4$

35. Do the points $(4, 3)$, $(2, 0)$, and $(-18, -12)$ lie on the same line? (*Hint:* Find the slopes between the points.)

36. Find k so that the line through $(4, -1)$ and $(k, 2)$ is

a. parallel to $2x + 3y = 6$,

b. perpendicular to $5x - 2y = -1$.

37. Use slopes to show that the quadrilateral with vertices at $(1, 3)$, $(-5/2, 2)$, $(-7/2, 4)$, and $(2, 1)$ is a parallelogram.

38. Use slopes to show that the square with vertices at $(-2, 5)$, $(4, 5)$, $(4, -1)$, and $(-2, -1)$ has diagonals that are perpendicular.

For the lines in Exercises 39 and 40, which of the following is closest to the slope of the line? (a) 1 (b) 2 (c) 3 (d) 21 (e) 22 (f) -3

39.

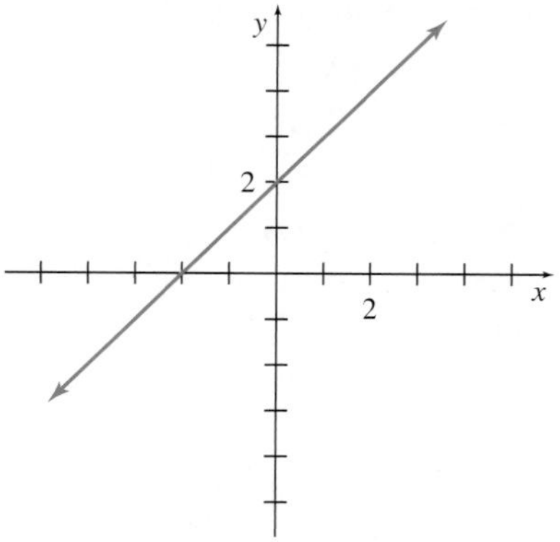

40.

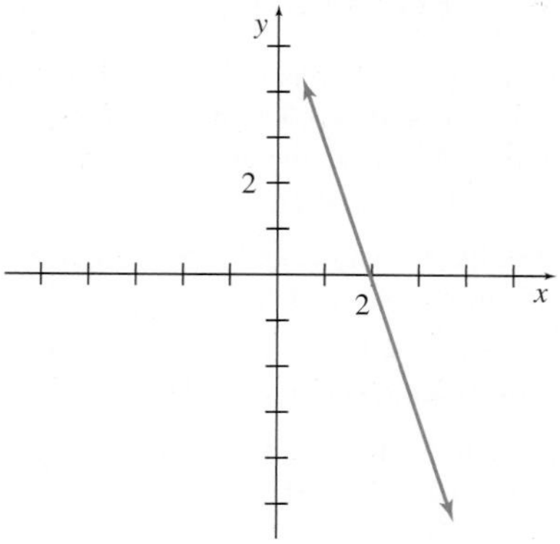

Estimate the slope of the lines in Exercises 41 and 42.

41.

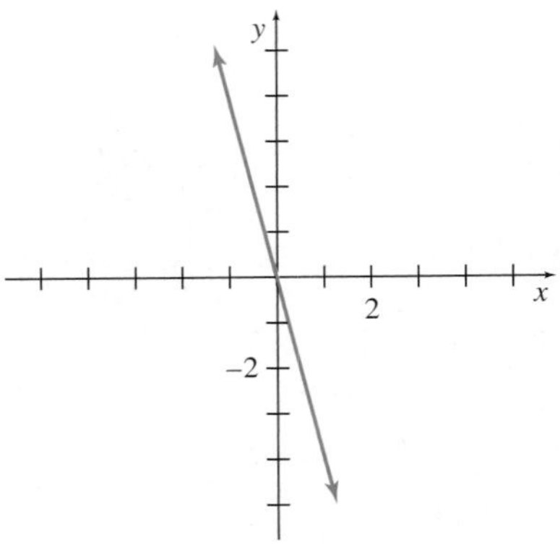

42.

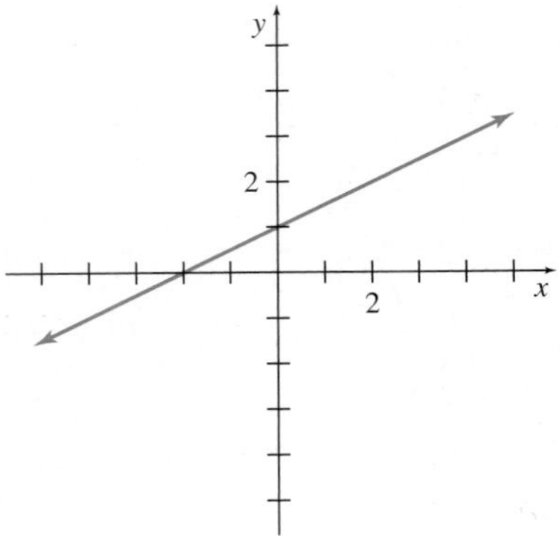

43. To show that two perpendicular lines, neither of which is vertical, have slopes with a product of −1, go through the following steps. Let line L_1 have equation $y = m_1x + b_1$, and let L_2 have equation $y = m_2x + b_2$. Assume that L_1 and L_2 are perpendicular, and use right triangle *MPN* shown in the figure. Prove each of the following statements.

a. *MQ* has length m_1.

b. *QN* has length $-m_2$.

c. Triangles *MPQ* and *PNQ* are similar.

d. $m_1/1 = 1/(-m_2)$ and $m_1m_2 = -1$

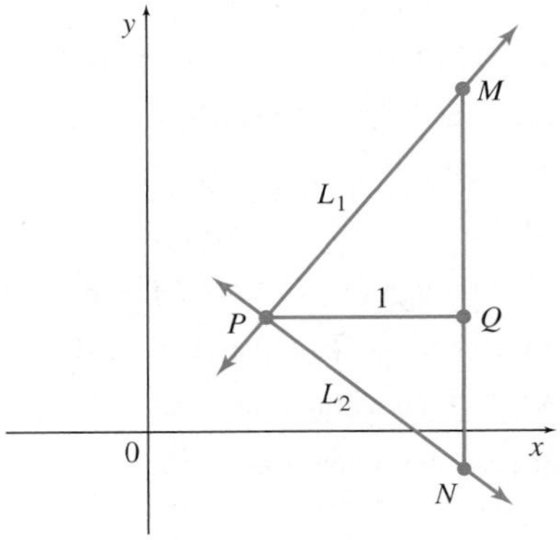

Graph each equation.

44. $y = x - 1$

45. $y = 2x + 3$

46. $y = -4x + 9$

47. $y = -6x + 12$

48. $2x - 3y = 12$

49. $3x - y = -9$

50. $3y + 4x = 12$

51. $4y + 5x = 10$

52. $y = -2$

53. $x = 4$

54. $x + 5 = 0$

55. $y - 4 = 0$

56. $y = 2x$

57. $y = -5x$

58. $x + 4y = 0$

59. $x - 3y = 0$

Applications

BUSINESS AND ECONOMICS

60. *Sales* The sales of a small company were \$27,000 in its second year of operation and \$63,000 in its fifth year. Let y represent sales in the xth year of operation. Assume that the data can be approximated by a straight line.

a. Find the slope of the sales line, and give an equation for the line in the form $y = mx + b$.

b. Use your answer from part a to find out how many years must pass before the sales surpass \$100,000.

61. *Federal Debt* The table lists the total federal debt (in trillions of dollars) from 1991 to 2001.*

Year	Federal Debt
1991	3.599
1992	4.002
1993	4.351
1994	4.644
1995	4.921
1996	5.182
1997	5.370
1998	5.479
1999	5.606
2000	5.629
2001	5.770

*Economic Research, Federal Reserve Bank of St. Louis, Fiscal Year 2002.

a. Plot the data by letting $x = 0$ correspond to 1990. Discuss any trends of the federal debt over this time period.

b. Find a linear equation that approximates the data, using the points $(1, 3.599)$ and $(11, 5.770)$. What does the slope of the graph represent? Graph the line and the data on the same coordinate axes.

c. Use the equation from part b to predict the federal debt in the year 2002. Compare your result to the actual value of 6.199 trillion dollars.

62. *Airline Passenger Growth* The following table estimates the growth in the number of airline passengers (in millions) at some of the fastest-growing airports in the United States between 1992 and 2005.*

Airport	1992	2005
Harrisburg Intl.	.7	1.4
Dayton Intl.	1.1	2.4
Austin Robert Mueller	2.2	4.7
Milwaukee Gen. Mitchell Intl.	2.2	4.4
Sacramento Metropolitan	2.6	5.0
Fort Lauderdale–Hollywood	4.1	8.1
Washington Dulles Intl.	5.3	10.9
Greater Cincinnati Airport	5.8	12.3

a. Determine a linear equation that approximates the data using the points $(.7, 1.4)$ and $(5.3, 10.9)$.

b. In 1992, 4.9 million passengers used Raleigh–Durham International Airport. Using the equation from part a, approximate the number of passengers using this airport in 2005 and compare it with the Federal Aviation Administration's estimate of 10.3 million passengers.

LIFE SCIENCES

63. *HIV Infection* The time interval between a person's initial infection with HIV and that person's eventual development of AIDS symptoms is an important issue. The method of infection with HIV affects the time interval before AIDS develops. One study of HIV patients who were infected by intravenous drug use found that 17% of the patients had AIDS after 4 years, and 33% had developed the disease after 7 years. The relationship between the time interval and the percentage of patients with AIDS can be modeled accurately with a linear equation.†

a. Write a linear equation $y = mx + b$ that models this data, using the ordered pairs $(4, 9.17)$ and $(7, 9.33)$.

b. Use your equation from part a to predict the number of years before half of these patients will have AIDS.

64. *Exercise Heart Rate* To achieve the maximum benefit for the heart when exercising, your heart rate (in beats per minute) should be in the target heart rate zone. The lower limit of this zone is found by taking 70% of the difference between 220 and your age. The upper limit is found by using 85%.‡

a. Find formulas for the upper and lower limits (u and l) as linear equations involving the age x.

b. What is the target heart rate zone for a 20-year-old?

c. What is the target heart rate zone for a 40-year-old?

d. Two women in an aerobics class stop to take their pulse, and are surprised to find that they have the same pulse. One woman is 36 years older than the other and is working at the upper limit of her target heart rate zone. The younger woman is working at the lower limit of her target heart rate zone. What are the ages of the two women, and what is their pulse?

e. Run for 10 minutes, take your pulse, and see if it is in your target heart rate zone. (After all, this is listed as an exercise!)

65. *Ponies Trotting* A 1991 study found that the peak vertical force on a trotting Shetland pony increased linearly with the pony's speed, and that when the force reached a critical level, the pony switched from a trot to a gallop.§ For one pony, the critical force was 1.16 times its body weight. It

*Federal Aviation Administration.

†Alcabes, P., A. Munoz, D. Vlahov, and G. Friedland, "Incubation Period of Human Immunodeficiency Virus," *Epidemiologic Review,* Vol. 15, No. 2, The Johns Hopkins University School of Hygiene and Public Health, 1993.

‡Hockey, Robert V., *Physical Fitness: The Pathway to Healthful Living,* Times Mirror/Mosby College Publishing, 1989, pp. 85–87.

§*Science*, July 19, 1991, pp. 306–308.

experienced a force of .75 times its body weight at a speed of 2 meters per second, and a force of .93 times its body weight at 3 meters per second. At what speed did the pony switch from a trot to a gallop?

66. *Life Span* Some scientists believe there is a limit to how long humans can live.* One supporting argument is that during the last century, life expectancy from age 65 has increased more slowly than life expectancy from birth, so eventually these two will be equal, at which point, according to these scientists, life expectancy should increase no further. In 1900, life expectancy at birth was 46 yr, and life expectancy at age 65 was 76. In 2000, these figures had risen to 76.9 and 82.9, respectively. In both cases, the increase in life expectancy has been linear. Using these assumptions and the data given, find the maximum life expectancy for humans.

67. *Deer Ticks* Deer ticks cause concern because they can carry Lyme disease. One study found a relationship between the density of acorns produced in the fall and the density of deer tick larvae the following spring.† The relationship can be approximated by the linear equation

$$y = 34x + 230,$$

where x is the number of acorns per square meter (m^2) in the fall, and y is the number of deer tick larvae per 400 m^2 the following spring. According to this formula, approximately how many acorns per square meter would result in 1000 deer tick larvae per 400 m^2?

SOCIAL SCIENCES

68. *Immigration* In 1974, there were 86,821 people from other countries who immigrated to the state of California. In 2000, the number of immigrants was 217,753.‡

a. If the change in foreign immigration to California is considered to be linear, write an equation expressing the number of immigrants, y, in terms of the number of years after 1974, x.

b. Use your result in part a to predict the foreign immigration to California in the year 2010.

69. *Cohabitation* The number of unmarried couples in the United States who are living together has been rising at a roughly linear rate in recent years. The number of cohabiting adults was 1.1 million in 1977 and 5.5 million in 2000.§

a. Write an equation expressing the number of cohabiting adults (in millions), y, in terms of the number of years after 1977, x.

b. Use your result in part a to predict the number of cohabiting adults in the year 2010.

70. *Older College Students* The percentage of college students who are age 35 and older has been increasing at roughly a linear rate. In 1970 the percentage was 9.6%. In 2001, an estimated 19.2% were 35 or older.‖

a. Find an equation giving the percentage of college students age 35 and older in terms of time t, where t represents the number of years since 1970.

b. If this linear trend continues, what percentage of college students will be 35 and over in 2010?

c. If this linear trend continues, in what year will the percentage of college students 35 and over reach 31%?

71. *Consumer Price Index* The Consumer Price Index (CPI) is a measure of the change in the cost of goods over time. If 1977 is used as the base year of comparison (CPI = 100 in 1977), then the CPI of 252.3 in 2000 would indicate that an item that cost $1.00 in 1977 would cost $2.52 in 2000. The CPI has been increasing at an approximately linear rate for the past 30 years.#

a. Use this information to determine a linear function for this data, letting x be the years since 1977.

b. Based on your function, what was the CPI in 1995? Compare this estimate with the actual CPI of 224.7.

c. How is the annual CPI changing?

PHYSICAL SCIENCES

72. *Global Warming* In 1990, the Intergovernmental Panel on Climate Change predicted that the average temperature on Earth would rise .3°C per decade in the absence of international controls on greenhouse emissions.** Let t measure the time in years since 1970, when the average global temperature was 15°C.

a. Find a linear equation giving the average global temperature in degrees Celsius in terms of t, the number of years since 1970.

b. Scientists have estimated that the sea level will rise by 65 cm if the average global temperature rises to 19°C.

**Science*, Nov. 15, 1991, pp. 936–938 and *The World Almanac and Book of Facts 2003,* p. 75.
†*Science*, Vol. 281, No. 5375, July 17, 1998, pp. 350–351.
‡*Legal Immigration to California in Federal Fiscal Year 1996,* State of California Demographic Research Unit, June 1999 and *The World Almanac and Book of Facts 2003,* p. 405.
§*The New York Times,* Feb. 15, 2000, p. F8 and U.S. Census Bureau Report, March 13, 2003.
‖National Center for Education Statistics.
#U.S. Census Bureau, Income 2000, http://www.census.gov/hhes/income/income00/cpiurs.html.
***Science News,* June 23, 1990, p. 391.

According to your answer to part a, when would this occur?

73. *Galactic Distance* The table lists the distances (in megaparsecs where 1 megaparsec $\approx 3.1 \cdot 10^{19}$ km) and velocities (in kilometers per second) of four galaxies moving rapidly away from Earth.*

Galaxy	Distance	Velocity
Virga	15	1600
Ursa Minor	200	15,000
Corona Borealis	290	24,000
Bootes	520	40,000

a. Plot the data points letting x represent distance and y represent velocity. Do the points lie in an approximately linear pattern?

b. Write a linear equation $y = mx$ to model this data, using the ordered pair $(520, 40{,}000)$.

c. The galaxy Hydra has a velocity of 60,000 km per sec. Use your equation to determine how far away it is from Earth.

d. The value of m in the equation is called the *Hubble constant*. The Hubble constant can be used to estimate the age of the universe A (in years) using the formula

$$A = \frac{9.5 \times 10^{11}}{m}.$$

Approximate A using your value of m.

GENERAL INTEREST

74. *Radio Stations* The graph shows the number of U.S. radio stations on the air along with the graph of a linear equation that models the data.†

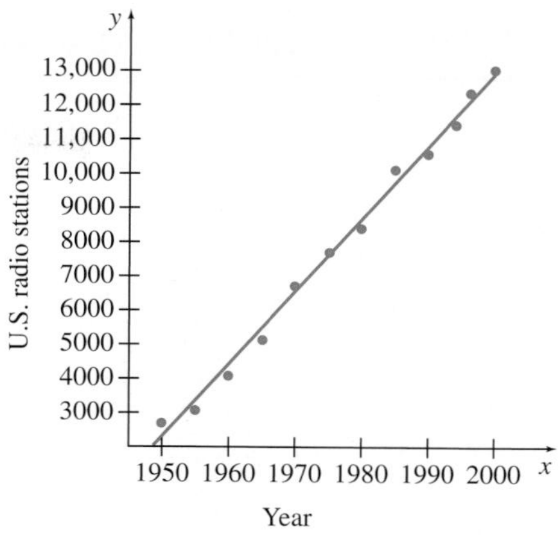

a. Use the two ordered pairs $(1950, 2773)$ and $(2000, 13{,}150)$ to find the approximate slope of the line shown. Interpret your answer.

b. Use the same two ordered pairs to write an equation of the line that models the data.

c. Estimate the year when it is expected that the number of stations will first exceed 15,000.

75. *Tuition* The table lists the average annual cost (in dollars) of tuition and fees at private four-year colleges for selected years.‡ See Example 15.

Year	Tuition and Fees
1990	9340
1992	10,449
1994	11,719
1996	12,994
1998	14,508
2000	16,332
2002	18,273

a. Sketch a graph of the data. Do the data appear to lie roughly along a straight line?

b. Let $x = 90$ correspond to 1990. Use the points $(90, 9340)$ and $(100, 16{,}332)$ to determine a linear equation that models the data. What does the slope of the graph of the equation indicate?

c. Discuss the accuracy of using this equation to estimate the cost of private college in 2020.

*Acker, A. and C. Jaschek, *Astronomical Methods and Calculations,* John Wiley & Sons, 1986; Karttunen, H. (editor), *Fundamental Astronomy,* Springer-Verlag, 1994.
†National Association of Broadcasters.
‡The College Board.

1.2 LINEAR FUNCTIONS AND APPLICATIONS

THINK ABOUT IT *How many units must be sold for a firm to break even?*

Later in this section, this question will be answered using a linear function.

As we saw in the previous section, many situations involve two variables related by a linear equation. For such a relationship, when we express the variable y in terms of x, we say that y is a **linear function** of x. This means that for any allowed value of x (the **independent variable**), we can use the equation to find the corresponding value of y (the **dependent variable**). Examples of linear functions include $y = 2x + 3$, $y = -5$, and $2x - 3y = 7$, which can be written as $y = (2/3)x - (7/3)$. Equations in the form $x = k$, where k is a constant, are not linear functions. All other linear equations define linear functions.

$f(x)$ Notation

Letters such as f, g, or h are often used to name functions. For example, f might be used to name the function

$$y = 5 - 3x.$$

To show that this function is named f, it is common to replace y with $f(x)$ (read "f of x") to get

$$f(x) = 5 - 3x.$$

By choosing 2 as a value of x, $f(x)$ becomes $5 - 3 \cdot 2 = 5 - 6 = -1$, written

$$f(2) = -1.$$

The corresponding ordered pair is $(2, -1)$. In a similar manner,

$$f(-4) = 5 - 3(-4) = 17, \qquad f(0) = 5, \qquad f(-6) = 23,$$

and so on.

EXAMPLE 1 *Function Notation*

Let $g(x) = -4x + 5$. Find $g(3)$, $g(0)$, and $g(-2)$.

Solution To find $g(3)$, substitute 3 for x.

$$g(3) = -4(3) + 5 = -12 + 5 = -7$$

Similarly,

$$g(0) = -4(0) + 5 = 0 + 5 = 5,$$

and

$$g(-2) = -4(-2) + 5 = 8 + 5 = 13.$$

We summarize the discussion below.

LINEAR FUNCTION

A relationship f defined by

$$y = f(x) = mx + b,$$

for real numbers m and b, is a **linear function.**

Supply and Demand

Linear functions are often good choices for **supply and demand curves.** Typically, as the price of an item increases, the demand for the item decreases, while the supply increases. On the other hand, when demand for an item increases, so does its price, causing the supply of the item to decrease.

For example, during the 1970s the price of gasoline increased rapidly. As the price continued to escalate, most buyers became more and more prudent in their use of gasoline in order to restrict their demand to an affordable amount. Consequently, the overall demand for gasoline decreased and the supply increased, to a point where there was an oversupply of gasoline. This caused prices to fall until supply and demand were approximately balanced. Many other factors were involved in the situation, but the relationship between price, supply, and demand was nonetheless typical. Some commodities, however, such as medical care, college education, and certain luxury items, may be exceptions to these typical relationships.

Although economists consider price to be the independent variable, they have the unfortunate habit of plotting price, usually denoted by p, on the vertical axis, while everyone else graphs the independent variable on the horizontal axis. This custom was started by the English economist Alfred Marshall (1842–1924). In order to abide by this custom, we will write p, the price, as a function of q, the quantity produced, and plot p on the vertical axis. But remember, it is really *price* that determines how much consumers demand and producers supply, not the other way around.

Supply and demand functions are not necessarily linear, the simplest kind of function. Yet most functions are approximately linear if a small enough piece of the graph is taken, allowing applied mathematicians to often use linear functions for simplicity. That approach will be taken in this chapter.

EXAMPLE 2 *Supply and Demand*

Suppose that Greg Tobin, an economist, has studied the supply and demand for vinyl siding and has determined that the price (in dollars) per square yard (yd^2), p, and the quantity demanded monthly (in thousands of square yards), q, are related by the linear function

$$p = D(q) = 60 - \frac{3}{4}q, \quad \text{Demand}$$

while the price p and the supply q are related by

$$p = S(q) = \frac{3}{4}q. \quad \text{Supply}$$

(a) Find the demand at a price of \$45 and at a price of \$18.

Solution Start with the demand function

$$p = 60 - \frac{3}{4}q,$$

> **FOR REVIEW**
>
> In the second-to-last step of the solution in Example 2(a), q was multiplied by $-3/4$, so both sides of the equation had to be divided by $-3/4$. This was done by multiplying both sides by $-4/3$. The way to divide by a fraction is to multiply by its reciprocal. In other words, to divide by the fraction a/b, multiply by b/a.

and replace p with 45.

$$45 = 60 - \frac{3}{4}q$$

$$-15 = -\frac{3}{4}q \qquad \text{Subtract 60 from both sides.}$$

$$20 = q \qquad \text{Multiply both sides by } -\tfrac{4}{3}.$$

Thus, at a price of \$45, the demand is 20,000 yd^2 per month.

Similarly, replace p with 18 to find the demand when the price is \$18. Verify that this leads to $q = 56$. When the price is lowered from \$45 to \$18, the demand increases from 20,000 yd^2 to 56,000 yd^2.

(b) Find the supply at a price of \$60 and at a price of \$12.

Solution Substitute 60 for p in the supply equation,

$$p = \frac{3}{4}q,$$

to find that $q = 80$, so the supply is 80,000 yd^2. Similarly, replacing p with 12 in the supply equation gives a supply of 16,000 yd^2. If the price decreases from \$60 to \$12, the supply also decreases, from 80,000 yd^2 to 16,000 yd^2.

(c) Graph both functions on the same axes.

Solution The results of part (a) are written as the ordered pairs $(20, 45)$ and $(56, 18)$. The line through those points is the graph of $p = 60 - (3/4)q$, shown in red in Figure 11(a). We used the ordered pairs $(80, 60)$ and $(16, 12)$ from the work in part (b) to get the supply graph shown in blue in Figure 11(a).

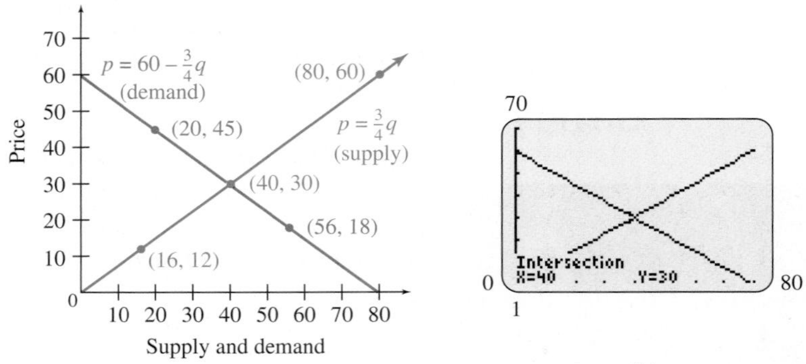

(a) (b)

FIGURE 11

A calculator-generated graph of the lines representing the supply and demand in Example 2 is shown in Figure 11(b). The equation of each line, using x and y instead of q and p, was entered along with an appropriate window to get this graph. A special menu choice gives the coordinates of the intersection point, shown at the bottom of the graph.

NOTE Not all supply/demand problems will have the same scale on both axes. It helps to consider the intercepts of both the supply and demand graphs to decide what scale to use. For example, in Figure 11, the y-intercept of the demand function is 60, so the scale should allow values from 0 to at least 60 on the vertical axis. The x-intercept of the demand function is 80, so values on the x-axis must go from 0 to 80. Letting each tick mark represent 10 gives a reasonable number of marks on each axis. ■

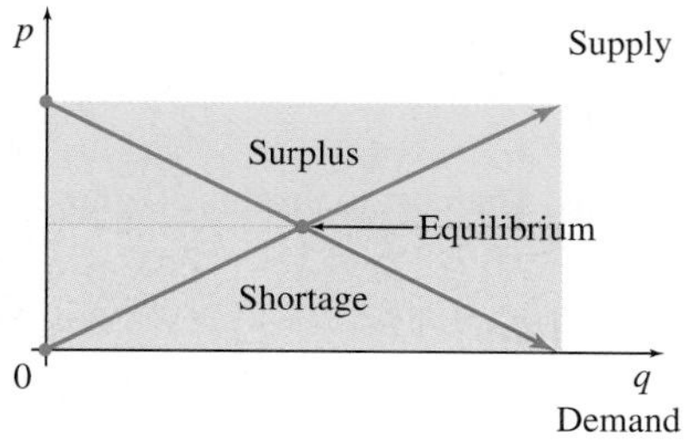

FIGURE 12

As shown in the graphs of Figure 11, both the supply and the demand graphs pass through the point $(40, 30)$. If the price of a square yard of siding is more than \$30, the supply will exceed the demand. At a price less than \$30, the demand will exceed the supply. Only at a price of \$30 will demand and supply be equal. For this reason, \$30 is called the *equilibrium price.* When the price is \$30, demand and supply both equal 40,000 yd^2, the *equilibrium quantity.* In general, the **equilibrium price** of a commodity is the price found at the point where the supply and demand graphs for that commodity intersect. The **equilibrium quantity** is the demand and the supply at that same point. Figure 12 illustrates a general supply and demand situation.

EXAMPLE 3 *Equilibrium Quantity*

Use algebra to find the equilibrium quantity for the vinyl siding in Example 2.

Solution The equilibrium quantity is found when the prices from both supply and demand are equal. Set the two expressions for p equal to each other and solve.

$$\begin{aligned} 60 - \frac{3}{4}q &= \frac{3}{4}q \\ 240 - 3q &= 3q && \text{Multiply both sides by 4.} \\ 240 &= 6q && \text{Add } 3q \text{ to both sides.} \\ 40 &= q \end{aligned}$$

The equilibrium quantity is 40,000 yd^2, the same answer found earlier.

You may prefer to find the equilibrium quantity by solving the equation with your calculator. Or, if your calculator has a TABLE feature, you can use it to find the value of q that makes the two expressions equal.

Another important issue is how, in practice, the equations of the supply and demand functions can be found. This issue is important for many problems involving linear functions in this section and the next. Data need to be collected, and if they lie perfectly along a line, then the equation can easily be found with any two points. What usually happens, however, is that the data are scattered, and there is no line that goes through all the points. In this case we must find a line that approximates the linear trend of the data as closely as possible (assuming the points lie approximately along a line) as in Example 15 in the previous section. This is usually done by the *method of least squares,* also referred to as *linear regression.* We will discuss this method in Section 1.3.

Cost Analysis The cost of manufacturing an item commonly consists of two parts. The first is a **fixed cost** for designing the product, setting up a factory, training workers, and so on. Within broad limits, the fixed cost is constant for a particular product and does not change as more items are made. The second part is a *cost per item* for labor, materials, packing, shipping, and so on. The total value of this second cost *does* depend on the number of items made.

EXAMPLE 4 *Cost Analysis*

Suppose that the cost of producing video games can be approximated by

$$C(x) = 12x + 100,$$

where $C(x)$ is the cost in dollars to produce x games. The cost to produce 0 games is

$$C(0) = 12(0) + 100 = 100,$$

or \$100. This sum, \$100, is the fixed cost.

Once the company has invested the fixed cost into the video game project, what will the additional cost per game be? As an example, let's compare the costs of making 5 games and 6 games.

$$C(5) = 12(5) + 100 = 160 \qquad \text{and} \qquad C(6) = 12(6) + 100 = 172,$$

or \$160 and \$172, respectively.

So the 6th game itself costs \$172 − \$160 = \$12 to produce. In the same way, the 81st game costs $C(81) - C(80) = \$1072 - \$1060 = \$12$ to produce. In fact, the $(n + 1)$st game costs

$$\begin{aligned} C(n+1) - C(n) &= [12(n+1) + 100] - (12n + 100) \\ &= 12, \end{aligned}$$

or \$12, to produce. The number 12 is also the slope of the graph of the cost function $C(x) = 12x + 100$; the slope gives us the cost to produce an additional item.

In economics, **marginal cost** is the rate of change of cost $C(x)$ at a level of production x and is equal to the slope of the cost function at x. It approximates the cost of producing one additional item. In fact, some books define the marginal cost to be the cost of producing one additional item. With *linear functions,* these two definitions are equivalent, and the marginal cost, which is equal to the slope of the cost function, is *constant.* For instance, in the video game example, the marginal cost of each game is \$12. For other types of functions, these two definitions are only approximately equal. Marginal cost is important to management in making decisions in areas such as cost control, pricing, and production planning.

The work in Example 4 can be generalized. Suppose the total cost to make x items is given by the linear cost function $C(x) = mx + b$. The fixed cost is found by letting $x = 0$:

$$C(0) = m \cdot 0 + b = b;$$

thus, the fixed cost is b dollars. The additional cost of the $(n + 1)$st item, the marginal cost, is m, the slope of the line $C(x) = mx + b$.

COST FUNCTION

In a cost function of the form $C(x) = mx + b$, the m represents the marginal cost per item and b the fixed cost. Conversely, if the fixed cost of producing an item is b and the marginal cost is m, then the **cost function** $C(x)$ for producing x items is $C(x) = mx + b$.

EXAMPLE 5 *Cost Function*

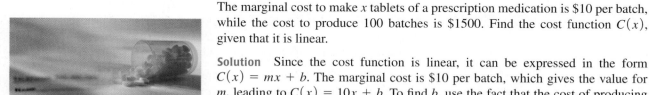

The marginal cost to make x tablets of a prescription medication is \$10 per batch, while the cost to produce 100 batches is \$1500. Find the cost function $C(x)$, given that it is linear.

Solution Since the cost function is linear, it can be expressed in the form $C(x) = mx + b$. The marginal cost is \$10 per batch, which gives the value for m, leading to $C(x) = 10x + b$. To find b, use the fact that the cost of producing 100 batches of tablets is \$1500, or $C(100) = 1500$. Substituting $C(x) = 1500$ and $x = 100$ into $C(x) = 10x + b$ gives

$$1500 = 10 \cdot 100 + b$$
$$1500 = 1000 + b$$
$$500 = b. \qquad \text{Subtract 1000 from both sides.}$$

The cost function is given by $C(x) = 10x + 500$, where the fixed cost is \$500.

Break-Even Analysis The **revenue** $R(x)$ from selling x units of an item is the product of the price per unit p and the number of units sold (demand) x, so that

$$R(x) = px.$$

The corresponding **profit** $P(x)$ is the difference between revenue $R(x)$ and cost $C(x)$. That is,

$$P(x) = R(x) - C(x).$$

A company can make a profit only if the revenue received from its customers exceeds the cost of producing and selling its goods and services. The number of units at which revenue just equals cost is the **break-even quantity;** the corresponding ordered pair gives the **break-even point.**

EXAMPLE 6 *Break-Even Analysis*

A firm producing poultry feed finds that the total cost $C(x)$ in dollars of producing and selling x units is given by

$$C(x) = 20x + 100.$$

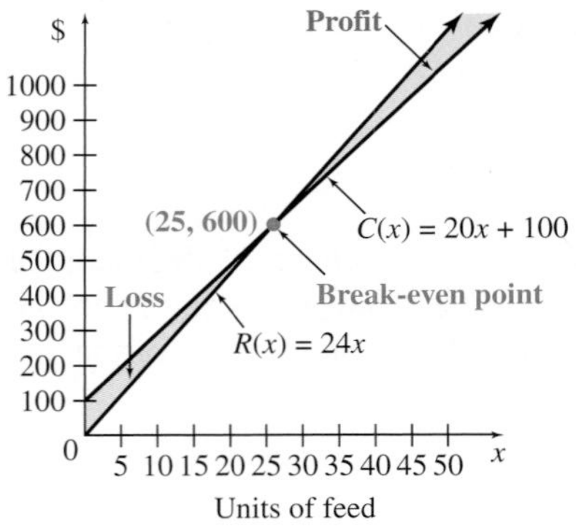

FIGURE 13

Management plans to charge \$24 per unit for the feed.

(a) How many units must be sold for the firm to break even?

Solution The firm will break even (no profit and no loss) as long as revenue just equals cost, or $R(x) = C(x)$. From the given information, since $R(x) = px$ and $p = \$24$,

$$R(x) = 24x.$$

Substituting for $R(x)$ and $C(x)$ in the equation $R(x) = C(x)$ gives

$$24x = 20x + 100,$$

from which $x = 25$. The firm breaks even by selling 25 units, which is the break-even quantity. The graphs of $C(x) = 20x + 100$ and $R(x) = 24x$ are shown in Figure 13. The break-even point (where $x = 25$) is shown on the graph. If the company sells more than 25 units (if $x > 25$), it makes a profit. If it sells less than 25 units, it loses money.

(b) What is the profit if 100 units of feed are sold?

Solution Use the formula for profit $P(x)$.

$$\begin{aligned} P(x) &= R(x) - C(x) \\ &= 24x - (20x + 100) \\ &= 4x - 100 \end{aligned}$$

Then $P(100) = 4(100) - 100 = 300$. The firm will make a profit of \$300 from the sale of 100 units of feed.

(c) How many units must be sold to produce a profit of \$900?

Solution Let $P(x) = 900$ in the equation $P(x) = 4x - 100$ and solve for x.

$$\begin{aligned} 900 &= 4x - 100 \\ 1000 &= 4x \\ x &= 250 \end{aligned}$$

Sales of 250 units will produce \$900 profit.

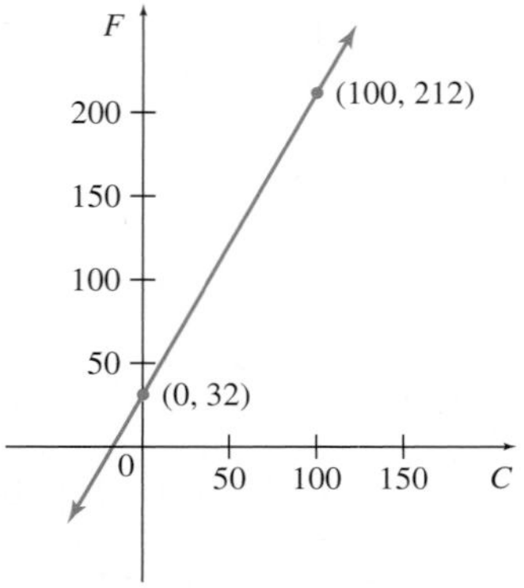

FIGURE 14

Temperature One of the most common linear relationships found in everyday situations deals with temperature. Recall that water freezes at 32° Fahrenheit and 0° Celsius, while it boils at 212° Fahrenheit and 100° Celsius.* The ordered pairs $(0, 32)$ and $(100, 212)$ are graphed in Figure 14 on axes showing Fahrenheit (F) as a function of Celsius (C). The line joining them is the graph of the function.

*Gabriel Fahrenheit (1686–1736), a German physicist, invented his scale with 0° representing the temperature of an equal mixture of ice and ammonium chloride (a type of salt), and 96° as the temperature of the human body. (It is often said, erroneously, that Fahrenheit set 100° as the temperature of the human body. Fahrenheit's own words are quoted in *A History of the Thermometer and Its Use in Meteorology* by W. E. Knowles, Middleton: The Johns Hopkins Press, 1966, p. 75.) The Swedish astronomer Anders Celsius (1701–1744) set 0° and 100° as the freezing and boiling points of water.

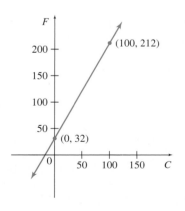

EXAMPLE 7 *Temperature*

Derive an equation relating F and C.

Solution To derive the required linear equation, first find the slope using the given ordered pairs, $(0, 32)$ and $(100, 212)$.

$$m = \frac{212 - 32}{100 - 0} = \frac{9}{5}$$

The F-intercept of the graph is 32, so by the slope-intercept form, the equation of the line is

$$F = \frac{9}{5}C + 32.$$

With simple algebra this equation can be rewritten to give C in terms of F:

$$C = \frac{5}{9}(F - 32).$$

1.2 EXERCISES

In Exercises 1–4, decide whether the statement is true or false.

1. To find the x-intercept of the graph of a linear function, we solve $y = f(x) = 0$, and to find the y-intercept, we evaluate $f(0)$.
2. The graph of $f(x) = -3$ is a vertical line.
3. The slope of the graph of a linear function cannot be undefined.
4. The graph of $f(x) = ax$ is a straight line that passes through the origin.
5. Describe what fixed costs and marginal costs mean to a company.
6. In a few sentences, explain why the price of a commodity not already at its equilibrium price should move in that direction.
7. Explain why a linear function may not be adequate for describing the supply and demand functions.

Write a linear cost function for each situation. Identify all variables used.

8. A chain saw rental firm charges \$12 plus \$1 per hour.
9. A trailer-hauling service charges \$45 plus \$2 per mile.
10. A parking garage charges 50 cents plus 35 cents per half-hour.
11. For a one-day rental, a car rental firm charges \$44 plus 28 cents per mile.

Assume that each situation can be expressed as a linear cost function. Find the cost function in each case.

12. Fixed cost: \$100; 50 items cost \$1600 to produce.
13. Fixed cost: \$400; 10 items cost \$650 to produce.
14. Marginal cost: \$90; 150 items cost \$16,000 to produce.
15. Marginal cost: \$120; 700 items cost \$96,500 to produce.
16. How is the average rate of change related to the graph of a function?
17. In your own words, describe the break-even quantity, how to find it, and what it indicates.

Applications

BUSINESS AND ECONOMICS

18. *Supply and Demand* Suppose that the demand and price for a certain model of electric can opener are related by

$$p = D(q) = 16 - \frac{5}{4}q,$$

where p is the price (in dollars) and q is the demand (in hundreds). Find the price at each level of demand.

a. 0 can openers **b.** 400 can openers

c. 800 can openers

Find the demand for the electric can opener at each price.

d. \$8 **e.** \$10 **f.** \$12

g. Graph $p = 16 - \frac{5}{4}q$.

Suppose the price and supply of the electric can opener are related by

$$p = S(q) = \frac{3}{4}q,$$

where p is the price (in dollars) and q is the supply (in hundreds) of can openers. Find the supply at each price.

h. \$0 **i.** \$10 **j.** \$20

k. Graph $p = \frac{3}{4}q$ on the same axes used for part g.

l. Find the equilibrium quantity and the equilibrium price.

19. *Supply and Demand* Let the supply and demand functions for strawberry-flavored licorice be given by

$$p = S(q) = \frac{3}{2}q \quad \text{and} \quad p = D(q) = 81 - \frac{3}{4}q,$$

where p is the price in dollars and q is the number of batches.

a. Graph these on the same axes.

b. Find the equilibrium quantity and the equilibrium price.

20. *Supply and Demand* Let the supply and demand functions for butter pecan ice cream be given by

$$p = S(q) = \frac{2}{5}q \quad \text{and} \quad p = D(q) = 100 - \frac{2}{5}q,$$

where p is the price in dollars and q is the number of 10-gallon tubs.

a. Graph these on the same axes.

b. Find the equilibrium quantity and the equilibrium price.

21. *Supply and Demand* Let the supply and demand functions for sugar be given by

$$p = S(q) = 1.4q - .6 \quad \text{and}$$
$$p = D(q) = -2q + 3.2,$$

where p is the price per pound and q is the quantity in thousands of pounds.

a. Graph these on the same axes.

b. Find the equilibrium quantity and the equilibrium price.

22. *T-Shirt Cost* Yoshi Yamamura sells silk-screened T-shirts at community festivals and crafts fairs. Her marginal cost to produce one T-shirt is \$3.50. Her total cost to produce 60 T-shirts is \$300, and she sells them for \$9 each.

a. Find the linear cost function for Yoshi's T-shirt production.

b. How many T-shirts must she produce and sell in order to break even?

c. How many T-shirts must she produce and sell to make a profit of \$500?

23. *Publishing Costs* Enrique Gonzales owns a small publishing house specializing in Latin American poetry. His fixed cost to produce a typical poetry volume is \$525, and his total cost to produce 1000 copies of the book is \$2675. His books sell for \$4.95 each.

a. Find the linear cost function for Enrique's book production.

b. How many poetry books must he produce and sell in order to break even?

c. How many books must he produce and sell to make a profit of \$1000?

24. *Marginal Cost of Coffee* The manager of a restaurant found that the cost to produce 100 cups of coffee is \$11.02, while the cost to produce 400 cups is \$40.12. Assume the cost $C(x)$ is a linear function of x, the number of cups produced.

a. Find a formula for $C(x)$.

b. What is the fixed cost?

c. Find the total cost of producing 1000 cups.

d. Find the total cost of producing 1001 cups.

e. Find the marginal cost of the 1001st cup.

f. What is the marginal cost of *any* cup and what does this mean to the manager?

25. *Marginal Cost of a New Plant* In deciding whether to set up a new manufacturing plant, company analysts have decided that a linear function is a reasonable estimation for the total cost $C(x)$ in dollars to produce x items. They estimate the cost to produce 10,000 items as \$547,500, and the cost to produce 50,000 items as \$737,500.

a. Find a formula for $C(x)$.

b. Find the fixed cost.

c. Find the total cost to produce 100,000 items.

d. Find the marginal cost of the items to be produced in this plant and what does this mean to the manager?

26. *Bread Sales* Bread Boutiques, which sell freshly baked bread with no preservatives, are located in many malls around the United States and are growing rapidly. The Saint Louis Bread Company (now called Panera Bread) claims a sales growth of 5000% in its first five years.*

a. Suppose sales were \$100,000 in 1991. What would they be in 1996 at that growth rate?

b. Let 1991 correspond to $x = 1$. Write two ordered pairs representing sales in 1991 and 1996.

c. Assuming sales increased linearly, write a linear sales function for this company.

d. If sales continue to increase at the same rate, when will they reach one billion dollars?

e. The actual sales were expected to be \$1 billion in 2003. Discuss the assumption that the growth rate has been linear.

27. *Break-Even Analysis* Producing x units of tacos costs $C(x) = 5x + 20$; revenue is $R(x) = 15x$, where $C(x)$ and $R(x)$ are in dollars.

a. What is the break-even quantity?

b. What is the profit from 100 units?

c. How many units will produce a profit of \$500?

28. *Break-Even Analysis* To produce x units of a religious medal costs $C(x) = 12x + 39$. The revenue is $R(x) = 25x$. Both $C(x)$ and $R(x)$ are in dollars.

a. Find the break-even quantity.

b. Find the profit from 250 units.

c. Find the number of units that must be produced for a profit of \$130.

Break-Even Analysis You are the manager of a firm. You are considering the manufacture of a new product, so you ask the accounting department for cost estimates and the sales department for sales estimates. After you receive the data, you must decide whether to go ahead with production of the new product. Analyze the data in Exercises 29–32 (find a break-even quantity) and then decide what you would do in each case. Also write the profit function.

29. $C(x) = 85x + 900$; $R(x) = 105x$; no more than 38 units can be sold.

30. $C(x) = 105x + 6000$; $R(x) = 250x$; no more than 400 units can be sold.

31. $C(x) = 70x + 500$; $R(x) = 60x$ (*Hint*: What does a negative break-even quantity mean?)

32. $C(x) = 1000x + 5000$; $R(x) = 900x$

PHYSICAL SCIENCES

33. *Temperature* Use the formula for conversion between Fahrenheit and Celsius derived in Example 7 to convert each temperature.

a. 58°F to Celsius

b. −20°F to Celsius

c. 50°C to Fahrenheit

34. *Body Temperature* You may have heard that the average temperature of the human body is 98.6°. Recent experiments show that the actual figure is closer to 98.2°.† The figure of 98.6 comes from experiments done by Carl Wunderlich in 1868. But Wunderlich measured the temperatures in degrees Celsius and rounded the average to the nearest degree, giving 37°C as the average temperature.‡

a. What is the Fahrenheit equivalent of 37°C?

b. Given that Wunderlich rounded to the nearest degree Celsius, his experiments tell us that the actual average human body temperature is somewhere between 36.5°C and 37.5°C. Find what this range corresponds to in degrees Fahrenheit.

35. *Temperature* Find the temperature at which the Celsius and Fahrenheit temperatures are numerically equal.

**The New York Times,* Nov. 18, 1995, pp. 19 and 21.
†*Science News,* Sept. 26, 1992, p. 195.
‡*Science News,* Nov. 7, 1992, p. 399.

1.3 THE LEAST SQUARES LINE

THINK ABOUT IT *How has the accidental death rate in the United States changed over time?*

In this section, we show how to answer such questions using the method of least squares. We use past data to find trends and to make tentative predictions about the future. The only assumption we make is that the data are related linearly—that is, if we plot pairs of data, the resulting points will lie close to some line. This method cannot give exact answers. The best we can expect is that, if we are careful, we will get a reasonable approximation.

The table lists the number of accidental deaths per 100,000 population in the United States through the past century.* If you were a manager at an insurance company, these data could be very important. You might need to make some predictions about how much you will pay out next year in accidental death benefits, and even a very tentative prediction based on past trends is better than no prediction at all.

Year	Death Rate
1910	84.4
1920	71.2
1930	80.5
1940	73.4
1950	60.3
1960	52.1
1970	56.2
1980	46.5
1990	36.9
2000	34.0

The first step is to draw a scatterplot, as we have done in Figure 15. Notice that the points lie approximately along a line, which means that a linear function may give a good approximation of the data. If we select two points and find the line that passes through them, as we did in Section 1.1, we will get a different line for each pair of points, and in some cases the lines will be very different. We want to draw one line that is simultaneously close to all the points on the graph, but many such lines are possible, depending upon how we define the phrase "simultaneously close to all the points." How do we decide on the best possible line? Before going on, you might want to try drawing the line you think is best on Figure 15.

The line used most often in applications is that in which the sum of the squares of the vertical distances from the data points to the line is as small as possible. Such a line is called the **least squares line.** The least squares line for the data in Figure 15 is drawn in Figure 16. How does the line compare with the one you drew on Figure 15? It may not be exactly the same, but should appear similar.

In Figure 16, the vertical distances from the points to the line are indicated by d_1, d_2, and so on, up through d_{10} (read "d-sub-one, d-sub-two, d-sub-three," and so on). For n points, corresponding to the n pairs of data, the least squares line is found by minimizing the sum $(d_1)^2 + (d_2)^2 + (d_3)^2 + \cdots + (d_n)^2$.

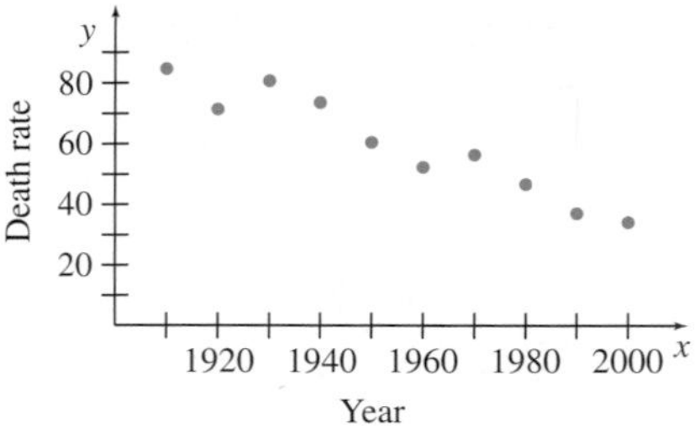

FIGURE 15

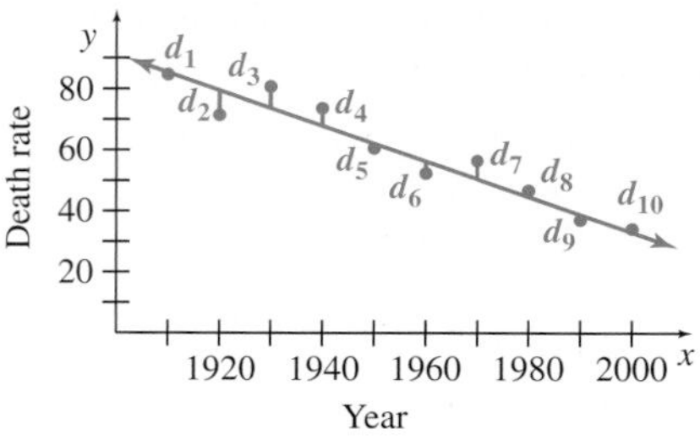

FIGURE 16

*U.S. Department of Health and Human Services, National Center for Health Statistics.

For the points $(x_1, y_1), (x_2, y_2), \ldots, (x_n, y_n)$, if the equation of the desired line is $Y = mx + b$, then

$$d_1 = |Y_1 - y_1| = |mx_1 + b - y_1|,$$
$$d_2 = |Y_2 - y_2| = |mx_2 + b - y_2|,$$

and so on. We use Y in the equation of the line instead of y to distinguish the predicted values (Y) from the y-values of the given data points. The sum to be minimized becomes

$$(mx_1 + b - y_1)^2 + (mx_2 + b - y_2)^2 + \cdots + (mx_n + b - y_n)^2,$$

where $(x_1, y_1), (x_2, y_2), \ldots, (x_n, y_n)$ are known and m and b are to be found.

The method of minimizing this sum requires advanced techniques and is not given here. The result gives equations that can be solved for the slope m and the y-intercept b of the least squares line.* In these equations, the symbol Σ, the Greek letter sigma, indicates "the sum of"; this notation is known as **summation notation.** For example, we write the sum $x_1 + x_2 + \cdots + x_n$, where n is the number of data points, as

$$x_1 + x_2 + \cdots + x_n = \Sigma x.$$

Similarly, Σxy means $x_1y_1 + x_2y_2 + \cdots + x_ny_n$, and so on.

CAUTION Note that Σx^2 means $x_1^2 + x_2^2 + \cdots + x_n^2$, which is *not* the same as squaring Σx. ■

LEAST SQUARES LINE

The **least squares line** $Y = mx + b$ that gives the best fit to the data points $(x_1, y_1), (x_2, y_2), \ldots, (x_n, y_n)$ has slope m and y-intercept b that satisfy the equations

$$nb + (\Sigma x)m = \Sigma y$$
$$(\Sigma x)b + (\Sigma x^2)m = \Sigma xy.$$

Method 1: Calculating by Hand

To find the least squares line for the given data, we first find the required sums. To reduce the size of the numbers, let x represent the years since 1900, so that, for example, $x = 10$ for the year 1910. Let y represent the death rate.

x	y	xy	x^2	y^2
10	84.4	844	100	7123.36
20	71.2	1424	400	5069.44
30	80.5	2415	900	6480.25
40	73.4	2936	1600	5387.56
50	60.3	3015	2500	3636.09
60	52.1	3126	3600	2714.41
70	56.2	3934	4900	3158.44
80	46.5	3720	6400	2162.25
90	36.9	3321	8100	1361.61
100	34.0	3400	10,000	1156.00
$\Sigma x = 550$	$\Sigma y = 595.5$	$\Sigma xy = 28{,}135$	$\Sigma x^2 = 38{,}500$	$\Sigma y^2 = 38{,}249.41$

*Equations for m and b are derived in Exercise 4.

(The column headed y^2 will be used later.) Now we can calculate m and b by solving a system of equations. Our method is to solve the first equation for b in terms of m, and substitute this into the second equation. We then solve the second equation for m. Once we have m, we put this back into the equation for b. Here, $n = 10$ (the number of data points).

$$nb + (\Sigma x)m = \Sigma y \qquad \text{First least squares equation.}$$

$$10b + 550m = 595.5 \qquad \text{Substitute from the table.}$$

$$10b = 595.5 - 550m \qquad \text{Subtract } 550m \text{ from both sides.}$$

$$b = (595.5 - 550m)/10 \qquad \text{Divide both sides by 10.}$$

$$(\Sigma x)b + (\Sigma x^2)m = \Sigma xy \qquad \text{Second least squares equation.}$$

$$550(595.5 - 550m)/10 + 38{,}500m = 28{,}135 \qquad \text{Substitute.}$$

$$32{,}752.5 - 30{,}250m + 38{,}500m = 28{,}135 \qquad \text{Multiply.}$$

$$8250m = -4617.5 \qquad \text{Combine terms.}$$

$$m \approx -.5596970 \approx -.560$$

The significance of m is that the death rate per 100,000 population is tending to drop (because of the negative) at a rate of .560 per year.

Now substitute the value of m into the equation for b.

$$b = \frac{595.5 - 550(-.5596970)}{10} \approx 90.333335 \approx 90.3$$

Substitute m and b into the least squares line equation, $Y = mx + b$; the least squares line that best fits the nine data points has equation $Y = -.560x + 90.3$. This gives a mathematical description of the relationship between the year and the number of accidental deaths per 100,000 population. The equation can be used to predict y from a given value of x, as we will show in Example 1. As we mentioned before, however, caution must be exercised when using the least squares equation to predict data points that are far from the range of points on which the equation was modeled.

CAUTION In computing m and b, we rounded the final answer to three digits because the original data were known only to three digits. It is important, however, *not* to round any of the intermediate results (such as Σx^2) because round-off error may have a detrimental effect on the accuracy of the answer. Similarly, it is important not to use a rounded-off value of m when computing b. ■

Method 2: Graphing Calculator

The calculations for finding the least squares line are often tedious, even with the aid of a calculator. Fortunately, many calculators can calculate the least squares line with just a few keystrokes. For purposes of illustration, we will show how the least squares line in the previous example is found with a TI-83/84 Plus graphing calculator.

We begin by entering the data into the calculator. We will be using the first two lists, called L_1 and L_2. Choosing the STAT menu, then choosing the fourth entry ClrList, we enter L_1, L_2, to indicate the lists to be cleared. Now we press STAT again and choose the first entry EDIT, which brings up the blank lists. As before, we will only use the last two digits of the year, putting the numbers in L_1. We put the death rate in L_2, giving the two screens shown in Figure 17.

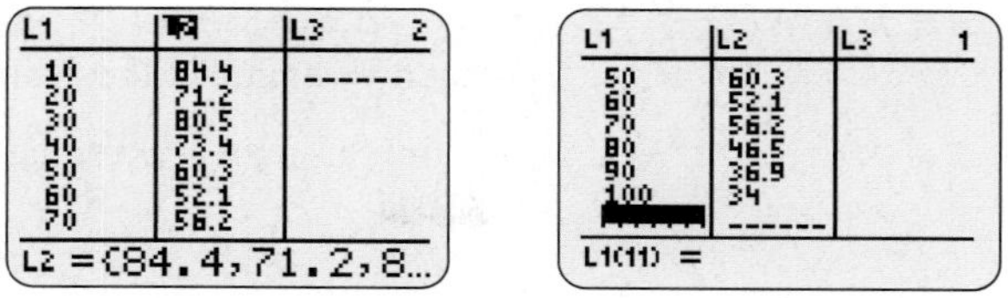

FIGURE 17

Press STAT again and choose CALC instead of EDIT. Then choose item 4 LinReg $(ax + b)$ to get the values of a (the slope) and b (the y-intercept) for the least squares line, as shown in Figure 18. With a and b rounded to three decimal places, the least squares line is $Y = -.560x + 90.3$. A graph of the data points and the line is shown in Figure 19.

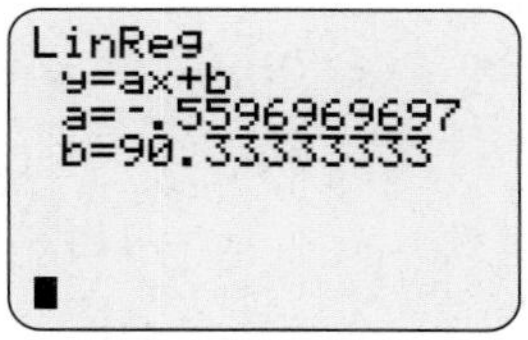

FIGURE 18

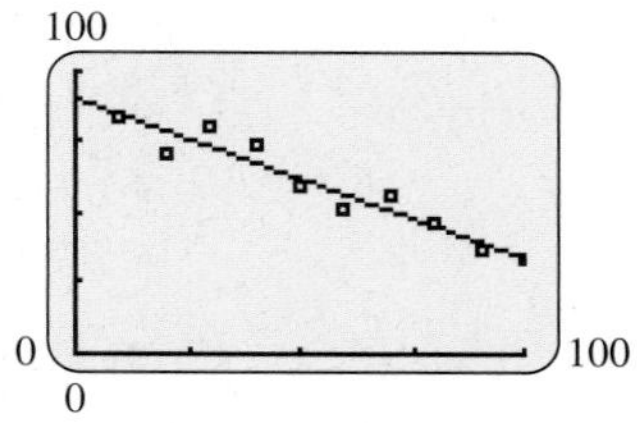

FIGURE 19

For more details on finding the least squares line with a graphing calculator, see *The Graphing Calculator Manual* that is available with this book.

Method 3: Spreadsheet

Many computer spreadsheet programs can also find the least squares line. Figure 20 shows the scatterplot and least squares line for the accidental death rate data using an Excel spreadsheet. The scatterplot was found using the XY(Scatter) command under Chart Wizard, and the line was found using the Add Trendline command under the Chart menu. For details, see *The Spreadsheet Manual* that is available with this book.

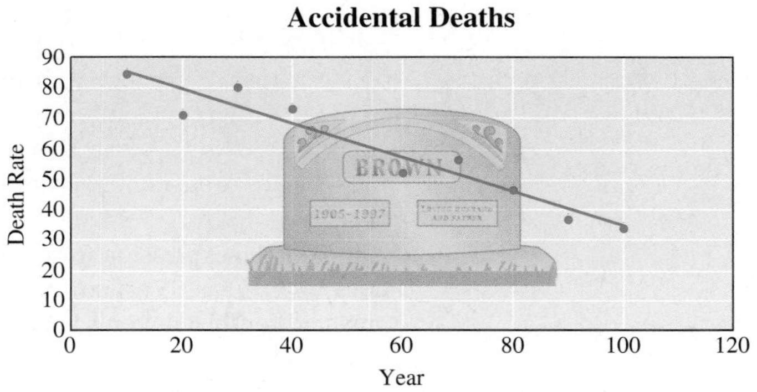

FIGURE 20

EXAMPLE 1 *Least Squares Line*

What do we predict the accidental death rate in 2001 to be?

Solution Use the least squares line equation given above with $x = 101$.

$$\begin{aligned} Y &= -.560x + 90.3 \\ &= -.560(101) + 90.3 \\ &= 33.74 \end{aligned}$$

The death rate in 2001 is predicted to be about 33.7 per 100,000 population. In this case, we have the actual value for 2001. It happens to be 34.3, which is close to the predicted value.

EXAMPLE 2 *Least Squares Line*

In what year is the death rate predicted to drop below 26 per 100,000 population?

Solution Let $Y = 26$ in the equation above and solve for x.

$$\begin{aligned} 26 &= -.560x + 90.3 \\ -64.3 &= -.560x && \text{Subtract 90.3 from both sides.} \\ x &= 114.8 && \text{Divide both sides by } -.560. \end{aligned}$$

This means that after 114 years, the rate will not have quite reached 26 per 100,000, so we must wait 115 years for this to happen. This corresponds to the year 2015 (115 years after 1900), when our equation predicts the death rate to be $-.560(115) + 90.3 = 25.9$ per 100,000 population.

Correlation Once an equation is found for the least squares line, we need to have some way of judging just how good the equation is for predictive purposes. If the points from the data fit the line quite closely, then we have more reason to expect future data pairs to do so. But if the points are widely scattered about even the best-fitting line, then predictions are not likely to be accurate.

In order to have a quantitative basis for confidence in our predictions, we need a measure of the "goodness of fit" of the original data to the prediction line. One such measure is called the **coefficient of correlation,** denoted r.

COEFFICIENT OF CORRELATION

$$r = \frac{n(\Sigma xy) - (\Sigma x)(\Sigma y)}{\sqrt{n(\Sigma x^2) - (\Sigma x)^2} \cdot \sqrt{n(\Sigma y^2) - (\Sigma y)^2}}$$

Although the expression for r looks daunting, remember that each of the summations, Σx, Σy, Σxy, and so on, are just the totals from a table like the one we prepared for the data on accidental deaths. Also, with a calculator, the arithmetic is no problem!

The coefficient of correlation r is always equal to or between 1 and -1. Values of exactly 1 or -1 indicate that the data points lie *exactly* on the least squares line. If $r = 1$, the least squares line has a positive slope; $r = -1$ gives a negative slope. If $r = 0$, there is no linear correlation between the data points (but some *nonlinear* function might provide an excellent fit for the data). A correlation of zero may also indicate that the data fit a horizontal line. To investigate what is happening, it is always helpful to sketch a scatterplot of the data. Some scatterplots that correspond to these values of r are shown in Figure 21.

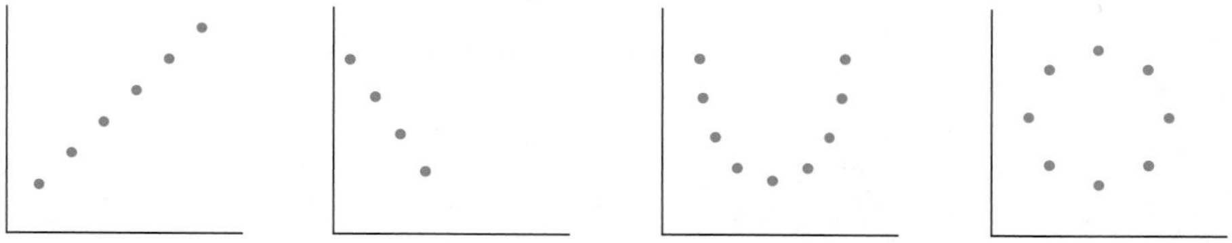

r close to 1 | r close to −1 | r close to 0 | r close to 0

FIGURE 21

A value of r close to 1 or -1 indicates the presence of a linear relationship. The exact value of r necessary to conclude that there is a linear relationship depends upon n, the number of data points, as well as how confident we want to be of our conclusion. For details, consult a text on statistics.*

EXAMPLE 3 *Coefficient of Correlation*

Find r for the data on accidental death rates.

Solution

Method 1: Calculating by Hand

From the table on page 31, $\sum x = 550$, $\sum y = 595.5$, $\sum xy = 28{,}135$, $\sum x^2 = 38{,}500$, and $\sum y^2 = 38{,}249.41$. Also, $n = 10$. Substituting these values into the formula for r gives

$$r = \frac{10(28{,}135) - (550)(595.5)}{\sqrt{10(38{,}500) - (550)^2} \cdot \sqrt{10(38{,}249.41) - (595.5)^2}}$$

$$= \frac{-46{,}175}{\sqrt{82{,}500} \cdot \sqrt{27{,}873.85}}$$

$$= -.962900585 \approx -.963.$$

This is a high correlation, which agrees with our observation that the data fit a line quite well.

Method 2: Graphing Calculator

Most calculators that give the least squares line will also give the coefficient of correlation. To do this on the TI-83/84 Plus, press the second function CATALOG and go down the list to the entry DiagnosticOn. Press ENTER at that point, then press STAT, CALC, and choose item 4 to get the display in Figure 22. The result is the same as we got by hand. The command DiagnosticOn need only be entered once, and the coefficient of correlation will always appear in the future.

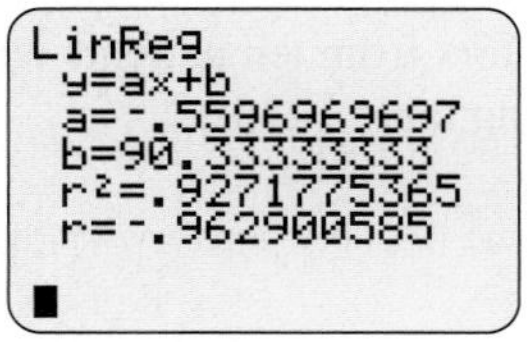

FIGURE 22

*For example, see *Introductory Statistics,* 6th edition, by Neil A. Weiss, Boston, Mass.: Addison-Wesley, 2002.

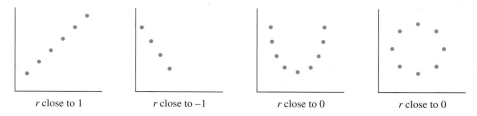

Method 3: Spreadsheet Many computer spreadsheet programs have a built-in command to find the coefficient of correlation. For example, in Excel, use the command "= CORREL(A1:A10,B1:B10)" to find the correlation of the 10 data points stored in columns A and B. For more details, see *The Spreadsheet Manual* that is available with this text.

EXAMPLE 4 *Airline Passengers*

The following table shows the number of airline passengers in the United States (in millions) from 1993 to 2002.*

Year	1993	1994	1995	1996	1997	1998	1999	2000	2001	2002
Passengers	488.5	528.8	547.8	581.2	594.7	612.9	636.0	666.2	622.1	611.9

Find the correlation coefficient, as well as the line that best fits the data.

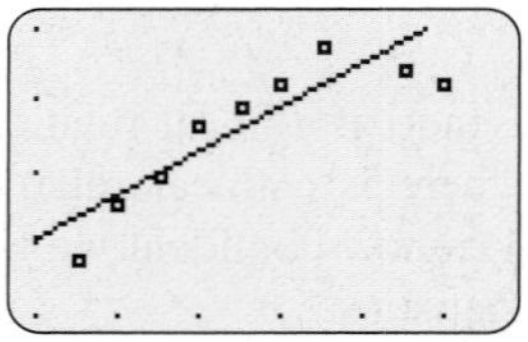

FIGURE 23

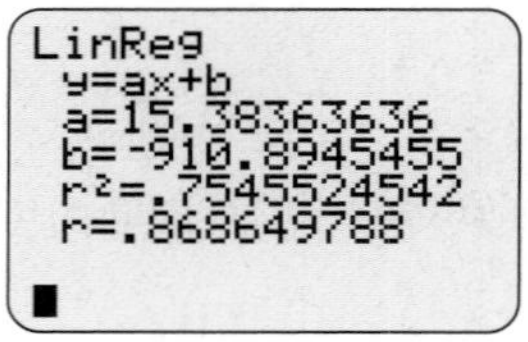

FIGURE 24

Solution A scatterplot of the data, along with the graph of the least squares line, is shown in Figure 23. Notice that the data appear to be linear for 1993 to 2000, but the last two points suggest that the number of passengers is no longer growing linearly. The most likely explanation for this downward spiral is the terrorist attacks on September 11, 2001. Figure 24 shows the result of the LinReg command on the TI-83/84 Plus. Although the correlation coefficient continues to reflect a linear trend, the scatterplot indicates that a significant event has occurred that has apparently stopped the increase, and in fact, has started a decline in the number of airline passengers. If this same analysis had been done in 2000 (that is, if we removed the last two data points), we would get $r = .99$ and $Y = 23.675x - 1702.625$. (Verify this on your calculator.) Therefore, the trend from 1993 to 2000 is very linear and many people would have been confident in predicting that there would have been about 712,225,000 airline passengers in 2002. Notice that the actual number is about 100 million fewer.

This example illustrates the hazards of using current trends to predict the future.

The final example is another illustration of why a plot of the data points is so important.

EXAMPLE 5 *Silicone Implants*

Silicones have long been used for fabricating medical devices on the presumption that they are biocompatible materials. This presumption is not entirely correct. Silicone prostheses, when implanted within the soft tissues of the breast, may evoke an inflammatory reaction. In response to silicone exposure, inflammatory mediator production was observed in experimental studies. After in vitro culture for 24 hours, the levels of four inflammatory mediators from ten patients with silicone breast implants were as shown in the following table.†

*Annual Traffic and Company, Air Transport Association, June 4, 2003, http://www.airlines.org/public/industry/display1.asp?nid=1032.

†Mena et al., "Inflammatory Intermediates Produced by Tissues Encasing Silicone Breast Prostheses," *Journal of Investigative Surgery*, Vol. 8, 1995, p. 33. Copyright © 1995. Reproduced by permission of Taylor & Francis, Inc., http://www.routledge-ny.com.

Patient	IL-2	TNF-α	IL-6	PGE_2
1	48	ND	231	68
2	219	78	308,287	2710
3	109	65	33,291	1804
4	2179	149	124,550	8053
5	219	451	17,075	7371
6	54	64	22,955	3418
7	6	79	95,102	9768
8	10	115	5649	441
9	42	618	840,585	9585
10	196	69	58,924	4536

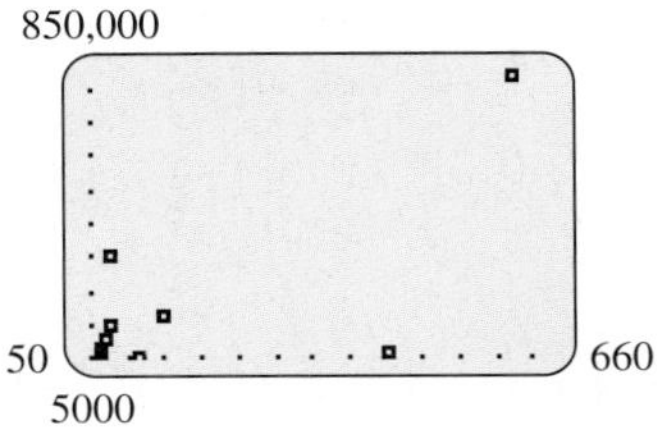

(a)

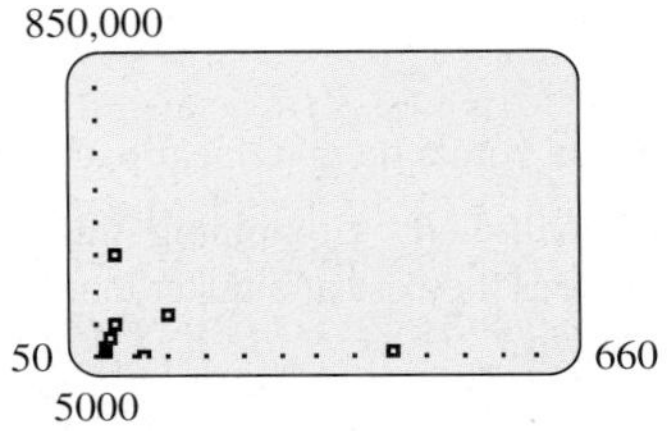

(b)

FIGURE 25

A graphing calculator was used to plot the last nine points in the table for TNF-$\alpha(x)$ and IL-6(y). (There is no data point for patient 1, since there is no TNF-α level.) The graph is shown in Figure 25(a). The correlation for these two mediators is .6936. Notice that one data point is way off by itself. As shown in Example 4, sometimes, by removing such a point from the graph, we can achieve a higher correlation.* Figure 25(b) shows the scatterplot with the remaining eight points. However, the correlation is now $-.2493$![†] With the ninth point, the points were closer to a line with a positive slope. Without the point, there is little linear correlation and it has become negative.

1.3 EXERCISES

1. Suppose a positive linear correlation is found between two quantities. Does this mean that one of the quantities increasing causes the other to increase? If not, what does it mean?

2. Given a set of points, the least squares line formed by letting x be the independent variable will not necessarily be the same as the least squares line formed by letting y be the independent variable. Give an example to show why this is true.

The following problem is reprinted from the November 1989 Actuarial Examination on Applied Statistical Methods.[‡]

3. You are given

X	6.8	7.0	7.1	7.2	7.4
Y	.8	1.2	.9	.9	1.5

Determine r^2, the coefficient of determination for the regression of Y on X. (*Note:* The coefficient of determination is defined as the square of the coefficient of correlation.)

a. .3 **b.** .4 **c.** .5 **d.** .6 **e.** .7

*Before discarding a point, we should investigate the reason it is an outlier.

†The observation that removing one point changes the correlation from positive to negative was made by Patrick Fleury, *Chance News* (an electronic newsletter), Vol. 4, No. 16, Dec. 1995.

‡"November 1989 Course 120 Examination Applied Statistical Methods" of the *Education and Examination Committee of The Society of Actuaries.* Reprinted by permission of The Society of Actuaries.

4. Follow the steps outlined in this section to solve the least squares line equations

$$nb + (\Sigma x)m = \Sigma y$$
$$(\Sigma x)b + (\Sigma x^2)m = \Sigma xy$$

for m and b to get

$$m = \frac{n\Sigma xy - (\Sigma x)(\Sigma y)}{n(\Sigma x^2) - (\Sigma x)^2}$$

$$b = \frac{\Sigma y - m(\Sigma x)}{n}.$$

Applications

BUSINESS AND ECONOMICS

5. *Recreation Spending* The U.S. Department of Commerce, Bureau of Economic Analysis, has reported the total U.S. expenditures on recreational goods (hobbies, music, sports, spectator admissions, etc.). From 1994 to 2001, expenditures have grown at an approximately linear rate. The results of the report, in which x represents the years since 1900 and y represents the total expenditures (in billions of dollars), provide the following summations.*

$$n = 8 \qquad \Sigma x^2 = 76{,}092$$
$$\Sigma x = 780 \qquad \Sigma xy = 374{,}850.8$$
$$\Sigma y = 3830.7 \qquad \Sigma y^2 = 1{,}878{,}286.33$$

a. Find an equation for the least squares line.

b. Predict the recreational expenditures in 2005.

c. If this growth continues linearly, when will recreational expenditures reach 750 billion dollars?

d. Find and interpret the coefficient of correlation.

6. *Decrease in Banks* The number of banks in the United States has dropped about 30% since 1992. The following data are from a survey in which x represents the years since 1900 and y corresponded to the number of banks, in thousands, in the United States.†

$$n = 10 \qquad \Sigma x^2 = 93{,}205$$
$$\Sigma x = 965 \qquad \Sigma xy = 9165.1$$
$$\Sigma y = 95.3 \qquad \Sigma y^2 = 920.47$$

a. Find an equation of the least squares line.

b. If the trend continues, how many banks will there be in 2004?

c. Find and interpret the coefficient of correlation.

7. *Air Fares* In January 2000, American Airlines ran an ad in *The New York Times* advertising one-way air fares from New York to various cities.‡ Fourteen of the cities are listed below, with the distances from New York to the cities added.

a. Plot the data. Do the data points lie in a linear pattern?

b. Find the correlation coefficient. Combining this with your answer to part a, does the cost of a ticket tend to go up with the distance flown?

c. Find the equation of the least squares line, and use it to find the approximate marginal cost per mile to fly.

d. For similar data in an October 1993 ad, the equation of the least squares line was $Y = 91.9 + .0313x$. Use this information and your answer to part b to compare the cost of flying American Airlines for these two time periods.

City	Distance (x) (miles)	Price (y) (dollars)
Boston	206	109
Chicago	802	124
Denver	1771	154
Kansas City	1198	144
Little Rock	1238	144
Los Angeles	2786	179
Minneapolis	1207	144
Nashville	892	144
Phoenix	2411	179
Portland	2885	179
Reno	2705	179
St. Louis	948	144
San Diego	2762	179
Seattle	2815	179

**The World Almanac and Book of Facts 2003,* p. 109.
†FDIC, Historical Statistics on Banking, http://www2.fdic.gov/hsob/hsobRpt.asp.
‡*The New York Times,* Jan. 7, 2000, p. A9.

Year (x)	95	96	97	98	99	00	01	02
Debt (y)	5832	6487	6900	7188	7564	8123	8367	8562

8. *Consumer Debt* Credit card debt has risen steadily over the years. The table above gives the average U.S. credit card debt (in dollars) per household. Years are represented as the number of years since 1900. (The table above includes all credit cards and U.S. households with at least one credit card.)*

a. Plot the data. Does the graph show a linear pattern?

b. Find the equation of the least squares line and graph it on the same axes. Does the line appear to be a good fit?

c. Find and interpret the coefficient of correlation.

d. If this linear trend continues, when will household debt reach $10,000?

9. *Used Car Sales* As cars are becoming more expensive, used car sales have increased at a faster rate since 1984 than new car sales.† Sales in millions from 1984 to 1996 are given in the table below.

Year	Sales	Year	Sales
84	12.3	91	12.3
85	13.2	92	12.8
86	13.6	93	13.9
87	13.2	94	15.0
88	14.5	95	14.7
89	14.5	96	14.6
90	13.8		

a. Find the equation of the least squares line and the coefficient of correlation.

b. Find the equation of the least squares line using only the data for every other year starting with 1985, 1987, and so on. Find the coefficient of correlation.

c. Compare your answers for parts a and b. What do you find? Why do you think this happens?

10. *Medical School Admissions* According to the American Association of Medical Colleges, the number of applications to medical schools in the United States began to decrease since 1996 as indicated in the following table.‡ Years are represented as the number of years since 1900 and applications are given in thousands.

Year (x)	94	95	96	97	98	99	00
Applications (y)	45.4	46.6	47.0	43.0	41.0	38.5	37.1

a. Plot the data. Do the data points lie in a linear pattern?

b. Determine the least squares line for this data and graph it on the same coordinate axes. Does the line fit the data reasonably well?

c. Find the coefficient of correlation. Does it agree with your estimate of the fit in part b?

d. Explain why the coefficient of correlation is close to 1, even though some of the data points do not appear to be linear.

11. *Bird Eggs* The average length and width of various bird eggs are given in the following table.§

Bird Name	Width (cm)	Length (cm)
Canada goose	5.8	8.6
Robin	1.5	1.9
Turtledove	2.3	3.1
Hummingbird	1.0	1.0
Raven	3.3	5.0

a. Plot the points, putting the length on the y-axis and the width on the x-axis. Do the data appear to be linear?

b. Find the least squares line, and plot it on the same graph as the data.

*MSN Money, *Is the Debt Binge Over?* http://money.msn.com/articles/smartbuy/basics/9526.asp.
†*The New York Times,* March 3, 1996.
‡Gabriel, B., "Medical School Applicant Pool Still Runs Deep," *New Room Reporter*, American Association of Medical Colleges, Vol. 6, No. 4, May 2003.
§www.nctm.org/wlme/wlme6/five.htm.

c. Suppose there are birds with eggs even smaller than those of hummingbirds. Would the equation found in part b continue to make sense for all positive widths, no matter how small? Explain.

d. Find the coefficient of correlation.

12. *Crickets Chirping* Biologists have observed a linear relationship between the temperature and the frequency with which a cricket chirps. The following data were measured for the striped ground cricket.*

Temperature °F (x)	Chirps per Second (y)
88.6	20.0
71.6	16.0
93.3	19.8
84.3	18.4
80.6	17.1
75.2	15.5
69.7	14.7
82.0	17.1
69.4	15.4
83.3	16.2
79.6	15.0
82.6	17.2
80.6	16.0
83.5	17.0
76.3	14.4

a. Find the equation for the least squares line for the data.

b. Use the results of part a to determine how many chirps per second you would expect to hear from the striped ground cricket if the temperature were 73°F.

c. Use the results of part a to determine what the temperature is when the striped ground crickets are chirping at a rate of 18 times per sec.

d. Find the coefficient of correlation.

SOCIAL SCIENCES

13. *Educational Expenditures* A 2000 report issued by the U.S. Department of Education listed the expenditure per pupil and the average mathematics proficiency in grade 8 for 39 states and the District of Columbia. Letting x equal the expenditure per pupil (ranging from \$4378 in Utah to \$10,107 in Washington, D.C.) and y equal the average mathematics proficiency score (ranging from 234 in Washington, D.C., to 288 in Minnesota), the data can be summarized as follows:†

$$\sum x = 266{,}947 \qquad \sum x^2 = 1{,}865{,}325{,}667$$
$$\sum y = 10{,}922 \qquad \sum y^2 = 2{,}986{,}764$$
$$\sum xy = 72{,}987{,}414 \qquad n = 40$$

Compute the coefficient of correlation for the given data. Does there appear to be a trend in the amount of money spent per pupil and the proficiency of eighth graders in mathematics?

14. *Poverty Levels* The following table lists how poverty level income cutoffs (in dollars) for a family of four have changed over time.‡

Year	Income
1970	3968
1975	5500
1980	8414
1985	10,989
1990	13,359
1995	15,569
2000	17,603

Let x be the year, with $x = 0$ corresponding to 1970, and y be the income in thousands of dollars. (*Note:* $\sum x = 105$, $\sum x^2 = 2275$, $\sum y = 75.402$, $\sum y^2 = 968.270792$, $\sum xy = 1460.97$.)

a. Plot the data. Do the data appear to lie along a straight line?

b. Calculate the coefficient of correlation. Does your result agree with your answer to part a?

c. Find the equation of the least squares line.

d. Use your answer from part c to predict the poverty level in the year 2015.

15. *SAT Scores* At Hofstra University, all students take the math SAT before entrance, and most students take a mathematics placement test before registration. Recently, one professor collected the data on the next page for 19 students in his Finite Mathematics class:

a. Find an equation for the least squares line. Let x be the math SAT and y be the placement test score.

*Pierce, George W., Data from *The Songs of Insects,* Cambridge, Mass., Harvard University Press, Copyright © 1948 by the President and Fellows of Harvard College.
†National Center for Educational Statistics.
‡U.S. Census Bureau, *Current Population Reports*.

Math SAT	Placement Test	Math SAT	Placement Test	Math SAT	Placement Test
540	20	580	8	440	10
510	16	680	15	520	11
490	10	560	8	620	11
560	8	560	13	680	8
470	12	500	14	550	8
600	11	470	10	620	7
540	10				

b. Use your answer from part a to predict the mathematics placement test score for a student with a math SAT score of 420.

c. Use your answer from part a to predict the mathematics placement test score for a student with a math SAT score of 620.

d. Calculate the coefficient of correlation.

e. Based on your answer to part d, what can you conclude about the relationship between a student's math SAT and mathematics placement test score?

PHYSICAL SCIENCES

16. *Air Conditioning* While shopping for an air conditioner, Adam Bryer consulted the following table giving a machine's BTUs and the square footage (ft^2) that it would cool.

ft^2 (x)	BTUs (y)
150	5000
175	5500
215	6000
250	6500
280	7000
310	7500
350	8000
370	8500
420	9000
450	9500

a. Find the equation for the least squares line for the data.

b. To check the fit of the data to the line, use the results from part a to find the BTUs required to cool a room of 150 ft^2, 280 ft^2, and 420 ft^2. How well does the actual data agree with the predicted values?

c. Suppose Adam's room measures 230 ft^2. Use the results from part a to decide how many BTUs it requires. If air conditioners are available only with the BTU choices in the table, which would Adam choose?

d. Why do you think the table gives ft^2 instead of ft^3, which would give the volume of the room?

17. *Length of a Pendulum* Grandfather clocks use pendulums to keep accurate time. The relationship between the length of a pendulum L and the time T for one complete oscillation can be determined from the data in the table.*

L (ft)	T (sec)
1.0	1.11
1.5	1.36
2.0	1.57
2.5	1.76
3.0	1.92
3.5	2.08
4.0	2.22

a. Plot the data from the table with L as the horizontal axis and T as the vertical axis.

b. Find the least squares line equation and graph it simultaneously, if possible, with the data points. Does it seem to fit the data?

c. Find the coefficient of correlation and interpret it. Does it confirm your answer to part b?†

GENERAL INTEREST

18. *Athletic Records* The table on the next page shows the men's and women's world records (in seconds) in the 800-m run.‡

a. Find the equation for the least squares line for the men's record (y) in terms of the year (x). Use 5 for 1905, 15 for 1915, and so on.

*Data provided by Gary Rockswold, Mankato State University, Minnesota.

†The actual relationship is $L = .81T^2$, which is not a linear equation. This illustrates that even if the relationship is not linear, a line can give a good approximation.

‡Whipp, Brian J. and Susan A. Ward, "Will Women Soon Outrun Men?" *Nature,* Vol. 355, Jan. 2, 1992, p. 25. The data are from Peter Matthews, *Track and Field Athletics: The Records,* Guinness, 1986, pp. 11, 44, and from Robert W. Schutz and Yuanlong Liu in *Statistics in Sport,* edited by Jay Bennett, Arnold, 1998, p. 189.

Year	Men's Record	Women's Record
1905	113.4	—
1915	111.9	—
1925	111.9	144.0
1935	109.7	135.6
1945	106.6	132.0
1955	105.7	125.0
1965	104.3	118.0
1975	103.7	117.48
1985	101.73	113.28
1995	101.11	113.28

b. Find the equation for the least squares line for the women's record.

c. Suppose the men's and women's records continue to improve as predicted by the equations found in parts a and b. In what year will the women's record catch up with the men's record? Do you believe that will happen? Why or why not?

d. Calculate the coefficient of correlation for both the men's and the women's record. What do these numbers tell you?

19. *Football* The following data give the expected points for a football team with first down and 10 yards to go from various points on the field.* (*Note:* $\Sigma x = 500$, $\Sigma x^2 = 33{,}250$, $\Sigma y = 20.668$, $\Sigma y^2 = 91.927042$, $\Sigma xy = 399.16$.)

a. Calculate the coefficient of correlation. Does there appear to be a linear correlation?

Yards from Goal (x)	Expected Points (y)
5	6.041
15	4.572
25	3.681
35	3.167
45	2.392
55	1.538
65	.923
75	.236
85	−.637
95	−1.245

b. Find the equation of the least squares line.

c. Use your answer from part a to predict the expected points when a team is at the 50-yd line.

20. *Baseball* Some baseball fans are concerned about the recent increase in time to complete the game. The following table shows the average time (in hours and minutes) to complete baseball games in recent years.[†]

Year	Average Completion Time	Year	Average Completion Time
1981	2:33	1989	2:46
1982	2:34	1990	2:48
1983	2:36	1991	2:49
1984	2:35	1992	2:49
1985	2:40	1993	2:48
1986	2:44	1994	2:54
1987	2:48	1995	2:57
1988	2:45		

Let x be the number of years since 1980, and let y be the number of minutes beyond 2 hours. (*Note:* $\Sigma x = 120$, $\Sigma x^2 = 1240$, $\Sigma y = 666$, $\Sigma xy = 5765$.)

a. Find the equation of the least squares line.

b. If the trend in the data continues, in what year will the average completion time be 3 hours and 15 minutes?

*Carter, Virgil and Robert E. Machol, *Operations Research,* Vol. 19, 1971, pp. 541–545.
[†]*The New York Times,* May 30, 1995, p. B9.

21. *Running* If you think a marathon is a long race, consider the Hardrock 100, a 101.7-mile running race held in southwestern Colorado. The following table lists the times that the 2000 winner, Kirk Apt, arrived at various mileage points along the way.*

a. What was Apt's average speed?

b. Graph the data, plotting time on the x-axis and distance on the y-axis. You will need to convert the time from hours and minutes into hours. Do the data appear to lie approximately on a straight line?

c. Find the equation for the least squares line, fitting distance as a linear function of time.

d. Calculate the coefficient of correlation. Does it indicate a good fit of the least squares line to the data?

e. Based on your answer to part d, what is a good value for Apt's average speed? Compare this with your answer to part a. Which answer do you think is better? Explain your reasoning.

Miles	Time (hr:min)
0	0
9.6	2:14
16.5	4:08
21.6	6:10
31.6	7:10
42.4	10:51
49.8	12:42
58.0	14:20
65.2	16:30
68.4	18:02
73.7	19:25
83.1	23:07
89.6	26:09
95.8	28:18
101.7	29:35

CHAPTER SUMMARY

In this chapter we have seen how to find the equation of a line, given a point and the slope or given two points. We have also seen how to express the result as a linear function. Equations of lines have a broad range of applications, as demonstrated in this chapter. They are used through the rest of this book, so fluency in their use is important. The method of least squares shows how mathematical models, such as many of those used throughout this book, are derived.

KEY TERMS

To understand the concepts presented in this chapter, you should know the meaning and use of the following terms. For easy reference, the section in the chapter where a word (or expression) was first used is provided.

mathematical model

1.1 **ordered pair**
Cartesian coordinate system
axes
origin
coordinates
quadrants
graph
intercepts
slope
linear equation
slope-intercept form
proportional
point-slope form
parallel
perpendicular
scatterplot

1.2 **linear function**
independent variable
dependent variable
supply curve
demand curve
equilibrium price
equilibrium quantity
fixed cost
marginal cost
cost function
break-even quantity
break-even point

1.3 **least squares line**
summation notation
coefficient of correlation

*Hardrock Hundred Mile Endurance Run, 2000 Hardrock Results Spreadsheet, http://www.run100s.com/HR/.

CHAPTER 1 REVIEW EXERCISES

1. What is marginal cost? Fixed cost?
2. What six quantities are needed to compute a coefficient of correlation?

Find the slope for each line that has a slope.

3. Through $(-2, 5)$ and $(4, 7)$
4. Through $(4, -1)$ and $(3, -3)$
5. Through the origin and $(11, -2)$
6. Through the origin and $(0, 7)$
7. $2x + 3y = 15$
8. $4x - y = 7$
9. $y + 4 = 9$
10. $3y - 1 = 14$
11. $y = -3x$
12. $x = 5y$

Find an equation in the form $y = mx + b$ (where possible) for each line.

13. Through $(5, -1)$; slope $= 2/3$
14. Through $(8, 0)$; slope $= -1/4$
15. Through $(5, -2)$ and $(1, 3)$
16. Through $(2, -3)$ and $(-3, 4)$
17. Through $(-1, 4)$; undefined slope
18. Through $(-2, 5)$; slope $= 0$
19. Through $(2, -1)$, parallel to $3x - y = 1$
20. Through $(0, 5)$, perpendicular to $8x + 5y = 3$
21. Through $(2, -10)$, perpendicular to a line with undefined slope
22. Through $(3, -5)$, parallel to $y = 4$
23. Through $(-7, 4)$, perpendicular to $y = 8$

Graph each linear equation defined as follows.

24. $y = 4x + 3$
25. $y = 6 - 2x$
26. $3x - 5y = 15$
27. $2x + 7y = 14$
28. $x + 2 = 0$
29. $y = 1$
30. $y = 2x$
31. $x + 3y = 0$

Applications

BUSINESS AND ECONOMICS

32. ***Profit*** To manufacture x thousand computer chips requires fixed expenditures of \$352 plus \$42 per thousand chips. Receipts from the sale of x thousand chips amount to \$130 per thousand.

a. Write an expression for expenditures.

b. Write an expression for receipts.

c. For profit to be made, receipts must be greater than expenditures. How many chips must be sold to produce a profit?

33. ***Supply and Demand*** The supply and demand for crabmeat in a local fish store are related by the equations

$$\text{Supply: } p = S(q) = 6q + 3$$

and

$$\text{Demand: } p = D(q) = 19 - 2q,$$

where p represents the price in dollars per pound and q represents the quantity of crabmeat in pounds per day. Find the supply and demand at each of the following prices.

a. \$10 **b.** \$15 **c.** \$18

d. Graph both the supply and the demand functions on the same axes.

e. Find the equilibrium price.

f. Find the equilibrium quantity.

34. *Supply* For a new diet pill, 60 pills will be supplied at a price of \$40, while 100 pills will be supplied at a price of \$60. Write a linear supply function for this product.

35. *Demand* The demand for the diet pills in Exercise 34 is 50 pills at a price of \$47.50 and 80 pills at a price of \$32.50. Determine a linear demand function for these pills.

36. *Supply and Demand* Find the equilibrium price and quantity for the diet pills in Exercises 34 and 35.

Cost Find a linear cost function in Exercises 37–40.

37. Eight units cost \$300; fixed cost is \$60.

38. Fixed cost is \$2000; 36 units cost \$8480.

39. Twelve units cost \$445; 50 units cost \$1585.

40. Thirty units cost \$1500; 120 units cost \$5640.

41. *Break-Even Analysis* The cost of producing x cartons of CDs is $C(x)$ dollars, where $C(x) = 200x + 1000$. The CDs sell for \$400 per carton.

a. Find the break-even quantity.

b. What revenue will the company receive if it sells just that number of cartons?

42. *Break-Even Analysis* The cost function for flavored coffee at an upscale coffeehouse is given in dollars by $C(x) = 3x + 160$, where x is in pounds. The coffee sells for \$7 per pound.

a. Find the break-even quantity.

b. What will the revenue be at that point?

43. *U.S. Imports from China* The U.S. is China's largest export market. Imports from China have grown from about 19 billion dollars in 1991 to 102 billion dollars in 2001.* This growth has been approximately linear. Use the given data pairs to write a linear equation that describes this growth in imports over the years. Let $x = 91$ represent 1991 and $x = 101$ represent 2001.

44. *U.S. Exports to China* U.S. exports to China have grown (although at a slower rate than imports) since 1991. In 1991, about 10 billion dollars of goods were exported to China. By 2001, this amount had grown to 19 billion dollars.* Write a linear equation describing the number of exports each year, with $x = 91$ representing 1991 and $x = 101$ representing 2001.

45. *Median Income* The U.S. Census Bureau reported that the median income for all U.S. households in 2000 was \$42,148. In 1993, the median income (in 2000 dollars) was \$36,746.† The median income is approximately linear and is a function of time. Find a formula for the median income, I, as a function of the year x, where x is the number of years since 1900.

46. *New Car Cost* The average new car cost for the years from 1975 to 2000 is given in the table where x is the number of years since 1900.‡

Year (x)	75	80	85	90	95	00
Cost (y)	6000	7500	12,000	16,000	20,400	24,900

a. Find an equation for the least squares line.

b. Use your equation from part a to predict the average cost of a new car in the year 2005 ($x = 105$).

c. Find and interpret the coefficient of correlation. Does it indicate that the line is a good fit for the data?

d. Plot the data. Does the scatterplot suggest the trend might not be linear?

LIFE SCIENCES

47. *World Health* In general, people tend to live longer in countries that have a greater supply of food. Listed below is the 1997 daily calorie supply and 2000 life expectancy at birth for 10 randomly selected countries.§

Country	Calories (x)	Life Expectancy (y)
Afghanistan	1523	43
Belize	2862	74
Cambodia	1974	56
France	3551	79
India	2415	64
Mexico	3137	73
New Zealand	3405	78
Peru	2310	70
Sweden	3160	80
United States	3642	78

a. Find the coefficient of correlation. Do the data seem to fit a straight line?

*International Trade Administration, Trade and Economy: Data and Analysis, Tables 55 and 56.
†U.S. Census Bureau, Current Population Survey, March 1994, 2000, 2001.
‡*Chicago Tribune,* Feb. 4, 1996, Sec. 5, p. 4 and NADA Industry Analysis Division 2002.
§*The New York Times 2003 Almanac,* pp. 479–481.

b. Draw a scatterplot of the data. Combining this with your results from part a, do the data seem to fit a straight line?

c. Find the equation for the least squares line.

d. Use your answer from part c to predict the life expectancy in the United Kingdom, which has a daily calorie supply of 3237. Compare your answer with the actual value of 78 years.

e. Briefly explain why countries with a higher daily calorie supply might tend to have a longer life expectancy.

f. (For the ambitious!) Find the coefficient of correlation and least squares line using the data for a larger sample of countries, as found in an almanac or other reference. Is the result in general agreement with the previous results?

48. *Blood Sugar and Cholesterol Levels* The following data show the connection between blood sugar levels and cholesterol levels for 8 different patients.

Patient	Blood Sugar Level (x)	Cholesterol Level (y)
1	130	170
2	138	160
3	142	173
4	159	181
5	165	201
6	200	192
7	210	240
8	250	290

For the data given in the preceding table, $\sum x = 1394$, $\sum y = 1607$, $\sum xy = 291{,}990$, $\sum x^2 = 255{,}214$, and $\sum y^2 = 336{,}155$.

a. Find the equation of the least squares line, $Y = mx + b$.

b. Predict the cholesterol level for a person whose blood sugar level is 190.

c. Find r.

SOCIAL SCIENCES

49. *Red Meat Consumption* The per capita consumption of red meat in the United States decreased from 129.5 lb in 1969 to 117.2 pounds in 1999.* Assume a linear function describes the decrease. Write a linear equation defining the function. Let x represent the number of years since 1900 and y represent the number of pounds of red meat consumed.

50. *Marital Status* More people are staying single longer in the United States. In 1990, the number of never-married adults, age 15 and over, was 52.6 million. By 2000, it was 59.9 million.[†] Assume the data increase linearly, and write an equation that defines a linear function for this data. Let x represent the number of years since 1900.

51. *Governors' Salaries* In general, the larger a state's population, the more its governor earns. Listed below are the estimated 2001 populations (in millions) and the salary of the governor (in thousands of dollars) for 8 randomly selected states.[‡]

a. Find the coefficient of correlation. Do the data seem to fit a straight line?

b. Draw a scatterplot of the data. Compare this with your answer from part a.

c. Find the equation for the least squares line.

d. Based on your answer to part c, how much does a governor's salary increase, on average, for each additional million in population?

e. Use your answer from part c to predict the governor's salary in your state. Based on your answers from parts a and b, would this prediction be very accurate? Compare with the actual salary, as listed in an almanac or other reference.

f. (For the ambitious!) Find the coefficient of correlation and least squares line using the data for all 50 states, as found in an almanac or other reference. Is the result in general agreement with the previous results?

State	AZ	DE	MD	MA	NY	PA	TN	WY
Population (x)	5.31	.80	5.38	6.38	19.01	12.29	5.74	.49
Governor's Salary (y)	95	114	120	135	179	142	85	95

**The World Almanac and Book of Facts 2003,* p. 100.
[†]U.S. Census Bureau, http://factfinder.census.gov/servlet/DTTable.
[‡]*The World Almanac and Book of Facts 2003,* pp. 364, 368.

EXTENDED APPLICATION: Using Extrapolation to Predict Life Expectancy

One reason for developing a mathematical model is to make predictions. If your model is a least squares line, you can predict the y value corresponding to some new x by substituting this x into an equation of the form $Y = mx + b$. (We use a capital Y to remind us that we're getting a predicted value rather than an actual data value.) Data analysts distinguish two very different kinds of prediction, *interpolation* and *extrapolation*. An interpolation uses a new x inside the x range of your original data. For example, if you have inflation data at five-year intervals from 1950 to 2000, estimating the rate of inflation in 1957 is an interpolation problem. But if you use the same data to estimate what the inflation rate was in 1920, or what it will be in 2020, you are extrapolating.

In general, interpolation is much safer than extrapolation, because data that are approximately linear over a short interval may be nonlinear over a larger interval. One way to detect nonlinearity is to look at *residuals,* which are the differences between the actual data values and the values predicted by the line of best fit. Here is a simple example:

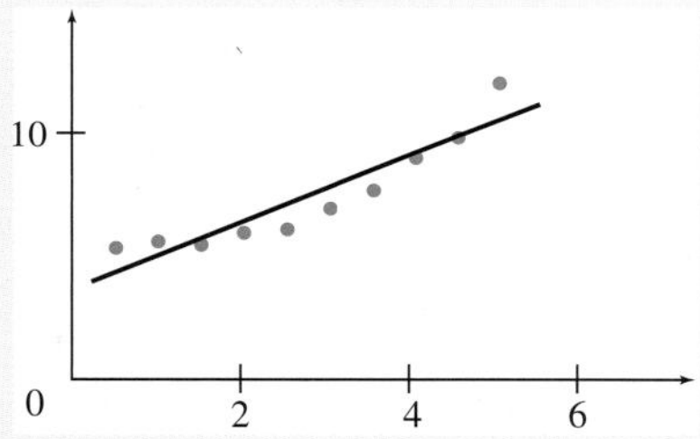

FIGURE 26

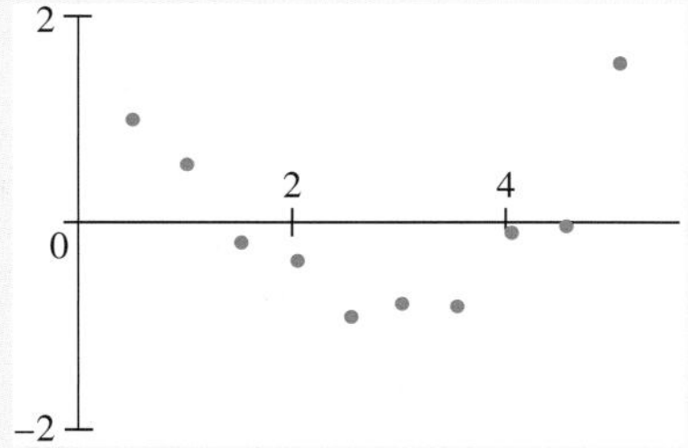

FIGURE 27

The regression equation for the linear fit on the top is $Y = 3.431 + 1.334x$. Since the r value for this regression line is .93, our linear model fits the data very well. But we might notice that the predictions are a bit low at the ends and high in the middle. We can get a better look at this pattern by plotting the residuals. To find them, we put each value of the independent variable into the regression equation, calculate the predicted value Y, and subtract it from the actual y value. The residual plot is below the linear fit graph, with the vertical axis rescaled to exaggerate the pattern. The residuals indicate that our data has a nonlinear, U-shaped component that is not captured by the linear fit. Extrapolating from this data set is probably not a good idea; our linear prediction for the value of y when x is 10 may be much too low.

Exercises

*The following table gives the life expectancy at birth of females born in the United States in various years from 1950 to 2000.**

Year of Birth	Life Expectancy (years)
1950	71.3
1960	73.1
1970	74.7
1980	77.4
1985	78.2
1990	78.8
1995	78.9
2000	79.5

1. Find an equation for the least squares line for this data, using year of birth as the independent variable.
2. Use your regression equation to guess a value for the life expectancy of females born in 1900.
3. Compare your answer with the actual life expectancy for females born in 1900, which was 48.3 years. Are you surprised?
4. Find the life expectancy predicted by your regression equation for each year in the table, and subtract it from the actual value in the second column. This gives you a table of residuals. Plot your residuals as points on a graph.
5. Now look at the residuals as a fresh data set and see if you can sketch the graph of a smooth function that fits the residuals well. How easy do you think it will be to predict the life expectancy at birth of females born in 2010?

**The World Almanac and Book of Facts 2003,* p. 75.

6. What will happen if you try linear regression on the *residuals?* If you're not sure, use your calculator or software to find the regression equation for the residuals. Why does this result make sense?
7. Since most of the females born in 1985 are still alive, how did the Public Health Service come up with a life expectancy of 78.2 years for these women?

Directions for Group Project

Assume that you and your group (3–5 students) are preparing a report for a local health agency that is interested in using linear regression to predict life expectancy. Using the questions above as a guide, write a report that addresses the spirit of each question and any issues related to that question. The report should be mathematically sound, grammatically correct, and professionally crafted. Provide recommendations as to whether the health agency should proceed with the linear equation or whether it should seek other means of making such predictions.

CHAPTER

2 Nonlinear Functions

There are fourteen mountain peaks over 8000 meters on the Earth's surface. At these altitudes climbers face the challenge of "thin air," since atmospheric pressure is about one third of the pressure at sea level. An exercise in Section 4 of this chapter shows how the change in atmospheric pressure with altitude can be modeled with an exponential function.

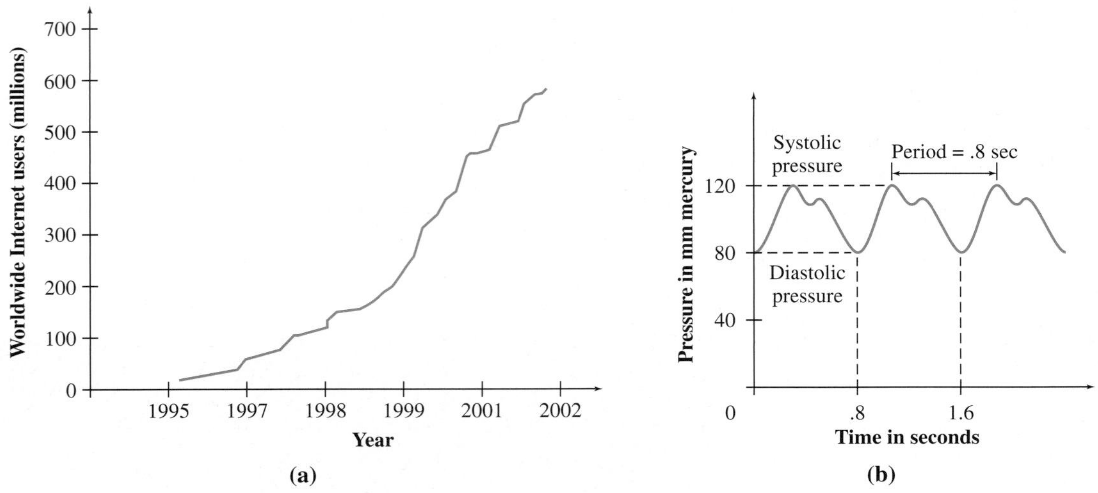

FIGURE 1

Figure 1(a) shows the estimated number of Internet users worldwide each year from 1995–2002.* For example, the figure shows in June 1998, there were about 120 million Internet users worldwide. Figure 1(b) shows the variation in blood pressure for a typical person.† (Systolic and diastolic pressures are the upper and lower limits in the periodic changes in pressure that produce the pulse. The length of time between peaks is called the period of the pulse.) After .8 second, the blood pressure is the same as its starting value, 80 millimeters of mercury.

Figures 1(a) and 1(b) illustrate functions that, unlike those studied in Chapter 1, are *nonlinear*. Their graphs are not straight lines. Linear functions are simple to study, and they can be used to approximate many functions over short intervals. But most functions exhibit behavior that, in the long run, does not follow a straight line. In this chapter we will study some of the most common nonlinear functions.

2.1 PROPERTIES OF FUNCTIONS

THINK ABOUT IT *How has the number of Internet users worldwide changed with time, and what does this indicate for future use of the Internet?*

After developing the concept of nonlinear functions, we will answer this question in one of the exercises.

As we saw in Chapter 1, the linear cost function $C(x) = 12x + 100$ for video games is related to the number of items produced. The number of games produced is the independent variable and the corresponding cost per game is the dependent variable because the cost of a game depends on the number produced. When a specific number of games (say 1000) is substituted for x, the cost $C(x)$ has one specific value $(12 \cdot 1000 + 100)$. Because of this, the variable $C(x)$ is said to be a *function* of x.

*Complied by NUA Internet Surveys.

†De Sapio, Rodolfo, *Calculus for the Life Sciences*. Copyright © 1976, 1978 by W. H. Freeman and Company. Reprinted by permission.

FUNCTION

A **function** is a rule that assigns to each element from one set exactly one element from another set.

In most cases in this book, the "rule" mentioned in the box is expressed as an equation, such as $C(x) = 12x + 100$. When an equation is given for a function, we say that the equation *defines* the function. Whenever x and y are used in this book to define a function, x represents the independent variable and y the dependent variable. Of course, letters other than x and y could be used and are often more meaningful. For example, if the independent variable represents the number of items sold for $4 each and the dependent variable represents revenue, we might write $R = 4s$.

The independent variable in a function can take on any value within a specified set of values called the *domain*.

DOMAIN AND RANGE

The set of all possible values of the independent variable in a function is called the **domain** of the function, and the resulting set of possible values of the dependent variable is called the **range.**

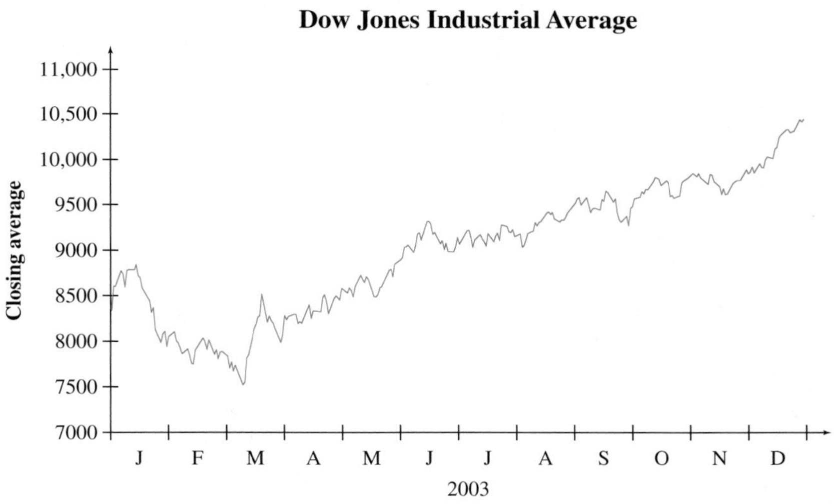

FIGURE 2

An important function to investors around the world is the Dow Jones industrial average, a performance measure of the stock market. Figure 2 shows how this average varied over the year 2003.* Let us label this function $y = f(x)$, where y is the Dow Jones industrial average and x is the time in days from the beginning of 2003. Notice that the function increases and decreases during the year, so it is not linear, although a linear function could be used as a very rough approximation. Such a function, whose graph is not a straight line, is called a *nonlinear function*.

*Data from Yahoo! Finance: www.yahoo.com.

The concepts you learned in the section on linear functions apply to this and other nonlinear functions as well. The independent variable here is x, the time in days; the dependent variable is y, the average at any time. The domain is $\{x \mid 0 \le x \le 365\}$, or $[0, 365]$; $x = 0$ corresponds to the beginning of the day on January 1, and $x = 365$ corresponds to the end of the day on December 31. By looking for the lowest and highest values of the function, we estimate the range to be approximately $\{y \mid 7500 \le y \le 9500\}$, or $[7500, 9500]$. As with linear functions, the domain is mapped along the horizontal axis and the range along the vertical axis.

We do not have a formula for $f(x)$. (If we had possessed such a formula at the beginning of 2003, we could have made a lot of money!) Instead, we can use the graph to estimate values of the function. To estimate $f(15)$, for example, we draw a vertical line from January 15, as shown in Figure 3. The y-coordinate seems to be roughly 8700, so we estimate $f(15) \approx 8700$. Similarly, if we wanted to solve the equation $f(x) = 8200$, we would look for points on the graph that have a y-coordinate of 8200. As Figure 4 shows, this occurs at several points. The first time is around January 24, and the last time is around April 11 (the 101st day of the year). Thus $f(x) = 8200$ when $x = 24$ and $x = 101$, as well as for a couple of values of x between 24 and 101.

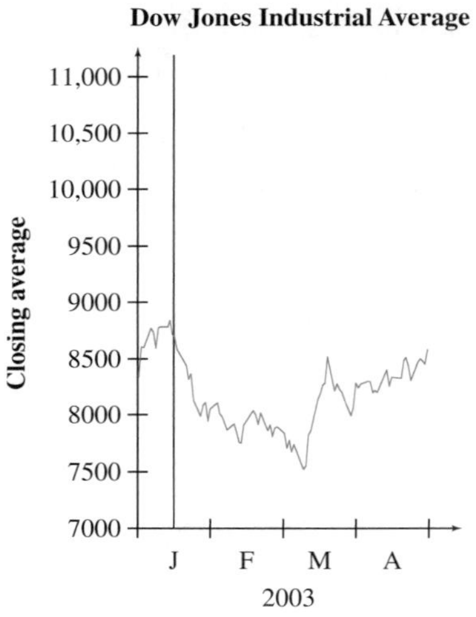

FIGURE 3

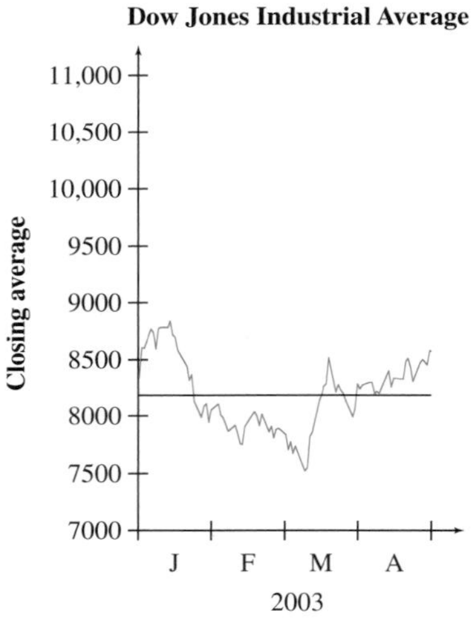

FIGURE 4

Dow Jones Industrial Average

Day (x)	Close (y)
0	8342.38
1	8342.38
2	8607.52
3	8601.69
4	8601.69
5	8601.69
6	8773.57
7	8740.59

This function can also be given as a table. The table in the margin shows the value of the function for several values of x.

Notice from the table that $f(0) = f(1) = 8342.38$. The stock market was closed for the first day of 2003, so the Dow Jones average did not change. This illustrates an important property of functions: several different values of the independent variable can have the same value for the dependent variable. On the other hand, we cannot have several different y-values corresponding to the same value of x; if we did, this would not be a function.

What is $f(5.5)$? We do not know. When the stock market opened on January 6, the Dow Jones industrial average was 8601.69. The closing value that day was 8773.57. We do not know what happened in between, although this information is recorded by the New York Stock Exchange.

Functions arise in numerous applications, and an understanding of them is critical for understanding calculus. The following example shows some of the ways functions can be represented and will help you in determining whether a relationship between two variables is a function or not.

EXAMPLE 1 *Functions*

Which of the following are functions?

(a)

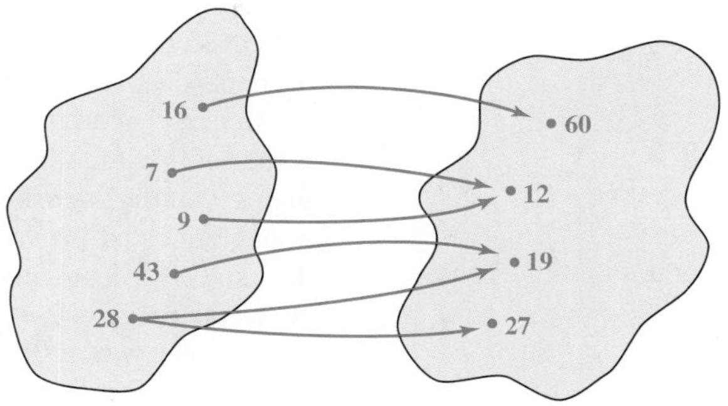

FIGURE 5

Solution Figure 5 shows that an x-value of 28 corresponds to *two* y-values, 19 and 27. In a function, each x must correspond to exactly one y, so this correspondence is not a function.

(b) The optical reader at the checkout counter in many stores that converts codes to prices

Solution For each code, the reader produces exactly one price, so this is a function.

(c) The x^2 key on a calculator

Solution This correspondence between input and output is a function because the calculator produces just one x^2 (one y-value) for each x-value entered. Notice also that two x-values, such as 3 and -3, produce the same y-value of 9, but this does not violate the definition of a function.

(d)

x	1	1	2	2	3	3
y	3	-3	5	-5	8	-8

Solution Since at least one x-value corresponds to more than one y-value, this table does not define a function.

(e) The set of ordered pairs with first elements mothers and second elements their children

Solution Here the mother is the independent variable and the child is the dependent variable. For a given mother, there may be several children, so this correspondence is not a function.

(f) The set of ordered pairs with first elements children and second elements their birth mothers

Solution In this case the child is the independent variable and the mother is the dependent variable. Since each child has only one birth mother, this is a function.

EXAMPLE 2 *Functions*

Decide whether each equation represents a function. (Assume that x represents the independent variable here, an assumption we shall make throughout this book.) Give the domain and range of any functions.

(a) $y = 11 - 4x^2$

Solution For a given value of x, calculating $11 - 4x^2$ produces exactly one value of y. (For example, if $x = -7$, then $y = 11 - 4(-7)^2 = -185$, so $f(-7) = -185$.) Since one value of the independent variable leads to exactly one value of the dependent variable, $y = 11 - 4x^2$ meets the definition of a function.

Because x can take on any real-number value, the domain of this function is the set of all real numbers. Finding the range is more difficult. One way to find it would be to ask what possible values of y could come out of this function. Notice that the value of y is 11 minus a quantity that is always 0 or positive, since $4x^2$ can never be negative. There is no limit to how large $4x^2$ can be, so the range is $(-\infty, 11]$.

Another way to find the range would be to examine the graph. Figure 6 shows a graphing calculator view of this function, and we can see that the function takes on y-values of 11 or less. The calculator cannot tell us, however, whether the function continues to go down past the viewing window, or turns back up. To find out, we need to study this type of function more carefully, as we will do in the next section.

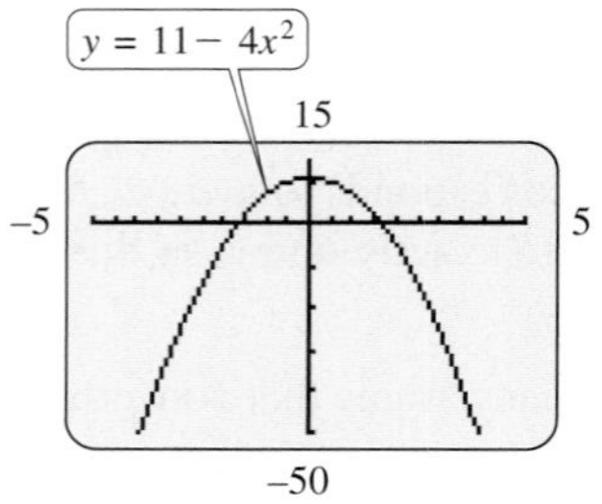

FIGURE 6

NOTE The actual graph of $y = 11 - 4x^2$ is a smooth curve. The jaggedness seen in Figure 6 is due to the limited resolution of the graphing calculator screen. ■

(b) $y^2 = x$

Solution Suppose $x = 36$. Then $y^2 = x$ becomes $y^2 = 36$, from which $y = 6$ or $y = -6$. Since one value of the independent variable can lead to two values of the dependent variable, $y^2 = x$ does not represent a function.

(c) $y = 7$

Solution No matter what the value of x, the value of y is always 7. This is indeed a function; it assigns exactly one element, 7, to each value of x. Such a function is known as a **constant function.** The domain is the set of all real numbers, and the range is the set $\{7\}$. Its graph is the horizontal line that intersects the y-axis at $y = 7$. Every constant function has a horizontal line for its graph.

The following agreement on domains is customary.

AGREEMENT ON DOMAINS

Unless otherwise stated, assume that the domain of all functions defined by an equation is the largest set of real numbers that are meaningful replacements for the independent variable.

For example, suppose

$$y = \frac{-4x}{2x - 3}.$$

Any real number can be used for x except $x = 3/2$, which makes the denominator equal 0. By the agreement on domains, the domain of this function is the set of all real numbers except 3/2, which we denote $\{x \mid x \neq 3/2\}$, $\{x \neq 3/2\}$, or $(-\infty, 3/2) \cup (3/2, \infty)$.*

CAUTION When finding the domain of a function, there are two operations to avoid: (1) dividing by zero; and (2) taking the square root (or any even root) of a negative number. Later chapters will present other functions, such as logarithms, which require further restrictions on the domain. For now, just remember these two restrictions on the domain. ■

EXAMPLE 3 *Domain and Range*

Find the domain and range for each function defined as follows.

(a) $y = x^2$

Solution Any number may be squared, so the domain is the set of all real numbers, written $(-\infty, \infty)$. Since $x^2 \geq 0$ for every value of x, the range is $[0, \infty)$.

(b) $y = \sqrt{6 - x}$

Solution For y to be a real number, $6 - x$ must be nonnegative. This happens only when $6 - x \geq 0$, or $6 \geq x$, making the domain $(-\infty, 6]$. The range is $[0, \infty)$ because $\sqrt{6 - x}$ is always nonnegative.

(c) $y = \sqrt{2x^2 + 5x - 12}$

Solution The domain includes only those values of x satisfying $2x^2 + 5x - 12 \geq 0$. Using the methods for solving a quadratic inequality produces the domain

$$(-\infty, -4] \cup [3/2, \infty).$$

As in part (b), the range is $[0, \infty)$.

FOR REVIEW

Section R.5 demonstrates the method for solving a quadratic inequality. To solve $2x^2 + 5x - 12 \geq 0$, factor the quadratic to get $(2x - 3)(x + 4) \geq 0$. Setting each factor equal to 0 gives $x = 3/2$ or $x = -4$, leading to the intervals $(-\infty, -4]$, $[-4, 3/2]$, and $[3/2, \infty)$. Testing a number from each interval shows that the solution is $(-\infty, -4] \cup [3/2, \infty)$.

(d) $y = \dfrac{1}{x + 3}$

Solution Since the denominator cannot be zero, $x \neq -3$ and the domain is

$$(-\infty, -3) \cup (-3, \infty).$$

Because the numerator can never be zero, $y \neq 0$. There are no other restrictions on y, so the range is $(-\infty, 0) \cup (0, \infty)$.

To understand how a function works, think of a function f as a machine—for example, a calculator or computer—that takes an input x from the domain and uses it to produce an output $f(x)$ (which represents the y-value), as shown in Figure 7 on the next page. In the Dow Jones example, when we put 4 into the machine, we get out 8601.69, since $f(4) = 8601.69$.

*The *union* of sets A and B, written $A \cup B$, is defined as the set of all elements in A or B or both.

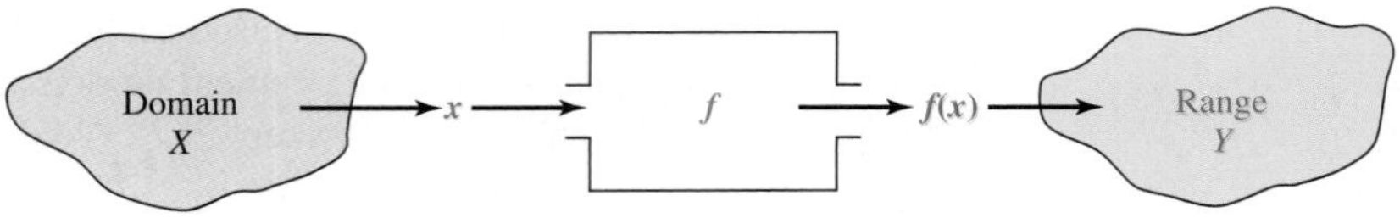

FIGURE 7

EXAMPLE 4 *Evaluating Functions*

Let $g(x) = -x^2 + 4x - 5$. Find the following.

(a) $g(3)$

Solution Replace x with 3.

$$g(3) = -3^2 + 4 \cdot 3 - 5 = -9 + 12 - 5 = -2$$

(b) $g(a)$

Solution Replace x with a to get $g(a) = -a^2 + 4a - 5$.

This replacement of one variable with another is important in later chapters.

(c)
$$\begin{aligned} g(x + h) &= -(x + h)^2 + 4(x + h) - 5 \\ &= -(x^2 + 2xh + h^2) + 4(x + h) - 5 \\ &= -x^2 - 2xh - h^2 + 4x + 4h - 5 \end{aligned}$$

(d) $g\left(\frac{2}{r}\right) = -\left(\frac{2}{r}\right)^2 + 4\left(\frac{2}{r}\right) - 5 = -\frac{4}{r^2} + \frac{8}{r} - 5$

(e) Find all values of x such that $g(x) = -12$.

Solution Set $g(x)$ equal to -12, and then add 12 to both sides to make one side equal to 0.

$$\begin{aligned} -x^2 + 4x - 5 &= -12 \\ -x^2 + 4x + 7 &= 0 \end{aligned}$$

This equation does not factor, but can be solved with the quadratic formula, which says that if $ax^2 + bx + c = 0$, where $a \neq 0$, then

$$x = \frac{-b \pm \sqrt{b^2 - 4ac}}{2a}.$$

In this case, with $a = -1$, $b = 4$, and $c = 7$, we have

$$\begin{aligned} x &= \frac{-4 \pm \sqrt{16 - 4(-1)7}}{2(-1)} \\ &= \frac{-4 \pm \sqrt{44}}{-2} \\ &= 2 \pm \sqrt{11} \\ &\approx -1.317 \quad \text{or} \quad 5.317. \end{aligned}$$

We can verify the results of parts (a) and (e) of the previous example using a graphing calculator. In Figure 8(a) on the next page, after graphing $f(x) = -x^2 + 4x - 5$, we have used the "value" feature on the TI-83/84 Plus

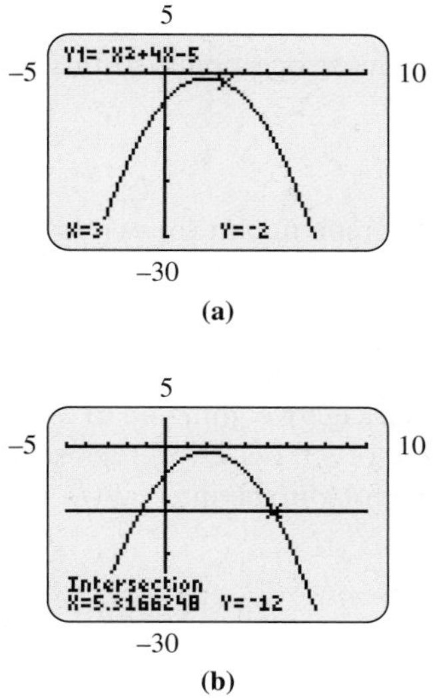

FIGURE 8

to support our answer from part (a). In Figure 8(b) we have used the "intersect" feature to find the intersection of $y = g(x)$ and $y = -12$. The result is $x = 5.3166248$, which is one of our two answers to part (e). The graph clearly shows that there is another answer on the opposite side of the y-axis.

CAUTION Notice from Example 4(c) that $g(x + h)$ is *not* the same as $g(x) + h$, which equals $-x^2 + 4x - 5 + h$. There is a significant difference between applying a function to the quantity $x + h$ and applying a function to x and adding h afterward. ■

If you tend to get confused when replacing x with $x + h$, as in Example 4(c), you might try replacing the x in the original function with a box, like this:

$$g\left(\boxed{}\right) = -\left(\boxed{}\right)^2 + 4\left(\boxed{}\right) - 5$$

Then, to compute $g(x + h)$, just enter $x + h$ into the box:

$$g\left(\boxed{x+h}\right) = -\left(\boxed{x+h}\right)^2 + 4\left(\boxed{x+h}\right) - 5$$

and proceed as in Example 4(c).

Notice in the Dow Jones example that to find the value of the function for a given value of x, we drew a vertical line from the value of x and found where it intersected the graph. If a graph is to represent a function, each value of x from the domain must lead to exactly one value of y. In the graph in Figure 9, the domain value x_1 leads to *two* y-values, y_1 and y_2. Since the given x-value corresponds to two different y-values, this is not the graph of a function. This example suggests the **vertical line test** for the graph of a function.

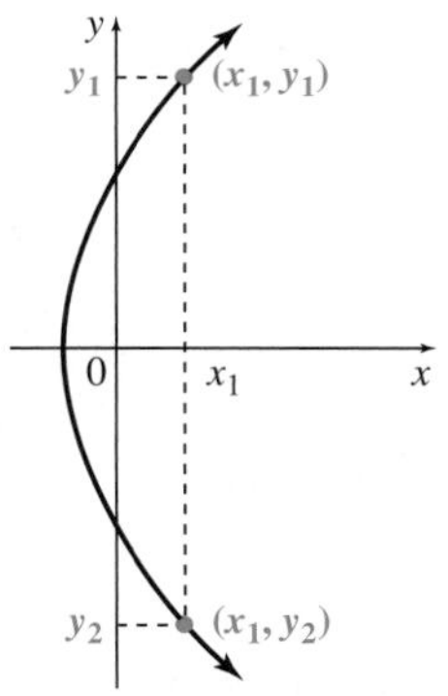

FIGURE 9

> **VERTICAL LINE TEST**
>
> If a vertical line intersects a graph in more than one point, the graph is not the graph of a function.

EXAMPLE 5 *Vertical Line Test*

Use the vertical line test to decide which of the graphs in Figure 10 are graphs of functions.

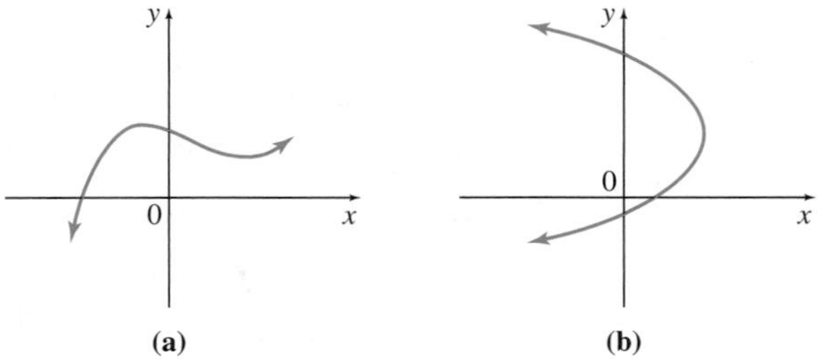

FIGURE 10

Solution

(a) Every vertical line intersects this graph in at most one point, so this is the graph of a function.

Solution

(b) It is possible for a vertical line to intersect the graph in part (b) twice. This is not the graph of a function.

EXAMPLE 6 *Delivery Charges*

An overnight delivery service charges \$25 for a package weighing up to 2 lb. For each additional pound there is an additional charge of \$3. Let $D(x)$ represent the cost to send a package weighing x lb. Graph $D(x)$ for x in the interval $(0, 6]$.

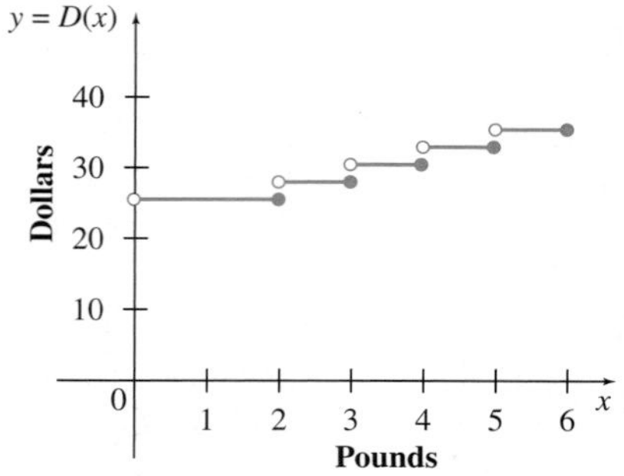

FIGURE 11

Solution For x in the interval $(0, 2]$, the shipping cost is $y = 25$. For x in $(2, 3]$, the shipping cost is $y = 25 + 3 = 28$. For x in $(3, 4]$, the shipping cost is $y = 28 + 3 = 31$, and so on. The graph is shown in Figure 11.

The function discussed in Example 6 is called a **step function.** Many real-life situations are best modeled by step functions. Additional examples are given in the exercises.

In Chapter 1 you saw several examples of linear models. Section 1.3 showed how the least squares line is used to find an equation to model data with a scatterplot that has an approximately linear pattern. In Example 7, we write an equation to model the area of a lot.

EXAMPLE 7 *Area*

A fence is to be built against a brick wall to form a rectangular lot, as shown in Figure 12. Only three sides of the fence need to be built, because the wall forms the fourth side. The contractor will use 200 m of fencing. Let the length of the wall be l and the width w, as shown in Figure 12.

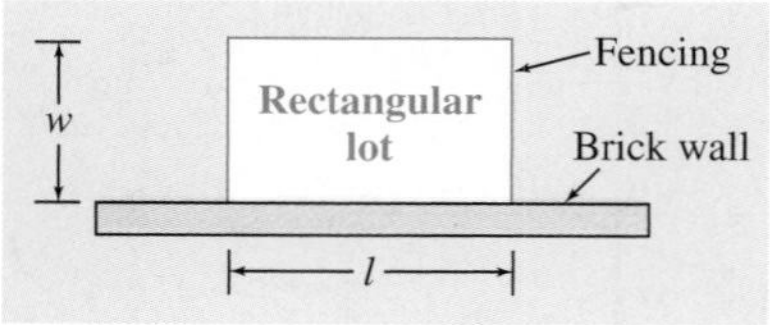

FIGURE 12

(a) Find the area of the lot as a function of the length l.

Solution The area formula for a rectangle is area = length × width, or

$$A = lw.$$

We want the area as a function of the length only, so we must eliminate the width. We use the fact that the total amount of fencing is the sum of the three sections, one length and two widths, so $200 = l + 2w$. Solve this for w:

$$\begin{aligned} 200 &= l + 2w \\ 200 - l &= 2w \qquad \text{Subtract } l \text{ from both sides.} \\ 100 - l/2 &= w. \qquad \text{Divide both sides by 2.} \end{aligned}$$

Substituting this into the formula for area gives

$$A = l(100 - l/2).$$

(b) Find the domain of the function in part (a).

Solution The length cannot be negative, so $l \geq 0$. Similarly, the width cannot be negative, so $100 - l/2 \geq 0$, from which we find $l \leq 200$. Therefore, the domain is $[0, 200]$.

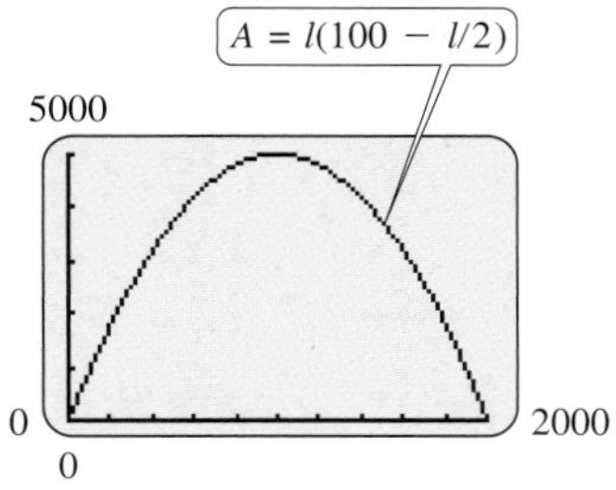

FIGURE 13

(c) Sketch a graph of the function in part (a).

Solution The result from a graphing calculator is shown in Figure 13. Notice that at the endpoints of the domain, when $l = 0$ and $l = 200$, the area is 0. This makes sense: If the length or width is 0, the area will be 0 as well. In between, as the length increases from 0 to 100 m, the area gets larger, and seems to reach a peak of 5000 m^2 when $l = 100$ m. After that, the area gets smaller as the length continues to increase because the width is becoming smaller.

In the next section, we will study this type of function in more detail and determine exactly where the maximum occurs.

2.1 EXERCISES

Which of the following rules define y as a function of x?

1.

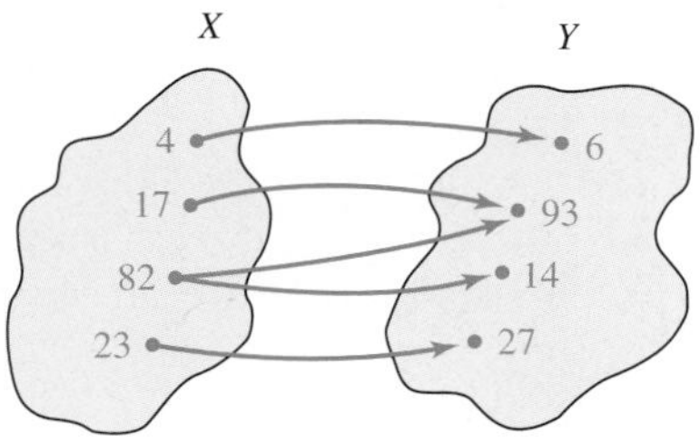

2.

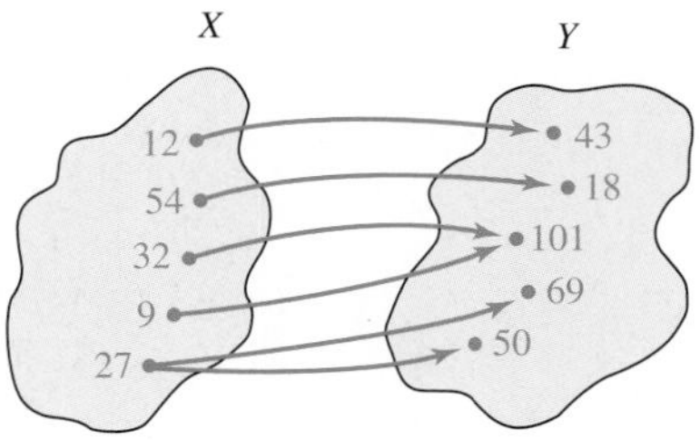

3.

x	y
3	9
2	4
1	1
0	0
−1	1
−2	4
−3	9

4.

x	y
9	3
4	2
1	1
0	0
1	−1
4	−2
9	−3

5. $y = x^3$

6. $y = \sqrt{x}$

7. $x = |y|$

8. $x = y^4 - 1$

List the ordered pairs obtained from each equation, given $\{-2, -1, 0, 1, 2, 3\}$ as the domain. Graph each set of ordered pairs. Give the range.

9. $y = 2x + 3$

10. $y = -4x + 9$

11. $2y - x = 5$

12. $6x - y = -3$

13. $y = x(x + 1)$

14. $y = (x - 2)(x - 3)$

15. $y = x^2$

16. $y = -2x^2$

17. $y = \dfrac{1}{x + 3}$

18. $y = \dfrac{-2}{x + 4}$

19. $y = \dfrac{3x - 3}{x + 5}$

20. $y = \dfrac{2x + 1}{x + 3}$

Give the domain of each function defined as follows.

21. $f(x) = 2x$

22. $f(x) = x + 2$

23. $f(x) = x^4$

24. $f(x) = (x - 2)^2$

25. $f(x) = \sqrt{16 - x^2}$

26. $f(x) = |x - 1|$

27. $f(x) = (x - 3)^{1/2}$

28. $f(x) = (3x + 5)^{1/2}$

29. $f(x) = \dfrac{2}{x^2 - 4}$

30. $f(x) = \dfrac{-8}{x^2 - 36}$

31. $f(x) = -\sqrt{\dfrac{2}{x^2 + 9}}$

32. $f(x) = -\sqrt{\dfrac{5}{x^2 + 36}}$

33. $f(x) = \sqrt{x^2 - 4x - 5}$

34. $f(x) = \sqrt{15x^2 + x - 2}$

35. $f(x) = \dfrac{1}{\sqrt{x^2 - 6x + 8}}$

36. $f(x) = \sqrt{\dfrac{x + 1}{x - 1}}$

Give the domain and the range of each function. Where arrows are drawn, assume the function continues in the indicated direction.

37.

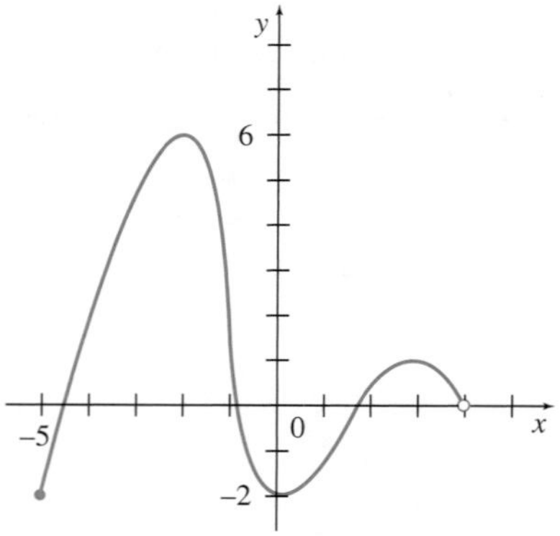

38.

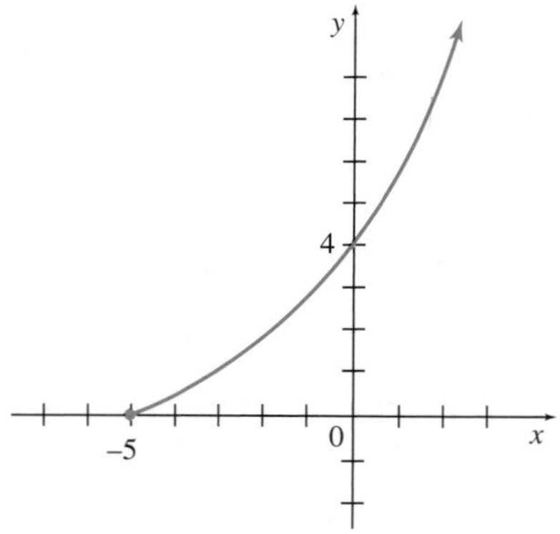

39.

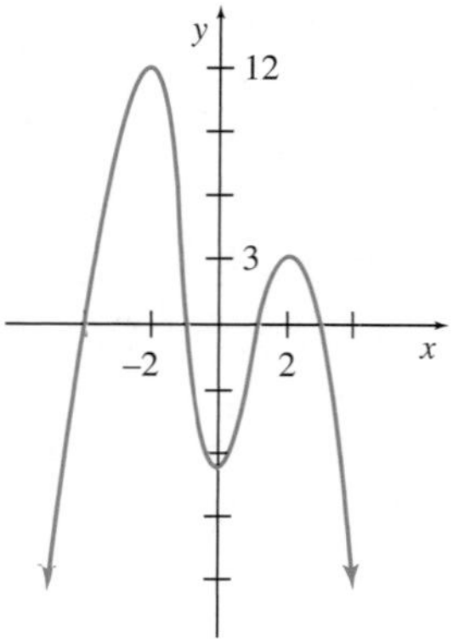

40.

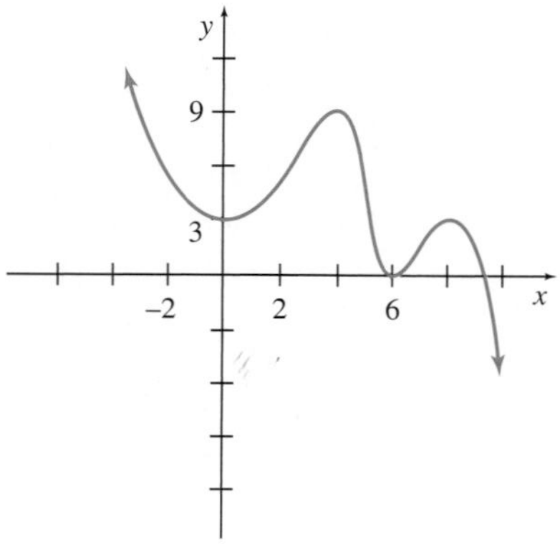

For each function, find **(a)** $f(4)$, **(b)** $f(-1/2)$, **(c)** $f(a)$, **(d)** $f(2/m)$, *and* **(e)** *any values of x such that* $f(x) = 1$.

41. $f(x) = -x^2 + 5x + 1$

42. $f(x) = (x + 3)(x - 4)$

43. $f(x) = \dfrac{2x + 1}{x - 2}$

44. $f(x) = \dfrac{3x - 5}{2x + 3}$

In Exercises 45–48, give the domain and range. Then, use each graph to find **(a)** $f(-2)$, **(b)** $f(0)$, **(c)** $f(1/2)$, *and* **(d)** *any values of x such that* $f(x) = 1$.

45.

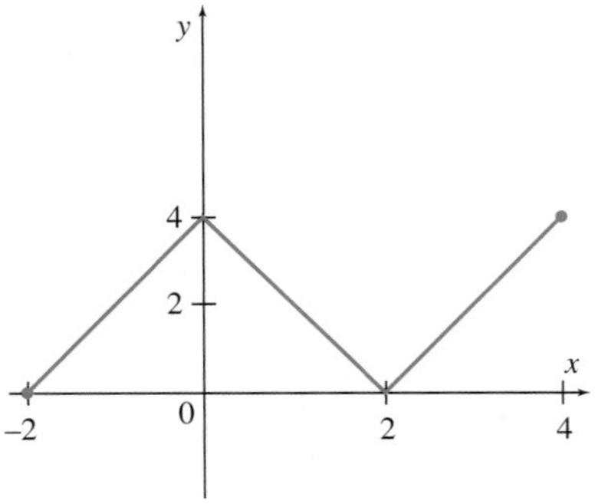

46.

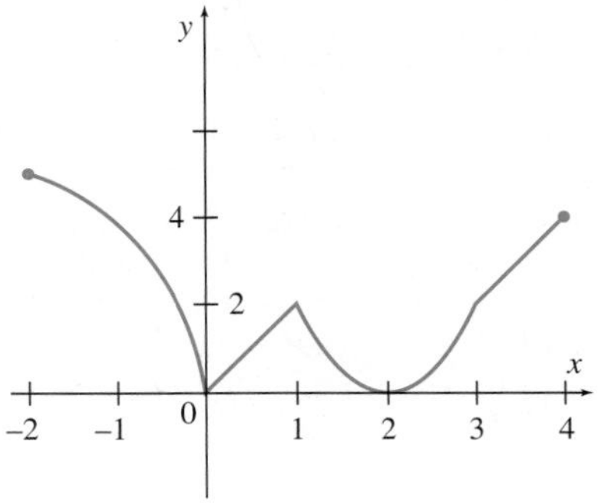

47.

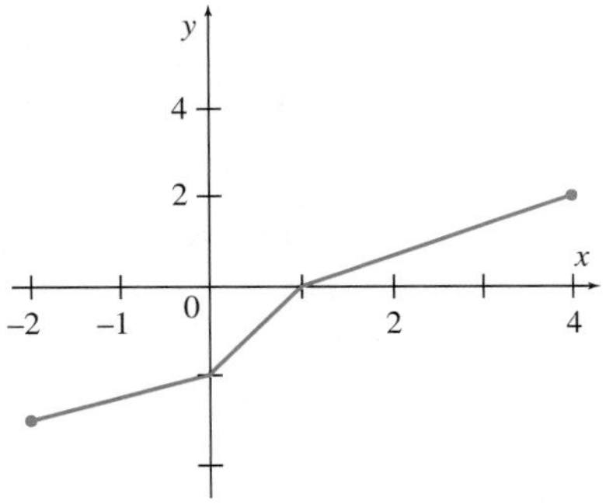

48.

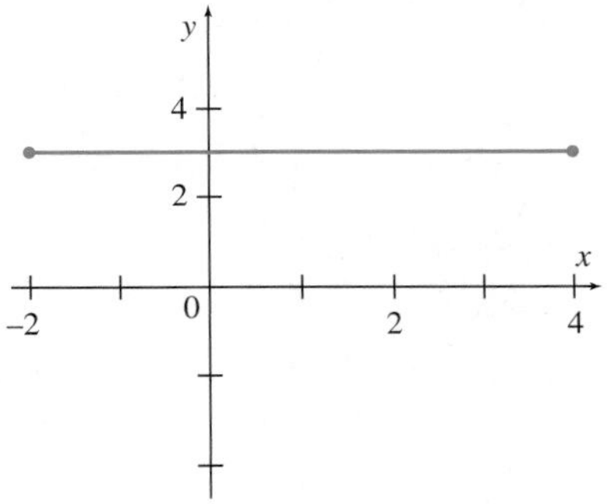

Let $f(x) = 6x^2 - 2$ *and* $g(x) = x^2 - 2x + 5$ *to find the following values.*

49. $f(m - 3)$

50. $f(2r - 1)$

51. $g(r + h)$

52. $g(z - p)$

53. $g\left(\dfrac{3}{q}\right)$

54. $g\left(-\dfrac{5}{z}\right)$

Decide whether each graph represents a function.

55.

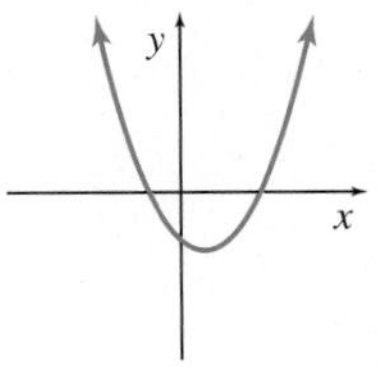

56.

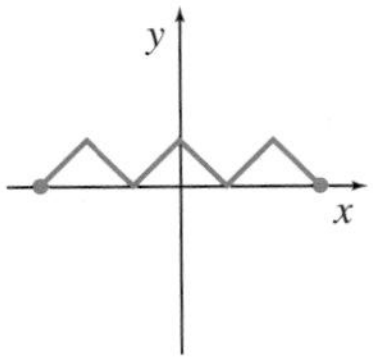

57.

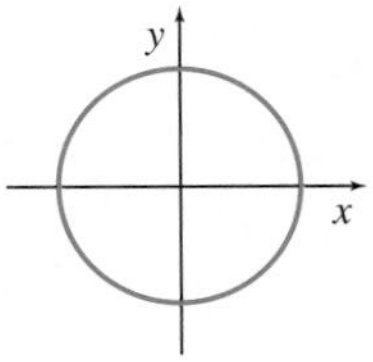

58.

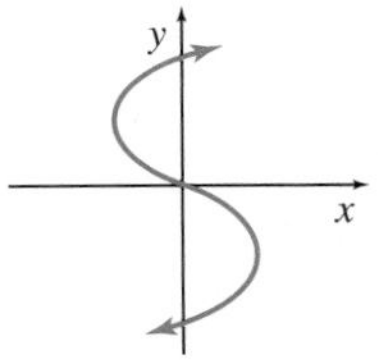

59.

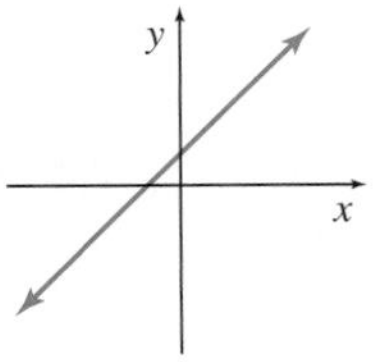

60.

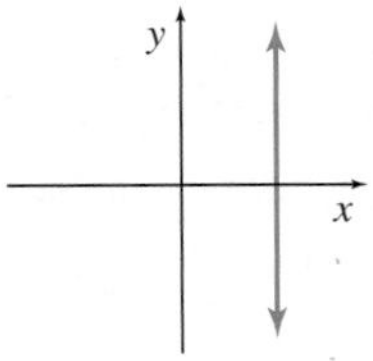

For each function defined as follows, find **(a)** $f(x + h)$, **(b)** $f(x + h) - f(x)$, *and* **(c)** $[f(x + h) - f(x)]/h$.

61. $f(x) = x^2 - 4$

62. $f(x) = 8 - 3x^2$

63. $f(x) = 2x^2 - 4x - 5$

64. $f(x) = -4x^2 + 3x + 2$

65. $f(x) = \dfrac{1}{x}$

66. $f(x) = -\dfrac{1}{x^2}$

Applications

BUSINESS AND ECONOMICS

67. *Internet Users* The following table* shows the estimated number of Internet users worldwide from 1995–2001. As we saw in Figure 1(a), these data define a function. However, the graph shows a function with a y-value for every x-value in the interval $1995 \le x \le 2001$, while the table shows a function with just 7 x-values. Let $y = f(x)$ represent the number of Internet users and x represent the years.

Worldwide Internet Users

Year	Millions of Users
1995	16
1996	36
1997	101
1998	150
1999	250
2000	451
2001	553

a. What is the independent variable?

b. What is the dependent variable?

c. Find $f(1997)$.

d. Give the domain and range of the function.

68. *Saw Rental* A chain-saw rental firm charges \$7 per day or fraction of a day to rent a saw, plus a fixed fee of \$4 for resharpening the blade. Let $S(x)$ represent the cost of renting a saw for x days. Find the following.

a. $S\left(\dfrac{1}{2}\right)$ **b.** $S(1)$ **c.** $S\left(1\dfrac{1}{4}\right)$

d. $S\left(3\dfrac{1}{2}\right)$ **e.** $S(4)$ **f.** $S\left(4\dfrac{1}{10}\right)$

g. What does it cost to rent a saw for $4\frac{9}{10}$ days?

h. A portion of the graph of $y = S(x)$ is shown here. Explain how the graph could be continued.

i. What is the independent variable?

j. What is the dependent variable?

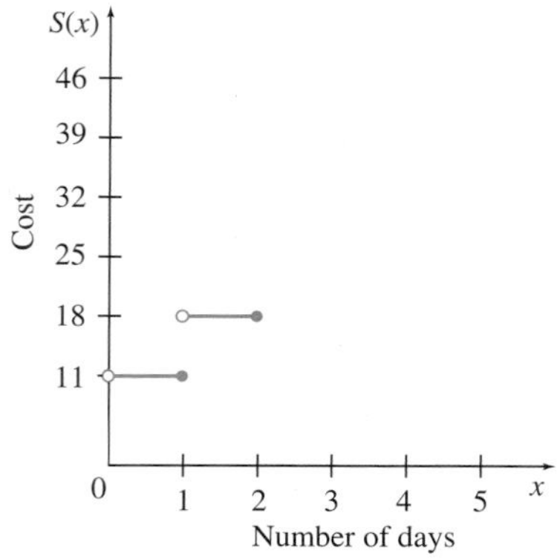

k. Write a sentence or two explaining what part f and its answer represent.

l. We have left $x = 0$ out of the graph. Discuss why it should or shouldn't be included. If it were included, how would you define $S(0)$?

69. *Rental Car Cost* The cost to rent a mid-size car is \$27 per day or fraction of a day. If the car is picked up in Lansing and dropped off in West Lafayette, there is a fixed \$25 drop-off charge. Let $C(x)$ represent the cost of renting the car for x days, taking it from Lansing to West Lafayette. Find the following.

a. $C(3/4)$ **b.** $C(9/10)$ **c.** $C(1)$ **d.** $C\left(1\dfrac{5}{8}\right)$

e. Find the cost of renting the car for 2.4 days.

f. Graph $y = C(x)$.

g. Is C a function?

h. Is C a linear function?

*NUA Internet Surveys.

LIFE SCIENCES

70. *Whales Diving* The figure shows the depth of a diving sperm whale as a function of time, as recorded by researchers at the Woods Hole Oceanographic Institution in Massachusetts.*

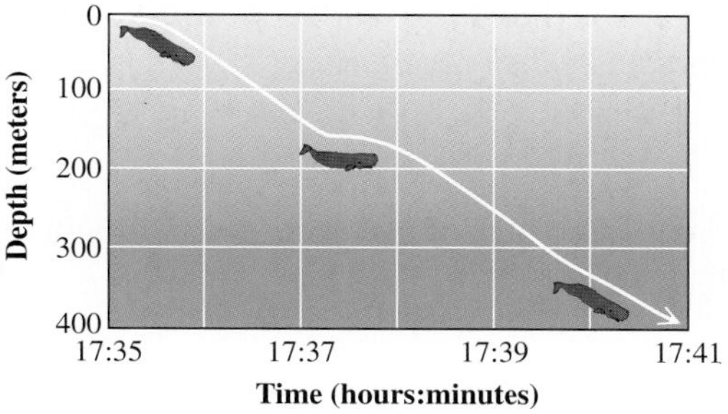

Find the depth of the whale at the following times.

a. 17 hours and 37 minutes

b. 17 hours and 39 minutes

71. *Metabolic Rate* The basal metabolic rate (in kcal/day) for large anteaters is given by

$$y = f(x) = 19.7x^{.753},$$

where x is the anteater's weight in kilograms.†‡

a. Find the basal metabolic rate for anteaters with the following weights.

i. 5 kg **ii.** 25 kg

b. Suppose the anteater's weight is given in pounds rather than kilograms. Given that 1 lb = .454 kg, find a function $x = g(z)$ giving the anteater's weight in kilograms if z is the animal's weight in pounds.

72. *Swimming Energy* The energy expenditure (in kcal/km) for animals swimming at the surface of the water is given by

$$y = f(x) = .01x^{.88},$$

where x is the animal's weight in grams.§

a. Find the energy for the following animals swimming at the surface of the water.

i. A muskrat weighing 800 g

ii. A sea otter weighing 20,000 g

b. Suppose the animal's weight is given in kilograms rather than grams. Given that 1 kg = 1000 g, find a function $x = g(z)$ giving the animal's weight in grams if z is the animal's weight in kilograms.

GENERAL INTEREST

73. *Energy Consumption* Over the last century, the world has shifted from using high-carbon sources of energy such as wood to lower carbon fuels such as oil and natural gas, as shown in the figure.‖ The rise in carbon emissions during this time has caused concern because of its suspected contribution to global warming.

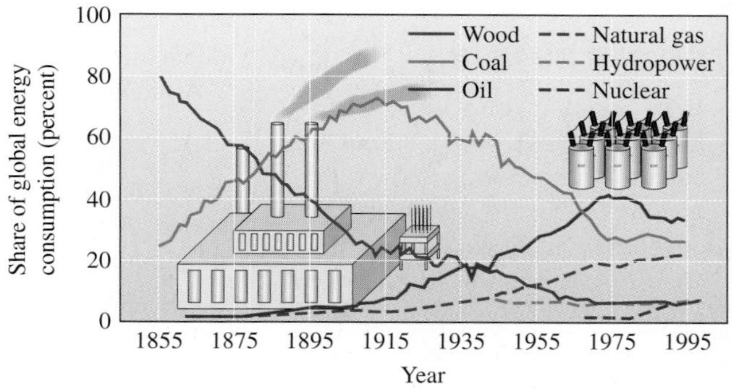

a. In what year were the percent of wood and coal use equal? What was the percent of each used in that year?

b. In what year were the percent of oil and coal use equal? What was the percent of each used in that year?

74. *Perimeter* A rectangular field is to have an area of 500 m^2.

a. Write the perimeter, P, of the field as a function of the width, w.

b. Find the domain of the function in part a.

c. Use a graphing calculator to sketch the graph of the function in part a.

d. Describe what the graph found in part c tells you about how the perimeter of the field varies with the width.

75. *Area* A rectangular field is to have a perimeter of 6000 ft.

a. Write the area, A, of the field as a function of the width, w.

b. Find the domain of the function in part a.

c. Use a graphing calculator to sketch the graph of the function in part a.

d. Describe what the graph found in part c tells you about how the area of the field varies with the width.

**Science*, Vol. 291, Jan. 26, 2001, p. 577. Courtesy of Woods Hole Oceanographic Institute.
†Robbins, Charles T., *Wildlife Feeding and Nutrition*, 2nd ed., Academic Press, 1993, p. 125.
‡Technically, kilograms are a measure of mass, not weight. Weight is a measure of the force of gravity, which varies with the distance from the center of Earth. For objects on the surface of Earth, weight and mass are often used interchangeably, and we will do so in this text.
§Robbins, Charles T., *Wildlife Feeding and Nutrition*, 2nd ed., Academic Press, 1993, p. 142.
‖*The New York Times*, Oct. 31, 1999, p. 38.

2.2 QUADRATIC FUNCTIONS; TRANSLATION AND REFLECTION

THINK ABOUT IT

How much should a company charge for its seminars? When Power and Money, Inc., charges $600 for a seminar on management techniques, it attracts 1000 people. For each $20 decrease in the fee, an additional 100 people will attend the seminar. The managers wonder how much to charge for the seminar to maximize their revenue.

FOR REVIEW

In this section you will need to know how to solve a quadratic equation by factoring and by the quadratic formula, which are covered in Sections R.2 and R.4. Factoring is usually easiest; when a polynomial is set equal to zero and factored, then a solution is found by setting any one factor equal to zero. But factoring is not always possible. The quadratic formula will provide the solution to *any* quadratic equation.

In this section we will see how knowledge of *quadratic functions* will help provide an answer to the question above.

A linear function is defined by

$$f(x) = ax + b,$$

for real numbers a and b. In a *quadratic function* the independent variable is squared. A quadratic function is an especially good model for many situations with a maximum or a minimum function value. Quadratic functions also may be used to describe supply and demand curves; cost, revenue, and profit; as well as other quantities. Next to linear functions, they are the simplest type of function, and well worth studying thoroughly.

QUADRATIC FUNCTION

A **quadratic function** is defined by

$$f(x) = ax^2 + bx + c,$$

where a, b, and c are real numbers, with $a \neq 0$.

The simplest quadratic function has $f(x) = x^2$, with $a = 1$, $b = 0$, and $c = 0$. This function describes situations where the dependent variable y is proportional to the *square* of the independent variable x. The function can be graphed on a graphing calculator as shown in Figure 14. This graph is called a **parabola.** Every quadratic function has a parabola as its graph. The lowest (or highest) point on a parabola is the **vertex** of the parabola. The vertex of the parabola in Figure 14 is $(0, 0)$.

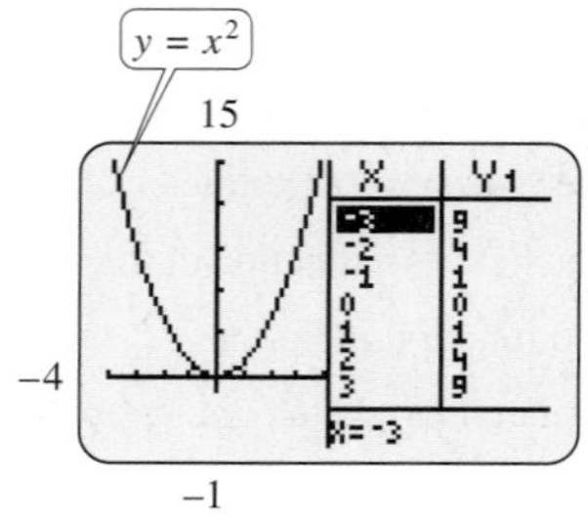

FIGURE 14

If the graph in Figure 14 were folded in half along the y-axis, the two halves of the parabola would match exactly. This means that the graph of a quadratic function is *symmetric* with respect to a vertical line through the vertex; this line is the **axis** of the parabola.

There are many real-world instances of parabolas. For example, cross sections of spotlight reflectors or radar dishes form parabolas. Also, a projectile thrown in the air follows a parabolic path.

EXAMPLE 1 *Graphing a Quadratic Function*

Graph $y = x^2 - 4$.

Solution Each value of y will be 4 less than the corresponding value of y in $y = x^2$. The graph of $y = x^2 - 4$ has the same shape as that of $y = x^2$ but is

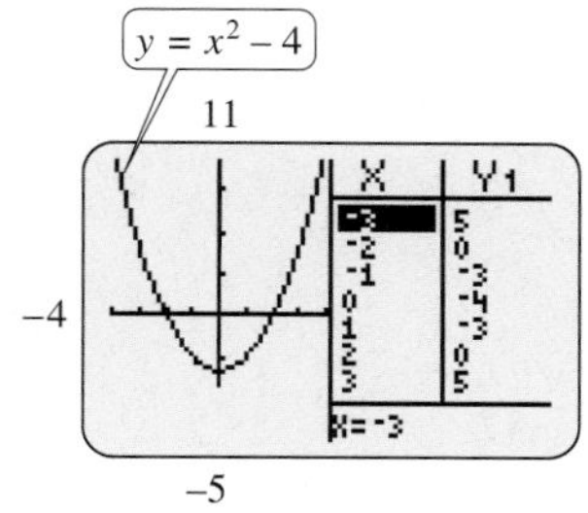

FIGURE 15

4 units lower. See Figure 15. The vertex of the parabola (on this parabola, the *lowest* point) is at $(0, -4)$. The x-intercepts can be found by letting $y = 0$ to get

$$0 = x^2 - 4,$$

from which $x = 2$ and $x = -2$ are the x-intercepts. The axis of the parabola is the vertical line $x = 0$.

Example 1 suggests that the effect of c in $ax^2 + bx + c$ is to lower the graph if c is negative and to raise the graph if c is positive. This is true for any function; the movement up or down is referred to as a **vertical translation** of the function.

EXAMPLE 2 *Graphing Quadratic Functions*

Graph $y = ax^2$ with $a = -.5$, $a = -1$, $a = -2$, and $a = -4$.

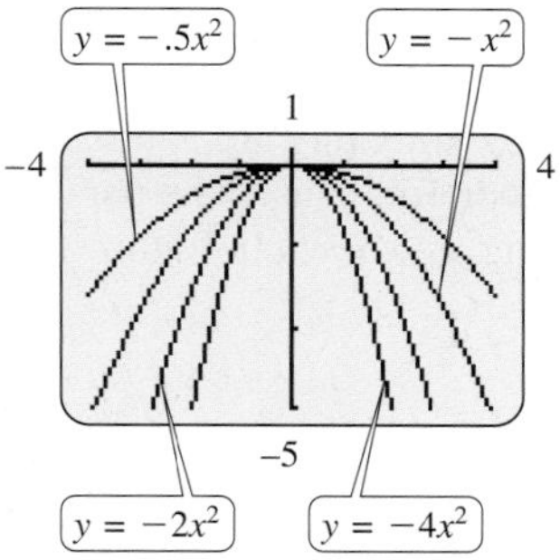

FIGURE 16

Solution Figure 16 shows all four functions plotted on the same axes. We see that since a is negative, the graph opens downward. When the magnitude of a is less than 1 (that is, when $a = -.5$), the graph is wider than the original graph, because the values of y are smaller in magnitude. On the other hand, when the magnitude of a is greater than 1, the graph is steeper.

Example 2 shows that the sign of a in $ax^2 + bx + c$ determines whether the parabola opens upward or downward. Multiplying $f(x)$ by a negative number flips the graph of f upside down. This is called a **vertical reflection** of the graph. The magnitude of a determines how steeply the graph increases or decreases.

EXAMPLE 3 *Graphing Quadratic Functions*

Graph $y = (x - h)^2$ for $h = 3, 0$, and -4.

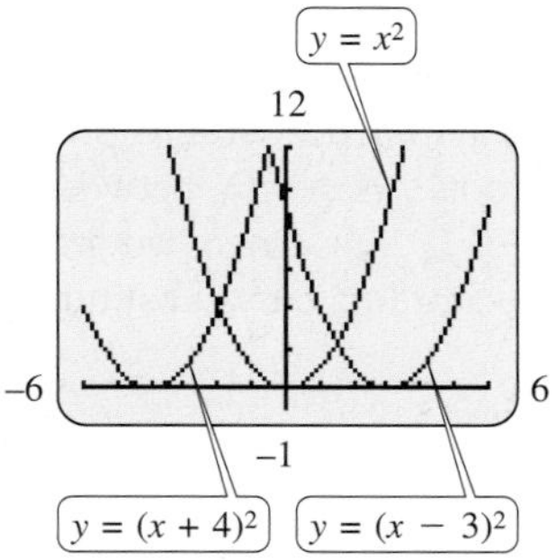

FIGURE 17

Solution Figure 17 shows a graphing calculator view of all three functions on the same axes. Notice that since the number is subtracted *before* the squaring occurs, the graph does not move up or down, but instead moves left or right. Evaluating $f(x) = (x - 3)^2$ at $x = 3$ gives the same result as evaluating $f(x) = x^2$ at $x = 0$. Therefore, when we subtract the positive number 3 from x, the graph shifts 3 units to the right, so the vertex is at $(3, 0)$. Similarly, when we subtract the negative number -4 from x—in other words, when the function becomes $f(x) = (x + 4)^2$—the graph shifts to the left 4 units.

The left or right shift of the graph illustrated in Figure 17 is called a **horizontal translation** of the function.

If a quadratic equation is given in the form $ax^2 + bx + c$, we can identify the translations and any vertical reflection by rewriting it in the form

$$y = a(x - h)^2 + k.$$

In this form, we can identify the vertex as (h, k). A quadratic equation not given in this form can be converted by a process called **completing the square.** The next example illustrates the process.

EXAMPLE 4 *Graphing a Quadratic Function*

Graph $f(x) = -3x^2 - 2x + 1$.

Solution To begin, factor -3 from the x-terms so the coefficient of x^2 is 1.

$$f(x) = -3\left(x^2 + \frac{2}{3}x\right) + 1 \qquad \text{Factor out } -3.$$

$$= -3\left(x^2 + \frac{2}{3}x + \frac{1}{9}\right) + 1 + 3\left(\frac{1}{9}\right) \qquad \text{Add and subtract } -3 \text{ times } (\tfrac{1}{2} \text{ the coefficient of } x)^2.$$

$$= -3\left(x + \frac{1}{3}\right)^2 + \frac{4}{3} \qquad \text{Factor and combine terms.}$$

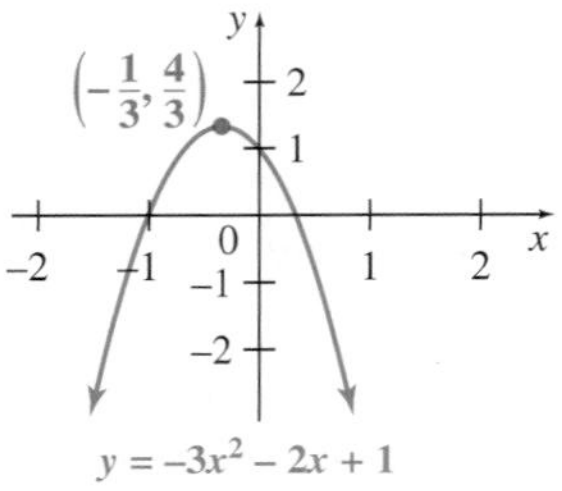

FIGURE 18

Note that in the second step we added 0. Now we can identify $a = -3$, $h = -1/3$, and $k = 4/3$, so the graph is the parabola $y = x^2$ translated $1/3$ unit to the left, flipped upside down, and translated $4/3$ units upward. This puts the vertex at $(-1/3, 4/3)$. The 3 in front of the squared term causes the parabola to be stretched vertically by a factor of 3. These results are shown in Figure 18.

Instead of completing the square to find the vertex of the graph of a quadratic function given in the form $y = ax^2 + bx + c$, we can develop a formula for the vertex. By the quadratic formula, if $ax^2 + bx + c = 0$, where $a \neq 0$, then

$$x = \frac{-b \pm \sqrt{b^2 - 4ac}}{2a}.$$

Notice that this is the same as

$$x = \frac{-b}{2a} \pm \frac{\sqrt{b^2 - 4ac}}{2a} = \frac{-b}{2a} \pm Q,$$

where $Q = \sqrt{b^2 - 4ac}/(2a)$. Since a parabola is symmetric with respect to its axis, the vertex is halfway between its two roots. Halfway between $x = -b/(2a) + Q$ and $x = -b/(2a) - Q$ is $x = -b/(2a)$. Once we have the x-coordinate of the vertex, we can easily find the y-coordinate by substituting the x-coordinate into the original equation.

GRAPH OF THE QUADRATIC FUNCTION

The graph of the quadratic function $f(x) = ax^2 + bx + c$ has its vertex at

$$\left(\frac{-b}{2a}, f\left(\frac{-b}{2a}\right)\right).$$

The graph opens upward if $a > 0$ and downward if $a < 0$.

A graphing calculator does not necessarily tell us the exact value of the vertex or the x- or y-intercepts, but it gives a sketch quickly, and it allows us to verify what we have found through algebra. In many examples, the exact value of the solutions given by the quadratic formula will be irrational numbers, and a calculator is required to approximate the solutions. Another situation that may arise is the absence of any x-intercepts, as in the next example.

EXAMPLE 5 *Graphing a Quadratic Function*

Graph $y = x^2 + 4x + 6$.

Solution This does not appear to factor, so we'll try the quadratic formula.

$$x = \frac{-b \pm \sqrt{b^2 - 4ac}}{2a} \qquad a = 1, b = 4, c = 6$$

$$= \frac{-4 \pm \sqrt{4^2 - 4(1)(6)}}{2(1)} = \frac{-4 \pm \sqrt{-8}}{2}$$

As soon as we see the negative under the square root sign, we know the solutions are complex numbers. Therefore, there are no x-intercepts. Nevertheless, the vertex is still at

$$x = \frac{-b}{2a} = \frac{-4}{2} = -2.$$

Substituting this into the equation gives

$$y = (-2)^2 + 4(-2) + 6 = 2.$$

The y-intercept is at $(0, 6)$, which is 2 units to the right of the parabola's axis $x = -2$. Using the symmetry of the figure, we can also plot the mirror image of this point on the opposite side of the parabola's axis: at $x = -4$ (2 units to the left of the axis), y is also equal to 6. Plotting the vertex, the y-intercept, and the point $(-4, 6)$ gives the graph in Figure 19.

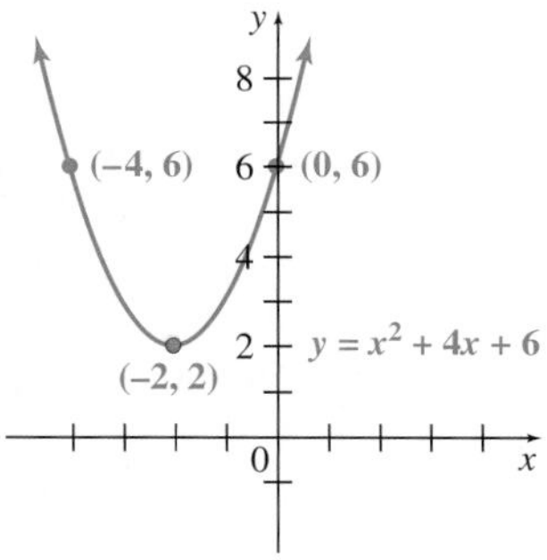

FIGURE 19

We now return to the question with which we started this section.

EXAMPLE 6 *Management Science*

When Power and Money, Inc., charges \$600 for a seminar on management techniques, it attracts 1000 people. For each \$20 decrease in the fee, an additional 100 people will attend the seminar. The managers are wondering how much to charge for the seminar to maximize their revenue.

Solution Let x be the number of \$20 decreases in the price. Then the price charged per person will be

$$\text{Price per person} = 600 - 20x,$$

and the number of people in the seminar will be

$$\text{Number of people} = 1000 + 100x.$$

The total revenue, $R(x)$, is given by the product of the price and the number of people attending, or

$$R(x) = (600 - 20x)(1000 + 100x)$$
$$= 600,000 + 40,000x - 2000x^2.$$

We see by the negative in the x^2-term that this defines a parabola opening downward, so the maximum revenue is at the vertex. The x-coordinate of the vertex is

$$x = \frac{-b}{2a} = \frac{-40,000}{2(-2000)} = 10.$$

The y-coordinate is then

$$y = 600,000 + 40,000(10) - 2000(10^2)$$
$$= 800,000.$$

Therefore, the maximum revenue is $800,000, which is achieved by charging $600 - 20x = 600 - 20(10) = \$400$ per person. ■

Notice in this last example that the maximum revenue was achieved by charging less than the current price of $600, which was more than made up for by the increase in sales. This is typical of many applications. Mathematics is a powerful tool for solving such problems, since the answer is not what one might have guessed intuitively.

The main difficulty in learning to solve such problems is deriving expressions for the price per person and the number of people. You will find this process easier if you notice that both expressions are linear functions of x. Notice also that you know the constant term in these functions because the original problem told what happens if $x = 0$ (i.e., if there is no $20 decrease). Finally, notice that the slope of these linear functions is just the amount of change each time x increases by 1. (In this example, an increase in x by 1 is equivalent to a $20 decrease in the price.) Once you understand these ideas (after getting some practice), you will be able to solve any such maximization problem.

One important observation from the previous example is that the maximum or minimum of a quadratic function occurs at the vertex. The concept of the maximum or minimum of a function is important in calculus, as we shall see in future chapters.

Section 1.3 showed how the equation of a line that closely approximates a set of data points is found using linear regression. Some graphing calculators with statistics capability perform other kinds of regression. For example, *quadratic regression* gives the coefficients of a quadratic equation that models a given set of points. (See Exercise 55.)

In the next example, we show how the calculation of profit can involve a quadratic function.

EXAMPLE 7 *Profit*
A deli owner has found that his revenue from producing x pounds of vegetable cream cheese is given by $R(x) = -x^2 + 30x$, while the cost in dollars is given by $C(x) = 5x + 100$.

(a) Find the minimum break-even quantity.

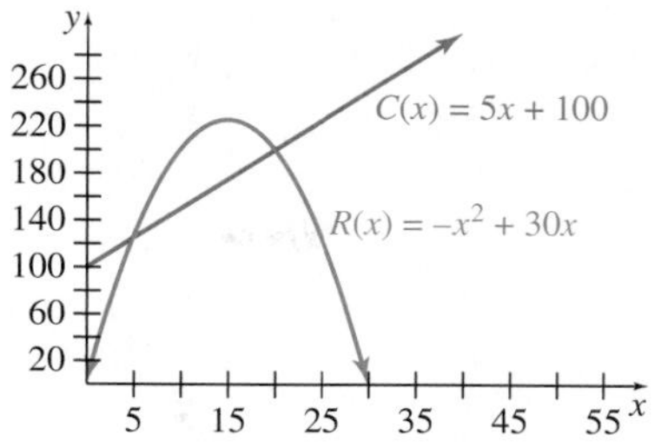

FIGURE 20

Solution Notice from the graph in Figure 20 that the revenue function is a parabola opening downward and the cost function is a linear function that crosses the revenue function at two points. To find the minimum break-even quantity, we find where the two functions are equal.

$$\begin{aligned} R(x) &= C(x) \\ -x^2 + 30x &= 5x + 100 \\ 0 &= x^2 - 25x + 100 && \text{Subtract } -x^2 + 30x \text{ from both sides.} \\ &= (x-5)(x-20) && \text{Factor.} \end{aligned}$$

The two graphs cross when $x = 5$ and $x = 20$. The minimum break-even point is at $x = 5$. The deli owner must sell at least 5 lb of cream cheese to break even.

(b) Find the maximum revenue.

Solution By factoring the revenue function, $R(x) = -x^2 + 30x = x(-x + 30)$, we can see that it has two roots, $x = 0$ and $x = 30$. The maximum is at the vertex, which has a value of x halfway between the two roots, or $x = 15$. (Alternatively, we could use the formula $x = -b/(2a) = -30/(-2) = 15$.) The maximum revenue is $R(15) = -15^2 + 30(15) = 225$, or \$225.

(c) Find the maximum profit.

Solution The profit is the difference between the revenue and the cost, or

$$\begin{aligned} P(x) &= R(x) - C(x) \\ &= (-x^2 + 30x) - (5x + 100) \\ &= -x^2 + 25x - 100. \end{aligned}$$

This is just the negative of the expression factored in part (a), where we found the roots to be $x = 5$ and $x = 20$. The value of x at the vertex is halfway between these two roots, or $x = (5 + 20)/2 = 12.5$. (Alternatively, we could use the formula $x = -b/(2a) = -25/(-2) = 12.5$.) The value of the function here is $P(12.5) = -12.5^2 + 25(12.5) - 100 = 56.25$. It is clear that this is a maximum, not only from Figure 20, but also because the profit function is a quadratic with a negative x^2-term. A maximum profit of \$56.25 is achieved by selling 12.5 lb of cream cheese.

On the next page, we provide guidelines for sketching graphs that involve translations and reflections.

TRANSLATIONS AND REFLECTIONS OF FUNCTIONS

Let f be any function, and let h and k be positive constants (Figure 21).
The graph of $y = f(x) + k$ is the graph of $y = f(x)$ translated upward by an amount k (Figure 22).

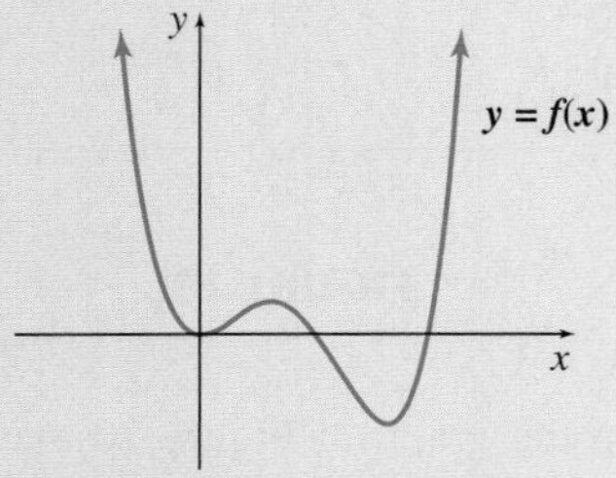

FIGURE 21

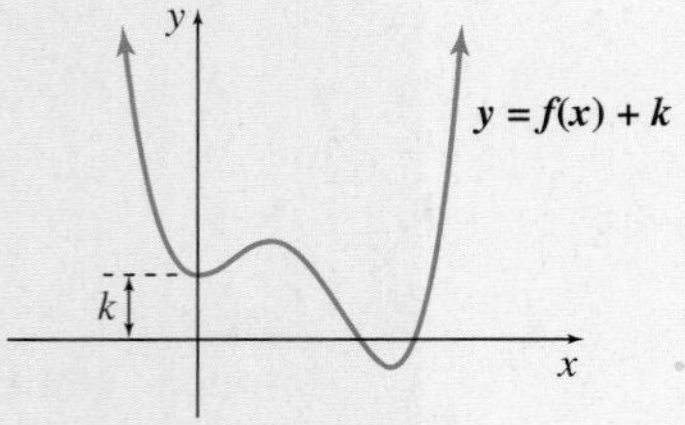

FIGURE 22

The graph of $y = f(x) - k$ is the graph of $y = f(x)$ translated downward by an amount k (Figure 23).
The graph of $y = f(x - h)$ is the graph of $y = f(x)$ translated to the right by an amount h (Figure 24).
The graph of $y = f(x + h)$ is the graph of $y = f(x)$ translated to the left by an amount h (Figure 25).

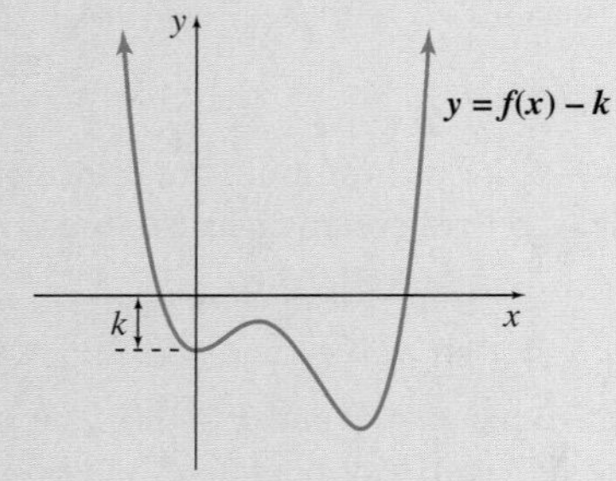

FIGURE 23

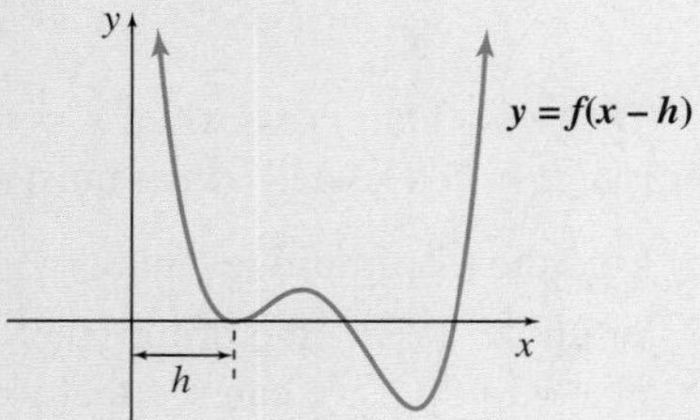

FIGURE 24

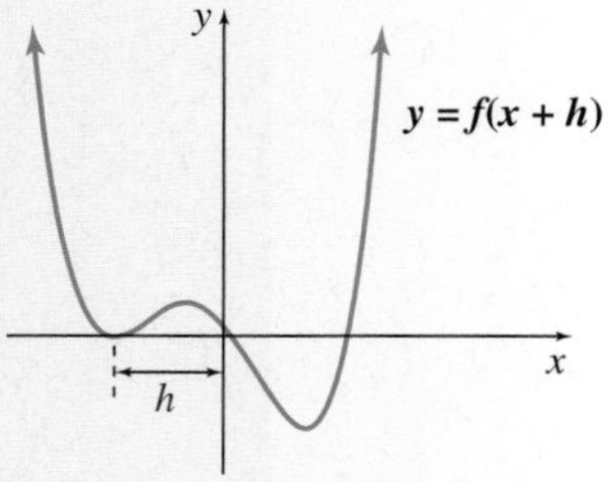

FIGURE 25

The graph of $y = -f(x)$ is the graph of $y = f(x)$ reflected vertically, that is, turned upside down (Figure 26).
The graph of $y = f(-x)$ is the graph of $y = f(x)$ reflected horizontally, that is, its mirror image (Figure 27).

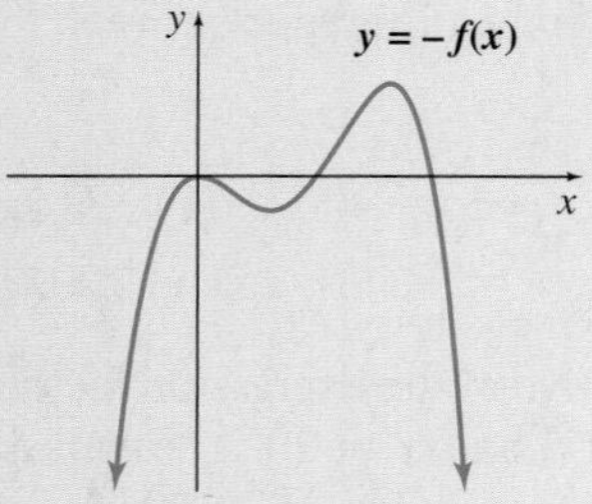

FIGURE 26

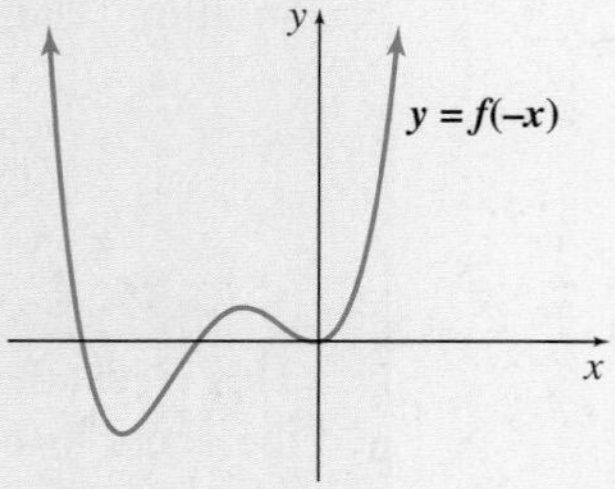

FIGURE 27

Multiplying x or $f(x)$ by a constant a, to get $y = f(ax)$ or $y = a \cdot f(x)$ does not change the general appearance of the graph, except to compress or stretch it. When a is negative, it also causes a reflection, as shown in the last two figures in the summary for $a = -1$.

EXAMPLE 8 *Translations and Reflections of a Graph*

Graph $f(x) = -\sqrt{4 - x} + 3$.

Solution Begin with the simplest possible function, then add each variation in turn. Start with the graph of $f(x) = \sqrt{x}$. As Figure 28 reveals, this is just one-half of the graph of $f(x) = x^2$ lying on its side.

Now add another component of the original function, the negative in front of the x, giving $f(x) = \sqrt{-x}$. This is a horizontal reflection of the $f(x) = \sqrt{x}$ graph, as shown in Figure 29. Next, include the 4 under the square root sign. To get $4 - x$ into the form $f(x - h)$ or $f(x + h)$, we need to factor out the negative: $\sqrt{4 - x} = \sqrt{-(x - 4)}$. Now the 4 is subtracted, so this function is a translation to the right of the function $f(x) = \sqrt{-x}$ by 4 units, as Figure 30 indicates.

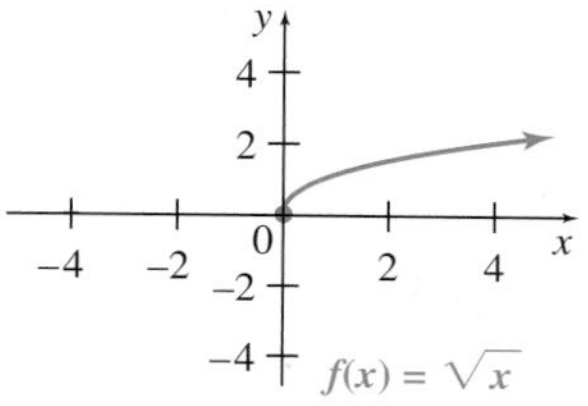

FIGURE 28

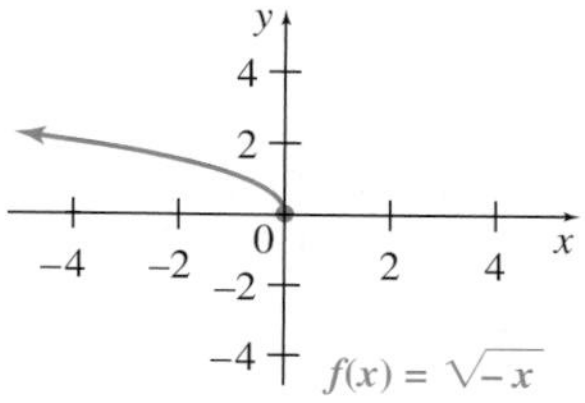

FIGURE 29

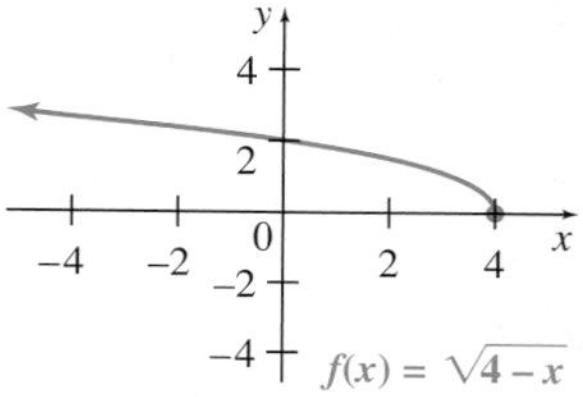

FIGURE 30

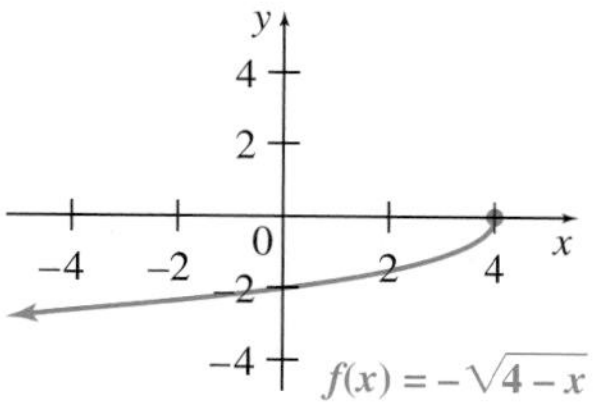

FIGURE 31

The effect of the negative in front of the radical is a vertical reflection, as in Figure 31, which shows the graph of $f(x) = -\sqrt{4 - x}$. Finally, adding the constant 3 raises the entire graph by 3 units, giving the graph of $f(x) = -\sqrt{4 - x} + 3$ in Figure 32(a).

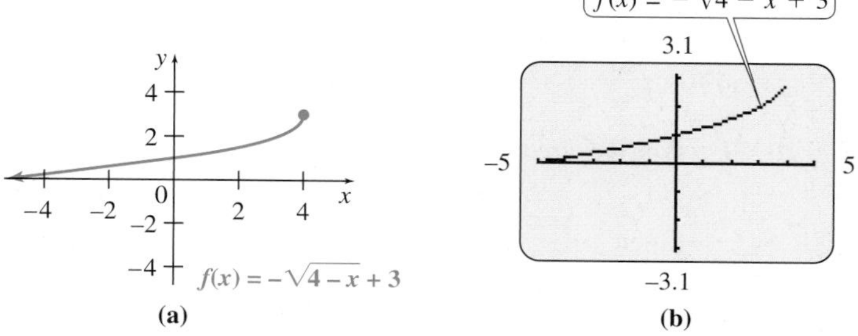

FIGURE 32

If you viewed a graphing calculator image such as Figure 32(b), you might think the function continues to go up and to the right. By realizing that $(4, 3)$ is the vertex of the sideways parabola, we see that this is the rightmost point on the graph. Another approach is to find the domain of f by setting $4 - x \geq 0$, from which we conclude that $x \leq 4$. This demonstrates the importance of knowing the algebraic techniques in order to interpret a graphing calculator image correctly.

2.2 EXERCISES

1. How does the value of a affect the graph of $y = ax^2$? Discuss the case for $a \geq 1$ and for $0 \leq a \leq 1$.

2. How does the value of a affect the graph of $y = ax^2$ if $a \leq 0$?

In Exercises 3–7, match the correct graph A–E to the function without using your calculator. Then, if you have a graphing calculator, use it to check your answers. Each graph in this group shows x and y in $[-10, 10]$.

3. $y = x^2 - 3$

4. $y = (x - 3)^2 + 2$

5. $y = (x + 3)^2 + 2$

6. $y = -(3 - x)^2 + 2$

7. $y = -(x + 3)^2 + 2$

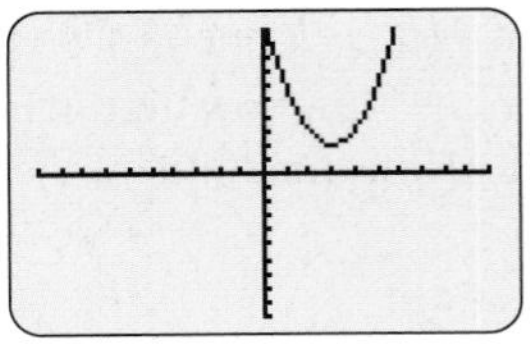

(a)

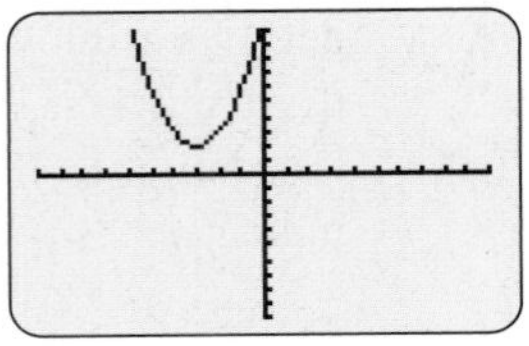

(b)

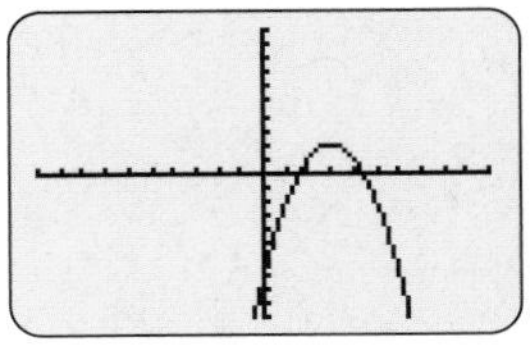

(c)

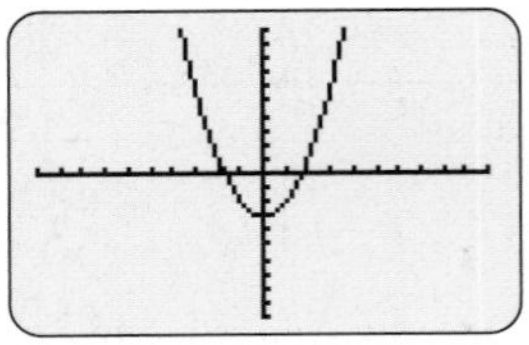

(d)

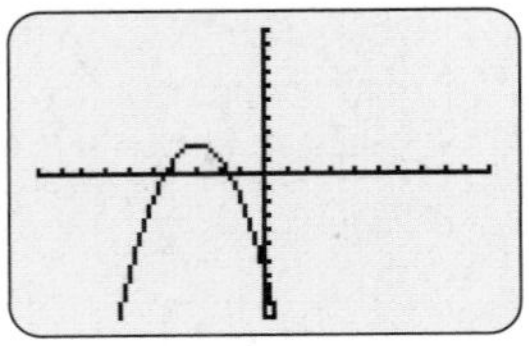

(e)

Graph each parabola and give its vertex, axis, x-intercepts, and y-intercept.

8. $y = x^2 + 6x + 5$

9. $y = x^2 - 10x + 21$

10. $y = -2x^2 - 12x - 16$

11. $y = -3x^2 + 12x - 11$

12. $f(x) = 2x^2 + 12x - 16$

13. $f(x) = -x^2 + 6x - 6$

14. $f(x) = 2x^2 - 4x + 5$

15. $f(x) = -3x^2 + 24x - 36$

16. $f(x) = -\frac{1}{3}x^2 + 2x + 4$

17. $f(x) = \frac{5}{2}x^2 + 10x + 8$

18. $f(x) = \frac{2}{3}x^2 - \frac{8}{3}x + \frac{5}{3}$

19. $f(x) = -\frac{1}{2}x^2 - x - \frac{7}{2}$

In Exercises 20–24, follow the directions for Exercises 3–7.

20. $y = \sqrt{x + 2} - 4$

21. $y = \sqrt{x - 2} - 4$

22. $y = \sqrt{-x + 2} - 4$

23. $y = \sqrt{-x - 2} - 4$

24. $y = -\sqrt{x + 2} - 4$

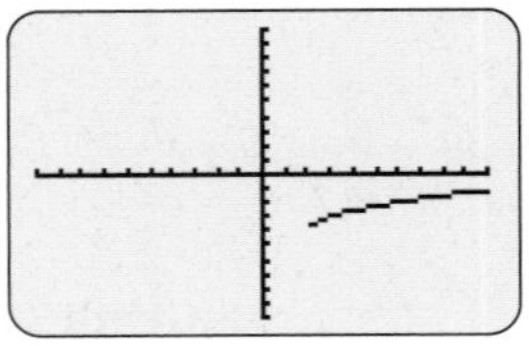

(a)

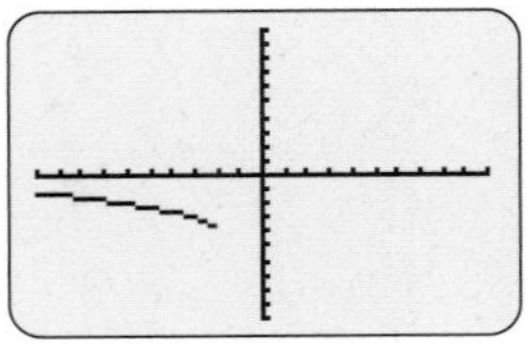

(b)

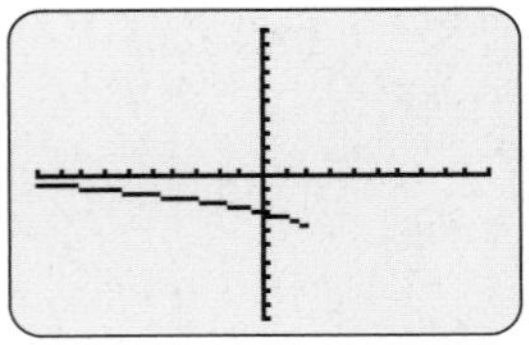

(c)

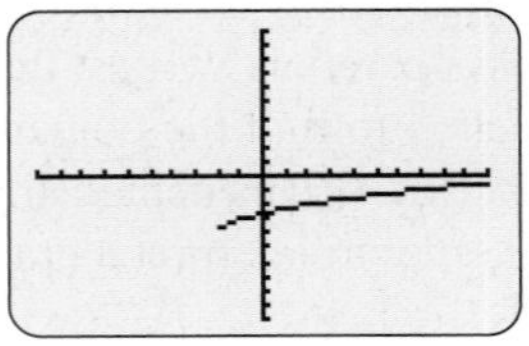

(d)

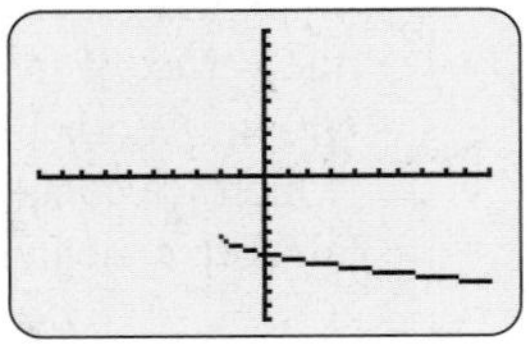

(e)

Given the following graph, sketch by hand the graph of the function described, indicating how the three points labeled on the original graph have been translated.

25. $y = -f(x)$

26. $y = f(x - 2) + 2$

27. $y = f(-x)$

28. $y = f(2 - x) + 2$

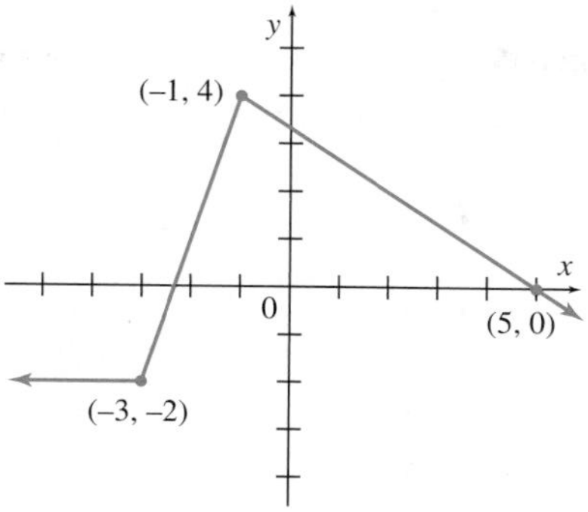

Use the ideas in this section to graph each function.

29. $f(x) = \sqrt{x - 1} + 3$

30. $f(x) = \sqrt{x + 1} - 4$

31. $f(x) = -\sqrt{-4 - x} - 2$

32. $f(x) = -\sqrt{2 - x} + 2$

Using the graph of $f(x)$ in Figure 21, show the graph of $f(ax)$ where a satisfies the given condition.

33. $0 < a < 1$

34. $1 < a$

35. $-1 < a < 0$

36. $a < -1$

Using the graph of $f(x)$ in Figure 21, show the graph of $af(x)$ where a satisfies the given condition.

37. $0 < a < 1$

38. $1 < a$

39. $-1 < a < 0$

40. $a < -1$

41. If r is an x-intercept of the graph of $y = f(x)$, what is an x-intercept of the graph of each of the following?

a. $y = -f(x)$ **b.** $y = f(-x)$ **c.** $y = -f(-x)$

42. If b is the y-intercept of the graph of $y = f(x)$, what is the y-intercept of the graph of each of the following?

a. $y = -f(x)$ **b.** $y = f(-x)$ **c.** $y = -f(-x)$

Applications

BUSINESS AND ECONOMICS

Profit In Exercises 43–46, let $C(x)$ be the cost to produce x widgets, and let $R(x)$ be the revenue. For each exercise, **(a)** *graph both functions,* **(b)** *find the minimum break-even quantity,* **(c)** *find the maximum revenue, and* **(d)** *find the maximum profit.*

43. $R(x) = -x^2 + 10x, \quad C(x) = 2x + 12$

44. $R(x) = -\dfrac{x^2}{2} + 8x, \quad C(x) = \dfrac{3}{2}x + 15$

45. $R(x) = -\dfrac{4}{5}x^2 + 10x, \quad C(x) = 2x + 15$

46. $R(x) = -4x^2 + 40x, \quad C(x) = 4x + 77$

47. *Maximizing Revenue* The revenue of a charter bus company depends on the number of unsold seats. If the revenue $R(x)$ is given by

$$R(x) = 5000 + 50x - x^2,$$

where x is the number of unsold seats, find the maximum revenue and the number of unsold seats that corresponds to maximum revenue.

48. *Maximizing Revenue* A charter flight charges a fare of \$200 per person plus \$4 per person for each unsold seat on the plane. The plane holds 100 passengers. Let x represent the number of unsold seats.

a. Find an expression for the total revenue received for the flight $R(x)$. (*Hint:* Multiply the number of people flying, $100 - x$, by the price per ticket.)

b. Graph the expression from part a.

c. Find the number of unsold seats that will produce the maximum revenue.

d. What is the maximum revenue?

e. Some managers might be concerned about the empty seats, arguing that it doesn't make economic sense to leave any seats empty. Write a few sentences explaining why this is not necessarily so.

49. *Maximizing Revenue* The demand for a certain type of cosmetic is given by

$$p = 500 - x,$$

where p is the price in dollars when x units are demanded.

a. Find the revenue $R(x)$ that would be obtained at a price p. (*Hint:* Revenue = Demand × Price)

b. Graph the revenue function $R(x)$.

c. From the graph of the revenue function, estimate the price that will produce maximum revenue.

d. What is the maximum revenue?

50. *Revenue* The manager of a peach orchard is trying to decide when to arrange for picking the peaches. If they are picked now, the average yield per tree will be 100 lb, which can be sold for 40¢ per pound. Past experience shows that the yield per tree will increase about 5 lb per week, while the price will decrease about 2¢ per pound per week.

a. Let x represent the number of weeks that the manager should wait. Find the income per pound.

b. Find the number of pounds per tree.

c. Find the total revenue from a tree.

d. When should the peaches be picked in order to produce maximum revenue?

e. What is the maximum revenue?

51. *Income* The manager of an 80-unit apartment complex is trying to decide what rent to charge. Experience has shown that at a rent of $600, all the units will be full. On the average, one additional unit will remain vacant for each $60 increase in rent.

a. Let x represent the number of $60 increases. Find an expression for the rent for each apartment.

b. Find an expression for the number of apartments rented.

c. Find an expression for the total revenue from all rented apartments.

d. What value of x leads to maximum revenue?

e. What is the maximum revenue?

52. *Advertising* A study done by an advertising agency reveals that when x thousands of dollars are spent on advertising, it results in a sales increase in thousands of dollars given by the function

$$S(x) = -\frac{1}{4}(x - 10)^2 + 40, \quad \text{for } 0 \le x \le 10.$$

a. Find the increase in sales when no money is spent on advertising.

b. Find the increase in sales when $10,000 is spent on advertising.

c. Sketch the graph of $S(x)$.

LIFE SCIENCES

53. *Length of Life* According to recent data from the Teachers Insurance and Annuity Association (TIAA), the survival function for life after 65 is approximately given by

$$S(x) = 1 - .058x - .076x^2,$$

where x is measured in decades. This function gives the probability that an individual who reaches the age of 65 will live at least x decades ($10x$ years) longer.*

a. Find the median length of life for people who reach 65, that is, the age for which the survival rate is .50.

b. Find the age beyond which virtually nobody lives. (There are, of course, exceptions.)

54. *Medicine* Between 1992 and 1998, the percent of college freshmen who planned to get a professional degree in a medical field can be modeled by

$$f(x) = -.2369x^2 + 1.425x + 6.905,$$

where $x = 0$ represents 1992.[†] Based on this model, in what year did the percent of freshmen planning to get a medical degree reach its maximum? What is the domain of $f(x)$?

55. *Tooth Length* The length (in mm) of the mesiodistal crown of the first molar for human fetuses can be approximated by

$$L(t) = -.01t^2 + .788t - 7.048,$$

where t is the number of weeks since conception.[‡]

a. What does this formula predict for the length at 14 weeks? 24 weeks?

*Exercise 53 is from Ralph DeMarr, University of New Mexico.

[†]*The American Freshmen: National Norms for Fall 1992–1998*, Higher Education Research Institute, UCLA.

[‡]Harris, Edward F., Joseph D. Hicks, and Betsy D. Barcroft, "Tissue Contributions to Sex and Race: Differences in Tooth Crown Size of Deciduous Molars," *American Journal of Physical Anthropology*, Vol. 115, 2001, pp. 223–237.

b. What does this formula predict for the maximum length, and when does that occur? Explain why the formula does not make sense past that time.

56. *Splenic Artery Resistance* Blood flow to the fetal spleen is of research interest because several diseases are associated with increased resistance in the splenic artery (the artery that goes to the spleen). Researchers have found that the index of splenic artery resistance in the fetus can be described by the function

$$y = .057x - .001x^2,$$

where x is the number of weeks of gestation.*

a. At how many weeks is the splenic artery resistance a maximum?

b. What is the maximum splenic artery resistance?

c. At how many weeks is the splenic artery resistance equal to 0, according to this formula? Is your answer reasonable for this function? Explain.

SOCIAL SCIENCE

57. *Head Start* The enrollment in Head Start for some recent years is included in the table.†

Year	Enrollment
1966	733,000
1970	477,400
1980	376,300
1990	540,930
1995	750,696
2000	857,664
2001	905,235
2002	912,345

a. Plot the data using 0 for 1960, and so on.

b. Would a linear or quadratic function best model this data? Explain.

c. If your graphing calculator has a regression feature, find the quadratic function that best fits the data. Graph this function on the same calculator window as the data. (On a TI-83/84 Plus calculator, press the STAT key, and then select the `CALC` menu. `QuadReg` is item 5. The command `QuadReg` L_1, L_2, Y_1 finds the quadratic regression equation for the data in L_1 and L_2 and stores the function in Y_1.)

d. Find a quadratic function defined by $f(x) = a(x - h)^2 + k$ that models the data using (20, 376,300) as the vertex and then choosing a second point such as (42, 912,345) to determine the value of a.

e. Graph the function from part d on the same calculator window as the data and function from part c. Do the graphs of the two functions differ by much?

58. *Highway Research* Since 1990 Congress has authorized more than \$700 million for research and development of "smart" highways, with the goal of enabling cars to drive themselves.‡ The spending per year is approximated by the function

$$f(x) = -19.321x^2 + 3608.7x - 168{,}310,$$

where $x = 90$ corresponds to 1990 and y is in millions of dollars.

a. In what year was the maximum amount spent?

b. What is the maximum amount spent?

c. Sketch a graph of the function.

59. *Accident Rate* According to data from the National Highway Traffic Safety Administration, the accident rate as a function of the age of the driver in years x can be approximated by the function

$$f(x) = 60.0 - 2.28x + .0232x^2$$

for $16 \le x \le 85$. Find the age at which the accident rate is a minimum and the minimum rate.§

PHYSICAL SCIENCES

60. *Maximizing the Height of an Object* If an object is thrown upward with an initial velocity of 32 ft/second, then its height after t seconds is given by

$$h = 32t - 16t^2.$$

a. Find the maximum height attained by the object.

b. Find the number of seconds it takes the object to hit the ground.

*Abuhamad, A. Z. et al., "Doppler Flow Velocimetry of the Splenic Artery in the Human Fetus: Is It a Marker of Chronic Hypoxia?" *American Journal of Obstetrics and Gynecology*, Vol. 172, No. 3, March 1995, pp. 820–825.

†http://www.acf.hhs.gov/programs/hsb/research/2003.htm.

‡IVHS America.

§Exercise 59 is from Ralph DeMarr, University of New Mexico.

61. *Stopping Distance* According to data from the National Traffic Safety Institute,* the stopping distance y in feet of a car traveling x mph can be described by the equation $y = .056057x^2 + 1.06657x$.

a. Find the stopping distance for a car traveling 25 mph.

b. How fast can you drive if you need to be certain of stopping within 150 ft?

GENERAL INTEREST

62. *Maximizing Area* Glenview Community College wants to construct a rectangular parking lot on land bordered on one side by a highway. It has 320 ft of fencing to use along the other three sides. What should be the dimensions of the lot if the enclosed area is to be a maximum? (*Hint:* Let x represent the width of the lot, and let $320 - 2x$ represent the length.)

63. *Maximizing Area* What would be the maximum area that could be enclosed by the college's 320 ft of fencing if it decided to close the entrance by enclosing all four sides of the lot? (See Exercise 62.)

In Exercises 64 and 65, draw a sketch of the arch or culvert on coordinate axes, with the horizontal and vertical axes through the vertex of the parabola. Use the given information to label points on the parabola. Then give the equation of the parabola and answer the question.

64. *Parabolic Arch* An arch is shaped like a parabola. It is 30 m wide at the base and 15 m high. How wide is the arch 10 m from the ground?

65. *Parabolic Culvert* A culvert is shaped like a parabola, 18 ft across the top and 12 ft deep. How wide is the culvert 8 ft from the top?

2.3 POLYNOMIAL AND RATIONAL FUNCTIONS

THINK ABOUT IT *How does the revenue collected by the government vary with the tax rate?*

In exercises in this section, we will explore this question using *polynomial* and *rational functions.*

Polynomial Functions

Earlier, we discussed linear and quadratic functions and their graphs. Both of these functions are special types of *polynomial functions.*

POLYNOMIAL FUNCTION

A **polynomial function** of degree n, where n is a nonnegative integer, is defined by

$$f(x) = a_n x^n + a_{n-1}x^{n-1} + \cdots + a_1 x + a_0,$$

where $a_n, a_{n-1}, \ldots, a_1$, and a_0 are real numbers, called **coefficients,** with $a_n \neq 0$. The number a_n is called the **leading coefficient.**

For $n = 1$, a polynomial function takes the form

$$f(x) = a_1 x + a_0,$$

a linear function. A linear function, therefore, is a polynomial function of degree 1. (Note, however, that a linear function of the form $f(x) = a_0$ for a real

**National Traffic Safety Institute Student Workbook*, 1993, p. 7.

number a_0 is a polynomial function of degree 0, the constant function.) A polynomial function of degree 2 is a quadratic function.

Accurate graphs of polynomial functions of degree 3 or higher require methods of calculus to be discussed later. Meanwhile, a graphing calculator is useful for obtaining such graphs, but care must be taken in choosing a viewing window that captures the significant behavior of the function.

The simplest polynomial functions of higher degree are those of the form $f(x) = x^n$. Such a function is known as a **power function.** Figure 33 below shows the graphs of $f(x) = x^3$ and $f(x) = x^5$. These functions are simple enough that they can be drawn by hand by plotting a few points and connecting them with a smooth curve. An important property of all polynomials is that their graphs are smooth curves.

The graphs of $f(x) = x^4$ and $f(x) = x^6$, shown in Figure 34, can be sketched in a similar manner. These graphs have symmetry about the y-axis, as does the graph of $f(x) = ax^2$ for a nonzero real number a. As with the graph of $f(x) = ax^2$, the value of a in $f(x) = ax^n$ affects the direction of the graph. When $a > 0$, the graph has the same general appearance as the graph of $f(x) = x^n$. However, if $a < 0$, the graph is reflected vertically.

$f(x) = x^3$

x	$f(x)$
-2	-8
-1	-1
0	0
1	1
2	8

$f(x) = x^5$

x	$f(x)$
-1.5	-7.6
-1	-1
0	0
1	1
1.5	7.6

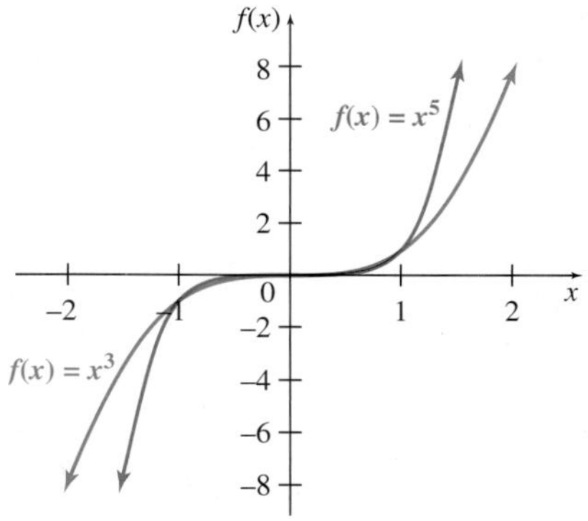

FIGURE 33

$f(x) = x^4$

x	$f(x)$
-2	16
-1	1
0	0
1	1
2	16

$f(x) = x^6$

x	$f(x)$
-1.5	11.4
-1	1
0	0
1	1
1.5	11.4

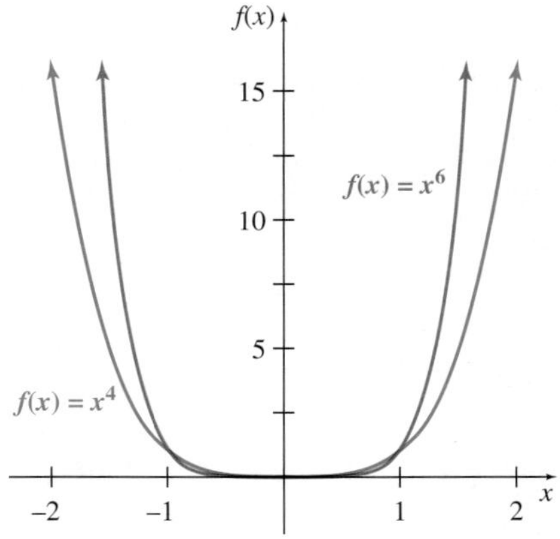

FIGURE 34

EXAMPLE 1 *Translations and Reflections*

Graph $f(x) = -(x - 2)^3 + 3$.

Solution Using the principles of translation and reflection from the previous section, we recognize that this is similar to the graph of $y = x^3$, but reflected vertically (because of the negative in front of $(x - 2)^3$), and with its center moved 2 units to the right and 3 units up. The result is shown in Figure 35.

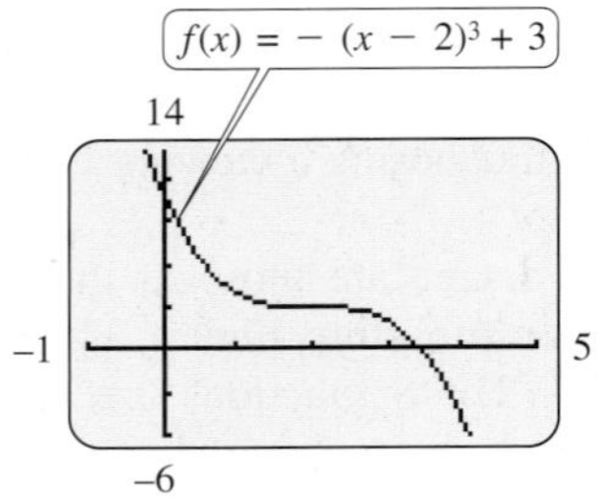

FIGURE 35

A polynomial of degree 3, such as that in the previous example and in the next, is known as a **cubic polynomial.** A polynomial of degree 4, such as that in Example 3, is known as a **quartic polynomial.**

EXAMPLE 2 *Graphing a Polynomial*

Graph $f(x) = 8x^3 - 12x^2 + 2x + 1$.

Solution Figure 36 shows the function graphed on the x- and y-intervals $[-.5, .6]$ and $[-2, 2]$. In this view, it appears similar to a parabola opening downward. Zooming out to $[-1, 2]$ by $[-8, 8]$, we see in Figure 37 that the graph goes upward as x gets large. There are also two **turning points** near $x = 0$ and $x = 1$. (In a later chapter, we will introduce another term for such turning points: *relative extrema*.) By zooming in with the graphing calculator, we can find these turning points to be at approximately $(.09175, 1.08866)$ and $(.90825, -1.08866)$.

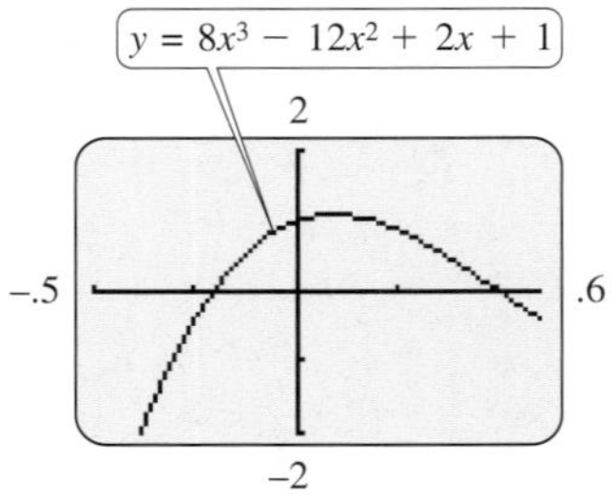

FIGURE 36

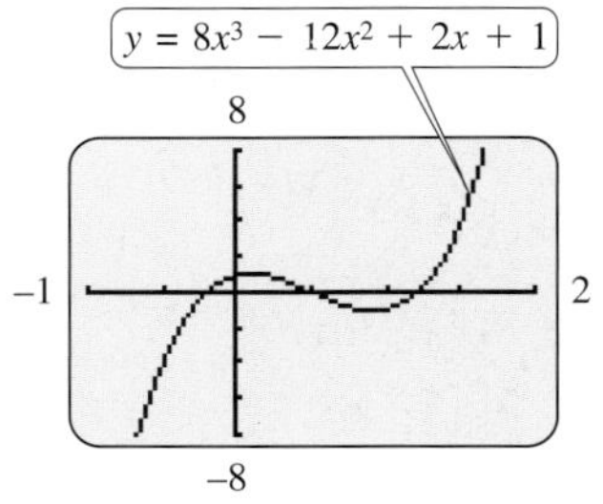

FIGURE 37

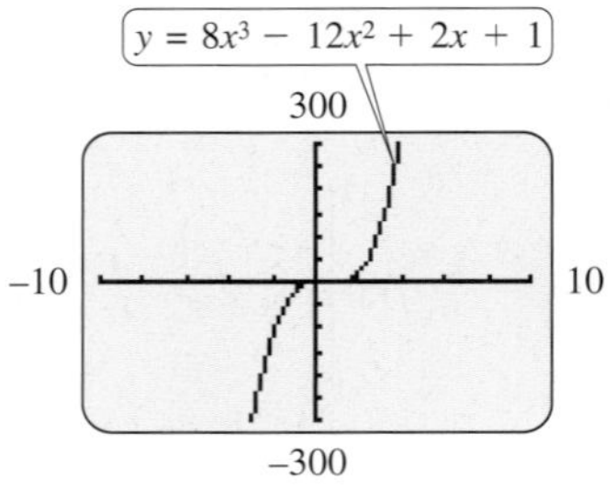

FIGURE 38

Zooming out still further, we see the function on $[-10, 10]$ by $[-300, 300]$ in Figure 38. From this viewpoint, we don't see the turning points at all, and the graph seems similar in shape to that of $y = x^3$. This is an important point: when x is large in magnitude, either positive or negative, $8x^3 - 12x^2 + 2x + 1$ behaves a lot like $8x^3$, because the other terms are small in comparison with the cubic term. So this viewpoint tells us something useful about the function, but it is less useful than the previous graph for determining the turning points.

After the previous example, you may wonder how to be sure you have the viewing window that exhibits all the important properties of a function. We will find an answer to this question in later chapters using the techniques of calculus. Meanwhile, let us consider one more example to get a better idea of what polynomials look like.

EXAMPLE 3 *Graphing a Polynomial*

Graph $f(x) = -3x^4 + 14x^3 - 54x + 3$.

Solution Figure 39 shows a graphing calculator view on $[-3, 5]$ by $[-50, 50]$. If you have a graphing calculator, we recommend that you experiment with various viewpoints and verify for yourself that this viewpoint captures the important behavior of the function. Notice that it has three turning points. Notice also that as $|x|$ gets large, the graph turns downward. This is because as $|x|$ becomes large, the x^4-term dominates the other terms, which are small in comparison, and the x^4-term has a negative coefficient.

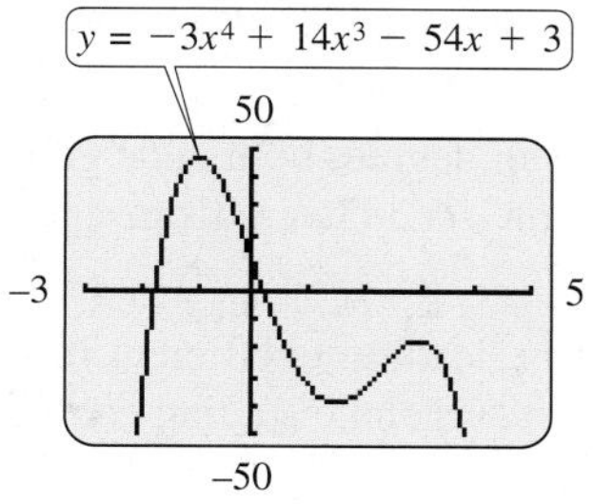

FIGURE 39

As suggested by the graphs above, the domain of a polynomial function is the set of all real numbers. The range of a polynomial function of odd degree is also the set of all real numbers. Some typical graphs of polynomial functions of odd degree are shown in Figure 40. These graphs suggest that for every polynomial function f of odd degree, there is at least one real value of x for which $f(x) = 0$. Such a value of x is called a **real zero** of f; these values are also the x-intercepts of the graph.

Polynomial functions of even degree have a range that takes either the form $(-\infty, k]$ or the form $[k, \infty)$ for some real number k. Figure 41 shows two typical graphs of polynomial functions of even degree.

A fifth-degree polynomial can have four turning points, as in the last graph of Figure 40, or no turning points, as in Figure 33. By examining the figures in this section, you may notice that the graph of a polynomial of degree n has at most $n - 1$ turning points. In a later chapter we will use calculus to see why this is true. Meanwhile, you can learn much about a polynomial by examining its graph. For example, if you are presented with the second graph in Figure 41 and told it is a polynomial, you know immediately that it is of even degree, because the range is of the form $(-\infty, k]$. It has five turning points, so it must be of degree 6 or higher. It could be of degree 8 or 10 or 12, etc., but you can't be sure from the graph alone. Because it goes down at the right, the leading coefficient must be negative.

These ideas about polynomial functions are summarized on the next page.

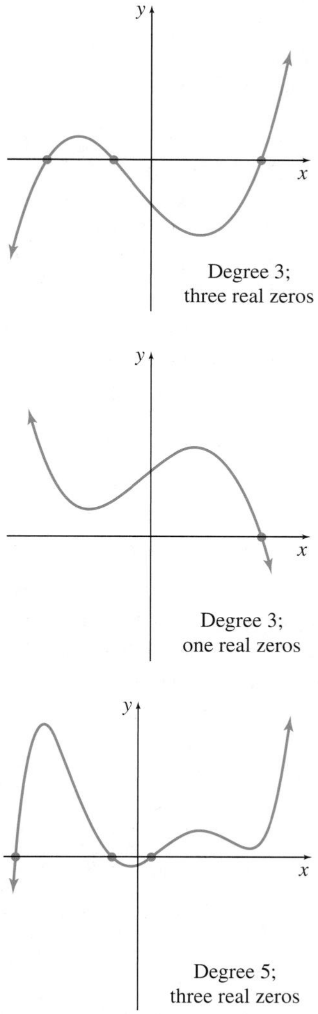

FIGURE 40

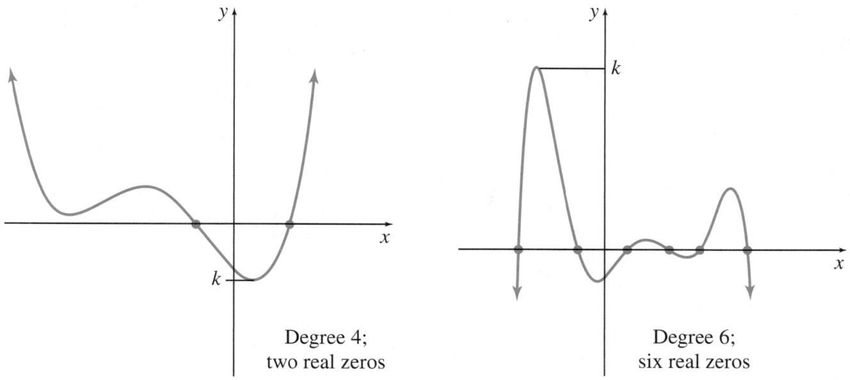

FIGURE 41

PROPERTIES OF POLYNOMIAL FUNCTIONS

1. A polynomial function of degree n can have at most $n - 1$ turning points. Conversely, if the graph of a polynomial function has n turning points, it must have degree at least $n + 1$.
2. In the graph of a polynomial function of even degree, both ends go up or both ends go down. For a polynomial function of odd degree, one end goes up and one end goes down.
3. If the graph goes up as x becomes large, the leading coefficient must be positive. If the graph goes down as x becomes large, the leading coefficient is negative.

Rational Functions

Many situations require mathematical models that are quotients. A common model for such situations is a *rational function.*

RATIONAL FUNCTION

A **rational function** is defined by

$$f(x) = \frac{p(x)}{q(x)},$$

where $p(x)$ and $q(x)$ are polynomial functions and $q(x) \neq 0$.

Since any values of x such that $q(x) = 0$ are excluded from the domain, a rational function often has a graph with one or more breaks.

EXAMPLE 4 *Graphing a Rational Function*

Graph $y = \dfrac{1}{x}$.

Solution This function is undefined for $x = 0$, since 0 is not allowed in the denominator of a fraction. For this reason, the graph of this function will not intersect the vertical line $x = 0$, which is the y-axis. Since x can take on any value except 0, the values of x can approach 0 as closely as desired from either side of 0.

x approaches 0. ↓

x	$-.5$	$-.2$	$-.1$	$-.01$	$.01$	$.1$	$.2$	$.5$
$y = \frac{1}{x}$	-2	-5	-10	-100	100	10	5	2

↑ $|y|$ gets larger and larger.

The table above suggests that as x gets closer and closer to 0, $|y|$ gets larger and larger. This is true in general: as the denominator gets smaller, the fraction gets larger. Thus, the graph of the function approaches the vertical line $x = 0$ (the y-axis) without ever touching it.

As $|x|$ gets larger and larger, $y = 1/x$ gets closer and closer to 0, as shown in the table below. This is also true in general: as the denominator gets larger, the fraction gets smaller.

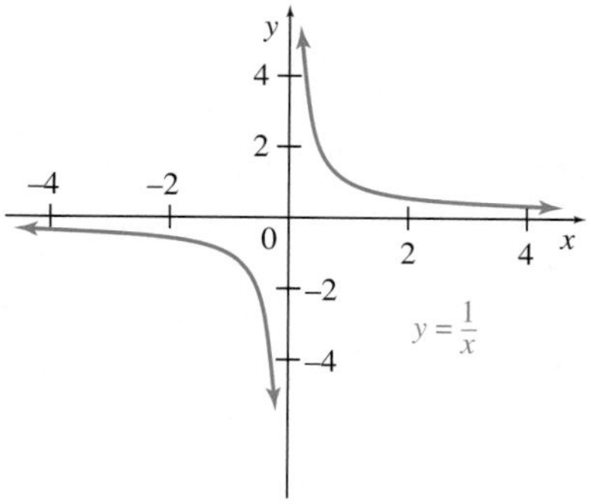

FIGURE 42

x	-100	-10	-4	-1	1	4	10	100
$y = \frac{1}{x}$	$-.01$	$-.1$	$-.25$	-1	1	.25	.1	.01

The graph of the function approaches the horizontal line $y = 0$ (the x-axis). The information from both tables supports the graph in Figure 42.

In Example 4, the vertical line $x = 0$ and the horizontal line $y = 0$ are *asymptotes*, defined as follows.

ASYMPTOTES

If a function gets larger and larger in magnitude without bound as x approaches the number k, then the line $x = k$ is a **vertical asymptote.**

If the values of y approach a number k as $|x|$ gets larger and larger, the line $y = k$ is a **horizontal asymptote.**

There is an easy way to find any vertical asymptotes of a rational function. If a number k makes the denominator 0 but does not make the numerator 0, then the line $x = k$ is a vertical asymptote. Thus, finding vertical asymptotes of a rational function amounts to finding the roots of the denominator. In the next chapter we will show another way to find asymptotes using the concept of a *limit*.

EXAMPLE 5 *Graphing a Rational Function*

Graph $y = \dfrac{3x + 2}{2x + 4}$.

Solution The value $x = -2$ makes the denominator 0, so the line $x = -2$ is a vertical asymptote. To find a horizontal asymptote, find y as x gets larger and larger, as in Figure 43 from a graphing calculator.

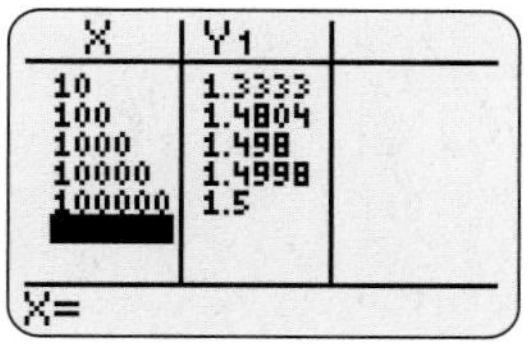

FIGURE 43

Figure 43 suggests that as x gets larger and larger, $(3x + 2)/(2x + 4)$ gets closer and closer to 1.5, or $3/2$. In fact, when $x = 100{,}000$, $(3x + 2)/(2x + 4)$ is equal to 1.5 *to the accuracy of the calculator*. Verify that the same behavior occurs with large negative values of x. Therefore, the line $y = 3/2$ is a horizontal asymptote, with the function approaching the asymptote as x becomes large in magnitude, either positive or negative.

The intercepts should also be noted. When $x = 0$, $y = 2/4 = 1/2$ (the y-intercept). To make a fraction 0, the numerator must be 0; so to make $y = 0$, it is necessary that $3x + 2 = 0$. Solve this for x to get $x = -2/3$ (the x-intercept). We can also use these values to determine where the function is positive and where it is negative. Using the techniques described in Chapter R, verify that the function is negative on $(-2, -2/3)$ and positive on $(-\infty, -2) \cup (-2/3, \infty)$. With this information, the two asymptotes to guide us, and the fact that there are only two intercepts, we suspect the graph is as shown in Figure 44. A graphing calculator can support this.

x	y
3	11/10
2	1
1	5/6
0	1/2
−1	−1/2
−3/2	−5/2
−5/2	11/2
−3	7/2
−4	5/2
−6	2

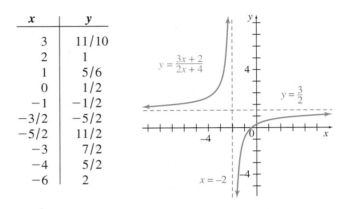

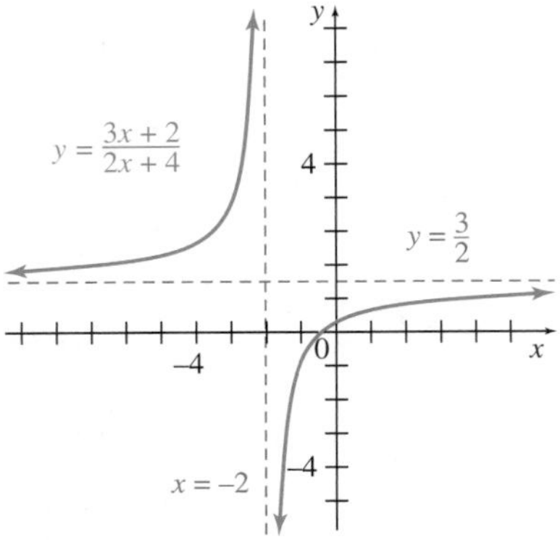

FIGURE 44

In Example 5, $y = 3/2$ was the horizontal asymptote for the rational function $y = (3x + 2)/(2x + 4)$. An equation for the horizontal asymptote can also be found by asking what happens when x gets large. If x is very large, then $3x + 2 \approx 3x$, because the 2 is very small by comparison. In other words, just keep the larger term ($3x$ in this case) and discard the smaller term. Similarly, the denominator is approximately equal to $2x$, so $y \approx 3x/2x = 3/2$. This means that the line $y = 3/2$ is a horizontal asymptote. A more precise way of approaching this idea will be seen later, when limits at infinity are discussed.

Rational functions occur often in practical applications. In many situations involving environmental pollution, much of the pollutant can be removed from the air or water at a fairly reasonable cost, but the last small part of the pollutant can be very expensive to remove. Cost as a function of the percentage of pollutant removed from the environment can be calculated for various percentages of removal, with a curve fitted through the resulting data points. This curve then leads to a mathematical model of the situation. Rational functions are often a good choice for these **cost–benefit models** because they rise rapidly as they approach a vertical asymptote.

EXAMPLE 6 *Cost–Benefit Analysis*

Suppose a cost–benefit model is given by

$$y = \frac{18x}{106 - x},$$

where y is the cost (in thousands of dollars) of removing x percent of a certain pollutant. The domain of x is the set of all numbers from 0 to 100 inclusive; any amount of pollutant from 0% to 100% can be removed. Find the cost to remove the following amounts of the pollutant: 100%, 95%, 90%, and 80%. Graph the function.

Solution Removal of 100% of the pollutant would cost

$$y = \frac{18(100)}{106 - 100} = 300,$$

or \$300,000. Check that 95% of the pollutant can be removed for \$155,000, 90% for \$101,000, and 80% for \$55,000. Using these points, as well as others obtained from the function, gives the graph shown in Figure 45.

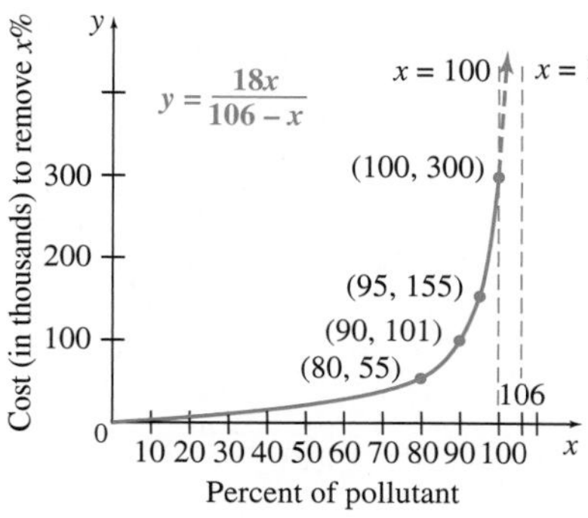

FIGURE 45

If a cost function has the form $C(x) = mx + b$, where x is the number of items produced, m is the marginal cost per item and b is the fixed cost, then the **average cost** per item is given by

$$\overline{C}(x) = \frac{C(x)}{x} = \frac{mx + b}{x}.$$

Notice that this is a rational function with a vertical asymptote at $x = 0$ and a horizontal asymptote at $y = m$. The vertical asymptote reflects the fact that, as the number of items produced approaches 0, the average cost per item becomes infinitely large, because the fixed costs are spread over fewer and fewer items. The horizontal asymptote shows that, as the number of items becomes large, the fixed costs are spread over more and more items, so most of the average cost per item is the marginal cost to produce each item. This is another example of how asymptotes give important information in real applications.

2.3 EXERCISES

1. Explain how translations and reflections can be used to graph $y = -(x - 1)^4 + 2$.
2. Describe an asymptote, and explain when a rational function will have (a) a vertical asymptote and (b) a horizontal asymptote.

In Exercises 3–6, use the principles of the previous section with the graphs of this section to sketch a graph of the given function.

3. $f(x) = (x + 2)^3 - 5$

4. $f(x) = (x - 4)^3 + 2$

5. $f(x) = -(x - 3)^4 + 1$

6. $f(x) = -(x + 1)^4 + 3$

In Exercises 7–15, match the correct graph A–I to the function without using your calculator. Then, after you have answered all of them, if you have a graphing calculator, use your calculator to check your answers. Each graph is plotted on $[-6, 6]$ by $[-50, 50]$.

7. $y = x^3 - 7x - 9$

8. $y = -x^3 + 4x^2 + 3x - 8$

9. $y = -x^3 - 4x^2 + x + 6$

10. $y = 2x^3 + 4x + 5$

11. $y = x^4 - 5x^2 + 7$

12. $y = x^4 + 4x^3 - 20$

13. $y = -x^4 + 2x^3 + 10x + 15$

14. $y = .7x^5 - 2.5x^4 - x^3 + 8x^2 + x + 2$

15. $y = -x^5 + 4x^4 + x^3 - 16x^2 + 12x + 5$

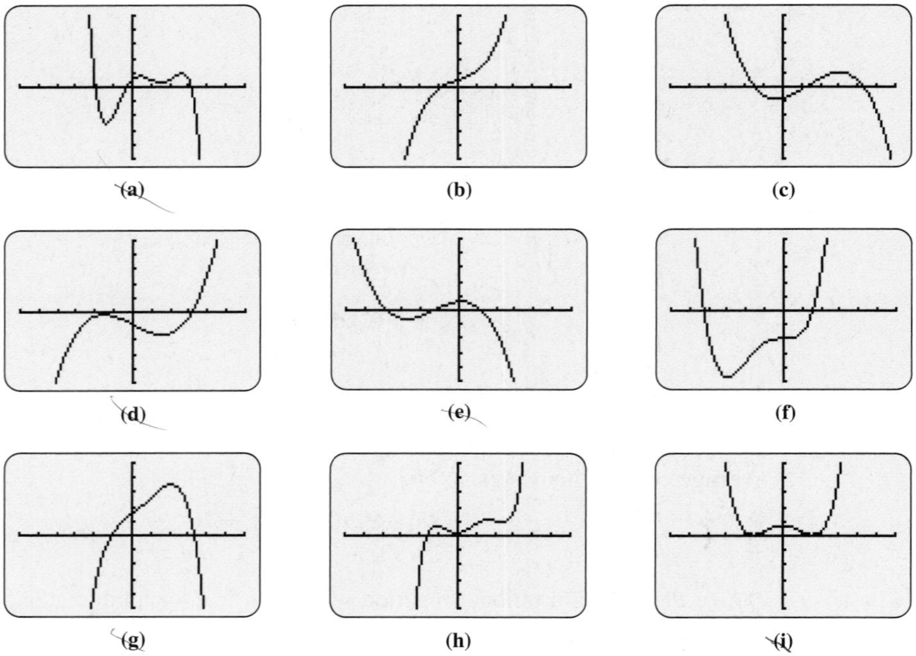

In Exercises 16–20, match the correct graph A–E to the function without using your calculator. Then, after you have answered all of them, if you have a graphing calculator, use your calculator to check your answers. Each graph in this group is plotted on $[-6, 6]$ by $[-6, 6]$. Hint: Consider the asymptotes. (These graphs are done using Dot mode. If you try graphing them in Connected mode, you will see some lines that are not part of the graph, but the result of the calculator connecting disconnected parts of the graph.)

16. $y = \dfrac{2x^2 + 3}{x^2 - 1}$

17. $y = \dfrac{2x^2 + 3}{x^2 + 1}$

18. $y = \dfrac{-2x^2 - 3}{x^2 - 1}$

19. $y = \dfrac{-2x^2 - 3}{x^2 + 1}$

20. $y = \dfrac{2x^2 + 3}{x^3 - 1}$

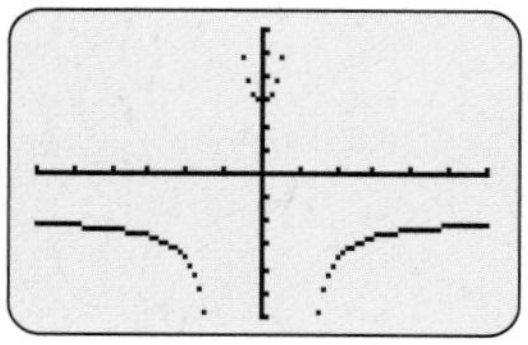

(a)

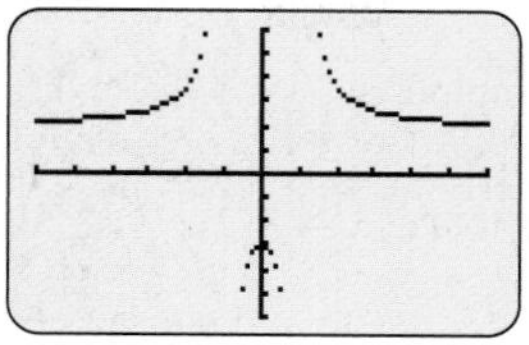

(b)

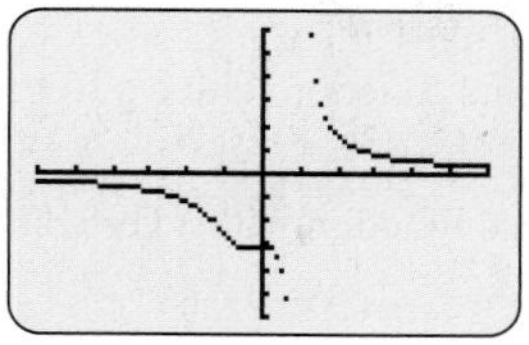

(c)

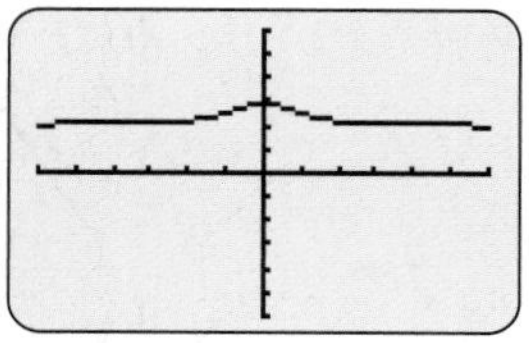

(d)

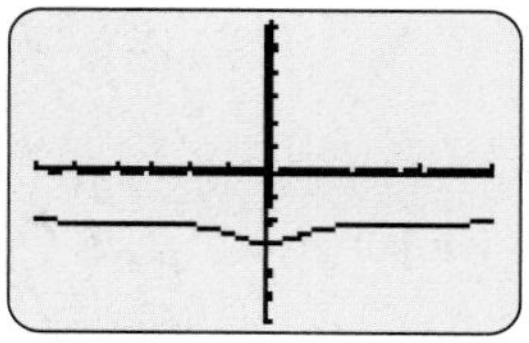

(e)

Each of the following is the graph of a polynomial function. Give the possible values for the degree of the polynomial, and give the sign (+ or −) for the x^n term.

21.

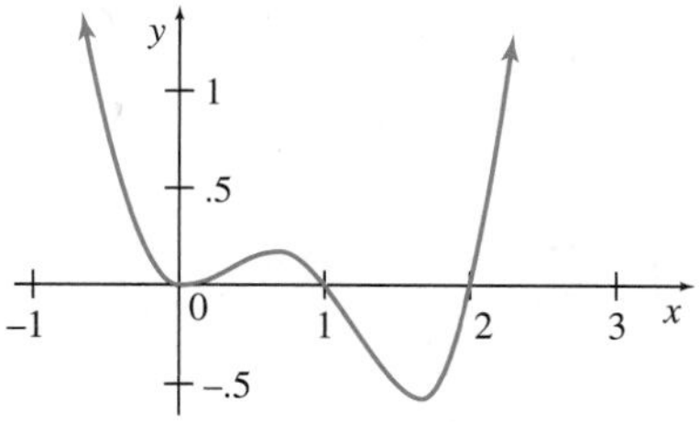

22.

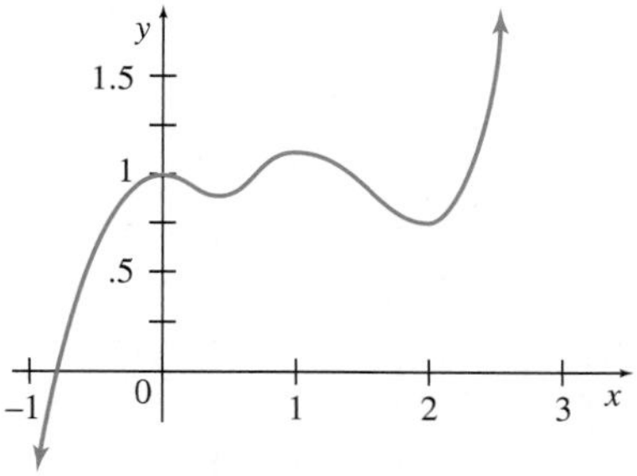

23.

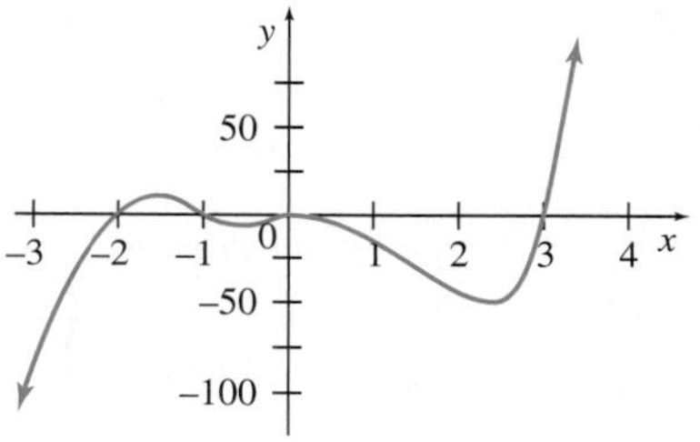

24.

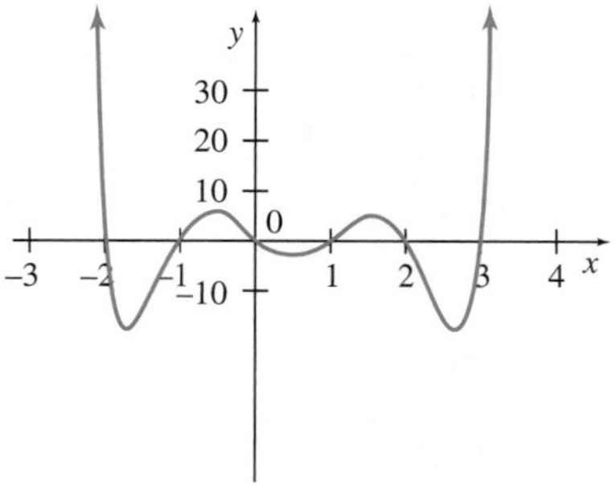

25.

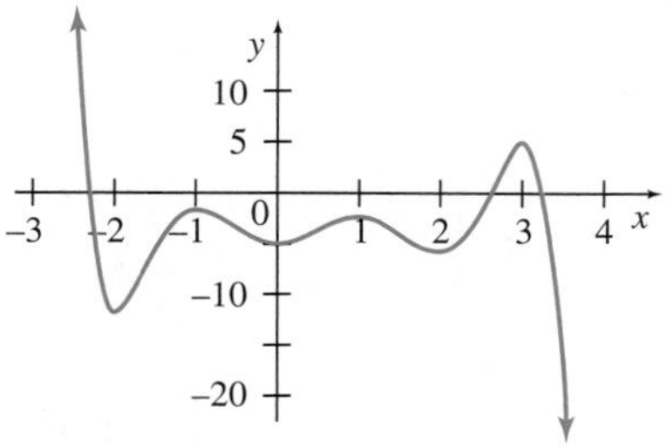

26.

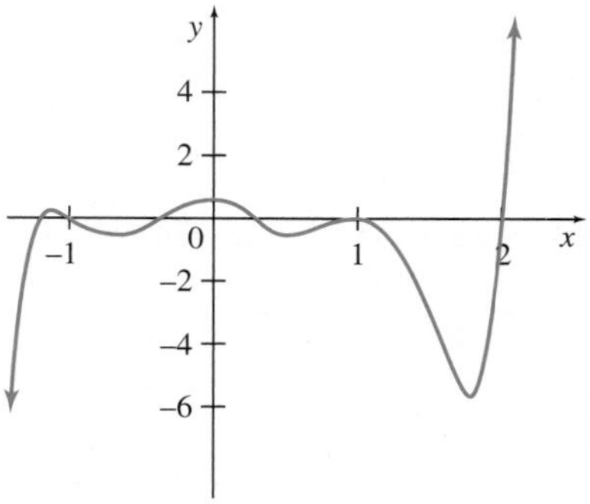

Find the horizontal and vertical asymptotes for each rational function. Draw the graph of each function, including any x- and y-intercepts.

27. $y = \dfrac{-4}{x - 3}$

28. $y = \dfrac{-1}{x + 3}$

29. $y = \dfrac{2}{3 + 2x}$

30. $y = \dfrac{4}{5 + 3x}$

31. $y = \dfrac{3x}{x - 1}$

32. $y = \dfrac{4x}{3 - 2x}$

33. $y = \dfrac{x + 1}{x - 4}$

34. $y = \dfrac{x - 3}{x + 5}$

35. $y = \dfrac{1 - 2x}{5x + 20}$

36. $y = \dfrac{6 - 3x}{4x + 12}$

37. $y = \dfrac{-x - 4}{3x + 6}$

38. $y = \dfrac{-x + 8}{2x + 5}$

39. Write an equation that defines a rational function with a vertical asymptote at $x = 1$ and a horizontal asymptote at $y = 2$.

40. Write an equation that defines a rational function with a vertical asymptote at $x = -2$ and a horizontal asymptote at $y = 0$.

41. Consider the polynomial functions defined by $f(x) = (x - 1)(x - 2)(x + 3)$, $g(x) = x^3 + 2x^2 - x - 2$, and $h(x) = 3x^3 + 6x^2 - 3x - 6$.

a. What is the value of $f(1)$?

b. For what values, other than 1, is $f(x) = 0$?

c. Verify that $g(-1) = g(1) = g(-2) = 0$.

d. Based on your answer from part c, what do you think is the factored form of $g(x)$? Verify your answer by multiplying it out and comparing with $g(x)$.

e. Using your answer from part d, what is the factored form of $h(x)$?

f. Based on what you have learned in this exercise, fill in the blank: If f is a polynomial and $f(a) = 0$ for some number a, then one factor of the polynomial is

________________.

42. Consider the function defined by

$$f(x) = \frac{x^7 - 4x^5 - 3x^4 + 4x^3 + 12x^2 - 12}{x^7}.*$$

a. Graph the function on $[-6, 6]$ by $[-6, 6]$. From your graph, estimate how many x-intercepts the function has and what their values are.

b. Now graph the function on $[-1.5, -1.4]$ by $[-10^{-4}, 10^{-4}]$ and also on $[1.4, 1.5]$ by $[-10^{-5}, 10^{-5}]$. From your graphs, estimate how many x-intercepts the function has and what their values are.

c. From your results in parts a and b, what advice would you give a friend on using a graphing calculator to find x-intercepts?

43. Consider the function defined by

$$f(x) = \frac{1}{x^5 - 2x^3 - 3x^2 + 6}.^{\dagger}$$

a. Graph the function on $[-3.4, 3.4]$ by $[-3, 3]$. From your graph, estimate how many vertical asymptotes the function has and where they are located.

b. Now graph the function on $[-1.5, -1.4]$ by $[-10, 10]$ and also on $[1.4, 1.5]$ by $[-1000, 1000]$. From your graphs, estimate how many vertical asymptotes the function has and where they are located.

c. From your results in parts a and b, what advice would you give a friend on using a graphing calculator to find vertical asymptotes?

*Donley, Edward and Elizabeth Ann George, "Hidden Behavior in Graphs," *Mathematics Teacher*, Vol. 86, No. 6, Sept. 1993.

†Ibid.

Applications

BUSINESS AND ECONOMICS

44. *Average Cost* Suppose the average cost per unit $\overline{C}(x)$, in dollars, to produce x units of margarine is given by

$$\overline{C}(x) = \frac{500}{x + 30}.$$

a. Find $\overline{C}(10)$, $\overline{C}(20)$, $\overline{C}(50)$, $\overline{C}(75)$, and $\overline{C}(100)$.

b. Which of the intervals $(0, \infty)$ and $[0, \infty)$ would be a more reasonable domain for $\overline{C}$? Why?

c. Give the equations of any asymptotes. Find any intercepts.

d. Graph $y = \overline{C}(x)$.

45. *Cost Analysis* In a recent year, the cost per ton, y, to build an oil tanker of x thousand deadweight tons was approximated by

$$\overline{C}(x) = \frac{110{,}000}{x + 225}$$

for $x > 0$.

a. Find $\overline{C}(25)$, $\overline{C}(50)$, $\overline{C}(100)$, $\overline{C}(200)$, $\overline{C}(300)$, and $\overline{C}(400)$.

b. Find any asymptotes.

c. Find any intercepts.

d. Graph $y = \overline{C}(x)$.

Tax Rates Exercises 46–48 refer to the Laffer curve, *originated by the economist Arthur Laffer. An idealized version of this curve is shown here. According to this curve, decreasing a tax rate, say from x_2 percent to x_1 percent on the graph, can actually lead to an increase in government revenue. The theory is that people will work harder and earn more money if they are taxed at a lower rate, so the government ends up with more revenue than it would at a higher tax rate. All economists agree on the endpoints—0 revenue at tax rates of both 0% and 100%—but there is much disagreement on the location of the tax rate x_1 that produces the maximum revenue.*

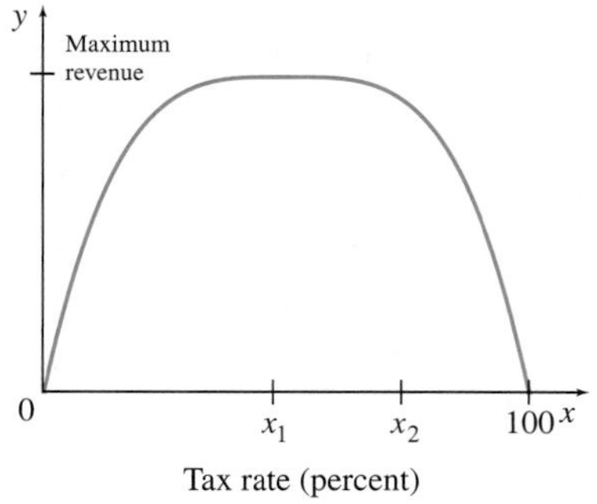

46. A function that might describe the entire Laffer curve is

$$y = x(100 - x)(x^2 + 500),$$

where y is government revenue in hundreds of thousands of dollars from a tax rate of x percent, with the function valid for $0 \le x \le 100$. Find the revenue from the following tax rates.

a. 10% **b.** 40% **c.** 50% **d.** 80%

e. Graph the function.

47. Find the equations of two quadratic functions that could describe the Laffer curve by having zeros at $x = 0$ and $x = 100$. Give the first a maximum of 100 and the second a maximum of 250, then multiply them together to get a new Laffer curve with a maximum of 25,000. Plot the resulting function.

48. An economist might argue that the models in the two previous exercises are unrealistic because they predict that a tax rate of 50% gives the maximum revenue, while the actual value is probably less than 50%. Consider the function

$$y = \frac{300x - 3x^2}{5x + 100},$$

where y is government revenue in millions of dollars from a tax rate of x percent, where $0 \le x \le 100$.*

a. Graph the function, and discuss whether the shape of the graph is appropriate.

b. Use a graphing calculator to find the tax rate that produces the maximum revenue. What is the maximum revenue?

49. *Cost–Benefit Model* Suppose a cost–benefit model is given by

$$y = \frac{6.7x}{100 - x},$$

where y is the cost in thousands of dollars of removing x percent of a given pollutant.

a. Find the cost of removing each percent of pollutants: 50%; 70%; 80%; 90%; 95%; 98%; 99%.

b. Is it possible, according to this function, to remove *all* the pollutant?

c. Graph the function.

50. *Cost–Benefit Model* Suppose a cost–benefit model is given by

$$y = \frac{6.5x}{102 - x},$$

*This exercise is from Dana Lee Ling.

where y is the cost in thousands of dollars of removing x percent of a certain pollutant.

a. Find the cost of removing each percent of pollutants: 0%; 50%; 80%; 90%; 95%; 99%; 100%.

b. Graph the function.

LIFE SCIENCES

51. *Contact Lenses* The strength of a contact lens is given in units known as diopters, as well as in mm of arc. The following is taken from a chart used by optometrists to convert diopters to mm of arc.*

Diopters	mm of Arc
36.000	9.37
36.125	9.34
36.250	9.31
36.375	9.27
36.500	9.24
36.625	9.21
36.750	9.18
36.875	9.15
37.000	9.12

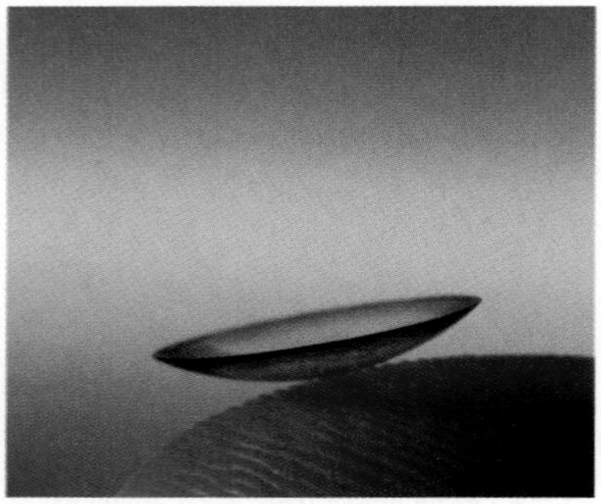

a. Notice that as the diopters increase, the mm of arc decrease. Find a value of k so the function $a = f(d) = k/d$ gives a, the mm of arc, as a function of d, the strength in diopters. (Round k to the nearest integer. For a more accurate answer, average all the values of k given by each pair of data.)

b. An optometrist wants to order 40.50 diopter lenses for a patient. The manufacturer needs to know the strength in mm of arc. What is the strength in mm of arc?

52. *Cardiac Output* A technique for measuring cardiac output depends on the concentration of a dye after a known amount is injected into a vein near the heart. In a normal heart, the concentration of the dye at time x (in seconds) is given by the function

$$g(x) = -.006x^4 + .140x^3 - .053x^2 + 1.79x.$$

a. Graph $g(x)$ on $[0, 6]$ by $[0, 20]$.

b. In your graph from part a, notice that the function initially increases. Considering the form of $g(x)$, do you think it can keep increasing forever? Explain.

c. Write a short paragraph about the extent to which the concentration of dye might be described by the function $g(x)$.

53. *Population Variation* During the early part of the twentieth century, the deer population of the Kaibab Plateau in Arizona experienced a rapid increase, because hunters had reduced the number of natural predators. The increase in population depleted the food resources and eventually caused the population to decline. For the period from 1905 to 1930, the deer population was approximated by

$$D(x) = -.125x^5 + 3.125x^4 + 4000,$$

where x is time in years from 1905.

a. Graph $D(x)$ on $0 \le x \le 25$.

b. From the graph, over what period of time (from 1905 to 1930) was the population increasing? Relatively stable? Decreasing?

54. *Alcohol Concentration* The polynomial function

$$A(x) = -.015x^3 + 1.058x$$

gives the approximate alcohol concentration (in tenths of a percent) in an average person's bloodstream x hours after drinking about eight ounces of 100-proof whiskey. The function is approximately valid for x in the interval $[0, 8]$.

a. Graph $A(x)$ on $0 \le x \le 9$.

b. Using the graph from part a, estimate the time of maximum alcohol concentration.

c. In many states, a person is legally drunk if the blood alcohol concentration exceeds .08%. Use the graph from part a to estimate the period in which this average person is legally drunk.

55. *Cancer* From 1930 to 1990, the rate of breast cancer was nearly constant at 30 cases per 100,000 females, whereas the rate of lung cancer in females over the same period increased. The number of lung cancer cases per 100,000 **e.**

*Data from Bausch & Lomb. The original chart gave all data to 2 decimal places.

females in the year t (where $t = 0$ corresponds to 1930) can be modeled using the function defined by

$$f(t) = 2.8 \times 10^{-4}t^3 - .011t^2 + .23t + .93.*$$

a. Graph the rates of breast and lung cancer on $0 \le t \le 60$.

b. Zoom in on the intersection of the two functions graphed in part a to determine the year when rates for lung cancer first exceeded those for breast cancer.

56. *Population Biology* The function

$$f(x) = \frac{\lambda x}{1 + (ax)^b}$$

is used in population models to give the size of the next generation $(f(x))$ in terms of the current generation (x).†

a. What is a reasonable domain for this function, considering what x represents?

b. Graph this function for $\lambda = a = b = 1$.

c. Graph this function for $\lambda = a = 1$ and $b = 2$.

d. What is the effect of making b larger?

57. *Growth Model* The function

$$f(x) = \frac{Kx}{A + x}$$

is used in biology to give the growth rate of a population in the presence of a quantity x of food. This is called Michaelis–Menten kinetics.‡

a. What is a reasonable domain for this function, considering what x represents?

b. Graph this function for $K = 5$ and $A = 2$.

c. Show that $y = K$ is a horizontal asymptote.

d. What do you think K represents?

e. Show that A represents the quantity of food for which the growth rate is half of its maximum.

58. *Brain Mass* The mass (in grams) of the human brain during the last trimester of gestation and the first two years after birth can be approximated by the function

$$m(c) = \frac{c^3}{100} - \frac{1500}{c},$$

where c is the circumference of the head in centimeters.§

a. Find the approximate mass of brains with a circumference of 30, 40, or 50 cm.

b. Clearly the formula is invalid for any values of c yielding negative values of w. For what values of c is this true?

c. Use a graphing calculator to sketch this graph on the interval $20 \le x \le 50$.

d. Suppose an infant brain has mass of 700 g. Use features on a graphing calculator to find what the circumference of the head is expected to be.

PHYSICAL SCIENCES

59. *Length of a Pendulum* A simple pendulum swings back and forth in regular time intervals. Grandfather clocks use pendulums to keep accurate time. The relationship between the length of a pendulum L and the period (time) T for one complete oscillation can be expressed by the function $L = kT^n$, where k is a constant and n is a positive integer to be determined. The data below were taken for different lengths of pendulums.‖

a. Find the value of k for $n = 1, 2$, and 3, using the data for the 4-ft pendulum.

b. Use a graphing calculator to plot the data in the table and to graph the function $L = kT^n$ for the three values of k (and their corresponding values of n) found in part a. Which function best fits the data?

T (sec)	*L* (ft)
1.11	1.0
1.36	1.5
1.57	2.0
1.76	2.5
1.92	3.0
2.08	3.5
2.22	4.0

c. Use the best-fitting function from part a to predict the period of a pendulum having a length of 5 ft.

d. If the length of pendulum doubles, what happens to the period?

*Valanis, B., *Epidemiology in Nursing and Health Care*, Appleton & Lange, 1992.
†Smith, J. Maynard, *Models in Ecology*, Oxford: Cambridge University Press, 1974.
‡Edelstein-Keshet, Leah, *Mathematical Models in Biology*, Random House, 1988.
§Dobbing, John and Jean Sands, "Head Circumference, Biparietal Diameter and Brain Growth in Fetal and Postnatal Life," *Early Human Development*, Vol. 2, No. 1, April 1978, pp. 81–87.
‖Data provided by Gary Rockswold, Mankato State University, Mankato, Minnesota. See Exercise 17, Section 1.3.

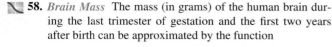

e. If you have a graphing calculator or computer program with a quadratic regression feature, use it to find a quadratic function that approximately fits the data. How does this answer compare with the answer to part b?

60. *Coal Consumption* The table gives U.S. coal consumption for selected years.*

Year	Millions of Short Tons
1950	494.1
1960	398.1
1970	523.2
1980	702.7
1985	818.0
1990	902.9
1995	962.1
1997	1029.5
1999	1038.6
2000	1084.1
2001	1060.3

a. Draw a scatterplot, letting $x = 0$ represent 1950.

b. Use the quadratic regression feature of a graphing calculator to get a quadratic function that approximates the data.

c. Graph the function from part b on the same window as the scatterplot.

d. Use cubic regression to get a cubic function that approximates the data.

e. Graph the cubic function from part d on the same window as the scatterplot.

f. Which of the two functions in parts b and d appears to be a better fit for the data? Explain your reasoning.

GENERAL INTEREST

61. *Abandoned Cars* The number of abandoned cars in New York City dropped during the 1990s because of reduced auto theft and increasing scrap metal prices.† The following table, based on estimates from a graph in *The New York Times*, gives the number of abandoned cars as a function of the year.

a. Plot the points from the table, using 0 for 1986, and so on.

b. If your graphing calculator has a cubic and quartic regression feature, find the third- and fourth-degree polynomials that best fit the data according to the least squares method. Plot these polynomials on the same calculator window as the data.

c. Discuss whether a third- or fourth-degree polynomial better fits the data.

Year	Abandoned Cars (1000s)
1986	82
1987	120
1988	140
1989	148
1990	135
1991	95
1992	75
1993	51
1994	38
1995	27
1996	20

2.4 EXPONENTIAL FUNCTIONS

THINK ABOUT IT *How much interest will an investment earn? What is the oxygen consumption of yearling salmon?*

Later in this section, in Examples 5 and 6, we will see that the answers to these questions depend on *exponential functions*.

In earlier sections we discussed functions involving expressions such as x^2, $(2x + 1)^3$, or x^{-1}, where the variable or variable expression is the base of an

**Annual Energy Review*, U.S. Department of Energy, 2001.
†*The New York Times*, Feb. 15, 1997, p. 25.

exponential expression, and the exponent is a constant. In an exponential function, the variable is in the exponent and the base is a constant.

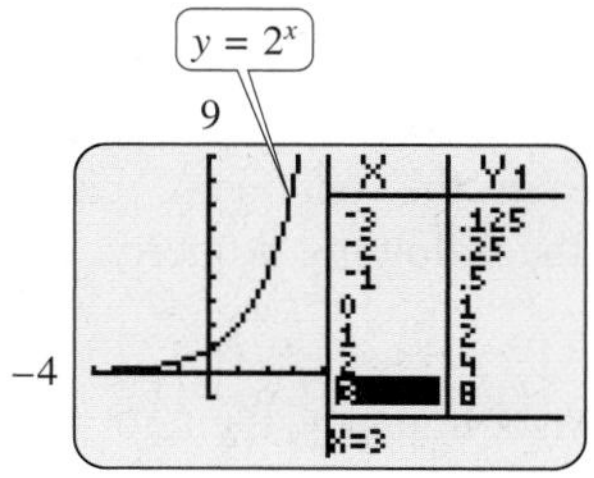

FIGURE 46

> **EXPONENTIAL FUNCTION**
>
> An **exponential function** with base a is defined as
>
> $$f(x) = a^x, \quad \text{where } a > 0 \text{ and } a \neq 1.$$

(If $a = 1$, the function is the constant function $f(x) = 1$.)

Exponential functions may be the single most important type of functions used in practical applications. They are used to describe growth and decay, which are important ideas in management, social science, and biology.

> **FOR REVIEW**
>
> Recall from the section on Quadratic Functions; Translation and Reflection, that the graph of $f(-x)$ is the reflection of the graph of $f(x)$ about the y-axis.

Figure 46 shows a graph of the exponential function defined by $f(x) = 2^x$ and a table of integer values from $x = -3$ to $x = 3$ of $f(x)$. This graph is typical of the graphs of exponential functions of the form $y = a^x$, where $a > 1$. Notice that as x gets larger and larger, the function also gets larger. As x gets more and more negative, the function becomes smaller and smaller, approaching but never reaching 0. Therefore, the x-axis is a horizontal asymptote, but the function only approaches the left side of the asymptote. In contrast, rational functions approach both the left and right sides of the asymptote. The graph suggests that the domain is the set of all real numbers and the range is the set of all positive numbers.

EXAMPLE 1 *Graphing an Exponential Function*

Graph $f(x) = 2^{-x}$.

Solution The graph, shown in Figure 47, is the horizontal reflection of the graph of $f(x) = 2^x$ given in Figure 46. Since $2^{-x} = 1/2^x = (1/2)^x$, this graph is typical of the graphs of exponential functions of the form $y = a^x$ where $0 < a < 1$. The domain includes all real numbers and the range includes all positive numbers. Notice that this function, with $f(x) = 2^{-x} = (1/2)^x$, is decreasing over its domain.

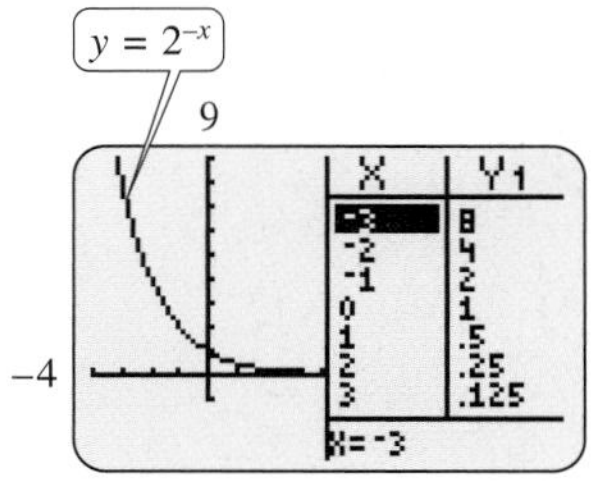

FIGURE 47

In the definition of an exponential function, notice that the base a is restricted to positive values, with negative or zero bases not allowed. For example, the function $y = (-4)^x$ could not include such numbers as $x = 1/2$ or $x = 1/4$ in the domain. The resulting graph would be at best a series of separate points having little practical use.

EXAMPLE 2 *Graphing an Exponential Function*

Graph $f(x) = -2^x + 3$.

Solution The graph of $y = -2^x$ is the vertical reflection of the graph of $y = 2^x$, so this is a decreasing function. (Notice that -2^x is not the same as $(-2)^x$. In -2^x, we raise 2 to the x power and then take the negative.) The 3 indicates that the graph should be translated vertically 3 units, as compared to the graph of $y = -2^x$. Since $y = -2^x$ would have y-intercept $(0, -1)$, this function has y-intercept $(0, 2)$, which is up 3 units. For negative values of x, the graph

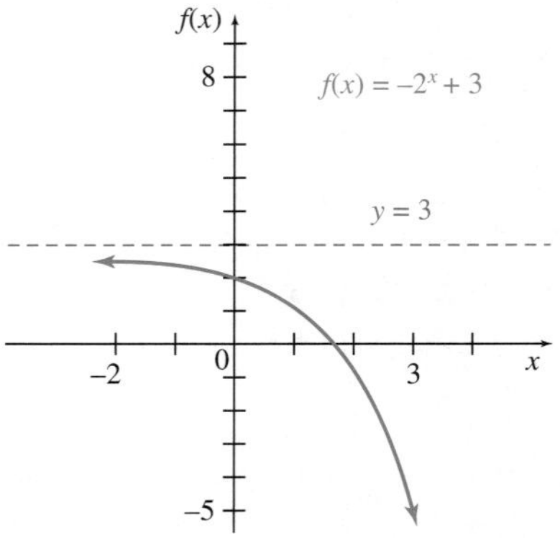

FIGURE 48

approaches the line $y = 3$, which is a horizontal asymptote. The graph is shown in Figure 48.

Exponential Equations In Figures 46 and 47, which are typical graphs of exponential functions, a given value of x leads to exactly one value of a^x. Because of this, an equation with a variable in the exponent, called an **exponential equation,** often can be solved using the following property.

> If $a > 0$, $a \neq 1$, and $a^x = a^y$, then $x = y$. Also, if $x = y$, then $a^x = a^y$.

(Both bases must be the same.) The value $a = 1$ is excluded since $1^2 = 1^3$, for example, even though $2 \neq 3$. To solve $2^{3x} = 2^7$ using this property, work as follows.

$$\begin{aligned} 2^{3x} &= 2^7 \\ 3x &= 7 \\ x &= \frac{7}{3} \end{aligned}$$

EXAMPLE 3 *Solving Exponential Equations*

(a) Solve $9^x = 27$.

Solution First rewrite both sides of the equation so the bases are the same. Since $9 = 3^2$ and $27 = 3^3$,

> **FOR REVIEW**
> Recall from the Algebra Reference that $(a^m)^n = a^{mn}$.

$$\begin{aligned} (3^2)^x &= 3^3 \\ 3^{2x} &= 3^3 \\ 2x &= 3 \\ x &= \frac{3}{2}. \end{aligned}$$

(b) Solve $32^{2x-1} = 128^{x+3}$.

Solution Since the bases must be the same, write 32 as 2^5 and 128 as 2^7, giving

$$\begin{aligned} 32^{2x-1} &= 128^{x+3} \\ (2^5)^{2x-1} &= (2^7)^{x+3} \\ 2^{10x-5} &= 2^{7x+21}. \quad \text{Multiply exponents.} \end{aligned}$$

Now use the property above to get

$$\begin{aligned} 10x - 5 &= 7x + 21 \\ 3x &= 26 \\ x &= \frac{26}{3}. \end{aligned}$$

Verify this solution in the original equation.

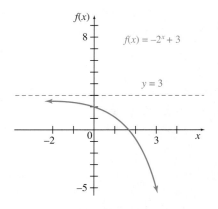

Compound Interest The calculation of compound interest is an important application of exponential functions. The cost of borrowing money or the return on an investment is called **interest.** The amount borrowed or invested is the **principal,** P. The **rate of interest** r is given as a percent per year, and t is the **time,** measured in years.

> **SIMPLE INTEREST**
>
> The product of the principal P, rate r, and time t gives **simple interest,** I:
>
> $$I = Prt.$$

With **compound interest,** interest is charged (or paid) on interest as well as on principal. To find a formula for compound interest, first suppose that P dollars, the principal, is deposited at a rate of interest r per year. The interest earned during the first year is found using the formula for simple interest.

$$\text{First-year interest} = P \cdot r \cdot 1 = Pr.$$

At the end of one year, the amount on deposit will be the sum of the original principal and the interest earned, or

$$P + Pr = P(1 + r). \tag{1}$$

If the deposit earns compound interest, the interest earned during the second year is found from the total amount on deposit at the end of the first year. Thus, the interest earned during the second year (again found by the formula for simple interest), is

$$[P(1 + r)](r)(1) = P(1 + r)r, \tag{2}$$

so the total amount on deposit at the end of the second year is the sum of amounts from (1) and (2) above, or

$$P(1 + r) + P(1 + r)r = P(1 + r)(1 + r) = P(1 + r)^2.$$

In the same way, the total amount on deposit at the end of three years is

$$P(1 + r)^3.$$

After t years, the total amount on deposit, called the *compound amount,* is $P(1 + r)^t$.

When interest is compounded more than once a year, the compound interest formula is adjusted. For example, if interest is to be paid quarterly (four times a year), $1/4$ of the interest rate is used each time interest is calculated, so the rate becomes $r/4$, and the number of compounding periods in t years becomes $4t$. Generalizing from this idea gives the following formula.

> **COMPOUND AMOUNT**
>
> If P dollars is invested at a yearly rate of interest r per year, compounded m times per year for t years, the **compound amount** is
>
> $$A = P\left(1 + \frac{r}{m}\right)^{tm} \text{ dollars.}$$

EXAMPLE 4 *Compound Interest*

Jessica Elbern invests a bonus of $9000 at 6% annual interest compounded semi-annually for 4 years. How much interest will she earn?

Solution Use the formula for compound interest with $P = 9000$, $r = .06$, $m = 2$, and $t = 4$.

$$A = P\left(1 + \frac{r}{m}\right)^{tm}$$

$$= 9000\left(1 + \frac{.06}{2}\right)^{4(2)}$$

$$= 9000(1.03)^8$$

$$\approx 11{,}400.93 \qquad \text{Use a calculator.}$$

The investment plus the interest is $11,400.93. The interest amounts to $11,400.93 − $9000 = $2400.93. ━━━

NOTE When using a calculator to compute the compound interest, store each partial result in the calculator and avoid rounding off until the final answer. ■

The Number *e* Perhaps the single most useful base for an exponential function is the number *e*, an irrational number that occurs often in practical applications. The letter *e* was chosen to represent this number in honor of the Swiss mathematician Leonhard Euler (pronounced "oiler") (1707–1783). To see how the number *e* occurs in an application, begin with the formula for compound interest,

$$P\left(1 + \frac{r}{m}\right)^{tm}.$$

Suppose that a lucky investment produces annual interest of 100%, so that $r = 1.00 = 1$. Suppose also that you can deposit only $1 at this rate, and for only one year. Then $P = 1$ and $t = 1$. Substituting these values into the formula for compound interest gives

$$P\left(1 + \frac{r}{m}\right)^{t(m)} = 1\left(1 + \frac{1}{m}\right)^{1(m)} = \left(1 + \frac{1}{m}\right)^{m}.$$

As interest is compounded more and more often, *m* gets larger and the value of this expression will increase. For example, if $m = 1$ (interest is compounded annually),

$$\left(1 + \frac{1}{m}\right)^{m} = \left(1 + \frac{1}{1}\right)^{1} = 2^1 = 2,$$

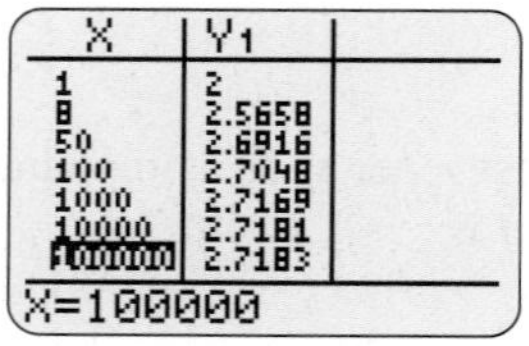

so that your $1 becomes $2 in one year. Using a graphing calculator, we produced Figure 49 (where *m* is represented by X and $(1 + 1/m)^m$ by Y_1) to see what happens as *m* becomes larger and larger. A spreadsheet can also be used to produce this table.

The table suggests that as *m* increases, the value of $(1 + 1/m)^m$ gets closer and closer to a fixed number, called *e*. As we shall see in the next chapter, this is an example of a limit.

X	Y1
1	2
8	2.5658
50	2.6916
100	2.7048
1000	2.7169
10000	2.7181
100000	2.7183

X=100000

FIGURE 49

DEFINITION OF *e*

As m becomes larger and larger, $\left(1 + \frac{1}{m}\right)^m$ becomes closer and closer to the number e, whose approximate value is 2.718281828.

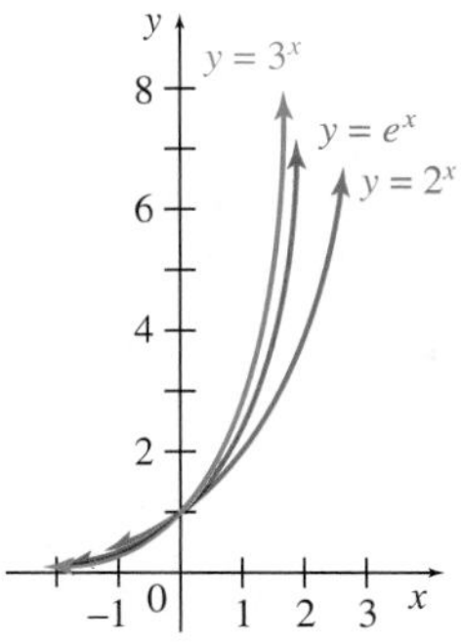

FIGURE 50

The value of e is approximated here to 9 decimal places. Euler approximated e to 23 decimal places using this definition. Many calculators give values of e^x, usually with a key labeled e^x. Some require two keys, either INV LN or 2nd LN. (We will define $\ln x$ in the next section.) In Figure 50, the functions $y = 2^x$, $y = e^x$, and $y = 3^x$ are graphed for comparison. Notice that e^x is between 2^x and 3^x, because e is between 2 and 3. For $x > 0$, the graphs show that $3^x > e^x > 2^x$. All three functions have y-intercept $(0, 1)$. It is difficult to see from the graph, but $3^x < e^x < 2^x$ when $x < 0$.

The number e is often used as the base in an exponential equation because it provides a good model for many natural, as well as economic, phenomena. In the exercises for this section, we will look at several examples of such applications.

Continuous Compounding

In economics, the formula for **continuous compounding** is a good example of an exponential growth function. Recall the formula for compound amount

$$A = P\left(1 + \frac{r}{m}\right)^{tm},$$

where m is the number of times annually that interest is compounded. As m becomes larger and larger, the compound amount also becomes larger, but not without bound. Recall that as m becomes larger and larger, $(1 + 1/m)^m$ becomes closer and closer to e. Similarly,

$$\left(1 + \frac{1}{(m/r)}\right)^{m/r}$$

becomes closer and closer to e. Let us rearrange the formula for compound amount to take advantage of this fact.

$$\begin{aligned} A &= P\left(1 + \frac{r}{m}\right)^{tm} \\ &= P\left(1 + \frac{1}{(m/r)}\right)^{tm} \\ &= P\left[\left(1 + \frac{1}{(m/r)}\right)^{m/r}\right]^{rt} \qquad \frac{m}{r} \cdot rt = tm \end{aligned}$$

This last expression becomes closer and closer to Pe^{rt} as m becomes larger and larger, which describes what happens when interest is compounded continuously. Essentially, the number of times annually that interest is compounded becomes infinitely large. We thus have the following formula for the compound amount when interest is compounded continuously.

CONTINUOUS COMPOUNDING

If a deposit of P dollars is invested at a rate of interest r compounded continuously for t years, the compound amount is

$$A = Pe^{rt} \text{ dollars.}$$

EXAMPLE 5 *Continuous Compound Interest*

Assuming continuous compounding, if the inflation rate averaged 6% per year for 5 years, how much would a $1 item cost at the end of the 5 years?

Solution In the formula for continuous compounding, let $P = 1$, $t = 5$, and $r = .06$ to get

$$A = 1e^{5(.06)} = e^{.3} \approx 1.34986.$$

An item that cost $1 at the beginning of the 5-year period would cost $1.35 at the end of the period, an increase of 35%, or about 1/3. ■

In situations that involve growth or decay of a population, the size of the population at a given time t often is determined by an exponential function of t. The next example illustrates a typical application of this kind.

EXAMPLE 6 *Oxygen Consumption*

Biologists studying salmon have found that the oxygen consumption of yearling salmon (in appropriate units) increases exponentially with the speed of swimming according to the function defined by

$$f(x) = 100e^{.6x},$$

where x is the speed in feet per second. Find the following.

(a) The oxygen consumption when the fish are still

Solution When the fish are still, their speed is 0. Substitute 0 for x:

$$f(0) = 100e^{(.6)(0)} = 100e^0$$
$$= 100 \cdot 1 = 100. \qquad e^0 = 1$$

When the fish are still, their oxygen consumption is 100 units.

(b) The oxygen consumption at a speed of 2 ft per second

Solution Find $f(2)$ as follows.

$$f(2) = 100e^{(.6)(2)} = 100e^{1.2} \approx 332$$

At a speed of 2 ft per second, oxygen consumption is about 332 units. ■

FOR REVIEW

Refer to the discussion on linear regression in Section 1.3. A similar process is used to fit data points to other types of functions. Many of the functions in this chapter's applications were determined in this way, including that given in Example 6.

NOTE In Example 6(b), we rounded the answer to the nearest integer. Because the function is only an approximation of the real situation, further accuracy is not realistic. ■

EXAMPLE 7 *Food Surplus*

A magazine article argued that the cause of the obesity epidemic in the United States is the decreasing cost of food (in real terms) due to the increasing surplus of food.* As one piece of evidence, the following table was provided, showing U.S. corn production (in billions of bushels) over the 20th century.

Year	Production (billions of bushels)
1930	1.757
1940	2.207
1950	2.764
1960	3.907
1970	4.152
1980	6.639
1990	7.934
2000	9.968

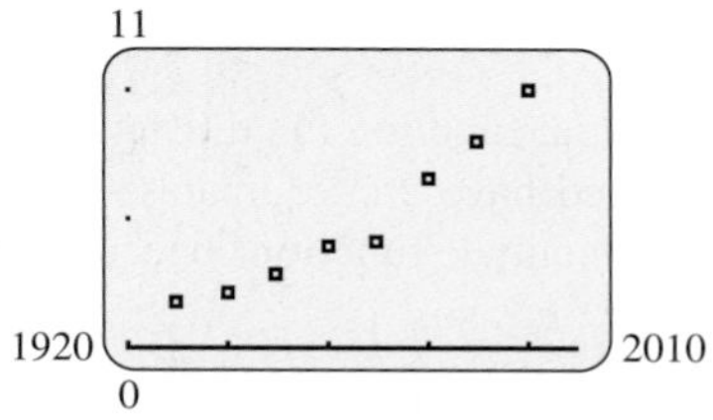

FIGURE 51

(a) Plot the data. Does the production appear to grow linearly or exponentially?

Solution Figure 51 shows a graphing calculator plot of the data, which suggests that corn production is growing exponentially.

(b) Find an exponential function in the form of $p(x) = p_0 a^{x-1930}$ that models this data, where x is the year and $p(x)$ is the production of corn. Use the data for 1930 and 2000.

Solution Since $p(1930) = p_0 a^0 = p_0$, we have $p_0 = 1.757$. Using $x = 2000$, we have

$$p(2000) = 1.757a^{2000-1930} = 1.757a^{70} = 9.968$$

$$a^{70} = \frac{9.968}{1.757} \qquad \text{Divide by } 1.757.$$

$$a = \left(\frac{9.968}{1.757}\right)^{1/70} \qquad \text{Take the 70th root.}$$

$$\approx 1.0251.$$

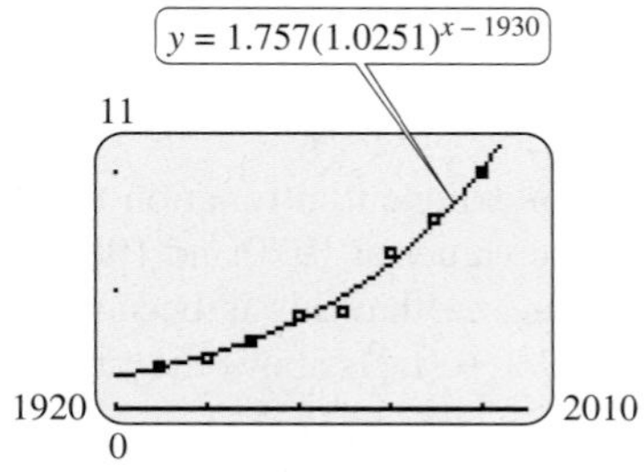

FIGURE 52

Thus $p(x) = 1.757(1.0251)^{x-1930}$. Figure 52 shows that this function fits the data well.

(c) Determine the expected annual percentage increase in corn production during this time period.

Solution Since a is 1.0251, the production of corn each year is 1.0251 times its value the previous year, for a rate of increase of $.0251 = 2.51\%$ per year.

(d) Graph p and estimate the year when corn production will be double what it was in 2000.

*Pollan, Michael, "The (Agri)Cultural Contradictions of Obesity," *The New York Times Magazine*, Oct. 12, 2003, p. 41.

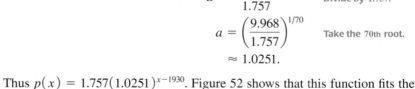

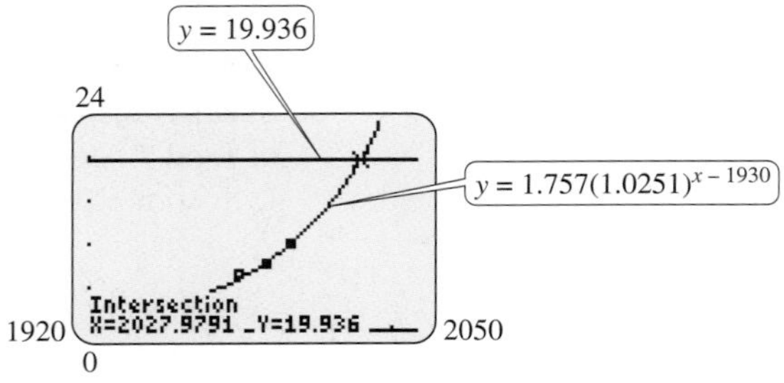

FIGURE 53

Solution Figure 53 shows the graphs of $p(x)$ and $y = 2 \cdot 9.968 = 19.936$ on the same coordinate axes. (Note that the scale in Figure 53 is different than the scale in Figures 51 and 52 so that larger values of x and $p(x)$ are visible.) Their graphs intersect at approximately 2028, which is thus the year when corn production will be double its 2000 level. In the next section, we will see another way to solve such problems that does not require the use of a graphing calculator.

NOTE Another way to check whether an exponential function fits the data is to see if points whose x-coordinates are equally spaced have y-coordinates with a constant ratio. This must be true for an exponential function because if $f(x) = a \cdot b^x$, then $f(x_1) = a \cdot b^{x_1}$ and $f(x_2) = a \cdot b^{x_2}$, so

$$\frac{f(x_2)}{f(x_1)} = \frac{a \cdot b^{x_2}}{a \cdot b^{x_1}} = b^{x_2 - x_1}.$$

This last expression is constant if $x_2 - x_1$ is constant, that is, if the x-coordinates are equally spaced.

In the previous example, all data points have x-coordinates 10 years apart, so we can compare the ratios of corn production for any pairs of years. Here are the ratios for the first pair of years and the last:

$$\frac{2.207}{1.757} = 1.256$$

$$\frac{9.968}{7.934} = 1.256$$

These ratios are identical to 3 decimal places, so an exponential function fits the data very well. Not all ratios are this close; using the values at 1970 and 1980, we have $6.639/4.152 = 1.599$. From Figure 52, we can see that this is because the 1970 value is below the exponential curve and the 1980 value is above the curve. ■

Another way to find an exponential function that fits a set of data is to use a graphing calculator or computer program with an exponential regression feature. This fits an exponential function through a set of points using the least squares method, introduced in Section 1.3 for fitting a line through a set of points. On a TI-83/84 Plus, for example, enter the year into the list L_1 and the corn production into L_2. For simplicity, subtract 1930 from each year, so that 1930 corresponds to

$x = 0$. Selecting `ExpReg` from the `STAT CALC` menu yields $y = 1.721(1.0256)^x$, which is close to the function we found in Example 7(b).

2.4 EXERCISES

*A ream of 20-lb paper contains 500 sheets and is about 2 in. high. Suppose you take one sheet, fold it in half, then fold it in half again, continuing in this way as long as possible.**

1. Complete the table.

Number of Folds	1	2	3	4	5	...	10	...	50
Layers of Paper									

2. After folding 50 times (if this were possible), what would be the height (in miles) of the folded paper?

For Exercises 3–11, match the correct graph A–F to the function without using your calculator. Notice that there are more functions than graphs; some of the functions are equivalent. After you have answered all of them, use a graphing calculator to check your answers. Each graph in this group is plotted on the window $[-2, 2]$ by $[-4, 4]$.

3. $y = 3^x$

4. $y = 3^{-x}$

5. $y = \left(\frac{1}{3}\right)^{1-x}$

6. $y = 3^{x+1}$

7. $y = 3(3)^x$

8. $y = \left(\frac{1}{3}\right)^x$

9. $y = 2 - 3^{-x}$

10. $y = -2 + 3^{-x}$

11. $y = 3^{x-1}$

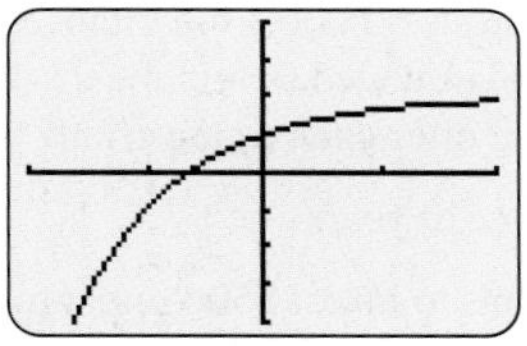
(a)

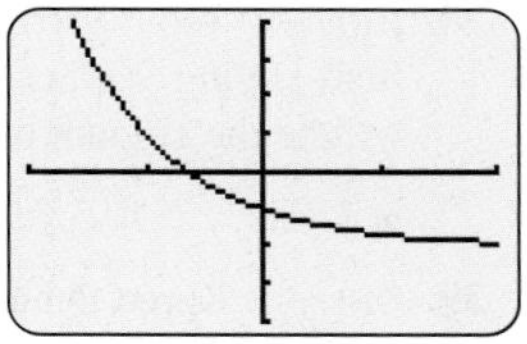
(b)

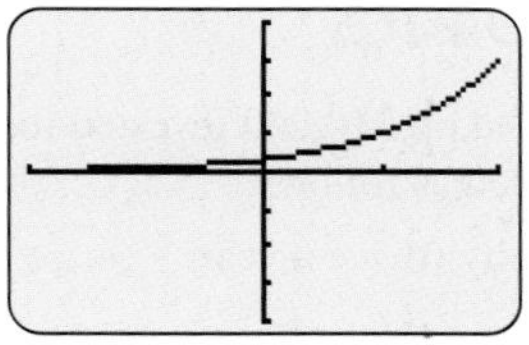
(c)

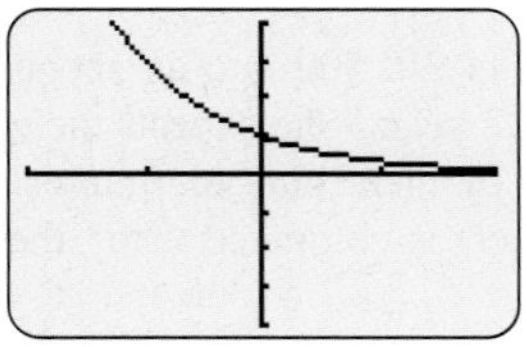
(d)

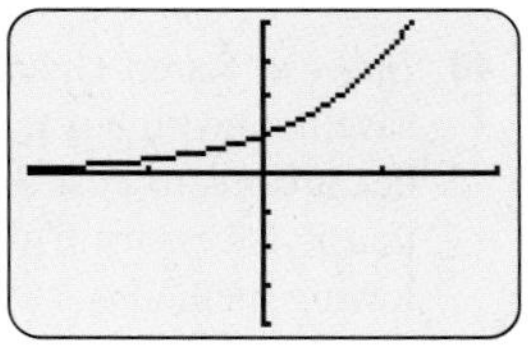
(e)

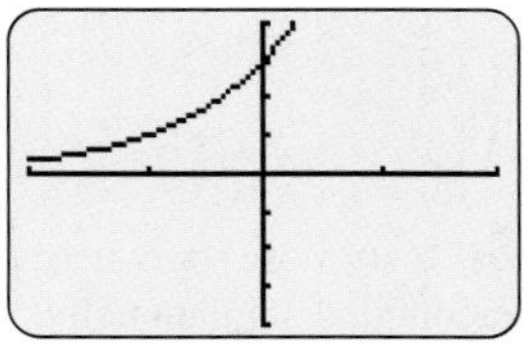
(f)

12. In Exercises 3–11, there were more formulas for functions than there were graphs. Explain how this is possible.

Solve each equation.

13. $2^x = \frac{1}{8}$

14. $4^x = 64$

15. $e^x = \frac{1}{e^2}$

16. $4^x = 8^{x+1}$

*Thomas, Jamie, "Exponential Functions," *The AMATYC Review*, Vol. 18, No. 2, Spring 1997.

17. $25^x = 125^{x-2}$ **18.** $16^{-x+1} = 8^x$ **19.** $16^{x+2} = 64^{2x-1}$ **20.** $(e^4)^{-2x} = e^{-x+1}$

21. $e^{-x} = (e^2)^{x+3}$ **22.** $2^{|x|} = 16$ **23.** $5^{-|x|} = \frac{1}{25}$ **24.** $2^{x^2-4x} = \left(\frac{1}{16}\right)^{x-4}$

25. $5^{x^2+x} = 1$ **26.** $8^{x^2} = 2^{5x+2}$ **27.** $9^x = 27^{x^2+x}$ **28.** $e^{x^2-3x+2} = 1$

29. In our definition of exponential function, we ruled out negative values of a. The author of a textbook on mathematical economics, however, obtained a "graph" of $y = (-2)^x$ by plotting the following points and drawing a smooth curve through them.

x	−4	−3	−2	−1	0	1	2	3
y	1/16	−1/8	1/4	−1/2	1	−2	4	−8

The graph oscillates very neatly from positive to negative values of y. Comment on this approach. (This exercise shows the dangers of point plotting when drawing graphs.)

30. Explain why the exponential equation $4^x = 6$ cannot be solved using the method described in Example 3.

31. Explain why $3^x > e^x > 2^x$ when $x > 0$, but $3^x < e^x < 2^x$ when $x < 0$.

32. A friend claims that as x becomes large, the expression $1 + 1/x$ gets closer and closer to 1, and 1 raised to any power is still 1. Therefore, $f(x) = (1 + 1/x)^x$ gets closer and closer to 1 as x gets larger. Use a graphing calculator to graph f on $.1 \le x \le 50$. How might you use this graph to explain to the friend why $f(x)$ does not approach 1 as x becomes large? What does it approach?

Applications

BUSINESS AND ECONOMICS

33. *Interest* Find the interest earned on \$10,000 invested for 5 years at 6% interest compounded as follows.

a. Annually **b.** Semiannually (twice a year)
c. Quarterly **d.** Monthly

34. *Interest* Suppose \$26,000 is borrowed for 3 years at 12% interest. Find the interest paid over this period if the interest is compounded as follows.

a. Annually **b.** Semiannually
c. Quarterly **d.** Monthly

35. *Interest* David Horwitz needs to choose between two investments: one pays 8% compounded semiannually, and the other pays $7\frac{1}{2}$% compounded monthly. If he plans to invest \$18,000 for $1\frac{1}{2}$ years, which investment should he choose? How much extra interest will he earn by making the better choice?

36. *Interest* Find the interest rate required for an investment of \$5000 to grow to \$8000 in 4 years if interest is compounded as follows.

a. Annually **b.** Quarterly

37. *Inflation* Assuming continuous compounding, what will it cost to buy a \$10 item in 3 years at the following inflation rates?

a. 3% **b.** 4% **c.** 5%

38. *Interest* Scott Perrine invests a \$25,000 inheritance in a fund paying 9% per year compounded continuously. What will be the amount on deposit after each time period?

a. 1 year **b.** 5 years **c.** 10 years

39. *Interest* Karen Panunzio plans to invest \$500 into a money market account. Find the interest rate that is needed for the money to grow to \$1200 in 14 years if the interest is compounded quarterly.

40. *Interest* Karen Guardino puts \$10,500 into an account to save money to buy a car in 12 years. She expects the car of her dreams to cost \$30,000 by then. Find the interest rate that is necessary if the interest is computed using the following methods.

a. Compounded quarterly

b. Compounded continuously

41. *Inflation* If money loses value at the rate of 8% per year, the value of \$1 in t years is given by

$$y = (1 - .08)^t = (.92)^t.$$

a. Use a calculator to help complete the following table.

t	0	1	2	3	4	5	6	7	8	9	10
y	1					.66					.43

b. Graph $y = (.92)^t$.

c. Suppose a house costs \$165,000 today. Use the results of part a to estimate the cost of a similar house in 10 years.

d. Find the cost of a \$50 textbook in 8 years.

42. *Internet Users* During the early days of the Internet, growth in the number of users worldwide could be approximated by an exponential function. The number of users was estimated to be 16 million in December 1995 and 451.04 million in December 2000.*

a. Find an exponential function with $f(x) = f_0 a^x$ that models the number of Internet users, where $x = 0$ (in years) corresponds to December 1995, and $f(x)$ gives the number of users in millions.

b. Graph the function on a graphing calculator and use the graph to estimate the number of users in December 1998. Is this value close to the estimate found using the graph at the beginning of this chapter?

c. Approximate the average yearly percent increase in users during this time period.

d. Use the function to estimate the number of users in December 2001. Compare with the actual number of 552.01 million. Does the exponential growth model continue to hold? Explain.

43. *Interest* On January 1, 1980, Jack deposited \$1000 into Bank X to earn interest at the rate of j per annum compounded semiannually. On January 1, 1985, he transferred his account to Bank Y to earn interest at the rate of k per annum compounded quarterly. On January 1, 1988, the balance at Bank Y was \$1990.76. If Jack could have earned interest at the rate of k per annum compounded quarterly from January 1, 1980, through January 1, 1988, his balance would have been \$2203.76. Which of the following represents the ratio k/j? (This exercise is reprinted from an actuarial examination.)†

a. 1.25 **b.** 1.30 **c.** 1.35 **d.** 1.40 **e.** 1.45

LIFE SCIENCES

44. *Population Growth* Since 1950, the growth in world population (in millions) closely fits the exponential function defined by

$$A(t) = 2600e^{.017t},$$

where t is the number of years since 1950.

a. World population was about 3700 million in 1970. How closely does the function approximate this value?

b. Use the function to approximate world population in 1990. (The actual 1990 population was about 5276 million.)

c. Estimate world population in the year 2010.

45. *Growth of Bacteria* Salmonella bacteria, found on almost all chicken and eggs, grow rapidly in a nice warm place. If just a few hundred bacteria are left on the cutting board when a chicken is cut up, and they get into the potato salad, the population begins compounding. Suppose the number present in the potato salad after x hours is given by

$$f(x) = 500 \cdot 2^{3x}.$$

a. If the potato salad is left out on the table, how many bacteria are present 1 hour later?

b. How many were present initially?

c. How often do the bacteria double?

d. How quickly will the number of bacteria increase to 32,000?

46. *Toronto's Jewish Population* The table gives the population of Toronto's Jewish community at various times.‡

Year	Population
1901	3103
1911	18,294
1921	34,770
1931	46,751
1941	52,798
1951	66,773
1961	85,000
1971	97,000
1981	128,650
1991	162,605

a. Plot the population on the y-axis against the year on the x-axis. Let x represent the years since 1900. Do the data appear to lie along a straight line?

b. Plot the natural logarithm of the population against the year. Does the graph appear to be more linear than the graph in part a?

c. Find an equation for the least squares line for the data plotted in part b.

d. If your graphing calculator has an exponential regression feature, find the exponential function that best fits the given data according to the least squares method.

*http://www.nua.com/surveys/how_many_online/world.html.

†Problem 5 from "November 1989 Course 140 Examination, Mathematics of Compound Interest" of the Education and Examination Committee of The Society of Actuaries. Reprinted by permission of The Society of Actuaries.

‡*The Globe and Mail*, Feb. 17, 1995. The data were quoted by Ron Lancaster and Charlie Marion, *Mathematics Teacher*, Vol. 90, No. 2, Feb. 1997.

e. Take the natural logarithm of the equation found in part d, and verify that the result is the same as the equation found in part c. (In the section on Solutions of Elementary and Separable Differential Equations, we will see another type of function that is a better fit to this data.)

PHYSICAL SCIENCES

47. *Chlorofluorocarbons* Chlorofluorocarbons (CFCs) are gases used in refrigeration units, foaming agents, and aerosols. They have great potential for destroying the ozone layer of the atmosphere. The following graph displays approximate concentrations (in parts per billion, or ppb) of CFC-11.*

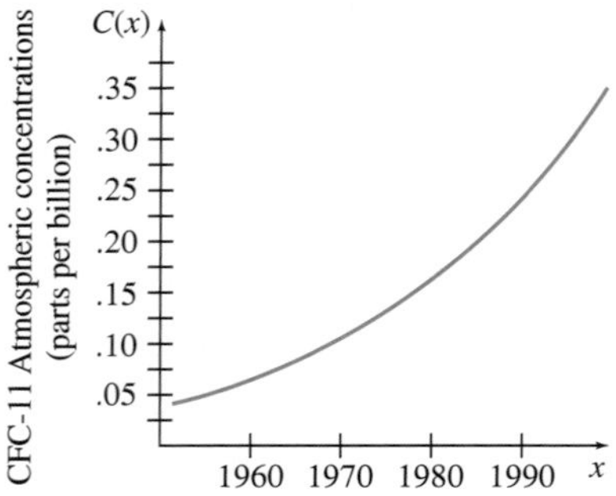

a. Letting $C(x)$ be the concentration of CFC-11 in the atmosphere from 1950 to 2000 in ppb, where x is the year, find $C(1975)/C(1950)$ and $C(2000)/C(1975)$. Based on these two ratios, does an exponential function seem to fit the data fairly well?

b. Use the values of the function at 1950 and 2000 to find an exponential function with $C(x) = C_0 a^{x-1950}$ that models the concentration of CFC-11 in the atmosphere from 1950 to 2000.

c. Graph $C(x)$, and describe how closely the graph resembles the original function. In particular, calculate $C(1975)$ and compare it with the actual concentration in 1975.

d. Approximate the average annual percentage increase of CFC-11 during this time period.

e. Use a graphing calculator to find the year in which the concentration would reach .50 ppb if the increase in CFC-11 were to continue.

48. *Radioactive Decay* Suppose the quantity (in grams) of a radioactive substance present at time t is

$$Q(t) = 1000(5^{-.3t}),$$

where t is measured in months.

a. How much will be present in 6 months?

b. How long will it take to reduce the substance to 8 g?

49. *Atmospheric Pressure* The atmospheric pressure (in millibars) at a given altitude (in meters) is listed in the table.†

Altitude	Pressure
0	1013
1000	899
2000	795
3000	701
4000	617
5000	541
6000	472
7000	411
8000	357
9000	308
10,000	265

a. Find functions of the form $P = ae^{kx}$, $P = mx + b$, and $P = 1/(ax + b)$ that fit the data at $x = 0$ and $x = 10{,}000$, where P is the pressure and x is the altitude.

b. Plot the data in the table and graph the three functions found in part a. Which function best fits the data?

c. Use the best-fitting function from part b to predict pressure at 1500 m and 11,000 m. Compare your answers to the true values of 846 millibars and 227 millibars, respectively.

d. If you have a graphing calculator or computer program with an exponential regression feature, use it to find an exponential function that approximately fits the data. How does this answer compare with the answer to part b?

50. *Computer Chips* The power of personal computers has increased dramatically as a result of the ability to place an increasing number of transistors on a single processor chip. The table on the next page lists the number of transistors on some popular computer chips made by Intel.‡

a. Let x be the year, where $x = 0$ corresponds to 1971, and y be the number of transistors. Find functions of the form $y = mx + b$, $y = ax^2 + b$, and $y = ab^x$ that fit the data at 1971 and 2000.

*Nilsson, A., *Greenhouse Earth*, New York: Wiley, 1992.
†Miller, A. and J. Thompson, *Elements of Meteorology*, Charles Merrill, 1975.
‡http://www.intel.com/research/silicon/mooreslaw.htm.

Year	Chip	Transistors
1971	4004	2250
1985	386	275,000
1989	486DX	1,180,000
1993	Pentium	3,100,000
1997	Pentium II	7,500,000
1999	Pentium III	24,000,000
2000	Pentium 4	42,000,000

b. Use a graphing calculator to plot the data in the table and to graph the three functions found in part a. Which function best fits the data?

c. Use the best-fitting function from part b to predict the number of transistors on a chip in the year 2008.

d. If you have a graphing calculator or computer program with an exponential regression feature, use it to find an exponential function that approximately fits the data. How does this answer compare with the answer to part b?

e. In 1965 Gordon Moore wrote a paper predicting how the power of computer chips would grow in the future. Moore's law says that the number of transistors that can be put on a chip doubles roughly every 18 months. Discuss the extent to which the data in this exercise confirms or refutes Moore's law.

2.5 LOGARITHMIC FUNCTIONS

THINK ABOUT IT

With an inflation rate averaging 5% per year, how long will it take for prices to double?

The number of years it will take for prices to double under given conditions is called the **years to double.** For \$1 to double (become \$2) in t years, assuming 5% annual compounding, means that

$$A = P\left(1 + \frac{r}{m}\right)^{mt}$$

becomes

$$2 = 1\left(1 + \frac{.05}{1}\right)^{1(t)}$$

or

$$2 = (1.05)^t.$$

This equation would be easier to solve if the variable were not in the exponent. **Logarithms** are defined for just this purpose. In Example 7, we will use logarithms to answer the question posed above.

LOGARITHM

For $a > 0$, $a \neq 1$, and $x > 0$,

$$\boldsymbol{y = \log_a x \quad \text{means} \quad a^y = x.}$$

(Read $y = \log_a x$ as "y is the logarithm of x to the base a.") For example, the exponential statement $2^4 = 16$ can be translated into the logarithmic statement $4 = \log_2 16$. Also, in the problem discussed above, $(1.05)^t = 2$ can be rewritten with this definition as $t = \log_{1.05} 2$. A logarithm is an exponent: $\log_a x$ is the exponent used with the base a to get x.

EXAMPLE 1 *Equivalent Expressions*

This example shows the same statements written in both exponential and logarithmic forms.

Exponential Form	*Logarithmic Form*
(a) $3^2 = 9$	$\log_3 9 = 2$
(b) $(1/5)^{-2} = 25$	$\log_{1/5} 25 = -2$
(c) $10^5 = 100{,}000$	$\log_{10} 100{,}000 = 5$
(d) $4^{-3} = 1/64$	$\log_4 1/64 = -3$
(e) $2^{-4} = 1/16$	$\log_2 1/16 = -4$
(f) $e^0 = 1$	$\log_e 1 = 0$

Logarithmic Functions For a given positive value of x, the definition of logarithm leads to exactly one value of y, so $y = \log_a x$ defines the *logarithmic function* of base a (the base a must be positive, with $a \neq 1$).

LOGARITHMIC FUNCTION

If $a > 0$ and $a \neq 1$, then the **logarithmic function** of base a is defined by

$$f(x) = \log_a x$$

for $x > 0$.

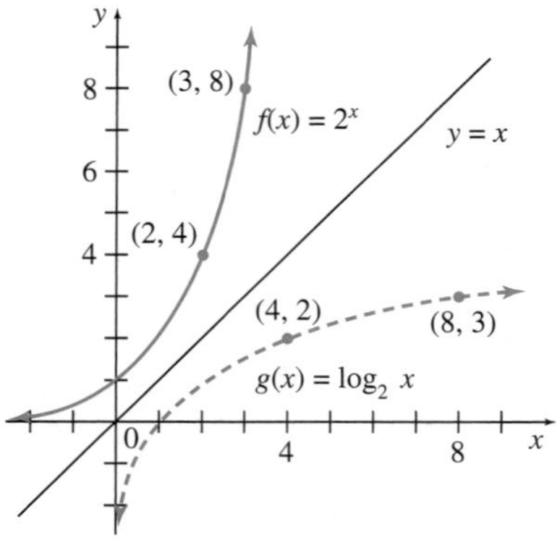

FIGURE 54

The graphs of the exponential function with $f(x) = 2^x$ and the logarithmic function with $g(x) = \log_2 x$ are shown in Figure 54. The graphs show that $f(3) = 2^3 = 8$, while $g(8) = \log_2 8 = 3$. Thus, $f(3) = 8$ and $g(8) = 3$. Also, $f(2) = 4$ and $g(4) = 2$. In fact, for any number m, if $f(m) = p$, then $g(p) = m$. Functions related in this way are called *inverses* of each other. The graphs also show that the domain of the exponential function (the set of real numbers) is the range of the logarithmic function. Also, the range of the exponential function (the set of positive real numbers) is the domain of the logarithmic function. Every logarithmic function is the inverse of some exponential function. This means that we can graph logarithmic functions by rewriting them as exponential functions using the definition of logarithm. The graphs in Figure 54 show a characteristic of a pair of inverse functions: their graphs are mirror images about the line $y = x$. Notice that because the exponential function has the x-axis as a horizontal asymptote, the logarithmic function has the y-axis as a vertical asymptote. A more complete discussion of inverse functions is given in most standard intermediate algebra and college algebra books.

The graph of $\log_2 x$ is typical of logarithms with bases $a > 1$. When $0 < a < 1$, the graph is the vertical reflection of the logarithm graph in Figure 54. Because logarithms with bases less than 1 are rarely used, we will not explore them here.

CAUTION The domain of $\log_a x$ includes all $x > 0$. In other words, you cannot take the logarithm of zero or a negative number. This also means that in a function such as $g(x) = \log_a(x - 2)$, the domain is given by $x - 2 > 0$, or $x > 2$. ■

Properties of Logarithms The usefulness of logarithmic functions depends in large part on the following **properties of logarithms.**

PROPERTIES OF LOGARITHMS

Let x and y be any positive real numbers and r be any real number. Let a be a positive real number, $a \neq 1$. Then

a. $\log_a xy = \log_a x + \log_a y$

b. $\log_a \frac{x}{y} = \log_a x - \log_a y$

c. $\log_a x^r = r \log_a x$

d. $\log_a a = 1$

e. $\log_a 1 = 0$

f. $\log_a a^r = r$.

To prove property (a), let $m = \log_a x$ and $n = \log_a y$. Then, by the definition of logarithm,

$$a^m = x \quad \text{and} \quad a^n = y.$$

Hence,

$$a^m a^n = xy.$$

By a property of exponents, $a^m a^n = a^{m+n}$, so

$$a^{m+n} = xy.$$

Now use the definition of logarithm to write

$$\log_a xy = m + n.$$

Since $m = \log_a x$ and $n = \log_a y$,

$$\log_a xy = \log_a x + \log_a y.$$

Proofs of properties (b) and (c) are left for the exercises. Properties (d) and (e) depend on the definition of a logarithm. Property (f) follows from properties (c) and (d).

EXAMPLE 2 *Properties of Logarithms*

If all the following variable expressions represent positive numbers, then for $a > 0$, $a \neq 1$, the statements in (a)–(c) are true.

(a) $\log_a x + \log_a(x - 1) = \log_a x(x - 1)$

(b) $\log_a \frac{x^2 - 4x}{x + 6} = \log_a(x^2 - 4x) - \log_a(x + 6)$

(c) $\log_a(9x^5) = \log_a 9 + \log_a(x^5) = \log_a 9 + 5 \cdot \log_a x$

Evaluating Logarithms The invention of logarithms is credited to John Napier (1550–1617), who first called logarithms "artificial numbers." Later he joined the Greek words *logos* (ratio) and *arithmos* (number) to form the word used today. The development of logarithms was motivated by a need for faster computation. Tables of logarithms and slide rule devices were developed by Napier, Henry Briggs (1561–1631), Edmund Gunter (1581–1626), and others.

For many years logarithms were used primarily to assist in involved calculations. Current technology has made this use of logarithms obsolete, but logarithmic functions play an important role in many applications of mathematics. Since our number system has base 10, logarithms to base 10 were most convenient for numerical calculations and so base 10 logarithms were called **common logarithms.** Common logarithms are still useful in other applications. For simplicity,

$$\log_{10} x \text{ is abbreviated } \log x.$$

Most practical applications of logarithms use the number e as base. (Recall that to 7 decimal places, $e = 2.7182818$.) Logarithms to base e are called **natural logarithms,** and

$$\log_e x \text{ is abbreviated } \ln x$$

(read "el-en x"). A graph of $f(x) = \ln x$ is given in Figure 55.

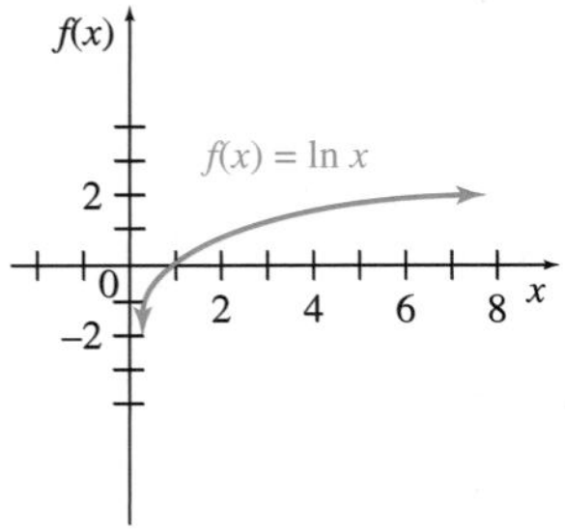

FIGURE 55

NOTE Keep in mind that $\ln x$ is a logarithmic function. Therefore, all of the properties of logarithms given previously are valid when $\log_a$ is replaced with ln and a is replaced with e. ■

Although common logarithms may seem more "natural" than logarithms to base e, there are several good reasons for using natural logarithms instead. The most important reason is discussed later, in the section on Derivatives of Logarithmic Functions.

A calculator can be used to find both common and natural logarithms. For example, using a calculator and 4 decimal places, we get the following values.

$$\log 2.34 = .3692, \quad \log 594 = 2.7738, \quad \text{and} \quad \log .0028 = -2.5528.$$

$$\ln 2.34 = .8502, \quad \ln 594 = 6.3869, \quad \text{and} \quad \ln .0028 = -5.8781.$$

Notice that logarithms of numbers less than 1 are negative when the base is greater than 1. A look at the graph of $y = \log_2 x$ or $y = \ln x$ will show why.

Sometimes it is convenient to use logarithms to bases other than 10 or e. For example, some computer science applications use base 2. In such cases, the following theorem is useful for converting from one base to another.

CHANGE-OF-BASE THEOREM FOR LOGARITHMS

If x is any positive number and if a and b are positive real numbers, $a \neq 1$, $b \neq 1$, then

$$\log_a x = \frac{\log_b x}{\log_b a}.$$

To prove this result, use the definition of logarithm to write $y = \log_a x$ as $x = a^y$ or $x = a^{\log_a x}$ (for positive x and positive a, $a \neq 1$). Now take base b logarithms of both sides of this last equation.

$$\log_b x = \log_b a^{\log_a x}$$

$$\log_b x = (\log_a x)(\log_b a), \qquad \log_a x^r = r \log_a x$$

$$\log_a x = \frac{\log_b x}{\log_b a} \qquad \text{Solve for } \log_a x.$$

If the base b is equal to e, then by the change-of-base theorem,

$$\log_a x = \frac{\log_e x}{\log_e a}.$$

Using $\ln x$ for $\log_e x$ gives the special case of the theorem using natural logarithms.

For any positive numbers a and x, $a \neq 1$,

$$\log_a x = \frac{\ln x}{\ln a}.$$

The change-of-base theorem is also useful when graphing $y = \log_b x$ on a graphing calculator for a base b other than e or 10. For example, to graph $y = \log_2 x$, let $y = \ln x / \ln 2$.

EXAMPLE 3 *Evaluating Logarithms*

Use natural logarithms to find each value. Round to the nearest hundredth.

(a) $\log_5 27$

Solution Let $x = 27$ and $a = 5$. Using the second form of the theorem gives

$$\log_5 27 = \frac{\ln 27}{\ln 5}.$$

Now use a calculator.

$$\log_5 27 \approx \frac{3.2958}{1.6094} \approx 2.05$$

To check, use a calculator, along with the definition of logarithm, to verify that $5^{2.05} \approx 27$.

(b) $\log_2 5.1$

Solution

$$\log_2 5.1 = \frac{\ln 5.1}{\ln 2} \approx -3.32$$

CAUTION As mentioned earlier, when using a calculator, do not round off intermediate results. Keep all numbers in the calculator until you have the final answer. In Example 3(a), we showed the rounded intermediate values of ln 27 and ln 5, but we used the unrounded quantities when doing the division. ■

Logarithmic Equations Equations involving logarithms are often solved by using the fact that exponential functions and logarithmic functions are inverses, so a logarithmic equation can be rewritten (with the definition of logarithm) as an exponential equation. In other cases, the properties of logarithms may be useful in simplifying a **logarithmic equation.**

EXAMPLE 4 *Solving Logarithmic Equations*

Solve each equation.

(a) $\log_x \frac{8}{27} = 3$

Solution Using the definition of logarithm, write the expression in exponential form. To solve for x, take the cube root on both sides.

$$x^3 = \frac{8}{27}$$

$$x = \frac{2}{3}$$

(b) $\log_4 x = \frac{5}{2}$

Solution In exponential form, the given statement becomes

$$4^{5/2} = x$$

$$(4^{1/2})^5 = x$$

$$2^5 = x$$

$$32 = x.$$

(c) $\log_2 x - \log_2(x - 1) = 1$

Solution By a property of logarithms,

$$\log_2 x - \log_2(x - 1) = \log_2 \frac{x}{x - 1},$$

so the original equation becomes

$$\log_2 \frac{x}{x - 1} = 1.$$

Now write this equation in exponential form, and solve.

$$\frac{x}{x - 1} = 2^1 = 2$$

Solve this equation.

$$\frac{x}{x-1}(x-1) = 2(x-1)$$
$$x = 2(x-1)$$
$$x = 2x - 2$$
$$-x = -2$$
$$x = 2$$

(d) $\log_3 \frac{1}{9} = x$

Solution Using the definition of logarithm, write the expression in exponential form. To solve for x, first rewrite in terms of 3 to some power. Thus, $3^x = 1/9 = (1/3)^2 = 3^{-2}$ and $x = -2$. It is important to check solutions when solving equations involving logarithms because $\log_a u$, where u is an expression in x, has domain given by $u > 0$.

Exponential Equations In the previous section exponential equations like $(1/3)^x = 81$ were solved by writing each side of the equation as a power of 3. That method cannot be used to solve an equation such as $3^x = 5$, however, since 5 cannot easily be written as a power of 3. Such equations can be solved approximately with a graphing calculator, but an algebraic method is also useful, particularly when the equation involves variables such as a and b rather than just numbers such as 3 and 5. A general method for solving these equations depends on the following property of logarithms, which is supported by the graphs of logarithmic functions (Figures 54 and 55).

> For $x > 0$, $y > 0$, $b > 0$, and $b \neq 1$,
>
> $$\text{if } x = y, \text{ then } \log_b x = \log_b y,$$
>
> and
>
> $$\text{if } \log_b x = \log_b y, \text{ then } x = y.$$

EXAMPLE 5 *Solving Exponential Equations*

Solve each equation.

(a) $3^{2x} = 4^{x+1}$

Solution Taking natural logarithms (logarithms to any base could be used) on both sides gives

$$\ln 3^{2x} = \ln 4^{x+1}$$
$$2x \ln 3 = (x+1)\ln 4 \qquad \ln u^r = r \ln u$$
$$(2\ln 3)x = (\ln 4)x + \ln 4$$
$$(2\ln 3)x - (\ln 4)x = \ln 4 \qquad \text{Subtract } (\ln 4)x \text{ from both sides.}$$
$$(2\ln 3 - \ln 4)x = \ln 4 \qquad \text{Factor } x.$$
$$x = \frac{\ln 4}{2\ln 3 - \ln 4} \qquad \text{Divide both sides by } 2\ln 3 - \ln 4.$$

Use a calculator to evaluate the logarithms, then divide, to get

$$x \approx \frac{1.3863}{2(1.0986) - 1.3863} \approx 1.710.$$

(b) $5e^{.01x} = 9$

Solution

$$e^{.01x} = \frac{9}{5} = 1.8 \qquad \text{Divide both sides by 5.}$$

$$\ln e^{.01x} = \ln 1.8 \qquad \text{Take natural logarithms on both sides.}$$

$$.01x = \ln 1.8 \qquad \ln e^u = u$$

$$x = \frac{\ln 1.8}{.01} \approx 58.779$$

Just as $\log_a x$ can be written as a base e logarithm, any exponential function $y = a^x$ can be written as an exponential function with base e. For example, there exists a real number k such that

$$2 = e^k.$$

Raising both sides to the power x gives

$$2^x = e^{kx},$$

so that powers of 2 can be found by evaluating appropriate powers of e. To find the necessary number k, solve the equation $2 = e^k$ for k by first taking logarithms on both sides.

$$2 = e^k$$

$$\ln 2 = \ln e^k$$

$$\ln 2 = k \ln e$$

$$\ln 2 = k \qquad \ln e = 1$$

Thus, $k = \ln 2$. In the section on Derivatives of Exponential Functions, we will see why this change of base is useful. A general statement can be drawn from this example.

CHANGE-OF-BASE THEOREM FOR EXPONENTIALS

For every positive real number a,

$$a^x = e^{(\ln a)x}.$$

Another way to see why the change-of-base theorem is true is to first observe that $e^{\ln a} = a$. Combining this with the fact that $e^{ab} = (e^a)^b$, we have $e^{(\ln a)x} = (e^{\ln a})^x = a^x$.

EXAMPLE 6 *Change-of-Base-Theorem*

(a) Write 7^x using base e rather than base 7.

Solution According to the change-of-base theorem,

$$7^x = e^{(\ln 7)x}.$$

Using a calculator to evaluate ln 7, we could also approximate this as $e^{1.9459x}$.

(b) Approximate the function $f(x) = e^{2x}$ as $f(x) = a^x$ for some base a.

Solution We do not need the change-of-base theorem here. Just use the fact that

$$e^{2x} = (e^2)^x \approx 7.389^x,$$

where we have used a calculator to approximate e^2.

EXAMPLE 7 *Doubling Time*

Complete the solution of the problem posed at the beginning of this section.

Solution Recall that if prices will double after t years at an inflation rate of 5%, compounded annually, t is given by the equation

$$2 = (1.05)^t.$$

We solve this equation by first taking natural logarithms on both sides.

$$\begin{aligned} \ln 2 &= \ln(1.05)^t \\ \ln 2 &= t \ln 1.05 && \ln x^r = r \ln x \\ t &= \frac{\ln 2}{\ln 1.05} \approx 14.2 \end{aligned}$$

It will take about 14 years for prices to double.

The problem solved in Example 7 can be generalized for the compound interest equation

$$A = P(1 + r)^t.$$

Solving for t as in Example 7 (with $A = 2$ and $P = 1$) gives the doubling time in years as

$$t = \frac{\ln 2}{\ln(1 + r)}.$$

It can be shown that for certain values of r,

$$t = \frac{\ln 2}{\ln(1 + r)} \approx \frac{.693}{r},$$

and

$$\frac{70}{100r} \le \frac{\ln 2}{\ln(1 + r)} \le \frac{72}{100r}.$$

The **rule of 70** says that for $.001 \le r \le .05$, the value of $70/100r$ gives a good approximation of t. The **rule of 72** says that for $.05 \le r \le .12$, the value of $72/100r$ approximates t quite well.

Figure 56 shows the three functions $y = \ln 2/\ln(1 + r)$, $y = 70/100r$, and $y = 72/100r$ graphed on the same axes with a graphing calculator. The three graphs are so close to each other that they appear to be one thick graph. In an exercise we will ask you to explore the relationship between these functions further.

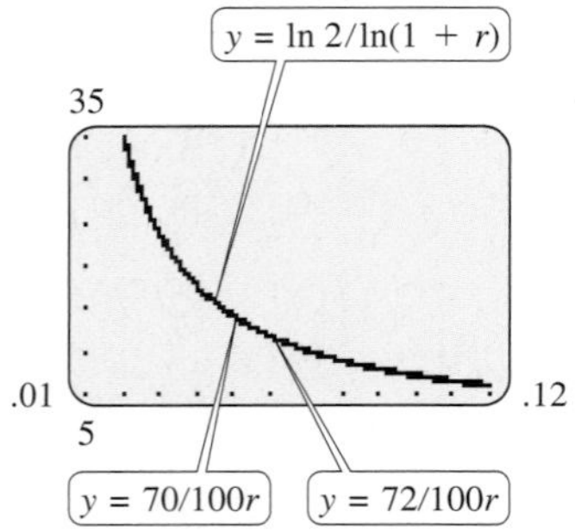

FIGURE 56

EXAMPLE 8 *Rules of 70 and 72*

Approximate the years to double at an interest rate of 6% using first the rule of 70, then the rule of 72.

Solution By the rule of 70, money will double at 6% interest after

$$\frac{70}{100r} = \frac{70}{100(.06)} = \frac{70}{6} = 11.67\left(\text{or } 11\frac{2}{3}\right)$$

years.

Using the rule of 72 gives

$$\frac{72}{100r} = \frac{72}{6} = 12$$

years doubling time. Since a more precise answer is given by

$$\frac{\ln 2}{\ln(1 + r)} = \frac{\ln 2}{\ln(1.06)} \approx \frac{.693}{.058} \approx 11.9,$$

the rule of 72 gives a better approximation than the rule of 70. This agrees with the statement that the rule of 72 works well for values of r where $.05 \le r \le .12$, since $r = .06$ falls into this category.

EXAMPLE 9 *Index of Diversity*

One measure of the diversity of the species in an ecological community is given by the **index of diversity** H, where

$$H = -[P_1 \ln P_1 + P_2 \ln P_2 + \cdots + P_n \ln P_n],$$

and $P_1, P_2, \ldots, P_n$ are the proportions of a sample belonging to each of n species found in the sample.* For example, in a community with two species, where

*Ludwig, John and James Reynolds, *Statistical Ecology: A Primer on Methods and Computing*, New York: Wiley, 1988, p. 92.

there are 90 of one species and 10 of the other, $P_1 = 90/100 = .9$ and $P_2 = 10/100 = .1$, with

$$H = -[.9 \ln .9 + .1 \ln .1].$$

Using a calculator, we find

$$\ln 5.9 \approx -.1054, \qquad \text{and} \qquad \ln 5.1 \approx -2.3026.$$

Therefore,

$$H \approx -[(.9)(-.1054) + (.1)(-2.3026)]$$
$$\approx .325.$$

Verify that $H \approx .673$ if there are 60 of one species and 40 of the other. As the proportions of n species get closer to $1/n$ each, the index of diversity increases to a maximum of $\ln n$.

2.5 EXERCISES

Write each exponential equation in logarithmic form.

1. $2^3 = 8$ **2.** $5^2 = 25$ **3.** $3^4 = 81$

4. $6^3 = 216$ **5.** $3^{-2} = \dfrac{1}{9}$ **6.** $\left(\dfrac{4}{3}\right)^{-2} = \dfrac{9}{16}$

Write each logarithmic equation in exponential form.

7. $\log_2 128 = 7$ **8.** $\log_3 81 = 4$ **9.** $\log_{25} \dfrac{1}{25} = -1$

10. $\log_2 \dfrac{1}{8} = -3$ **11.** $\log 10{,}000 = 4$ **12.** $\log 5.00001 = -5$

Evaluate each logarithm without using a calculator.

13. $\log_5 25$ **14.** $\log_9 81$ **15.** $\log_4 64$ **16.** $\log_6 216$

17. $\log_2 \dfrac{1}{4}$ **18.** $\log_3 \dfrac{1}{27}$ **19.** $\log_2 \sqrt[3]{\dfrac{1}{4}}$ **20.** $\log_8 \sqrt[4]{\dfrac{1}{2}}$

21. $\ln e$ **22.** $\ln e^2$ **23.** $\ln e^{5/3}$ **24.** $\ln 1$

25. Is the "logarithm to the base 3 of 4" written as $\log_4 3$ or $\log_3 4$?

26. Write a few sentences describing the relationship between e^x and $\ln x$.

Use the properties of logarithms to write each expression as a sum, difference, or product.

27. $\log_9(7m)$ **28.** $\log_5(8p)$ **29.** $\log_3 \dfrac{3p}{5k}$

30. $\log_7 \dfrac{11p}{13y}$ **31.** $\ln \dfrac{5\sqrt{2}}{\sqrt[4]{7}}$ **32.** $\ln \dfrac{9\sqrt[3]{5}}{\sqrt[4]{3}}$

Suppose $\log_b 2 = a$ *and* $\log_b 3 = c$*. Use the properties of logarithms to find the following.*

33. $\log_b 8$ **34.** $\log_b 24$ **35.** $\log_b(72b)$ **36.** $\log_b(4b^2)$

Use natural logarithms to evaluate each logarithm to the nearest hundredth.

37. $\log_5 20$ **38.** $\log_{12} 170$ **39.** $\log_{1.2} 5.55$ **40.** $\log_{2.8} 5.12$

Solve each equation in Exercises 41–61. Round decimal answers to the nearest hundredth.

41. $\log_x 25 = -2$ **42.** $\log_9 27 = m$ **43.** $\log_8 4 = z$

44. $\log_y 8 = \frac{3}{4}$ **45.** $\log_r 7 = \frac{1}{2}$ **46.** $\log_3(5x + 1) = 2$

47. $\log_5(9x - 4) = 1$ **48.** $\log_4 x - \log_4(x + 3) = -1$ **49.** $\log_9 m - \log_9(m - 4) = -2$

50. $\log (x + 5) + \log(x + 2) = 1$ **51.** $\log_3(x - 2) + \log_3(x + 6) = 2$ **52.** $\log_3(x^2 + 17) - \log_3(x + 5) = 1$

53. $\log_2(x^2 - 1) - \log_2(x + 1) = 2$ **54.** $3^x = 5$ **55.** $4^x = 12$

56. $e^{k-1} = 4$ **57.** $e^{2y} = 12$ **58.** $2e^{5a+12} = 8$

59. $10e^{3z-7} = 5$ **60.** $5(.10)^x = 4(.12)^x$ **61.** $1.5(1.05)^x = 2(1.01)^x$

Find the domain of each function.

62. $f(x) = \log(3 - x)$ **63.** $f(x) = \ln(x^2 - 4)$

64. Lucky Larry was faced with solving

$$\log(2x + 1) - \log(3x - 1) = 0.$$

Larry just dropped the logs and proceeded:

$$(2x + 1) - (3x - 1) = 0$$
$$-x + 2 = 0$$
$$x = 2.$$

Although Lucky Larry is wrong in dropping the logs, his procedure will always give the correct answer to an equation of the form

$$\log A - \log B = 0,$$

where A and B are any two expressions in x. Prove that this last equation leads to the equation $A - B = 0$, which is what you get when you drop the logs.*

65. Prove: $\log_a\left(\frac{x}{y}\right) = \log_a x - \log_a y.$

66. Prove: $\log_a x^r = r \log_a x.$

Applications

BUSINESS AND ECONOMICS

67. *Inflation* Assuming annual compounding, find the time it would take for the general level of prices in the economy to double at the following annual inflation rates.

a. 3% **b.** 6% **c.** 8%

d. Check your answers using either the rule of 70 or the rule of 72, whichever applies.

68. *Interest* Lori Hales invests $15,000 in an account paying 6% per year compounded annually.

a. How many years are required for the compound amount to at least double? (Note that interest is only paid at the end of each year.)

b. In how many years will the amount at least triple?

c. Check your answer to part a using the rule of 72.

*Based on Lucky Larry #16 by Joan Page, *The AMATYC Review*, Vol. 16, No. 1, Fall 1994, p. 67.

69. *Interest* Karen Panunzio plans to invest $500 into a money market account. Find the interest rate that is needed for the money to grow to $1200 in 14 years if the interest is compounded continuously. (Compare with Exercise 39 in the previous section.)

70. *Rule of 72* Complete the following table, and use the results to discuss when the rule of 70 gives a better approximation for the doubling time, and when the rule of 72 gives a better approximation.

r (sec)	.001	.02	.05	.08	.12
$\ln 2/\ln(1+r)$					
$70/100r$					
$72/100r$					

71. *Pay Increases* You are offered two jobs starting July 1, 2007. Humongous Enterprises offers you $45,000 a year to start, with a raise of 4% every July 1. At Crabapple Inc. you start at $30,000, with an annual increase of 6% every July 1. On July 1 of what year would the job at Crabapple Inc. pay more than the job at Humongous Enterprises? Use the algebra of logarithms to solve this problem, and support your answer by using a graphing calculator to see where the two salary functions intersect.

LIFE SCIENCES

72. *Insect Species* An article in *Science* stated that the number of insect species of a given mass is proportional to $m^{-.6}$, where m is the mass in grams.* A graph accompanying the article shows the common logarithm of the mass on the horizontal axis and the common logarithm of the number of species on the vertical axis. Explain why the graph is a straight line. What is the slope of the line?

Index of Diversity For Exercises 73–75, refer to Example 9.

73. Suppose a sample of a small community shows two species with 50 individuals each.

a. Find the index of diversity H.

b. What is the maximum value of the index of diversity for two species?

c. Does your answer for part a equal ln 2? Explain why.

74. A virgin forest in northwestern Pennsylvania has 4 species of large trees with the following proportions of each: hemlock, .521; beech, .324; birch, .081; maple, .074. Find the index of diversity H.

75. Find the value of the index of diversity for populations with n species and $1/n$ of each if

a. $n = 3$; **b.** $n = 4$.

c. Verify that your answers for parts a and b equal ln 3 and ln 4, respectively.

76. *Allometric Growth* The allometric formula is used to describe a wide variety of growth patterns. It says that $y = nx^m$, where x and y are variables, and n and m are constants. For example, the famous biologist J. S. Huxley used this formula to relate the weight of the large claw of the fiddler crab to the weight of the body without the claw.† Show that if x and y are given by the allometric formula, then $X = \log_b x$, $Y = \log_b y$, and $N = \log_b n$ are related by the linear equation

$$Y = mX + N.$$

77. *Drug Concentration* When a pharmaceutical drug is injected into the bloodstream, its concentration at time t can be approximated by $C(t) = C_0 e^{-kt}$, where C_0 is the concentration at $t = 0$. Suppose the drug is ineffective below a concentration C_1 and harmful above a concentration C_2. Then it can be shown that the drug should be given at intervals of time T, where

$$T = \frac{1}{k}\ln\frac{C_2}{C_1}.‡$$

A certain drug is harmful at a concentration five times the concentration below which it is ineffective. At noon an injection of the drug results in a concentration of 2 mg per liter of blood. Three hours later the concentration is down to 1 mg per liter. How often should the drug be given?

**Science*, Vol. 284, June 18, 1999, p. 1937.
†Huxley, J. S., *Problems of Relative Growth*, Dover, 1968.
‡Horelick, Brindell, and Sinan Koont, "Applications of Calculus to Medicine: Prescribing Safe and Effective Dosage," *UMAP Module 202*, 1977.

The graph for Exercise 78 is plotted on a* logarithmic scale *where differences between successive measurements are not always the same. Data that do not plot in a linear pattern on the usual Cartesian axes often form a linear pattern when plotted on a logarithmic scale. Notice that on the vertical scale, the distance from* 1 *to* 2 *is not the same as the distance from* 2 *to* 3, *and so on. This is characteristic of a graph drawn on logarithmic scales.*

78. *Oxygen Consumption* The accompanying graph gives the rate of oxygen consumption for resting guinea pigs of various sizes. This rate is proportional to body mass raised to the power .67.

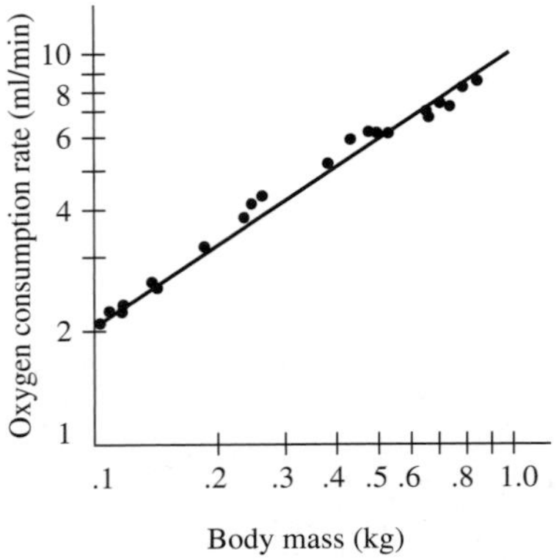

a. Estimate the oxygen consumption for a guinea pig with body mass of .3 kg. Do the same for one with body mass of .7 kg.

b. Verify that if the relationship between x and y is of the form $y = ax^b$, then there will be a linear relationship between $\ln x$ and $\ln y$. (*Hint:* Apply ln to both sides of $y = ax^b$.)

c. If a function of the form $y = ax^b$ contains the points (x_1, y_1) and (x_2, y_2), then values for a and b can be found by dividing $y_1 = ax_1^b$ by $y_2 = ax_2^b$, solving the resulting equation for b, and putting the result back into either equation to solve for a. Use this procedure and the results from part a to find an equation of the form $y = ax^b$ that gives the oxygen consumption rate as a function of body mass.

d. Use the result of part c to predict the oxygen consumption of a guinea pig whose body mass is .5 kg.

79. *Population Growth* In July 1994 the population of New York state was 18.2 million and increasing at a rate of .1% per year. The population of Florida was 14.0 million and increasing at a rate of 1.7% per year.† If this trend continues, estimate in what year the population of Florida will exceed the population of New York. Use the algebra of logarithms to solve this problem, and verify your answer by using a graphing calculator or spreadsheet to see where the two population functions intersect.

SOCIAL SCIENCES

80. *Evolution of Languages* The number of years $N(r)$ since two independently evolving languages split off from a common ancestral language is approximated by

$$N(r) = -5000 \ln r,$$

where r is the proportion of the words from the ancestral language that are common to both languages now. Find the following.

a. $N(.9)$ **b.** $N(.5)$ **c.** $N(.3)$

d. How many years have elapsed since the split if 70% of the words of the ancestral language are common to both languages today?

e. If two languages split off from a common ancestral language about 1000 years ago, find r.

PHYSICAL SCIENCES

81. According to the Shannon–Hartley theorem, the capacity of a communications channel in bits per second is given by

$$C = B \log_2\left(\frac{s}{n} + 1\right),$$

where B is the frequency bandwidth of the channel in hertz and s/n is its signal-to-noise ratio.‡ It is physically impossible to exceed this limit. Solve the equation for the signal-to-noise ratio s/n.

For Exercises 82–85, recall that log x *represents the common (base* 10*) logarithm of* x.

82. *Intensity of Sound* The loudness of sounds is measured in a unit called a *decibel*. To do this, a very faint sound, called the *threshold sound*, is assigned an intensity I_0. If a particular sound has intensity I, then the decibel rating of this louder sound is

$$10 \log \frac{I}{I_0}.$$

Find the decibel ratings of the following sounds having intensities as given. Round answers to the nearest whole number.

*McMahon, Thomas A. and John Tyler Bonner, *On Size and Life*, Copyright (c) 1983 by Thomas A. McMahon and John Tyler Bonner. Reprinted by permission of W. H. Freeman and Company.
†U.S. Census Bureau.
‡*Scientific American*, Oct. 1999, p. 103.

a. Whisper, $115I_0$

b. Busy street, $9{,}500{,}000I_0$

c. Heavy truck, 20 m away, $1{,}200{,}000{,}000I_0$

d. Rock music concert, $895{,}000{,}000{,}000I_0$

e. Jetliner at takeoff, $109{,}000{,}000{,}000{,}000I_0$

f. In a noise ordinance instituted in Stamford, Connecticut, the threshold sound I_0 was defined as .0002 microbars.* Use this definition to express the sound levels in parts c and d in microbars.

83. *Intensity of Sound* A story on the National Public Radio program *All Things Considered* on May 7, 2002, discussed a proposal to lower the noise limit in Austin, Texas, from 85 decibels to 75 decibels. A manager for a restaurant was quoted as saying, "If you cut from 85 to 75, . . . you're basically cutting the sound down in half." Is this correct? If not, to what fraction of its original level is the sound being cut?

84. *Earthquake Intensity* The magnitude of an earthquake, measured on the Richter scale, is given by

$$R(I) = \log \frac{I}{I_0},$$

where I is the amplitude registered on a seismograph located 100 km from the epicenter of the earthquake, and I_0 is the amplitude of a certain small size earthquake. Find the Richter scale ratings of earthquakes with the following amplitudes.

a. $1{,}000{,}000I_0$ **b.** $100{,}000{,}000I_0$

c. On June 16, 1999, the city of Puebla in central Mexico was shaken by an earthquake that measured 6.7 on the Richter scale. Express this reading in terms of I_0.†

d. On September 19, 1985, Mexico's largest recent earthquake, measuring 8.1 on the Richter scale, killed about 9500 people. Express the magnitude of an 8.1 reading in terms of I_0.†

e. Compare your answers to parts c and d. How much greater was the force of the 1985 earthquake than the 1999 earthquake?

f. The relationship between the energy E of an earthquake and the magnitude on the Richter scale is given by

$$R(E) = \frac{2}{3} \log \frac{E}{E_0},$$

where E_0 is the energy of a certain small earthquake. Compare the energies of the 1999 and 1985 earthquakes.

g. According to a newspaper article, "Scientists say such an earthquake of magnitude 7.5 could release 15 times as much energy as the magnitude 6.7 trembler that struck the Northridge section of Los Angeles"‡ in 1994. Using the formula from part f, verify this quote by computing the magnitude of an earthquake with 15 times the energy of a magnitude 6.7 earthquake.

85. *Acidity of a Solution* A common measure for the acidity of a solution is its pH. It is defined by pH $= -\log[H^+]$, where H^+ measures the concentration of hydrogen ions in the solution. The pH of pure water is 7. Solutions that are more acidic than pure water have a lower pH, while solutions that are less acidic (referred to as basic solutions) have a higher pH.

a. Acid rain sometimes has a pH as low as 4. How much greater is the concentration of hydrogen ions in such rain than in pure water?

b. A typical mixture of laundry soap and water for washing clothes has a pH of about 11, while black coffee has a pH of about 5. How much greater is the concentration of hydrogen ions in black coffee than in the laundry mixture?

2.6 APPLICATIONS: GROWTH AND DECAY; MATHEMATICS OF FINANCE

THINK ABOUT IT

What interest rate will cause \$5000 to grow to \$7250 in 4 years if money is compounded continuously?

This is one of many situations that occur in biology, economics, and the social sciences, in which a quantity changes at a rate proportional to the amount of the quantity present. In such cases the amount present at time t is a function of t, called the **exponential growth and decay function.** (The derivation of this equation is presented in a later section on Differential Equations.)

**The New York Times*, June 6, 1999, p. 41.

†*Times Picayune*.

‡*The New York Times*, Jan. 13, 1995.

EXPONENTIAL GROWTH AND DECAY FUNCTION

Let y_0 be the amount or number of some quantity present at time $t = 0$. Then, under certain conditions, the amount present at any time t is given by

$$y = y_0 e^{kt},$$

where k is a constant.

If $k > 0$, then k is called the **growth constant;** if $k < 0$, then k is called the **decay constant.** A common example is the growth of bacteria in a culture. The more bacteria present, the faster the population increases.

EXAMPLE 1 *Yeast Production*

Yeast in a sugar solution is growing at a rate such that 1 g becomes 1.5 g after 20 hours. Find the growth function, assuming exponential growth.

Solution The values of y_0 and k in the exponential growth function $y = y_0 e^{kt}$ must be found. Since y_0 is the amount present at time $t = 0$, $y_0 = 1$. To find k, substitute $y = 1.5$, $t = 20$, and $y_0 = 1$ into the equation.

$$y = y_0 e^{kt}$$
$$1.5 = 1e^{k(20)}$$

Now take natural logarithms on both sides and use the power rule for logarithms and the fact that $\ln e = 1$.

$$1.5 = e^{20k}$$
$$\ln 1.5 = \ln e^{20k} \quad \text{Take ln of both sides.}$$
$$\ln 1.5 = 20k \quad \ln e^x = x$$
$$\frac{\ln 1.5}{20} = k \quad \text{Divide both sides by 20.}$$
$$k \approx .02 \text{ (to the nearest hundredth)}$$

The exponential growth function is $y = e^{.02t}$, where y is the number of grams of yeast present after t hours.

The decline of a population or decay of a substance may also be described by the exponential growth function. In this case the decay constant k is negative, since an increase in time leads to a decrease in the quantity present. Radioactive substances provide a good example of exponential decay. By definition, the **half-life** of a radioactive substance is the time it takes for exactly half of the initial quantity to decay.

EXAMPLE 2 *Carbon Dating*

Carbon 14 is a radioactive form of carbon that is found in all living plants and animals. After a plant or animal dies, the carbon 14 disintegrates. Scientists determine the age of the remains by comparing its carbon 14 with the amount found in living plants and animals. The amount of carbon 14 present after t years is given by the exponential equation

$$A(t) = A_0 e^{kt},$$

with $k = -[(\ln 2)/5600]$.

(a) Find the half-life of carbon 14.

Solution Let $A(t) = (1/2)A_0$ and $k = -[(\ln 2)/5600]$.

$$\frac{1}{2}A_0 = A_0 e^{-[(\ln 2)/5600]t}$$

$$\frac{1}{2} = e^{-[(\ln 2)/5600]t} \qquad \text{Divide by } A_0.$$

$$\ln\frac{1}{2} = \ln e^{-[(\ln 2)/5600]t} \qquad \text{Take logarithms of both sides.}$$

$$\ln\frac{1}{2} = -\frac{\ln 2}{5600}t \qquad \ln e^x = x$$

$$-\frac{5600}{\ln 2}\ln\frac{1}{2} = t \qquad \text{Multiply by } -\frac{5600}{\ln 2}.$$

$$-\frac{5600}{\ln 2}(\ln 1 - \ln 2) = t \qquad \ln\tfrac{x}{y} = \ln x - \ln y$$

$$-\frac{5600}{\ln 2}(-\ln 2) = t \qquad \ln 1 = 0$$

$$5600 = t$$

The half-life is 5600 years.

(b) Charcoal from an ancient fire pit on Java had $1/4$ the amount of carbon 14 found in a living sample of wood of the same size. Estimate the age of the charcoal.

Solution Let $A(t) = (1/4)A_0$ and $k = -[(\ln 2)/5600]$.

$$\frac{1}{4}A_0 = A_0 e^{-[(\ln 2)/5600]t}$$

$$\frac{1}{4} = e^{-[(\ln 2)/5600]t}$$

$$\ln\frac{1}{4} = \ln e^{-[(\ln 2)/5600]t}$$

$$\ln\frac{1}{4} = -\frac{\ln 2}{5600}t$$

$$-\frac{5600}{\ln 2}\ln\frac{1}{4} = t$$

$$-\frac{5600}{\ln 2}(\ln 2^{-2}) = t$$

$$-\frac{5600}{\ln 2}(-2\ln 2) = t$$

$$t = 11{,}200$$

The charcoal is about 11,200 years old.

By following the steps in Example 2, we get the general equation giving the half-life T in terms of the decay constant k as

$$T = -\frac{\ln 2}{k}.$$

For example, the decay constant for potassium 40, where t is in billions of years, is approximately $-.5545$, so its half-life is

$$T = -\frac{\ln 2}{(-.5545)}$$

$$\approx 1.25 \text{ billion years.}$$

We can rewrite the growth and decay function as

$$y = y_0 e^{kt} = y_0(e^k)^t = y_0 a^t,$$

where $a = e^k$. This is sometimes a helpful way to look at an exponential growth or decay function.

EXAMPLE 3 *Radioactive Decay*

Rewrite the function for radioactive decay of carbon 14 in the form $A(t) = A_0 a^{f(t)}$.

Solution From the previous example, we have

$$A(t) = A_0 e^{kt} = A_0 e^{-[(\ln 2)/5600]t}$$

$$= A_0(e^{\ln 2})^{-t/5600}$$

$$= A_0 2^{-t/5600} = A_0 (2^{-1})^{t/5600} = A_0 \left(\frac{1}{2}\right)^{t/5600}.$$

This last expression shows clearly that every time t increases by 5600 years, the amount of carbon 14 decreases by a factor of $1/2$. ■

EXAMPLE 4 *Insurance Boycott*

A major insurance company canceled a textbook author's homeowners' and umbrella liability insurance because of the risk the company believed that the author incurred by owning an American Staffordshire terrier (a.k.a. a pit bull). Some participants in a dog-loving Internet newsgroup urged a boycott of the insurer. Two days after the original posting, two people had written to the news group that they supported a boycott. One day later, the number of boycott supporters had risen to five. How many days from the original posting would it take for the number of boycott supporters to reach 1 million if these numbers continued to grow exponentially?

Solution If we assume exponential growth, the number of boycott supporters can be given by $y = y_0 e^{kt}$. We have $y = 2$ when $t = 2$ and $y = 5$ when $t = 3$. This gives the two equations

$$2 = y_0 e^{k2}$$

and

$$5 = y_0 e^{k3}.$$

To solve equations of this type, eliminate y_0 by dividing the second equation by the first, giving

$$\frac{5}{2} = \frac{y_0 e^{k3}}{y_0 e^{k2}} = e^{3k-2k} = e^k.$$

Taking the logarithms of both sides yields

$$k = \ln(5/2) \approx .91629.$$

Putting this into the first of the two equations above gives us

$$2 = y_0 e^{.91629 \cdot 2},$$

or

$$y_0 = 2/e^{.91629 \cdot 2} = .32.$$

We now have the growth equation

$$y = .32e^{.91629t}.$$

Letting $y = 1{,}000{,}000$, we solve for t as we did in Example 2.

$$\begin{aligned} 1{,}000{,}000 &= .32e^{.91629t} \\ 3{,}125{,}000 &= e^{.91629t} && \text{Divide by } .32. \\ \ln 3{,}125{,}000 &= \ln e^{.91629t} && \text{Take logarithms of both sides.} \\ \ln 3{,}125{,}000 &= .91629t && \ln e^x = x \\ t &= \frac{\ln 3{,}125{,}000}{.91629} \approx 16.32 \end{aligned}$$

It would take approximately 16 days for the 1 million number to be reached.

Effective Rate We could use a calculator to see that \$1 at 8% interest (per year) compounded semiannually is $1(1.04)^2 = 1.0816$ or \$1.0816. The actual increase of \$.0816 is 8.16% rather than the 8% that would be earned with interest compounded annually. To distinguish between these two amounts, 8% (the annual interest rate) is called the **nominal** or **stated** interest rate, and 8.16% is called the **effective** interest rate. We will continue to use r to designate the stated rate and we will use r_E for the effective rate.

EFFECTIVE RATE FOR COMPOUND INTEREST

If r is the annual stated rate of interest and m is the number of compounding periods per year, the effective rate of interest is

$$r_E = \left(1 + \frac{r}{m}\right)^m - 1.$$

Effective rate is sometimes called *annual yield.*

With continuous compounding, \$1 at 8% for 1 year becomes $(1)e^{1(.08)} = e^{.08} = 1.0833$. The increase is 8.33% rather than 8%, so a stated interest rate of 8% produces an effective rate of 8.33%.

EFFECTIVE RATE FOR CONTINUOUS COMPOUNDING

If interest is compounded continuously at an annual stated rate of r, the effective rate of interest is

$$r_E = e^r - 1.$$

EXAMPLE 5 *Effective Rate*

Find the effective rate corresponding to each stated rate.

(a) 6% compounded quarterly

Solution Using the formula, we get

$$\left(1 + \frac{.06}{4}\right)^4 - 1 = (1.015)^4 - 1 = .0614.$$

The effective rate is 6.14%.

(b) 6% compounded continuously

Solution The formula for continuous compounding gives

$$e^{.06} - 1 = .0618,$$

so the effective rate is 6.18%.

The formula for interest compounded m times a year, $A = P(1 + r/m)^{tm}$, has five variables: A, P, r, m, and t. If the values of any four are known, then the value of the fifth can be found. In particular, if A, the amount of money we wish to end up with, is given as well as r, m, and t, then P can be found. Here P is the amount that should be deposited today to produce A dollars in t years. The amount P is called the **present value** of A dollars.

EXAMPLE 6 *Present Value*

Tom Shaffer has a balloon payment of $100,000 due in 3 years. What is the present value of that amount if the money earns interest at 12% annually?

Solution Here P in the compound interest formula is unknown, with $A = 100{,}000$, $r = .12$, $t = 3$, and $m = 1$. Substitute the known values into the formula to get $100{,}000 = P(1.12)^3$. Solve for P, using a calculator to find $(1.12)^3$.

$$P = \frac{100{,}000}{(1.12)^3} = 71{,}178.02$$

The present value of $100,000 in 3 years at 12% per year is $71,178.02.

In general, to find the present value for an interest rate r compounded m times per year for t years, solve the equation

$$A = P\left(1 + \frac{r}{m}\right)^{tm}$$

for the variable P. To find the present value for an interest rate r compounded continuously for t years, solve the equation

$$A = Pe^{rt}$$

for the variable P.

EXAMPLE 7 *Continuous Compound Interest*

Find the interest rate that will cause \$5000 to grow to \$7250 in 4 years if the money is compounded continuously.

Solution Use the formula for continuous compounding, $A = Pe^{rt}$, with $A = 7250$, $P = 5000$, and $t = 4$. Solve first for e^{rt}, then for r.

$$\begin{aligned} A &= Pe^{rt} \\ 7250 &= 5000e^{4r} \\ 1.45 &= e^{4r} && \text{Divide by 5000.} \\ \ln 1.45 &= \ln e^{4r} && \text{Take logarithms of both sides.} \\ \ln 1.45 &= 4r && \ln e^x = x \\ r &= \frac{\ln 1.45}{4} \\ r &\approx .093 \end{aligned}$$

The required interest rate is 9.3%.

Limited Growth Functions The exponential growth functions discussed so far all continued to grow without bound. More realistically, many populations grow exponentially for a while, but then the growth is slowed by some external constraint that eventually limits the growth. For example, an animal population may grow to the point where its habitat can no longer support the population and the growth rate begins to dwindle until a stable population size is reached. Models that reflect this pattern are called **limited growth functions.** The next example discusses a function of this type that occurs in industry.

EXAMPLE 8 *Employee Turnover*

Assembly-line operations tend to have a high turnover of employees, forcing companies to spend much time and effort in training new workers. It has been found that a worker new to a task on the line will produce items according to the function defined by

$$P(x) = 25 - 25e^{-.3x},$$

where $P(x)$ items are produced by the worker on day x.

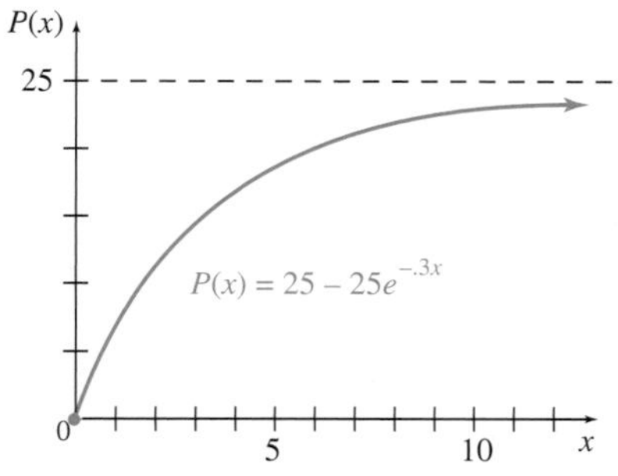

FIGURE 57

(a) What happens to the number of items a worker can produce as x gets larger and larger?

Solution As x gets larger, $e^{-.3x}$ becomes closer to 0, so $P(x)$ approaches 25. This represents the limit on the number of items a worker can produce. Note that this limit represents a horizontal asymptote on the graph of P, shown in Figure 57.

(b) How many days will it take for a new worker to produce 20 items?

Solution Let $P(x) = 20$ and solve for x.

$$P(x) = 25 - 25e^{-.3x}$$
$$20 = 25 - 25e^{-.3x}$$
$$-5 = -25e^{-.3x}$$
$$.2 = e^{-.3x}$$

Now take natural logarithms of both sides and use properties of logarithms.

$$\ln .2 = \ln e^{-.3x}$$
$$\ln .2 = -.3x \qquad \ln e^u = u$$
$$x = \frac{\ln .2}{-.3} \approx 5.4$$

In about $5\frac{1}{2}$ days on the job a new worker will be producing 20 items.

Graphs such as the one in Figure 57 are called **learning curves.** According to such a graph, a new worker tends to learn quickly at first; then learning tapers off and approaches some upper limit. This is characteristic of the learning of certain types of skills involving the repetitive performance of the same task.

2.6 EXERCISES

1. What is the difference between stated interest rate and effective interest rate?
2. In the exponential growth or decay function $y = y_0e^{kt}$, what does y_0 represent? What does k represent?
3. In the exponential growth or decay function, explain the circumstances that cause k to be positive or negative.
4. What is meant by the half-life of a quantity?
5. Show that if a radioactive substance has a half-life of T, then the corresponding constant k in the exponential decay function is given by $k = -(\ln 2)/T$.
6. Show that if a radioactive substance has a half-life of T, then the corresponding exponential decay function can be written as $y = y_0(1/2)^{(t/T)}$.

Applications

BUSINESS AND ECONOMICS

Effective Rate Find the effective rate corresponding to each nominal rate of interest.

7. 15% compounded quarterly
8. 18% compounded monthly
9. 11% compounded continuously
10. 7% compounded continuously

Present Value Find the present value of each deposit.

11. \$10,000 at 10% compounded quarterly for 8 years
12. \$45,678.93 if interest is 12.6% compounded monthly for 11 months
13. \$7300 at 11% compounded continuously for 3 years
14. \$25,000 at 9% compounded continuously for 8 years
15. *Effective Rate* Becky Anderson bought a television set with money borrowed from the bank at 9% interest com-

pounded semiannually. What effective interest rate did she pay?

16. *Effective Rate* A firm deposits some funds in a special account at 7.2% compounded quarterly. What effective rate will they earn?

17. *Effective Rate* Michelle Renda deposits \$7500 of lottery winnings in an account paying 6% interest compounded monthly. What effective rate does the account earn?

18. *Present Value* Frank Steed must make a balloon payment of \$20,000 in 4 years. Find the present value of the payment if it includes annual interest of 8%.

19. *Present Value* A company must pay a \$307,000 settlement in 3 years.

a. What amount must be deposited now at 6% compounded semiannually to have enough money for the settlement?

b. How much interest will be earned?

c. Suppose the company can deposit only \$200,000 now. How much more will be needed in 3 years?

20. *Present Value* A couple wants to have \$20,000 in 5 years for a down payment on a new house.

a. How much could they deposit today, at 8% compounded quarterly, to have the required amount in 5 years?

b. How much interest will be earned?

c. If they can deposit only \$10,000 now, how much more will they need to complete the \$20,000 after 5 years?

21. *Interest* Christine O'Brien, who is self-employed, wants to invest \$60,000 in a pension plan. One investment offers 10% compounded quarterly. Another offers 9.75% compounded continuously.

a. Which investment will earn the most interest in 5 years?

b. How much more will the better plan earn?

c. What is the effective rate in each case?

d. If Ms. O'Brien chooses the plan with continuous compounding, how long will it take for her \$60,000 to grow to \$80,000?

e. How long will it take for her \$60,000 to grow to at least \$80,000 if she chooses the plan with quarterly compounding? (Be careful; interest is only added to the account every quarter.)

22. *Interest* Greg Tobin wishes to invest a \$5000 bonus check into a savings account that pays 6.3% interest. Find how many years it will take for the \$5000 to grow to at least \$11,000 if interest is compounded

a. quarterly. (Be careful; interest is only added to the account every quarter.)

b. continuously.

23. *Rent Increase* A newspaper article contained the following quotes: "Next year's . . . legal rent increase has been pegged at 4.9%. Coming on the heels of this year's 6 percent increase and 5.4 percent in 1991, it means rents have gone up 16.3 percent in three years."* Is this figure correct? If not, what is the correct figure?

24. *Electric Rates* In 1987, a New York State Assemblywoman wrote a letter to her constituents about the electric rates of the Long Island Lighting Company (LILCO). She wrote. "If the [Public Service] Commission grants this increase, it will mean that between 1970 and 1987, LILCO's electric rates will have increased by more than 478 percent. That's an average rise of over 27 percent a year!" Is this correct? If not, what annual percent increase over 17 years leads to a total increase of 478 percent?

25. *Sales* Sales of a new model of compact disc player are approximated by the function $S(x) = 1000 - 800e^{-x}$, where $S(x)$ is in appropriate units and x represents the number of years the disc player has been on the market.

a. Find the sales during year 0.

b. In how many years will sales reach 500 units?

c. Will sales ever reach 1000 units?

d. Is there a limit on sales for this product? If so, what is it?

26. *Sales* Sales of a new model of word processor are approximated by

$$S(x) = 5000 - 4000e^{-x},$$

where x represents the number of years that the word processor has been on the market, and $S(x)$ represents sales in thousands of dollars.

a. Find the sales in year 0.

b. When will sales reach \$4,500,000?

c. Find the limit on sales.

LIFE SCIENCES

27. *Population Growth* The population of the world in the year 1650 was 470 million, and in the year 1999 was 5996 million.†

a. Assuming that the population of the world grows exponentially, find the equation for the population $P(t)$ in millions in the year t.

**Toronto Star*, Aug. 2, 1992.
†*Statistical Abstract of the United States*, 1999.

b. Use your answer from part a to find the population of the world in the year 1.

c. Is your answer to part b reasonable? What does this tell you about how the population of the world grows?

28. *Giardia* When a person swallows giardia cysts, stomach acids and pancreatic enzymes cause the cysts to release trophozoites, which divide every 12 hours.*

a. Suppose the number of trophozoites at time $t = 0$ is y_0. Write a function giving the number after t hours.

b. The article quoted above said that a single trophozoite can multiply to a million in just 10 days and a billion in 15 days. Verify this fact.

29. *Melanoma* An article on skin cancer said that the number of diagnosed melanoma cases is growing by 4% a year.† It also said that the chance of a U.S. resident getting cancer increased from 1 in 250 in 1980 to 1 in 84 in 1997. Is this a 4% increase annually? If not, what percent increase annually is this?

30. *Growth of Bacteria* A culture contains 25,000 bacteria, with the population increasing exponentially. The culture contains 40,000 bacteria after 10 hours.

a. Write an exponential equation to express the growth function y in terms of time t in hours.

b. How long will it be until there are 60,000 bacteria?

31. *Decrease in Bacteria* When an antibiotic is introduced into a culture of 50,000 bacteria, the number of bacteria decreases exponentially. After 9 hours, there are only 20,000 bacteria.

a. Write an exponential equation to express the growth function y in terms of time t in hours.

b. In how many hours will half the number of bacteria remain?

32. *Growth of Bacteria* The growth of bacteria in food products makes it necessary to time-date some products (such as milk) so that they will be sold and consumed before the bacteria count is too high. Suppose for a certain product that the number of bacteria present is given by

$$f(t) = 500e^{.1t},$$

under certain storage conditions, where t is time in days after packing of the product and the value of $f(t)$ is in millions.

a. If the product cannot be safely eaten after the bacteria count reaches 3000 million, how long will this take?

b. If $t = 0$ corresponds to January 1, what date should be placed on the product?

33. *Cancer Research* An article on cancer treatment contains the following statement: A 37% 5-year survival rate for women with ovarian cancer yields an estimated annual mortality rate of .1989.‡ The authors of this article assume that the number of survivors is described by the exponential decay function given at the beginning of this section, where y is the number of survivors and k is the mortality rate. Verify that the given survival rate leads to the given mortality rate.

34. *Chromosomal Abnormality* The graph below shows how the risk of chromosomal abnormality in a child rises with the age of the mother.§

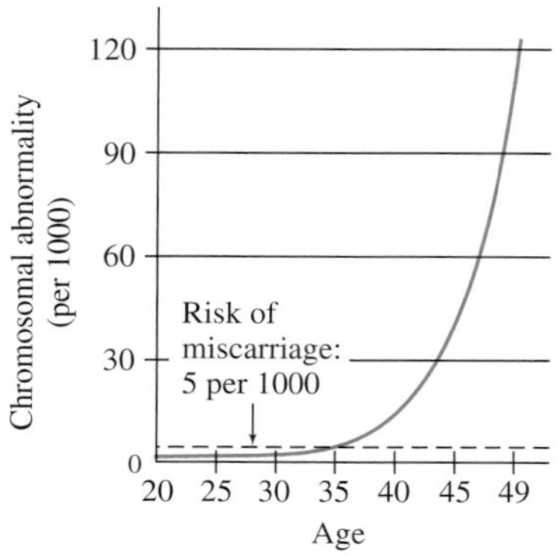

a. Read from the graph the risk of chromosomal abnormality (per 1000) at ages 20, 35, 42, and 49.

b. Assuming the graph to be of the form $y = Ce^{kt}$, find k using $t = 20$ and $t = 35$.

c. Still assuming the graph to be of the form $y = Ce^{kt}$, find k using $t = 42$ and $t = 49$.

d. Based on your results from parts a–c, is it reasonable to assume the graph is of the form $y = Ce^{kt}$? Explain.

e. In situations such as parts a–c, where an exponential function does not fit because different data points give different values for the growth constant k, it is often appropriate to describe the data using an equation of the form $y = Ce^{kt^n}$. Parts b and c show that $n = 1$ results in a smaller constant using the interval $[20, 35]$ than using the interval $[42, 49]$. Repeat parts b and c using $n = 2, 3$, etc., until the interval $[20, 35]$ yields a larger value of k than the interval $[42, 49]$, and then estimate what n should be.

*Brody, Jane, "Sly Parasite Menaces Pets and their Owners," *The New York Times*, Dec. 21, 1999, p. F7.
†Seppa, N., "Skin Cancer Makes Unexpected Reappearance," *Science News*, June 21, 1997, p. 383.
‡Speroff, Theodore et al., "A Risk-Benefit Analysis of Elective Bilateral Oophorectomy: Effect of Changes in Compliance with Estrogen Therapy on Outcome," *American Journal of Obstetrics and Gynecology*, Vol. 164, Jan. 1991, pp. 165–74.
§*The New York Times*, Feb. 5, 1994, p. 24. Reprinted with permission.

PHYSICAL SCIENCES

35. *Carbon Dating* Refer to Example 2. A sample from a refuse deposit near the Strait of Magellan had 60% of the carbon 14 found in a contemporary living sample. How old was the sample?

Half-Life Find the half-life of each radioactive substance. See Example 2.

36. Plutonium 241; $A(t) = A_0e^{-.053t}$

37. Radium 226; $A(t) = A_0e^{-.00043t}$

38. *Half-Life* The half-life of plutonium 241 is approximately 13 years.

a. How much of a sample weighing 2 g will remain after 100 years?

b. How much time is necessary for a sample weighing 2 g to decay to .1 g?

39. *Half-Life* The half-life of radium 226 is approximately 1620 years.

a. How much of a sample weighing 2 g will remain after 100 years?

b. How much time is necessary for a sample weighing 2 g to decay to .1 g?

40. *Radioactive Decay* 500 g of iodine 131 is decaying exponentially. After 3 days 386 g of iodine 131 is left.

a. Write an exponential equation to express the decay function y in terms of t in days.

b. Use your answer from part a to find the half-life of iodine 131.

41. *Radioactive Decay* 25 g of polonium 210 is decaying exponentially. After 50 days 19.5 g of polonium 210 is left.

a. Write an exponential equation to express the decay function y in terms of t in days.

b. Use your answer from part a to find the half-life of polonium 210.

42. *Nuclear Energy* Nuclear energy derived from radioactive isotopes can be used to supply power to space vehicles. The output of the radioactive power supply for a certain satellite is given by the function $y = 40e^{-.004t}$, where y is in watts and t is the time in days.

a. How much power will be available at the end of 180 days?

b. How long will it take for the amount of power to be half of its original strength?

c. Will the power ever be completely gone? Explain.

43. *Botany* A group of Tasmanian botanists have claimed that a King's holly shrub, the only one of its species in the world, is also the oldest living plant.* Using carbon 14 dating of charcoal found along with fossilized leaf fragments, they arrived at an age of 43,000 years for the plant, whose exact location in southwest Tasmania is being kept a secret. What percent of the original carbon 14 in the charcoal was present?

44. *Decay of Radioactivity* A large cloud of radioactive debris from a nuclear explosion has floated over the Pacific Northwest, contaminating much of the hay supply. Consequently, farmers in the area are concerned that the cows who eat this hay will give contaminated milk. (The tolerance level for radioactive iodine in milk is 0.) The percent of the initial amount of radioactive iodine still present in the hay after t days is approximated by $P(t)$, which is given by the mathematical model

$$P(t) = 100e^{-.1t}.$$

a. Find the percent remaining after 4 days.

b. Find the percent remaining after 10 days.

c. Some scientists feel that the hay is safe after the percent of radioactive iodine has declined to 10% of the original amount. Solve the equation $10 = 100e^{-.1t}$ to find the number of days before the hay may be used.

d. Other scientists believe that the hay is not safe until the level of radioactive iodine has declined to only 1% of the original level. Find the number of days that this would take.

45. *Chemical Dissolution* The amount of chemical that will dissolve in a solution increases exponentially as the temperature is increased. At 0°C 10 g of the chemical dissolves, and at 10°C 11 g dissolves.

a. Write an equation to express the amount of chemical dissolved, y, in terms of temperature, t, in degrees Celsius.

b. At what temperature will 15 g dissolve?

Newton's Law of Cooling Newton's law of cooling says that the rate at which a body cools is proportional to the difference in temperature between the body and an environment into which it is introduced. This leads to an equation where the temperature $f(t)$ of the body at time t after being introduced into an environment having constant temperature T_0 is

$$f(t) = T_0 + Ce^{-kt},$$

where C and k are constants. Use this result in Exercises 46–48.

46. Find the temperature of an object when $t = 9$ if $T_0 = 18$, $C = 5$, and $k = .6$.

47. If $C = 100$, $k = .1$, and t is time in minutes, how long will it take a hot cup of coffee to cool to a temperature of 25°C in a room at 20°C?

48. If $C = -14.6$ and $k = .6$ and t is time in hours, how long will it take a frozen pizza to thaw to 10°C in a room at 18°C?

**Science*, Vol. 277, July 25, 1997, p. 483.

CHAPTER SUMMARY

In this chapter we have studied the general properties of functions. By investigating the special nature of specific families of functions, such as quadratic, polynomial, rational, exponential, and logarithmic functions, we are now able to use this knowledge to analyze and understand a large number of real-life applications. We now know when the use of one function is more appropriate than another, and we can predict how the dependent variable will react to changes in the independent variable. In particular, we have applied this knowledge to the study of growth and decay and the mathematics of finance. In the next chapters, we will build on this knowledge.

EXPONENTIAL AND LOGARITHMIC FUNCTIONS SUMMARY

Exponential Functions

Exponential Growth and Decay Function If y_0 is the number present at time $t = 0$,

$$y = y_0 e^{kt}.$$

Math of Finance Formulas If I is the interest, P is the principal or present value, r is the annual interest rate, t is time in years, and m is the number of compounding periods per year:

Simple interest: $I = Prt$

	Compounded m Times per Year	Compounded Continuously
Compound amount	$A = P\left(1 + \dfrac{r}{m}\right)^{tm}$	$A = Pe^{rt}$
Effective rate	$r_E = \left(1 + \dfrac{r}{m}\right)^{m} - 1$	$r_E = e^r - 1$

Change-of-Base Theorem for Exponentials $a^x = e^{(\ln a)x}$

Logarithmic Functions

Properties of Logarithms Let x and y be any positive real numbers and r be any real number. Let a be a positive real number, $a \neq 1$. Then

a. $\log_a xy = \log_a x + \log_a y$

b. $\log_a \dfrac{x}{y} = \log_a x - \log_a y$

c. $\log_a x^r = r \log_a x$

d. $\log_a a = 1$

e. $\log_a 1 = 0$

f. $\log_a a^r = r.$

Change-of-Base Theorem for Logarithms $\log_a x = \dfrac{\log_b x}{\log_b a} = \dfrac{\ln x}{\ln a}$

Graphs of Basic Functions

Quadratic

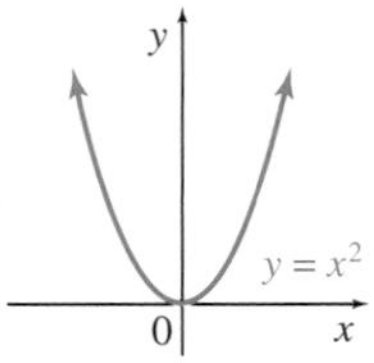

Absolute Value

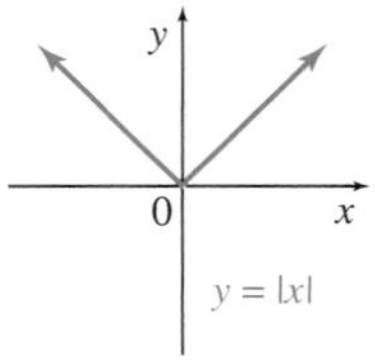

Square Root

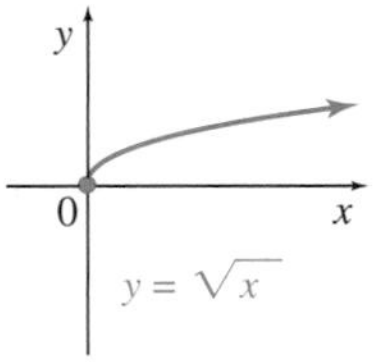

Rational

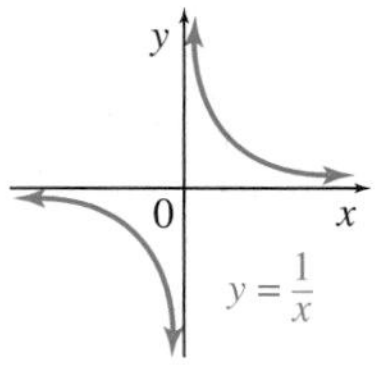

Exponential

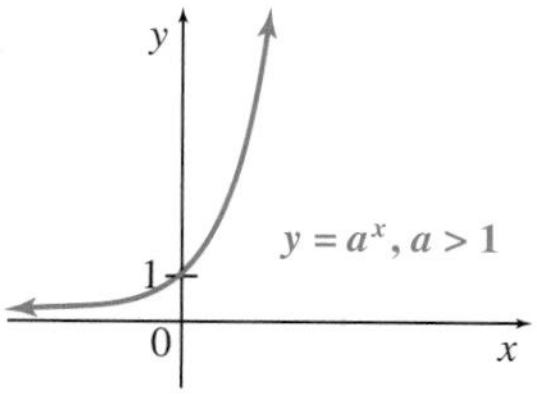

Logarithmic

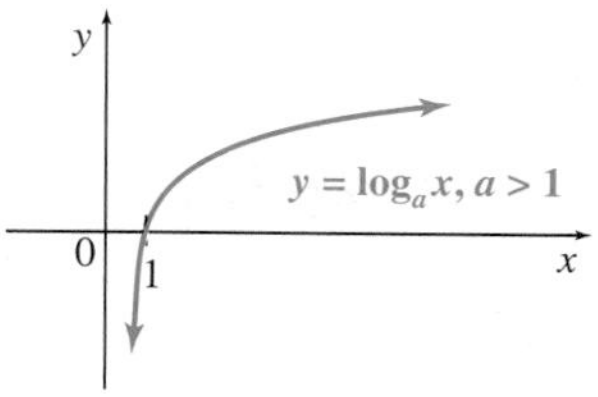

KEY TERMS

To understand the concepts presented in this chapter, you should know the meaning and use of the following terms. For easy reference, the section in the chapter where a word (or expression) was first used is provided.

2.1 **function**
domain
range
constant function
vertical line test
step function

2.2 **quadratic function**
parabola
vertex
axis
vertical translation
horizontal translation
vertical reflection
completing the square

2.3 **polynomial function**
coefficient
leading coefficient
power function
cubic polynomial
quartic polynomial
turning point
real zero
rational function
vertical asymptote
horizontal asymptote
cost–benefit model
average cost

2.4 **exponential function**
exponential equation
simple interest
compound interest
rate of interest
time
principal
compound amount
e
continuous compounding

2.5 **years to double**
logarithm
logarithmic function
properties of logarithms
common logarithms
natural logarithms
change-of-base theorem for logarithms
logarithmic equation
change-of-base theorem for exponentials
rule of 70
rule of 72
index of diversity

2.6 **exponential growth and decay function**
growth constant
decay constant
half-life
effective rate
nominal (stated) rate
present value
limited growth function
learning curve

CHAPTER 2 REVIEW EXERCISES

1. What is a function? A linear function? A quadratic function? A rational function?

2. How do you find a vertical asymptote? A horizontal asymptote?

3. What can you tell about the graph of a polynomial function of degree n before you plot any points?

4. Describe in words what a logarithm is.

List the ordered pairs obtained from the following if the domain of x for each exercise is $\{-3, -2, -1, 0, 1, 2, 3\}$. *Graph each set of ordered pairs. Give the range.*

5. $y = (2x + 1)(x - 1)$

6. $y = \dfrac{2}{x^2 + 1}$

In Exercises 7 and 8, find **(a)** $f(6)$, **(b)** $f(-2)$, **(c)** $f(-4)$, and **(d)** $f(r + 1)$.

7. $f(x) = -x^2 + 2x - 4$

8. $f(x) = 8 - x - x^2$

9. Let $f(x) = 5x^2 - 3$ and $g(x) = -x^2 + 4x + 1$. Find the following.

a. $f(-2)$ **b.** $g(3)$ **c.** $f(-k)$ **d.** $g(3m)$

e. $f(x + h)$ **f.** $g(x + h)$ **g.** $\dfrac{f(x + h) - f(x)}{h}$ **h.** $\dfrac{g(x + h) - g(x)}{h}$

Find the domain of each function defined as follows.

10. $y = \frac{\sqrt{x-2}}{2x+3}$ **11.** $y = \ln(x^2 - 9)$ **12.** $y = \frac{3x-4}{x}$

Graph the following.

13. $y = 3x^2 + 6x - 2$ **14.** $y = -\frac{1}{4}x^2 + x + 2$ **15.** $y = x^2 - 4x + 2$ **16.** $y = -3x^2 - 12x - 1$

17. $f(x) = x^3 + 5$ **18.** $f(x) = 1 - x^4$ **19.** $y = -(x-1)^3 + 4$ **20.** $y = -(x+2)^4 - 2$

21. $f(x) = \frac{8}{x}$ **22.** $f(x) = \frac{2}{3x-1}$ **23.** $f(x) = \frac{4x-2}{3x+1}$ **24.** $f(x) = \frac{6x}{x+2}$

25. $y = 5^x$ **26.** $y = 5^{-x} + 1$ **27.** $y = \left(\frac{1}{5}\right)^{2x-3}$ **28.** $y = \left(\frac{1}{2}\right)^{x-1}$

29. $y = \log_2(x-1)$ **30.** $y = 1 + \log_3 x$ **31.** $y = -\log_3 x$ **32.** $y = 2 - \ln x^2$

Solve each equation.

33. $2^{3x} = \frac{1}{8}$ **34.** $\left(\frac{9}{16}\right)^x = \frac{3}{4}$ **35.** $9^{2y-1} = 27^y$ **36.** $\frac{1}{2} = \left(\frac{b}{4}\right)^{1/4}$

Write each equation using logarithms.

37. $2^6 = 64$ **38.** $3^{1/2} = \sqrt{3}$ **39.** $e^{.09} = 1.09417$ **40.** $10^{1.07918} = 12$

Write each equation using exponents.

41. $\log_2 32 = 5$ **42.** $\log_{10} 100 = 2$ **43.** $\ln 82.9 = 4.41763$ **44.** $\log 15.46 = 1.18921$

Evaluate each expression without using a calculator. Then support your work using a calculator and the change-of-base theorem for logarithms.

45. $\log_3 81$ **46.** $\log_{32} 16$ **47.** $\log_{25} 5$ **48.** $\log_{100} 1000$

Simplify each expression using the properties of logarithms.

49. $\log_5 3k + \log_5 7k^3$ **50.** $\log_3 2y^3 - \log_3 8y^2$ **51.** $2\log_2 x - 3\log_2 m$ **52.** $5\log_4 r - 3\log_4 r^2$

Solve each equation. If necessary, round each answer to the nearest thousandth.

53. $8^p = 19$ **54.** $3^{z-2} = 11$ **55.** $2^{1-m} = 7$ **56.** $15^{-k} = 9$

57. $e^{-5-2x} = 5$ **58.** $e^{3x-1} = 12$ **59.** $\left(1 + \frac{m}{3}\right)^5 = 10$ **60.** $\left(1 + \frac{2p}{5}\right)^2 = 3$

61. $\log_k 64 = 6$ **62.** $\log_3(2x+5) = 2$ **63.** $\log(4p+1) + \log p = \log 3$

64. $\log_2(5m-2) - \log_2(m+3) = 2$

65. Give the following properties of the exponential function $f(x) = a^x$; $a > 0$, $a \neq 1$.

a. Domain **b.** Range **c.** y-intercept

d. Discontinuities **e.** Asymptote(s) **f.** Increasing if a is ______

g. Decreasing if a is ______

66. Give the following properties of the logarithmic function $f(x) = \log_a x$; $a > 0$, $a \neq 1$.

a. Domain **b.** Range **c.** x-intercept

d. Discontinuities **e.** Asymptote(s) **f.** Increasing if a is ______

g. Decreasing if a is ______

67. Compare your answers for Exercises 65 and 66. What similarities do you notice? What differences?

Applications

BUSINESS AND ECONOMICS

68. *Car Rental* To rent a mid-size car from one agency costs \$40 per day or fraction of a day. If you pick up the car in Boston and drop it off in Utica, there is a fixed \$40 charge. Let $C(x)$ represent the cost of renting the car for x days and taking it from Boston to Utica. Find the following.

a. $C\left(\frac{3}{4}\right)$ **b.** $C\left(\frac{9}{10}\right)$ **c.** $C(1)$

d. $C\left(1\frac{5}{8}\right)$ **e.** $C\left(2\frac{1}{9}\right)$

f. Graph the function defined by $y = C(x)$ for $0 < x \le 5$.

g. What is the independent variable?

h. What is the dependent variable?

69. *Pollution* The cost to remove x percent of a pollutant is

$$y = \frac{7x}{100 - x},$$

in thousands of dollars. Find the cost of removing the following percents of the pollutant.

a. 80% **b.** 50% **c.** 90%

d. Graph the function.

e. Can all of the pollutant be removed?

Interest *Find the amount of interest earned by each deposit.*

70. \$6902 at 12% compounded semiannually for 8 years

71. \$2781.36 at 8% compounded quarterly for 6 years

72. How long will it take for \$1000 deposited at 6% compounded semiannually to double? To triple?

73. How long will it take for \$2100 deposited at 4% compounded quarterly to double? To triple?

Interest *Find the compound amount if* \$12,104 *is invested at 8% compounded continuously for each period.*

74. 2 years 75. 4 years

Interest *Find the compound amounts for the following deposits if interest is compounded continuously.*

76. \$1500 at 10% for 9 years

77. \$12,000 at 5% for 8 years

Effective Rate *Find the effective rate to the nearest hundredth for each nominal interest rate.*

78. 7% compounded quarterly

79. 9% compounded monthly

80. 9% compounded continuously

Present Value *Find the present value of each amount.*

81. \$2000 at 6% interest compounded annually for 5 years

82. \$10,000 at 8% interest compounded semiannually for 6 years

83. *Interest* To help pay for college expenses, Tiffany Lockman borrowed \$10,000 at 10% interest compounded semiannually for 8 years. How much will she owe at the end of the 8-year period?

84. *Interest* How long will it take for \$1 to triple at an annual inflation rate of 8% compounded continuously?

85. *Interest* Find the interest rate needed for \$6000 to grow to \$8000 in 3 years with continuous compounding.

86. *Present Value* Michael Garbin wants to open a camera shop. How much must he deposit now at 6% interest compounded monthly to have \$25,000 at the end of 3 years?

87. *Revenue* A concert promoter finds she can sell 1000 tickets at \$20 each. She will not sell the tickets for less than \$20, but she finds that for every \$1 increase in the ticket price above \$20, she will sell 10 fewer tickets.

a. Express n, the number of tickets sold, as a function of p, the price.

b. Express R, the revenue, as a function of p, the price.

c. Find the domain of the function found in part b.

d. Express R, the revenue, as a function of n, the number sold.

e. Find the domain of the function found in part d.

f. Find the price that produces the maximum revenue.

g. Find the number of tickets sold that produces the maximum revenue.

h. Find the maximum revenue.

i. Sketch the graph of the function found in part b.

j. Describe what the graph found in part i tells you about how the revenue varies with price.

88. *Cost* Suppose the cost in dollars to produce x posters is given by

$$C(x) = \frac{5x + 3}{x + 1}.$$

a. Sketch a graph of $C(x)$.

b. Find a formula for $C(x+1) - C(x)$, the cost to produce an additional poster when x posters are already produced.

c. Find a formula for $A(x)$, the average cost per poster.

d. Find a formula for $A(x+1) - A(x)$, the change in the average cost per poster when one additional poster is produced. (This quantity is approximately equal to the marginal average cost, which will be discussed in the chapter on the derivative.)

89. *Cost* Suppose the cost in dollars to produce x hundreds of nails is given by

$$C(x) = x^2 + 4x + 7.$$

a. Sketch a graph of $C(x)$.

b. Find a formula for $C(x+1) - C(x)$, the cost to produce an additional hundred nails when x hundred are already produced. (This quantity is approximately equal to the marginal cost.)

c. Find a formula for $A(x)$, the average cost per hundred nails.

d. Find a formula for $A(x+1) - A(x)$, the change in the average cost per nail when one additional batch of 100 nails is produced. (This quantity is approximately equal to the marginal average cost, which will be discussed in the chapter on the derivative.)

LIFE SCIENCES

90. *Fever* A certain viral infection causes a fever that typically lasts 6 days. A model of the fever (in °F) on day x, $1 \le x \le 6$, is

$$F(x) = -\frac{2}{3}x^2 + \frac{14}{3}x + 96.$$

According to the model, on what day should the maximum fever occur? What is the maximum fever?

91. *Sunscreen* An article in a medical journal says that a sunscreen with a sun protection factor (SPF) of 2 provides 50% protection against ultraviolet B (UVB) radiation, an SPF of 4 provides 75% protection, and an SPF of 8 provides 87.5% protection (which the article rounds to 87%).*

a. 87.5% protection means that 87.5% of the UVB radiation is screened out. Write as a fraction the amount of radiation that is let in, and then describe how this fraction, in general, relates to the SPF rating.

b. Plot UVB percent protection (y) against x, where $x = 1/\text{SPF}$.

c. Based on your graph from part b, give an equation relating UVB protection to SPF rating.

d. An SPF of 8 has double the chemical concentration of an SPF 4. Find the increase in the percent protection.

e. An SPF of 30 has double the chemical concentration of an SPF 15. Find the increase in the percent protection.

f. Based on your answers from parts d and e, what happens to the increase in the percent protection as the SPF continues to double?

92. *AIDS Deaths* The following table lists the number of AIDS deaths in the United States in recent years.†

Year	Deaths
1990	31,836
1991	37,106
1992	41,849
1993	45,733
1994	50,657
1995	51,414
1996	38,074
1997	21,846
1998	18,148
1999	16,762
2000	14,499
2001	8,998

a. Plot the data on a graphing calculator, letting $x = 0$ correspond to the year 1990.

b. Using the regression feature on your calculator, find a quadratic, a cubic, and a quartic function that models this data.

c. Plot the three functions with the data on the same coordinate axes. Which function or functions best capture the behavior of the data over the years plotted?

d. Find the number of deaths predicted by all three functions for 2002. Explain why none of these predictions is realistic.

93. *Respiratory Rate* Researchers have found that the 95th percentile (the value at which 95% of the data is at or

Family Practice, May 17, 1993, p. 55.

†*The New York Times 2003 Almanac*, p. 375.

below) for respiratory rates (in breaths per minute) during the first 3 years of infancy are given by

$$y = 10^{1.82411 - .0125995x + .00013401x^2}$$

for awake infants and

$$y = 10^{1.72858 - .0139928x + .00017646x^2}$$

for sleeping infants, where x is the age in months.*

a. What is the domain for each function?

b. For each respiratory rate, is the rate decreasing or increasing over the first 3 years of life? (*Hint:* Is the graph of the quadratic in the exponent opening upward or downward? Where is the vertex?)

c. Verify your answer to part b using a graphing calculator.

d. For a 1-year-old infant in the 95th percentile, how much higher is the waking respiratory rate than the sleeping respiratory rate?

94. *Polar Bear Mass* One formula for estimating the mass (in kg) of a polar bear is given by

$$m(g) = e^{.02 + .062g - .000165g^2},$$

where g is the axillary girth in centimeters.† It seems reasonable that as girth increases, so does the mass. What is the largest girth for which this formula gives a reasonable answer? What is the predicted mass of a polar bear with this girth?

95. *Population Growth* A population of 15,000 small deer in a specific region has grown exponentially to 17,000 in 4 years.

a. Write an exponential equation to express the population growth y in terms of time t in years.

b. At this rate, how long will it take for the population to reach 45,000?

96. *Population Growth* In 1960 in an article in *Science* magazine, H. Van Forester, P. M. Mora, and W. Amiot predicted that world population would be infinite in the year 2026. Their projection was based on the rational function defined by

$$p(t) = \frac{1.79 \cdot 10^{11}}{(2026.87 - t)^{.99}},$$

where $p(t)$ gives population in year t.‡ This function has provided a relatively good fit to the population so far.

a. Estimate world population in 1999 using this function, and compare it with the estimate given in *Statistical Abstract of the United States 1999*, of 5.996 billion.

b. What does the function predict for world population in 2020? 2025?

c. Discuss why this function is not realistic, despite its good fit to past data.

97. *Intensity of Light* The intensity of light (in appropriate units) passing through water decreases exponentially with the depth it penetrates beneath the surface according to the function

$$I(x) = 10e^{-.3x},$$

where x is the depth in meters. A certain water plant requires light of an intensity of 1 unit. What is the greatest depth of water in which it will grow?

98. *Drug Concentration* The concentration of a certain drug in the bloodstream at time t (in minutes) is given by

$$c(t) = e^{-t} - e^{-2t}.$$

Use a graphing calculator to find the maximum concentration and the time when it occurs.

99. *Glucose Concentration* When glucose is infused into a person's bloodstream at a constant rate of c grams per minute, the glucose is converted and removed from the bloodstream at a rate proportional to the amount present. The amount of glucose in the bloodstream at time t (in minutes) is given by

$$g(t) = \frac{c}{a} + \left(g_0 - \frac{c}{a}\right)e^{-at},$$

where a is a positive constant. Assume $g_0 = .08$, $c = .1$, and $a = 1.3$.

a. At what time is the amount of glucose a maximum? What is the maximum amount of glucose in the bloodstream?

b. When is the amount of glucose in the bloodstream .1 g?

c. What happens to the amount of glucose in the bloodstream after a very long time?

PHYSICAL SCIENCES

100. *Oil Production* The production of an oil well has decreased exponentially from 128,000 barrels per year 5 years ago to 100,000 barrels per year at present.

*Rusconi, Franca et al., "Reference Values for Respiratory Rate in the First 3 Years of Life," *Pediatrics*, Vol. 94, No. 3, Sept. 1994, pp. 350–355.

†Cattet, Marc R. L. et al., "Predicting Body Mass in Polar Bears: Is Morphometry Useful?" *Journal of Wildlife Management*, Vol. 61. No. 4, 1997, pp. 1083–1090.

‡*Chance News* 7.01, Jan. 1998.

a. Letting $t = 0$ represent the present time, write an exponential equation for production y in terms of time t in years.

b. Find the time it will take for production to fall to 70,000 barrels per year.

101. *Dating Rocks* Geologists sometimes measure the age of rocks by using "atomic clocks." By measuring the amounts of potassium 40 and argon 40 in a rock, the age t of the specimen (in years) is found with the formula

$$t = (1.26 \times 10^9)\frac{\ln[1 + 8.33(A/K)]}{\ln 2},$$

where A and K, respectively, are the numbers of atoms of argon 40 and potassium 40 in the specimen.

a. How old is a rock in which $A = 0$ and $K > 0$?

b. The ratio A/K for a sample of granite from New Hampshire is .212. How old is the sample?

c. Let $A/K = r$. What happens to t as r gets larger? Smaller?

102. *Average Speed* Suppose a plane flies from one city to another that is a distance d miles away. The plane flies at a constant speed v relative to the wind, but there is a constant wind speed of w. Therefore, the speed in one direction is $v + w$, and the speed in the other direction is $v - w$.*

a. Using the formula distance = rate × time, show that the time to make the round trip is

$$\frac{d}{v+w} + \frac{d}{v-w}.$$

b. Using the result from part a, show that the average speed for the round trip is

$$v_{aver} = \frac{2d}{\dfrac{d}{v+w} + \dfrac{d}{v-w}}.$$

c. Simplify your result from part b to get

$$v_{aver} = v - \frac{w^2}{v}.$$

d. Consider v in the result in part c to be a constant. What wind speed results in the greatest average speed? Explain why your result makes sense.

103. *Average Speed* Suppose the plane in the previous exercise makes the trip one way at a speed of v, and the return trip at a speed xv.†

a. Explain in words what $x = .9$ and $x = 1.1$ represent.

b. Show that the average velocity can be written as

$$v_{aver} = \left(\frac{2x}{x+1}\right)v.$$

(*Hint:* Use the steps in parts a–c of the previous exercise.)

c. With v held constant, the equation in part b defines v_{aver} as a function of x. Discuss the behavior of this function. In particular, consider the horizontal asymptote and what it says about the average velocity.

104. *Carbon Monoxide* The total amount of carbon monoxide (in millions of short tons) in the United States emitted by fuel combustion, industrial processes, and transportation between the years 1983 and 2002 can be approximated by

$$f(t) = .01408t^3 - .4168t^2 - .421t + 170,$$

where t is the number of years since 1983.‡

a. Using a graphing calculator sketch of the function, determine the trend in carbon monoxide emissions from 1983 to 2002.

b. Use a graphing calculator to estimate the year that emissions were 125 million short tons.

105. *Planets* The following table contains the average distance D from the sun for the first eight planets and their period P of revolution around the sun in years.§

Planet	Distance (D)	Period (P)
Mercury	.39	.24
Venus	.72	.62
Earth	1	1
Mars	1.52	1.89
Jupiter	5.20	11.9
Saturn	9.54	29.5
Uranus	19.2	84.0
Neptune	30.1	164.8

The distances are given in astronomical units (A.U.); 1 A.U. is the average distance from Earth to the sun. For exam-

*This exercise is based on a letter by Tom Blazey to *Mathematics Teacher*, Vol. 86, No. 2, Feb. 1993, p. 178.

†This exercise is based on the article "Problems Whose Solutions Lie on a Hyperbola," by Steven Schwartzman, *The AMATYC Review*, Vol. 14, No. 2, Spring 1993, pp. 27–36.

‡*Latest Findings on National Air Quality: 2002 Status and Trends*, Environmental Protection Agency, p. 16.

§Ronan, C., *The Natural History of the Universe*, Macmillan, 1991.

ple, since Jupiter's distance is 5.2 A.U., its distance from the sun is 5.2 times farther than Earth's.

a. Find functions of the form $P = kD^n$ for $n = 1, 1.5$, and 2 that fit the data at Neptune.

b. Use a graphing calculator to plot the data in the table and to graph the three functions found in part a. Which function best fits the data?

c. Use the best-fitting function from part b to predict the period of the planet Pluto, which has a distance from the sun of 39.5 A.U. Compare your answer to the true value of 248.5 years.

d. If you have a graphing calculator or computer program with a power regression feature, use it to find a power function (a function of the form $P = kD^n$) that approximately fits the data. How does this answer compare with the answer to part b?

GENERAL INTEREST

106. *Automobiles* The following graph shows the horsepower and torque for a 1991 Porsche 911 Turbo as a function of the engine speed.*

a. Find the engine speed giving the maximum horsepower.

b. Find the maximum horsepower.

c. Find the torque when the horsepower is at its maximum.

911 Turbo Porsche Performance and Torque

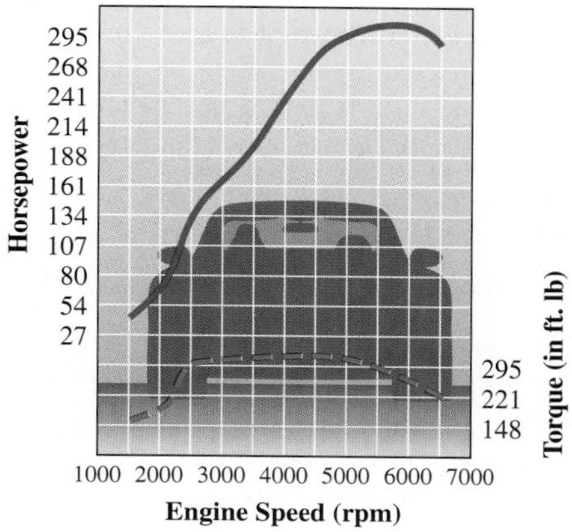

107. *Consumer Price Index* The U.S. consumer price index (CPI, or cost of living index) has risen over the years, as shown in the table below, using an index in which the average over the years 1982 to 1984 is set to 100.†

Year	CPI
1960	29.6
1970	38.8
1980	82.4
1990	130.7
1995	152.4
2000	172.2
2002	179.9

a. Letting t be the years since 1960, write an exponential function in the form $y = a^t$ that fits the data at 1960 and 2002.

b. If your calculator has an exponential regression feature, find the best fitting exponential function for the data.

c. Use a graphing calculator to plot the answers to parts a and b on the same axes as the data. Are the answers to parts a and b close to each other?

*Porsche product information kit for the 1991 Turbo Performance and Torque 1991 model year. Reprinted by permission of Porsche Cars North America, Inc.

*http://data.bls.gov/cgi-bin/surveymost?cu.

d. If your calculator has a quadratic and cubic regression feature, find the best fitting quadratic and cubic functions for the data.

e. Use a graphing calculator to plot the answers to parts b and d on the same window as the data. Discuss the extent to which any one of these functions models the data better than the others.

108. *Pace of Life* In an attempt to measure how the pace of city life is related to the size of the city, two researchers measured the mean speed of pedestrians in 15 cities by measuring the mean time it took them to walk 50 ft.*

a. Plot the original pairs of numbers. The pattern should be nonlinear.

b. Compute the coefficient of correlation for the data.

c. Plot y against $\log x$, using a calculator to compute $\log x$. Are the data more linear now than in part a?

d. Compute the coefficient of correlation for y against $\log x$. Is r closer to 1 than in part b?

e. Compute the least squares line for y against $\log x$.

City	Population (x)	Speed (ft/sec) (y)
Brno, Czechoslovakia	341,948	4.81
Prague, Czechoslovakia	1,092,759	5.88
Corte, France	5491	3.31
Bastia, France	49,375	4.90
Munich, Germany	1,340,000	5.62
Psychro, Crete	365	2.67
Itea, Greece	2500	2.27
Iráklion, Greece	78,200	3.85
Athens, Greece	867,023	5.21
Safed, Israel	14,000	3.70
Dimona, Israel	23,700	3.27
Netanya, Israel	70,700	4.31
Jerusalem, Israel	304,500	4.42
New Haven, Conn., U.S.A	138,000	4.39
Brooklyn, N.Y., U.S.A	2,602,000	5.05

Nature, Feb. 19, 1976, pp. 557–559.

EXTENDED APPLICATION: CHARACTERISTICS OF THE MONKEYFACE PRICKLEBACK*

The monkeyface prickleback (*Cebidichthys violaceus*), known to anglers as the monkeyface "eel," is found in rocky intertidal and subtidal habitats ranging from San Quintin Bay, Baja California, to Brookings, Oregon. Pricklebacks are prime targets of the few sports anglers who "poke pole" in the rocky intertidal zone at low tide. Little is known about the life history of this species. The results of a study of the length, weight, and age of this species are discussed in this case.

Data on standard length (*SL*) and total length (*TL*) were collected. Early in the study only *TL* was measured, so a conversion to *SL* was necessary. The equation relating the two lengths, calculated from 177 observations for which both lengths had been measured, is

$$SL = TL(.931) + 1.416.$$

Ages (determined by standard aging techniques) were used to estimate parameters of the von Bertanfany growth model

$$L_t = L_x(1 - e^{-kt}) \qquad (1)$$

where

L_t = length at age t,
L_x = asymptotic age of the species,
k = growth completion rate, and
t_0 = theoretical age at zero length.

The constants a and b in the model

$$W = aL^b \qquad (2)$$

where

W = weight in grams,
L = standard length in centimeters,

were determined using 139 fish ranging from 27 cm and 145 g to 60 cm and 195 g.

Growth curves giving length as a function of age are shown in Figure 58. For the data marked opercle, the lengths were computed from the ages using equation (1).

Estimated length from Equation (1) at a given age was larger for males than females after age 8. See the table. Weight/length relationships found with Equation (2) are shown in Figure 59, along with data from other studies.

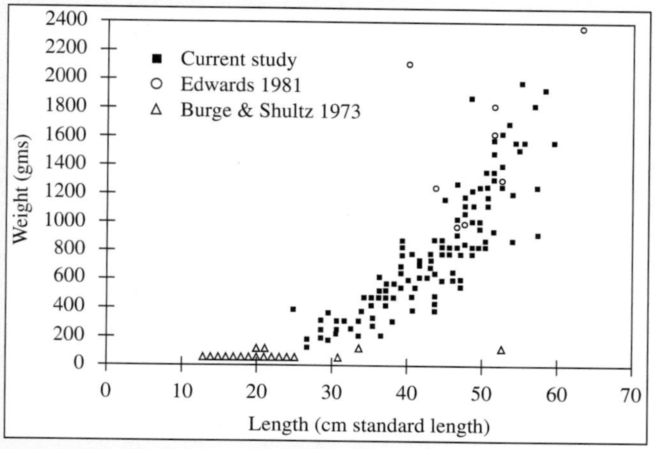

FIGURE 59

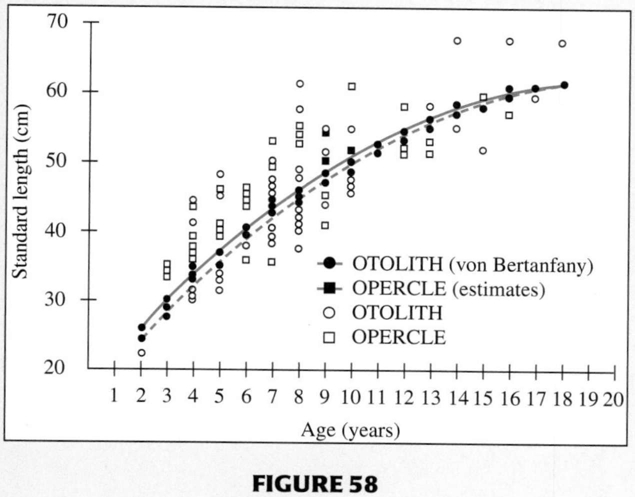

FIGURE 58

Structure/ Sex	Age (yr)	Length (cm)	L_x	k	t_0	n
Otolith						
Est.	2–18	23–67	72	.10	−1.89	91
S.D.			8	.03	1.08	
Opercle						
Est.	2–18	23–67	71	.10	−2.63	91
S.D.			8	.04	1.31	
Opercle-Females						
Est.	0–18	15–62	62	.14	−1.95	115
S.D.			2	.02	.28	
Opercle-Males						
Est.	0–18	13–67	70	.12	−1.91	74
S.D.			5	.02	.29	

*From Marshall, William H. and Tina Wyllie Echeverria, "Characteristics of the Monkeyface Prickleback," *California Fish & Game*, Vol. 78, No. 2, Spring 1992. Copyright 1992, American Association for the Advancement of Science.

Exercises

1. Use Equation (1) to estimate the lengths at ages 4, 11, and 17. Let $L_x = 71.5$ and $k = .1$. Compare your answers with the results in Figure 58. What do you find?
2. Use Equation (2) with $a = .01289$ and $b = 2.9$ to estimate the weights for lengths of 25 cm, 40 cm, and 60 cm. Compare with the results in Figure 59. Are your answers reasonable compared to the curve?

Directions for Group Project

Suppose that your group (3–5 students) is preparing an article for a fishing magazine. Your goal is to introduce and expose "poke poling" and pricklebacks to sports anglers. Use the information given above and the answers to Exercises 1 and 2 as a guide, along with other information you can find on the subject. The article should be professionally developed with pictures and charts.

CHAPTER

3 The Derivative

Controlling environmental pollution means controlling rates like kilograms of sulfur emitted per day by a power plant or micrograms per liter of pollutant flowing into a river. These rates are examples of derivatives, the subject of this chapter. Rates also govern the economics of pollution, as we will see in an exercise in Section 3 exploring the rate of increase in fines paid by polluters.

In earlier chapters we began to see how mathematics can be used to study a wide range of situations. We saw how to answer questions like the following:

- How much would 100 items cost?
- In how many years will the population reach 10,000?
- How wide is the arch 10 m from the ground?

In this chapter, we will build upon this knowledge by introducing calculus. With calculus, we will be able to solve problems that involve changing situations.

- How fast is the car moving after 20 seconds?
- At what rate is the population growing?
- What is the rate of change of profit when sales reach $1000?

Each of these problems can be solved by calculating a *derivative*. Since the definition of the derivative involves the idea of a *limit*, we begin with limits.

3.1 LIMITS

THINK ABOUT IT

What happens to the oxygen concentration in a pond over the long run?

In this section we will find an answer to this question using the concept of limit.

We can find the value of the function defined by

$$f(x) = \frac{x^2 - 4}{x - 2}$$

when $x = 1$ by substitution:

$$f(1) = \frac{1^2 - 4}{1 - 2} = 3.$$

FOR REVIEW

Evaluating function notation, such as $f(1) = 3$, was discussed in Section 1.2. Verify that for

$$f(x) = \frac{x^2 - 4}{x - 2},$$

$f(0) = 2$, $f(3) = 5$, and $f(-1) = 1$.

It is also true that when x is a number *very close* to 1 (on either side of 1), then $f(x)$ is a number *very close* to 3, as the following table shows.

x approaches 1 from the left. → ↓ ← x approaches 1 from the right.

x	.8	.9	.99	.9999	**1**	1.0000001	1.0001	1.001	1.01	1.05	1.1
$f(x)$	2.8	2.9	2.99	2.9999	**3**	3.0000001	3.0001	3.001	3.01	3.05	3.1

$f(x)$ approaches 3. → ↑ ← $f(x)$ approaches 3.

At $x = 2$ the situation is different: $f(2)$ is not defined because the denominator is 0 when $x = 2$. But we can still ask, what happens to $f(x)$ when x is a number *very close* to (but not equal to) 2? The table on the next page provides an answer.

				x approaches 2 from the left.	↓	*x approaches 2 from the right.*				
x	1.8	1.9	1.99	1.9999	**2**	2.0000001	2.00001	2.001	2.05	2.1
$f(x)$	3.8	3.9	3.99	3.9999	**4**	4.0000001	4.00001	4.001	4.05	4.1
				f(x) approaches 4.	↑	*f(x) approaches 4.*				

The table suggests that, as x gets closer and closer to 2 from either side, $f(x)$ gets closer and closer to 4. In fact, by experimenting with a calculator you can convince yourself that the values of $f(x)$ can be made as close as you want to 4 by taking values of x close enough to 2. In such a case, we say "the limit of $f(x)$ as x approaches 2 is 4," which is written as

$$\lim_{x\to 2} f(x) = 4.$$

In the first example, we found that

$$\lim_{x\to 1} f(x) = 3,$$

because the values of $f(x)$ got closer and closer to 3 as x got closer and closer to 1, *from either side* of 1.

The phrase "x approaches 1 from the left" is written $x \longrightarrow 1^-$. Similarly, "x approaches 1 from the right" is written $x \longrightarrow 1^+$. These expressions are used to write **one-sided limits.** The **limit from the left** (as x approaches 1 from the negative direction) is written

$$\lim_{x\to 1^-} f(x) = 3,$$

and the **limit from the right** (as x approaches 1 from the positive direction) is written

$$\lim_{x\to 1^+} f(x) = 3.$$

A **two-sided limit,** such as

$$\lim_{x\to 1} f(x) = 3,$$

exists only if both one-sided limits exist and are the same; that is, if $f(x)$ approaches the same number as x approaches a given number from *either* side.

The examples suggest the following informal definition.

LIMIT OF A FUNCTION

Let f be a function and let a and L be real numbers. If

1. as x takes values closer and closer (but not equal) to a on both sides of a, the corresponding values of $f(x)$ get closer and closer (and perhaps equal) to L; and
2. the value of $f(x)$ can be made as close to L as desired by taking values of x close enough to a;

then L is the **limit** of $f(x)$ as x approaches a, written

$$\lim_{x\to a} f(x) = L.$$

This definition is informal because the expressions "closer and closer to" and "as close as desired" have not been defined. A more formal definition would be needed to prove the rules for limits given later in this section.*

NOTE The definition of a limit describes what happens to $f(x)$ when x is near a. It is not affected by whether $f(a)$ is defined. Also, the definition implies that the function values cannot approach two different numbers, so that if a limit exists, it is unique. ■

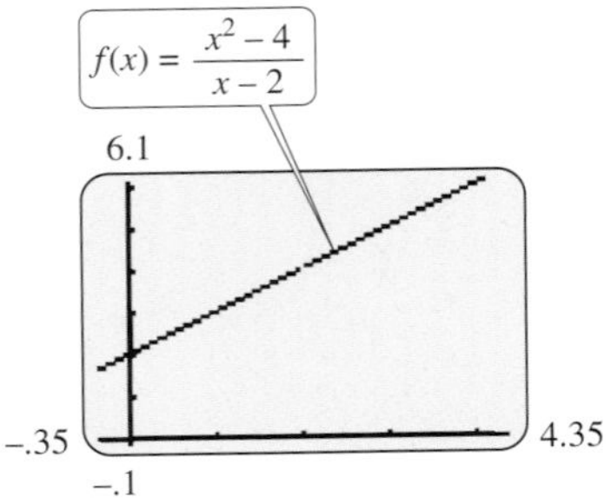

FIGURE 1

Figure 1 shows the graph of the function in the previous example drawn with a graphing calculator. Notice that the function has a small gap at the point $(2, 4)$, which agrees with our previous observation that the function is undefined at $x = 2$, where the limit is 4. (Due to the limitations of the graphing calculator, this gap may vanish when the viewing window is changed very slightly.) Furthermore, notice that for other values of x, the graph of the function appears to be a straight line. This is because when $x \neq 2$, the function can be simplified:

$$f(x) = \frac{x^2 - 4}{x - 2}$$

$$f(x) = \frac{(x-2)(x+2)}{x-2} \qquad \text{Using } a^2 - b^2 = (a-b)(a+b)$$

$$= x + 2.$$

The graph of $y = x + 2$ is a straight line, and it seems reasonable that as x gets closer and closer to 2, the value of $x + 2$ gets closer and closer to 4.

The previous discussion shows a second way of finding a limit: Use algebra to simplify the expression so the limit is easy to find.

A third way of finding a limit is to use the TRACE feature on a graphing calculator. For the example above, the result after pressing the TRACE key is shown in Figure 2. The cursor is already located at $x = 2$; if it were not, we could use the right or left arrow key to move the cursor there. The calculator does not give a y-value because the function is undefined at $x = 2$. Moving the cursor back a step gives $x = 1.95$, $y = 3.95$. Moving the cursor forward two steps gives $x = 2.05$, $y = 4.05$. It seems that as x approaches 2, y approaches 4, or at least something close to 4. Zooming in on the point $(2, 4)$ (such as using $[1.9, 2.1]$ by $[3.9, 4.1]$) allows the limit to be estimated more accurately and helps ensure that the graph has no unexpected behavior very close to $x = 2$.

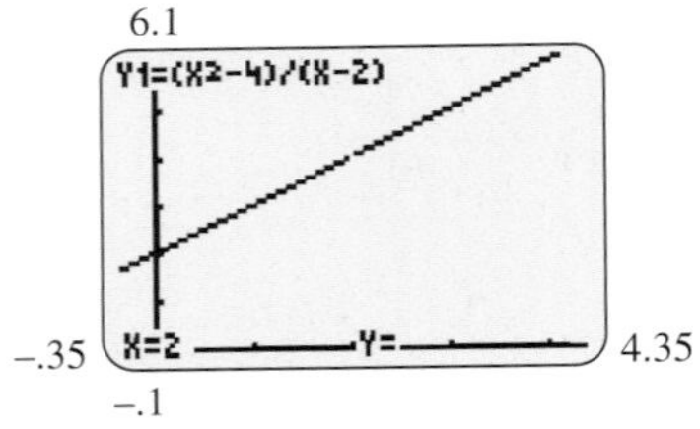

FIGURE 2

EXAMPLE 1 *Finding a Limit*

Find $\lim_{x \to 2} g(x)$, where $g(x) = \frac{x^2 + 4}{x - 2}$.

Solution Using the TABLE feature on a TI-83/84 Plus, we produce the table of numbers shown in Figure 3 on the next page, where Y_1 represents the function

*The limit is the key concept from which all the ideas of calculus flow. Calculus was independently discovered by the English mathematician Isaac Newton (1642–1727) and the German mathematician Gottfried Wilhelm Leibniz (1646–1716). For the next century, supporters of each accused the other of plagiarism, resulting in a lack of communication between mathematicians in England and on the European continent. Neither Newton nor Leibniz developed a mathematically rigorous definition of the limit (and we have no intention of doing so here). More than 100 years passed before the French mathematician Augustin-Louis Cauchy (1789–1857) accomplished this feat.

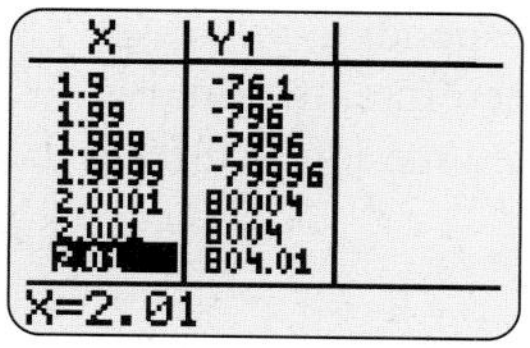

FIGURE 3

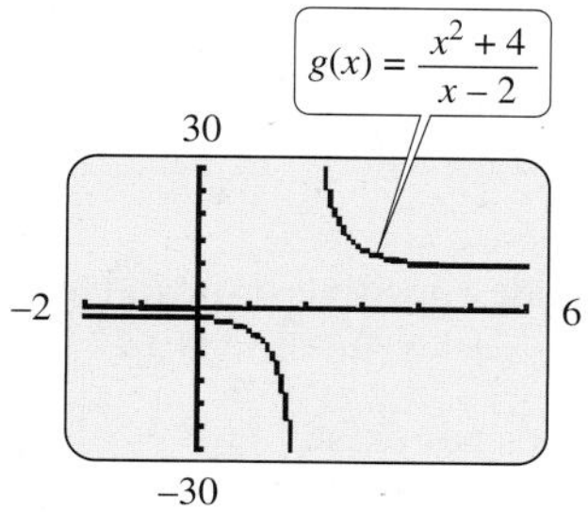

FIGURE 4

$g(x)$. Figure 4 shows the graph of the function. Both the table and the corresponding graph can be easily generated using a spreadsheet. Consult *The Spreadsheet Manual* that is available with this text for details.

Both the table and the graph suggest that as $x \to 2$ from the left, $g(x)$ gets more and more negative, becoming larger and larger in magnitude. This is indicated by writing

$$\lim_{x \to 2^-} g(x) = -\infty.$$

Because $-\infty$ is not a real number, the limit in this case does not exist. The symbol $-\infty$ simply indicates that as $x \to 2^-$, $g(x)$ becomes more and more negative without bound. Whenever we say that a limit has the value of ∞ or $-\infty$, this implies that the limit does not exist.

In the same way, the behavior of the function as $x \to 2$ from the right is indicated by writing

$$\lim_{x \to 2^+} g(x) = \infty.$$

Since there is no real number that $g(x)$ approaches as $x \to 2$ (from either side), nor does $g(x)$ approach either ∞ or $-\infty$, we simply say

$$\lim_{x \to 2} \frac{x^2 + 4}{x - 2} \text{ does not exist.}$$

We have shown three methods for determining limits: (1) using a table of numbers, (2) using algebraic simplification, and (3) tracing the graph on a graphing calculator. Which method you choose depends on the complexity of the function and the accuracy required by the application. Algebraic simplification gives the exact answer, but it can be difficult or even impossible to use in some situations. Calculating a table of numbers or tracing the graph may be easier when the function is complicated, but be careful, because the results could be inaccurate, inconclusive, or misleading. A graphing calculator does not tell us what happens between or beyond the points that are plotted.

EXAMPLE 2 *Finding a Limit*

Find $\lim_{x \to 0} \frac{|x|}{x}$.

Solution

Method 1: Algebraic Approach

The function $f(x) = |x|/x$ is not defined when $x = 0$. When $x > 0$, the definition of absolute value shows that $f(x) = |x|/x = x/x = 1$. When $x < 0$, then $|x| = -x$ and $f(x) = -x/x = -1$.

Method2: Graphing Calculator Approach

A calculator graph of f is shown in Figure 5 on the next page.

As x approaches 0 from the right, x is always positive and the corresponding value of $f(x)$ is 1, so

$$\lim_{x \to 0^+} f(x) = 1.$$

But as x approaches 0 from the left, x is always negative and the corresponding value of $f(x)$ is -1, so

$$\lim_{x \to 0^-} f(x) = -1.$$

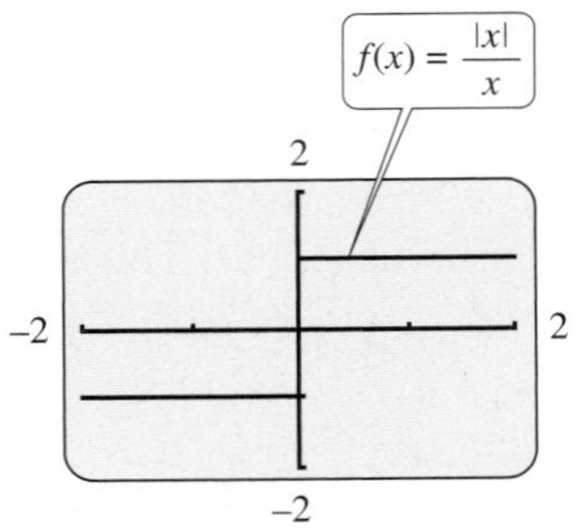

FIGURE 5

Thus, as x approaches 0 from either side, the corresponding values of $f(x)$ do not get closer and closer to a *single* real number. Therefore, the limit of $|x|/x$ as x approaches 0 does not exist because the limits from the left and from the right are not equal.

The discussion up to this point can be summarized as follows.

EXISTENCE OF LIMITS

The limit of f as x approaches a may not exist.

1. If $f(x)$ becomes infinitely large in magnitude (positive or negative) as x approaches the number a from either side, we write $\lim_{x\to a} f(x) = \infty$ or $\lim_{x\to a} f(x) = -\infty$. In either case, the limit does not exist.
2. If $f(x)$ becomes infinitely large in magnitude (positive) as x approaches a from one side and infinitely large in magnitude (negative) as x approaches a from the other side, then $\lim_{x\to a} f(x)$ does not exist.
3. If $\lim_{x\to a^-} f(x) = L$ and $\lim_{x\to a^+} f(x) = M$, and $L \neq M$, then $\lim_{x\to a} f(x)$ does not exist.

Figure 6 illustrates these three facts.

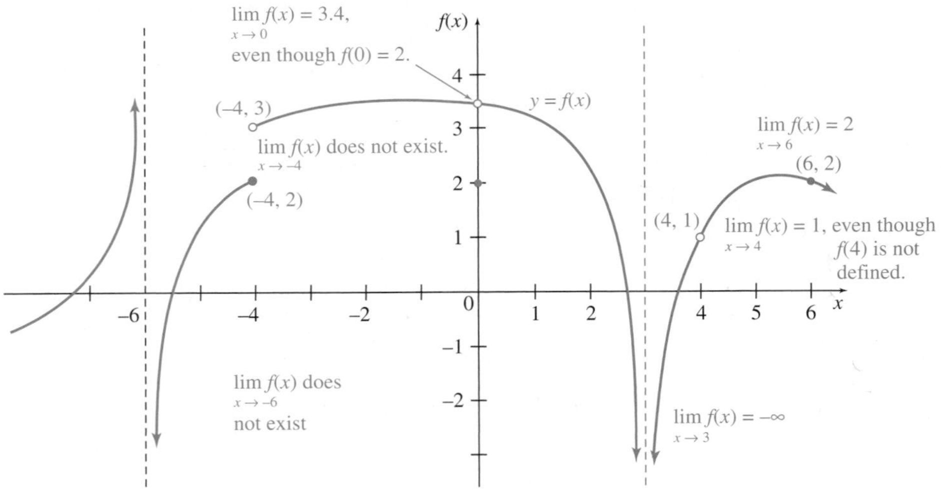

FIGURE 6

Rules for Limits As shown by the preceding examples, tables and graphs can be used to find limits. However, it is usually more efficient to use the rules for limits given on the next page. (Proofs of these rules require a formal definition of limit, which we have not given.)

RULES FOR LIMITS

Let a, A, and B be real numbers, and let f and g be functions such that

$$\lim_{x\to a} f(x) = A \qquad \text{and} \qquad \lim_{x\to a} g(x) = B.$$

1. If k is a constant, then $\lim_{x\to a} k = k$ and $\lim_{x\to a}[k \cdot f(x)] = k \cdot \lim_{x\to a} f(x) = k \cdot A$.
2. $\lim_{x\to a}[f(x) \pm g(x)] = \lim_{x\to a} f(x) \pm \lim_{x\to a} g(x) = A \pm B$
   (The limit of a sum or difference is the sum or difference of the limits.)
3. $\lim_{x\to a}[f(x) \cdot g(x)] = [\lim_{x\to a} f(x)] \cdot [\lim_{x\to a} g(x)] = A \cdot B$
   (The limit of a product is the product of the limits.)
4. $\lim_{x\to a} \dfrac{f(x)}{g(x)} = \dfrac{\lim_{x\to a} f(x)}{\lim_{x\to a} g(x)} = \dfrac{A}{B}$ if $B \neq 0$
   (The limit of a quotient is the quotient of the limits, provided the limit of the denominator is not zero.)
5. If $p(x)$ is a polynomial, then $\lim_{x\to a} p(x) = p(a)$.
6. For any real number k, $\lim_{x\to a}[f(x)]^k = [\lim_{x\to a} f(x)]^k = A^k$, provided this limit exists.*
7. $\lim_{x\to a} f(x) = \lim_{x\to a} g(x)$ if $f(x) = g(x)$ for all $x \neq a$.
8. For any real number $b > 0$, $\lim_{x\to a} b^{f(x)} = b^{[\lim_{x\to a} f(x)]} = b^A$.
9. For any real number b such that $0 < b < 1$ or $1 < b$,
   $\lim_{x\to a}[\log_b f(x)] = \log_b[\lim_{x\to a} f(x)] = \log_b A$ if $A > 0$.

This list may seem imposing, but these limit rules, once understood, agree with common sense. For example, Rule 3 says that if $f(x)$ becomes close to A as x approaches a, and if $g(x)$ becomes close to B, then $f(x) \cdot g(x)$ should become close to $A \cdot B$, which seems plausible.

EXAMPLE 3 *Rules for Limits*

Suppose $\lim_{x\to 2} f(x) = 3$ and $\lim_{x\to 2} g(x) = 4$. Use the limit rules to find the following limits.

(a) $\lim_{x\to 2}[f(x) + 5g(x)]$

Solution

$$\begin{aligned}\lim_{x\to 2}[f(x) + 5g(x)] &= \lim_{x\to 2} f(x) + \lim_{x\to 2} 5g(x) && \text{Rule 2}\\ &= \lim_{x\to 2} f(x) + 5\lim_{x\to 2} g(x) && \text{Rule 1}\\ &= 3 + 5(4)\\ &= 23\end{aligned}$$

*This limit does not exist, for example, when $A < 0$ and $k = 1/2$, or when $A = 0$ and $k \leq 0$.

(b) $\displaystyle\lim_{x \to 2} \frac{[f(x)]^2}{\ln g(x)}$

Solution

$$\lim_{x \to 2} \frac{[f(x)]^2}{\ln g(x)} = \frac{\lim_{x \to 2} [f(x)]^2}{\lim_{x \to 2} \ln g(x)} \qquad \text{Rule 4}$$

$$= \frac{[\lim_{x \to 2} f(x)]^2}{\ln[\lim_{x \to 2} g(x)]} \qquad \text{Rule 6 and Rule 9}$$

$$= \frac{3^2}{\ln 4}$$

$$\approx \frac{9}{1.38629} \approx 6.4921$$

EXAMPLE 4 *Finding a Limit*

Find $\displaystyle\lim_{x \to 3} \frac{x^2 - x - 1}{\sqrt{x + 1}}$.

Solution

$$\lim_{x \to 3} \frac{x^2 - x - 1}{\sqrt{x + 1}} = \frac{\lim_{x \to 3} (x^2 - x - 1)}{\lim_{x \to 3} \sqrt{x + 1}} \qquad \text{Rule 4}$$

$$= \frac{3^2 - 3 - 1}{\sqrt{\lim_{x \to 3} (x + 1)}} \qquad \text{Rule 5 and Rule 6 } (\sqrt{a} = a^{1/2})$$

$$= \frac{5}{\sqrt{4}} \qquad \text{Rule 5}$$

$$= \frac{5}{2}$$

As Examples 3 and 4 suggest, the rules for limits actually mean that many limits can be found simply by evaluation. This process is valid for polynomials, rational functions, exponential functions, logarithmic functions, and roots and powers, so long as this does not involve an illegal operation, such as division by 0 or taking the logarithm of a negative number. Division by 0 presents particular problems that can often be solved by algebraic simplification, as the following example shows.

EXAMPLE 5 *Finding a Limit*

Find $\displaystyle\lim_{x \to 2} \frac{x^2 + x - 6}{x - 2}$.

Solution Rule 4 cannot be used here, since

$$\lim_{x \to 2} (x - 2) = 0.$$

The numerator also approaches 0 as x approaches 2, and $0/0$ is meaningless. For $x \neq 2$, we can, however, simplify the function by rewriting the fraction as

$$\frac{x^2 + x - 6}{x - 2} = \frac{(x + 3)(x - 2)}{x - 2} = x + 3.$$

Now Rule 7 can be used.

$$\lim_{x \to 2} \frac{x^2 + x - 6}{x - 2} = \lim_{x \to 2}(x + 3) = 2 + 3 = 5$$

EXAMPLE 6 *Finding a Limit*

Find $\lim_{x \to 4} \frac{\sqrt{x} - 2}{x - 4}$.

Solution As $x \to 4$, the numerator approaches 0 and the denominator also approaches 0, giving the meaningless expression 0/0. Algebra can be used to rationalize the numerator by multiplying both the numerator and the denominator by $\sqrt{x} + 2$. This gives

$$\frac{\sqrt{x} - 2}{x - 4} \cdot \frac{\sqrt{x} + 2}{\sqrt{x} + 2} = \frac{(\sqrt{x})^2 - 2^2}{(x - 4)(\sqrt{x} + 2)}$$

$(a - b)(a + b) = a^2 - b^2$

$$= \frac{x - 4}{(x - 4)(\sqrt{x} + 2)} = \frac{1}{\sqrt{x} + 2}$$

if $x \neq 4$. Now use the rules for limits.

$$\lim_{x \to 4} \frac{\sqrt{x} - 2}{x - 4} = \lim_{x \to 4} \frac{1}{\sqrt{x} + 2} = \frac{1}{\sqrt{4} + 2} = \frac{1}{2 + 2} = \frac{1}{4}$$

You can support this result by using a graphing calculator to make a table of values as we did in Example 1, or by graphing the function and using the TRACE feature to investigate that near $x = 4$, y is close to .25, although at $x = 4$, the function is undefined.

Examples 5 and 6 suggest the following principle: **To calculate the limit of $f(x)/g(x)$ as x approaches a, where $f(a) = g(a) = 0$, you should attempt to factor $x - a$ from both the numerator and the denominator.**

CAUTION Simply because the expression in a limit is approaching 0/0, as in Examples 5 and 6, does *not* mean that the limit is 0 or that the limit does not exist. For such a limit, try to simplify the expression using algebra. ■

EXAMPLE 7 *Finding a Limit*

Find $\lim_{x \to 1} \frac{x + 1}{x^2 - 1}$.

Solution

Method 1: Algebraic Approach

Again, Rule 4 cannot be used since $\lim_{x \to 1} x^2 - 1 = 0$. If $x \neq 1$, the function can be rewritten as

$$\frac{x + 1}{x^2 - 1} = \frac{x + 1}{(x + 1)(x - 1)} = \frac{1}{x - 1}.$$

Then

$$\lim_{x\to1}\frac{x+1}{x^2-1}=\lim_{x\to1}\frac{1}{x-1}$$

by Rule 7. None of the rules can be used to find

$$\lim_{x\to1}\frac{1}{x-1},$$

but as x approaches 1, the denominator approaches 0 while the numerator stays at 1, making the result larger and larger in magnitude. If $x > 1$, both the numerator and denominator are positive, so $\lim_{x\to1^+} 1/(x-1)=\infty$. If $x < 1$, the denominator is negative, so $\lim_{x\to1^-} 1/(x-1)=-\infty$. Therefore,

$$\lim_{x\to1}\frac{1}{x-1}\text{ does not exist.}$$

Another way to understand the behavior of this function near $x = 1$ is to recall from the section on Polynomial and Rational Functions that a rational function often has a vertical asymptote at a value of x where the denominator is 0, although it may not if the numerator there is also 0. In this example, we see after simplifying that the function has a vertical asymptote at $x = 1$ because that would make the denominator of $1/(x-1)$ equal to 0, while the numerator is 1.

Method 2: Graphing Calculator Approach

Figure 7 shows a graphing calculator view of $y = 1/(x-1)$ on $[0, 2]$ by $[-10, 10]$. The behavior of the function indicates a vertical asymptote at $x = 1$, with the limit approaching $-\infty$ from one side and ∞ from the other, so

$$\lim_{x\to1}\frac{1}{x-1}\text{ does not exist.}$$

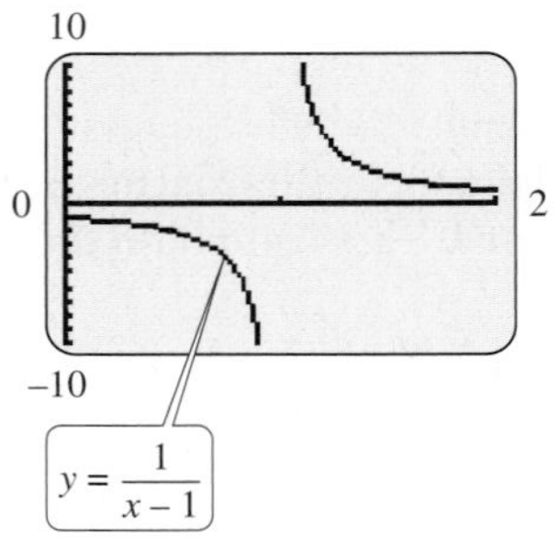

FIGURE 7

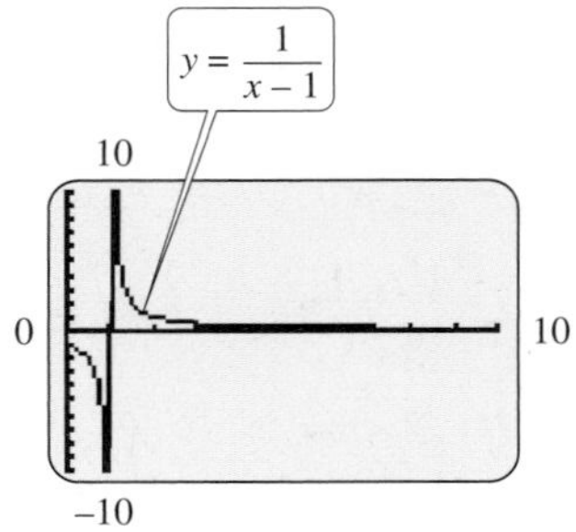

FIGURE 8

CAUTION A graphing calculator can give a deceptive view of a function. Figure 8 shows the result if we graph the previous function on $[0, 10]$ by $[-10, 10]$. Near $x = 1$, the graph appears to be a steep line connecting the two pieces. The graph in Figure 7 is more representative of the function near $x = 1$. When using a graphing calculator, you may need to experiment with the viewing window, guided by what you have learned about functions and limits, to get a good picture of a function. On many calculators, extraneous lines connecting parts of the graph can be avoided by using Dot mode rather than Connected mode. ■

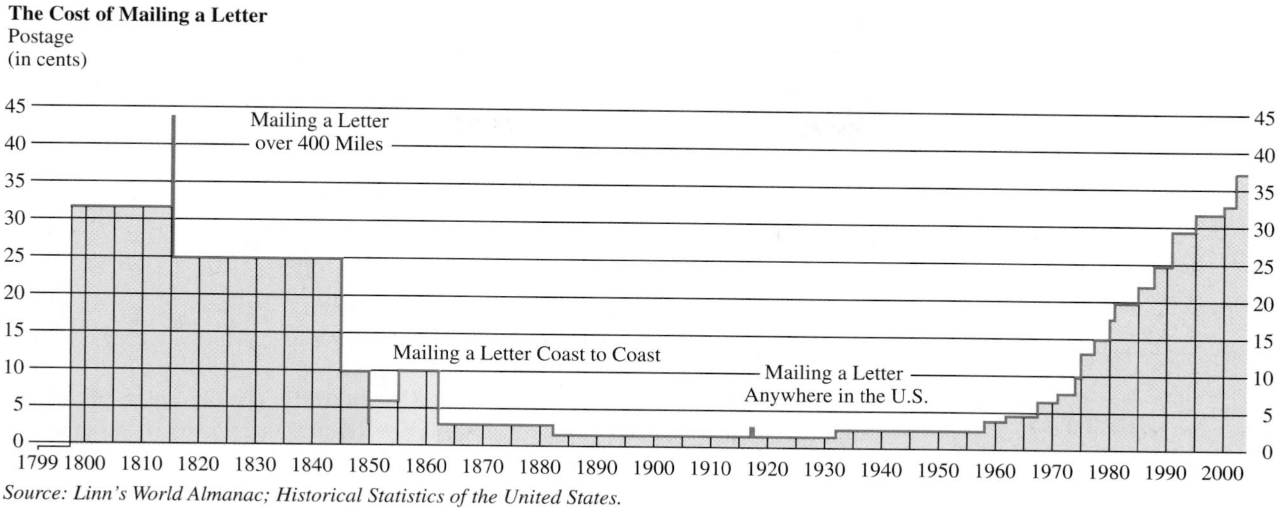

FIGURE 9

EXAMPLE 8 *Postage*

Figure 9 shows how the postage required to mail a letter in the United States has changed with time.*

Let $C(t)$ be the cost of a letter in the year t. Consider the behavior of the function near $t = 2002$, when postage jumped from 34¢ to 37¢. The graph as drawn does not pass the vertical line test, so we have redrawn this section of the graph in Figure 10 using the notation in this text. Notice that $\lim_{t \to 2002^-} C(t) = 34$ and $\lim_{t \to 2002^+} C(t) = 37$, so $\lim_{t \to 2002} C(t)$ does not exist. Notice also that $C(2002)$ does exist and is equal to 37.

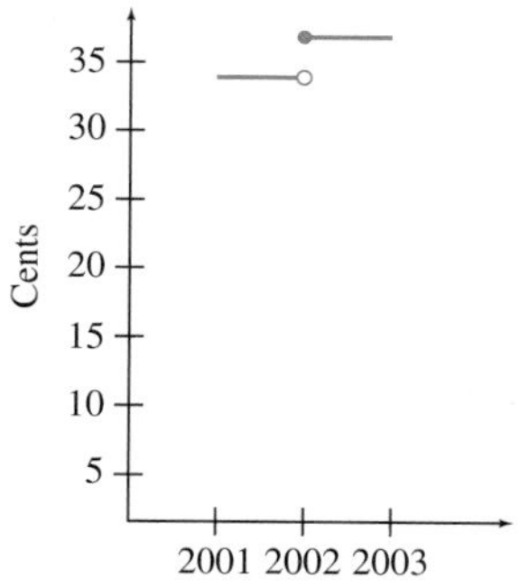

FIGURE 10

Limits at Infinity Sometimes it is useful to examine the behavior of the values of $f(x)$ as x gets larger and larger (or smaller and smaller). For example, suppose a small pond normally contains 12 units of dissolved oxygen in a fixed volume of water. Suppose also that at time $t = 0$ a quantity of organic waste is introduced into the pond, with the oxygen concentration t weeks later given by

$$f(t) = \frac{12t^2 - 15t + 12}{t^2 + 1}.$$

As time goes on, what will be the ultimate concentration of oxygen? Will it return to 12 units?

After 2 weeks, the pond contains

$$f(2) = \frac{12 \cdot 2^2 - 15 \cdot 2 + 12}{2^2 + 1} = \frac{30}{5} = 6$$

units of oxygen, and after 4 weeks, it contains

$$f(4) = \frac{12 \cdot 4^2 - 15 \cdot 4 + 12}{4^2 + 1} \approx 8.5$$

**The New York Times,* March 13, 1994, p. 2, updated. Reprinted with permission.

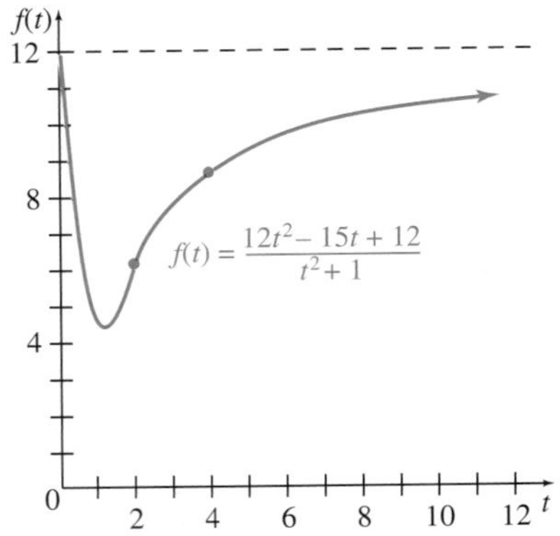

FIGURE 11

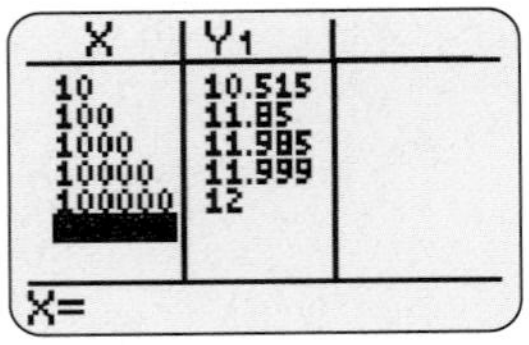

FIGURE 12

> **FOR REVIEW**
> In the section on Polynomial and Rational Functions, we saw a way to find horizontal asymptotes by considering the behavior of the function as x (or t) gets large. For large t, $12t^2 - 15t + 12 \approx 12t^2$, because the t-term and the constant term are small compared with the t^2-term when t is large. Similarly, $t^2 + 1 \approx t^2$. Thus, for large t, $f(t) = \dfrac{12t^2 - 15t + 12}{t^2 + 1} \approx \dfrac{12t^2}{t^2} = 12$. Thus the function f has a horizontal asymptote at $y = 12$.

units. Choosing several values of t and finding the corresponding values of $f(t)$, or using a graphing calculator or computer, leads to the graph in Figure 11.

The graph suggests that, as time goes on, the oxygen level gets closer and closer to the original 12 units. If so, the line $y = 12$ is a horizontal asymptote. We can use the TABLE feature on a graphing calculator to investigate the behavior for large values of t. Figure 12 shows the result using a TI-83/84 Plus.

The table suggests that

$$\lim_{t \to \infty} f(t) = 12,$$

where $t \to \infty$ means that t increases without bound. (Similarly, $t \to -\infty$ means that t *decreases* without bound; that is, t becomes more and more negative.) Thus, the oxygen concentration will approach 12, but it will never be *exactly* 12.

The preceding example illustrates a **limit at infinity.** The phrase "t approaches infinity" (symbolically, $t \to \infty$) is simply convenient shorthand to express the fact that t becomes larger and larger without bound. Similarly, the phrase "t approaches negative infinity" (symbolically, $t \to -\infty$) means that t becomes more and more negative without bound (such as -10, -1000, $-10{,}000$, etc.).

As we saw in the previous example, limits at infinity or negative infinity, if they exist, correspond to horizontal asymptotes of the graph of the function. In the previous chapter, we saw one way to find horizontal asymptotes. We will now show a more precise way, based upon some simple limits at infinity. The graphs of $f(x) = 1/x$ (in red) and $g(x) = 1/x^2$ (in blue) shown in Figure 13, as well as the table there, indicate that $\lim_{x \to \infty} 1/x = 0$, $\lim_{x \to -\infty} 1/x = 0$, $\lim_{x \to \infty} 1/x^2 = 0$, and $\lim_{x \to -\infty} 1/x^2 = 0$, suggesting the following rule.

x	$\frac{1}{x}$	$\frac{1}{x^2}$
-100	$-.01$	.0001
-10	$-.1$	.01
-1	-1	1
1	1	1
10	.1	.01
100	.01	.0001

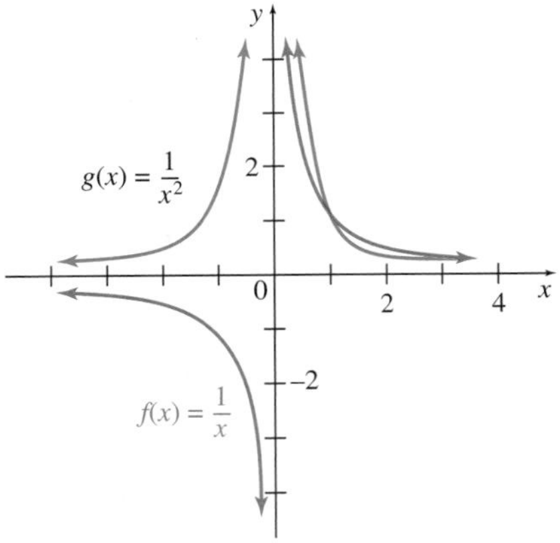

FIGURE 13

> **LIMITS AT INFINITY**
> For any positive real number n,
>
> $$\lim_{x \to \infty} \frac{1}{x^n} = 0 \quad \text{and} \quad \lim_{x \to -\infty} \frac{1}{x^n} = 0.^*$$

*If x is negative, x^n does not exist for certain values of n, so the second limit is undefined.

The rules for limits given earlier remain unchanged when a is replaced with ∞ or $-\infty$.

To evaluate the limit at infinity of a rational function, divide the numerator and denominator by the largest power of the variable that appears in the denominator, t^2 here, and then use these results. In the previous example, we find that

$$\lim_{t\to\infty}\frac{12t^2-15t+12}{t^2+1}=\lim_{t\to\infty}\frac{\frac{12t^2}{t^2}-\frac{15t}{t^2}+\frac{12}{t^2}}{\frac{t^2}{t^2}+\frac{1}{t^2}}$$

$$=\lim_{t\to\infty}\frac{12-15\cdot\frac{1}{t}+12\cdot\frac{1}{t^2}}{1+\frac{1}{t^2}}.$$

Now apply the limit rules and the fact that $\lim_{t\to\infty} 1/t^n = 0$.

$$\frac{\lim_{t\to\infty}\left(12-15\cdot\frac{1}{t}+12\cdot\frac{1}{t^2}\right)}{\lim_{t\to\infty}\left(1+\frac{1}{t^2}\right)}$$

$$=\frac{\lim_{t\to\infty}12-\lim_{t\to\infty}15\cdot\frac{1}{t}+\lim_{t\to\infty}12\cdot\frac{1}{t^2}}{\lim_{t\to\infty}1+\lim_{t\to\infty}\frac{1}{t^2}} \qquad \text{Rules 4 and 2}$$

$$=\frac{12-15\left(\lim_{t\to\infty}\frac{1}{t}\right)+12\left(\lim_{t\to\infty}\frac{1}{t^2}\right)}{1+\lim_{t\to\infty}\frac{1}{t^2}} \qquad \text{Rule 1}$$

$$=\frac{12-15\cdot 0+12\cdot 0}{1+0}=12. \qquad \text{Limits at infinity}$$

EXAMPLE 9 *Limits at Infinity*

Find each limit.

(a) $\lim_{x\to\infty}\frac{8x+6}{3x-1}$

Solution We can use the rule $\lim_{x\to\infty} 1/x^n = 0$ to find this limit by first dividing numerator and denominator by x, as follows.

$$\lim_{x\to\infty}\frac{8x+6}{3x-1}=\lim_{x\to\infty}\frac{\frac{8x}{x}+\frac{6}{x}}{\frac{3x}{x}-\frac{1}{x}}=\lim_{x\to\infty}\frac{8+6\cdot\frac{1}{x}}{3-\frac{1}{x}}=\frac{8+0}{3-0}=\frac{8}{3}$$

(b) $$\lim_{x\to\infty}\frac{3x+2}{4x^3-1}=\lim_{x\to\infty}\frac{3\cdot\frac{1}{x^2}+2\cdot\frac{1}{x^3}}{4-\frac{1}{x^3}}=\frac{0+0}{4-0}=\frac{0}{4}=0$$

Here, the highest power of x in the denominator is x^3, which is used to divide each term in the numerator and denominator.

(c) $$\lim_{x\to\infty} \frac{3x^2 + 2}{4x - 3} = \lim_{x\to\infty} \frac{3x + \frac{2}{x}}{4 - \frac{3}{x}}$$

The highest power of x in the denominator is x (to the first power). There is a higher power of x in the numerator, but we don't divide by this. Notice that the denominator approaches 4, while the numerator becomes infinitely large, so

$$\lim_{x\to\infty} \frac{3x^2 + 2}{4x - 3} = \infty.$$

(d) $$\lim_{x\to\infty} \frac{5x^2 - 4x^3}{3x^2 + 2x - 1} = \lim_{x\to\infty} \frac{3 - 4x}{3 + \frac{2}{x} - \frac{1}{x}}$$

The highest power of x in the denominator is x^2. The denominator approaches 3, while the numerator becomes a negative number that is larger and larger in magnitude, so

$$\lim_{x\to\infty} \frac{5x^2 - 4x^3}{3x^2 + 2x - 1} = -\infty.$$

The method used in Example 9 is a useful way to rewrite expressions with fractions so that the rules for limits at infinity can be used.

FINDING LIMITS AT INFINITY

If $f(x) = p(x)/q(x)$, for polynomials $p(x)$ and $q(x)$, $q(x) \neq 0$, $\lim_{x\to-\infty} f(x)$ and $\lim_{x\to\infty} f(x)$ can be found as follows.

1. Divide $p(x)$ and $q(x)$ by the highest power of x in $q(x)$.
2. Use the rules for limits, including the rules for limits at infinity,

$$\lim_{x\to\infty} \frac{1}{x^n} = 0 \quad \text{and} \quad \lim_{x\to-\infty} \frac{1}{x^n} = 0,$$

to find the limit of the result from step 1.

3.1 EXERCISES

In Exercises 1–4, choose the best answer for each limit.

1. If $\lim_{x\to 2^-} f(x) = 5$ and $\lim_{x\to 2^+} f(x) = 6$, then $\lim_{x\to 2} f(x)$

a. is 5.

b. is 6.

c. does not exist.

d. is infinite.

2. If $\lim_{x\to 2^-} f(x) = \lim_{x\to 2^+} f(x) = -1$, but $f(2) = 1$, then $\lim_{x\to 2} f(x)$

a. is -1.

b. does not exist.

c. is infinite.

d. is 1.

3. If $\lim_{x\to 4^-} f(x) = \lim_{x\to 4^+} f(x) = 6$, but $f(4)$ does not exist, then $\lim_{x\to 4} f(x)$

a. does not exist.

b. is 6.

c. is $-\infty$.

d. is ∞.

4. If $\lim_{x\to 1^-} f(x) = -\infty$ and $\lim_{x\to 1^+} f(x) = -\infty$, then $\lim_{x\to 1} f(x)$

a. is ∞.

b. is $-\infty$.

c. does not exist.

d. is 1.

Decide whether each limit exists. If a limit exists, find its value.

5. $\lim_{x\to 3} f(x)$

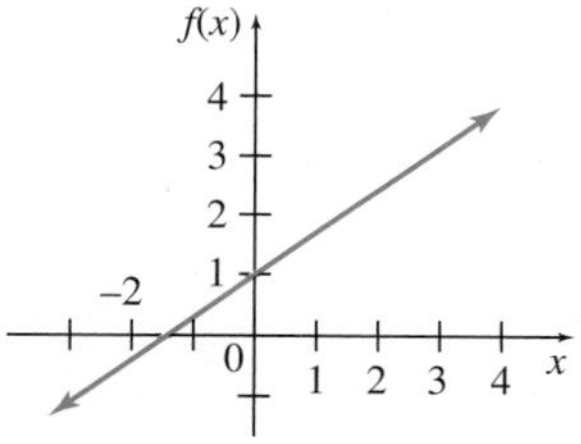

6. $\lim_{x\to 2} F(x)$

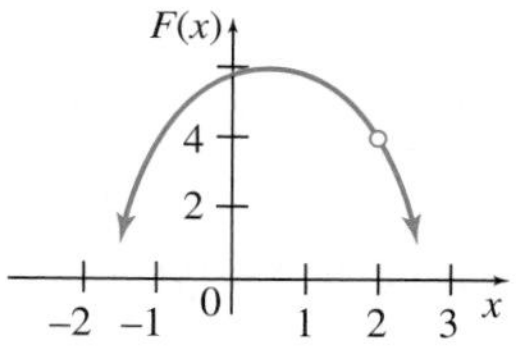

7. $\lim_{x\to 0} f(x)$

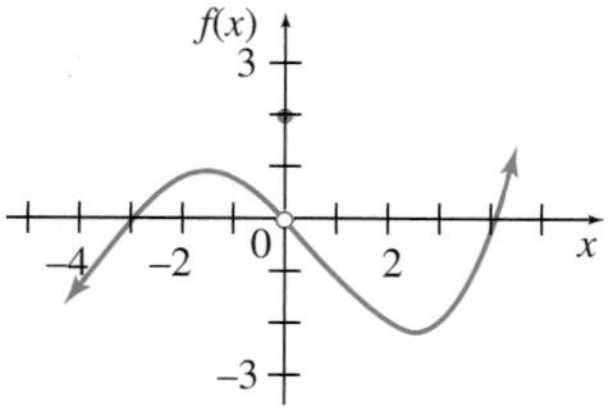

8. $\lim_{x\to 3} g(x)$

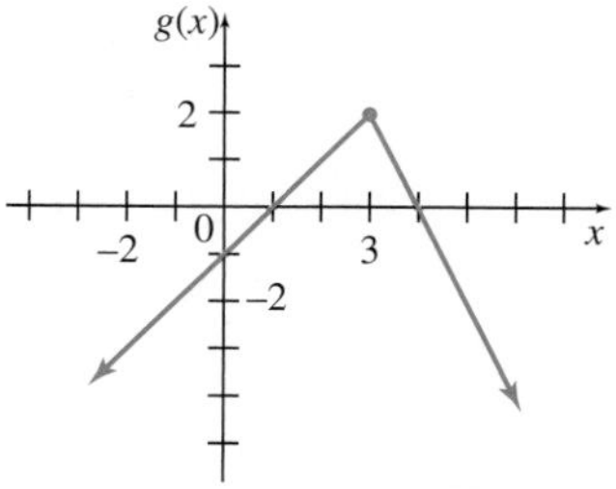

In Exercises 9 and 10, use the graph to find **(i)** $\lim_{x\to a^-} f(x)$, **(ii)** $\lim_{x\to a^+} f(x)$, **(iii)** $\lim_{x\to a} f(x)$, *and* **(iv)** $f(a)$ *if it exists.*

9. a. $a = -2$ **b.** $a = -1$

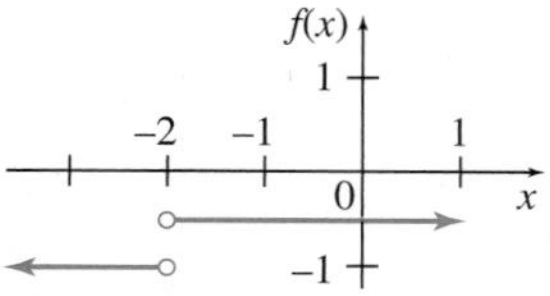

10. a. $a = 1$ **b.** $a = 2$

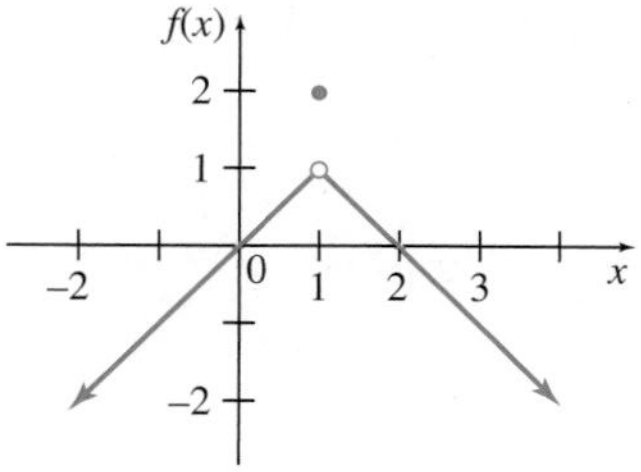

Decide whether each limit exists. If a limit exists, find its value.

11. $\lim_{x \to \infty} f(x)$

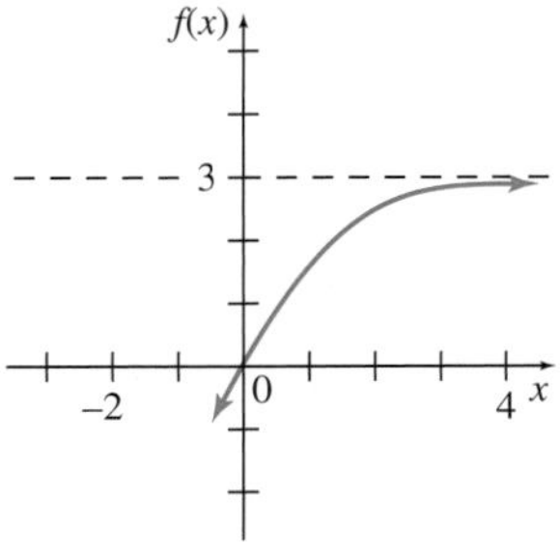

12. $\lim_{x \to -\infty} g(x)$

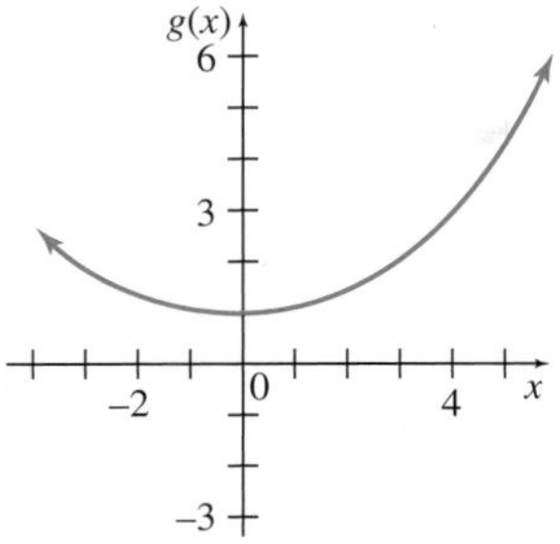

13. Explain why $\lim_{x \to 2} F(x)$ in Exercise 6 exists, but $\lim_{x \to -2} f(x)$ in Exercise 9 does not.

14. In Exercise 10, why does $\lim_{x \to 1} f(x) = 1$, even though $f(1) = 2$?

15. Use the table of values to estimate $\lim_{x \to 1} f(x)$.

x	.9	.99	.999	.9999	1.0001	1.001	1.01	1.1
$f(x)$	1.9	1.99	1.999	1.9999	2.0001	2.001	2.01	2.1

Complete the tables and use the results to find the indicated limits.

16. If $f(x) = 2x^2 - 4x + 3$, find $\lim_{x \to 1} f(x)$.

x	.9	.99	.999	1.001	1.01	1.1
$f(x)$			1.000002	1.000002		

17. If $k(x) = \dfrac{x^3 - 2x - 4}{x - 2}$, find $\lim_{x \to 2} k(x)$.

x	1.9	1.99	1.999	2.001	2.01	2.1
$k(x)$						

18. If $f(x) = \dfrac{2x^3 + 3x^2 - 4x - 5}{x + 1}$, find $\lim_{x \to -1} f(x)$.

x	−1.1	−1.01	−1.001	−.999	−.99	−.9
$f(x)$						

19. If $h(x) = \dfrac{\sqrt{x} - 2}{x - 1}$, find $\lim_{x \to 1} h(x)$.

x	.9	.99	.999	1.001	1.01	1.1
$h(x)$						

20. If $f(x) = \dfrac{\sqrt{x} - 3}{x - 3}$, find $\lim_{x \to 3} f(x)$.

x	2.9	2.99	2.999	3.001	3.01	3.1
$f(x)$						

Let $\lim_{x \to 4} f(x) = 16$ *and* $\lim_{x \to 4} g(x) = 8$. *Use the limit rules to find each limit.*

21. $\lim_{x \to 4} [f(x) - g(x)]$

22. $\lim_{x \to 4} [g(x) \cdot f(x)]$

23. $\lim_{x \to 4} \dfrac{f(x)}{g(x)}$

24. $\lim_{x \to 4} \log_2 f(x)$

25. $\lim_{x \to 4} \sqrt{f(x)}$

26. $\lim_{x \to 4} \sqrt[3]{g(x)}$

27. $\lim_{x \to 4} 2^{g(x)}$

28. $\lim_{x \to 4} [1 + f(x)]^2$

29. $\lim_{x \to 4} \dfrac{f(x) + g(x)}{2g(x)}$

30. $\lim_{x \to 4} \dfrac{5g(x) + 2}{1 - f(x)}$

Use the properties of limits to help decide whether each limit exists. If a limit exists, find its value.

31. $\lim_{x\to 3} \frac{x^2 - 9}{x - 3}$

32. $\lim_{x\to -2} \frac{x^2 - 4}{x + 2}$

33. $\lim_{x\to 1} \frac{5x^2 - 7x + 2}{x^2 - 1}$

34. $\lim_{x\to -3} \frac{x^2 - 9}{x^2 + x - 6}$

35. $\lim_{x\to -2} \frac{x^2 - x - 6}{x + 2}$

36. $\lim_{x\to 5} \frac{x^2 - 3x - 10}{x - 5}$

37. $\lim_{x\to 0} \frac{[1/(x + 3)] - 1/3}{x}$

38. $\lim_{x\to 0} \frac{[-1/(x + 2)] + 1/2}{x}$

39. $\lim_{x\to 25} \frac{\sqrt{x} - 5}{x - 25}$

40. $\lim_{x\to 36} \frac{\sqrt{x} - 6}{x - 36}$

41. $\lim_{h\to 0} \frac{(x + h)^2 - x^2}{h}$

42. $\lim_{h\to 0} \frac{(x + h)^3 - x^3}{h}$

43. $\lim_{x\to \infty} \frac{3x}{5x - 1}$

44. $\lim_{x\to -\infty} \frac{8x + 2}{2x - 5}$

45. $\lim_{x\to \infty} \frac{x^2 + 2x}{2x^2 - 2x + 1}$

46. $\lim_{x\to \infty} \frac{x^2 + 2x - 5}{3x^2 + 2}$

47. $\lim_{x\to \infty} \frac{3x^3 + 2x - 1}{2x^4 - 3x^3 - 2}$

48. $\lim_{x\to \infty} \frac{2x^2 - 1}{3x^4 + 2}$

49. $\lim_{x\to \infty} \frac{2x^3 - x - 3}{6x^2 - x - 1}$

50. $\lim_{x\to \infty} \frac{x^4 - x^3 - 3x}{7x^2 + 9}$

51. $\lim_{x\to \infty} \frac{2x^2 - 7x^4}{4x^2 + 5x - 6}$

52. $\lim_{x\to \infty} \frac{-5x^3 - 4x^2 + 8}{6x^2 + 3x + 2}$

53. Let $F(x) = \frac{3x}{(x + 2)^3}$.

a. Find $\lim_{x\to -2} F(x)$.

b. Find the vertical asymptote of the graph of $F(x)$.

c. Compare your answers for parts a and b. What can you conclude?

54. Let $G(x) = \frac{-6}{(x - 4)^2}$.

a. Find $\lim_{x\to 4} G(x)$.

b. Find the vertical asymptote of the graph of $G(x)$.

c. Compare your answers for parts a and b. Are they related? How?

55. How can you tell that the graph in Figure 7 is more representative of the function $f(x) = 1/(x - 1)$ than the graph in Figure 8?

56. A friend who is confused about limits wonders why you investigate the value of a function closer and closer to a point, instead of just finding the value of a function at the point. How would you respond?

57. Use a graph of $f(x) = e^x$ to answer the following questions.

a. Find $\lim_{x\to -\infty} e^x$.

b. Where does the function e^x have a horizontal asymptote?

58. Use a graphing calculator to answer the following questions.

a. From a graph of $y = xe^{-x}$, what do you think is the value of $\lim_{x\to\infty} xe^{-x}$? Support this by evaluating the function for several large values of x.

b. Repeat part a, this time using the graph of $y = x^2e^{-x}$.

c. Based on your results from parts a and b, what do you think is the value of $\lim_{x\to\infty} x^n e^{-x}$, where n is a positive integer? Support this by experimenting with other positive integers n.

59. Use a graph of $f(x) = \ln x$ to answer the following questions.

a. Find $\lim_{x \to 0^+} \ln x$.

b. Where does the function $\ln x$ have a vertical asymptote?

60. Use a graphing calculator to answer the following questions.

a. From a graph of $y = x \ln x$, what do you think is the value of $\lim_{x \to 0^+} x \ln x$? Support this by evaluating the function for several small values of x.

b. Repeat part a, this time using the graph of $y = x(\ln x)^2$.

c. Based on your results from parts a and b, what do you think is the value of $\lim_{x \to 0^+} x(\ln x)^n$, where n is a positive integer? Support this by experimenting with other positive integers n.

61. Explain in your own words why the rules for limits at infinity should be true.

62. Explain in your own words what Rule 4 for limits means.

Find each of the following limits **(a)** *by investigating values of the function near the x-value where the limit is taken, and* **(b)** *using a graphing calculator to view the function near that value of x.*

63. $\lim_{x \to 1} \dfrac{x^4 + 4x^3 - 9x^2 + 7x - 3}{x - 1}$

64. $\lim_{x \to 2} \dfrac{x^4 + x - 18}{x^2 - 4}$

65. $\lim_{x \to -1} \dfrac{x^{1/3} + 1}{x + 1}$

66. $\lim_{x \to 4} \dfrac{x^{3/2} - 8}{x + x^{1/2} - 6}$

Use a graphing calculator to graph the function. **(a)** *Determine the limit from the graph.* **(b)** *Explain how your answer could be determined from the expression for $f(x)$.*

67. $\lim_{x \to \infty} \dfrac{\sqrt{9x^2 + 5}}{2x}$

68. $\lim_{x \to -\infty} \dfrac{\sqrt{9x^2 + 5}}{2x}$

69. $\lim_{x \to -\infty} \dfrac{\sqrt{36x^2 + 2x + 7}}{3x}$

70. $\lim_{x \to \infty} \dfrac{\sqrt{36x^2 + 2x + 7}}{3x}$

71. $\lim_{x \to \infty} \dfrac{(1 + 5x^{1/3} + 2x^{5/3})^3}{x^5}$

72. $\lim_{x \to -\infty} \dfrac{(1 + 5x^{1/3} + 2x^{5/3})^3}{x^5}$

73. Explain why the following rules can be used to find $\lim_{x \to \infty} p(x)/q(x)$:

a. If the degree of $p(x)$ is less than the degree of $q(x)$, the limit is 0.

b. If the degree of $p(x)$ is equal to the degree of $q(x)$, the limit is A/B, where A and B are the leading coefficients of $p(x)$ and $q(x)$, respectively.

c. If the degree of $p(x)$ is greater than the degree of $q(x)$, the limit is ∞ or $-\infty$.

Applications

BUSINESS AND ECONOMICS

74. *Consumer Demand* When the price of an essential commodity (such as gasoline) rises rapidly, consumption drops slowly at first. If the price continues to rise, however, a "tipping" point may be reached, at which consumption takes a sudden substantial drop. Suppose the accompanying graph shows the consumption of gasoline, $G(t)$, in millions of gallons, in a certain area. We assume that the price is rising rapidly. Here t is time in months after the price began rising. Use the graph to find the following.

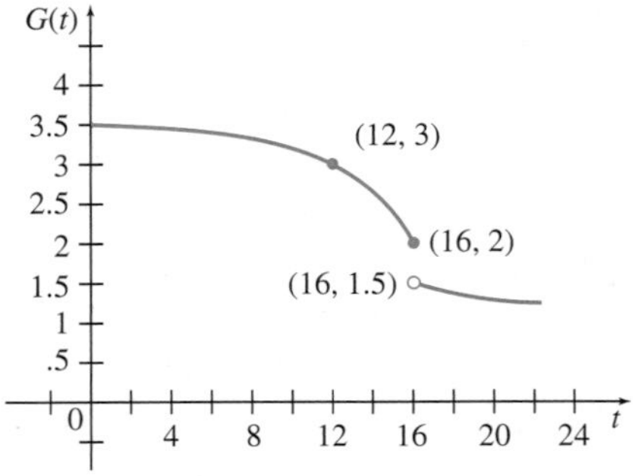

a. $\lim_{t \to 12} G(t)$

b. $\lim_{t \to 16} G(t)$

c. $G(16)$

d. The tipping point (in months)

75. *Sales Tax* Officials in California tend to raise the sales tax in years in which the state faces a budget deficit, and then cut the tax when the state has a surplus. The graph below shows the California state sales tax since it was first established in 1933. Let $T(x)$ represent the sales tax in year x. Find the following.

a. $\lim_{x \to 98} T(x)$ **b.** $\lim_{x \to 02^-} T(x)$

c. $\lim_{x \to 02^+} T(x)$ **d.** $\lim_{x \to 02} T(x)$

e. $T(02)$

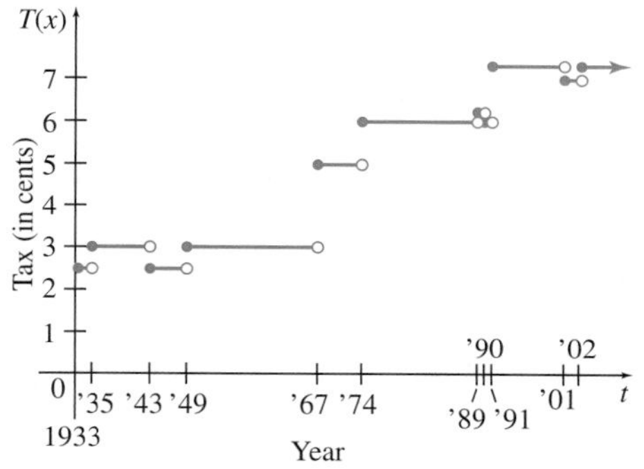

76. *Average Cost* The cost (in dollars) for manufacturing a particular DVD is

$$C(x) = 15{,}000 + 6x,$$

where x is the number of DVDs produced. The average cost per DVD, denoted by $\overline{C}(x)$, is found by dividing $C(x)$ by x. Find and interpret $\lim_{x \to \infty} \overline{C}(x)$.

77. *Employee Productivity* A company training program has determined that, on the average, a new employee produces $P(s)$ items per day after s days of on-the-job training, where

$$P(s) = \frac{75s}{s + 8}.$$

Find and interpret $\lim_{s \to \infty} P(s)$.

78. *Preferred Stock* In business finance, an annuity is a series of equal payments received at equal intervals for a finite period of time. The *present value* of an n-period annuity takes the form

$$P = R\left[\frac{1 - (1 + i)^{-n}}{i}\right],$$

where R is the amount of the periodic payment and i is the fixed interest rate per period. Many corporations raise money by issuing preferred stock. Holders of the preferred stock, called a *perpetuity*, receive payments that take the form of an annuity in that the amount of the payment never changes. However, normally the payments for preferred stock do not end but theoretically continue forever. Find the limit of this present value equation as n approaches infinity to derive a formula for the present value of a share of preferred stock paying a periodic dividend R.*

79. *Growing Annuities* For some annuities encountered in business finance, called *growing annuities*, the amount of the periodic payment is not constant but grows at a constant periodic rate. Leases with escalation clauses can be examples of growing annuities. The present value of a growing annuity takes the form

$$P = \frac{R}{i - g}\left[1 - \left(\frac{1 + g}{1 + i}\right)^n\right],$$

where

R = amount of the next annuity payment,

g = expected constant annuity growth rate,

i = required periodic return at the time the annuity is evaluated,

n = number of periodic payments.

A corporation's common stock may be thought of as a claim on a growing annuity where the annuity is the company's annual dividend. However, in the case of common stock, these payments have no contractual end but theoretically continue forever. Compute the limit of the expression above as n approaches infinity to derive the Gordon–Shapiro Dividend Model popularly used to estimate the value of common stock. Make the reasonable assumption that $i > g$.* (*Hint:* What happens to a^n as $n \to \infty$ if $0 < a < 1$?)

LIFE SCIENCES

80. *Alligator Teeth* Researchers have developed a mathematical model that can be used to estimate the number of teeth

*Exercises 78 and 79 were contributed by Robert D. Campbell of the Frank G. Zarb School of Business at Hofstra University.

$N(t)$ at time t (days of incubation) for *Alligator mississippiensis*,* where

$$N(t) = 71.8e^{-8.96e^{-.0685t}}.$$

a. Find $N(65)$, the number of teeth of an alligator that hatched after 65 days.

b. Find $\lim_{t\to\infty} N(t)$ and use this value as an estimate of the number of teeth of a newborn alligator. Does this estimate differ significantly from the estimate of part a?

81. *Sediment* To develop strategies to manage water quality in polluted lakes, biologists must determine the depths of sediments and the rate of sedimentation. It has been determined that the depth of sediment $D(t)$ (in centimeters) with respect to time (in years before 1990) for Lake Coeur d'Alene, Idaho, can be estimated by the equation

$$D(t) = 155(1 - e^{-.0133t}).$$ [†]

a. Find $D(20)$ and interpret.

b. Find $\lim_{t\to\infty} D(t)$ and interpret.

82. *Drug Concentration* The concentration of a drug in a patient's bloodstream h hours after it was injected is given by

$$A(h) = \frac{.17h}{h^2 + 2}.$$

Find and interpret $\lim_{h\to\infty} A(h)$.

SOCIAL SCIENCES

83. *Legislative Voting* Members of a legislature often must vote repeatedly on the same bill. As time goes on, members may change their votes. Suppose that p_0 is the probability that an individual legislator favors an issue before the first roll call vote, and suppose that p is the probability of a change in position from one vote to the next. Then the probability that the legislator will vote "yes" on the nth roll call is given by

$$p_n = \frac{1}{2} + \left(p_0 - \frac{1}{2}\right)(1 - 2p)^n.$$ [‡]

For example, the chance of a "yes" on the third roll call vote is

$$p_3 = \frac{1}{2} + \left(p_0 - \frac{1}{2}\right)(1 - 2p)^3.$$

Suppose that there is a chance of $p_0 = .7$ that Congressman Stephens will favor the budget appropriation bill before the first roll call, but only a probability of $p = .2$ that he will change his mind on the subsequent vote. Find and interpret the following.

a. p_2

b. p_4

c. p_8

d. $\lim_{n\to\infty} p_n$

3.2 CONTINUITY

THINK ABOUT IT

When the minimum wage is raised, how does the minimum wage change when adjusted for inflation?

Figure 14 on the next page shows how the federal minimum wage has changed with time before 1996, when it was raised.[§] (It was raised again in 1997.) The nominal wage is how much workers were actually paid per hour. The real wage is measured in 1994 dollars, adjusted for inflation. Notice that whenever the minimum wage changes, both graphs go straight up, so they appear not to be functions at all because they do not satisfy the vertical line test. To clarify the behavior at these points, we have drawn a portion of the constant dollar graph around the beginning of 1979 in Figure 15 on the next page, using the notation of our text. Let us denote this function by $f(t)$, where t is the year.

*Kulesa, P., G. Cruywagen et al. "On a Model Mechanism for the Spatial Patterning of Teeth Primordia in the Alligator," *Journal of Theoretical Biology,* Vol. 180, 1996, pp. 287–296.

†Nord, Gail and John Nord, "Sediment in Lake Coeur d'Alene, Idaho," *Mathematics Teacher,* Vol. 91, No. 4, April 1998, pp. 292–295.

‡Bishir, John W. and Donald W. Drewes, *Mathematics in the Behavioral and Social Sciences,* New York: Harcourt Brace Jovanovich, 1970, p. 538.

§*The New York Times,* Jan. 12, 1994, p. D1. Reprinted with permission.

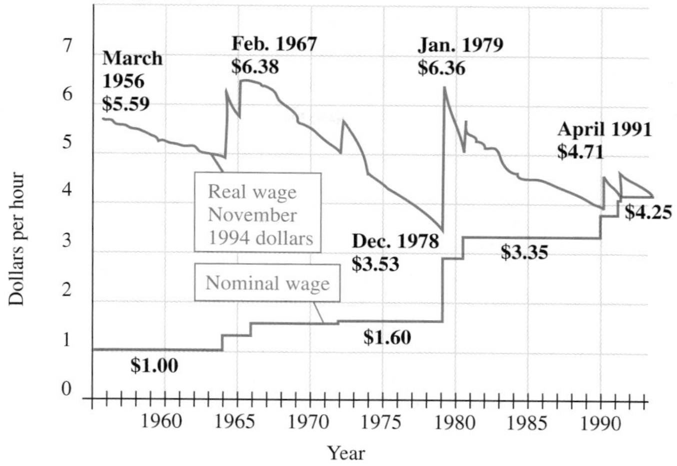

FIGURE 14

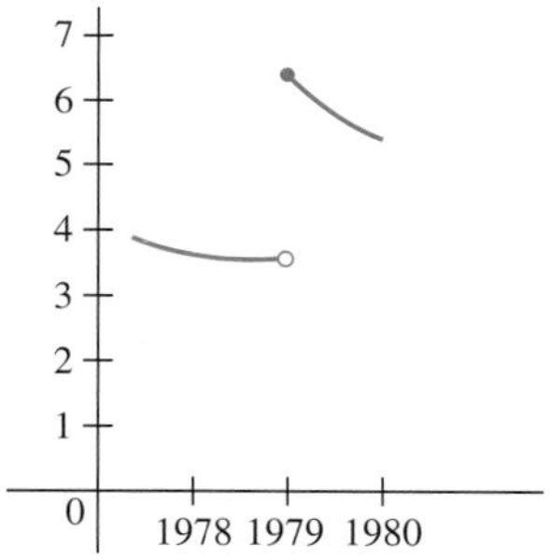

FIGURE 15

Notice from the graph that $\lim_{t \to 1979^-} f(t) = 3.53$ and that $\lim_{t \to 1979^+} f(t) = 6.36$, so $\lim_{t \to 1979} f(t)$ does not exist. Notice also that $f(1979) = 6.36$. A point such as this, where a function has a sudden sharp break, is a point where the function is *discontinuous*.

Intuitively speaking, a function is *continuous* at a point if you can draw the graph of the function in the vicinity of that point without lifting your pencil from the paper. As we already mentioned, this would not be possible in Figure 14 if it were drawn correctly; there would be a break in the graph at $t = 1979$, for example. Conversely, a function is discontinuous at any x-value where the pencil *must* be lifted from the paper in order to draw the graph on both sides of the point. A more precise definition is as follows.

CONTINUITY AT $x = c$

A function f is **continuous** at $x = c$ if the following three conditions are satisfied:

1. $f(c)$ is defined,
2. $\lim_{x \to c} f(x)$ exists, and
3. $\lim_{x \to c} f(x) = f(c)$.

If f is not continuous at c, it is **discontinuous** there.

The following example shows the various ways a function can be discontinuous.

EXAMPLE 1 *Continuity*

Tell why each function is discontinuous at the indicated x-value.

(a) $f(x)$ in Figure 16 on the next page at $x = 3$

Solution The open circle on the graph of Figure 16 at the point where $x = 3$ means that $f(3)$ is not defined. Because of this, part 1 of the definition fails.

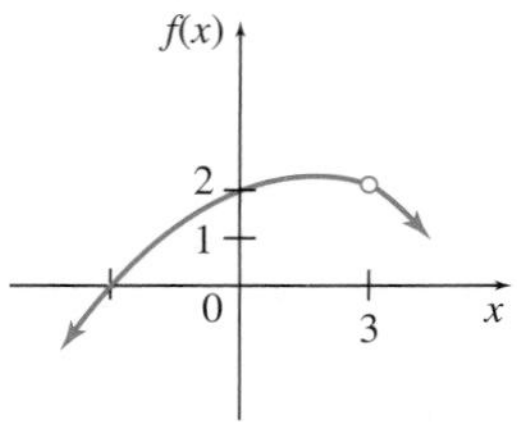

FIGURE 16

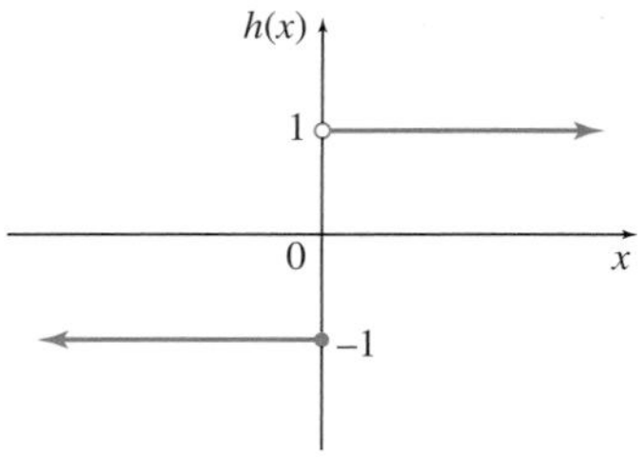

FIGURE 17

(b) $h(x)$ in Figure 17 at $x = 0$

Solution According to the graph of Figure 17, $h(0) = -1$. Also, as x approaches 0 from the left, $h(x)$ is -1. As x approaches 0 from the right, however, $h(x)$ is 1. In other words,

$$\lim_{x\to 0^-} h(x) = -1,$$

while

$$\lim_{x\to 0^+} h(x) = 1.$$

Since no single number is approached by the values of $h(x)$ as x approaches 0, the limit $\lim_{x\to 0} h(x)$ does not exist, and part 2 of the definition fails.

(c) $g(x)$ in Figure 18 at $x = 4$

Solution In Figure 18, the heavy dot above 4 shows that $g(4)$ is defined. In fact, $g(4) = 1$. The graph also shows, however, that

$$\lim_{x\to 4} g(x) = -2,$$

so $\lim_{x\to 4} g(x) \neq g(4)$, and part 3 of the definition fails.

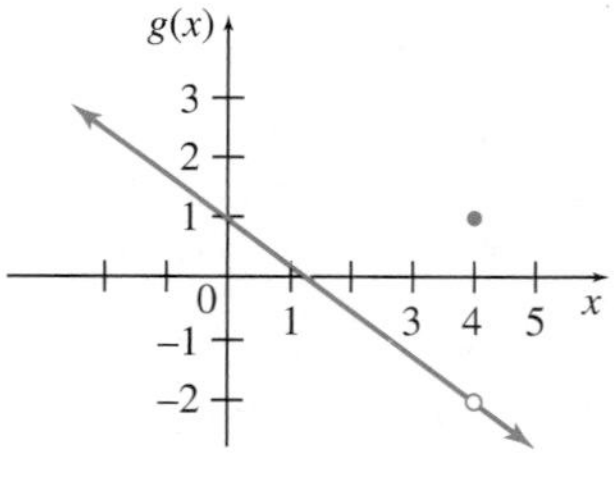

FIGURE 18

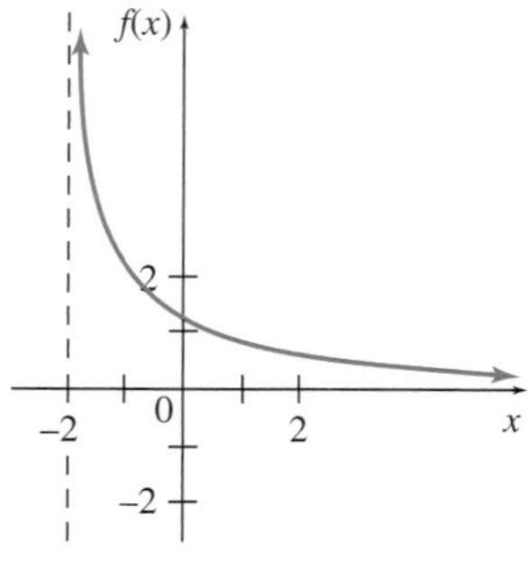

FIGURE 19

(d) $f(x)$ in Figure 19 at $x = -2$

Solution The function f graphed in Figure 19 is not defined at $x = -2$, and $\lim_{x\to -2} f(x)$ does not exist there. Either of these reasons is sufficient to show that f is not continuous at -2. (Function f is continuous at any value of x greater than -2, however.)

Notice that the function in part (a) of the previous example could be made continuous simply by defining $f(3) = 2$. Similarly, the function in part (c) could be made continuous by defining $g(4) = -2$. In such cases, when the function can be made continuous simply by defining or redefining it at a single point, the function is said to have a **removable discontinuity.**

A function is said to be **continuous on an open interval** if it is continuous at every x-value in the interval. Continuity on a closed interval is slightly more complicated because we must decide what to do with the endpoints. We will say that a function f is **continuous from the right** at $x = c$ if $\lim_{x \to c^+} f(x) = f(c)$. A function f is **continuous from the left** at $x = c$ if $\lim_{x \to c^-} f(x) = f(c)$. With these ideas, we can now define continuity on a closed interval.

CONTINUITY ON A CLOSED INTERVAL

A function is **continuous on a closed interval** $[a, b]$ if

1. it is continuous on the open interval (a, b),
2. it is continuous from the right at $x = a$, and
3. it is continuous from the left at $x = b$.

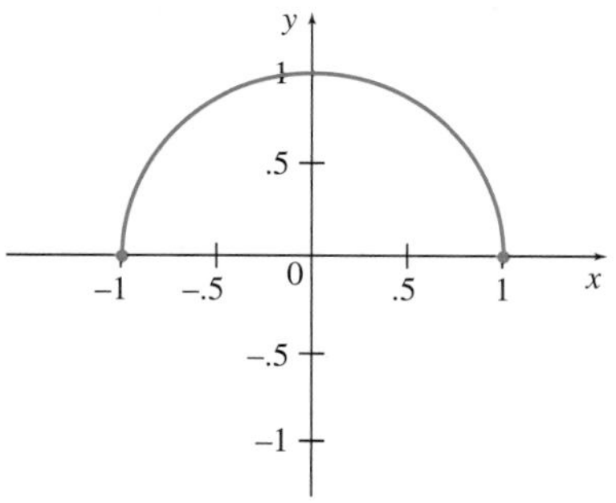

FIGURE 20

For example, the function $f(x) = \sqrt{1 - x^2}$, shown in Figure 20, is continuous on the closed interval $[-1, 1]$. By defining continuity on a closed interval in this way, we need not worry about the fact that $\sqrt{1 - x^2}$ does not exist to the left of $x = -1$ or to the right of $x = 1$.

The table on the next page lists some key functions and tells where each is continuous.

Continuous functions are nice to work with because finding $\lim_{x \to c} f(x)$ is simple if f is continuous: just evaluate $f(c)$.

When a function is given by a graph, any discontinuities are clearly visible. When a function is given by a formula, it is usually continuous at all x-values except those where the function is undefined, or possibly where there is a change in the defining formula for the function, as in the following example.

EXAMPLE 2 *Continuity*

Find all values of x where the following function is discontinuous.

$$f(x) = \begin{cases} x + 1 & \text{if } x < 1 \\ x^2 - 3x + 4 & \text{if } 1 \le x \le 3. \\ 5 - x & \text{if } x > 3 \end{cases}$$

Solution A function defined by two or more cases is called a *piecewise function*. The only x-values where f might be discontinuous here are 1 and 3. We investigate at $x = 1$ first. From the left,

$$\lim_{x \to 1^-} f(x) = \lim_{x \to 1^-} (x + 1) = 1 + 1 = 2.$$

From the right,

$$\lim_{x \to 1^+} f(x) = \lim_{x \to 1^+} (x^2 - 3x + 4) = 1^2 - 3 + 4 = 2.$$

Continuous Functions

Type of Function	Where It Is Continuous	Graphic Example
Polynomial Function $y = a_n x^n + a_{n-1}x^{n-1} + \cdots + a_1 x + a_0$, where a_n, $a_{n-1}, \ldots, a_1, a_0$ are real numbers, not all 0	For all x	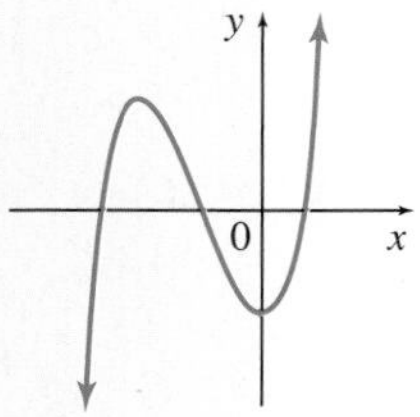
Rational Function $y = \dfrac{p(x)}{q(x)}$, where $p(x)$ and $q(x)$ are polynomials, with $q(x) \neq 0$	For all x where $q(x) \neq 0$	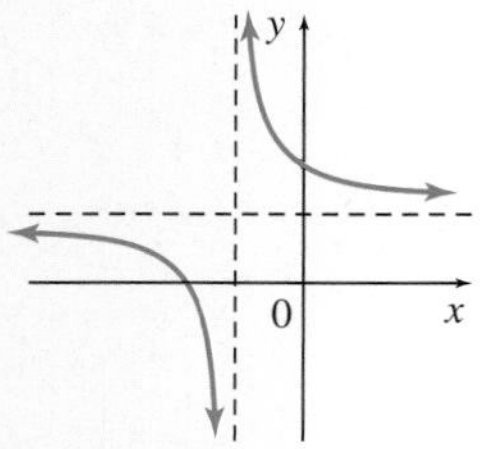
Root Function $y = \sqrt{ax + b}$, where a and b are real numbers, with $a \neq 0$ and $ax + b \geq 0$	For all x where $ax + b \geq 0$	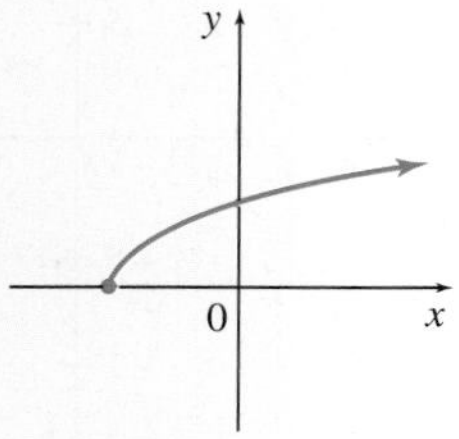
Exponential Function $y = a^x$ where $a > 0$	For all x	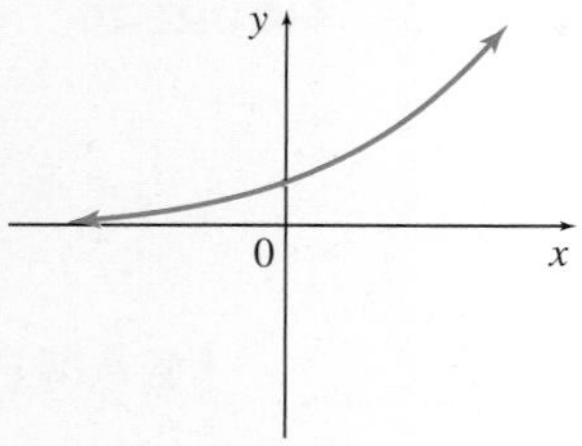
Logarithmic Function $y = \log_a x$ where $a > 0$	For all $x > 0$	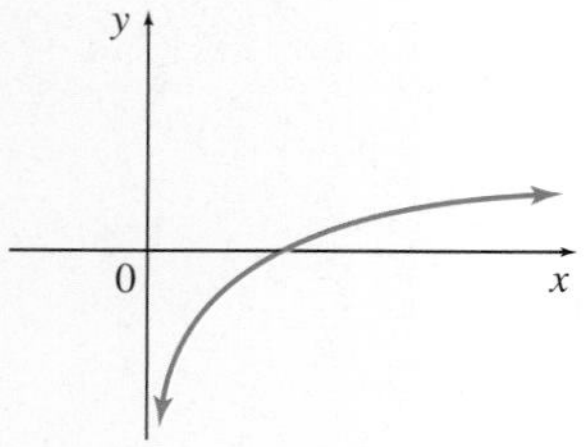

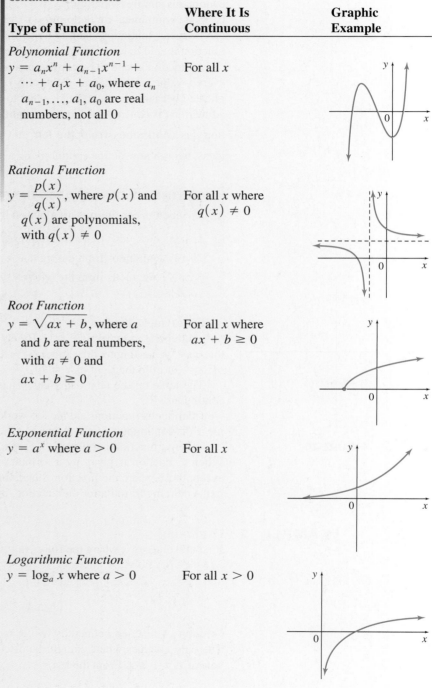

Furthermore, $f(1) = 1^2 - 3 + 4 = 2$, so $\lim_{x\to 1} f(x) = f(1) = 2$. Thus f is continuous at $x = 1$.

Now let us investigate $x = 3$. From the left,

$$\lim_{x\to 3^-} f(x) = \lim_{x\to 3^-} (x^2 - 3x + 4) = 3^2 - 3(3) + 4 = 4.$$

From the right,

$$\lim_{x\to 3^+} f(x) = \lim_{x\to 3^+} (5 - x) = 5 - 3 = 2.$$

Because $\lim_{x\to 3^-} f(x) \neq \lim_{x\to 3^+} f(x)$, the limit $\lim_{x\to 3} f(x)$ does not exist, so f is discontinuous at $x = 3$, regardless of the value of $f(3)$.

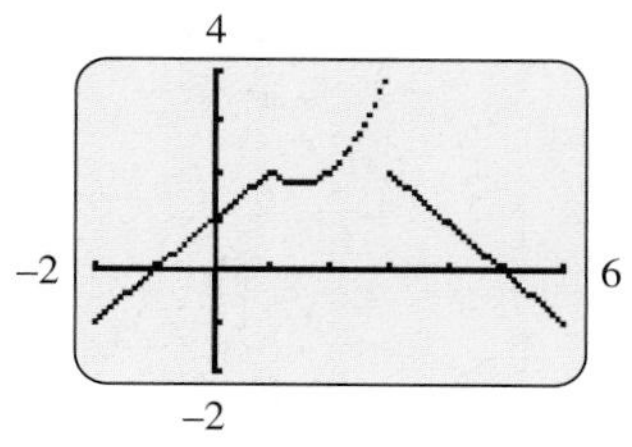

FIGURE 21

The graph of the function in Example 2 can be drawn by considering each of the three parts separately. For example, for the first part, the line $y = x + 1$ is drawn using the techniques of Chapter 1, but including only the section of the line to the left of $x = 1$. The other two parts are drawn similarly.

Alternatively, some graphing calculators have the ability to draw piecewise functions. On the TI-83/84 Plus, letting `Y1=(X + 1)(X < 1)+(X² − 3X + 4)(1 ≤ X)(X ≤ 3)+(5 − X)(X > 3)` produces the graph shown in Figure 21. It is important here that the graphing mode be set on Dot rather than Connected. Otherwise, the calculator will show a line segment at $x = 3$ connecting the parabola to the line, although such a segment does not really exist.

EXAMPLE 3 *Cost Analysis*

A trailer rental firm charges a flat \$8 to rent a hitch. The trailer itself is rented for \$22 per day or fraction of a day. Let $C(x)$ represent the cost of renting a hitch and trailer for x days.

(a) Graph C.

Solution The charge for one day is \$8 for the hitch and \$22 for the trailer, or \$30. In fact, if $0 < x \leq 1$, then $C(x) = 30$. To rent the trailer for more than one day, but not more than two days, the charge is $8 + 2 \cdot 22 = 52$ dollars. For any value of x satisfying $1 < x \leq 2$, the cost is $C(x) = 52$. Also, if $2 < x \leq 3$, then $C(x) = 74$. These results lead to the graph in Figure 22.

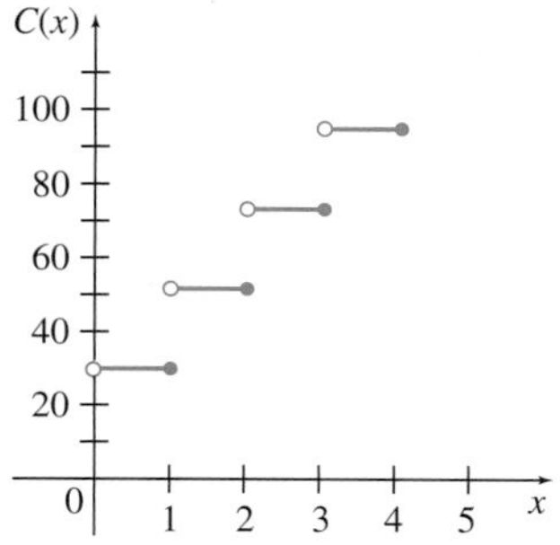

FIGURE 22

(b) Find any values of x where C is discontinuous.

Solution As the graph suggests, C is discontinuous at $x = 1, 2, 3, 4$, and all other positive integers.

One application of continuity is the **Intermediate Value Theorem,** which says that if a function is continuous on a closed interval $[a, b]$, the function takes on every value between $f(a)$ and $f(b)$. For example, if $f(1) = -3$ and $f(2) = 5$, then f must take on every value between -3 and 5 as x varies over the interval $[1, 2]$. In particular (in this case), there must be a value of x in the interval $(1, 2)$ such that $f(x) = 0$. If f were discontinuous, however, this conclusion would not necessarily be true. Before searching for a solution to $f(x) = 0$ in $[1, 2]$, we would like to know that a solution exists.

3.2 EXERCISES

In Exercises 1–6, find all values $x = a$ where the function is discontinuous. For each point of discontinuity, give **(a)** $\lim_{x \to a^-} f(x)$, **(b)** $\lim_{x \to a^+} f(x)$, **(c)** $\lim_{x \to a} f(x)$, *and* **(d)** $f(a)$ *if it exists.*

1.

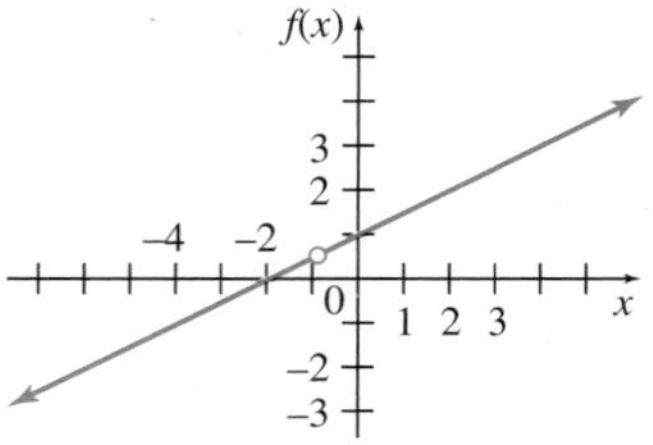

2.

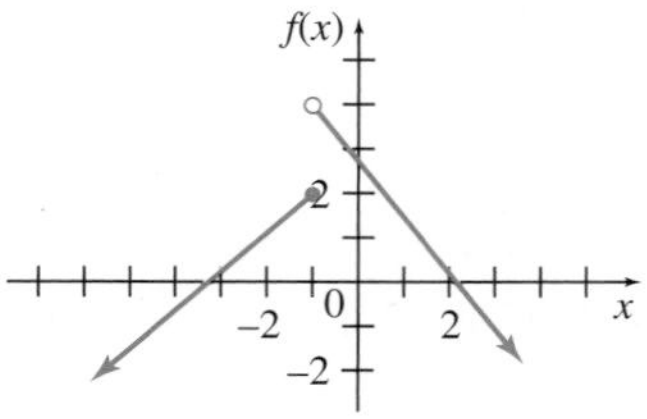

3.

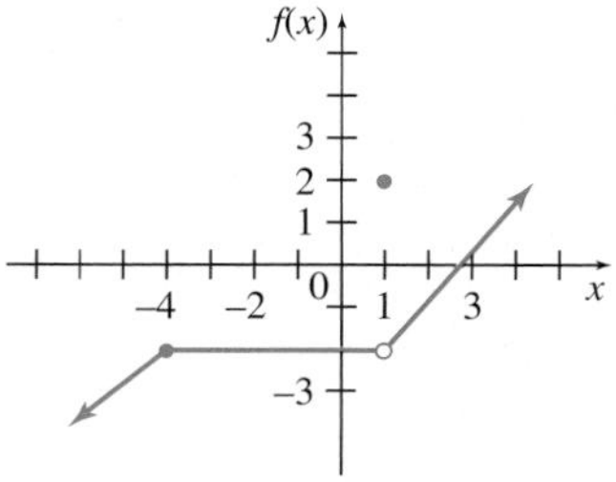

4.

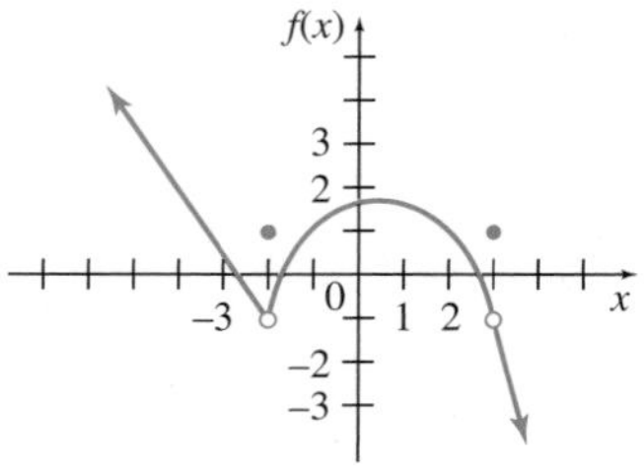

5.

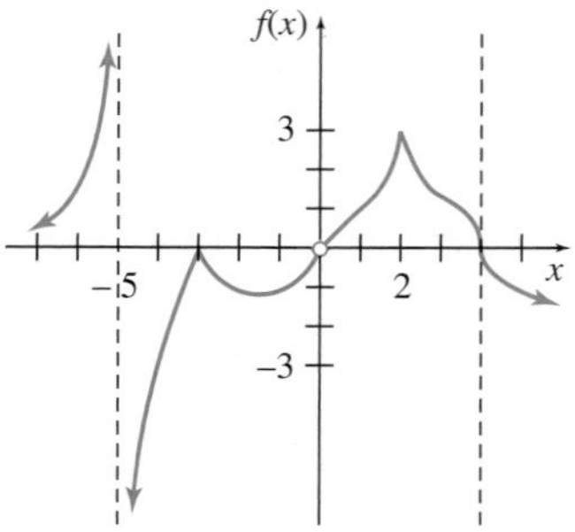

6.

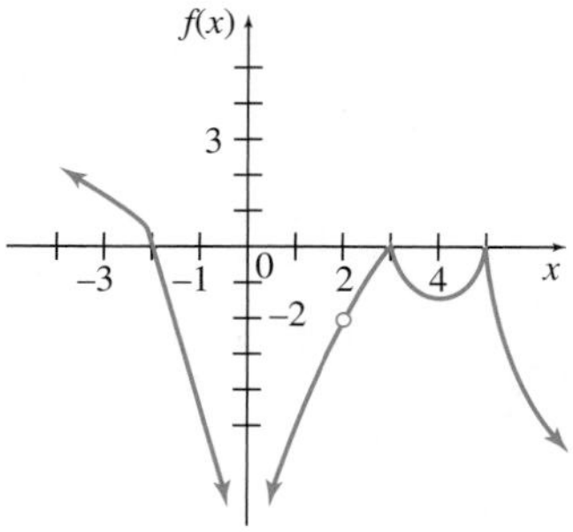

Find all values $x = a$ where the function is discontinuous. For each value of x, give the limit of the function as x approaches a.

7. $f(x) = \dfrac{5 + x}{x(x - 2)}$

8. $f(x) = \dfrac{-2x}{(2x + 1)(3x + 6)}$

9. $f(x) = \dfrac{x^2 - 4}{x - 2}$

10. $f(x) = \dfrac{x^2 - 25}{x + 5}$

11. $p(x) = x^2 - 4x + 11$

12. $q(x) = -3x^3 + 2x^2 - 4x + 1$

13. $p(x) = \dfrac{|x + 2|}{x + 2}$

14. $r(x) = \dfrac{|5 - x|}{x - 5}$

15. $k(x) = e^{\sqrt{x-1}}$

16. $j(x) = e^{1/x}$

17. $r(x) = \ln\left(\dfrac{x}{x - 1}\right)$

18. $j(x) = \ln\left(\dfrac{x + 2}{x - 1}\right)$

In Exercises 19–24, **(a)** *graph the given function,* **(b)** *find all values of x where the function is discontinuous, and* **(c)** *find the limit from the left and from the right at any values of x found in part b.*

19. $f(x) = \begin{cases} 1 & \text{if } x < 2 \\ x + 3 & \text{if } 2 \le x \le 4 \\ 7 & \text{if } x > 4 \end{cases}$

20. $f(x) = \begin{cases} x - 1 & \text{if } x < 1 \\ 0 & \text{if } 1 \le x \le 4 \\ x - 2 & \text{if } x > 4 \end{cases}$

21. $g(x) = \begin{cases} 11 & \text{if } x < -1 \\ x^2 + 2 & \text{if } -1 \le x \le 3 \\ 11 & \text{if } x > 3 \end{cases}$

22. $g(x) = \begin{cases} 0 & \text{if } x < 0 \\ x^2 - 5x & \text{if } 0 \le x \le 5 \\ 5 & \text{if } x > 5 \end{cases}$

23. $h(x) = \begin{cases} 4x + 4 & \text{if } x \le 0 \\ x^2 - 4x + 4 & \text{if } x > 0 \end{cases}$

24. $h(x) = \begin{cases} x^2 + x - 12 & \text{if } x \le 1 \\ 3 - x & \text{if } x > 1 \end{cases}$

In Exercises 25–28, find the value of the constant k that makes the function continuous.

25. $f(x) = \begin{cases} kx^2 & \text{if } x \le 2 \\ x + k & \text{if } x > 2 \end{cases}$

26. $g(x) = \begin{cases} x^3 + k & \text{if } x \le 3 \\ kx - 5 & \text{if } x > 3 \end{cases}$

27. $g(x) = \begin{cases} \dfrac{2x^2 - x - 15}{x - 3} & \text{if } x = 3 \\ kx - 1 & \text{if } x = 3 \end{cases}$

28. $h(x) = \begin{cases} \dfrac{3x^2 + 2x - 8}{x + 2} & \text{if } x \ne -2 \\ 3x + k & \text{if } x = -2 \end{cases}$

29. Explain in your own words what the Intermediate Value Theorem says and why it seems plausible.

30. Explain why $\lim_{x \to 2} (3x^2 + 8x)$ can be evaluated by substituting $x = 3$.

In Exercises 31–32, **(a)** *use a graphing calculator to tell where the rational function $P(x)/Q(x)$ is discontinuous, and* **(b)** *verify your answer from part (a) by using the graphing calculator to plot $Q(x)$ and determine where $Q(x) = 0$. You will need to choose the viewing window carefully.*

31. $f(x) = \dfrac{x^2 + x + 2}{x^3 - .9x^2 + 4.14x - 5.4}$

32. $f(x) = \dfrac{x^2 + 3x - 2}{x^3 - .9x^2 + 4.14x + 5.4}$

Applications

BUSINESS AND ECONOMICS

33. *Production* The graph shows the profit from the daily production of x thousand kilograms of an industrial chemical. Use the graph to find the following limits.

a. $\lim_{x \to 6} P(x)$ **b.** $\lim_{x \to 10^-} P(x)$ **c.** $\lim_{x \to 10^+} P(x)$

d. $\lim_{x \to 10} P(x)$

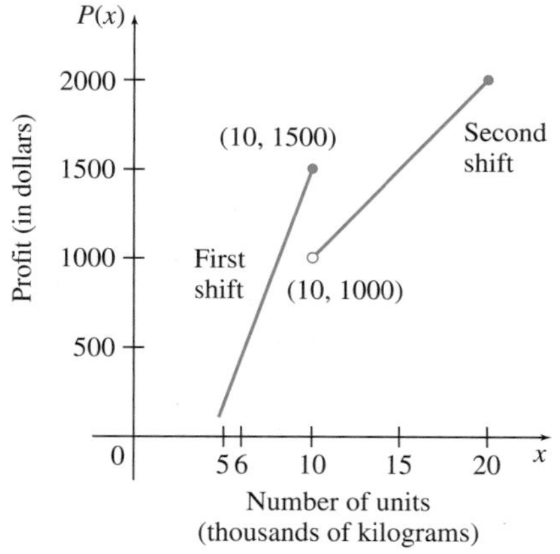

e. Where is the function discontinuous? What might account for such a discontinuity?

f. Use the graph to estimate the number of units of the chemical that must be produced before the second shift is as profitable as the first.

34. *Cost Analysis* The cost to transport a mobile home depends on the distance, x, in miles that the home is moved. Let $C(x)$ represent the cost to move a mobile home x miles. One firm charges as follows.

Cost per Mile	Distance in Miles
\$4.00	$0 < x \le 150$
\$3.00	$150 < x \le 400$
\$2.50	$400 < x$

Find the cost to move a mobile home the following distances.

a. 130 miles **b.** 150 miles **c.** 210 miles

d. 400 miles **e.** 500 miles

f. Where is C discontinuous?

35. *Cost Analysis* A company charges \$1.20 per lb for a certain fertilizer on all orders not over 100 lb, and \$1 per lb for orders over 100 lb. Let $F(x)$ represent the cost for buying x lb of the fertilizer. Find the cost of buying the following.

a. 80 lb **b.** 150 lb **c.** 100 lb

d. Where is F discontinuous?

36. *Car Rental* Recently, a car rental firm charged \$30 per day or portion of a day to rent a car for a period of 1 to 5 days. Days 6 and 7 were then free, while the charge for days 8 through 12 was again \$30 per day. Let $A(t)$ represent the average cost to rent the car for t days, where $0 < t \le 12$.

Find the average cost of a rental for the following number of days.

a. 4 **b.** 5 **c.** 6 **d.** 7 **e.** 8

f. Find $\lim_{x\to 5^-} A(t)$. **g.** Find $\lim_{x\to 5^+} A(t)$.

h. Where is A discontinuous on the given interval?

37. *Postage* To send an international airmail letter from the United States to Japan, Australia, or New Zealand in 2003, it cost \$.80 for the first ounce and \$.90 for each additional ounce, up to a total of 8 oz. Let $C(x)$ represent the postage for a letter weighing x oz. Find the following.

a. $\lim_{x\to 3^-} C(x)$ **b.** $\lim_{x\to 3^+} C(x)$

c. $\lim_{x\to 3} C(x)$ **d.** $C(3)$

e. Find all values on the interval $(0, 8)$ where the function C is discontinuous.

f. Sketch the graph of $y = C(x)$ on the interval $(0, 8]$.

LIFE SCIENCES

38. *Pregnancy* During pregnancy, a woman's weight naturally increases during the course of the event. When she delivers, her weight immediately decreases by the approximate weight of the child. Suppose that a 120-lb woman gains 27 lb during pregnancy, delivers a 7-lb baby, and then, through diet and exercise, loses the remaining weight during the next 20 weeks.

a. Graph the weight gain and loss during the pregnancy and the 20 weeks following the birth of the baby. Assume that the pregnancy lasts 40 weeks, that delivery occurs immediately after this time interval, and that the weight gain/loss before and after birth is linear.

b. Is this a continuous function? If not, then find the value(s) of t where the function is discontinuous.

39. *Poultry Farming* Researchers at Iowa State University, and the University of Arkansas have developed a piecewise function that can be used to estimate the body weight (in grams) of a male broiler during the first 56 days of life according to

$$W(t) = \begin{cases} 48 + 3.64t + .6363t^2 + .00963t^3 & \text{if } 1 \le t \le 28, \\ -1004 + 65.8t & \text{if } 28 < t \le 56, \end{cases}$$

where t is the age of the chicken (in days).*

a. Determine the weight of a male broiler that is 25 days old.

b. Is $W(t)$ a continuous function?

c. Use a graphing calculator to graph $W(t)$ on $[1, 56]$ by $[0, 3000]$. Comment on the accuracy of the graph.

d. Comment on why researchers would use two different types of functions to estimate the weight of a chicken at various ages.

3.3 RATES OF CHANGE

THINK ABOUT IT

How does the manufacturing cost of a DVD change as the number of DVDs manufactured changes?

This question will be answered in Example 4 of this section as we develop a method for finding the rate of change of one variable with respect to a unit change in another variable.

Average Rate of Change One of the main applications of calculus is determining how one variable changes in relation to another. A marketing manager wants to know how profit changes with respect to the amount spent on advertising, while a physician wants to know how a patient's reaction to a drug changes with respect to the dose.

For example, suppose we take a trip from San Francisco driving south. Every half-hour we note how far we have traveled, with the following results for the first three hours.

Time in Hours	0	.5	1	1.5	2	2.5	3
Distance in Miles	0	20	48	80	104	126	150

*Xin, H., I. Berry, T. Barton, and G. Tabler, "Feed and Water Consumption, Growth, and Mortality of Male Broiler," *Poultry Science,* Vol. 73, No. 5, May 1994, pp. 610–616.

If s is the function whose rule is

$$s(t) = \text{Distance from San Francisco at time } t,$$

then the table shows, for example, that $s(0) = 0$, $s(1) = 48$, $s(2.5) = 126$, and so on. The distance traveled during, say, the second hour can be calculated by $s(2) - s(1) = 104 - 48 = 56$ miles.

Distance equals time multiplied by rate (or speed); so the distance formula is $d = rt$. Solving for rate gives $r = d/t$, or

$$\text{Average speed} = \frac{\text{Distance}}{\text{Time}}.$$

For example, the average speed over the time interval from $t = 0$ to $t = 3$ is

$$\text{Average speed} = \frac{s(3) - s(0)}{3 - 0} = \frac{150 - 0}{3} = 50,$$

or 50 mph. We can use this formula to find the average speed for any interval of time during the trip, as shown below.

FOR REVIEW

Recall the formula for the slope of a line through two points (x_1, y_1) and (x_2, y_2):

$$\frac{y_2 - y_1}{x_2 - x_1}.$$

Find the slopes of the lines through the following points.

$(.5, 20)$	and	$(1, 48)$
$(.5, 20)$	and	$(1.5, 80)$
$(1, 48)$	and	$(2, 104)$

Compare your answers to the average speeds shown in the table.

Time Interval	**Average Speed** $= \dfrac{\textbf{Distance}}{\textbf{Time}}$
$t = .5$ to $t = 1$	$\dfrac{s(1) - s(.5)}{1 - .5} = \dfrac{28}{.5} = 56$
$t = .5$ to $t = 1.5$	$\dfrac{s(1.5) - s(.5)}{1.5 - .5} = \dfrac{60}{1} = 60$
$t = 1$ to $t = 2$	$\dfrac{s(2) - s(1)}{2 - 1} = \dfrac{56}{1} = 56$
$t = 1$ to $t = 3$	$\dfrac{s(3) - s(1)}{3 - 1} = \dfrac{102}{2} = 51$
$t = a$ to $t = b$	$\dfrac{s(b) - s(a)}{b - a}$

The analysis of the average speed or *average rate of change* of distance s with respect to t can be applied to any function defined by $f(x)$ to get a formula for the average rate of change of f with respect to x.

AVERAGE RATE OF CHANGE

The **average rate of change** of $f(x)$ with respect to x for a function f as x changes from a to b is

$$\frac{f(b) - f(a)}{b - a}.$$

NOTE The formula for the average rate of change is the same as the formula for the slope of the line through $(a, f(a))$ and $(b, f(b))$. This connection between slope and rate of change will be examined more closely in the next section. ■

We will sometimes refer to the quantity $(f(b) - f(a))/(b - a)$ as the **difference quotient.**

EXAMPLE 1 *Elderly Employment*

The percentage of men aged 65 and older in the workforce has been declining over the last century.* The percent can be approximated by the function

$$f(x) = 68.7(.986)^x,$$

where x is the number of years since 1900. Find the average rate of change of this percent from 1960 to 2000.

Solution On the interval from $x = 60$ to $x = 100$, the average rate of change is

$$\begin{aligned}\frac{f(100) - f(60)}{100 - 60} &= \frac{68.7(.986)^{100} - 68.7(.986)^{60}}{40}\\ &= \frac{16.774 - 29.483}{40} = \frac{-12.709}{40}\\ &\approx -.32.\end{aligned}$$

Therefore, the percentage of men aged 65 and older in the workforce decreased on average at a rate of about .32% (about one-third of 1%) per year between 1960 and 2000.

EXAMPLE 2 *Drug Control*

The graph in Figure 23 shows federal spending $S(t)$, measured in millions of dollars, as a function of time t, measured in years, on drug control programs.† The average rate of change of spending with respect to time on some interval is defined as the change in spending divided by the change in time. From the graph, we see that \$7.6 billion was spent in 1998 and \$11.0 billion in 2002. Find the

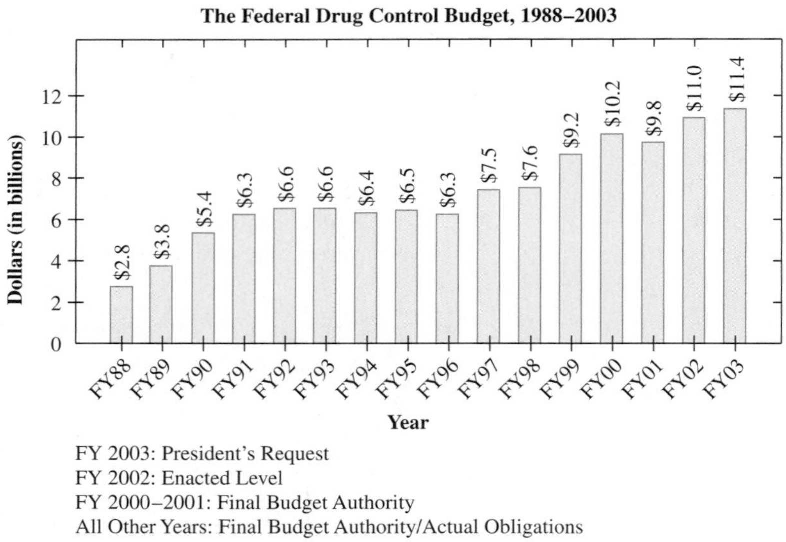

FIGURE 23

**The World Almanac and Book of Facts 2003,* p. 143.
†http://www.whitehousedrugpolicy.gov/drugfact/october2002.ppt.

average rate of change in the amount of federal drug control spending with respect to time from 1998 to 2002.

Solution The average rate of change is

$$\frac{S(2002) - S(1998)}{2002 - 1998} = \frac{11.0 - 7.6}{4} = \frac{3.4}{4} = .85.$$

On average, the amount the federal government spent to control drugs increased by .85 billion dollars, or 850 million dollars, per year during this time interval.

Instantaneous Rate of Change Finding the average rate of change of a function over a large interval can sometimes lead to answers that are not very helpful. The results often are more useful if the average rate of change is found over a fairly narrow interval. Finding the exact rate of change at a particular x-value requires a continuous function. For example, suppose a car starts from a stop sign and moves along a straight road. Assume that the distance in feet traveled in t seconds is given by $s(t) = t^2 + 3t$. After 5 seconds, the car has traveled $s(5) = 5^2 + 3(5) = 40$ ft. The speed of the car at exactly 10 seconds after starting can be estimated by finding the average speed over shorter and shorter time intervals.

Interval	Average Speed
$t = 10$ to $t = 10.1$	$\frac{s(10.1) - s(10)}{10.1 - 10} = \frac{132.31 - 130}{.1} = 23.1$
$t = 10$ to $t = 10.01$	$\frac{s(10.01) - s(10)}{10.01 - 10} = \frac{130.2301 - 130}{.01} = 23.01$
$t = 10$ to $t = 10.001$	$\frac{s(10.001) - s(10)}{10.001 - 10} = \frac{130.023001 - 130}{.001} = 23.001$

The results in the table suggest that the exact speed at $t = 10$ seconds is 23 ft per second.

This example can be easily generalized to any function f. Let a be a specific x-value, such as 10 in the example. Let h be a (small) number. The average rate of change of f as x changes from a to $a + h$ is

$$\frac{f(a + h) - f(a)}{(a + h) - a} = \frac{f(a + h) - f(a)}{h}.$$

The exact rate of change of f at $x = a$, called the *instantaneous rate of change of f at $x = a$*, is the limit of this quotient.

INSTANTANEOUS RATE OF CHANGE

The **instantaneous rate of change** for a function f when $x = a$ is

$$\lim_{h \to 0} \frac{f(a + h) - f(a)}{h},$$

provided this limit exists.

CAUTION Remember that $f(x + h) \neq f(x) + f(h)$. To find $f(x + h)$, replace x with $x + h$ in the expression for $f(x)$. For example, if $f(x) = x^2$,

$$f(x + h) = (x + h)^2 = x^2 + 2xh + h^2,$$

but
$$f(x) + f(h) = x^2 + h^2. \qquad ▪$$

For the next example, we need to make a subtle distinction between speed, the term we have been using to describe how fast someone is moving, and velocity, which also incorporates the direction of the motion. In any motion along a straight line, one direction is arbitrarily labeled as positive, so the other direction is negative. Velocity in the negative direction is considered negative. Speed is always positive; it is the absolute value of velocity.

EXAMPLE 3 *Velocity*
The distance in feet of an object from a starting point is given by $s(t) = 2t^2 - 5t + 40$, where t is time in seconds.

(a) Find the average velocity of the object from 2 seconds to 4 seconds.

   Solution The average velocity is
$$\frac{s(4) - s(2)}{4 - 2} = \frac{52 - 38}{2} = \frac{14}{2} = 7$$

ft per second.

(b) Find the instantaneous velocity at 4 seconds.

   Solution For $t = 4$, the instantaneous velocity is
$$\lim_{h \to 0} \frac{s(4 + h) - s(4)}{h}$$

ft per second.

$$\begin{aligned} s(4 + h) &= 2(4 + h)^2 - 5(4 + h) + 40 \\ &= 2(16 + 8h + h^2) - 20 - 5h + 40 \\ &= 32 + 16h + 2h^2 - 20 - 5h + 40 \\ &= 2h^2 + 11h + 52 \end{aligned}$$

Also,
$$s(4) = 2(4)^2 - 5(4) + 40 = 52.$$

Therefore, the instantaneous velocity at $t = 4$ is
$$\lim_{h \to 0} \frac{(2h^2 + 11h + 52) - 52}{h} = \lim_{h \to 0} \frac{2h^2 + 11h}{h}$$
$$= \lim_{h \to 0} (2h + 11) = 11,$$

or 11 ft per second.

EXAMPLE 4 *Manufacturing*

A company determines that the cost in dollars to manufacture x cases of the DVD "Mathematicians Caught in Embarrassing Moments" is given by

$$C(x) = 100 + 15x - x^2 \quad (0 \le x \le 7).$$

(a) Find the average rate of change of cost per case for manufacturing between 1 and 5 cases.

Solution Use the formula for average rate of change. The cost to manufacture 1 case is

$$C(1) = 100 + 15(1) - 1^2 = 114,$$

or \$114. The cost to manufacture 5 cases is

$$C(5) = 100 + 15(5) - 5^2 = 150,$$

or \$150. The average rate of change of cost is

$$\frac{C(5) - C(1)}{5 - 1} = \frac{150 - 114}{4} = 9.$$

Thus, on the average, the cost increases at the rate of \$9 per case when production increases from 1 to 5 cases.

(b) Find the instantaneous rate of change of cost with respect to the number of cases produced when just one case is produced.

Solution The instantaneous rate of change for $x = 1$ is given by

$$\lim_{h \to 0} \frac{C(1 + h) - C(1)}{h}$$

$$= \lim_{h \to 0} \frac{[100 + 15(1 + h) - (1 + h)^2] - [100 + 15(1) - 1^2]}{h}$$

$$= \lim_{h \to 0} \frac{100 + 15 + 15h - 1 - 2h - h^2 - 114}{h}$$

$$= \lim_{h \to 0} \frac{13h - h^2}{h} \qquad \text{Combine terms.}$$

$$= \lim_{h \to 0} \frac{h(13 - h)}{h} \qquad \text{Factor.}$$

$$= \lim_{h \to 0} (13 - h) \qquad \text{Divide by } h.$$

$$= 13. \qquad \text{Calculate the limit.}$$

When 1 case is manufactured, the cost is increasing at the rate of \$13 per case. The instantaneous rate of change of cost represents the approximate cost of manufacturing one additional case. As we mentioned in Chapter 1, this rate of change of cost is called the *marginal cost*.

(c) Find the instantaneous rate of change of cost when 5 cases are made.

Solution The instantaneous rate of change for $x = 5$ is given by

$$\lim_{h \to 0} \frac{C(5 + h) - C(5)}{h}$$

$$= \lim_{h\to 0} \frac{[100 + 15(5 + h) - (5 + h)^2] - [100 + 15(5) - 5^2]}{h}$$

$$= \lim_{h\to 0} \frac{100 + 75 + 15h - 25 - 10h - h^2 - 150}{h}$$

$$= \lim_{h\to 0} \frac{5h - h^2}{h} \qquad \text{Combine terms.}$$

$$= \lim_{h\to 0} \frac{h(5 - h)}{h} \qquad \text{Factor.}$$

$$= \lim_{h\to 0} (5 - h) \qquad \text{Divide by } h.$$

$$= 5. \qquad \text{Calculate the limit.}$$

When 5 cases are manufactured, the cost is increasing at the rate of \$5 per case; that is, the marginal cost when $x = 5$ is \$5. Notice that as the number of items produced goes up, the marginal cost goes down, as might be expected.

EXAMPLE 5 *Elderly Employment*

Estimate the instantaneous rate of change in 1990 in the percentage of men aged 65 and older in the workforce.

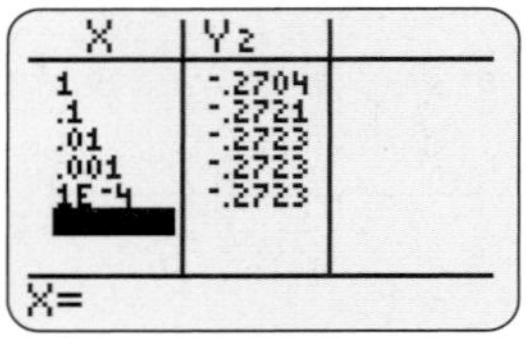

FIGURE 24

Solution We saw in Example 1 that the percentage is approximately given by $f(x) = 68.7(.986)^x$, where x is the number of years since 1900. Unlike the previous example, in which the function was a polynomial, the function in this example is an exponential, making it harder to compute the limit directly using the formula for instantaneous rate of change. Instead, we will approximate the instantaneous rate of change at $x = 90$ by using smaller and smaller values of h. This can be done using the TABLE feature on a TI-83/84 Plus graphing calculator by entering Y_1 as the function above, and $Y_2 = (Y_1(90 + X) - Y_1(90))/X$. (The graphing calculator requires us to use X in place of h in the formula for instantaneous rate of change.) The result is shown in Figure 24. This table can also be generated using a spreadsheet.

The limit seems to be approaching $-.2723$. Thus, the instantaneous rate of change in the percentage of men aged 65 and older in the work force in 1990 seems to be about $-.27$ percent per year, that is, decreasing at a rate of about one-quarter of 1% per year.

EXAMPLE 6 *Velocity*

One day Musk, the friendly pit bull, escaped from the yard and ran across the street to see a neighbor, who was 50 ft away. An estimate of the distance Musk ran as a function of time is given by the following table.

t (sec)	0	1	2	3	4
s (ft)	0	10	25	42	50

(a) Find Musk's average velocity during her 4-second trip.

Solution The total distance she traveled is 50 ft, and the total time is 4 seconds, so her average velocity is $50/4 = 12.5$ ft per second.

(b) Estimate Musk's velocity at 2 seconds.

Solution We could estimate her velocity by taking the short time interval from 2 to 3 seconds, for which the velocity is

$$\frac{42 - 25}{1} = 17 \text{ ft per second.}$$

Alternatively, we could estimate her velocity by taking the short time interval from 1 to 2 seconds, for which the velocity is

$$\frac{25 - 10}{1} = 15 \text{ ft per second.}$$

A better estimate is found by averaging these two values, to get

$$\frac{17 + 15}{2} = 16 \text{ ft per second.}$$

Another way to get this same answer is to take the time interval from 1 to 3 seconds, for which the velocity is

$$\frac{42 - 10}{2} = 16 \text{ ft per second.}$$

This answer is reasonable if we assume Musk's velocity changes at a fairly steady rate, and does not increase or decrease drastically from one second to the next. It is impossible to calculate Musk's exact velocity without knowing her position at times arbitrarily close to 2 seconds, or without a formula for her position as a function of time, or without a radar gun or speedometer on her. (In any case, she was very happy when she reached the neighbor.)

3.3 EXERCISES

Find the average rate of change for each function over the given interval.

1. $y = x^2 + 2x$ between $x = 0$ and $x = 3$

2. $y = -4x^2 - 6$ between $x = 2$ and $x = 5$

3. $y = 2x^3 - 4x^2 + 6x$ between $x = -1$ and $x = 1$

4. $y = -3x^3 + 2x^2 - 4x + 1$ between $x = 0$ and $x = 1$

5. $y = \sqrt{x}$ between $x = 1$ and $x = 4$

6. $y = \sqrt{3x - 2}$ between $x = 1$ and $x = 2$

7. $y = \dfrac{1}{x - 1}$ between $x = -2$ and $x = 0$

8. $y = \dfrac{-5}{2x - 3}$ between $x = 2$ and $x = 4$

Suppose the position of an object moving in a straight line is given by $s(t) = t^2 + 5t + 2$. *Find the instantaneous velocity at each time.*

9. $t = 6$

10. $t = 1$

Suppose the position of an object moving in a straight line is given by $s(t) = t^3 + 2t + 9$. Find the instantaneous velocity at each time.

11. $t = 1$

12. $t = 4$

Find the instantaneous rate of change for each function at the given value.

13. $f(x) = x^2 + 2x$ at $x = 0$

14. $s(t) = -4t^2 - 6$ at $t = 2$

15. $g(t) = 1 - t^2$ at $t = -1$

16. $F(x) = x^2 + 2$ at $x = 0$

Use the formula for instantaneous rate of change, approximating the limit by using smaller and smaller values of h, to find the instantaneous rate of change for each function at the given value.

17. $f(x) = x^x$ at $x = 2$

18. $f(x) = x^x$ at $x = 3$

19. $f(x) = x^{\ln x}$ at $x = 2$

20. $f(x) = x^{\ln x}$ at $x = 3$

21. Explain the difference between the average rate of change of y as x changes from a to b, and the instantaneous rate of change of y at $x = a$.

22. If the instantaneous rate of change of $f(x)$ with respect to x is positive when $x = 1$, is f increasing or decreasing there?

Applications

BUSINESS AND ECONOMICS

23. *Medicare Trust Fund* The graph shows the money remaining in the Medicare Trust Fund at the end of the calendar year, adjusted for inflation in 2000 dollars.*

Medicare Trust Fund

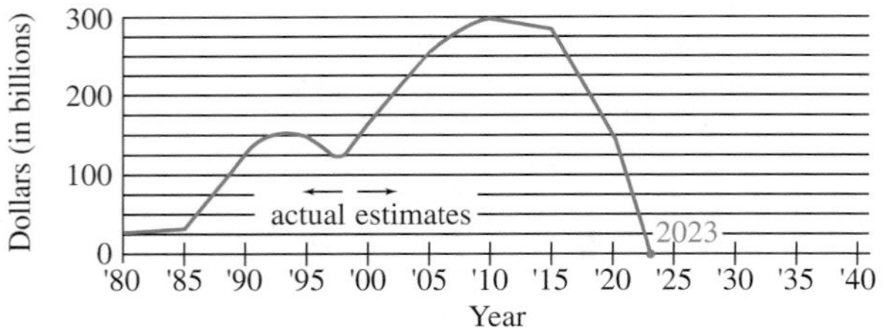

Using the Consumer Price Index for Urban Wage Earners and Clerical Workers

Find the approximate average rate of change in the trust fund for each time period.

a. From 1994 (the peak) to 1998 (the low point)

b. From 1998 to the estimated value for 2010

c. From 1990 to 1998

24. *Imported Cars* The percentage of U.S. automobile sales consisting of imports is shown in the following figure.†

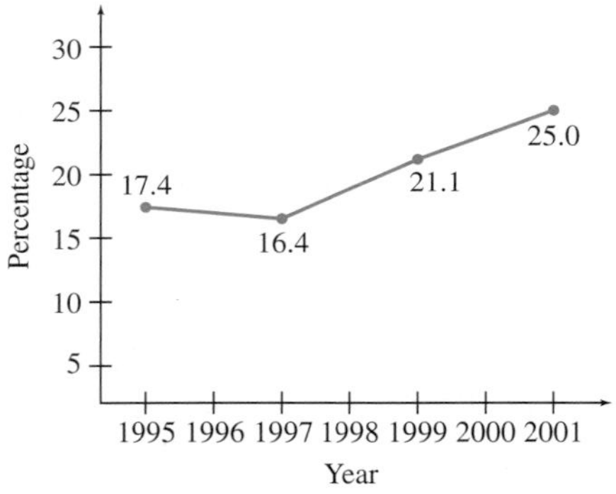

a. Use the points on the graph to find the slope of the line segment from 1995 to 1997. Find and interpret the rate of change in the percentage of automobiles imported for this period.

b. Use the points on the graph to find the slope of the line segment from 1999 to 2001. Find and interpret the rate of change in the percentage of automobiles imported for this period.

c. According to our definition, does the instantaneous rate of change in the percentage of automobiles imported exist for all values of x for the function in the graph? Explain your answer.

*Social Security Administration; Department of Health and Human Services.
†*The World Almanac and Book of Facts 2003*, p. 225.

25. *Sales* The graph shows annual sales (in thousands of dollars) of a Nintendo game. Find the average annual rate of change in sales for the following changes in years.

a. 1 to 4 **b.** 4 to 7 **c.** 7 to 12

d. What do your answers for parts a–c tell you about the sales of this product?

e. Give an example of another product that might have such a sales curve.

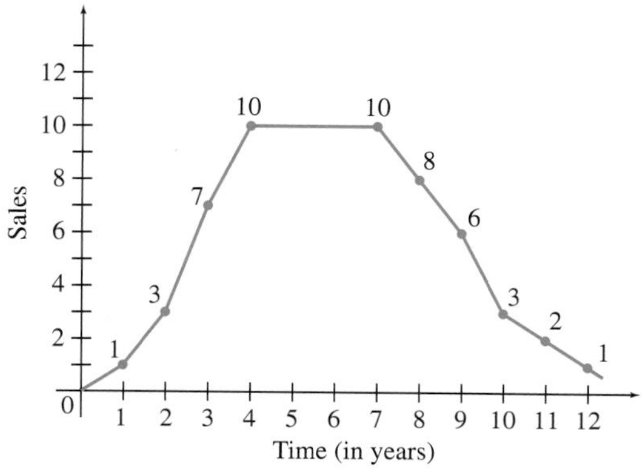

26. *New Technology versus Old* The graph at the top of the next column shows the ratio of users of new technology to users of old technology for two new technologies—Internet messages and PCs.*

a. Which curve has a greater rate of change from 1991 to 1997? Interpret your answer.

b. The graphs indicate that one curve is approximately linear, while the other could be modeled by an exponential function. Which ratio has an increasing, rather than constant, rate of change?

27. *Profit* Suppose that the total profit in hundreds of dollars from selling x items is given by

$$P(x) = 2x^2 - 5x + 6.$$

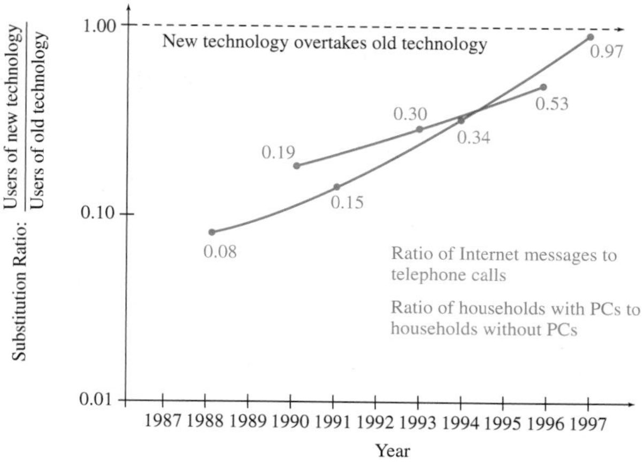

Find the average rate of change of profit for the following changes in x.

a. 2 to 4 **b.** 2 to 3

c. What is the instantaneous rate of change of profit with respect to the number of items produced when $x = 2$? (This number, called the *marginal profit* at $x = 2$, is the approximate profit from producing the third item.)

d. Find the marginal profit at $x = 4$.

28. *Revenue* The revenue (in thousands of dollars) from producing x units of an item is

$$R(x) = 10x - .002x^2.$$

a. Find the average rate of change of revenue when production is increased from 1000 to 1001 units.

b. Find the instantaneous rate of change of revenue with respect to the number of items produced when 1000 units are produced. (This number is called the *marginal revenue* at $x = 1000$.)

c. Find the additional revenue if production is increased from 1000 to 1001 units.

d. Compare your answers for parts a and b. What do you find?

29. *Demand* Suppose customers in a hardware store are willing to buy $N(p)$ boxes of nails at p dollars per box, as given by

$$N(p) = 80 - 5p^2, \quad 1 \le p \le 4.$$

a. Find the average rate of change of demand for a change in price from \$2 to \$3.

b. Find the instantaneous rate of change of demand when the price is \$2.

*Data from a May 1999 survey by Mercer Management Consulting in Washington, D.C.

c. Find the instantaneous rate of change of demand when the price is \$3.

d. As the price is increased from \$2 to \$3, how is demand changing? Is the change to be expected?

LIFE SCIENCES

30. *Flu Epidemic* Epidemiologists in College Station, Texas, estimate that t days after the flu begins to spread in town, the percent of the population infected by the flu is approximated by

$$p(t) = t^2 + t$$

for $0 \le t \le 5$.

a. Find the average rate of change of p with respect to t over the interval from 1 to 4 days.

b. Find the instantaneous rate of change of p with respect to t at $t = 3$.

31. *World Population Growth* The future size of the world population depends on how soon it reaches replacement-level fertility, the point at which each woman bears on average about 2.1 children. The graph shows projections for reaching that point in different years.*

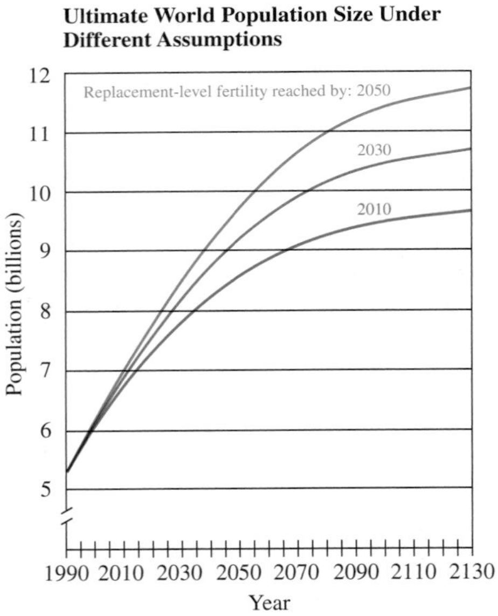

a. Estimate the average rate of change in population for each projection from 1990 to 2050. Which projection shows the smallest rate of change in world population?

b. Estimate the average rate of change in population from 2090 to 2130 for each projection. Interpret your answer.

32. *Bacteria Population* The graph shows the population in millions of bacteria t minutes after an antibiotic is introduced into a culture. Find and interpret the average rate of change of population with respect to time for the following time intervals.

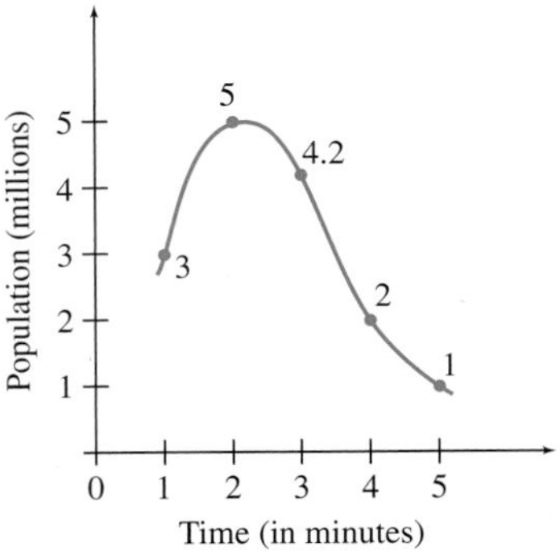

a. 1 to 2 **b.** 2 to 3

c. 3 to 4 **d.** 4 to 5

e. How long after the antibiotic was introduced did the population begin to decrease?

f. At what time did the rate of decrease of the population slow down?

33. *Molars* The mesiodistal crown length (as shown below) of deciduous mandibular first molars in fetuses is related to the postconception age of the tooth as

$$L(t) = -.01t^2 + .788t - 7.048,$$

where $L(t)$ is the crown length, in millimeters, of the molar t weeks after conception.†

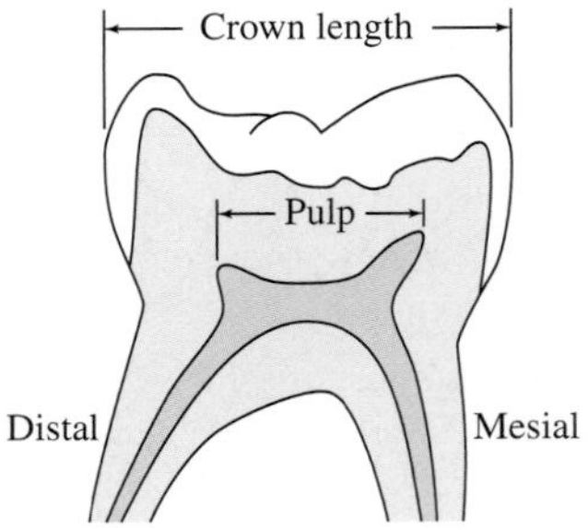

a. Find the average rate of growth in mesiodistal crown length during weeks 22 through 28.

*Carl Haub, Population Reference Bureau, 2000.

†Harris, E. F., J. D. Hicks, and B. D. Barcroft, "Tissue Contributions to Sex and Race: Differences in Tooth Crown Size of Deciduous Molars," *American Journal of Physical Anthropology,* Vol. 115, 2001, pp. 223–237.

b. Find the instantaneous rate of growth in mesiodistal crown length when the tooth is exactly 22 weeks of age.

c. Graph the function on $[0, 50]$ by $[0, 9]$. Does a function that increases and then begins to decrease make sense for this particular application? What do you suppose is happening during the first 11 weeks? Does this function accurately model crown length during those weeks?

34. *Thermic Effect of Food* The metabolic rate of a person who has just eaten a meal tends to go up and then, after some time has passed, returns to a resting metabolic rate. This phenomenon is known as the thermic effect of food. Researchers have indicated that the thermic effect of food (in kJ/hr) for a particular person is

$$F(t) = -10.28 + 175.9te^{-t/1.3},$$

where t is the number of hours that have elapsed since eating a meal.*

a. Graph the function on $[0, 6]$ by $[-20, 100]$.

b. Find the average rate of change of the thermic effect of food during the first hour after eating.

c. Use a graphing calculator to find the instantaneous rate of change of the thermic effect of food exactly 1 hour after eating.

d. Use a graphing calculator to estimate when the function stops increasing and begins to decrease.

35. *Mass of Bighorn Yearlings* The body mass of yearling bighorn sheep on Ram Mountain in Alberta, Canada, can be estimated by

$$M(t) = 27.5 + .3t - .001t^2$$

where $M(t)$ is measured in kilograms and t is days since May 25.†

a. Find the average rate of change of the weight of a bighorn yearling between 105 and 115 days past May 25.

b. Find the instantaneous rate of change of weight for a bighorn yearling sheep whose age is 105 days past May 25.

c. Graph the function $M(t)$ on $[5, 125]$ by $[25, 65]$.

d. Does the behavior of the function past 125 days accurately model the mass of the sheep? Why or why not?

SOCIAL SCIENCES

36. *Drug Trade* The U.S. government estimates of the worldwide potential net drug production (in metric tons) for the years 1994–1998 are given in the following graph.‡

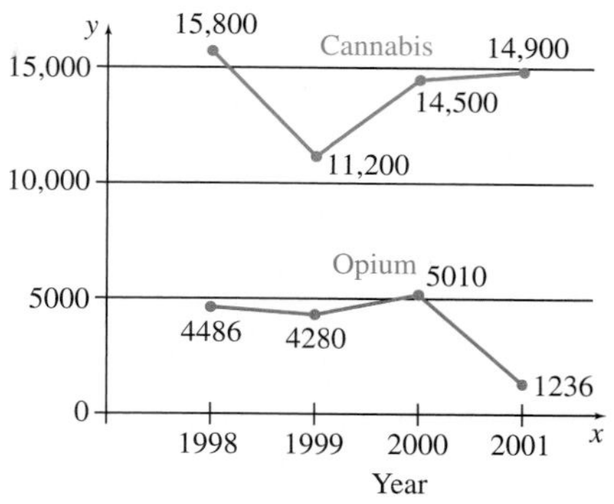

a. Estimate the average rate of change in the net production of cannabis and opium from 1998 to 2000 and from 2000 to 2001.

b. Which drug had the greatest change in net production during this time?

c. List some possible reasons for the observed behavior in marijuana production and opium production during this time.

37. *High School Drug Use* The U.S. government estimates of the percentage of high school seniors who have tried illegal drugs in the past month are shown in the following graph.§

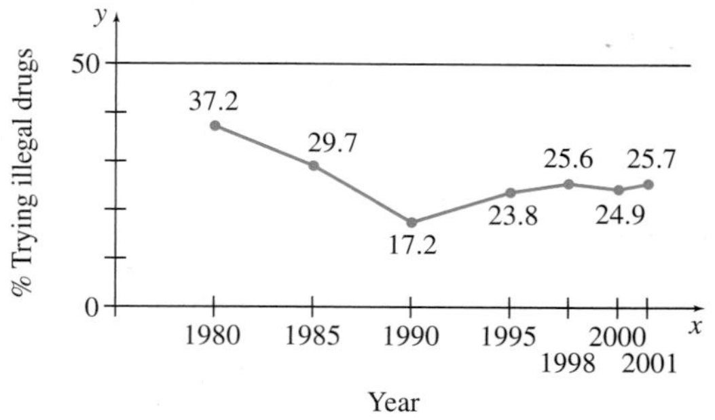

*Reed, G. and J. Hill, "Measuring the Thermic Effect of Food," *American Journal of Clinical Nutrition,* Vol. 63, 1996, pp. 164–169.

†Jorgenson, J., M. Festa-Bianchet, M. Lucherini, and W. Wishart, "Effects of Body Size, Population Density, and Maternal Characteristics on Age at First Reproduction of Bighorn Ewes," *Canadian Journal of Zoology,* Vol. 71, No. 12, Dec. 1993, pp. 2509–2517.

‡"ONDCP Drug Policy Information Clearinghouse, Drug Data Summary," Executive Office of the President, Office of National Drug Control Policy, NCJ-191351, March 2003, p. 5.

§www.whitehousedrugpolicy.gov/publications/factsht/druguse/index.html.

a. Estimate the average rate of change in the percentage of high school seniors who have tried illegal drugs from 1980 to 1990 and from 1990 to 2000.

b. Estimate the average rate of change in the percentage of high school seniors who have tried illegal drugs from 1980 to 2000 and compare your answer to those found in part a.

c. List some possible reasons for the observed changes in the percentage of high school seniors who have tried illegal drugs.

PHYSICAL SCIENCES

38. *Temperature* The graph shows the temperature T in degrees Celsius as a function of the altitude h in feet when an inversion layer is over Southern California. (An inversion layer is formed when air at a higher altitude, say 3000 ft, is warmer than air at sea level, even though air normally is cooler with increasing altitude.) Estimate and interpret the average rate of change in temperature for the following changes in altitude.

a. 1000 to 3000 ft **b.** 1000 to 5000 ft

c. 3000 to 9000 ft **d.** 1000 to 9000 ft

e. At what altitude at or below 7000 ft is the temperature highest? Lowest? How would your answer change if 7000 ft is changed to 10,000 ft?

f. At what altitude is the temperature the same as it is at 1000 ft?

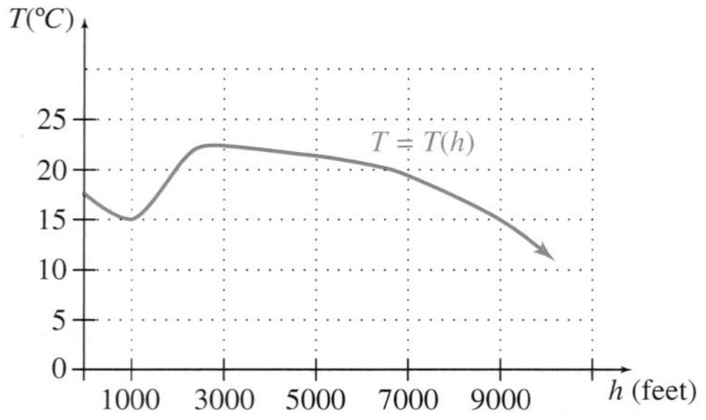

39. *Velocity* A car is moving along a straight test track. The position in feet of the car, $s(t)$, at various times t is measured, with the following results.

t (sec)	0	2	4	6	8	10
$s(t)$ (ft)	0	10	14	20	30	36

Find and interpret the average velocities for the following changes in t.

a. 0 to 2 seconds **b.** 2 to 4 seconds

c. 4 to 6 seconds **d.** 6 to 8 seconds

e. Estimate the instantaneous velocity at 4 seconds

i. by using the formula for estimating instantaneous rate (with $h = 2$), and

ii. by averaging the answers for the average velocity in the two seconds before and the two seconds after (that is, the answers to parts b and c).

f. Estimate the instantaneous velocity at 6 seconds using the two methods in part e.

g. Notice in parts e and f that your two answers are the same. Discuss whether this will always be the case, and why or why not.

40. *Velocity* Consider the example at the beginning of this section regarding the car traveling from San Francisco.

a. Estimate the instantaneous velocity at 1 hour. Assume that the velocity changes at a steady rate from one half-hour to the next.

b. Estimate the instantaneous velocity at 2 hours.

41. *Velocity* The distance of a particle from some fixed point is given by

$$s(t) = t^2 + 5t + 2,$$

where t is time measured in seconds. Find the average velocity of the particle over the following intervals.

a. 4 to 6 seconds

b. 4 to 5 seconds

c. Find the instantaneous velocity of the particle when $t = 4$.

3.4 DEFINITION OF THE DERIVATIVE

THINK ABOUT IT

How does the risk of chromosomal abnormality in a child change with the mother's age?

We will answer this question in Example 3, using the concept of the derivative.

In the previous section, the formula

$$\lim_{h \to 0} \frac{f(a+h) - f(a)}{h}$$

was used to calculate the instantaneous rate of change of a function f at the point where $x = a$. Now we will give a geometric interpretation of this limit.

The Tangent Line In geometry, a *tangent line* to a circle is defined as a line that touches the circle at only one point, as at the point P in Figure 25 (which shows the top half of a circle). If you think of this half-circle as part of a curving road on which you are driving at night, then the tangent line indicates the direction of the light beam from your headlights as you pass through the point P. Intuitively, the tangent line to an arbitrary curve at a point P on the curve should touch the curve at P, but not at any points nearby, and should indicate the direction of the curve. In Figure 26, for example, the lines through P_1 and P_3 are tangent lines, while the lines through P_2 and P_5 are not. The tangent lines just touch the curve, while the other lines pass through it. To decide about the line at P_4, we need to define the idea of a tangent line to the graph of a function more carefully.

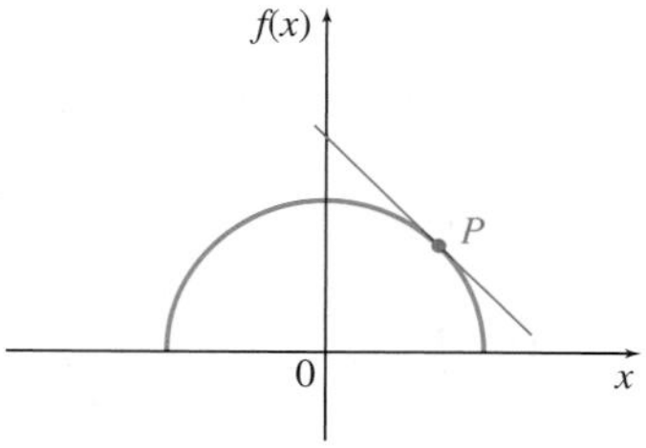

FIGURE 25

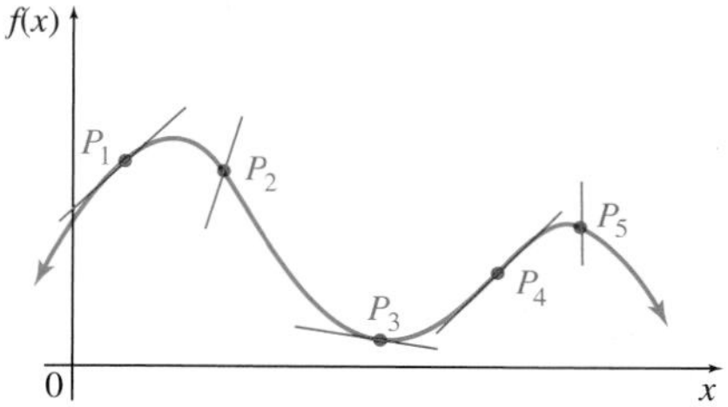

FIGURE 26

To see how we might define the slope of a line tangent to the graph of a function f at a given point, let R be a fixed point with coordinates $(a, f(a))$ on the graph of a function $y = f(x)$, as in Figure 27. Choose a different point S on the graph and draw the line through R and S; this line is called a **secant line.** If S has coordinates $(a + h, f(a + h))$, then by the definition of slope, the slope of the secant line RS is given by

$$\text{Slope of secant} = \frac{\Delta y}{\Delta x} = \frac{f(a + h) - f(a)}{a + h - a} = \frac{f(a + h) - f(a)}{h}.$$

This slope corresponds to the average rate of change of y with respect to x over the interval from a to $a + h$. As h approaches 0, point S will slide along the curve, getting closer and closer to the fixed point R. See Figure 28, which shows

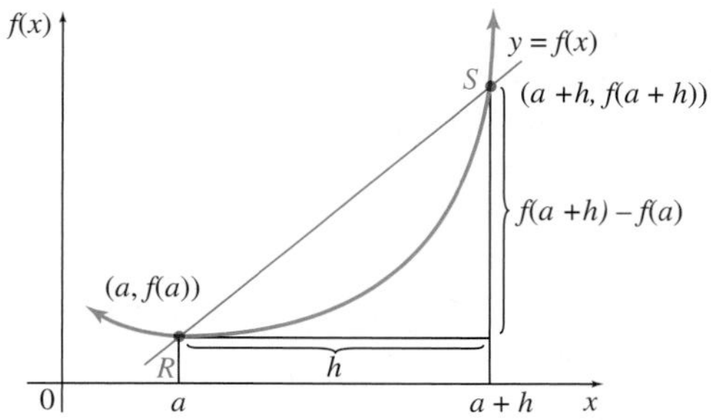

FIGURE 27

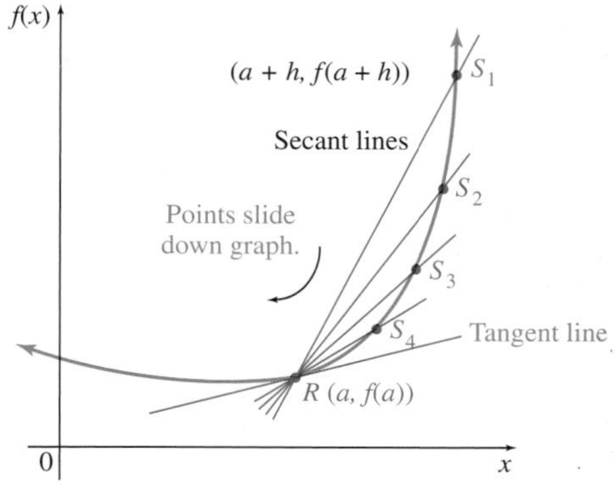

FIGURE 28

successive positions S_1, S_2, S_3, and S_4 of the point S. If the slopes of the corresponding secant lines approach a limit as h approaches 0, then this limit is defined to be the slope of the tangent line at point R.

SLOPE OF THE TANGENT LINE

The **tangent line** of the graph of $y = f(x)$ at the point $(a, f(a))$ is the line through this point having slope

$$\lim_{h \to 0} \frac{f(a+h) - f(a)}{h},$$

provided this limit exists. If this limit does not exist, then there is no tangent at the point.

The slope of this line at a point is also called the **slope of the curve** at the point and corresponds to the instantaneous rate of change of y with respect to x at the point.

In certain applications of mathematics it is necessary to determine the equation of a line tangent to the graph of a function at a given point, as in the next example.

EXAMPLE 1 *Tangent Line*

Find the slope of the tangent line to the graph of $f(x) = x^2 + 2$ at $x = -1$. Find the equation of the tangent line.

Solution Use the definition given above, with $f(x) = x^2 + 2$ and $a = -1$. The slope of the tangent line is given by

$$\begin{aligned}
\text{Slope of tangent} &= \lim_{h \to 0} \frac{f(a+h) - f(a)}{h} \\
&= \lim_{h \to 0} \frac{[(-1+h)^2 + 2] - [(-1)^2 + 2]}{h} \\
&= \lim_{h \to 0} \frac{[1 - 2h + h^2 + 2] - [1 + 2]}{h} \\
&= \lim_{h \to 0} \frac{-2h + h^2}{h} \\
&= \lim_{h \to 0} (-2 + h) = -2.
\end{aligned}$$

The slope of the tangent line at $(-1, f(-1)) = (-1, 3)$ is -2.

The equation of the tangent line can be found with the point-slope form of the equation of a line from Chapter 1.

$$\begin{aligned}
y - y_1 &= m(x - x_1) \\
y - 3 &= -2[x - (-1)] \\
y - 3 &= -2(x + 1) \\
y - 3 &= -2x - 2 \\
y &= -2x + 1
\end{aligned}$$

FOR REVIEW

In Section 1.1, we saw that the equation of a line can be found with the point-slope form $y - y_1 = m(x - x_1)$, if the slope m and the coordinates (x_1, y_1) of a point on the line are known. Use the point-slope form to find the equation of the line with slope 3 that goes through the point $(-1, 4)$.

Let $m = 3$, $x_1 = -1$, $y_1 = 4$. Then

$$\begin{aligned} y - y_1 &= m(x - x_1) \\ y - 4 &= 3(x - (-1)) \\ y - 4 &= 3x + 3 \\ y &= 3x + 7. \end{aligned}$$

Figure 29 shows a graph of $f(x) = x^2 + 2$, along with a graph of the tangent line at $x = -1$.

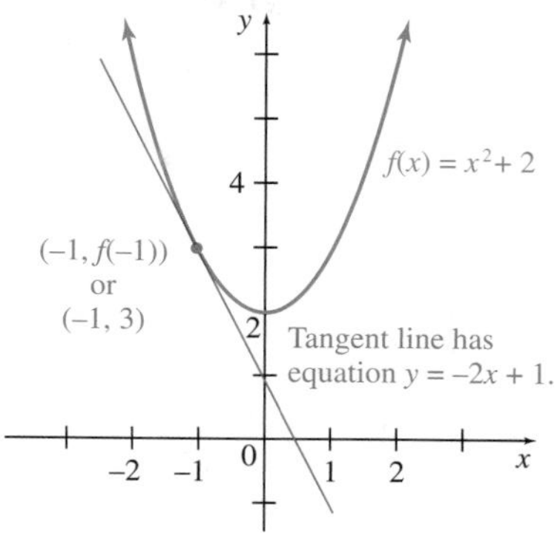

FIGURE 29

NOTE Some graphing calculators have a TANLN feature that draws the tangent line and gives its equation. If your calculator has this feature, use it to duplicate Figure 29. ■

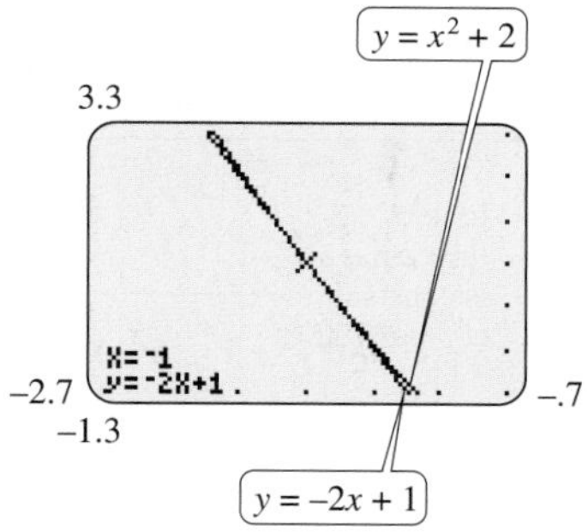

FIGURE 30

Figure 30 shows the result of using a graphing calculator to zoom in on the point $(-1, 3)$ in Figure 29. Notice that in this closeup view, the graph and its tangent line appear virtually identical. This gives us another interpretation of the tangent line. Suppose, as we zoom in on a function, the graph appears to become a straight line. Then this line is the tangent line to the graph at that point. In other words, the tangent line captures the behavior of the function very close to the point under consideration. (This assumes, of course, that the function when viewed close up is approximately a straight line. As we will see later in this section, this may not occur.)

If it exists, the tangent line at $x = a$ is a good approximation of the graph of a function near $x = a$.

Consequently, another way to approximate the slope of the curve is to zoom in on the function until it appears to be a straight line (the tangent line). Then find the slope using any two points on that line.

EXAMPLE 2 *Slope*

Use a graphing calculator to find the slope of the graph of $f(x) = x^x$ at $x = 1$.

Solution The slope would be challenging to evaluate algebraically using the limit definition. Instead, using a graphing calculator on the window [0, 2] by [0, 2], we see the graph in Figure 31 on the next page. Zooming in gives the view in Figure 32. Using the TRACE key, we find two points on the line to be $(1, 1)$ and $(1.0021277, 1.0021322)$. Therefore, the slope is approximately

$$\frac{1.0021322 - 1}{1.0021277 - 1} \approx 1.$$

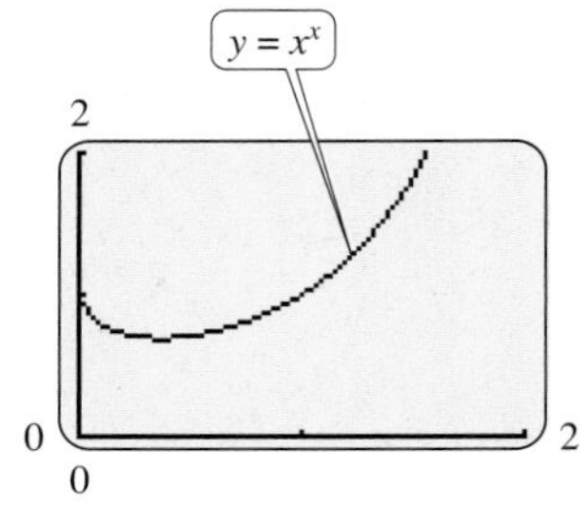

FIGURE 31

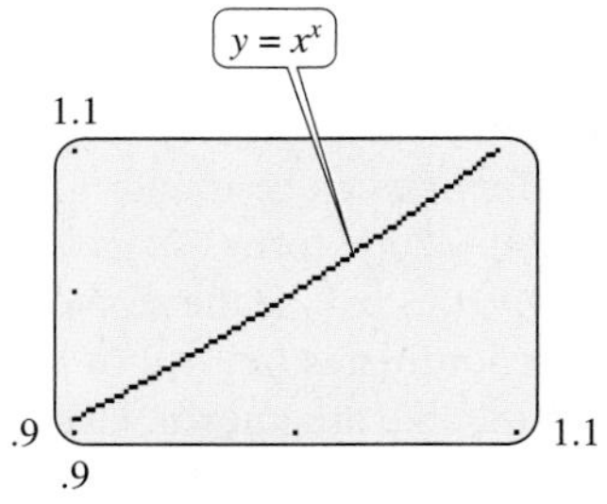

FIGURE 32

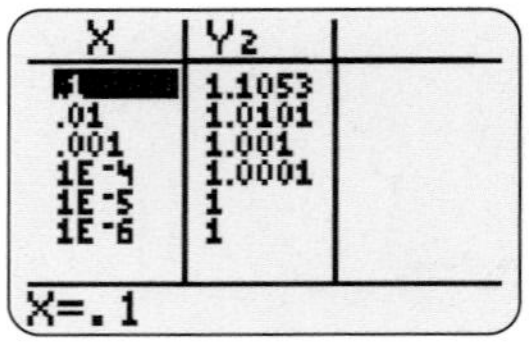

FIGURE 33

Rather than using a graph, we could use the graphing calculator to create a table, as we did in the previous section to estimate the instantaneous rate of change. Letting $\mathtt{Y_1} = \mathtt{X \hat{} X}$ and $\mathtt{Y_2=(Y_1(1+X)-Y_1(1))/X}$ results in the table shown in Figure 33. Based on this table, we estimate that the slope of the graph of $f(x) = x^x$ at $x = 1$ is 1.

EXAMPLE 3 *Genetics*

Figure 34 shows how the risk of chromosomal abnormality in a child increases with the age of the mother.* Find the rate that the risk is rising when the mother is 40 years old.

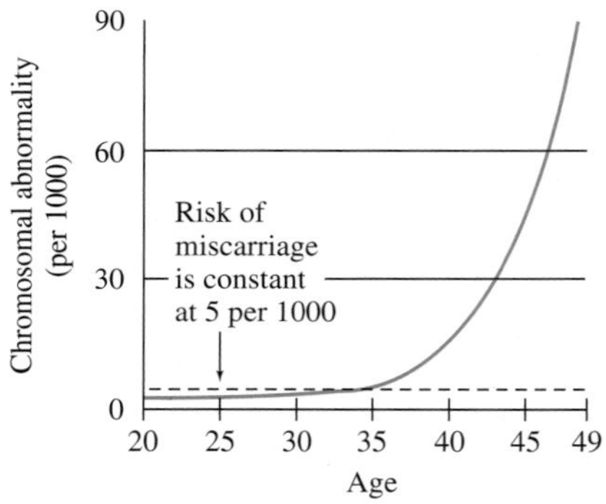

FIGURE 34

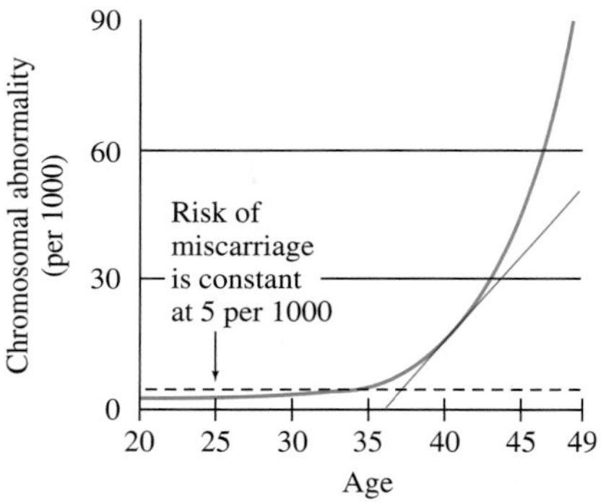

FIGURE 35

Solution In Figure 35, we have added the tangent line to the graph at the point where the age of the mother is 40. At that point, the risk is approximately 15 per 1000. Extending the line, we estimate that when the age is 45, the y-coordinate of the line is roughly 35. Thus, the slope of the line is

$$\frac{35 - 15}{45 - 40} = \frac{20}{5} = 4.$$

Therefore, at the age of 40, the risk of chromosomal abnormality in the child is increasing at the rate of about 4 per 1000 for each additional year of the mother's age.

*_The New York Times,_ Feb. 5, 1994, p. 24.

The Derivative If $y = f(x)$ is a function and a is a number in its domain, then we shall use the symbol $f'(a)$ to denote the special limit

$$\lim_{h \to 0} \frac{f(a+h) - f(a)}{h},$$

provided that it exists. This means that for each number a we can assign the number $f'(a)$ found by calculating this limit. This assignment defines an important new function.

DERIVATIVE

The **derivative** of the function f at x is defined as

$$f'(x) = \lim_{h \to 0} \frac{f(x+h) - f(x)}{h},$$

provided this limit exists.

The notation $f'(x)$ is read "f-prime of x." The function $f'(x)$ is called the derivative of f with respect to x. If x is a value in the domain of f and if $f'(x)$ exists, then f is **differentiable** at x. The process that produces f' is called **differentiation.**

NOTE The derivative is a *function of* x, since $f'(x)$ varies as x varies. This differs from both the slope of the tangent line and the instantaneous rate of change, either of which is represented by the number $f'(a)$ that corresponds to a number a. ■

The derivative function has several interpretations, two of which we have discussed.

1. The function f' represents the *instantaneous rate of change* of y with respect to x. This instantaneous rate of change could be interpreted as marginal cost, revenue, or profit (if the original function represented cost, revenue, or profit) or velocity (if the original function described displacement along a line). From now on we will use *rate of change* to mean *instantaneous* rate of change.
2. The function f' represents the *slope* of the graph of $f(x)$ at any point x. If the derivative is evaluated at the point $x = a$, then it represents the slope of the curve at that point.

The following table compares the different interpretations of the difference quotient and the derivative.

The Difference Quotient and the Derivative

Difference Quotient	Derivative
$\frac{f(b) - f(a)}{b - a}$	$\lim_{h \to 0} \frac{f(x+h) - f(x)}{h}$
■ Slope of the secant line	■ Slope of the tangent line
■ Average rate of change	■ Instantaneous rate of change
■ Average velocity	■ Instantaneous velocity

The next few examples show how to use the definition to find the derivative of a function by means of a four-step procedure.

EXAMPLE 4 *Derivative*

Let $f(x) = x^2$.

(a) Find the derivative.

Solution By definition, for all values of x where the following limit exists, the derivative is given by

$$f'(x) = \lim_{h \to 0} \frac{f(x+h) - f(x)}{h}.$$

Use the following sequence of steps to evaluate this limit.

Step 1 Find $f(x+h)$.
Replace x with $x + h$ in the equation for $f(x)$. Simplify the result.

$$\begin{aligned} f(x) &= x^2 \\ f(x+h) &= (x+h)^2 \\ &= x^2 + 2xh + h^2 \end{aligned}$$

(Note that $f(x+h) \neq f(x) + h$, since $f(x) + h = x^2 + h$.)

Step 2 Find $f(x+h) - f(x)$.
Since $f(x) = x^2$,

$$f(x+h) - f(x) = (x^2 + 2xh + h^2) - x^2 = 2xh + h^2.$$

Step 3 Find and simplify the quotient $\dfrac{f(x+h) - f(x)}{h}$.

$$\frac{f(x+h) - f(x)}{h} = \frac{2xh + h^2}{h} = \frac{h(2x+h)}{h} = 2x + h$$

Step 4 Finally, find the limit as h approaches 0. In this step, h is the variable and x is fixed.

$$\begin{aligned} f'(x) &= \lim_{h \to 0} \frac{f(x+h) - f(x)}{h} \\ &= \lim_{h \to 0}(2x + h) \\ &= 2x + 0 = 2x \end{aligned}$$

(b) Calculate and interpret $f'(3)$. Use the function defined by $f'(x) = 2x$.

Solution

Method 1: Algebraic Method

$$f'(3) = 2 \cdot 3 = 6$$

The number 6 is the slope of the tangent line to the graph of $f(x) = x^2$ at the point where $x = 3$, that is, at $(3, f(3)) = (3, 9)$. See Figure 36(a) on the next page.

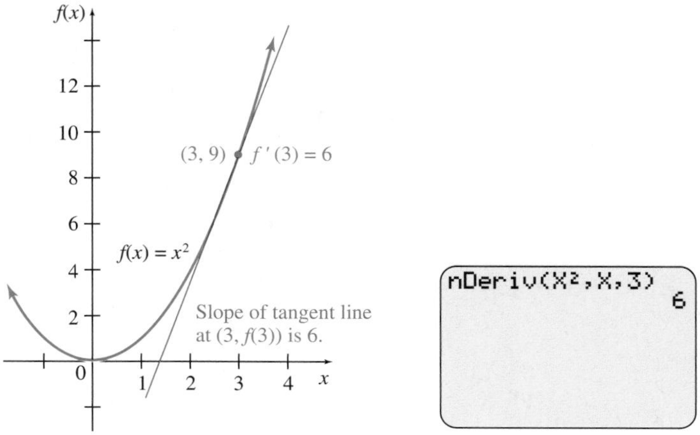

(a) (b)

FIGURE 36

Method 2: Graphing Calculator Some graphing calculators can calculate the value of the derivative at a given x-value. For example, the TI-83/84 Plus uses the `nDeriv` command as shown in Figure 36(b), with the expression for $f(x)$, the variable, and the value of a entered in the parentheses, to find $f'(3)$ for $f(x) = x^2$.

CAUTION

1. In Example 4(a) notice that $f(x + h)$ is *not* equal to $f(x) + h$. In fact,

$$f(x + h) = (x + h)^2 = x^2 + 2xh + h^2,$$

but

$$f(x) + h = x^2 + h.$$

2. In Example 4(b), do not confuse $f(3)$ and $f'(3)$. The value $f(3)$ is the y-value that corresponds to $x = 3$. It is found by substituting 3 for x in $f(x)$; $f(3) = 3^2 = 9$. On the other hand, $f'(3)$ is the slope of the tangent line to the curve at $x = 3$; as Example 4(b) shows, $f'(3) = 2 \cdot 3 = 6$. ■

FINDING $f'(x)$ FROM THE DEFINITION OF DERIVATIVE

The four steps used to find the derivative $f'(x)$ for a function $y = f(x)$ are summarized here.

1. Find $f(x + h)$.
2. Find and simplify $f(x + h) - f(x)$.
3. Divide by h to get $\dfrac{f(x + h) - f(x)}{h}$.
4. Let $h \to 0$; $f'(x) = \lim_{h \to 0} \dfrac{f(x + h) - f(x)}{h}$ if this limit exists.

EXAMPLE 5 *Derivative*

Let $f(x) = 2x^3 + 4x$. Find $f'(x)$, $f'(2)$, and $f'(-3)$.

Solution Go through the four steps to find $f'(x)$.

Step 1 Find $f(x+h)$ by replacing x with $x+h$.

$$\begin{aligned} f(x+h) &= 2(x+h)^3 + 4(x+h) \\ &= 2(x^3 + 3x^2h + 3xh^2 + h^3) + 4(x+h) \\ &= 2x^3 + 6x^2h + 6xh^2 + 2h^3 + 4x + 4h \end{aligned}$$

Step 2
$$\begin{aligned} f(x+h) - f(x) &= 2x^3 + 6x^2h + 6xh^2 + 2h^3 + 4x + 4h \\ &\quad - 2x^3 - 4x \\ &= 6x^2h + 6xh^2 + 2h^3 + 4h \end{aligned}$$

Step 3
$$\begin{aligned} \frac{f(x+h) - f(x)}{h} &= \frac{6x^2h + 6xh^2 + 2h^3 + 4h}{h} \\ &= \frac{h(6x^2 + 6xh + 2h^2 + 4)}{h} \\ &= 6x^2 + 6xh + 2h^2 + 4 \end{aligned}$$

Step 4 Now use the rules for limits to get

$$\begin{aligned} f'(x) &= \lim_{h\to 0} \frac{f(x+h) - f(x)}{h} \\ &= \lim_{h\to 0} (6x^2 + 6xh + 2h^2 + 4) \\ &= 6x^2 + 6x(0) + 2(0)^2 + 4 \\ f'(x) &= 6x^2 + 4. \end{aligned}$$

Use this result to find $f'(2)$ and $f'(-3)$.

$$\begin{aligned} f'(2) &= 6 \cdot 2^2 + 4 = 28 \\ f'(-3) &= 6 \cdot (-3)^2 + 4 = 58 \end{aligned}$$

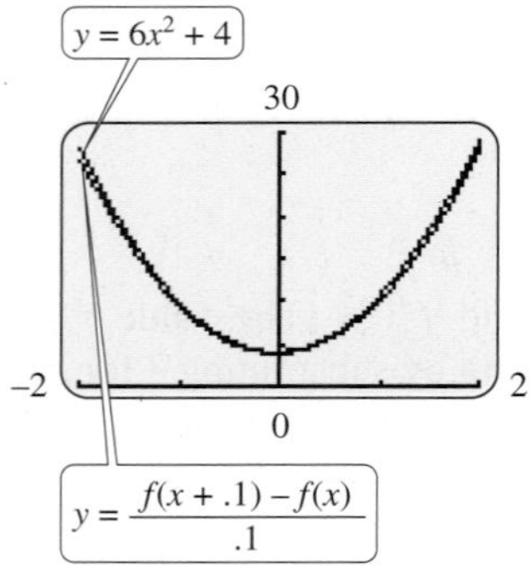

FIGURE 37

One way to support this result is to plot $[f(x+h) - f(x)]/h$ on a graphing calculator with a small value of h. Figure 37 shows a graphing calculator screen of $y = [f(x+.1) - f(x)]/.1$, where f is the function $f(x) = 2x^3 + 4x$, and $y = 6x^2 + 4$, which was just found to be the derivative of f. The two functions, plotted on the window $[-2, 2]$ by $[0, 30]$, appear virtually identical. If $h = .01$ had been used, the two functions would be indistinguishable.

EXAMPLE 6 *Derivative*

Let $f(x) = \dfrac{4}{x}$. Find $f'(x)$.

Solution

Step 1 $f(x+h) = \dfrac{4}{x+h}$

Step 2 $f(x+h) - f(x) = \dfrac{4}{x+h} - \dfrac{4}{x}$

$= \dfrac{4x - 4(x+h)}{x(x+h)}$ Find a common denominator.

$= \dfrac{4x - 4x - 4h}{x(x+h)}$ Simplify the numerator.

$= \dfrac{-4h}{x(x+h)}$

Step 3 $\dfrac{f(x+h) - f(x)}{h} = \dfrac{\frac{-4h}{x(x+h)}}{h}$

$= \dfrac{-4h}{x(x+h)} \cdot \dfrac{1}{h}$ Invert and multiply.

$= \dfrac{-4}{x(x+h)}$

Step 4 $f'(x) = \lim_{h \to 0} \dfrac{f(x+h) - f(x)}{h}$

$= \lim_{h \to 0} \dfrac{-4}{x(x+h)}$

$= \dfrac{-4}{x(x+0)}$

$f'(x) = \dfrac{-4}{x(x)} = \dfrac{-4}{x^2}$

Notice that in Example 6 neither $f(x)$ nor $f'(x)$ is defined when $x = 0$. Look at a graph of $f(x) = 4/x$ to see why this is true.

EXAMPLE 7 *Weight Gain*

A mathematics professor found that, after introducing his dog Django to a new brand of food, Django's weight began to increase. After x weeks on the new food, Django's weight (in pounds) was approximately given by $w(x) = \sqrt{x} + 40$ for $0 \le x \le 6$. Find the rate of change of Django's weight after x weeks.

Solution

Step 1 $w(x+h) = \sqrt{x+h} + 40$

Step 2 $w(x+h) - w(x) = \sqrt{x+h} + 40 - (\sqrt{x} + 40)$

$= \sqrt{x+h} - \sqrt{x}$

Step 3 $\dfrac{w(x+h) - w(x)}{h} = \dfrac{\sqrt{x+h} - \sqrt{x}}{h}$

At this point, in order to be able to divide by h, multiply both numerator and denominator by $\sqrt{x+h}+\sqrt{x}$; that is, rationalize the *numerator*.

$$\frac{w(x+h)-w(x)}{h} = \frac{\sqrt{x+h}-\sqrt{x}}{h}\cdot\frac{\sqrt{x+h}+\sqrt{x}}{\sqrt{x+h}+\sqrt{x}}$$

$$= \frac{(\sqrt{x+h})^2-(\sqrt{x})^2}{h(\sqrt{x+h}+\sqrt{x})} \qquad (a-b)(a+b)=a^2-b^2.$$

$$= \frac{x+h-x}{h(\sqrt{x+h}+\sqrt{x})}$$

$$= \frac{1}{\sqrt{x+h}+\sqrt{x}} \qquad \text{Simplify.}$$

Step 4 $$w'(x) = \lim_{h\to 0}\frac{1}{\sqrt{x+h}+\sqrt{x}} = \frac{1}{\sqrt{x}+\sqrt{x}} = \frac{1}{2\sqrt{x}}$$

This tells us, for example, that after 4 weeks, when Django's weight is $w(4)=\sqrt{4}+40=42$ lb, her weight is increasing at a rate of $w'(4)=1/(2\sqrt{4})=1/4$ lb per week.

EXAMPLE 8 *Cost Analysis*

The cost in dollars to manufacture x graphing calculators is given by $C(x)=-.005x^2+20x+150$ when $0\le x\le 2000$. Find the rate of change of cost with respect to the number manufactured when 100 calculators are made and when 1000 calculators are made.

Solution The rate of change of cost is given by the derivative of the cost function,

$$C'(x)=\lim_{h\to 0}\frac{C(x+h)-C(x)}{h}.$$

Going through the steps for finding $C'(x)$ gives

$$C'(x)=-.01x+20.$$

When $x=100$,

$$C'(100)=-.01(100)+20=19.$$

This rate of change of cost per calculator gives the marginal cost at $x=100$, which means the approximate cost of producing the 101st calculator is \$19.

When 1000 calculators are made, the marginal cost is

$$C'(1000)=-.01(1000)+20=10,$$

or \$10.

We can use the notation for the derivative to write the equation of the tangent line. Using the point-slope form, $y-y_1=m(x-x_1)$, and letting $y_1=f(x_1)$ and $m=f'(x_1)$, we have the following formula.

EQUATION OF THE TANGENT LINE

The tangent line to the graph of $y = f(x)$ at the point $(x_1, f(x_1))$ is given by the equation

$$y - f(x_1) = f'(x_1)(x - x_1),$$

provided $f'(x)$ exists.

EXAMPLE 9 *Tangent Line*

Find the equation of the tangent line to the graph of $f(x) = 4/x$ at $x = 2$.

Solution From the answer to Example 6, we have $f'(x) = -4/x^2$, so $f'(x_1) = f'(2) = -4/2^2 = -1$. Also $f(x_1) = f(2) = 4/2 = 2$. Then the equation of the tangent line is

$$y - 2 = (-1)(x - 2),$$

or

$$y = -x + 4$$

after simplifying.

Existence of the Derivative The definition of the derivative included the phrase "provided this limit exists." If the limit used to define the derivative does not exist, then of course the derivative does not exist. For example, a derivative cannot exist at a point where the function itself is not defined. If there is no function value for a particular value of x, there can be no tangent line for that value. This was the case in Example 6—there was no tangent line (and no derivative) when $x = 0$.

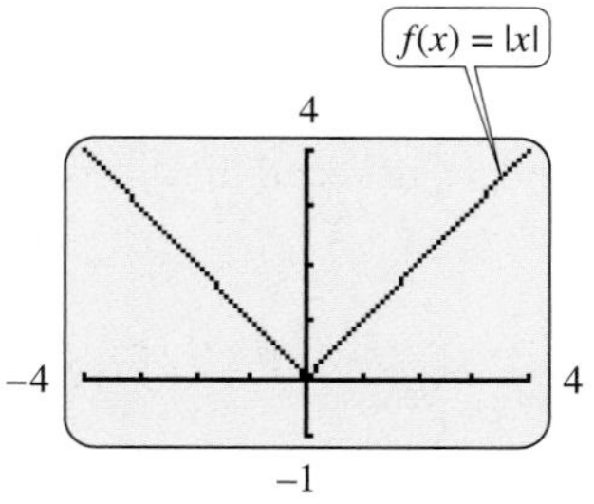

FIGURE 38

Derivatives also do not exist at "corners" or "sharp points" on a graph. For example, the function graphed in Figure 38 is the *absolute value function*, defined previously as

$$f(x) = \begin{cases} x & \text{if } x \geq 0 \\ -x & \text{if } x < 0, \end{cases}$$

and written $f(x) = |x|$. By the definition of derivative, the derivative at any value of x is given by

$$f'(x) = \lim_{h \to 0} \frac{f(x + h) - f(x)}{h},$$

provided this limit exists. To find the derivative at 0 for $f(x) = |x|$, replace x with 0 and $f(x)$ with $|0|$ to get

$$f'(0) = \lim_{h \to 0} \frac{|0 + h| - |0|}{h} = \lim_{h \to 0} \frac{|h|}{h}.$$

In Example 2 in the first section of this chapter, we showed that

$$\lim_{h \to 0} \frac{|h|}{h} \text{ does not exist;}$$

therefore, the derivative does not exist at 0. However, the derivative does exist for all values of x other than 0.

In Figure 39, we have zoomed in on the origin in Figure 38. Notice that the graph looks essentially the same. The corner is still sharp, and the graph does not resemble a straight line any more than it originally did. As we observed earlier, the derivative only exists at a point when the function more and more resembles a straight line as we zoom in on the point.

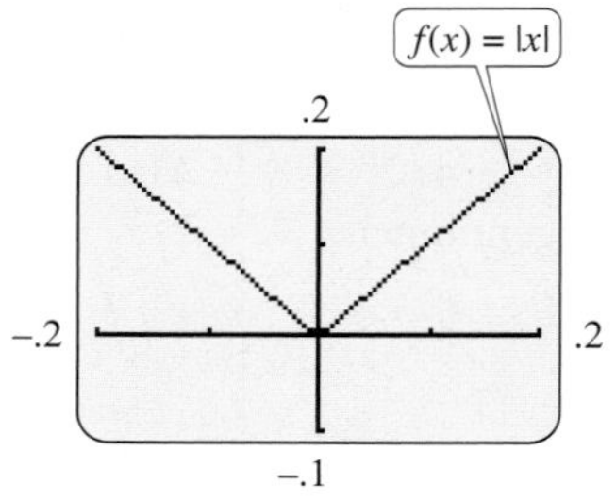

FIGURE 39

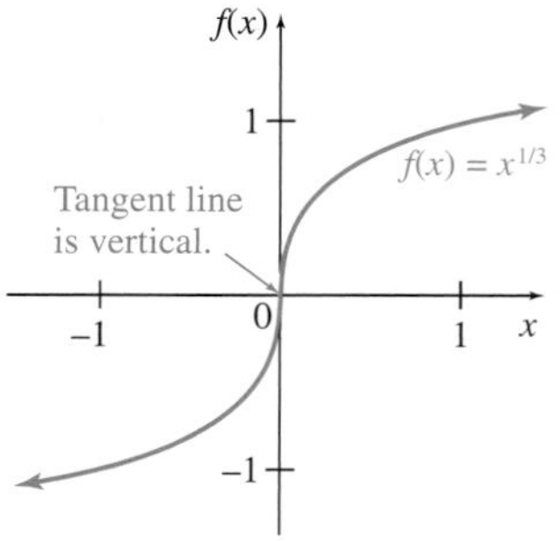

FIGURE 40

A graph of the function $f(x) = x^{1/3}$ is shown in Figure 40. As the graph suggests, the tangent line is vertical when $x = 0$. Since a vertical line has an undefined slope, the derivative of $f(x) = x^{1/3}$ cannot exist when $x = 0$. Use the fact that $\lim_{h\to 0} h^{1/3}/h = \lim_{h\to 0} 1/h^{2/3}$ does not exist and the definition of the derivative to verify that $f'(0)$ does not exist for $f(x) = x^{1/3}$.

Figure 41 summarizes the various ways that a derivative can fail to exist. Notice in Figure 41 that at a point where the function is discontinuous, such as x_3, x_4, and x_6, the derivative does not exist. A function must be continuous at a point for the derivative to exist there. But just because a function is continuous at a point does not mean the derivative necessarily exists; for example, observe the behavior of the function in Figure 41 at x_1, x_2, and x_5, or recall that $f(x) = |x|$ is continuous at $x = 0$ but $f'(0)$ does not exist.

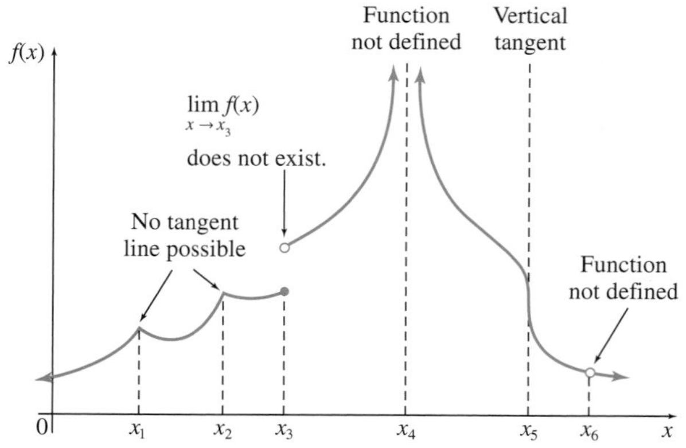

FIGURE 41

EXAMPLE 10 *Astronomy*

A nova is a star whose brightness suddenly increases and then gradually fades. The cause of the sudden increase in brightness is thought to be an explosion of some kind. The intensity of light emitted by a nova as a function of time is shown in Figure 42.* Notice that although the graph is a continuous curve, it is not differentiable at the point of the explosion.

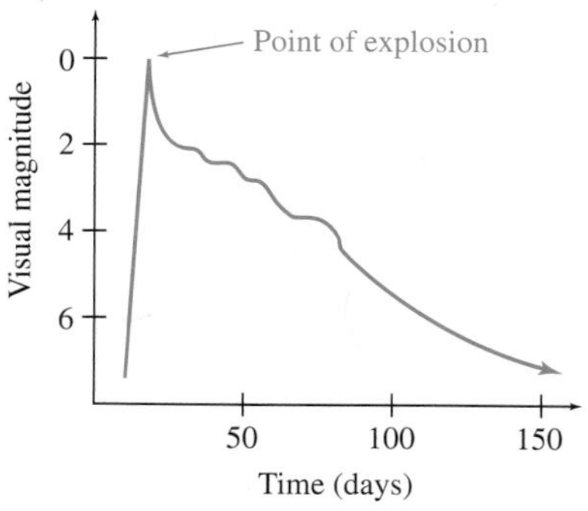

FIGURE 42

3.4 EXERCISES

1. By considering, but not calculating, the slope of the tangent line, give the derivative of the following.

   a. $f(x) = 5$ b. $f(x) = x$ c. $f(x) = -x$ d. The line $x = 3$

   e. The line $y = mx + b$

2. a. Suppose $g(x) = \sqrt[3]{x}$. Use the graph of $g(x)$ to find $g'(0)$.

   b. Explain why the derivative of a function does not exist at a point where the tangent line is vertical.

3. If $f(x) = \dfrac{x^2 - 1}{x + 2}$, where is f not differentiable?

4. If the rate of change of $f(x)$ is zero when $x = a$, what can be said about the tangent line to the graph of $f(x)$ at $x = a$?

Estimate the slope of the tangent line to each curve at the given point (x, y).

5.

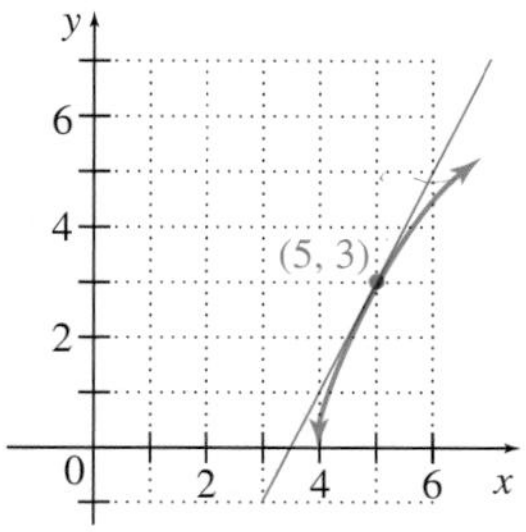

6.

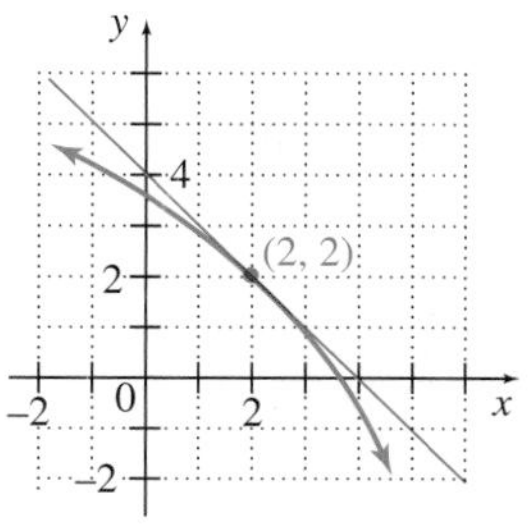

7.

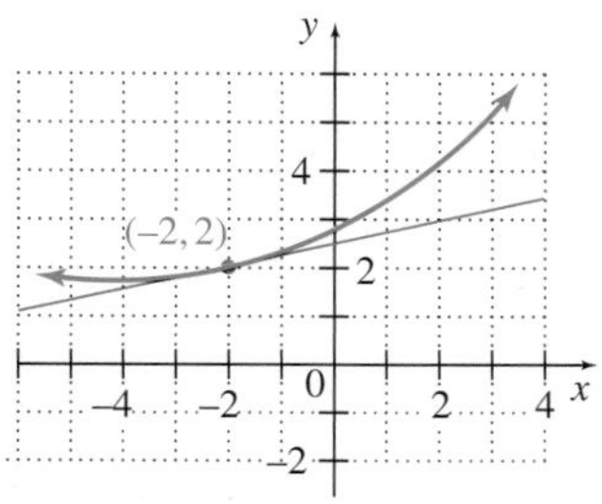

*Kaufmann III, William J., "The Light Curve of a Nova," *Astronomy: The Structure of the Universe*, New York: Macmillan, 1977. Reprinted by permission of William J. Kaufmann III.

8.

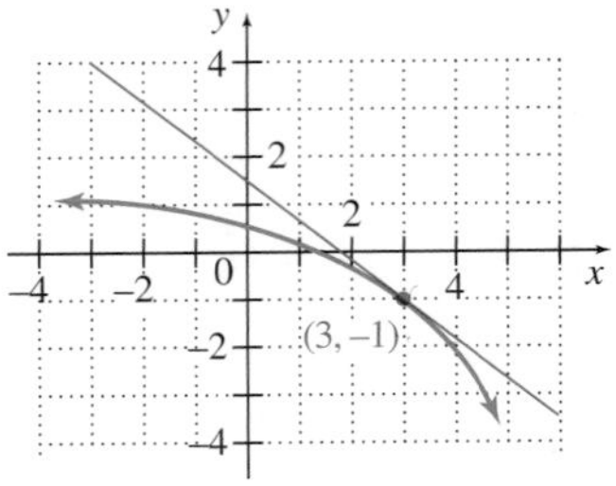

9.

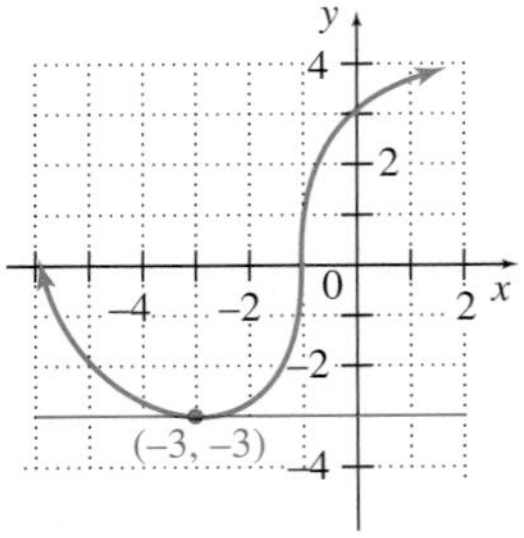

10.

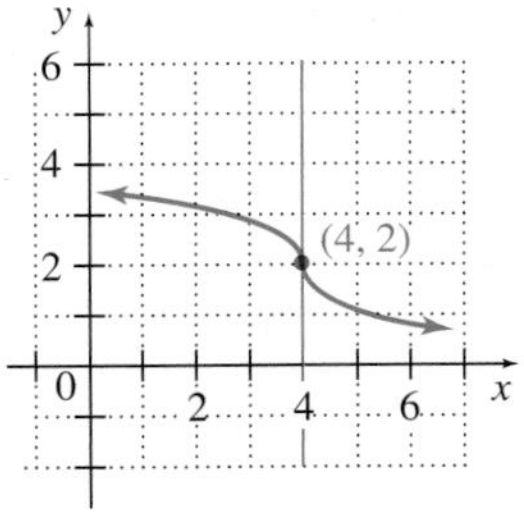

Using the definition of the derivative, find $f'(x)$. Then find $f'(-2)$, $f'(0)$, and $f'(3)$. (Hint for Exercises 15 and 16: *In Step 3, multiply numerator and denominator by* $\sqrt{x+h} + \sqrt{x}$.)

11. $f(x) = -4x^2 + 11x$

12. $f(x) = 6x^2 - 4x$

13. $f(x) = -2/x$

14. $f(x) = 6/x$

15. $f(x) = \sqrt{x}$

16. $f(x) = -3\sqrt{x}$

Find the equation of the tangent line to each curve when x has the given value.

17. $f(x) = x^2 + 2x;\quad x = 3$

18. $f(x) = 6 - x^2;\quad x = -1$

19. $f(x) = 5/x;\quad x = 2$

20. $f(x) = -3/(x + 1);\quad x = 1$

21. $f(x) = 4\sqrt{x};\quad x = 9$

22. $f(x) = \sqrt{x};\quad x = 25$

Use a graphing calculator to find $f'(2)$, $f'(16)$, and $f'(-3)$ for the following.

23. $f(x) = -4x^2 + 11x$

24. $f(x) = 6x^2 - 4x$

25. $f(x) = 8x + 6$

26. $f(x) = -9x - 5$

27. $f(x) = -\dfrac{2}{x}$

28. $f(x) = \dfrac{6}{x}$

29. $f(x) = \sqrt{x}$

30. $f(x) = -3\sqrt{x}$

Find the x-values where the following do not have derivatives.

31.

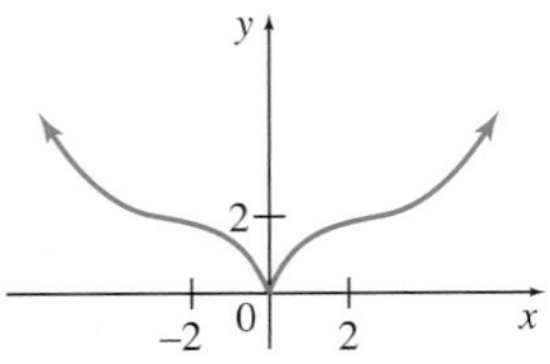

32.

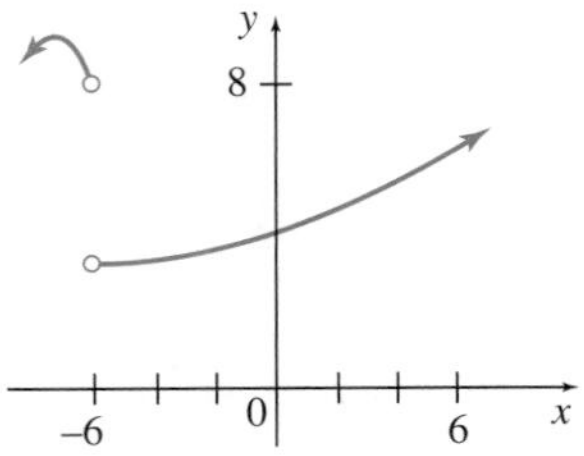

33.

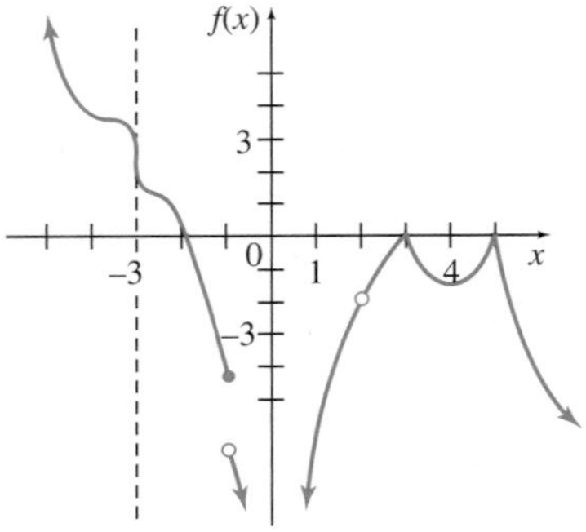

34.

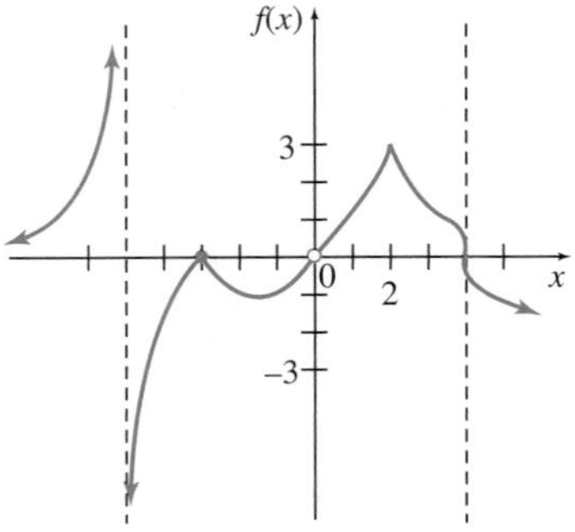

35. For the function shown in the sketch, give the intervals or points on the x-axis where the rate of change of $f(x)$ with respect to x is

a. positive; **b.** negative; **c.** zero.

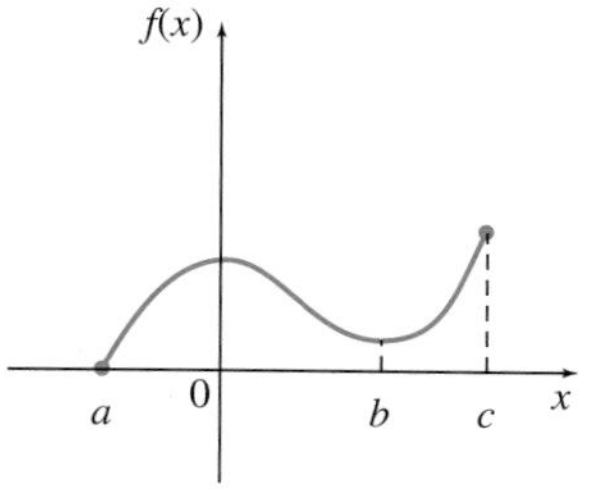

In Exercises 36 and 37, tell which graph, **a** *or* **b**, *represents velocity and which represents distance from a starting point.* (Hint: *Consider where the derivative is zero, positive, or negative.*)

36. a.

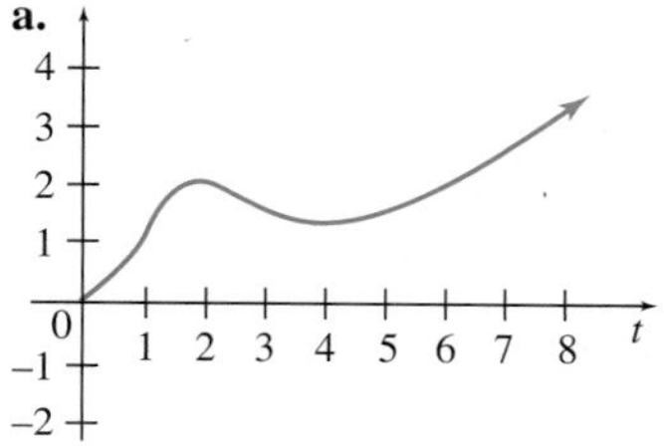

b.

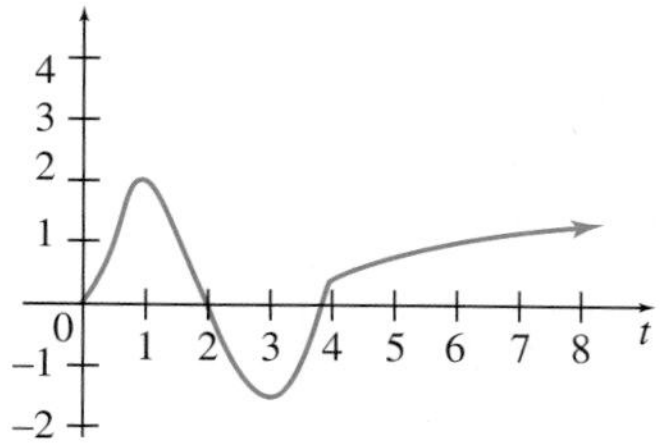

37. a.

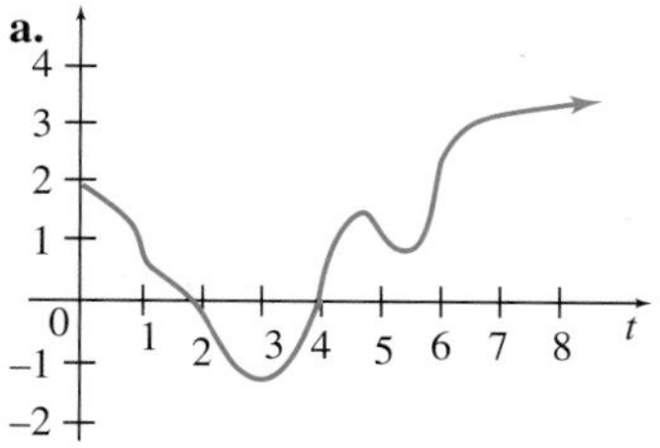

b.

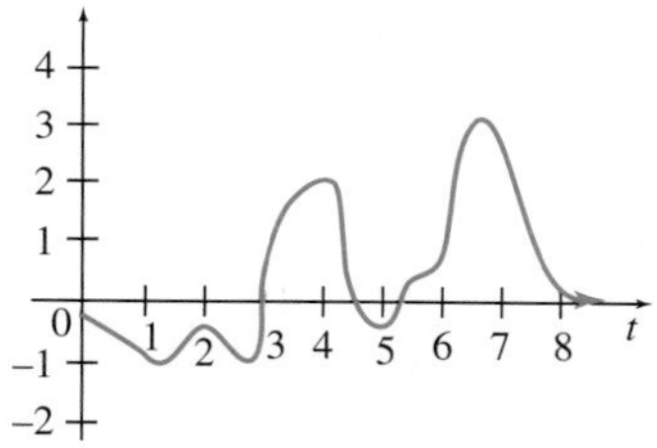

In Exercises 38–41, find the derivative of the function at the given point.

a. *Approximate the definition of the derivative with small values of h.*

b. *Use a graphing calculator to zoom in on the function until it appears to be a straight line, and then find the slope of that line.*

38. $f(x) = x^x$; $a = 2$

39. $f(x) = x^x$; $a = 3$

40. $f(x) = x^{1/x}$; $a = 2$

41. $f(x) = x^{1/x}$; $a = 3$

42. For each function in Column A, graph $[f(x + h) - f(x)]/h$ for a small value of h on the window $[-2, 2]$ by $[-2, 8]$. Then graph each function in Column B on the same window. Compare the first set of graphs with the second set to choose from Column B the derivative of each of the functions in Column A.

Column A	Column B
$\ln \lvert x \rvert$	e^x
e^x	$3x^2$
x^3	$\frac{1}{x}$

43. Explain why

$$\frac{f(x+h)-f(x-h)}{2h}$$

should give a reasonable approximation of $f'(x)$ when h is small.

44. a. For the function $f(x) = -4x^2 + 11x$, find the value of $f'(3)$, as well as the approximation using

$$\frac{f(x+h)-f(x)}{h}$$

and using the formula in Exercise 43 with $h = .1$.

b. Repeat part a using $h = .01$.

c. Repeat part a using the function $f(x) = -2/x$ and $h = .1$.

d. Repeat part c using $h = .01$.

e. Repeat part a using the function $f(x) = \sqrt{x}$ and $h = .1$.

f. Repeat part f using $h = .01$.

g. Using the results of parts a through f, discuss which approximation formula seems to give better accuracy.

Applications

BUSINESS AND ECONOMICS

45. *Demand* Suppose the demand for a certain item is given by $D(p) = -2p^2 - 4p + 300$, where p represents the price of the item in dollars.

a. Find the rate of change of demand with respect to price.

b. Find and interpret the rate of change of demand when the price is \$10.

46. *Profit* The profit (in dollars) from the expenditure of x thousand dollars on advertising is given by $P(x) = 1000 + 32x - 2x^2$. Find the marginal profit at the following expenditures. In each case, decide whether the firm should increase the expenditure.

a. \$8000 **b.** \$6000 **c.** \$12,000 **d.** \$20,000

47. *Revenue* The revenue in dollars generated from the sale of x picnic tables is given by $R(x) = 20x - \dfrac{x^2}{500}$.

a. Find the marginal revenue when 1000 tables are sold.

b. Estimate the revenue from the sale of the 1001st table by finding $R'(1000)$.

c. Determine the actual revenue from the sale of the 1001st table.

d. Compare your answers for parts b and c. What do you find?

48. *Cost* The cost of producing x tacos is $C(x) = 1000 + .24x^2$, $0 \le x \le 30{,}000$.

a. Find the marginal cost.

b. Find and interpret the marginal cost at a production level of 100 tacos.

c. Find the exact cost to produce the 101st taco.

d. Compare the answers to parts b and c. How are they related?

49. *Social Security Assets* The table on the next page gives actual and projected year-end assets in Social Security trust funds, in billions of current dollars, where Year represents the number of years since 1990.*

The polynomial function defined by

$$f(x) = -.0142x^4 + .6698x^3 - 6.113x^2 + 84.05x + 203.9$$

models these data quite well.

a. To verify the fit of the model, find $f(10)$, $f(20)$, and $f(30)$.

b. Use a graphing calculator with a command such as `nDeriv` to find the slope of the tangent line to the graph of f at the following x-values: 10, 20, 30, 35.

c. Use your results in part b to describe the graph of f and interpret the corresponding changes in Social Security assets.

*Social Security Administration.

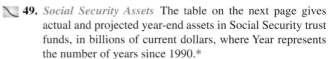

Year	Billions of Dollars
0	214
6	550
8	800
10	1000
20	2500
30	3800
40	250

LIFE SCIENCES

50. *Flight Speed* The graph in the next column shows the relationship between the speed of the Arctic tern in flight and the required power expended by its flight muscles.* Several significant flight speeds are indicated on the curve.

a. The speed V_{mp} minimizes energy costs per unit of time. What is the slope of the line tangent to the curve at the point corresponding to V_{mp}? What is the physical significance of the slope at that point?

b. The speed V_{mr} minimizes the energy costs per unit of distance covered. Estimate the slope of the curve at the point corresponding to V_{mr}. Give the significance of the slope at that point.

c. The speed V_{opt} minimizes the total duration of the migratory journey. Estimate the slope of the curve at the point corresponding to V_{opt}. Relate the significance of this slope to the slopes found in parts a and b.

d. By looking at the shape of the curve, describe how the power level decreases and increases for various speeds.

e. Notice that the slope of the lines found in parts a–c represents the power divided by speed. Power is energy per unit time, and speed is distance per unit time, so the slope represents energy per unit distance. If a line is drawn from the origin to a point on the graph, at which point is the slope of the line (representing energy per unit distance) smallest? How does this compare with your answers to parts a–c?

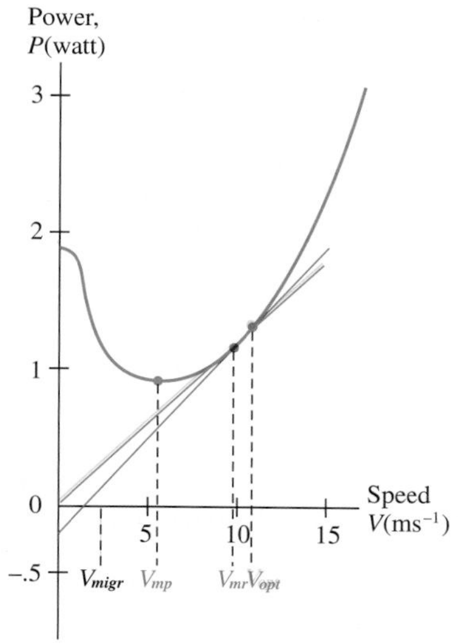

51. *Shellfish Population* In one research study, the population of a certain shellfish in an area at time t was closely approximated by the following graph. Estimate and interpret the derivative at each of the marked points.

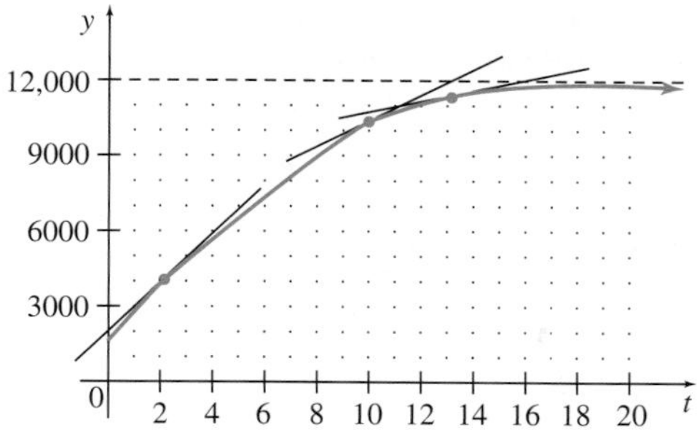

52. *Eating Behavior* The eating behavior of a typical human during a meal can be described by

$$I(t) = 27 + 72t - 1.5t^2,$$

where t is the number of minutes since the meal began, and $I(t)$ represents the amount (in grams) that the person has eaten at time t.†

*Alerstam, Thomas, "Bird Flight and Optimal Migration," *Trends in Ecology and Evolution,* July 1991. Copyright © 1991 by Elsevier Trends Journals. Reprinted by permission of Elsevier Trends Journals and Thomas Alerstam.

†Kissileff, H. R. and J. L. Guss, "Microstructure of Eating Behavior in Humans," *Appetite,* Vol. 36, No. 1, Feb. 2001, pp. 70–78.

a. Find the rate of change of the intake of food for this particular person 5 minutes into a meal and interpret.

b. Verify that the rate in which food is consumed is zero 24 minutes after the meal starts.

c. Comment on the assumptions and usefulness of this function after 24 minutes. Given this fact, determine a logical range for this function.

53. *Quality Control of Cheese* It is often difficult to evaluate the quality of products that undergo a ripening or maturation process. Researchers have successfully used ultrasonic velocity to determine the maturation time of Mahon cheese. The age can be determined by

$$M(v) = .0312443v^2 - 101.39v + 82{,}264, \quad v \geq 1620,$$

where $M(v)$ is the estimated age of the cheese (in days) for a velocity v (m per second).*

a. If Mahon cheese ripens in 150 days, determine the velocity of the ultrasound that one would expect to measure. (*Hint:* Set $M(v) = 150$ and solve for v.)

b. Determine the derivative of this function when $v = 1700$ m per second and interpret.

PHYSICAL SCIENCES

54. *Temperature* The graph shows the temperature in degrees Celsius as a function of the altitude h in feet when an inversion layer is over Southern California. (See Exercise 38 in the previous section.) Estimate and interpret the derivatives of $T(h)$ at the marked points.

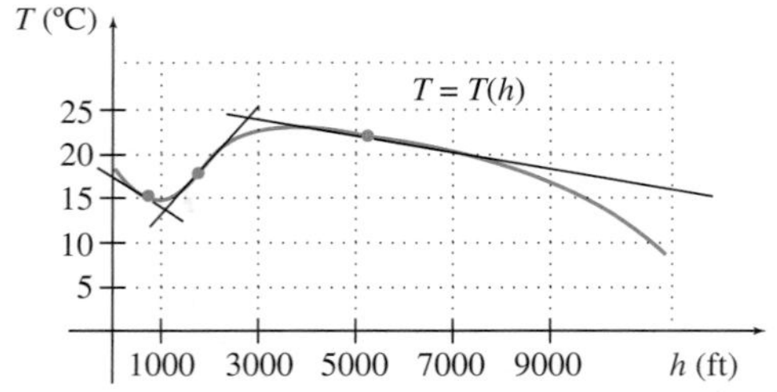

55. *Oven Temperature* The graph shows the temperature in an oven during a self-cleaning cycle.[†] (*Note:* The circles on the graph do not represent points of discontinuity, but merely the times when the thermal door lock turns on and off.) Let $T(x)$ be the temperature after x hours.

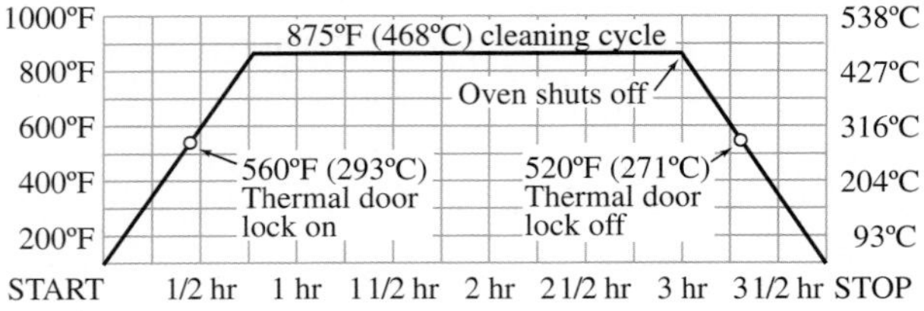

a. Find all x-values where the derivative does not exist.

b. Find and interpret $T'(.5)$.

c. Find and interpret $T'(2)$.

d. Find and interpret $T'(3.5)$.

56. *Baseball* The graph shows how the distance a baseball travels varies with the weight of the bat.[‡] Estimate and interpret the derivative for a 24-oz and a 51-oz bat under the assumption that the bat plus the batter have a constant energy.

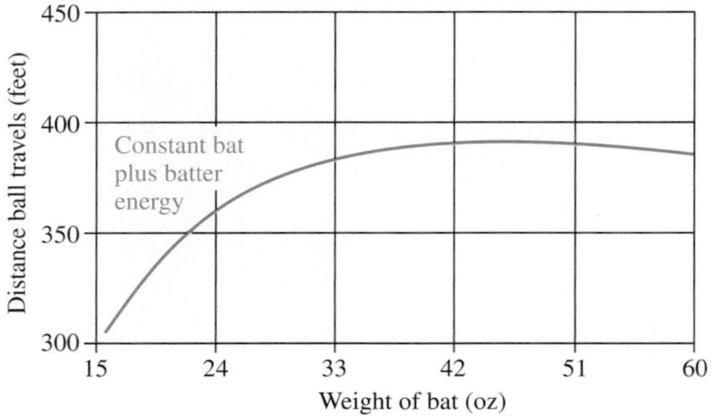

57. *Baseball* The graph shows how the velocity of the hands and the baseball bat vary with the time of the swing. Estimate and interpret the derivative for the hands and the bat at the time when the velocity of the two are equal. (*Hint:* The rate of change of velocity is called acceleration.)

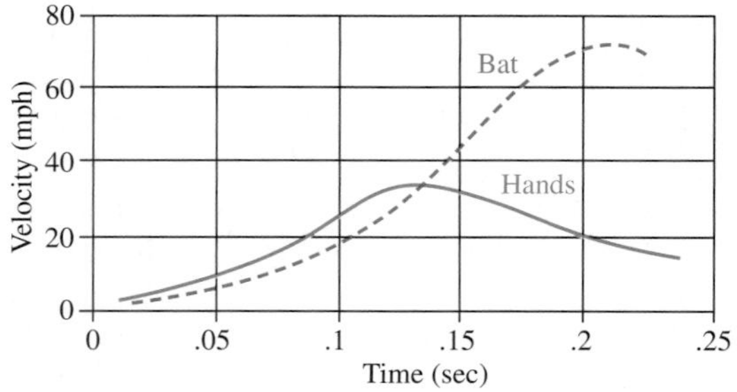

*Benedito, J., J. Carcel, M. Gisbert, and A. Mulet, "Quality Control of Cheese Maturation and Defects Using Ultrasonics," *Journal of Food Science,* Vol. 66, No. 1, 2001, pp. 100–104.

[†]*Whirlpool Use and Care Guide, Self-Cleaning Electric Range,* Whirlpool Corporation. Reprinted with permission.

[‡]Exercises 56 and 57 are from Adair, Robert K., *The Physics of Baseball,* HarperCollins, © 1990, p. 82. © 1994 by Robert K. Adair. Reprinted by permission of HarperCollins Publishers, Inc.

3.5 GRAPHICAL DIFFERENTIATION

THINK ABOUT IT *Given a graph of the cost function, how can we find the graph of the marginal cost function?*

To understand how cost and marginal cost are affected by production, it is helpful to draw graphs of both functions. In Figure 43(a), from an economics text, the graph shows total production, measured in bushels of wheat per year, as a function of the labor used, measured in workers hired.* The graph in Figure 43(b), labeled MP_L, shows the derivative of the top function, namely, the marginal production curve.

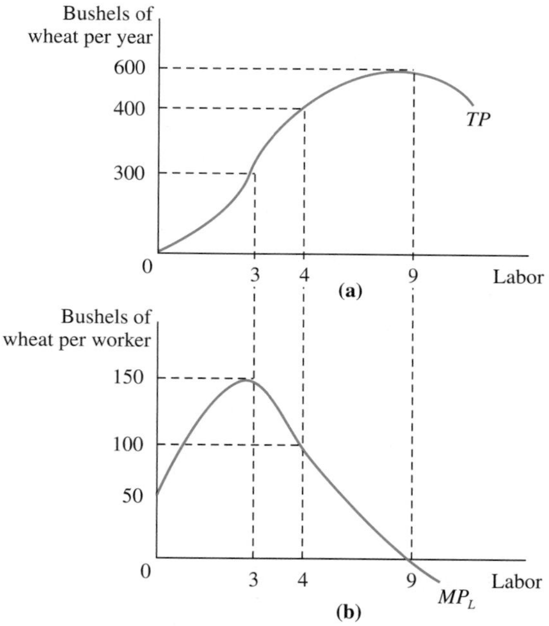

FIGURE 43

In the previous section, we estimated the derivative at various points of a graph by estimating the slope of the tangent line at those points. We will now extend this process to show how to sketch the graph of the derivative given the graph of the original function. This is important because, in many applications, a graph is all we have, and it is easier to find the derivative graphically than to find a formula that fits the graph and take the derivative of that formula.

In the economics example above, let q refer to the quantity of labor. We begin by choosing a point where estimating the derivative is simple. Observe that when $q = 9$, TP has a horizontal tangent line, so its derivative is 0. This explains why

*Browning, Edgar K. and Jacquelene M. Browning, *Microeconomic Theory and Applications,* 4th ed., HarperCollins, 1992, p. 184. Copyright © 1992. This material is used by permission of John Wiley & Sons, Inc.

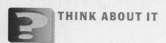

the graph of MP_L equals 0 when $q = 9$. Observe that when $q < 9$, the slope of TP is positive (that is, the tangent line is going up as q goes from left to right), and the slope is steepest when $q = 3$. (See Figure 43.) This means the derivative should be positive for $q < 9$, and largest when $q = 3$. Verify that the graph of MP_L has this property. Finally, as Figure 43 shows, the graph of TP has a negative slope when $q > 9$, so its derivative, represented by the graph of MP_L, should also be negative there.

In this example, we have seen how the general shape of the graph of the derivative can be found from the graph of the original function. To get a more accurate graph of the derivative, we need to estimate the slope of the tangent line at various points, as we did in the previous section.

EXAMPLE 1 *Temperature*

Figure 44 is from Exercise 54 in the previous section. It gives the temperature in degrees Celsius as a function of the altitude h in feet when an inversion layer is over Southern California. The exercise asks for an estimate of the derivative of $T(h)$ at the marked points. If you have not done that exercise yet, do the estimation now by finding two points on each tangent line and computing the slope between the two points.

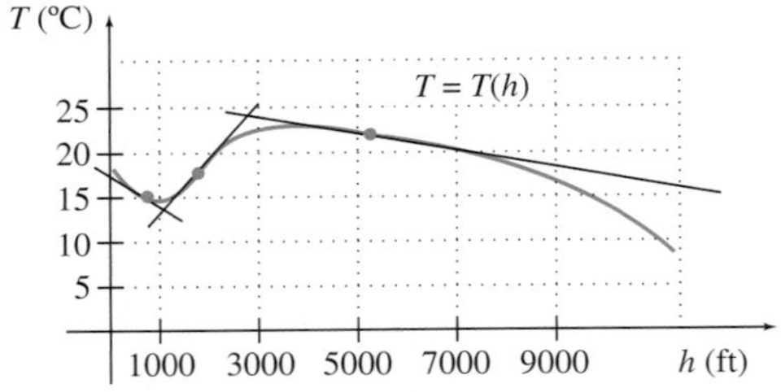

FIGURE 44

Your answers should be roughly $T'(500) = -.005$, $T'(1500) = .008$, and $T'(5000) = -.00125$. Your answers may be slightly different, since estimation from a picture can be inexact. We will also add the estimate $T'(3500) = 0$, because the tangent line is horizontal there. Figure 45 on the next page shows a graph of these values of T' found so far.

Notice from the graph of $T(h)$ that the slope is largest at $h = 1500$, and that the slope becomes more negative as we move to the left from 500. Also, as we move to the right from $h = 5000$, the slope becomes more and more negative. Using these facts, we connect the points in the graph $T'(h)$ smoothly, with the result shown in Figure 46.

CAUTION Remember that when you graph the derivative, you are graphing the *slope* of the original function. Do not confuse the slope of the original function with the y-value of the original function. ■

t	T'
500	$-.005$
1500	.008
3500	0
5000	$-.00125$

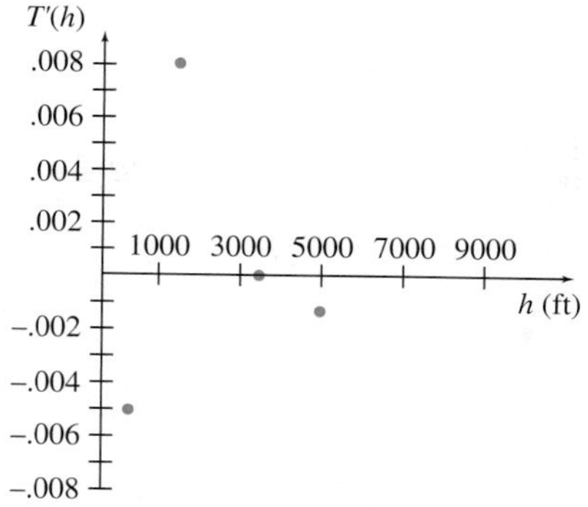

FIGURE 45

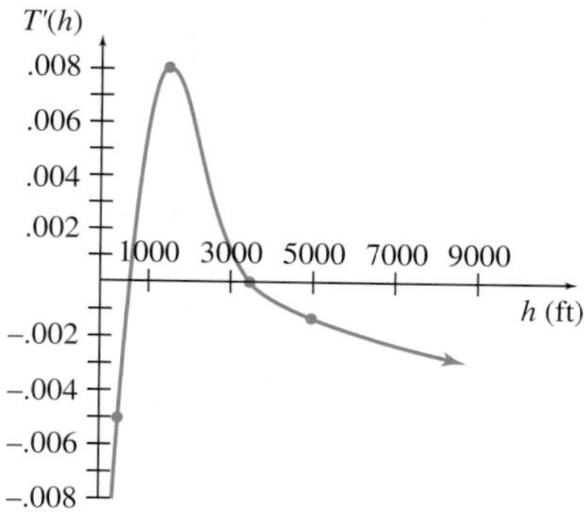

FIGURE 46

Sometimes the original function is not smooth or even continuous, so the graph of the derivative may also be discontinuous.

EXAMPLE 2 *Graphing a Derivative*

Sketch the graph of the derivative of the function shown in Figure 47.

Solution Notice that when $x < -2$, the slope is 1, and when $-2 < x < 0$, the slope is -1. At $x = -2$, the derivative does not exist due to the sharp corner in the graph. The derivative also does not exist at $x = 0$ because the function is discontinuous there. Using this information, the graph of $f'(x)$ on $x < 0$ is shown in Figure 48.

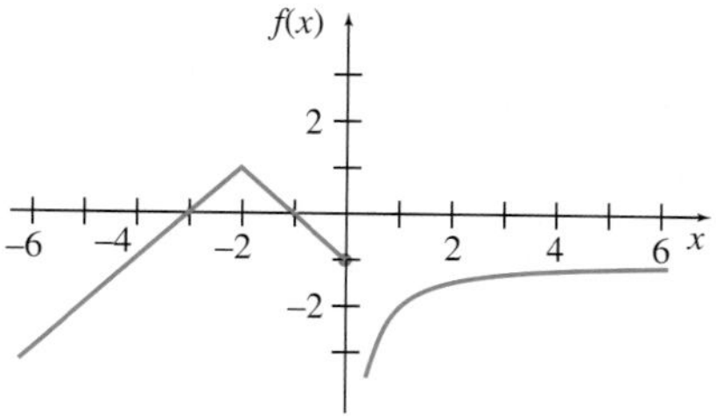

FIGURE 47

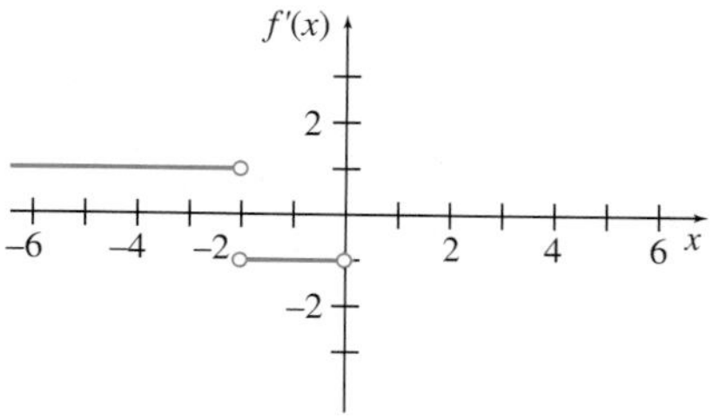

FIGURE 48

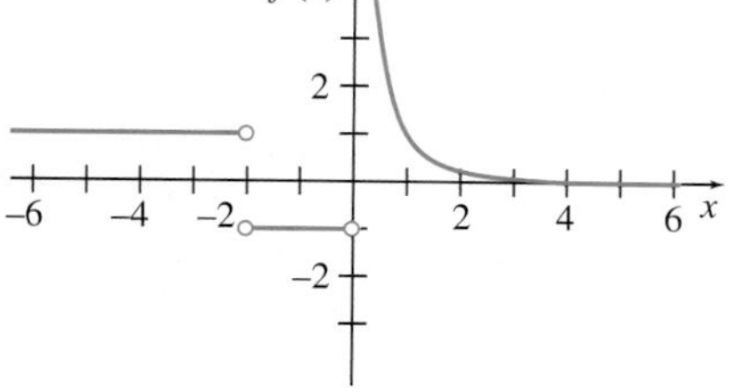

FIGURE 49

For $x > 0$, the derivative is positive. If you draw a tangent line at $x = 1$, you should find that the slope of this line is roughly 1. As x approaches 0 from the right, the derivative becomes larger and larger. As x approaches infinity, the derivative approaches 0. The resulting sketch of the graph of $y = f'(x)$ is shown in Figure 49.

Finding the derivative graphically may seem difficult at first, but with practice you should be able to quickly sketch the derivative of any function graphed. Your answers to the exercises may not look exactly like those in the back of the book, because estimating the slope accurately can be difficult, but your answers should have the same general shape.

Figures 50(a), (b), and (c) show the graphs of $y = x^2$, $y = x^4$, and $y = x^{4/3}$ on a graphing calculator. When finding the derivative graphically, all three seem to have the same behavior: negative derivative for $x < 0$, 0 derivative at $x = 0$, and positive derivative for $x > 0$. Beyond these general features, however, the derivatives look quite different, as you can see from Figures 51(a), (b), and (c), which show the graphs of the derivatives. When finding derivatives graphically, detailed information can only be found by very carefully measuring the slope of the tangent line at a large number of points.

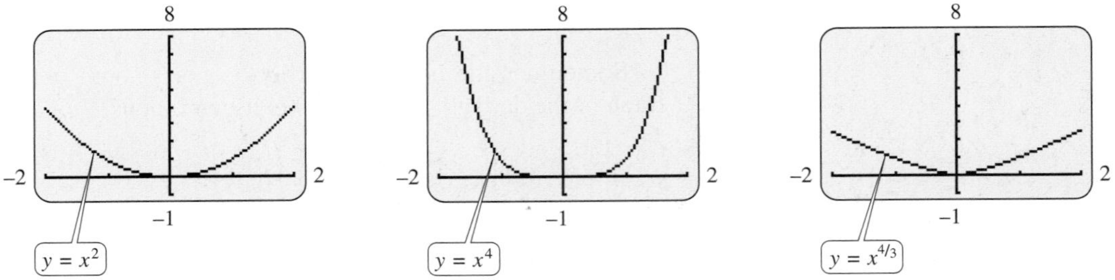

(a) (b) (c)

FIGURE 50

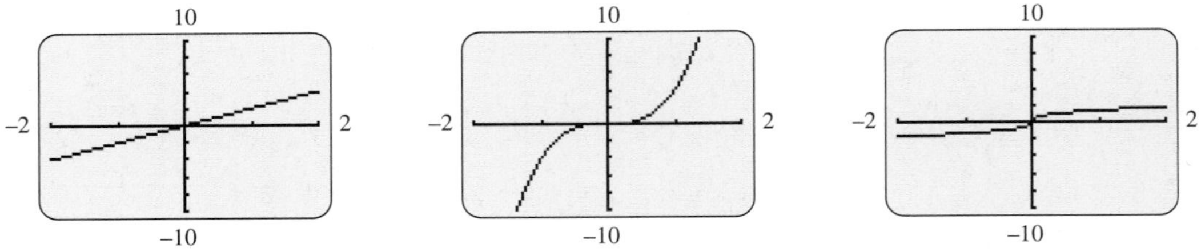

(a) (b) (c)

FIGURE 51

NOTE On many calculators, the graph of the derivative can be plotted if a formula for the original function is known. For example, the graphs in Figure 51 were drawn on a TI-83/84 Plus by defining Y_2 = `nDeriv(Y₁, x, x)` after entering the original function into Y_1. You can use this feature to practice finding the derivative graphically. Enter a function into Y_1, sketch the graph on the graphing calculator, and use it to draw by hand the graph of the derivative. Then use `nDeriv` to draw the graph of the derivative, and compare it with your sketch. ■

EXAMPLE 3 *Graphical Differentiation*

Figure 52 shows the graph of a function f and its derivative function f'. Use slopes to decide which graph is that of f and which is the graph of f'.

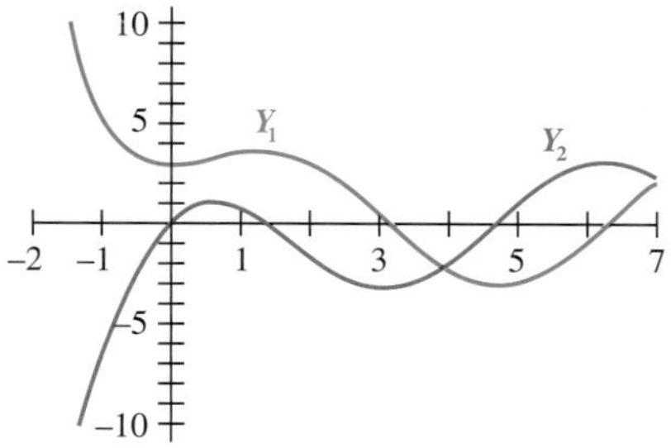

FIGURE 52

Solution Look at the places where each graph crosses the x-axis; that is, the x-intercepts, since x-intercepts occur on the graph of f' whenever the graph of f has a horizontal tangent line or slope of zero. Also, a decreasing graph corresponds to negative slope or a negative derivative, while an increasing graph corresponds to positive slope or a positive derivative. Y_1 has zero slope near $x = 0$, $x = 1$, and $x = 5$; Y_2 has x-intercepts near these values of x. Y_1 decreases on $(-2, 0)$ and $(1, 5)$; Y_2 is negative on those intervals. Y_1 increases on $(0, 1)$ and $(5, 7)$; Y_2 is positive there. Thus, Y_1 is the graph of f and Y_2 is the graph of f'.

3.5 EXERCISES

1. Explain how to graph the derivative of a function given the graph of the function.

2. Explain how to graph a function given the graph of the derivative function.

Each graphing calculator window shows the graph of a function f and its derivative function f'. Decide which is the graph of the function and which is the graph of the derivative.

3.

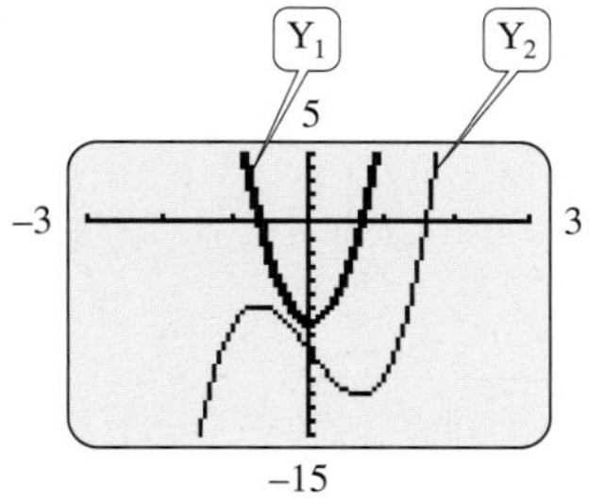

4.

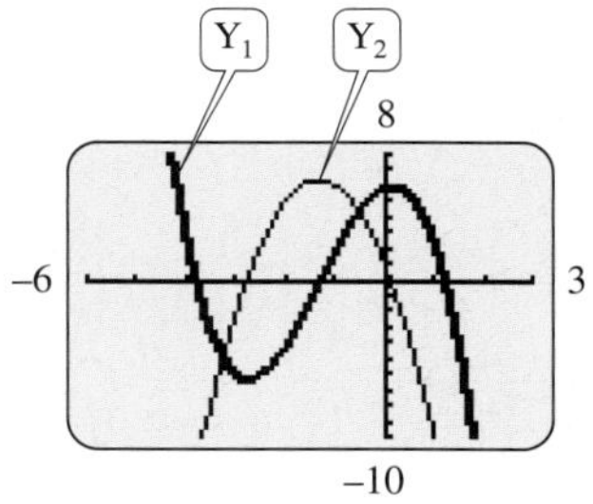

5.

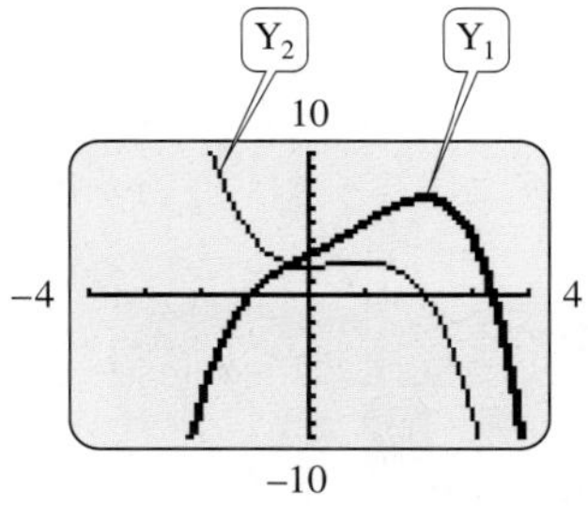

6.

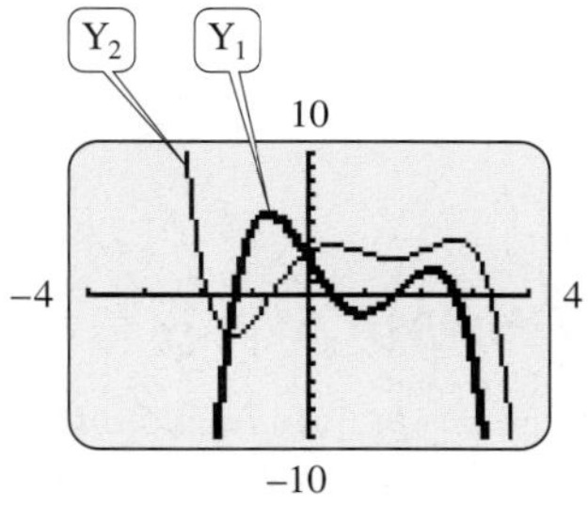

Sketch the graph of the derivative for each function shown.

7.

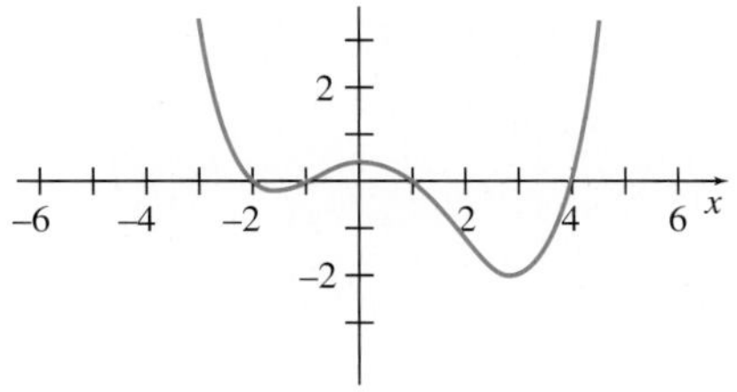

8.

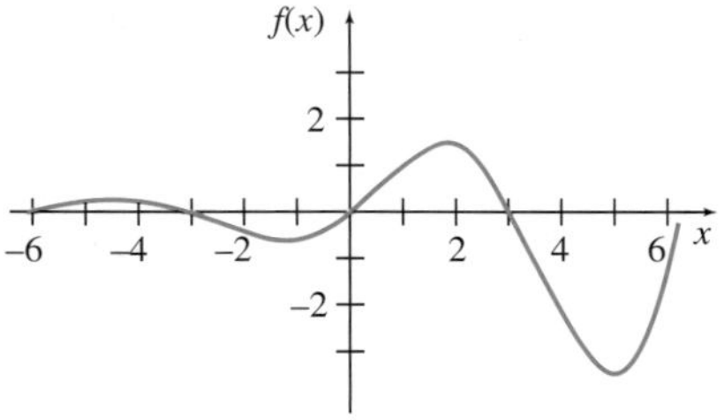

9.

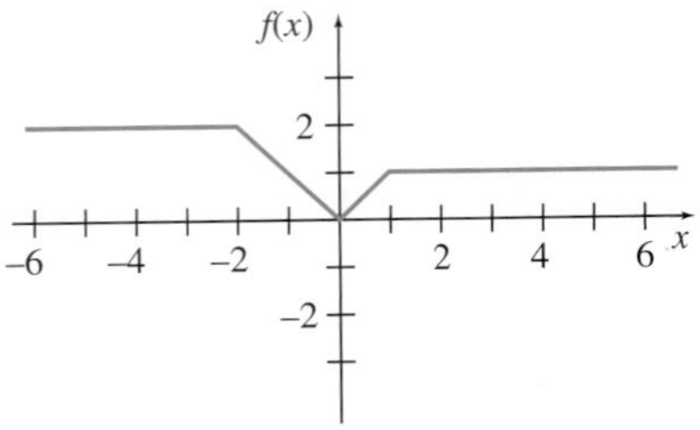

10.

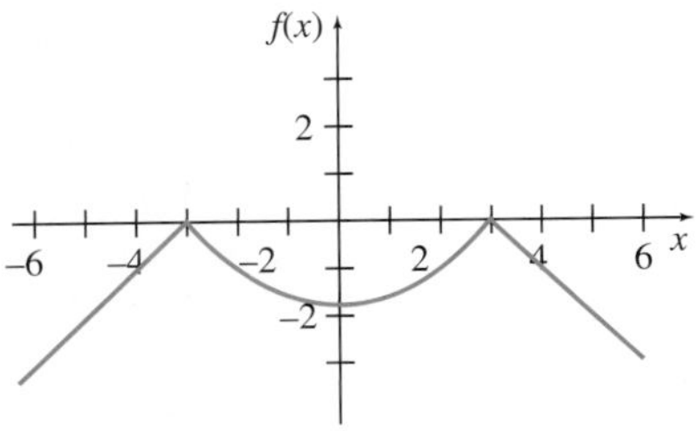

11.

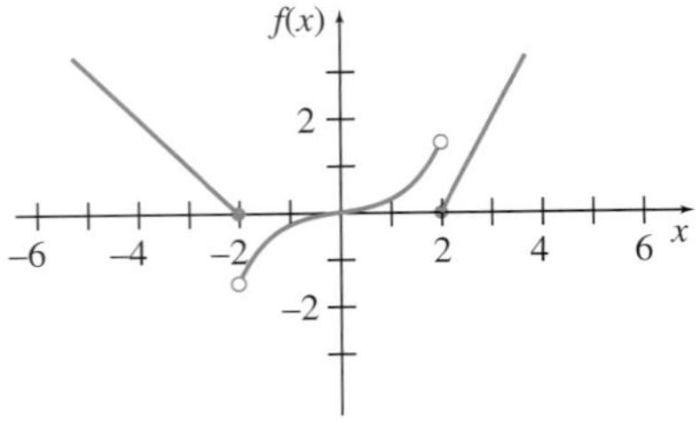

12.

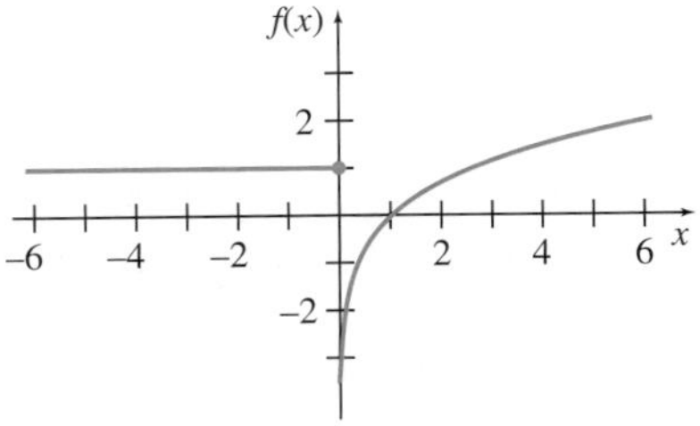

13.

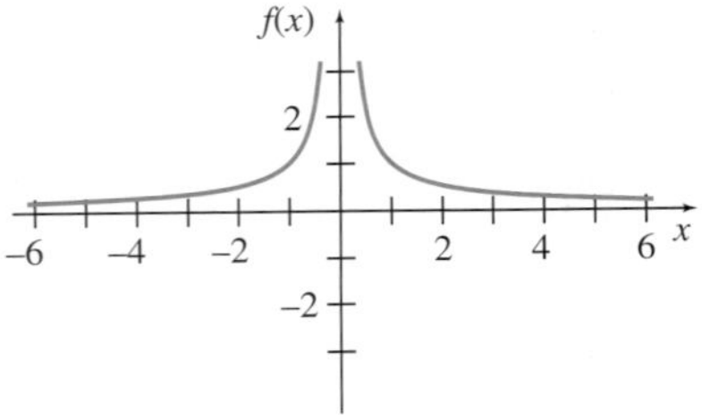

14.

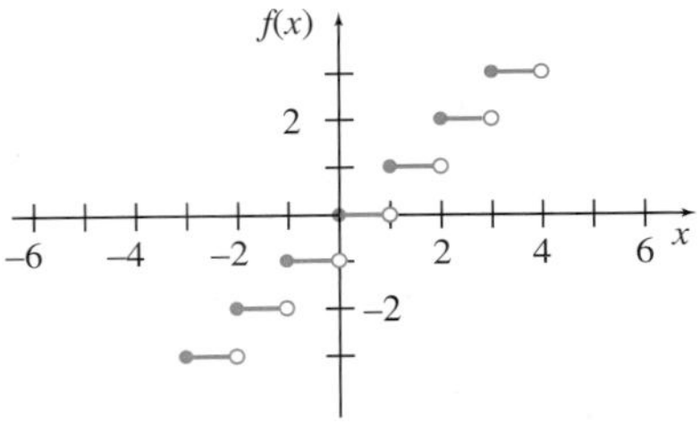

Applications

BUSINESS AND ECONOMICS

Mortgage Interest Rates *The graphs in Exercises 15 and 16 on the next page show average mortgage interest rates for 30-year fixed rate mortgages in New York during two different annual periods.* Sketch a graph of the rate of change of interest rates with respect to time.*

*HSH Associates, NJ.

15.

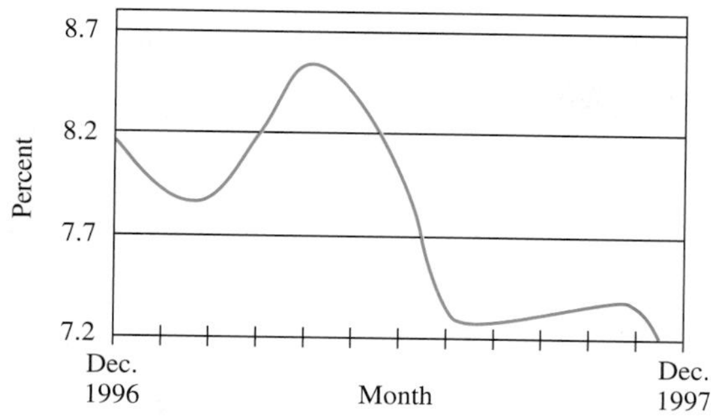

16.

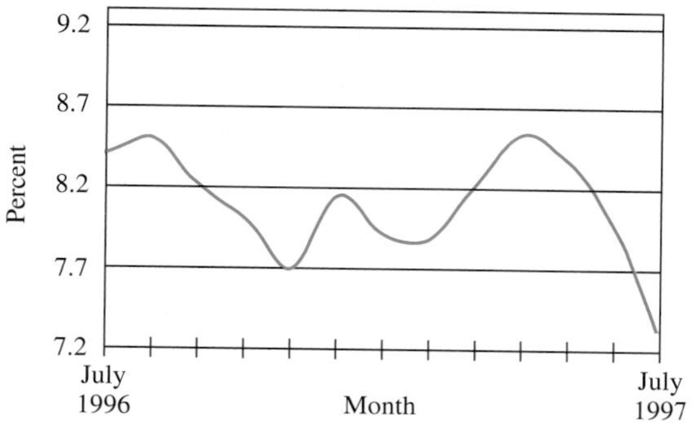

LIFE SCIENCES

17. *Shellfish Population* In one research study, the population of a certain shellfish in an area at time t was closely approximated by the following graph. Sketch a graph of the growth rate of the population.

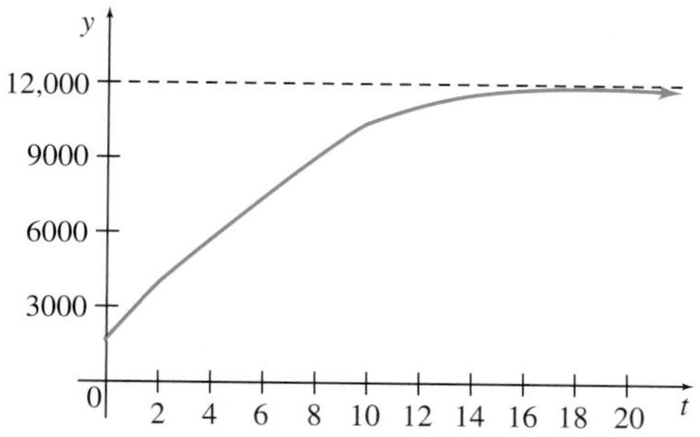

18. *Flight Speed* The graph shows the relationship between the speed of the Arctic tern in flight and the required power expended by flight muscles.* Sketch the graph of the rate of change of the power as a function of the speed.

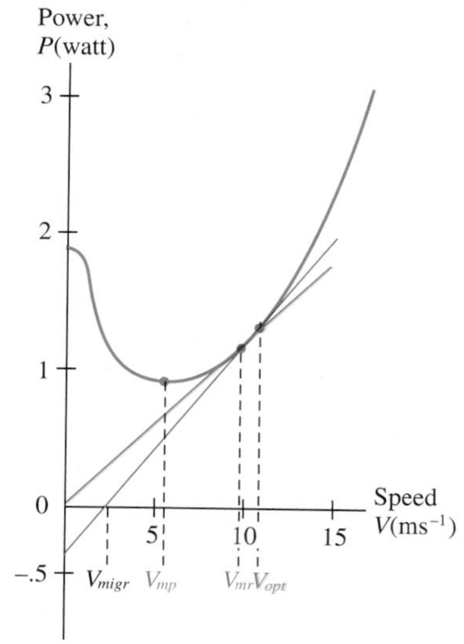

19. *Human Growth* The growth remaining in sitting height at consecutive skeletal age levels is indicated below for boys.† Sketch a graph showing the rate of change of growth remaining for the indicated years. Use the graph and your

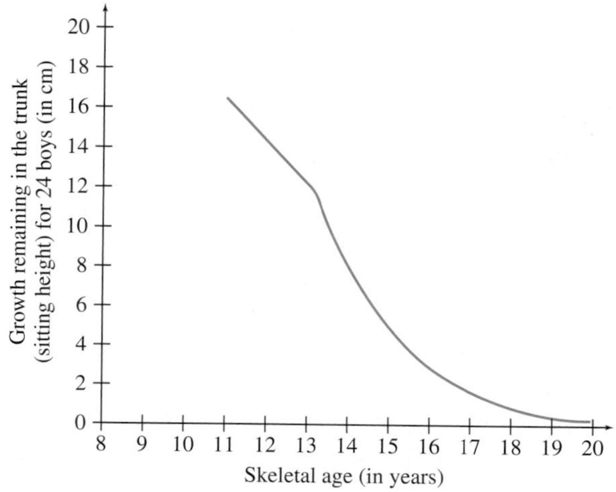

*Alerstam, Thomas, "Bird Flight and Optimal Migration," *Ecology and Evolution,* July 1991. Copyright © 1991 by Elsevier Trends Journals. Reprinted by permission of Elsevier Trends Journals and Thomas Alerstam.

†Hensinger, Robert, *Standards in Pediatric Orthopedics: Tables, Charts, and Graphs Illustrating Growth,* New York: Raven Press, 1986, p. 192.

sketch to estimate the remaining growth and the rate of change of remaining growth for a 14-year-old boy.

20. *Weight Gain* The graph to the right shows the typical weight (in kilograms) of an English boy for his first 18 years of life.* Sketch the graph of the rate of change of weight with respect to time.

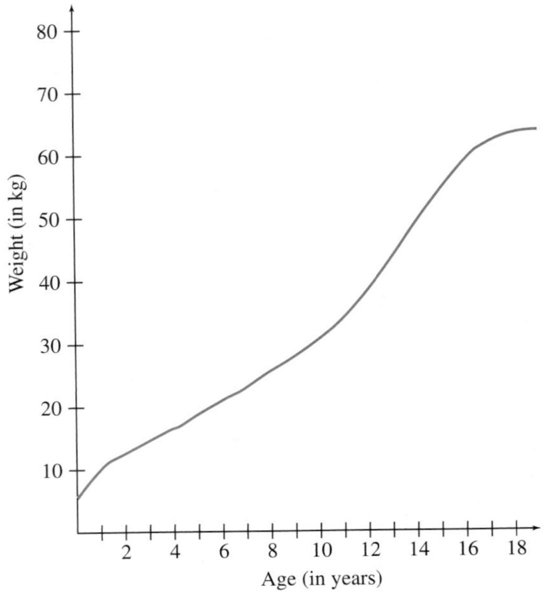

CHAPTER SUMMARY

In this chapter we introduced the ideas of limit and continuity of functions and then used these ideas to discover calculus. We investigated the relationship between instantaneous rate of change, velocity, and the definition of a derivative. We also learned how to estimate the value of the derivative using graphical differentiation. In the next chapter, we will take a closer look at the definition of a derivative to develop a set of rules in which the derivative of a wide range of functions can be quickly and easily calculated without having to directly apply the definition of a derivative each time.

KEY TERMS

To understand the concepts presented in this chapter, you should know the meaning and use of the following terms. For easy reference, the section in the chapter where a word (or expression) was first used is provided.

3.1 **limit**
limit from the left/right
one-/two-sided limit
limit at infinity

3.2 **continuous**
discontinuous
removable discontinuity
continuous from the right/left
continuous on an open/closed interval
Intermediate Value Theorem

3.3 **average rate of change**
difference quotient
instantaneous rate of change
derivative
differentiable
differentiation

3.4 **secant line**
tangent line
slope of a curve

*Sinclair, David, *Human Growth After Birth,* New York: Oxford University Press, 1985.

CHAPTER 3 REVIEW EXERCISES

1. Is a derivative always a limit? Is a limit always a derivative? Explain.

2. Is every continuous function differentiable? Is every differentiable function continuous? Explain.

3. Describe how to tell when a function is discontinuous at the real number $x = a$.

4. Give two applications of the derivative

$$f'(x) = \lim_{h \to 0} \frac{f(x+h) - f(x)}{h}.$$

Decide whether the limits in Exercises 5–22 exist. If a limit exists, find its value.

5. a. $\lim_{x \to -3^-} f(x)$ **b.** $\lim_{x \to -3^+} f(x)$ **c.** $\lim_{x \to -3} f(x)$ **d.** $f(-3)$

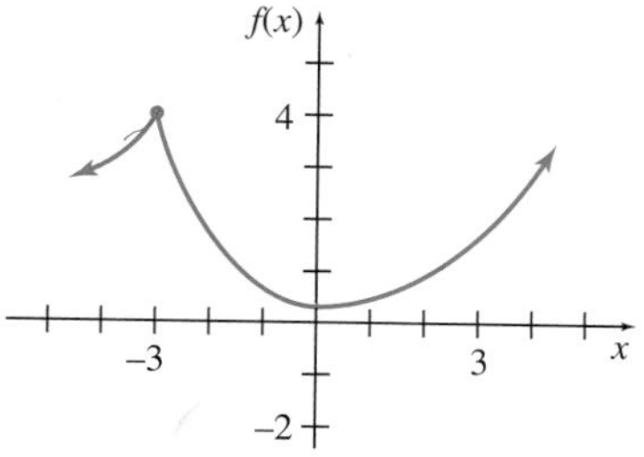

6. a. $\lim_{x \to -1^-} g(x)$ **b.** $\lim_{x \to -1^+} g(x)$ **c.** $\lim_{x \to -1} g(x)$ **d.** $g(-1)$

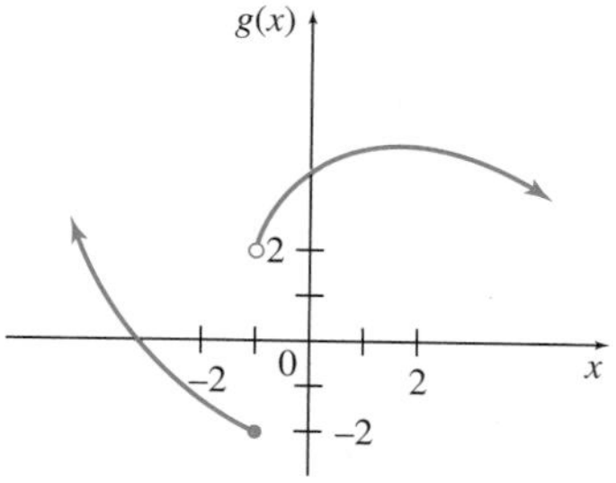

7. a. $\lim_{x \to 4^-} f(x)$ **b.** $\lim_{x \to 4^+} f(x)$ **c.** $\lim_{x \to 4} f(x)$ **d.** $f(4)$

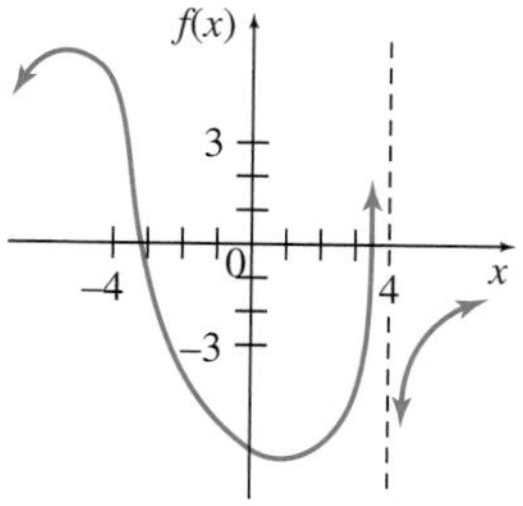

8. a. $\lim_{x \to 2^-} h(x)$ **b.** $\lim_{x \to 2^+} h(x)$ **c.** $\lim_{x \to 2} h(x)$ **d.** $h(2)$

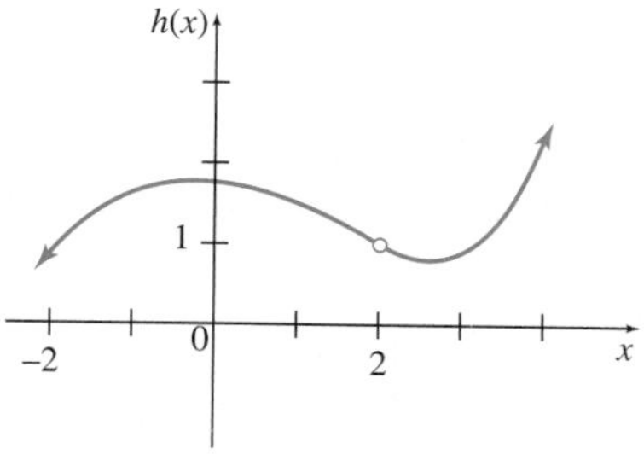

9. $\lim_{x \to -\infty} g(x)$

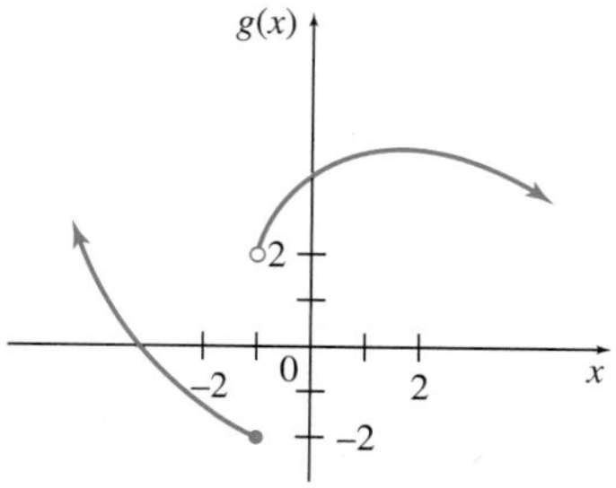

10. $\lim_{x \to \infty} f(x)$

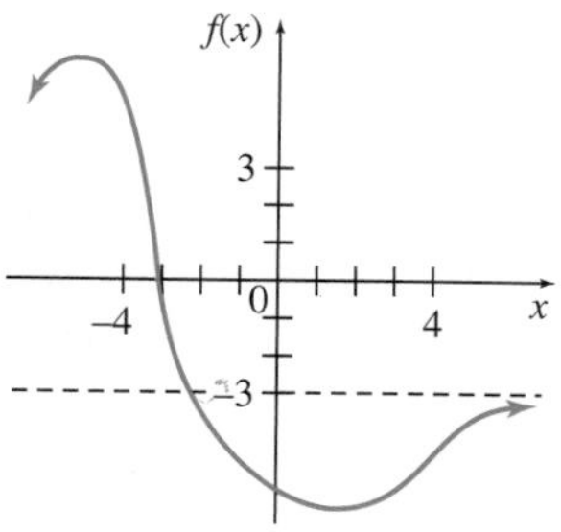

11. $\lim_{x \to 6} \frac{2x + 5}{x - 3}$

12. $\lim_{x \to 3} \frac{2x + 5}{x - 3}$

13. $\lim_{x \to 4} \frac{x^2 - 16}{x - 4}$

14. $\lim_{x \to 2} \frac{x^2 + 3x - 10}{x - 2}$

15. $\lim_{x \to -4} \frac{2x^2 + 3x - 20}{x + 4}$

16. $\lim_{x \to 3} \frac{3x^2 - 2x - 21}{x - 3}$

17. $\lim_{x \to 9} \frac{\sqrt{x} - 3}{x - 9}$

18. $\lim_{x \to 16} \frac{\sqrt{x} - 4}{x - 16}$

19. $\lim_{x \to \infty} \frac{x^2 + 5}{5x^2 - 1}$

20. $\lim_{x \to \infty} \frac{x^2 + 6x + 8}{x^3 + 2x + 1}$

21. $\lim_{x \to -\infty} \left(\frac{3}{4} + \frac{2}{x} - \frac{5}{x^2} \right)$

22. $\lim_{x \to -\infty} \left(\frac{9}{x^4} + \frac{1}{x^2} - 3 \right)$

Identify the x-values where f is discontinuous.

23.

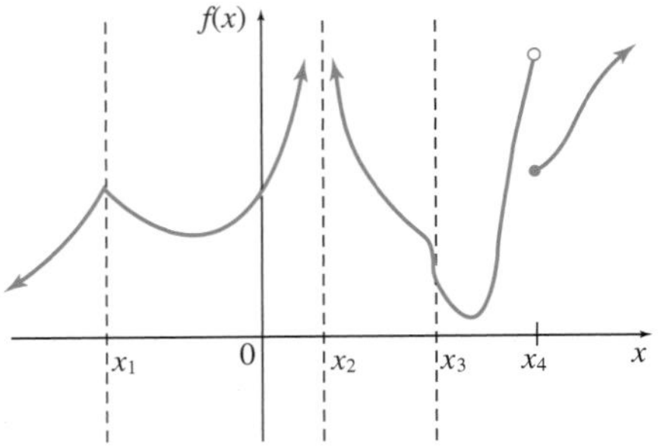

24.

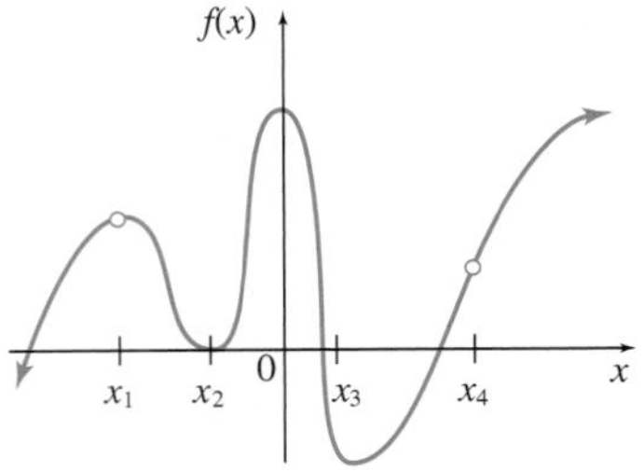

Find all x-values where the function is discontinuous. For each such value, give $f(a)$ and $\lim_{x \to a} f(x)$.

25. $f(x) = \frac{-5}{3x(2x - 1)}$

26. $f(x) = \frac{2 - 3x}{(1 + x)(2 - x)}$

27. $f(x) = \frac{x - 6}{x + 5}$

28. $f(x) = \frac{x^2 - 9}{x + 3}$

29. $f(x) = x^2 + 3x - 4$

30. $f(x) = 2x^2 - 5x - 3$

*In Exercises 31 and 32, **(a)** graph the given function, **(b)** find all values of x where the function is discontinuous, and **(c)** find the limit from the left and from the right at any values of x found in part b.*

31. $f(x) = \begin{cases} 1 - x & \text{if } x < 1 \\ 2 & \text{if } 1 \le x \le 2 \\ 4 - x & \text{if } x > 2 \end{cases}$

32. $f(x) = \begin{cases} 2 & \text{if } x < 0 \\ -x^2 + x + 2 & \text{if } 0 \le x \le 2 \\ 1 & \text{if } x > 2 \end{cases}$

*Find each limit **(a)** by investigating values of the function near the point where the limit is taken, and **(b)** using a graphing calculator to view the function near the point.*

33. $\lim_{x \to 1} \frac{x^4 + 2x^3 + 2x^2 - 10x + 5}{x^2 - 1}$

34. $\lim_{x \to -2} \frac{x^4 + 3x^3 + 7x^2 + 11x + 2}{x^3 + 2x^2 - 3x - 6}$

Find the average rate of change for the following on the given interval. Then find the instantaneous rate of change at the first x-value.

35. $y = 6x^2 + 2$ from $x = 1$ to $x = 4$

36. $y = -2x^3 - x^2 + 5$ from $x = -2$ to $x = 6$

37. $y = \frac{-6}{3x - 5}$ from $x = 4$ to $x = 9$

38. $y = \frac{x + 4}{x - 1}$ from $x = 2$ to $x = 5$

Use the definition of the derivative to find the derivative of the following.

39. $y = 4x + 3$

40. $y = 5x^2 + 6x$

In Exercises 41 and 42, find the derivative of the function at the given point **(a)** *by approximating the definition of the derivative with small values of h, and* **(b)** *by using a graphing calculator to zoom in on the function until it appears to be a straight line, and then finding the slope of that line.*

41. $f(x) = (\ln x)^x; \quad x_0 = 2$

42. $f(x) = x^{\ln x}; \quad x_0 = 3$

Sketch the graph of the derivative for each function shown.

43.

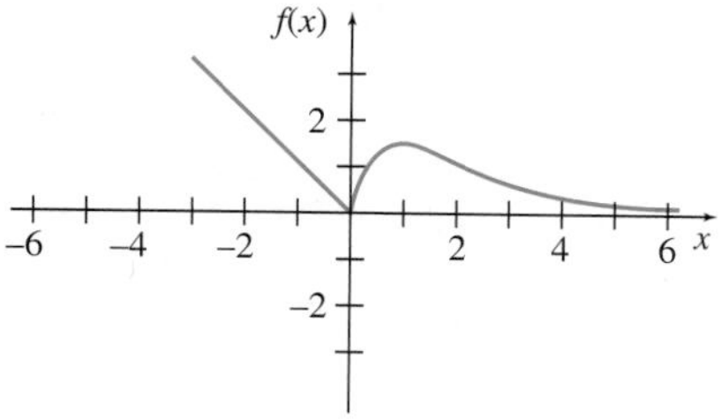

44.

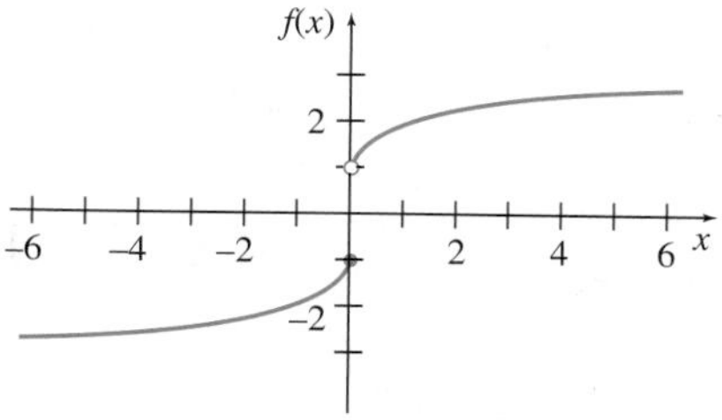

Applications

BUSINESS AND ECONOMICS

45. *Productivity* Average real hourly earnings in private nonagricultural industry in dollars for 1980–1999 are shown in the graph.* Sketch a graph showing the rate of change of earnings for this period. Use the given graph and your sketch to estimate the earnings and rate of change of earning in 1997.

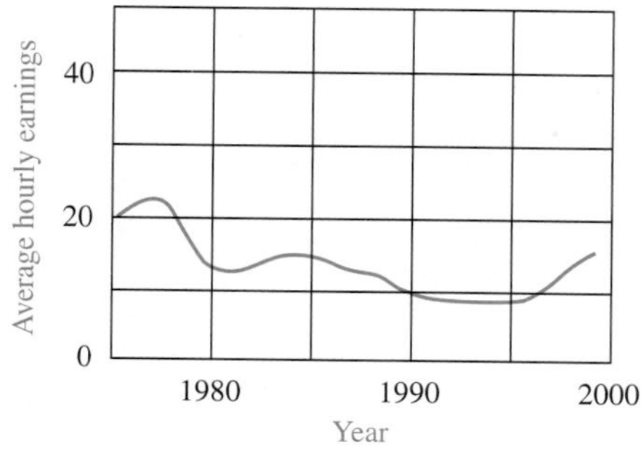

46. *Revenue* Waverly Products has found that its revenue is related to advertising expenditures by the function

$$R(x) = 5000 + 16x - 3x^2,$$

where $R(x)$ is the revenue in dollars when x hundred dollars are spent on advertising.

a. Find the marginal revenue function.

b. Find and interpret the marginal revenue when \$1000 is spent on advertising.

47. *Cost Analysis* A company charges \$1.50 per lb when a certain chemical is bought in lots of 125 lb or less, with a price per pound of \$1.35 if more than 125 lb are purchased. Let $C(x)$ represent the cost of x lb. Find the cost for the following numbers of pounds.

a. 100 **b.** 125 **c.** 140

d. Graph $y = C(x)$.

e. Where is C discontinuous?

Find the average cost per pound if the following number of pounds are bought.

f. 100 **g.** 125 **h.** 140

Find and interpret the marginal cost (that is, the instantaneous rate of change of the cost) for the following numbers of pounds.

i. 100 **j.** 140

48. *Marginal Analysis* Suppose the profit (in cents) from selling x lb of potatoes is given by

$$P(x) = 15x + 25x^2.$$

*U.S. Bureau of Labor Statistics.

Find the average rate of change in profit from selling each of the following amounts.

a. 6 lb to 7 lb **b.** 6 lb to 6.5 lb

c. 6 lb to 6.1 lb

Find the marginal profit (that is, the instantaneous rate of change of the profit) from selling the following amounts.

d. 6 lb **e.** 20 lb **f.** 30 lb

g. What is the domain of x?

h. Is it possible for the marginal profit to be negative here? What does this mean?

i. Find the average profit function. (Recall that average profit is given by total profit divided by the number produced, or $\overline{P}(x) = P(x)/x$.)

j. Find the marginal average profit function (that is, the function giving the instantaneous rate of change of the average profit function).

k. Is it possible for the marginal average profit to vary here? What does this mean?

l. Discuss whether this function describes a realistic situation.

49. *Average Cost* The graph shows the total cost $C(x)$ to produce x tons of cement. (Recall that average cost is given by total cost divided by the number produced, or $\overline{C}(x) = C(x)/x$.)

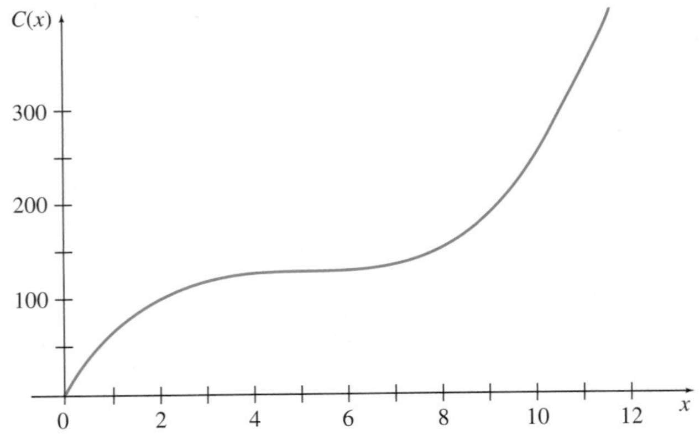

a. Draw a line through $(0, 0)$ and $(5, C(5))$. Explain why the slope of this line represents the average cost per ton when 5 tons of cement are produced.

b. Find the value of x for which the average cost is smallest.

c. What can you say about the marginal cost at the point where the average cost is smallest?

50. *Tax Rates* A simplified income tax considered in the U.S. Senate in 1986 had two tax brackets.* Married couples earning \$29,300 or less would pay 15% of their income in taxes. Those earning more than \$29,300 would pay \$4350 plus 27% of the income over \$29,300 in taxes. Let $T(x)$ be the amount of taxes paid by someone earning x dollars in a year.

a. Find $\lim_{x\to 29{,}300^-} T(x)$. **b.** Find $\lim_{x\to 29{,}300^+} T(x)$.

c. Find $\lim_{x\to 29{,}300} T(x)$. **d.** Sketch a graph of $T(x)$.

e. Identify any x-values where T is discontinuous.

f. Let $A(x) = T(x)/x$ be the average tax rate, that is, the amount paid in taxes divided by the income. Find a formula for $A(x)$. (*Note:* The formula will have two parts: one for $x \le 29{,}300$ and one for $x > 29{,}300$.)

g. Find $\lim_{x\to 29{,}300^-} A(x)$. **h.** Find $\lim_{x\to 29{,}300^+} A(x)$.

i. Find $\lim_{x\to 29{,}300} A(x)$. **j.** Find $\lim_{x\to\infty} A(x)$.

k. Sketch the graph of $A(x)$.

LIFE SCIENCES

51. *Cholesterol* The graph shows how the risk of coronary heart attack rises as blood cholesterol increases.† Estimate and interpret the derivative when blood cholesterol is as follows.

a. 100 mg/dL **b.** 200 mg/dL

c. Find the average rate of change of the risk of coronary heart attack as blood cholesterol goes from 100 to 300 mg/dL.

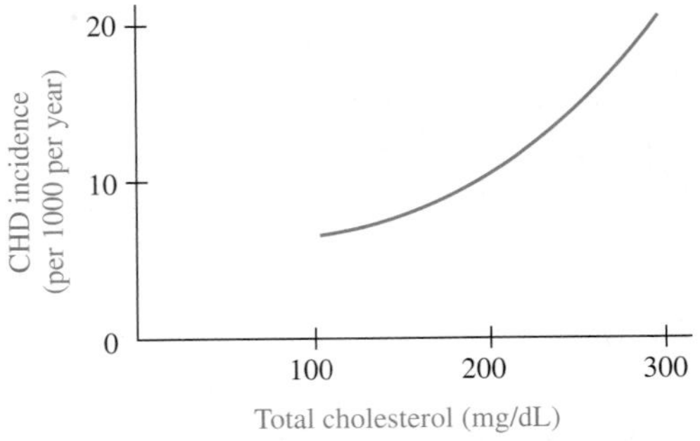

*Murray, Alan, "Winners? Losers? Estimates Show How Impact of Tax Proposal Varies," *Wall Street Journal,* May 9, 1986, p. 29.

†LaRosa, John C. et al., "The Cholesterol Facts: A Joint Statement by the American Heart Association and the National Heart, Lung, and Blood Institute," *Circulation,* Vol. 81, No. 5, May 1990, p. 1722. Reprinted with permission.

52. *Spread of a Virus* The spread of a virus is modeled by

$$V(t) = -t^2 + 6t - 4,$$

where $V(t)$ is the number of people (in hundreds) with the virus and t is the number of weeks since the first case was observed.

a. Graph $V(t)$.

b. What is a reasonable domain of t for this problem?

c. When does the number of cases reach a maximum? What is the maximum number of cases?

d. Find the rate of change function.

e. What is the rate of change in the number of cases at the maximum?

f. Give the sign (+ or −) of the rate of change up to the maximum and after the maximum.

53. *Whales Diving* The following figure, already shown in the section on Properties of Functions, shows the depth of a sperm whale as a function of time, recorded by researchers at the Woods Hole Oceanographic Institution in Massachusetts.*

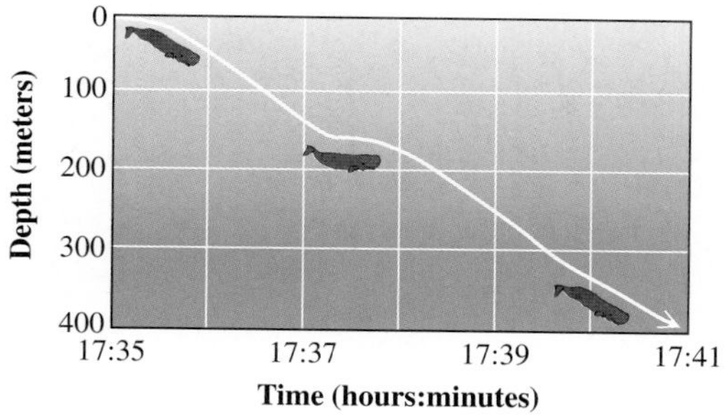

a. Find the rate that the whale was descending at the following times.

i. 17 hours and 37 minutes

ii. 17 hours and 39 minutes

b. Sketch a graph of the rate the whale was descending as a function of time.

54. *Human Growth* The growth remaining in sitting height at consecutive skeletal age levels is indicated below for girls.† Sketch a graph showing the rate of change of growth remaining for the indicated years. Use the graph and your sketch to estimate the remaining growth and the rate of change of remaining growth for a 10-year-old girl.

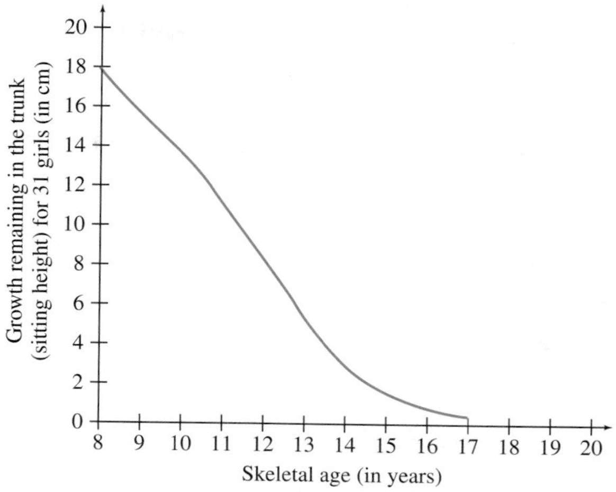

PHYSICAL SCIENCES

55. *Baseball* When a batter hits a baseball, the bat may not hit the center of the ball, but might hit over or under the center by various amounts. The graph shows the trajectories of balls struck by a bat swung under the ball by various amounts.‡ Estimate and interpret the derivative for a bat swung under the ball by 1.5 in. when the ball has traveled the following distances.

a. 100 ft **b.** 200 ft

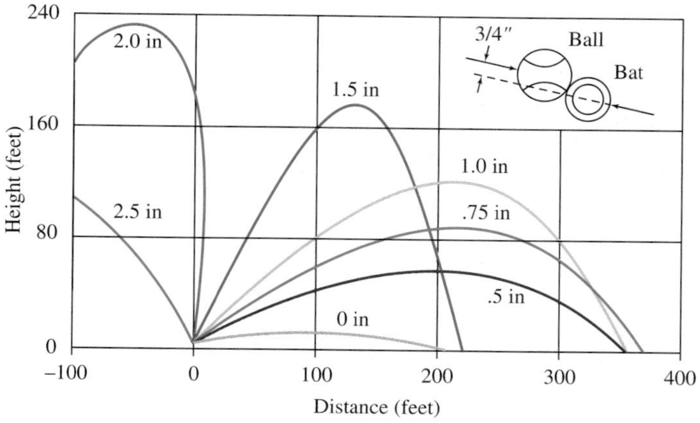

Science, Vol. 291, Jan. 26, 2001, p. 577. Courtesy of Woods Hole Oceanographic Institute.

†Hensinger, Robert, *Standards in Pediatric Orthopedics: Tables, Charts, and Graphs Illustrating Growth*, New York: Raven Press, 1986, p. 193.

‡Adair, Robert K., *The Physics of Baseball*, HarperCollins, © 1990, p. 83. Reprinted with permission.

56. *Temperature* Suppose a gram of ice is at a temperature of −100°C. The graph shows the temperature of the ice as increasing numbers of calories of heat are applied. It takes 80 calories to melt one gram of ice at 0°C into water, and 540 calories to boil one gram of water at 100°C into steam.

a. Where is this graph discontinuous?

b. Where is this graph not differentiable?

c. Sketch the graph of the derivative.

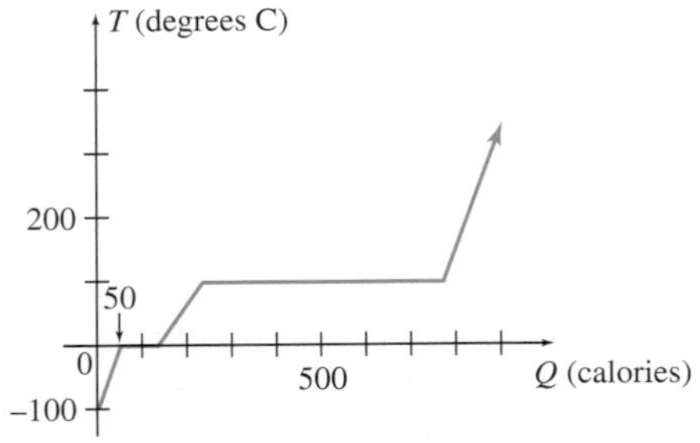

CHAPTER

4 Calculating the Derivative

By differentiating the function defining a mathematical model we can see how the model's output changes with the input. In an exercise in Section 2 we explore a rational-function model for the length of the rest period needed to recover from vigorous exercise such as riding a bike. The derivative indicates how the rest required changes with the work expended in kilocalories per minute.

In the previous chapter, we found the derivative to be a useful tool for describing the rate of change, velocity, and the slope of a curve. Taking the derivative by using the definition, however, can be difficult. To take full advantage of the power of the derivative, we need faster ways of calculating the derivative. That is the goal of this chapter.

4.1 TECHNIQUES FOR FINDING DERIVATIVES

THINK ABOUT IT

How can a manager determine the best production level if the relationship between profit and production is known? How fast is the number of Americans who are expected to be over 100 years old growing?

These questions can be answered by finding the derivative of an appropriate function. We shall return to them at the end of this section in Examples 8 and 9.

Using the definition to calculate the derivative of a function is a very involved process even for simple functions. In this section we develop rules that make the calculation of derivatives much easier. Keep in mind that even though the process of finding a derivative will be greatly simplified with these rules, *the interpretation of the derivative will not change.* But first, a few words about notation are in order.

Several alternative notations for the derivative are used. In the previous chapter the symbol $f'(x)$ was used to represent the derivative of $y = f(x)$. Sometimes it is important to show that the derivative is taken with respect to a particular variable; for example, if y is a function of x, the notation

$$\frac{dy}{dx} \quad \text{or} \quad D_x y$$

(both read "the derivative of y with respect to x") can be used for the derivative of y with respect to x. The dy/dx notation for the derivative is sometimes referred to as *Leibniz notation,* named after one of the co-inventors of the calculus, Gottfried Wilhelm von Leibniz (1646–1716). (The other was Sir Isaac Newton, 1642–1727.)

With the above notation, the derivative of $f(x) = 2x^3 + 4x$, for example, which was found in Example 5 of the section on the Definition of the Derivative to be $f'(x) = 6x^2 + 4$, would be written

$$\frac{dy}{dx} = \frac{d}{dx}(2x^3 + 4x) = 6x^2 + 4, \qquad \text{or} \qquad D_x(2x^3 + 4x) = 6x^2 + 4.$$

Either $(d/dx)[f(x)]$ or $D_x[f(x)]$ represents the derivative of the function f with respect to x.

NOTATIONS FOR THE DERIVATIVE

The derivative of $y = f(x)$ may be written in any of the following ways:

$$f'(x), \quad \frac{dy}{dx}, \quad \frac{d}{dx}[f(x)], \quad \text{or} \quad D_x[f(x)].$$

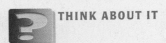

A variable other than x is often used as the independent variable. For example, if $y = f(t)$ gives population growth as a function of time, then the derivative of y with respect to t could be written

$$f'(t), \qquad \frac{dy}{dt}, \qquad \frac{d}{dt}[f(t)], \qquad \text{or} \qquad D_t[f(t)].$$

Other variables also may be used to name the function, as in $g(x)$ or $h(t)$.

Now we will use the definition

$$f'(x) = \lim_{h \to 0} \frac{f(x+h) - f(x)}{h}$$

to develop some rules for finding derivatives more easily than by the four-step process given in the previous chapter.

The first rule tells how to find the derivative of a constant function defined by $f(x) = k$, where k is a constant real number. Since $f(x+h)$ is also k, by definition $f'(x)$ is

$$\begin{aligned} f'(x) &= \lim_{h \to 0} \frac{f(x+h) - f(x)}{h} \\ &= \lim_{h \to 0} \frac{k-k}{h} = \lim_{h \to 0} \frac{0}{h} = \lim_{h \to 0} 0 = 0, \end{aligned}$$

establishing the following rule.

CONSTANT RULE

If $f(x) = k$, where k is any real number, then

$$\boldsymbol{f'(x) = 0.}$$

(The derivative of a constant is 0.)

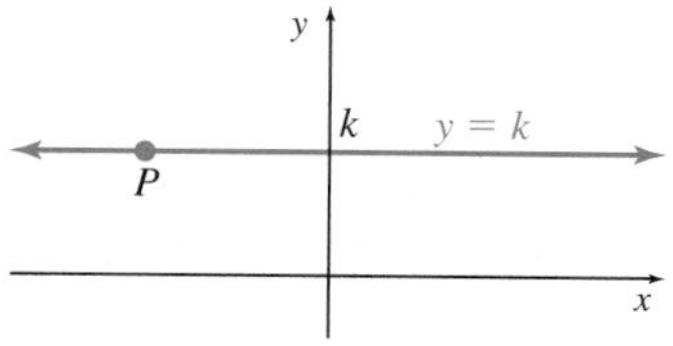

FIGURE 1

Figure 1 illustrates this constant rule geometrically; it shows a graph of the horizontal line $y = k$. At any point P on this line, the tangent line at P is the line itself. Since a horizontal line has a slope of 0, the slope of the tangent line is 0. This agrees with the result above: The derivative of a constant is 0.

EXAMPLE 1 *Derivative of a Constant*

(a) If $f(x) = 9$, then $f'(x) = 0$.

(b) If $h(t) = \pi$, then $D_t[h(t)] = 0$.

(c) If $y = 2^3$, then $dy/dx = 0$.

Functions of the form $y = x^n$, where n is a real number, are very common in applications. To obtain a rule for finding the derivative of such a function, we can use the definition to work out the derivatives for various special values of n. This was done in the section on the Definition of the Derivative in Example 4 to show that for $f(x) = x^2$, $f'(x) = 2x$.

For $f(x) = x^3$, the derivative is found as follows.

$$\begin{aligned} f'(x) &= \lim_{h\to 0} \frac{f(x+h) - f(x)}{h} \\ &= \lim_{h\to 0} \frac{(x+h)^3 - x^3}{h} \\ &= \lim_{h\to 0} \frac{(x^3 + 3x^2h + 3xh^2 + h^3) - x^3}{h} \end{aligned}$$

The binomial theorem (discussed in most intermediate and college algebra texts) was used to expand $(x + h)^3$ in the last step. Now, the limit can be determined.

$$\begin{aligned} f'(x) &= \lim_{h\to 0} \frac{3x^2h + 3xh^2 + h^3}{h} \\ &= \lim_{h\to 0} (3x^2 + 3xh + h^2) \\ &= 3x^2 \end{aligned}$$

The results in the following table were found in a similar way, using the definition of the derivative. (These results are modifications of some of the examples and exercises from the previous chapter.)

Function	n	Derivative
$f(x) = x^2$	2	$f'(x) = 2x = 2x^1$
$f(x) = x^3$	3	$f'(x) = 3x^2$
$f(x) = x^4$	4	$f'(x) = 4x^3$
$f(x) = x^{-1}$	-1	$f'(x) = -1 \cdot x^{-2} = \dfrac{-1}{x^2}$
$f(x) = x^{1/2}$	$1/2$	$f'(x) = \dfrac{1}{2}x^{-1/2} = \dfrac{1}{2x^{1/2}}$

These results suggest the following rule.

POWER RULE

If $f(x) = x^n$ for any real number n, then

$$f'(x) = nx^{n-1}.$$

(The derivative of $f(x) = x^n$ is found by multiplying by the exponent n and decreasing the exponent on x by 1.)

While the power rule is true for every real-number value of n, a proof is given here only for positive integer values of n. This proof follows the steps used above in finding the derivative of $f(x) = x^3$.

For any real numbers p and q, by the binomial theorem,

$$(p + q)^n = p^n + np^{n-1}q + \frac{n(n-1)}{2}p^{n-2}q^2 + \cdots + npq^{n-1} + q^n.$$

Replacing p with x and q with h gives

$$(x+h)^n = x^n + nx^{n-1}h + \frac{n(n-1)}{2}x^{n-2}h^2 + \cdots + nxh^{n-1} + h^n,$$

from which

$$(x+h)^n - x^n = nx^{n-1}h + \frac{n(n-1)}{2}x^{n-2}h^2 + \cdots + nxh^{n-1} + h^n.$$

Dividing each term by h yields

$$\frac{(x+h)^n - x^n}{h} = nx^{n-1} + \frac{n(n-1)}{2}x^{n-2}h + \cdots + nxh^{n-2} + h^{n-1}.$$

Use the definition of derivative, and the fact that each term except the first contains h as a factor and thus approaches 0 as h approaches 0, to get

$$\begin{aligned} f'(x) &= \lim_{h\to 0}\frac{(x+h)^n - x^n}{h} \\ &= nx^{n-1} + \frac{n(n-1)}{2}x^{n-2}0 + \cdots + nx0^{n-2} + 0^{n-1} \\ &= nx^{n-1}. \end{aligned}$$

This shows that the derivative of $f(x) = x^n$ is $f'(x) = nx^{n-1}$, proving the power rule for positive integer values of n.

EXAMPLE 2 *Power Rule*

(a) If $y = x^6$, find $D_x y$.

Solution $D_x y = 6x^{6-1} = 6x^5$

(b) If $y = t = t^1$, find $\dfrac{dy}{dt}$.

Solution $\dfrac{dy}{dt} = 1t^{1-1} = t^0 = 1$

FOR REVIEW

At this point you may wish to turn back to the Algebra Reference for a review of negative exponents and rational exponents. The relationship between powers, roots, and rational exponents is explained there.

(c) If $y = 1/x^3$, find dy/dx.

Solution Use a negative exponent to rewrite this equation as $y = x^{-3}$; then

$$\frac{dy}{dx} = -3x^{-3-1} = -3x^{-4} \quad \text{or} \quad \frac{-3}{x^4}.$$

(d) Find $D_x(x^{4/3})$.

Solution $D_x(x^{4/3}) = \dfrac{4}{3}x^{4/3-1} = \dfrac{4}{3}x^{1/3}$

(e) If $y = \sqrt{z}$, find dy/dz.

Solution Rewrite this as $y = z^{1/2}$; then

$$\frac{dy}{dz} = \frac{1}{2}z^{1/2-1} = \frac{1}{2}z^{-1/2} \quad \text{or} \quad \frac{1}{2z^{1/2}} \quad \text{or} \quad \frac{1}{2\sqrt{z}}.$$

The next rule shows how to find the derivative of the product of a constant and a function.

CONSTANT TIMES A FUNCTION

Let k be a real number. If $f'(x)$ exists, then

$$D_x[kf(x)] = kf'(x).$$

(The derivative of a constant times a function is the constant times the derivative of the function.)

This rule is proved with the definition of the derivative and rules for limits.

$$\begin{aligned} D_x[kf(x)] &= \lim_{h\to 0} \frac{kf(x+h) - kf(x)}{h} \\ &= \lim_{h\to 0} k\frac{[f(x+h) - f(x)]}{h} && \text{Factor out } k. \\ &= k \lim_{h\to 0} \frac{f(x+h) - f(x)}{h} && \text{Limit rule 1} \\ &= kf'(x) && \text{Definition of derivative} \end{aligned}$$

EXAMPLE 3 *Derivative of a Constant Times a Function*

(a) If $y = 8x^4$, find $\dfrac{dy}{dx}$.

Solution $\dfrac{dy}{dx} = 8(4x^3) = 32x^3$

(b) If $y = -\dfrac{3}{4}x^{12}$, find dy/dx.

Solution $\dfrac{dy}{dx} = -\dfrac{3}{4}(12x^{11}) = -9x^{11}$

(c) Find $D_t(-8t)$.

Solution $D_t(-8t) = -8(1) = -8$

(d) Find $D_p(10p^{3/2})$.

Solution $D_p(10p^{3/2}) = 10\left(\dfrac{3}{2}p^{1/2}\right) = 15p^{1/2}$

(e) If $y = \dfrac{6}{x}$, find $\dfrac{dy}{dx}$.

Solution Rewrite this as $y = 6x^{-1}$; then

$$\frac{dy}{dx} = 6(-1x^{-2}) = -6x^{-2} \quad \text{or} \quad \frac{-6}{x^2}.$$

EXAMPLE 4 *Beagles*

Researchers have determined that the daily energy requirements of female beagles who are at least 1 year old change with respect to age according to the function

$$E(t) = 753t^{-.1321},$$

where $E(t)$ is the daily energy requirements (in $\text{kJ}/\text{W}^{.67}$) for a dog that is t years old.*

(a) Find $E'(t)$.

Solution Using the rules of differentiation we find that

$$E'(t) = -99.4713t^{-1.1321}.$$

(b) Determine the rate of change of the daily energy requirements of a 2-year-old female beagle.

Solution $E'(2) = -99.4713(2)^{-1.1321} \approx -45.4$

Thus, the daily energy requirements of a 2-year-old female beagle are decreasing at the rate of 45.4 $\text{kJ}/\text{W}^{.67}$ per year.

The final rule in this section is for the derivative of a function that is a sum or difference of terms.

SUM OR DIFFERENCE RULE

If $f(x) = u(x) \pm v(x)$, and if $u'(x)$ and $v'(x)$ exist, then

$$f'(x) = u'(x) \pm v'(x).$$

(The derivative of a sum or difference of functions is the sum or difference of the derivatives.)

The proof of the sum part of this rule is as follows: If $f(x) = u(x) + v(x)$, then

$$\begin{aligned} f'(x) &= \lim_{h \to 0} \frac{[u(x+h) + v(x+h)] - [u(x) + v(x)]}{h} \\ &= \lim_{h \to 0} \frac{[u(x+h) - u(x)] + [v(x+h) - v(x)]}{h} \\ &= \lim_{h \to 0} \left[\frac{u(x+h) - u(x)}{h} + \frac{v(x+h) - v(x)}{h} \right] \\ &= \lim_{h \to 0} \frac{u(x+h) - u(x)}{h} + \lim_{h \to 0} \frac{v(x+h) - v(x)}{h} \\ &= u'(x) + v'(x). \end{aligned}$$

A similar proof can be given for the difference of two functions.

*Finke, M., "Energy Requirements of Adult Female Beagles," *Journal of Nutrition,* Vol. 124, 1994, pp. 2604s–2608s.

EXAMPLE 5 *Derivative of a Sum*

Find the derivative of each function.

(a) $y = 6x^3 + 15x^2$

Solution Let $u(x) = 6x^3$ and $v(x) = 15x^2$; then $y = u(x) + v(x)$. Since $u'(x) = 18x^2$ and $v'(x) = 30x$,

$$\frac{dy}{dx} = 18x^2 + 30x.$$

(b) $p(t) = 12t^4 - 6\sqrt{t} + \frac{5}{t}$

Solution Rewrite $p(t)$ as $p(t) = 12t^4 - 6t^{1/2} + 5t^{-1}$; then

$$p'(t) = 48t^3 - 3t^{-1/2} - 5t^{-2}.$$

Also, $p'(t)$ may be written as $p'(t) = 48t^3 - 3/\sqrt{t} - 5/t^2$.

(c) $f(x) = \dfrac{x^3 + 3\sqrt{x}}{x}$

Solution Rewrite $f(x)$ as $f(x) = \dfrac{x^3}{x} + \dfrac{3x^{1/2}}{x} = x^2 + 3x^{-1/2}$. Then,

$$D_x[f(x)] = 2x - \frac{3}{2}x^{-3/2},$$

or

$$D_x[f(x)] = 2x - \frac{3}{2\sqrt{x^3}}.$$

(d) $f(x) = (4x^2 - 3x)^2$

Solution Rewrite $f(x)$ as $f(x) = 16x^4 - 24x^3 + 9x^2$; then, using the fact that $(a - b)^2 = a^2 - 2ab + b^2$,

$$f'(x) = 64x^3 - 72x^2 + 18x.$$

NOTE Some computer programs and calculators have built-in methods for taking derivatives symbolically, which is what we have been doing in this section, as opposed to approximating the derivative numerically by using a small number for h in the definition of the derivative. In the computer program Maple, we would do part (a) of Example 5 by entering

```
> diff(6*x^3+15*x^2,x);
```

and Maple would respond with

```
18*x^2+30*x.
```

Similarly, on the TI-89, we would enter `d(6x^3+15x^2,x)` and the calculator would give "`18·x²+30·x.`"

Other graphing calculators, such as the TI-83/84 Plus, do not have built-in methods for taking derivatives symbolically. As we saw in the last chapter, however, they do have the ability to calculate the derivative of a function at a particular point and to simultaneously graph a function and its derivative.

Recall that, on the TI-83/84 Plus, we could enter `nDeriv(6X^3+15X^2,X,1)`. The number `48` will appear on the screen of the calculator, indicating the value of the derivative when $x = 1$. Figure 2 and Figure 3 indicate how to input the functions into the calculator and the corresponding graphs of both the function and its derivative. Consult *The Graphing Calculator Manual* that is available with this book for assistance.

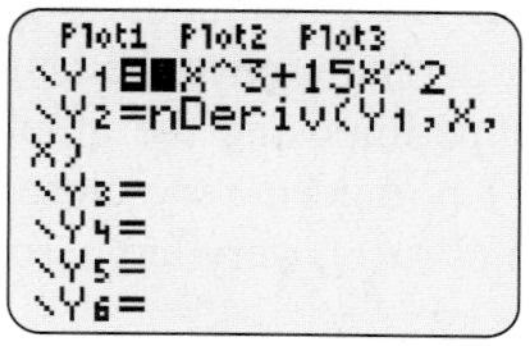

FIGURE 2   **FIGURE 3**   ■

The rules developed in this section make it possible to find the derivative of a function more directly, so that applications of the derivative can be dealt with more effectively. The following examples illustrate some business applications.

Marginal Analysis   In previous sections we discussed the concepts of marginal cost, marginal revenue, and marginal profit. These concepts of **marginal analysis** are summarized here.

In business and economics the rates of change of such variables as cost, revenue, and profit are important considerations. Economists use the word *marginal* to refer to rates of change. For example, *marginal cost* refers to the rate of change of cost. Since the derivative of a function gives the rate of change of the function, a marginal cost (or revenue, or profit) function is found by taking the derivative of the cost (or revenue, or profit) function. Roughly speaking, the marginal cost at some level of production x is the cost to produce the $(x + 1)$st item. (Similar statements could be made for revenue or profit.)

To see why it is reasonable to say that the marginal cost function is approximately the cost of producing one more unit, look at Figure 4, where $C(x)$ represents the cost of producing x units of some item. Then the cost of producing $x + 1$ units is $C(x + 1)$. The cost of the $(x + 1)$st unit is therefore $C(x + 1) - C(x)$. This quantity is shown in the graph in Figure 4.

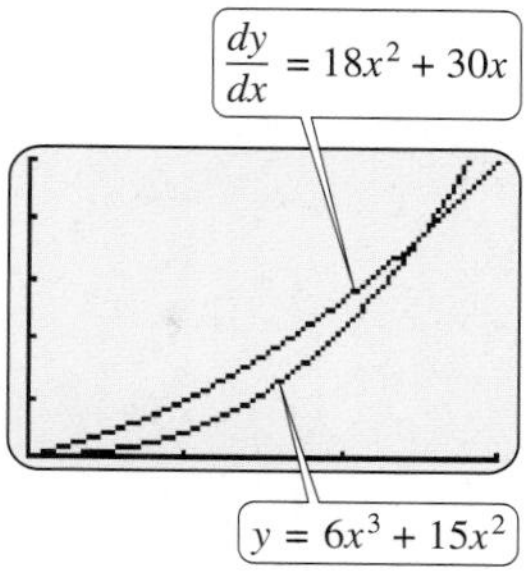

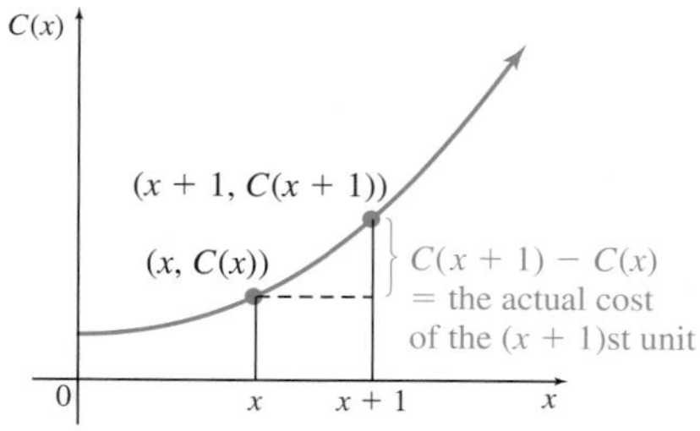

FIGURE 4

Now if $C(x)$ is the cost function, then the marginal cost $C'(x)$ represents the slope of the tangent line at any point $(x, C(x))$. The graph in Figure 5 shows the cost function $C(x)$ and the tangent line at a point $(x, C(x))$. Remember what it means for a line to have a given slope. If the slope of the line is $C'(x)$, then

$$\frac{\Delta y}{\Delta x} = C'(x) = \frac{C'(x)}{1},$$

and beginning at any point on the line and moving 1 unit to the right requires moving $C'(x)$ units up to get back to the line again. The vertical distance from the horizontal line to the tangent line shown in Figure 5 is therefore $C'(x)$.

Superimposing the graphs from Figures 4 and 5 as in Figure 6 shows that $C'(x)$ is indeed very close to $C(x + 1) - C(x)$. The two values are closest when $C'(x)$ is very large, so that 1 unit is relatively small.

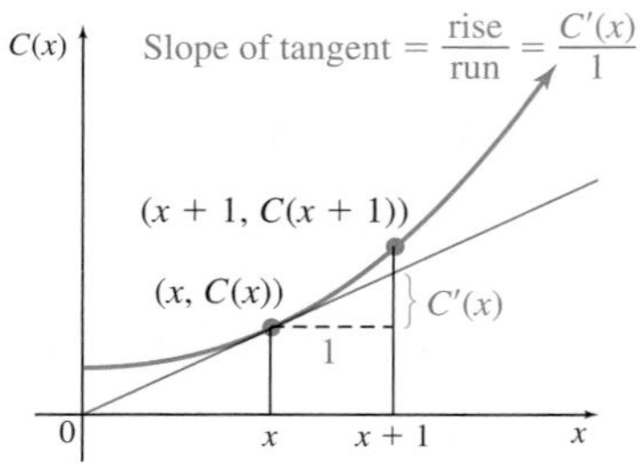

FIGURE 5

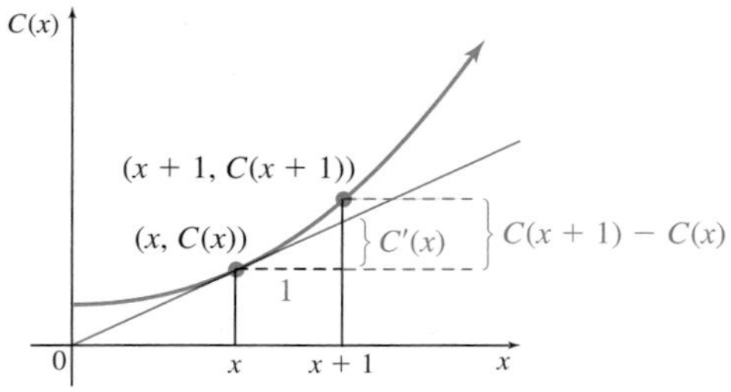

FIGURE 6

EXAMPLE 6 *Marginal Cost*

Suppose that the total cost in hundreds of dollars to produce x thousand barrels of a beverage is given by

$$C(x) = 4x^2 + 100x + 500.$$

Find the marginal cost for the following values of x.

(a) $x = 5$

Solution To find the marginal cost, first find $C'(x)$, the derivative of the total cost function.

$$C'(x) = 8x + 100$$

When $x = 5$,

$$C'(5) = 8(5) + 100 = 140.$$

After 5 thousand barrels of the beverage have been produced, the cost to produce one thousand more barrels will be *approximately* 140 hundred dollars, or \$14,000.

The *actual* cost to produce one thousand more barrels is $C(6) - C(5)$:

$$\begin{aligned} C(6) - C(5) &= (4 \cdot 6^2 + 100 \cdot 6 + 500) - (4 \cdot 5^2 + 100 \cdot 5 + 500) \\ &= 1244 - 1100 = 144, \end{aligned}$$

144 hundred dollars, or \$14,400.

(b) $x = 30$

Solution After 30 thousand barrels have been produced, the cost to produce one thousand more barrels will be approximately

$$C'(30) = 8(30) + 100 = 340,$$

or $34,000. Notice that the cost to produce an additional thousand barrels of beverage has increased by approximately $20,000 at a production level of 30,000 barrels compared to a production level of 5000 barrels. Management must be careful to keep track of marginal costs. If the marginal cost of producing an extra unit exceeds the revenue received from selling it, then the company will lose money on that unit. ■

Demand Functions The demand function, defined by $p = f(x)$, relates the number of units x of an item that consumers are willing to purchase to the price p. (Demand functions were also discussed in Chapter 1.) The total revenue $R(x)$ is related to price per unit and the amount demanded (or sold) by the equation

$$R(x) = xp = x \cdot f(x).$$

EXAMPLE 7 *Marginal Revenue*
The demand function for a certain product is given by

$$p = \frac{50,000 - x}{25,000}.$$

Find the marginal revenue when $x = 10,000$ units and p is in dollars.

Solution From the given function for p, the revenue function is given by

$$R(x) = xp$$
$$= x\left(\frac{50,000 - x}{25,000}\right)$$
$$= \frac{50,000x - x^2}{25,000}$$
$$= 2x - \frac{1}{25,000}x^2.$$

The marginal revenue is

$$R'(x) = 2 - \frac{2}{25,000}x.$$

When $x = 10,000$, the marginal revenue is

$$R'(10,000) = 2 - \frac{2}{25,000}(10,000) = 1.2,$$

or $1.20 per unit. Thus, the next item sold (at sales of 10,000) will produce additional revenue of about $1.20 per unit. ■

EXAMPLE 8 *Marginal Profit*

Suppose that the cost function for the product in Example 7 is given by

$$C(x) = 2100 + .25x, \qquad \text{where } 0 \le x \le 30{,}000.$$

Find the marginal profit from the production of the following numbers of units.

(a) 15,000

Solution From Example 6, the revenue from the sale of x units is

$$R(x) = 2x - \frac{1}{25{,}000}x^2.$$

Since profit, P, is given by $P = R - C$,

$$\begin{aligned} P(x) &= R(x) - C(x) \\ &= \left(2x - \frac{1}{25{,}000}x^2\right) - (2100 + .25x) \\ &= 2x - \frac{1}{25{,}000}x^2 - 2100 - .25x \\ &= 1.75x - \frac{1}{25{,}000}x^2 - 2100. \end{aligned}$$

See Figure 7.

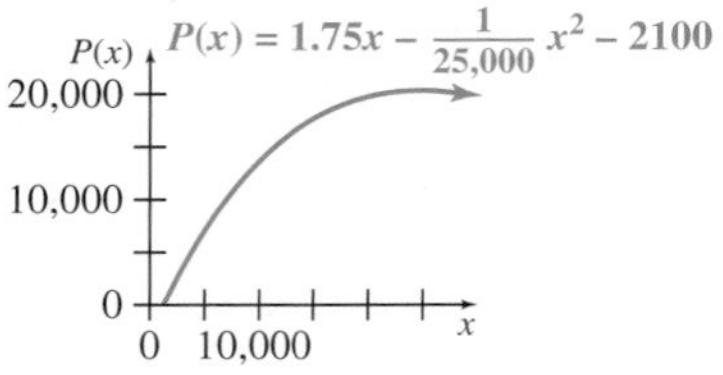

FIGURE 7

The marginal profit from the sale of x units is

$$P'(x) = 1.75 - \frac{2}{25{,}000}x = 1.75 - \frac{1}{12{,}500}x.$$

At $x = 15{,}000$ the marginal profit is

$$P'(15{,}000) = 1.75 - \frac{1}{12{,}500}(15{,}000) = .55,$$

or \$.55 per unit.

(b) 21,875

Solution When $x = 21{,}875$, the marginal profit is

$$P'(21{,}875) = 1.75 - \frac{1}{12{,}500}(21{,}875) = 0.$$

(c) 25,000

Solution When $x = 25{,}000$, the marginal profit is

$$P'(25{,}000) = 1.75 - \frac{1}{12{,}500}(25{,}000) = -.25,$$

or $-$\$.25 per unit.

As shown by parts (b) and (c), if more than 21,875 units are sold, the marginal profit is negative. This indicates that increasing production beyond that level will *reduce* profit.

The final example shows an application of the derivative to a problem of demography.

EXAMPLE 9 *Centenarians*

The number of Americans (in thousands) who are expected to be over 100 years old can be approximated by the function

$$f(x) = .4018x^2 + 2.039x + 50.071,^*$$

where x is the year, with $x = 0$ corresponding to 1994. This formula is based on estimates from 1994–2004.

(a) Find a formula giving the rate of change of the number of Americans over 100 years old.

Solution Using the techniques for finding the derivative, we have

$$f'(x) = .8036x + 2.039.$$

This tells us that the number of Americans over 100 years old is expected to grow at a linear rate.

(b) Find the rate of change in the number of Americans who were expected to be over 100 years old in the year 2003.

Solution The year 2003 corresponds to $x = 9$.

$$f'(9) = .8036(9) + 2.039 \approx 9.27$$

The number of Americans over 100 years old was expected to grow at a rate of 9.27 thousand, or about 9300, per year in the year 2003.

4.1 EXERCISES

Find the derivative of each function defined as follows.

1. $y = 10x^3 - 9x^2 + 6x + 5$

2. $y = 3x^3 - x^2 - \frac{x}{12}$

3. $y = x^4 - 5x^3 + \frac{x^2}{9} + 5$

4. $y = 3x^4 + 11x^3 + 2x^2 - 4x$

5. $f(x) = 6x^{1.5} - 4x^{.5}$

6. $f(x) = -2x^{2.5} + 8x^{.5}$

7. $y = 8\sqrt{x} + 6x^{3/4}$

8. $y = -100\sqrt{x} - 11x^{2/3}$

9. $g(x) = 6x^{-5} - x^{-1}$

10. $y = 10x^{-2} + 3x^{-4} - 6x$

11. $y = x^{-5} - x^{-2} + 5x^{-1}$

12. $f(t) = \frac{6}{t} - \frac{8}{t^2}$

13. $f(t) = \frac{4}{t} + \frac{2}{t^3} + \sqrt{2}$

14. $y = \frac{9}{x^4} - \frac{8}{x^3} + \frac{2}{x}$

15. $y = \frac{3}{x^6} + \frac{1}{x^5} - \frac{7}{x^2}$

16. $p(x) = -10x^{-1/2} + 8x^{-3/2}$

17. $h(x) = x^{-1/2} - 14x^{-3/2}$

18. $y = \frac{6}{\sqrt[4]{x}}$

19. $y = \frac{-2}{\sqrt[3]{x}}$

20. $f(x) = \frac{x^2 + 2}{x}$

21. $g(x) = \frac{x^2 - 2x}{\sqrt{x}}$

22. $g(x) = (8x^2 - 4x)^2$

23. $h(x) = (x^2 - 1)^3$

24. Which of the following describes the derivative function f' of a quadratic function f?

a. Quadratic **b.** Linear **c.** Constant **d.** Cubic (third degree)

25. Explain the relationship between the slope and the derivative of $f(x)$ at $x = a$.

26. Which of the following does *not* equal $\frac{d}{dx}(4x^3 - 6x^{-2})$?

a. $\frac{12x^2 + 12}{x^3}$ **b.** $\frac{12x^5 + 12}{x^3}$ **c.** $12x^2 + \frac{12}{x^3}$ **d.** $12x^3 + 12x^{-3}$

*U.S. Census Bureau.

Find each derivative.

27. $D_x\left[9x^{-1/2} + \dfrac{2}{x^{3/2}}\right]$

28. $D_x\left[\dfrac{8}{\sqrt[4]{x}} - \dfrac{3}{\sqrt{x^3}}\right]$

29. $f'(-2)$ if $f(x) = \dfrac{x^2}{6} - 4x$

30. $f'(3)$ if $f(x) = \dfrac{x^3}{9} - 8x^2$

In Exercises 31–34, find the slope of the tangent line to the graph of the given function at the given value of x. Find the equation of the tangent line in Exercises 31 and 32.

31. $y = x^4 - 5x^3 + 2;\quad x = 2$

32. $y = -2x^5 - 7x^3 + 8x^2;\quad x = 1$

33. $y = -2x^{1/2} + x^{3/2};\quad x = 4$

34. $y = -x^{-3} + x^{-2};\quad x = 1$

35. Find all points on the graph of $f(x) = 9x^2 - 8x + 4$ where the slope of the tangent line is 0.

In Exercises 36–39, for each function find all values of x where the tangent line is horizontal.

36. $f(x) = 2x^3 + 9x^2 - 60x + 4$

37. $f(x) = x^3 + 15x^2 + 63x - 10$

38. $f(x) = x^3 - 4x^2 - 7x + 8$

39. $f(x) = x^3 - 5x^2 + 6x + 3$

40. At what points on the graph of $f(x) = 6x^2 + 4x - 9$ is the slope of the tangent line -2?

41. At what points on the graph of $f(x) = 2x^3 - 9x^2 - 12x + 5$ is the slope of the tangent line 12?

42. At what points on the graph of $f(x) = x^3 + 6x^2 + 21x + 2$ is the slope of the tangent line 9?

43. If $g'(5) = 10$ and $h'(5) = -4$, find $f'(5)$ for $f(x) = 3g(x) - 2h(x) + 3$.

44. If $g'(2) = 3$ and $h'(2) = 6$, find $f'(2)$ for $f(x) = \dfrac{1}{2}g(x) + \dfrac{1}{4}h(x)$.

45. Use the information given in the figure to find the following values.

a. $f(1)$ **b.** $f'(1)$ **c.** The domain of f **d.** The range of f

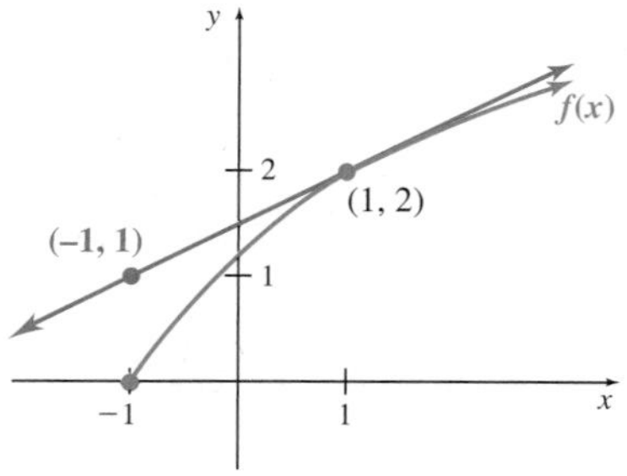

46. Explain the concept of marginal cost. How does it relate to cost? How is it found?

47. In Exercises 37–40 of the section on Quadratic Functions; Translation and Reflection, the effect of a when graphing $y = af(x)$ was discussed. Now describe how this relates to the fact that $D_x[af(x)] = af'(x)$.

48. Show that, for any constant k,

$$\frac{d}{dx}\left[\frac{f(x)}{k}\right] = \frac{f'(x)}{k}.$$

49. Use the differentiation feature on your graphing calculator to solve the problems (to 2 decimal places) below, where $f(x)$ is defined as follows:

$$f(x) = 1.25x^3 + .01x^2 - 2.9x + 1.$$

a. Find $f'(4)$.

b. Find all values of x where $f'(x) = 0$.

Applications

BUSINESS AND ECONOMICS

50. *Revenue* Assume that a demand equation is given by $x = 5000 - 100p$. Find the marginal revenue for the following production levels (values of x). (*Hint:* Solve the demand equation for p and use $R(x) = xp$.)

a. 1000 units **b.** 2500 units **c.** 3000 units

51. *Profit* Suppose that for the situation in Exercise 50 the cost of producing x units is given by $C(x) = 3000 - 20x + .03x^2$. Find the marginal profit for the following production levels.

a. 500 units **b.** 815 units **c.** 1000 units

52. *Sales* Often sales of a new product grow rapidly at first and then level off with time. This is the case with the sales represented by the function

$$S(t) = 100 - 100t^{-1},$$

where t represents time in years. Find the rate of change of sales for the following numbers of years.

a. 1 **b.** 10

53. *Revenue* If the price in dollars of a stereo system is given by

$$p(x) = \frac{1000}{x^2} + 1000,$$

where x represents the demand for the product, find the marginal revenue when the demand is 10.

54. *Profit* Suppose that for the situation in Exercise 53 the cost in dollars of producing x stereo systems is given by $C(x) = .2x^2 + 6x + 50$. Find the marginal profit when the demand is 10.

55. *Profit* An analyst has found that a company's costs and revenues in dollars for one product are given by

$$C(x) = 2x \quad \text{and} \quad R(x) = 6x - \frac{x^2}{1000},$$

respectively, where x is the number of items produced.

a. Find the marginal cost function.

b. Find the marginal revenue function.

c. Using the fact that profit is the difference between revenue and costs, find the marginal profit function.

d. What value of x makes marginal profit equal 0?

e. Find the profit when the marginal profit is 0.

(As we shall see in the next chapter, this process is used to find *maximum* profit.)

56. *Postal Rates* U.S. postal rates have steadily increased since 1932. Using data depicted in the table for the years 1932–2002, the cost to mail a single letter can be modeled using a quadratic formula as follows:

$$C(x) = .0102x^2 - .2164x + 3.23,$$

where x is the number of years since 1932.*

Year	Cost	Year	Cost
1932	3	1981	20
1958	4	1985	22
1963	5	1988	25
1968	6	1991	29
1971	8	1995	32
1974	10	1999	33
1975	13	2001	34
1978	15	2002	37
1981	18		

a. Find the predicted cost of mailing a letter in 1982 and 2002 and compare these estimates with the actual rates.

b. Find the rate of change of the postage cost for the years ending 1982 and 2002 and interpret your results.

c. Critically evaluate possible limitations of this model and make suggestions for improvements.

57. *Money* The total amount of money in circulation for the years 1915–2002 can be closely approximated by

$$M(x) = 3.044x^3 - 379.6x^2 + 14{,}274.5x - 139{,}433,$$

*U.S. Postal Service.

where x represents the number of years since 1900 and $M(x)$ is in millions of dollars.* Find the derivative of $M(x)$ and use it to find the rate of change of money in circulation in the following years.

a. 1920 **b.** 1960 **c.** 1980 **d.** 2000

e. What do your answers to parts a–d tell you about the amount of money in circulation in those years?

LIFE SCIENCES

58. *Cancer* Insulation workers who were exposed to asbestos and employed before 1960 experienced an increased likelihood of lung cancer. If a group of insulation workers has a cumulative total of 100,000 years of work experience with their first date of employment t years ago, then the number of lung cancer cases occurring within the group can be modeled using the function

$$N(t) = .00437t^{3.2}.^{\dagger}$$

Find the rate of growth of the number of workers with lung cancer in a group as described by the following first dates of employment.

a. 5 years ago **b.** 10 years ago

59. *Blood Sugar Level* Insulin affects the glucose, or blood sugar, level of some diabetics according to the function

$$G(x) = -.2x^2 + 450,$$

where $G(x)$ is the blood sugar level 1 hour after x units of insulin are injected. (This mathematical model is only approximate, and it is valid only for values of x less than about 40.) Find the blood sugar level after the following numbers of units of insulin are injected.

a. 0 **b.** 25

Find the rate of change of blood sugar level after injection of the following numbers of units of insulin.

c. 10 **d.** 25

60. *Bighorn Sheep* The cumulative horn volume for certain types of bighorn rams, found in the Rocky Mountains, can be described by the quadratic function

$$V(t) = -2159 + 1313t - 60.82t^2,$$

where $V(t)$ is the horn volume (in cm^3) and t is the year of growth, $2 \le t \le 9$.‡

a. Find the horn volume for a 3-year-old ram.

b. Find the rate at which the horn volume of a 3-year-old ram is changing.

61. *Brain Mass* The brain mass of a human fetus during the last trimester can be accurately estimated from the circumference of the head by

$$m(c) = \frac{c^3}{100} - \frac{1500}{c},$$

where $m(c)$ is the mass of the brain (in grams) and c is the circumference (in centimeters) of the head.§

a. Estimate the brain mass of a fetus that has a head circumference of 30 cm.

b. Find the rate of change of the brain mass for a fetus that has a head circumference of 30 cm and interpret your results.

62. *Velocity of Marine Organism* The typical velocity (in centimeters per second) of a marine organism of length l (in centimeters) is given by $v = 2.69l^{1.86}$.‖ Find the rate of change of the velocity with respect to the length of the organism.

The World Almanac and Book of Facts 2003, p. 114.

†Walker, A., *Observation and Inference: An Introduction to the Methods of Epidemiology,* Epidemiology Resources, Inc., 1991.

‡Fitzsimmons, N., S. Buskirk, and M. Smith, "Population History, Genetic Variability, and Horn Growth in Bighorn Sheep," *Conservation Biology,* Vol. 9, No. 2, April 1995, pp. 314–323.

§Dobbing, John and Jean Sands, "Head Circumference, Biparietal Diameter and Brain Growth in Fetal and Postnatal Life," *Early Human Development,* Vol. 2, No. 1, April 1978, pp. 81–87.

‖Okubo, Akira, "Fantastic Voyage into the Deep: Marine Biofluid Mechancis," in *Mathematical Topics in Population Biology Morphogenesis and Neurosciences,* edited by E. Teramoto and M. Yamaguti, Springer-Verlag, 1987, pp. 32–47.

63. *Heart* The left ventricular length (viewed from the front of the heart) of a fetus that is at least 18 weeks old can be estimated by

$$l(x) = -2.318 + .2356x - .002674x^2,$$

where $l(x)$ is the ventricular length (in centimeters) and x is the age (in weeks) of the fetus.*

a. Determine a meaningful domain for this function.

b. Find $l'(x)$.

c. Find $l'(25)$.

64. *Track and Field* In 1906 Kennelly developed a simple formula for predicting an upper limit on the fastest time that humans could ever run distances from 100 yards to 10 miles. His formula is given by

$$t = .0588s^{1.125},$$

where s is the distance in meters and t is the time to run that distance in seconds.†

a. Find Kennelly's estimate for the fastest mile. (*Hint:* 1 mile ≈ 1609 meters.)

b. Find dt/ds when $s = 100$ and interpret your answer.

c. Compare this and other estimates to the current world records. Have these estimates been surpassed?

65. *Human Cough* To increase the velocity of the air flowing through the trachea when a human coughs, the body contracts the windpipe, producing a more effective cough. Tuchinsky formulated that the velocity of air that is flowing through the trachea during a cough is

$$V = C(R_0 - R)R^2,$$

where C is a constant based on individual body characteristics, R_0 is the radius of the windpipe before the cough, and R is the radius of the windpipe during the cough.‡ It can be shown that the maximum velocity of the cough occurs when $dV/dR = 0$. Find the value of R that maximizes the velocity.§

66. *Body Mass Index* The body mass index (BMI) is a number that can be calculated for any individual as follows: Multiply weight (lb) by 703 and divide by the person's height (in.) squared. That is,

$$BMI = \frac{703w}{h^2},$$

where w is in pounds and h is in inches. The National Heart, Lung, and Blood Institute uses the BMI to determine whether a person is "overweight" $(25 \leq BMI < 30)$ or "obese" $(BMI \geq 30)$.

a. Calculate the BMI for a person who weighs 220 lb and is 6′2″ tall.

b. How much weight would the person in part a have to lose until he reaches a BMI of 24.9 and is no longer "overweight"?

c. For a 125-lb female, what is the rate of change of BMI with respect to height? (*Hint:* Take the derivative of the function: $f(h) = 703(125)/h^2$.)

d. Calculate and interpret the meaning of $f'(65)$.

e. Use the TABLE feature on your graphing calculator to construct a table for BMI for various weights and heights.

PHYSICAL SCIENCES

Velocity We saw in the previous chapter that if a function $s(t)$ gives the position of an object at time t, the derivative gives the velocity, that is, $v(t) = s'(t)$. For each position function in Exercises 64–67, find **(a)** *$v(t)$ and* **(b)** *the velocity when $t = 0$, $t = 5$, and $t = 10$.*

67. $s(t) = 11t^2 + 4t + 2$

68. $s(t) = 25t^2 - 9t + 8$

69. $s(t) = 4t^3 + 8t^2$

70. $s(t) = -2t^3 + 4t^2 - 1$

71. *Velocity* If a rock is dropped from a 144-ft building, its position (in feet above the ground) is given by $s(t) = -16t^2 + 144$, where t is the time in seconds since it was dropped.

a. What is its velocity 1 second after being dropped? 2 seconds after being dropped?

b. When will it hit the ground?

c. What is its velocity upon impact?

*Tan J., N. Silverman, J. Hoffman, M. Villegas, and K. Schmidt, "Cardiac Dimensions Determined by Cross-Sectional Echocardiography in the Normal Human Fetus From 18 Weeks to Term," *American Journal of Cardiology,* Vol. 70, No. 18, Dec. 1, 1992, pp. 1459–1467.

†Kennelly, A., "An Approximate Law of Fatigue in Speeds of Racing Animals," *Proceedings of the American Academy of Arts and Sciences,* Vol. 42, 1906, pp. 275–331.

‡Tuchinsky, Philip, "The Human Cough," *UMAP Module 211,* Lexington, MA, COMAP, Inc., 1979, pp. 1–9.

§Interestingly, Tuchinsky also states that X-rays indicate that the body naturally contracts the windpipe to this radius during a cough.

72. *Velocity* A ball is thrown vertically upward from the ground at a velocity of 64 ft per second. Its distance from the ground at t seconds is given by $s(t) = -16t^2 + 64t$.

a. How fast is the ball moving 2 seconds after being thrown? 3 seconds after being thrown?

b. How long after the ball is thrown does it reach its maximum height?

c. How high will it go?

73. *Dead Sea* Researchers who have been studying the alarming rate in which the level of the Dead Sea has been dropping have shown that the density $d(x)$ (in g per cm^3) of the Dead Sea brine during evaporation can be estimated by the function

$$d(x) = 1.66 - .90x + .47x^2,$$

where x is the fraction of the remaining brine, $0 \le x \le 1$.*

a. Estimate the density of the brine when 50% of the brine remains.

b. Find and interpret the instantaneous rate of change of the density when 50% of the brine remains.

74. *Dog's Human Age* From the data printed in the following table from the *Minneapolis Star Tribune* on September 20, 1998, a dog's age when compared to a human's age can be modeled using either a linear formula or a quadratic formula as follows:

$$y_1 = 4.13x + 14.63$$
$$y_2 = -0.033x^2 + 4.647x + 13.347,$$

where y_1 and y_2 represent a dog's human age for each formula and x represents a dog's actual age.†

Dog Age	Human Age
1	16
2	24
3	28
5	36
7	44
9	52
11	60
13	68
15	76

a. Find y_1 and y_2 when $x = 5$.

b. Find dy_1/dx and dy_2/dx when $x = 5$ and interpret your answers.

c. If the first three points are eliminated from the table, find the equation of a line that perfectly fits the reduced set of data. Interpret your findings.

d. Of the three formulas, which do you prefer?

4.2 DERIVATIVES OF PRODUCTS AND QUOTIENTS

THINK ABOUT IT *A manufacturer of small motors wants to make the average cost per motor as small as possible. How can this be done?*

We show how the derivative is used to solve a problem like this in Example 5, later in this section. In the previous section we developed several rules for finding derivatives. We develop two additional rules in this section, again using the definition of the derivative.

*Yechieli, Yoseph, Ittai Gavrieli, Brian Berkowitz, and Daniel Ronen, "Will the Dead Sea Die?" *Geology,* Vol. 26, No. 8, Aug. 1998, pp. 755–758. These researchers have predicted that the Dead Sea will not die but reach an equilibrium level.

†Vennebush, Patrick, "Media Clips: A Dog's Human Age," *Mathematics Teacher,* Vol. 92, 1999, pp. 710–712.

The derivative of a sum of two functions is found from the sum of the derivatives. What about products? Is the derivative of a product equal to the product of the derivatives? For example, if

$$u(x) = 2x + 3 \qquad \text{and} \qquad v(x) = 3x^2,$$

then

$$u'(x) = 2 \qquad \text{and} \qquad v'(x) = 6x.$$

Let $f(x)$ be the product of u and v; that is, $f(x) = (2x + 3)(3x^2) = 6x^3 + 9x^2$. By the rules of the preceding section, $f'(x) = 18x^2 + 18x$. On the other hand, $u'(x) = 2$ and $v'(x) = 6x$, with the product $u'(x) \cdot v'(x) = 2(6x) = 12x \neq f'(x)$. In this example, the derivative of a product is *not* equal to the product of the derivatives, nor is this usually the case.

The rule for finding derivatives of products is now developed.

PRODUCT RULE

If $f(x) = u(x) \cdot v(x)$, and if $u'(x)$ and $v'(x)$ both exist, then

$$\boldsymbol{f'(x) = u(x) \cdot v'(x) + v(x) \cdot u'(x).}$$

(The derivative of a product of two functions is the first function times the derivative of the second, plus the second function times the derivative of the first.)

FOR REVIEW

This proof uses several of the rules for limits given in the first section of the previous chapter. You may want to review them at this time.

To sketch the method used to prove the product rule, let

$$f(x) = u(x) \cdot v(x).$$

Then $f(x + h) = u(x + h) \cdot v(x + h)$, and, by definition, $f'(x)$ is given by

$$\begin{aligned} f'(x) &= \lim_{h \to 0} \frac{f(x + h) - f(x)}{h} \\ &= \lim_{h \to 0} \frac{u(x + h) \cdot v(x + h) - u(x) \cdot v(x)}{h}. \end{aligned}$$

Now subtract and add $u(x + h) \cdot v(x)$ in the numerator, giving

$$\begin{aligned} f'(x) &= \lim_{h \to 0} \frac{u(x + h) \cdot v(x + h) - u(x + h) \cdot v(x) + u(x + h) \cdot v(x) - u(x) \cdot v(x)}{h} \\ &= \lim_{h \to 0} \frac{u(x + h)[v(x + h) - v(x)] + v(x)[u(x + h) - u(x)]}{h} \\ &= \lim_{h \to 0} u(x + h)\left[\frac{v(x + h) - v(x)}{h}\right] + \lim_{h \to 0} v(x)\left[\frac{u(x + h) - u(x)}{h}\right] \\ &= \lim_{h \to 0} u(x + h) \cdot \lim_{h \to 0} \frac{v(x + h) - v(x)}{h} + \lim_{h \to 0} v(x) \cdot \lim_{h \to 0} \frac{u(x + h) - u(x)}{h}. \qquad (1) \end{aligned}$$

If u' and v' both exist, then

$$\lim_{h \to 0} \frac{u(x + h) - u(x)}{h} = u'(x) \qquad \text{and} \qquad \lim_{h \to 0} \frac{v(x + h) - v(x)}{h} = v'(x).$$

The fact that u' exists can be used to prove

$$\lim_{h \to 0} u(x+h) = u(x),$$

and since no h is involved in $v(x)$,

$$\lim_{h \to 0} v(x) = v(x).$$

Substituting these results into Equation (1) gives

$$f'(x) = u(x) \cdot v'(x) + v(x) \cdot u'(x),$$

the desired result.

To help see why the product rule is true, consider the special case in which u and v are positive functions. Then $u(x) \cdot v(x)$ represents the area of a rectangle, as shown in Figure 8. If we assume that u and v are increasing, then $u(x+h) \cdot v(x+h)$ represents the area of a slightly larger rectangle when h is a small positive number, as shown in the figure. The change in the area of the rectangle is given by the pink rectangle, with an area of $u(x)$ times the amount v has changed, plus the blue rectangle, with an area of $v(x)$ times the amount u has changed, plus the small green rectangle. As h becomes smaller and smaller, the green rectangle becomes negligibly small, and the change in the area is essentially $u(x)$ times the change in v, plus $v(x)$ times the change in u.

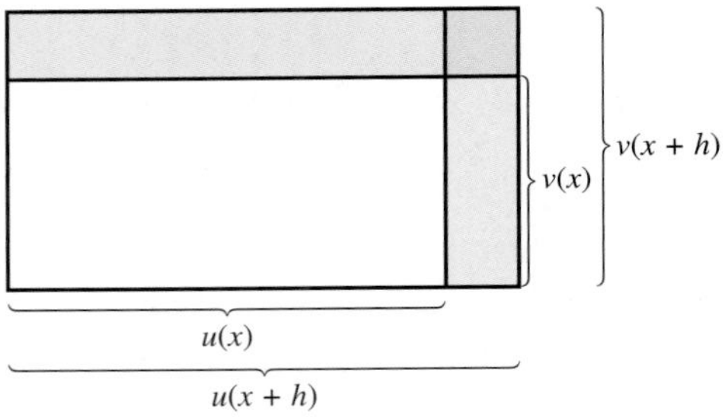

FIGURE 8

EXAMPLE 1 *Product Rule*

Let $f(x) = (2x+3)(3x^2)$. Use the product rule to find $f'(x)$.

Solution Here f is given as the product of $u(x) = 2x + 3$ and $v(x) = 3x^2$. By the product rule and the fact that $u'(x) = 2$ and $v'(x) = 6x$,

$$\begin{aligned} f'(x) &= u(x) \cdot v'(x) + v(x) \cdot u'(x) \\ &= (2x+3)(6x) + (3x^2)(2) \\ &= 12x^2 + 18x + 6x^2 = 18x^2 + 18x. \end{aligned}$$

This result is the same as that found at the beginning of the section.

EXAMPLE 2 *Product Rule*

Find the derivative of $y = (\sqrt{x} + 3)(x^2 - 5x)$.

Solution Let $u(x) = \sqrt{x} + 3 = x^{1/2} + 3$, and $v(x) = x^2 - 5x$. Then

$$\frac{dy}{dx} = u(x) \cdot v'(x) + v(x) \cdot u'(x)$$

$$= (x^{1/2} + 3)(2x - 5) + (x^2 - 5x)\left(\frac{1}{2}x^{-1/2}\right).$$

Simplify by multiplying and combining terms.

$$\frac{dy}{dx} = (2x)(x^{1/2}) + 6x - 5x^{1/2} - 15 + (x^2)\left(\frac{1}{2}x^{-1/2}\right) - (5x)\left(\frac{1}{2}x^{-1/2}\right)$$

$$= 2x^{3/2} + 6x - 5x^{1/2} - 15 + \frac{1}{2}x^{3/2} - \frac{5}{2}x^{1/2}$$

$$= \frac{5}{2}x^{3/2} + 6x - \frac{15}{2}x^{1/2} - 15$$

We could have found the derivatives above by multiplying out the original functions. The product rule then would not have been needed. In the next section, however, we shall see products of functions where the product rule is essential.

What about *quotients* of functions? To find the derivative of the quotient of two functions, use the next rule.

QUOTIENT RULE

If $f(x) = u(x)/v(x)$, if all indicated derivatives exist, and if $v(x) \neq 0$, then

$$f'(x) = \frac{v(x) \cdot u'(x) - u(x) \cdot v'(x)}{[v(x)]^2}.$$

(The derivative of a quotient is the denominator times the derivative of the numerator, minus the numerator times the derivative of the denominator, all divided by the square of the denominator.)

The proof of the quotient rule is similar to that of the product rule and is left for the exercises. (See Exercises 32 and 33.)

CAUTION Just as the derivative of a product is *not* the product of the derivatives, the derivative of a quotient is *not* the quotient of the derivatives. If you are asked to take the derivative of a product or a quotient, it is essential that you recognize that the function contains a product or quotient and then use the appropriate rule. ■

EXAMPLE 3 *Quotient Rule*

Find $f'(x)$ if $f(x) = \dfrac{2x - 1}{4x + 3}$.

FOR REVIEW

You may want to consult the Rational Expressions section of the Algebra Reference chapter (Section 3) to help you work with the fractions in the section.

Solution Let $u(x) = 2x - 1$, with $u'(x) = 2$. Also, let $v(x) = 4x + 3$, with $v'(x) = 4$. Then, by the quotient rule,

$$\begin{aligned} f'(x) &= \frac{v(x) \cdot u'(x) - u(x) \cdot v'(x)}{[v(x)]^2} \\ &= \frac{(4x + 3)(2) - (2x - 1)(4)}{(4x + 3)^2} \\ &= \frac{8x + 6 - 8x + 4}{(4x + 3)^2} \\ &= \frac{10}{(4x + 3)^2}. \end{aligned}$$

CAUTION In the second step of Example 3, we had the expression

$$\frac{(4x + 3)(2) - (2x - 1)(4)}{(4x + 3)^2}.$$

Students often incorrectly "cancel" the $4x + 3$ in the numerator with one factor of the denominator. Because the numerator is a *difference* of two products, however, you must multiply and combine terms *before* looking for common factors in the numerator and denominator. ■

EXAMPLE 4 *Product and Quotient Rules*

Find $D_x\left[\frac{(3 - 4x)(5x + 1)}{7x - 9}\right]$.

Solution This function has a product within a quotient. Instead of multiplying the factors in the numerator first (which is an option), we can use the quotient rule together with the product rule, as follows. Use the quotient rule first to get

$$D_x\left[\frac{(3 - 4x)(5x + 1)}{7x - 9}\right] = \frac{(7x - 9)D_x[(3 - 4x)(5x + 1)] - [(3 - 4x)(5x + 1)D_x(7x - 9)]}{(7x - 9)^2}.$$

Now use the product rule to find $D_x[(3 - 4x)(5x + 1)]$ in the numerator.

$$\begin{aligned} &= \frac{(7x - 9)[(3 - 4x)5 + (5x + 1)(-4)] - (3 + 11x - 20x^2)(7)}{(7x - 9)^2} \\ &= \frac{(7x - 9)(15 - 20x - 20x - 4) - (21 + 77x - 140x^2)}{(7x - 9)^2} \\ &= \frac{(7x - 9)(11 - 40x) - 21 - 77x + 140x^2}{(7x - 9)^2} \\ &= \frac{-280x^2 + 437x - 99 - 21 - 77x + 140x^2}{(7x - 9)^2} \\ &= \frac{-140x^2 + 360x - 120}{(7x - 9)^2} \end{aligned}$$

Average Cost Suppose $y = C(x)$ gives the total cost to manufacture x items. As mentioned earlier, the average cost per item is found by dividing the total cost by the number of items. The rate of change of average cost, called the *marginal average cost*, is the derivative of the average cost.

MARGINAL AVERAGE COST

If the total cost to manufacture x items is given by $C(x)$, then the average cost per item is $\overline{C}(x) = C(x)/x$. The **marginal average cost** is the derivative of the average cost function, $\overline{C}'(x)$.

Similarly, the marginal average revenue function, $\overline{R}'(x)$, is defined as the derivative of the average revenue function, $\overline{R}(x) = R(x)/x$, and the marginal average profit function, $\overline{P}'(x)$, is defined as the derivative of the average profit function, $\overline{P}(x) = P(x)/x$.

A company naturally would be interested in making the average cost as small as possible. The next chapter will show that this can be done by using the derivative of $C(x)/x$. This derivative often can be found by means of the quotient rule, as in the next example.

EXAMPLE 5 *Minimum Average Cost*

Suppose the cost in dollars of manufacturing x hundred small motors is given by

$$C(x) = \frac{3x^2 + 120}{2x + 1}.$$

(a) Find the average cost per hundred motors.

Solution The average cost is defined by

$$\overline{C}(x) = \frac{C(x)}{x} = \frac{3x^2 + 120}{2x + 1} \cdot \frac{1}{x} = \frac{3x^2 + 120}{2x^2 + x}.$$

(b) Find the marginal average cost.

Solution The marginal average cost is given by

$$\begin{aligned}\frac{d}{dx}[\overline{C}(x)] &= \frac{(2x^2 + x)(6x) - (3x^2 + 120)(4x + 1)}{(2x^2 + x)^2}\\ &= \frac{12x^3 + 6x^2 - 12x^3 - 480x - 3x^2 - 120}{(2x^2 + x)^2}\\ &= \frac{3x^2 - 480x - 120}{(2x^2 + x)^2}.\end{aligned}$$

(c) Average cost is generally minimized when the marginal average cost is zero. Find the level of production that minimizes average cost.

Solution Set the derivative $\overline{C}'(x) = 0$ and solve for x.

$$\frac{3x^2 - 480x - 120}{(2x^2 + x)^2} = 0$$

$$3x^2 - 480x - 120 = 0$$

$$3(x^2 - 160x - 40) = 0$$

Use the quadratic formula to solve this quadratic equation. Discarding the negative solution leaves $x = (160 + \sqrt{(160)^2 + 160})/2 \approx 160$ as the solution. Since x is in hundreds, production of 160 hundred or 16,000 motors will minimize average cost.

4.2 EXERCISES

Use the product rule to find the derivative of the following. (Hint for Exercises 3–6: Write the quantity as a product.)

1. $y = (3x^2 + 2)(2x - 1)$

2. $y = (5x^2 - 1)(4x + 3)$

3. $y = (2x - 5)^2$

4. $y = (7x - 6)^2$

5. $k(t) = (t^2 - 1)^2$

6. $g(t) = (3t^2 + 2)^2$

7. $y = (x + 1)(\sqrt{x} + 2)$

8. $y = (2x - 3)(\sqrt{x} - 1)$

9. $p(y) = (y^{-1} + y^{-2})(2y^{-3} - 5y^{-4})$

10. $q(x) = (x^{-2} - x^{-3})(3x^{-1} + 4x^{-4})$

Use the quotient rule to find the derivative of the following.

11. $f(x) = \dfrac{7x + 1}{3x + 8}$

12. $f(x) = \dfrac{6x - 11}{8x + 1}$

13. $y = \dfrac{5 - 3t}{4 + t}$

14. $y = \dfrac{9 - 7t}{1 - t}$

15. $y = \dfrac{x^2 + x}{x - 1}$

16. $y = \dfrac{x^2 - 4x}{x + 3}$

17. $f(t) = \dfrac{4t + 11}{t^2 - 3}$

18. $y = \dfrac{-x^2 + 6x}{4x^2 + 1}$

19. $g(x) = \dfrac{x^2 - 4x + 2}{x + 3}$

20. $k(x) = \dfrac{x^2 + 7x - 2}{x - 2}$

21. $p(t) = \dfrac{\sqrt{t}}{t - 1}$

22. $r(t) = \dfrac{\sqrt{t}}{2t + 3}$

23. $y = \dfrac{5x + 6}{\sqrt{x}}$

24. $h(z) = \dfrac{z^{2.2}}{z^{3.2} + 5}$

25. $g(y) = \dfrac{y^{1.4} + 1}{y^{2.5} + 2}$

26. $f(x) = \dfrac{(3x^2 + 1)(2x - 1)}{5x + 4}$

27. If $g(3) = 4$, $g'(3) = 5$, $f(3) = 7$, and $f'(3) = 6$, find $h'(3)$ when $h(x) = f(x)g(x)$.

28. If $g(3) = 4$, $g'(3) = 5$, $f(3) = 7$, and $f'(3) = 6$, find $h'(3)$ when $h(x) = f(x)/g(x)$.

29. Find the error in the following work.

$$D_x\left(\frac{2x + 5}{x^2 - 1}\right) = \frac{(2x + 5)(2x) - (x^2 - 1)2}{(x^2 - 1)^2} = \frac{4x^2 + 10x - 2x^2 + 2}{(x^2 - 1)^2}$$
$$= \frac{2x^2 + 10x + 2}{(x^2 - 1)^2}$$

30. Find the error in the following work.

$$D_x\left(\frac{x^2 - 4}{x^3}\right) = x^3(2x) - (x^2 - 4)(3x^2) = 2x^4 - 3x^4 + 12x^2$$
$$= -x^4 + 12x^2$$

31. Find an equation of the line tangent to the graph of $f(x) = x/(x - 2)$ at $(3, 3)$.

32. Following the steps used to prove the product rule for derivatives, prove the quotient rule for derivatives.

33. Use the fact that $f(x) = u(x)/v(x)$ can be rewritten as $f(x)v(x) = u(x)$ and the product rule for derivatives to verify the quotient rule for derivatives. (*Hint:* After applying the product rule, substitute $u(x)/v(x)$ for $f(x)$ and simplify.)

For each function, find the value(s) of x in which $f'(x) = 0$, to 3 decimal places.

34. $f(x) = (x^2 - 2)(x^2 - \sqrt{2})$

35. $f(x) = \dfrac{x - 2}{x^2 + 4}$

Applications

BUSINESS AND ECONOMICS

36. *Average Cost* The total cost (in hundreds of dollars) to produce x units of perfume is

$$C(x) = \frac{3x + 2}{x + 4}.$$

Find the average cost for each production level.

a. 10 units **b.** 20 units **c.** x units

d. Find the marginal average cost function.

37. *Average Profit* The total profit (in tens of dollars) from selling x self-help books is

$$P(x) = \frac{5x - 6}{2x + 3}.$$

Find the average profit from each sales level.

a. 8 books **b.** 15 books **c.** x books

d. Find the marginal average profit function.

e. Is this a reasonable function for profit? Why or why not?

38. *Employee Training* A company that manufactures bicycles has determined that a new employee can assemble $M(d)$ bicycles per day after d days of on-the-job training, where

$$M(d) = \frac{100d^2}{3d^2 + 10}.$$

a. Find the rate of change function for the number of bicycles assembled with respect to time.

b. Find and interpret $M'(2)$ and $M'(5)$.

39. *Marginal Revenue* Suppose that the demand function is given by $p = D(q)$, where q is the quantity that consumers demand when the price is p. Show that the marginal revenue is given by

$$R'(q) = D(q) + qD'(q).$$

40. *Marginal Average Cost* Suppose that the marginal cost function is given by $\overline{C}(x) = C(x)/x$, where x is the number of items produced. Show that the marginal average cost function is given by

$$\overline{C}'(x) = \frac{xC'(x) - C(x)}{x^2}.$$

LIFE SCIENCES

41. *Muscle Reaction* When a certain drug is injected into a muscle, the muscle responds by contracting. The amount of contraction, s (in millimeters) is related to the concentration of the drug, x (in milliliters) by

$$s(x) = \frac{x}{m + nx},$$

where m and n are constants.

a. Find $s'(x)$.

b. Find the rate of contraction when the concentration of the drug is 50 ml, $m = 10$, and $n = 3$.

42. *Growth Models* In Exercise 57 of the section on polynomial and rational functions, the formula for the growth rate of a population in the presence of a quantity x of food was given as

$$f(x) = \frac{Kx}{A + x}.$$

This was referred to as Michaelis–Menten kinetics.

a. Find the rate of change of the growth rate with respect to the amount of food.

b. The quantity A in the formula for $f(x)$ represents the quantity of food for which the growth rate is half of its maximum. Using your answer from part a, find the rate of change of the growth rate when $x = A$.

43. *Bacteria Population* Assume that the total number (in millions) of bacteria present in a culture at a certain time t (in hours) is given by

$$N(t) = (t - 10)^2(2t) + 50.$$

a. Find $N'(t)$.

Find the rate at which the population of bacteria is changing at the following times.

b. 8 hours **c.** 11 hours

d. The answer in part b is negative, and the answer in part c is positive. What does this mean in terms of the population of bacteria?

44. *Work/Rest Cycles* Murrell's formula for calculating the total amount of rest, in minutes, required after performing a particular type of work activity for 30 minutes is given by the formula

$$R(w) = \frac{30(w - 4)}{w - 1.5},$$

where w is the work expended in kilocalories per minute, kcal/min.*

a. A value of 5 for w indicates light work, such as riding a bicycle on a flat surface at 10 mph. Find $R(5)$.

b. A value of 7 for w indicates moderate work, such as mowing grass with a pushmower on level ground. Find $R(7)$.

c. Find $R'(5)$ and $R'(7)$ and compare your answers. Explain whether these answers make sense.

45. *Optimal Foraging* Using data collected by zoologist Reto Zach, the work done by a crow to break open a whelk (large marine snail) can be estimated by the function

$$W = \left(1 + \frac{20}{H - 0.93}\right)H,$$

where H is the height (in meters) of the whelk when it is dropped.†

a. Find dW/dH.

b. One can show that the amount of work is minimized when $dW/dH = 0$. Find the value of H that minimizes W.

c. Interestingly, Zach observed the crows dropping the whelks from an average height of 5.23 m. What does this imply?

SOCIAL SCIENCES

46. *Memory Retention* Some psychologists contend that the number of facts of a certain type that are remembered after t hours is given by

$$f(t) = \frac{90t}{99t - 90}.$$

Find the rate at which the number of facts remembered is changing after the following numbers of hours.

a. 1 **b.** 10

GENERAL INTEREST

47. *Vehicle Waiting Time* The average number of vehicles waiting in a line to enter a parking ramp can be modeled by the function

$$f(x) = \frac{x^2}{2(1 - x)},$$

where x is a quantity between 0 and 1 known as the traffic intensity.‡ Find the rate of change of the number of vehicles in line with respect to the traffic intensity for the following values of the intensity.

a. $x = .1$ **b.** $x = .6$

4.3 THE CHAIN RULE

THINK ABOUT IT

Suppose we know how fast the radius of a circular oil slick is growing, and we know how much the area of the oil slick is growing per unit of change in the radius. How fast is the area growing?

The answer to this question involves the chain rule for derivatives. Before discussing the chain rule, we consider the composition of functions. Many of the most useful functions for modeling are created by combining simpler functions. Viewing complex functions as combinations of simpler functions often makes them easier to understand and use.

*Sanders, Mark and Ernest McCormick, *Human Factors in Engineering and Design,* 7th ed., New York: McGraw-Hill, 1993, pp. 243–246.

†Kellar, Brian and Heather Thompson, "Whelk-come to Mathematics," *Mathematics Teacher,* Vol. 92, No. 6, September 1999, pp. 475–481.

‡Mannering, F. and W. Kilareski, *Principles of Highway Engineering and Traffic Control,* New York: Wiley, 1990.

Composition of Functions Suppose a function f assigns to each element x in set X some element $y = f(x)$ in set Y. Suppose also that a function g takes each element in set Y and assigns to it a value $z = g[f(x)]$ in set Z. By using both f and g, an element x in X is assigned to an element z in Z, as illustrated in Figure 9. The result of this process is a new function called the *composition* of functions g and f and defined as follows.

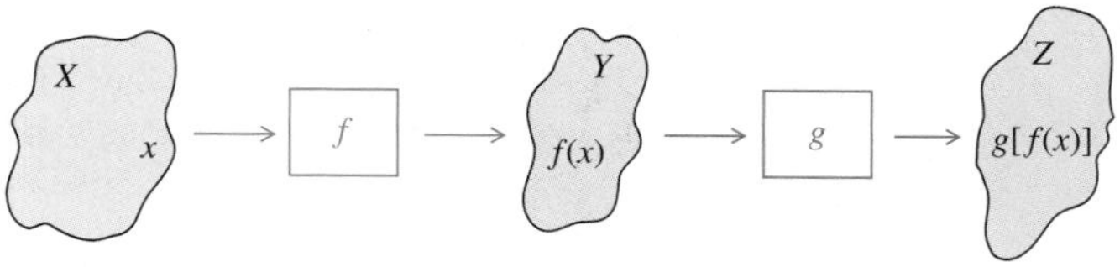

FIGURE 9

COMPOSITE FUNCTION

Let f and g be functions. The **composite function,** or **composition,** of g and f is the function whose values are given by $g[f(x)]$ for all x in the domain of f such that $f(x)$ is in the domain of g. (Read $g[f(x)]$ as "g of f of x".)

EXAMPLE 1 *Composite Functions*

Let $f(x) = 2x - 1$ and $g(x) = \sqrt{3x + 5}$. Find the following.

FOR REVIEW

You may want to review how to find the domain of a function. Domain was discussed in the section on Properties of Functions.

(a) $g[f(4)]$

Solution Find $f(4)$ first.

$$f(4) = 2 \cdot 4 - 1 = 8 - 1 = 7$$

Then

$$g[f(4)] = g[7] = \sqrt{3 \cdot 7 + 5} = \sqrt{26}.$$

(b) $f[g(4)]$

Solution Since $g(4) = \sqrt{3 \cdot 4 + 5} = \sqrt{17}$,

$$f[g(4)] = 2 \cdot \sqrt{17} - 1 = 2\sqrt{17} - 1.$$

(c) $f[g(-2)]$ does not exist since -2 is not in the domain of g.

EXAMPLE 2 *Composition of Functions*

Let $f(x) = 2x^2 + 5x$ and $g(x) = 4x + 1$. Find the following.

(a) $f[g(x)]$

Solution Using the given functions, we have

$$\begin{aligned} f[g(x)] &= f[4x + 1] \\ &= 2(4x + 1)^2 + 5(4x + 1) \\ &= 2(16x^2 + 8x + 1) + 20x + 5 \\ &= 32x^2 + 16x + 2 + 20x + 5 \\ &= 32x^2 + 36x + 7. \end{aligned}$$

(b) $g[f(x)]$

Solution By the definition above, with f and g interchanged,

$$\begin{aligned} g[f(x)] &= g[2x^2 + 5x] \\ &= 4(2x^2 + 5x) + 1 \\ &= 8x^2 + 20x + 1. \end{aligned}$$

As Example 2 shows, it is not always true that $f[g(x)] = g[f(x)]$. In fact, it is rare to find two functions f and g such that $f[g(x)] = g[f(x)]$. The domain of both composite functions given in Example 2 is the set of all real numbers.

EXAMPLE 3 *Composition of Functions*

Write each function as the composition of two functions f and g so that $h(x) = f[g(x)]$.

(a) $h(x) = 2(4x + 1)^2 + 5(4x + 1)$

Solution Let $f(x) = 2x^2 + 5x$ and $g(x) = 4x + 1$. Then $f[g(x)] = f(4x + 1) = 2(4x + 1)^2 + 5(4x + 1)$. Notice that $h(x)$ here is the same as $f[g(x)]$ in Example 2(a).

(b) $h(x) = \sqrt{1 - x^2}$

Solution One way to do this is to let $f(x) = \sqrt{x}$ and $g(x) = 1 - x^2$. Another choice is to let $f(x) = \sqrt{1 - x}$ and $g(x) = x^2$. Verify that with either choice, $f[g(x)] = \sqrt{1 - x^2}$. For the purposes of this section, the first choice is better; it is useful to think of f as being the function on the outer layer and g as the function on the inner layer. With this function h, we see a square root on the outer layer, and when we peel that away we see $1 - x^2$ on the inside.

The Chain Rule Suppose $f(x) = x^2$ and $g(x) = 5x^3 + 2$. What is the derivative of $h(x) = f[g(x)] = (5x^3 + 2)^2$? At first you might think the answer is just $h'(x) = 2(5x^3 + 2) = 10x^3 + 4$ by using the power rule. You can check this answer by multiplying out $h(x) = (5x^3 + 2)^2 = 25x^6 + 20x^3 + 4$. Now calculate $h'(x) = 150x^5 + 60x^2$. The guess using the power rule was clearly wrong! The error is that the power rule applies to x raised to a power, not to some other function of x raised to a power.

How, then, could we take the derivative of $p(x) = (5x^3 + 2)^{20}$? This seems far too difficult to multiply out. Fortunately, there is a way. Notice from the previous paragraph that $h'(x) = 150x^5 + 60x^2 = 2(5x^3 + 2)15x^2$. So the original guess was almost correct, except it was missing the factor of $15x^2$, which just happens to be $g'(x)$. This is not a coincidence. To see why the derivative of $f[g(x)]$ involves taking the derivative of f and then multiplying by the derivative of g, let us consider a realistic example, the question from the beginning of this section.

A leaking oil well off the Gulf Coast is spreading a circular film of oil over the water surface. At any time t (in minutes) after the beginning of the leak, the radius of the circular oil slick is given by

$$r(t) = 4t, \qquad \text{with} \qquad \frac{dr}{dt} = 4,$$

where dr/dt is the rate of change in radius over time. The area of the oil slick is given by

$$A(r) = \pi r^2, \qquad \text{with} \qquad \frac{dA}{dr} = 2\pi r,$$

where dA/dr is the rate of change in area per unit change in radius.

As these derivatives show, the radius is increasing 4 times as fast as the time t, and the area is increasing $2\pi r$ times as fast as the radius r. It seems reasonable, then, that the area is increasing $2\pi r \cdot 4 = 8\pi r$ times as fast as time. That is,

$$\frac{dA}{dt} = \frac{dA}{dr} \cdot \frac{dr}{dt} = 2\pi r \cdot 4 = 8\pi r.$$

Since $r = 4t$,

$$\frac{dA}{dt} = 8\pi r = 8\pi(4t) = 32\pi t.$$

To check this, use the fact that $r = 4t$ and $A = \pi r^2$ to get the same result:

$$A = \pi(4t)^2 = 16\pi t^2, \qquad \text{with} \qquad \frac{dA}{dt} = 32\pi t.$$

(Notice that because A is a function of r, which is a function of t, A as a function of t is a composition of two functions.)

The product used above,

$$\frac{dA}{dt} = \frac{dA}{dr} \cdot \frac{dr}{dt},$$

is an example of the **chain rule,** which is used to find the derivative of a composite function.

CHAIN RULE

If y is a function of u, say $y = f(u)$, and if u is a function of x, say $u = g(x)$, then $y = f(u) = f[g(x)]$, and

$$\frac{dy}{dx} = \frac{dy}{du} \cdot \frac{du}{dx}.$$

One way to remember the chain rule is to pretend that dy/du and du/dx are fractions, with du "canceling out." The proof of the chain rule requires advanced concepts and therefore is not given here.

EXAMPLE 4 *Chain Rule*

Find dy/dx if $y = (3x^2 - 5x)^{1/2}$.

Solution Let $y = u^{1/2}$, and $u = 3x^2 - 5x$. Then

$$\frac{dy}{dx} = \frac{dy}{du} \cdot \frac{du}{dx}$$

$$= \frac{1}{2}u^{-1/2} \cdot (6x - 5).$$

Replacing u with $3x^2 - 5x$ gives

$$\frac{dy}{dx} = \frac{1}{2}(3x^2 - 5x)^{-1/2}(6x - 5) = \frac{6x - 5}{2(3x^2 - 5x)^{1/2}}.$$

The following alternative version of the chain rule is stated in terms of composite functions.

CHAIN RULE (ALTERNATIVE FORM)

If $y = f[g(x)]$, then

$$\frac{dy}{dx} = f'[g(x)] \cdot g'(x).$$

(To find the derivative of $f[g(x)]$, find the derivative of $f(x)$, replace each x with $g(x)$, and then multiply the result by the derivative of $g(x)$.)

EXAMPLE 5 *Chain Rule*

Use the chain rule to find $D_x(x^2 + 5x)^8$.

Solution Let $f(x) = x^8$ and $g(x) = x^2 + 5x$. Then $(x^2 + 5x)^8 = f[g(x)]$ and

$$D_x(x^2 + 5x)^8 = f'[g(x)]g'(x).$$

Here $f'(x) = 8x^7$, with $f'[g(x)] = 8[g(x)]^7 = 8(x^2 + 5x)^7$ and $g'(x) = 2x + 5$.

$$\begin{aligned} D_x(x^2 + 5x)^8 &= f'[g(x)]g'(x) \\ &= 8[g(x)]^7 g'(x) \\ &= 8(x^2 + 5x)^7(2x + 5) \end{aligned}$$

CAUTION

(a) A common error is to forget to multiply by $g'(x)$ when using the chain rule. Remember, the derivative must involve a "chain," or product, of derivatives.

(b) Another common mistake is to write the derivative as $f'[g'(x)]$. Remember to leave $g(x)$ unchanged in $f'[g(x)]$, and then to multiply by $g'(x)$.

One way to avoid both of the errors described above is to remember that the chain rule is a two-step process. In Example 5, the first step was taking the derivative of the power, and the second step was multiplying by $g'(x)$. Forgetting to multiply by $g'(x)$ would be an erroneous one-step process. The other erroneous one-step process is to take the derivative inside the power, getting $f'[g'(x)]$, or $8(2x + 5)^7$ in Example 5. ■

Sometimes both the chain rule and either the product or quotient rule are needed to find a derivative, as the next examples show.

EXAMPLE 6 *Derivative Rules*

Find the derivative of $y = 4x(3x + 5)^5$.

Solution Write $4x(3x + 5)^5$ as the product

$$(4x) \cdot (3x + 5)^5.$$

To find the derivative of $(3x + 5)^5$, let $g(x) = 3x + 5$, with $g'(x) = 3$. Now use the product rule and the chain rule.

$$\frac{dy}{dx} = 4x\overbrace{[5(3x + 5)^4 \cdot 3]}^{\text{Derivative of } (3x+5)^5} + (3x + 5)^5\overbrace{(4)}^{\text{Derivative of } 4x}$$

$$= 60x(3x + 5)^4 + 4(3x + 5)^5$$

$$= 4(3x + 5)^4[15x + (3x + 5)^1]$$ Factor out the greatest common factor, $4(3x + 5)^4$.

$$= 4(3x + 5)^4(18x + 5)$$ Simplify inside brackets.

EXAMPLE 7 *Derivative Rules*

Find $D_x\left[\frac{(3x + 2)^7}{x - 1}\right]$.

Solution Use the quotient rule and the chain rule.

$$D_x\left[\frac{(3x + 2)^7}{x - 1}\right] = \frac{(x - 1)[7(3x + 2)^6 \cdot 3] - (3x + 2)^7(1)}{(x - 1)^2}$$

$$= \frac{21(x - 1)(3x + 2)^6 - (3x + 2)^7}{(x - 1)^2}$$

$$= \frac{(3x + 2)^6[21(x - 1) - (3x + 2)]}{(x - 1)^2}$$ Factor out the greatest common factor, $(3x + 2)^6$.

$$= \frac{(3x + 2)^6[21x - 21 - 3x - 2]}{(x - 1)^2}$$ Simplify inside brackets.

$$= \frac{(3x + 2)^6(18x - 23)}{(x - 1)^2}$$

Some applications requiring the use of the chain rule are illustrated in the next examples.

EXAMPLE 8 *City Revenue*

The revenue realized by a small city from the collection of fines from parking tickets is given by

$$R(n) = \frac{8000n}{n + 2},$$

where n is the number of work-hours each day that can be devoted to parking patrol. At the outbreak of a flu epidemic, 30 work-hours are used daily in parking patrol, but during the epidemic that number is decreasing at the rate of 6 work-hours per day. How fast is revenue from parking fines decreasing during the epidemic?

Solution We want to find dR/dt, the change in revenue with respect to time. By the chain rule,

$$\frac{dR}{dt} = \frac{dR}{dn} \cdot \frac{dn}{dt}.$$

First find dR/dn, as follows.

$$\frac{dR}{dn} = \frac{(n + 2)(8000) - 8000n(1)}{(n + 2)^2} = \frac{16{,}000}{(n + 2)^2}$$

Since 30 work-hours were used at the outbreak of the epidemic, $n = 30$, so $dR/dn = 16{,}000/32^2 = 15.625$. Also, $dn/dt = -6$. Thus,

$$\frac{dR}{dt} = (15.625)(-6) = -93.75.$$

Revenue is being lost at the rate of about $94 per day.

EXAMPLE 9 *Compound Interest*

Suppose a sum of $500 is deposited in an account with an interest rate of r percent per year compounded monthly. At the end of 10 years, the balance in the account (as illustrated in Figure 10) is given by

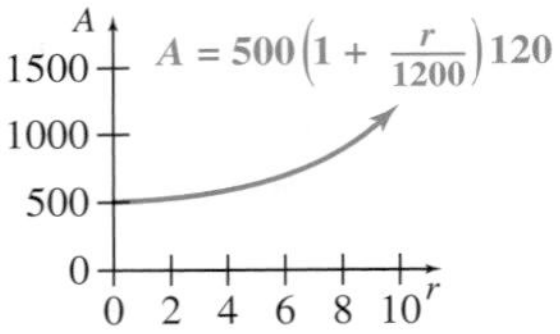

FIGURE 10

$$A = 500\left(1 + \frac{r}{1200}\right)^{120}.$$

Find the rate of change of A with respect to r if $r = 5$ or 7.*

Solution First find dA/dr using the chain rule.

$$\frac{dA}{dr} = (120)(500)\left(1 + \frac{r}{1200}\right)^{119}\left(\frac{1}{1200}\right)$$

$$= 50\left(1 + \frac{r}{1200}\right)^{119}$$

If $r = 5$,

$$\frac{dA}{dr} = 50\left(1 + \frac{5}{1200}\right)^{119}$$

$$\approx 82.01,$$

*Notice that r is given here as an integer percent, rather than as a decimal, which is why the formula for compound interest has 1200 where you would expect to see 12. This leads to a simpler interpretation of the derivative.

or \$82.01 per percentage point. If $r = 7$,

$$\frac{dA}{dr} = 50\left(1 + \frac{7}{1200}\right)^{119}$$

$$\approx 99.90,$$

or \$99.90 per percentage point.

NOTE One lesson to learn from this section is that a derivative is always with respect to some variable. In the oil slick example, notice that the derivative of the area with respect to the radius is $2\pi r$, while the derivative of the area with respect to time is $8\pi r$. As another example, consider the velocity of a conductor walking at 2 mph on a train car. Her velocity with respect to the ground may be 50 mph, but the earth on which the train is running is moving about the sun at 1.6 million mph. The derivative of her position function might be 2, 50, or 1.6 million mph, depending on what variable it is with respect to.

4.3 EXERCISES

Let $f(x) = 4x^2 - 2x$ and $g(x) = 8x + 1$. Find the following.

1. $f[g(2)]$ **2.** $f[g(-5)]$ **3.** $g[f(2)]$ **4.** $g[f(-5)]$ **5.** $f[g(k)]$ **6.** $g[f(5z)]$

Find $f[g(x)]$ and $g[f(x)]$ in the following.

7. $f(x) = \frac{x}{8} + 12;\quad g(x) = 3x - 1$

8. $f(x) = -6x + 9;\quad g(x) = \frac{x}{5} + 7$

9. $f(x) = \frac{1}{x};\quad g(x) = x^2$

10. $f(x) = \frac{2}{x^4};\quad g(x) = 2 - x$

11. $f(x) = \sqrt{x + 2};\quad g(x) = 8x^2 - 6$

12. $f(x) = 9x^2 - 11x;\quad g(x) = 2\sqrt{x + 2}$

13. $f(x) = \sqrt{x + 1};\quad g(x) = \frac{-1}{x}$

14. $f(x) = \frac{8}{x};\quad g(x) = \sqrt{3 - x}$

15. In your own words, explain how to form the composition of two functions.

Write each function as the composition of two functions. (There may be more than one way to do this.)

16. $y = (3x - 7)^{1/3}$

17. $y = (5 - x)^{2/5}$

18. $y = \sqrt{9 - 4x}$

19. $y = -\sqrt{13 + 7x}$

20. $y = (x^{1/2} - 3)^2 + (x^{1/2} - 3) + 5$

21. $y = (x^2 + 5x)^{1/3} - 2(x^2 + 5x)^{2/3} + 7$

Find the derivative of each function defined as follows.

22. $y = (2x^3 + 9x)^5$

23. $y = (8x^4 - 3x^2)^3$

24. $f(x) = -8(3x^4 + 2)^3$

25. $k(x) = -2(12x^2 + 5)^6$

26. $s(t) = 12(2t^4 + 5)^{3/2}$

27. $s(t) = 45(3t^3 - 8)^{3/2}$

28. $f(t) = 8\sqrt{4t^2 + 7}$

29. $g(t) = -3\sqrt{7t^3 - 1}$

30. $r(t) = 4t(2t^5 + 3)^2$

31. $m(t) = -6t(5t^4 - 1)^2$

32. $y = (x^3 + 2)(x^2 - 1)^2$

33. $y = (3x^4 + 1)^2(x^3 + 4)$

34. $p(z) = z(6z + 1)^{4/3}$

35. $q(y) = 4y^2(y^2 + 1)^{5/4}$

36. $y = \frac{1}{(3x^2 - 4)^5}$

37. $y = \dfrac{-5}{(2x^3 + 1)^2}$

38. $p(t) = \dfrac{(2t + 3)^3}{4t^2 - 1}$

39. $r(t) = \dfrac{(5t - 6)^4}{3t^2 + 4}$

40. $y = \dfrac{x^2 + 4x}{(3x^3 + 2)^4}$

41. $y = \dfrac{3x^2 - x}{(2x - 1)^5}$

42. The generalized power rule says that if $g(x)$ is a function of x and $y = [g(x)]^n$ for any real number n, then

$$\frac{dy}{dx} = n \cdot [g(x)]^{n-1} \cdot g'(x).$$

Explain why the generalized power rule is a consequence of the chain rule and the power rule.

Consider the following table of values of the functions f and g and their derivatives at various points.

x	1	2	3	4
$f(x)$	2	4	1	3
$f'(x)$	−6	−7	−8	−9
$g(x)$	2	3	4	1
$g'(x)$	2/7	3/7	4/7	5/7

Find the following using the table.

43. a. $D_x(f[g(x)])$ at $x = 1$ **b.** $D_x(f[g(x)])$ at $x = 2$

44. a. $D_x(g[f(x)])$ at $x = 1$ **b.** $D_x(g[f(x)])$ at $x = 2$

In Exercises 45–48, find the equation of the tangent line to the graph of the given function at the given value of x.

45. $f(x) = \sqrt{x^2 + 16}$; $x = 3$

46. $f(x) = (x^3 + 7)^{2/3}$; $x = 1$

47. $f(x) = x(x^2 - 4x + 5)^4$; $x = 2$

48. $f(x) = x^2\sqrt{x^4 - 12}$; $x = 2$

In Exercises 49 and 50, find all values of x for the given function where the tangent line is horizontal.

49. $f(x) = \sqrt{x^3 - 6x^2 + 9x + 1}$

50. $f(x) = \dfrac{x}{(x^2 + 4)^4}$

51. Katie and Sarah are working on taking the derivative of

$$f(x) = \frac{2x}{3x + 4}.$$

Katie uses the quotient rule to get

$$f'(x) = \frac{(3x + 4)2 - 2x(3)}{(3x + 4)^2} = \frac{8}{(3x + 4)^2}.$$

Sarah converts it into a product and uses the product rule and the chain rule:

$$\begin{aligned} f(x) &= 2x(3x + 4)^{-1} \\ f'(x) &= 2x(-1)(3x + 4)^{-2}(3) + 2(3x + 4)^{-1} \\ &= 2(3x + 4)^{-1} - 6x(3x + 4)^{-2}. \end{aligned}$$

Explain the discrepancies between the two answers. Which procedure do you think is preferable?

52. Margy and Nate are working on taking the derivative of

$$f(x) = \frac{2}{(3x+1)^4}.$$

Margy uses the quotient rule as follows:

$$f'(x) = \frac{(3x+1)^4 \cdot 0 - 2 \cdot 4(3x+1)^3 \cdot 3}{(3x+1)^8}$$

$$= \frac{-24(3x+1)^3}{(3x+1)^8} = \frac{-24}{(3x+1)^5}.$$

Nate rewrites the function and uses the chain rule as follows:

$$f(x) = 2(3x+1)^{-4}$$

$$f'(x) = (-4)2(3x+1)^{-5} \cdot 3 = \frac{-24}{(3x+1)^5}.$$

Compare the two procedures. Which procedure do you think is preferable?

Applications

BUSINESS AND ECONOMICS

53. *Demand* Suppose the demand for a certain brand of vacuum cleaner is given by

$$D(p) = \frac{-p^2}{100} + 500,$$

where p is the price in dollars. If the price, in terms of the cost c, is expressed as

$$p(c) = 2c - 10,$$

find the demand in terms of the cost.

54. *Revenue* Assume that the total revenue from the sale of x television sets is given by

$$R(x) = 1000\left(1 - \frac{x}{500}\right)^2.$$

Find the marginal revenue when the following numbers of sets are sold.

a. 400 **b.** 500 **c.** 600

d. Find the average revenue from the sale of x sets.

e. Find the marginal average revenue.

f. Write a paragraph covering the following questions. When does the revenue begin to decrease? What sales produce zero marginal average revenue? How should a manager use this information?

55. *Interest* A sum of \$1500 is deposited in an account with an interest rate of r percent per year, compounded daily. At the end of 5 years, the balance in the account is given by

$$A = 1500\left(1 + \frac{r}{36{,}500}\right)^{1825}.$$

Find the rate of change of A with respect to r for the following interest rates.

a. 6% **b.** 8% **c.** 9%

56. *Demand* Suppose a demand function is given by

$$q = D(p) = 30\left(5 - \frac{p}{\sqrt{p^2+1}}\right),$$

where q is the demand for a product and p is the price per item in dollars. Find the rate of change in the demand for the product per unit change in price (i.e., find dq/dp).

57. *Depreciation* A certain truck depreciates according to the formula

$$V = \frac{6000}{1 + .3t + .1t^2},$$

where t is time measured in years and $t = 0$ represents the time of purchase (in years). Find the rate at which the value of the truck is changing at the following times.

a. 2 years **b.** 4 years

58. *Cost* Suppose the cost in dollars of manufacturing x items is given by

$$C = 2000x + 3500,$$

and the demand equation is given by

$$x = \sqrt{15{,}000 - 1.5p}.$$

In terms of the demand x,

a. find an expression for the revenue R;

b. find an expression for the profit P;

c. find an expression for the marginal profit.

d. Determine the value of the marginal profit when the price is \$5000.

LIFE SCIENCES

59. *Fish Population* Suppose the population P of a certain species of fish depends on the number x (in hundreds) of a smaller fish that serves as its food supply, so that

$$P(x) = 2x^2 + 1.$$

Suppose, also, that the number of the smaller species of fish depends on the amount a (in appropriate units) of its food supply, a kind of plankton. Specifically,

$$x = f(a) = 3a + 2.$$

A biologist wants to find the relationship between the population P of the large fish and the amount a of plankton available, that is, $P[f(a)]$. What is the relationship?

60. *Oil Pollution* An oil well off the Gulf Coast is leaking, with the leak spreading oil over the surface as a circle. At any time t (in minutes) after the beginning of the leak, the radius of the circular oil slick on the surface is $r(t) = t^2$ Let $A(r) = \pi r^2$ feet. represent the area of a circle of radius r.

a. Find and interpret $A[r(t)]$.

b. Find and interpret $D_t A[r(t)]$ when $t = 100$.

61. *Thermal Inversion* When there is a thermal inversion layer over a city (as happens often in Los Angeles), pollutants cannot rise vertically but are trapped below the layer and must disperse horizontally. Assume that a factory smokestack begins emitting a pollutant at 8 A.M. Assume that the pollutant disperses horizontally, forming a circle. If t represents the time (in hours) since the factory began emitting pollutants ($t = 0$ represents 8 A.M.), assume that the radius of the circle of pollution is $r(t) = 2t$ miles. Let $A(r) = \pi r^2$ represent the area of a circle of radius r.

a. Find and interpret $A[r(t)]$.

b. Find and interpret $D_t A[r(t)]$ when $t = 4$.

62. *Bacteria Population* The total number of bacteria (in millions) present in a culture is given by

$$N(t) = 2t(5t + 9)^{1/2} + 12,$$

where t represents time (in hours) after the beginning of an experiment. Find the rate of change of the population of bacteria with respect to time for the following numbers of hours.

a. 0 **b.** 7/5 **c.** 8

63. *Calcium Usage* To test an individual's use of calcium, a researcher injects a small amount of radioactive calcium into the person's bloodstream. The calcium remaining in the bloodstream is measured each day for several days. Suppose the amount of the calcium remaining in the bloodstream (in milligrams per cubic centimeter) t days after the initial injection is approximated by

$$C(t) = \frac{1}{2}(2t + 1)^{-1/2}.$$

Find the rate of change of the calcium level with respect to time for the following numbers of days.

a. 0 **b.** 4 **c.** 7.5

d. Is C always increasing or always decreasing? How can you tell?

64. *Drug Reaction* The strength of a person's reaction to a certain drug is given by

$$R(Q) = Q\left(C - \frac{Q}{3}\right)^{1/2},$$

where Q represents the quantity of the drug given to the patient and C is a constant.

a. The derivative $R'(Q)$ is called the *sensitivity* to the drug. Find $R'(Q)$.

b. Find the sensitivity to the drug if $C = 59$ and a patient is given 87 units of the drug.

c. Is the patient's sensitivity to the drug increasing or decreasing when $Q = 87$?

GENERAL INTEREST

65. *Candy* The volume and surface area of a "jawbreaker" for any radius is given by the formulas

$$V(r) = \frac{4}{3}\pi r^3 \quad \text{and} \quad S(r) = 4\pi r^2,$$

respectively. Roger Guffey estimates the radius of a jaw-breaker while in a person's mouth to be

$$r(t) = 6 - \frac{3}{17}t,$$

where $r(t)$ is in millimeters and t is in minutes.*

a. What is the life expectancy of a jawbreaker?

b. Find dV/dt and dS/dt when $t = 17$ and interpret your answer.

c. Construct an analogous experiment using some other type of food or verify the results of this experiment.

4.4 DERIVATIVES OF EXPONENTIAL FUNCTIONS

 THINK ABOUT IT

Given a new product whose rate of growth is rapid at first and then slows, how can we find the rate of growth?

We will use a derivative to answer this question in Example 5 at the end of this section.

We can find the derivative of the exponential function by using the definition of the derivative. Thus

$$\frac{d(e^x)}{dx} = \lim_{h \to 0} \frac{e^{x+h} - e^x}{h}$$

$$= \lim_{h \to 0} \frac{e^x e^h - e^x}{h} \qquad \text{Property 1 of exponents}$$

$$= e^x \lim_{h \to 0} \frac{e^h - 1}{h}. \qquad \text{Property 1 of limits}$$

In the last step, since e^x does not involve h, we were able to bring e^x in front of the limit. The result says that the derivative of e^x is e^x times a constant, namely, $\lim_{h \to 0}(e^h - 1)/h$. To investigate this limit, we will use a graphing calculator to evaluate the expression for smaller and smaller values of h. Figure 11 shows the result, in which X represents h and Y_1 represents $\lim_{h \to 0}(e^h - 1)/h$. Based on this evidence, it appears that $\lim_{h \to 0}(e^h - 1)/h = 1$. This is proven rigorously in more advanced courses. We therefore have the following formula.

X	Y₁
.1	1.0517
.01	1.005
.001	1.0005
1E-4	1.0001
1E-5	1

X=

FIGURE 11

DERIVATIVE OF e^x

$$D_x[e^x] = e^x$$

To find the derivative of the exponential function with a base other than e, use the change-of-base theorem for exponentials to rewrite a^x as $e^{(\ln a)x}$. Thus

$$\frac{d(a^x)}{dx} = \frac{d[e^{(\ln a)x}]}{dx} \qquad \text{Change-of-base theorem for exponentials}$$

$$= e^{(\ln a)x} \ln a \qquad \text{Chain rule}$$

$$= (\ln a)a^x. \qquad \text{Change-of-base theorem again}$$

*Guffey, Roger, "The Life Expectancy of a Jawbreaker: An Application of the Composition of Functions," *Mathematics Teacher,* Vol. 92, No. 2, Feb. 1999, pp. 125–127.

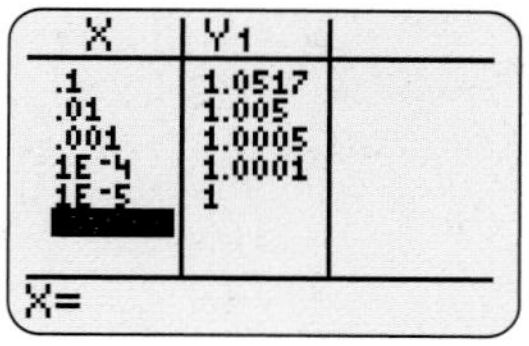

DERIVATIVE OF a^x

$$D_x[a^x] = (\ln a)a^x$$

We now see why e is the best base to work with: It has the simplest derivative of all the exponential functions. Even if we choose a different base, e appears in the derivative anyway through the $\ln a$ term. (Recall that $\ln a$ is the logarithm of a to the base e.) In fact, of all the functions we have studied, e^x is the simplest to differentiate, because its derivative is just itself.*

The chain rule can be used to find the derivative of the more general exponential function $y = a^{g(x)}$. Let $y = f(u) = a^u$ and $u = g(x)$, so that $f[g(x)] = a^{g(x)}$. Then

$$f'[g(x)] = f'(u) = (\ln a)a^u = (\ln a)a^{g(x)},$$

and by the chain rule,

$$\begin{aligned}\frac{dy}{dx} &= f'[g(x)] \cdot g'(x)\\ &= (\ln a)a^{g(x)} \cdot g'(x).\end{aligned}$$

As before, this formula becomes simpler when we use natural logarithms because $\ln e = 1$. We summarize these results next.

DERIVATIVE OF $a^{g(x)}$ AND $e^{g(x)}$

$$D_x[a^{g(x)}] = (\ln a)a^{g(x)}g'(x)$$

and

$$D_x[e^{g(x)}] = e^{g(x)}g'(x)$$

CAUTION Notice the difference between the derivative of a variable to a constant power, such as $D_x x^3 = 3x^2$, and a constant to a variable power, like $D_x e^x = e^x$. Remember, $D_x e^x \neq xe^{x-1}$. ■

EXAMPLE 1 *Derivatives of Exponential Functions*

Find derivatives of each function.

(a) $y = e^{5x}$

Solution Let $g(x) = 5x$, with $g'(x) = 5$. Then

$$\frac{dy}{dx} = 5e^{5x}.$$

(b) $s = 3^t$

Solution

$$\frac{ds}{dt} = (\ln 3)3^t$$

*There is a joke about a deranged mathematician who frightened other inmates at an insane asylum by screaming at them, "I'm going to differentiate you!" But one inmate remained calm and simply responded, "I don't care; I'm e^x."

(c) $y = 10e^{3x^2}$

Solution $$\frac{dy}{dx} = 10(e^{3x^2})(6x) = 60xe^{3x^2}$$

(d) $s = 8 \cdot 10^{1/t}$

Solution $$\frac{ds}{dt} = 8(\ln 10)10^{1/t}\left(\frac{-1}{t^2}\right)$$

$$= \frac{-8(\ln 10)10^{1/t}}{t^2}$$

EXAMPLE 2 *Derivative of an Exponential Function*

Let $y = e^x\sqrt{x}$. Find $\frac{dy}{dx}$.

Solution Use the product rule.

$$\frac{dy}{dx} = e^x \cdot \frac{1}{2\sqrt{x}} + \sqrt{x}e^x$$

$$= \frac{e^x}{2\sqrt{x}} + e^x\sqrt{x} \cdot \frac{2\sqrt{x}}{2\sqrt{x}}$$ Get a common denominator.

$$= \frac{e^x}{2\sqrt{x}}(1 + 2x)$$

EXAMPLE 3 *Derivative of an Exponential Function*

Let $f(x) = \frac{100{,}000}{1 + 100e^{-.3x}}$. Find $f'(x)$.

Solution Use the quotient rule.

$$f'(x) = \frac{(1 + 100e^{-.3x})(0) - 100{,}000(-30e^{-.3x})}{(1 + 100e^{-.3x})^2}$$

$$= \frac{3{,}000{,}000e^{-.3x}}{(1 + 100e^{-.3x})^2}$$

NOTE In the previous example, we could also have taken the derivative by writing $f(x) = 100{,}000(1 + 100e^{-.3x})^{-1}$, from which we have $f'(x) = -100{,}000(1 + 100e^{-.3x})^{-2}100e^{-.3x}(-.3)$. This simplifies to the same expression as in Example 3. ■

EXAMPLE 4 *Radioactivity*

The amount in grams in a sample of Uranium 239 after t years is given by

$$A(t) = 100e^{-.362t}.$$

Find the rate of change of the amount present after 3 years.

Solution The rate of change is given by the derivative dA/dt.

$$\frac{dA}{dt} = 100(e^{-.362t})(-.362) = -36.2e^{-.362t}$$

After 3 years ($t = 3$), the rate of change is

$$\frac{dA}{dt} = -36.2e^{-.362(3)} = -36.2e^{-1.086} \approx -12.2$$

grams per year.

Frequently a population, or the sales of a certain product, will start growing slowly, then grow more rapidly, and then gradually level off. Such growth can often be approximated by a mathematical model of the form

$$f(x) = \frac{b}{1 + ae^{kx}}$$

for appropriate constants a, b, and k.

EXAMPLE 5 *Product Sales*

Suppose that the sales of a new product can be approximated for its first few years on the market by the function

$$S(x) = \frac{100{,}000}{1 + 100e^{-.3x}},$$

where x is time in years since the introduction of the product.

(a) Find the rate of change of sales after 4 years.

Solution The derivative of this sales function, which gives the rate of change of sales, was found in Example 3. Using that derivative,

$$S'(4) = \frac{3{,}000{,}000e^{-.3(4)}}{[1 + 100e^{-.3(4)}]^2} = \frac{3{,}000{,}000e^{-1.2}}{(1 + 100e^{-1.2})^2}.$$

Using a calculator, $e^{-1.2} \approx .3012$, and

$$S'(4) \approx \frac{3{,}000{,}000(.3012)}{[1 + 100(.3012)]^2} \approx \frac{903{,}600}{(1 + 30.12)^2} \approx \frac{903{,}600}{968.5} \approx 933.$$

The rate of change of sales after 4 years is about 933 units per year. The positive number indicates that sales are increasing at this time.

(b) What happens to sales in the long run?

Solution As time increases, $x \to \infty$, and

$$e^{-.3x} = \frac{1}{e^{.3x}} \to 0.$$

Thus,

$$\lim_{x\to\infty} S(x) = \frac{100{,}000}{1 + 100(0)} = 100{,}000.$$

Sales approach a horizontal asymptote of 100,000.

The graph of the function in Example 5 is shown in Figure 12. This is an example of the *logistic function*

$$G(t) = \frac{mG_0}{G_0 + (m - G_0)e^{-kmt}}.$$

x	y
0	990
5	4300
10	17,000
15	47,000
20	80,000
30	99,000

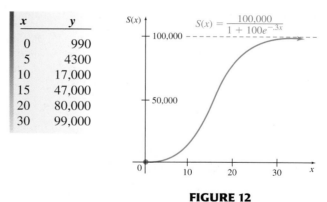

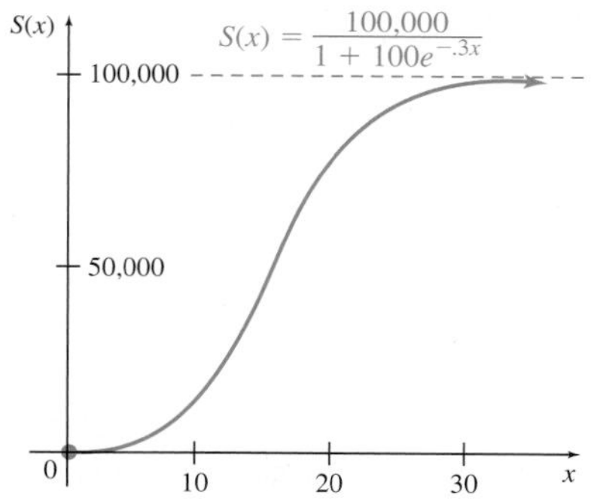

FIGURE 12

The **logistic function** is useful for modeling environments where the population growth is limited, perhaps by a shortage of food or other resources. In this type of application, t represents time in appropriate units, G_0 is the initial number present, m is the maximum possible size of the population, k is a positive constant, and $G(t)$ is the population at time t. Notice that the logistic function is a special case of the model discussed before Example 5.

4.4 EXERCISES

Find derivatives of the functions defined as follows.

1. $y = e^{4x}$
2. $y = e^{-2x}$
3. $y = -8e^{2x}$
4. $y = .2e^{5x}$
5. $y = -16e^{x+1}$
6. $y = -4e^{-.1x}$
7. $y = e^{x^2}$
8. $y = e^{-x^2}$
9. $y = 3e^{2x^2}$
10. $y = -5e^{4x^3}$
11. $y = 4e^{2x^2-4}$
12. $y = -3e^{3x^2+5}$
13. $y = xe^{x}$
14. $y = x^2e^{-2x}$
15. $y = (x - 3)^2e^{2x}$

16. $y = (3x^2 - 4x)e^{-3x}$

17. $y = \dfrac{x^2}{e^x}$

18. $y = \dfrac{e^x}{2x + 1}$

19. $y = \dfrac{e^x + e^{-x}}{x}$

20. $y = \dfrac{e^x - e^{-x}}{x}$

21. $p = \dfrac{10{,}000}{9 + 4e^{-.2t}}$

22. $p = \dfrac{500}{12 + 5e^{-.5t}}$

23. $f(z) = (2z + e^{-z^2})^2$

24. $y = 8^{5x}$

25. $y = 2^{-x}$

26. $y = 3 \cdot 4^{x^2+2}$

27. $y = -10^{3x^2-4}$

28. $s = 2 \cdot 3^{\sqrt{t}}$

29. $s = 5 \cdot 7^{\sqrt{t-2}}$

30. Prove that if $y = y_0 e^{kt}$, then $dy/dt = ky$. (This says that for exponential growth and decay, the rate of change of the population is proportional to the size of the population, and the constant of proportionality is the growth or decay constant.)

31. Use a graphing calculator to sketch the graph of $y = [f(x + h) - f(x)]/h$ using $f(x) = e^x$ and $h = .0001$. Compare it with the graph of $y = e^x$ and discuss what you observe.

32. Use graphical differentiation to verify that $de^x/dx = e^x$.

Applications

BUSINESS AND ECONOMICS

33. *Sales* The sales of a new personal computer (in thousands) are given by

$$S(t) = 100 - 90e^{-.3t},$$

where t represents time in years. Find the rate of change of sales at each time.

a. After 1 year **b.** After 5 years

c. What is happening to the rate of change of sales as time goes on?

d. Does the rate of change of sales ever equal zero?

34. *Cost* The cost in dollars to produce x computer diskettes can be approximated by

$$C(x) = \sqrt{900 - 800 \cdot 1.1^{-x}}.$$

Find the marginal cost when the following quantities are made.

a. 0 **b.** 20

c. What happens to the marginal cost as the number produced becomes larger and larger?

35. *Product Awareness* After the introduction of a new product for tanning without sun exposure, the percent of the public that is aware of the product is approximated by

$$A(t) = 10t^2 2^{-t},$$

where t is the time in months. Find the rate of change of the percent of the public that is aware of the product after the following numbers of months.

a. 2 **b.** 4

c. Notice that the answer to part a is positive and the answer to part b is negative. What does this tell you about how public awareness of the product has changed?

36. *Internet Users* In Exercise 42 of the section on Exponential Functions, we found that the number of Internet users, in millions, is approximated by

$$f(x) = 16(1.950)^x,$$

where $x = 0$ (in years) corresponds to December 1995. Find the instantaneous rate of change in the number of Internet users in December of the following years.

a. 1998 **b.** 2000

37. *Product Durability* Using data in a car magazine, we constructed the mathematical model

$$y = 100e^{-.03045t}$$

for the percent of cars of a certain type still on the road after t years. Find the percent of cars on the road after the following numbers of years.

a. 0 **b.** 2 **c.** 4 **d.** 6

Find the rate of change of the percent of cars still on the road after the following numbers of years.

e. 0 **f.** 2

g. Interpret your answers to parts e and f.

LIFE SCIENCES

38. *Population Growth* In Exercise 79 of the section on Logarithmic Functions, we found that the population of Florida (in millions) could be approximated by

$$p(t) = 14.0(1.017)^t,$$

where $t = 0$ (in years) corresponds to July 1994. Find the instantaneous rate of change in the population of Florida at the following times.

a. July 1998 **b.** January 2005

39. *Insect Growth* The growth of a population of rare South American beetles is given by the logistic function with $k = .00001$ and t in months. Assume that there are 200 beetles initially and that the maximum population size is 10,000.

a. Find the growth function $G(t)$ for these beetles.

Find the population and rate of growth of the population after the following times.

b. 6 months **c.** 3 years **d.** 7 years

e. What happens to the rate of growth over time?

40. *Clam Population* The population of a bed of clams in the Great South Bay off Long Island is described by the logistic function with $k = .0001$ and t in years. Assume that there are 400 clams initially and that the maximum population size is 5200.

a. Find the growth function $G(t)$ for the clams.

Find the population and rate of growth of the population after the following times.

b. 1 year **c.** 4 years **d.** 10 years

e. What happens to the rate of growth over time?

41. *Pollution Concentration* The concentration of pollutants (in grams per liter) in the east fork of the Big Weasel River is approximated by

$$P(x) = .04e^{-4x},$$

where x is the number of miles downstream from a paper mill that the measurement is taken. Find the following values.

a. The concentration of pollutants .5 mile downstream

b. The concentration of pollutants 1 mile downstream

c. The concentration of pollutants 2 miles downstream

Find the rate of change of concentration with respect to distance for the following distances.

d. .5 mile **e.** 1 mile **f.** 2 miles

42. *Breast Cancer* It has been observed that the following formula accurately models the relationship between the size of a breast tumor and the amount of time that it has been growing.

$$V(t) = 1100[1023e^{-.02415t} + 1]^{-4},$$

where t is in months and $V(t)$ is measured in cubic centimeters.*

a. Find the tumor volume at 240 months.

b. Assuming that the shape of a tumor is spherical, find the radius of the tumor from part a. (*Hint:* The volume of a sphere is given by the formula $V = (4/3)\pi r^3$.)

c. If a tumor of size .5 cm^3 is detected, according to the formula, how long has it been growing? What does this imply?

d. Find $\lim_{t \to \infty} V(t)$ and interpret this value. Explain whether this makes sense.

e. Calculate the rate of change of tumor volume at 240 months and interpret.

43. *Mortality* The percentage of people of any particular age group that will die in a given year may be approximated by the formula

$$P(t) = .00239e^{.0957t},$$

where t is the age of the person in years.†

a. Find $P(25)$, $P(50)$, and $P(75)$.

b. Find $P'(25)$, $P'(50)$, and $P'(75)$.

c. Interpret your answers for parts a and b. Are there any limitations of this formula?

44. *Dialysis* One measure of whether a dialysis patient has been adequately dialyzed is by the urea reduction ratio (URR). It is generally agreed that a patient has been adequately dialyzed when URR exceeds a value of .65. The value of URR can be calculated for a particular patient using the following formula by Gotch:

$$URR = 1 - \left\{(.96)^{.14t-1} + \frac{8t}{126t + 900}[1 - (.96)^{.14t-1}]\right\},$$

where t is measured in minutes.‡

a. Find the value of URR after a patient receives dialysis for 180 minutes. Has the patient received adequate dialysis?

b. Find the value of URR after a patient receives dialysis for 240 minutes. Has the patient received adequate dialysis?

c. Calculate the instantaneous rate of change of URR when time on dialysis is 240 minutes and interpret.

*Spratt, John et al., "Decelerating Growth and Human Breast Cancer," *Cancer,* Vol. 71, No. 6, 1993, pp. 2013–2019.

†U.S. Vital Statistics, 1995.

‡Kessler, Edward and Nathan Ritchey et al., "Urea Reduction Ratio and Urea Kinetic Modeling: A Mathematical Analysis of Changing Dialysis Parameters," *American Journal of Nephrology,* Vol. 18, 1998, pp. 471–477.

45. *Medical Literature* It has been observed that there has been an increase in the proportion of medical research papers that use the word "novel" in the title or abstract, and that this proportion can be accurately modeled by the function

$$p(x) = .001131e^{.1268x},$$

where x is the number of years since 1970.*

a. Find $p(25)$.

b. If this phenomenon continues, estimate the year in which every medical article will contain the word "novel" in its title or abstract.

c. Estimate the rate of increase in the proportion of medical papers using this word in the year 2002.

d. Explain some factors that may be contributing to researchers using this word.

46. *Arctic Foxes* The age/weight relationship of female Arctic foxes caught in Svalbard, Norway, can be estimated by the function

$$M(t) = 3102e^{-e^{-.022(t-56)}},$$

where t is the age of the fox in days and $M(t)$ is the weight of the fox in grams.†

a. Estimate the weight of a female fox that is 200 days old.

b. Use $M(t)$ to estimate the largest size that a female fox can attain. (*Hint:* Find $\lim_{t\to\infty} M(t)$.)

c. Estimate the age of a female fox when it has reached 80% of its maximum weight.

d. Estimate the rate of change in weight of an Arctic fox that is 200 days old. (*Hint:* Recall that $D_t[e^{f(t)}] = f'(t)e^{f(t)}$.)

e. Use a graphing calculator to graph $M(t)$ and then describe the growth pattern.

f. Use the table function on a graphing calculator or a spreadsheet to develop a chart that shows the estimated weight and growth rate of female foxes for days 50, 100, 150, 200, 250, and 300.

47. *Cutlassfish* The cutlassfish is one of the most important resources of the commercial marine fishing industry in China. Researchers have developed a von Bertalanffy growth model that uses the age of a certain species of cutlassfish to estimate length such that

$$L(t) = 589[1 - e^{-.168(t+2.682)}],$$

where $L(t)$ is the length of the fish (in millimeters) at time t (in years).‡

a. What happens to the length of the average cutlassfish of this species over time?

b. Determine the age of a fish that has grown to 95% of its maximum length.

c. Find $L'(4)$ and interpret the result.

d. Graph the function on $[0, 20]$ by $[0, 600]$.

48. *Beef Cattle* Researchers have compared two models that are used to predict the weight of beef cattle of various ages,

$$W_1(t) = 509.7(1 - .941e^{-.00181t})$$

and

$$W_2(t) = 498.4(1 - .889e^{-.00219t})^{1.25},$$

where $W_1(t)$ and $W_2(t)$ represent the weight (in kilograms) of a t-day-old beef cow.§

a. What is the maximum weight predicted by each function for the average beef cow? Is this difference significant?

b. According to each function, find the age that the average beef cow reaches 90% of its maximum weight.

c. Find $W_1'(750)$ and $W_2'(750)$. Compare your results.

d. Graph the two functions on $[0, 2500]$ by $[0, 525]$ and comment on the differences in growth patterns for each of these functions.

e. Graph the derivative of these two functions on $[0, 2500]$ by $[0, 1]$ and comment on any differences you notice between these functions.

SOCIAL SCIENCES

49. *Habit Strength* According to work by the psychologist C. L. Hull, the strength of a habit is a function of the number of times the habit is repeated. If N is the number of repetitions and $H(N)$ is the strength of the habit, then

$$H(N) = 1000(1 - e^{-kN}),$$

where k is a constant. Find $H'(N)$ if $k = .1$ and the number of times the habit is repeated is as follows.

a. 10 **b.** 100 **c.** 1000

d. Show that $H'(N)$ is always positive. What does this mean?

*Friedman, Simon H. and Jens O. Karlsson, "A Novel Paradigm," *Nature,* Vol. 385, No. 6616, Feb. 6, 1997, p. 480.

†Prestrud, Pal and Kjell Nilssen, "Growth, Size, and Sexual Dimorphism in Arctic Foxes," *Journal of Mammalogy,* Vol. 76, No. 2, May 1995, pp. 522–530.

‡Kwok, K. and I-Hsun Ni, "Age and Growth of Cutlassfishes, *Trichiurus* spp., from the South China Sea." *Fish Bulletin,* Vol. 98, No. 4, Oct. 2000, pp. 748–758.

§DeNise, R. and J. Brinks, "Genetic and Environmental Aspects of the Growth Curve Parameters in Beef Cows," *Journal of Animal Science,* Vol. 61, No. 6, 1985, pp. 1431–1440.

50. *Population* A recent report by the U.S. Census Bureau predicts that the Latino-American population will increase from 26.7 million in 1995 to 96.5 million in 2050.* Assuming an exponential growth pattern, the population can be approximated by

$$P(t) = 26.7e^{.023t},$$

where t is the number of years since 1995.

a. Estimate the Latino-American population for the year 2005.

b. What is the instantaneous rate of change of the Latino-American population when $t = 10$? Interpret your answer.

PHYSICAL SCIENCES

51. *Radioactive Decay* The amount (in grams) of a sample of lead 214 present after t years is given by

$$A(t) = 500e^{-.31t}.$$

Find the rate of change of the quantity present after each of the following years.

a. 4 **b.** 6 **c.** 10

d. What is happening to the rate of change of the amount present as the number of years increases?

e. Will the substance ever be gone completely?

52. *Electricity* In a series resistance-capacitance DC circuit, the instantaneous charge Q on the capacitor as a function of time (where $t = 0$ is the moment the circuit is energized by closing a switch) is given by the equation

$$Q(t) = CV(1 - e^{-t/RC}),$$

where C, V, and R are constants. Further, the instantaneous charging current I_C is the rate of change of charge on the capacitor, or $I_C = dQ/dt$.†

a. Find the expression for I_C as a function of time.

b. If $C = 10^{-5}$ farads, $R = 10^7$ ohms, and $V = 10$ volts, what is the charging current after 200 seconds? (*Hint:* When placed into the function in part a the units can be combined into amps.)

GENERAL INTEREST

53. *Track and Field* In 1958, L. Lucy developed a method for predicting the world record for any given year that a human could run a distance of 1 mile. His formula is given as follows:

$$t(n) = 218 + 31(.933)^n,$$

where $t(n)$ is the world record (in seconds) for the mile run in year $1950 + n$. Thus, $n = 5$ corresponds to the year 1955.‡

a. Find the estimate for the world record in the year 2005.

b. Calculate the instantaneous rate of change for the world record at the end of year 2005 and interpret.

c. Find $\lim_{n\to\infty} t(n)$ and interpret. How does this compare with the current world record?

4.5 DERIVATIVES OF LOGARITHMIC FUNCTIONS

THINK ABOUT IT

How does the average velocity of pedestrians in a city vary with the population size?

In an exercise from an earlier chapter, we found a logarithmic relationship between the average velocity of pedestrians and the population of the city. In this section, we will find the derivative of logarithmic functions and use the result to answer the question above.

To find the derivative of $\ln x$, use the definition of the derivative.

$$\frac{d(\ln x)}{dx} = \lim_{h\to 0} \frac{\ln(x+h) - \ln x}{h} \qquad \text{Definition of the derivative}$$

$$= \lim_{h\to 0} \frac{\ln\left(\dfrac{x+h}{x}\right)}{h} \qquad \text{Property b of logarithms}$$

*Population Projections of the United States by Age, Race, and Hispanic origin: 1995–2050, U.S. Census Bureau.

†Problem submitted by Kevin Friedrich, Sharon, PA.

‡Bennett, Jay, "Statistical Modeling in Track and Field," *Statistics in Sports,* Arnold, 1998, p. 179.

$$= \lim_{h \to 0} \frac{1}{h} \ln\left(1 + \frac{h}{x}\right) \qquad \text{Algebra}$$

$$= \lim_{h \to 0} \ln\left(1 + \frac{h}{x}\right)^{1/h} \qquad \text{Property c of logarithms}$$

The last step in the derivation was motivated by a limit that was introduced in the chapter on nonlinear functions, before the concept of limit had been explicitly mentioned:

$$\lim_{m \to \infty} \left(1 + \frac{1}{m}\right)^m = e.$$

The expression inside the natural logarithm in the derivation of $d(\ln x)/dx$ looks a lot like this limit. To make the expression the same, let $h/x = 1/m$, so $1/h = m/x$. As $h \to 0$, $m \to \infty$ (Do you see why?) so that

$$\frac{d(\ln x)}{dx} = \lim_{m \to \infty} \ln\left(1 + \frac{1}{m}\right)^{m/x}.$$

Because $\ln x$ is a continuous function, we can bring the limit inside of the natural logarithm to get

$$\begin{aligned} \frac{d(\ln x)}{dx} &= \ln \lim_{m \to \infty} \left(1 + \frac{1}{m}\right)^{m/x} \\ &= \ln e^{1/x} \\ &= 1/x. \end{aligned}$$

This observation can be further justified geometrically. Notice what happens to the slope of the line $y = 2x + 4$ if the x-axis and y-axis are switched. That is, if we replace x with y and y with x, then the resulting line $x = 2y + 4$ or $y = x/2 - 2$ is a reflection of the line $y = 2x + 4$ across the line $y = x$, as seen in Figure 13. Furthermore, the slope of the new line is the reciprocal of the original line. In fact, the reciprocal property holds for all lines.

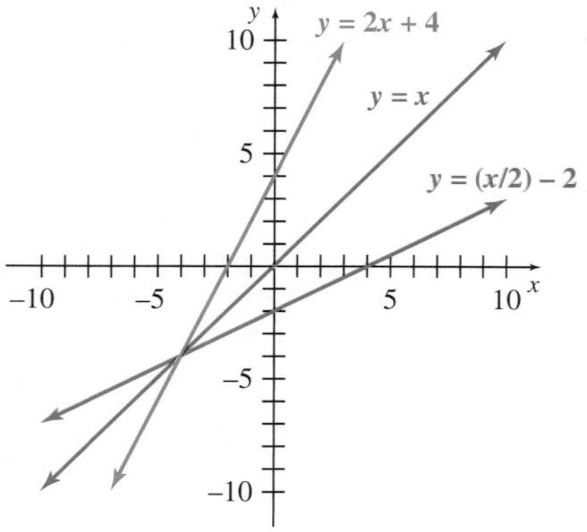

FIGURE 13

In the section on Logarithmic Functions, we showed that switching the x and y variables changes the exponential graph into a logarithmic graph, a defining property of functions that are inverses of each other. We also showed in the previous section that the slope of the tangent line of e^x at any point is e^x—that is, the

y-coordinate itself. So, if we switch the x and y variables, the new slope of the tangent line will be $1/y$, except that it is no longer y, it is x. Thus, the slope of the tangent line of $y = \ln x$ must be $1/x$ and hence $D_x \ln x = 1/x$.

We can also prove this fact rigorously by using a technique for finding the derivative of a function when you already know the derivative of its inverse function. Returning to the general logarithmic function, $f(x) = \log_a x$, we solve for x.

$$f(x) = \log_a x$$

$$a^{f(x)} = x \qquad \text{Definition of the logarithm}$$

Now consider the left and right sides of the last equation as functions of x that are equal, so their derivatives with respect to x should also be equal. Notice in the first step that we need to use the chain rule when differentiating $a^{f(x)}$.

$$(\ln a)a^{f(x)}f'(x) = 1 \qquad \text{Derivative of the exponential function}$$

$$(\ln a)xf'(x) = 1 \qquad \text{Substitute } a^{f(x)} = x.$$

Finally, divide both sides of this equation by $(\ln a)x$ to get

$$f'(x) = \frac{1}{(\ln a)x}.$$

DERIVATIVE OF $\log_a x$

$$D_x[\log_a x] = \frac{1}{(\ln a)x}$$

As with the exponential function, this formula becomes particularly simple when we let $a = e$, because of the fact that $\ln e = 1$.

DERIVATIVE OF $\ln x$

$$D_x[\ln x] = \frac{1}{x}$$

EXAMPLE 1 *Derivatives of Logarithmic Functions*

Find the derivative of each function.

(a) $f(x) = \ln 6x$

Solution Use the properties of logarithms and the rules for derivatives.

$$f'(x) = \frac{d}{dx}(\ln 6x)$$

$$= \frac{d}{dx}(\ln 6 + \ln x)$$

$$= \frac{d}{dx}(\ln 6) + \frac{d}{dx}(\ln x) = 0 + \frac{1}{x} = \frac{1}{x}$$

(b) $y = \log x$

Solution Recall that when the base is not specified, we assume that the logarithm is a common logarithm, which has a base of 10.

$$\frac{dy}{dx} = \frac{1}{(\ln 10)x}$$

Applying the chain rule to the formulas for the derivative of logarithmic functions gives us

$$\frac{d}{dx} \log_a g(x) = \frac{1}{\ln a} \cdot \frac{g'(x)}{g(x)}$$

and

$$\frac{d}{dx} \ln g(x) = \frac{g'(x)}{g(x)}.$$

EXAMPLE 2 *Derivatives of Logarithmic Functions*

Find the derivative of each function.

(a) $f(x) = \ln(x^2 + 1)$

Solution Here $g(x) = x^2 + 1$ and $g'(x) = 2x$. Thus,

$$f'(x) = \frac{g'(x)}{g(x)} = \frac{2x}{x^2 + 1}.$$

(b) $y = \log_2(3x^2 - 4x)$

Solution

$$\frac{dy}{dx} = \frac{1}{\ln 2} \cdot \frac{6x - 4}{3x^2 - 4x}$$

$$= \frac{6x - 4}{(\ln 2)(3x^2 - 4x)}$$

If $y = \ln(-x)$, where $x < 0$, the chain rule with $g(x) = -x$ and $g'(x) = -1$ gives

$$\frac{dy}{dx} = \frac{g'(x)}{g(x)} = \frac{-1}{-x} = \frac{1}{x}.$$

The derivative of $y = \ln(-x)$ is the same as the derivative of $y = \ln x$. For this reason, these two results can be combined into one rule using the absolute value of x. A similar situation holds true for $y = \ln[g(x)]$ and $y = \ln[-g(x)]$, as well as for $y = \log_a[g(x)]$ and $y = \log_a[-g(x)]$. These results are summarized as follows.

DERIVATIVE OF $\log_a|x|$, $\log_a|g(x)|$, $\ln|x|$, AND $\ln|g(x)|$

$$D_x[\log_a|x|] = \frac{1}{(\ln a)x} \qquad D_x[\log_a|g(x)|] = \frac{1}{\ln a} \cdot \frac{g'(x)}{g(x)}$$

$$D_x[\ln|x|] = \frac{1}{x} \qquad D_x[\ln|g(x)|] = \frac{g'(x)}{g(x)}$$

You need not remember four separate formulas. If you know the derivative of $y = \log_a x$ as well as the chain rule, all the other formulas follow. An absolute value inside of a logarithm has no effect on the derivative, other than making the result valid for more values of x.

EXAMPLE 3 *Derivatives of Logarithmic Functions*

Find the derivative of each function.

(a) $y = \ln|5x|$

Solution Let $g(x) = 5x$, so that $g'(x) = 5$. From the formula above,

$$\frac{dy}{dx} = \frac{g'(x)}{g(x)} = \frac{5}{5x} = \frac{1}{x}.$$

Notice that the derivative of $\ln|5x|$ is the same as the derivative of $\ln|x|$. Also, in Example 1, the derivative of $\ln 6x$ was the same as that for $\ln x$. This suggests that for any constant a,

$$\frac{d}{dx}\ln|ax| = \frac{d}{dx}\ln|x| = \frac{1}{x}.$$

Exercise 41 asks for a proof of this result.

(b) $f(x) = 3x \ln x^2$

Solution This function is the product of the two functions $3x$ and $\ln x^2$, so use the product rule.

$$f'(x) = (3x)\left[\frac{d}{dx}\ln x^2\right] + (\ln x^2)\left[\frac{d}{dx}3x\right]$$
$$= 3x\left(\frac{2x}{x^2}\right) + (\ln x^2)(3)$$
$$= 6 + 3\ln x^2$$

By the power rule for logarithms,

$$f'(x) = 6 + \ln(x^2)^3$$
$$= 6 + \ln x^6.$$

Alternatively, write the answer as $f'(x) = 6 + 6\ln x$.

(c) $s(t) = 6 \log_8 t^{3/2}$

Solution

$$s'(t) = 6 \cdot \frac{1}{\ln 8} \cdot \frac{(3/2)t^{1/2}}{t^{3/2}}$$

$$= \frac{9}{(\ln 8)t} \qquad \text{Simplify with algebra.}$$

Alternatively, we could have used one of the rules of logarithms to simplify the function to $s = 6(3/2)\log_8 t = 9 \log_8 t$ and then taken the derivative.

EXAMPLE 4 *Pedestrian Speed*

In the last of the review exercises for the chapter on Nonlinear Functions, we found that the average speed of pedestrians depends on the population (x) of the city in which they are walking by the function

$$f(x) = .873 \log x - .0255.$$

Find and interpret $f(1{,}000{,}000)$ and $f'(1{,}000{,}000)$.

Solution Recognizing this function as a common (base 10) logarithm, we have $f(1{,}000{,}000) = .873 \log 1{,}000{,}000 - .0255 = 5.2125$. In a city of 1,000,000 people, pedestrians walk, on average, about 5.2 ft per second.

$$f'(x) = \frac{.873}{(\ln 10)x},$$

so $f'(1{,}000{,}000) \approx 3.791 \cdot 10^{-7}$. This means that for each additional person in a city of 1,000,000, the average pedestrian speed increases by about $3.8 \cdot 10^{-7}$ ft per second. Perhaps a simpler interpretation is that for a city of 1,000,000, every additional 10,000 people causes an increase in the average pedestrian velocity of about $3.8 \cdot 10^{-7} \cdot 10{,}000 = .0038$ ft per second, a very small quantity. The increase in speed with larger populations is only noticeable when comparing cities that vary greatly in size.

4.5 EXERCISES

Find the derivative of each function.

1. $y = \ln(8x)$

2. $y = \ln(-4x)$

3. $y = \ln(3 - x)$

4. $y = \ln(1 + x^2)$

5. $y = \ln|2x^2 - 7x|$

6. $y = \ln|-8x^2 + 6x|$

7. $y = \ln\sqrt{x + 5}$

8. $y = \ln\sqrt{2x + 1}$

9. $y = \ln(x^4 + 5x^2)^{3/2}$

10. $y = \ln(5x^3 - 2x)^{3/2}$

11. $y = -3x \ln(x + 2)$

12. $y = (3x + 1)\ln(x - 1)$

13. $s = t^2 \ln|t|$

14. $y = x \ln|2 - x^2|$

15. $y = \dfrac{2\ln(x + 3)}{x^2}$

16. $v = \dfrac{\ln u}{u^3}$

17. $y = \dfrac{\ln x}{4x + 7}$

18. $y = \dfrac{-2\ln x}{3x - 1}$

19. $y = \dfrac{3x^2}{\ln x}$

20. $y = \dfrac{x^3 - 1}{2\ln x}$

21. $y = (\ln|x + 1|)^4$

22. $y = \sqrt{\ln|x - 3|}$

23. $y = \ln|\ln x|$

24. $y = (\ln 4)(\ln|3x|)$

25. $y = e^{x^2} \ln x$ **26.** $y = e^{2x-1} \ln(2x - 1)$ **27.** $y = \dfrac{e^x}{\ln x}$ **28.** $p(y) = \dfrac{\ln y}{e^y}$

29. $g(z) = (e^{2z} + \ln z)^3$ **30.** $y = \log(6x)$ **31.** $y = \log(2x - 3)$ **32.** $y = \log|1 - x|$

33. $y = \log|3x|$ **34.** $y = \log_5\sqrt{5x + 2}$ **35.** $y = \log_7\sqrt{2x - 3}$ **36.** $y = \log_3(x^2 + 2x)^{3/2}$

37. $y = \log_2(2x^2 - x)^{5/2}$ **38.** $w = \log_8(6^p - 1)$ **39.** $z = 10^y \log y$

40. Why do we use the absolute value of x or of $g(x)$ in the derivative formulas for the natural logarithm?

41. Prove $\dfrac{d}{dx}\ln|ax| = \dfrac{d}{dx}\ln|x|$ for any constant a.

42. A friend concludes that because $y = \ln 6x$ and $y = \ln x$ have the same derivative, namely $dy/dx = 1/x$, these two functions must be the same. Explain why this is incorrect.

43. Use a graphing calculator to sketch the graph of $y = [f(x + h) - f(x)]/h$ using $f(x) = \ln|x|$ and $h = .0001$. Compare it with the graph of $y = 1/x$ and discuss what you observe.

44. Using the fact that

$$\ln[u(x)v(x)] = \ln u(x) + \ln v(x),$$

use the chain rule and the formula for the derivative of $\ln x$ to derive the product rule. In other words, find $[u(x)v(x)]'$ without assuming the product rule.

45. Using the fact that

$$\ln\frac{u(x)}{v(x)} = \ln u(x) - \ln v(x),$$

use the chain rule and the formula for the derivative of $\ln x$ to derive the quotient rule. In other words, find $[u(x)/v(x)]'$ without assuming the quotient rule.

46. Use graphical differentiation to verify that $d \ln x/dx = 1/x$.

47. Use the fact that $d \ln x/dx = 1/x$, as well as the change-of-base theorem for logarithms, to prove that

$$\frac{d \log_a x}{dx} = \frac{1}{x \ln a}.$$

Applications

BUSINESS AND ECONOMICS

48. *Profit* Assume that the total revenue received from the sale of x items is given by

$$R(x) = 30 \ln(2x + 1),$$

while the total cost to produce x items is $C(x) = x/2$. Find the number of items that should be manufactured so that profit, $R(x) - C(x)$, is a maximum. (*Hint:* Set the derivative of the profit function equal to 0.)

49. *Revenue* Suppose the demand function for x units of a certain item is

$$p = 100 + \frac{50}{\ln x}, \quad x > 1,$$

where p is in dollars.

a. Find the marginal revenue.

b. Approximate the revenue from one more unit when 8 units are sold.

c. How might a manager use the information from part b?

50. *Profit* If the cost function in dollars for x units of the item in Exercise 49 is $C(x) = 100x + 100$, find the following.

a. The marginal cost

b. The profit function $P(x)$

c. The profit from one more unit when 8 units are sold

d. How might a manager use the information from part c?

51. *Marginal Average Cost* Suppose the cost in dollars to make x oboe reeds is given by

$$C(x) = 5 \log_2 x + 10.$$

Find the marginal average cost when the following numbers of reeds are sold.

a. 10 **b.** 20

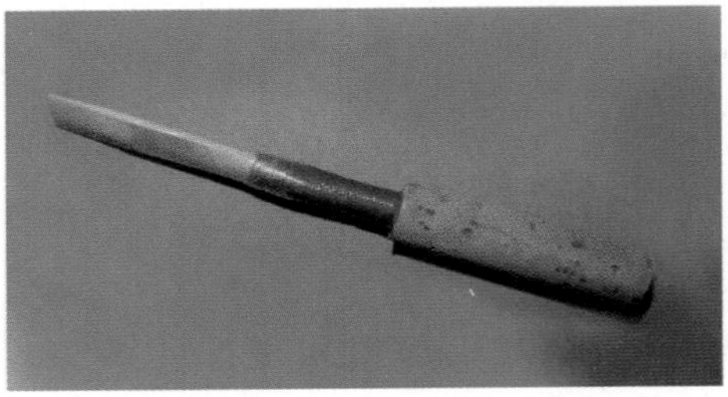

LIFE SCIENCES

52. *Body Surface Area* There is a mathematical relationship between an infant's weight and total body surface area (BSA), given by

$$A(w) = 4.688w^{.8168 - .0154 \log_{10} w},$$

where w is the weight (in grams) and $A(w)$ is the BSA in square centimeters.*

a. Find the BSA for an infant who weighs 4000 g.

b. Find $A'(4000)$ and interpret your answer.

c. Use a graphing calculator to graph $A(w)$ on $[2000, 10{,}000]$ by $[0, 6000]$.

53. *Bologna Sausage* Scientists in Italy have developed a modified Gompertz model to predict the growth of *Enterococcus faecium* in bologna sausage at 32°C. Another way of representing the number of bacteria is given by

$$\ln\left(\frac{N(t)}{N_0}\right) = 9.8901e^{-e^{2.54197-.2167t}},$$

where N_0 is the number of bacteria present at the beginning of the experiment and $N(t)$ is the number of bacteria present at time t (in hours).†

a. Use the properties of logarithms to find an expression for $N(t)$. Assume that $N_0 = 1000$.

b. Use a graphing calculator to estimate the derivative of $N(t)$ when $t = 20$ and interpret.

c. Let $S(t) = \ln(N(t)/N_0)$. Graph $S(t)$ on $[0, 35]$ by $[0, 12]$.

d. Graph $N(t)$ on $[0, 35]$ by $[0, 20{,}000{,}000]$ and compare the graphs from parts c and d.

e. Find $\lim_{t\to\infty} S(t)$ and then use this limit to find $\lim_{t\to\infty} N(t)$.

54. *Pronghorn Fawns* The field metabolic rate (FMR), or the total energy expenditure per day in excess of growth, can be calculated for pronghorn fawns using Nagy's formula,

$$F(x) = .774 + .727 \log x,$$

where x is the mass (in grams) of the fawn and $F(x)$ is the energy expenditure (in kJ/day).‡

a. Determine the total energy expenditure per day in excess of growth for a pronghorn fawn that weighs 25,000 g.

b. Find $F'(25{,}000)$ and interpret the result.

c. Graph the function on $[5000, 30{,}000]$ by $[3, 5]$.

55. *Fruit Flies* A study of the relation between the rate of reproduction in *Drosophila* (fruit flies) bred in bottles and the density of the mated population found that the number of imagoes (sexually mature adults) per mated female per day (y) can be approximated by

$$\log y = 1.54 - .008x - .658 \log x,$$

where x is the mean density of the mated population (measured as flies per bottle) over a 16-day period.§

a. Show that the above equation is equivalent to

$$y = 34.7(1.0186)^{-x}x^{-.658}.$$

b. Using your answer from part a, find the number of imagoes per mated female per day when the density is

i. 20 flies per bottle;

ii. 40 flies per bottle.

*Sharkey, I. et al., "Body Surface Area Estimation in Children Using Weight Alone: Application in Pediatric Oncology," *British Journal of Cancer,* Vol. 85, No. 1, 2001, pp. 23–28.

†Zanoni, B., C. Garzaroli, S. Anselmi, and G. Rondinini, "Modeling the Growth of *Enterococcus faecium* in Bologna Sausage," *Applied and Environmental Microbiology,* Vol. 59, No. 10, Oct. 1993, pp. 3411–3417.

‡Miller, Michelle N. and John A. Byers, "Energetic Cost of Locomotor Play in Pronghorn Fawns," *Animal Behavior,* Vol. 41, 1991, pp. 1007–1013.

§Pearl, R. and S. Parker, *Proc. Natl. Acad. Sci.,* Vol. 8, 1922, p. 212, quoted in *Elements of Mathematical Biology* by Alfred J. Lotka, Dover Publications, 1956, pp. 308–311.

c. Using your answer from part a, find the rate of change in the number of imagoes per mated female per day with respect to the density when the density is

i. 20 flies per bottle;

ii. 40 flies per bottle.

56. *Insect Mating* Consider an experiment in which equal numbers of male and female insects of a certain species are permitted to intermingle. Assume that

$$M(t) = (.1t + 1)\ln\sqrt{t}$$

represents the number of matings observed among the insects in an hour, where t is the temperature in degrees Celsius. (*Note:* The formula is an approximation at best and holds only for specific temperature intervals.)

a. Find the number of matings when the temperature is 15°C.

b. Find the number of matings when the temperature is 25°C.

c. Find the rate of change of the number of matings when the temperature is 15°C.

57. *Population Growth* Suppose that the population of a certain collection of rare Brazilian ants is given by

$$P(t) = (t + 100)\ln(t + 2),$$

where t represents the time in days. Find the rates of change of the population on the second day and on the eighth day.

PHYSICAL SCIENCES

58. *Richter Scale* The Richter scale provides a measure of the magnitude of an earthquake. In fact, the largest Richter number M ever recorded for an earthquake was 8.9 from the 1933 earthquake in Japan. The following formula shows a relationship between the amount of energy released and the Richter number.

$$M = \frac{2}{3}\log\frac{E}{.007},$$

where E is measured in kilowatt-hours.*

a. For the 1933 earthquake in Japan, what value of E gives a Richter number $M = 8.9$?

b. If the average household uses 247 kWh per month, how many months would the energy released by an earthquake of this magnitude power 10 million households?

c. Find the rate of change of the Richter number M with respect to energy when $E = 70{,}000$ kWh.

d. What happens to dM/dE as E increases?

GENERAL INTEREST

59. *Street Crossing* Consider a child waiting at a street corner for a gap in traffic that is large enough so that he can safely cross the street. A mathematical model for traffic shows that if the expected waiting time for the child is to be at most 1 minute, then the maximum traffic flow, in cars per hour, is given by

$$f(x) = \frac{29{,}000(2.322 - \log x)}{x},$$

where x is the width of the street in feet.† Find the maximum traffic flow and the rate of change of the maximum traffic flow with respect to street width for the following values of the street width.

a. 30 ft **b.** 40 ft

CHAPTER SUMMARY

In this chapter we used the definition of the derivative to develop techniques for finding derivatives of several types of functions. With the help of the rules that were developed, such as the power rule, product rule, quotient rule, and chain rule, we can now directly compute the derivative of a large variety of functions. In particular, we developed rules for finding derivatives of exponential and logarithmic functions. We also began to see the wide range of applications that these functions have in business, life sciences, social sciences, and the physical sciences. In the next chapter we will apply these techniques to study the behavior of certain functions and we will learn that differentiation can be used to find maximum and minimum values of continuous functions.

*Bradley, Christopher, "Media Clips," *Mathematics Teacher,* Vol. 93, No. 4, April 2000, pp. 300–303.
†Bender, Edward, *An Introduction to Mathematical Modeling,* New York: Wiley, 1978, p. 213.

Rules for Derivatives Summary

Assume all indicated derivatives exist.

Constant Function If $f(x) = k$, where k is any real number, then $f'(x) = 0$.

Power Rule If $f(x) = x^n$, for any real number n, then $f'(x) = n \cdot x^{n-1}$.

Constant Times a Function Let k be a real number. Then the derivative of $y = k \cdot f(x)$ is $dy/dx = k \cdot f'(x)$.

Sum or Difference Rule If $y = u(x) \pm v(x)$, then $\dfrac{dy}{dx} = u'(x) \pm v'(x)$.

Product Rule If $f(x) = u(x) \cdot v(x)$, then

$$f'(x) = u(x) \cdot v'(x) + v(x) \cdot u'(x).$$

Quotient Rule If $f(x) = \dfrac{u(x)}{v(x)}$, then

$$f'(x) = \frac{v(x) \cdot u'(x) - u(x) \cdot v'(x)}{[v(x)]^2}.$$

Chain Rule If y is a function of u, say $y = f(u)$, and if u is a function of x, say $u = g(x)$, then $y = f(u) = f[g(x)]$, and

$$\frac{dy}{dx} = \frac{dy}{du} \cdot \frac{du}{dx}.$$

Chain Rule (Alternative Form) Let $y = f[g(x)]$. Then $\dfrac{dy}{dx} = f'[g(x)] \cdot g'(x)$.

Exponential Functions

$$D_x(e^x) = e^x \qquad D_x(a^x) = (\ln a)a^x$$

$$D_x(e^{g(x)}) = e^{g(x)}g'(x) \qquad D_x(a^{g(x)}) = (\ln a)a^{g(x)}g'(x)$$

Logarithmic Functions

$$D_x(\ln |x|) = \frac{1}{x} \qquad D_x(\log_a |x|) = \frac{1}{(\ln a)x}$$

$$D_x(\ln |g(x)|) = \frac{g'(x)}{g(x)} \qquad D_x(\log_a |g(x)|) = \frac{1}{\ln a} \cdot \frac{g'(x)}{g(x)}$$

KEY TERMS

To understand the concepts presented in this chapter, you should know the meaning and use of the following terms. For easy reference, the section in the chapter where a word (or expression) was first used is provided.

4.1 **marginal analysis**

4.2 **marginal average cost**

4.3 **composite function**
chain rule

4.4 **logistic function**

CHAPTER 4 REVIEW EXERCISES

Use the rules for derivatives to find the derivative of each function defined as follows.

1. $y = 5x^2 - 7x - 9$

2. $y = x^3 - 4x^2$

3. $y = 6x^{7/3}$

4. $y = -3x^{-2}$

5. $f(x) = x^{-3} + \sqrt{x}$

6. $f(x) = 6x^{-1} - 2\sqrt{x}$

7. $k(x) = \dfrac{3x}{x + 5}$

8. $r(x) = \dfrac{-8}{2x + 1}$

9. $y = \dfrac{x^2 - x + 1}{x - 1}$

10. $y = \dfrac{2x^3 - 5x^2}{x + 2}$

11. $f(x) = (3x - 2)^4$

12. $k(x) = (5x - 1)^6$

13. $y = \sqrt{2t - 5}$

14. $y = -3\sqrt{8t - 1}$

15. $y = 3x(2x + 1)^3$

16. $y = 4x^2(3x - 2)^5$

17. $r(t) = \dfrac{5t^2 - 7t}{(3t + 1)^3}$

18. $s(t) = \dfrac{t^3 - 2t}{(4t - 3)^4}$

19. $p(t) = t^2(t^2 + 1)^{5/2}$

20. $g(t) = t^3(t^4 + 5)^{7/2}$

21. $y = -6e^{2x}$

22. $y = 8e^{.5x}$

23. $y = e^{-2x^3}$

24. $y = -4e^{x^2}$

25. $y = 5xe^{2x}$

26. $y = -7x^2e^{-3x}$

27. $y = \ln(2 + x^2)$

28. $y = \ln(5x + 3)$

29. $y = \dfrac{\ln |3x|}{x - 3}$

30. $y = \dfrac{\ln |2x - 1|}{x + 3}$

31. $y = \dfrac{xe^x}{\ln(x^2 - 1)}$

32. $y = \dfrac{(x^2 + 1)e^{2x}}{\ln x}$

33. $s = (t^2 + e^t)^2$

34. $q = (e^{2p+1} - 2)^4$

35. $y = 3 \cdot 10^{-x^2}$

36. $y = 10 \cdot 2^{\sqrt{x}}$

37. $g(z) = \log_2(z^3 + z + 1)$

38. $h(z) = \log(1 + e^z)$

39. Why is e a convenient base for exponential and logarithmic functions?

Find the slope of the tangent line to the given curve at the given value of x. Find the equation of each tangent line.

40. $y = x^2 - 6x; \quad x = 2$

41. $y = 8 - x^2; \quad x = 1$

42. $y = \dfrac{3}{x - 1}; \quad x = -1$

43. $y = \dfrac{x}{x^2 - 1}; \quad x = 2$

44. $y = \sqrt{6x - 2}; \quad x = 3$

45. $y = -\sqrt{8x + 1}; \quad x = 3$

46. $y = e^x; \quad x = 0$

47. $y = xe^x; \quad x = 1$

48. $y = \ln x; \quad x = 1$

49. $y = x \ln x; \quad x = e$

*The following exercise is from the 1991 examination for applicants to the Economics Division of Shiga University in Japan.**

50. Consider the graphs of the function $y = \sqrt{2x - 1}$ and the straight line $y = x + k$. Discuss the number of points of intersection versus the change in the value of k.

51. a. Verify that

$$\frac{d \ln f(x)}{dx} = \frac{f'(x)}{f(x)}.$$

*"Japanese University Entrance Examination Problems in Mathematics," edited by Ling-Erl Eileen T. Wu, published by the Mathematical Association of America, copyright 1993, pp. 18–19.

This expression is called the *relative rate of change*. It expresses the rate of change of f relative to the size of f. Stephen B. Maurer denotes this expression by $\hat{f}$ and notes that economists commonly work with relative rates of change.*

b. Verify that

$$\widehat{fg} = \hat{f} + \hat{g}$$

Interpret this equation in terms of relative rates of change.

c. In his article, Maurer uses the result of part b to solve the following problem:

"Last year, the population grew by 1% and the average income per person grew by 2%. By what approximate percent did the national income grow?"

Explain why the result from part b implies that the answer to this question is approximately 3%.

52. Suppose that the student body in your college grows by 2% and the tuition goes up by 3%. Use the result from the previous exercise to calculate the approximate amount that the total tuition collected goes up, and compare this with the actual amount.

Applications

BUSINESS AND ECONOMICS

Marginal Average Cost *Find the marginal average cost function of each function defined as follows.*

53. $C(x) = \sqrt{x + 1}$

54. $C(x) = \sqrt{3x + 2}$

55. $C(x) = (x^2 + 3)^3$

56. $C(x) = (4x + 3)^4$

57. $C(x) = 10 - e^{-x}$

58. $C(x) = \ln(x + 5)$

59. *Sales* The sales of a company are related to its expenditures on research by

$$S(x) = 1000 + 50\sqrt{x} + 10x,$$

where $S(x)$ gives sales in millions when x thousand dollars is spent on research. Find and interpret dS/dx if the following amounts are spent on research.

a. \$9000 **b.** \$16,000 **c.** \$25,000

d. As the amount spent on research increases, what happens to sales?

60. *Profit* Suppose that the profit (in hundreds of dollars) from selling x units of a product is given by

$$P(x) = \frac{x^2}{x - 1}, \quad \text{where } x > 1.$$

Find and interpret the marginal profit when the following numbers of units are sold.

a. 4 **b.** 12 **c.** 20

d. What is happening to the marginal profit as the number sold increases?

e. Find and interpret the marginal average profit when 4 units are sold.

61. *Costs* A company finds that its costs are related to the amount spent on training programs by

$$T(x) = \frac{1000 + 50x}{x + 1},$$

where $T(x)$ is costs in thousands of dollars when x hundred dollars are spent on training. Find and interpret $T'(x)$ if the following amounts are spent on training.

a. \$900 **b.** \$1900

c. Are costs per dollar spent on training always increasing or decreasing?

62. *Compound Interest* If a sum of \$1000 is deposited into an account that pays r% interest compounded quarterly, the balance after 12 years is given by

$$A = 1000\left(1 + \frac{r}{400}\right)^{48}.$$

Find and interpret $\frac{dA}{dr}$ when $r = 5$.

*Maurer, Stephen B., "Hat Derivatives," *The College Mathematics Journal,* Vol. 33, No. 2, Jan. 2002, pp. 32–37.

63. *Continuous Compounding* If a sum of $1000 is deposited into an account that pays $r\%$ interest compounded continuously, the balance after 12 years is given by

$$A = 1000e^{12r/100}.$$

Find and interpret $\frac{dA}{dr}$ when $r = 5$.

64. *Doubling Time* If a sum of money is deposited into an account that pays $r\%$ interest compounded annually, the doubling time (in years) is given by

$$T = \frac{\ln 2}{\ln(1 + r/100)}.$$

Find and interpret dT/dr when $r = 5$.

LIFE SCIENCES

65. *Exponential Growth* Suppose a population is growing exponentially with an annual growth constant $k = .05$. How fast is the population growing when it is 1,000,000? Use the derivative to calculate your answer, and then explain how the answer can be obtained without using the derivative.

66. *Logistic Growth* Suppose a population is growing logistically with $k = 5 \cdot 10^{-6}$, $m = 30{,}000$, and $G_0 = 2000$. Assume time is measured in years.

a. Find the growth function $G(t)$ for this population.

b. Find the population and rate of growth of the population after 6 years.

67. *Fish* The length of the monkeyface prickleback, a West Coast game fish, can be approximated by

$$L = 71.5(1 - e^{-.1t})$$

and the weight by

$$W = .01289 \cdot L^{2.9},$$

where L is the length in centimeters, t is the age in years, and W is the weight in grams.* (See the Extended Application for the chapter on Nonlinear Functions.)

a. Find the approximate length of a 5-year-old monkeyface.

b. Find how fast the length of a 5-year-old monkeyface is growing.

c. Find the approximate weight of a 5-year-old monkeyface. (*Hint:* Use your answer from part a.)

d. Find the rate of change of the weight with respect to length for a 5-year-old monkeyface.

e. Using the chain rule and your answers to parts b and d, find how fast the weight of a 5-year-old monkeyface is growing.

68. *Arctic Foxes* The age/weight relationship of male Arctic foxes caught in Svalbard, Norway, can be estimated by the function

$$M(t) = 3583e^{-e^{-.020(t-66)}},$$

where t is the age of the fox in days and $M(t)$ is the weight of the fox in grams.†

a. Estimate the weight of a male fox that is 250 days old.

b. Use $M(t)$ to estimate the largest size that a male fox can attain. (*Hint:* Find $\lim_{t\to\infty} M(t)$.)

c. Estimate the age of a male fox when it has reached 50% of its maximum weight.

d. Estimate the rate of change in weight of a male Arctic fox that is 250 days old. (*Hint:* Recall that $D_t e^{f(t)} = f'(t)e^{f(t)}$.)

e. Use a graphing calculator to graph $M(t)$ and then describe the growth pattern.

f. Use the table function on a graphing calculator or a spreadsheet to develop a chart that shows the estimated weight and growth rate of male foxes for days 50, 100, 150, 200, 250, and 300.

PHYSICAL SCIENCES

69. *Chlorofluorocarbons* In Exercise 47 in the section on Exponential Functions, we found that the concentration (in ppb) of the chlorofluorocarbon CFC-11 in the atmosphere could be approximated by

$$C(x) = .05(1.040)^{x-1950},$$

where x is the year. Find and interpret $C'(1998)$.

70. *Cats* The distance from Lisa Wunderle's cat, Belmar, to a piece of string he is stalking is given in feet by

$$f(t) = \frac{8}{t+1} + \frac{20}{t^2+1},$$

where t is the time in seconds since he begins.

a. Find Belmar's average velocity between 1 second and 3 seconds.

b. Find Belmar's instantaneous velocity at 3 seconds.

*Marshall, William H. and Tina Wylie Echeverria, "Characteristics of the Monkeyface Prickleback," *California Fish and Game,* Vol. 78, No. 2, Spring 1992.

†Pestrud, Pal and Kjell Nilssen, "Growth, Size, and Sexual Dimorphism in Arctic Foxes," *Journal of Mammalogy,* Vol. 76, No. 2, May 1995, pp. 522–530.

GENERAL INTEREST

71. *Food Surplus* In Example 7 in the section on Exponential Functions, we found that the production of corn (in billions of bushels) in the United States since 1930 could be approximated by

$$p(x) = 1.757(1.0251)^{x-1930},$$

where x is the year. Find and interpret $p'(2000)$.

72. *Dating a Language* Over time, the number of original basic words in a language tends to decrease as words become obsolete or are replaced with new words. Linguists have used calculus to study this phenomenon and have developed a methodology for dating a language, called *glottochronology*. Experiments have indicated that a good estimate of the number of words that remain in use at a given time is given by

$$N(t) = N_0 e^{-.217t},$$

where $N(t)$ is the number of words in a particular language, t is measured in the number of millennium, and N_0 is the original number of words in the language.*

a. In 1950, C. Feng and M. Swadesh established that of the original 210 basic ancient Chinese words from 950 A.D., 167 were still being used. Letting $t = 0$ correspond to 1950, with $N_0 = 167$, find the number of words predicted to have been in use in 950 A.D., and compare it with the actual number in use.

b. Estimate the number of words that will remain in the year 2050.

c. Find $N'(2)$ and interpret your answer.

73. *Driving Fatalities* A study by the National Highway Traffic Safety Administration found that driver fatalities rates were highest for the youngest and oldest drivers.† The rates per 1000 licensed drivers for every 100 million miles may be approximated by the function

$$f(x) = k(x - 49)^6 + .8,$$

where x is the driver's age in years and k is the constant $3.8 \cdot 10^{-9}$. Find and interpret the rate of change of the fatality rate when the driver is

a. 20 years old; **b.** 60 years old.

*Lo Bello, Anthony and Maurice Weir, "Glottochronology: An Application of Calculus to Linguistics," *The UMAP Journal,* Vol. 3, No. 1, Spring 1982, pp. 85–99.

†www-nrd.nhtsa.dot.gov/pdf/nrd-30/NCSA/RNotes/1998/AgeSex96.pdf.

CHAPTER

5 Graphs and the Derivative

Derivatives provide useful information about the behavior of functions and the shapes of their graphs. The first derivative describes the rate of increase or decrease, while the second derivative indicates the degree of *nonlinearity* in the function. In an exercise at the end of this chapter we will see what changes in the sign of the second derivative tell us about the shape of a weightlifter's age versus performance graph.

The graph in Figure 1 shows the relationship between the number of sleep-related accidents and traffic density during a 24-hour period.* The blue line indicates the hourly distribution of sleep-related accidents. The green line indicates the hourly distribution of traffic density. The red line indicates the relative risk of sleep-related accidents. For example, the relative risk graph shows us that a person is nearly seven times as likely to have an accident at 4:00 A.M. than at 10:00 P.M.

Given a graph like the one in Figure 1, we can often locate maximum and minimum values simply by looking at the graph. It is difficult to get *exact* values or *exact* locations of maxima and minima from a graph, however, and many functions are difficult to graph without the aid of technology. In the chapter on Nonlinear Functions we saw how to find exact maximum and minimum values for quadratic functions by identifying the vertex. A more general approach is to use the derivative of a function to determine precise maximum and minimum values of the function. The procedure for doing this is described in this chapter, which begins with a discussion of increasing and decreasing functions.

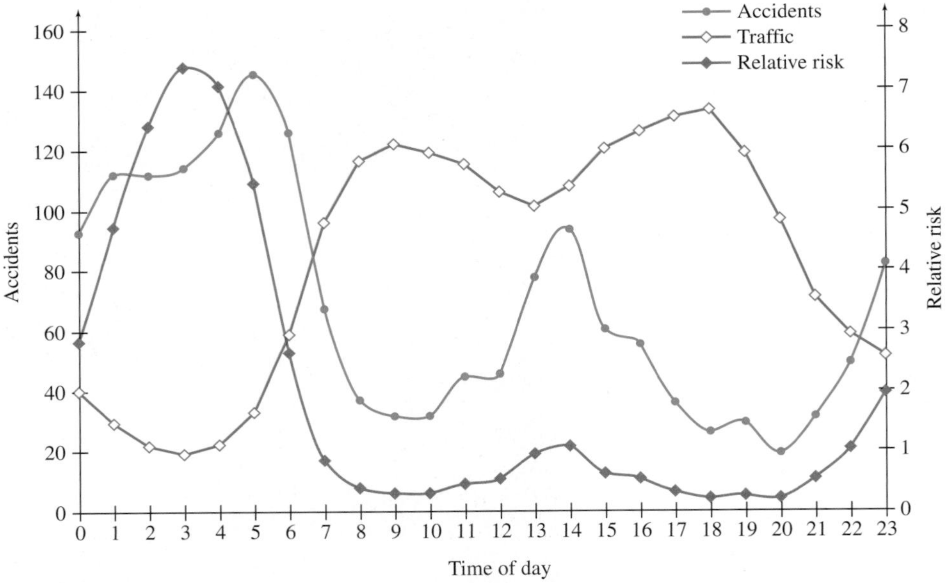

FIGURE 1

5.1 INCREASING AND DECREASING FUNCTIONS

THINK ABOUT IT *How long is it profitable to increase production?*

We will answer this question in Example 3 after further investigating increasing and decreasing functions.

*Garbarino, S., L. Nobili, M. Beelke, F. Phy, and F. Ferrillo, "The Contribution Role of Sleepiness in Highway Vehicle Accidents," *Sleep,* Vol. 24, No. 2, 2001, pp. 203–206. © Copyright 2001 by the American Academy of Sleep Medicine. Reproduced with permission of the American Academy of Sleep Medicine via Copyright Clearance Center.

The graph of a typical function may increase on some intervals and decrease on others as shown in Figure 2. How can we tell from the equation that defines a function where the graph increases and where it decreases? The derivative can be used to answer this question. Remember that the derivative of a function at a point gives the slope of the line tangent to the function at that point. Recall also that a line with a positive slope rises from left to right and a line with a negative slope falls from left to right.

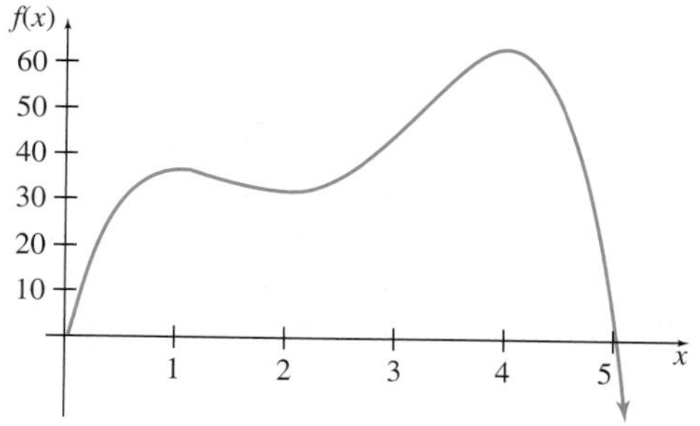

FIGURE 2

Think of the graph of f in Figure 2 as a roller coaster track moving from left to right along the graph. Now, picture one of the cars on the roller coaster. As shown in Figure 3, when the car is on level ground or parallel to level ground, its floor is horizontal, but as the car moves up the slope, its floor tilts upward. When the car reaches a peak, its floor is again horizontal, but it then begins to tilt downward (very steeply) as the car rolls downhill. The floor of the car as it moves from left to right along the track represents the tangent line at each point. Using this analogy, we can see that the slope of the tangent line will be *positive* when the car travels uphill and f is *increasing*, and the slope of the tangent line will be *negative* when the car travels downhill and f is *decreasing*. (In this case it is also true that the slope of the tangent line will be zero at "peaks" and "valleys.")

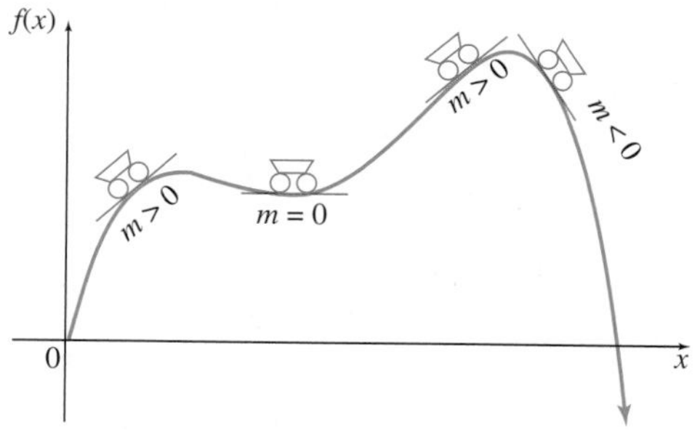

FIGURE 3

Thus, on intervals where $f'(x) > 0$, the function $f(x)$ will increase, and on intervals where $f'(x) < 0$, $f(x)$ will decrease. We can determine where $f(x)$ peaks by finding the intervals on which it increases and decreases.

Summarizing, a function is *increasing* if the graph goes *up* from left to right and *decreasing* if its graph goes *down* from left to right. Examples of increasing functions are shown in Figures 4(a)–(c), and examples of decreasing functions in Figures 4(d)–(f).

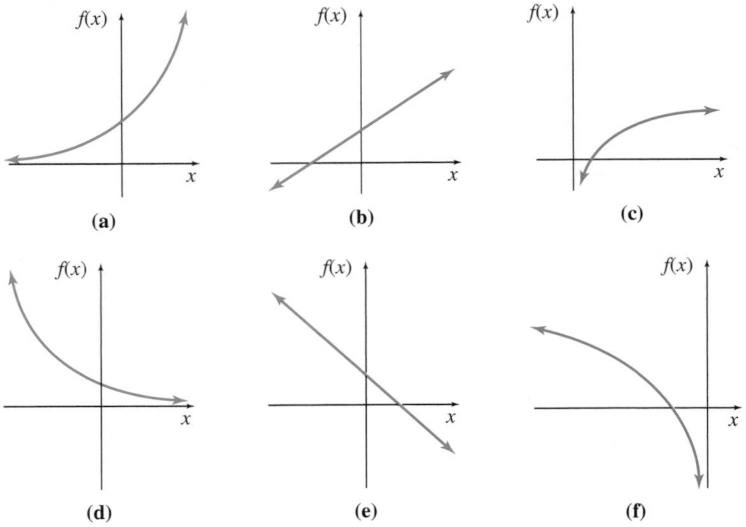

FIGURE 4

INCREASING AND DECREASING FUNCTIONS

Let f be a function defined on some interval. Then for any two numbers x_1 and x_2 in the interval, f is **increasing** on the interval if

$$f(x_1) < f(x_2) \quad \text{whenever} \quad x_1 < x_2,$$

and f is **decreasing** on the interval if

$$f(x_1) > f(x_2) \quad \text{whenever} \quad x_1 < x_2.$$

EXAMPLE 1 *Increasing and Decreasing*

Where is the function graphed in Figure 5 increasing? Where is it decreasing?

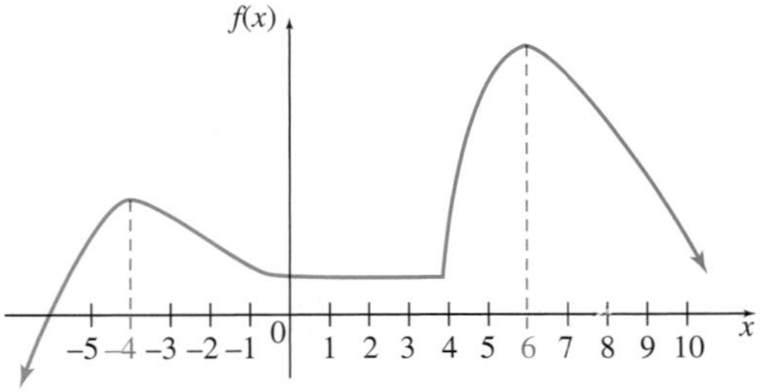

FIGURE 5

Solution Moving from left to right, the function is increasing up to -4, then decreasing from -4 to 0, constant (neither increasing nor decreasing) from 0 to 4, increasing from 4 to 6, and decreasing from 6 onward. In interval notation, the function is increasing on $(-\infty, -4)$ and $(4, 6)$, decreasing on $(-4, 0)$ and $(6, \infty)$, and constant on $(0, 4)$.

Our discussion suggests the following test.

TEST FOR INTERVALS WHERE $f(x)$ IS INCREASING AND DECREASING

Suppose a function f has a derivative at each point in an open interval; then
if $f'(x) > 0$ for each x in the interval, f is *increasing* on the interval;
if $f'(x) < 0$ for each x in the interval, f is *decreasing* on the interval;
if $f'(x) = 0$ for each x in the interval, f is *constant* on the interval.

The derivative $f'(x)$ can change signs from positive to negative (or negative to positive) at points where $f'(x) = 0$, and also at points where $f'(x)$ does not exist. The values of x where this occurs are called *critical numbers.*

CRITICAL NUMBERS

The **critical numbers** for a function f are those numbers c in the domain of f for which $f'(c) = 0$ or $f'(c)$ does not exist. A **critical point** is a point whose x-coordinate is the critical number c, and whose y-coordinate is $f(c)$.

It is shown in more advanced classes that if the critical numbers of a polynomial function are used to determine open intervals on a number line, then the sign of the derivative at any point in an interval will be the same as the sign of the derivative at any other point in the interval. This suggests that the test for increasing and decreasing functions be applied as follows (assuming that no open intervals exist where the function is constant).

FOR REVIEW

The method for finding where a function is increasing and decreasing is similar to the method introduced in Section R.5 for solving quadratic inequalities.

APPLYING THE TEST

1. Locate the critical numbers for f on a number line, as well as any points where f is undefined. These points determine several open intervals.
2. Choose a value of x in each of the intervals determined in Step 1. Use these values to decide whether $f'(x) > 0$ or $f'(x) < 0$ in that interval.
3. Use the test above to decide whether f is increasing or decreasing on the interval.

EXAMPLE 2 *Increasing and Decreasing*

Find the intervals in which the following functions are increasing or decreasing. Locate all points where the tangent line is horizontal. Graph the function.

FOR REVIEW

In this chapter you will need all of the rules for derivatives you learned in the previous chapter. If any of these are still unclear, go over the Derivative Summary at the end of that chapter and practice some of the Review Exercises before proceeding.

(a) $f(x) = x^3 + 3x^2 - 9x + 4$

Solution Here $f'(x) = 3x^2 + 6x - 9$. To find the critical numbers, set this derivative equal to 0 and solve the resulting equation by factoring.

$$3x^2 + 6x - 9 = 0$$
$$3(x^2 + 2x - 3) = 0$$
$$3(x + 3)(x - 1) = 0$$
$$x = -3 \quad \text{or} \quad x = 1$$

The tangent line is horizontal at $x = -3$ or $x = 1$. Since there are no values of x where $f'(x)$ fails to exist, the only critical numbers are -3 and 1. To determine where the function is increasing or decreasing, locate -3 and 1 on a number line, as in Figure 6. (Be sure to place the values on the number line in numerical order.) These points determine three intervals: $(-\infty, -3)$, $(-3, 1)$, and $(1, \infty)$.

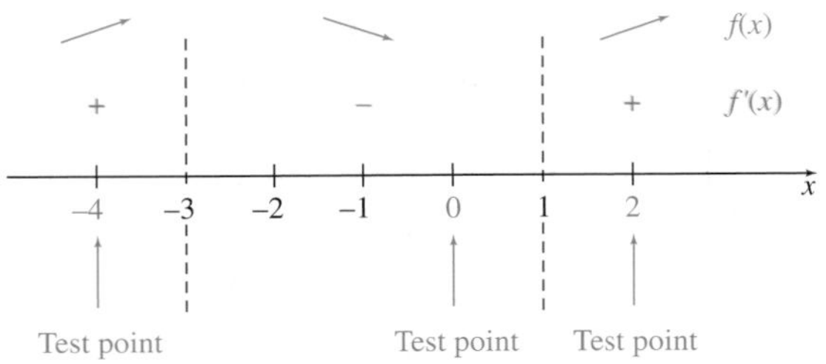

FIGURE 6

Now choose any value of x in the interval $(-\infty, -3)$. Choosing $x = -4$ gives

$$f'(-4) = 3(-4)^2 + 6(-4) - 9 = 15,$$

which is positive. Since one value of x in this interval makes $f'(x) > 0$, all values will do so, and therefore f is increasing on $(-\infty, -3)$. Selecting 0 from the middle interval gives $f'(0) = -9$, so f is decreasing on $(-3, 1)$. Finally, choosing 2 in the right-hand region gives $f'(2) = 15$, with f increasing on $(1, \infty)$. The arrows in each interval in Figure 6 indicate where f is increasing or decreasing.

Up to now our only method of graphing most functions has been by plotting points that lie on the graph, either by hand or using a graphing calculator or computer. Now an additional tool is available: the test for determining where a function is increasing or decreasing. (Other tools are discussed in the next few sections.) To graph the function, plot a point at each of the critical numbers by finding $f(-3) = 31$ and $f(1) = -1$. Also plot points for $x = -4, 0$, and 2, the test values of each interval. Use these points along with the information about where the function is increasing and decreasing to get the graph in Figure 7 on the next page.

CAUTION Be careful to use $f(x)$, not $f'(x)$, to find the y-values of the points to plot. ■

(b) $f(x) = \dfrac{x-1}{x+1}$

Solution Use the quotient rule to find $f'(x)$.

$$f'(x) = \frac{(x+1)(1) - (x-1)(1)}{(x+1)^2}$$

$$= \frac{x + 1 - x + 1}{(x+1)^2} = \frac{2}{(x+1)^2}$$

This derivative is never 0, but it fails to exist at $x = -1$, where the function is undefined. This divides the number line into two intervals: $(-\infty, -1)$ and

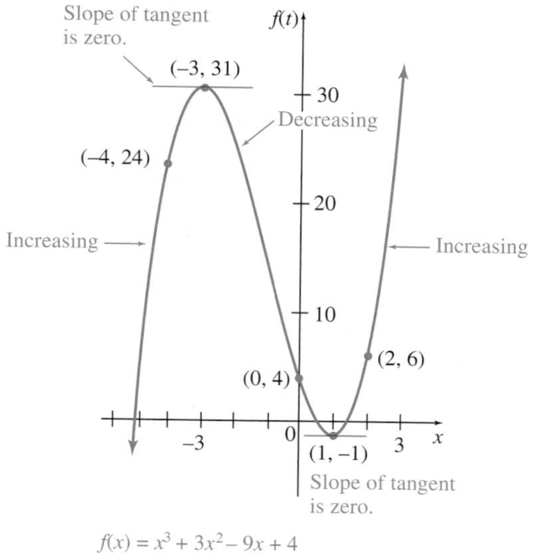

FIGURE 7

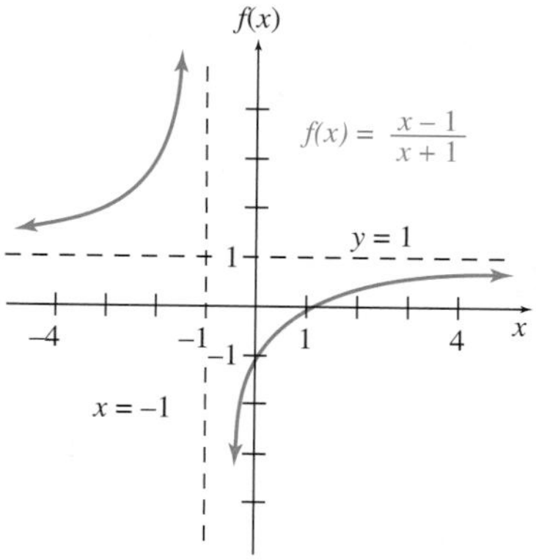

FIGURE 8

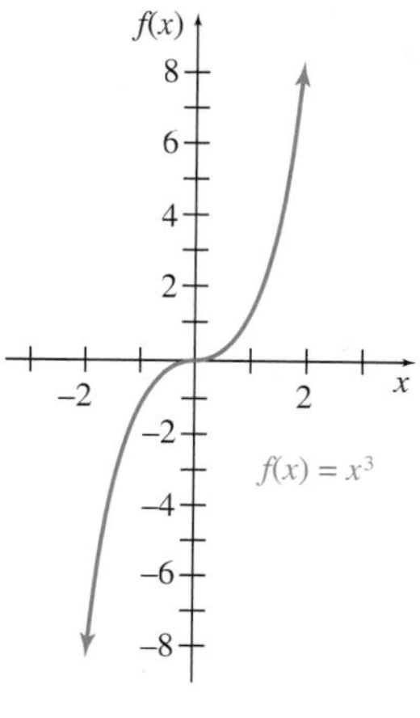

FIGURE 9

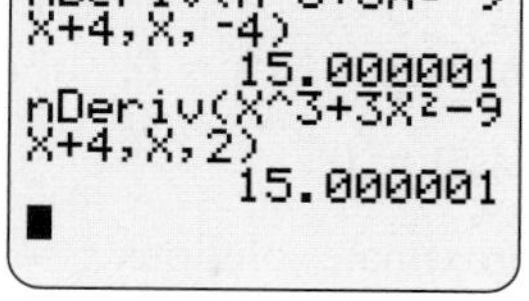

FIGURE 10

$(-1, \infty)$. Draw a number line for f', and use a test point in each of these intervals to find that $f'(x) > 0$ for all x except -1. (This can also be seen by observing that $f'(x)$ is the quotient of 2, which is positive, and $(x + 1)^2$, which is always positive or 0.) This means that the function f is increasing on both $(-\infty, -1)$ and $(-1, \infty)$. Note that $x = -1$ is not a critical number since -1 is not in the domain of f.

We also note that this function has a horizontal asymptote:

$$\lim_{x\to\infty} \frac{x-1}{x+1} = \lim_{x\to\infty} \frac{1 - 1/x}{1 + 1/x} \qquad \text{Divide numerator and denominator by } x.$$
$$= 1.$$

We get the same limit as x approaches $-\infty$, so the graph has the line $y = 1$ as a horizontal asymptote. Verify that at $x = -1$ the graph has a vertical asymptote. Using this information, as well as the intercept $y = 0$ when $x = 1$, gives the graph in Figure 8.

CAUTION It is important to note that the reverse of the test for increasing and decreasing functions is not true—it is possible for a function to be increasing on an interval even though the derivative is not positive at every point in the interval. A good example is given by $f(x) = x^3$, which is increasing on every interval, even though $f'(x) = 0$ when $x = 0$. See Figure 9.

Similarly, it is incorrect to assume that the sign of the derivative in regions separated by critical numbers must alternate between + and −. If this were always so, it would lead to a simple rule for finding the sign of the derivative: just check one test point, and then make the other regions alternate in sign. But this is not true if one of the factors in the derivative is raised to an even power. In the function $f(x) = x^3$ just considered, $f'(x) = 3x^2$ is positive on both sides of the critical number $x = 0$. ■

A graphing calculator can be used to find the derivative of a function at a particular x-value. The screen in Figure 10 supports our results in Example 2(a) for the test values, -4 and 2. The results are not exact because the calculator uses a numerical method to approximate the derivative at the given x-value.

Some graphing calculators can find where a function changes from increasing to decreasing by finding a maximum or minimum. The calculator windows in Figure 11 show this feature for the function in Example 2(a). Notice that these, too, are approximations. This concept will be explored further in the next section.

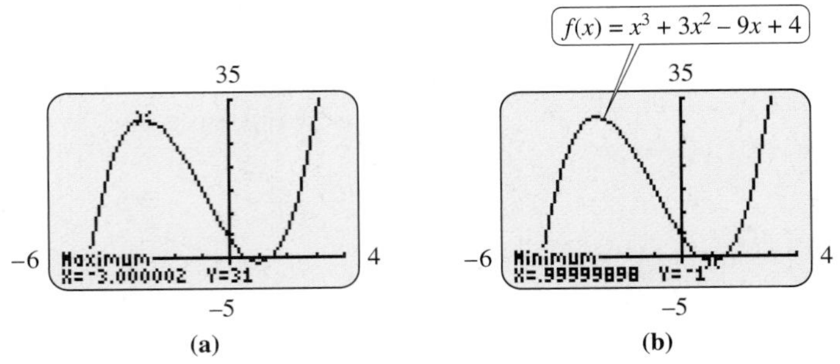

FIGURE 11

Knowing the intervals where a function is increasing or decreasing can be important in applications, as shown by the next examples.

EXAMPLE 3 *Profit Analysis*

A company selling computers finds that the cost per computer decreases linearly with the number sold monthly, decreasing from \$1000 when none are sold to \$800 when 1000 are sold. Thus, the average cost function has a y-intercept of 1000 and a slope of $-200/1000 = -.2$, so it is given by the formula

$$\overline{C}(x) = 1000 - .2x, \quad 0 \leq x \leq 1000,$$

where x is the number of computers sold monthly. Since $\overline{C}(x) = C(x)/x$, the cost function is given by

$$\begin{aligned} C(x) = x\overline{C}(x) &= x(1000 - .2x) \\ &= 1000x - .2x^2, \quad 0 \leq x \leq 1000. \end{aligned}$$

Suppose the revenue function can be approximated by

$$R(x) = .0008x^3 - 2.4x^2 + 2400x, \quad 0 \leq x \leq 1000.$$

Determine any intervals on which the profit function is increasing.

Solution First find the profit function $P(x)$.

$$\begin{aligned} P(x) &= R(x) - C(x) \\ &= (.0008x^3 - 2.4x^2 + 2400x) - (1000x - .2x^2) \\ &= .0008x^3 - 2.2x^2 + 1400x \end{aligned}$$

To find any intervals where this function is increasing, set $P'(x) = 0$.

$$P'(x) = .0024x^2 - 4.4x + 1400 = 0$$

Solving this with the quadratic formula gives the approximate solutions $x = 409.8$ and $x = 1423.6$. The latter number is outside of the domain. Use $x = 409.8$ to

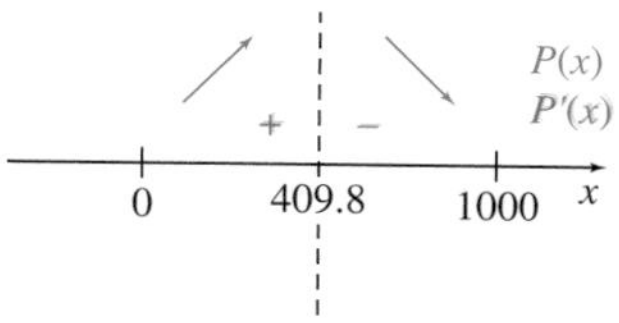

FIGURE 12

determine two intervals on a number line, as shown in Figure 12. Choose $x = 0$ and $x = 1000$ as test points.

$$P'(0) = .0024(0^2) - 4.4(0) + 1400 = 1400$$
$$P'(1000) = .0024(1000^2) - 4.4(1000) + 1400 = -600$$

This means that when no computers are sold monthly, the profit is going up at a rate of \$1400 per computer. When 1000 computers are sold monthly, the profit is going down at a rate of \$600 per computer. The test points show that the function increases on $(0, 409.8)$ and decreases on $(409.8, 1000)$. See Figure 12. Thus, the profit is increasing when 409 computers or fewer are sold, and decreasing when 410 or more are sold, as shown in Figure 13.

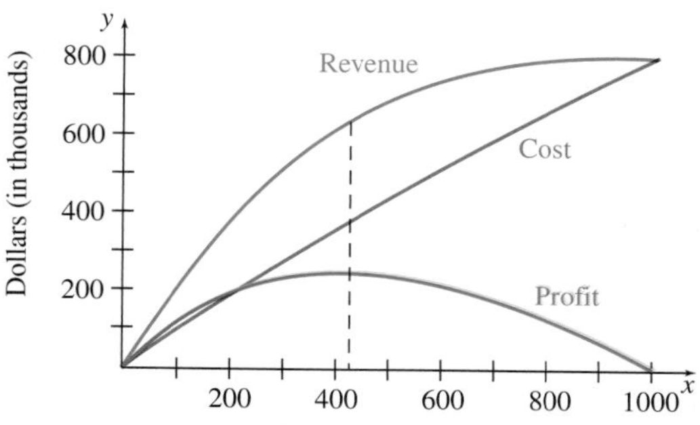

FIGURE 13

As the graph in Figure 13 shows, the profit will increase as long as the revenue function increases faster than the cost function. That is, increasing production will produce more profit as long as the marginal revenue is greater than the marginal cost.

EXAMPLE 4 *Recollection of Facts*

In the exercises in the previous chapter, the function

$$f(t) = \frac{90t}{99t - 90}$$

gave the number of facts recalled after t hours for $t > 10/11$. Find the intervals in which $f(t)$ is increasing or decreasing.

Solution First find the derivative, $f'(t)$.

$$f'(t) = \frac{(99t - 90)(90) - 90t(99)}{(99t - 90)^2}$$
$$= \frac{8910t - 8100 - 8910t}{(99t - 90)^2} = \frac{-8100}{(99t - 90)^2}$$

Since $(99t - 90)^2$ is positive everywhere in the domain of the function and since the numerator is a negative constant, $f'(t) < 0$ for all t in the domain of $f(t)$. Thus $f(t)$ always decreases and, as expected, the number of words recalled decreases steadily over time.

The next example illustrates the case where a function has a critical number at c because the derivative does not exist at c.

EXAMPLE 5 *Increasing and Decreasing*

Find the critical numbers and decide where f is increasing and decreasing if $f(x) = (x - 1)^{2/3}$.

Solution We find $f'(x)$ first, using the power rule and the chain rule.

$$f'(x) = \frac{2}{3}(x - 1)^{-1/3}(1) = \frac{2}{3(x - 1)^{1/3}}$$

We need to find any values of x that make $f'(x) = 0$, but here $f'(x)$ is never 0. To find the numbers where $f'(x)$ does not exist, set the denominator equal to 0 and solve.

$$3(x - 1)^{1/3} = 0$$
$$x - 1 = 0$$
$$x = 1$$

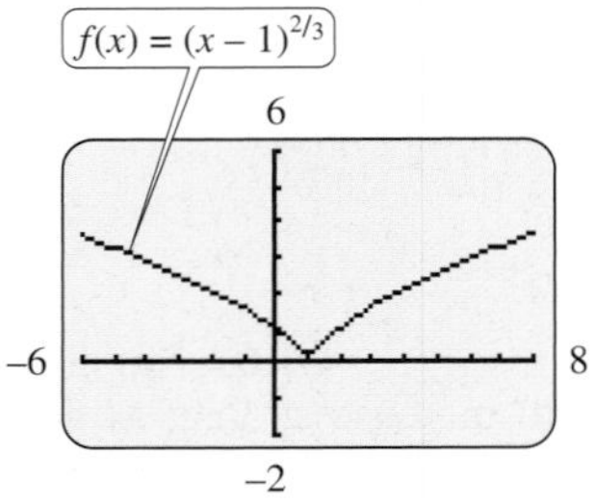

FIGURE 14

Since $f'(1)$ does not exist but $f(1)$ is defined, $x = 1$ is a critical number, the only critical number. Now use the first derivative test to find where f is increasing and decreasing.

$$f'(0) = \frac{2}{3(0 - 1)^{1/3}} = \frac{2}{-3} = -\frac{2}{3}$$

$$f'(2) = \frac{2}{3(2 - 1)^{1/3}} = \frac{2}{3}$$

Since f is defined for all x, these results show that f is decreasing on $(-\infty, 1)$ and increasing on $(1, \infty)$. A calculator graph of f is shown in Figure 14.

5.1 EXERCISES

Find the open intervals where the functions graphed as follows are **(a)** *increasing, or* **(b)** *decreasing.*

1.
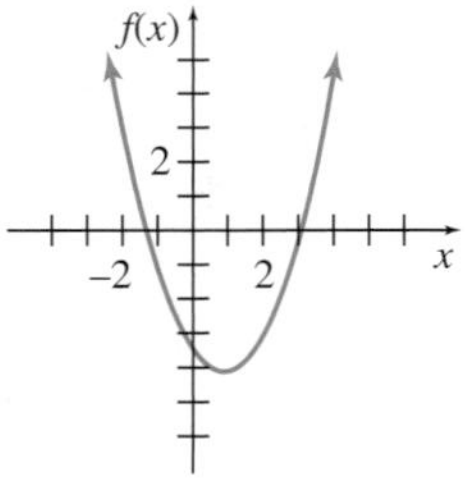

2.
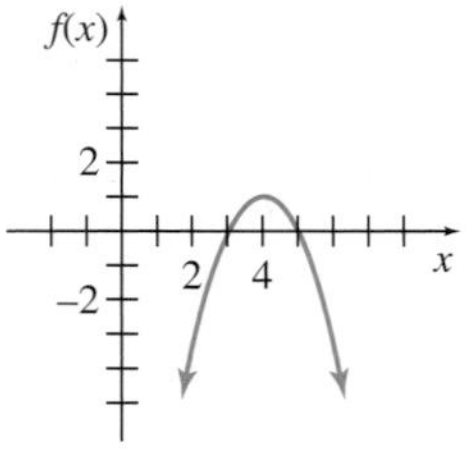

3.
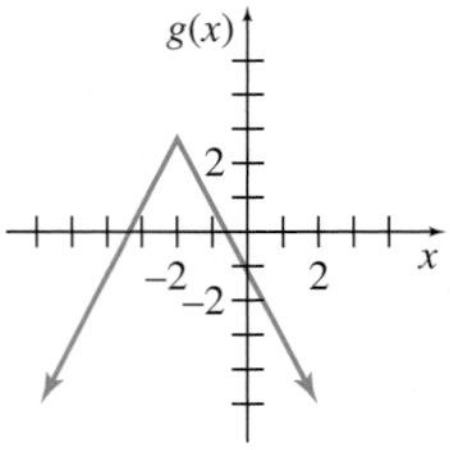

4.
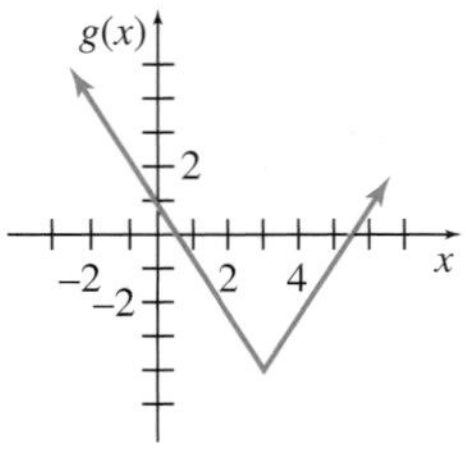

5.
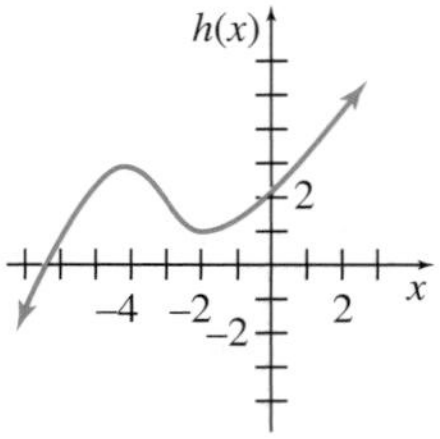

6.
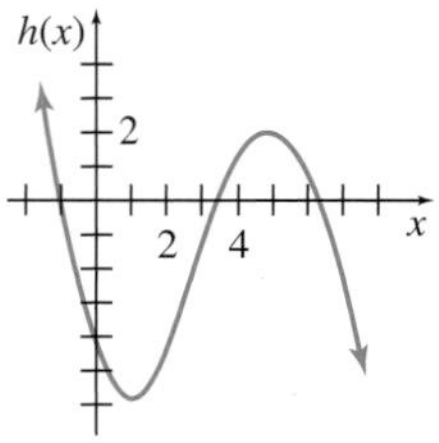

7.

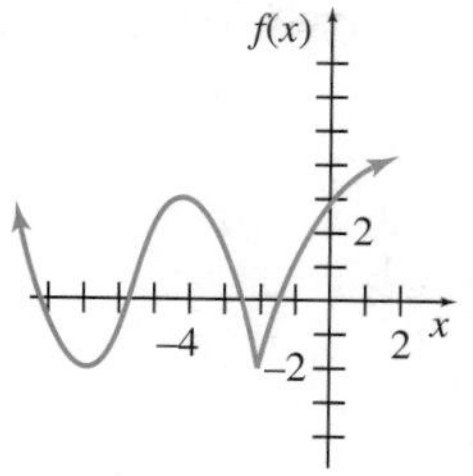

8.

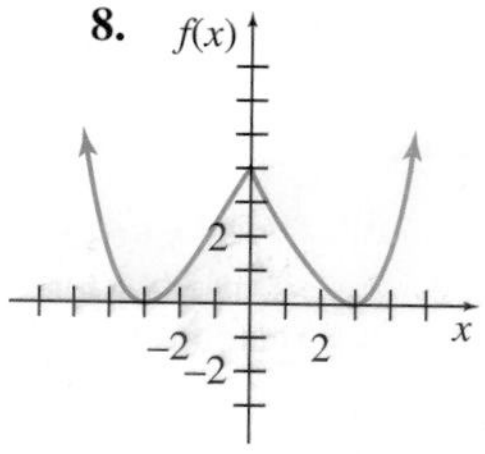

For each function, find **(a)** *the critical numbers;* **(b)** *the open intervals where the function is increasing; and* **(c)** *the open intervals where it is decreasing.*

9. $y = 2 + 3.6x - 1.2x^2$

10. $y = .3 + .4x - .2x^2$

11. $f(x) = \frac{2}{3}x^3 - x^2 - 24x - 4$

12. $f(x) = \frac{2}{3}x^3 - x^2 - 4x + 2$

13. $f(x) = 4x^3 - 15x^2 - 72x + 5$

14. $f(x) = 4x^3 - 9x^2 - 30x + 6$

15. $f(x) = x^4 + 4x^3 + 4x^2 + 1$

16. $f(x) = 3x^4 + 8x^3 - 18x^2 + 5$

17. $y = -3x + 6$

18. $y = 6x - 9$

19. $f(x) = \dfrac{x + 2}{x + 1}$

20. $f(x) = \dfrac{x + 3}{x - 4}$

21. $y = \sqrt{x^2 + 1}$

22. $y = x\sqrt{9 - x^2}$

23. $f(x) = x^{2/3}$

24. $f(x) = (x + 1)^{4/5}$

25. $y = x - 4\ln(3x - 9)$

26. $y = xe^{x^2 - 3x}$

27. $y = x^{2/3} - x^{5/3}$

28. $y = x^{1/3} + x^{4/3}$

29. A friend looks at the graph of $y = x^2$ and observes that if you start at the origin, the graph increases whether you go to the right or the left, so the graph is increasing everywhere. Explain why this reasoning is incorrect.

30. Use the techniques of this chapter to find the vertex and intervals where f is increasing and decreasing, given

$$f(x) = ax^2 + bx + c,$$

where we assume $a > 0$. Verify that this agrees with what we found in the chapter on Nonlinear Functions.

31. Repeat Exercise 30 under the assumption $a < 0$.

32. Where is the function defined by $f(x) = e^x$ increasing? Decreasing? Where is the tangent line horizontal?

33. Repeat Exercise 32 with the function defined by $f(x) = \ln x$.

Applications

BUSINESS AND ECONOMICS

34. *Housing Starts* A county realty group estimates that the number of housing starts per year over the next three years will be

$$H(r) = \frac{300}{1 + .03r^2},$$

where r is the mortgage rate (in percent).

a. Where is $H(r)$ increasing?

b. Where is $H(r)$ decreasing?

35. *Cost* Suppose the total cost $C(x)$ (in dollars) to manufacture a quantity x of weed killer (in hundreds of liters) is given by

$$C(x) = x^3 - 2x^2 + 8x + 50.$$

a. Where is $C(x)$ decreasing?

b. Where is $C(x)$ increasing?

36. *Profit* A manufacturer sells video games with the following cost and revenue functions, where x is the number of games sold.

$$C(x) = 4.8x - .0004x^2, \quad 0 \le x \le 2250$$
$$R(x) = 8.4x - .002x^2, \quad 0 \le x \le 2250$$

Determine the interval(s) on which the profit function is increasing.

37. *Profit* A manufacturer of CD players has determined that the profit $P(x)$ (in thousands of dollars) is related to the quantity x of CD players produced (in hundreds) per month by

$$P(x) = -(x - 4)e^x - 4, \qquad 0 < x \leq 3.9$$

as long as the number of CD players produced is fewer than 390 per month.

a. At what production levels is the profit increasing?

b. At what levels is it decreasing?

LIFE SCIENCES

38. *Air Pollution* The graph shows the amount of air pollution removed by trees in the Chicago urban region for each month of the year.* From the graph we see, for example, that the ozone level starting in May increases up to June, and then abruptly decreases.

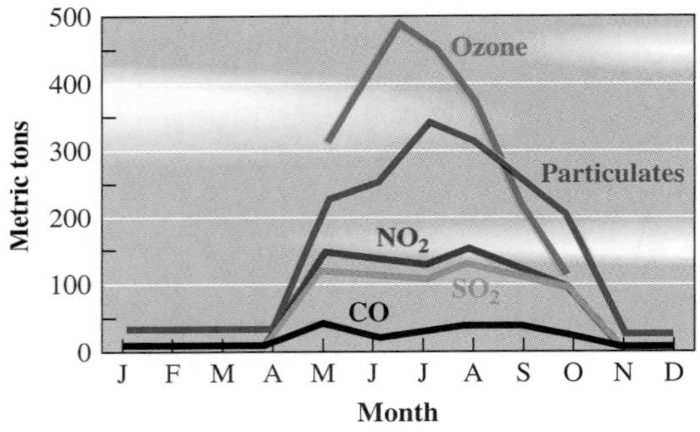

a. Are these curves the graphs of functions?

b. Look at the graph for particulates. Where is the function increasing? Decreasing? Constant?

c. On what intervals do all four lower graphs indicate that the corresponding functions are constant? Why do you think the functions are constant on those intervals?

39. *Spread of Infection* The number of people $P(t)$ (in hundreds) infected t days after an epidemic begins is approximated by

$$P(t) = \frac{10 \ln(.19t + 1)}{.19t + 1}.$$

When will the number of people infected start to decline?

40. *Alcohol Concentration* In an earlier exercise set we gave the function defined by

$$A(x) = -.015x^3 + 1.058x$$

as the approximate alcohol concentration (in tenths of a percent) in an average person's bloodstream x hours after drinking 8 oz of 100-proof whiskey. The function applies only for the interval $[0, 8]$.

a. On what time intervals is the alcohol concentration increasing?

b. On what intervals is it decreasing?

41. *Drug Concentration* The percent of concentration of a drug in the bloodstream x hours after the drug is administered is given by

$$K(x) = \frac{4x}{3x^2 + 27}.$$

a. On what time intervals is the concentration of the drug increasing?

b. On what intervals is it decreasing?

42. *Drug Concentration* Suppose a certain drug is administered to a patient, with the percent of concentration of the drug in the bloodstream t hours later given by

$$K(t) = \frac{5t}{t^2 + 1}.$$

a. On what time intervals is the concentration of the drug increasing?

b. On what intervals is it decreasing?

43. *Cardiology* The aortic pressure-diameter relation in a particular patient who underwent cardiac catheterization can be modeled by the polynomial

$$D(p) = .000002p^3 - .0008p^2 + .1141p + 16.683, \qquad 55 \leq p \leq 130,$$

where $D(p)$ is the aortic diameter (in millimeters) and p is the aortic pressure (in mmHg).† Determine where this func-

*National Arbor Day Foundation, 100 Arbor Ave., Nebraska City, NE 68410. Ad in *Chicago Tribune,* Feb. 4, 1996, Sec. 2, p. 11. Used with permission of the National Arbor Day Foundation.
†Stefanadis, C., J. Dernellis et al., "Assessment of Aortic Line of Elasticity Using Polynomial Regression Analysis," *Circulation,* Vol. 101, No. 15, April 18, 2000, pp. 1819–1825.

tion is increasing and where it is decreasing within the interval given above.

44. *Thermic Effect of Food* The metabolic rate of a person who has just eaten a meal tends to go up and then, after some time has passed, returns to a resting metabolic rate. This phenomenon is known as the thermic effect of food. Researchers have indicated that the thermic effect of food for one particular person is

$$F(t) = -10.28 + 175.9te^{-t/1.3},$$

where $F(t)$ is the thermic effect of food (in kJ/hr) and t is the number of hours that have elapsed since eating a meal.*

a. Find $F'(t)$

b. Determine where this function is increasing and where it is decreasing. Interpret your answers.

45. *Holstein Dairy Cattle* Researchers have used the Richardson model to develop the following function that can be used to accurately predict the weight of Holstein cows (females) of various ages:

$$W_1(t) = 619(1 - .905e^{-.002t})^{1.2386},$$

where $W_1(t)$ is the weight of the Holstein cow (in kilograms) that is t days old.† Where is this function increasing?

SOCIAL SCIENCES

46. *Population* The standard normal probability function is used to describe many different populations. Its graph is the well-known normal curve. This function is defined by

$$f(x) = \frac{1}{\sqrt{2\pi}}e^{-x^2/2}.$$

Give the intervals where the function is increasing and decreasing.

47. *Nuclear Arsenals* The figure shows estimated totals of nuclear warhead stockpiles for the United States and the Soviet Union (and its successor states) from 1945 to 2002.‡

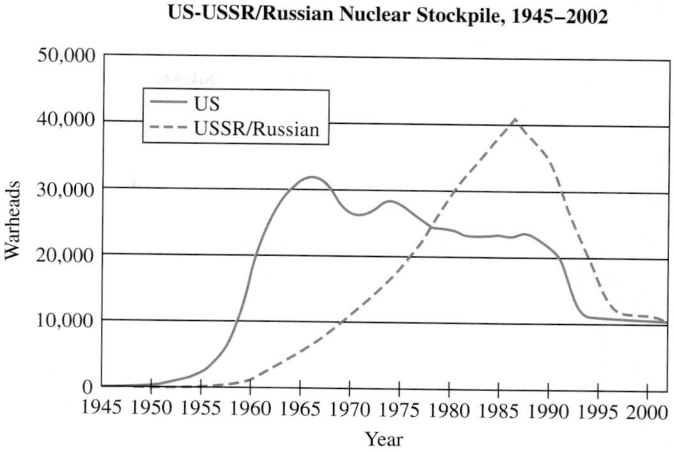

a. On what intervals were the stockpiles of both countries increasing?

b. On what intervals were the stockpiles of both countries decreasing?

GENERAL INTEREST

48. *Sports Cars* The following graph shows the horsepower and torque as a function of the engine speed for a 1991 Porsche 928 GT.§

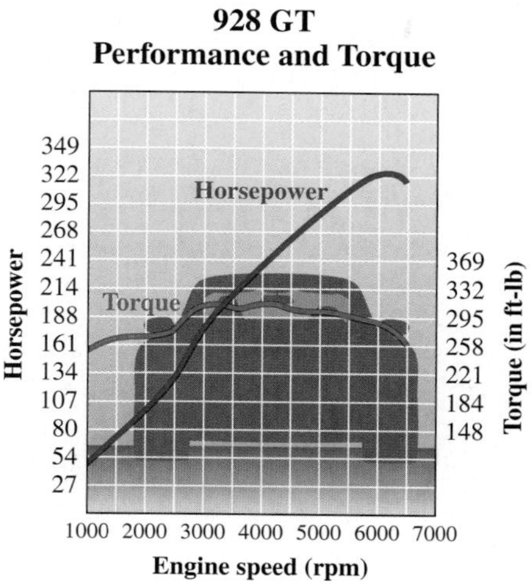

a. On what intervals is the horsepower increasing with engine speed?

b. On what intervals is the horsepower decreasing with engine speed?

*Reed, George and James Hill, "Measuring the Thermic Effect of Food," *American Journal of Clinical Nutrition,* Vol. 63, 1996, pp. 164–169.

†Perotto, D., R. Cue, and A. Lee, "Comparison of Nonlinear Functions of Describing the Growth Curve of Three Genotypes of Dairy Cattle," *Canadian Journal of Animal Science,* Vol. 73, Dec. 1992, pp. 773–782.

‡http://www.nrdc.org/nuclear/nudb/dafig11.asp.

§Porsche 928 and the distinctive shapes of Porsche automobiles are registered trademarks of Dr. Ing h. c. F. Porsche AG. Used with permission of Porsche Cars North America, Inc. and Porsche AG. Copyrighted by Porsche Cars North America.

c. On what intervals is the torque increasing with engine speed?

d. On what intervals is the torque decreasing with engine speed?

49. *Automobile Mileage* As a mathematics professor loads more weight in the back of his Mazda, the mileage goes down. Let x be the amount of weight (in pounds) that he adds, and let $y = f(x)$ be the mileage (in mpg).

a. Is $f'(x)$ positive or negative? Explain.

b. What are the units of $f'(x)$?

5.2 RELATIVE EXTREMA

THINK ABOUT IT *In a 30-second commercial, when is the best time to present the sales message?*

Suppose that the manufacturer of a diet soft drink is disappointed by sales after airing a new series of 30-second television commercials. The company's market research analysts hypothesize that the problem lies in the timing of the commercial's message, Drink Sparkling Light. Either it comes too early in the commercial, before the viewer has become involved; or it comes too late, after the viewer's attention has faded. After extensive experimentation, the research group finds that the percent of full attention that a viewer devotes to a commercial is a function of time (in seconds) since the commercial began, where

$$\text{Viewer's attention} = f(t) = -\frac{3}{20}t^2 + 6t + 20, \quad 0 \le t \le 30.$$

When is the best time to present the commercial's sales message?

Clearly, the message should be delivered when the viewer's attention is at a maximum. To find this time, find $f'(t)$.

$$f'(t) = -\frac{3}{10}t + 6 = -.3t + 6$$

The derivative $f'(t)$ is greater than 0 when $-.3t + 6 > 0$, $-3t > -60$, or $t < 20$. Similarly, $f'(t) < 0$ when $-.3t + 6 < 0$, $-3t < -60$, or $t > 20$. Thus, attention increases for the first 20 seconds and decreases for the last 10 seconds. The message should appear about 20 seconds into the commercial. At that time the viewer will devote $f(20) = 80\%$ of his attention to the commercial.

The maximum level of viewer attention (80%) in the example above is a *relative maximum*, defined as follows.

RELATIVE MAXIMUM OR MINIMUM

Let c be a number in the domain of a function f. Then $f(c)$ is a **relative** (or **local**) **maximum** for f if there exists an open interval (a, b) containing c such that

$$f(x) \le f(c)$$

for all x in (a, b), and $f(c)$ is a **relative** (or **local**) **minimum** for f if there exists an open interval (a, b) containing c such that

$$f(x) \ge f(c)$$

for all x in (a, b).

A function has a **relative** (or **local**) **extremum** (plural: **extrema**) at c if it has either a relative maximum or a relative minimum there.

If c is an endpoint of the domain of f, we only consider x in the half-open interval that is in the domain.*

The intuitive idea is that a relative maximum is the greatest value of the function in some region right around the point, although there may be greater values elsewhere. For example, the highest value of the Dow Jones industrial average this week is a relative maximum, although the Dow may have reached a higher value earlier this year. Similarly, a relative minimum is the least value of a function in some region around the point.

EXAMPLE 1 *Relative Extrema*

Identify the x-values of all points where the graph in Figure 15 has relative extrema.

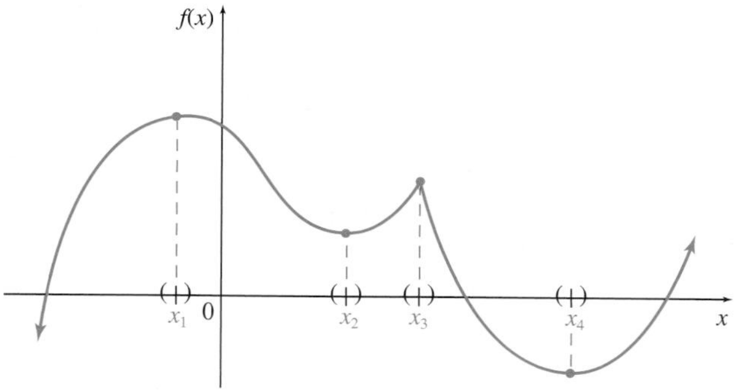

FIGURE 15

Solution The parentheses around x_1 show an open interval containing x_1 such that $f(x) \leq f(x_1)$, so there is a relative maximum of $f(x_1)$ at $x = x_1$. Notice that many other open intervals would work just as well. Similar intervals around x_2, x_3, and x_4 can be used to find a relative maximum of $f(x_3)$ at $x = x_3$ and relative minima of $f(x_2)$ at $x = x_2$ and $f(x_4)$ at $x = x_4$.

The function graphed in Figure 16 on the next page has relative maxima when $x = x_1$ or $x = x_3$ and relative minima when $x = x_2$ or $x = x_4$. The tangent lines at the points having x-values x_1 and x_2 are shown in the figure. Both tangent lines

*There is disagreement on calling an endpoint a maximum or minimum. We define it this way because this is an applied calculus book, and in an application it would be considered a maximum or minimum value of the function.

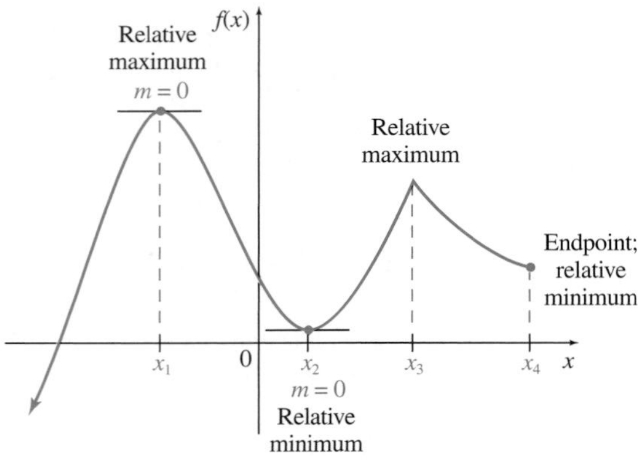

FIGURE 16

are horizontal and have slope 0. There is no single tangent line at the point where $x = x_3$.

Since the derivative of a function gives the slope of a line tangent to the graph of the function, to find relative extrema we first identify all critical numbers and endpoints. A relative extremum *may* exist at a critical number. (A rough sketch of the graph of the function near a critical number often is enough to tell whether an extremum has been found.) These facts about extrema are summarized below.

> If a function f has a relative extremum at c, then c is a critical number or c is an endpoint of the domain.

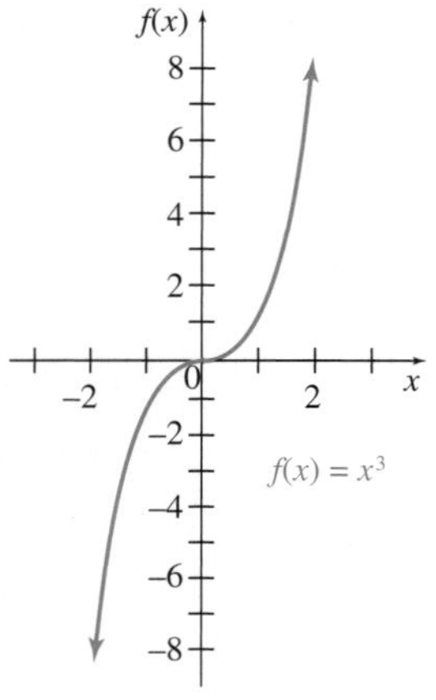

FIGURE 17

CAUTION Be very careful not to get this result backward. It does *not* say that a function has relative extrema at all critical numbers of the function. For example, Figure 17 shows the graph of $f(x) = x^3$. The derivative, $f'(x) = 3x^2$, is 0 when $x = 0$, so that 0 is a critical number for that function. However, as suggested by the graph of Figure 17, $f(x) = x^3$ has neither a relative maximum nor a relative minimum at $x = 0$ (or anywhere else, for that matter). A critical number is a candidate for the location of a relative extremum, but only a candidate. ■

First Derivative Test Suppose all critical numbers have been found for some function f. How is it possible to tell from the equation of the function whether these critical numbers produce relative maxima, relative minima, or neither? One way is suggested by the graph in Figure 18.

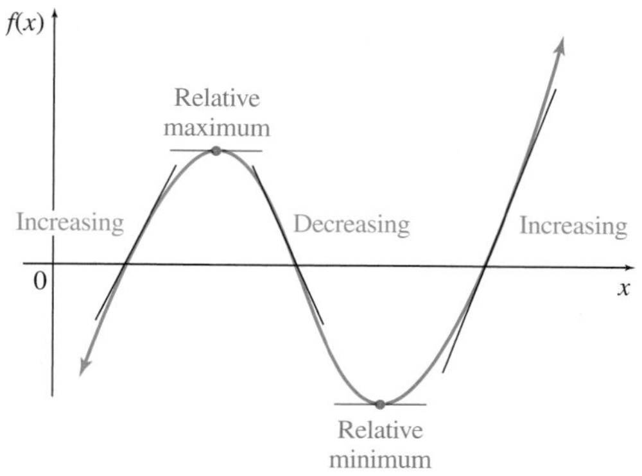

FIGURE 18

As shown in Figure 18, on the left of a relative maximum the tangent lines to the graph of a function have positive slopes, indicating that the function is increasing. At the relative maximum, the tangent line is horizontal. On the right of the relative maximum the tangent lines have negative slopes, indicating that the function is decreasing. Around a relative minimum the opposite occurs. As shown by the tangent lines in Figure 18, the function is decreasing on the left of the relative minimum, has a horizontal tangent at the minimum, and is increasing on the right of the minimum.

Putting this together with the methods from Section 1 for identifying intervals where a function is increasing or decreasing gives the following **first derivative test** for locating relative extrema.

FIRST DERIVATIVE TEST

Let c be a critical number for a function f. Suppose that f is continuous on (a, b) and differentiable on (a, b) except possibly at c, and that c is the only critical number for f in (a, b).

1. $f(c)$ is a relative maximum of f if the derivative $f'(x)$ is positive in the interval (a, c) and negative in the interval (c, b).
2. $f(c)$ is a relative minimum of f if the derivative $f'(x)$ is negative in the interval (a, c) and positive in the interval (c, b).

The sketches in the following table show how the first derivative test works. Assume the same conditions on a, b, and c for the table as those given for the first derivative test.

$f(x)$ has:	Sign of f' in (a, c)	Sign of f' in (c, b)	Sketches
Relative maximum	+	−	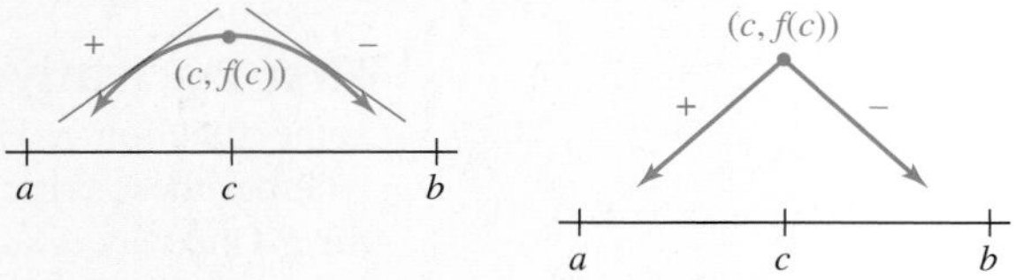
Relative minimum	−	+	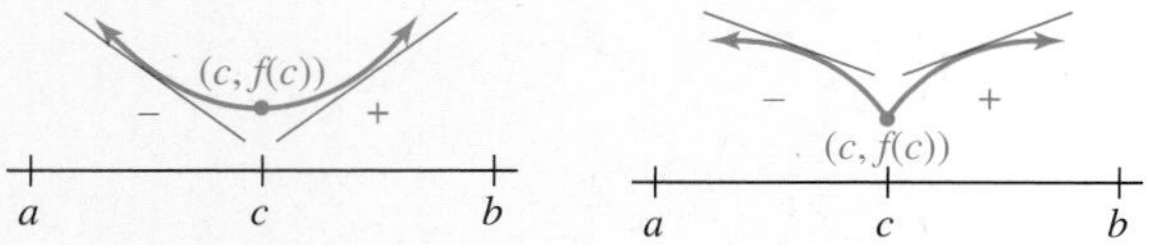
No relative extrema	+	+	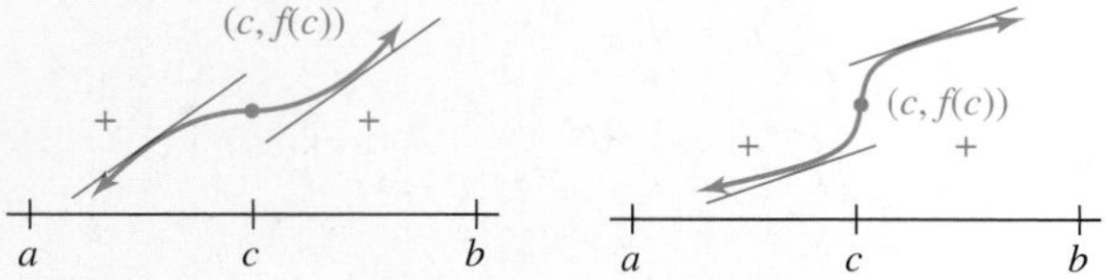
No relative extrema	−	−	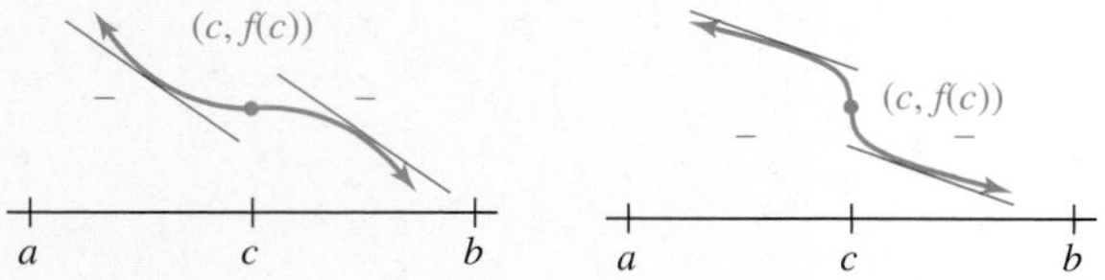

EXAMPLE 2 *Relative Extrema*

Find all relative extrema for the following functions, as well as where each function is increasing and decreasing.

(a) $f(x) = 2x^3 - 3x^2 - 72x + 15$

Method 1: First Derivative Test

Solution The derivative is $f'(x) = 6x^2 - 6x - 72$. There are no points where $f'(x)$ fails to exist, so the only critical numbers will be found where the derivative equals 0. Setting the derivative equal to 0 gives

$$6x^2 - 6x - 72 = 0$$
$$6(x^2 - x - 12) = 0$$
$$6(x - 4)(x + 3) = 0$$
$$x - 4 = 0 \quad \text{or} \quad x + 3 = 0$$
$$x = 4 \quad \text{or} \quad x = -3.$$

As in the previous section, the critical numbers 4 and -3 are used to determine the three intervals $(-\infty, -3)$, $(-3, 4)$, and $(4, \infty)$ shown on the number line in Figure 19 on the next page. Any number from each of the three intervals can be used as a test point to find the sign of f' in each interval. Using -4, 0, and 5 gives the following information.

$$f'(-4) = 6(-8)(-1) > 0$$
$$f'(0) = 6(-4)(3) < 0$$
$$f'(5) = 6(1)(8) > 0$$

Thus, the derivative is positive on $(-\infty, -3)$, negative on $(-3, 4)$, and positive on $(4, \infty)$. By Part 1 of the first derivative test, this means that the function has a

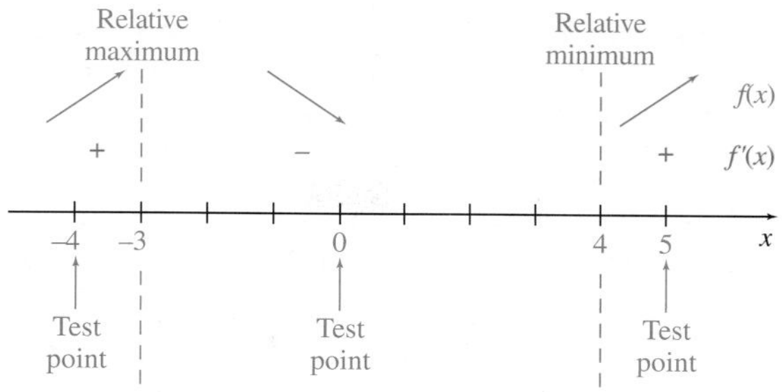

FIGURE 19

relative maximum of $f(-3) = 150$ when $x = -3$; by Part 2, f has a relative minimum of $f(4) = -193$ when $x = 4$. The function is increasing on $(-\infty, -3)$ and $(4, \infty)$ and decreasing on $(-3, 4)$. The graph is shown in Figure 20.

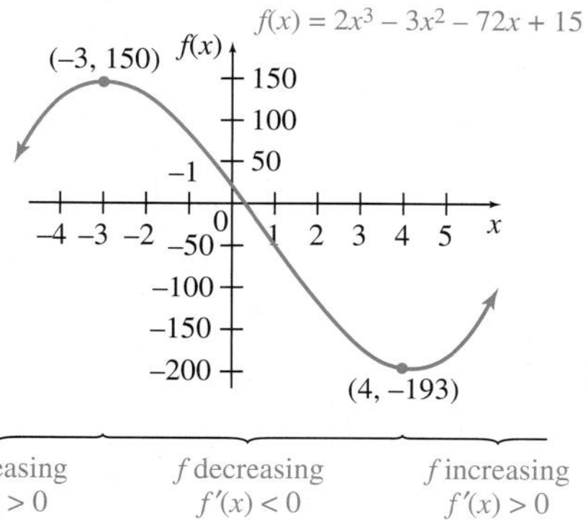

FIGURE 20

Method 2: Graphing Calculator

Many graphing calculators can locate a relative extremum when supplied with an interval containing the extremum. For example, after graphing the function $f(x) = 2x^3 - 3x^2 - 72x + 15$ on a TI-83/84 Plus, we selected "maximum" from the CALC menu and entered a left bound of -4 and a right bound of 0. The calculator asks for an initial guess, but in this example it doesn't matter what we enter. The result of this process, as well as a similar process for finding the relative minimum, is shown in Figure 21.

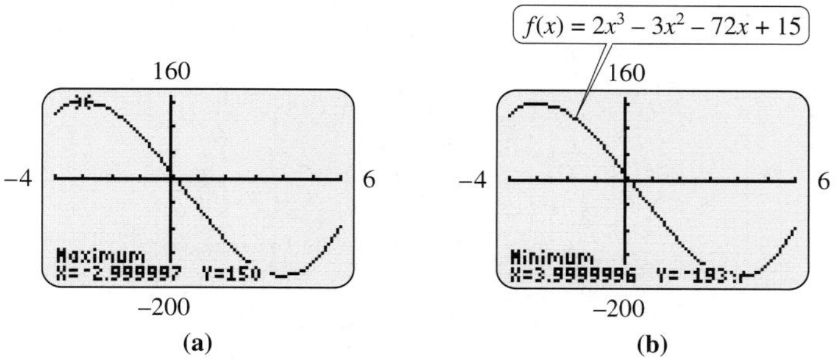

FIGURE 21

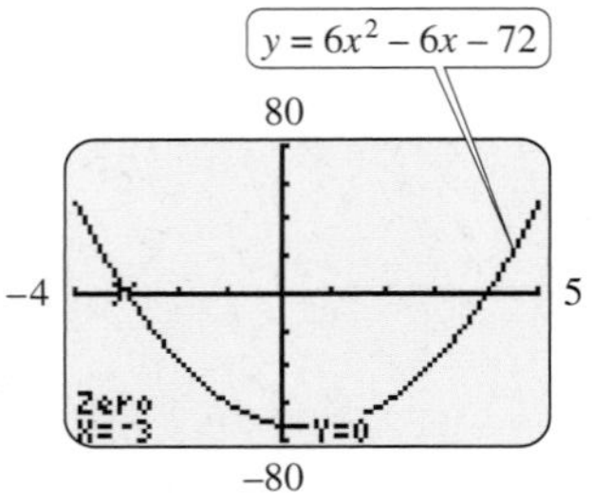

FIGURE 22

Another way to verify the extrema with a graphing calculator is to graph $y = f'(x)$ and find where the graph crosses the x-axis. Figure 22 shows the result of this approach for finding the relative minimum of the previous function.

(b) $f(x) = 6x^{2/3} - 4x$

Solution Find $f'(x)$.

$$f'(x) = 4x^{-1/3} - 4 = \frac{4}{x^{1/3}} - 4$$

The derivative fails to exist when $x = 0$, but the function itself is defined when $x = 0$, making 0 a critical number for f. To find other critical numbers, set $f'(x) = 0$.

$$f'(x) = 0$$
$$\frac{4}{x^{1/3}} - 4 = 0$$
$$\frac{4}{x^{1/3}} = 4$$
$$4 = 4x^{1/3}$$
$$1 = x^{1/3}$$
$$1 = x$$

The critical numbers 0 and 1 are used to locate the intervals $(-\infty, 0)$, $(0, 1)$, and $(1, \infty)$ on a number line as in Figure 23. Evaluating $f'(x)$ at the test points -1, $1/2$, and 2 and using the first derivative test shows that f has a relative maximum at $x = 1$; the value of this relative maximum is $f(1) = 2$. Also, f has a relative minimum at $x = 0$; this relative minimum is $f(0) = 0$. The function is increasing on $(0, 1)$ and decreasing on $(-\infty, 0)$ and $(1, \infty)$. Notice that the graph, shown in Figure 24, has a sharp point at the critical number where the derivative does not exist. In the last section of this chapter we will show how to verify other features of the graph.

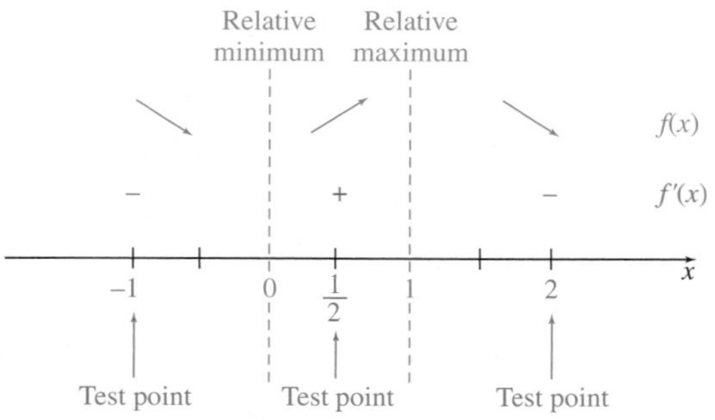

FIGURE 23

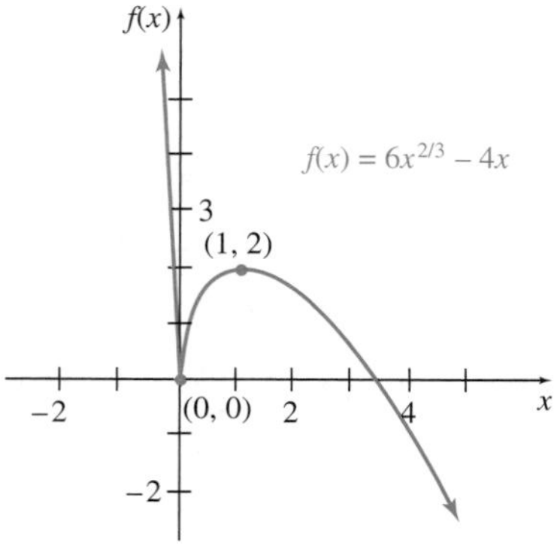

FIGURE 24

(c) $f(x) = xe^{2-x^2}$

Solution The derivative is

$$\begin{aligned} f'(x) &= x(-2x)e^{2-x^2} + e^{2-x^2} \\ &= e^{2-x^2}(-2x^2+1). \end{aligned}$$

This expression exists for all x in the domain of f. The derivative is 0 when

$$\begin{aligned} -2x^2 + 1 &= 0 \\ 1 &= 2x^2 \\ \frac{1}{2} &= x^2 \\ x &= \pm\sqrt{1/2} \\ x &= \pm\frac{1}{\sqrt{2}} \approx \pm .707. \end{aligned}$$

There are two critical points, $-1/\sqrt{2}$ and $1/\sqrt{2}$. Using test points of -1, 0, and 1 gives the results shown in Figure 25.

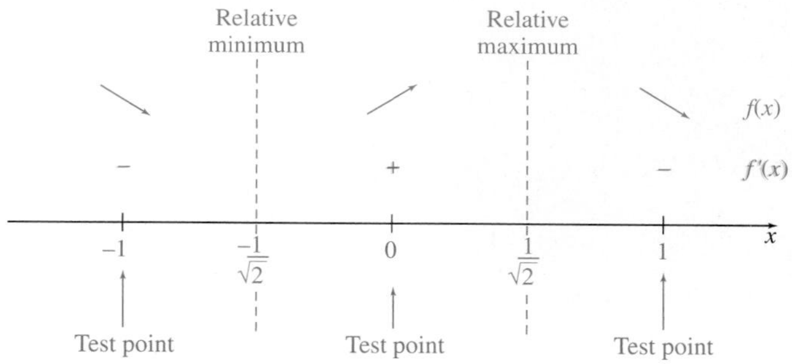

FIGURE 25

The function has a relative minimum at $-1/\sqrt{2}$ of $f(-1/\sqrt{2}) \approx -3.17$, and a relative maximum at $1/\sqrt{2}$ of $f(1/\sqrt{2}) \approx 3.17$. It is decreasing on the interval $(-\infty, -1/\sqrt{2})$, increasing on the interval $(-1/\sqrt{2}, 1/\sqrt{2})$, and decreasing on the interval $(1/\sqrt{2}, \infty)$. The graph is shown in Figure 26.

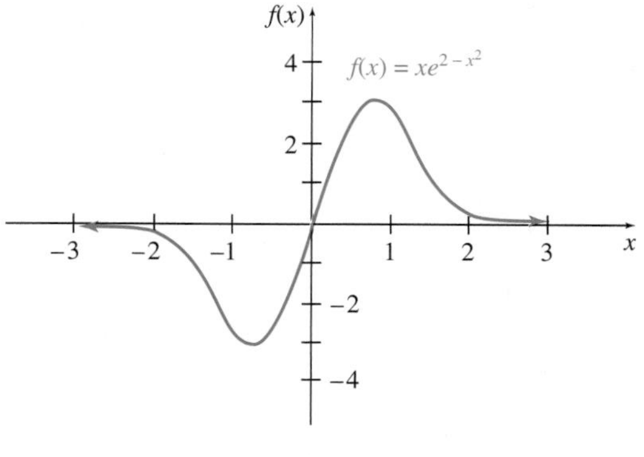

FIGURE 26

CAUTION A critical number must be in the domain of the function. For example, the derivative of $f(x) = x/(x - 4)$ is $f'(x) = -4/(x - 4)^2$, which fails to exist when $x = 4$. But $f(4)$ does not exist, so 4 is not a critical number, and the function has no relative extrema. ■

As mentioned at the beginning of this section, finding the maximum or minimum value of a quantity is important in applications of mathematics. The final example gives a further illustration.

EXAMPLE 3 *Bicycle Sales*

A small company manufactures and sells bicycles. The production manager has determined that the cost and demand functions for x $(x \geq 0)$ bicycles per week are

$$C(x) = 10 + 5x + \frac{1}{60}x^3 \qquad \text{and} \qquad p = 90 - x,$$

where p is the price per bicycle.

(a) Find the maximum weekly revenue.

Solution The revenue each week is given by

$$R(x) = xp = x(90 - x) = 90x - x^2.$$

To maximize $R(x) = 90x - x^2$, find $R'(x)$. Then find the critical numbers.

$$\begin{aligned} R'(x) = 90 - 2x &= 0 \\ 90 &= 2x \\ x &= 45 \end{aligned}$$

Since $R'(x)$ exists for all x, 45 is the only critical number. To verify that $x = 45$ will produce a *maximum,* evaluate the derivative on either side of $x = 45$.

$$R'(40) = 10 \qquad \text{and} \qquad R'(50) = -10$$

This shows that $R(x)$ is increasing up to $x = 45$, then decreasing, so there is a maximum value at $x = 45$ of $R(45) = 2025$. The maximum revenue will be \$2025 and will occur when 45 bicycles are produced and sold each week.

(b) Find the maximum weekly profit.

Solution Since profit equals revenue minus cost, the profit is given by

$$\begin{aligned} P(x) &= R(x) - C(x) \\ &= (90x - x^2) - \left(10 + 5x + \frac{1}{60}x^3\right) \\ &= -\frac{1}{60}x^3 - x^2 + 85x - 10. \end{aligned}$$

Find the derivative and set it equal to 0 to find the critical numbers. (The derivative exists for all x.)

$$P'(x) = -\frac{1}{20}x^2 - 2x + 85 = 0$$

Solving this equation by the quadratic formula gives the solutions $x \approx 25.8$ and $x \approx -65.8$. Since x cannot be negative, the only critical number of concern is 25.8. Determine whether $x = 25.8$ produces a maximum by testing a value on either side of 25.8 in $P'(x)$.

$$P'(0) = 85 \qquad \text{and} \qquad P'(40) = -75$$

These results show that $P(x)$ increases to $x = 25.8$ and then decreases. Since x must be an integer, verify that $x = 26$ produces a maximum value of $P(26) = 1231.07$. (We should also check $P(25) = 1229.58$.) Thus, the maximum profit of \$1231.07 occurs when 26 bicycles are produced and sold each week. Notice that this is not the same as the number that should be produced to yield maximum revenue.

(c) Find the price the company should charge to realize maximum profit.

Solution As shown in part (b), 26 bicycles per week should be produced and sold to get the maximum profit of \$1231.07 per week. Since the price is given by

$$p = 90 - x,$$

if $x = 26$, then $p = 64$. The manager should charge \$64 per bicycle and produce and sell 26 bicycles per week to get the maximum profit of \$1231.07 per week. Figure 27 shows the graphs of the functions used in this example. Notice that the slopes of the revenue and cost functions are the same at the point where the maximum profit occurs. Why is this true?

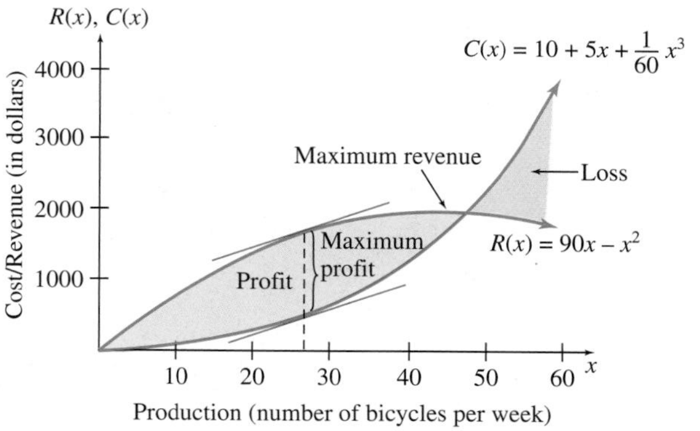

FIGURE 27

CAUTION Be careful to give the y-value of the point where an extremum occurs. Although we solve the equation $f'(x) = 0$ for x to find the extremum, the maximum or minimum value of the function is the corresponding y-value. Thus, in Example 3(a), we found that at $x = 45$, the maximum weekly revenue is \$2025 (not \$45). ■

The examples in this section involving the maximization of a quadratic function, such as the opening example and the bicycle revenue example, could be solved by the methods described in the chapter on Nonlinear Functions. But those involving more complicated functions, such as the bicycle profit example, are difficult to analyze without the tools of calculus.

Finding extrema for realistic problems requires an accurate mathematical model of the problem. In particular, it is important to be aware of restrictions on the values of the variables. For example, if $T(x)$ closely approximates the number of items that can be manufactured daily on a production line when x is the number of employees on the line, x must certainly be restricted to the positive integers, or perhaps to a few common fractional values. (We can imagine half-time workers, but not $1/49$-time workers.)

On the other hand, to apply the tools of calculus to obtain an extremum for some function, the function must be defined and be meaningful at every real number in some interval. Because of this, the answer obtained from a mathematical model might be a number that is not feasible in the actual problem.

Usually, the requirement that a continuous function be used, rather than one that can take on only certain selected values, is of theoretical interest only. In most cases, the methods of calculus give acceptable results as long as the assumptions of continuity and differentiability are not totally unreasonable. If they lead to the conclusion, say, that $80\sqrt{2}$ workers should be hired, it is usually only necessary to investigate acceptable values close to $80\sqrt{2}$. This was done in Example 3.

5.2 EXERCISES

Find the locations and values of all relative extrema for the functions with graphs as follows. Compare with Exercises 1–8 in the preceding section.

1.

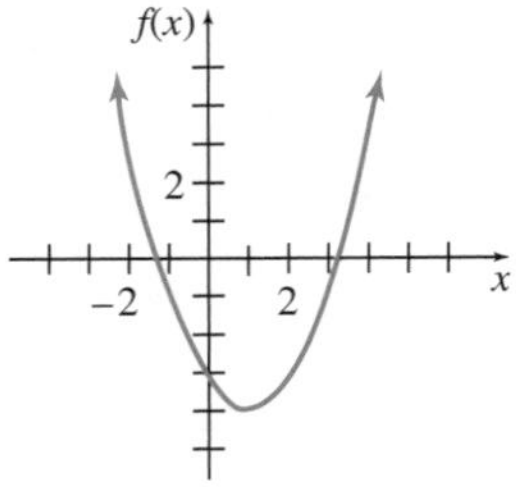

2.

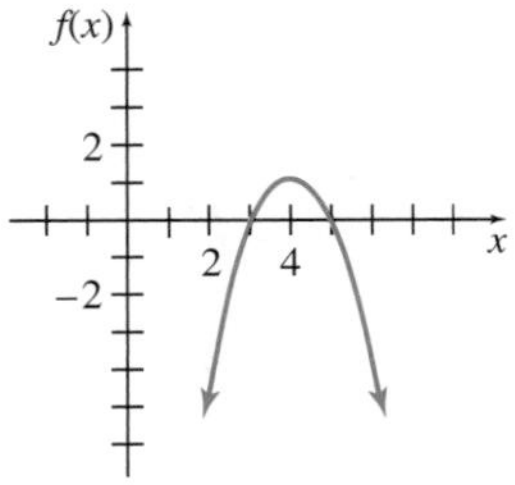

3.

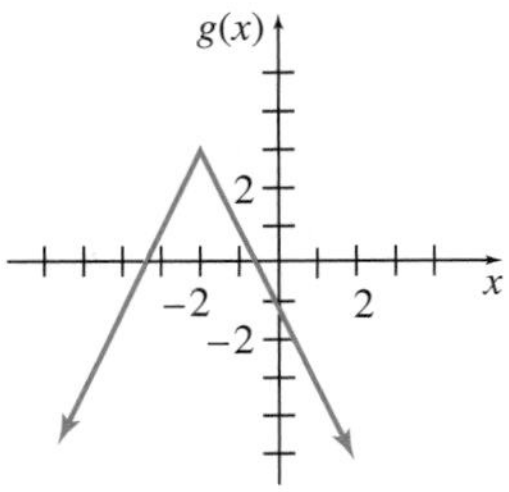

4.

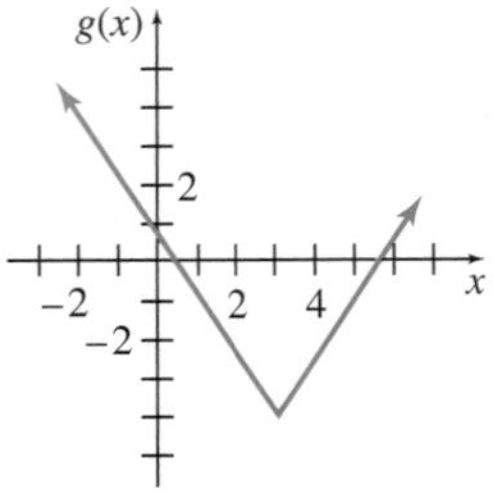

5.

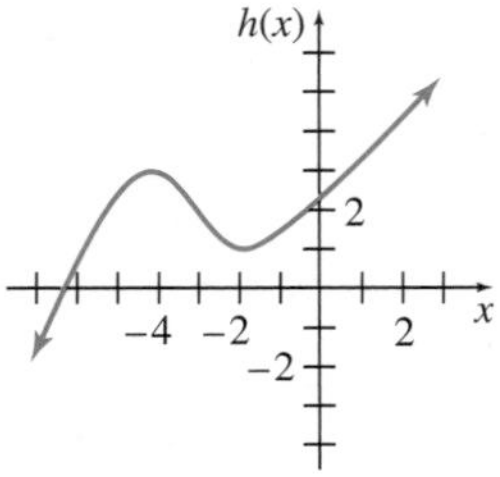

6.

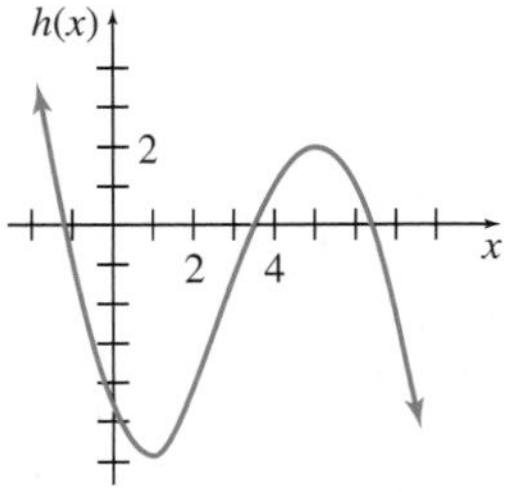

7.

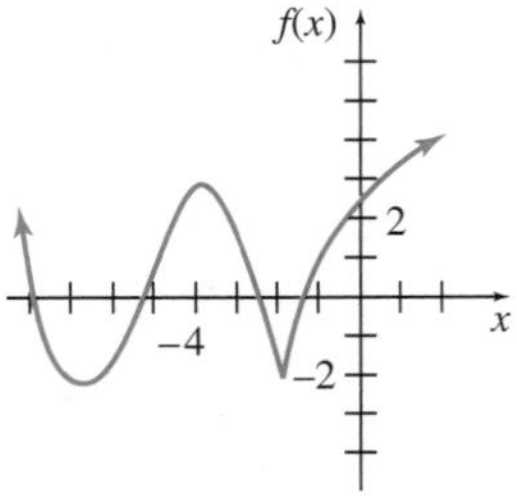

8.

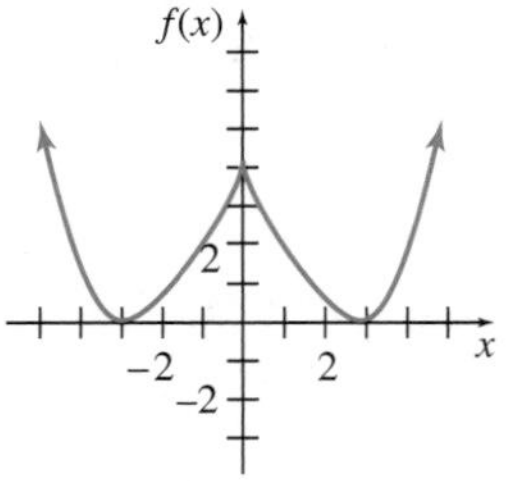

Find the x-value of all points where the functions defined as follows have any relative extrema. Find the value(s) of any relative extrema.

9. $f(x) = x^2 + 12x - 8$

10. $f(x) = x^2 - 4x + 6$

11. $f(x) = x^3 + 6x^2 + 9x - 8$

12. $f(x) = x^3 + 3x^2 - 24x + 2$

13. $f(x) = -\frac{4}{3}x^3 - \frac{21}{2}x^2 - 5x + 8$

14. $f(x) = -\frac{2}{3}x^3 - \frac{1}{2}x^2 + 3x - 4$

15. $f(x) = x^4 - 18x^2 - 4$

16. $f(x) = x^4 - 8x^2 + 9$

17. $f(x) = -(8 - 5x)^{2/3}$

18. $f(x) = (2 - 9x)^{2/3}$

19. $f(x) = 2x + 3x^{2/3}$

20. $f(x) = 3x^{5/3} - 15x^{2/3}$

21. $f(x) = x - \frac{1}{x}$

22. $f(x) = x^2 + \frac{1}{x}$

23. $f(x) = \frac{x^2 - 2x + 1}{x - 3}$

24. $f(x) = \frac{x^2 - 6x + 9}{x + 2}$

25. $f(x) = x^2e^x - 3$

26. $f(x) = 3xe^x + 2$

27. $f(x) = 2x + \ln x$

28. $f(x) = \frac{x^2}{\ln x}$

Use the derivative to find the vertex of each parabola.

29. $y = -2x^2 + 8x - 1$

30. $y = ax^2 + bx + c$

Graph each function on a graphing calculator, and then use the graph to find all relative extrema (to three decimal places). Then confirm your answer by finding the derivative and using the calculator to solve the equation $f'(x) = 0$.

31. $f(x) = x^5 - x^4 + 4x^3 - 30x^2 + 5x + 6$

32. $f(x) = -x^5 - x^4 + 2x^3 - 25x^2 + 9x + 12$

33. Graph $f(x) = 2|x + 1| + 4|x - 5| - 20$ with a graphing calculator in the window $[-10, 10]$ by $[-15, 30]$. Use the graph and the function to determine the x-values of all extrema.

34. Consider the function*

$$g(x) = \frac{1}{x^{12}} - 2\left(\frac{1000}{x}\right)^6.$$

a. Using a graphing calculator, try to find any local minima, or tell why finding a local minimum is difficult for this function.

b. Find any local minima using the techniques of calculus.

c. Based on your results in parts a and b, describe circumstances under which relative extrema are easier to find using the techniques of calculus than using a graphing calculator.

Applications

BUSINESS AND ECONOMICS

Profit *In Exercises 35–37, find* **(a)** *the number, x, of units that produces maximum profit;* **(b)** *the price, p, per unit that produces maximum profit; and* **(c)** *the maximum profit, P.*

35. $C(x) = 75 + 10x;\quad p = 70 - 2x$

36. $C(x) = 25x + 5000;\quad p = 80 - .01x$

37. $C(x) = 100 + 20xe^{-.01x};\quad p = 40e^{-.01x}$

*Dubinsky, Ed, "Is Calculus Obsolete?" *Mathematics Teacher,* Vol. 88, No. 2, Feb. 1995, pp. 146–148.

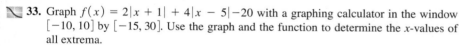

38. *Profit* The total profit $P(x)$ (in thousands of dollars) from the sale of x units of a certain prescription drug is given by

$$P(x) = \ln(-x^3 + 3x^2 + 72x + 1)$$

for x in $[0, 10]$.

a. Find the number of units that should be sold in order to maximize the total profit.

b. What is the maximum profit?

39. *Revenue* The demand equation for telephones at one store is

$$p = D(q) = 200e^{-.1q},$$

where p is the price (in dollars) and q is the quantity of telephones sold per week. Find the values of q and p that maximize revenue.

40. *Revenue* The demand equation for one type of computer networking system is

$$p = D(q) = 500qe^{-.0016q^2},$$

where p is the price (in dollars) and q is the quantity of servers sold per month. Find the values of q and p that maximize revenue.

41. *Cost* Suppose that the cost function for a product is given by $C(x) = .002x^3 - 9x + 4000$. Find the production level (i.e., value of x) that will produce the minimum average cost per unit $\overline{C}(x)$.

LIFE SCIENCES

42. *Activity Level* In the summer the activity level of a certain type of lizard varies according to the time of day. A biologist has determined that the activity level is given by the function

$$a(t) = .008t^3 - .288t^2 + 2.304t + 7,$$

where t is the number of hours after 12 noon. When is the activity level highest? When is it lowest?

43. *Milk Consumption* The average individual daily milk consumption for herds of Charolais, Angus, and Hereford calves can be described by the function

$$M(t) = 6.281t^{.242}e^{-.025t}, \qquad 1 \le t \le 26,$$

where $M(t)$ is the milk consumption (in kilograms) and t is the age of the calf (in weeks).*

a. Find the time in which the maximum daily consumption occurs and the maximum daily consumption.

b. If the general formula for this model is given by

$$M(t) = at^b e^{-ct},$$

find the time where the maximum consumption occurs and the maximum consumption. (*Hint:* Express your answer in terms of a, b, and c.)

44. *Alaskan Moose* The mathematical relationship between the age of a captive female moose and its mass can be described by the function

$$M(t) = 369(.93)^t t^{.36}, \quad t \le 12,$$

where $M(t)$ is the mass of the moose (in kilograms) and t is the age (in years) of the moose.† Find the age at which the mass of a female moose is maximized. What is the maximum mass?

45. *Thermic Effect of Food* As we saw in the last section, the metabolic rate after a person eats a meal tends to go up and then, after some time has passed, returns to a resting metabolic rate. This phenomenon is known as the thermic effect of food and can be described for a particular individual as

$$F(t) = -10.28 + 175.9te^{-t/1.3},$$

where $F(t)$ is the thermic effect of food (in kJ/hr), and t is the number of hours that have elapsed since eating a meal.‡ Find the time after the meal when the thermic effect of the food is maximized.

SOCIAL SCIENCES

46. *Training Accidents* Many American troops die each year as a result of training accidents. As a result, the military has been redoubling its efforts to increase safety standards and prevent violence in the ranks. The number of casualties (in hundreds) each year is approximated by

$$f(x) = .198x^3 - 1.56x^2 + 2.03x + 14.8,$$

where x is the number of years since 1990.§ The function is accurate only on the finite domain $\{0, 1, 2, 3, 4, 5\}$. Find the location and value of any extrema for this function. (*Hint:* Refer to the last two paragraphs in the section.) Interpret your results.

*Mezzadra, C., R. Paciaroni, S. Vulich, E. Villarreal, and L. Melucci, "Estimation of Milk Consumption Curve Parameters for Different Genetic Groups of Bovine Calves," *Animal Production,* Vol. 49, 1989, pp. 83–87.

†Schwartz, C. and Kris Hundertmark, "Reproductive Characteristics of Alaskan Moose," *Journal of Wildlife Management,* Vol. 57, No. 3, July 1993, pp. 454–468.

‡Reed, George and James Hill, "Measuring the Thermic Effects of Food," *American Journal of Clinical Nutrition,* Vol. 63, 1996, pp. 164–169.

§*The New York Times,* Dec. 25, 1995, p. 8.

47. *Attitude Change* Social psychologists have found that as the discrepancy between the views of a speaker and those of an audience increases, the attitude change in the audience also increases to a point, but decreases when the discrepancy becomes too large, particularly if the communicator is viewed by the audience as having low credibility.* Suppose that the degree of change can be approximated by the function

$$D(x) = -x^4 + 8x^3 + 80x^2,$$

where x is the discrepancy between the views of the speaker and those of the audience, as measured by scores on a questionnaire. Find the amount of discrepancy the speaker should aim for to maximize the attitude change in the audience.

48. *Film Length* A group of researchers found that people prefer training films of moderate length; shorter films contain too little information, while longer films are boring. For a training film on the care of exotic birds, the researchers determined that the ratings people gave for the film could be approximated by

$$R(t) = \frac{20t}{t^2 + 100},$$

where t is the length of the film (in minutes). Find the film length that received the highest rating.

PHYSICAL SCIENCES

49. *Height* After a great deal of experimentation, two Atlantic Institute of Technology senior physics majors determined that when a bottle of French champagne is shaken several times, held upright, and uncorked, its cork travels according to

$$s(t) = -16t^2 + 64t + 3,$$

where s is its height (in feet) above the ground t seconds after being released.

a. How high will it go?

b. How long is it in the air?

5.3 HIGHER DERIVATIVES, CONCAVITY, AND THE SECOND DERIVATIVE TEST

THINK ABOUT IT

Just because the price of a stock is increasing, does that alone make it a good investment?

The following discussion addresses this question.

To understand the behavior of a function on an interval, it is important to know the *rate* at which the function is increasing or decreasing. For example, suppose that your friend, a finance major, has studied a young company and is trying to get you to invest in its stock. He shows you the following function, which represents the price $P(t)$ of the company's stock since it became available in January two years ago:

$$P(t) = 17 + t^{1/2},$$

*Eagly, A. H. and K. Telaak, "Width of the Latitude of Acceptance as a Determinant of Attitude Change," *Journal of Personality and Social Psychology*, Vol. 23, 1972, pp. 388–397.

where t is the number of months since the stock became available. He points out that the derivative of the function is always positive, so the price of the stock is always increasing. He claims that you cannot help but make a fortune on it. Should you take his advice and invest?

It is true that the price function increases for all t. The derivative is

$$P'(t) = \frac{1}{2}t^{-1/2} = \frac{1}{2\sqrt{t}},$$

which is always positive because $\sqrt{t}$ is positive for $t > 0$. The catch lies in *how fast* the function is increasing. The derivative $P'(t) = 1/(2\sqrt{t})$ tells how fast the price is increasing at any number of months, t, since the stock became available. For example, when $t = 1$, $P'(t) = 1/2$, and the price is increasing at the rate of $1/2$ dollar, or 50 cents, per month. When $t = 4$, $P'(t) = 1/4$; the stock is increasing at 25 cents per month. At $t = 9$ months, $P'(t) = 1/6$, or about 17 cents per month. By the time you could buy in at $t = 24$ months, the price is increasing at 10 cents per month, and the *rate of increase* looks as though it will continue to decrease.

The rate of increase in P' is given by the derivative of $P'(t)$, called the **second derivative** of P and denoted by $P''(t)$. Since $P'(t) = (1/2)t^{-1/2}$,

$$P''(t) = -\frac{1}{4}t^{-3/2} = -\frac{1}{4\sqrt{t^3}}.$$

$P''(t)$ is negative for all positive values of t and therefore confirms the suspicion that the *rate* of increase in price does indeed decrease for all $t \geq 0$. The price of the company's stock will not drop, but the amount of return will certainly not be the fortune that your friend predicts. For example, at $t = 24$ months, when you would buy, the price would be \$21.90. A year later, it would be \$23.00 a share. If you were rich enough to buy 100 shares for \$2190, they would be worth \$2300 in a year. The increase of \$110 is about 5% of the investment—similar to the return that you could get in many types of savings accounts. The only investors to make a lot of money on this stock would be those who bought early, when the rate of increase was much greater.

As mentioned earlier, the second derivative of a function f, written f'', gives the rate of change of the *derivative* of f. Before continuing to discuss applications of the second derivative, we need to introduce some additional terminology and notation.

Higher Derivatives If a function f has a derivative f', then the derivative of f', if it exists, is the second derivative of f, written f''. The derivative of f'', if it exists, is called the **third derivative** of f, and so on. By continuing this process, we can find **fourth derivatives** and other higher derivatives. For example, if $f(x) = x^4 + 2x^3 + 3x^2 - 5x + 7$, then

$$f'(x) = 4x^3 + 6x^2 + 6x - 5, \quad \text{First derivative of } f$$

$$f''(x) = 12x^2 + 12x + 6, \quad \text{Second derivative of } f$$

$$f'''(x) = 24x + 12, \quad \text{Third derivative of } f$$

and

$$f^{(4)}(x) = 24. \quad \text{Fourth derivative of } f$$

NOTATION FOR HIGHER DERIVATIVES

The second derivative of $y = f(x)$ can be written using any of the following notations:

$$f''(x), \quad \frac{d^2y}{dx^2}, \quad \text{or} \quad D_x^2[f(x)].$$

The third derivative can be written in a similar way. For $n \geq 4$, the nth derivative is written $f^{(n)}(x)$.

CAUTION Notice the difference in notation between $f^{(4)}$, which indicates the fourth derivative of f, and f^4, which indicates f raised to the fourth power. ■

EXAMPLE 1 *Second Derivative*

Let $f(x) = x^3 + 6x^2 - 9x + 8$.

(a) Find $f''(x)$.

Solution To find the second derivative of $f(x)$, find the first derivative, and then take its derivative.

$$f'(x) = 3x^2 + 12x - 9$$
$$f''(x) = 6x + 12$$

(b) Find $f''(0)$.

Solution Since $f''(x) = 6x + 12$,

$$f''(0) = 6(0) + 12 = 12.$$

EXAMPLE 2 *Second Derivative*

Find the second derivative for the functions defined as follows.

(a) $f(x) = (x^2 - 1)^2$

Solution Here

$$f'(x) = 2(x^2 - 1)(2x) = 4x(x^2 - 1).$$

Use the product rule to find $f''(x)$.

$$\begin{aligned} f''(x) &= 4x(2x) + (x^2 - 1)(4) \\ &= 8x^2 + 4x^2 - 4 \\ &= 12x^2 - 4 \end{aligned}$$

(b) $g(x) = 4x(\ln x)$

Solution Use the product rule.

$$g'(x) = 4x \cdot \frac{1}{x} + (\ln x) \cdot 4 = 4 + 4(\ln x)$$

$$g''(x) = 0 + 4 \cdot \frac{1}{x} = \frac{4}{x}$$

(c) $h(x) = \dfrac{x}{e^x}$

Solution Here, we need the quotient rule.

$$h'(x) = \frac{e^x - xe^x}{(e^x)^2} = \frac{e^x(1 - x)}{(e^x)^2} = \frac{1 - x}{e^x}$$

$$h''(x) = \frac{e^x(-1) - (1 - x)e^x}{(e^x)^2} = \frac{e^x(-1 - 1 + x)}{(e^x)^2} = \frac{-2 + x}{e^x}$$

Earlier, we saw that the first derivative of a function represents the rate of change of the function. The second derivative, then, represents the rate of change of the first derivative. If a function describes the position of a vehicle (along a straight line) at time t, then the first derivative gives the velocity of the vehicle. That is, if $y = s(t)$ describes the position (along a straight line) of the vehicle at time t, then $v(t) = s'(t)$ gives the velocity at time t.

We also saw that *velocity* is the rate of change of distance with respect to time. Recall, the difference between velocity and speed is that velocity may be positive or negative, whereas speed is always positive. A negative velocity indicates travel in a negative direction (backing up) with regard to the starting point; positive velocity indicates travel in the positive direction (going forward) from the starting point.

The instantaneous rate of change of velocity is called **acceleration.** Since instantaneous rate of change is the same as the derivative, acceleration is the derivative of velocity. Thus if $a(t)$ represents the acceleration at time t, then

$$a(t) = \frac{d}{dt}v(t) = s''(t).$$

If the velocity is positive and the acceleration is positive, the velocity is increasing, so the vehicle is speeding up. If the velocity is positive and the acceleration is negative, the vehicle is slowing down. A negative velocity and a positive acceleration mean the vehicle is backing up and slowing down. If both the velocity and acceleration are negative, the vehicle is speeding up in the negative direction.

EXAMPLE 3 *Velocity and Acceleration*

Suppose a car is moving in a straight line, with its position from a starting point (in feet) at time t (in seconds) given by

$$s(t) = t^3 - 2t^2 - 7t + 9.$$

Find the following.

(a) The velocity at any time t

Solution The velocity is given by

$$v(t) = s'(t) = 3t^2 - 4t - 7$$

feet per second.

(b) The acceleration at any time t

Solution Acceleration is given by

$$a(t) = v'(t) = s''(t) = 6t - 4$$

feet per second per second.

(c) The time intervals (for $t \geq 0$) when the car is going forward or backing up

Solution We first find when the velocity is 0, that is, when the car is stopped.

$$v(t) = 3t^2 - 4t - 7 = 0$$
$$(3t - 7)(t + 1) = 0$$
$$t = 7/3 \quad \text{or} \quad t = -1$$

We are interested in $t \geq 0$. Choose a value of t in each of the intervals $(0, 7/3)$ and $(7/3, \infty)$ to see that the velocity is negative in $(0, 7/3)$ and positive in $(7/3, \infty)$. The car is backing up for the first 7/3 seconds, then going forward.

(d) The time intervals (for $t \geq 0$) when the car is speeding up or slowing down

Solution The car will speed up when the velocity and acceleration are the same sign and slow down when they have opposite signs. Here, the acceleration is positive when $6t - 4 > 0$, that is, $t > 2/3$ seconds, and negative for $t < 2/3$ seconds. Since the velocity is negative in $(0, 7/3)$ and positive in $(7/3, \infty)$, the car is speeding up for $0 \leq t \leq 2/3$ seconds, slowing down for $2/3 \leq t \leq 7/3$ seconds, and speeding up again for $t > 7/3$ seconds. See the sign graphs.

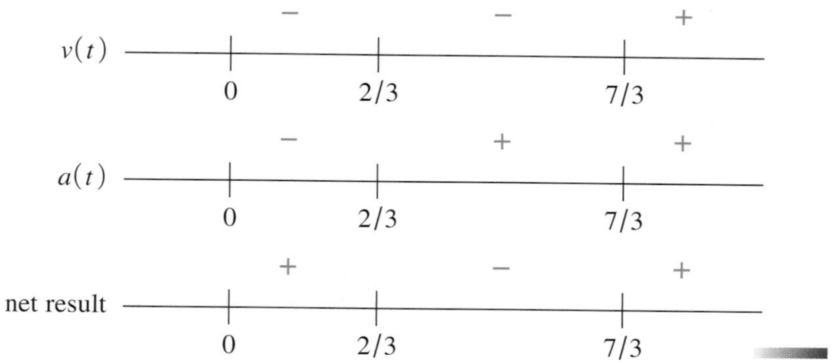

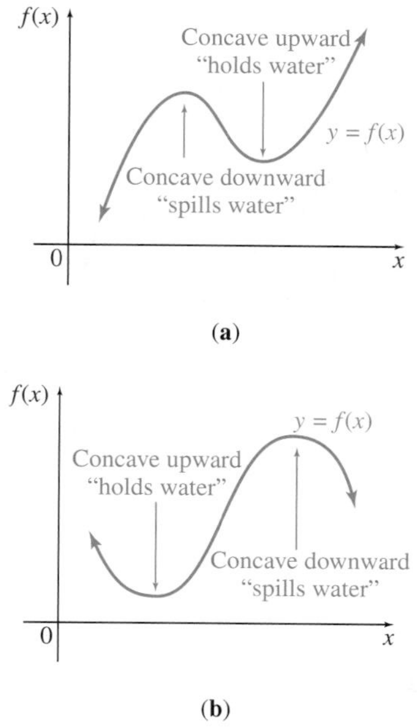

FIGURE 28

Concavity of a Graph The first derivative has been used to show where a function is increasing or decreasing and where the extrema occur. The second derivative gives the rate of change of the first derivative; it indicates *how fast* the function is increasing or decreasing. The rate of change of the derivative (the second derivative) affects the *shape* of the graph. Intuitively, we say that a graph is *concave upward* on an interval if it "holds water" and *concave downward* if it "spills water." See Figure 28.

More precisely, a function is **concave upward** on an interval (a, b) if the graph of the function lies above its tangent line at each point of (a, b). A function is **concave downward** on (a, b) if the graph of the function lies below its tangent line at each point of (a, b). A point where a graph changes **concavity** is called a **point of inflection.** See Figure 29 on the next page.

Users of soft contact lenses recognize concavity as the way to tell if a lens is inside out. As Figure 30 on the next page shows, a correct contact lens has a profile that is entirely concave upward. The profile of an inside-out lens has points of inflection near the edges, where the profile begins to turn concave downward very slightly.

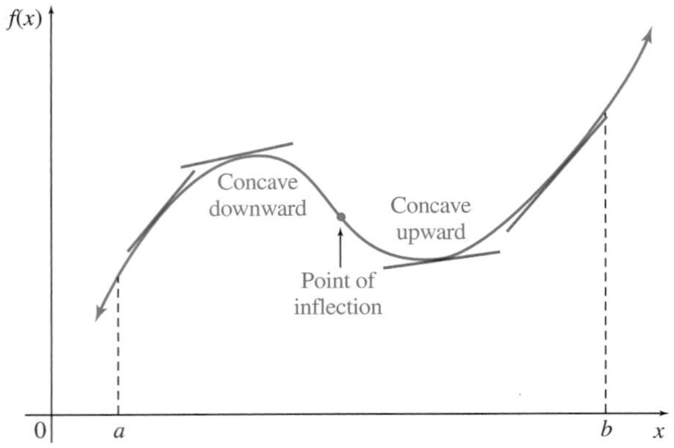

FIGURE 29

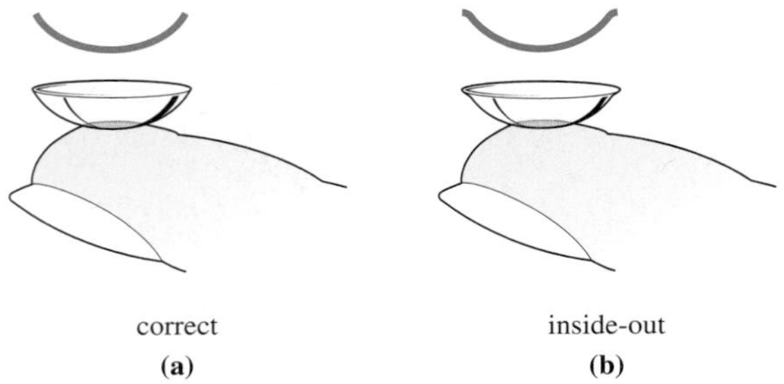

FIGURE 30

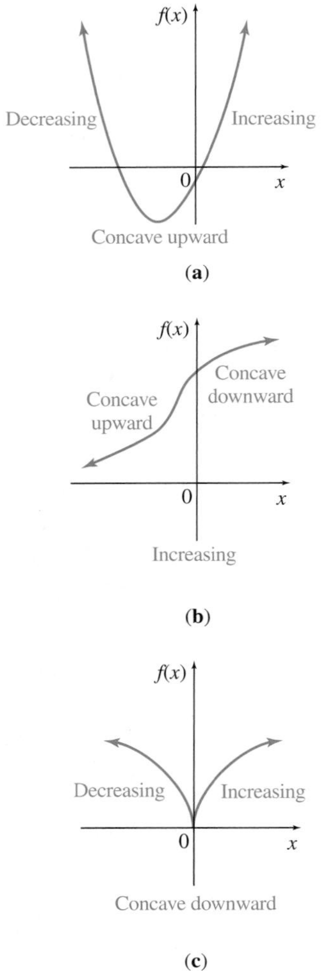

FIGURE 31

Just as a function can be either increasing or decreasing on an interval, it can be either concave upward or concave downward on an interval. Examples of various combinations are shown in Figure 31.

Figure 32 shows two functions that are concave upward on an interval (a, b) Several tangent lines are also shown. In Figure 32(a), the slopes of the tangent lines (moving from left to right) are first negative, then 0, and then positive. In Figure 32(b), the slopes are all positive, but they get larger.

In both cases, the slopes are *increasing*. The slope at a point on a curve is given by the derivative. Since a function is increasing if its derivative is positive, its slope is increasing if the derivative of the slope function is positive. Since the

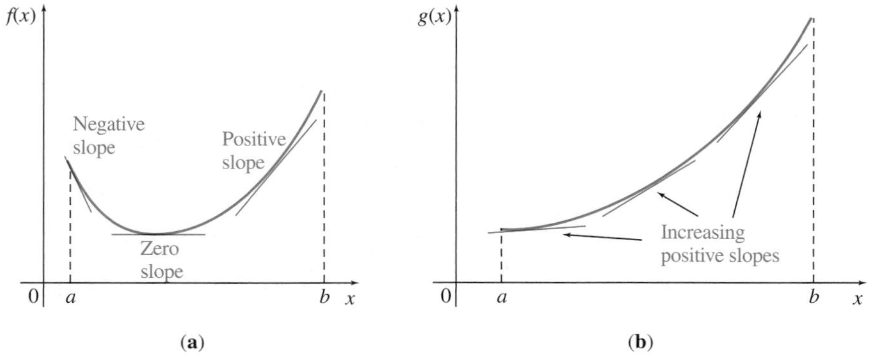

FIGURE 32

derivative of a derivative is the second derivative, a function is concave upward on an interval if its second derivative is positive at each point of the interval.

A similar result is suggested by Figure 33 for functions whose graphs are concave downward. In both graphs, the slopes of the tangent lines are *decreasing* as we move from left to right. Since a function is decreasing if its derivative is negative, a function is concave downward on an interval if its second derivative is negative at each point of the interval. These observations suggest the following test.

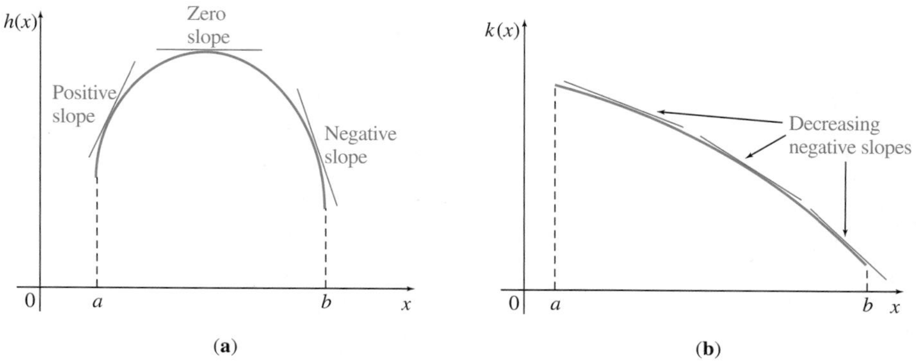

FIGURE 33

> **TEST FOR CONCAVITY**
>
> Let f be a function with derivatives f' and f'' existing at all points in an interval (a, b). Then f is concave upward on (a, b) if $f''(x) > 0$ for all x in (a, b), and concave downward on (a, b) if $f''(x) < 0$ for all x in (a, b).

An easy way to remember this test is by the faces shown in Figure 34. When the second derivative is positive at a point $(+\ +)$, the graph is concave upward $(\smile)$. When the second derivative is negative at a point $(-\ -)$, the graph is concave downward $(\frown)$.

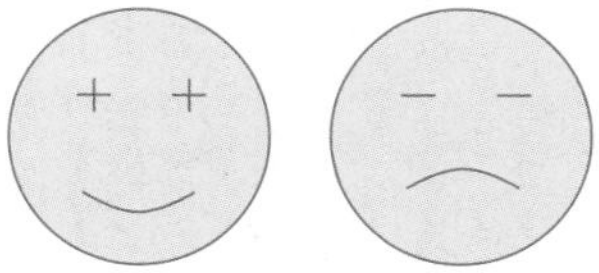

FIGURE 34

EXAMPLE 4 *Concavity*

Find all intervals where $f(x) = x^4 - 8x^3 + 18x^2$ is concave upward or downward, and find all inflection points.

Solution The first derivative is $f'(x) = 4x^3 - 24x^2 + 36x$, and the second derivative is $f''(x) = 12x^2 - 48x + 36$. We factor $f''(x)$ as $12(x - 1)(x - 3)$, and then create a number line for $f''(x)$ as we did in the previous two sections for $f'(x)$.

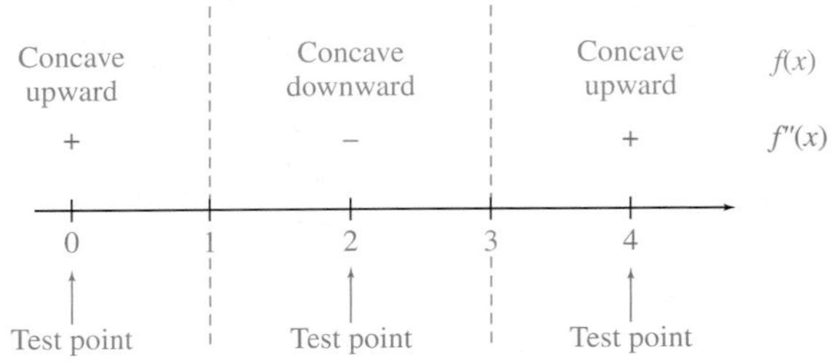

FIGURE 35

We see from Figure 35 that $f''(x) > 0$ on the intervals $(-\infty, 1)$ and $(3, \infty)$, so f is concave upward on these intervals. Also, $f''(x) < 0$ on the interval $(1, 3)$, so f is concave downward on this interval.

Finally, we have inflection points where f'' changes sign, namely, at $x = 1$ and $x = 3$. Since $f(1) = 11$ and $f(3) = 27$, the inflection points are $(1, 11)$ and $(3, 27)$. The function is graphed in Figure 36.

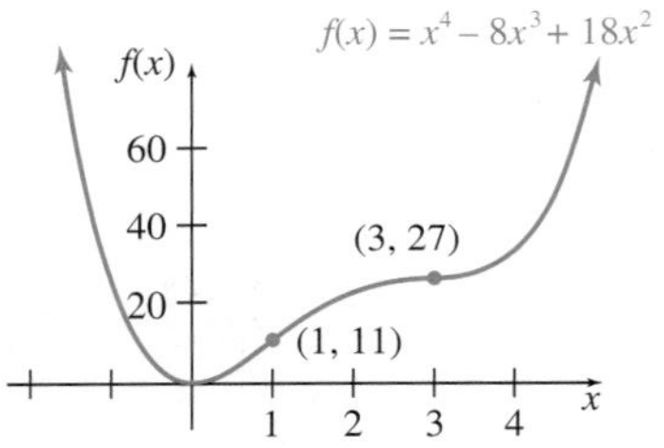

FIGURE 36

Example 4 suggests the following result.

> At a point of inflection for a function f, the second derivative is 0 or does not exist.

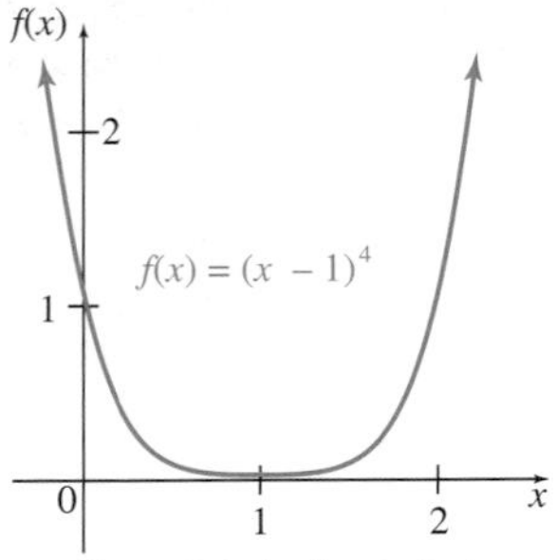

Second derivative is 0 at $x = 1$, but $(1, f(1))$ is not a point of inflection.

FIGURE 37

CAUTION Be careful with the previous statement. Finding a value of x where $f''(x) = 0$ does not mean that a point of inflection has been located. For example, if $f(x) = (x - 1)^4$, then $f''(x) = 12(x - 1)^2$, which is 0 at $x = 1$. The graph of $f(x) = (x - 1)^4$ is always concave upward, however, so it has no point of inflection. See Figure 37. ■

Second Derivative Test The idea of concavity can often be used to decide whether a given critical number produces a relative maximum or a relative minimum. This test, an alternative to the first derivative test, is based on the fact that a curve with a horizontal tangent at a point c and concave downward on an open interval containing c also has a relative maximum at c. A relative minimum occurs when a graph has a horizontal tangent at a point d and is concave upward on an open interval containing d. See Figure 38 on the next page.

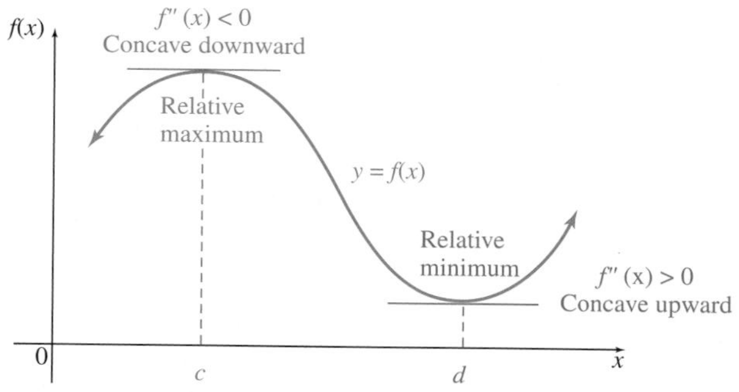

FIGURE 38

A function f is concave upward on an interval if $f''(x) > 0$ for all x in the interval, while f is concave downward on an interval if $f''(x) < 0$ for all x in the interval. These ideas lead to the **second derivative test** for relative extrema.

SECOND DERIVATIVE TEST

Let f'' exist on some open interval containing c, and let $f'(c) = 0$.

1. If $f''(c) > 0$, then $f(c)$ is a relative minimum.
2. If $f''(c) < 0$, then $f(c)$ is a relative maximum.
3. If $f''(c) = 0$, then the test gives no information about extrema.

EXAMPLE 5 *Second Derivative Test*

Find all relative extrema for

$$f(x) = 4x^3 + 7x^2 - 10x + 8.$$

Solution First, find the points where the derivative is 0. Here $f'(x) = 12x^2 + 14x - 10$. Solve the equation $f'(x) = 0$ to get

$$\begin{aligned} 12x^2 + 14x - 10 &= 0 \\ 2(6x^2 + 7x - 5) &= 0 \\ 2(3x + 5)(2x - 1) &= 0 \end{aligned}$$

$$\begin{aligned} 3x + 5 &= 0 &\quad \text{or} \quad 2x - 1 &= 0 \\ 3x &= -5 & 2x &= 1 \\ x &= -\frac{5}{3} & x &= \frac{1}{2}. \end{aligned}$$

Now use the second derivative test. The second derivative is $f''(x) = 24x + 14$. Evaluate $f''(x)$ first at $-5/3$, getting

$$f''\left(-\frac{5}{3}\right) = 24\left(-\frac{5}{3}\right) + 14 = -40 + 14 = -26 < 0,$$

so that by Part 2 of the second derivative test, $-5/3$ leads to a relative maximum of $f(-5/3) = 691/27$. Also, when $x = 1/2$,

$$f''\left(\frac{1}{2}\right) = 24\left(\frac{1}{2}\right) + 14 = 12 + 14 = 26 > 0,$$

with $1/2$ leading to a relative minimum of $f(1/2) = 21/4$.

The second derivative test works only for those critical numbers c that make $f'(c) = 0$. This test does not work for critical numbers c for which $f'(c)$ does not exist (since $f''(c)$ would not exist either). Also, the second derivative test does not work for critical numbers c that make $f''(c) = 0$. In both of these cases, use the first derivative test.

EXAMPLE 6 *Catfish Farming*

The graph in Figure 39 shows the population of catfish in a commercial catfish farm as a function of time. As the graph shows, the population increases rapidly up to a point and then increases at a slower rate. We saw in the section on Derivatives of Exponential Functions that the logistic function produces such a graph. The horizontal dashed line shows that the population will approach some upper limit determined by the capacity of the farm. The point at which the rate of population growth starts to slow is the point of inflection for the graph.

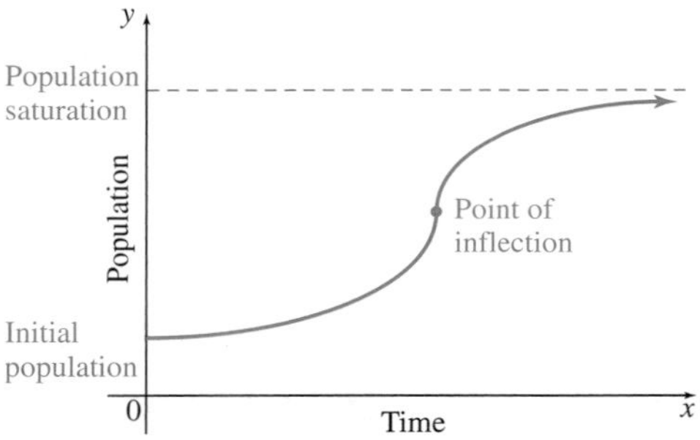

FIGURE 39

To produce the maximum yield of catfish, harvesting should take place at the point of fastest possible growth of the population; here, this is at the point of inflection. The rate of change of the population, given by the first derivative, is increasing up to the inflection point (on the interval where the second derivative is positive) and decreasing past the inflection point (on the interval where the second derivative is negative).

The *law of diminishing returns* in economics is related to the idea of concavity. The function graphed in Figure 40 on the next page gives the output y from a given input x. If the input were advertising costs for some product, for example, the output might be the corresponding revenue from sales.

The graph in Figure 40 shows an inflection point at $(c, f(c))$. For $x < c$, the graph is concave upward, so the rate of change of the slope is increasing. This indicates that the output y is increasing at a faster rate with each additional dollar

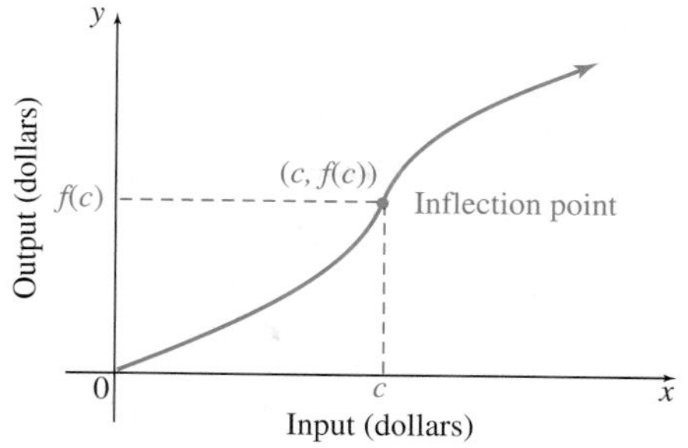

FIGURE 40

spent. When $x > c$, however, the graph is concave downward, the rate of change of the slope is decreasing, and the increase in y is smaller with each additional dollar spent. Thus, further input beyond c dollars produces diminishing returns. The point of inflection at $(c, f(c))$ is called the **point of diminishing returns.** Beyond this point there is a smaller and smaller return for each dollar invested.

As another example of diminishing returns from agriculture, with a fixed amount of land, machinery, fertilizer, and so on, adding workers increases production a lot at first, then less and less with each additional worker.

EXAMPLE 7 *Point of Diminishing Returns*

The revenue $R(x)$ generated from sales of a certain product is related to the amount x spent on advertising by

$$R(x) = \frac{1}{15{,}000}(600x^2 - x^3), \quad 0 \le x \le 600,$$

where x and $R(x)$ are in thousands of dollars. Is there a point of diminishing returns for this function? If so, what is it?

Solution Since a point of diminishing returns occurs at an inflection point, look for an x-value that makes $R''(x) = 0$. Write the function as

$$R(x) = \frac{600}{15{,}000}x^2 - \frac{1}{15{,}000}x^3 = \frac{1}{25}x^2 - \frac{1}{15{,}000}x^3.$$

Now find $R'(x)$ and then $R''(x)$.

$$R'(x) = \frac{2x}{25} - \frac{3x^2}{15{,}000} = \frac{2}{25}x - \frac{1}{5000}x^2$$

$$R''(x) = \frac{2}{25} - \frac{1}{2500}x$$

Set $R''(x)$ equal to 0 and solve for x.

$$\frac{2}{25} - \frac{1}{2500}x = 0$$

$$-\frac{1}{2500}x = -\frac{2}{25}$$

$$x = \frac{5000}{25} = 200$$

Test a number in the interval $(0, 200)$ to see that $R''(x)$ is positive there. Then test a number in the interval $(200, 600)$ to find $R''(x)$ negative in that interval. Since the sign of $R''(x)$ changes from positive to negative at $x = 200$, the graph changes from concave upward to concave downward at that point, and there is a point of diminishing returns at the inflection point $(200, 1066\frac{2}{3})$. Investments in advertising beyond \$200,000 return less and less for each dollar invested. Verify that $R'(200) = 8$. This means that when \$200,000 is invested, another \$1000 invested returns approximately \$8000 in additional revenue. Thus it may still be economically sound to invest in advertising beyond the point of diminishing returns.

5.3 EXERCISES

Find $f''(x)$ for each function. Then find $f''(0)$ and $f''(2)$.

1. $f(x) = 3x^3 - 4x + 5$

2. $f(x) = x^3 + 4x^2 + 2$

3. $f(x) = 3x^4 - 5x^3 + 2x^2$

4. $f(x) = -x^4 + 2x^3 - x^2$

5. $f(x) = 3x^2 - 4x + 8$

6. $f(x) = 8x^2 + 6x + 5$

7. $f(x) = \dfrac{x^2}{1 + x}$

8. $f(x) = \dfrac{-x}{1 - x^2}$

9. $f(x) = \sqrt{x + 4}$

10. $f(x) = \sqrt{2x + 9}$

11. $f(x) = 5x^{3/5}$

12. $f(x) = -2x^{2/3}$

13. $f(x) = 5e^{-x^2}$

14. $f(x) = .5e^{x^2}$

15. $f(x) = \dfrac{\ln x}{4x}$

16. $f(x) = \ln x + \dfrac{1}{x}$

Find $f'''(x)$, the third derivative of f, and $f^{(4)}(x)$, the fourth derivative of f, for each function.

17. $f(x) = -x^4 + 2x^2 + 8$

18. $f(x) = 2x^4 - 3x^3 + x^2$

19. $f(x) = 4x^5 + 6x^4 - x^2 + 2$

20. $f(x) = 3x^5 - x^4 + 2x^3 - 7x$

21. $f(x) = \dfrac{x - 1}{x + 2}$

22. $f(x) = \dfrac{x + 1}{x}$

23. $f(x) = \dfrac{3x}{x - 2}$

24. $f(x) = \dfrac{x}{2x + 1}$

25. Let $f(x) = \ln x$.

a. Compute $f'(x)$, $f''(x)$, $f'''(x)$, $f^{(4)}(x)$, and $f^{(5)}(x)$.

b. Guess a formula for $f^{(n)}(x)$, where n is any positive integer.

26. For $f(x) = e^x$, find $f''(x)$ and $f'''(x)$. What is the nth derivative of f with respect to x?

In Exercises 27–44, find the open intervals where the functions are concave upward or concave downward. Find any points of inflection.

27.

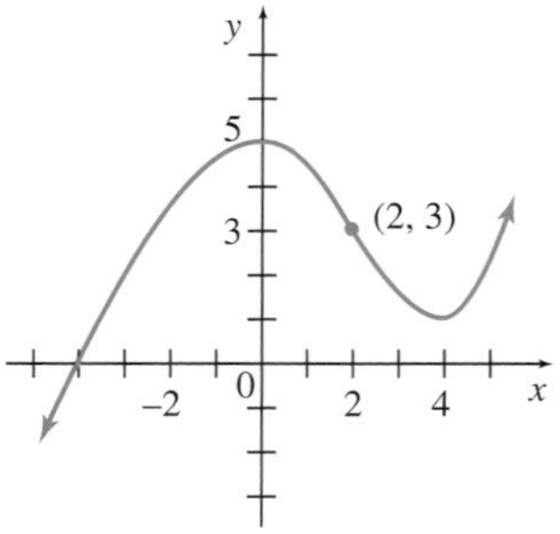

28.

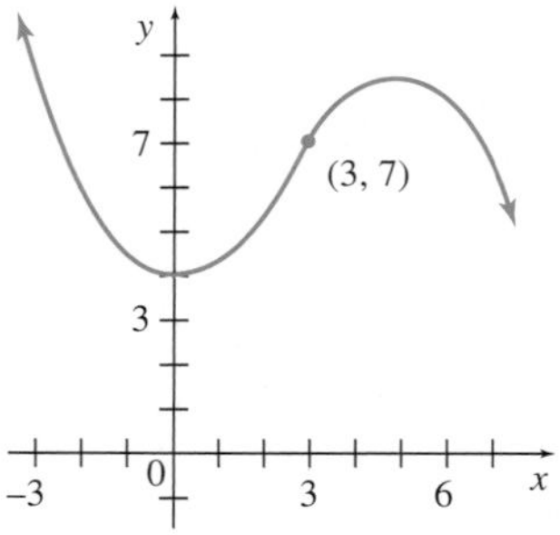

29.

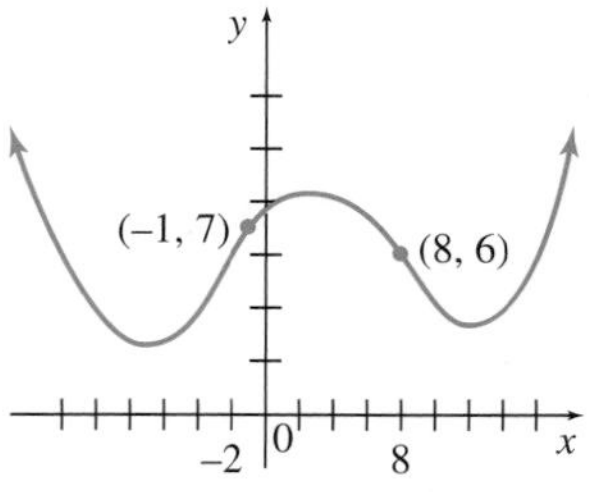

30.

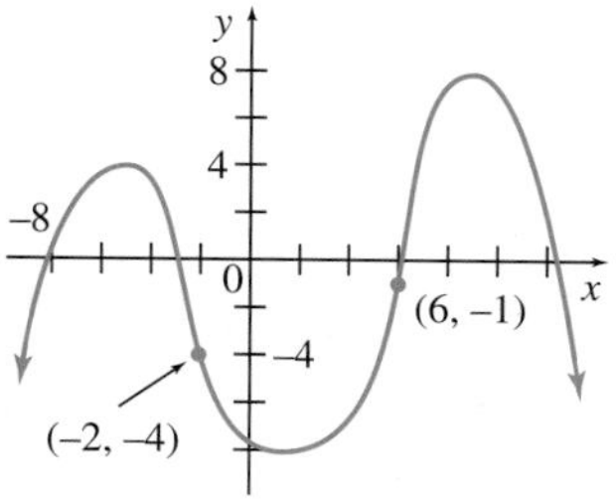

31.

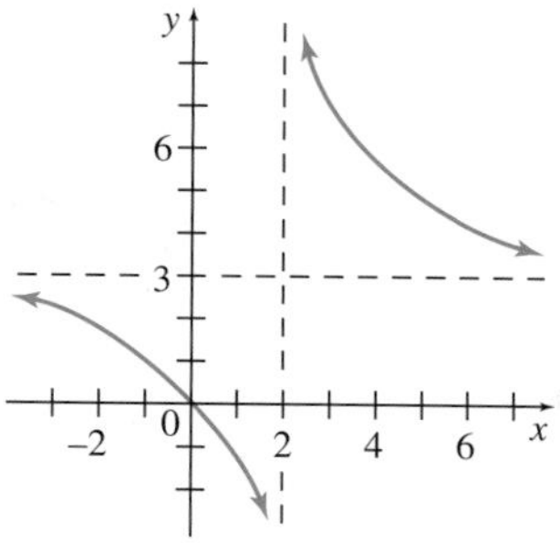

32.

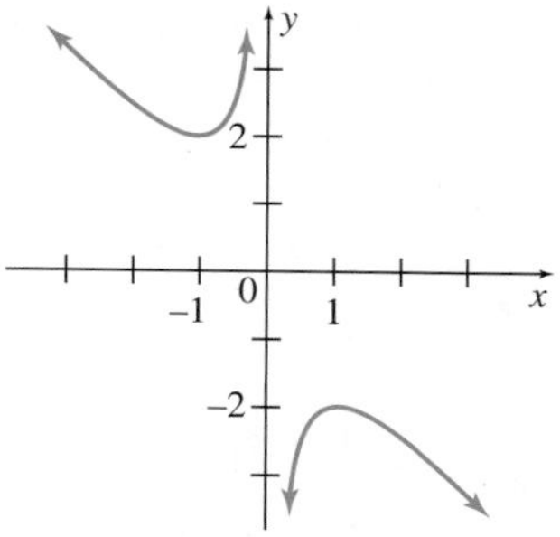

33. $f(x) = x^2 + 10x - 9$

34. $f(x) = 8 - 6x - x^2$

35. $f(x) = -2x^3 + 9x^2 + 168x - 3$

36. $f(x) = -x^3 - 12x^2 - 45x + 2$

37. $f(x) = \dfrac{3}{x - 5}$

38. $f(x) = \dfrac{-2}{x + 1}$

39. $f(x) = x(x + 5)^2$

40. $f(x) = -x(x - 3)^2$

41. $f(x) = 18x - 18e^{-x}$

42. $f(x) = 2e^{-x^2}$

43. $f(x) = x^{8/3} - 4x^{5/3}$

44. $f(x) = x^{7/3} + 56x^{4/3}$

45. **a.** Graph the two functions $f(x) = x^{7/3}$ and $g(x) = x^{5/3}$ on the window $[-2, 2]$ by $[-2, 2]$.

b. Verify that both f and g have an inflection point at $(0, 0)$.

c. How is the value of $f''(0)$ different from $g''(0)$?

d. Based on what you have seen so far in this exercise, is it always possible to tell the difference between a point where the second derivative is 0 or undefined based on the graph? Explain.

46. Describe the slope of the tangent line to the graph of $f(x) = e^x$ for the following.

a. $x \to -\infty$ **b.** $x \to 0$

47. What is true about the slope of the tangent line to the graph of $f(x) = \ln x$ as $x \to \infty$? As $x \to 0$?

Find any critical numbers for f in Exercises 48–53 and then use the second derivative test to decide whether the critical numbers lead to relative maxima or relative minima. If $f''(c) = 0$ for a critical number c, then the second derivative test gives no information. In this case, use the first derivative test instead.

48. $f(x) = -x^2 - 10x - 25$

49. $f(x) = x^2 - 12x + 36$

50. $f(x) = 3x^3 - 3x^2 + 1$

51. $f(x) = 2x^3 - 4x^2 + 2$

52. $f(x) = (x + 3)^4$

53. $f(x) = x^3$

54. Suppose a friend makes the following argument. A function f is increasing and concave downward. Therefore, f' is positive and decreasing, so it eventually becomes 0 and then negative, at which point f decreases. Show that your friend is wrong by giving an example of a function that is always increasing and concave downward.

Sometimes the derivative of a function is known, but not the function. We will see more of this later in the book. For each function f' defined as shown, find $f''(x)$, then use a graphing calculator to graph f' and f'' in the indicated window. Use the graph to do the following.

a. *Give the (approximate) x-values where f has a maximum or minimum.*

b. *By considering the sign of $f'(x)$, give the (approximate) intervals where $f(x)$ is increasing and decreasing.*

c. *Give the (approximate) x-values of any inflection points.*

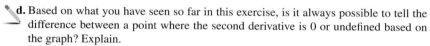

d. *By considering the sign of $f''(x)$, give the intervals where f is concave upward or concave downward.*

55. $f'(x) = x^3 - 6x^2 + 7x + 4$; $[-5, 5]$ by $[-5, 15]$

56. $f'(x) = 10x^2(x - 1)(5x - 3)$; $[-1, 1.5]$ by $[-20, 20]$

57. $f'(x) = \dfrac{1 - x^2}{(x^2 + 1)^2}$; $[-3, 3]$ by $[-1.5, 1.5]$

Applications

BUSINESS AND ECONOMICS

58. *Product Life Cycle* The accompanying figure shows the *product life cycle* graph, with typical products marked on it. It illustrates the fact that a new product is often purchased at a faster and faster rate as people become familiar with it. In time, saturation is reached and the purchase rate stays constant until the product is made obsolete by newer products, after which it is purchased less and less.*

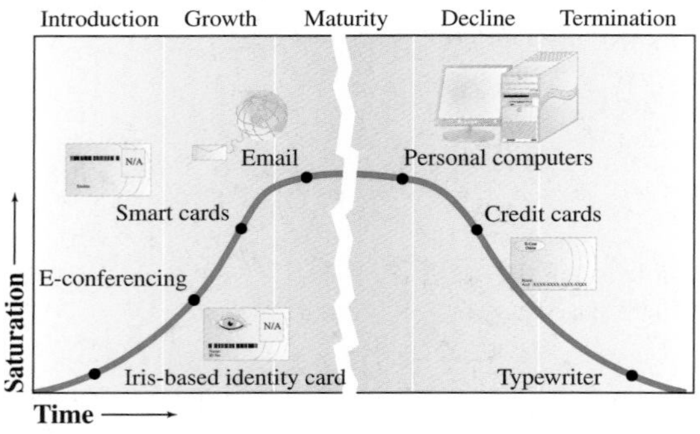

a. Which products on the left side of the graph are closest to the left-hand point of inflection? What does the point of inflection mean here?

b. Which product on the right side of the graph is closest to the right-hand point of inflection? What does the point of inflection mean here?

c. Discuss where portable DVD players, fax machines, and other new technologies should be placed on the graph.

Point of Diminishing Returns In Exercises 59–62, find the point of diminishing returns (x, y) for the given functions, where $R(x)$ represents revenue (in thousands of dollars) and x represents the amount spent on advertising (in thousands of dollars).

59. $R(x) = 10{,}000 - x^3 + 42x^2 + 800x, \quad 0 \le x \le 20$

60. $R(x) = \dfrac{4}{27}(-x^3 + 66x^2 + 1050x - 400), \quad 0 \le x \le 25$

61. $R(x) = -.3x^3 + x^2 + 11.4x, \quad 0 \le x \le 6$

62. $R(x) = -.6x^3 + 3.7x^2 + 5x, \quad 0 \le x \le 6$

63. *Risk Aversion* In economics, an index of *absolute risk aversion* is defined as

$$I(M) = \frac{-U''(M)}{U'(M)},$$

where M measures how much of a commodity is owned and $U(M)$ is a *utility function,* which measures the ability of quantity M of a commodity to satisfy a consumer's wants. Find $I(M)$ for $U(M) = \sqrt{M}$ and for $U(M) = M^{2/3}$, and determine which indicates a greater aversion to risk.

64. *Demand Function* The authors of an article† in an economics journal state that if $D(q)$ is the demand function, then the inequality

$$qD''(q) + D'(q) < 0$$

is equivalent to saying that the marginal revenue declines more quickly than does the price. Prove that this equivalence is true.

LIFE SCIENCES

65. *Population Growth* When a hardy new species is introduced into an area, the population often increases as shown. Explain the significance of the following function values on the graph.

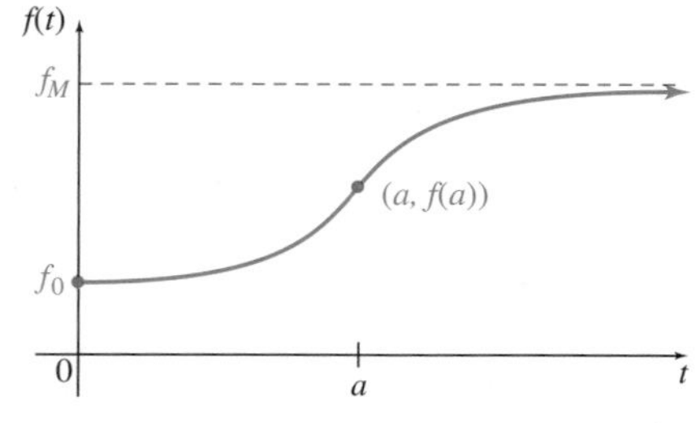

a. f_0 **b.** $f(a)$ **c.** f_M

*http://www.tutor2u.net/business/marketing/products_lifecycle.asp.

†Fudenberg, Drew and Jean Tirole, "Learning by Doing and Market Performance," *Bell Journal of Economics,* Vol. 14, 1983.

66. *Bacteria Population* Assume that the number of bacteria $R(t)$ (in millions) present in a certain culture at time t (in hours) is given by

$$R(t) = t^2(t-18) + 96t + 1000.$$

a. At what time before 8 hours will the population be maximized?

b. Find the maximum population.

67. *Ozone Depletion* According to an article in *The New York Times*, "Government scientists reported last week that they had detected a slowdown in the rate at which chemicals that deplete the earth's protective ozone layer are accumulating in the atmosphere."* Letting $c(t)$ be the amount of ozone-depleting chemicals at time t, what does this statement tell you about $c(t)$, $c'(t)$, and $c''(t)$?

68. *Drug Concentration* The percent of concentration of a certain drug in the bloodstream x hours after the drug is administered is given by

$$K(x) = \frac{3x}{x^2 + 4}.$$

For example, after 1 hour the concentration is given by

$$K(1) = \frac{3(1)}{1^2 + 4} = \frac{3}{5}\% = .6\% = .006.$$

a. Find the time at which concentration is a maximum.

b. Find the maximum concentration.

69. *Drug Concentration* The percent of concentration of a drug in the bloodstream x hours after the drug is administered is given by

$$K(x) = \frac{4x}{3x^2 + 27}.$$

a. Find the time at which the concentration is a maximum.

b. Find the maximum concentration.

The next two exercises are a continuation of exercises first given in the section on Derivatives of Exponential Functions. Find the inflection point of the graph of each logistic function. This is the point at which the growth rate begins to decline.

70. *Insect Growth* The growth function for a population of beetles is given by

$$G(t) = \frac{10{,}000}{1 + 49e^{-.1t}}.$$

71. *Clam Population Growth* The population of a bed of clams is described by

$$G(t) = \frac{5200}{1 + 12e^{-.5t}}.$$

Hints for Exercises 72 and 73: Leave B, c, and k as constants until you are ready to calculate your final answer. Recall that $d(e^{g(t)})/dt = g'(t)e^{g(t)}$.

72. *Clam Growth* Researchers used a version of the Gompertz curve to model the growth of razor clams during the first seven years of the clams' lives with the equation

$$L(t) = Be^{-ce^{-kt}},$$

where $L(t)$ gives the length (in centimeters) after t years, $B = 14.3032$, $c = 7.267963$, and $k = .670840$.† Find the inflection point and describe what it signifies.

73. *Breast Cancer Growth* Researchers used a version of the Gompertz curve to model the growth of breast cancer tumors with the equation

$$N(t) = e^{c(1 - e^{-kt})},$$

where $N(t)$ is the number of cancer cells after t days, $c = 27.3$, and $k = .011$.‡ Find the inflection point and describe what it signifies.

74. *Popcorn* Researchers have determined that the amount of moisture present in a kernel of popcorn affects the volume of the popped corn and can be modeled for certain sizes of kernels by the function

$$v(x) = -35.98 + 12.09x - .4450x^2,$$

where x is moisture content (%, wet basis) and $v(x)$ is the expansion volume (in cm^3/gram).§ Describe the concavity of this function.

75. *Alligator Teeth* Researchers have developed a mathematical model that can be used to estimate the number of teeth $N(t)$ at time t (days of incubation) for *Alligator mississippiensis*,‖ where

$$N(t) = 71.8e^{-8.96e^{-.0685t}}.$$

The New York Times, Aug. 29, 1993, p. E2.

†Weymouth, F. W., H. C. McMillin, and Willis H. Rich, "Latitude and Relative Growth in the Razor Clam," *Journal of Experimental Biology*, Vol. 8, 1931, pp. 228–249.

‡Speer, John F. et al., "A Stochastic Numerical Model of Breast Cancer Growth That Simulates Clinical Data," *Cancer Research*, Vol. 44, Sept. 1984, pp. 4124–4130.

§Song, A. and S. Eckhoff, "Optimum Popping Moisture Content for Popcorn Kernels of Different Sizes," *Cereal Chemistry*, Vol. 71, No. 5, 1994, pp. 458–460.

‖Kulesa, P. et al., "On a Model Mechanism for the Spatial Performing of Teeth Primordia in the Alligator," *Journal of Theoretical Biology*, Vol. 180, 1996, pp. 287–296.

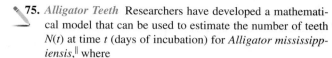

Find the inflection point and describe its importance to this research.

SOCIAL SCIENCES

76. *Crime* In 1995, the rate of violent crimes in New York City continued to decrease, but at a slower rate than in previous years.* Letting $f(t)$ be the rate of violent crime as a function of time, what does this tell you about $f(t)$, $f'(t)$, and $f''(t)$?

PHYSICAL SCIENCES

77. *Chemical Reaction* An autocatalytic chemical reaction is one in which the product being formed causes the rate of formation to increase. The rate of a certain autocatalytic reaction is given by

$$V(x) = 12x(100 - x),$$

where x is the quantity of the product present and 100 represents the quantity of chemical present initially. For what value of x is the rate of the reaction a maximum?

78. *Velocity and Acceleration* When an object is dropped straight down, the distance (in feet) that it travels in t seconds is given by

$$s(t) = -16t^2.$$

Find the velocity at each of the following times.

a. After 3 seconds

b. After 5 seconds

c. After 8 seconds

d. Find the acceleration. (The answer here is a constant—the acceleration due to the influence of gravity alone near the surface of Earth.)

79. *Baseball* Roger Clemens, former ace pitcher for the New York Yankees and Boston Red Sox, is standing on top of the 37-ft-high "Green Monster" left-field wall in Boston's Fenway Park, to which he has returned for a visit. We have asked him to fire his famous 95 mph (140 ft per second) fastball straight up. The position equation, which gives the height of the ball at any time t, in seconds, is given by $s(t) = -16t^2 + 140t + 37$.† Find the following.

a. The maximum height of the ball

b. The time and velocity when the ball hits the ground

80. *Height of a Ball* If a cannonball is shot directly upward with a velocity of 256 ft per second, its height above the ground after t seconds is given by $s(t) = 256t - 16t^2$. Find the velocity and the acceleration after t seconds. What is the maximum height the cannonball reaches? When does it hit the ground?

81. *Velocity and Acceleration of a Car* A car rolls down a hill. Its distance (in feet) from its starting point is given by $s(t) = 1.5t^2 + 4t$, where t is in seconds.

a. How far will the car move in 10 seconds?

b. What is the velocity at 5 seconds? At 10 seconds?

c. How can you tell from $v(t)$ that the car will not stop?

d. What is the acceleration at 5 seconds? At 10 seconds?

e. What is happening to the velocity and the acceleration as t increases?

5.4 CURVE SKETCHING

The test for concavity, the test for increasing and decreasing functions, and the concept of limits at infinity help us sketch the graphs and describe the behavior of a variety of functions. This process, called **curve sketching,** has decreased somewhat in importance in recent years due to the widespread use of graphing calculators. We believe, however, that this topic is worth studying for the following reasons.

For one thing, a graphing calculator picture can be misleading, particularly if important points lie outside the viewing window. Even if all important features

**The New York Times,* Dec. 17, 1995, p. 49.

†This exercise provided by Frederick Russell of College of Southern Maryland.

are within the viewing windows, there is still the problem that the calculator plots and connects points, and misses what goes on between those points. As an example of the difficulty in choosing an appropriate window without a knowledge of calculus, see Exercise 34 in the second section of this chapter.

Furthermore, curve sketching may be the best way to learn the material in the previous three sections. You may feel confident that you understand what increasing and concave upward mean, but using those concepts in a graph will put your understanding to the test.

Curve sketching may be done with the following steps.

CURVE SKETCHING

To sketch the graph of a function *f*:

1. Consider the domain of the function, and note any restrictions. (Avoid dividing by zero or taking a square root of a negative number.)
2. Find the *y*-intercept (if it exists) by substituting $x = 0$ into $f(x)$. Find any *x*-intercepts by solving $f(x) = 0$ if this is not too difficult.
3. **a.** If *f* is a rational function, find any vertical asymptotes by investigating where the denominator is 0, and find any horizontal asymptotes by finding the limits as $x \to \infty$ and $x \to -\infty$.

   b. If *f* is an exponential function, find any horizontal asymptotes; if *f* is a logarithmic function, find any vertical asymptotes.
4. Find f'. Locate any critical points by solving the equation $f'(x) = 0$ and determining where f' does not exist, but *f* does. Find any relative extrema and determine where *f* is increasing or decreasing.
5. Find f''. Locate potential points of inflection by solving the equation $f''(x) = 0$ and determining where $f''(x)$ does not exist. Determine where *f* is concave upward or concave downward.
6. Plot the intercepts, the critical points, the inflection points, the asymptotes, and other points as needed.
7. Connect the points with a smooth curve using the correct concavity, being careful not to connect points where the function is not defined.
8. Check your graph using a graphing calculator. If the picture looks very different from what you've drawn, see in what ways the picture differs and use that information to help find your mistake.

There are four possible combinations for a function to be increasing or decreasing and concave up or concave down, as shown in the following table.

$f'(x)$ / $f''(x)$	**+ (Function Is Increasing)**	**− (Function Is Decreasing)**
+ (function is concave up)		
− (function is concave down)		

EXAMPLE 1 *Polynomial Graph*

Graph $f(x) = 2x^3 - 3x^2 - 12x + 1$.

Solution The domain is $(-\infty, \infty)$. The y-intercept is located at $y = f(0) = 1$. Finding the x-intercepts requires solving the equation $f(x) = 0$. But this is a third-degree equation; since we have not covered a procedure for solving such equations, we will skip this step. This is not a rational function, so we also skip step 3.

To find the intervals where the function is increasing or decreasing, find the first derivative.

$$f'(x) = 6x^2 - 6x - 12$$

This derivative is 0 when

$$6(x^2 - x - 2) = 0$$
$$6(x - 2)(x + 1) = 0$$
$$x = 2 \quad \text{or} \quad x = -1.$$

These critical numbers divide the number line in Figure 41 into three regions. Testing a number from each region in $f'(x)$ shows that f is increasing on $(-\infty, -1)$ and $(2, \infty)$ and decreasing on $(-1, 2)$. This is shown with the arrows in Figure 41. By the first derivative test, f has a relative maximum when $x = -1$ and a relative minimum when $x = 2$. The relative maximum is $f(-1) = 8$, while the relative minimum is $f(2) = -19$.

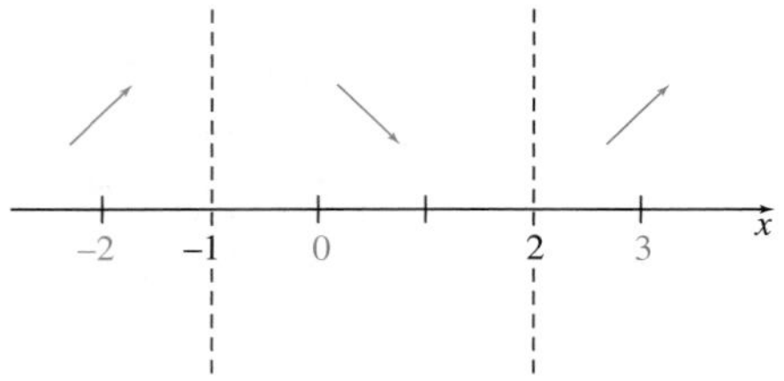

FIGURE 41

Now use the second derivative to find the intervals where the function is concave upward or downward. Here

$$f''(x) = 12x - 6,$$

which is 0 when $x = 1/2$. Testing a point with x less than $1/2$, and one with x greater than $1/2$, shows that f is concave downward on $(-\infty, 1/2)$ and concave upward on $(1/2, \infty)$. The graph has an inflection point at $(1/2, f(1/2))$, or $(1/2, -11/2)$. This information is summarized in the following table.

Interval	$(-\infty, -1)$	$(-1, 1/2)$	$(1/2, 2)$	$(2, \infty)$
Sign of f'	+	−	−	+
Sign of f''	−	−	+	+
f Increasing or Decreasing	Increasing	Decreasing	Decreasing	Increasing
Concavity of f	Downward	Downward	Upward	Upward
Shape of Graph				

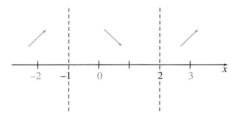

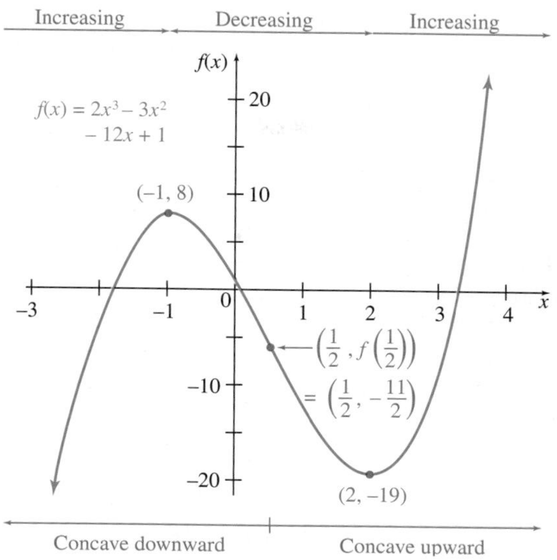

FIGURE 42

Use this information and the critical points to get the graph shown in Figure 42.

A graphing calculator picture of the function in Figure 42 on the arbitrarily chosen window $[-3, 3]$ by $[-7, 7]$ gives a misleading picture, as Figure 43(a) shows. Knowing where the turning points lie tells us that a better window would be $[-3, 4]$ by $[-20, 20]$, with the results shown in Figure 43(b).

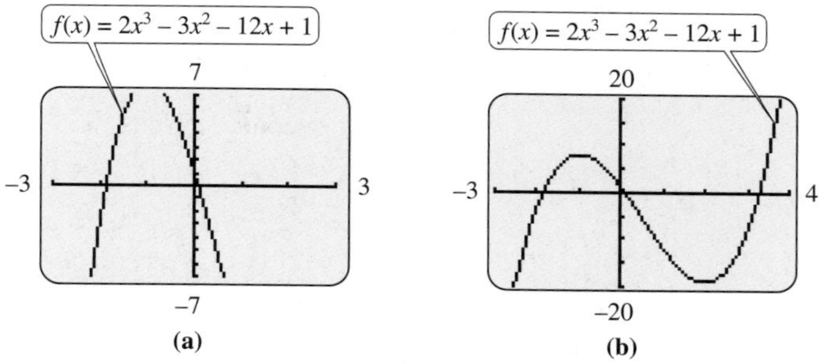

FIGURE 43

EXAMPLE 2 *Rational Function Graph*

Graph $f(x) = x + 1/x$.

Solution Notice that $x = 0$ is not in the domain of the function, so there is no y-intercept. To find the x-intercept, solve $f(x) = 0$.

$$x + \frac{1}{x} = 0$$

$$x = -\frac{1}{x}$$

$$x^2 = -1$$

Since x^2 is always positive, there is also no x-intercept.

The function is a rational function, but it is not written in the usual form of one polynomial over another. It can be rewritten in that form:

$$f(x) = x + \frac{1}{x} = \frac{x^2 + 1}{x}.$$

FOR REVIEW

Asymptotes were discussed in the section on Polynomial and Rational Functions. You may wish to refer back to that section to review. To review limits, refer to the first section in the chapter titled The Derivative.

Because $x = 0$ makes the denominator (but not the numerator) 0, the line $x = 0$ is a vertical asymptote. To find any horizontal asymptotes, we investigate

$$\lim_{x \to \infty} \frac{x^2 + 1}{x} = \lim_{x \to \infty} \left(\frac{x^2}{x} + \frac{1}{x} \right) = \lim_{x \to \infty} \left(x + \frac{1}{x} \right).$$

The second term, $1/x$, approaches 0 as $x \to \infty$, but the first term, x, becomes infinitely large, so the limit does not exist. Verify that $\lim_{x \to -\infty} f(x)$ also does not exist, so there are no horizontal asymptotes.

Observe that as x gets very large, the second term $(1/x)$ in $f(x)$ gets very small, so $f(x) = x + (1/x) \approx x$. The graph gets closer and closer to the straight line $y = x$ as x becomes larger and larger. This is what is known as an **oblique asymptote.**

Here $f'(x) = 1 - (1/x^2)$, which is 0 when

$$\frac{1}{x^2} = 1$$

$$x^2 = 1$$

$$x = 1 \qquad \text{or} \qquad x = -1.$$

The derivative fails to exist at 0, where the vertical asymptote is located. Evaluating $f'(x)$ in each of the regions determined by the critical numbers and the asymptote shows that f is increasing on $(-\infty, -1)$ and $(1, \infty)$ and decreasing on $(-1, 0)$ and $(0, 1)$. By the first derivative test, f has a relative maximum of $y = f(-1) = -2$ when $x = -1$, and a relative minimum of $y = f(1) = 2$ when $x = 1$.

The second derivative is

$$f''(x) = \frac{2}{x^3},$$

which is never equal to 0 and does not exist when $x = 0$. (The function itself also does not exist at 0.) Because of this, there may be a change of concavity, but not an inflection point, when $x = 0$. The second derivative is negative when x is negative, making f concave downward on $(-\infty, 0)$. Also, $f''(x) > 0$ when $x > 0$, making f concave upward on $(0, \infty)$.

Interval	$(-\infty, -1)$	$(-1, 0)$	$(0, 1)$	$(1, \infty)$
Sign of f'	+	−	−	+
Sign of f''	−	−	+	+
f Increasing or Decreasing	Increasing	Decreasing	Decreasing	Increasing
Concavity of f	Downward	Downward	Upward	Upward
Shape of Graph				

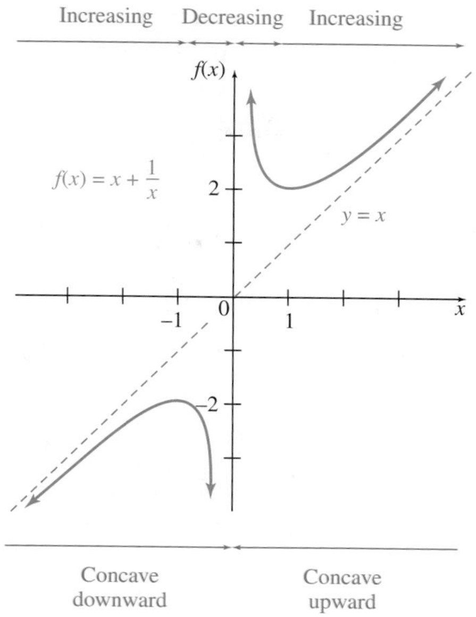

FIGURE 44

Use this information, the asymptotes, and the critical points to get the graph shown in Figure 44.

EXAMPLE 3 *Rational Function Graph*

Graph $f(x) = \dfrac{3x^2}{x^2 + 5}$.

Solution The y-intercept is located at $y = f(0) = 0$. Verify that this is also the only x-intercept. There is no vertical asymptote, because $x^2 + 5 \neq 0$ for any value of x. Find any horizontal asymptote by calculating $\lim_{x \to \infty} f(x)$ and $\lim_{x \to -\infty} f(x)$. First, divide both the numerator and the denominator of $f(x)$ by x^2.

$$\lim_{x \to \infty} \frac{3x^2}{x^2 + 5} = \lim_{x \to \infty} \frac{\dfrac{3x^2}{x^2}}{\dfrac{x^2}{x^2} + \dfrac{5}{x^2}} = \frac{3}{1 + 0} = 3$$

Verify that the limit of $f(x)$ as $x \to -\infty$ is also 3. Thus, the horizontal asymptote is $y = 3$.

We now compute $f'(x)$:

$$f'(x) = \frac{(x^2 + 5)(6x) - (3x^2)(2x)}{(x^2 + 5)^2}.$$

Notice that $6x$ can be factored out of each term in the numerator:

$$f'(x) = \frac{(6x)[(x^2 + 5) - x^2]}{(x^2 + 5)^2}$$

$$= \frac{(6x)(5)}{(x^2 + 5)^2} = \frac{30x}{(x^2 + 5)^2}.$$

From the numerator, $x = 0$ is a critical number. The denominator is always positive. (Why?) Evaluating $f'(x)$ in each of the regions determined by $x = 0$ shows that f is decreasing on $(-\infty, 0)$ and increasing on $(0, \infty)$. By the first derivative test, f has a relative minimum when $x = 0$.

The second derivative is

$$f''(x) = \frac{(x^2 + 5)^2(30) - (30x)(2)(x^2 + 5)(2x)}{(x^2 + 5)^4}.$$

Factor $30(x^2 + 5)$ out of the numerator:

$$f''(x) = \frac{30(x^2 + 5)[(x^2 + 5) - (x)(2)(2x)]}{(x^2 + 5)^4}.$$

Divide a factor of $(x^2 + 5)$ out of the numerator and denominator, and simplify the numerator:

$$\begin{aligned} f''(x) &= \frac{30[(x^2 + 5) - (x)(2)(2x)]}{(x^2 + 5)^3} \\ &= \frac{30[(x^2 + 5) - (4x^2)]}{(x^2 + 5)^3} \\ &= \frac{30(5 - 3x^2)}{(x^2 + 5)^3}. \end{aligned}$$

The numerator of $f''(x)$ is 0 when $x = \pm\sqrt{5/3} \approx \pm 1.29$. Testing a point in each of the three intervals defined by these points shows that f is concave downward on $(-\infty, -1.29)$ and $(1.29, \infty)$, and concave upward on $(-1.29, 1.29)$. The graph has inflection points at $(\pm\sqrt{5/3}, f(\pm\sqrt{5/3})) \approx (\pm 1.29, y.75)$.

Interval	$(-\infty, -1.29)$	$(-1.29, 0)$	$(0, 1.29)$	$(1.29, \infty)$
Sign of f'	−	−	+	+
Sign of f''	−	+	+	−
f Increasing or Decreasing	Decreasing	Decreasing	Increasing	Increasing
Concavity of f	Downward	Upward	Upward	Downward
Shape of Graph				

Use this information, the asymptote, the critical point, and the inflection points to get the graph shown in Figure 45 on the next page.

EXAMPLE 4 *Graph with Logarithm*

Graph $f(x) = \dfrac{\ln x}{x^2}$.

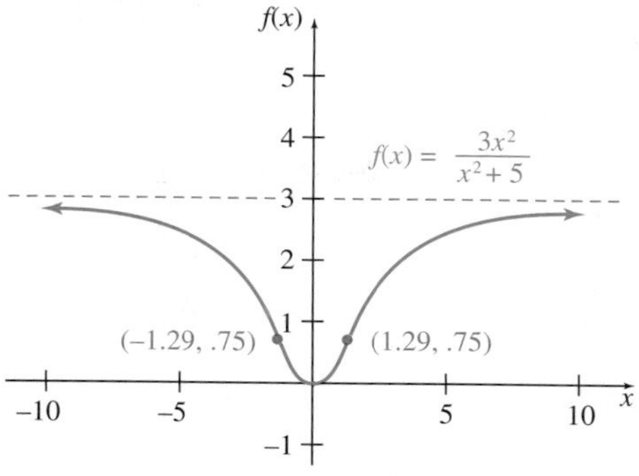

FIGURE 45

Solution The domain is $x > 0$, so there is no y-intercept. The x-intercept is 1, because $\ln 1 = 0$. We know that $y = \ln x$ has a vertical asymptote at $x = 0$, because $\lim_{x \to 0^+} \ln x = -\infty$. Dividing by x^2 when x is small makes $\ln x/x^2$ even more negative than $\ln x$. The first derivative is

$$f'(x) = \frac{x^2 \cdot \frac{1}{x} - 2x \ln x}{(x^2)^2} = \frac{x(1 - 2\ln x)}{x^4} = \frac{1 - 2\ln x}{x^3}$$

by the quotient rule. Setting the numerator equal to 0 and solving for x gives

$$\begin{aligned} 1 - 2\ln x &= 0 \\ 1 &= 2\ln x \\ \ln x &= .5 \\ x &= e^{.5} \approx 1.65. \end{aligned}$$

Since $f'(1)$ is positive and $f'(2)$ is negative, f increases on $(0, 1.65)$, then decreases on $(1.65, \infty)$, with a maximum value of $f(1.65) \approx .18$.

To find any inflection points, we set $f''(x) = 0$.

$$\begin{aligned} f''(x) &= \frac{x^3\left(-2 \cdot \frac{1}{x}\right) - (1 - 2\ln x) \cdot 3x^2}{(x^3)^2} \\ &= \frac{-2x^2 - 3x^2 + 6x^2 \ln x}{x^6} = \frac{-5 + 6\ln x}{x^4} \end{aligned}$$

$$\frac{-5 + 6\ln x}{x^4} = 0$$

$$-5 + 6\ln x = 0 \qquad \text{Set the numerator equal to 0.}$$

$$6\ln x = 5 \qquad \text{Add 5 to both sides.}$$

$$\ln x = 5/6 \qquad \text{Divide both sides by 6.}$$

$$x = e^{5/6} \approx 2.3 \qquad e^{\ln x} = x.$$

There is an inflection point at $(2.3, f(2.3)) \approx (2.3,y.16)$. Verify that $f''(1)$ is negative and $f''(3)$ is positive, so the graph is concave downward on $(1, 2.3)$ and upward on $(2.3, \infty)$. This information is summarized in the following table and

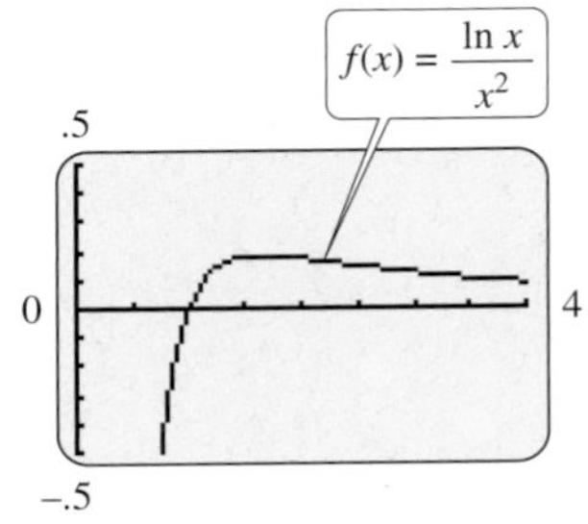

FIGURE 46

could be used to sketch the graph. A calculator graph of the function is shown in Figure 46.

Interval	(0, 1.65)	(1.65, 2.3)	(2.3, ∞)
Sign of f′	+	–	–
Sign of f″	–	–	+
f Increasing or Decreasing	Increasing	Decreasing	Decreasing
Concavity of f	Downward	Downward	Upward
Shape of Graph			

As we saw earlier, a graphing calculator, when used with care, can be helpful in studying the behavior of functions. This section has illustrated that calculus is also a great help. The techniques of calculus show where the important points of a function, such as the relative extrema and the inflection points, are located. Furthermore, they tell how the function behaves between and beyond the points that are graphed, something a graphing calculator cannot always do.

5.4 EXERCISES

1. By sketching a graph of the function or by investigating values of the function near 0, find $\lim_{x \to 0} x \ln |x|$. (This result will be useful in Exercise 21.)

2. Describe how you would find the equation of the horizontal asymptote for the graph of
$$f(x) = \frac{3x^2 - 2x}{2x^2 + 5}.$$

*Graph each function, considering the domain, critical points, regions where the function is increasing or decreasing, points of inflection, regions where the function is concave upward or concave downward, intercepts where possible, and asymptotes where applicable. (*Hint: *In Exercise 21, use the result of Exercise 1. In Exercises 25–27, recall from Exercise 58 in the section on Limits that* $\lim_{x \to \infty} x^n e^{-x} = 0$.*)*

3. $f(x) = -2x^3 - 9x^2 + 108x - 10$

4. $f(x) = x^3 - \frac{15}{2}x^2 - 18x - 1$

5. $f(x) = -3x^3 + 6x^2 - 4x - 1$

6. $f(x) = x^3 - 6x^2 + 12x - 11$

7. $f(x) = x^4 - 18x^2 + 5$

8. $f(x) = x^4 - 8x^2$

9. $f(x) = x^4 - 2x^3$

10. $f(x) = x^5 - 15x^3$

11. $f(x) = x + \frac{2}{x}$

12. $f(x) = 2x + \frac{8}{x}$

13. $f(x) = \frac{x - 1}{x + 1}$

14. $f(x) = \frac{x}{1 + x}$

15. $f(x) = \frac{1}{x^2 + x - 2}$

16. $f(x) = \frac{-2}{x^2 - x - 6}$

17. $f(x) = \frac{x}{x^2 + 1}$

18. $f(x) = \frac{1}{x^2 + 1}$

19. $f(x) = \frac{1}{x^2 - 4}$

20. $f(x) = \frac{x}{x^2 - 1}$

21. $y = x \ln |x|$

22. $y = x - \ln |x|$

23. $y = \dfrac{\ln x}{x}$

24. $y = \dfrac{\ln x^2}{x^2}$

25. $y = xe^{-x}$

26. $y = x^2 e^{-x}$

27. $y = (x - 1)e^{-x}$

28. $y = e^x + e^{-x}$

29. $y = x^{2/3} - x^{5/3}$

30. $y = x^{1/3} + x^{4/3}$

31. The default window on many calculators is $[-10, 10]$ by $[-10, 10]$. For the odd exercises between 3 and 15, tell which would give a poor representation in this window. (*Note:* Your answers may differ from ours, depending on what you consider "poor.")

32. Repeat Exercise 31 for the even exercises between 4 and 16.

33. Repeat Exercise 31 for the odd exercises between 17 and 29.

34. Repeat Exercise 31 for the even exercises between 18 and 30.

In Exercises 35–39, sketch the graph of a single function that has all of the properties listed.

35. **a.** Continuous and differentiable everywhere except at $x = 1$, where it has a vertical asymptote

b. $f'(x) < 0$ everywhere it is defined

c. A horizontal asymptote at $y = 2$

d. $f''(x) < 0$ on $(-\infty, 1)$ and $(2, 4)$

e. $f''(x) > 0$ on $(1, 2)$ and $(4, \infty)$

36. **a.** Continuous for all real numbers

b. $f'(x) < 0$ on $(-\infty, -6)$ and $(1, 3)$

c. $f'(x) > 0$ on $(-6, 1)$ and $(3, \infty)$

d. $f''(x) > 0$ on $(-\infty, -6)$ and $(3, \infty)$

e. $f''(x) < 0$ on $(-6, 3)$

f. A y-intercept at $(0, 2)$

37. **a.** Continuous and differentiable for all real numbers

b. $f'(x) > 0$ on $(-\infty, -3)$ and $(1, 4)$

c. $f'(x) < 0$ on $(-3, 1)$ and $(4, \infty)$

d. $f''(x) < 0$ on $(-\infty, -1)$ and $(2, \infty)$

e. $f''(x) > 0$ on $(-1, 2)$

f. $f'(-3) = f'(4) = 0$

g. $f''(x) = 0$ at $(-1, 3)$ and $(2, 4)$

38. **a.** Continuous for all real numbers

b. $f'(x) > 0$ on $(-\infty, -2)$ and $(0, 3)$

c. $f'(x) < 0$ on $(-2, 0)$ and $(3, \infty)$

d. $f''(x) < 0$ on $(-\infty, 0)$ and $(0, 5)$

e. $f''(x) > 0$ on $(5, \infty)$

f. $f'(-2) = f'(3) = 0$

g. $f'(0)$ doesn't exist

h. Differentiable everywhere except at $x = 0$

i. An inflection point at $(5, 1)$

39. a. Continuous for all real numbers

b. Differentiable everywhere except at $x = 4$

c. $f(1) = 5$

d. $f'(1) = 0$ and $f'(3) = 0$

e. $f'(x) > 0$ on $(-\infty, 1)$ and $(4, \infty)$

f. $f'(x) < 0$ on $(1, 3)$ and $(3, 4)$

g. $\lim_{x \to 4^-} f'(x) = -\infty$ and $\lim_{x \to 4^+} f'(x) = \infty$

h. $f''(x) > 0$ on $(2, 3)$

i. $f''(x) < 0$ on $(-\infty, 2)$, $(3, 4)$, and $(4, \infty)$

40. On many calculators, graphs of rational functions produce lines at vertical asymptotes. For example, graphing $y = (x - 1)/(x + 1)$ on the window $[-4.9, 4.9]$ by $[-4.9, 4.9]$ produces such a line at $x = -1$ on the TI-83/84 Plus and TI-86. But with the window $[-4.7, 4.7]$ by $[-4.7, 4.7]$ on a TI-83/84 Plus; or $[-6.3, 6.3]$ by $[-6.3, 6.3]$ on a TI-86, the spurious line does not appear. Experiment with this function on your calculator, trying different windows, and try to figure out an explanation for this phenomenon. (*Hint:* Consider the number of pixels on the calculator screen.)

CHAPTER SUMMARY

In this chapter we have explored various concepts related to the graph of a function: increasing and decreasing, relative maxima and minima, concavity, critical points, and inflection points. The first and second derivative tests provide ways to locate relative extrema. The last section brings all these concepts together. Also, we investigated two applications of the second derivative: acceleration and the point of diminishing returns.

KEY TERMS

5.1 **increasing function**
decreasing function
critical number
critical point

5.2 **relative (or local) maximum**
relative (or local) minimum
relative (or local) extremum
first derivative test

5.3 **second derivative**
third derivative
fourth derivative
acceleration
concavity
concave upward and downward
point of inflection
second derivative test
point of diminishing returns

5.4 **oblique asymptote**

CHAPTER 5 REVIEW EXERCISES

1. When given the equation for a function, how can you determine where it is increasing and where it is decreasing?

2. When given the equation for a function, how can you determine where the relative extrema are located? Give two ways to test whether a relative extremum is a minimum or a maximum.

3. Does a relative maximum of a function always have the largest y-value in the domain of the function? Explain your answer.

4. What information about a graph can be found from the second derivative?

Find the open intervals where f is increasing or decreasing.

5. $f(x) = x^2 - 5x + 3$

6. $f(x) = -2x^2 - 3x + 4$

7. $f(x) = -x^3 - 5x^2 + 8x - 6$

8. $f(x) = 4x^3 + 3x^2 - 18x + 1$

9. $f(x) = \dfrac{6}{x - 4}$

10. $f(x) = \dfrac{5}{2x + 1}$

11. $f(x) = \ln(x^2 - 1)$

12. $f(x) = 3xe^{2x}$

Find the locations and values of all relative maxima and minima.

13. $f(x) = -x^2 + 4x - 8$

14. $f(x) = x^2 - 6x + 4$

15. $f(x) = 2x^2 - 8x + 1$

16. $f(x) = -3x^2 + 2x - 5$

17. $f(x) = 2x^3 + 3x^2 - 36x + 20$

18. $f(x) = 2x^3 + 3x^2 - 12x + 5$

19. $f(x) = \dfrac{xe^x}{x - 1}$

20. $f(x) = \dfrac{\ln(3x)}{2x^2}$

Find the second derivative of each function, and then find $f''(1)$ and $f''(-3)$.

21. $f(x) = 3x^4 - 5x^2 - 11x$

22. $f(x) = 9x^3 + \dfrac{1}{x}$

23. $f(x) = \dfrac{5x - 1}{2x + 3}$

24. $f(x) = \dfrac{4 - 3x}{x + 1}$

25. $f(t) = \sqrt{t^2 + 1}$

26. $f(t) = -\sqrt{5 - t^2}$

Graph each function, considering the domain, critical points, regions where the function is increasing or decreasing, points of inflection, regions where the function is concave up or concave down, intercepts where possible, and asymptotes where applicable.

27. $f(x) = -2x^3 - \dfrac{1}{2}x^2 + x - 3$

28. $f(x) = -\dfrac{4}{3}x^3 + x^2 + 30x - 7$

29. $f(x) = x^4 - \dfrac{4}{3}x^3 - 4x^2 + 1$

30. $f(x) = -\dfrac{2}{3}x^3 + \dfrac{9}{2}x^2 + 5x + 1$

31. $f(x) = \dfrac{x - 1}{2x + 1}$

32. $f(x) = \dfrac{2x - 5}{x + 3}$

33. $f(x) = -4x^3 - x^2 + 4x + 5$

34. $f(x) = x^3 + \dfrac{5}{2}x^2 - 2x - 3$

35. $f(x) = x^4 + 2x^2$

36. $f(x) = 6x^3 - x^4$

37. $f(x) = \dfrac{x^2 + 4}{x}$

38. $f(x) = x + \dfrac{8}{x}$

39. $f(x) = \dfrac{2x}{3 - x}$

40. $f(x) = \dfrac{-4x}{1 + 2x}$

In Exercises 41 and 42, sketch the graph of a single function that has all of the properties listed.

41. **a.** Continuous everywhere except at $x = -4$, where there is a vertical asymptote
b. A y-intercept at $y = -2$
c. x-intercepts at $x = -3$, 1, and 4
d. $f'(x) < 0$ on $(-\infty, -5)$, $(-4, -1)$, and $(2, \infty)$
e. $f'(x) > 0$ on $(-5, -4)$ and $(-1, 2)$
f. $f''(x) > 0$ on $(-\infty, -4)$ and $(-4, -3)$
g. $f''(x) < 0$ on $(-3, -1)$ and $(-1, \infty)$
h. Differentiable everywhere except at $x = -4$ and $x = -1$

42. **a.** Continuous and differentiable everywhere except at $x = -3$, where it has a vertical asymptote
b. A horizontal asymptote at $y = 1$
c. An x-intercept at $x = -2$

d. A y-intercept at $y = 4$

e. $f'(x) > 0$ on the intervals $(-\infty, -3)$ and $(-3, 2)$

f. $f'(x) < 0$ on the interval $(2, \infty)$

g. $f''(x) > 0$ on the intervals $(-\infty, -3)$ and $(4, \infty)$

h. $f''(x) < 0$ on the interval $(-3, 4)$

i. $f'(2) = 0$

j. An inflection point at $(4, 3)$

Applications

BUSINESS AND ECONOMICS

Stock Prices In Exercises 43 and 44, $P(t)$ is the price of a certain stock at time t during a particular day.

43. a. If the price of the stock is falling faster and faster, are $P'(t)$ and $P''(t)$ positive or negative?

b. Explain your answer.

44. a. When the stock reaches its highest price of the day, are $P'(t)$ and $P''(t)$ positive, zero, or negative?

b. Explain your answer.

45. *Cat Brushes* The cost function to produce x electric cat brushes is given by $C(x) = -10x^2 + 250x$. The demand equation is given by $p = -x^2 - 3x + 299$, where p is the price in dollars.

a. Find and simplify the profit function.

b. Find the number of brushes that will produce the maximum profit.

c. Find the price that produces the maximum profit.

d. Find the maximum profit.

e. Find the point of diminishing returns for the profit function.

LIFE SCIENCES

46. *Weightlifting* An abstract for an article states, "We tentatively conclude that Olympic weightlifting ability in trained subjects undergoes a nonlinear decline with age, in which the second derivative of the performance versus age curve repeatedly changes sign."*

a. What does this quote tell you about the first derivative of the performance versus age curve?

b. Describe what you know about the performance versus age curve based on the information in the quote.

47. *Scaling Laws* Many biological variables depend on body mass, with a functional relationship of the form

$$Y = Y_0 M^b,$$

where M represents body mass, b is a multiple of $1/4$, and Y_0 is a constant.† For example, when Y represents metabolic rate, $b = 3/4$. When Y represents heartbeat, $b = -1/4$. When Y represents life span, $b = 1/4$.

a. Determine which of metabolic rate, heartbeat, and life span are increasing or decreasing functions of mass. Also determine which have graphs that are concave upward and which have graphs that are concave downward.

b. Verify that all functions of the form given above satisfy the equation

$$\frac{dY}{dM} = \frac{b}{M}Y.$$

This means that the rate of change of Y is proportional to Y and inversely proportional to body mass.

48. *Thoroughbred Horses* The association between velocity during exercise and blood lactate concentration after submaximal 800-m exercise of thoroughbred racehorses on sand and grass tracks has been studied. The lactate-velocity relationship can be described by the functions.

$$l_1(v) = .08e^{.33v} \quad \text{and}$$
$$l_2(v) = -.87v^2 + 28.17v - 211.41,$$

where $l_1(v)$ and $l_2(v)$ are the lactate concentrations (in mmol/L) and v is the velocity (in m/sec) of the horse during workout on sand and grass tracks, respectively.§ Sketch the graph of both functions for $13 \le v \le 17$.

*Meltzer, David E., "Age Dependence of Olympic Weightlifting," *Medicine and Science in Sports and Exercise,* Vol. 26, No. 8, Aug. 1994, p. 1053.

†West, Geoffrey B., James H. Brown, and Brian J. Enquist, "A General Model for the Origin of Allometric Scaling Laws in Biology," *Science,* Vol. 276, April 4, 1997, pp. 122–126.

§Davie, A. and D. Evans, "Blood Lactate Responses to Submaximal Field Exercise Tests in Thoroughbred Horses," *The Veterinary Journal,* Vol. 159, 2000, pp. 252–258.

49. *Neuron Communications* In the FitzHugh-Nagumo model of how neurons communicate, the rate of change of the electric potential v with respect to time is given as a function of v by $f(v) = v(a - v)(v - 1)$, where a is a positive constant.[*] Sketch a graph of this function when $a = .25$ and $0 \le v \le 1$.

50. *Fruit Flies* The number of imagoes (sexually mature adult fruit flies) per mated female per day (y) can be approximated by

$$y = 34.7(1.0186)^{-x}x^{-.658},$$

where x is the mean density of the mated population (measured as flies per bottle) over a 16-day period.[†] Sketch the graph of the function.

51. *Blood Volume* A formula proposed by Hurley[‡] for the red cell volume (RCV) in milliliters for males is

$$RCV = 1486S^2 - 4106S + 4514,$$

where S is the surface area (in square meters). A formula given by Pearson et al.,[§] is

$$RCV = 1486S - 825.$$

a. For the value of S which the RCV values given by the two formulas are closest, find the rate of change of RCV with respect to S for both formulas. What does this number represent?

b. The formula for plasma volume for males given by Hurley is

$$PV = 995e^{.6085S},$$

while the formula given by Pearson et al., is

$$PV = 1578S,$$

where PV is measured in millimeters and S in square meters. Find the value of S for which the PV values given by the two formulas are the closest. Then find the value of PV that each formula gives for this value of S.

c. For the value of S found in part b, find the rate of change of PV with respect to S for both formulas. What does this number represent?

d. Notice in parts a and c that both formulas give the same instantaneous rate of change at the value of S for which the function values are closest. Prove that if two functions f and g are differentiable and never cross but are closest together when $x = x_0$, then $f'(x_0) = g'(x_0)$.

SOCIAL SCIENCES

52. *Murder Rate* The number of murders in Chicago from 1990–2002 is approximated by $f(x) = .817x^3 - 15.4x^2 + 49.4x + 874$, where x corresponds to the number of years after 1990—that is, 1991 corresponds to $x = 1$, 1992 corresponds to $x = 2$, and so on.[||]

a. In what years did a relative maximum or a relative minimum number of murders occur? How many murders occurred in those years?

b. In what year did the rate of decrease in murders start to slow down?

53. *Learning* Researchers used a version of the Gompertz curve to model the rate that children learn with the equation

$$y(t) = A^{c^t},$$

where $y(t)$ is the portion of children of age t years passing a certain mental test, $A = .3982 \cdot 10^{-291}$, and $c = .4152$.[#] Find the inflection point and describe what it signifies. (*Hint:* Leave A and c as constants until you are ready to calculate your final answer. If A is too small for your calculator to handle, use common logarithms and properties of logarithms to calculate $(\log A)/(\log e)$. Recall that $d(a^{g(t)})/dt = (\ln a)g'(t)a^{g(t)}$.)

54. *Nuclear Weapons* The graph on the next page shows the stockpile of nuclear weapons held by the United States and by the Soviet Union and its successor states from 1945 to 2002.[**] (See Exercise 47 in the first section of this chapter.)

a. In what years was the U.S. stockpile of weapons at a relative maximum?

[*]Murray, J. D., *Mathematical Biology,* Springer–Verlag, 1989, p.163.
[†]Pearl, R. and S. Parker, *Proc. Natl. Acad. Sci.,* Vol. 8, 1922, p. 212, quoted in *Elements of Mathematical Biology* by Alfred J. Lotka, Dover Publications, 1956, pp. 308–311.
[‡]Hurley, Peter J., "Red Cell Plasma Volumes in Normal Adults," *Journal of Nuclear Medicine,* Vol. 16, 1975, pp. 46–52.
[§]Pearson, T. C. et al. "Interpretation of Measured Red Cell Mass and Plasma Volume in Adults," *British Journal of Haematology,* Vol. 89, 1995, pp. 748–756.
[||]http://qrc.depaul.edu/djabon/Articles/ChicagoCrime20030101.htm.
[#]Courtis, S. A., "Maturation Units for the Measurement of Growth," *School and Society,* Vol. 30, 1929, pp. 683–690.
[**]http://www.nrdc.org/nuclear/nudb/dafig11.asp.

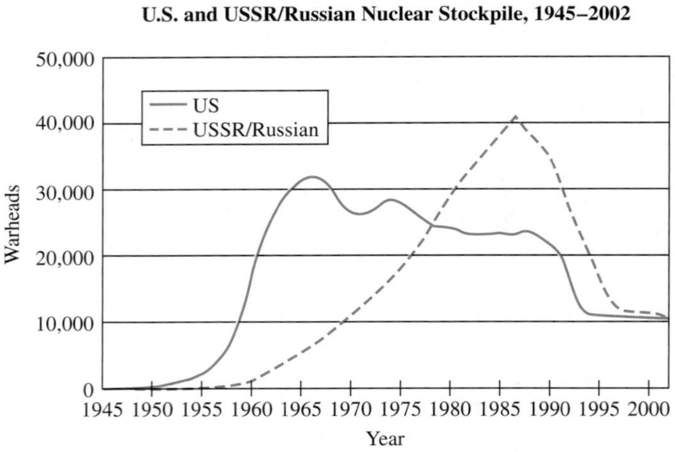

b. When the U.S. stockpile of weapons was at the largest relative maximum, is the graph for the Soviet stockpile concave up or concave down? What does this mean?

PHYSICAL SCIENCES

55. *Velocity and Acceleration* A projectile is shot straight up with an initial velocity of 512 ft per second. Its height above the ground after t seconds is given by $s(t) = 512t - 16t^2$.

a. Find the velocity and acceleration after t seconds.

b. What is the maximum height attained?

c. When does the projectile hit the ground and with what velocity?

CHAPTER

6 Applications of the Derivative

When several variables are related by a single equation, their rates of change are also related. For example, the height and horizontal distance of a kite are related to the length of the string holding the kite. In an exercise in Section 5 we differentiate this relationship to discover how fast the kite flier must let out the string to maintain the kite at a constant height and constant horizontal speed.

What do aluminum cans, shipments of antibiotics, elasticity of demand, and a melting icicle have in common? All involve applications of the derivative. The previous chapter included examples in which we used the derivative to find the maximum or minimum value of a function. This problem is ubiquitous; consider the efforts people expend trying to maximize their income, or to minimize their costs or the time required to complete a task. In this chapter we will treat the topic of optimization in greater depth.

The derivative is applicable in far wider circumstances, however. In roughly 500 B.C., Heraclitus said, "Nothing endures but change," and his observation has relevance here. If change is continuous, rather than in sudden jumps, the derivative can be used to describe the rate of change. This explains why calculus has been applied to so many fields.

6.1 ABSOLUTE EXTREMA

THINK ABOUT IT

How can the maximum number of people who are below poverty level in a given period be found?

We will answer this question later in this section.

If a function has more than one relative maximum, in a practical situation it is often important to know if one function value is larger than any other. In other cases we may want to know whether one function value is smaller than any other. For example, in Figure 1, $f(x_1) \geq f(x)$ for all x in the domain. There is no function value that is smaller than all others, however, because $f(x) \to -\infty$ as $x \to \infty$ or as $x \to -\infty$.

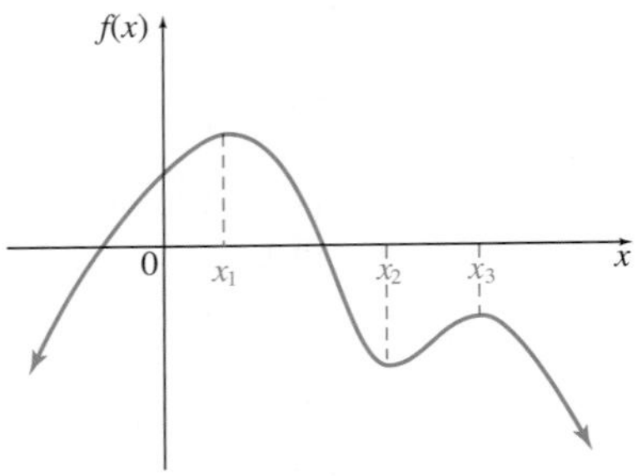

FIGURE 1

The largest possible value of a function is called the *absolute maximum* and the smallest possible value of a function is called the *absolute minimum*. As Figure 1 shows, one or both of these may not exist on the domain of the function, $(-\infty, \infty)$ here. Absolute extrema often coincide with relative extrema, as with $f(x_1)$ in Figure 1. Although a function may have several relative maxima or relative minima, it never has more than one *absolute maximum* or *absolute minimum*.

ABSOLUTE MAXIMUM OR MINIMUM

Let f be a function defined on some interval. Let c be a number in the interval. Then $f(c)$ is the **absolute maximum** of f on the interval if

$$f(x) \leq f(c)$$

for every x in the interval, and $f(c)$ is the **absolute minimum** of f on the interval if

$$f(x) \geq f(c)$$

for every x in the interval.

A function has an **absolute extremum** (plural: **extrema**) at c if it has either an absolute maximum or an absolute minimum there.

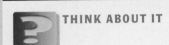

CAUTION Notice that, just like a relative extremum, an absolute extremum is a y-value, not an x-value. ■

Now look at Figure 2, which shows three functions defined on closed intervals. In each case there is an absolute maximum value and an absolute minimum value. These absolute extrema may occur at the endpoints or at relative extrema. As the graphs in Figure 2 show, an absolute extremum is either the largest or the smallest function value occurring on a closed interval, while a relative extremum is the largest or smallest function value in some (perhaps small) open interval.

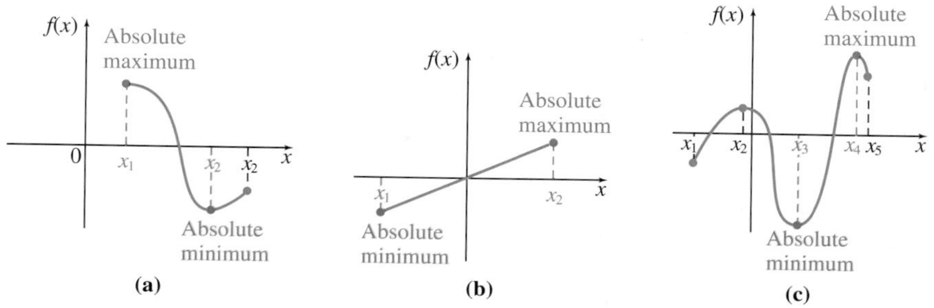

FIGURE 2

Although a function can have only one absolute minimum value and only one absolute maximum value, it can have many points where these values occur. (Note that the absolute maximum value and absolute minimum value are numbers, not points.) As an extreme example, consider the function $f(x) = 2$. The absolute minimum value of this function is clearly 2, as is the absolute maximum value. Both the absolute minimum and the absolute maximum occur at every real number x.

One of the main reasons for the importance of absolute extrema is given by the **extreme value theorem** (which is proved in more advanced courses).

> **EXTREME VALUE THEOREM**
>
> A function f that is continuous on a closed interval $[a, b]$ will have both an absolute maximum and an absolute minimum on the interval.

As Figure 2 shows, an absolute extremum *may* occur on an open interval. See x_2 in part (a) and x_3 in part (c), for example. The extreme value theorem says that a function *must* have both an absolute maximum and an absolute minimum on a closed interval. The conditions that f be *continuous* and on a *closed* interval are very important. In Figure 3, f is discontinuous at $x = b$. Since $f(x)$ gets larger and larger as $x \to b$, there is no absolute (or relative) maximum on (a, b). Also, as $x \to a^+$, the values of $f(x)$ get smaller and smaller approaching $f(a)$, but there is no smallest value of $f(x)$ on the *open* interval (a, b), since there is no endpoint.

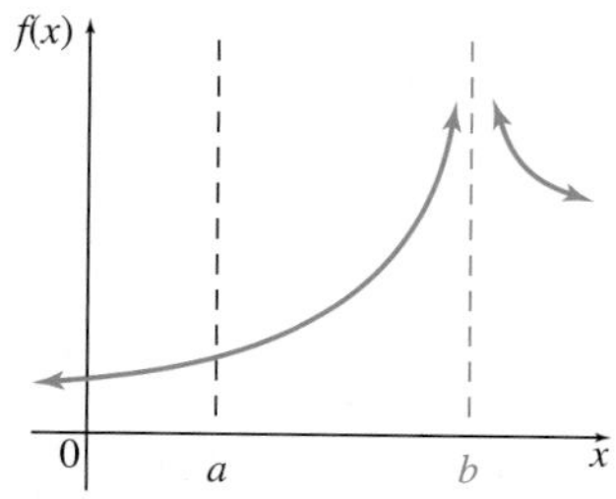

The extreme value theorem guarantees the existence of absolute extrema. To find these extrema, use the following steps.

FIGURE 3

FINDING ABSOLUTE EXTREMA

To find absolute extrema for a function f continuous on a closed interval $[a, b]$:

1. Find all critical numbers for f in (a, b).
2. Evaluate f for all critical numbers in (a, b).
3. Evaluate f for the endpoints a and b of the interval $[a, b]$.
4. The largest value found in Step 2 or 3 is the absolute maximum for f on $[a, b]$, and the smallest value found is the absolute minimum for f on $[a, b]$.

EXAMPLE 1 *Absolute Extrema*

Find the absolute extrema of the function

$$f(x) = x^{8/3} - 16x^{2/3}$$

on the interval $[-1, 8]$.

Solution First look for critical numbers in the interval $(-1, 8)$.

$$\begin{aligned} f'(x) &= \frac{8}{3}x^{5/3} - \frac{32}{3}x^{-1/3} \\ &= \frac{8}{3}\left(x^{5/3} - \frac{4}{x^{1/3}}\right) \\ &= \frac{8}{3}\left(x^{5/3} \cdot \frac{x^{1/3}}{x^{1/3}} - \frac{4}{x^{1/3}}\right) && \text{Factor and find a common denominator.} \\ &= \frac{8}{3}\left(\frac{x^2 - 4}{x^{1/3}}\right) \end{aligned}$$

Set $f'(x) = 0$ and solve for x. Notice that $f'(x) = 0$ at $x = 2$ and $x = -2$, but -2 is not in the interval $(-1, 8)$, so we ignore it. The derivative is undefined at $x = 0$, but the function is defined there, so 0 is also a critical number.

Evaluate the function at the critical numbers and the endpoints.

x-Value	Value of Function
-1	-15
0	0
2	-19.05
8	192

The absolute maximum, 192, occurs when $x = 8$, and the absolute minimum, approximately -19.05, occurs when $x = 2$. A graph of f is shown in Figure 4 on the next page.

In Example 1, a graphing calculator that gives the maximum and minimum values of a function on an interval, such as the `fMax` or `fMin` feature of the TI-83/84 Plus, could replace the table. Alternatively, we could first graph the function on the given interval and then select the feature that gives the maximum or minimum value of the graph of the function instead of completing the table.

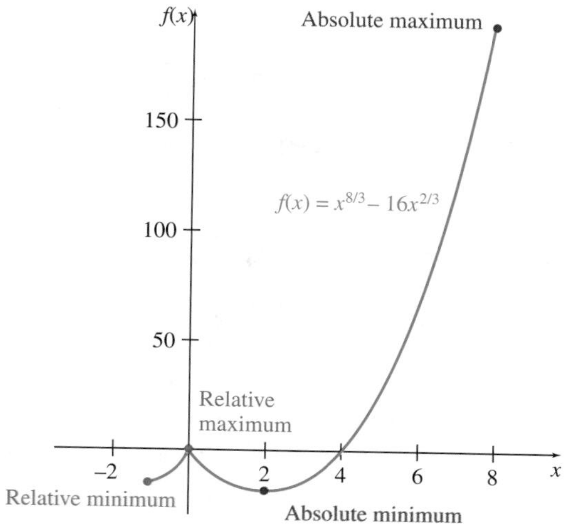

FIGURE 4

EXAMPLE 2 *Absolute Extrema*

Find the locations of the absolute extrema, if they exist, for the function

$$f(x) = 3x^4 - 4x^3 - 12x^2 + 2.$$

Solution In this example, the extreme value theorem does not apply since the domain is an open interval, $(-\infty, \infty)$, rather than a closed interval. Begin as before by finding any critical numbers.

$$f'(x) = 12x^3 - 12x^2 - 24x = 0$$
$$12x(x^2 - x - 2) = 0$$
$$12x(x + 1)(x - 2) = 0$$
$$x = 0 \quad \text{or} \quad x = -1 \quad \text{or} \quad x = 2$$

There are no values of x where $f'(x)$ does not exist. Evaluate the function at the critical numbers.

x-Value	Value of Function
-1	3
0	2
2	-30

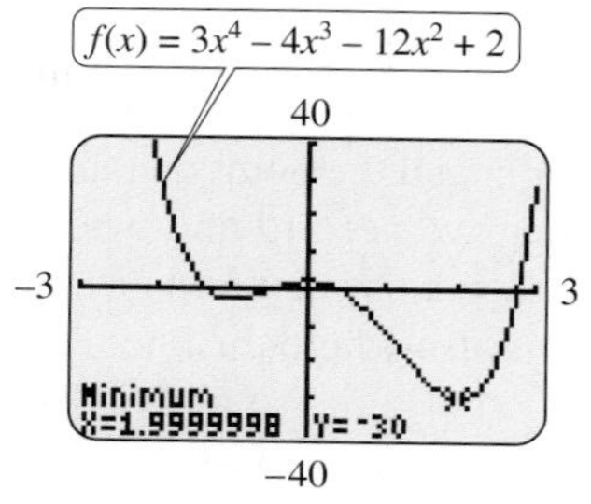

FIGURE 5

For an open interval, rather than evaluating the function at the endpoints, we evaluate the limit of the function when the endpoints are approached. Because the positive x^4-term dominates the other terms as x becomes large,

$$\lim_{x \to \infty} (3x^4 - 4x^3 - 12x^2 + 2) = \infty.$$

The limit is also ∞ as x approaches $-\infty$. Since the function can be made arbitrarily large, it has no absolute maximum. The absolute minimum, -30, occurs at $x = 2$. This result can be confirmed with a graphing calculator, as shown in Figure 5.

EXAMPLE 3 *Poverty*

Based on information provided by the U.S. Census Bureau, the number y of people who were considered to be below the poverty level between 1990 and 2002 can be modeled by the function

$$y = .0399x^3 - .7863x^2 + 3.7794x + 33.2615,$$

where $x = 0$ corresponds to the year 1990. In what year during this period did the number of people living below the poverty level reach its absolute maximum? Based on this model, what was the maximum number of people who were below the poverty level during this time period?

Solution The function is defined on the interval $[0, 12]$. We first look for critical numbers in this interval. Here $f'(x) = .1197x^2 - 1.5726x + 3.7794$. We set this derivative equal to zero and use the quadratic formula to solve for x.

$$.1197x^2 - 1.5726x + 3.7794 = 0$$

$$x = \frac{1.5726 \pm \sqrt{(-1.5726)^2 - 4(.1197)(3.7794)}}{2(.1197)}$$

$$x = 3.17 \quad \text{or} \quad x = 9.97$$

Both x-values are in the interval $[0, 12]$. Now evaluate the function at the critical numbers and the endpoints, 0 and 12.

x-Value	Value of Function	
0	33.3	
3.17	38.6	← Absolute maximum
9.97	32.3	
12	34.3	

During the third year, that is, during 1993, the poverty level reached its absolute maximum in the given period, with almost 39 million people considered to be below poverty level. It is also worth noting that in 2000, the number of people living below poverty level reached its minimum during this time period.

Graphical Optimization Figure 6, from an economics textbook, shows how the total production output for a product might vary with the hours of labor used.* A manager may want to know how many hours of labor to use in order to maximize the output per hour of labor. For any point on the curve, the y-coordinate measures the output and the x-coordinate measures the hours of labor, so the y-coordinate divided by the x-coordinate gives the output per hour of labor. This quotient is also the slope of the line through the origin and the point on the curve. Therefore, to maximize the output per hour of labor, we need to find where this slope is greatest. As shown in Figure 6, on the next page, this occurs when approximately 30 hours of labor are used. Notice that this is also where the line from the

*Browning, Edgar K. and Jacquelene M. Browning, *Microeconomic Theory and Applications,* 4th ed., New York: HarperCollins, 1992, p. 208. This material is used by permission of John Wiley & Sons, Inc.

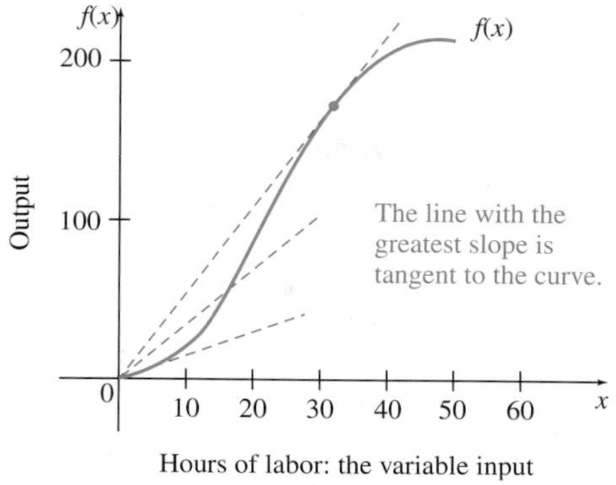

FIGURE 6

origin to the curve is tangent to the curve. Another way of looking at this is to say the point on the curve where the tangent line passes through the origin is the point that maximizes the output per hour of labor.

We can show that, in general, when $y = f(x)$ represents the output as a function of input, the maximum output per unit input occurs when the line from the origin to a point on the graph of the function is tangent to the function. Our goal is to maximize

$$g(x) = \frac{\text{output}}{\text{input}} = \frac{f(x)}{x}.$$

Taking the derivative and setting it equal to 0 gives

$$g'(x) = \frac{xf'(x) - f(x)}{x^2} = 0$$

$$xf'(x) = f(x)$$

$$f'(x) = \frac{f(x)}{x}.$$

Notice that $f'(x)$ gives the slope of the tangent line at the point, and $f(x)/x$ gives the slope of the line from the origin to the point. When these are equal, as in Figure 6, the output per input is maximized. In other examples, the point on the curve where the tangent line passes through the origin gives a minimum. For a life science example of this, see Exercise 50 in the section on the Definition of the Derivative.

6.1 EXERCISES

Find the locations of any absolute extrema for the functions with graphs as follows.

1.

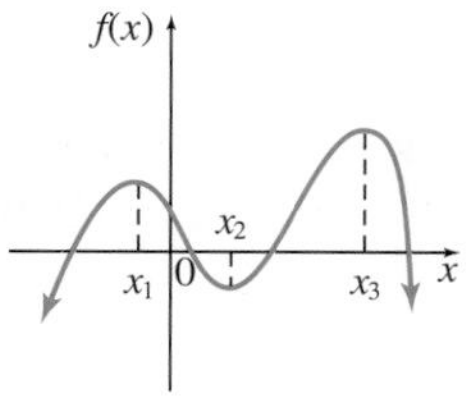

2.

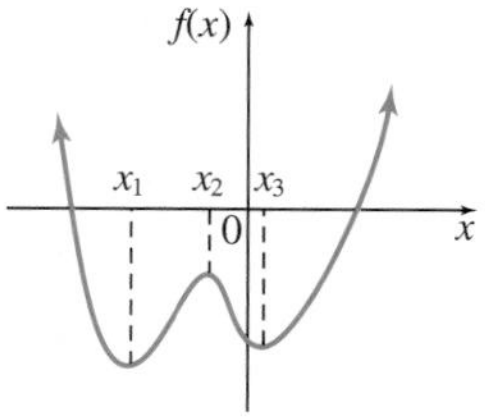

3.

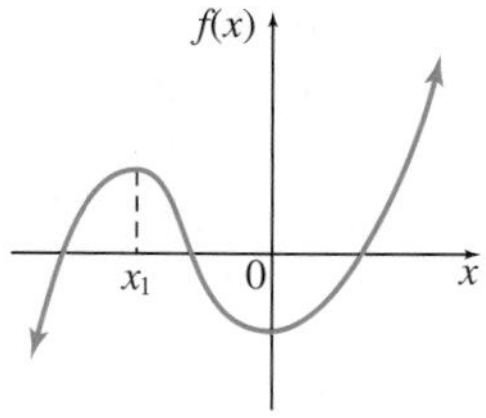

4.

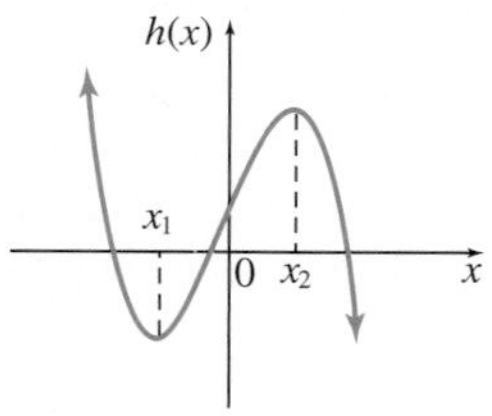

5.

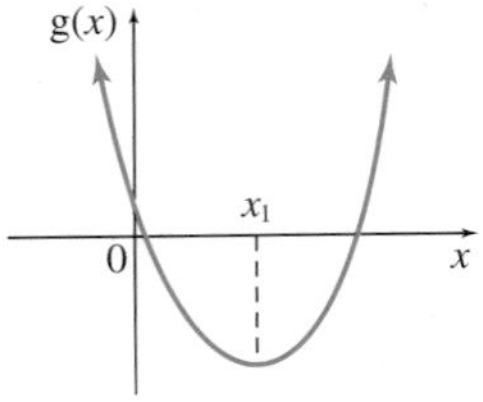

6.

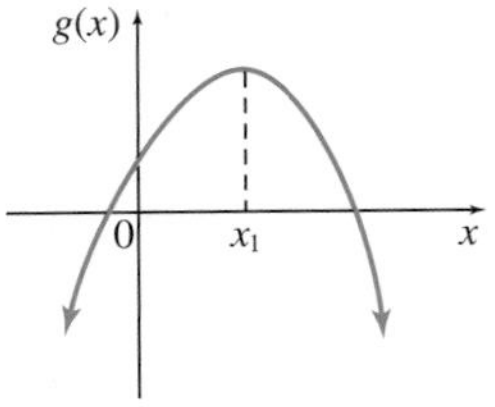

7.

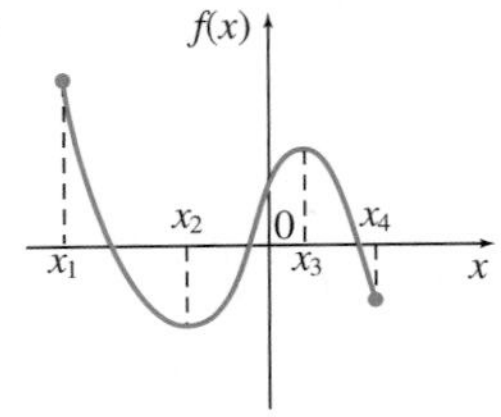

8.

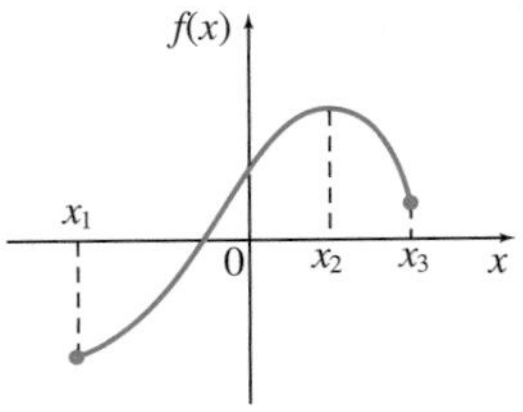

9. What is the difference between a relative extremum and an absolute extremum? Can a relative extremum be an absolute extremum? Is a relative extremum necessarily an absolute extremum?

Find the absolute extrema if they exist, as well as all values of x where they occur, for each function, and specified domain. If you have one, use a graphing calculator to verify your answers.

10. $f(x) = x^3 - 3x^2 - 24x + 5;\ [-3, 6]$

11. $f(x) = x^3 - 6x^2 + 9x - 8;\ [0, 5]$

12. $f(x) = \frac{1}{3}x^3 - \frac{1}{2}x^2 - 6x + 3;\ [-4, 4]$

13. $f(x) = \frac{1}{3}x^3 + \frac{3}{2}x^2 - 4x + 1;\ [-5, 2]$

14. $f(x) = x^4 - 32x^2 - 7;\ [-5, 6]$

15. $f(x) = x^4 - 18x^2 + 1;\ [-4, 4]$

16. $f(x) = \frac{8 + x}{8 - x};\ [4, 6]$

17. $f(x) = \frac{1 - x}{3 + x};\ [0, 3]$

18. $f(x) = \frac{x}{x^2 + 2};\ [0, 4]$

19. $f(x) = \frac{x - 1}{x^2 + 1};\ [1, 5]$

20. $f(x) = (x^2 + 18)^{2/3};\ [-3, 3]$

21. $f(x) = (x^2 + 4)^{1/3};\ [-2, 2]$

22. $f(x) = (x + 1)(x + 2)^2;\ [-4, 0]$

23. $f(x) = (x - 3)(x - 1)^3;\ [-2, 3]$

Graph each function on the indicated domain, and use the capabilities of your calculator to find the location of the absolute extrema.

24. $f(x) = \frac{x^3 + 2x + 5}{x^4 + 3x^3 + 10};\ [-3, 0]$

25. $f(x) = \frac{-5x^4 + 2x^3 + 3x^2 + 9}{x^4 - x^3 + x^2 + 7};\ [-1, 1]$

Find the absolute extrema if they exist, as well as all values of x where they occur.

26. $f(x) = 12 - x - 9/x,\quad x > 0$

27. $f(x) = 2x + 8/x^2 + 1,\quad x > 0$

28. $f(x) = x^4 - 4x^3 + 4x^2 + 1$

29. $f(x) = -3x^4 + 8x^3 + 18x^2 + 2$

30. $f(x) = \frac{x}{x^2 + 1}$

31. $f(x) = \frac{x - 1}{x^2 + 2x + 6}$

32. Find the absolute maximum and minimum of $f(x) = 2x - 3x^{2/3}$ **(a)** on the interval $[-1, 5.5]$; **(b)** on the interval $[.5, 2]$.

Applications

BUSINESS AND ECONOMICS

33. *Bank Robberies* According to the FBI, the number of bank robbery incidents in the United States for the years 1990–2001 is given in the figure on the next page.* Consider the closed interval [1990, 2001].

a. Give all approximate relative maxima and minima and when they occur on the interval.

b. Give the approximate absolute maximum and minimum values and when they occur on the interval. Interpret your results.

*http://www.fbi.gov.

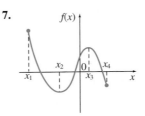

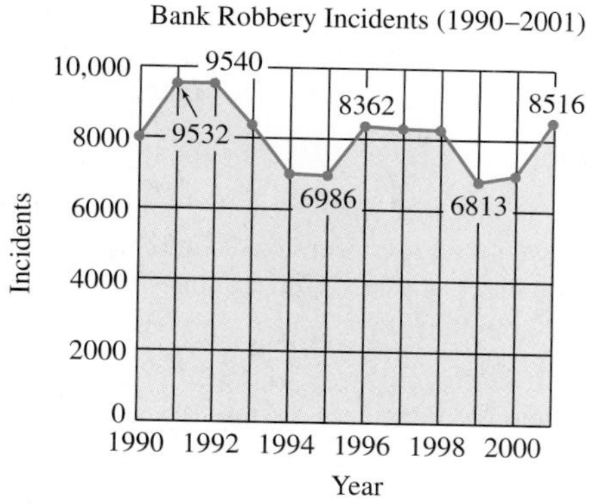

34. *Automobile Theft* According to the FBI, the rate (per 100,000 inhabitants) at which automobiles were stolen each year during the years 1984–2002 are given in the figure. Consider the closed interval [1984, 2002].*

a. Give all relative maxima and minima on the interval and when they occur.

b. Give the absolute maximum and minimum on the interval and when they occur.

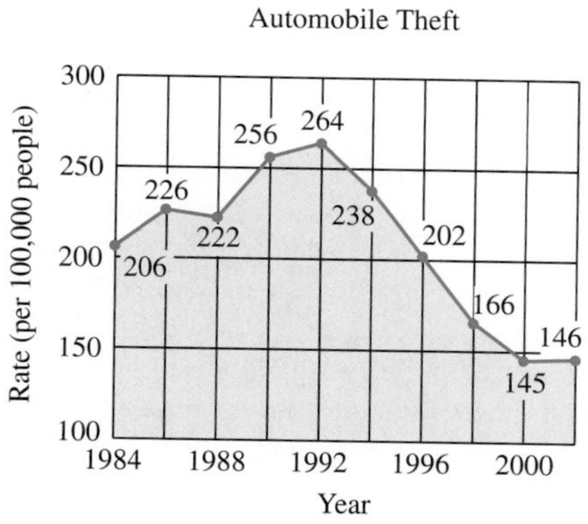

35. *Profit* A company has found that its weekly profit from the sale of x units of an auto part is given by

$$P(x) = -.02x^3 + 600x - 20{,}000.$$

Production bottlenecks limit the number of units that can be made per week to no more than 150, while a long-term contract requires that at least 50 units be made each week. Find the maximum possible weekly profit that the firm can make.

36. *Profit* The total profit $P(x)$ (in thousands of dollars) from the sale of x hundred thousand automobile tires is approximated by

$$P(x) = -x^3 + 9x^2 + 120x - 400, \quad x \geq 5.$$

Find the number of hundred thousands of tires that must be sold to maximize profit. Find the maximum profit.

Average Cost Find the minimum value of the average cost for the given cost function on the given intervals.

37. $C(x) = 81x^2 + 17x + 324$ on the following intervals.

a. $1 \leq x \leq 10$ **b.** $10 \leq x \leq 20$

38. $C(x) = x^3 + 37x + 250$ on the following intervals.

a. $1 \leq x \leq 10$ **b.** $10 \leq x \leq 20$

Cost Each graph gives the cost as a function of production level. Use the method of graphical optimization to estimate the production level that results in the minimum cost per item produced.

39.

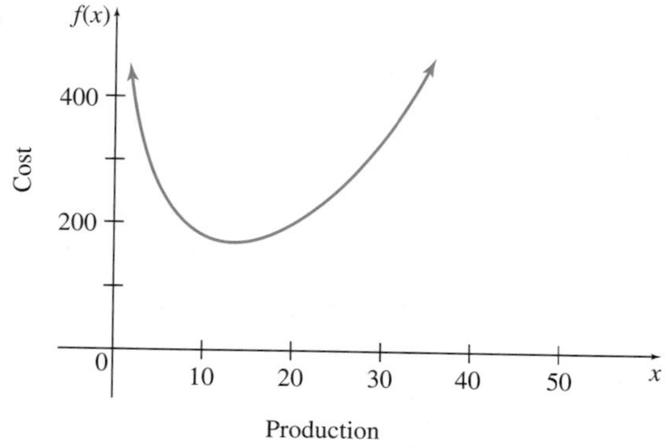

40.

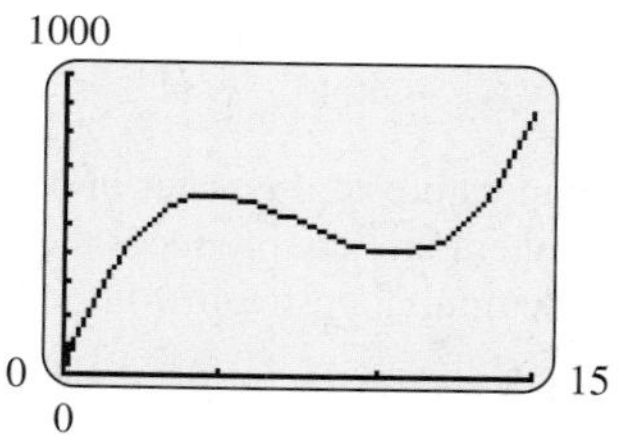

Profit Each graph gives the profit as a function of production level. Use graphical optimization to estimate the production level that gives the maximum profit per item produced.

41.

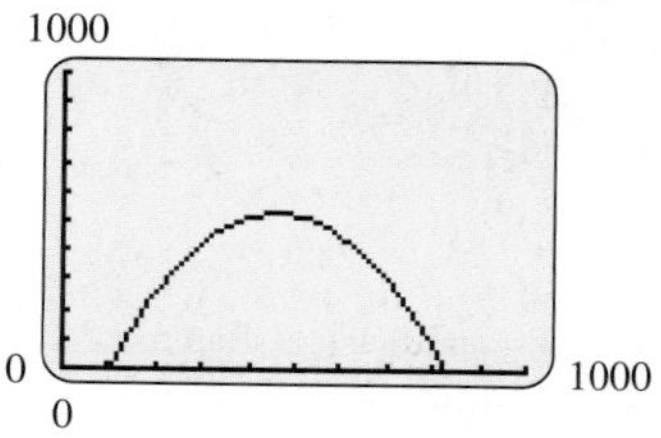

*http://www.fbi.gov.

42. 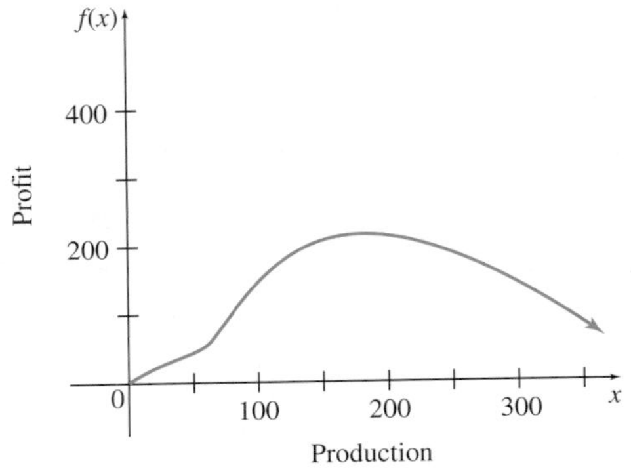

LIFE SCIENCES

43. *Pollution* A marshy region used for agricultural drainage has become contaminated with selenium. It has been determined that flushing the area with clean water will reduce the selenium for a while, but it will then begin to build up again. A biologist has found that the percent of selenium in the soil x months after the flushing begins is given by

$$f(x) = \frac{x^2 + 36}{2x}, \quad 1 \le x \le 12.$$

When will the selenium be reduced to a minimum? What is the minimum percent?

44. *Salmon Spawning* The number of salmon swimming upstream to spawn is approximated by

$$S(x) = -x^3 + 3x^2 + 360x + 5000, \quad 6 \le x \le 20,$$

where x represents the temperature of the water in degrees Celsius. Find the water temperature that produces the maximum number of salmon swimming upstream.

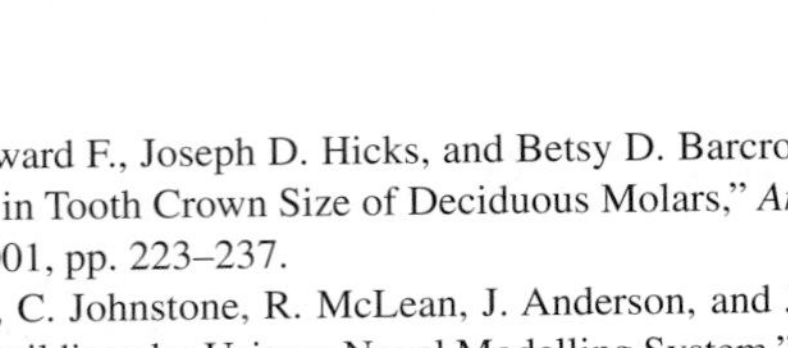

45. *Molars* Researchers have determined that the mesiodistal crown length of deciduous mandibular first molars is related to the postconception age of the tooth as

$$L(t) = -.01t^2 + .788t - 7.048,$$

where $L(t)$ is the crown length (in millimeters) of the molar t weeks after conception.* Find the maximum length in mesiodistal crown of mandibular first molars during weeks 22 through 28.

46. *Fungal Growth* Because of the time that many people spend indoors, there is a concern about the health risk of being exposed to harmful fungi that thrive in buildings. The risk appears to increase in damp environments. Researchers have discovered that by controlling both the temperature and the relative humidity in a building, the growth of the fungus *A. versicolor* can be limited. The relationship between temperature and relative humidity, which limits growth, can be described by

$$R(T) = -.00007T^3 + .0401T^2 - 1.6572T + 97.086, \quad 15 \le T \le 46,$$

where $R(T)$ is the relative humidity (in percent) and T is the temperature (in degrees Celsius).† Find the temperature at which the relative humidity is minimized.

PHYSICAL SCIENCES

47. *Gasoline Mileage* From information given in a recent business publication, we constructed the mathematical model

$$M(x) = -\frac{1}{45}x^2 + 2x - 20, \quad 30 \le x \le 65,$$

to represent the miles per gallon used by a certain car at a speed of x mph. Find the absolute maximum miles per gallon and the absolute minimum.

48. *Gasoline Mileage* For a certain compact car,

$$M(x) = -.018x^2 + 1.24x + 6.2, \quad 30 \le x \le 60,$$

represents the miles per gallon obtained at a speed of x mph. Find the absolute maximum miles per gallon and the absolute minimum.

GENERAL INTEREST

Area A piece of wire 12 ft long is cut into two pieces. (See the figure.) One piece is made into a circle and the other piece is

*Harris, Edward F., Joseph D. Hicks, and Betsy D. Barcroft, "Tissue Contributions to Sex and Race: Differences in Tooth Crown Size of Deciduous Molars," *American Journal of Physical Anthropology*, Vol. 115, 2001, pp. 223–237.

†Rowan, N., C. Johnstone, R. McLean, J. Anderson, and J. Clarke, "Prediction of Toxigenic Fungal Growth in Buildings by Using a Novel Modelling System," *Applied and Environmental Microbiology*, Vol. 65, No. 11, Nov. 1999, pp. 4814–4821.

made into a square. Let the piece of length x be formed into a circle. We allow x to equal 0 or 12, so all the wire is used for the square or for the circle.

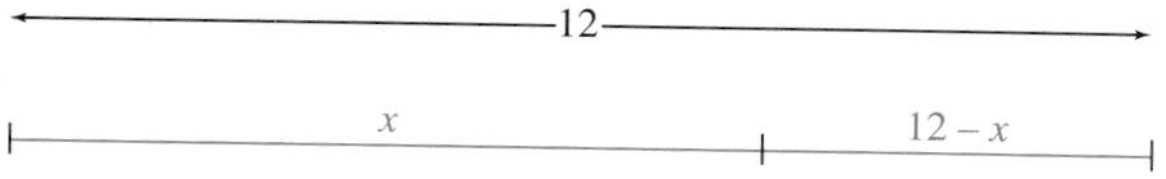

$$\text{Radius of circle} = \frac{x}{2\pi} \qquad \text{Area of circle} = \pi\left(\frac{x}{2\pi}\right)^2$$

$$\text{Side of square} = \frac{12 - x}{4} \qquad \text{Area of square} = \left(\frac{12 - x}{4}\right)^2$$

49. Where should the cut be made in order to minimize the sum of the areas enclosed by both figures?

50. Where should the cut be made in order to make the sum of the areas maximum? (*Hint*: Remember to use the endpoints of a domain when looking for absolute maxima and minima.)

51. *Information Content* Suppose dots and dashes are transmitted over a telegraph line so that dots occur a fraction p of the time (where $0 < p < 1$) and dashes occur a fraction $1 - p$ of the time. The *information content* of the telegraph line is given by $I(p)$, where

$$I(p) = -p \ln p - (1 - p)\ln(1 - p).$$

a. Show that $I'(p) = -\ln p + \ln(1 - p)$.

b. Set $I'(p) = 0$ and find the value of p that maximizes the information content.

c. How might the result in part b be used?

6.2 APPLICATIONS OF EXTREMA

THINK ABOUT IT *How should boxes and cans be designed to minimize the material needed to construct them?*

In Examples 3 and 4 we will use the techniques of calculus to find an answer to this question.

In this section, we give several examples showing applications of calculus to maximum and minimum problems. To solve these examples, go through the following steps.

SOLVING AN APPLIED EXTREMA PROBLEM

1. Read the problem carefully. Make sure you understand what is given and what is unknown.
2. If possible, sketch a diagram. Label the various parts.
3. Decide on the variable that must be maximized or minimized. Express that variable as a function of *one* other variable. Be sure to find the domain of the function.
4. Find the critical points for the function from Step 3.
5. If the domain is a closed interval, evaluate the function at the endpoints and the critical points to see which yields the absolute maximum or minimum. If the domain is an open interval, find the limit as the endpoints are approached, as in Example 2 of the previous section, to determine if an absolute maximum or minimum exists at one of the critical points.

CAUTION Do not skip Step 5 in the preceding box. If a problem asks you to maximize a quantity and you find a critical point at Step 4, do not automatically assume the maximum occurs there, for it may occur at an endpoint, as in Exercise 50 of the previous section, or it may not exist at all.

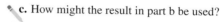

An infamous case of such an error occurred in a 1945 study of "flying wing" aircraft designs similar to the Stealth bomber. In seeking to maximize the range of the aircraft (how far it can fly on a tank of fuel), the study's authors found that a critical point occurred when almost all of the volume of the plane was in the wing. They claimed that this critical point was a maximum. But another engineer later found that this critical point, in fact, *minimized* the range of the aircraft!* ■

EXAMPLE 1 *Maximization*

Find two nonnegative numbers x and y for which $2x + y = 30$, such that xy^2 is maximized.

Solution First we must decide what is to be maximized and assign a variable to that quantity. Here, xy^2 is to be maximized, so let

$$M = xy^2.$$

Now, express M in terms of just *one* variable. Use the equation $2x + y = 30$ to do that. Solve $2x + y = 30$ for either x or y. Solving for y gives

$$\begin{aligned} 2x + y &= 30 \\ y &= 30 - 2x. \end{aligned}$$

Substitute for y in the expression for M to get

$$\begin{aligned} M &= x(30 - 2x)^2 \\ &= x(900 - 120x + 4x^2) \\ &= 900x - 120x^2 + 4x^3. \end{aligned}$$

Note that x must be at least 0. Since y must also be at least 0, we require $30 - 2x \geq 0$, so $x \leq 15$. Thus x is confined to the interval $[0, 15]$.

Find the critical points for M by finding dM/dx, then solving the equation $dM/dx = 0$ for x.

$$\begin{aligned} \frac{dM}{dx} = 900 - 240x + 12x^2 &= 0 \\ 12(75 - 20x + x^2) &= 0 \\ (5 - x)(15 - x) &= 0 \quad \text{Divide both sides by 12 and factor} \\ x = 5 \quad \text{or} \quad x &= 15 \end{aligned}$$

x	M	
0	0	
5	2000	← Maximum
15	0	

Find M for the critical numbers $x = 5$ and $x = 15$ as well as for $x = 0$, one endpoint of the domain. (The other endpoint has already been included as a critical number.)

We see in the table that the maximum value of the function occurs when $x = 5$. Since $y = 30 - 2(5) = 20$, the values that maximize xy^2 are $x = 5$ and $y = 20$.

EXAMPLE 2 *Minimizing Time*

A math professor participating in the sport of orienteering must get to a specific tree in the woods as fast as possible. He can get there by traveling east along the

*Biddle, Wayne, "Skeleton Alleged in the Stealth Bomber's Closet," *Science,* Vol. 244, May 12, 1989.

trail for 300 m, and then north through the woods for 800 m. He can run 160 m per minute along the trail, but only 70 m per minute through the woods. Running directly through the woods toward the tree minimizes the distance, but he will be going slowly the whole time. He could instead run 300 m along the trail before entering the woods, maximizing the total distance but minimizing the time in the woods. Perhaps the fastest route is a combination, as shown in Figure 7. Find the path that will get him to the tree in the minimum time.

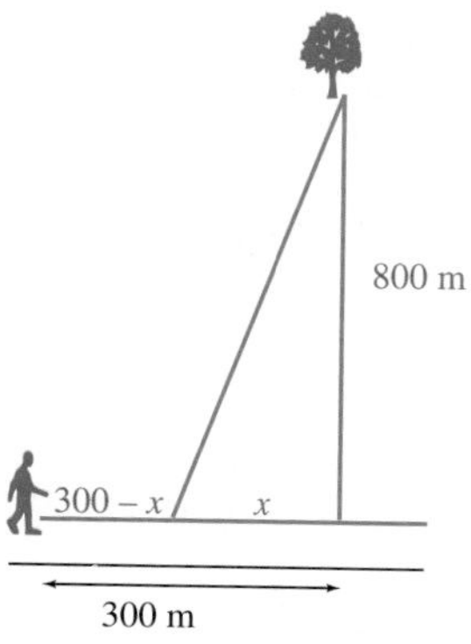

FIGURE 7

Solution Let x be the distance shown in Figure 7, so the distance he runs on the trail is $300 - x$. By the Pythagorean theorem, the distance he runs through the woods is $\sqrt{800^2 + x^2}$. The total time is the sum of the time on the trail and the time through the woods. Since time = distance/speed, the total time is

$$T(x) = \frac{\sqrt{800^2 + x^2}}{70} + \frac{300 - x}{160},$$

where $0 \le x \le 300$. Set the derivative equal to 0. Since $\sqrt{800^2 + x^2} = (800^2 + x^2)^{1/2}$,

$$T'(x) = \frac{1}{70}\left(\frac{1}{2}\right)(800^2 + x^2)^{-1/2}(2x) - \frac{1}{160} = 0.$$

$$\frac{x}{70\sqrt{800^2 + x^2}} = \frac{1}{160}$$

$$16x = 7\sqrt{800^2 + x^2} \qquad \text{Cross multiply.}$$

$$256x^2 = 49(800^2 + x^2) = (49 \cdot 800^2) + 49x^2 \qquad \text{Square both sides.}$$

$$207x^2 = 49 \cdot 800^2$$

$$x^2 = \frac{49 \cdot 800^2}{207}$$

$$x = \frac{7 \cdot 800}{\sqrt{207}} \approx 389$$

x	$T(x)$	
0	13.93	
300	12.83	← Minimum

Since 389 is not in the interval $(0, 300)$, the minimum time must occur at one of the endpoints.

From the table we see that the time is minimized when $x = 300$, that is, when the professor heads straight for the tree.

EXAMPLE 3 *Maximizing Volume*

An open box is to be made by cutting a square from each corner of a 12-in. by 12-in. piece of metal and then folding up the sides. What size square should be cut from each corner to produce a box of maximum volume?

Solution Let x represent the length of a side of the square that is cut from each corner, as shown in Figure 8(a) on the next page. The width of the box is $12 - 2x$, with the length also $12 - 2x$. As shown in Figure 8(b), the depth of the box will be x inches. The volume of the box is given by the product of the length, width, and height. In this example, the volume, $V(x)$, depends on x:

$$V(x) = x(12 - 2x)(12 - 2x) = 144x - 48x^2 + 4x^3.$$

Clearly, $0 \le x$, and since neither the length nor the width can be negative, $0 \le 12 - 2x$, so $x \le 6$. Thus, the domain of V is the interval $[0, 6]$.

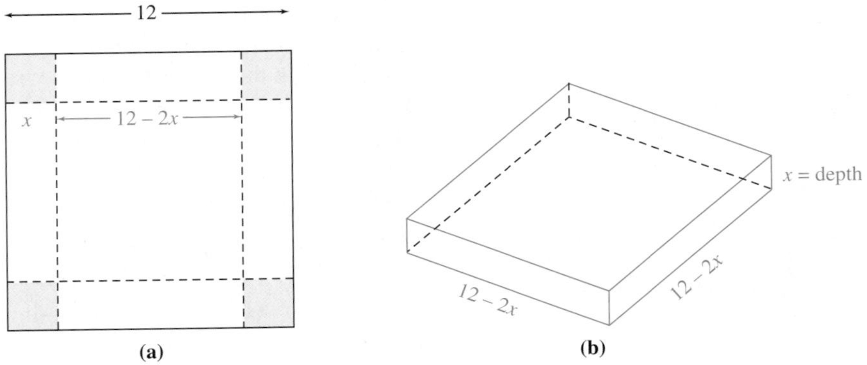

FIGURE 8

The derivative is $V'(x) = 144 - 96x + 12x^2$. Set this derivative equal to 0.

$$12x^2 - 96x + 144 = 0$$
$$12(x^2 - 8x + 12) = 0$$
$$12(x - 2)(x - 6) = 0$$
$$x - 2 = 0 \quad \text{or} \quad x - 6 = 0$$
$$x = 2 \qquad\qquad x = 6$$

x	$V(x)$	
0	0	
2	128	← Maximum
6	0	

Find $V(x)$ for x equal to 0, 2, and 6 to find the depth that will maximize the volume. The table indicates that the box will have maximum volume when $x = 2$ and that the maximum volume will be 128 in^3.

EXAMPLE 4 *Minimizing Area*

A company wants to manufacture cylindrical aluminum cans with a volume of 1000 cm^3 (1 liter). What should the radius and height of the can be to minimize the amount of aluminum used?

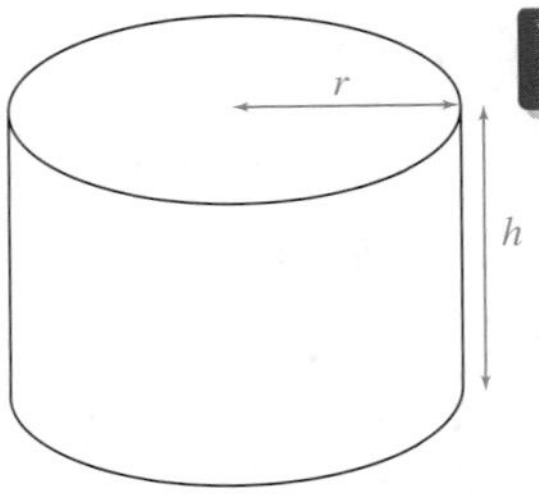

FIGURE 9

Solution The two variables in this problem are the radius and the height of the can, which we shall label r and h, as in Figure 9. Minimizing the amount of aluminum used requires minimizing the surface area of the can, which we will designate S. The surface area consists of a top and a bottom, each of which is a circle with an area πr^2, plus the side. If the side was sliced vertically and unrolled, it would form a rectangle with height h and width equal to the circumference of the can, which is $2\pi r$. Thus the surface area is given by

$$S = 2\pi r^2 + 2\pi rh.$$

The right side of the equation involves two variables. We need to get a function of a single variable. We can do this by using the information about the volume of the can:

$$V = \pi r^2 h = 1000.$$

(Here we have used the formula for the volume of a cylinder.) Solve this for h:

$$h = \frac{1000}{\pi r^2}.$$

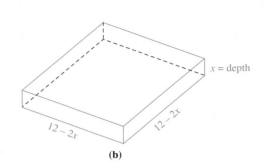

(Solving for r would have involved a square root and a more complicated function.)

We now substitute this expression for h into the equation for S to get

$$S = 2\pi r^2 + 2\pi r\frac{1000}{\pi r^2} = 2\pi r^2 + \frac{2000}{r}.$$

There are no restrictions on r other than that it be a positive number, so the domain of S is $(0, \infty)$.

Find the critical points for S by finding dS/dr, then solving the equation $dS/dr = 0$ for r.

$$\frac{dS}{dr} = 4\pi r - \frac{2000}{r^2} = 0$$

$$4\pi r^3 = 2000$$

$$r^3 = \frac{500}{\pi}$$

Take the cube root of both sides to get

$$r = \left(\frac{500}{\pi}\right)^{1/3} \approx 5.419$$

centimeters. Substitute this expression into the equation for h to get

$$h = \frac{1000}{\pi 5.419^2} \approx 10.84$$

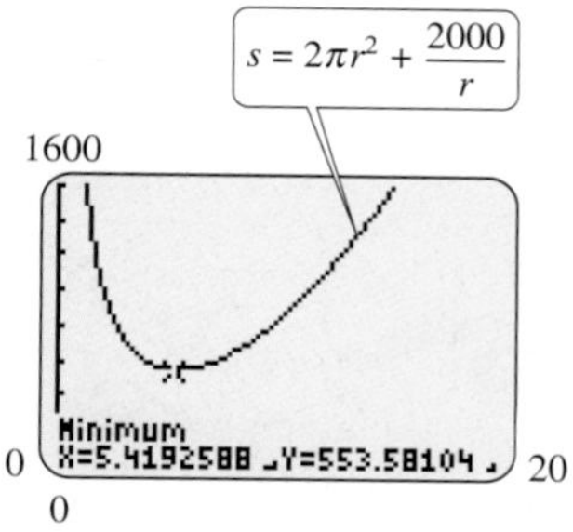

FIGURE 10

centimeters. Notice that the height of the can is twice its radius. But we are not yet assured that the critical number we have found is indeed the minimum. Since the domain is an open interval, find the limit as the endpoints are approached.

$$\lim_{r\to 0} S = \lim_{r\to\infty} S = \infty$$

The surface area becomes arbitrarily large as r approaches the endpoints of the domain, so the critical number must give the absolute minimum surface area. The graphing calculator screen in Figure 10 confirms this result.

Notice that if the previous example had asked for the height and radius that maximize the amount of aluminum used, the problem would have no answer. There is no maximum for a function that can be made arbitrarily large.

In Example 4, we could use the first derivative test to observe that since $dS/dr < 0$ when $0 < r < 5.419$, and $dS/dr > 0$ when $r > 5.419$, there is a relative minimum at $r = 5.419$. The second derivative test also leads to this conclusion because $d^2S/dr^2 = 4\pi + 4000/r^3 > 0$ when $r > 0$. Because there is only one relative extremum, the relative minimum must be the absolute minimum in this case. Be careful when using the first or second derivative test, however—in general, a relative extremum is not necessarily an absolute extremum.

For most living things, reproduction is *seasonal*—it can take place only at selected times of the year.* Large whales, for example, reproduce every two years during a relatively short time span of about two months. Shown on the time

*From Cullen, Michael R., *Mathematics for the Biosciences*. Copyright © 1983 PWS Publishers. Reprinted by permission.

axis in Figure 11 are the reproductive periods. Let S = number of adults present during the reproductive period and let R = number of adults that return the next season to reproduce.

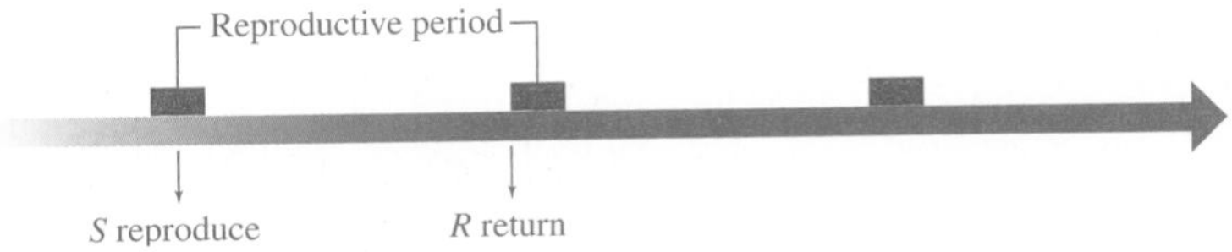

FIGURE 11

If we find a relationship between R and S, $R = f(S)$, then we have formed a **spawner-recruit** function or **parent-progeny** function. These functions are notoriously hard to develop because of the difficulty of obtaining accurate counts and because of the many hypotheses that can be made about the life stages. We will simply suppose that the function f takes various forms.

If $R > S$, we can presumably harvest

$$H = R - S = f(S) - S$$

individuals, leaving S to reproduce. Next season, $R = f(S)$ will return and the harvesting process can be repeated, as shown in Figure 12.

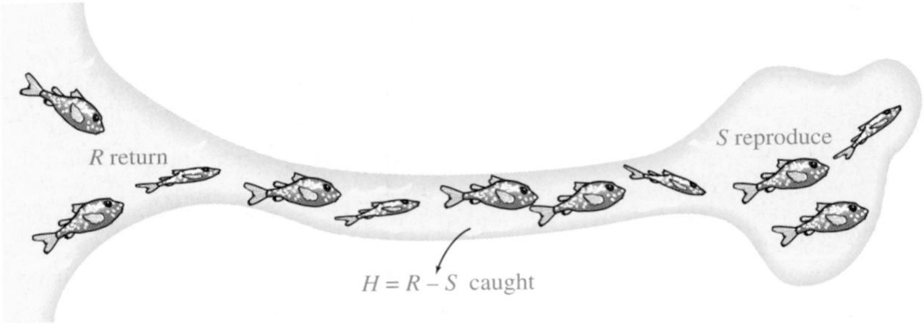

FIGURE 12

Let S_0 be the number of spawners that will allow as large a harvest as possible without threatening the population with extinction. Then $H(S_0)$ is called the **maximum sustainable harvest.**

EXAMPLE 5 *Maximum Sustainable Harvest*

Suppose the spawner-recruit function for Idaho rabbits is $f(S) = 2.17\sqrt{S}\ln(S+1)$, where S is measured in thousands of rabbits. Find S_0 and the maximum sustainable harvest, $H(S_0)$.

Solution S_0 is the value of S that maximizes H. Since

$$\begin{aligned} H(S) &= f(S) - S \\ &= 2.17\sqrt{S}\ln(S+1) - S, \end{aligned}$$

$$H'(S) = 2.17\left(\frac{\ln(S+1)}{2\sqrt{S}} + \frac{\sqrt{S}}{S+1}\right) - 1.$$

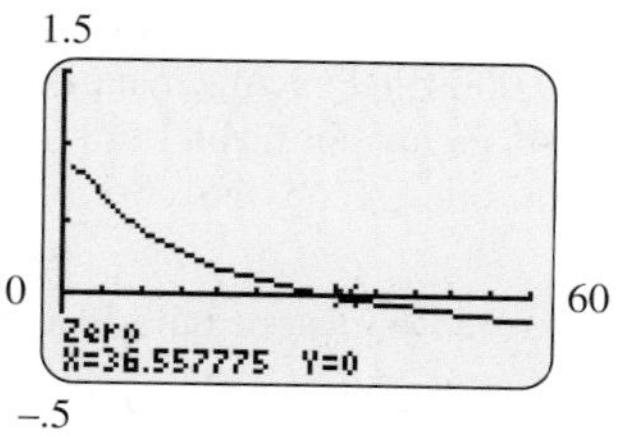

FIGURE 13

Now we want to set this derivative equal to 0 and solve for S.

$$0 = 2.17\left(\frac{\ln(S+1)}{2\sqrt{S}} + \frac{\sqrt{S}}{S+1}\right) - 1.$$

This equation cannot be solved analytically, so we will graph $H'(S)$ with a graphing calculator and find any S-values where $H'(S)$ is 0. (An alternative approach is to use the equation solver some graphing calculators have.) The graph with the value where $H'(S)$ is 0 is shown in Figure 13.

From the graph we see that $H'(S) = 0$ when $S = 36.557775$, so the number of rabbits needed to sustain the population is about 36,600. A graph of H will show that this is a maximum. From the graph, using the capability of the calculator, we find that the harvest is $H(36.557775) \approx 11.015504$. These results indicate that after one reproductive season, a population of 36,600 rabbits will have increased to 47,600. Of these, 11,000 may be harvested, leaving 36,600 to regenerate the population. Any harvest larger than 11,000 will threaten the future of the rabbit population, while a harvest smaller than 11,000 will allow the population to grow larger each season. Thus 11,000 is the maximum sustainable harvest for this population.

6.2 EXERCISES

In Exercises 1–4, use the steps shown in Exercise 1 to find nonnegative numbers x and y that satisfy the given requirements. Give the optimum value of the indicated expression.

1. $x + y = 100$ and the product $P = xy$ is as large as possible.

a. Solve $x + y = 100$ for y.

b. Substitute the result from part a into $P = xy$, the equation for the variable that is to be maximized.

c. Find the domain of the function P found in part b.

d. Find dP/dx. Solve the equation $dP/dx = 0$.

e. Evaluate P at any solutions found in part d, as well as the endpoints of the domain found in part c.

f. Give the maximum value of P, as well as the two numbers x and y whose product is that value.

2. The sum of x and y is 200 and the sum of the squares of x and y is minimized.

3. $x + y = 150$ and x^2y is maximized.

4. $x + y = 45$ and xy^2 is maximized.

Applications

BUSINESS AND ECONOMICS

Average Cost *In Exercises 5 and 6, determine the average cost function. Use a graphing calculator to find where the average cost is smallest by taking the derivative, then finding where the derivative is 0. Check your work by finding the minimum from the graph of the function.*

5. $C(x) = \frac{1}{2}x^3 + 2x^2 - 3x + 35$

6. $C(x) = 10 + 20x^{1/2} + 16x^{3/2}$

7. *Revenue* If the price charged for a candy bar is $p(x)$ cents, then x thousand candy bars will be sold in a certain city, where

$$p(x) = 100 - \frac{x}{10}.$$

a. Find an expression for the total revenue from the sale of x thousand candy bars.

b. Find the value of x that leads to maximum revenue.

c. Find the maximum revenue.

8. *Revenue* The sale of compact disks of "lesser" performers is very sensitive to price. If a CD manufacturer charges $p(x)$ dollars per CD, where

$$p(x) = 6 - \frac{x}{8},$$

then x thousand CDs will be sold.

a. Find an expression for the total revenue from the sale of x thousand CDs.

b. Find the value of x that leads to maximum revenue.

c. Find the maximum revenue.

9. *Area* A farmer has 1200 m of fencing. He wants to enclose a rectangular field bordering a river, with no fencing needed along the river. (See the sketch.) Let x represent the width of the field.

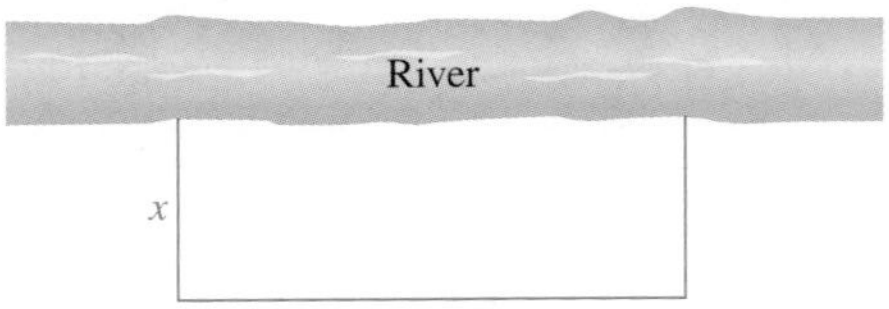

a. Write an expression for the length of the field.

b. Find the area of the field (area = length × width).

c. Find the value of x leading to the maximum area.

d. Find the maximum area.

10. *Area* Find the dimensions of the rectangular field of maximum area that can be made from 200 m of fencing material. (This fence has four sides.)

11. *Area* An ecologist is conducting a research project on breeding pheasants in captivity. She first must construct suitable pens. She wants a rectangular area with two additional fences across its width, as shown in the sketch. Find the maximum area she can enclose with 3600 m of fencing.

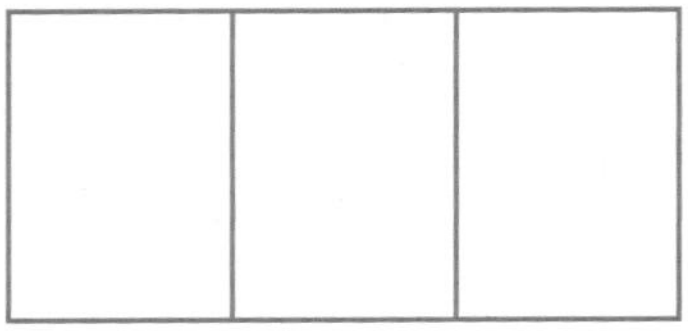

12. *Cost with Fixed Area* A fence must be built to enclose a rectangular area of 20,000 ft^2. Fencing material costs \$3 per foot for the two sides facing north and south, and \$6 per foot for the other two sides. Find the cost of the least expensive fence.

13. *Cost with Fixed Area* A fence must be built in a large field to enclose a rectangular area of 15,625 m^2. One side of the area is bounded by an existing fence; no fence is needed there. Material for the fence costs \$2 per meter for the two ends, and \$4 per meter for the side opposite the existing fence. Find the cost of the least expensive fence.

14. *Profit* In planning a restaurant, it is estimated that a profit of \$5 per seat will be made if the number of seats is between 60 and 80, inclusive. On the other hand, the profit on each seat will decrease by 5¢ for each seat above 80.

a. Find the number of seats that will produce the maximum profit.

b. What is the maximum profit?

15. *Timing Income* A local group of scouts has been collecting aluminum cans for recycling. The group has already collected 12,000 lb of cans, for which they could currently receive \$4 per hundred pounds. The group can continue to collect cans at the rate of 400 lb per day. However, a glut in the aluminum market has caused the recycling company to announce that it will lower its price, starting immediately, by \$.10 per hundred pounds per day. The scouts can make only one trip to the recycling center. Find the best time for the trip. What total income will be received?

16. *Packaging Design* A television manufacturing firm needs to design an open-topped box with a square base. The box must hold 32 in^3. Find the dimensions of the box that can be built with the minimum amount of materials. (See the figure.)

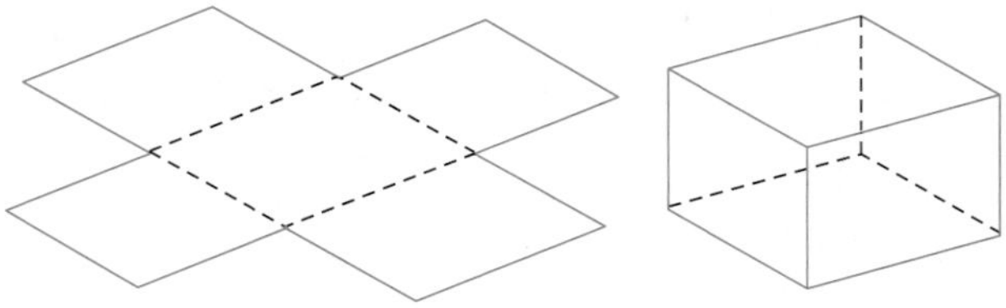

17. *Revenue* A local club is arranging a charter flight to Hawaii. The cost of the trip is \$425 each for 75 passengers, with a refund of \$5 per passenger for each passenger in excess of 75.

a. Find the number of passengers that will maximize the revenue received from the flight.

b. Find the maximum revenue.

18. *Packaging Design* A company wishes to manufacture a box with a volume of 36 ft^3 that is open on top and is twice as long as it is wide. Find the dimensions of the box produced from the minimum amount of material.

19. *Packaging Cost* A closed box with a square base is to have a volume of 16,000 cm^3. The material for the top and bottom of the box costs \$3 per square centimeter, while the material for the sides costs \$1.50 per square centimeter. Find the dimensions of the box that will lead to the minimum total cost. What is the minimum total cost?

20. *Packaging Design* A cylindrical box will be tied up with ribbon as shown in the figure. The longest piece of ribbon available is 130 cm long, and 10 cm of that are required for the bow. Find the radius and height of the box with the largest possible volume.

21. *Can Design*

a. For the can problem in Example 4, the minimum surface area required that the height be twice the radius. Show that this is true for a can of arbitrary volume V.

b. Do many cans in grocery stores have a height that is twice the radius? If not, discuss why this may be so.

22. *Container Design* Your company needs to design cylindrical metal containers with a volume of 16 cubic feet. The top and bottom will be made of a sturdy material that costs \$2 per square foot, while the material for the sides costs \$1 per square foot. Find the radius, height, and cost of the least expensive container.

23. *Container Design* An open box will be made by cutting a square from each corner of a 3-ft by 8-ft piece of cardboard and then folding up the sides. What size square should be cut from each corner in order to produce a box of maximum volume?

24. *Container Design* Consider the problem of cutting corners out of a rectangle and folding up the sides to make a box. Specific examples of this problem are discussed in Example 3 and Exercise 23.

a. In the solution to Example 3, compare the area of the base of the box with the area of the walls.

b. Repeat part a for the solution to Exercise 23.

c. Make a conjecture about the area of the base compared with the area of the walls for the box with the maximum volume.

25. *Use of Materials* A mathematics book is to contain 36 in^2 of printed matter per page, with margins of 1 in. along the sides and $1\frac{1}{2}$ in. along the top and bottom. Find the dimensions of the page that will require the minimum amount of paper. (See the figure.)

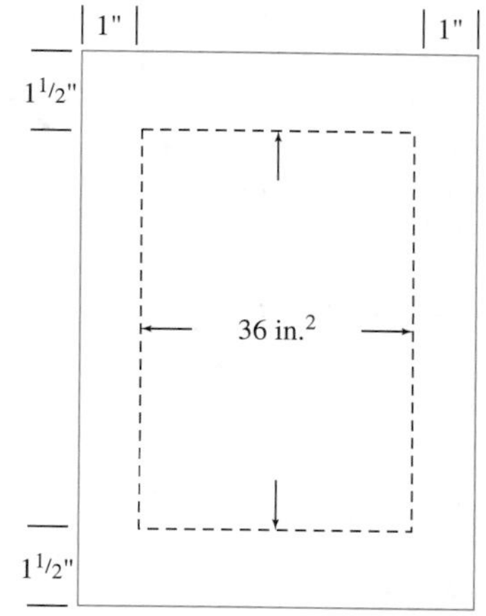

26. *Cost* A company wishes to run a utility cable from point A on the shore (see the figure) to an installation at point B on the island. The island is 6 miles from the shore. It costs \$400 per mile to run the cable on land and \$500 per mile underwater. Assume that the cable starts at A and runs along the shoreline, then angles and runs underwater to the island. Find the point at which the line should begin to angle in order to yield the minimum total cost.

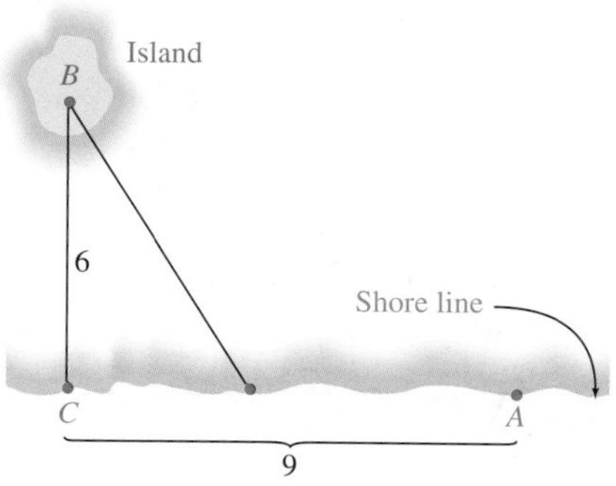

27. *Cost* Repeat Exercise 26, but make point A 7 miles from point C.

28. *Pricing* Decide what you would do if your assistant presented the following contract for your signature:

> Your firm offers to deliver 300 tables to a dealer, at \$90 per table, and to reduce the price per table on the entire order by 25¢ for each additional table over 300.

Find the dollar total involved in the largest possible transaction between the manufacturer and the dealer; then find the smallest possible dollar amount.

In Exercises 29–31, use a graphing calculator to determine where the derivative is equal to zero.

29. *Can Design* Modify the can problem in Example 4 so the cost must be minimized. Assume that aluminum costs 3¢ per square centemeter, and that there is an additional cost of 2¢ per cm times the perimeter of the top, and a similar cost for the bottom, to seal the top and bottom of the can to the side.

30. *Can Design* In this modification of the can problem in Example 4, the cost must be minimized. Assume that aluminum costs 3¢ per square centemeter, and that there is an additional cost of 1¢ per cm times the height of the can to make a vertical seam on the side.

31. *Can Design* This problem is a combination of Exercises 29 and 30. We will again minimize the cost of the can, assuming that aluminum costs 3¢ per square centemeter. In addition, there is a cost of 2¢ per cm to seal the top and bottom of the can to the side, plus 1¢ per cm to make a vertical seam.

LIFE SCIENCES

32. *Disease* Epidemiologists have found a new communicable disease running rampant in College Station, Texas. They estimate that t days after the disease is first observed in the community, the percent of the population infected by the disease is approximated by

$$p(t) = \frac{20t^3 - t^4}{1000}$$

for $0 \le t \le 20$.

a. After how many days is the percent of the population infected a maximum?

b. What is the maximum percent of the population infected?

33. *Pollution* A lake polluted by bacteria is treated with an antibacterial chemical. After t days, the number N of bacteria per milliliter of water is approximated by

$$N(t) = 20\left(\frac{t}{12} - \ln\left(\frac{t}{12}\right)\right) + 30$$

for $1 \le t \le 15$.

a. When during this time will the number of bacteria be a minimum?

b. What is the minimum number of bacteria during this time?

c. When during this time will the number of bacteria be a maximum?

d. What is the maximum number of bacteria during this time?

34. *Disease* Another disease hits the chronically ill town of College Station, Texas. This time the percent of the population infected by the disease t days after it hits town is approximated by $p(t) = 10te^{-t/8}$ for $0 \le t \le 40$.

a. After how many days is the percent of the population infected a maximum?

b. What is the maximum percent of the population infected?

Maximum Sustainable Harvest Find the maximum sustainable harvest in Exercises 35 and 36. See Example 5.

35. $f(S) = 12S^{.25}$ **36.** $f(S) = \dfrac{25S}{S + 2}$

37. *Maximum Sustainable Harvest* The population of salmon next year is given by $f(S) = Se^{r(1 - S/P)}$, where S is this year's salmon population, P is the equilibrium population, and r is a constant that depends upon how fast the population grows.* The number of salmon that can be fished next year while keeping the population the same is $H(S) = f(S) - S$. The maximum value of $H(S)$ is the maximum sustainable harvest.

a. Show that the maximum sustainable harvest occurs when $f'(S) = 1$. (*Hint:* To maximize, set $H'(S) = 0$.)

b. Let the value of S found in part a be denoted by S_0. Show that the maximum sustainable harvest is given by

$$S_0\left(\frac{1}{1 - rS_0/P} - 1\right).$$

(*Hint:* Set $f'(S_0) = 1$ and solve for $e^{r(1 - S_0/P)}$. Then find $H(S_0)$ and substitute the expression for $e^{r(1 - S_0/P)}$.)

Maximum Sustainable Harvest In Exercises 38 and 39, refer to Exercise 37. Find $f'(S_0)$ and solve the equation $f'(S_0) = 1$, using a calculator to find the intersection of the graphs of $f'(S_0)$ and $y = 1$.

38. Find the maximum sustainable harvest if $r = .1$ and $P = 100$.

39. Find the maximum sustainable harvest if $r = .4$ and $P = 500$.

40. *Pigeon Flight* Homing pigeons avoid flying over large bodies of water, preferring to fly around them instead. (One possible explanation is the fact that extra energy is required to fly over water because air pressure drops over water in the daytime.) Assume that a pigeon released from a boat 1 mile from the shore of a lake (point B in the figure) flies

*Ricker, W. E., "Stock and Recruitment," *Journal of the Fisheries Research Board of Canada*, Vol. 11, 1957, pp. 559–623.

first to point P on the shore and then along the straight edge of the lake to reach its home at L. If L is 2 miles from point A, the point on the shore closest to the boat, and if a pigeon needs $4/3$ as much energy per mile to fly over water as over land, find the location of point P, which minimizes energy used.

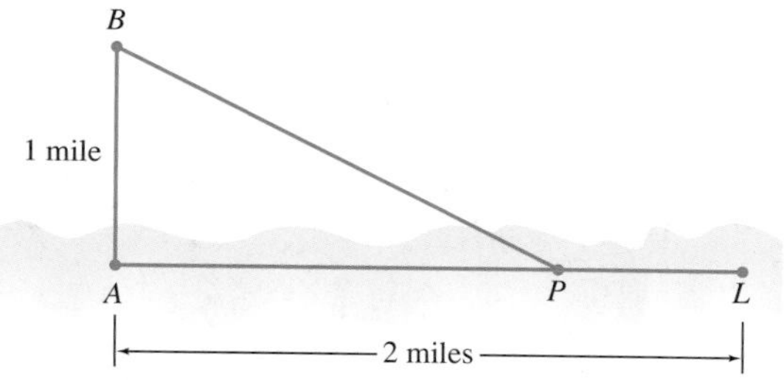

41. *Pigeon Flight* Repeat Exercise 40, but assume a pigeon needs $10/9$ as much energy to fly over water as over land.

42. *Harvesting Cod* A recent article described the population $f(S)$ of cod in the North Sea next year as a function of this year's population S (in thousands of tons) by various mathematical models.*

$$\text{Shepherd:}\quad f(S) = \frac{aS}{1 + (S/b)^c};$$

$$\text{Ricker:}\quad f(S) = aSe^{-bS};$$

$$\text{Beverton–Holt:}\quad f(S) = \frac{aS}{1 + (S/b)},$$

where a, b, and c are constants.

a. Find a replacement of variables in the Ricker model above that will make it the same as another form of the Ricker model described in Exercise 37 of this section, $f(S) = Se^{r(1-S/P)}$.

b. Find $f'(S)$ for all three models.

c. Find $f'(0)$ for all three models. From your answer, describe in words the geometric meaning of the constant a.

d. The values of a, b, and c reported in the article for the Shepherd model are 3.026, 248.72, and 3.24, respectively. Find the value of this year's population that maximizes next year's population using the Shepherd model.

e. The values of a and b reported in the article for the Ricker model are 4.151 and .0039, respectively. Find the value of this year's population that maximizes next year's population using the Ricker model.

f. Explain why, for the Beverton–Holt model, there is no value of this year's population that maximizes next year's population.

43. *Bird Migration* Suppose a migrating bird flies at a velocity v, and suppose the amount of time the bird can fly depends on its velocity according to the function $T(v)$.†

a. If E is the bird's initial energy, then the bird's effective power is given by kE/T, where k is the fraction of the power that can be converted into mechanical energy. According to principles of aerodynamics,

$$\frac{kE}{T} = aSv^3 + I,$$

where a is a constant, S is the wind speed, and I is the induced power, or rate of working against gravity. Using this result and the fact that distance is velocity multiplied by time, show that the distance that the bird can fly is given by

$$D(v) = \frac{kEv}{aSv^3 + I}.$$

b. Show that the migrating bird can fly a maximum distance by flying at a velocity

$$v = \left(\frac{I}{2aS}\right)^{1/3}.$$

GENERAL INTEREST

44. *Postal Regulations* The U.S. Postal Service stipulates that any boxes sent through the mail must have a length plus girth totaling no more than 108 in. (See the figure.) Find the dimensions of the box with maximum volume that can be sent through the U.S. mail, assuming that the width and the height of the box are equal.

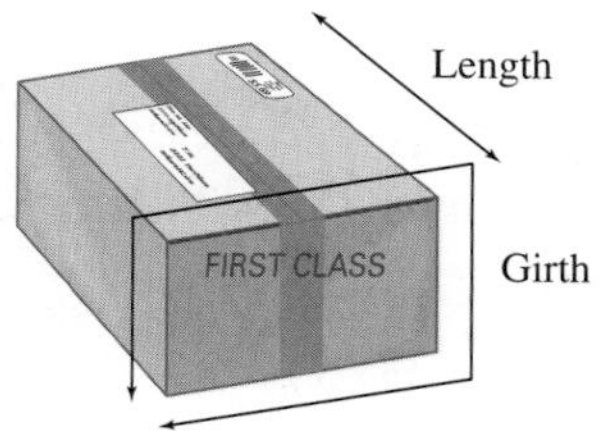

*Cook, R. M., A. Sinclair, and G. Stefánsson, "Potential Collapse of North Sea Cod Stocks," *Nature*, Vol. 385, Feb. 6, 1997, pp. 521–522.

†This exercise is based on an example in *A Concrete Approach to Mathematical Modelling* by Michael Mesterton-Gibbons, Wiley-Interscience, 1995, pp. 93–96.

45. *Travel Time* A hunter is at a point along a river bank. He wants to get to his cabin, located 3 miles north and 8 miles west. (See the figure.) He can travel 5 mph along the river but only 2 mph on this very rocky land. How far upriver should he go in order to reach the cabin in minimum time?

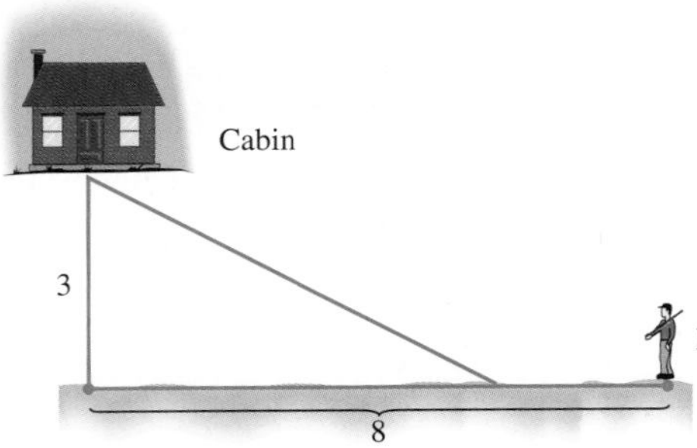

46. *Travel Time* Repeat Exercise 45, but assume the cabin is 19 miles north and 8 miles west.

6.3 FURTHER BUSINESS APPLICATIONS: ECONOMIC LOT SIZE; ECONOMIC ORDER QUANTITY; ELASTICITY OF DEMAND

THINK ABOUT IT *How many batches of primer should a paint company produce per year to minimize its costs while meeting its customers' demand?*

We will answer this question in Example 1 using the concept of *economic lot size*.

In this section we introduce three common business applications of calculus. The first two, *economic lot size* and *economic order quantity*, are related. A manufacturer must determine the production lot (or batch) size that will result in minimum production and storage costs, while a purchaser must decide what quantity of an item to order in an effort to minimize reordering and storage costs. The third application, *elasticity of demand*, deals with the sensitivity of demand for a product to changes in the price of the product.

Economic Lot Size Suppose that a company manufactures a constant number of a product per year and that the product can be manufactured in several batches of equal size throughout the year. If the company were to produce a large number of items in each batch, resulting in a small number of batches each year, it would minimize setup costs but incur high warehouse costs. On the other hand, if it produced a small number of items in each batch, it would have to produce a batch more often, which would increase setup costs. Calculus can be used to find the number that should be manufactured in each batch in order to minimize the total cost. This number is called the **economic lot size.**

Figure 14 on the next page shows several possibilities for a product having an annual demand of 12,000 units. The top graph shows the results if all 12,000 units are made in one batch per year. In this case an average of 6000 units will be held in a warehouse. If 3000 units are made in each batch, four batches will be made at equal time intervals during the year, and the average number of units in the warehouse falls to only 1500. If 1000 units are made in each of twelve batches, an average of 500 units will be in the warehouse.

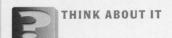

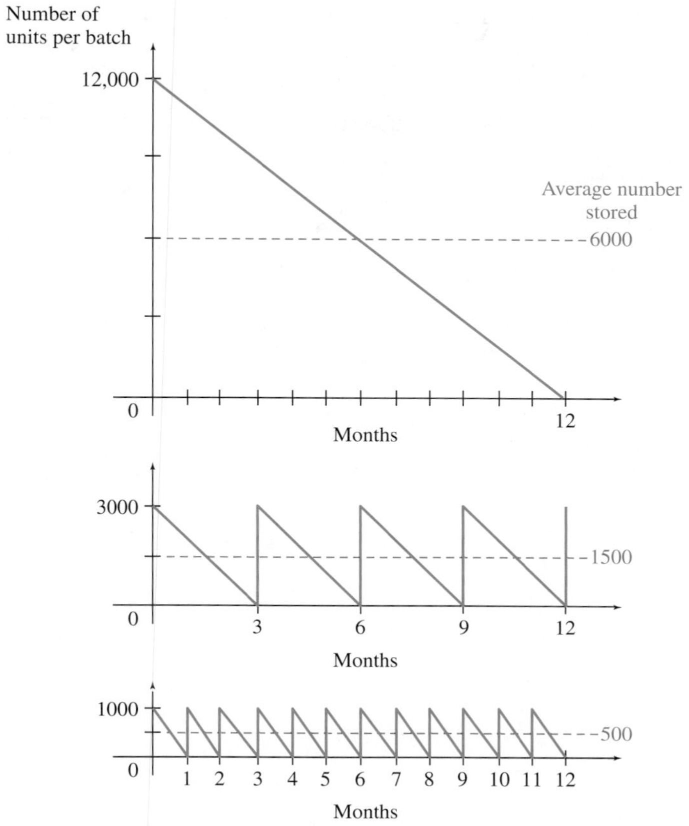

FIGURE 14

The variable in our discussion of economic lot size will be

$$q = \text{number of units in each batch.}$$

In addition, we have the following constants:

$$k = \text{cost of storing one unit of the product for one year;}$$
$$f = \text{fixed setup cost to manufacture the product;}$$
$$g = \text{cost of manufacturing a single unit of the product;}$$
$$M = \text{total number of units produced annually.}$$

The company has two types of costs: a cost associated with manufacturing the item and a cost associated with storing the finished product. Because q units are produced in each batch, and each batch has a fixed cost f and a variable cost g per unit, the manufacturing cost per batch is

$$f + gq.$$

The number of units produced in a year is M, so the number of batches per year must be M/q. Therefore, the total annual manufacturing cost is

$$(f + gq)\frac{M}{q} = \frac{fM}{q} + gM. \quad \textbf{(1)}$$

Since demand is constant, the inventory goes down linearly from q to 0, as in Figure 14, with an average inventory of $q/2$ units per year. The cost of storing one unit of the product for a year is k, so the total storage cost is

$$k\left(\frac{q}{2}\right) = \frac{kq}{2}. \tag{2}$$

The total production cost is the sum of the manufacturing and storage costs, or the sum of expressions (1) and (2). If $T(q)$ is the total cost of producing M units in batches of size q,

$$T(q) = \frac{fM}{q} + gM + \frac{kq}{2}.$$

Since the only constraint on q is that it be a positive number, the domain of T is $(0, \infty)$. To find the value of q that will minimize $T(q)$, remember that f, g, k, and M are constants and find $T'(q)$.

$$T'(q) = \frac{-fM}{q^2} + \frac{k}{2}$$

Set this derivative equal to 0.

$$\frac{-fM}{q^2} + \frac{k}{2} = 0$$

$$\frac{k}{2} = \frac{fM}{q^2}$$

$$q^2\frac{k}{2} = fM$$

$$q^2 = \frac{2fM}{k}$$

$$q = \sqrt{\frac{2fM}{k}} \tag{3}$$

Examination of the limits as q approaches 0 and ∞ can be used to show that $\sqrt{(2fM)/k}$ is the economic lot size that minimizes total production costs. (See Exercise 1.)

This application is referred to as the *inventory problem* and is treated in more detail in management science courses. Please note that Equation (3) was derived under very specific assumptions. If the assumptions are changed slightly, a different conclusion might be reached, and it would not necessarily be valid to use Equation (3).

In some examples Equation (3) may not give an integer value, in which case we must investigate the next integer smaller than q and the next integer larger to see which gives the minimum cost.

EXAMPLE 1 *Lot Size*

A paint company has a steady annual demand for 24,500 cans of automobile primer. The comptroller for the company says that it costs \$2 to store one can of paint for 1 year and \$500 to set up the plant for the production of the primer. Find the number of cans of primer that should be produced in each batch, as well as the number of batches per year, in order to minimize total production costs.

Solution Use Equation (3), with $k = 2$, $M = 24{,}500$, and $f = 500$.

$$q = \sqrt{\frac{2fM}{k}} = \sqrt{\frac{2(500)(24{,}500)}{2}} = \sqrt{12{,}250{,}000} = 3500$$

The company should make 3500 cans of primer in each batch to minimize production costs. The number of batches per year is $M/q = 24{,}500/3500 = 7$.

Economic Order Quantity We can extend our previous discussion to the problem of reordering an item that is used at a constant rate throughout the year. Here, the company using a product must decide how often to order and how many units to request each time an order is placed; that is, it must identify the **economic order quantity.** In this case, the variable is

$$q = \text{number units to order each time.}$$

We also have the following constants:

$$\begin{aligned} k &= \text{cost of storing one unit for one year} \\ f &= \text{fixed cost to place an order} \\ M &= \text{total units needed per year} \end{aligned}$$

The goal is to minimize the total cost of ordering over a year's time, where

$$\text{Total cost} = \text{Storage cost} + \text{Reorder cost.}$$

Again assume an average inventory of $q/2$, so the yearly storage cost is $kq/2$. The number of orders placed annually is M/q. The reorder cost is the product of this quantity and the cost per order, f. Thus, the reorder cost is fM/q, and the total cost is

$$T(q) = \frac{fM}{q} + \frac{kq}{2}.$$

This is almost the same formula we derived for the inventory problem, which also had a constant term gM. Since a constant does not affect the derivative, Equation (3) is also valid for the economic order quantity problem. As before, the number of orders placed annually is M/q. This illustrates how two different applications might have the same mathematical structure, so a solution to one applies to both.

EXAMPLE 2 *Order Quantity*

A large pharmacy has an annual need for 480 units of a certain antibiotic. It costs \$3 to store one unit for one year. The fixed cost of placing an order (clerical time, mailing, and so on) amounts to \$31. Find the number of units to order each time, and how many times a year the antibiotic should be ordered.

Solution Here $k = 3$, $M = 480$, and $f = 31$. We have

$$q = \sqrt{\frac{2fM}{k}} = \sqrt{\frac{2(31)(480)}{3}} = \sqrt{9920} \approx 99.6$$

$T(99) = 298.803$ and $T(100) = 298.800$, so ordering 100 units of the drug each time minimizes the annual cost. The drug should be ordered $M/q = 480/100 = 4.8$ times a year, or about once every $2\frac{1}{2}$ months.

Elasticity of Demand Anyone who sells a product or service is concerned with how a change in price affects demand. The sensitivity of demand to changes in price varies with different items. Luxury items tend to be more sensitive to price than essentials. For items such as milk, heating fuel, and light bulbs, relatively small percentage changes in price will not change the demand for the item much, so long as the price is not far from its normal range. For cars, home loans, jewelry, and concert tickets, however, small percentage changes in price can have a significant effect on demand.

> **FOR REVIEW**
> Recall from Chapter 1 that the Greek letter Δ, pronounced *delta*, is used in mathematics to mean "change."

One way to measure the sensitivity of demand to changes in price is by the relative change—the ratio of percent change in demand to percent change in price. If q represents the quantity demanded and p the price, this ratio can be written as

$$\frac{\Delta q/q}{\Delta p/p},$$

where Δq represents the change in q and Δp represents the change in p. This ratio is always negative, because q and p are positive, while Δq and Δp have opposite signs. (An *increase* in price causes a *decrease* in demand.) If the absolute value of this ratio is large, it suggests that a relatively small increase in price causes a relatively large drop (decrease) in demand.

This ratio can be rewritten as

$$\frac{\Delta q/q}{\Delta p/p} = \frac{\Delta q}{q} \cdot \frac{p}{\Delta p} = \frac{p}{q} \cdot \frac{\Delta q}{\Delta p}.$$

Suppose $q = f(p)$. (Note that this is the inverse of the way our demand functions have been expressed so far; previously we had $p = D(q)$.) Then $\Delta q = f(p + \Delta p) - f(p)$, and

$$\frac{\Delta q}{\Delta p} = \frac{f(p + \Delta p) - f(p)}{\Delta p}.$$

As $\Delta p \to 0$, this quotient becomes

$$\lim_{\Delta p \to 0} \frac{\Delta q}{\Delta p} = \lim_{\Delta p \to 0} \frac{f(p + \Delta p) - f(p)}{\Delta p} = \frac{dq}{dp},$$

and

$$\lim_{\Delta p \to 0} \frac{p}{q} \cdot \frac{\Delta q}{\Delta p} = \frac{p}{q} \cdot \frac{dq}{dp}.$$

The negative of this last quantity is called the *elasticity of demand* (E) and measures the instantaneous responsiveness of demand to price.

> **ELASTICITY OF DEMAND**
> Let $q = f(p)$, where q is demand at a price p. The **elasticity of demand** is
> $$E = -\frac{p}{q} \cdot \frac{dq}{dp}.$$
> Demand is inelastic if $E < 1$.
> Demand is elastic if $E > 1$.
> Demand has unit elasticity if $E = 1$.

For example, E may be .2 for medical services, but may be 1.2 for stereo equipment. The demand for essential medical services is much less responsive to price changes than is the demand for nonessential commodities, such as stereo equipment.

If $E < 1$, the relative change in demand is less than the relative change in price, and the demand is called *inelastic*. If $E > 1$, the relative change in demand is greater than the relative change in price, and the demand is called *elastic*. When $E = 1$, the percentage changes in price and demand are relatively equal and the demand is said to have **unit elasticity.**

EXAMPLE 3 *Elasticity*

Terrence Wales described the demand for distilled spirits as

$$q = f(p) = -.00375p + 7.87,$$

where p represents the retail price of a case of liquor in dollars per case.* Here q represents the average number of cases purchased per year by a consumer. Calculate and interpret the elasticity of demand when $p = \$118.30$ per case.

Solution From $q = -.00375p + 7.87$, $dq/dp = -.00375$. Now we find E.

$$\begin{aligned} E &= -\frac{p}{q} \cdot \frac{dq}{dp} \\ &= -\frac{p}{-.00375p + 7.87}(-.00375) \\ &= \frac{.00375p}{-.00375p + 7.87} \end{aligned}$$

Let $p = 118.30$ to get

$$E = \frac{.00375(118.30)}{-.00375(118.30) + 7.87} \approx .0597.$$

Since $.0597 < 1$, the demand is inelastic, and a percentage change in price will result in a smaller percentage change in demand. Thus an increase in price will increase revenue. For example, a 10% increase in price will cause an approximate decrease in demand of $(.0597)(.10) = .00597$ or about .6%.

EXAMPLE 4 *Elasticity*

The demand for beer was modeled by Hogarty and Elzinga with the function given by $q = f(p) = 1/p$.† The price was expressed in dollars per can of beer, and the quantity sold in cans per day per adult. Calculate and interpret the elasticity of demand.

*Wales, Terrence J., "Distilled Spirits and Interstate Consumption Efforts," *The American Economic Review*, Vol. 57, No. 4, 1968, pp. 853–863.

†Hogarty, T. F. and K. G. Elzinga, "The Demand for Beer," *The Review of Economics and Statistics*, Vol. 54, No. 2, 1972, pp. 195–198.

Solution Since $q = 1/p$,

$$\frac{dq}{dp} = \frac{-1}{p^2}, \quad \text{and}$$

$$E = -\frac{p}{q} \cdot \frac{dq}{dp} = -\frac{p}{1/p} \cdot \frac{-1}{p^2} = 1.$$

Here, the elasticity is 1, unit elasticity, at every (positive) price. Revenues remain constant when the price changes.

Elasticity can be related to the total revenue, R, by considering the derivative of R. Since revenue is given by price times sales (demand),

$$R = pq.$$

Differentiate with respect to p using the product rule.

$$\begin{aligned}
\frac{dR}{dp} &= p \cdot \frac{dq}{dp} + q \cdot 1 \\
&= \frac{q}{q} \cdot p \cdot \frac{dq}{dp} + q \qquad \text{Multiply by } \frac{q}{q} \text{ (or 1).} \\
&= q\left(\frac{p}{q} \cdot \frac{dq}{dp}\right) + q \\
&= q(-E) + q \\
&= q(-E + 1) \\
&= q(1 - E)
\end{aligned}$$

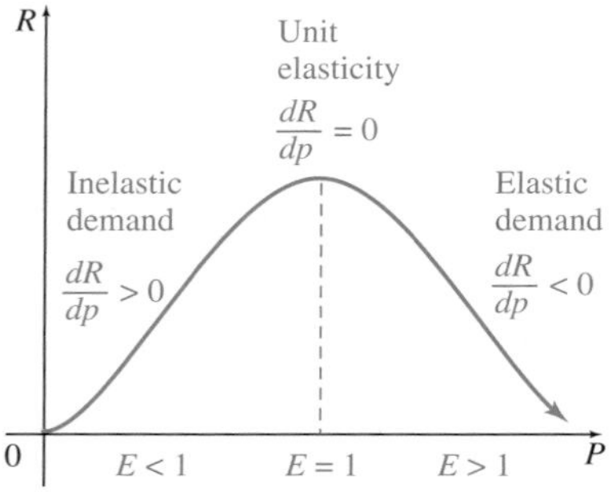

FIGURE 15

Total revenue R is increasing, optimized, or decreasing depending on whether $dR/dp > 0$, $dR/dp = 0$, or $dR/dp < 0$. These three situations correspond to $E < 1$, $E = 1$, or $E > 1$. See Figure 15.

In summary, total revenue is related to elasticity as follows.

REVENUE AND ELASTICITY

1. If the demand is inelastic, total revenue increases as price increases.
2. If the demand is elastic, total revenue decreases as price increases.
3. Total revenue is maximized at the price where demand has unit elasticity.

EXAMPLE 5 *Elasticity*

Assume that the demand for a product is $q = 216 - 2p^2$, where p is the price.

(a) Find the price intervals where demand is elastic and where demand is inelastic.

Solution Since $q = 216 - 2p^2$, $dq/dp = -4p$, and

$$\begin{aligned}
E &= -\frac{p}{q} \cdot \frac{dq}{dp} \\
&= -\frac{p}{216 - 2p^2}(-4p) \\
&= \frac{4p^2}{216 - 2p^2}.
\end{aligned}$$

To decide where $E < 1$ or $E > 1$, solve the corresponding *equation*.

$$E = 1$$
$$\frac{4p^2}{216 - 2p^2} = 1$$
$$4p^2 = 216 - 2p^2$$
$$6p^2 = 216$$
$$p^2 = 36$$
$$p = 6$$

Substitute a test number on either side of 6 in the expression for E to see which values make $E < 1$ and which make $E > 1$.

$$\text{Let } p = 1\text{: } E = \frac{4(1)^2}{216 - 2(1)^2} = \frac{4}{214} < 1.$$
$$\text{Let } p = 10\text{: } E = \frac{4(10)^2}{216 - 2(10)^2} = \frac{400}{216 - 200} > 1.$$

Demand is inelastic when $E < 1$. This occurs when $p < 6$. Demand is elastic when $E > 1$; that is, when $p > 6$.

(b) What prices will cause revenue to increase or decrease?

Solution Revenue will increase when demand is inelastic, so keeping prices below \$6 per item will keep demand high enough to continue to increase revenue. When demand is elastic, revenue is decreasing, so any price over \$6 per item will cause demand to decrease to the point where revenue also decreases.

6.3 EXERCISES

1. In the discussion of economic lot size, use the limits as q approaches 0 and ∞ to show that $\sqrt{(2fM)/k}$ is the economic lot size that minimizes total production costs.

2. Why do you think that the cost g does not appear in the equation for q [Equation (3)]?

3. *Choose the correct answer.**

The economic order quantity formula assumes that

a. Purchase costs per unit differ due to quantity discounts.

b. Costs of placing an order vary with quantity ordered.

c. Periodic demand for the goods is known.

d. Erratic usage rates are cushioned by safety stocks.

Applications

BUSINESS AND ECONOMICS

4. *Lot Size* Suppose 100,000 lamps are to be manufactured annually. It costs \$1 to store a lamp for 1 year, and it costs \$500 to set up the factory to produce a batch of lamps. Find the number of lamps to produce in each batch.

5. *Lot Size* A manufacturer has a steady annual demand for 16,800 cases of sugar. It costs \$3 to store 1 case for 1 year and it costs \$7 to produce each batch. Find the number of cases per batch that should be produced.

*Material from the Uniform CPA Examination Questions and Unofficial Answers, Copyright © 1991 by the American Institute of Certified Public Accountants, Inc. Reprinted (or adapted) with permission.

6. *Lot Size* Find the number of batches of lamps that should be manufactured annually in Exercise 4.

7. *Lot Size* Find the number of batches of sugar that should be manufactured annually in Exercise 5.

8. *Order Quantity* A bookstore has an annual demand for 100,000 copies of a best-selling book. It costs $.50 to store 1 copy for 1 year, and it costs $60 to place an order. Find the optimum number of copies per order.

9. *Order Quantity* A restaurant has an annual demand for 900 bottles of a California wine. It costs $1 to store 1 bottle for 1 year, and it costs $5 to place a reorder. Find the optimum number of bottles per order.

10. *Lot Size* Suppose that in the inventory problem, the storage cost depends on the maximum inventory size, rather than the average. This would be more realistic if, for example, the company had to build a warehouse large enough to hold the maximum inventory, and the cost of storage was the same no matter how full or empty the warehouse was. Show that in this case the number of units that should be ordered or manufactured to minimize the total cost is

$$q = \sqrt{\frac{fM}{k}}.$$

11. *Lot Size* A book publisher wants to know how many times a year a print run should be scheduled. Suppose it costs $1000 to set up the printing process, and the subsequent cost per book is so low it can be ignored. Suppose further that the annual warehouse cost is $6 times the maximum number of books stored. Assuming 5000 copies of the book are needed per year, how many books should be printed in each print run? (See Exercise 10.)

12. *Lot Size* Suppose that in the inventory problem, the storage cost is a combination of the cost described in the text and the cost described in Exercise 10. In other words, suppose there is an annual cost, k_1, for storing a single unit, plus an annual cost per unit, k_2, that must be paid for each unit up to the maximum number of units stored. Show that the number of units that should be ordered or manufactured to minimize the total cost in this case is

$$q = \sqrt{\frac{2fM}{k_1 + 2k_2}}.$$

13. *Lot Size* Every year, Karen Panunzio sells 30,000 cases of her Famous Spaghetti Sauce. It costs her $1 per year in electricity to store a case, plus she must pay annual warehouse fees of $2 per case for the maximum number of cases she will store. If it costs her $750 to set up a production run, plus $8 per case to manufacture a single case, how many production runs should she have each year to minimize her total costs? (See Exercise 12.)

Elasticity For each of the following demand functions, find **(a)** *E, and* **(b)** *values of q (if any) at which total revenue is maximized.*

14. $q = 25{,}000 - 50p$

15. $q = 50 - \frac{p}{4}$

16. $q = 48{,}000 - 10p^2$

17. $q = 37{,}500 - 5p^2$

18. $q = 10 - \ln p$

19. $p = 400e^{-.2q}$

Elasticity Find the elasticity of demand (E) for the given demand function at the indicated values of p. Is the demand elastic, inelastic, or neither at the indicated values? Interpret your results.

20. $q = 300 - 2p$

a. $p = \$100$ **b.** $p = \$50$

21. $q = 400 - .2p^2$

a. $p = \$20$ **b.** $p = \$40$

22. *Elasticity* Research has indicated that the demand for heroin is given by $q = 100p^{-.17}$.[*]

a. Find E.

b. Is the demand for heroin elastic or inelastic?

c. Discuss possible reasons for your answer in part b.

23. *Elasticity* The retail price of heroin has been related to the crime rate in Detroit by a function defined by $c = 3351p^{.287}$.[†]

a. Find the elasticity of c with respect to p.

b. By approximately what percentage does the rate of property crimes c rise if the price of heroin p rises by 5%?

*Brown, George F. Jr., and Lester R. Silverman, *The Retail Price of Heroin: Estimation and Applications*, Washington, D.C.: The Drug Abuse Council, Inc., 1973.

†Nievergelt, Yves, UMAP Module 674, *Price Elasticity of Demand: Gambling, Heroin, Marijuana, Whiskey, Prostitution, and Fish*, COMAP, Inc., 1987.

24. *Elasticity* The demand for marijuana (in dollars per ounce) among UCLA students was found to be given by $q = -.225p + 3.74$, where p represents the price (in dollars per ounce) and q represents the average number of ounces purchased monthly per consumer.*

a. Find E.

b. Compute the elasticity at a price of \$10 (the prevailing price at the time of the research).

c. Determine the price at which the demand for marijuana is unit elastic.

25. *Elasticity* Suppose that a demand function is linear—that is, $q = m - np$ for $0 \le p \le m/n$, where m and n are positive constants. Show that $E = 1$ at the midpoint of the demand curve on the interval $0 \le p \le m/n$; that is, at $p = m/(2n)$.

26. *Elasticity* What must be true about the demand function if $E = 0$?

27. *Elasticity* Suppose the demand function is of the form $q = Cp^{-k}$, where C and k are positive constants.

a. Find the elasticity E.

b. If $0 < k < 1$, what does your answer from part a say about how prices should be set to maximize the revenue?

c. If $k > 1$, what does your answer from part a say about how prices should be set to maximize the revenue?

d. If $k = 1$, what does your answer from part a tell you about setting prices to maximize revenue?

e. Based on your answers above, is a demand function of the form $q = Cp^{-k}$ realistic? Explain your answer.

6.4 IMPLICIT DIFFERENTIATION

In almost all of the examples and applications so far, all functions have been defined in the form

$$y = f(x),$$

with y given **explicitly** in terms of x, or as an **explicit function** of x. For example,

$$y = 3x - 2, \qquad y = x^2 + x + 6, \qquad \text{and} \qquad y = -x^3 + 2$$

are all explicit functions of x. The equation $4xy - 3x = 6$ can be expressed as an explicit function of x by solving for y. This gives

$$4xy - 3x = 6$$
$$4xy = 3x + 6$$
$$y = \frac{3x + 6}{4x}.$$

On the other hand, some equations in x and y cannot be readily solved for y, and some equations cannot be solved for y at all. For example, while it would be possible (but tedious) to use the quadratic formula to solve for y in the equation $y^2 + 2yx + 4x^2 = 0$, it is not possible to solve for y in the equation $y^5 + 8y^3 + 6y^2x^2 + 2yx^3 + 6 = 0$. In equations such as these last two, y is said to be given **implicitly** in terms of x.

In such cases, it may still be possible to find the derivative dy/dx by a process called **implicit differentiation.** In doing so, we assume that there exists some function or functions f, which we may or may not be able to find, such that

*Nisbet, C. T. and F. Vakil, "Some Estimates of Price and Expenditure Elasticities of Demand for Marijuana among UCLA Students," *The Review of Economics and Statistics*, Vol. 54, No. 4, 1972, pp. 473–475.

$y = f(x)$ and dy/dx exists. It is useful to use dy/dx here rather than $f'(x)$ to make it clear which variable is independent and which is dependent.

FOR REVIEW

In Chapter 1, we pointed out that when y is given as a function of x, x is the independent variable and y is the dependent variable. We later defined the derivative dy/dx when y is a function of x. In an equation such as $3xy + 4y^2 = 10$, either variable can be considered the independent variable. If a problem asks for dy/dx, consider x the independent variable; if it asks for dx/dy, consider y the independent variable. A similar rule holds when other variables are used.

EXAMPLE 1 *Implicit Differentiation*

Find dy/dx if $3xy + 4y^2 = 10$.

Solution Differentiate with respect to x on both sides of the equation.

$$3xy + 4y^2 = 10$$

$$\frac{d}{dx}(3xy + 4y^2) = \frac{d}{dx}(10) \tag{1}$$

Now differentiate each term on the left side of the equation. Think of $3xy$ as the product $(3x)(y)$ and use the product rule and the chain rule. Since

$$\frac{d}{dx}(3x) = 3 \quad \text{and} \quad \frac{d}{dx}(y) = \frac{dy}{dx},$$

the derivative of $(3x)(y)$ is

$$(3x)\frac{dy}{dx} + (y)3 = 3x\frac{dy}{dx} + 3y.$$

To differentiate the second term, $4y^2$, use the chain rule, since y is assumed to be some function of x.

$$\frac{d}{dx}(4y^2) = 4\overbrace{(2y^1)\frac{dy}{dx}}^{\text{Derivative of } y^2} = 8y\frac{dy}{dx}$$

On the right side of Equation (1), the derivative of 10 is 0. Taking the indicated derivatives in Equation (1) term by term gives

$$3x\frac{dy}{dx} + 3y + 8y\frac{dy}{dx} = 0.$$

Now solve this result for dy/dx.

$$(3x + 8y)\frac{dy}{dx} = -3y$$

$$\frac{dy}{dx} = \frac{-3y}{3x + 8y}$$

NOTE Because we are treating y as a function of x, notice that each time an expression has y in it, we use the chain rule. ■

EXAMPLE 2 *Implicit Differentiation*

Find dy/dx for $x + \sqrt{x}\sqrt{y} = y^2$.

Solution Take the derivative on both sides.

$$\frac{d}{dx}(x + \sqrt{x}\sqrt{y}) = \frac{d}{dx}(y^2)$$

Since $\sqrt{x} \cdot \sqrt{y} = x^{1/2} \cdot y^{1/2}$, use the product rule and the chain rule as follows.

$$\underbrace{1}_{\text{Derivative of } x} + \underbrace{x^{1/2}\left(\frac{1}{2}y^{-1/2} \cdot \frac{dy}{dx}\right) + y^{1/2}\left(\frac{1}{2}x^{-1/2}\right)}_{\text{Derivative of } x^{1/2}y^{1/2}} = \underbrace{2y\frac{dy}{dx}}_{\text{Derivative of } y^2}$$

$$1 + \frac{x^{1/2}}{2y^{1/2}} \cdot \frac{dy}{dx} + \frac{y^{1/2}}{2x^{1/2}} = 2y\frac{dy}{dx}$$

Multiply both sides by $2x^{1/2} \cdot y^{1/2}$.

$$2x^{1/2} \cdot y^{1/2} + x\frac{dy}{dx} + y = 4x^{1/2} \cdot y^{3/2} \cdot \frac{dy}{dx}$$

Combine terms and solve for dy/dx.

$$2x^{1/2} \cdot y^{1/2} + y = (4x^{1/2} \cdot y^{3/2} - x)\frac{dy}{dx}$$

$$\frac{dy}{dx} = \frac{2x^{1/2} \cdot y^{1/2} + y}{4x^{1/2} \cdot y^{3/2} - x}$$

EXAMPLE 3 *Tangent Line*

The graph of $x^3 + y^3 = 9xy$, shown in Figure 16, is a *folium of Descartes*.* Find the equation of the tangent line at the point $(2, 4)$, shown in Figure 16.

Solution Since this is not the graph of a function, y is not a function of x, and dy/dx is not defined. But if we restrict the curve to the vicinity of $(2, 4)$, as shown in Figure 17, the curve does represent the graph of a function, and we can calculate dy/dx by implicit differentiation.

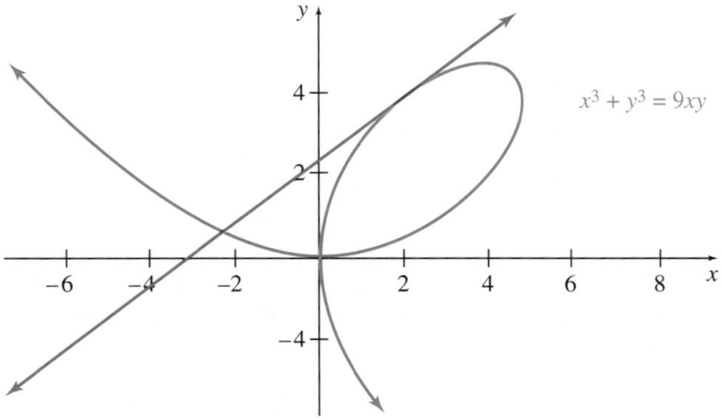

FIGURE 16

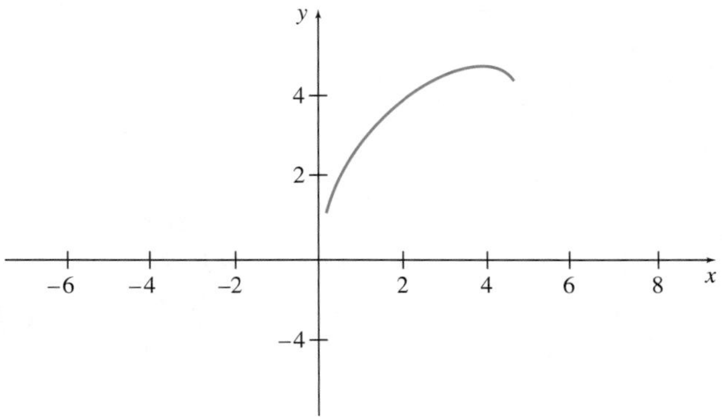

FIGURE 17

*Information on this curve and others is available on the Famous Curves section of the MacTutor History of Mathematics Archive Web site at www-history.mcs.st-and.ac.uk/~history. See Exercises 30–33 for more curves.

$$3x^2 + 3y^2 \cdot \frac{dy}{dx} = 9x\frac{dy}{dx} + 9y \qquad \text{Chain rule and product rule}$$

$$3y^2 \cdot \frac{dy}{dx} - 9x\frac{dy}{dx} = 9y - 3x^2 \qquad \text{Move all } dy/dx \text{ terms to the same side of the equation.}$$

$$\frac{dy}{dx}(3y^2 - 9x) = 9y - 3x^2 \qquad \text{Factor.}$$

$$\frac{dy}{dx} = \frac{9y - 3x^2}{3y^2 - 9x}$$

$$= \frac{3(3y - x^2)}{3(y^2 - 3x)} = \frac{3y - x^2}{y^2 - 3x}$$

To find the slope of the tangent line at the point $(2, 4)$, let $x = 2$ and $y = 4$. The slope is

$$m = \frac{3y - x^2}{y^2 - 3x} = \frac{3(4) - 2^2}{4^2 - 3(2)} = \frac{8}{10} = \frac{4}{5}.$$

The equation of the tangent line is then found by using the point-slope form of the equation of a line.

$$y - y_1 = m(x - x_1)$$

$$y - 4 = \frac{4}{5}(x - 2)$$

$$y - 4 = \frac{4}{5}x - \frac{8}{5}$$

$$y = \frac{4}{5}x + \frac{12}{5}$$

The tangent line is graphed in Figure 16.

NOTE In Example 3, we could have substituted $x = 2$ and $y = 4$ immediately after taking the derivative implicitly. You may find that such a substitution makes solving the equation for dy/dx easier.

The steps used in implicit differentiation can be summarized as follows.

IMPLICIT DIFFERENTIATION

To find dy/dx for an equation containing x and y:

1. Differentiate on both sides of the equation with respect to x, keeping in mind that y is assumed to be a function of x.
2. Place all terms with dy/dx on one side of the equals sign, and all terms without dy/dx on the other side.
3. Factor out dy/dx, and then solve for dy/dx.

When an applied problem involves an equation that is not given in explicit form, implicit differentiation can be used to locate maxima and minima or to find rates of change.

EXAMPLE 4 *Demand*

The demand function for a certain commodity is given by

$$p = \frac{500{,}000}{2q^3 + 400q + 5000},$$

where p is the price in dollars and q is the demand in hundreds of units. Find the rate of change of demand with respect to price when $q = 100$ (that is, find dq/dp).

Solution Rewrite the equation as

$$2q^3 + 400q + 5000 = 500{,}000p^{-1}.$$

Use implicit differentiation to find dq/dp.

$$6q^2\frac{dq}{dp} + 400\frac{dq}{dp} = -500{,}000p^{-2}$$

$$(6q^2 + 400)\frac{dq}{dp} = \frac{-500{,}000}{p^2}$$

$$\frac{dq}{dp} = \frac{-500{,}000}{p^2(6q^2 + 400)}$$

When $q = 100$,

$$p = \frac{500{,}000}{2(100)^3 + 400(100) + 5000} \approx .244,$$

and

$$\frac{dq}{dp} = \frac{-500{,}000}{(.244)^2[6(100)^2 + 400]} \approx -139.$$

This means that when demand (q) is 100 hundreds, or 10,000, demand is decreasing at the rate of 139 hundred, or 13,900, units per dollar change in price.

6.4 EXERCISES

Find dy/dx by implicit differentiation for the following.

1. $4x^2 + 3y^2 = 6$

2. $2x^2 - 5y^2 = 4$

3. $6x^2 + 8xy + y^2 = 6$

4. $8x^2 = 6y^2 + 2xy$

5. $x^3 = y^2 + 4$

6. $x^3 - 6y^2 = 10$

7. $3x^2 = \frac{2 - y}{2 + y}$

8. $2y^2 = \frac{5 + x}{5 - x}$

9. $\sqrt{x} + \sqrt{y} = 4$

10. $2\sqrt{x} - \sqrt{y} = 1$

11. $x^4y^3 + 4x^{3/2} = 6y^{3/2} + 5$

12. $(xy)^{4/3} + x^{1/3} = y^6 + 1$

13. $e^{x^2y} = 5x + 4y + 2$

14. $x^2e^y + y = x^3$

15. $x + \ln y = x^2y^3$

16. $y \ln x + 2 = x^{3/2}y^{5/2}$

Find the equation of the tangent line at the given point on each curve.

17. $x^2 + y^2 = 25;\quad (-3, 4)$

18. $x^2 + y^2 = 100;\quad (8, -6)$

19. $x^2y^2 = 1;\quad (-1, 1)$

20. $x^2y^3 = 8;\quad (-1, 2)$

21. $2y^2 - \sqrt{x} = 4;\quad (16, 2)$

22. $y + \frac{\sqrt{x}}{y} = 3;\quad (4, 2)$

Find the equation of the tangent line at the given value of x on each curve.

23. $y^3 + xy - y = 8x^4;\quad x = 1$

24. $y^3 + 2x^2y - 8y = x^3 + 19;\quad x = 2$

25. $y^3 + xy^2 + 1 = x + 2y^2;\quad x = 2$

26. $y^4(1 - x) + xy = 2;\quad x = 1$

27. $2y^3(x - 3) + x\sqrt{y} = 3;\quad x = 3$

28. $\frac{y}{18}(x^2 - 64) + x^{2/3}y^{1/3} = 12;\quad x = 8$

29. The graph of $x^2 + y^2 = 100$ is a circle having center at the origin and radius 10.

a. Write the equations of the tangent lines at the points where $x = 6$.

b. Graph the circle and the tangent lines.

Information on curves in Exercises 30–33, as well as many other curves, is available on the Famous Curves section of the MacTutor History of Mathematics Archive Web site at www-history.mcs.st-and.ac.uk/~history.

30. The graph of $x^{2/3} + y^{2/3} = 2$, shown in the figure, is an *astroid*. Find the equation of the tangent line at the point $(1, 1)$.

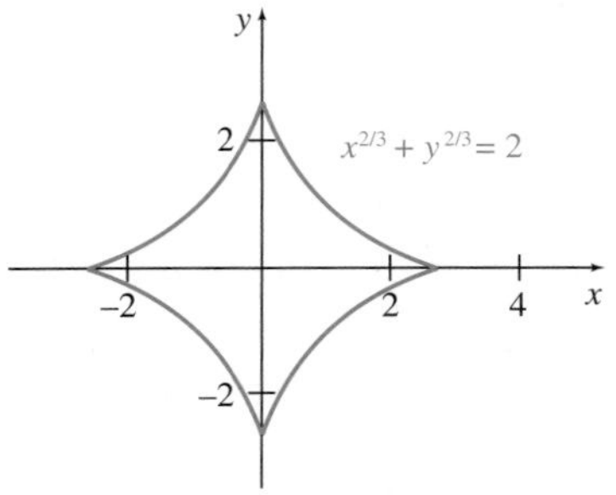

31. The graph of $3(x^2 + y^2)^2 = 25(x^2 - y^2)$, shown in the figure, is a *lemniscate of Bernoulli*. Find the equation of the tangent line at the point $(2, 1)$.

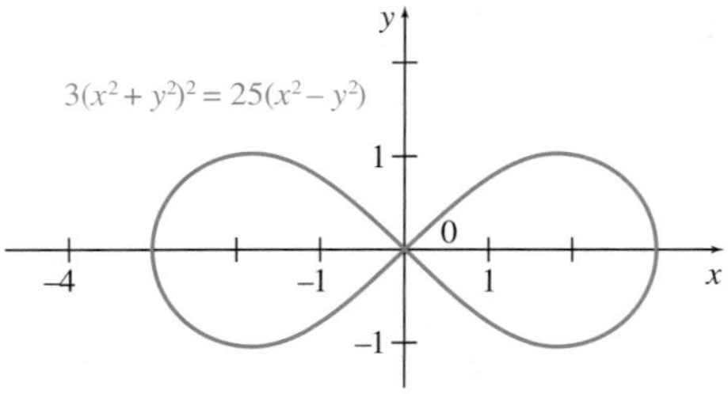

32. The graph of $y^2(x^2 + y^2) = 20x^2$, shown in the figure, is a *kappa curve*. Find the equation of the tangent line at the point $(1, 2)$.

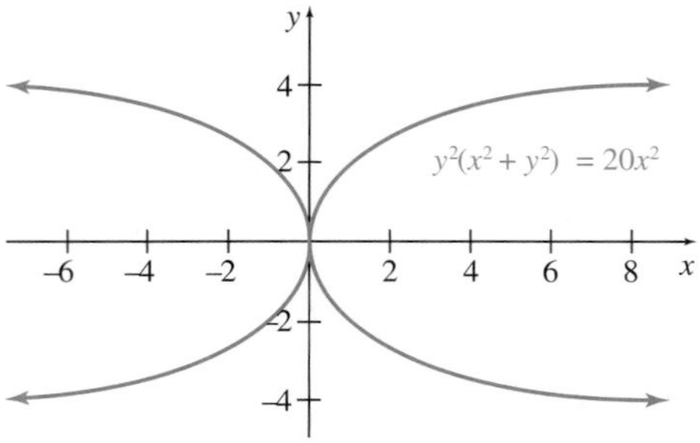

33. The graph of $2(x^2 + y^2)^2 = 25xy^2$, shown in the figure, is a *double folium*. Find the equation of the tangent line at the point $(2, 1)$.

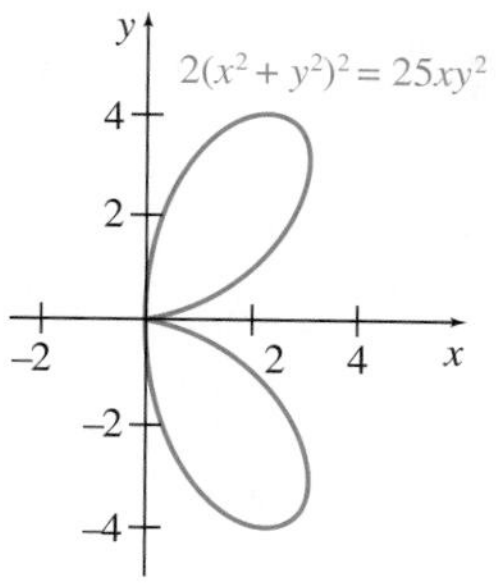

34. Suppose $x^2 + y^2 + 1 = 0$. Use implicit differentiation to find dy/dx. Then explain why the result you got is meaningless. (*Hint:* Can $x^2 + y^2 + 1$ equal 0?)

Let $\sqrt{u} + \sqrt{2v + 1} = 5$. *Find each derivative.*

35. $\dfrac{du}{dv}$

36. $\dfrac{dv}{du}$

Applications

BUSINESS AND ECONOMICS

37. *Demand* The demand equation for a certain product is $2p^2 + q^2 = 1600$, where p is the price per unit in dollars and q is the number of units demanded.

a. Find and interpret dq/dp.

b. Find and interpret dp/dq.

38. *Cost and Revenue* For a certain product, cost C and revenue R are given as follows, where x is the number of units sold (in hundreds)

$$\text{Cost: } C^2 = x^2 + 100\sqrt{x} + 50$$

$$\text{Revenue: } 900(x - 5)^2 + 25R^2 = 22{,}500$$

a. Find and interpret the marginal cost dC/dx at $x = 5$.

b. Find and interpret the marginal revenue dR/dx at $x = 5$.

LIFE SCIENCES

39. *Respiratory Rate* Researchers have found a correlation between respiratory rate and body mass in the first three years of life. This correlation can be expressed by the function

$$\log R(w) = 1.83 - .43 \log(w),$$

where w is the body weight (in kilograms) and $R(w)$ is the respiratory rate (in breaths per minute).*

a. Find $R'(w)$ using implicit differentiation.

b. Find $R'(w)$ by first solving the equation for $R(w)$.

c. Discuss the two procedures. Is there a situation when you would want to use one method over another?

40. *Biochemical Reaction* A simple biochemical reaction with three molecules has solutions that oscillate toward a steady state when positive constants a and b are below the curve $b - a = (b + a)^3$.† Find the largest possible value of a for which the reaction has solutions that oscillate toward a steady state. (*Hint:* Find where $da/db = 0$. Derive values for $a + b$ and $a - b$, and then solve the equations in two unknowns.)

41. *Species* The relationship between the number of species in a genus (x) and the number of genera (y) comprising x species is given by

$$xy^a = k,$$

where a and k are constants.‡ Find dy/dx.

PHYSICAL SCIENCES

Velocity The position of a particle at time t is given by s. Find the velocity ds/dt.

42. $s^3 - 4st + 2t^3 - 5t = 0$

43. $2s^2 + \sqrt{st} - 4 = 3t$

*Gagliardi, L. and F. Rusconi, "Respiratory Rate and Body Mass in the First Three Years of Life," *Archives of Disease in Children*, Vol. 76, 1997, pp. 151–154.

†Murray, J. D., *Mathematical Biology*, New York: Springer-Verlag, 1989, pp. 156–158.

‡Lotka, Alfred J., *Elements of Mathematical Biology*, Dover Publications, 1956, p. 313.

6.5 RELATED RATES

THINK ABOUT IT *When a skier's blood vessels contract because of the cold, how fast is the velocity of blood changing?*

We use related rates to answer this question in Example 5 of this section.

It is common for variables to be functions of time; for example, sales of an item may depend on the season of the year, or a population of animals may be increasing at a certain rate several months after being introduced into an area. Time is often present implicitly in a mathematical model, meaning that derivatives with respect to time must be found by the method of implicit differentiation discussed in the previous section.

EXAMPLE 1 *Area*

A small rock is dropped into a lake. Circular ripples spread over the surface of the water, with the radius of each circle increasing at the rate of 3/2 ft per second. Find the rate of change of the area inside the circle formed by a ripple at the instant the radius is 4 ft.

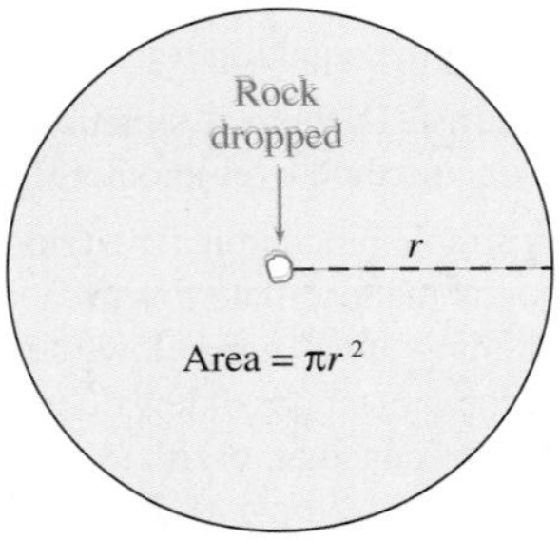

FIGURE 18

Solution As shown in Figure 18, the area A and the radius r are related by

$$A = \pi r^2.$$

Take the derivative of both sides with respect to time.

$$\frac{d}{dt}(A) = \frac{d}{dt}(\pi r^2)$$

$$\frac{dA}{dt} = 2\pi r \cdot \frac{dr}{dt} \quad \textbf{(1)}$$

Since the radius is increasing at the rate of 3/2 ft per second,

$$\frac{dr}{dt} = \frac{3}{2}.$$

The rate of change of area at the instant $r = 4$ is given by dA/dt evaluated at $r = 4$. Substituting into Equation (1) gives

$$\frac{dA}{dt} = 2\pi \cdot 4 \cdot \frac{3}{2}$$

$$\frac{dA}{dt} = 12\pi \approx 37.7 \text{ ft}^2 \text{ per second.}$$

In Example 1, the derivatives (or rates of change) dA/dt and dr/dt are related by Equation (1); for this reason they are called **related rates.** As suggested by Example 1, four basic steps are involved in solving problems about related rates.

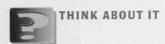

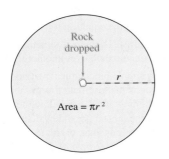

SOLVING A RELATED RATE PROBLEM

1. Identify all given quantities, as well as the quantities to be found. Draw a sketch when possible.
2. Write an equation relating the variables of the problem.
3. Use implicit differentiation to find the derivative of both sides of the equation in Step 2 with respect to time.
4. Solve for the derivative giving the unknown rate of change and substitute the given values.

CAUTION Differentiate *first*, and *then* substitute values for the variables. If the substitutions were performed first, differentiating would not lead to useful results. ■

EXAMPLE 2 *Sliding Ladder*

A 50-ft ladder is placed against a large building. The base of the ladder is resting on an oil spill, and it slips (to the right in Figure 19) at the rate of 3 ft per minute. Find the rate of change of the height of the top of the ladder above the ground at the instant when the base of the ladder is 30 ft from the base of the building.

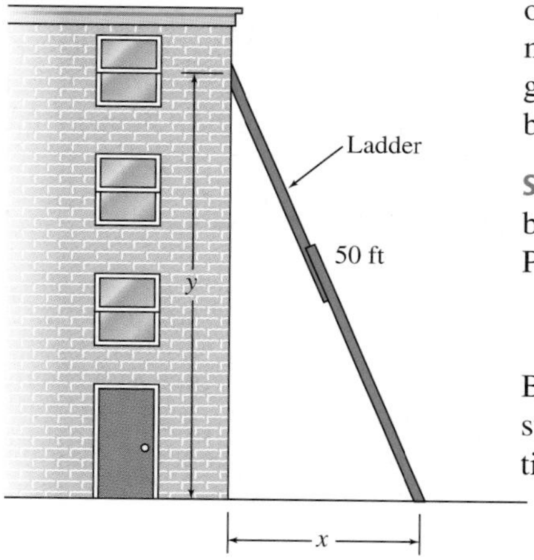

FIGURE 19

Solution Let y be the height of the top of the ladder above the ground, and let x be the distance of the base of the ladder from the base of the building. By the Pythagorean theorem,

$$x^2 + y^2 = 50^2. \tag{2}$$

Both x and y are functions of time t in minutes after the moment that the ladder starts slipping. Take the derivative of both sides of Equation (2) with respect to time, getting

$$\frac{d}{dt}(x^2 + y^2) = \frac{d}{dt}(50^2)$$

$$2x\frac{dx}{dt} + 2y\frac{dy}{dt} = 0. \tag{3}$$

Since the base is sliding at the rate of 3 ft per minute,

$$\frac{dx}{dt} = 3.$$

Also, the base of the ladder is 30 ft from the base of the building. Use this to find y.

$$50^2 = 30^2 + y^2$$
$$2500 = 900 + y^2$$
$$1600 = y^2$$
$$y = 40$$

In summary, $y = 40$ when $x = 30$. Also, the rate of change of x over time t is $dx/dt = 3$. Substituting these values into Equation (3) to find the rate of change of y over time gives

$$2(30)(3) + 2(40)\frac{dy}{dt} = 0$$
$$180 + 80\frac{dy}{dt} = 0$$
$$80\frac{dy}{dt} = -180$$
$$\frac{dy}{dt} = \frac{-180}{80} = \frac{-9}{4} = -2.25.$$

At the instant when the base of the ladder is 30 ft from the base of the building, the top of the ladder is sliding down the building at the rate of 2.25 ft per minute. (The minus sign shows that the ladder is sliding *down*, so the distance y is *decreasing*.)*

EXAMPLE 3 *Icicle*

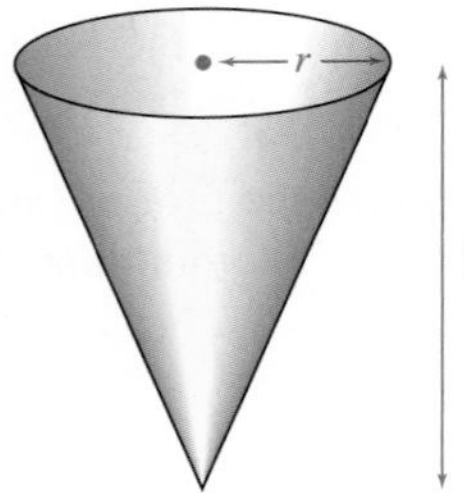

FIGURE 20

A cone-shaped icicle is dripping from the roof. The radius of the icicle is decreasing at a rate of .2 cm per hour, while the length is increasing at a rate of .8 cm per hour. If the icicle is currently 4 cm in radius and 20 cm long, is the volume of the icicle increasing or decreasing, and at what rate?

Solution For this problem we need the formula for the volume of a cone:

$$V = \frac{1}{3}\pi r^2 h, \qquad (4)$$

where r is the radius of the cone and h is the height of the cone, which in this case is the length of the icicle, as in Figure 20.

In this problem, both r and h are functions of the time t in hours. Taking the derivative of both sides of Equation (4) with respect to time yields

$$\frac{dV}{dt} = \frac{1}{3}\pi\left[r^2\frac{dh}{dt} + (h)(2r)\frac{dr}{dt}\right]. \qquad (5)$$

Since the radius is decreasing at a rate of .2 cm per hour and the length is increasing at a rate of .8 cm per hour,

$$\frac{dr}{dt} = -.2 \qquad \text{and} \qquad \frac{dh}{dt} = .8.$$

Substituting these, as well as $r = 4$ and $h = 20$, into Equation (5) yields

$$\frac{dV}{dt} = \frac{1}{3}\pi[4^2(.8) + (20)(8)(-.2)]$$
$$= \frac{1}{3}\pi(-19.2) \approx -20.$$

*The model in Example 2 breaks down as the top of the ladder nears the ground. As y approaches 0, dy/dt becomes infinitely large. In reality, the ladder loses contact with the wall before y reaches 0.

Because the sign of dV/dt is negative, the volume of the icicle is decreasing at a rate of 20 cm^3 per hour.

EXAMPLE 4 *Revenue*

A company is increasing production of peanuts at the rate of 50 cases per day. All cases produced can be sold. The daily demand function is given by

$$p = 50 - \frac{q}{200},$$

where q is the number of units produced (and sold) and p is price in dollars. Find the rate of change of revenue with respect to time (in days) when the daily production is 200 units.

Solution The revenue function,

$$R = qp = q\left(50 - \frac{q}{200}\right) = 50q - \frac{q^2}{200},$$

relates R and q. The rate of change of q over time (in days) is $dq/dt = 50$. The rate of change of revenue over time, dR/dt, is to be found when $q = 200$. Differentiate both sides of the equation

$$R = 50q - \frac{q^2}{200}$$

with respect to t.

$$\frac{dR}{dt} = 50\frac{dq}{dt} - \frac{1}{100}q\frac{dq}{dt} = \left(50 - \frac{1}{100}q\right)\frac{dq}{dt}$$

Now substitute the known values for q and dq/dt.

$$\frac{dR}{dt} = \left[50 - \frac{1}{100}(200)\right](50) = 2400$$

Thus revenue is increasing at the rate of \$2400 per day.

EXAMPLE 5 *Blood Flow*

Blood flows faster the closer it is to the center of a blood vessel. According to Poiseuille's laws, the velocity V of blood is given by

$$V = k(R^2 - r^2),$$

where R is the radius of the blood vessel, r is the distance of a layer of blood flow from the center of the vessel, and k is a constant, assumed here to equal 375. See Figure 21. Suppose a skier's blood vessel has radius $R = .08$ mm and that cold

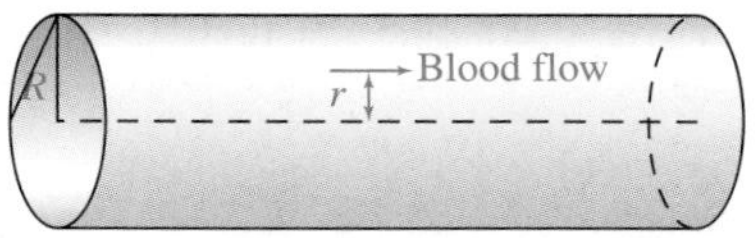

FIGURE 21

weather is causing the vessel to contract at a rate of $dR/dt = -.01$ mm per minute. How fast is the velocity of blood changing?

Solution Find dV/dt. Treat r as a constant. Assume the given units are compatible.

$$V = 375(R^2 - r^2)$$

$$\frac{dV}{dt} = 375\left(2R\frac{dR}{dt} - 0\right) \qquad r \text{ is a constant.}$$

$$\frac{dV}{dt} = 750R\frac{dR}{dt}$$

Here $R = .08$ and $dR/dt = -.01$, so

$$\frac{dV}{dt} = 750(.08)(-.01) = -.6.$$

That is, the velocity of the blood is decreasing at a rate of $-.6$ mm per minute each minute. The units (mm/min^2) indicate that this is a deceleration (negative acceleration), since it gives the rate of change of velocity.

6.5 EXERCISES

Assume x and y are functions of t. Evaluate dy/dt for each of the following.

1. $y^2 - 5x^2 = -1; \quad \dfrac{dx}{dt} = -3, x = 1, y = 2$

2. $8y^3 + x^2 = 1; \quad \dfrac{dx}{dt} = 2, x = 3, y = -1$

3. $xy - 5x + 2y^3 = -70; \quad \dfrac{dx}{dt} = -5, x = 2, y = -3$

4. $4x^3 - 9xy^2 + y = -80; \quad \dfrac{dx}{dt} = 4, x = -3, y = 1$

5. $\dfrac{x^2 + y}{x - y} = 9; \quad \dfrac{dx}{dt} = 2, x = 4, y = 2$

6. $\dfrac{y^3 - x^2}{x + 2y} = \dfrac{17}{7}; \quad \dfrac{dx}{dt} = 1, x = -3, y = -2$

7. $xe^y = 2 - \ln 2 + \ln x; \quad \dfrac{dx}{dt} = 6, x = 2, y = 0$

8. $y \ln x + xe^y = 6; \quad \dfrac{dx}{dt} = 5, x = 1, y = 0$

Applications

BUSINESS AND ECONOMICS

9. *Cost* A manufacturer of handcrafted wine racks has determined that the cost to produce x units per month is given by $C = .1x^2 + 10{,}000$. How fast is cost per month changing when production is changing at the rate of 10 units per month and the production level is 100 units?

10. *Cost/Revenue* The manufacturer in Exercise 9 has found that the cost C and revenue R (in dollars) from the production and sale of x units are related by the equation

$$C = \frac{R^2}{400{,}000} + 10{,}000.$$

Find the rate of change of revenue per unit when the cost per unit is changing by \$10 and the revenue is \$20,000.

11. *Revenue/Cost/Profit* Given the revenue and cost functions $R = 50x - .4x^2$ and $C = 5x + 15$, where x is the daily production (and sales), find the following when 40 units are produced daily and the rate of change of production is 10 units per day.

a. The rate of change of revenue with respect to time

b. The rate of change of cost with respect to time

c. The rate of change of profit with respect to time

12. *Revenue/Cost/Profit* Repeat Exercise 11, given that 200 units are produced daily and the rate of change of production is 50 units per day.

13. *Demand* The demand function for a certain product is determined by the fact that the product of the price and the quantity demanded equals 8000. The product currently sells for \$3.50 per unit. Suppose manufacturing costs are increasing over time at a rate of 15% and the company plans to increase the price p at this rate as well. Find the rate of change of demand over time.

14. *Revenue* A company is increasing production at the rate of 25 units per day. The daily demand function is determined by the fact that the price (in dollars) is a linear function of q. At a price of \$70, the demand is 0, and 100 items will be demanded at a price of \$60. Find the rate of change of revenue with respect to time (in days) when the daily production (and sales) is 20 items.

LIFE SCIENCES

15. *Blood Velocity* A cross-country skier has a history of heart problems. She takes nitroglycerin to dilate blood vessels, thus avoiding angina (chest pain) due to blood vessel contraction. Use Poiseuille's law with $k = 555.6$ to find the rate of change of the blood velocity when $R = .02$ mm and R is changing at .003 mm per minute. Assume r is constant.

16. *Allometric Growth* Suppose x and y are two quantities that vary with time according to the allometric formula $y = nx^m$. (See Exercise 76 in the section on Logarithmic Functions.) Show that the derivatives of x and y are related by the formula

$$\frac{1}{y}\frac{dy}{dt} = m\frac{1}{x}\frac{dx}{dt}.$$

(*Hint:* Take natural logarithms of both sides before taking the derivatives.)

17. *Brain Mass* The brain mass of a fetus can be estimated using the total mass of the fetus by the function

$$b = .22m^{.87},$$

where m is the mass of the fetus (in grams) and b is the brain mass (in grams).* Suppose the brain mass of a 25-g fetus is changing at a rate of .25 g per day. Use this to estimate the rate of change of the total mass of the fetus, dm/dt.

18. *Birds* The energy cost of bird flight as a function of body mass is given by

$$E = 429m^{-.35},$$

where m is the mass of the bird (in grams) and E is the energy expenditure (in calories per gram per hour).† Suppose that the mass of a 10-g bird is increasing at a rate of .001 g per hour. Find the rate at which the energy expenditure is changing with respect to time.

19. *Metabolic Rate* The average daily metabolic rate for captive animals from weasels to elk can be expressed as a function of mass by

$$r = 140.2m^{.75},$$

where m is the mass of the animal (in kilograms) and r is the metabolic rate (in kcal per day).‡

a. Suppose that the mass of a weasel is changing with respect to time at a rate dm/dt. Find dr/dt.

b. Determine dr/dt for a 250-kg elk that is gaining mass at a rate of 2 kg per day.

20. *Lizards* The energy cost of horizontal locomotion as a function of the body mass of a lizard is given by

$$E = 26.5m^{-.34},$$

where m is the mass of the lizard (in kilograms) and E is the energy expenditure (in kcal/kg/km).§ Suppose that the mass of a 5-kg lizard is increasing at a rate of .05 kg per day. Find the rate at which the energy expenditure is changing with respect to time.

SOCIAL SCIENCES

21. *Crime Rate* Sociologists have found that crime rates are influenced by temperature. In a midwestern town of 100,000 people, the crime rate has been approximated as

$$C = \frac{1}{10}(T - 60)^2 + 100,$$

where C is the number of crimes per month and T is the average monthly temperature in degrees Fahrenheit. The average temperature for May was 76°, and by the end of May the temperature was rising at the rate of 8° per month. How fast is the crime rate rising at the end of May?

22. *Learning Skills* It is estimated that a person learning a certain assembly-line task takes

$$T(x) = \frac{2 + x}{2 + x^2}$$

minutes to perform the task after x repetitions. Find dT/dt if dx/dt is 4, and 4 repetitions of the task have been performed.

*Wanderley, S., M. Costa-Neves, and R. Rega, "Relative Growth of the Brain in Human Fetuses: First Gestational Trimester," *Archives d'anatomie, d'histologie et d'embryologie*, Vol. 73, 1990, pp. 43–46.
†Robbins, C., *Wildlife Feeding and Nutrition*, New York: Academic Press, 1983, p. 119.
‡Ibid., p. 133.
§Ibid., p. 114.

23. *Memorization Skills* Under certain conditions, a person can memorize W words in t minutes, where

$$W(t) = \frac{-.02t^2 + t}{t + 1}.$$

Find dW/dt when $t = 5$.

PHYSICAL SCIENCES*

24. *Sliding Ladder* A 25-ft ladder is placed against a building. The base of the ladder is slipping away from the building at a rate of 4 ft per minute. Find the rate at which the top of the ladder is sliding down the building at the instant when the bottom of the ladder is 7 ft from the base of the building.

25. *Distance*

a. One car leaves a given point and travels north at 30 mph. Another car leaves the same point at the same time and travels west at 40 mph. At what rate is the distance between the two cars changing at the instant when the cars have traveled 2 hours?

b. Suppose that, in part a, the second car left 1 hour later than the first car. At what rate is the distance between the two cars changing at the instant when the second car has traveled 1 hour?

26. *Area* A rock is thrown into a still pond. The circular ripples move outward from the point of impact of the rock so that the radius of the circle formed by a ripple increases at the rate of 2 ft per minute. Find the rate at which the area is changing at the instant the radius is 4 ft.

27. *Volume* A spherical snowball is placed in the sun. The sun melts the snowball so that its radius decreases 1/4 in. per hour. Find the rate of change of the volume with respect to time at the instant the radius is 4 in.

28. *Volume* A sand storage tank used by the highway department for winter storms is leaking. As the sand leaks out, it forms a conical pile. The radius of the base of the pile increases at the rate of 1 in. per minute. The height of the pile is always twice the radius of the base. Find the rate at which the volume of the pile is increasing at the instant the radius of the base is 5 in.

29. *Shadow Length* A man 6 ft tall is walking away from a lamp post at the rate of 50 ft per minute. When the man is 8 ft from the lamp post, his shadow is 10 ft long. Find the rate at which the length of the shadow is increasing when he is 25 ft from the lamp post. (See the figure.)

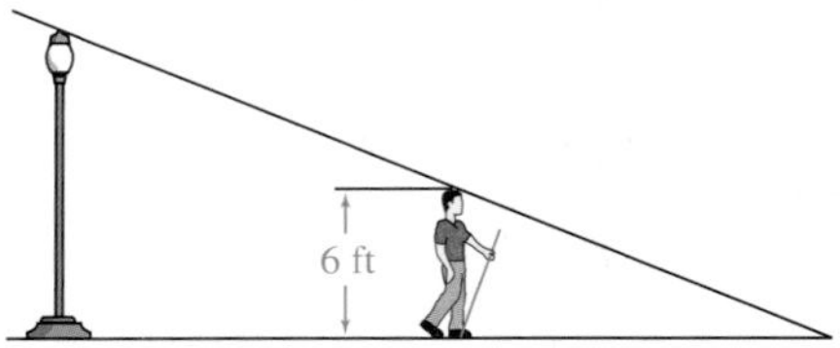

30. *Water Level* A trough has a triangular cross section. The trough is 6 ft across the top, 6 ft deep, and 16 ft long. Water is being pumped into the trough at the rate of 4 ft^3 per minute. Find the rate at which the height of the water is increasing at the instant that the height is 4 ft.

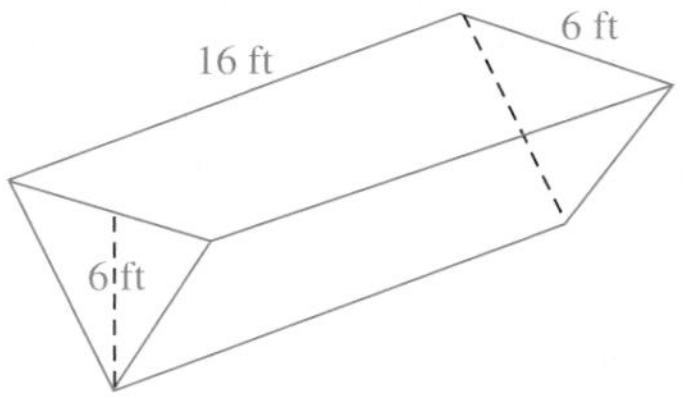

31. *Velocity* A pulley is on the edge of a dock, 8 ft above the water level. (See the figure.) A rope is being used to pull in a boat. The rope is attached to the boat at water level. The rope is being pulled in at the rate of 1 ft per second. Find the rate at which the boat is approaching the dock at the instant the boat is 8 ft from the dock.

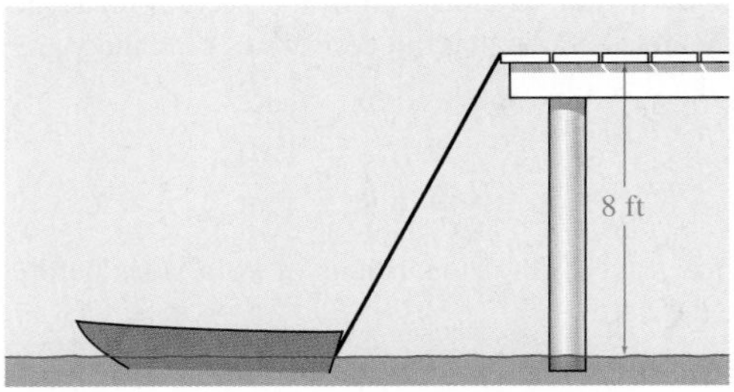

32. *Kite Flying* Christine O'Brien is flying her kite in a wind that is blowing it east at a rate of 50 ft per minute. She has already let out 200 ft of string, and the kite is flying 100 ft above her hand. How fast must she let out string at this moment to keep the kite flying with the same speed and altitude?

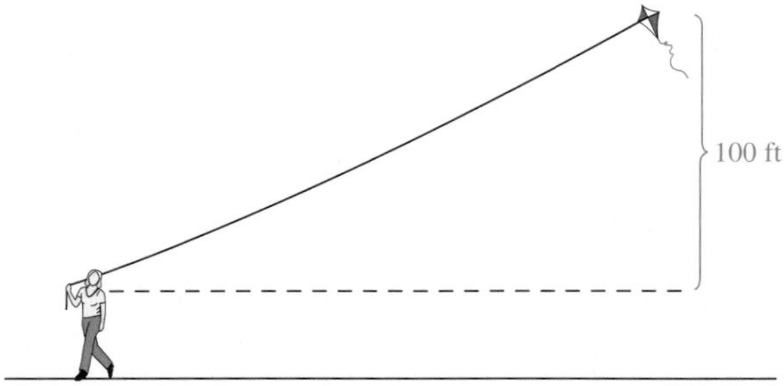

*You may wish to refer to the Appendix of Formulas from Geometry for some of these exercises.

6.6 DIFFERENTIALS: LINEAR APPROXIMATION

THINK ABOUT IT *If the estimated sales of microwave ovens turns out to be inaccurate, approximately how much are profits affected?*

Using differentials, we will answer this question in Example 4.

As mentioned earlier, the symbol Δx represents a change in the variable x. Similarly, Δy represents a change in y. An important problem that arises in many applications is to determine Δy given specific values of x and Δx. This quantity is often difficult to evaluate. In this section we show a method of approximating Δy that uses the derivative dy/dx. In essence, we use the tangent line at a particular value of x to approximate $f(x)$ for values close to x.

For values x_1 and x_2,

$$\Delta x = x_2 - x_1.$$

Solving for x_2 gives

$$x_2 = x_1 + \Delta x.$$

For a function $y = f(x)$, the symbol Δy represents a change in y:

$$\Delta y = f(x_2) - f(x_1).$$

Replacing x_2 with $x_1 + \Delta x$ gives

$$\Delta y = f(x_1 + \Delta x) - f(x_1).$$

If Δx is used instead of h, the derivative of a function f at x_1 could be defined as

$$\frac{dy}{dx} = \lim_{\Delta x \to 0} \frac{\Delta y}{\Delta x}.$$

If the derivative exists, then

$$\frac{dy}{dx} \approx \frac{\Delta y}{\Delta x}$$

as long as Δx is close to 0. Multiplying both sides by Δx (assume $\Delta x \neq 0$) gives

$$\Delta y \approx \frac{dy}{dx} \cdot \Delta x.$$

Until now, dy/dx has been used as a single symbol representing the derivative of y with respect to x. In this section, separate meanings for dy and dx are introduced in such a way that their quotient, when $dx \neq 0$, is the derivative of y with respect to x. These meanings of dy and dx are then used to find an approximate value of Δy.

To define dy and dx, look at Figure 22 on the next page, which shows the graph of a function $y = f(x)$. The tangent line to the graph has been drawn at the point P. Let Δx be any nonzero real number (in practical problems, Δx is a small number) and locate the point $x + \Delta x$ on the x-axis. Draw a vertical line through $x + \Delta x$. Let this vertical line cut the tangent line at M and the graph of the function at Q.

Define the new symbol dx to be the same as Δx. Define the new symbol dy to equal the length MR. The slope of PM is $f'(x)$. By the definition of slope, the slope of PM is also dy/dx, so that

$$f'(x) = \frac{dy}{dx},$$

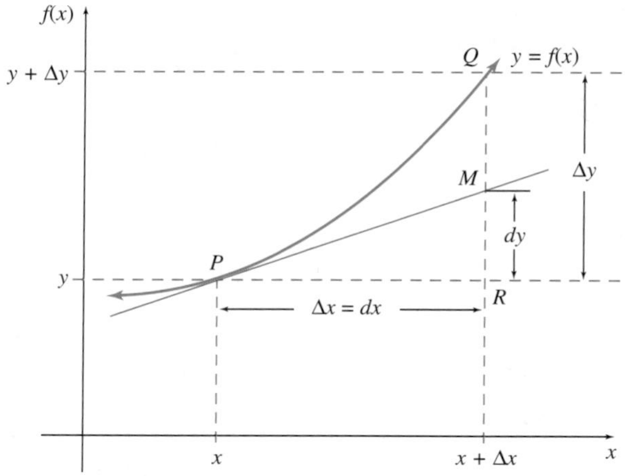

FIGURE 22

or

$$dy = f'(x)dx.$$

In summary, the definitions of the symbols dy and dx are as follows.

DIFFERENTIALS

For a function $y = f(x)$ whose derivative exists, the **differential** of x, written dx, is an arbitrary real number (usually small compared with x); the **differential** of y, written dy, is the product of $f'(x)$ and dx, or

$$\boldsymbol{dy = f'(x)dx.}$$

The usefulness of the differential is suggested by Figure 22. As dx approaches 0, the value of dy gets closer and closer to that of Δy, so that for small nonzero values of dx

$$dy \approx \Delta y,$$

or

$$\Delta y \approx f'(x)dx.$$

EXAMPLE 1 *Differential*

Find dy for the following functions.

(a) $y = 6x^2$

Solution The derivative is $dy/dx = 12x$ so

$$dy = 12x\,dx.$$

(b) $y = 800x^{-3/4}$, $x = 16$, $dx = .01$

Solution

$$\begin{aligned} dy &= -600x^{-7/4}dx \\ &= -600(16)^{-7/4}(.01) \\ &= -600\left(\frac{1}{2^7}\right)(.01) = -.046875 \end{aligned}$$

Differentials can be used to approximate function values for a given x-value (in the absence of a calculator or computer). As discussed above,

$$\Delta y = f(x + \Delta x) - f(x).$$

For small nonzero values of Δx, $\Delta y \approx dy$, so that

$$dy \approx f(x + \Delta x) - f(x),$$

or

$$f(x) + dy \approx f(x + \Delta x).$$

Replacing dy with $f'(x)dx$ gives the following result.

LINEAR APPROXIMATION

Let f be a function whose derivative exists. For small nonzero values of Δx,

$$\boldsymbol{dy \approx \Delta y,}$$

and

$$\boldsymbol{f(x + \Delta x) \approx f(x) + dy = f(x) + f'(x)dx.}$$

EXAMPLE 2 *Approximation*

Approximate $\sqrt{50}$.

Solution We know that $\sqrt{49} = 7$, so we let $f(x) = \sqrt{x}$, $x = 49$, $\Delta x = dx = 1$, and use dy to approximate $\Delta y = \sqrt{50} - \sqrt{49}$. Since $\sqrt{x} = x^{1/2}$,

$$f'(x) = \frac{1}{2}x^{-1/2} = \frac{1}{2x^{1/2}},$$

so

$$dy = \frac{1}{2x^{1/2}}dx.$$

Substituting $x = 49$ and $dx = 1$ gives

$$dy = \frac{1}{2 \cdot 49^{1/2}} \cdot 1 = \frac{1}{14}.$$

Thus,

$$\begin{aligned} \sqrt{50} = f(x + \Delta x) &\approx f(x) + dy \\ &= \sqrt{x} + dy \\ &= \sqrt{49} + \frac{1}{14} = 7\frac{1}{14}. \end{aligned}$$

A calculator gives $7\frac{1}{4} \approx 7.07143$ and $\sqrt{50} \approx 7.07107$. Our approximation of $7\frac{1}{4}$ is close to the true answer and does not require a calculator. ■

While calculators have made differentials less important, the approximation of functions, including linear approximation, is still important in the branch of mathematics known as numerical analysis.

Marginal Analysis Differentials are used to find an approximate value of the change in a value of the dependent variable corresponding to a given change in the independent variable. When the concept of marginal cost (or profit or revenue) was used to approximate the change in cost for nonlinear functions, the same idea was developed. Thus the differential dy approximates Δy in much the same way as the marginal quantities approximate changes in functions.

For example, for a cost function $C(x)$,

$$dC = C'(x)dx = C'(x)\Delta x.$$

Since $\Delta C \approx dC$,

$$\Delta C \approx C'(x)\Delta x.$$

If the change in production, Δx, is equal to 1, then

$$C(x + 1) - C(x) = \Delta C$$
$$\approx C'(x)\Delta x$$
$$= C'(x),$$

which shows that marginal cost $C'(x)$ approximates the cost of the next unit produced, as mentioned earlier.

EXAMPLE 3 *Cost*
Let $C(x) = 2x^3 + 300$.

(a) Find ΔC and $C'(x)$ when $\Delta x = 1$ and $x = 3$.

Solution
$$\Delta C = C(4) - C(3) = 428 - 354 = 74$$
$$C'(x) = 6x^2$$
$$C'(3) = 54$$

Here, the approximation of $C'(3)$ for ΔC is poor, since $\Delta x = 1$ is large relative to $x = 3$.

(b) Find ΔC and $C'(x)$ when $\Delta x = 1$ and $x = 50$.

Solution
$$\Delta C = C(51) - C(50) = 265{,}602 - 250{,}300 = 15{,}302$$
$$C'(50) = 6(2500) = 15{,}000$$

This approximation is quite good since $\Delta x = 1$ is small compared to $x = 50$. ■

EXAMPLE 4 *Profit*

An analyst for a manufacturer of small appliances estimates that the profit (in dollars) from the sale of x microwave ovens is given by

$$P(x) = 4000 \ln x.$$

In a report to management, the analyst projected sales for the coming year to be 1000 ovens, for a total profit of about \$27,630. He now realizes that his sales estimate may have been as much as 100 ovens too high. Approximately how far off is his profit estimate?

Solution Differentials can be used to find the approximate change in P resulting from decreasing x by 100. This change can be approximated by $dP = P'(x)dx$ where $x = 1000$ and $dx = -100$. Since $P'(x) = 4000/x$,

$$\begin{aligned} \Delta P \approx dp &= \frac{4000}{x}dx \\ &= \frac{4000}{1000}(-100) \\ &= -400. \end{aligned}$$

Thus the profit estimate may have been as much as \$400 too high. Computing the actual difference with a calculator gives $4000 \ln 900 - 4000 \ln 1000 \approx -421$, which is close to our approximation.

Error Estimation The final example in this section shows how differentials are used to estimate errors that might enter into measurements of a physical quantity.

EXAMPLE 5 *Error Estimation*

In a precision manufacturing process, ball bearings must be made with a radius of .6 mm, with a maximum error in the radius of $\pm .015$ mm. Estimate the maximum error in the volume of the ball bearing.

Solution The formula for the volume of a sphere is

$$V = \frac{4}{3}\pi r^3.$$

If an error of Δr is made in measuring the radius of the sphere, the maximum error in the volume is

$$\Delta V = \frac{4}{3}\pi(r + \Delta r)^3 - \frac{4}{3}\pi r^3.$$

Rather than calculating ΔV, approximate ΔV with dV, where

$$dV = 4\pi r^2 dr.$$

Replacing r with .6 and $dr = \Delta r$ with $\pm .015$ gives

$$\begin{aligned} dV &= 4\pi(.6)^2(\pm .015) \\ &\approx \pm .0679. \end{aligned}$$

The maximum error in the volume is about.07 mm^3.

6.6 EXERCISES

For Exercises 1–8, find dy for the given values of x and Δx.

1. $y = 2x^2 - 5x;\quad x = -2, \Delta x = .2$

2. $y = x^2 - 3x;\quad x = 3, \Delta x = .1$

3. $y = x^3 - 2x^2 + 3;\quad x = 1, \Delta x = -.1$

4. $y = 2x^3 + x^2 - 4x;\quad x = 2, \Delta x = -.2$

5. $y = \sqrt{3x};\quad x = 1, \Delta x = .15$

6. $y = \sqrt{4x - 1};\quad x = 5, \Delta x = .08$

7. $y = \dfrac{2x - 5}{x + 1};\quad x = 2, \Delta x = -.03$

8. $y = \dfrac{6x - 3}{2x + 1};\quad x = 3, \Delta x = -.04$

Use the differential to approximate each quantity. Then use a calculator to approximate the quantity, and give the absolute value of the difference in the two results to 4 decimal places.

9. $\sqrt{145}$

10. $\sqrt{23}$

11. $\sqrt{.99}$

12. $\sqrt{17.02}$

13. $e^{.01}$

14. $e^{-.002}$

15. ln 1.05

16. ln5.98

Applications

BUSINESS AND ECONOMICS

17. *Demand* The demand for grass seed (in thousands of pounds) at a price of x dollars is

$$d(x) = -5x^3 - 2x^2 + 1500.$$

Use the differential to approximate the changes in demand for the following changes in x.

a. \$2 to \$2.50 **b.** \$6 to \$6.30

18. *Average Cost* The average cost to manufacture x dozen marking pencils is

$$A(x) = .04x^3 + .1x^2 + .5x + 6.$$

Use the differential to approximate the changes in the average cost for the following changes in x.

a. 3 to 4 **b.** 5 to 6

19. *Revenue* A company estimates that the revenue (in dollars) from the sale of x doghouses is given by

$$R(x) = 625 + .03x + .0001x^2.$$

Use the differential to approximate the change in revenue from the sale of one more doghouse when 1000 doghouses are sold.

20. *Profit* The profit function for the company in Exercise 19 is

$$P(x) = -390 + 24x + 5x^2 - \frac{1}{3}x^3,$$

where x represents the demand for the product. Find the approximate change in profit for a 1-unit change in demand when demand is at a level of 1000 doghouses. Use the differential.

21. *Material Requirement* A cube 4 in. on an edge is given a protective coating .1 in. thick. About how much coating should a production manager order for 1000 such cubes?

22. *Material Requirement* Beach balls 1 ft in diameter have a thickness of .03 in. How much material would be needed to make 5000 beach balls?

LIFE SCIENCES

23. *Alcohol Concentration* The mathematical model

$$y = A(x) = \frac{-7}{480}x^3 + \frac{127}{120}x, \quad 0 \le x < 9,$$

gives the approximate alcohol concentration (in tenths of a percent) in an average person's bloodstream x hours after drinking 8 oz of 100-proof whiskey.

a. Approximate the change in alcohol concentration from 3 to 3.5 hours.

b. Approximate the change in alcohol concentration from 6 to 6.25 hours.

24. *Drug Concentration* The concentration of a certain drug in the bloodstream x hours after being administered is approximately

$$C(x) = \frac{5x}{9 + x^2}.$$

Use the differential to approximate the changes in concentration for the following changes in x.

a. 1 to 1.5 **b.** 2 to 2.25

25. *Bacteria Population* The population of bacteria (in millions) in a certain culture x hours after an experimental nutrient is introduced into the culture is

$$P(x) = \frac{25x}{8 + x^2}.$$

Use the differential to approximate the changes in population for the following changes in x.

a. 2 to 2.5 **b.** 3 to 3.25

26. *Area of a Blood Vessel* The radius of a blood vessel is 1.7 mm. A drug causes the radius to change to 1.6 mm. Find the approximate change in the area of a cross section of the vessel.

27. *Volume of a Tumor* A tumor is approximately spherical in shape. If the radius of the tumor changes from 14 mm to 16 mm, find the approximate change in volume.

28. *Area of an Oil Slick* An oil slick is in the shape of a circle. Find the approximate increase in the area of the slick if its radius increases from 1.2 miles to 1.4 miles.

29. *Area of a Bacteria Colony* The shape of a colony of bacteria on a Petri dish is circular. Find the approximate increase in its area if the radius increases from 20 mm to 22 mm.

30. *Gray Wolves* Accurate methods of estimating the age of gray wolves are important to scientists who study wolf population dynamics. One method of estimating the age of a gray wolf is to measure the percent closure of the pulp cavity of a canine tooth and estimate age by

$$A(p) = \frac{1.181p}{94.359 - p},$$

where p is the percent closure and $A(p)$ is the age of the wolf (in years).*

a. What is a sensible domain for this function?

b. Use differentials to estimate how long it will take for a gray wolf that first measures a 60% closure to obtain a 65% closure. Compare this with the actual value of about .55 years.

31. *Pigs* Researchers have observed that the mass of a female (gilt) pig can be estimated by the function

$$M(t) = -3.5 + 197.5e^{-e^{-.01394(t-108.4)}},$$

where t is the age of the pig (in days) and $M(t)$ is the mass of the pig (in kilograms).†

a. If a particular gilt is 80 days old, use differentials to estimate how much it will gain before it is 90 days old.

b. What is the actual gain in mass?

PHYSICAL SCIENCES

32. *Volume* A spherical balloon is being inflated. Find the approximate change in volume if the radius increases from 4 cm to 4.2 cm.

33. *Volume* A spherical snowball is melting. Find the approximate change in volume if the radius decreases from 3 cm to 2.8 cm.

GENERAL INTEREST

34. *Measurement Error* The edge of a square is measured as 3.45 in., with a possible error of $\pm .002$ in. Estimate the maximum error in the area of the square.

35. *Measurement Error* The radius of a circle is measured as 4.87 in., with a possible error of $\pm .040$ in. Estimate the maximum error in the area of the circle.

36. *Measurement Error* A sphere has a radius of 5.81 in., with a possible error of $\pm .003$ in. Estimate the maximum error in the volume of the sphere.

37. *Measurement Error* A cone has a known height of 7.284 in. The radius of the base is measured as 1.09 in., with a possible error of $\pm .007$ in. Estimate the maximum error in the volume of the cone. (The volume of the cone is given by $V = (1/3)\pi r^2 h$.)

*Landon, D., C. Waite, R. Peterson, and L. Mech, "Evaluation of Age Determination Techniques for Gray Wolves," *Journal of Wildlife Management*, Vol. 62, No. 2, 1998, pp. 674–682.

†Van Lunen, T. and D. Cole, "Growth and Body Composition of Highly Selected Boars and Gilts," *Animal Science*, Vol. 67, 1998, pp. 107–116.

CHAPTER SUMMARY

In this chapter, after discussing how to find an absolute maximum or minimum, we saw a diversity of applications with that as the goal. Two more applications, economic lot size and economic order quantity, were covered in a separate section, which also applied the derivative to the economic concept of elasticity of demand. Implicit differentiation is more a technique than an application, but it underlies related rate problems, in which one or more rates are given and another is to be found. Finally, we studied the differential as a way to find linear approximations to functions.

KEY TERMS

6.1 **absolute maximum**
absolute minimum
absolute extremum (or extrema)
extreme value theorem
graphical optimization

6.2 **spawner-recruit function**
parent-progeny function
maximum sustainable harvest

6.3 **economic lot size**
economic order quantity
elasticity of demand
unit elasticity

6.4 **explicit function**
implicit differentiation

6.5 **related rates**

6.6 **differential**

CHAPTER 6 REVIEW EXERCISES

Find the absolute extrema if they exist, and all values of x where they occur on the given intervals.

1. $f(x) = -x^2 + 5x + 1;\quad [1, 4]$

2. $f(x) = 4x^2 - 8x - 3;\quad [-1, 2]$

3. $f(x) = x^3 + 2x^2 - 15x + 3;\quad [-4, 2]$

4. $f(x) = -2x^3 - 2x^2 + 2x - 1;\quad [-3, 1]$

5. When solving applied extrema problems, why is it necessary to check the endpoints of the domain?

6. What is elasticity of demand (in words; no mathematical symbols allowed)? Why is the derivative used to describe elasticity?

7. Find the absolute maximum and minimum of $f(x) = \dfrac{2 \ln x}{x^2}$

a. on the interval $[1, 4]$;

b. on the interval $[2, 5]$.

8. When is it necessary to use implicit differentiation?

9. When a term involving y is differentiated in implicit differentiation, it is multiplied by dy/dx. Why? Why aren't terms involving x multiplied by dx/dx?

Find dy/dx.

10. $x^2y^3 + 4xy = 2$

11. $\dfrac{x}{y} - 4y = 3x$

12. $9\sqrt{x} + 4y^3 = \dfrac{2}{x}$

13. $2\sqrt{y - 1} = 8x^{2/3}$

14. $\dfrac{x + 2y}{x - 3y} = y^{1/2}$

15. $\dfrac{6 + 5x}{2 - 3y} = \dfrac{1}{5x}$

16. Find the equation of the line tangent to the graph of $\sqrt{2x} - 4yx = -22$ at the point $(2, 3)$.

17. What is the difference between a related rate problem and an applied extremum problem?

18. Why is implicit differentiation used in related rate problems?

Find dy/dt.

19. $y = 8x^3 - 7x^2;\quad \dfrac{dx}{dt} = 4,\ x = 2$

20. $y = \dfrac{9 - 4x}{3 + 2x};\quad \dfrac{dx}{dt} = -1,\ x = -3$

21. $y = \dfrac{1 + \sqrt{x}}{1 - \sqrt{x}};\quad \dfrac{dx}{dt} = -4,\ x = 4$

22. $\dfrac{x^2 + 5y}{x - 2y} = 2;\quad \dfrac{dx}{dt} = 1,\ x = 2,\ y = 0$

23. What is a differential? What is it used for?

Evaluate dy.

24. $y = 8 - x^2 + x^3$; $x = -1, \Delta x = .02$

25. $y = \dfrac{3x - 7}{2x + 1}$; $x = 2, \Delta x = .003$

26. **a.** Suppose x and y are related by the equation

$$-12x + x^3 + y + y^2 = 4.$$

Find all critical points on the curve.

b. Determine whether the critical points found in part a are relative maxima or relative minima by taking values of x nearby and solving for the corresponding values of y.

c. Is there an absolute maximum or minimum for x and y in the relationship given in part a? Why or why not?

27. In Exercise 26, implicit differentiation was used to find the relative extrema. The exercise was contrived to avoid various difficulties that could have arisen. Discuss some of the difficulties that might be encountered in such problems, and how these difficulties might be resolved.

Applications

BUSINESS AND ECONOMICS

28. *Profit* The total profit (in tens of dollars) from the sale of x hundred boxes of candy is given by

$$P(x) = -x^3 + 10x^2 - 12x.$$

a. Find the number of boxes of candy that should be sold in order to produce maximum profit.

b. Find the maximum profit.

29. *Packaging Design* The packaging department of a corporation is designing a box with a square base and no top. The volume is to be 32 m^3. To reduce cost, the box is to have minimum surface area. What dimensions (height, length, and width) should the box have?

30. *Packaging Design* Fruit juice will be packaged in cylindrical cans with a volume of 40 in^3 each. The top and bottom of the can cost 4¢ per in^2, while the sides cost 3¢ per in^2. Find the radius and height of the can of minimum cost.

31. *Packaging Design* A company plans to package its product in a cylinder that is open at one end. The cylinder is to have a volume of 27π in^3. What radius should the circular bottom of the cylinder have to minimize the cost of the material?

32. *Order Quantity* A store sells 980,000 cases of a product annually. It costs \$2 to store 1 case for 1 year and \$20 to place a reorder. Find the number of cases that should be ordered each time.

33. *Order Quantity* A very large camera store sells 320,000 rolls of film annually. It costs 10¢ to store 1 roll for 1 year and \$10 to place a reorder. Find the number of rolls that should be ordered each time.

34. *Lot Size* In 1 year, a health food manufacturer produces and sells 240,000 cases of vitamins. It costs \$2 to store 1 case for 1 year and \$15 to produce each batch. Find the number of batches that should be produced annually.

35. *Lot Size* A company produces 128,000 cases of soft drink annually. It costs \$1 to store 1 case for 1 year and \$10 to produce 1 lot. Find the number of lots that should be produced annually.

36. *Elasticity* Suppose the demand function for a product is given by $q = A/p^k$, where A and k are positive constants. For what values of k is the demand elastic? Inelastic?

LIFE SCIENCES

37. *Pollution* A circle of pollution is spreading from a broken underwater waste disposal pipe, with the radius increasing at the rate of 4 ft per minute. Find the rate of change of the area of the circle when the radius is 7 ft.

38. *Logistic Growth* Many populations grow according to the logistic equation

$$\frac{dx}{dt} = rx(N - x),$$

where r is a constant involving the rate of growth and N is the carrying capacity of the environment, beyond which the population decreases. Show that the graph of x has an inflection point where $x = N/2$. (*Hint:* Use implicit differentiation. Then set $d^2x/dt^2 = 0$, and factor)

39. *Dentin Growth* The dentinal formation of molars in mice has been studied by researchers in Copenhagen. They determined that the growth curve that best fits dentinal formation for the first molar is

$$M(t) = 1.3386309 - .4321173t + .0564512t^2$$
$$-.0020506t^3 + .0000315t^4 - .0000001785t^5,$$
$$5 \le t \le 51,$$

where t is the age of the mouse (in days), and $M(t)$ is the cumulative dentin volume (in 10^{-1} mm^3).*

a. Use a graphing calculator to sketch the graph of this function on $[5, 51]$ by $[0, 7.5]$.

b. Find the time in which the dentin formation is growing most rapidly. (*Hint:* Find the maximum value of the derivative of this function.)

40. *Human Skin Surface* The surface of the skin is made up of a network of intersecting lines which form polygons. Researchers have discovered a functional relationship between the age of a female and the number of polygons per area of skin according to

$$P(t) = 237.09 - 8.0398t + .20813t^2 - .0027563t^3 + .000013016t^4, \quad 0 \le t \le 95,$$

where t is the age of the person (in years), and $P(t)$ is the number of polygons for a particular surface area of skin.†

a. Use a graphing calculator to sketch a graph of $P(t)$ on $[0, 95]$ by $[0, 300]$.

b. Find the maximum and minimum number of polygons per area predicted by the model.

c. Discuss the accuracy of this model for older people.

PHYSICAL SCIENCES

41. *Sliding Ladder* A 50-ft ladder is placed against a building. The top of the ladder is sliding down the building at the rate of 2 ft per minute. Find the rate at which the base of the ladder is slipping away from the building at the instant that the base is 30 ft from the building.

42. *Spherical Radius* A large weather balloon is being inflated with air at the rate of 1.2 ft^3 per minute. Find the rate of change of the radius when the radius is 1.2 ft.

43. *Water Level* A water trough 2 ft across, 4 ft long, and 1 ft deep has ends in the shape of isosceles triangles. (See the figure.) It is being filled with 3.5 ft^3 of water per minute. Find the rate at which the depth of water in the tank is changing when the water is $1/3$ ft deep.

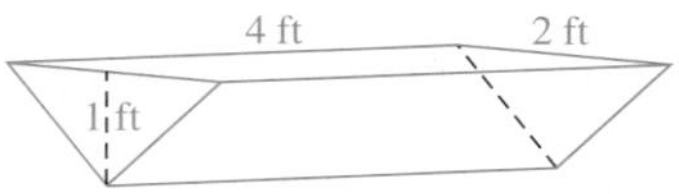

GENERAL INTEREST

44. *Volume* Approximate the volume of coating on a sphere of radius 4 in. if the coating is .02 in. thick.

45. *Area* A square has an edge of 9.2 in., with a possible error in the measurement of $\pm .04$ in. Estimate the possible error in the area of the square.

46. *Package Dimensions* UPS has the following rule regarding package dimensions. The length can be no more than 108 in., and the length plus the girth (twice the sum of the width and the height) can be no more than 130 in. If the width of a package is 4 in. more than its height and it has the maximum girth, find the length that produces maximum volume.

47. *Pursuit* A boat moves north at a constant speed. A second boat, moving at the same speed, pursues the first boat in such a way that it always points directly at the first boat. When the first boat is at the point $(0, 1)$, the second boat is at the point $(6, 2.5)$, with the positive y-axis pointing north. It can then be shown that the curve traced by the second boat, known as a pursuit curve, is given by

$$y = \frac{x^2}{16} - 2 \ln x + \frac{1}{4} + 2 \ln 6.^{‡}$$

Find the y-coordinate of the southernmost point of the second boat's path.

48. *Playground Area* The city park department is planning an enclosed play area in a new park. One side of the area will be against an existing building, with no fence needed there. Find the dimensions of the maximum rectangular area that can be made with 900 m of fence.

*Matsumoto, B., K. Nonaka, and M. Nakata, "A Genetic Study of Dentin Growth in the Mandibular Second and Third Molars of Male Mice," *Journal of Craniofacial Genetics and Developmental Biology*, Vol. 16, No. 3, July–Sept. 1996, pp. 137–147.

†Voros, E., C. Robert, and A. Robert, "Age-Related Changes of the Human Skin Surface Microrelief," *Gerontology*, Vol. 36, 1990, pp. 276–285.

‡For example, see *Differential Equations: Theory and Applications*, Ray Redheffer, Jones and Bartlett Publishers, 1991, pp. 107–108.

EXTENDED APPLICATION: A Total Cost Model for a Training Program*

In this application, we set up a mathematical model for determining the total costs in setting up a training program. Then we use calculus to find the time interval between training programs that produces the minimum total cost. The model assumes that the demand for trainees is constant and that the fixed cost of training a batch of trainees is known. Also, it is assumed that people who are trained, but for whom no job is readily available, will be paid a fixed amount per month while waiting for a job to open up.

The model uses the following variables.

D = demand for trainees per month

N = number of trainees per batch

C_1 = fixed cost of training a batch of trainees

C_2 = marginal cost of training per trainee per month

C_3 = salary paid monthly to a trainee who has not yet been given a job after training

m = time interval in months between successive batches of trainees

t = length of training program in months

$Z(m)$ = total monthly cost of program

The total cost of training a batch of trainees is given by $C_1 + NtC_2$. However, $N = mD$, so that the total cost per batch is $C_1 + mDtC_2$.

After training, personnel are given jobs at the rate of D per month. Thus, $N - D$ of the trainees will not get a job the first month, $N - 2D$ will not get a job the second month, and so on. The $N - D$ trainees who do not get a job the first month produce total costs of $(N - D)C_3$, those not getting jobs during the second month produce costs of $(N - 2D)C_3$, and so on. Since $N = mD$, the costs during the first month can be written as

$$(N - D)C_3 = (mD - D)C_3 = (m - 1)DC_3,$$

while the costs during the second month are $(m - 2)DC_3$, and so on. The total cost for keeping the trainees without a job is thus

$$(m - 1)DC_3 + (m - 2)DC_3 + (m - 3)DC_3 + \cdots + 2DC_3 + DC_3,$$

which can be factored to give

$$DC_3[(m - 1) + (m - 2) + (m - 3) + \cdots + 2 + 1].$$

The expression in brackets is the sum of the terms of an arithmetic sequence, discussed in most algebra texts. Using formulas for arithmetic sequences, the expression in brackets can be shown to equal $m(m - 1)/2$, so that we have

$$DC_3\left[\frac{m(m - 1)}{2}\right] \tag{1}$$

as the total cost for keeping jobless trainees.

The total cost per batch is the sum of the training cost per batch, $C_1 + mDtC_2$, and the cost of keeping trainees without a proper job, given by Equation (1). Since we assume that a batch of trainees is trained every m months, the total cost per month, $Z(m)$, is given by

$$Z(m) = \frac{C_1 + mDtC_2}{m} + \frac{DC_3\left[\frac{m(m - 1)}{2}\right]}{m}$$

$$= \frac{C_1}{m} + DtC_2 + DC_3\left(\frac{m - 1}{2}\right).$$

Exercises

1. Find $Z'(m)$.
2. Solve the equation $Z'(m) = 0$.

As a practical matter, it is usually required that m be a whole number. If m does not come out to be a whole number, then m^+ and m^-, the two whole numbers closest to m, must be chosen. Calculate both $Z(m^+)$ and $Z(m^-)$; the smaller of the two provides the optimum value of Z.

*Based on "A Total Cost Model for a Training Program" by P. L. Goyal and S. K. Goyal, Faculty of Commerce and Administration, Concordia University.

3. Suppose a company finds that its demand for trainees is 3 per month, that a training program requires 12 months, that the fixed cost of training a batch of trainees is \$15,000, that the marginal cost per trainee per month is \$100, and that trainees are paid \$900 per month after training but before going to work. Use your result from Exercise 2 and find m.
4. Since m is not a whole number, find m^+ and m^-.
5. Calculate $Z(m^+)$ and $Z(m^-)$.
6. What is the optimum time interval between successive batches of trainees? How many trainees should be in a batch?

Directions for Group Project

Suppose you have read an article in the paper announcing that a new high-tech company is locating in your town. Given that the company is manufacturing very specialized equipment, you realize that it must develop a program to train all new employees. Because you would like to get an internship at this new company, use the information above to develop a hypothetical training program that optimizes the time interval between successive batches of trainees and the number of trainees that should be in each session. Assume that you know the new CEO because you and three of your friends served her pizza at various times at the local pizza shop (your current jobs) and that she is willing to listen to a proposal that describes your training program. Prepare a presentation for your interview that will describe your training program. Use presentation software such as Microsoft Powerpoint.

CHAPTER

7 Integration

If we know the rate at which a quantity is changing, we can find the total change over a period of time by integrating. An exercise in Section 2 illustrates this process with an exponential model for the rate at which the number of Mexican vehicle passengers crossing the border into the United States is changing. The integral of the exponential rate function helps us predict how long it will be until the number of such border crossings doubles.

In previous chapters, we studied the mathematical process of finding the derivative of a function, and we considered various applications of derivatives. The material that was covered belongs to the branch of calculus called *differential calculus*. In this chapter and the next we will study another branch of calculus, called *integral calculus*. Like the derivative of a function, the definite integral of a function is a special limit with many diverse applications. Geometrically, the derivative is related to the slope of the tangent line to a curve, while the definite integral is related to the area under a curve. As this chapter will show, differential and integral calculus are connected by the Fundamental Theorem of Calculus.

7.1 ANTIDERIVATIVES

THINK ABOUT IT *If an object is thrown from the top of the Sears Tower in Chicago, how fast is it going when it hits the ground?*

Using *antiderivatives*, we can answer this question.

Functions used in applications in previous chapters have provided information about a *total amount* of a quantity, such as cost, revenue, profit, temperature, gallons of oil, or distance. Working with these functions provided information about the rate of change of these quantities and allowed us to answer important questions about the extrema of the functions. It is not always possible to find ready-made functions that provide information about the total amount of a quantity, but it is often possible to collect enough data to come up with a function that gives the *rate of change* of a quantity. Since we know that derivatives give the rate of change when the total amount is known, is it possible to reverse the process and use a known rate of change to get a function that gives the total amount of a quantity? The answer is yes; this reverse process, called *antidifferentiation*, is the topic of this section. The *antiderivative* of a function is defined as follows.

ANTIDERIVATIVE

If $F'(x) = f(x)$, then $F(x)$ is an **antiderivative** of $f(x)$.

EXAMPLE 1 *Antiderivative*

(a) If $F(x) = 10x$, then $F'(x) = 10$, so $F(x) = 10x$ is an antiderivative of $f(x) = 10$.

(b) For $F(x) = x^2$, $F'(x) = 2x$, making $F(x) = x^2$ an antiderivative of $f(x) = 2x$.

EXAMPLE 2 *Antiderivative*

Find an antiderivative of $f(x) = 5x^4$.

Solution To find a function $F(x)$ whose derivative is $5x^4$, work backwards. Recall that the derivative of x^n is nx^{n-1}. If

$$nx^{n-1} \text{ is } 5x^4,$$

then $n - 1 = 4$ and $n = 5$, so x^5 is an antiderivative of $5x^4$.

EXAMPLE 3 *Population*

Suppose a population is growing at a rate given by $f(x) = e^x$, where x is time in years from some initial date. Find a function giving the population at time x.

Solution Let the population function be $F(x)$. Then

$$f(x) = F'(x) = e^x.$$

The derivative of the function defined by $F(x) = e^x$ is $F'(x) = e^x$, so one possible population function with the given growth rate is $F(x) = e^x$.

The function from Example 1(b), defined by $F(x) = x^2$, is not the only function whose derivative is $f(x) = 2x$; for example, both $G(x) = x^2 + 2$ and $H(x) = x^2 - 7$ have $f(x) = 2x$ as a derivative. In fact, for any real number C, the function $F(x) = x^2 + C$ has $f(x) = 2x$ as its derivative. This means that there is a *family* or *class* of functions having $2x$ as an antiderivative. As the next theorem states, if two functions $F(x)$ and $G(x)$ are antiderivatives of $f(x)$, then $F(x)$ and $G(x)$ can differ only by a constant.

If $F(x)$ and $G(x)$ are both antiderivatives of a function $f(x)$ on an interval, then there is a constant C such that

$$F(x) - G(x) = C.$$

(Two antiderivatives of a function can differ only by a constant.) The arbitrary real number C is called an integration constant.

For example,

$$F(x) = x^2 + 2, \qquad G(x) = x^2, \qquad \text{and} \qquad H(x) = x^2 - 4$$

are all antiderivatives of $f(x) = 2x$, and any two of them differ only by a constant. The derivative of a function gives the slope of the tangent line at any x-value. The fact that these three functions have the same derivative, $f(x) = 2x$, means that their slopes at any particular value of x are the same, as shown in Figure 1. Thus, each graph can be obtained from another by a vertical shift of $|C|$ units, where C is any constant. We will represent this family of antiderivatives of $f(x)$ by $F(x) + C$.

The family of all antiderivatives of the function f is indicated by

$$\int f(x)\,dx.$$

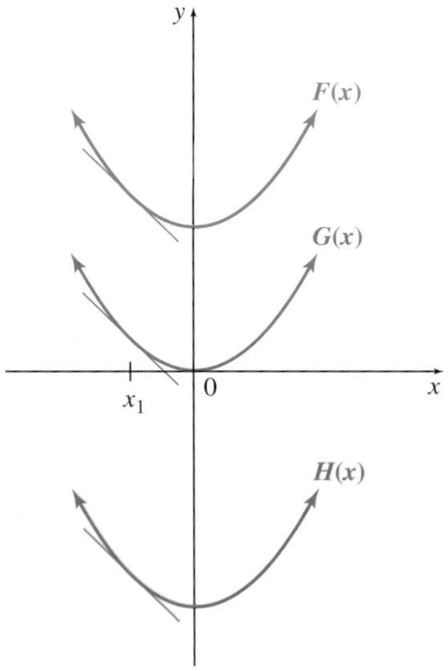

Slopes of the tangent lines at $x = x_1$ are the same.

FIGURE 1

The symbol $\int$ is the **integral sign,** $f(x)$ is the **integrand,** and $\int f(x)\,dx$ is called an **indefinite integral,** the most general antiderivative of f.

INDEFINITE INTEGRAL

If $F'(x) = f(x)$, then

$$\int f(x)\,dx = F(x) + C,$$

for any real number C.

For example, using this notation,

$$\int 2x\,dx = x^2 + C.$$

The dx in the indefinite integral indicates that $\int f(x)\,dx$ is the "integral of $f(x)$ *with respect* to x" just as the symbol dy/dx denotes the "derivative of y with respect to x." For example, in the indefinite integral $\int 2ax\,dx$, dx indicates that a is to be treated as a constant and x as the variable, so that

$$\int 2ax\,dx = \int a(2x)dx = ax^2 + C.$$

On the other hand,

$$\int 2ax\,da = a^2x + C = xa^2 + C.$$

A more complete interpretation of dx will be discussed later.

The symbol $\int f(x)\,dx$ was created by G. W. Leibniz (1646–1716) in the latter part of the seventeenth century. The $\int$ is an elongated S from *summa,* the Latin word for *sum*. The word *integral* as a term in the calculus was coined by Jakob Bernoulli (1654–1705), a Swiss mathematician who corresponded frequently with Leibniz. The relationship between sums and integrals will be clarified in a later section.

The method of working backwards, as above, to find an antiderivative is not very satisfactory. Some rules for finding antiderivatives are needed. Since the process of finding an indefinite integral is the inverse of the process of finding a derivative, each formula for derivatives leads to a rule for indefinite integrals.

FOR REVIEW

Recall that $\frac{d}{dx}x^n = nx^{n-1}$.

As mentioned in Example 2, the derivative of x^n is found by multiplying x by n and reducing the exponent on x by 1. To find an indefinite integral—that is, to undo what was done—*increase* the exponent by 1 and *divide* by the new exponent, $n + 1$.

POWER RULE

For any real number $n \neq -1$,

$$\int x^n dx = \frac{x^{n+1}}{n+1} + C.$$

This rule can be verified by differentiating the expression on the right above:

$$\frac{d}{dx}\left(\frac{x^{n+1}}{n+1} + C\right) = \frac{n+1}{n+1}x^{(n+1)-1} + 0 = x^n.$$

(If $n = -1$, the expression in the denominator is 0, and the above rule cannot be used. Finding an antiderivative for this case is discussed later.)

EXAMPLE 4 *Power Rule*

Use the power rule to find each indefinite integral.

(a) $\int t^3 dt$

Solution Use the power rule with $n = 3$.

$$\int t^3\,dt = \frac{t^{3+1}}{3+1} + C = \frac{t^4}{4} + C$$

(b) $\int \frac{1}{t^2}\,dt$

Solution First, write $1/t^2$ as t^{-2}. Then

$$\int \frac{1}{t^2}\,dt = \int t^{-2}dt = \frac{t^{-1}}{-1} + C = -\frac{1}{t} + C.$$

(c) $\int \sqrt{u}\,du$

Solution Since $\sqrt{u} = u^{1/2}$,

$$\int \sqrt{u}\,du = \int u^{1/2}\,du = \frac{u^{3/2}}{1/2 + 1} + C = \frac{2}{3}u^{3/2} + C.$$

To check this, differentiate $(2/3)u^{3/2} + C$; the derivative is $u^{1/2}$, the original function.

(d) $\int dx$

Solution Write dx as $1 \cdot dx$ and use the fact that $x^0 = 1$ for any nonzero number x to get

$$\int dx = \int 1\,dx = \int x^0\,dx = \frac{x^1}{1} + C = x + C.$$

FOR REVIEW

Recall that $\frac{d}{dx}[f(x) \pm g(x)] = [f'(x) \pm g'(x)]$ and $\frac{d}{dx}[kf(x)] = kf'(x)$.

As shown earlier, the derivative of the product of a constant and a function is the product of the constant and the derivative of the function. A similar rule applies to indefinite integrals. Also, since derivatives of sums or differences are found term by term, indefinite integrals also can be found term by term.

CONSTANT MULTIPLE RULE AND SUM OR DIFFERENCE RULE

If all indicated integrals exist,

$$\int k \cdot f(x)\,dx = k\int f(x)\,dx, \qquad \text{for any real number } k,$$

and

$$\int [f(x) \pm g(x)]\,dx = \int f(x)\,dx \pm \int g(x)\,dx.$$

EXAMPLE 5 *Rules of Integration*

Use the rules to find each integral.

(a) $\int 2v^3\,dv$

Solution By the constant multiple rule and the power rule,

$$\int 2v^3\,dv = 2\int v^3\,dv = 2\left(\frac{v^4}{4}\right) + C = \frac{v^4}{2} + C.$$

(b) $\int \frac{12}{z^5}\,dz$

Solution Use negative exponents.

$$\int \frac{12}{z^5}\,dz = \int 12z^{-5}\,dz = 12\int z^{-5}\,dz = 12\left(\frac{z^{-4}}{-4}\right) + C$$

$$= -3z^{-4} + C = \frac{-3}{z^4} + C$$

(c) $\int (3z^2 - 4z + 5)\,dz$

Solution Using the rules in this section,

$$\int (3z^2 - 4z + 5)\,dz = 3\int z^2\,dz - 4\int z\,dz + 5\int dz$$

$$= 3\left(\frac{z^3}{3}\right) - 4\left(\frac{z^2}{2}\right) + 5z + C$$

$$= z^3 - 2z^2 + 5z + C.$$

Only one constant C is needed in the answer; the three constants from integrating term by term are combined. In Example 5(a), C represents any real number, so it is not necessary to multiply it by 2 in the next-to-last step.

EXAMPLE 6 *Rules of Integration*

Use the rules to find each integral.

(a) $\int \frac{x^2 + 1}{\sqrt{x}}\,dx$

Solution First rewrite the integrand as follows.

$$\int \frac{x^2 + 1}{\sqrt{x}}\,dx = \int \left(\frac{x^2}{\sqrt{x}} + \frac{1}{\sqrt{x}}\right) dx$$

$$= \int \left(\frac{x^2}{x^{1/2}} + \frac{1}{x^{1/2}}\right) dx$$

$$= \int (x^{3/2} + x^{-1/2})\,dx$$

Now find the antiderivative.

$$\int (x^{3/2} + x^{-1/2})\,dx = \frac{x^{5/2}}{5/2} + \frac{x^{1/2}}{1/2} + C$$

$$= \frac{2}{5}x^{5/2} + 2x^{1/2} + C$$

(b) $\int (x^2 - 1)^2 \, dx$

Solution Square the binomial first, and then find the antiderivative.

$$\int (x^2 - 1)^2 \, dx = \int (x^4 - 2x^2 + 1) \, dx$$
$$= \frac{x^5}{5} - \frac{2x^3}{3} + x + C$$

To check integration, take the derivative of the result. For instance, in Example 5(c) check that $z^3 - 2z^2 + 5z + C$ is the required indefinite integral by taking the derivative

$$\frac{d}{dz}(z^3 - 2z^2 + 5z + C) = 3z^2 - 4z + 5,$$

which agrees with the original information.

It was shown earlier that the derivative of $f(x) = e^x$ is $f'(x) = e^x$, and the derivative of $f(x) = a^x$ is $f'(x) = (\ln a)a^x$. Also, the derivative of $f(x) = e^{kx}$ is $f'(x) = k \cdot e^{kx}$, and the derivative of $f(x) = a^{kx}$ is $f'(x) = k(\ln a)a^{kx}$. These results lead to the following formulas for indefinite integrals of exponential functions.

INDEFINITE INTEGRALS OF EXPONENTIAL FUNCTIONS

$$\int e^x \, dx = e^x + C$$
$$\int e^{kx} \, dx = \frac{e^{kx}}{k} + C, \quad k \neq 0$$
$$\int a^x \, dx = \frac{a^x}{\ln a} + C$$
$$\int a^{kx} \, dx = \frac{a^{kx}}{k(\ln a)} + C, \quad k \neq 0$$

EXAMPLE 7 *Exponential Functions*

(a) $\int 9e^t \, dt = 9\int e^t \, dt = 9e^t + C$

(b) $\int e^{9t} \, dt = \frac{e^{9t}}{9} + C$

(c) $\int 3e^{(5/4)u} \, du = 3\left(\frac{e^{(5/4)u}}{5/4}\right) + C$

$$= 3\left(\frac{4}{5}\right)e^{(5/4)u} + C$$
$$= \frac{12}{5}e^{(5/4)u} + C$$

(d) $\int 2^{-5x} \, dx = \frac{2^{-5x}}{-5(\ln 2)} + C = -\frac{2^{-5x}}{5(\ln 2)} + C$

The restriction $n \neq -1$ was necessary in the formula for $\int x^n\,dx$ since $n = -1$ made the denominator of $1/(n + 1)$ equal to 0. To find $\int x^n\,dx$ when $n = -1$, that is, to find $\int x^{-1}\,dx$, recall the differentiation formula for the logarithmic function: The derivative of $f(x) = \ln|x|$, where $x \neq 0$, is $f'(x) = 1/x = x^{-1}$. This formula for the derivative of $f(x) = \ln|x|$ gives a formula for $\int x^{-1}\,dx$.

INDEFINITE INTEGRAL OF x^{-1}

$$\int x^{-1}\,dx = \int \frac{1}{x}\,dx = \ln|x| + C$$

CAUTION Don't neglect the absolute value sign in the natural logarithm when integrating x^{-1}. If x can take on a negative value, $\ln x$ will be undefined there. Note, however, that the absolute value is redundant (but harmless) in an expression such as $\ln|x^2 + 1|$, since $x^2 + 1$ can never be negative. ■

EXAMPLE 8 *Integrals*

(a) $\displaystyle\int \frac{4}{x}\,dx = 4\int \frac{1}{x}\,dx = 4\ln|x| + C$

(b) $\displaystyle\int\left(-\frac{5}{x} + e^{-2x}\right)dx = -5\ln|x| - \frac{1}{2}e^{-2x} + C$

In all these examples, the antiderivative family of functions was found. In many applications, however, the given information allows us to determine the value of the integration constant C. The next examples illustrate this idea.

EXAMPLE 9 *Cost*

Suppose a publishing company has found that the marginal cost at a level of production of x thousand books is given by

$$C'(x) = \frac{50}{\sqrt{x}}$$

and that the fixed cost (the cost before the first book can be produced) is \$25,000. Find the cost function $C(x)$.

Solution Write $50/\sqrt{x}$ as $50/x^{1/2}$ or $50x^{-1/2}$, and then use the indefinite integral rules to integrate the function.

$$\int \frac{50}{\sqrt{x}}\,dx = \int 50x^{-1/2}\,dx = 50(2x^{1/2}) + k = 100x^{1/2} + k$$

(Here k is used instead of C to avoid confusion with the cost function $C(x)$.) To find the value of k, use the fact that $C(0)$ is 25,000.

$$C(x) = 100x^{1/2} + k$$
$$25{,}000 = 100 \cdot 0 + k$$
$$k = 25{,}000$$

With this result, the cost function is $C(x) = 100x^{1/2} + 25{,}000$.

EXAMPLE 10 *Demand*

Suppose the marginal revenue from a product is given by $50 - 3x - x^2$. Find the demand function for the product.

Solution The marginal revenue is the derivative of the revenue function

$$\frac{dR}{dx} = 50 - 3x - x^2,$$

so

$$R = \int (50 - 3x - x^2)\, dx$$
$$= 50x - \frac{3x^2}{2} - \frac{x^3}{3} + C.$$

If $x = 0$, then $R = 0$ (no items sold means no revenue), and

$$0 = 50(0) - \frac{3(0)^2}{2} - \frac{(0)^3}{3} + C$$
$$0 = C.$$

Thus,

$$R = 50x - \frac{3x^2}{2} - \frac{x^3}{3}$$

gives the revenue function. Now, recall that $R = xp$, where p is the demand function. Then

$$50x - \frac{3}{2}x^2 - \frac{x^3}{3} = xp$$
$$50 - \frac{3}{2}x - \frac{x^2}{3} = p,$$

which gives the demand function.

In the next example integrals are used to find the position of a particle when the acceleration of the particle is given.

EXAMPLE 11 *Velocity and Acceleration*

Recall that if the function $s(t)$ gives the position of a particle at time t, then its velocity $v(t)$ and its acceleration $a(t)$ are given by

$$v(t) = s'(t) \quad \text{and} \quad a(t) = v'(t) = s''(t).$$

(a) Suppose the velocity of an object is $v(t) = 6t^2 - 8t$ and that the object is at 5 when time is 0. Find $s(t)$.

Solution Since $v(t) = s'(t)$, the function $s(t)$ is an antiderivative of $v(t)$:

$$s(t) = \int v(t)\,dt = \int (6t^2 - 8t)\,dt$$
$$= 2t^3 - 4t^2 + C$$

for some constant C. Find C from the given information that $s = 5$ when $t = 0$.

$$s(t) = 2t^3 - 4t^2 + C$$
$$5 = 2(0)^3 - 4(0)^2 + C$$
$$5 = C$$
$$s(t) = 2t^3 - 4t^2 + 5$$

(b) Many experiments have shown that when an object is dropped, its acceleration (ignoring air resistance) is constant. This constant has been found to be approximately 32 ft per second every second; that is,

$$a(t) = -32.$$

The negative sign is used because the object is falling. Suppose an object is thrown down from the top of the 1100-ft-tall Sears Tower in Chicago. If the initial velocity of the object is -20 ft per second, find $s(t)$, the distance of the object from the ground at time t.

Solution First find $v(t)$ by integrating $a(t)$:

$$v(t) = \int (-32)\,dt = -32t + k.$$

When $t = 0$, $v(t) = -20$:

$$-20 = -32(0) + k$$
$$-20 = k$$

and

$$v(t) = -32t - 20.$$

Be sure to evaluate the constant of integration k before integrating again to get $s(t)$. Now integrate $v(t)$ to find $s(t)$.

$$s(t) = \int (-32t - 20)\,dt = -16t^2 - 20t + C$$

Since $s(t) = 1100$ when $t = 0$, we can substitute these values into the equation for $s(t)$ to get $C = 1100$ and

$$s(t) = -16t^2 - 20t + 1100$$

as the distance of the object from the ground after t seconds.

(c) Use the equations derived in (b) to find the velocity of the object when it hit the ground and how long it took to strike the ground.

Solution When the object strikes the ground, $s = 0$, so

$$0 = -16t^2 - 20t + 1100.$$

To solve this equation for t, factor out the common factor of -4 and then use the quadratic formula.

$$0 = -4(4t^2 + 5t - 275)$$
$$t = \frac{-5 \pm \sqrt{25 + 4400}}{8} \approx \frac{-5 \pm 66.5}{8}$$

Only the positive value of t is meaningful here: $t \approx 7.69$. It takes the object about 7.69 seconds to strike the ground. From the velocity equation, with $t = 7.69$, we find.

$$v(t) = -32t - 20$$
$$v(7.69) = -32(7.69) - 20 \approx -266,$$

so the object was falling (as indicated by the negative sign) at about 266 ft per second when it hit the ground.

EXAMPLE 12 *Slope*

Find a function f whose graph has slope $f'(x) = 6x^2 + 4$ and goes through the point $(1, 1)$.

Solution Since $f'(x) = 6x^2 + 4$,

$$f(x) = \int (6x^2 + 4)\,dx = 2x^3 + 4x + C.$$

The graph of f goes through $(1, 1)$, so C can be found by substituting 1 for x and 1 for $f(x)$.

$$1 = 2(1)^3 + 4(1) + C$$
$$1 = 6 + C$$
$$C = -5$$

Finally, $f(x) = 2x^3 + 4x - 5$.

7.1 EXERCISES

1. What must be true of $F(x)$ and $G(x)$ if both are antiderivatives of $f(x)$?
2. How is the antiderivative of a function related to the function?
3. In your own words, describe what is meant by an integrand.
4. Explain why the restriction $n \neq -1$ is necessary in the rule $\int x^n\,dx = \frac{x^{n+1}}{n+1} + C$.

Find the following.

5. $\int 6\,dk$

6. $\int 2\,dy$

7. $\int (2z + 3)\,dz$

8. $\int (3x - 5)\,dx$

9. $\int (t^2 - 4t + 5)\,dt$

10. $\int (5x^2 - 6x + 3)\,dx$

11. $\int (4z^3 + 3z^2 + 2z - 6)\,dz$

12. $\int (12y^3 + 6y^2 - 8y + 5)\,dy$

13. $\int 5\sqrt{z}\,dz$

14. $\int t^{1/4}\,dt$

15. $\int x(x^2 - 3)\,dx$

16. $\int x^2(x^4 + 4x + 3)\,dx$

17. $\int (4\sqrt{v} - 3v^{3/2})\,dv$

18. $\int (15x\sqrt{x} + 2\sqrt{x})\,dx$

19. $\int (10u^{3/2} - 14u^{5/2})\,du$

20. $\int (56t^{5/2} + 18t^{7/2})\,dt$

21. $\int \left(\frac{1}{z^2}\right)dz$

22. $\int \left(\frac{4}{x^3}\right)dx$

23. $\int \left(\frac{1}{y^3} - \frac{1}{\sqrt{y}}\right)dy$

24. $\int \left(\sqrt{u} + \frac{1}{u^2}\right)du$

25. $\int (-9t^{-2} - 2t^{-1})\,dt$

26. $\int (8x^{-3} + 4x^{-1})\,dx$

27. $\int \frac{1}{3x^2}\,dx$

28. $\int \frac{2}{3x^4}\,dx$

29. $\int 3e^{-.2x}\,dx$

30. $\int -4e^{.2v}\,dv$

31. $\int \left(\frac{3}{x} + 4e^{-.5x}\right)dx$

32. $\int \left(\frac{9}{x} - 3e^{-.4x}\right)dx$

33. $\int \frac{1 + 2t^3}{t}\,dt$

34. $\int \frac{2y^{1/2} - 3y^2}{y}\,dy$

35. $\int (e^{2u} + 4u)\,du$

36. $\int (v^2 - e^{3v})\,dv$

37. $\int (x + 1)^2\,dx$

38. $\int (2y - 1)^2\,dy$

39. $\int \frac{\sqrt{x} + 1}{\sqrt[3]{x}}\,dx$

40. $\int \frac{1 - 2\sqrt[3]{z}}{\sqrt[3]{z}}\,dz$

41. $\int 10^x\,dx$

42. $\int 3^{2x}\,dx$

43. Find an equation of the curve whose tangent line has a slope of
$$f'(x) = x^{2/3},$$
given that the point $(1, 3/5)$ is on the curve.

44. The slope of the tangent line to a curve is given by
$$f'(x) = 6x^2 - 4x + 3.$$
If the point $(0, 1)$ is on the curve, find an equation of the curve.

Applications

BUSINESS AND ECONOMICS

Cost *Find the cost function for each marginal cost function.*

45. $C'(x) = 4x - 5$; fixed cost is \$8

46. $C'(x) = .2x^2 + 5x$; fixed cost is \$10

47. $C'(x) = .03e^{.01x}$; fixed cost is \$8

48. $C'(x) = x^{1/2}$; 16 units cost \$45

49. $C'(x) = x^{2/3} + 2$; 8 units cost \$58

50. $C'(x) = x + 1/x^2$; 2 units cost \$5.50

51. $C'(x) = 5x - 1/x$; 10 units cost \$94.20

52. $C'(x) = 1.2^x(\ln 1.2)$; 2 units cost \$9.44

Demand *Find the demand function for each marginal revenue function. Recall that if no items are sold, the revenue is 0.*

53. $R'(x) = 175 - .02x - .03x^2$

54. $R'(x) = 50 - x$

55. $R'(x) = 500 - .15\sqrt{x}$

56. $R'(x) = 600 - 5e^{.0002x}$

57. *Cellular Telephones* The approximate rate of change in the number (in millions) of subscribers to cellular telephone plans is given by

$$f'(x) = 1.659x + .743,$$

where x represents the number of years since 1990.* In 1990 $(x = 0)$ there were approximately 4.4 million subscribers.

a. Find the function that gives the total number of cellular telephone subscribers in year x.

b. According to this function, how many subscribers were there in 2003? Compare this with the actual number of approximately 150 million.

58. *Profit* The marginal profit of a small fast-food stand is given by

$$P'(x) = 2x + 20,$$

where x is the sales volume in thousands of hamburgers. The "profit" is $-\$50$ when no hamburgers are sold. Find the profit function.

59. *Profit* Suppose the marginal profit from the sale of x hundred items is

$$P'(x) = 4 - 6x + 3x^2,$$

and the profit on 0 items is $-\$40$.

a. Find the profit function.

b. Find the profit from selling 800 items.

LIFE SCIENCES

60. *Biochemical Excretion* If the rate of excretion of a biochemical compound is given by

$$f'(t) = .01e^{-.01t},$$

the total amount excreted by time t (in minutes) is $f(t)$.

a. Find an expression for $f(t)$.

b. If 0 units are excreted at time $t = 0$, how many units are excreted in 10 minutes?

61. *Flour Beetles* A model for describing the population of adult flour beetles involves evaluating the integral

$$\int \frac{g(x)}{x}\,dx,$$

where $g(x)$ is the per-unit-abundance growth rate for a population of size x.† The researchers consider the simple case in which $g(x) = a - bx$ for positive constants a and b. Find the integral in this case.

62. *Concentration of a Solute* According to Fick's law, the diffusion of a solute across a cell membrane is given by

$$c'(t) = \frac{kA}{V}[C - c(t)], \quad (1)$$

where A is the area of the cell membrane, V is the volume of the cell, $c(t)$ is the concentration inside the cell at time t, C is the concentration outside the cell, and k is a constant. If c_0 represents the concentration of the solute inside the cell when $t = 0$, then it can be shown that

$$c(t) = (c_0 - C)e^{-kAt/V} + M. \quad (2)$$

a. Use the last result to find $c'(t)$.

b. Substitute back into Equation (1) to show that (2) is indeed the correct antiderivative of (1).

SOCIAL SCIENCES

63. *Seat Belt Usage* In California, the use of seat belts has risen steadily since the enactment of a seat belt law in 1986. The rate of change (in percent) of drivers using seat belts in year x, where 1985 corresponds to $x = 0$, is modeled by

$$f'(x) = .072x^2 - 1.672x + 11.503.^{\ddagger}$$

In 1985 (year 0), 26% of drivers used seat belts.

a. Find the function that gives the percent of drivers using seat belts in year x.

b. According to this function, what percent of drivers used seat belts in 2000?

c. Does the function found in part a give reasonable results past 2002? Explain.

64. *Vehicle-Related Deaths* Vehicle-related deaths in California have declined since 1986 when the California seat belt law was enacted.§ The rate of change in deaths per 100 million miles of vehicle travel is modeled by

$$g'(x) = -.00097x^3 + .022x^2 - .137x + .0989,$$

where $x = 0$ corresponds to 1985 and so on. There were 2.4 deaths per 100 million miles driven in 1985.

a. Find the function giving the number of deaths per 100 million miles driven in year x.

b. How many deaths were there in 2000?

c. Does the function found in part a give reasonable results past 2002? Explain.

*Cellular Telecommunications & Internet Association.
†Dennis, Brian and Robert F. Costantino, "Analysis of Steady-State Populations with the Gamma Abundance Model: Application to *Tribolium*," *Ecology,* Vol. 69, No. 4, Aug. 1988, pp. 1200–1213.
‡California Highway Patrol, May 2001.
§Ibid.

PHYSICAL SCIENCES

Exercises 65–69 refer to Example 11 in this section.

65. *Velocity* For a particular object, $a(t) = t^2 + 1$ and $v(0) = 6$. Find $v(t)$.

66. *Distance* Suppose $v(t) = 6t^2 - 2/t^2$ and $s(1) = 8$. Find $s(t)$.

67. *Time* An object is dropped from a small plane flying at 6400 ft. Assume that $a(t) = -32$ ft per second and $v(0) = 0$, and find $s(t)$. How long will it take the object to hit the ground?

68. *Distance* Suppose $a(t) = 18t + 8$, $v(1) = 15$, and $s(1) = 19$. Find $s(t)$.

69. *Distance* Suppose $a(t) = (15/2)\sqrt{t} + 3e^{-t}$, $v(0) = -3$, and $s(0) = 4$. Find $s(t)$.

70. *Rocket Science* In the 1999 movie *October Sky*, Homer Hickum was accused of launching a rocket that started a forest fire. Homer proved his innocence by showing that his rocket could not have flown far enough to reach where the fire started. He used the following reasoning.

a. Using the fact that $a(t) = -32$ (see Example 11(b)), find $v(t)$ and $s(t)$, given $v(0) = v_0$ and $s(0) = 0$. (The initial velocity was unknown, and the initial height was 0 ft.)

b. Homer estimated that the rocket was in the air for 14 seconds. Use $s(14) = 0$ to find v_0.

c. If the rocket left the ground at a 45° angle, the velocity in the horizontal direction would be equal to v_0, the velocity in the vertical direction, so the distance traveled horizontally would be v_0t. (The rocket left the ground at a steeper angle, so this would overestimate the distance from starting to landing point.) Find the distance the rocket would travel horizontally during its 14-second flight.

7.2 SUBSTITUTION

THINK ABOUT IT *If a formula for the marginal revenue is known, how can a formula for the total revenue be found?*

Using the method of substitution, this question will be answered in an exercise in this section.

We saw how to integrate a few simple functions in the previous section. More complicated functions can sometimes be integrated by *substitution*. The substitution technique depends on the idea of a differential, discussed in an earlier chapter. If $u = f(x)$, the *differential* of u, written du, is defined as

$$du = f'(x)\,dx.$$

For example, if $u = 6x^4$, then $du = 24x^3dx$.

Recall the chain rule for derivatives as used in the following example:

$$\frac{d}{dx}(x^2 - 1)^5 = 5(x^2 - 1)^4(2x) = 10x(x^2 - 1)^4.$$

> **FOR REVIEW**
>
> The chain rule, discussed in detail in the chapter on Calculating the Derivative, states that
>
> $$\frac{d}{dx}[f(g(x))] = f'(g(x)) \cdot g'(x).$$

As in this example, the result of using the chain rule is often a product of two functions. Because of this, functions formed by the product of two functions can sometimes be integrated by using the chain rule in reverse. In the example above, working backwards from the derivative gives

$$\int 10x(x^2 - 1)^4\,dx = (x^2 - 1)^5 + C.$$

To find an antiderivative involving products, it often helps to make a substitution: let $u = x^2 - 1$, so that $du = 2x\,dx$. Now substitute u for $x^2 - 1$ and du for $2x\,dx$ in the indefinite integral above.

$$\begin{aligned}\int 10x(x^2-1)^4\,dx &= \int 5 \cdot 2x(x^2-1)^4\,dx\\ &= 5\int (x^2-1)^4(2x\,dx)\\ &= 5\int u^4\,du\end{aligned}$$

With substitution we have changed a complicated integral into a simple one. This last integral can now be found by the power rule.

$$5\int u^4\,du = 5 \cdot \frac{u^5}{5} + C = u^5 + C$$

Finally, substitute $x^2 - 1$ for u.

$$\int 10x(x^2-1)^4\,dx = (x^2-1)^5 + C$$

This method of integration is called **integration by substitution.** As shown above, it is simply the chain rule for derivatives in reverse. The results can always be verified by differentiation.

This discussion leads to the following integration formula, which is sometimes called the general power rule for integrals.

GENERAL POWER RULE FOR INTEGRALS

For $u = f(x)$ and $du = f'(x)\,dx$,

$$\int u^n\,du = \frac{u^{n+1}}{n+1} + C.$$

EXAMPLE 1 *General Power Rule*

Find $\int 6x(3x^2 + 4)^4\,dx$.

Solution A certain amount of trial and error may be needed to decide on the expression to set equal to u. The integrand must be written as two factors, one of which is the derivative of the quantity chosen for u. In this example, if $u = 3x^2 + 4$, then $du = 6x\,dx$ and the integrand can be written as the product of $(3x^2 + 4)^4$ and $6x\,dx$. Now substitute.

$$\int 6x(3x^2+4)^4\,dx = \int (3x^2+4)^4(6x\,dx) = \int u^4\,du$$

Find this last indefinite integral.

$$\int u^4\,du = \frac{u^5}{5} + C$$

Now replace u with $3x^2 + 4$.

$$\int 6x(3x^2 + 4)^4\, dx = \frac{u^5}{5} + C = \frac{(3x^2 + 4)^5}{5} + C$$

To verify this result, find the derivative.

$$\frac{d}{dx}\left[\frac{(3x^2 + 4)^5}{5} + C\right] = \frac{5}{5}(3x^2 + 4)^4(6x) + 0 = (3x^2 + 4)^4\,(6x)$$

The derivative is the original function, as required.

EXAMPLE 2 *General Power Rule*

Find $\int x^2\sqrt{x^3 + 1}\, dx$.

Solution An expression raised to a power is usually a good choice for u, so because of the square root or $1/2$ power, let $u = x^3 + 1$; then $du = 3x^2\, dx$. The integrand does not contain the constant 3, which is needed for du. To take care of this, multiply by $3/3$, placing 3 inside the integral sign and $1/3$ outside.

$$\int x^2\sqrt{x^3 + 1}\, dx = \frac{1}{3}\int 3x^2\sqrt{x^3 + 1}\, dx = \frac{1}{3}\int \sqrt{x^3 + 1}\,(3x^2\, dx)$$

Now substitute u for $x^3 + 1$ and du for $3x^2\, dx$, and then integrate.

$$\frac{1}{3}\int \sqrt{x^3 + 1}\, 3x^2 dx = \frac{1}{3}\int \sqrt{u}\, du = \frac{1}{3}\int u^{1/2}\, du$$

$$= \frac{1}{3}\cdot\frac{u^{3/2}}{3/2} + C = \frac{2}{9}u^{3/2} + C$$

Since $u = x^3 + 1$,

$$\int x^2\sqrt{x^3 + 1}\, dx = \frac{2}{9}(x^3 + 1)^{3/2} + C.$$

The substitution method given in the examples above *will not always work.* For example, you might try to find

$$\int x^3\sqrt{x^3 + 1}\, dx$$

by substituting $u = x^3 + 1$, so that $du = 3x^2\, dx$. However, there is no *constant* that can be inserted inside the integral sign to give $3x^2$. This integral, and a great many others, cannot be evaluated by substitution.

With practice, choosing u will become easy if you keep two principles in mind.

1. u should equal some expression in the integral that, when replaced with u, tends to make the integral simpler.
2. u must be an expression whose derivative—disregarding any constant multiplier, such as the 3 in $3x^2$—is also present in the integral.

The substitution should include as much of the integral as possible, as long as its derivative is still present. In Example 1, we could have chosen $u = 3x^2$, but

$u = 3x^2 + 4$ is better, because it has the same derivative as $3x^2$ and captures more of the original integral. If we carry this reasoning further, we might try $u = (3x^2 + 4)^4$, but this is a poor choice, for $du = 4(3x^2 + 4)^3(6x)\,dx$, an expression not present in the original integral.

EXAMPLE 3 *Substitution*

Find $\displaystyle\int \frac{x + 3}{(x^2 + 6x)^2}\,dx$.

Solution Let $u = x^2 + 6x$, so that $du = (2x + 6)\,dx = 2(x + 3)\,dx$. The integral is missing the 2, so multiply by $2 \cdot (1/2)$, putting 2 inside the integral sign and $1/2$ outside.

$$\int \frac{x + 3}{(x^2 + 6x)^2}\,dx = \frac{1}{2}\int \frac{2(x + 3)}{(x^2 + 6x)^2}\,dx$$
$$= \frac{1}{2}\int \frac{du}{u^2} = \frac{1}{2}\int u^{-2}\,du = \frac{1}{2} \cdot \frac{u^{-1}}{-1} + C = \frac{-1}{2u} + C$$

Substituting $x^2 + 6x$ for u gives

$$\int \frac{x + 3}{(x^2 + 6x)^2}\,dx = \frac{-1}{2(x^2 + 6x)} + C.$$

Recall the formula for $\frac{d}{dx}(e^u)$, where $u = f(x)$:

$$\frac{d}{dx}(e^u) = e^u \frac{du}{dx}.$$

For example, if $u = x^2$, then $\frac{du}{dx} = \frac{d}{dx}(x^2) = 2x$, and

$$\frac{d}{dx}(e^{x^2}) = e^{x^2} \cdot 2x.$$

Working backwards, if $u = x^2$, then $du = 2x\,dx$, so

$$\int e^{x^2} \cdot 2x\,dx = \int e^u\,du = e^u + C = e^{x^2} + C.$$

The work above suggests the following rule for the indefinite integral of e^u, where $u = f(x)$.

INDEFINITE INTEGRAL OF e^u

If $u = f(x)$, then $du = f'(x)\,dx$ and

$$\int e^u\,du = e^u + C.$$

EXAMPLE 4 *Substitution*

Find $\int x^2 e^{x^3}\,dx$.

Solution Let $u = x^3$, the exponent on e. Then $du = 3x^2\,dx$. Multiplying by $3/3$ gives

$$\int x^2 e^{x^3}\,dx = \frac{1}{3}\int e^{x^3}(3x^2\,dx)$$
$$= \frac{1}{3}\int e^u\,du = \frac{1}{3}e^u + C = \frac{1}{3}e^{x^3} + C.$$

Recall that the antiderivative of $f(x) = 1/x$ is $\ln|x|$. The next example uses the fact that $\int x^{-1}\,dx = \ln|x| + C$, together with the method of substitution.

EXAMPLE 5 *Substitution*

Find $\displaystyle\int \frac{(2x-3)\,dx}{x^2-3x}$.

Solution Let $u = x^2 - 3x$, so that $du = (2x-3)\,dx$. Then

$$\int \frac{(2x-3)\,dx}{x^2-3x} = \int \frac{du}{u} = \ln|u| + C = \ln|x^2-3x| + C.$$

Generalizing from the results of Example 5 suggests the following rule for finding the indefinite integral of u^{-1}, where $u = f(x)$.

INDEFINITE INTEGRAL OF u^{-1}

If $u = f(x)$, then $du = f'(x)\,dx$ and

$$\int u^{-1}\,du = \int \frac{du}{u} = \ln|u| + C.$$

The next example shows a more complicated integral evaluated by the method of substitution.

EXAMPLE 6 *Substitution*

Find $\int x\sqrt{1-x}\,dx$.

Solution Let $u = 1 - x$. To get the x outside the radical in terms of u, solve $u = 1 - x$ for x to get $x = 1 - u$. Then $dx = -du$ and we can substitute as follows.

$$\int x\sqrt{1-x}\,dx = \int (1-u)\sqrt{u}(-du) = \int (u-1)u^{1/2}\,du$$
$$= \int (u^{3/2} - u^{1/2})\,du = \frac{2}{5}u^{5/2} - \frac{2}{3}u^{3/2} + C$$
$$= \frac{2}{5}(1-x)^{5/2} - \frac{2}{3}(1-x)^{3/2} + C$$

The substitution method is useful if the integral can be written in one of the following forms, where u is some function of x.

SUBSTITUTION METHOD

In general, for the types of problems we are concerned with, there are three cases. We choose u to be one of the following:

1. the quantity under a root or raised to a power;
2. the exponent on e;
3. the quantity in the denominator.

Remember that some integrands may need to be rearranged to fit one of these cases.

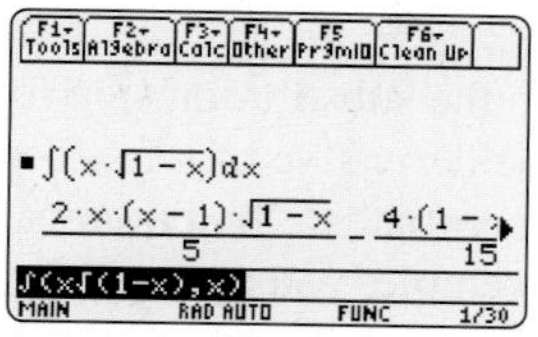

FIGURE 2

Some calculators, such as the TI-89, can find indefinite integrals automatically. Many computer algebra systems also do this. Figure 2 shows the integral in Example 6 performed on a TI-89. The answer looks different but is algebraically equivalent to the answer found in Example 6.

EXAMPLE 7 *Demand*

The research department for a hardware chain has determined that at one store the marginal price of x boxes per week of a particular type of nails is

$$p'(x) = \frac{-4000}{(2x + 15)^3}.$$

Find the demand equation if the weekly demand for this type of nails is 10 boxes when the price of a box of nails is \$4.

Solution To find the demand function, first integrate $p'(x)$ as follows.

$$p(x) = \int p'(x)\,dx = \int \frac{-4000}{(2x + 15)^3}\,dx$$

Let $u = 2x + 15$. Then $du = 2\,dx$, and

$$\begin{aligned} p(x) &= -2000\int (2x + 15)^{-3}\,2\,dx \\ &= -2000\int u^{-3}\,du \\ &= (-2000)\frac{u^{-2}}{-2} + C \\ &= \frac{1000}{u^2} + C \end{aligned}$$

$$p(x) = \frac{1000}{(2x + 15)^2} + C. \qquad (1)$$

Find the value of C by using the given information that $p = 4$ when $x = 10$.

$$4 = \frac{1000}{(2 \cdot 10 + 15)^2} + C$$

$$4 = \frac{1000}{35^2} + C$$

$$4 = .82 + C$$

$$3.18 = C$$

Replacing C with 3.18 in Equation (1) gives the demand function,

$$p(x) = \frac{1000}{(2x + 15)^2} + 3.18.$$

With a little practice, you will find you can skip the substitution step for integrals such as that shown in Example 7, in which the derivative of u is a constant. Recall from the chain rule that when you differentiate a function, such as $p(x) = 1000/(2x + 15)^2 + 3.18$ in the previous example, you multiply by 2, the derivative of $(2x + 15)$. So when taking the antiderivative, simply divide by 2:

$$\int -4000(2x + 15)^{-3}\,dx = \frac{-4000}{2} \cdot \frac{(2x + 15)^{-2}}{-2} + C$$
$$= \frac{1000}{(2x + 15)^2} + C.$$

CAUTION This procedure is valid because of the constant multiple rule presented in the previous section, which says that constant multiples can be brought into or out of integrals, just as they can with derivatives. This procedure is *not* valid with any expression other than a constant. ■

EXAMPLE 8 *Popularity Index*

To determine the top 100 popular songs of each year since 1956, Jim Quirin and Barry Cohen developed a function that represents the rate of change on the charts of *Billboard* magazine required for a song to earn a "star" on the *Billboard* "Hot 100" survey.* They developed the function

$$f(x) = \frac{A}{B + x},$$

where $f(x)$ represents the rate of change in position on the charts, x is the position on the "Hot 100" survey, and A and B are constants. The function

$$F(x) = \int f(x)\,dx$$

is defined as the "Popularity Index." Find $F(x)$.

*Formulas for determining "Popularity Index" from Quirin, Jim, and Barry Cohen, *Chartmasters' Rock 100,* 5th ed. Copyright 1992 by Chartmasters. Reprinted by permission.

Solution Integrating $f(x)$ gives

$$F(x) = \int f(x)\,dx = \int \frac{A}{B + x}\,dx = A\int \frac{1}{B + x}\,dx.$$

Let $u = B + x$, so that $du = dx$. Then

$$F(x) = A\int \frac{1}{u}\,du = A \ln u + C = A \ln(B + x) + C.$$

(The absolute value bars are not necessary, since $B + x$ is always positive here.)

7.2 EXERCISES

1. Integration by substitution is related to what differentiation method? What type of integrand suggests using integration by substitution?

2. The following integrals may be solved using substitution. Choose a function u that may be used to solve each problem. Then find du.

a. $\int (3x^2 - 5)^4\, 2x\,dx$ **b.** $\int \sqrt{1 - x}\,dx$ **c.** $\int \frac{x^2}{2x^3 + 1}\,dx$ **d.** $\int 4x^3 e^{x^4}\,dx$

Use substitution to find each indefinite integral.

3. $\int 4(2x + 3)^4\,dx$

4. $\int (-4t + 1)^3\,dt$

5. $\int \frac{2\,dm}{(2m + 1)^3}$

6. $\int \frac{3\,du}{\sqrt{3u - 5}}$

7. $\int \frac{2x + 2}{(x^2 + 2x - 4)^4}\,dx$

8. $\int \frac{6x^2\,dx}{(2x^3 + 7)^{3/2}}$

9. $\int z\sqrt{z^2 - 5}\,dz$

10. $\int r\sqrt{r^2 + 2}\,dr$

11. $\int -4e^{2p}\,dp$

12. $\int 5e^{-.3g}\,dg$

13. $\int 3x^2 e^{2x^3}\,dx$

14. $\int re^{-r^2}\,dr$

15. $\int (1 - t)e^{2t - t^2}\,dt$

16. $\int (x^2 - 1)e^{x^3 - 3x}\,dx$

17. $\int \frac{e^{1/z}}{z^2}\,dz$

18. $\int \frac{e^{\sqrt{y}}}{2\sqrt{y}}\,dy$

19. $\int (x^3 + 2x)(x^4 + 4x^2 + 7)^8\,dx$

20. $\int \frac{t^2 + 2}{t^3 + 6t + 3}\,dt$

21. $\int \frac{2x + 1}{(x^2 + x)^3}\,dx$

22. $\int \frac{B^3 - 1}{(2B^4 - 8B)^{3/2}}\,dB$

23. $\int p(p + 1)^5\,dp$

24. $\int 4r\sqrt{8 - r}\,dr$

25. $\int \frac{u}{\sqrt{u - 1}}\,du$

26. $\int \frac{2x}{(x + 5)^6}\,dx$

27. $\int (\sqrt{x^2 + 12x})(x + 6)\,dx$

28. $\int (\sqrt{x^2 - 6x})(x - 3)\,dx$

29. $\int \frac{t}{t^2 + 2}\,dt$

30. $\int \frac{-4x}{x^2 + 3}\,dx$

31. $\int \frac{(1 + \ln x)^2}{x}\,dx$

32. $\int \frac{\sqrt{2 + \ln x}}{x}\,dx$

33. $\int \frac{e^{2x}}{e^{2x} + 5}\,dx$

34. $\int \frac{1}{x(\ln x)}\,dx$

35. Stan and Ollie work on the integral

$$\int 3x^2 e^{x^3}\,dx.$$

Stan lets $u = x^3$ and proceeds to get

$$\int e^u\,du = e^u + C = e^{x^3} + C.$$

Ollie tries $u = e^{x^3}$ and proceeds to get

$$\int du = u + C = e^{x^3} + C.$$

Discuss which procedure you prefer, and why.

36. Stan and Ollie work on the integral

$$\int 2x(x^2 + 2)\,dx.$$

Stan lets $u = x^2 + 2$ and proceeds to get

$$\int u\,du = \frac{u^2}{2} + C = \frac{(x^2 + 2)^2}{2} + C.$$

Ollie multiplies out the function under the integral and gets

$$\int (2x^3 + 4x)\,dx = \frac{x^4}{2} + 2x^2 + C.$$

How can they both be right?

Applications

BUSINESS AND ECONOMICS

37. *Revenue* Suppose the marginal revenue (in dollars) from the sale of x jet planes is

$$R'(x) = 2x(x^2 + 50)^2.$$

a. Find the total revenue function if the revenue from 3 planes is \$206,379.

b. How many planes must be sold for a revenue of at least \$450,000?

38. *Maintenance* The rate of expenditure for maintenance of a particular machine is given by

$$M'(x) = \sqrt{x^2 + 12x}(2x + 12),$$

where x is time measured in years. Total maintenance costs through the fourth year are \$612.

a. Find the total maintenance function.

b. How many years must pass before the total maintenance costs reach \$2000?

39. *Cost* A company has found that the marginal cost of a new production line (in thousands) is

$$C'(x) = \frac{60x}{5x^2 + e},$$

where x is the number of years the line is in use.

a. Find the total cost function for the production line. The fixed cost is \$10,000.

b. The company will add the new line if the total cost is reduced to \$20,000 within 5 years. Should they add the new line?

40. *Profit* The rate of growth of the profit (in millions of dollars) from a new technology is approximated by

$$P'(x) = xe^{-x^2},$$

where x represents time measured in years. The total profit in the third year that the new technology is in operation is \$10,000.

a. Find the total profit function.

b. What happens to the total amount of profit in the long run?

LIFE SCIENCES

41. *Cell Growth* Under certain conditions, the number of cancer cells $N(t)$ at time t increases at a rate

$$N'(t) = Ae^{kt},$$

where A is the rate of increase at time 0 (in cells per day) and k is a constant.

a. Suppose $A = 50$, and at 5 days, the cells are growing at a rate of 250 per day. Find a formula for the number of cells after t days, given that 300 cells are present at $t = 0$.

b. Use your answer from part a to find the number of cells present after 12 days.

42. *Blood Pressure* The rate of change of the volume $V(t)$ of blood in the aorta at time t is given by

$$V'(t) = -kP(t),$$

where $P(t)$ is the pressure in the aorta at time t and k is a constant that depends upon properties of the aorta. The pressure in the aorta is given by

$$P(t) = P_0 e^{-mt},$$

where P_0 is the pressure at time $t = 0$ and m is another constant. Letting V_0 be the volume at time $t = 0$, find a formula for $V(t)$.

GENERAL INTEREST

43. *Border Crossing* According to U.S. Customs, the number of Mexicans who cross by vehicle into the United States has been increasing since 1997. The rate of change in the number of Mexicans crossing the border by vehicle (in millions) is given by

$$S'(t) = 8.93e^{.07t},$$

where t is the number of years since 1990.

a. Find $S(t)$ if there were 210.3 million such border crossings in 1997 ($t = 7$).

b. In how many years past 1997 will the number of Mexicans crossing the border by vehicle double?

7.3 AREA AND THE DEFINITE INTEGRAL

THINK ABOUT IT

If the rate of change of annual maintenance charges is known, how can the total amount paid for maintenance over a 10-year period be estimated?

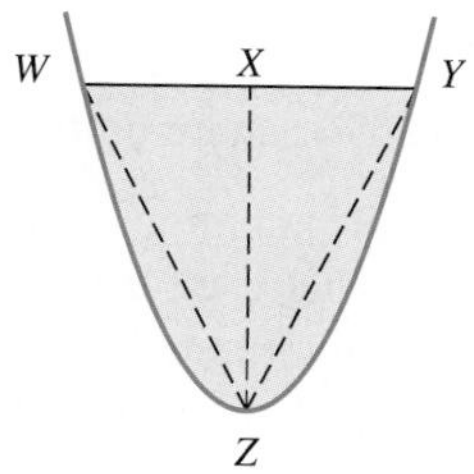

Area of parabolic segment
$= \frac{4}{3}$ (area of triangle WYZ)

FIGURE 3

This section introduces a method for answering such questions.

To calculate the areas of geometric figures such as rectangles, squares, triangles, and circles, we use specific formulas. In this section we consider the problem of finding the area of a figure or region that is bounded by curves, such as the shaded region in Figure 3.

The brilliant Greek mathematician Archimedes (about 287 B.C.–212 B.C.) is considered one of the greatest mathematicians of all time. His development of a rigorous method known as *exhaustion* to derive results was a forerunner of the ideas of integral calculus. Archimedes used a method that would later be verified by the theory of integration. His method involved viewing a geometric figure as a sum of other figures. For example, he thought of a plane surface area as a figure consisting of infinitely many parallel line segments. Among the results established by Archimedes' method was the fact that the area of a segment of a parabola (shown in color in Figure 3) is equal to $4/3$ the area of a triangle with the same base and the same height.

Under certain conditions the area of a region can be thought of as a sum of parts. Figure 4 shows the region bounded by the y-axis, the x-axis, and the graph of

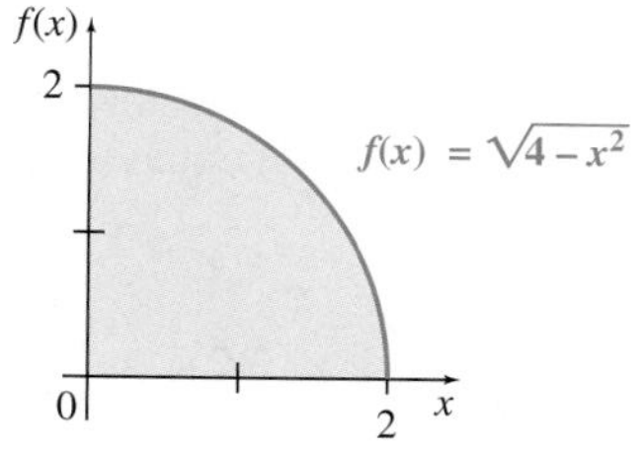

FIGURE 4

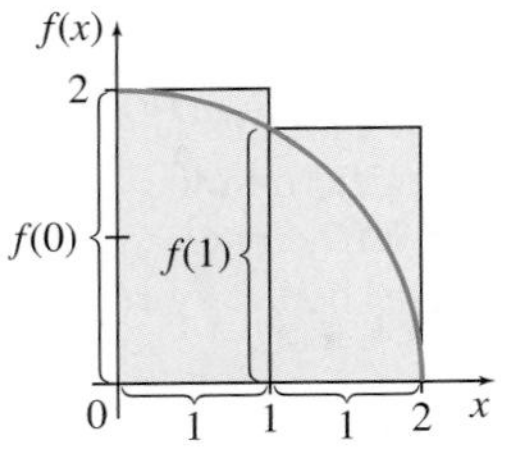

FIGURE 5

$f(x) = \sqrt{4 - x^2}$. A very rough approximation of the area of this region can be found by using two inscribed rectangles as in Figure 5. The height of the rectangle on the left is $f(0) = 2$ and the height of the rectangle on the right is $f(1) = \sqrt{3}$. The width of each rectangle is 1, making the total area of the two rectangles

$$1 \cdot f(0) + 1 \cdot f(1) = 2 + \sqrt{3} \approx 3.7321 \text{ square units.}$$

In this example, the function is decreasing, and we will overestimate the area when we evaluate the function at the left endpoint to determine the height of the rectangle in that interval. If we use the right endpoint, the answer will be too small. For example, using the right endpoint, the area of the two rectangles is

$$1 \cdot f(1) + 1 \cdot f(2) = \sqrt{3} + 0 \approx 1.7321 \text{ square units.}$$

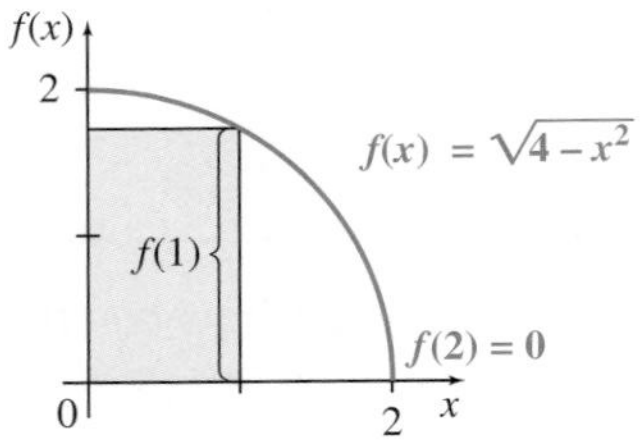

FIGURE 6

See Figure 6.

If the left endpoint gives an answer too big, and the right endpoint an answer too small, it seems reasonable to average the two answers. This produces the method called the *trapezoidal rule*, discussed in more detail later in this chapter. In this example, we get

$$\frac{3.7321 + 1.7321}{2} = 2.7321.$$

Another way to get an improved answer would be to use the midpoint of each interval, rather than the left endpoint or the right endpoint. This is called the **midpoint rule.** See Figure 7. In this example, it gives

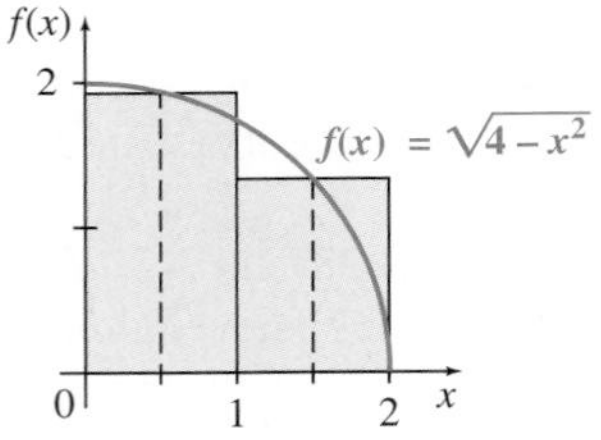

FIGURE 7

$$1 \cdot f(.5) + 1 \cdot f(1.5) = \sqrt{3.75} + \sqrt{1.75} \approx 3.2594 \text{ square units.}$$

To improve the accuracy of all of the previous approximations, we could divide the interval from $x = 0$ to $x = 2$ into more parts. The result using the left endpoint again with four parts, each of width $1/2$, is shown in Figure 8 on the next page. This approximation is greater than the actual area. As before, the height of each rectangle is given by the value of f at the left side of the rectangle, and its area is the width, $1/2$, multiplied by the height. The total area of the four rectangles is

$$\frac{1}{2} \cdot f(0) + \frac{1}{2} \cdot f\left(\frac{1}{2}\right) + \frac{1}{2} \cdot f(1) + \frac{1}{2} \cdot f\left(1\frac{1}{2}\right)$$
$$= \frac{1}{2}(2) + \frac{1}{2}\left(\frac{\sqrt{15}}{2}\right) + \frac{1}{2}(\sqrt{3}) + \frac{1}{2}\left(\frac{\sqrt{7}}{2}\right)$$
$$= 1 + \frac{\sqrt{15}}{4} + \frac{\sqrt{3}}{2} + \frac{\sqrt{7}}{4} \approx 3.4957 \text{ square units.}$$

This approximation looks better, but it is still greater than the actual area. To improve the approximation, divide the interval from $x = 0$ to $x = 2$ into 8 parts with equal widths of $1/4$ (see Figure 9). The total area of all these rectangles is

$$\frac{1}{4} \cdot f(0) + \frac{1}{4} \cdot f\left(\frac{1}{4}\right) + \frac{1}{4} \cdot f\left(\frac{1}{2}\right) + \frac{1}{4} \cdot f\left(\frac{3}{4}\right) + \frac{1}{4} \cdot f(1)$$
$$+ \frac{1}{4} \cdot f\left(\frac{5}{4}\right) + \frac{1}{4} \cdot f\left(\frac{3}{2}\right) + \frac{1}{4} \cdot f\left(\frac{7}{4}\right)$$
$$\approx 3.3398 \text{ square units.}$$

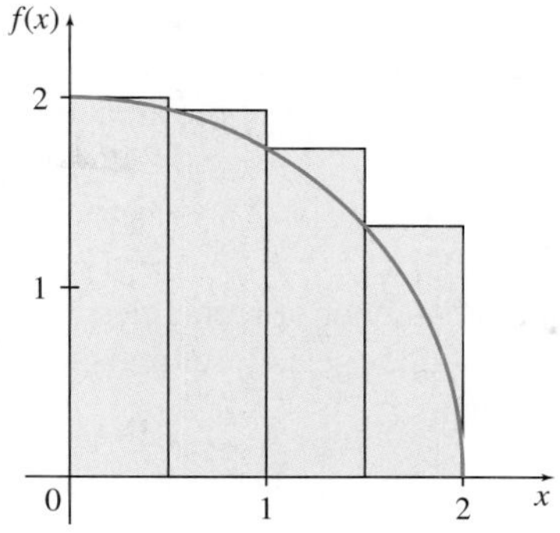

FIGURE 8

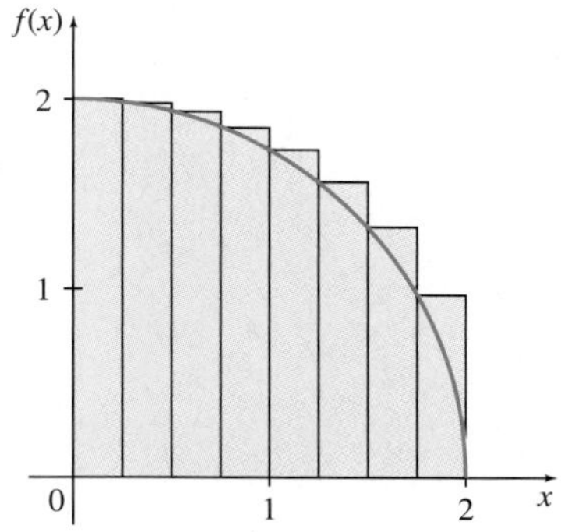

FIGURE 9

This process of approximating the area under a curve by using more and more rectangles to get a better and better approximation can be generalized. To do this, divide the interval from $x = 0$ to $x = 2$ into n equal parts. Each of these n intervals has width

$$\frac{2-0}{n} = \frac{2}{n},$$

so each rectangle has width $2/n$ and height determined by the function value at the left side of the rectangle, or the right side, or the midpoint. We could also average the left and right side values as before. Using a computer or graphing calculator to find approximations to the area for several values of n gives the results in the following table.

n	Left Sum	Right Sum	Trapezoidal	Midpoint
2	3.7321	1.7321	2.7321	3.2594
4	3.4957	2.4957	2.9957	3.1839
8	3.3398	2.8398	3.0898	3.1567
10	3.3045	2.9045	3.1045	3.1524
20	3.2285	3.0285	3.1285	3.1454
50	3.1783	3.0983	3.1383	3.1426
100	3.1604	3.1204	3.1404	3.1419
500	3.1455	3.1375	3.1415	3.1416

The numbers in the last four columns of this table represent approximations to the area under the curve, above the x-axis, and between the lines $x = 0$ and $x = 2$. As n becomes larger and larger, all four approximations become better and better, getting closer to the actual area. In this example, the exact area can be found by a formula from plane geometry. Write the given function as

$$y = \sqrt{4 - x^2},$$

then square both sides to get

$$y^2 = 4 - x^2$$
$$x^2 + y^2 = 4,$$

the equation of a circle centered at the origin with radius 2. The region in Figure 4 is the quarter of this circle that lies in the first quadrant. The actual area of this region is one-quarter of the area of the entire circle, or

$$\frac{1}{4}\pi(2)^2 = \pi \approx 3.1416.$$

As the number of rectangles increases without bound, the sum of the areas of these rectangles gets closer and closer to the actual area of the region, π. This can be written as

$$\lim_{n\to\infty}(\text{sum of areas of } n \text{ rectangles}) = \pi.$$

(The value of π was originally found by a process similar to this.)*

Notice in the above example that for a particular value of n, the midpoint rule gave the best answer (the one closest to the true value of 3.1416), followed by the trapezoidal rule, followed by the left and right sums. In fact, the midpoint rule with $n = 20$ gives a value (3.1454) that is slightly more accurate than the left sum with $n = 500$ (3.1455).

Now we can generalize to get a method of finding the area bounded by the curve $y = f(x)$, the x-axis, and the vertical lines $x = a$ and $x = b$, as shown in Figure 10. To approximate this area, we could divide the region under the curve first into 10 rectangles (Figure 10(a)) and then into 20 rectangles (Figure 10(b)). The sum of the areas of the rectangles give approximations to the area under the curve.

To develop a process that would yield the *exact* area, begin by dividing the interval from a to b into n pieces of equal width, using each of these n pieces as the base of a rectangle (see Figure 11). Let x_1 be an arbitrary point in the first

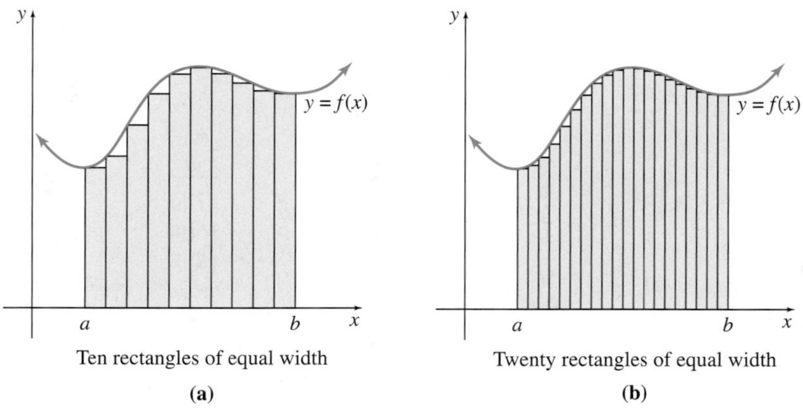

FIGURE 10

*The number π is the ratio of the circumference of a circle to its diameter. It is an example of an *irrational number*, and as such it cannot be expressed as a terminating or repeating decimal. Many approximations have been used for π over the years. A passage in the Bible (1 Kings 7:23) indicates a value of 3. The Egyptians used the value 3.16, and Archimedes showed that its value must be between 22/7 and 223/71. A Hindu writer, Brahmagupta, used $\sqrt{10}$ as its value in the seventh century. The search for the digits of π has continued into modern times. Yasumasa Kanada and his coworkers at the University of Tokyo recently computed the value to over 1.2 trillion places.

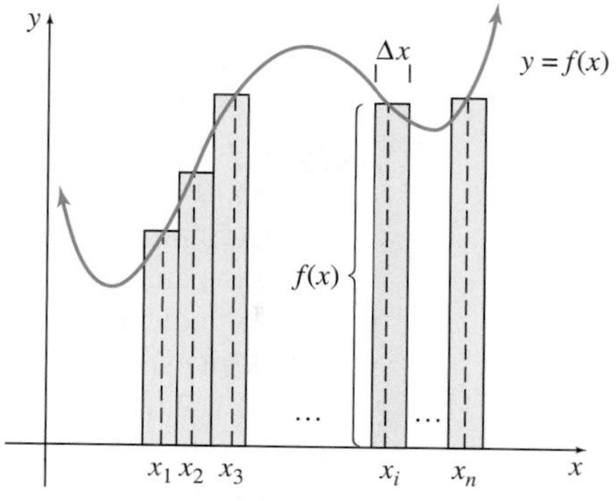

FIGURE 11

> **FOR REVIEW**
>
> Recall from Chapter 1 that the symbol Σ (sigma) indicates "the sum of." Here, we use $\sum_{i=1}^{n}$ to indicate the sum $f(x_1)\Delta x + f(x_2)\Delta x + f(x_3)\Delta x + \cdots + f(x_n)\Delta x$, where we replace i with 1 in the first term, 2 in the second term, and so on, ending with n replacing i in the last term.

interval, x_2 be an arbitrary point in the second interval, and so on, up to the nth interval. In the graph of Figure 11, the symbol Δx is used to represent the width of each of the intervals. The pink rectangle is an arbitrary rectangle called the ith rectangle. Its area is the product of its length and width. Since the width of the ith rectangle is Δx and the length of the ith rectangle is given by the height $f(x_i)$,

$$\text{Area of the } i\text{th rectangle} = f(x_i) \cdot \Delta x.$$

The total area under the curve is approximated by the sum of the areas of all n of the rectangles. With sigma notation, the approximation to the total area becomes

$$\text{Area of all } n \text{ rectangles} = \sum_{i=1}^{n} f(x_i) \cdot \Delta x.$$

The exact area is defined to be the limit of this sum (if the limit exists) as the number of rectangles increases without bound:

$$\text{Exact area} = \lim_{n\to\infty} \sum_{i=1}^{n} f(x_i)\Delta x.$$

This limit is called the *definite integral* of $f(x)$ from a to b and is written as follows.

> **THE DEFINITE INTEGRAL**
>
> If f is defined on the interval $[a, b]$, the **definite integral** of f from a to b is given by
>
> $$\int_a^b f(x)\,dx = \lim_{n\to\infty} \sum_{i=1}^{n} f(x_i)\Delta x,$$
>
> provided the limit exists, where $\Delta x = (b - a)/n$ and x_i is *any* value of x in the ith interval.

The definite integral can be approximated by

$$\sum_{i=1}^{n} f(x_i)\Delta x.$$

If $f(x) \geq 0$ on the interval $[a, b]$, the definite integral gives the area under the curve between $x = a$ and $x = b$. In the midpoint rule, x_i is the midpoint of the ith interval. We may also let x_i be the left endpoint, the right endpoint, or any other point in the ith interval.

Some calculators have a built-in function for evaluating the definite integral. For example, on the TI-83/84 Plus, the command `fnInt(√(4 − X²), X, 0, 2)` gives the answer 3.141593074, with an error of approximately .0000004.

As indicated in this definition, although the left endpoint of the ith interval has been used to find the height of the ith rectangle, any number in the ith interval can be used. (A more general definition is possible in which the rectangles do not necessarily all have the same width.) The b above the integral sign is called the **upper limit** of integration, and the a is the **lower limit** of integration. This use of the word *limit* has nothing to do with the limit of the sum; it refers to the limits, or boundaries, on x.

The sum in the definition of the definite integral is an example of a Riemann sum, named for the German mathematician Georg Riemann (1826–1866), who at the age of 20 changed his field of study from theology and the classics to mathematics. Twenty years later he died of tuberculosis while traveling in Italy in search of a cure. The concepts of *Riemann sum* and *Riemann integral* are still studied in rigorous calculus textbooks.

In the example at the beginning of this section, the area bounded by the x-axis, the curve $y = \sqrt{4 - x^2}$, and the lines $x = 0$ and $x = 2$ could be written as the definite integral

$$\int_0^2 \sqrt{4 - x^2}\, dx = \pi.$$

NOTE Notice that unlike the indefinite integral, which is a set of *functions*, the definite integral represents a *number*. The next section will show how antiderivatives are used in finding the definite integral and thus the area under a curve. ■

Keep in mind that finding the definite integral of a function can be thought of as a mathematical process that gives the sum of an infinite number of individual parts (within certain limits). The definite integral represents area only if the function involved is *nonnegative* $(f(x) \geq 0)$ at every x-value in the interval $[a, b]$. There are many other interpretations of the definite integral, and all of them involve this idea of approximation by appropriate sums.

EXAMPLE 1 *Approximation of Area*

Approximate $\int_0^4 2x\, dx$, the area of the region under the graph of $f(x) = 2x$, above the x-axis, and between $x = 0$ and $x = 4$, by using four rectangles of equal width whose heights are the values of the function at the midpoint of each subinterval.

Solution

Method 1: Calculating by Hand

We want to find the area of the shaded region in Figure 12 on the next page. The heights of the four rectangles given by $f(x_i)$ for $i = 1, 2, 3,$ and 4 are as follows.

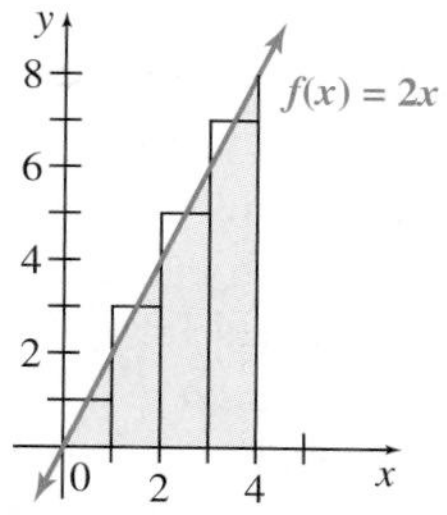

FIGURE 12

i	x_i	$f(x_i)$
1	$x_1 = .5$	$f(.5) = 1.0$
2	$x_2 = 1.5$	$f(1.5) = 3.0$
3	$x_3 = 2.5$	$f(2.5) = 5.0$
4	$x_4 = 3.5$	$f(3.5) = 7.0$

The width of each rectangle is $\Delta x = (4 - 0)/4 = 1$. The sum of the areas of the four rectangles is

$$\begin{aligned}\sum_{i=1}^{4} f(x_i)\Delta x &= f(x_1)\Delta x + f(x_2)\Delta x + f(x_3)\Delta x + f(x_4)\Delta x \\ &= f(.5)\Delta x + f(1.5)\Delta x + f(2.5)\Delta x + f(3.5)\Delta x \\ &= 1(1) + 3(1) + 5(1) + 7(1) \\ &= 16.\end{aligned}$$

Using the formula for the area of a triangle, $A = (1/2)bh$, with b, the length of the base, equal to 4 and h, the height, equal to 8, gives

$$A = \frac{1}{2}bh = \frac{1}{2}(4)(8) = 16,$$

the exact value of the area. The approximation equals the exact area in this case because our use of the midpoints of each subinterval distributed the error evenly above and below the graph.

Method 2: Graphing Calculator

A graphing calculator can be used to organize the information in this example. For example, the `seq` feature in the LIST OPS menu of the TI-83/84 Plus calculator can be used to store the values of i in the list L_1. Using the STAT EDIT menu, the entries for x_i can be generated by entering the formula $-.5 + L_1$ as the heading of L_2. Similarly, entering the formula for $f(x_i)$, $2 * L_2$, at the top of list L_3 will generate the values of $f(x_i)$ in L_3. (The entries are listed automatically when the formula is entered.) Then the `sum` feature in the LIST MATH menu can be used to add the values in L_3. The resulting screens are shown in Figure 13.

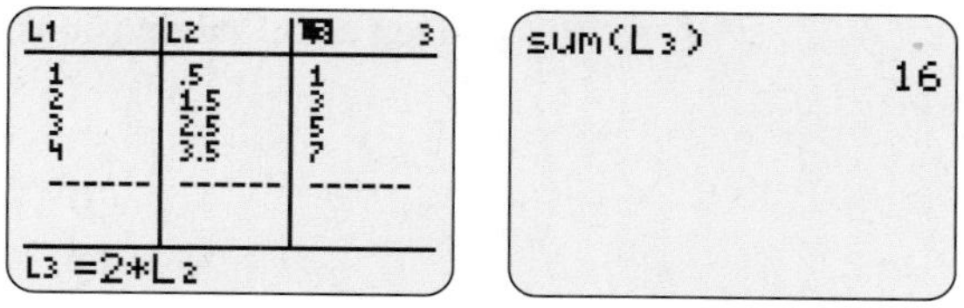

FIGURE 13

Method 3: Spreadsheet

The calculations in this example can also be done on a spreadsheet. In Microsoft Excel, for example, store the values of i in column A. Put the command "=A1−.5" into B1; copying this formula into the rest of column B gives the values of x_i. Similarly, use the formula for $f(x_i)$ to fill column C. Column D is the product of Column C and Δx. Sum column D to get the answer. For more details, see *The Spreadsheet Manual* that is available with this book.

Total Change Suppose the function $f(x) = x^2 + 20$ gives the marginal cost of some item at a particular x-value. Then $f(2) = 24$ gives the rate of change of cost at $x = 2$. That is, a unit change in x (at this point) will produce a change of 24 units in the cost function. Also, $f(3) = 29$ means that each unit of change in x (when $x = 3$) will produce a change of 29 units in the cost function.

To find the *total* change in the cost function as x changes from 2 to 3, we could divide the interval from 2 to 3 into n equal parts, using each part as the base of a rectangle as we did above. The area of each rectangle would approximate the change in cost at the x-value that is the left endpoint of the base of the rectangle. Then the sum of the areas of these rectangles would approximate the net total change in cost from $x = 2$ to $x = 3$. The limit of this sum as $n \to \infty$ would give the exact total change.

This result produces another application of the definite integral: the area of the region under the graph of the marginal cost function $f(x)$ that is above the x-axis and between $x = a$ and $x = b$ gives the *net total change in the cost* as x goes from a to b.

TOTAL CHANGE IN $F(x)$

If $f(x)$ gives the rate of change of $F(x)$ for x in $[a, b]$, then the **total change** in $F(x)$ as x goes from a to b is given by

$$\lim_{n\to\infty} \sum_{i=1}^{n} f(x_i)\Delta x = \int_a^b f(x)\,dx.$$

In other words, the total change in a quantity can be found from the function that gives the rate of change of the quantity, using the same methods used to approximate the area under a curve.

EXAMPLE 2 *Total Maintenance Charges*

Figure 14 shows the rate of change of the annual maintenance charges for a certain machine. To approximate the total maintenance charges over the 10-year life of the machine, use approximating rectangles, dividing the interval from 0 to 10 into ten equal subdivisions. Each rectangle has width 1; using the left endpoint of

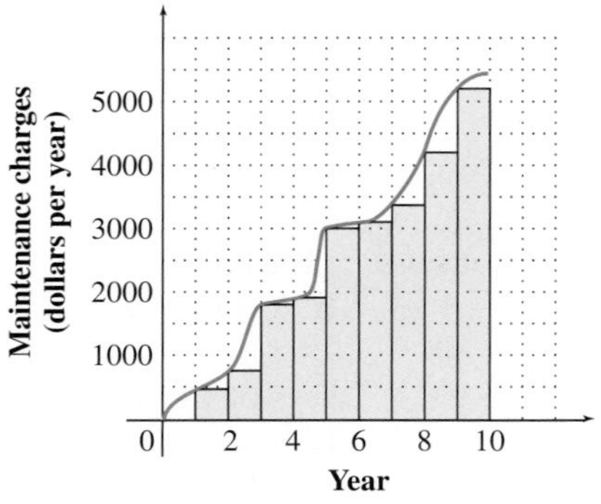

FIGURE 14

each rectangle to determine the height of the rectangle, the approximation becomes

$$1 \cdot 0 + 1 \cdot 500 + 1 \cdot 750 + 1 \cdot 1800 + 1 \cdot 1900 + 1 \cdot 3000 + 1 \cdot 3100 + 1 \cdot 3400 + 1 \cdot 4200 + 1 \cdot 5200 = 23{,}850.$$

About \$23,850 will be spent on maintenance over the 10-year life of the machine.

Recall, velocity is the rate of change in distance from time a to time b. Thus the area under the velocity function defined by $v(t)$ from $t = a$ to $t = b$ gives the distance traveled in that time period.

EXAMPLE 3 *Total Distance*

A driver traveling on a business trip checks the speedometer each hour. The table shows the driver's velocity at several times.
Approximate the total distance traveled during the 3-hour period using the left endpoint of each interval, then the right endpoint.

Time (hr)	0	1	2	3
Velocity (mph)	0	52	58	60

Solution Using left endpoints, the total distance is

$$0 \cdot 1 + 52 \cdot 1 + 58 \cdot 1 = 110.$$

With right endpoints, we get

$$52 \cdot 1 + 58 \cdot 1 + 60 \cdot 1 = 170.$$

Again, left endpoints give a total that is too small, while right endpoints give a total that is too large. The average, 140 miles, is a better estimate of the total distance traveled.

Before discussing further applications of the definite integral, we need a more efficient method for evaluating it. This method will be developed in the next section.

7.3 EXERCISES

1. Explain the difference between an indefinite integral and a definite integral.

2. Complete the following statement.

$$\int_0^3 (x^2 + 2)\,dx = \lim_{n \to \infty} \_\_\_\_\_\_, \text{ where } \Delta x = \_\_\_\_\_\_, \text{ and } x_i \text{ is } \_\_\_\_\_\_.$$

3. Let $f(x) = 2x + 1$, $x_1 = 0$, $x_2 = 2$, $x_3 = 4$, $x_4 = 6$, and $\Delta x = 2$.

a. Find $\sum_{i=1}^{4} f(x_i)\Delta x$.

b. The sum in part a approximates a definite integral using rectangles. The height of each rectangle is given by the value of the function at the left endpoint. Write the definite integral that the sum approximates.

4. Let $f(x) = 1/x$, $x_1 = 1/2$, $x_2 = 1$, $x_3 = 3/2$, $x_4 = 2$, and $\Delta x = 1/2$.

a. Find $\sum_{i=1}^{4} f(x_i)\Delta x$.

b. The sum in part a approximates a definite integral using rectangles. The height of each rectangle is given by the value of the function at the left endpoint. Write the definite integral that the sum approximates.

5. The booklet *All About Lawns** published by Ortho Books gives the following instructions for measuring the area of an irregularly shaped region.

Irregular Shapes
(within 5% accuracy)
Measure a long (L) axis of the area. Every 10 feet along the length line, measure the width at right angles to the length line. Total widths and multiply by 10.

Area $= (\overline{A_1A_2} + \overline{B_1B_2} + \overline{C_1C_2}$ **etc.**$) \times 10$

A $= (40' + 60' + 32') \times 10$

A $= 132' \times 10'$

A $= 1320$ square feet

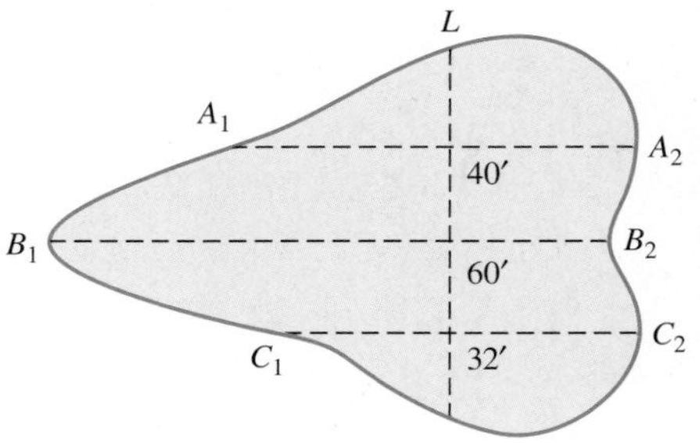

How does this method relate to the discussion in this section?

In Exercises 6–13, approximate the area under the graph of $f(x)$ and above the x-axis using the following methods with $n = 4$. **(a)** *Use left endpoints.* **(b)** *Use right endpoints.* **(c)** *Average the answers in parts a and b.* **(d)** *Use midpoints.*

6. $f(x) = 3x + 2$ from $x = 1$ to $x = 5$

7. $f(x) = x + 5$ from $x = 2$ to $x = 4$

8. $f(x) = x^2$ from $x = 1$ to $x = 5$

9. $f(x) = -x^2 + 4$ from $x = -2$ to $x = 2$

10. $f(x) = e^x - 1$ from $x = 0$ to $x = 4$

11. $f(x) = e^x + 1$ from $x = -2$ to $x = 2$

12. $f(x) = \frac{1}{x}$ from $x = 1$ to $x = 5$

13. $f(x) = \frac{2}{x}$ from $x = 1$ to $x = 9$

14. Consider the region below $f(x) = x/2$, above the x-axis, and between $x = 0$ and $x = 4$. Let x_i be the midpoint of the ith subinterval.

a. Approximate the area of the region using four rectangles.

b. Find $\int_0^4 f(x)\,dx$ by using the formula for the area of a triangle.

15. Find $\int_0^5 (5 - x)\,dx$ by using the formula for the area of a triangle.

16. Find $\int_0^4 f(x)\,dx$ for each graph of $y = f(x)$.

a.

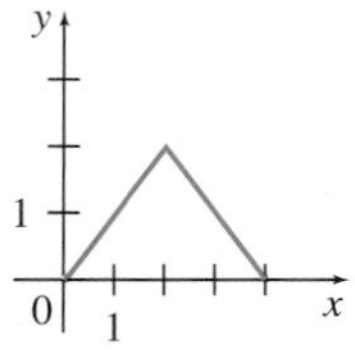

b.

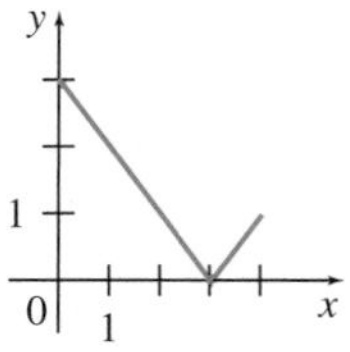

Find the exact value of each integral using formulas from geometry.

17. $\int_{-3}^{3} \sqrt{9 - x^2}\,dx$

18. $\int_{-4}^{0} \sqrt{16 - x^2}\,dx$

19. $\int_{1}^{3} (5 - x)\,dx$

20. $\int_{2}^{5} (1 + 2x)\,dx$

*MacLaskey, Michael, *All About Lawns,* ed. by Alice Mace, Ortho Information Services, © 1980, p. 108.

21. In this exercise, we investigate the value of $\int_0^1 x^2\,dx$ using larger and larger values of n in the definition of the definite integral.

a. First let $n = 10$, so $\Delta x = .1$. Fill a list on your calculator with values of x^2 as x goes from .1 to 1. (On a TI-83/84 Plus, use the command `seq(X^2,X,.1,1,.1)→L1`.)

b. Sum the values in the list formed in part a, and multiply by .1, to estimate $\int_0^1 x^2\,dx$ with $n = 10$. (On a TI-83/84 Plus, use the command `.1*sum(L1)`.)

c. Repeat parts a and b with $n = 100$.

d. Repeat parts a and b with $n = 500$.

e. Based on your answers to parts b through d, what do you estimate the value of $\int_0^1 x^2\,dx$ to be?

22. Repeat Exercise 21 for $\int_0^1 x^3\,dx$.

Applications

In Exercises 23–28, estimate the area under each curve by summing the area of rectangles. Use the left endpoints, then the right endpoints, then give the average of those answers.

BUSINESS AND ECONOMICS

23. *Oil Consumption* The graph at the bottom of the page shows U.S. oil production and consumption rates (in millions of barrels per day) for the years 1950–2001.*

a. Estimate the amount of oil produced in the United States (domestic supply) between 1990 and 2000. Use rectangles with widths of 5 years.

b. Estimate the amount of oil imported in the United States between 1990 and 2000. Use rectangles with widths of 5 years.

c. Estimate the amount of oil consumed in the United States between 1990 and 2000 and then compare it to the total of parts a and b. Comment on the accuracy of these calculations.

24. *Electricity Consumption* The following graph shows the rate of use of electrical energy (in kilowatts) in a certain city on a very hot day. Estimate the total usage of electricity on that day. Let the width of each rectangle be 2 hours.

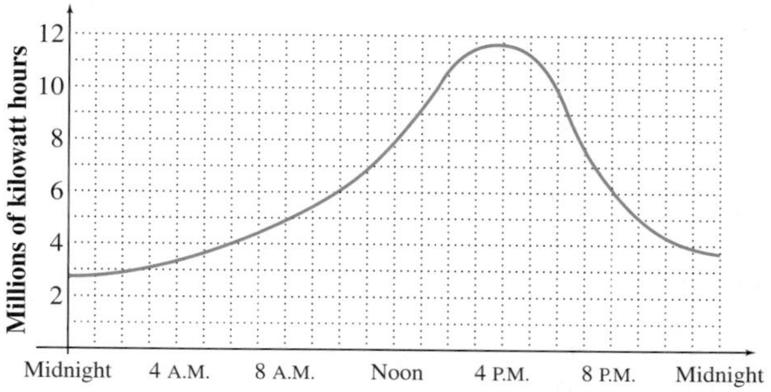

25. *Wages* The graph at the top of the next page shows the average manufacturing hourly wage in the United States and Canada for 1992–2001.[†] All figures are in U.S. dollars.

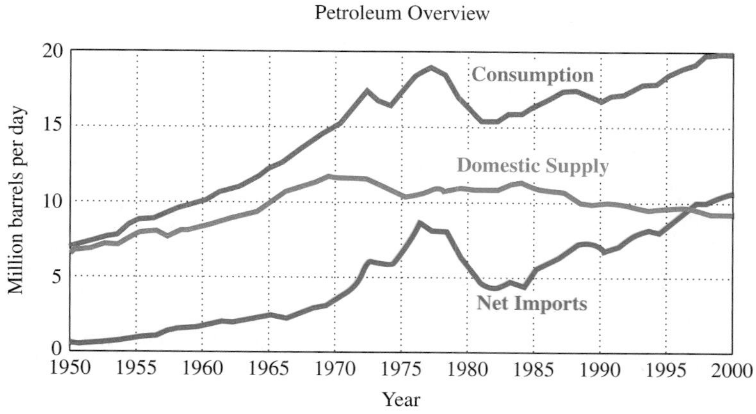

*U.S. Department of Energy, www.eia.doe.gov.
[†]U.S. Bureau of Labor Statistics, August 2003.

Average Manufacturing Hourly Wage

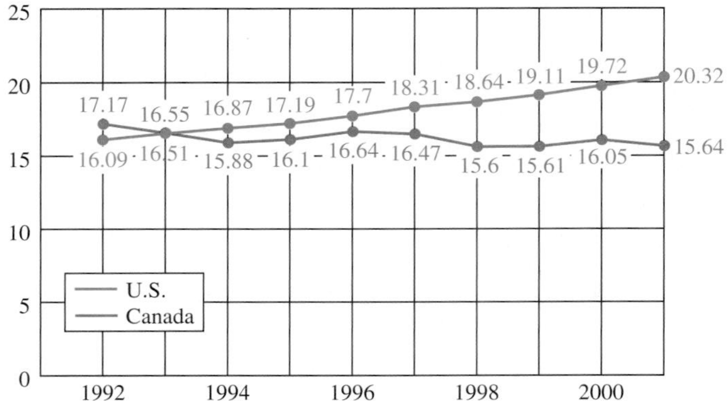

Assume that the average employee works 2000 hours per year.

a. Estimate the total amount earned by an average U.S. worker during the 4-year period from the beginning of 1998 to the end of 2001. Use rectangles with widths of 1 year.

b. Estimate the total amount earned (in U.S. dollars) by an average Canadian worker during the 4-year period from the beginning of 1998 to the end of 2001. Use rectangle widths of 1 year.

LIFE SCIENCES

26. *Alcohol Concentration* The following graph shows the approximate concentration of alcohol in a person's bloodstream t hours after drinking 2 oz of alcohol. Estimate the total amount of alcohol in the bloodstream by estimating the area under the curve. Use rectangles with widths of 1 hour.

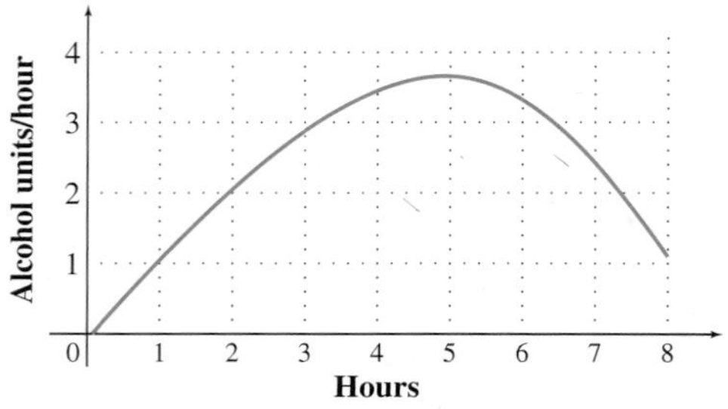

27. *Oxygen Inhalation* The following graph shows the rate of inhalation of oxygen (in liters per minute) by a person riding a bicycle very rapidly for 10 minutes. Estimate the total volume of oxygen inhaled in the first 20 minutes after the beginning of the ride. Use rectangles with widths of 1 minute.

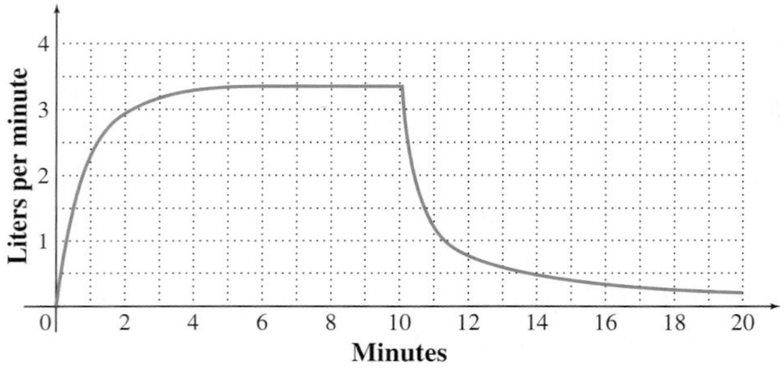

28. *Foot-and-Mouth Epidemic* In 2001, the United Kingdom suffered an epidemic of foot-and-mouth disease. The graph shows the number of reported cases each day since February 18, as well as the number of cases epidemiologists project would have occurred had they culled all livestock on infected farms within 24 hours, and all livestock on neighboring farms within 48 hours of the infection.*

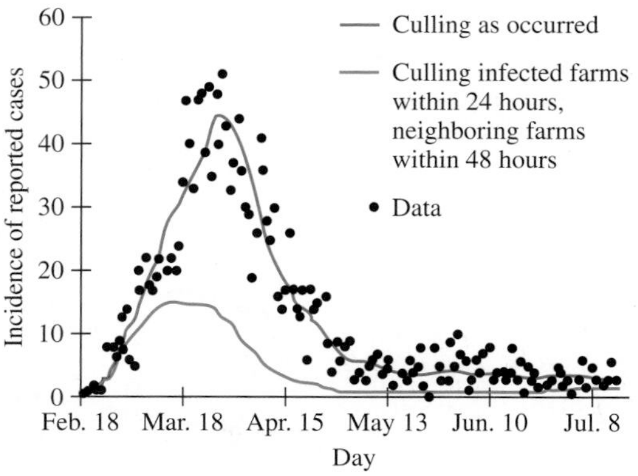

*Science, Vol. 294, Oct. 5, 2001, p. 26. Peter Morrison/AP. Reprinted by permission of the Associated Press.

a. Estimate the total number of cases that occurred from February 18 through May 13. Use rectangles with widths of 14 days.

b. Estimate the total number of cases that would have occurred from February 18 through May 13 using the more aggressive culling plan. Use rectangles with widths of 14 days.

PHYSICAL SCIENCES

Distance *The next two graphs are from* Road & Track *magazine.* The curve shows the velocity at t seconds after the car accelerates from a dead stop. To find the total distance traveled by the car in reaching* 100 *mph, we must estimate the definite integral*

$$\int_0^T v(t)\,dt,$$

where T represents the number of seconds it takes for the car to reach 100 *mph.*

Use the graphs to estimate this distance by adding the areas of rectangles with widths of 5 *seconds. (The last rectangle will have a width of* 4 *seconds or* 3 *seconds.) Use the midpoint rule. To adjust your answer to miles per hour, divide by* 3600 *(the number of seconds in an hour). You then have the number of miles that the car traveled in reaching* 100 *mph. Finally, multiply by* 5280 *ft per mile to convert the answer to feet.*

29. Estimate the distance traveled by the Porsche 928, using the graph below.

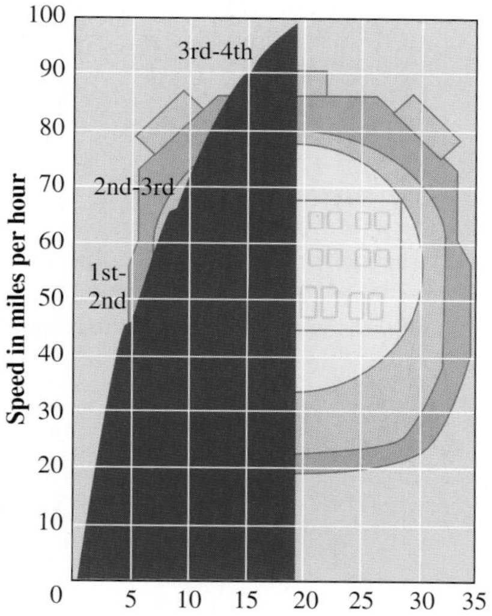

Porsche 928

30. Estimate the distance traveled by the BMW 733i, using the graph below.

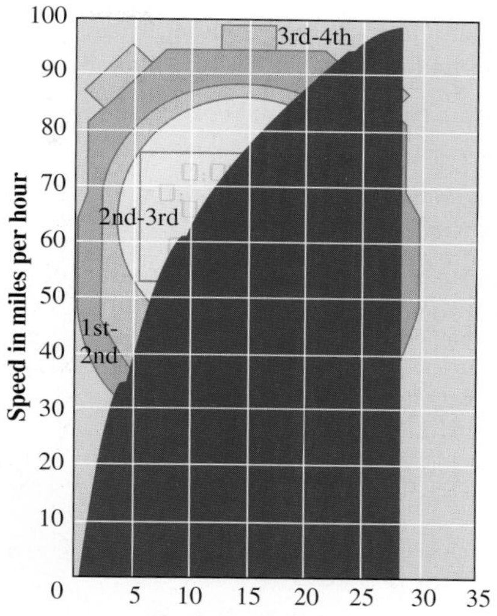

BMW 733i

Heat Gain *The following graphs*[†] *show the typical heat gain, in BTUs per hour per square foot, for a window facing east and one facing south, with plain glass and with a black Shade-Screen. Estimate the total heat gain per square foot by summing*

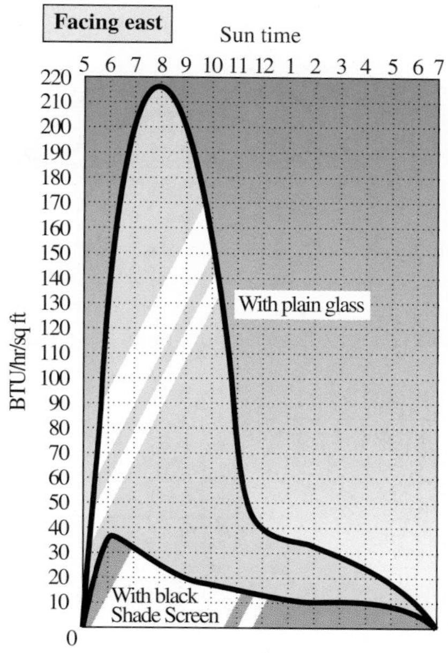

**Road & Track,* April and May, 1978. Reprinted with permission of *Road & Track.*

[†]Graphs courtesy of Phifer Wire Products. Reprinted by permission of Phifer Wire Products.

the areas of rectangles. Use rectangles with widths of 2 hours, and let the function value at the midpoint of the subinterval give the height of the rectangle.

31. a. Estimate the total heat gain per square foot for a plain glass window facing east.

b. Estimate the total heat gain per square foot for a window facing east with a ShadeScreen.

32. a. Estimate the total heat gain per square foot for a plain glass window facing south.

b. Estimate the total heat gain per square foot for a window facing south with a ShadeScreen.

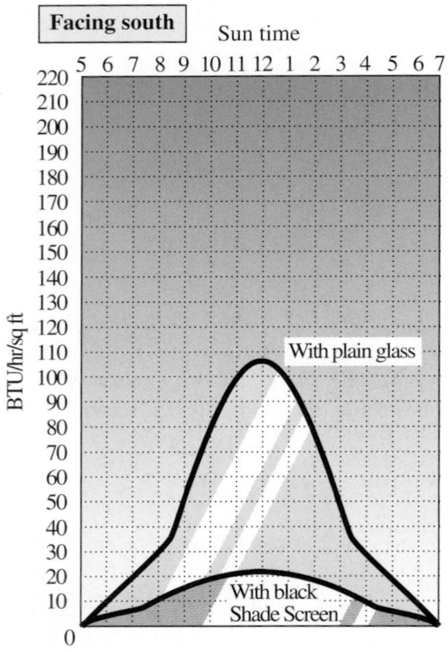

33. *Automobile Velocity* Two cars start from rest at a traffic light and accelerate for several minutes. The graph shows their velocities (in feet per second) as a function of time (in seconds). Car A is the one that initially has greater velocity.*

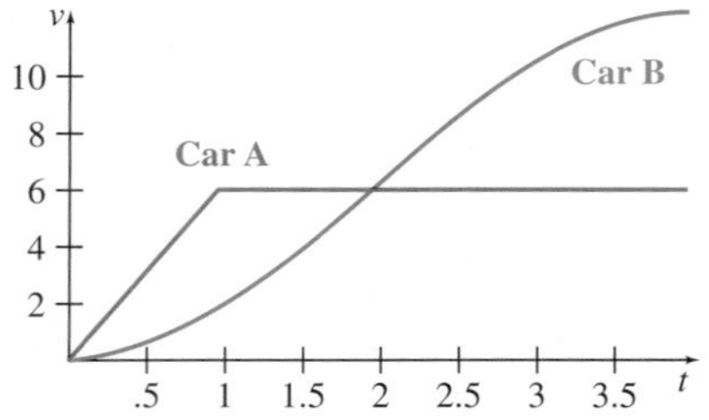

a. How far has car A traveled after 2 seconds? (*Hint*: Use formulas from geometry.)

b. When is car A farthest ahead of car B?

c. Estimate the farthest that car A gets ahead of car B. For car A, use formulas from geometry. For car B, use $n = 4$ and the value of the function at the midpoint of each interval.

d. Give a rough estimate of when car B catches up with car A.

34. *Distance* Musk the friendly pit bull has escaped again! Here is her velocity during the first 4 seconds of her romp.

t (sec)	0	1	2	3	4
v (ft/sec)	0	8	13	17	18

Give two estimates for the total distance Musk traveled during her 4-second trip, one using the left endpoint of each interval and one using the right endpoint.

35. *Distance* The speed of a particle in a test laboratory was noted every second for 3 seconds. The results are shown in the following table. Use the left endpoints and then the right endpoints to estimate the total distance the particle moved in the first three seconds.

t (sec)	0	1	2	3
v (ft/sec)	10	6.5	6	5.5

36. *Running* In 1987, Canadian Ben Johnson set a world record in the 100-m sprint. (The record was later taken away when he was found to have used an anabolic steroid

*Based on an example given by Stephen Monk of the University of Washington.

to enhance his performance.) His speed at various times in the race is given in the following table.*

Time (sec)	Speed (mph)
0	0
1.84	12.9
3.80	23.8
6.38	26.3
7.23	26.3
8.96	26.0
9.83	25.7

a. Use the information in the table and left endpoints to estimate the distance that Johnson ran in miles. You will first need to calculate Δt for each interval. At the end, you will need to divide by 3600 (the number of seconds in an hour), since the speed is in miles per hour.

b. Repeat part a, using right endpoints.

c. Wait a minute; we know that the distance Johnson ran is 100 m. Divide this by 1609, the number of meters in a mile, to find how far Johnson ran in miles. Is your answer from part a or part b closer to the true answer? Briefly explain why you think this answer should be more accurate.

7.4 THE FUNDAMENTAL THEOREM OF CALCULUS

THINK ABOUT IT *If we know how the rate of consumption of natural gas varies over time, how can we compute the total amount of natural gas used?*

This section introduces a powerful theorem for answering such questions.

The work from the last two sections can now be put together. We have seen that, if $f(x) > 0$,

$$\int_a^b f(x)\,dx$$

gives the area between the graph of $f(x)$ and the x-axis, from $x = a$ to $x = b$. We can find this definite integral by using the antiderivatives discussed earlier. The definite integral was defined and evaluated in the previous section using the limit of a sum. In that section, we also saw that if $f(x)$ gives the rate of change of $F(x)$, the definite integral $\int_a^b f(x)\,dx$ gives the total change of $F(x)$ as x changes from a to b. If $f(x)$ gives the rate of change of $F(x)$, then $F(x)$ is an antiderivative of $f(x)$. Writing the total change in $F(x)$ from $x = a$ to $x = b$ as $F(b) - F(a)$ shows the connection between antiderivatives and definite integrals. This relationship is called the **Fundamental Theorem of Calculus.**

*Wildbur, Peter, *Information Graphics,* Van Nostrand Reinhold, 1989, pp. 126–127. The world record of 9.78 seconds is currently held by Tim Montgomery.

FUNDAMENTAL THEOREM OF CALCULUS

Let f be continuous on the interval $[a, b]$, and let F be *any* antiderivative of f. Then

$$\int_a^b f(x)\,dx = F(b) - F(a) = F(x)\Big|_a^b.$$

The symbol $F(x)|_a^b$ is used to represent $F(b) - F(a)$. It is important to note that the Fundamental Theorem does not require $f(x) > 0$. The condition $f(x) > 0$ is necessary only when using the Fundamental Theorem to find area. Also, note that the Fundamental Theorem does not *define* the definite integral; it just provides a method for evaluating it.

EXAMPLE 1 *Fundamental Theorem of Calculus*

First find $\int 4t^3\,dt$ and then find $\int_1^2 4t^3\,dt$.

Solution By the rules given earlier,

$$\int 4t^3\,dt = t^4 + C.$$

By the Fundamental Theorem, the value of $\int_1^2 4t^3\,dt$ is found by evaluating $t^4|_1^2$, with no constant C required.

$$\int_1^2 4t^3\,dt = t^4\Big|_1^2 = 2^4 - 1^4 = 15$$

NOTE No constant C is needed, as it is for the indefinite integral, because if C were added to the antiderivative F, it would be eliminated in the final answer:

$$\begin{aligned}\int_a^b f(x)\,dx &= (F(x) + C)\Big|_a^b \\ &= (F(b) + C) - (F(a) + C) \\ &= F(b) - F(a).\end{aligned}$$

In other words, any antiderivative will give the same answer, so for simplicity, we choose the one with $C = 0$. ■

Example 1 illustrates the difference between the definite integral and the indefinite integral. A definite integral is a real number; an indefinite integral is a family of functions in which all the functions are antiderivatives of a function f.

To see why the Fundamental Theorem of Calculus is true for $f(x) > 0$ when f is continuous, look at Figure 15. Define the function $A(x)$ as the area between the x-axis and the graph of $y = f(x)$ from a to x. We first show that A is an antiderivative of f; that is $A'(x) = f(x)$.

To do this, let h be a small positive number. Then $A(x + h) - A(x)$ is the shaded area in Figure 15. This area can be approximated with a rectangle having width h and height $f(x)$. The area of the rectangle is $h \cdot f(x)$, and

$$A(x + h) - A(x) \approx h \cdot f(x).$$

Dividing both sides by h gives

$$\frac{A(x + h) - A(x)}{h} \approx f(x).$$

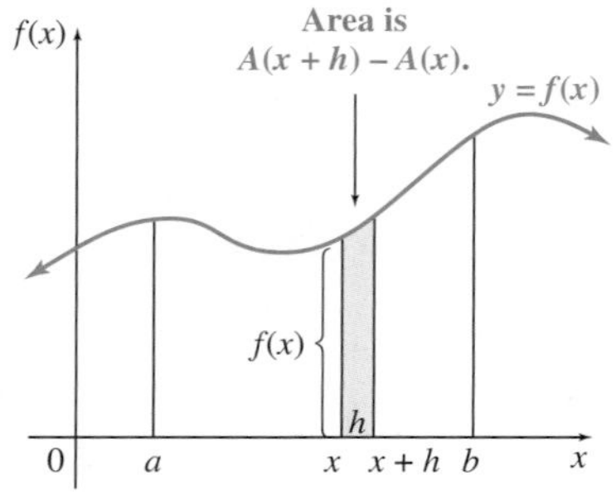

FIGURE 15

This approximation improves as h gets smaller and smaller. Taking the limit on the left as h approaches 0 gives an exact result.

$$\lim_{h \to 0} \frac{A(x+h) - A(x)}{h} = f(x)$$

This limit is simply $A'(x)$, so

$$A'(x) = f(x).$$

This result means that A is an antiderivative of f, as we set out to show.

$A(b)$ is the area under the curve from a to b, and $A(a) = 0$, so the area under the curve can be written as $A(b) - A(a)$. From the previous section, we know that the area under the curve is also given by $\int_a^b f(x)\,dx$. Putting these two results together gives

$$\int_a^b f(x)\,dx = A(b) - A(a)$$
$$= A(x)\Big|_a^b$$

where A is an antiderivative of f. From the note after Example 1, we know that any antiderivative will give the same answer, which proves the Fundamental Theorem of Calculus.

The Fundamental Theorem of Calculus certainly deserves its name, which sets it apart as the most important theorem of calculus. It is the key connection between differential calculus and integral calculus, which originally were developed separately without knowledge of this connection between them.

The variable used in the integrand does not matter; each of the following definite integrals represents the number $F(b) - F(a)$.

$$\int_a^b f(x)\,dx = \int_a^b f(t)\,dt = \int_a^b f(u)\,du$$

Key properties of definite integrals are listed below. Some of them are just restatements of properties from Section 1.

PROPERTIES OF DEFINITE INTEGRALS

If all indicated definite integrals exist,

1. $\int_a^a f(x)\,dx = 0;$
2. $\int_a^b k \cdot f(x)\,dx = k \cdot \int_a^b f(x)\,dx$ for any real constant k (constant multiple of a function);
3. $\int_a^b [f(x) \pm g(x)]\,dx = \int_a^b f(x)\,dx \pm \int_a^b g(x)\,dx$ (sum or difference of functions);
4. $\int_a^b f(x)\,dx = \int_a^c f(x)\,dx + \int_c^b f(x)\,dx$ for any real number c;
5. $\int_a^b f(x)\,dx = -\int_b^a f(x)\,dx.$

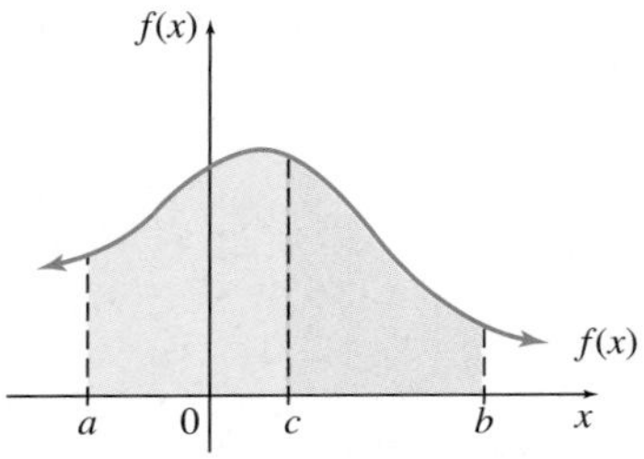

FIGURE 16

For $f(x) > 0$, since the distance from a to a is 0, the first property says that the "area" under the graph of f bounded by $x = a$ and $x = a$ is 0. Also, since $\int_a^c f(x)\,dx$ represents the blue region in Figure 16 and $\int_c^b f(x)\,dx$ represents the pink region,

$$\int_a^b f(x)\,dx = \int_a^c f(x)\,dx + \int_c^b f(x)\,dx,$$

as stated in the fourth property. While the figure shows $a < c < b$, the property is true for any value of c where both $f(x)$ and $F(x)$ are defined.

An algebraic proof is given here for the third property; proofs of the other properties are left for the exercises. If $F(x)$ and $G(x)$ are antiderivatives of $f(x)$ and $g(x)$, respectively,

$$\begin{aligned}\int_a^b [f(x) + g(x)]\,dx &= [F(x) + G(x)]\Big|_a^b \\ &= [F(b) + G(b)] - [F(a) + G(a)] \\ &= [F(b) - F(a)] + [G(b) - G(a)] \\ &= \int_a^b f(x)\,dx + \int_a^b g(x)\,dx.\end{aligned}$$

EXAMPLE 2 *Fundamental Theorem of Calculus*

Find $\int_2^5 (6x^2 - 3x + 5)\,dx$.

Solution Use the properties above and the Fundamental Theorem, along with properties from Section 1.

$$\begin{aligned}\int_2^5 (6x^2 - 3x + 5)\,dx &= 6\int_2^5 x^2\,dx - 3\int_2^5 x\,dx + 5\int_2^5 dx \\ &= 2x^3\Big|_2^5 - \frac{3}{2}x^2\Big|_2^5 + 5x\Big|_2^5 \\ &= 2(5^3 - 2^3) - \frac{3}{2}(5^2 - 2^2) + 5(5 - 2) \\ &= 2(125 - 8) - \frac{3}{2}(25 - 4) + 5(3) \\ &= 234 - \frac{63}{2} + 15 = \frac{435}{2}\end{aligned}$$

EXAMPLE 3 *Fundamental Theorem of Calculus*

$$\begin{aligned}\int_1^2 \frac{dy}{y} = \ln|y|\Big|_1^2 &= \ln|2| - \ln|1| \\ &= \ln 2 - \ln 1 \approx .6931 - 0 = .6931\end{aligned}$$

EXAMPLE 4 *Substitution*

Evaluate $\int_0^5 x\sqrt{25 - x^2}\,dx$.

Solution Use substitution. Let $u = 25 - x^2$, so that $du = -2x\,dx$. With a definite integral, the limits should be changed, too. The new limits on u are found as follows.

$$\text{If } x = 5, \text{ then } u = 25 - 5^2 = 0.$$
$$\text{If } x = 0, \text{ then } u = 25 - 0^2 = 25.$$

Then

$$\begin{aligned}\int_0^5 x\sqrt{25 - x^2}\,dx &= -\frac{1}{2}\int_0^5 \sqrt{25 - x^2}(-2x\,dx)\\ &= -\frac{1}{2}\int_{25}^0 \sqrt{u}\,du \qquad \text{Substitute and change limits.}\\ &= -\frac{1}{2}\int_{25}^0 u^{1/2}\,du\\ &= -\frac{1}{2}\cdot\frac{u^{3/2}}{3/2}\bigg|_{25}^0\\ &= -\frac{1}{2}\cdot\frac{2}{3}[0^{3/2} - 25^{3/2}]\\ &= -\frac{1}{3}(-125) = \frac{125}{3}.\end{aligned}$$

CAUTION When substitution is used on a definite integral, it is best to not revert from u back to the original variable after antidifferentiation, as we did for the indefinite integral. Notice in Example 4 that after changing from the old limits on x to the new limits on u, we never returned to x or its limits. We recommend the practice of labeling the limits when doing a substitution, so the substitution in Example 4 becomes

$$\int_{x=0}^{x=5} x\sqrt{25 - x^2}\,dx = -\frac{1}{2}\int_{u=25}^{u=0} \sqrt{u}\,du.$$ ■

Area In the previous section we saw that, if $f(x) > 0$ in $[a, b]$, the definite integral $\int_a^b f(x)\,dx$ gives the area below the graph of the function $y = f(x)$, above the x-axis, and between the lines $x = a$ and $x = b$.

To see how to work around the requirement that $f(x) > 0$, look at the graph of $f(x) = x^2 - 4$ in Figure 17. The area bounded by the graph of f, the x-axis, and the vertical lines $x = 0$ and $x = 2$ lies below the x-axis. Using the Fundamental Theorem to find this area gives

$$\begin{aligned}\int_0^2 (x^2 - 4)\,dx &= \left(\frac{x^3}{3} - 4x\right)\bigg|_0^2\\ &= \left(\frac{8}{3} - 8\right) - (0 - 0) = -\frac{16}{3}.\end{aligned}$$

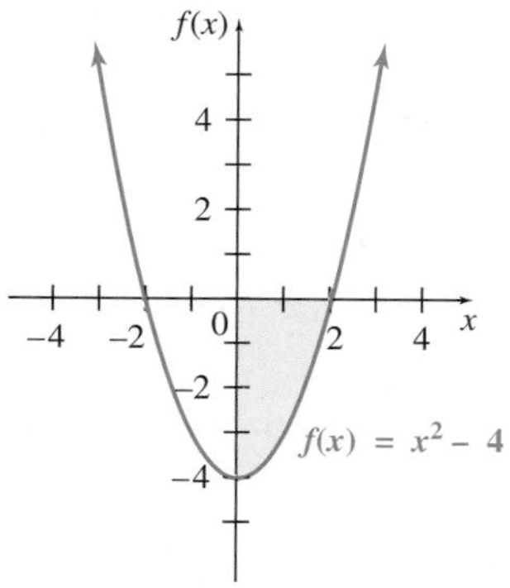

FIGURE 17

The result is a negative number because $f(x)$ is negative for values of x in the interval $[0, 2]$. Since Δx is always positive, if $f(x) < 0$ the product $f(x)\cdot\Delta x$ is negative, so $\int_0^2 f(x)\,dx$ is negative. Since area is nonnegative, the required area is given by $|-16/3|$ or $16/3$. Using a definite integral, the area could be written as

$$\left|\int_0^2 (x^2 - 4)\,dx\right| = \left|-\frac{16}{3}\right| = \frac{16}{3}.$$

EXAMPLE 5 *Area*

Find the area of the region between the x-axis and the graph of $f(x) = x^2 - 3x$ from $x = 1$ to $x = 3$.

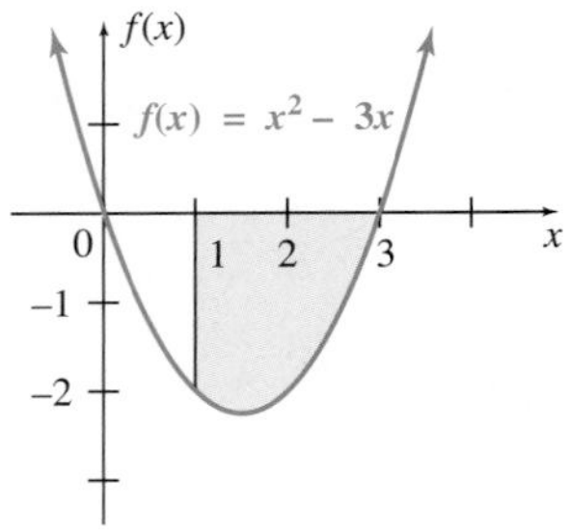

FIGURE 18

Solution The region is shown in Figure 18. Since the region lies below the x-axis, the area is given by

$$\left|\int_1^3 (x^2 - 3x)\,dx\right|.$$

By the Fundamental Theorem,

$$\int_1^3 (x^2 - 3x)\,dx = \left(\frac{x^3}{3} - \frac{3x^2}{2}\right)\Bigg|_1^3 = \left(\frac{27}{3} - \frac{27}{2}\right) - \left(\frac{1}{3} - \frac{3}{2}\right) = -\frac{10}{3}.$$

The required area is $|-10/3| = 10/3$.

EXAMPLE 6 *Area*

Find the area between the x-axis and the graph of $f(x) = x^2 - 4$ from $x = 0$ to $x = 4$.

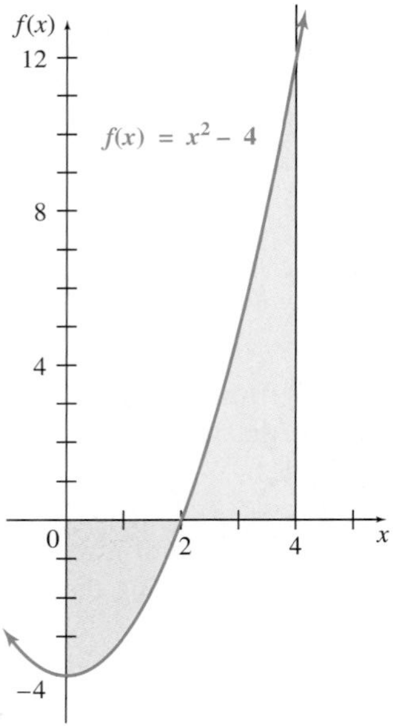

FIGURE 19

Solution Figure 19 shows the required region. Part of the region is below the x-axis. The definite integral over that interval will have a negative value. To find the area, integrate the negative and positive portions separately and take the absolute value of the first result before combining the two results to get the total area. Start by finding the point where the graph crosses the x-axis. This is done by solving the equation

$$x^2 - 4 = 0.$$

The solutions of this equation are 2 and -2. The only solution in the interval $[0, 4]$ is 2. The total area of the region in Figure 19 is

$$\begin{aligned}\left|\int_0^2 (x^2 - 4)\,dx\right| + \int_2^4 (x^2 - 4)\,dx &= \left|\left(\frac{1}{3}x^3 - 4x\right)\Bigg|_0^2\right| + \left(\frac{1}{3}x^3 - 4x\right)\Bigg|_2^4 \\ &= \left|\frac{8}{3} - 8\right| + \left(\frac{64}{3} - 16\right) - \left(\frac{8}{3} - 8\right) \\ &= 16.\end{aligned}$$

Incorrectly using one integral over the entire interval to find the area in Example 6 would have given

$$\int_0^4 (x^2 - 4)\,dx = \left(\frac{x^3}{3} - 4x\right)\Bigg|_0^4 = \left(\frac{64}{3} - 16\right) - 0 = \frac{16}{3},$$

which is not the correct area. This definite integral does not represent any area, but is just a real number.

For instance, if $f(x)$ in Example 6 represents the annual rate of profit of a company, then $16/3$ represents the total profit for the company over a 4-year period. The integral between 0 and 2 is $-16/3$; the negative sign indicates a loss for the first two years. The integral between 2 and 4 is $32/3$, indicating a profit. The overall profit is $32/3 - 16/3 = 16/3$, although the total shaded area is $32/3 + |-16/3| = 16$.

FINDING AREA

In summary, to find the area bounded by $f(x)$, $x = a$, $x = b$, and the x-axis, use the following steps.

1. Sketch a graph.
2. Find any x-intercepts of $f(x)$ in $[a, b]$. These divide the total region into subregions.
3. The definite integral will be *positive* for subregions above the x-axis and *negative* for subregions below the x-axis. Use separate integrals to find the (positive) areas of the subregions.
4. The total area is the sum of the areas of all of the subregions.

In the last section, we saw that the area under a rate of change function $f'(x)$ from $x = a$ to $x = b$ gives the total value of $f(x)$ on $[a, b]$. Now we can use the definite integral to solve these problems.

EXAMPLE 7 *Natural Gas Consumption*

The yearly rate of consumption of natural gas (in trillions of cubic feet) for a certain city is

$$C'(t) = t + e^{.01t},$$

where t is time in years and $t = 0$ corresponds to 1990. At this consumption rate, what was the total amount the city used in the 10-year period of the 1990s?

Solution To find the consumption over the 10-year period from 1990 through 1999, use the definite integral.

$$\begin{aligned}\int_0^{10} (t + e^{.01t})\,dt &= \left(\frac{t^2}{2} + \frac{e^{.01t}}{.01}\right)\Bigg|_0^{10} \\ &= (50 + 100e^{.1}) - (0 + 100) \\ &\approx -50 + 100(1.10517) \approx 60.5\end{aligned}$$

Therefore, a total of 60.5 trillion ft^3 of natural gas was used during the 1990s at this consumption rate.

7.4 EXERCISES

Evaluate each definite integral.

1. $\int_{-2}^{4} (-1)\, dp$

2. $\int_{-4}^{1} 6x\, dx$

3. $\int_{-1}^{2} (3t - 1)\, dt$

4. $\int_{-2}^{2} (4z + 3)\, dz$

5. $\int_{0}^{2} (5x^2 - 4x + 2)\, dx$

6. $\int_{-2}^{3} (-x^2 - 3x + 5)\, dx$

7. $\int_{0}^{2} 3\sqrt{4u + 1}\, du$

8. $\int_{3}^{9} \sqrt{2r - 2}\, dr$

9. $\int_{0}^{1} 2(t^{1/2} - t)\, dt$

10. $\int_{0}^{4} -(3x^{3/2} + x^{1/2})\, dx$

11. $\int_{1}^{4} (5y\sqrt{y} + 3\sqrt{y})\, dy$

12. $\int_{4}^{9} (4\sqrt{r} - 3r\sqrt{r})\, dr$

13. $\int_{4}^{6} \frac{2}{(x - 3)^2}\, dx$

14. $\int_{1}^{4} \frac{-3}{(2p + 1)^2}\, dp$

15. $\int_{1}^{5} (5n^{-2} + n^{-3})\, dn$

16. $\int_{2}^{3} (3x^{-2} - x^{-4})\, dx$

17. $\int_{2}^{3} \left(2e^{-.1A} + \frac{3}{A}\right) dA$

18. $\int_{1}^{2} \left(\frac{-1}{B} + 3e^{.2B}\right) dB$

19. $\int_{1}^{2} \left(e^{5u} - \frac{1}{u^2}\right) du$

20. $\int_{.5}^{1} (p^3 - e^{4p})\, dp$

21. $\int_{-1}^{0} y(2y^2 - 3)^5\, dy$

22. $\int_{0}^{3} m^2(4m^3 + 2)^3\, dM$

23. $\int_{1}^{64} \frac{\sqrt{z} - 2}{\sqrt[3]{z}}\, dz$

24. $\int_{1}^{8} \frac{3 - y^{1/3}}{y^{2/3}}\, dy$

25. $\int_{1}^{2} \frac{\ln x}{x}\, dx$

26. $\int_{1}^{3} \frac{\sqrt{\ln x}}{x}\, dx$

27. $\int_{0}^{8} x^{1/3}\sqrt{x^{4/3} + 9}\, dx$

28. $\int_{1}^{2} \frac{3}{x(1 + \ln x)}\, dx$

29. $\int_{0}^{1} \frac{e^t}{(3 + e^t)^2}\, dt$

30. $\int_{0}^{1} \frac{e^{2z}}{\sqrt{1 + e^{2z}}}\, dz$

In Exercises 31–40, use the definite integral to find the area between the x-axis and $f(x)$ over the indicated interval. Check first to see if the graph crosses the x-axis in the given interval.

31. $f(x) = 2x + 3;\quad [8, 10]$

32. $f(x) = 4x - 7;\quad [5, 10]$

33. $f(x) = 2 - 2x^2;\quad [0, 5]$

34. $f(x) = 9 - x^2;\quad [0, 6]$

35. $f(x) = x^3;\quad [-1, 3]$

36. $f(x) = x^3 - 2x;\quad [-2, 4]$

37. $f(x) = e^x - 1;\quad [-1, 2]$

38. $f(x) = 1 - e^{-x};\quad [-1, 2]$

39. $f(x) = \frac{1}{x};\quad [1, e]$

40. $f(x) = \frac{1}{x};\quad [e, e^2]$

Find the area of each shaded region.

41.

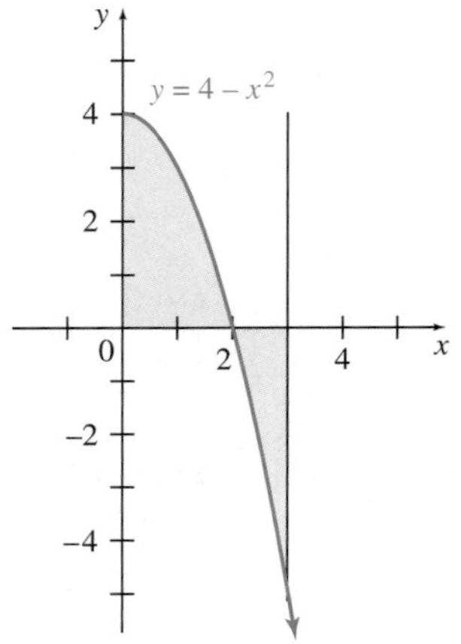

42.

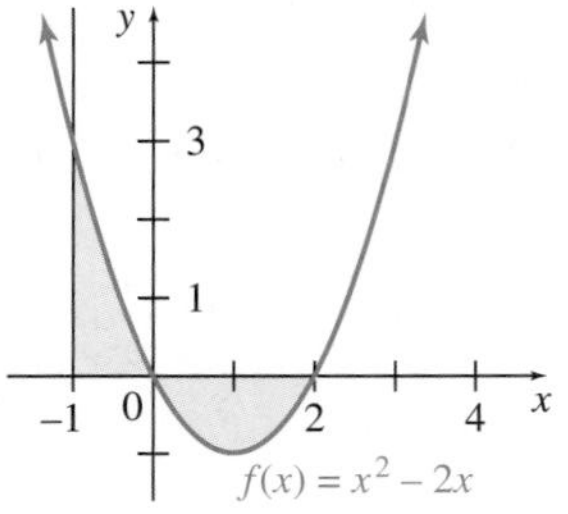

43.

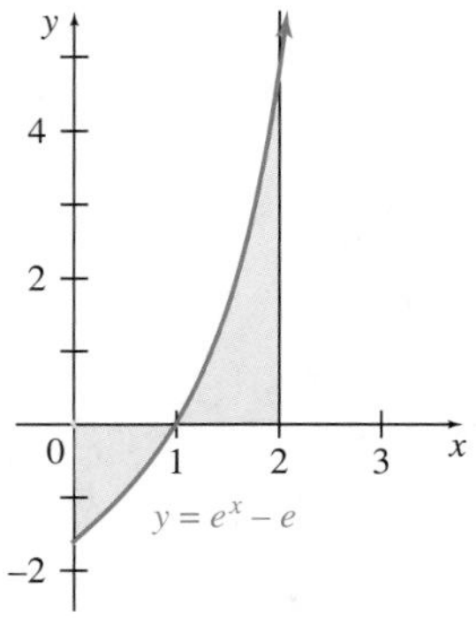

44. Assume $f(x)$ is continuous for $g \leq x \leq c$ as shown in the figure. Write an equation relating the three quantities

$$\int_a^b f(x)\,dx, \qquad \int_a^c f(x)\,dx, \qquad \int_b^c f(x)\,dx.$$

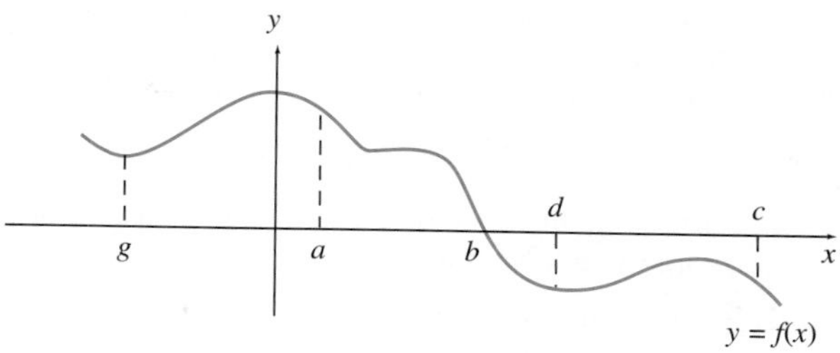

45. Is the equation you wrote for Exercise 44 still true

a. if b is replaced by d?

b. if b is replaced by g?

46. The graph of $f(x)$, shown here, consists of two straight line segments and two quarter circles. Find the value of $\int_0^{16} f(x)\,dx$.

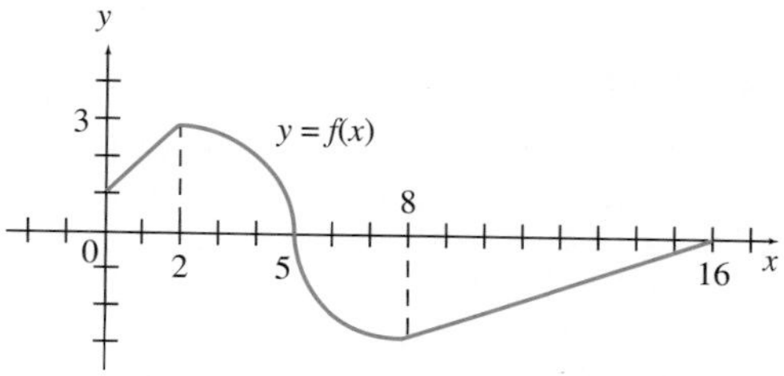

Use the Fundamental Theorem to show that the following are true.

47. $\displaystyle\int_a^b kf(x)\,dx = k\int_a^b f(x)\,dx$

48. $\displaystyle\int_a^b f(x)\,dx = \int_a^c f(x)\,dx + \int_c^b f(x)\,dx$

49. $\displaystyle\int_a^b f(x)\,dx = -\int_b^a f(x)\,dx$

50. Use Exercise 48 to find $\int_{-1}^{4} f(x)\,dx$, given

$$f(x) = \begin{cases} 2x + 3 & \text{if } x \leq 0 \\ -\dfrac{x}{4} - 3 & \text{if } x > 0. \end{cases}$$

51. You are given $\int_0^1 e^{x^2}\,dx; = 1.46265$ and $\int_0^2 e^{x^2}\,dx = 16.45263$. Use this information to find

a. $\displaystyle\int_{-1}^{1} e^{x^2}\,dx;$ **b.** $\displaystyle\int_{1}^{2} e^{x^2}\,dx.$

52. Let $g(t) = t^4$ and define $f(x) = \displaystyle\int_c^x g(t)\,dt$ with $c = 1$.

a. Find a formula for $f(x)$.

b. Verify that $f'(x) = g(x)$. The fact that $\dfrac{d}{dx}\displaystyle\int_c^x g(t)\,dt = g(x)$ is true for all continuous functions g is an alternative version of the Fundamental Theorem of Calculus.

c. Let us verify the result in part b for a function whose antiderivative cannot be found. Let $g(t) = e^{t^2}$ and let $c = 0$. Use the integration feature on a graphing calculator to find $f(x)$ for $x = 1$ and $x = 1.01$. Then use the definition of the derivative with $h = .01$ to approximate $f'(1)$, and compare it with $g(1)$.

Applications

BUSINESS AND ECONOMICS

53. *Profit* Karla Harby Communications, a small company of science writers, found that its rate of profit (in thousands of dollars) after t years of operation is given by

$$P'(t) = (3t + 3)(t^2 + 2t + 2)^{1/3}.$$

a. Find the total profit in the first three years.

b. Find the profit in the fourth year of operation.

c. What is happening to the annual profit over the long run?

54. *Worker Efficiency* A worker new to a job will improve his efficiency with time so that it takes him fewer hours to produce an item with each day on the job, up to a certain point.

Suppose the rate of change of the number of hours it takes a worker in a certain factory to produce the xth item is given by

$$H'(x) = 20 - 2x.$$

a. What is the total number of hours required to produce the first 5 items?

b. What is the total number of hours required to produce the first 10 items?

LIFE SCIENCES

55. *Pollution* Pollution from a factory is entering a lake. The rate of concentration of the pollutant at time t is given by

$$P'(t) = 140t^{5/2},$$

where t is the number of years since the factory started introducing pollutants into the lake. Ecologists estimate that the lake can accept a total level of pollution of 4850 units before all the fish life in the lake ends. Can the factory operate for 4 years without killing all the fish in the lake?

56. *Spread of an Oil Leak* An oil tanker is leaking oil at the rate given (in barrels per hour) by

$$L'(t) = \frac{80\ln(t + 1)}{t + 1},$$

where t is the time (in hours) after the tanker hits a hidden rock (when $t = 0$).

a. Find the total number of barrels that the ship will leak on the first day.

b. Find the total number of barrels that the ship will leak on the second day.

c. What is happening over the long run to the amount of oil leaked per day?

57. *Tree Growth* After long study, tree scientists conclude that a eucalyptus tree will grow at the rate of $.2 + 4t^{-4}$ ft per year, where t is time (in years).

a. Find the number of feet that the tree will grow in the second year.

b. Find the number of feet the tree will grow in the third year.

58. *Growth of a Substance* The rate at which a substance grows is given by

$$R'(x) = 200e^{.2x},$$

where x is the time (in days). What is the total accumulated growth during the first 2.5 days?

59. *Drug Reaction* For a certain drug, the rate of reaction in appropriate units is given by

$$R'(t) = \frac{5}{t} + \frac{2}{t^2},$$

where t is time (in hours) after the drug is administered. Find the total reaction to the drug over the following time periods.

a. From $t = 1$ to $t = 12$ **b.** From $t = 12$ to $t = 24$

60. *Human Mortality* If $f(x)$ is the instantaneous death rate for members of a population at time t, then the number of individuals who survive to age T is given by

$$F(T) = \int_0^T f(x)\,dx.$$

In 1825 the biologist Benjamin Gompertz proposed that $f(x) = kb^x$.* Find a formula for $F(T)$.

61. *Cell Division* Let the expected number of cells in a culture that have an x percent probability of undergoing cell division during the next hour be denoted by $n(x)$.

a. Explain why $\int_{20}^{30} n(x)\,dx$ approximates the total number of cells with a 20 to 30% chance of dividing during the next hour.

b. Give an integral representing the number of cells that have less than a 60% chance of dividing during the next hour.

c. Let $n(x) = \sqrt{5x + 1}$ give the expected number of cells (in millions) with x percent probability of dividing during the next hour. Find the number of cells with a 5 to 10% chance of dividing.

62. *Bacterial Growth* A population of *E. coli* bacteria will grow at a rate given by

$$w'(t) = (4t + 1)^{1/3},$$

where w is the weight (in milligrams) after t hours. Find the change in weight of the population from $t = 0$ to $t = 3$.

63. *Blood Flow* In an example from an earlier chapter, the velocity v of the blood in a blood vessel was given as

$$v = k(R^2 - r^2),$$

where R is the (constant) radius of the blood vessel, r is the distance of the flowing blood from the center of the blood vessel, and k is a constant. Total blood flow (in millimeters per minute) is given by

$$Q(R) = \int_0^R 2\pi vr\,dr.$$

*Gompertz, Benjamin, "On the Nature of the Function Expressive of the Law of Human Mortality," *Philosophical Transactions of the Royal Society of London,* 1825.

a. Find the general formula for Q in terms of R by evaluating the definite integral given above.

b. Evaluate $Q(.4)$.

64. *Rams' Horns* The average annual increment in the horn length (in centimeters) of bighorn rams born since 1986 can be approximated by

$$y = .1762x^2 - 3.986x + 22.68,$$

where x is the ram's age (in years) for x between 3 and 9.* Integrate to find the total increase in the length of a ram's horn during this time.

65. *Beagles* The daily energy requirements of female beagles who are at least 1 year old change with respect to time according to the function

$$E(t) = 753t^{-.1321},$$

where $E(t)$ is the daily energy requirement (in $\text{kJ}/W^{.67}$), where W is the dog's weight (in kilograms) for a beagle that is t years old.†

a. Assuming 365 days in a year, show that the energy requirement for a female beagle that is t days old is given by

$$E(t) = 1642t^{-.1321}.$$

b. Using the formula from part a, determine the total energy requirements (in $\text{kJ}/W^{.67}$) for a female beagle between her first and third birthday.

66. *Sediment* The density of sediment (in grams per cubic centimeter) at the bottom of Lake Coeur d'Alene, Idaho, is given by

$$p(x) = p_0e^{.0133x},$$

where x is the depth (in centimeters) and p_0 is the density at the surface.‡ The total mass of a square-centimeter column of sediment above a depth of h cm is given by

$$\int_0^h p(x)\,dx.$$

If $p_0 = .85$ g per cm^3, find the total mass above a depth of 100 cm.

SOCIAL SCIENCES

67. *Age Distribution* The 2000 U.S. census gives us an age distribution that is approximately given (in millions) by the function

$$f(x) = 40.1 + 2.03x - .741x^2,$$

where x varies from 0 to 9 decades.§ The population of a given age group can be found by integrating this function over the interval for that age group.

a. Find the integral over the interval $[0, 9]$. What does this integral represent?

b. Baby boomers are those born between 1945 and 1965, that is, those in the range of 3.5 to 5.5 decades in 2000. Find the number of baby boomers.

68. *Income Distribution* Based on 2000 census data, an approximate income distribution for the United States is given by the function

$$f(x) = .058x^3 - 1.08x^2 + 4.81x + 6.26$$

where x is annual income in units of \$10,000, $0 \le x \le 10$.§ For example, $x = .5$ represents an annual income of \$5000. (*Note:* This function does not give a good representation for incomes over \$100,000.) The percent of the population with an income in a given range can be found by integrating this function over that range. Find the percentage of the population with an income between \$25,000 and \$50,000.

PHYSICAL SCIENCES

69. *Oil Consumption* Suppose that the rate of consumption of a natural resource is $c'(t)$, where

$$c'(t) = ke^{rt}.$$

Here t is time in years, r is a constant, and k is the consumption in the year when $t = 0$. In 2003, an oil company sold 1.2 billion barrels of oil. Assume that $r = .04$.

a. Write $c'(t)$ for the oil company, letting $t = 0$ represent 2003.

*Jorgenson, Jon T., et al., "Effects of Population Density on Horn Development in Bighorn Rams," *Journal of Wildlife Management,* Vol. 62, No. 3, 1998, pp. 1011–1020.

†Finke, M., "Energy Requirements of Adult Female Beagles," *Journal of Nutrition,* Vol. 124, 1994, pp. 2604s–2608s.

‡Nord, Gail and John Nord, "Sediment in Lake Coeur d'Alene, Idaho," *Mathematics Teacher,* Vol. 91, No. 4, April 1998, pp. 292–295.

§Exercises 67 and 68 were originally contributed by Ralph DeMarr, University of New Mexico.

b. Set up a definite integral for the amount of oil that the company will sell in the next 10 years.

c. Evaluate the definite integral of part b.

d. The company has about 20 billion barrels of oil in reserve. To find the number of years that this amount will last, solve the equation

$$\int_0^T 1.2e^{.04t}\,dt = 20.$$

e. Rework part d, assuming that $r = .02$.

70. *Oil Consumption* The rate of consumption of oil (in billions of barrels) by the company in Exercise 69 was given as

$$1.2e^{.04t},$$

where $t = 0$ corresponds to 2003. Find the total amount of oil used by the company from 2003 to year T. At this rate, how much will be used in 5 years?

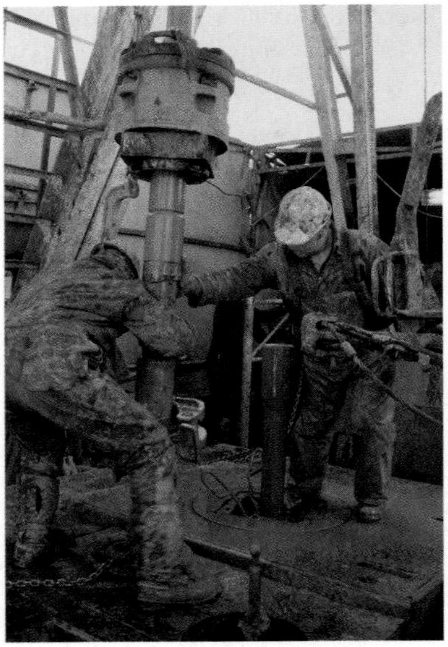

7.5 THE AREA BETWEEN TWO CURVES

THINK ABOUT IT *If an executive knows how the savings from a new manufacturing process decline over time and how the costs of that process will increase, how can she compute when the net savings will cease and what the total savings will be?*

This section shows a method for answering such questions.

Many important applications of integrals require finding the area between two graphs. The method used in previous sections to find the area between the graph of a function and the x-axis from $x = a$ to $x = b$ can be generalized to find such an area. For example, the area between the graphs of $f(x)$ and $g(x)$ from $x = a$ to $x = b$ in Figure 20(a) is the same as the difference of the area from a to b between $f(x)$ and the x-axis, shown in Figure 20(b), and the area from a to b

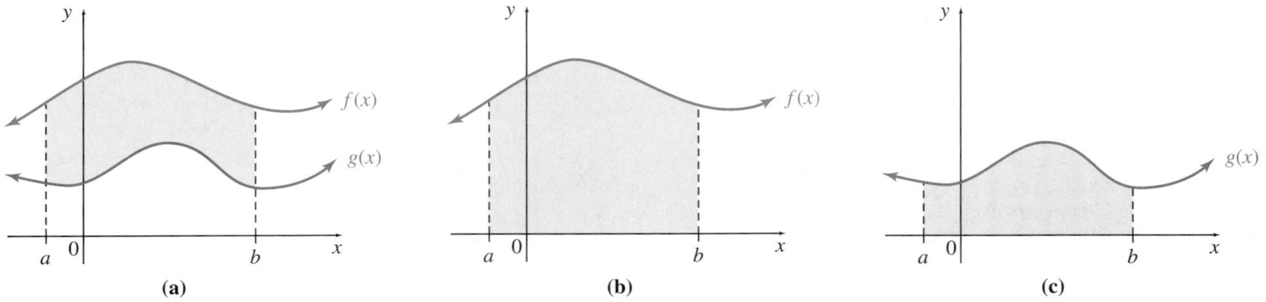

FIGURE 20

between $g(x)$ and the x-axis (see Figure 20(c)). That is, the area between the graphs is given by

$$\int_a^b f(x)\,dx - \int_a^b g(x)\,dx,$$

which can be written as

$$\int_a^b [f(x) - g(x)]\,dx.$$

AREA BETWEEN TWO CURVES

If f and g are continuous functions and $f(x) \geq g(x)$ on $[a, b]$, then the area between the curves $f(x)$ and $g(x)$ from $x = a$ to $x = b$ is given by

$$\int_a^b [f(x) - g(x)]\,dx.$$

EXAMPLE 1 *Area*

Find the area bounded by $f(x) = -x^2 + 1$, $g(x) = 2x + 4$, $x = -1$, and $x = 2$.

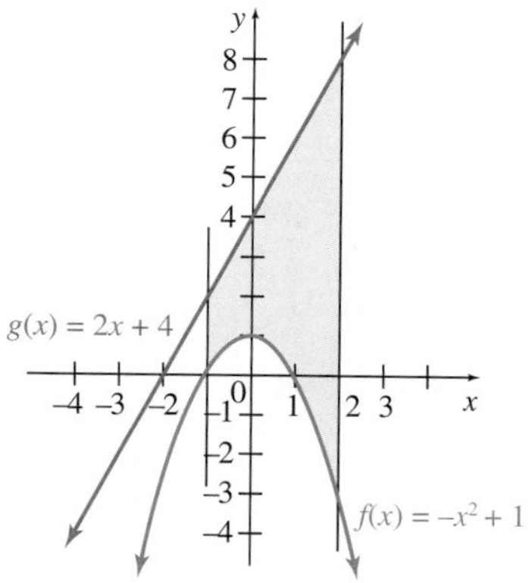

FIGURE 21

Solution A sketch of the four equations is shown in Figure 21. In general, it is not necessary to spend time drawing a detailed sketch, but only to know whether the two functions intersect, and which function is greater between the intersections. To find out, set the two functions equal.

$$-x^2 + 1 = 2x + 4$$
$$0 = x^2 + 2x + 3$$

Verify by the quadratic formula that this equation has no roots. Since the graph of f is a parabola opening downward that does not cross the graph of g (a line), the parabola must be entirely under the line, as shown in Figure 21. Therefore $g(x) \geq f(x)$ for x in the interval $[-1, 2]$, and the area is given by

$$\begin{aligned}
\int_{-1}^{2} [g(x) - f(x)]\,dx &= \int_{-1}^{2} (2x + 4) - (-x^2 + 1)\,dx \\
&= \int_{-1}^{2} (2x + 4 + x^2 - 1)\,dx \\
&= \int_{-1}^{2} (x^2 + 2x + 3)\,dx \\
&= \frac{x^3}{3} + x^2 + 3x \Big|_{-1}^{2} \\
&= \left(\frac{8}{3} + 4 + 6\right) - \left(\frac{-1}{3} + 1 - 3\right) \\
&= \frac{8}{3} + 10 + \frac{1}{3} + 2 \\
&= 15.
\end{aligned}$$

NOTE It is not necessary to draw the graphs to determine which function is greater. Since the functions in the previous example do not intersect, we can evaluate them at *any* point to make this determination. For example, $f(0) = 1$ and $g(0) = 4$. Because $g(x) > f(x)$ at $x = 4$, and the two functions are continuous and never intersect, $g(x) > f(x)$ for all x. ■

EXAMPLE 2 *Area*

Find the area between the curves $y = x^{1/2}$ and $y = x^3$.

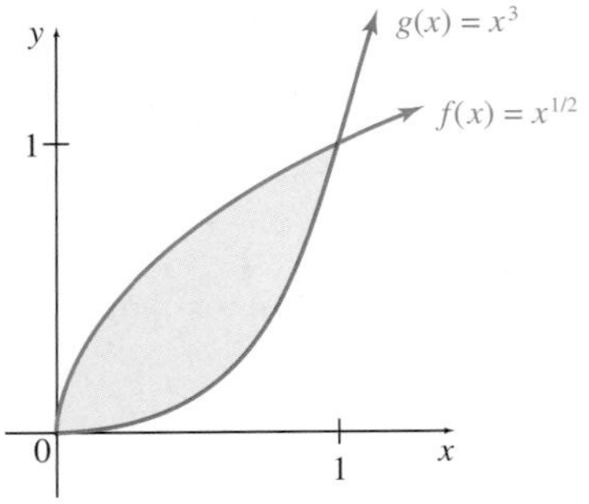

FIGURE 22

Solution Let $f(x) = x^{1/2}$ and $g(x) = x^3$. As before, set the two equal to find where they intersect.

$$x^{1/2} = x^3$$
$$0 = x^3 - x^{1/2}$$
$$0 = x^{1/2}(x^{5/2} - 1)$$

The only solutions are $x = 0$ and $x = 1$. Verify that the graph of f is concave downward, while the graph of g is concave upward, so the graph of f must be greater between 0 and 1. (This may also be verified by taking a point between 0 and 1, such as .5, and verifying that $.5^{1/2} > .5^3$.) The graph is shown in Figure 22.

The area between the two curves is given by

$$\int_a^b [f(x) - g(x)]\, dx = \int_0^1 (x^{1/2} - x^3)\, dx.$$

Using the Fundamental Theorem,

$$\begin{aligned}\int_0^1 (x^{1/2} - x^3)\, dx &= \left(\frac{x^{3/2}}{3/2} - \frac{x^4}{4}\right)\Bigg|_0^1 \\ &= \left(\frac{2}{3}x^{3/2} - \frac{x^4}{4}\right)\Bigg|_0^1 \\ &= \frac{2}{3}(1) - \frac{1}{4} \\ &= \frac{5}{12}.\end{aligned}$$

A graphing calculator is very useful in approximating solutions of problems involving the area between two curves. First, it can be used to graph the functions and identify any intersection points. Then it can be used to approximate the definite integral that represents the area. (A function that gives the numerical integral is located in the MATH menu of a TI-83/84 Plus calculator.) Figure 23 on the next page shows the results of using these steps for Example 2. The second window shows that the area closely approximates $5/12$.

The difference between two integrals can be used to find the area between the graphs of two functions even if one graph lies below the x-axis. In fact, if $f(x) \geq g(x)$ for all values of x in the interval $[a, b]$, then the area between the two graphs is always given by

$$\int_a^b [f(x) - g(x)]\, dx.$$

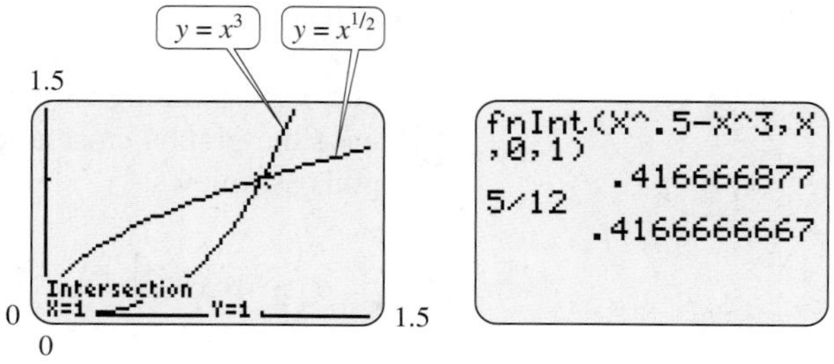

FIGURE 23

To see this, look at the graphs in Figure 24(a), where $f(x) \geq g(x)$ for x in $[a, b]$. Suppose a constant C is added to both functions, with C large enough so that both graphs lie above the x-axis, as in Figure 24(b). The region between the graphs is not changed. By the work above, this area is given by $\int_a^b [f(x) - g(x)]dx$ regardless of where the graphs of $f(x)$ and $g(x)$ are located. As long as $f(x) \geq g(x)$ on $[a, b]$, then the area between the graphs from $x = a$ to $x = b$ will equal $\int_a^b [f(x) - g(x)]dx$.

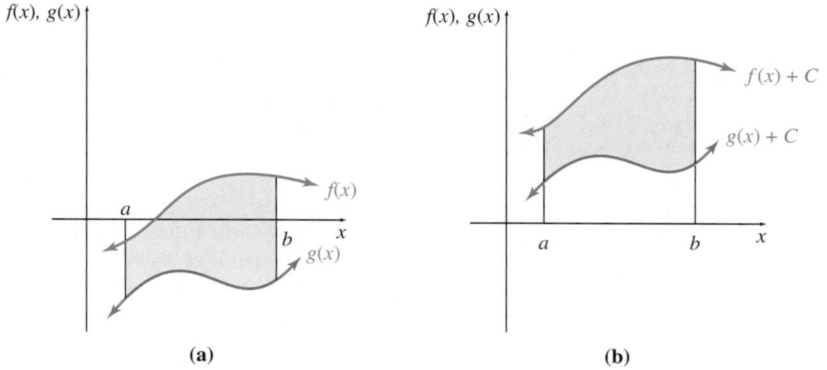

FIGURE 24

EXAMPLE 3 *Area*

Find the area of the region enclosed by $y = x^2 - 2x$ and $y = x$ on $[0, 4]$.

Solution Verify that the two graphs cross at $x = 0$ and $x = 3$. Because the first graph is a parabola opening upward, the parabola must be below the line between 0 and 3 and above the line between 3 and 4. See Figure 25. (The greater function

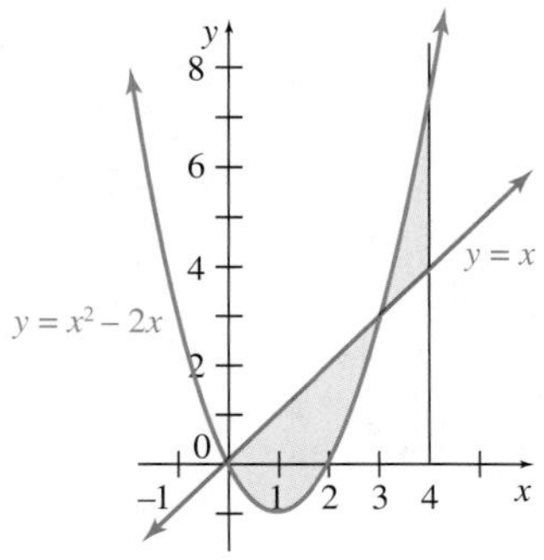

FIGURE 25

could also be identified by checking a point between 0 and 3, such as 1, and a point between 3 and 4, such as 3.5. For each of these values of x, we could calculate the corresponding value of y for the two functions and see which is greater.) Because the graphs cross at $x = 3$, the area is found by taking the sum of two integrals as follows.

$$\begin{aligned}
\text{Area} &= \int_0^3 [x - (x^2 - 2x)]\,dx + \int_3^4 [(x^2 - 2x) - x]\,dx \\
&= \int_0^3 (-x^2 + 3x)\,dx + \int_3^4 (x^2 - 3x)\,dx \\
&= \left(\frac{-x^3}{3} + \frac{3x^2}{2}\right)\Bigg|_0^3 + \left(\frac{x^3}{3} - \frac{3x^2}{2}\right)\Bigg|_3^4 \\
&= \left(-9 + \frac{27}{2} - 0\right) + \left(\frac{64}{3} - 24 - 9 + \frac{27}{2}\right) \\
&= \frac{19}{3}
\end{aligned}$$

In the remainder of this section we will consider some typical applications that require finding the area between two curves.

EXAMPLE 4 *Savings*

A company is considering a new manufacturing process in one of its plants. The new process provides substantial initial savings, with the savings declining with time x (in years) according to the rate-of-savings function

$$S(x) = 100 - x^2,$$

where $S(x)$ is in thousands of dollars per year. At the same time, the cost of operating the new process increases with time x (in years), according to the rate-of-cost function (in thousands of dollars per year)

$$C(x) = x^2 + \frac{14}{3}x.$$

(a) For how many years will the company realize savings?

Solution Figure 26 shows the graphs of the rate-of-savings and rate-of-cost functions. The rate of cost (marginal cost) is increasing, while the rate of savings (marginal savings) is decreasing. The company should use this new process until the difference between these quantities is zero; that is, until the time at which these graphs intersect. The graphs intersect when

$$C(x) = S(x),$$

or

$$x^2 + \frac{14}{3}x = 100 - x^2.$$

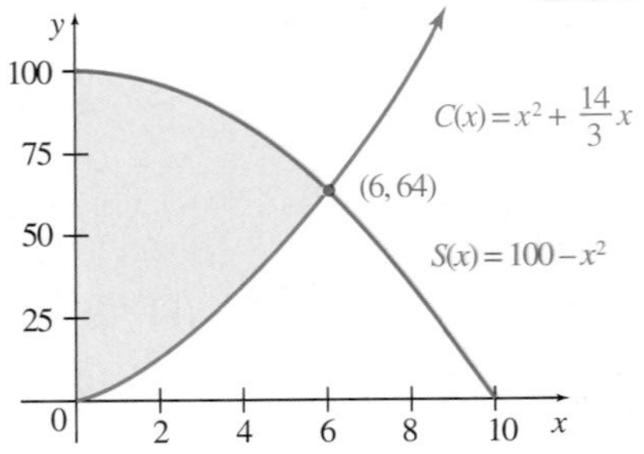

FIGURE 26

Solve this equation as follows.

$$0 = 2x^2 + \frac{14}{3}x - 100$$

$$0 = 3x^2 + 7x - 150 \qquad \text{Multiply by } \tfrac{3}{2}.$$

$$= (x - 6)(3x + 25) \qquad \text{Factor.}$$

Set each factor equal to 0 separately to get

$$x = 6 \qquad \text{or} \qquad x = -25/3.$$

Only 6 is a meaningful solution here. The company should use the new process for 6 years.

(b) What will be the net total savings during this period?

Solution Since the total savings over the 6-year period is given by the area under the rate-of-savings curve and the total additional cost by the area under the rate-of-cost curve, the net total savings over the 6-year period is given by the area between the rate-of-cost and the rate-of-savings curves and the lines $x = 0$ and $x = 6$. This area can be evaluated with a definite integral as follows.

$$\text{Total savings} = \int_0^6 \left[(100 - x^2) - \left(x^2 + \frac{14}{3}x \right) \right] dx$$

$$= \int_0^6 \left(100 - \frac{14}{3}x - 2x^2 \right) dx$$

$$= \left(100x - \frac{7}{3}x^2 - \frac{2}{3}x^3 \right) \Bigg|_0^6$$

$$= 100(6) - \frac{7}{3}(36) - \frac{2}{3}(216) = 372$$

The company will save a total of \$372,000 over the 6-year period.

The answer to a problem will not always be an integer. Suppose in solving the quadratic equation in Example 4 we found the solutions to be $x = 6.7$ and $x = -7.3$. It may not be realistic to use a new process for 6.7 years; it may be necessary to choose between 6 years and 7 years. Since the mathematical model produces a result that is not in the domain of the function in this case, it is necessary to find the total savings after 6 years and after 7 years and then select the best result.

Consumers' Surplus The market determines the price at which a product is sold. As indicated earlier, the point of intersection of the demand curve and the supply curve for a product gives the equilibrium price. At the equilibrium price, consumers will purchase the same amount of the product that the manufacturers want to sell. Some consumers, however, would be willing to spend more for an item than the equilibrium price. The total of the differences between the equilibrium price of the item and the higher prices that individuals would be willing to

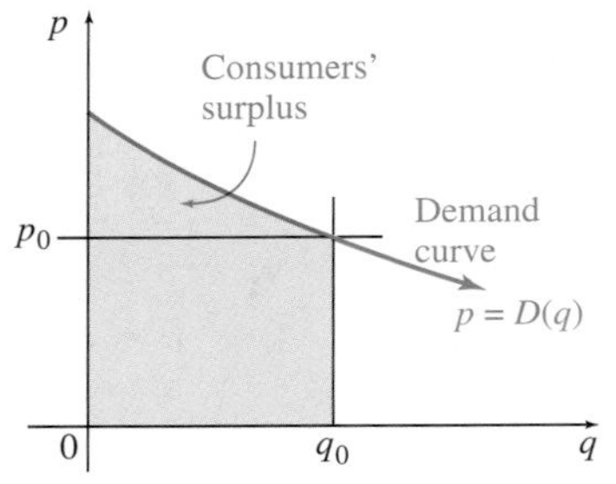

FIGURE 27

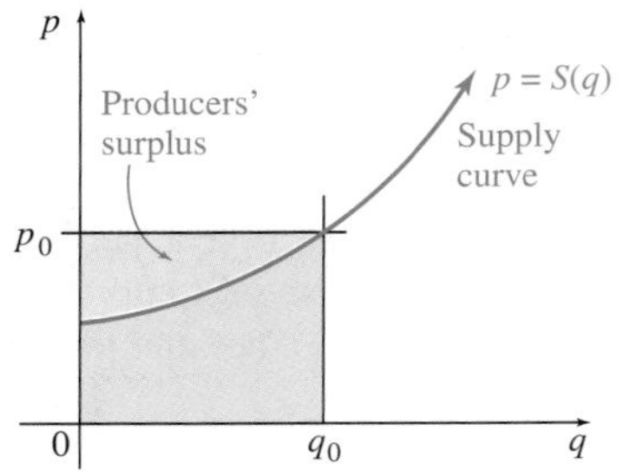

FIGURE 28

pay is thought of as savings realized by those individuals and is called the **consumers' surplus.**

In Figure 27, the area under the demand curve is the total amount consumers are willing to spend for q_0 items. The green shaded area under the line $y = p_0$ shows the total amount consumers actually will spend at the equilibrium price of p_0. The pink shaded area represents the consumers' surplus. As the figure suggests, the consumers' surplus is given by an area between the two curves $p = D(q)$ and $p = p_0$, so its value can be found with a definite integral as follows.

CONSUMERS' SURPLUS

If $D(q)$ is a demand function with equilibrium price p_0 and equilibrium demand q_0, then

$$\textbf{Consumers' surplus} = \int_0^{q_0} [D(q) - p_0]\, dq.$$

Similarly, if some manufacturers would be willing to supply a product at a price *lower* than the equilibrium price p_0, the total of the differences between the equilibrium price and the lower prices at which the manufacturers would sell the product is considered added income for the manufacturers and is called the **producers' surplus.** Figure 28 shows the (green shaded) total area under the supply curve from $q = 0$ to $q = q_0$, which is the minimum total amount the manufacturers are willing to realize from the sale of q_0 items. The total area under the line $p = p_0$ is the amount actually realized. The difference between these two areas, the producers' surplus, is also given by a definite integral.

PRODUCERS' SURPLUS

If $S(q)$ is a supply function with equilibrium price p_0 and equilibrium supply q_0, then

$$\textbf{Producers' surplus} = \int_0^{q_0} [p_0 - S(q)]\, dq.$$

EXAMPLE 5 *Consumers' and Producers' Surplus*

Suppose the price (in dollars per ton) for oat bran is

$$D(q) = 400 - e^{q/2},$$

when the demand for the product is q tons. Also, suppose the function

$$S(q) = e^{q/2} - 1$$

gives the price (in dollars per ton) when the supply is q tons. Find the consumers' surplus and the producers' surplus.

Solution Begin by finding the equilibrium quantity. This is done by setting the two equations equal.

$$\begin{aligned} e^{q/2} - 1 &= 400 - e^{q/2} \\ 2e^{q/2} &= 401 \\ e^{q/2} &= \frac{401}{2} \\ q/2 &= \ln\left(\frac{401}{2}\right) \\ q &= 2\ln\left(\frac{401}{2}\right) \approx 10.60163 \end{aligned}$$

The result can be further rounded to 10.60 tons as long as this rounded value is not used in future calculations. At the equilibrium point where the supply and demand are both 10.60 tons, the price is

$$S(10.60163) = e^{10.60163/2} - 1 = 199.50,$$

or \$199.50. Verify that this same answer is found by computing $D(10.60163)$. The consumers' surplus, represented by the area shown in Figure 29, is

$$\int_0^{10.60163} [(400 - e^{q/2}) - 199.50]\,dq = \int_0^{10.60163} [200.5 - e^{q/2}]\,dq.$$

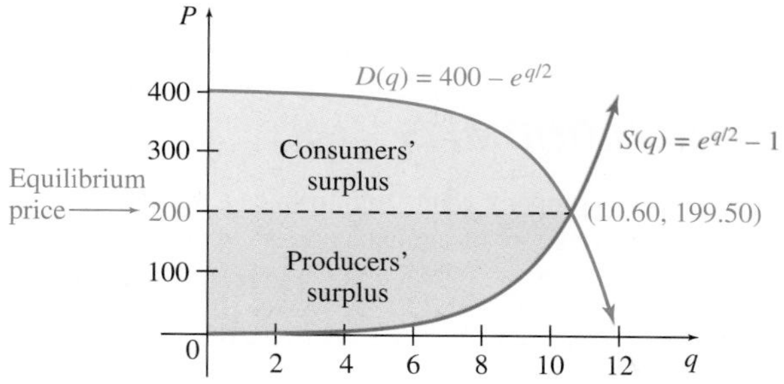

FIGURE 29

Evaluating the definite integral gives

$$\begin{aligned} (200.5q - 2e^{q/2})\Big|_0^{10.60163} &= [200.5(10.60163) - 2e^{10.60163/2}] - (0 - 2) \\ &\approx 1726.63. \end{aligned}$$

Here the consumers' surplus is \$1726.63. The producers' surplus, also shown in Figure 29, is given by

$$\int_0^{10.60163} [199.50 - (e^{q/2} - 1)]\,dq = \int_0^{10.60163} [200.5 - e^{q/2}]\,dq,$$

which is exactly the same as the expression found for the consumers' surplus, so the producers' surplus is also \$1726.63.

NOTE In general, the producers' surplus and consumers' surplus are not the same, as they are in Example 5. ■

7.5 EXERCISES

Find the area between the curves in Exercises 1–22.

1. $x = -2, \quad x = 1, \quad y = x^2 + 4, \quad y = 0$

2. $x = 1, \quad x = 2, \quad y = x^3, \quad y = 0$

3. $x = -3, \quad x = 1, \quad y = x + 1, \quad y = 0$

4. $x = -2, \quad x = 0, \quad y = 1 - x^2, \quad y = 0$

5. $x = -2, \quad x = 1, \quad y = 2x, \quad y = x^2 - 3$

6. $x = 0, \quad x = 6, \quad y = 5x, \quad y = 3x + 10$

7. $y = x^2 - 30, \quad y = 10 - 3x$

8. $y = x^2 - 18, \quad y = x - 6$

9. $y = x^2, \quad y = 2x$

10. $y = x^2, \quad y = x^3$

11. $x = 1, \quad x = 6, \quad y = \frac{1}{x}, \quad y = -1$

12. $x = 0, \quad x = 4, \quad y = \frac{1}{x + 1}, \quad y = \frac{x - 1}{2}$

13. $x = -1, \quad x = 1, \quad y = e^x, \quad y = 3 - e^x$

14. $x = -1, \quad x = 2, \quad y = e^{-x}, \quad y = e^x$

15. $x = 1, \quad x = 2, \quad y = e^x, \quad y = \frac{1}{x}$

16. $x = 2, \quad x = 4, \quad y = \frac{x}{2} + 3, \quad y = \frac{1}{x - 1}$

17. $y = x^3 - x^2 + x + 1, \quad y = 2x^2 - x + 1$

18. $y = 2x^3 + x^2 + x + 5, \quad y = x^3 + x^2 + 2x + 5$

19. $y = x^4 + \ln(x + 10), \quad y = x^3 + \ln(x + 10)$

20. $y = x^5 - 2\ln(x + 5), \quad y = x^3 - 2\ln(x + 5)$

21. $y = x^{4/3}, \quad y = 2x^{1/3}$

22. $y = \sqrt{x}, \quad y = x\sqrt{x}$

In Exercises 23 and 24, use a graphing calculator to find the values of x where the curves intersect, and then to find the area between the two curves.

23. $y = e^x, \quad y = -x^2 - 2x$

24. $y = \ln x, \quad y = x^3 - 5x^2 + 6x - 1$

Applications

BUSINESS AND ECONOMICS

25. *Net Savings* Suppose a company wants to introduce a new machine that will produce a rate of annual savings (in dollars) given by

$$S(x) = 150 - x^2,$$

where x is the number of years of operation of the machine, while producing a rate of annual costs (in dollars) of

$$C(x) = x^2 + \frac{11}{4}x.$$

a. For how many years will it be profitable to use this new machine?

b. What are the net total savings during the first year of use of the machine?

c. What are the net total savings over the entire period of use of the machine?

26. *Net Savings* A new smog-control device will reduce the output of sulfur oxides from automobile exhausts. It is estimated that the rate of savings to the community from the use of this device will be approximated by

$$S(x) = -x^2 + 4x + 8,$$

where $S(x)$ is the rate of savings (in millions of dollars) after x years of use of the device. The new device cuts down on the production of sulfur oxides, but it causes an increase in the production of nitrous oxides. The rate of additional costs (in millions) to the community after x years is approximated by

$$C(x) = \frac{3}{25}x^2.$$

a. For how many years will it pay to use the new device?

b. What will be the net savings over this period of time?

27. *Profit* De Win Enterprises had an expenditure rate of $E(x) = e^{.1x}$ dollars per day and an income rate of $I(x) = 98.8 - e^{.1x}$ dollars per day on a particular job, where x was the number of days from the start of the job. The company's profit on that job will equal total income less total expenditures. Profit will be maximized if the job ends at the optimum time, which is the point where the two curves meet. Find the following.

a. The optimum number of days for the job to last

b. The total income for the optimum number of days

c. The total expenditures for the optimum number of days

d. The maximum profit for the job

28. *Net Savings* A factory of Marisa Raffaele Industries has installed a new process that will produce an increased rate of revenue (in thousands of dollars per year) of

$$R(t) = 104 - .4e^{t/2},$$

where t is time measured in years. The new process produces additional costs (in thousands of dollars per year) at the rate of

$$C(t) = .3e^{t/2}.$$

a. When will it no longer be profitable to use this new process?

b. Find the net total savings.

29. *Producers' Surplus* Find the producers' surplus if the supply function for pork bellies is given by

$$S(q) = q^{5/2} + 2q^{3/2} + 50.$$

Assume supply and demand are in equilibrium at $q = 16$.

30. *Producers' Surplus* Suppose the supply function for concrete is given by

$$S(q) = 100 + 3q^{3/2} + q^{5/2},$$

and that supply and demand are in equilibrium at $q = 9$. Find the producers' surplus.

31. *Consumers' Surplus* Find the consumers' surplus if the demand function for grass seed is given by

$$D(q) = \frac{100}{(3q + 1)^2},$$

assuming supply and demand are in equilibrium at $q = 3$.

32. *Consumers' Surplus* Find the consumers' surplus if the demand function for extra virgin olive oil is given by

$$D(q) = \frac{16{,}000}{(2q + 8)^3},$$

and if supply and demand are in equilibrium at $q = 6$.

33. *Consumers' and Producers' Surplus* Suppose the supply function for oil is given (in dollars) by

$$S(q) = q^2 + 10q,$$

and the demand function is given (in dollars) by

$$D(q) = 900 - 20q - q^2.$$

a. Graph the supply and demand curves.

b. Find the point at which supply and demand are in equilibrium.

c. Find the consumers' surplus.

d. Find the producers' surplus.

34. *Consumers' and Producers' Surplus* Suppose the supply function for a certain item is given by

$$S(q) = (q + 1)^2,$$

and the demand function is given by

$$D(q) = \frac{1000}{q + 1}.$$

a. Graph the supply and demand curves.

b. Find the point at which supply and demand are in equilibrium.

c. Find the consumers' surplus.

d. Find the producers' surplus.

35. *Consumers' and Producers' Surplus* Suppose that with the supply and demand for oil as in Exercise 33, the government sets the price at \$264 per unit.

a. Use the supply function to calculate the quantity that will be produced at the new price.

b. Find the consumers' surplus for the new price, using the quantity found in part a in place of the equilibrium quantity. How much larger is this than the consumers' surplus in Exercise 33?

c. Find the producers' surplus for the new price, using the quantity found in part a in place of the equilibrium quantity. How much smaller is this than the producers' surplus in Exercise 33?

d. Calculate the difference between the total of the consumers' and producers' surplus under the equilibrium price and under the government price. Economists refer to this loss as the *welfare cost* of the government's setting the price.

e. Because of the welfare cost calculated in part d, many economists argue that it is bad economics for the government to set prices. Others point to the increase in the consumers' surplus, calculated in part b, as a justification for such government action. Discuss the pros and cons of this issue.

36. *Fuel Economy* In an article in the December 1994 *Scientific American* magazine, the authors estimated future gas use.* Without a change in U.S. policy, auto fuel use is forecasted to rise along the projection shown at the right in the figure on the next page. The shaded band predicts gas use if the technologies for increased fuel economy are phased in by the year 2010. The moderate estimate (center curve) corresponds to an average of 46 mpg for all cars on the road. Discuss the interpretation of the shaded area and other regions of the graph that pertain to the topic in this section.

*DeCicco, John and Marc Ross, "Improving Automotive Efficiency," *Scientific American,* Vol. 271, No. 6, Dec. 1994, p. 56.

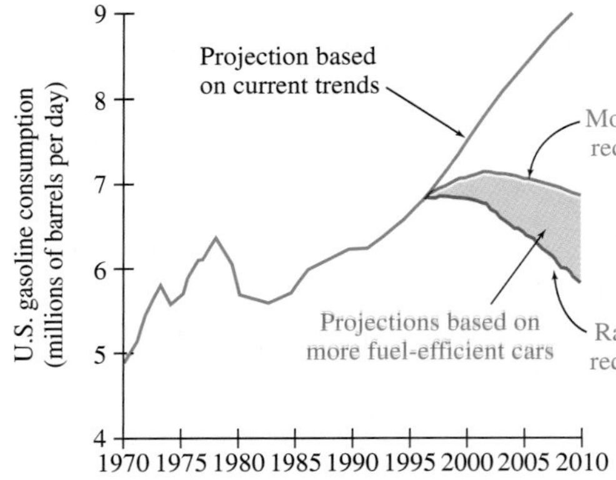

LIFE SCIENCES

37. *Pollution* Pollution begins to enter a lake at time $t = 0$ at a rate (in gallons per hour) given by the formula

$$f(t) = 10(1 - e^{-.5t}),$$

where t is the time (in hours). At the same time, a pollution filter begins to remove the pollution at a rate

$$g(t) = .4t$$

as long as pollution remains in the lake.

a. How much pollution is in the lake after 12 hours?

b. Use a graphing calculator to find the time when the rate that pollution enters the lake equals the rate the pollution is removed.

c. Find the amount of pollution in the lake at the time found in part b.

d. Use a graphing calculator to find the time when all the pollution has been removed from the lake.

38. *Pollution* Repeat the steps of Exercise 37, using the functions

$$f(t) = 15(1 - e^{-.05t})$$

and

$$g(t) = .3t.$$

SOCIAL SCIENCES

39. *Distribution of Income* Suppose that all the people in a country are ranked according to their incomes, starting at the bottom. Let x represent the fraction of the community making the lowest income $(0 \leq x \leq 1)$; $x = .4$, therefore, represents the lower 40% of all income producers. Let $I(x)$ represent the proportion of the total income earned by the lowest x of all people. Thus, $I(.4)$ represents the fraction of total income earned by the lowest 40% of the population. The curve described by this function is known as a *Lorenz curve*. Suppose

$$I(x) = .9x^2 + .1x.$$

Find and interpret the following.

a. $I(.1)$ **b.** $I(.4)$

If income were distributed uniformly, we would have $I(x) = x$. The area under this line of complete equality is $1/2$. As $I(x)$ dips further below $y = x$, there is less equality of income distribution. This inequality can be quantified by the ratio of the area between $I(x)$ and $y = x$ to $1/2$. This ratio is called the *Gini index of income inequality* and equals $2\int_0^1[x - I(x)]\,dx$.

c. Graph $I(x) = x$ and $I(x) = .9x^2 + .1x$, for $0 \leq x \leq 1$, on the same axes.

d. Find the area between the curves.

e. For U.S. families, the Gini index was .4 in 1968 and .46 in 2001.* Describe how the distribution of family incomes has changed over this time.

PHYSICAL SCIENCES

40. *Metal Plate* A worker sketches the curves $y = \sqrt{x}$ and $y = x/2$ on a sheet of metal and cuts out the region between the curves to form a metal plate. Find the area of the plate.

7.6 NUMERICAL INTEGRATION

THINK ABOUT IT

If the velocity of a vehicle is known only at certain points in time, how can the total distance traveled by the vehicle be estimated?

Using numerical integration, we will answer this question in Example 3 of this section.

*http://www.wsws.org/articles/2001/nov2001/inco-n09.shtml.

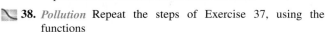

Some integrals cannot be evaluated by any technique. One solution to this problem was presented in Section 3 of this chapter, in which the area under a curve was approximated by summing the areas of rectangles. This method is seldom used in practice because better methods exist that are more accurate for the same amount of work. These methods are referred to as **numerical integration** methods. We discuss two such methods here: the trapezoidal rule and Simpson's rule.

Trapezoidal Rule Recall, the trapezoidal rule was mentioned briefly in Section 3, where we found approximations with it by averaging the sums of rectangles found by using left endpoints and then using right endpoints. In this section we derive an explicit formula for the trapezoidal rule in terms of function values.* To illustrate the derivation of the trapezoidal rule, consider the integral

$$\int_1^5 \frac{1}{x}\,dx.$$

The shaded region in Figure 30 shows the area representing that integral, the area under the graph $f(x) = 1/x$, above the x-axis, and between the lines $x = 1$ and $x = 5$. As shown in the figure, if the area under the curve is approximated with trapezoids rather than rectangles, the approximation should be improved.

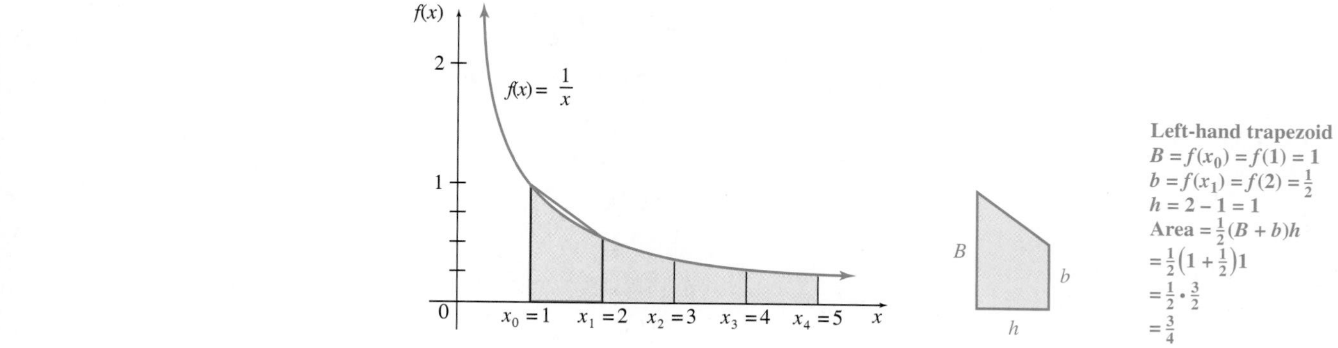

FIGURE 30

Since $\int(1/x)\,dx = \ln|x| + C$,

$$\int_1^5 \frac{1}{x}\,dx = \ln|x|\Big|_1^5 = \ln 5 - \ln 1 = \ln 5 - 0 = \ln 5 \approx 1.609438.$$

As in earlier work, to approximate this area we divide the interval $[1, 5]$ into subintervals of equal widths. To get a first approximation to ln 5 by the trapezoidal rule, find the sum of the areas of the four trapezoids shown in Figure 30. From geometry, the area of a trapezoid is half the product of the sum of the bases and the altitude. Each of the trapezoids in Figure 30 has altitude 1. (In this case,

*In American English a trapezoid is a four-sided figure with two parallel sides, contrasted with a trapezium, which has no parallel sides. In British English, however, it is just the opposite. What Americans call a trapezoid is called a trapezium in Great Britain.

the bases of the trapezoid are vertical and the altitudes are horizontal.) Adding the areas gives

$$\ln 5 = \int_1^5 \frac{1}{x}\,dx \approx \frac{1}{2}\left(\frac{1}{1}+\frac{1}{2}\right)(1) + \frac{1}{2}\left(\frac{1}{2}+\frac{1}{3}\right)(1) + \frac{1}{2}\left(\frac{1}{3}+\frac{1}{4}\right)(1) + \frac{1}{2}\left(\frac{1}{4}+\frac{1}{5}\right)(1)$$

$$= \frac{1}{2}\left(\frac{3}{2}+\frac{5}{6}+\frac{7}{12}+\frac{9}{20}\right) \approx 1.68333.$$

To get a better approximation, divide the interval $[1, 5]$ into more subintervals. Generally speaking, the larger the number of subintervals, the better the approximation. The results for selected values of n are shown below to 5 decimal places.

n	$\int_1^5 \frac{1}{x}\,dx = \ln 5 \approx 1.609438$
6	1.64360
8	1.62897
10	1.62204
20	1.61263
100	1.60957
1000	1.60944

When $n = 1000$, the approximation agrees with the true value to 5 decimal places.

Generalizing from this example, let f be a continuous function on an interval $[a, b]$. Divide the interval from a to b into n equal subintervals by the points $a = x_0, x_1, x_2, \ldots, x_n = b$, as shown in Figure 31. Use the subintervals to make trapezoids that approximately fill in the region under the curve. The approximate value of the definite integral $\int_a^b f(x)\,dx$ is given by the sum of the areas of the trapezoids, or

$$\int_a^b f(x)\,dx \approx \frac{1}{2}[f(x_0)+f(x_1)]\left(\frac{b-a}{n}\right) + \frac{1}{2}[f(x_1)+f(x_2)]\left(\frac{b-a}{n}\right)$$

$$+ \cdots + \frac{1}{2}[f(x_{n-1})+f(x_n)]\left(\frac{b-a}{n}\right)$$

$$= \left(\frac{b-a}{n}\right)\left[\frac{1}{2}f(x_0)+\frac{1}{2}f(x_1)+\frac{1}{2}f(x_1)+\frac{1}{2}f(x_2)+\frac{1}{2}f(x_2)+\cdots+\frac{1}{2}f(x_{n-1})+\frac{1}{2}f(x_n)\right]$$

$$= \left(\frac{b-a}{n}\right)\left[\frac{1}{2}f(x_0)+f(x_1)+f(x_2)+\cdots+f(x_{n-1})+\frac{1}{2}f(x_n)\right].$$

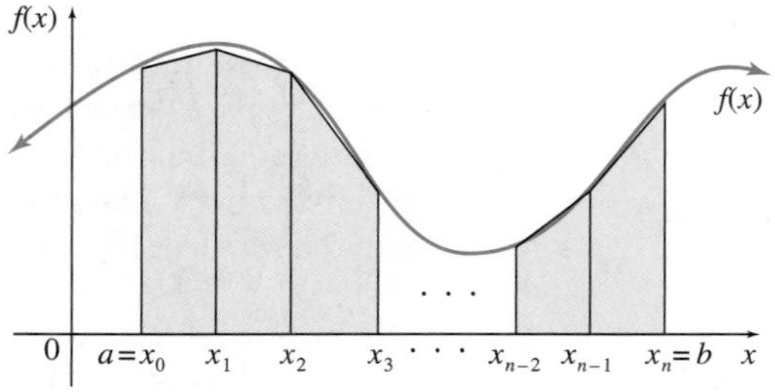

FIGURE 31

This result gives the following rule.

TRAPEZOIDAL RULE

Let f be a continuous function on $[a, b]$ and let $[a, b]$ be divided into n equal subintervals by the points $a = x_0, x_1, x_2, \ldots, x_n = b$. Then, by the **trapezoidal rule,**

$$\int_a^b f(x)\,dx \approx \left(\frac{b-a}{n}\right)\left[\frac{1}{2}f(x_0) + f(x_1) + \cdots + f(x_{n-1}) + \frac{1}{2}f(x_n)\right].$$

EXAMPLE 1 *Trapezoidal Rule*

Use the trapezoidal rule with $n = 4$ to approximate

$$\int_0^2 \sqrt{x^2+1}\,dx.$$

Solution

Method 1: Calculating by Hand

Here $a = 0$, $b = 2$, and $n = 4$, with $(b - a)/n = (2 - 0)/4 = 1/2$ as the altitude of each trapezoid. Then $x_0 = 0$, $x_1 = 1/2$, $x_2 = 1$, $x_3 = 3/2$, and $x_4 = 2$. Now find the corresponding function values. The work can be organized into a table, as follows.

i	x_i	$f(x_i)$
0	0	$\sqrt{0^2+1} = 1$
1	1/2	$\sqrt{(1/2)^2+1} \approx 1.11803$
2	1	$\sqrt{1^2+1} \approx 1.41421$
3	3/2	$\sqrt{(3/2)^2+1} \approx 1.80278$
4	2	$\sqrt{2^2+1} \approx 2.23607$

Substitution into the trapezoidal rule gives

$$\int_0^2 \sqrt{x^2+1}\,dx$$
$$\approx \frac{2-0}{4}\left[\frac{1}{2}(1) + 1.11803 + 1.41421 + 1.80278 + \frac{1}{2}(2.23607)\right]$$
$$\approx 2.97653.$$

The approximation 2.97653 found above using the trapezoidal rule with $n = 4$ differs from the true value of 2.95789 by .01864. As mentioned above, this error would be reduced if larger values were used for n. For example, if $n = 8$, the trapezoidal rule gives an answer of 2.96254, which differs from the true value by .00465. Techniques for estimating such errors are considered in more advanced courses.

Method 2: Graphing Calculator Just as we used a graphing calculator to approximate area using rectangles, we can also use it for the trapezoidal rule. As before, put the values of i in L_1 and the values of x_i in L_2. In the heading for L_3, put $\sqrt{(L_2^2+1)}$. Using the fact that $(b-a)/n = (2-0)/4 = .5$, the command `.5*(.5*L3(1)+sum(L3,2,4)+.5*L3(5))` gives the result 2.976528589. For more details, see *The Graphing Calculator Manual* that is available with this book.

Method 3: Spreadsheet The trapezoidal rule can also be done on a spreadsheet. In Microsoft Excel, for example, store the values of 0 through n in column A. After putting the left endpoint in E1 and Δx in E2, put the command "`=$E$1+A1*$E$2`" into B1; copying this formula into the rest of column B gives the values of x_i. Similarly, use the formula for $f(x_i)$ to fill column C. Using the fact that $n = 5$ in this example, the command "`$E$2*(.5*C1+sum(C2:C4)+.5*C5)`" gives the result 2.976529. For more details, see *The Spreadsheet Manual* that is available with this book.

Simpson's Rule Another numerical method, *Simpson's rule*, approximates consecutive portions of the curve with portions of parabolas rather than the line segments of the trapezoidal rule. Simpson's rule usually gives a better approximation than the trapezoidal rule for the same number of subintervals. As shown in Figure 32, one parabola is fitted through points A, B, and C, another through C, D, and E, and so on. Then the sum of the areas under these parabolas will approximate the area under the graph of the function. Because of the way the parabolas overlap, it is necessary to have an even number of intervals, and therefore an odd number of points, to apply Simpson's rule.

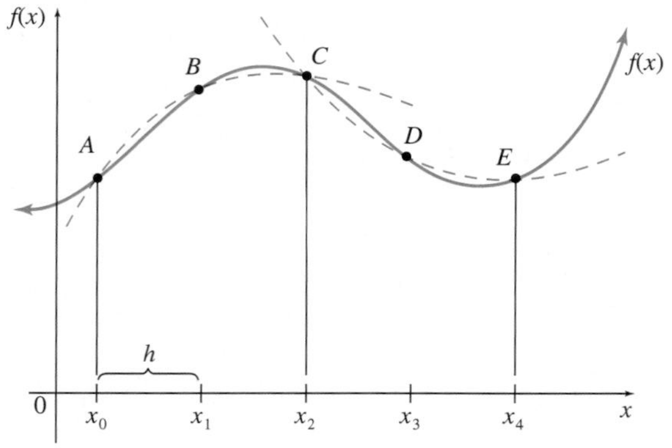

FIGURE 32

If h, the length of each subinterval, is $(b-a)/n$, the area under the parabola through points A, B, and C can be found by a definite integral. The details are omitted; the result is

$$\frac{h}{3}[f(x_0) + 4f(x_1) + f(x_2)].$$

Similarly, the area under the parabola through points C, D, and E is

$$\frac{h}{3}[f(x_2) + 4f(x_3) + f(x_4)].$$

When these expressions are added, the last term of one expression equals the first term of the next. For example, the sum of the two areas given above is

$$\frac{h}{3}[f(x_0) + 4f(x_1) + 2f(x_2) + 4f(x_3) + f(x_4)].$$

This illustrates the origin of the pattern of the terms in the following rule.

SIMPSON'S RULE

Let f be a continuous function on $[a, b]$ and let $[a, b]$ be divided into an even number n of equal subintervals by the points $a = x_0, x_1, x_2, \ldots, x_n = b$. Then by **Simpson's rule,**

$$\int_a^b f(x)\,dx \approx \left(\frac{b-a}{3n}\right)[f(x_0) + 4f(x_1) + 2f(x_2) + 4f(x_3) + \cdots + 2f(x_{n-2}) + 4f(x_{n-1}) + f(x_n)].$$

Thomas Simpson (1710–1761), a British mathematician, wrote texts on many branches of mathematics. Some of these texts went through as many as ten editions. His name became attached to this numerical method of approximating definite integrals even though the method preceded his work.

CAUTION In Simpson's rule, n (the number of subintervals) must be even. ■

EXAMPLE 2 *Simpson's Rule*

Use Simpson's rule with $n = 4$ to approximate

$$\int_0^2 \sqrt{x^2 + 1}\,dx,$$

which was approximated by the trapezoidal rule in Example 1.

Solution As in Example 1, $a = 0$, $b = 2$, and $n = 4$, and the endpoints of the four intervals are $x_0 = 0$, $x_1 = 1/2$, $x_2 = 1$, $x_3 = 3/2$, and $x_4 = 2$. The table of values is also the same.

i	x_i	$f(x_i)$
0	0	1
1	1/2	1.11803
2	1	1.41421
3	3/2	1.80278
4	2	2.23607

Since $(b - a)/(3n) = 2/12 = 1/6$, substituting into Simpson's rule gives

$$\int_0^2 \sqrt{x^2 + 1}\, dx \approx \frac{1}{6}[1 + 4(1.11803) + 2(1.41421) + 4(1.80278) + 2.23607] \approx 2.95796.$$

This differs from the true value by .00007, which is less than the trapezoidal rule with $n = 8$. If $n = 8$ for Simpson's rule, the approximation is 2.95788, which differs from the true value by only .00001.

NOTE

1. Just as we can use a graphing calculator or a spreadsheet for the trapezoidal rule, we can also use such technology for Simpson's rule. For more details, see *The Graphing Calculator Manual* and *The Spreadsheet Manual* that are available with this book.
2. Let M represent the midpoint rule approximation and T the trapezoidal rule approximation, using n subintervals in each. Then the formula $S = (2M + T)/3$ gives the Simpson's rule approximation with $2n$ subintervals.

Numerical methods make it possible to approximate

$$\int_a^b f(x)\, dx$$

even when $f(x)$ is not known. The next example shows how this is done.

EXAMPLE 3 *Total Distance*

As mentioned earlier, the velocity $v(t)$ gives the rate of change of distance $s(t)$ with respect to time t. Suppose a vehicle travels an unknown distance. The passengers keep track of the velocity at 10-minute intervals (every 1/6 of an hour) with the following results.

Time in Hours, t	1/6	2/6	3/6	4/6	5/6	1	7/6
Velocity in Miles per Hour, v(t)	45	55	52	60	64	58	47

What is the total distance traveled in the 60-minute period from $t = 1/6$ to $t = 7/6$?

Solution The distance traveled in t hours is $s(t)$, with $s'(t) = v(t)$. The total distance traveled between $t = 1/6$ and $t = 7/6$ is given by

$$\int_{1/6}^{7/6} v(t)\, dt.$$

Even though this integral cannot be evaluated since we do not have an expression for $v(t)$, either the trapezoidal rule or Simpson's rule can be used to approximate its value and give the total distance traveled. In either case, let $n = 6$, $a = t_0 = 1/6$, and $b = t_6 = 7/6$. By the trapezoidal rule,

$$\int_{1/6}^{7/6} v(t)\, dt \approx \frac{7/6 - 1/6}{6}\left[\frac{1}{2}(45) + 55 + 52 + 60 + 64 + 58 + \frac{1}{2}(47)\right]$$

$$\approx 55.83.$$

By Simpson's rule,

$$\int_{1/6}^{7/6} v(t)\,dt \approx \frac{7/6 - 1/6}{3(6)}[45 + 4(55) + 2(52) + 4(60) + 2(64) + 4(58) + 47]$$

$$= \frac{1}{18}(45 + 220 + 104 + 240 + 128 + 232 + 47) \approx 56.44.$$

The distance traveled in the 1-hour period was about 56 miles.

As already mentioned, Simpson's rule generally gives a better approximation than the trapezoidal rule. As n increases, the two approximations get closer and closer. For the same accuracy, however, a smaller value of n generally can be used with Simpson's rule so that less computation is necessary. Simpson's rule is the method used by many calculators that have a built-in integration feature.

The branch of mathematics that studies methods of approximating definite integrals (as well as many other topics) is called *numerical* analysis. Numerical integration is useful even with functions whose antiderivatives can be determined if the antidifferentiation is complicated and a computer or calculator programmed with Simpson's rule is handy. You may want to program your calculator for both the trapezoidal rule and Simpson's rule. For some calculators, these programs are in *The Graphing Calculator Manual* that is available with this book.

7.6 EXERCISES

In Exercises 1–10, use $n = 4$ to approximate the value of the given integrals by the following methods: **(a)** *the trapezoidal rule, and* **(b)** *Simpson's rule.* **(c)** *Find the exact value by integration.*

1. $\int_0^2 x^2\,dx$

2. $\int_0^2 (2x + 1)\,dx$

3. $\int_{-1}^3 \frac{1}{4 - x}\,dx$

4. $\int_1^5 \frac{1}{x + 1}\,dx$

5. $\int_{-2}^2 (2x^2 + 1)\,dx$

6. $\int_0^3 (2x^2 + 1)\,dx$

7. $\int_1^5 \frac{1}{x^2}\,dx$

8. $\int_2^4 \frac{1}{x^3}\,dx$

9. $\int_0^1 xe^{x^2}\,dx$

10. $\int_0^4 \sqrt{2x + 1}\,dx$

11. Find the area under the semicircle $y = \sqrt{4 - x^2}$ and above the x-axis by using $n = 8$ with the following methods.

a. The trapezoidal rule **b.** Simpson's rule

c. Compare the results with the area found by the formula for the area of a circle. Which of the two approximation techniques was more accurate?

12. Find the area between the x-axis and the ellipse $4x^2 + 9y^2 = 36$ by using $n = 12$ with the following methods.

a. The trapezoidal rule **b.** Simpson's rule

(*Hint:* Solve the equation for y and find the area of the semiellipse.)

c. Compare the results with the actual area, $3\pi \approx 9.4248$ (which can be found by methods not considered in this text). Which approximation technique was more accurate?

13. Suppose that $f(x) > 0$ and $f''(x) > 0$ for all x between a and b, where $a < b$. Which of the following cases is true of a trapezoidal approximation T for the integral $\int_a^b f(x)\,dx$? Explain.

a. $T < \int_a^b f(x)dx$ **b.** $T > \int_a^b f(x)dx$ **c.** Can't say which is larger

14. Refer to Exercise 13. Which of the three cases applies to these functions?

a. $f(x) = x^2$; $[0, 3]$ **b.** $f(x) = \sqrt{x}$; $[0, 9]$

c.

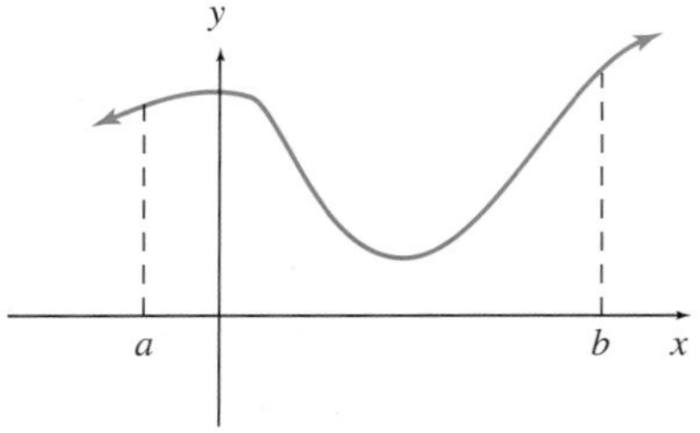

Exercises 15–18 require both the trapezoidal rule and Simpson's rule. They can be worked without calculator programs if such programs are not available, although they require more calculation than the other problems in this exercise set.

Error Analysis *The difference between the true value of an integral and the value given by the trapezoidal rule or Simpson's rule is known as the error. In numerical analysis, the error is studied to determine how large n must be for the error to be smaller than some specified amount. For both rules, the error is inversely proportional to a power of n, the number of subdivisions. In other words, the error is roughly k/n^p, where k is a constant that depends on the function and the interval, and p is a power that depends only on the method used. With a little experimentation, you can find out what the power p is for the trapezoidal rule and for Simpson's rule.*

15. a. Find the exact value of $\int_0^1 x^4\,dx$.

b. Approximate the integral in part a using the trapezoidal rule with $n = 4, 8, 16$, and 32. For each of these answers, find the absolute value of the error by subtracting the trapezoidal rule answer from the exact answer found in part a.

c. If the error is k/n^p, then the error times n^p should be approximately a constant. Multiply the errors in part b times n^p for $p = 1, 2$, etc., until you find a power p yielding the same answer for all four values of n.

16. Based on the results of Exercise 15, what happens to the error in the trapezoidal rule when the number of intervals is doubled?

17. Repeat Exercise 15 using Simpson's rule.

18. Based on the results of Exercise 17, what happens to the error in Simpson's rule when the number of intervals is doubled?

19. For the integral in Exercise 7, apply the midpoint rule with $n = 4$ and Simpson's rule with $n = 8$ to verify the formula $S = (2M + T)/3$.

20. Repeat the instructions of Exercise 19 using the integral in Exercise 8.

Applications

BUSINESS AND ECONOMICS

21. *Total Sales* A sales manager presented the following results at a sales meeting.

Year, x	1	2	3	4	5	6	7
Rate of Sales, $f(x)$	.4	.6	.9	1.1	1.3	1.4	1.6

Find the total sales over the given period as follows.

a. Plot these points. Connect the points with line segments.

b. Use the trapezoidal rule to find the area bounded by the broken line of part a, the x-axis, the line $x = 1$, and the line $x = 7$.

c. Find the same area using Simpson's rule.

22. *Total Cost* A company's marginal costs (in hundreds of dollars per year) were as follows over a certain period.

Year, x	1	2	3	4	5	6	7
Marginal Cost, $f(x)$	9.0	9.2	9.5	9.4	9.8	10.1	10.5

Repeat parts a–c of Exercise 21 for these data to find the total cost over the given period.

LIFE SCIENCES

23. *Drug Reaction Rate* The reaction rate to a new drug is given by

$$y = e^{-t^2} + \frac{1}{t},$$

where t is time (in hours) after the drug is administered. Find the total reaction to the drug from $t = 1$ to $t = 9$ by letting $n = 8$ and using the following methods.

a. The trapezoidal rule **b.** Simpson's rule

24. *Growth Rate* The growth rate of a certain tree (in feet) is given by

$$y = \frac{2}{t} + e^{-t^2/2},$$

where t is time (in years). Find the total growth from $t = 1$ to $t = 7$ by using $n = 12$ with the following methods.

a. The trapezoidal rule **b.** Simpson's rule

Blood Level Curves In the study of bioavailability in pharmacy, a drug is given to a patient. The level of concentration of the drug is then measured periodically, producing blood level curves such as the ones shown in the figure. The areas under the curves give the total amount of the drug available to the patient for each milliliter of blood. Use the trapezoidal rule with $n = 10$ to find the following areas.*

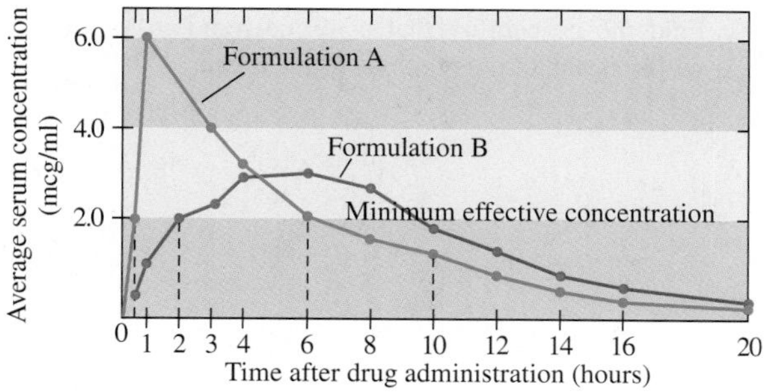

25. Find the total area under the curve for Formulation A. What does this area represent?

26. Find the total area under the curve for Formulation B. What does this area represent?

27. Find the area between the curve for Formulation A and the minimum effective concentration line. What does your answer represent?

28. Find the area between the curve for Formulation B and the minimum effective concentration line. What does this area represent?

29. *Calves* The daily milk consumption (in kilograms) for calves can be approximated by the function

$$y = b_0 w^{b_1} e^{-b_2 w},$$

where w is the age of the calf (in weeks) and b_0, b_1, and b_2 are constants.†

a. The age in days is given by $t = 7w$. Use this fact to convert the function above to a function in terms of t.

b. For a group of Angus calves, $b_0 = 5.955$, $b_1 = .233$, and $b_2 = .027$. Use the trapezoidal rule with $n = 10$, and then Simpson's rule with $n = 10$, to find the total amount of milk consumed by one of these calves over the first 25 weeks of life.

c. For a group of Nelore calves, $b_0 = 8.409$, $b_1 = .143$, and $b_2 = .037$. Use the trapezoidal rule with $n = 10$, and then Simpson's rule with $n = 10$, to find the total amount of milk consumed by one of these calves over the first 25 weeks of life.

30. *Foot-and-Mouth Epidemic* In 2001, the United Kingdom suffered an epidemic of foot-and-month disease. The graph on the next page shows the number of reported cases each day since Feb. 18, as well as the number of cases epidemiologists project would have occurred had they culled all livestock on infected farms within 24 hours, and all livestock on neighboring farms within 48 hours of infection.‡ In the section on Area and the Definite Integral, we estimated the number of cases using rectangles.

a. Estimate the total number of cases that occurred from Feb. 18 through May 13. Use Simpson's rule with interval widths of 14 days.

b. Estimate the total number of cases that would have occurred from Feb. 18 through May 13 using the more aggressive culling plan. Use Simpson's rule with interval widths of 14 days.

*These graphs are from D. J. Chodos and A. R. DeSantos, *Basics of Bioavailability*. Copyright 1978 by the Upjohn Company.

†Mezzadra, C., R., Paciaroni, S. Vulich, E. Villarreal, and L. Melucci, "Estimation of Milk Consumption Curve Parameters for Different Genetic Groups of Bovine Calves," *Animal Production,* Vol. 49, 1989, pp. 83–87.

‡*Science,* Vol. 294, Oct. 5, 2001, p. 26. Peter Morrison/AP. Reprinted by permission of the Associated Press.

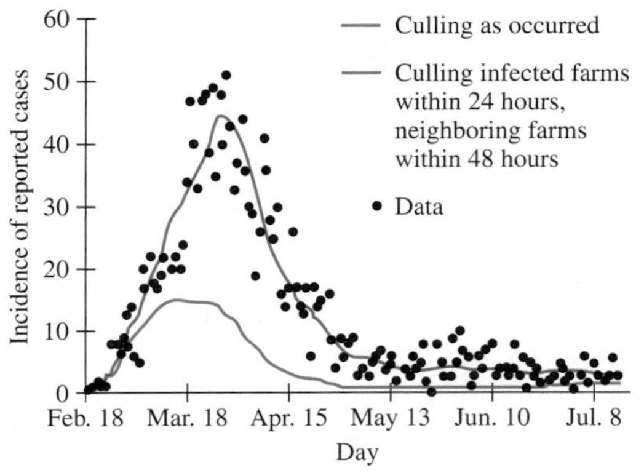

SOCIAL SCIENCES

31. *Educational Psychology* The results from a research study in psychology were as follows.

Number of Hours of Study, x	1	2	3	4	5	6	7
Rate of Extra Points Earned on a Test, $f(x)$	4	7	11	9	15	16	23

Repeat parts a–c of Exercise 21 for these data.

PHYSICAL SCIENCES

32. *Chemical Formation* The following table shows the results from a chemical experiment.

Concentration of Chemical A, x	1	2	3	4	5	6	7
Rate of Formation of Chemical B, $f(x)$	12	16	18	21	24	27	32

Repeat parts a–c of Exercise 21 for these data.

If you have a program for Simpson's rule in your graphing calculator, use it with $n = 20$ *for Exercises 33–35.*

33. *Total Revenue* An electronics company analyst has determined that the rate per month at which revenue comes in from the calculator division is given by

$$R(x) = 105e^{.01x} + 32,$$

where x is the number of months the division has been in operation. Find the total revenue between the 12th and 36th months.

34. *Blood Pressure* Blood pressure in an artery changes rapidly over a very short time for a healthy young adult, from a high of about 120 to a low of about 80. Suppose the blood pressure function over an interval of 1.5 seconds is given by

$$f(x) = .2x^5 - .68x^4 + .8x^3 - .39x^2 + .055x + 100,$$

where x is the time in seconds after a peak reading. The area under the curve for one cycle is important in some blood pressure studies. Find the area under $f(x)$ from .1 second to 1.1 seconds.

35. *Probability* The most important function in probability and statistics is the density function for the standard normal distribution, which is the familiar bell-shaped curve. The function is

$$f(x) = \frac{1}{\sqrt{2\pi}}e^{-x^2/2}.$$

a. The area under this curve between $x = -1$ and $x = 1$ represents the probability that a normal random variable is within 1 standard deviation of the mean. Find this probability.

b. Find the area under this curve between $x = -2$ and $x = 2$, which represents the probability that a normal random variable is within 2 standard deviations of the mean.

c. Find the probability that a normal random variable is within 3 standard deviations of the mean.

CHAPTER SUMMARY

Earlier chapters dealt with the derivative, one of the two main ideas of calculus. This chapter deals with integration, the second main idea. There are two aspects of integration. The first is indefinite integration, or finding an antiderivative; the second is definite integration, which can be used to find the area under a curve. The Fundamental Theorem of Calculus unites these two ideas by showing that the way to find the area under a curve is to use the antiderivative. Substitution is a technique for finding antiderivatives. Numerical integration can be used to find the definite integral when finding an antiderivative is not feasible. The idea of the definite integral can also be applied to finding the area between two curves.

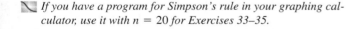

INTEGRATION SUMMARY

Antidifferentiation Formulas

Power Rule

$$\int x^n\,dx = \frac{x^{n+1}}{n+1} + C,\ n \neq -1$$

Constant Multiple Rule

$$\int k \cdot f(x)\,dx = k\int f(x)\,dx, \quad \text{for any real number } k$$

Sum or Difference Rule

$$\int [f(x) \pm g(x)]\,dx = \int f(x)\,dx \pm \int g(x)\,dx$$

Integration of Exponential Functions

$$\int e^{kx}\,dx = \frac{e^{kx}}{k} + C, \quad k \neq 0$$

Integration of x^{-1}

$$\int x^{-1}\,dx = \ln|x| + C$$

Substitution Method

Choose u to be one of the following:

1. the quantity under a root or raised to a power;
2. the exponent on e;
3. the quantity in the denominator.

Definite Integrals

Definition of the Definite Integral

$\int_a^b f(x)\,dx = \lim_{n\to\infty} \sum_{i=1}^{n} f(x_i)\Delta x$, where $\Delta x = (b-a)/n$ and x_i is any value of x in the ith interval. If $f(x)$ gives the rate of change of $F(x)$ for x in $[a, b]$, then this represents the total change in $F(x)$ as x goes from a to b.

Properties of Definite Integrals

1. $\int_a^a f(x)\,dx = 0$
2. $\int_a^b k \cdot f(x)\,dx = k\int_a^b f(x)\,dx$, for any real number k.
3. $\int_a^b [f(x) \pm g(x)]\,dx = \int_a^b f(x)\,dx \pm \int_a^b g(x)\,dx$
4. $\int_a^b f(x)\,dx = \int_a^c f(x)\,dx + \int_c^b f(x)\,dx$, for any real number c
5. $\int_a^b f(x)\,dx = -\int_b^a f(x)\,dx$

Fundamental Theorem of Calculus

$\int_a^b f(x)\,dx = F(x)\big|_a^b = F(b) - F(a)$, where f is continuous on $[a, b]$ and F is any antiderivative of f.

Area Between Two Curves

$\int_a^b [f(x) - g(x)]\,dx$, where f and g are continuous functions and $f(x) \geq g(x)$ on $[a, b]$

Consumers' Surplus

$\int_0^{q_0} [D(q) - p_0]\,dq$, where D is the demand function and p_0 and q_0 are the equilibrium price and demand

Producers' Surplus $\int_0^{q_0}[p_0 - S(q)]\,dq$, where S is the supply function and p_0 and q_0 are the equilibrium price and supply

Trapezoidal Rule $$\int_a^b f(x)\,dx \approx \left(\frac{b-a}{n}\right)\left[\frac{1}{2}f(x_0) + f(x_1) + \cdots + f(x_{n-1}) + \frac{1}{2}f(x_n)\right]$$

Simpson's Rule $$\int_a^b f(x)\,dx \approx \left(\frac{b-a}{3n}\right)[f(x_0) + 4f(x_1) + 2f(x_2) + \cdots + 4f(x_{n-1}) + f(x_n)]$$

KEY TERMS

7.1 antiderivative
integral sign
integrand
indefinite integral

7.2 integration by substitution

7.3 definite integral
limits of integration
total change

7.4 Fundamental Theorem of Calculus

7.5 consumers' surplus
producers' surplus

7.6 numerical integration
trapezoidal rule
Simpson's rule

CHAPTER 7 REVIEW EXERCISES

1. Explain the differences between an indefinite integral and a definite integral.

2. Explain under what circumstances substitution is useful in integration.

3. Explain why the limits of integration are changed when u is substituted for an expression in x in a definite integral.

4. Describe the type of integral for which numerical integration is useful.

Find each indefinite integral.

5. $\int (2x + 3)\,dx$

6. $\int (5x - 1)\,dx$

7. $\int (x^2 - 3x + 2)\,dx$

8. $\int (6 - x^2)\,dx$

9. $\int 3\sqrt{x}\,dx$

10. $\int \frac{\sqrt{x}}{2}\,dx$

11. $\int (x^{1/2} + 3x^{-2/3})\,dx$

12. $\int (2x^{4/3} + x^{-1/2})\,dx$

13. $\int \frac{-4}{x^3}\,dx$

14. $\int \frac{5}{x^4}\,dx$

15. $\int -3e^{2x}\,dx$

16. $\int 5e^{-x}\,dx$

17. $\int xe^{3x^2}\,dx$

18. $\int 2xe^{x^2}\,dx$

19. $\int \frac{3x}{x^2 - 1}\,dx$

20. $\int \frac{-x}{2 - x^2}\,dx$

21. $\int \frac{x^2\,dx}{(x^3 + 5)^4}$

22. $\int (x^2 - 5x)^4(2x - 5)\,dx$

23. $\int \frac{x^3}{e^{3x^4}}\,dx$

24. $\int e^{3x^2+4}x\,dx$

25. Let $f(x) = 3x + 1$, $x_1 = -1$, $x_2 = 0$, $x_3 = 1$, $x_4 = 2$, and $x_5 = 3$. Find $\sum_{i=1}^{5} f(x_i)$.

26. Find $\int_0^4 f(x)\,dx$ for each graph of $y = f(x)$.

a.

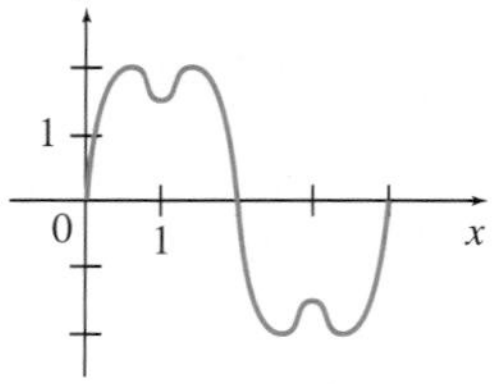

b.

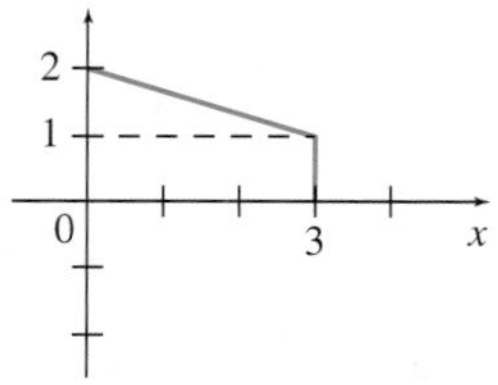

27. Approximate the area under the graph of $f(x) = 2x + 3$ and above the x-axis from $x = 0$ to $x = 4$ using four rectangles. Let the height of each rectangle be the function value on the left side.

28. Find $\int_0^4 (2x + 3)\,dx$ by using the formula for the area of a trapezoid: $A = (1/2)(B + b)h$, where B and b are the lengths of the parallel sides and h is the distance between them. Compare with Exercise 27.

29. In Exercises 29 and 30 of the section on Area and the Definite Integral, you calculated the distance that a car traveled by estimating the integral $\int_0^T v(t)\,dt$.

a. Let $s(t)$ represent the mileage reading on the odometer. Express the distance traveled between $t = 0$ and $t = T$ using the function $s(t)$.

b. Since your answer to part a and the original integral both represent the distance traveled by the car, the two can be set equal. Explain why the resulting equation is a statement of the Fundamental Theorem of Calculus.

30. What does the Fundamental Theorem of Calculus state?

Find each definite integral.

31. $\int_1^2 (3x^2 + 5)\,dx$

32. $\int_1^6 (2x^2 + x)\,dx$

33. $\int_1^5 (3x^{-2} + x^{-3})\,dx$

34. $\int_0^1 x\sqrt{5x^2 + 4}\,dx$

35. $\int_1^3 2x^{-1}\,dx$

36. $\int_1^6 8x^{-1}\,dx$

37. $\int_0^4 2e^x\,dx$

38. $\int_1^6 \frac{5}{2}e^{4x}\,dx$

39. Use the substitution $u = 4x^2$ and the equation of a semicircle to evaluate

$$\int_0^{1/2} x\sqrt{1 - 16x^4}\,dx.$$

Find the area between the x-axis and $f(x)$ over each of the given intervals.

40. $f(x) = \sqrt{x - 1}$; $[1, 10]$

41. $f(x) = (x + 2)^6$; $[-2, 0]$

42. $f(x) = e^x$; $[0, 2]$

43. $f(x) = 1 + e^{-x}$; $[0, 4]$

Find the area of the region enclosed by each group of curves.

44. $f(x) = 5 - x^2$, $g(x) = x^2 - 3$

45. $f(x) = x^2 - 4x$, $g(x) = x - 6$

46. $f(x) = x^2 - 4x$, $g(x) = x + 1$, $x = 2$, $x = 4$

47. $f(x) = 5 - x^2$, $g(x) = x^2 - 3$, $x = 0$, $x = 4$

Use the trapezoidal rule with $n = 4$ to approximate the value of each integral. Then find the exact value and compare the two answers.

48. $\int_1^3 \frac{\ln x}{x}\,dx$

49. $\int_2^{10} \frac{x\,dx}{x - 1}$

50. $\int_0^1 e^x\sqrt{e^x + 1}\,dx$

Use Simpson's rule with $n = 4$ to approximate the value of each integral. Compare your answers with the answers to Exercises 48–50.

51. $\int_1^3 \frac{\ln x}{x}\,dx$

52. $\int_2^{10} \frac{x\,dx}{x - 1}$

53. $\int_0^1 e^x\sqrt{e^x + 1}\,dx$

54. Find the area of the region between the graphs of $y = \sqrt{x - 1}$ and $2y = x - 1$ from $x = 1$ to $x = 5$ in three ways.

a. Use antidifferentiation.

b. Use the trapezoidal rule with $n = 4$.

c. Use Simpson's rule with $n = 4$.

55. Let $y = [x(x-1)(x+1)(x-2)(x+2)]^2$.

a. Find $\int_{-2}^{2} f(x)\,dx$ using the trapezoidal rule with $n = 4$.

b. Find $\int_{-2}^{2} f(x)\,dx$ using Simpson's rule with $n = 4$.

c. Without evaluating $\int_{-2}^{2} f(x)\,dx$, explain why your answers to parts a and b cannot possibly be correct.

d. Explain why the trapezoidal rule and Simpson's rule with $n = 4$ give incorrect answers for $\int_{-2}^{2} f(x)\,dx$ with this function.

Applications

BUSINESS AND ECONOMICS

Cost *Find the cost function for each of the marginal cost functions in Exercises 56 and 57.*

56. $C'(x) = 3\sqrt{2x-1}$; 13 units cost \$270.

57. $C'(x) = \dfrac{1}{x+1}$; fixed cost is \$18.

58. *Investment* The curve shown below gives the rate that an investment accumulates income (in dollars per year). Use rectangles of width 2 units and height determined by the function value at the midpoint to find the total income accumulated over 10 years.

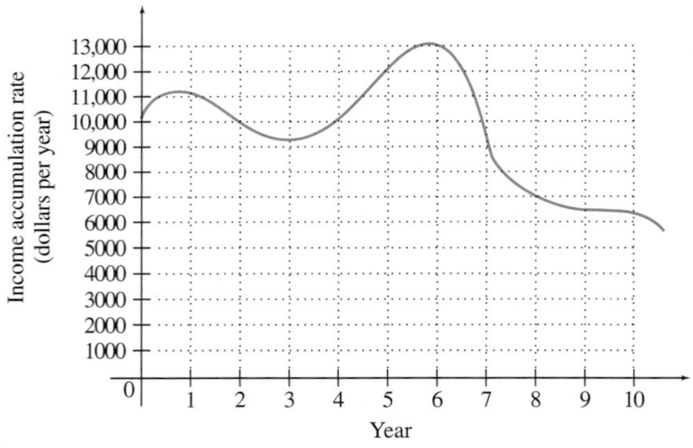

59. *Utilization of Reserves* A manufacturer of electronic equipment requires a certain rare metal. He has a reserve supply of 4,000,000 units that he will not be able to replace. If the rate at which the metal is used is given by

$$f(t) = 100{,}000e^{.03t},$$

where t is time (in years), how long will it be before he uses up the supply? (*Hint*: Find an expression for the total amount used in t years and set it equal to the known reserve supply.)

60. *Sales* The rate of change of sales of a new brand of tomato soup (in thousands per month) is given by

$$S'(x) = \sqrt{x} + 2,$$

where x is the time (in months) that the new product has been on the market. Find the total sales after 9 months.

61. *Productivity* The function defined by $f'(x) = .28x + .54$ approximates marginal U.S. nonfarm productivity from 1990–2002.* Productivity is measured as total output per hour compared to a measure of 100 for 1992, and x represents the end of the year with 1990 corresponding to $x = 0$, 1991 corresponding to $x = 1$, and so on.

a. Give the function that describes total productivity in year x.

b. Use your function from part a to find productivity at the end of 2002. In 2002, productivity actually measured 124.7. How does your value using the function compare with this?

62. *Producers' and Consumers' Surplus* Suppose that the supply function for some commodity is

$$S(q) = q^2 + 5q + 100$$

and the demand function for the commodity is

$$D(q) = 350 - q^2.$$

a. Find the producers' surplus.

b. Find the consumers' surplus.

63. *Net Savings* A company has installed new machinery that will produce a savings rate (in thousands of dollars per year) of

$$S'(x) = 225 - x^2,$$

*Bureau of Labor Statistics.

where x is the number of years the machinery is to be used. The rate of additional costs (in thousands of dollars per year) to the company due to the new machinery is expected to be

$$C'(x) = x^2 + 25x + 150.$$

For how many years should the company use the new machinery? Find the net savings (in thousands of dollars) over this period.

LIFE SCIENCES

64. *Population Growth* The rate of change of the population of a rare species of Australian spider is given by

$$f(t) = 100 - \sqrt{2.4t + 1},$$

where $f(t)$ is the number of spiders present at time t (in months). Find the total number of additional spiders in the first 10 months.

65. *Infection Rate* The rate of infection of a disease (in people per month) is given by the function

$$I'(t) = \frac{100t}{t^2 + 1},$$

where t is the time (in months) since the disease broke out. Find the total number of infected people over the first four months of the disease.

66. *Insect Cannibalism* In certain species of flour beetles, the larvae cannibalize the unhatched eggs. In calculating the population cannibalism rate per egg, researchers needed to evaluate the integral

$$\int_0^A c(x)\,dx,$$

where A is the length of the larval stage and $c(x)$ is the cannibalism rate per egg per larva of age x.* The minimum value of A for the flour beetle *Tribolium castaneum* is 17.6 days, which is the value we will use. The function $c(x)$ starts at day 0 with a value of 0, increases linearly to the value .024 at day 12, and then stays constant. Find the values of the integral using

a. formulas from geometry;

b. the Fundamental Theorem of Calculus.

67. *Insulin in Sheep* A research group studied the effect of a large injection of glucose in sheep fed a normal diet compared with sheep that were fasting.† A graph of the plasma insulin levels (in pM—pico molars, or 10^{-12} of a molar) for both groups is shown below. The red circles designate the fasting sheep and the green circles the sheep fed a normal diet. The researchers compared the area under the curves for the two groups.

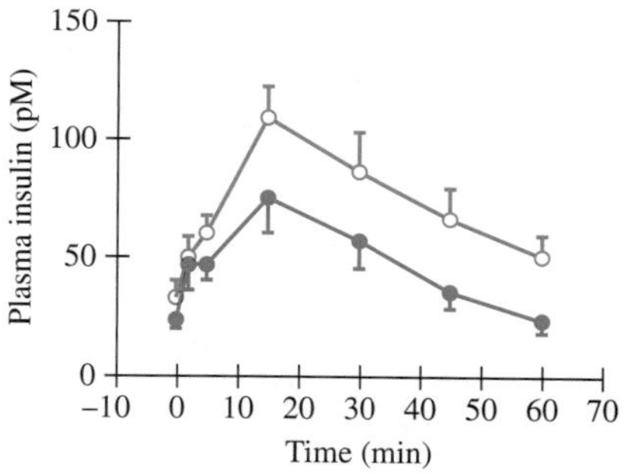

a. For the fasting sheep, estimate the area under the curve using rectangles, first by using the left endpoints, then the right endpoints, and then averaging the two. Note that the width of the rectangles will vary.

b. Repeat part a for the sheep fed a normal diet.

c. How much higher is the area under the curve for the fasting sheep compared with the normal sheep?

*Hastings, Alan and Robert F. Costantino, "Oscillations in Population Numbers: Age-Dependent Cannibalism," *Journal of Animal Ecology,* Vol. 60, No. 2, June 1991, pp. 471–482.

†Oliver, M. H. et al., "Material Undernutrition During the Periconceptual Period Increases Plasma Taurine Levels and Insulin Response to Glucose but not Arginine in the Late Gestation Fetal Sheep," *Endocrinology,* Vol 14, No. 10, 2001, pp. 4576–4579. Reprinted with permission from The Endocrine Society.

68. *Risk in Fisheries* The maximum sustainable harvest is an important quantity in managing a fishery. Suppose that θ is the maximum sustainable harvest, but that we do not know it exactly, and must instead use an estimator, T, plus a correction factor, x. There is a risk that our estimator is too high or too low. The expected risk of the estimator, $T + x$, is

$$E = \int_{-\infty}^{\infty}\left(\frac{|t + x - \theta|}{\theta} + \frac{a(t + x - \theta)^2}{\theta^2}\right)h(t;\theta)\,dt,$$

where $h(t;\theta)$ is a probability density function of t that depends on the parameter θ, and a is a constant.* Our goal is to find the value of x that minimizes the risk.

a. For simplicity, let $\theta = 1$. The simplest function h is a constant on an interval, set so that the integral of h over that interval is 1. Let

$$h(t;\theta) = \begin{cases} 1 & \text{if } .5 \le t \le 1.5 \\ 0 & \text{otherwise.} \end{cases}$$

Show that

$$E = x^2 + \frac{1}{4} + \frac{a}{3}[(1.5 + x)^3 - (.5 + x)^3]$$

(*Hint:* To integrate, note that $t + x - 1 < 0$ if $.5 \le t \le 1 - x$, in which case $|t + x - 1| = -(t + x - 1)$, and $t + x - 1 \ge 0$ if $1 - x \le t \le 1.5$.)

b. We can minimize E by setting $dE/dx = 0$. (How do we know this is a minimum and not a maximum?) Show that

$$x = \frac{-a}{1 + a},$$

that is, the expected risk is minimized using the estimator $T - a/(1 + a)$.

c. Suppose we leave θ as an unknown parameter in part a, and let

$$h(t;\theta) = \begin{cases} 1/\theta & \text{if } .5\theta \le t \le 1.5\theta \\ 0 & \text{otherwise.} \end{cases}$$

Show that in this case, the expected risk is minimized using the estimator $T - a\theta/(1 + a)$.

SOCIAL SCIENCES

69. *Crime* Based on data from the Chicago Police Department,† the homicide rate in Chicago between 1992 and 2002 can be approximated by

$$f(t) = .817x^3 - 15.4x^2 + 49.4x + 874,$$

where t is the number of years since 1992. Find the total number of homicides during the 10-year period from the beginning of 1992 to the beginning of 2002.

PHYSICAL SCIENCES

70. *Linear Motion* A particle is moving along a straight line with velocity $v(t) = t^2 - 2t$. Its distance from the starting point after 3 seconds is 8 cm. Find $s(t)$, the distance of the particle from the starting point after t seconds.

71. *Weather* The following graph shows 2003 weather statistics for New York City, as well as the normal high and low temperatures.‡ The amount of cold weather in a year is

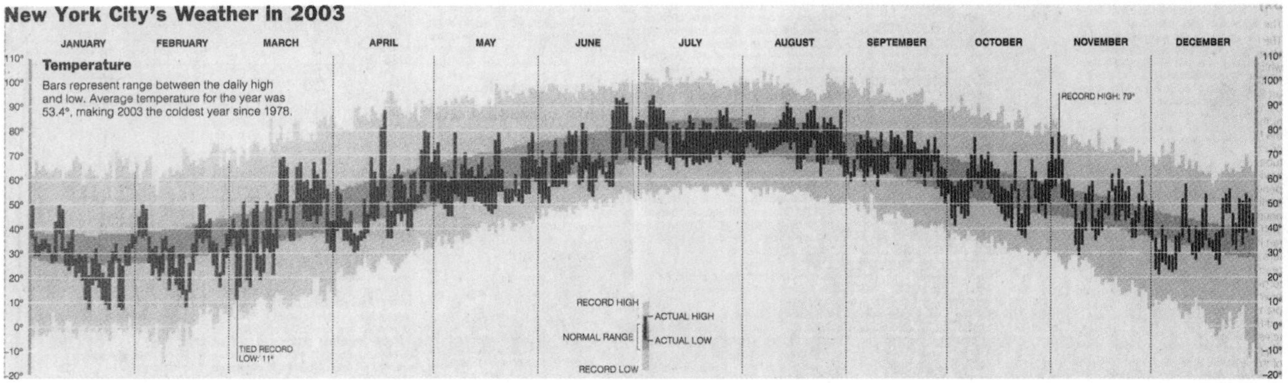

*Kirkwood, G. P., "Risks in Setting Catch Limits," *Mathematical Biosciences,* Vol. 53, No. 1/2, Feb. 1981, pp. 119–129.

†Bureau of Justice Statistics.

‡*The New York Times,* Jan. 7, 2001, p. 29. Reprinted with permission.

measured in heating degree-days, where 1 degree-day is added to the total for each degree that a day's average falls below 65°F. For example, if the average temperature on November 15 is 50°F, 15 degree-days are added to the years' total. Estimate the total number of heating degree-days in an average New York City year, using rectangles of width 1 month, with the height determined by the average temperature at the middle of the rectangle. Assume that the normal average is halfway between the normal high and low (at the center line of the dark band). Use the actual number of days in each month.

EXTENDED APPLICATION: Estimating Depletion Dates for Minerals

It is becoming more and more obvious that the earth contains only a finite quantity of minerals. The "easy and cheap" sources of minerals are being used up, forcing an ever more expensive search for new sources. For example, oil from the North Slope of Alaska would never have been used in the United States during the 1930s because a great deal of Texas and California oil was readily available.

We said in an earlier chapter that population tends to follow an exponential growth curve. Mineral usage also follows such a curve. Thus, if q represents the rate of consumption of a certain mineral at time t, while q_0 represents consumption when $t = 0$, then

$$q = q_0 e^{kt},$$

where k is the growth constant. For example, the world consumption of petroleum in 1970 was 17,100 million barrels. During this period energy use was growing rapidly, and by 1975 annual world consumption had risen to 20,500 million barrels. We can use these two values to make a rough estimate of the constant k, and we find that over this 5-year span the average value of k was about .037, representing 3.7% annual growth. If we let $t = 0$ correspond to the base year 1970, then

$$q = 17{,}100e^{.037t}$$

is the rate of consumption at time t, assuming that all the trends of the early 1970s have continued. In 1970 a reasonable guess would have put the total amount of oil in provable reserves or likely to be discovered in the future at 1,500,000 million barrels. At the 1970–1975 rate of consumption, in how many years after 1970 would you expect the world's reserves to be depleted? We can use the integral calculus of this chapter to find out.

To begin, we need to know the total quantity of petroleum that would be used between time $t = 0$ and some future time $t = T$. Figure 33 shows a typical graph of the function $q = q_0 e^{kt}$.

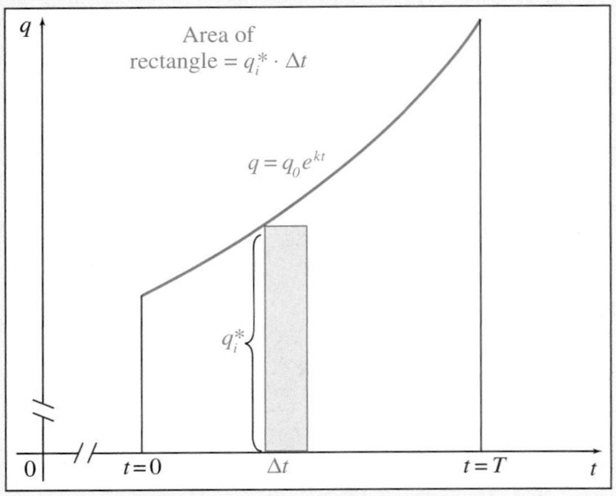

FIGURE 33

Following the work we did in Section 3, divide the time interval from $t = 0$ to $t = T$ into n subintervals. Let each subinterval have width Δt. Let the rate of consumption for the ith subinterval be approximated by q_i^*. Thus, the approximate total consumption for the subinterval is given by

$$q_i^* \cdot \Delta t,$$

and the total consumption over the interval from time $t = 0$ to $t = T$ is approximated by

$$\sum_{i=1}^{n} q_i^* \cdot \Delta t.$$

The limit of this sum as Δt approaches 0 gives the total consumption from time $t = 0$ to $t = T$. That is,

$$\text{Total consumption} = \lim_{\Delta t \to 0} \sum q_i^* \cdot \Delta t.$$

We have seen, however, that this limit is the definite integral of the function $q = q_0 e^{kt}$ from $t = 0$ to $t = T$, or

$$\text{Total consumption} = \int_0^T q_0 e^{kt}\, dt.$$

We can now evaluate this definite integral.

$$\begin{aligned}
\int_0^T q_0 e^{kt}\, dt &= q_0 \int_0^T e^{kt}\, dt = q_0 \left(\frac{e^{kt}}{k} \right) \Bigg|_0^T \\
&= \frac{q_0}{k} e^{kt} \Bigg|_0^T = \frac{q_0}{k} e^{kT} - \frac{q_0}{k} e^0 \\
&= \frac{q_0}{k} e^{kT} - \frac{q_0}{k}(1) \\
&= \frac{q_0}{k}(e^{kT} - 1) \qquad \textbf{(1)}
\end{aligned}$$

Now let us return to the numbers we gave for petroleum. We said that $q_0 = 17{,}100$ million barrels, where q_0 represents consumption in the base year of 1970. We have $k = .037$ with total petroleum reserves estimated at 1,500,000 million barrels. Thus, using Equation (1) we have

$$1{,}500{,}000 = \frac{17{,}100}{.037}(e^{.037T} - 1).$$

Multiply both sides of the equation by .037.

$$55{,}500 = 17{,}100(e^{.037T} - 1)$$

Divide both sides of the equation by 17,100.

$$3.2 = e^{.037T} - 1$$

Add 1 to both sides.

$$4.2 = e^{.037T}$$

Take natural logarithms of both sides.

$$\begin{aligned}
\ln 4.2 &= \ln e^{.037T} \\
&= .037T
\end{aligned}$$

Finally,

$$T = \frac{\ln 4.2}{.037} \approx 39.$$

By this result, petroleum reserves should last until 39 years after 1970, that is, until about 2009.

In fact, in the early 1970s some analysts were predicting that reserves would be exhausted before the end of the century, and this was a reasonable guess. But two things have happened since then: The growth in consumption has slowed, and more reserves have been discovered. One way to refine our model is to look at the historical data over a longer time span. The following table gives average world annual petroleum consumption in millions of barrels at 5-year intervals from 1970 to 2000.*

Year	World Consumption (in millions of barrels)
1970	17,100
1975	20,500
1980	23,000
1985	21,900
1990	24,100
1995	25,400
2000	28,100

The first step in comparing this data with our exponential model is to estimate a value for the growth constant k. One simple way of doing this is to solve the equation

$$28{,}100 = 17{,}100 \cdot e^{k \cdot 30}.$$

Using natural logarithms just as we did in estimating the time to depletion for $k = .037$, we find that

$$k = \frac{\ln\left(\dfrac{28{,}100}{17{,}100}\right)}{30} \approx .017.$$

So the data from the Bureau of Transportation Statistics suggests a growth constant of about 1.7%. We can check the fit by plotting the function $17{,}100 \cdot e^{.017t}$ along with a bar graph of the consumption data, shown in Figure 34. The fit looks reasonably good, but over this short range of 30 years, the exponential model is close to a linear model, and the growth in consumption is certainly not smooth.

The exponential model rests on the assumption of a constant growth rate. As already noted, we might expect instead that the growth rate would change as the world comes closer to exhausting its reserves. In particular, scarcity might drive up the price of oil and thus reduce consumption. We can use integration to explore an alternative model in which the factor k changes over time, so that k becomes $k(t)$, a function of time.

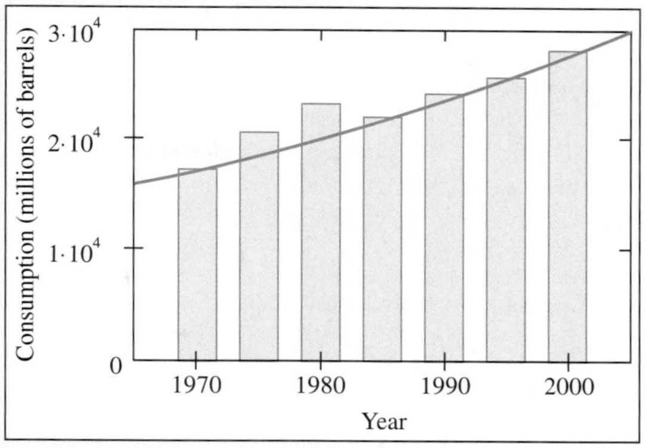

FIGURE 34

As an illustration, we explore a model in which the growth constant k declines toward 0 over time. We'll use 1970 as our base year, so the variable t will count years since 1970. We need a simple positive function $k(t)$ that tends toward 0 as t gets large. To get some numbers to work with, assume that the growth rate was 2% in 1970 and declined to 1% by 1995. There are many possible choices for the function $k(t)$, but a convenient one is

$$k(t) = \frac{.5}{t + 25}.$$

Using integration to turn the instantaneous rate of consumption into the total consumption up to time T, we can write

$$\text{Total consumption} = 17{,}100\int_0^T e^{k(t)\cdot t}\,dt$$
$$= 17{,}100\int_0^T e^{.5t/(t+25)}\,dt.$$

We'd like to find out when the world will use up its estimated reserves, but as just noted, the estimates have increased since the 1970s. A paper presented by the U.S. Geological Survey at the 1998 World Petroleum Conference estimates the *current* global petroleum reserves at 2,300,000 million barrels.† So we need to solve

$$2{,}300{,}000 = 17{,}100\int_0^T e^{.5t/(t+25)}\,dt \qquad \textbf{(2)}$$

But this problem is much harder to solve than the corresponding problem for constant growth, because *there is no formula for evaluating this definite integral!* The function

$$g(t) = e^{.5t/(t+25)}$$

doesn't have an antiderivative that we can write down in terms of functions that we know how to compute.

*Bureau of Transportation Statistics of the U.S. Department of Transportation.
†http://energy.er.usgs.gov/products/papers/WPC/14/.

Here the numerical integration techniques discussed in Section 6 come to the rescue. We can use one of the integration rules to *approximate* the integral numerically for various values of T, and with some trial and error we can estimate how long the reserves will last. If you have a calculator or computer algebra system that does numerical integration, you can pick some T values and evaluate the right-hand side of Equation (2). Here are the results produced by one computer algebra system:

For $T = 90$ the integral is about 2,062,000.

For $T = 100$ the integral is about 2,316,000.

For $T = 110$ the integral is about 2,572,000.

So using this model we would estimate that starting in 1970 the petroleum reserves would last for about 100 years, that is, until 2070.

Our integration tools are essential in building and exploring models of resource use, but the difference in our two predictions (39 years vs. 100 years) illustrates the difficulty of making accurate predictions. A model that performs well on historical data may not take the changing dynamics of resource use into account, leading to forecasts that are either unduly gloomy or too optimistic.

Exercises

1. Find the number of years that the estimated petroleum reserves would last if used at the same rate as in the base year.
2. How long would the estimated petroleum reserves last if the growth constant was only 2% instead of 3.7%?

Estimate the length of time until depletion for each mineral.

3. Bauxite (the ore from which aluminum is obtained): estimated reserves in base year 15,000,000 thousand tons; rate of consumption in base year 63,000 thousand tons; growth constant 6%
4. Bituminous coal: estimated world reserves 2,000,000 million tons; rate of consumption in base year 2200 million tons; growth constant 4%
5. **a.** Verify that the function $k(t)$ defined on the previous page has the right values at $k = 0$ and $k = 25$.
   b. Find a similar function that has $k(0) = .03$ and $k(25) = .02$.
6. **a.** Use the function you defined in Exercise 5 b to write an integral for world petroleum consumption from 1970 until T years after 1970.
   b. If you have access to a numerical integrator, compute some values of your integral and estimate the time required to exhaust the reserve of 2,300,000 million barrels.
7. A reasonable assumption is that over time scarcity might drive up the price of oil and thus reduce consumption. Comment on the fact that the rate of oil consumption actually increased in 2002, connecting current events and economic forecasts to the short-term possibility of a reduction in consumption.

Directions for Group Project

Suppose that you and three other students are spending a summer as interns for a local congresswoman. During your internship your realize that the information contained in your calculus class could be used to help with a new bill under consideration. The primary purpose of the bill is to require, by law, that all cars manufactured after a certain date get at least 60 miles per gallon of gasoline. Prepare a report that uses the information above to make a case for or against a bill of this nature.

CHAPTER

8 Further Techniques and Applications of Integration

It might seem that definite integrals with infinite limits have only theoretical interest, but in fact these *improper* integrals provide answers to many practical questions. An example in Section 4 models an environmental cleanup process in which the amount of pollution entering a stream decreases by a constant fraction each year. An improper integral gives the total amount of pollutant that will ever enter the river.

In the previous chapter we discussed indefinite and definite integrals, and presented rules for finding the antiderivatives of several types of functions. We showed how numerical methods can be used for functions that cannot be integrated by the techniques presented there. In this chapter we develop additional methods of integrating functions. We also show how to evaluate an integral that has one or both limits at infinity. These new techniques allow us to consider additional applications of integration, such as volumes of solids of revolution, the average value of a function, and continuous money flow.

8.1 INTEGRATION BY PARTS

THINK ABOUT IT *If we know the rate of growth of a patch of moss, how can we calculate the area the moss covers?*

The technique of *integration by parts* often makes it possible to reduce a complicated integral to a simpler integral. We know that if u and v are both differentiable functions, then uv is also differentiable and, by the product rule for derivatives,

$$\frac{d(uv)}{dx} = u\frac{dv}{dx} + v\frac{du}{dx}.$$

This expression can be rewritten, using differentials, as

$$d(uv) = u\,dv + v\,du.$$

Integrating both sides of this last equation gives

$$\int d(uv) = \int u\,dv + \int v\,du,$$

or

$$uv = \int u\,dv + \int v\,du.$$

Rearranging terms gives the following formula.

INTEGRATION BY PARTS

If u and v are differentiable functions, then

$$\int u\,dv = uv - \int v\,du.$$

The process of finding integrals by this formula is called **integration by parts.** There are two ways to do integration by parts: the standard method and column integration. Both methods are illustrated in the following example.

EXAMPLE 1 *Integration by Parts*

Find $\int xe^{5x}\,dx$.

Method 1: Standard Method

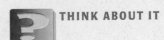

Solution Although this integral cannot be found by using any method studied so far, it can be found with integration by parts. First write the expression $xe^{5x}\,dx$ as a product of two functions u and dv in such a way that $\int dv$ can be found. One way to do this is to choose the two functions x and e^{5x}. Both x and e^{5x} can be integrated, but $\int x\,dx$, which is $x^2/2$, is more complicated than x itself, while the derivative of x is 1, which is simpler than x. Since e^{5x} remains the same (except for the coefficient) whether it is integrated or differentiated, it is best here to choose

$$dv = e^{5x}\,dx \qquad \text{and} \qquad u = x.$$

Then

$$du = dx,$$

and v is found by integrating dv:

$$v = \int dv = \int e^{5x}\,dx = \frac{e^{5x}}{5}.$$

We need not introduce the constant of integration until the last step, because only one constant is needed. Now substitute into the formula for integration by parts and complete the integration.

$$\int u\,dv = uv - \int v\,du$$

$$\int \underbrace{x}_{u}\underbrace{e^{5x}\,dx}_{dv} = \underbrace{x}_{u}\underbrace{\left(\frac{e^{5x}}{5}\right)}_{v} - \int \underbrace{\frac{e^{5x}}{5}}_{v}\underbrace{dx}_{du}$$

$$= \frac{xe^{5x}}{5} - \frac{e^{5x}}{25} + C$$

The constant C was added in the last step. As before, check the answer by taking its derivative.

FOR REVIEW

In the section on Substitution, we pointed out that when the chain rule is used to find the derivative of the function e^{kx}, we multiply by k, so when finding the antiderivative of e^{kx}, we divide by k. Thus $\int e^{5x}\,dx = e^{5x}/5 + C$. Keeping this technique in mind makes integration by parts simpler.

Method 2: Column Integration

A technique called **column integration,** or *tabular integration*, is equivalent to integration by parts but helps in organizing the details.* We begin by creating two columns. The first column, labeled D, contains u, the part to be differentiated in the original integral. The second column, labeled I, contains the rest of the integral: that is, the part to be integrated, but without the dx. To create the remainder of the first column, write the derivative of the function in the first row underneath it in the second row. Now write the derivative of the function in the second row underneath it in the third row. Proceed in this manner down the first column, taking derivatives until you get a 0. Form the second column in a similar manner, except take an antiderivative at each row, until the second column has the same number of rows as the first.

To illustrate this process, consider our goal of finding $\int xe^{5x}\,dx$. Here $u = x$, so e^{5x} is left for the second column. Taking derivatives down the first column and antiderivatives down the second column results in the following table.

D	***I***
x	e^{5x}
1	$e^{5x}/5$
0	$e^{5x}/25$

*This technique appeared in the 1988 movie *Stand and Deliver*.

Next, draw a diagonal line from each term (except the last) in the left column to the term in the row below it in the right column. Label the first such line with "+", the next with "−", and continue alternating the signs as shown.

D		I
x	$+$	e^{5x}
1	$-$	$e^{5x}/5$
0		$e^{5x}/25$

Then multiply the terms on opposite ends of each diagonal line. Finally, sum up the products just formed, adding the "+" terms and subtracting the "−" terms.

$$\int xe^{5x}\,dx = x(e^{5x}/5) - 1(e^{5x}/25) + C$$

$$= \frac{xe^{5x}}{5} - \frac{e^{5x}}{25} + C$$

Compare these steps with those of Method 1 and convince yourself that the process is the same.

CONDITIONS FOR INTEGRATION BY PARTS

Integration by parts can be used only if the integrand satisfies the following conditions.

1. The integrand can be written as the product of two factors, u and dv.
2. It is possible to integrate dv to get v and to differentiate u to get du.
3. The integral $\int v\,du$ can be found.

EXAMPLE 2 *Integration by Parts*

Find $\int \ln x\,dx$ for $x > 0$.

Method 1: Standard Method

Solution No rule has been given for integrating $\ln x$, so choose

$$dv = dx \qquad \text{and} \qquad u = \ln x.$$

Then

$$v = x \qquad \text{and} \qquad du = \frac{1}{x}\,dx,$$

and, since $uv = vu$, we have

$$\int u \cdot dv = v \cdot u - \int v \cdot du$$

$$\int \overbrace{\ln x}\,\overbrace{dx} = \overbrace{x}\,\overbrace{\ln x} - \int \overbrace{x} \cdot \overbrace{\frac{1}{x}\,dx}$$

$$= x \ln x - \int dx$$

$$= x \ln x - x + C.$$

Method 2: Column Integration

Column integration works a little differently here. As in Method 1, choose $\ln x$ as the part to differentiate. The part to be integrated must be 1. (Think of $\ln x$ as

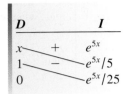

$1 \cdot \ln x$.) No matter how many times $\ln x$ is differentiated, the result is never 0. In this case, stop as soon as the natural logarithm is gone.

D	*I*
$\ln x$	1
$1/x$	x

Draw diagonal lines with alternating + and − as before. On the last line, because the left column does not contain a 0, draw a horizontal line.

D		*I*
$\ln x$	+	1
$1/x$	−	x

The presence of a horizontal line indicates that the product is to be integrated, just as the original integral was represented by the first row of the two columns.

$$\begin{aligned}\int \ln x\,dx &= (\ln x)x - \int \frac{1}{x}\cdot x\,dx\\ &= x\ln x - \int dx\\ &= x\ln x - x + C.\end{aligned}$$

Note that when setting up the columns, a horizontal line is drawn only when a 0 does not eventually appear in the left column.

Sometimes integration by parts must be applied more than once, as in the next example.

EXAMPLE 3 *Integration by Parts*

Find $\int 2x^2e^{-3x}\,dx$.

Method 1: Standard Method

Solution Choose

$$dv = e^{-3x}\,dx \qquad \text{and} \qquad u = 2x^2.$$

Then

$$v = \frac{-e^{-3x}}{3} \qquad \text{and} \qquad du = 4x\,dx.$$

Substitute these values into the formula for integration by parts.

$$\int u\,dv = uv - \int v\,du$$

$$\begin{aligned}\int 2x^2e^{-3x}\,dx &= 2x^2\left(\frac{-e^{-3x}}{3}\right) - \int\left(\frac{-e^{-3x}}{3}\right)4x\,dx\\ &= -\frac{2}{3}x^2e^{-3x} + \frac{4}{3}\int xe^{-3x}\,dx\end{aligned}$$

Now apply integration by parts to the last integral, letting

$$dv = e^{-3x}\,dx \qquad \text{and} \qquad u = x,$$

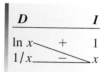

so

$$v = \frac{-e^{-3x}}{3} \quad \text{and} \quad du = dx.$$

$$\begin{aligned}\int 2x^2e^{-3x}\,dx &= -\frac{2}{3}x^2e^{-3x} + \frac{4}{3}\int xe^{-3x}\,dx \\ &= -\frac{2}{3}x^2e^{-3x} + \frac{4}{3}\left[x\left(\frac{-e^{-3x}}{3}\right) - \int\left(\frac{-e^{-3x}}{3}\right)dx\right] \\ &= -\frac{2}{3}x^2e^{-3x} + \frac{4}{3}\left[-\frac{x}{3}e^{-3x} - \frac{e^{-3x}}{9}\right] + C \\ &= -\frac{2}{3}x^2e^{-3x} - \frac{4}{9}xe^{-3x} - \frac{4}{27}e^{-3x} + C\end{aligned}$$

Method 2: Column Integration

Choose $2x^2$ as the part to be differentiated, and put e^{-3x} in the integration column.

D		*I*
$2x^2$	$+$	e^{-3x}
$4x$	$-$	$-e^{-3x}/3$
4	$+$	$e^{-3x}/9$
0		$-e^{-3x}/27$

Multiplying and adding as before yields

$$\begin{aligned}\int 2x^2e^{-3x}\,dx &= 2x^2(-e^{-3x}/3) - 4x(e^{-3x}/9) + 4(-e^{-3x}/27) + C \\ &= -\frac{2}{3}x^2e^{-3x} - \frac{4}{9}xe^{-3x} - \frac{4}{27}e^{-3x} + C.\end{aligned}$$

With the functions discussed so far in this book, choosing u and dv (or the parts to be differentiated and integrated) is relatively simple. Before trying integration by parts, first see if the integration can be performed using substitution. If substitution does not work, see if $\ln x$ is in the integral. If it is, set $u = \ln x$ and dv equal to the rest of the integral. (Equivalently, put $\ln x$ in the *D* column and the rest of the function in the *I* column.) If $\ln x$ is not present, see if x^k is present, where k is any positive integer. If it is present, set $u = x^k$ and dv equal to the rest of the integral. (Equivalently, put x^k in the *D* column and the rest of the function in the *I* column.)

EXAMPLE 4 *Definite Integral*

Find $\displaystyle\int_1^e \frac{\ln x}{x^2}\,dx.$

Solution First find the indefinite integral using integration by parts by the standard method. (You may wish to verify this using column integration.) Whenever $\ln x$ is present, it is selected as u, so let

$$u = \ln x \quad \text{and} \quad dv = \frac{1}{x^2}\,dx.$$

FOR REVIEW

Recall that $\int x^n\,dx = x^{n+1}/(n+1) + C, n \neq -1$, so $\int 1/x^2\,dx = \int x^{-2}\,dx = x^{-1}/(-1) + C = -1/x + C$.

Then

$$du = \frac{1}{x}\,dx \quad \text{and} \quad v = -\frac{1}{x}.$$

Substitute these values into the formula for integration by parts, and integrate the second term on the right.

$$\int u\,dv = uv - \int v\,du$$

$$\begin{aligned}\int \frac{\ln x}{x^2}\,dx &= (\ln x)\frac{-1}{x} - \int\left(-\frac{1}{x}\cdot\frac{1}{x}\right)dx\\ &= -\frac{\ln x}{x} + \int \frac{1}{x^2}\,dx\\ &= -\frac{\ln x}{x} - \frac{1}{x} + C\\ &= \frac{-\ln x - 1}{x} + C\end{aligned}$$

Now find the definite integral.

$$\begin{aligned}\int_1^e \frac{\ln x}{x^2}\,dx &= \left.\frac{-\ln x - 1}{x}\right|_1^e\\ &= \left(\frac{-1-1}{e}\right) - \left(\frac{0-1}{1}\right)\\ &= \frac{-2}{e} + 1 \approx .2642411177\end{aligned}$$

Definite integrals can be found with a graphing calculator using the function integral feature or by finding the area under the graph of the function between the limits. For example, using the `fnInt` feature of the TI-83/84 Plus calculator to find the integral in Example 4 gives .2642411177. Using the area under the graph approach gives .26424112, the same result rounded.

Many integrals cannot be found by the methods presented so far. For example, consider the integral

$$\int \frac{1}{4 - x^2}\,dx.$$

Substitution of $u = 4 - x^2$ will not help, because $du = -2x\,dx$, and there is no x in the numerator of the integral. We could try integration by parts, using $dv = dx$ and $u = (4 - x^2)^{-1}$. Integration gives $v = x$ and differentiation gives $du = 2x\,dx/(4 - x^2)^2$, with

$$\int \frac{1}{4 - x^2}\,dx = \frac{x}{4 - x^2} - \int \frac{2x^2}{(4 - x^2)^2}\,dx.$$

The integral on the right is more complicated than the original integral, however. A second use of integration by parts on the new integral would only make matters worse. Since we cannot choose $dv = (4 - x^2)^{-1}\,dx$ because it cannot be integrated by the methods studied so far, integration by parts is not possible for this problem.

This integration can be performed using one of the many techniques of integration beyond the scope of this text.* Tables of integrals can also be used, but technology is rapidly making such tables obsolete, and even reducing the impor-

*For example, see Thomas, George B., Ross L. Finney, Maurice D. Weir, and Frank Giordano, *Thomas' Calculus*, 10th ed., Addison Wesley Longman, 2000.

tance of techniques of integration. The following example shows how the table of integrals given in the appendix of this book may be used.

EXAMPLE 5 *Tables of Integrals*

Find $\int \frac{1}{4 - x^2}\,dx$.

Solution Using formula 7 in the table of integrals in the appendix, with $a = 2$, gives

$$\int \frac{1}{4 - x^2}\,dx = \frac{1}{4} \cdot \ln \left|\frac{2 + x}{2 - x}\right| + C.$$

We mentioned in the previous chapter how computer algebra systems and some calculators can perform integration. Using a TI-89, the answer to the above integral is

$$\frac{\ln\left(\frac{|x + 2|}{|x - 2|}\right)}{4}.$$

(The C is not included.) Verify that this is equivalent to the answer given in Example 5.

If you don't have a calculator or computer program that integrates symbolically, there is a Web site (http://integrals.wolfram.com), as of this writing, that finds indefinite integrals using the computer algebra system Mathematica. It includes instructions on how to enter your function. When the previous integral was entered, it returned the answer

$$-\frac{1}{4}\text{Log}[-2 + x] + \frac{1}{4}\text{Log}[2 + x].$$

Note that Mathematica does not include the C or the absolute value, and that natural logarithms are written as Log. Verify that this answer is equivalent to the answer given by the TI-89 and the answer given in Example 5.

Unfortunately, there are integrals that cannot be antidifferentiated by any technique, in which case numerical integration must be used. (See the last section of the previous chapter.) In this book, for simplicity, all integrals to be antidifferentiated can be done with substitution or by parts, except for Exercise 23–28 in this section.

8.1 EXERCISES

Use integration by parts to find the integrals in Exercises 1–10.

1. $\int xe^x\,dx$

2. $\int (x + 1)e^x\,dx$

3. $\int (5x - 9)e^{-3x}\,dx$

4. $\int (6x + 3)e^{-2x}\,dx$

5. $\int_0^1 \frac{2x + 1}{e^x}\,dx$

6. $\int_0^1 \frac{1 - x}{3e^x}\,dx$

7. $\int_1^4 \ln 2x\,dx$

8. $\int_1^2 \ln 5x\,dx$

9. $\int x \ln x\,dx$

10. $\int x^2 \ln x\,dx$

11. Find the area between $y = (x - 2)e^x$ and the x-axis from $x = 2$ to $x = 4$.

12. Find the area between $y = xe^x$ and the x-axis from $x = 0$ to $x = 1$.

Exercises 13–22 are mixed—some require integration by parts, while others can be integrated by using techniques discussed in the chapter on Integration.

13. $\int x^2 e^{2x}\, dx$

14. $\int_1^2 (1 - x^2)e^{2x}\, dx$

15. $\int_0^5 x\sqrt[3]{x^2 + 2}\, dx$

16. $\int (2x - 1)\ln(3x)\, dx$

17. $\int (8x + 7)\ln(5x)\, dx$

18. $\int xe^{x^2}\, dx$

19. $\int x^2\sqrt{x + 2}\, dx$

20. $\int_0^1 \frac{x^2\, dx}{2x^3 + 1}$

21. $\int_0^1 \frac{x^3\, dx}{\sqrt{3 + x^2}}$

22. $\int \frac{x^2\, dx}{2x^3 + 1}$

Use the table of integrals, or a computer or calculator with symbolic integration capabilities, to find each indefinite integral.

23. $\int \frac{9}{\sqrt{x^2 + 9}}\, dx$

24. $\int \frac{6}{x^2 - 9}\, dx$

25. $\int \frac{3}{x\sqrt{121 - x^2}}\, dx$

26. $\int \frac{2}{3x(3x - 5)}\, dx$

27. $\int \frac{-3}{x(4x + 3)^2}\, dx$

28. $\int \sqrt{x^2 + 10}\, dx$

29. What rule of differentiation is related to integration by parts?

30. Explain why the two methods of solving Example 2 are equivalent.

31. Use integration by parts to derive the following formula from the table of integrals.

$$\int x^n \cdot \ln|x|\, dx = x^{n+1}\left[\frac{\ln|x|}{n + 1} - \frac{1}{(n + 1)^2}\right] + C, \quad n \neq -1$$

32. Use integration by parts to derive the following formula from the table of integrals.

$$\int x^n e^{ax}\, dx = \frac{x^n e^{ax}}{a} - \frac{n}{a}\int x^{n-1}e^{ax}\, dx + C, \quad a \neq 0$$

33. a. One way to integrate $\int x\sqrt{x + 1}\, dx$ is to use integration by parts. Do so to find the antiderivative.

b. Another way to evaluate the integral in part a is by using the substitution $u = x + 1$. Do so to find the antiderivative.

c. Compare the results from the two methods. If they do not look the same, explain how this can happen. Discuss the advantages and disadvantages of each method.

34. Using integration by parts,

$$\begin{aligned}\int \frac{1}{x}\, dx &= \int \frac{1}{x} \cdot 1\, dx \\ &= \frac{1}{x} \cdot x - \int\left(-\frac{1}{x^2}\right)x\, dx \\ &= 1 + \int \frac{1}{x}\, dx.\end{aligned}$$

Subtracting $\int \frac{1}{x}\, dx$ from both sides we conclude that $0 = 1$. What is wrong with this logic?*

*Problem submitted by Sam Northshield, Plattsburgh State University.

Applications

BUSINESS AND ECONOMICS

35. *Rate of Change of Revenue* The rate of change of revenue (in dollars per calculator) from the sale of x small desk calculators is

$$R'(x) = (x + 1)\ln(x + 1).$$

Find the total revenue from the sale of the first 12 calculators. (*Hint:* In this exercise, it simplifies matters to write an antiderivative of $x + 1$ as $(x + 1)^2/2$ rather than $x^2/2 + x$.)

LIFE SCIENCES

36. *Reaction to a Drug* The rate of reaction to a drug is given by

$$r'(x) = 2x^2e^{-x},$$

where x is the number of hours since the drug was administered. Find the total reaction to the drug from $x = 1$ to $x = 6$.

37. *Growth of a Population* The rate of growth of a microbe population is given by

$$m'(x) = 30xe^{2x},$$

where x is time in days. What is the total accumulated growth during the first 3 days?

38. *Rate of Growth* The area covered by a patch of moss is growing at a rate of

$$A'(t) = \sqrt{t}\ln t$$

cm^2 per day, for $t \geq 1$. Find the additional amount of area covered by the moss between 4 and 9 days.

39. *Thermic Effect of Food* As we saw in an earlier chapter, a person's metabolic rate tends to go up after eating a meal and then, after some time has passed, it returns to a resting metabolic rate. This phenomenon is known as the thermic effect of food, and the effect (in kJ per hour) for one individual is

$$F(t) = -10.28 + 175.9te^{-t/1.3},$$

where t is the number of hours that have elapsed since eating a meal.* Find the total thermic energy of a meal for the next six hours after a meal by integrating the thermic effect function between $t = 0$ and $t = 6$.

40. *Rumen Fermentation* The rumen is the first division of the stomach of a ruminant, or cud-chewing animal. An article on the rumen microbial system reports that the fraction of the soluble material passing from the rumen without being fermented during the first hour after its ingestion could be calculated by the integral

$$\int_0^1 ke^{-kt}(1 - t)dt,$$

where k measures the rate that the material is fermented.†

a. Determine the above integral, and evaluate it for the following values of k used in the article: $1/12$, $1/24$, and $1/48$ hour.

b. The fraction of intermediate material left in the rumen at 1 hour that escapes digestion by passage between 1 and 6 hours is given by

$$\int_1^6 ke^{-kt}(6 - t)/5\,dt.$$

Determine this integral, and evaluate it for the values of k given in part a.

*Reed, George and James Hill, "Measuring the Thermic Effect of Food," *American Journal of Clinical Nutrition*, Vol. 63, 1996, pp. 164–169.

†Hungate, R. E., "The Rumen Microbial Ecosystem," *Annual Review of Ecology and Systematics*, Vol. 6, 1975, pp. 39–66.

8.2 VOLUME AND AVERAGE VALUE

THINK ABOUT IT

If we have a formula giving the price of a common stock as a function of time, how can we find the average price of the stock over a certain period of time?

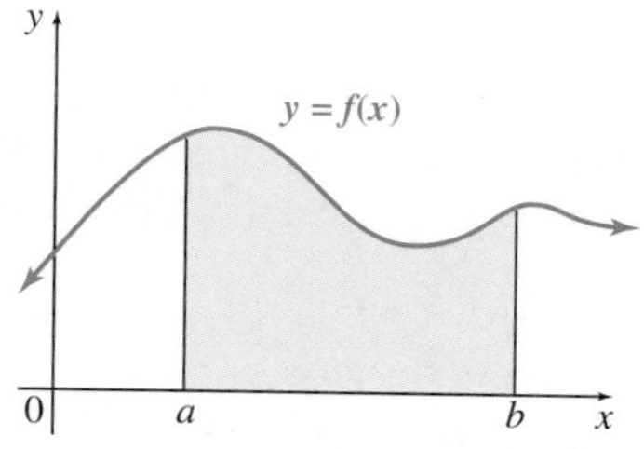

FIGURE 1

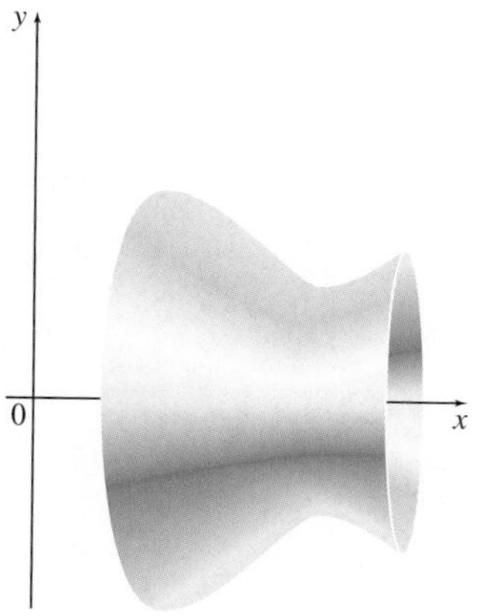

FIGURE 2

In this section, we will discover how to find the average value of a function, as well as how to compute the volume of a solid.

Volume Figure 1 shows the region below the graph of some function $y = f(x)$, above the x-axis, and between $x = a$ and $x = b$. We have seen how to use integrals to find the area of such a region. Now, suppose this region is revolved about the x-axis as shown in Figure 2. The resulting figure is called a **solid of revolution.** In many cases, the volume of a solid of revolution can be found by integration.

To begin, divide the interval $[a, b]$ into n subintervals of equal width Δx by the points $a = x_0, x_1, x_2, \dots, x_i, \dots, x_n = b$. Then think of slicing the solid into n slices of equal thickness Δx, as shown in Figure 3. If the slices are thin enough, each slice is very close to being a right circular cylinder. The formula for the volume of a right circular cylinder is $\pi r^2 h$, where r is the radius of the circular base and h is the height of the cylinder. As shown in Figure 4, the height of each slice is Δx. (The height is horizontal here, since the cylinder is on its side.) The radius of the circular base of each slice is $f(x_i)$. Thus, the volume of the slice is closely approximated by $\pi[f(x_i)]^2\Delta x$. The volume of the solid of revolution will be approximated by the sum of the volumes of the slices:

$$V \approx \sum_{i=1}^{n} \pi[f(x_i)]^2 \Delta x.$$

By definition, the volume of the solid of revolution is the limit of this sum as the thickness of the slices approaches 0, or

$$V = \lim_{\Delta x \to 0} \sum_{i=1}^{n} \pi[f(x_i)]^2 \Delta x.$$

This limit, like the one discussed earlier for area, is a definite integral.

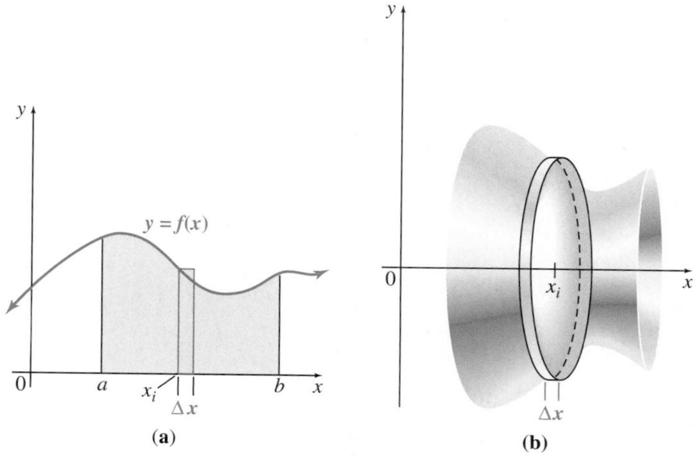

FIGURE 3

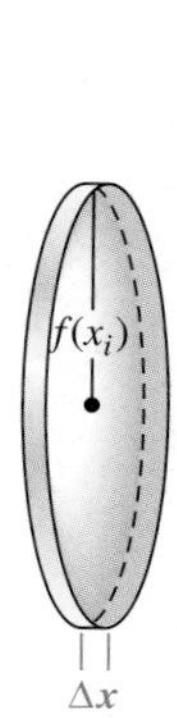

FIGURE 4

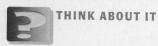

VOLUME OF A SOLID OF REVOLUTION

If $f(x)$ is nonnegative and R is the region between $f(x)$ and the x-axis from $x = a$ to $x = b$, the volume of the solid formed by rotating R about the x-axis is given by

$$V = \lim_{\Delta x \to 0} \sum_{i=1}^{n} \pi[f(x_i)]^2 \Delta x = \int_a^b \pi[f(x)]^2\,dx.$$

The technique of summing disks to approximate volumes was originated by Johannes Kepler (1571–1630), a famous German astronomer who discovered three laws of planetary motion. He estimated volumes of wine casks used at his wedding by means of solids of revolution.

EXAMPLE 1 *Volume*

Find the volume of the solid of revolution formed by rotating about the x-axis the region bounded by $y = x + 1$, $y = 0$, $x = 1$, and $x = 4$.

Solution The region and the solid are shown in Figure 5. Use the formula given above for the volume, with $a = 1$, $b = 4$, and $f(x) = x + 1$.

$$V = \int_1^4 \pi(x+1)^2\,dx = \pi\left[\frac{(x+1)^3}{3}\right]\Bigg|_1^4$$

$$= \frac{\pi}{3}(5^3 - 2^3) = \frac{117\pi}{3} = 39\pi$$

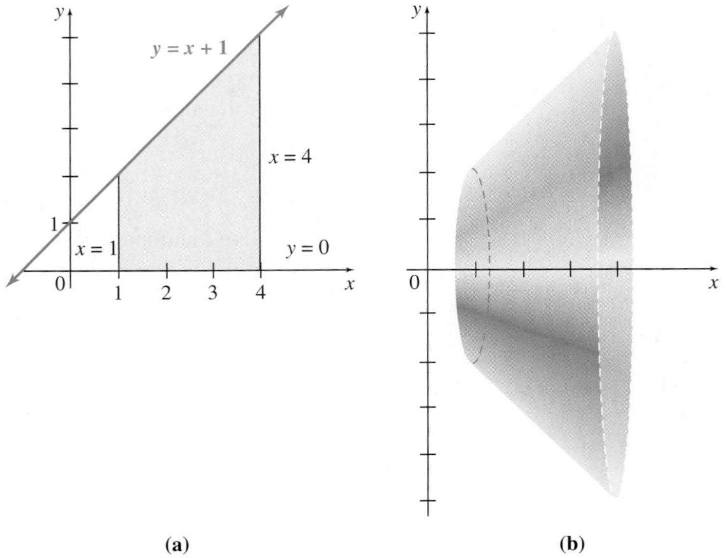

FIGURE 5

EXAMPLE 2 *Volume*

Find the volume of the solid of revolution formed by rotating about the x-axis the area bounded by $f(x) = 4 - x^2$ and the x-axis.

Solution The region and the solid are shown in Figure 6. Find a and b from the x-intercepts. If $y = 0$, then $x = 2$ or $x = -2$, so that $a = -2$ and $b = 2$. The volume is

$$V = \int_{-2}^{2} \pi(4 - x^2)^2\,dx$$

$$= \int_{-2}^{2} \pi(16 - 8x^2 + x^4)\,dx$$

$$= \pi\left(16x - \frac{8x^3}{3} + \frac{x^5}{5}\right)\Bigg|_{-2}^{2} = \frac{512\pi}{15}.$$

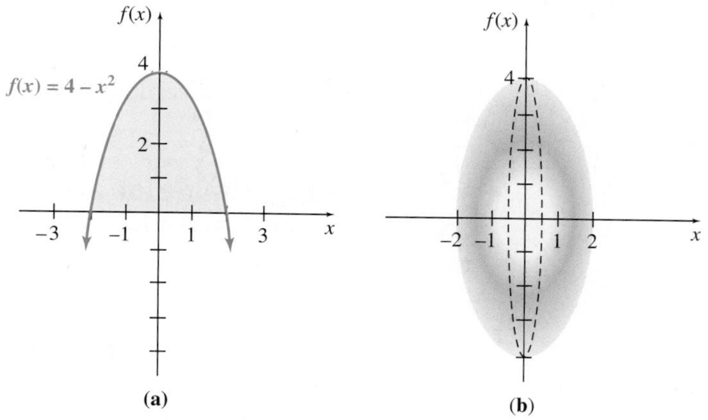

FIGURE 6

A graphing calculator with the fnInt feature gives the value as 107.2330292, which agrees with the approximation of $512\pi/15$ to the 7 decimal places shown.

EXAMPLE 3 *Volume*

Find the volume of a right circular cone with height h and base radius r.

Solution Figure 7(a) shows the required cone, while Figure 7(b) shows an area that could be rotated about the x-axis to get such a cone. The cone formed by

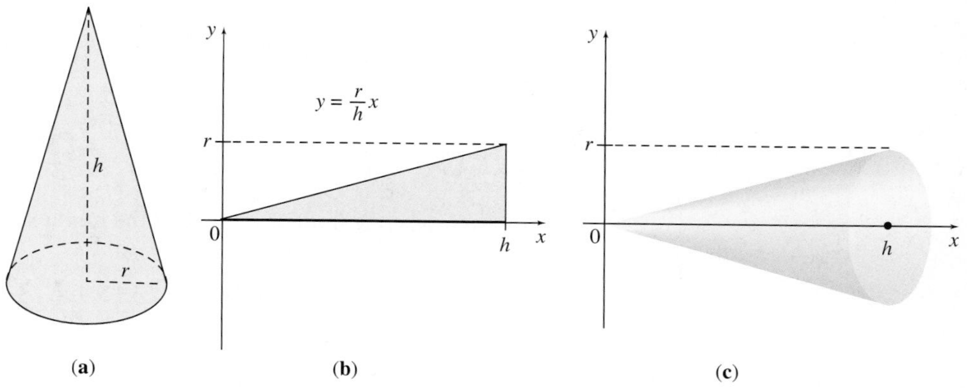

FIGURE 7

rotation is shown in Figure 7(c). Here $y = f(x)$ is the equation of the line through $(0, 0)$ and (h, r). The slope of this line is r/h, and since the y-intercept is 0, the equation of the line is

$$y = \frac{r}{h}x.$$

Then the volume is

$$\begin{aligned} V &= \int_0^h \pi\left(\frac{r}{h}x\right)^2 dx = \pi\int_0^h \frac{r^2x^2}{h^2}\,dx \\ &= \pi\frac{r^2x^3}{3h^2}\bigg|_0^h \qquad \text{Since } r \text{ and } h \text{ are constants} \\ &= \frac{\pi r^2 h}{3}. \end{aligned}$$

This is the familiar formula for the volume of a right circular cone.

Average Value of a Function The average of the n numbers $v_1, v_2, v_3, \ldots, v_i, \ldots, v_n$ is given by

$$\frac{v_1 + v_2 + v_3 + \cdots + v_n}{n} = \frac{\sum_{i=1}^{n} v_i}{n}.$$

For example, to compute an average temperature, we could take readings at equally spaced intervals and average the readings.

The average value of a function f on $[a, b]$ can be defined in a similar manner; divide the interval $[a, b]$ into n subintervals, each of width Δx. Then choose an x-value, x_i, in each subinterval, and find $f(x_i)$. The average function value for the n subintervals and the given choices of x_i is

$$\frac{f(x_1) + f(x_2) + \cdots + f(x_n)}{n} = \frac{\sum_{i=1}^{n} f(x_i)}{n}.$$

Since $(b - a)/n = \Delta x$, multiply the expression on the right side of the equation by $(b - a)/(b - a)$ and rearrange the expression to get

$$\begin{aligned} \frac{b-a}{b-a}\cdot\frac{\sum_{i=1}^{n} f(x_i)}{n} &= \frac{b-a}{n}\cdot\frac{\sum_{i=1}^{n} f(x_i)}{b-a} = \Delta x\cdot\frac{\sum_{i=1}^{n} f(x_i)}{b-a} \\ &= \frac{1}{b-a}\sum_{i=1}^{n} f(x_i)\Delta x. \end{aligned}$$

Now, take the limit as $n \to \infty$. If the limit exists, then

$$\lim_{n\to\infty}\frac{1}{b-a}\sum_{i=1}^{n} f(x_i)\Delta x = \frac{1}{b-a}\lim_{n\to\infty}\sum_{i=1}^{n} f(x_i)\Delta x = \frac{1}{b-a}\int_a^b f(x)\,dx.$$

The following definition summarizes this discussion.

AVERAGE VALUE OF A FUNCTION

The **average value of a function** f on the interval $[a, b]$ is

$$\frac{1}{b-a}\int_a^b f(x)\,dx,$$

provided the indicated definite integral exists.

In Figure 8 the quantity $\bar{y}$ represents the average height of the irregular region. The average height can be thought of as the height of a rectangle with base $b - a$. For $f(x) \geq 0$, this rectangle has area $\bar{y}(b - a)$, which equals the area under the graph of $f(x)$ from $x = a$ to $x = b$, so that

$$\bar{y}(b-a) = \int_a^b f(x)\,dx.$$

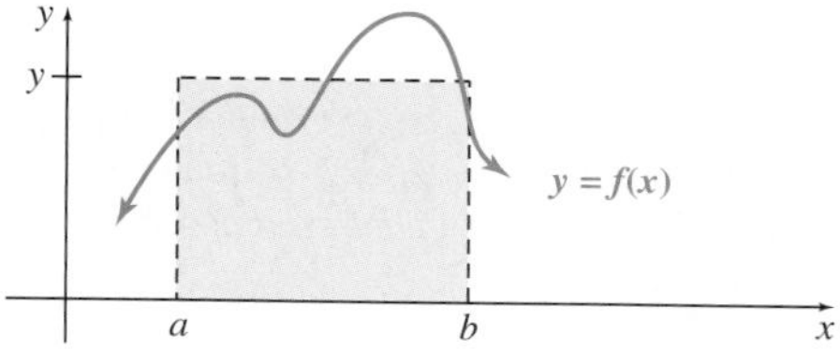

FIGURE 8

EXAMPLE 4 *Average Price*

A stock analyst plots the price per share of a certain common stock as a function of time and finds that it can be approximated by the function

$$S(t) = 25 - 5e^{-.01t},$$

where t is the time (in years) since the stock was purchased. Find the average price of the stock over the first six years.

Solution Use the formula for average value with $a = 0$ and $b = 6$. The average price is

$$\frac{1}{6-0}\int_0^6 (25 - 5e^{-.01t})\,dt = \frac{1}{6}\left(25t - \frac{5}{-.01}e^{-.01t}\right)\Bigg|_0^6$$

$$= \frac{1}{6}(25t + 500e^{-.01t})\Big|_0^6$$

$$= \frac{1}{6}(150 + 500e^{-.06} - 500)$$

$$= 20.147,$$

or approximately \$20.15.

Alternatively, using a graphing calculator with the `fnInt` feature to evaluate the integral, we get 120.88268. Dividing that result by 6 gives the average price as $20.14704447 \approx 20.15$.

8.2 EXERCISES

Find the volume of the solid of revolution formed by rotating about the x-axis each region bounded by the given curves.

1. $f(x) = x, \quad y = 0, \quad x = 0, \quad x = 2$

2. $f(x) = 2x, \quad y = 0, \quad x = 0, \quad x = 3$

3. $f(x) = 2x + 1, \quad y = 0, \quad x = 0, \quad x = 4$

4. $f(x) = x - 4, \quad y = 0, \quad x = 4, \quad x = 10$

5. $f(x) = \frac{1}{3}x + 2, \quad y = 0, \quad x = 1, \quad x = 3$

6. $f(x) = \frac{1}{2}x + 4, \quad y = 0, \quad x = 0, \quad x = 5$

7. $f(x) = \sqrt{x}, \quad y = 0, \quad x = 1, \quad x = 2$

8. $f(x) = \sqrt{x + 1}, \quad y = 0, \quad x = 0, \quad x = 3$

9. $f(x) = \sqrt{2x + 1}, \quad y = 0, \quad x = 1, \quad x = 4$

10. $f(x) = \sqrt{3x + 2}, \quad y = 0, \quad x = 1, \quad x = 2$

11. $f(x) = e^x, \quad y = 0, \quad x = 0, \quad x = 2$

12. $f(x) = 2e^x, \quad y = 0, \quad x = -2, \quad x = 1$

13. $f(x) = \frac{1}{\sqrt{x}}, \quad y = 0, \quad x = 1, \quad x = 4$

14. $f(x) = \frac{1}{\sqrt{x + 1}}, \quad y = 0, \quad x = 0, \quad x = 2$

15. $f(x) = x^2, \quad y = 0, \quad x = 1, \quad x = 5$

16. $f(x) = \frac{x^2}{2}, \quad y = 0, \quad x = 0, \quad x = 4$

17. $f(x) = 1 - x^2, \quad y = 0$

18. $f(x) = 2 - x^2, \quad y = 0$

The function defined by $y = \sqrt{r^2 - x^2}$ has as its graph a semicircle of radius r with center at $(0, 0)$ (see the figure). In Exercise 19–21, find the volume that results when each semicircle is rotated about the x-axis. (The result of Exercise 21 gives a formula for the volume of a sphere with radius r.)

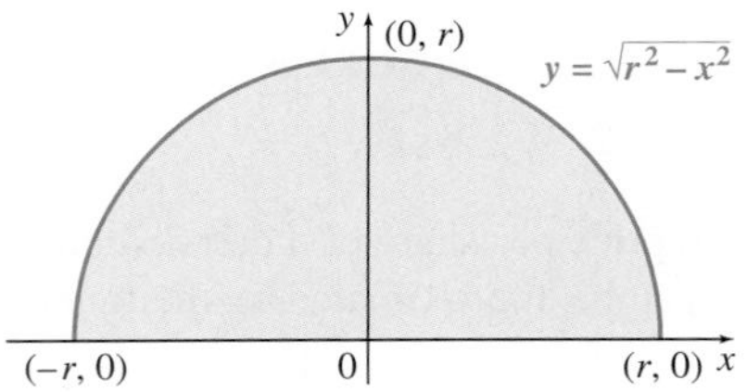

19. $f(x) = \sqrt{1 - x^2}$

20. $f(x) = \sqrt{16 - x^2}$

21. $f(x) = \sqrt{r^2 - x^2}$

22. Find a formula for the volume of an ellipsoid. See Exercise 19–21 and the following figures.

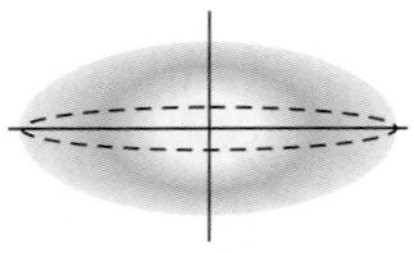

Ellipsoid

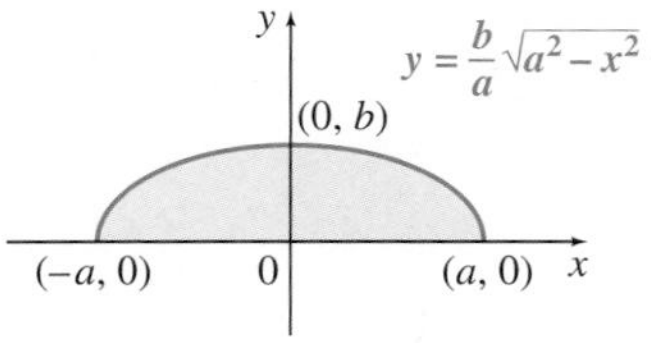

23. Use the methods of this section to find the volume of a cylinder with height h and radius r.

Find the average value of each function on the given interval.

24. $f(x) = 3 - 2x^2; \quad [1, 9]$

25. $f(x) = x^2 - 2; \quad [0, 5]$

26. $f(x) = (2x - 1)^{1/2}; \quad [1, 13]$

27. $f(x) = \sqrt{x + 1}; \quad [3, 8]$

28. $f(x) = e^{.1x}; \quad [0, 10]$

29. $f(x) = e^{x/5}; \quad [0, 5]$

30. $f(x) = x \ln x; \quad [1, e]$

31. $f(x) = x^2 e^{2x}; \quad [0, 2]$

In Exercise 32 and 33, use the integration feature on a graphing calculator to find the volume of the solid of revolution by rotating about the x-axis each region bounded by the given curves.

32. $f(x) = \dfrac{1}{1 + x^2}$, $y = 0$, $x = -1$, $x = 1$

33. $f(x) = e^{-x^2}$, $y = 0$, $x = -2$, $x = 2$

Applications

BUSINESS AND ECONOMICS

34. *Average Price* Otis Taylor plots the price per share of a stock that he owns as a function of time and finds that it can be approximated by the function

$$S(t) = t(25 - 5t) + 18,$$

where t is the time (in years) since the stock was purchased. Find the average price of the stock over the first five years.

35. *Average Price* A stock analyst plots the price per share of a certain common stock as a function of time and finds that it can be approximated by the function

$$S(t) = 15 + 2e^{-.02t},$$

where t is the time (in years) since the stock was purchased. Find the average price of the stock over the first six years.

36. *Average Inventory* The Carter Fenton Fragrance Company (CFFC) receives a shipment of 400 cases of specialty perfume early Monday morning of every week. CFFC sells the perfume to retail outlets in California at a rate of about 80 cases per day during each business day (Monday through Friday). What is the average daily inventory for CFFC? (*Hint:* Find a function that represents the inventory for any given business day and then integrate.)

37. *Average Inventory* The DeMarco Pasta Company receives 600 cases of imported San Marzano tomato sauce every 30 days. The number of cases of sauce on inventory t days after the shipment arrives is

$$N(t) = 600 - 20\sqrt{30t}.$$

Find the average daily inventory.

LIFE SCIENCES

38. *Blood Flow* The figure shows the blood flow in a small artery of the body. The flow of blood is *laminar* (in layers), with the velocity very low near the artery walls and highest in the center of the artery. In this model of blood flow, we calculate the total flow in the artery by thinking of the flow as being made up of many layers of concentric tubes sliding one on the other.

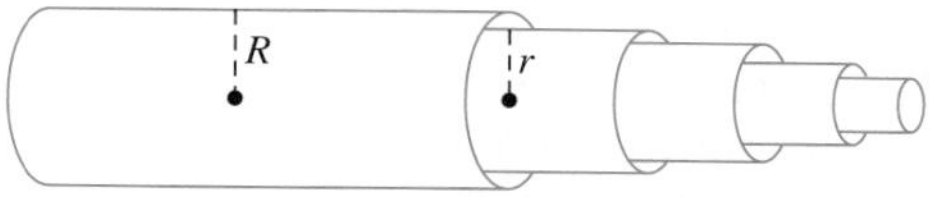

Suppose R is the radius of an artery and r is the distance from a given layer to the center. Then the velocity of blood in a given layer can be shown to equal

$$v(r) = k(R^2 - r^2),$$

where k is a numerical constant.

Since the area of a circle is $A = \pi r^2$, the change in the area of the cross section of one of the layers, corresponding to a small change in the radius, Δr, can be approximated by differentials. For $dr = \Delta r$, the differential of the area A is

$$dA = 2\pi r\, dr = 2\pi r\, \Delta r,$$

where Δr is the thickness of the layer. The total flow in the layer is defined to be the product of velocity and cross-section area, or

$$F(r) = 2\pi rk(R^2 - r^2)\Delta r.$$

a. Set up a definite integral to find the total flow in the artery.

b. Evaluate this definite integral.

39. *Drug Reaction* The intensity of the reaction to a certain drug, in appropriate units, is given by

$$R(t) = te^{-.1t},$$

where t is time (in hours) after the drug is administered. Find the average intensity during the following hours.

a. Second hour

b. Twelfth hour

c. Twenty-fourth hour

40. *Bird Eggs* The average length and width of various bird eggs are given in the table on the following page.*

*www.nctm.org/wlme/wlme6/five.htm.

Bird Name	Length (cm)	Width (cm)
Canada goose	8.6	5.8
Robin	1.9	1.5
Turtledove	3.1	2.3
Hummingbird	1.0	1.0
Raven	5.0	3.3

a. Assume for simplicity that a bird's egg is roughly the shape of an ellipsoid. Use the result of Exercise 22 to estimate the volume of an egg of each bird.

i. Canada goose

ii. Robin

iii. Turtledove

iv. Hummingbird

v. Raven

b. In Exercise 11 of Section 1.3, we showed that the average length (in centimeters) of an egg of width w cm is given by

$$l = 1.585w - .487.$$

Using this result and the ideas in part a, show that the average volume of an egg of width w centimeters is given by

$$V = \pi(1.585w^3 - .487w^2)/6.$$

Use this formula to calculate the average volume for the bird eggs in part a, and compare with your results from part a.

SOCIAL SCIENCES

41. *Production Rate* Suppose the number of items a new worker on an assembly line produces daily after t days on the job is given by

$$I(t) = 35\ln(t + 1).$$

Find the average number of items produced daily by this employee after the following numbers of days.

a. 5 **b.** 10 **c.** 15

42. *Typing Speed* The function $W(t) = -3.75t^2 + 30t + 40$ describes a typist's speed (in words per minute) over a time interval $[0, 5]$.

a. Find $W(0)$.

b. Find the maximum W value and the time t when it occurs.

c. Find the average speed over $[0, 5]$.

8.3 CONTINUOUS MONEY FLOW

THINK ABOUT IT *Given a changing rate of annual income and a certain rate of interest, how can we find the present value of the income?*

In an earlier chapter we looked at the concepts of present value and future value when a lump sum of money is deposited in an account and allowed to accumulate interest. In some situations, however, money flows into and out of an account almost continuously over a period of time. Examples include income in a store, bank receipts and payments, and highway tolls. Although the flow of money in such cases is not exactly continuous, it can be treated as though it were continuous, with useful results.

EXAMPLE 1 *Total Income*

The income from a soda machine (in dollars per year) is growing exponentially. When the machine was first installed, it was producing income at a rate of $500 per year. By the end of the first year, it was producing income at a rate of \$510.10 per year. Find the total income produced by the machine during its first 3 years of operation.

Solution Let t be the time (in years) since the installation of the machine. The assumption of exponential growth, coupled with the initial value of 500, implies that the rate of change of income is of the form

$$f(t) = 500e^{kt},$$

where k is some constant. To find k, use the value at the end of the first year.

$$\begin{aligned} f(1) = 500e^{k(1)} &= 510.10 \\ e^k &= 1.0202 && \text{Divide by 500.} \\ k &= \ln 1.0202 && \text{Take logarithms of both sides.} \\ &\approx .02 && \text{Round to the nearest hundredth.} \end{aligned}$$

We therefore have

$$f(t) = 500e^{.02t}.$$

Since the rate of change of incomes is given, the total income can be determined by using the definite integral.

$$\begin{aligned} \text{Total income} &= \int_0^3 500e^{.02t}\,dt \\ &= \frac{500}{.02}e^{.02t}\Big|_0^3 \\ &= 25{,}000e^{.02t}\Big|_0^3 = 25{,}000(e^{.06} - 1) = 1545.91 \end{aligned}$$

Thus, the soda machine will produce \$1545.91 total income in its first three years of operation.

The money in Example 1 is not received as a one-time lump sum payment of \$1545.91. Instead, it comes in on a regular basis, perhaps daily, weekly, or monthly. In discussions of such problems it is usually assumed that the income is received continuously over a period of time.

Total Money Flow Let the continuous function $f(x)$ represent the rate of flow of money per unit of time. If x is in years and $f(x)$ is in dollars per year, the area under $f(x)$ between two points in time gives the total dollar flow over the given time interval.

The function $f(x) = 2000$, shown in Figure 9 on the next page, represents a uniform rate of money flow of \$2000 per year. The graph of this money flow is a horizontal line; the *total money flow* over a specified time t is given by the rectangular area below the graph of $f(x)$ and above the x-axis between $x = 0$ and $x = t$. For example, the total money flow over $t = 5$ years would be $2000(5) = 10{,}000$, or \$10,000.

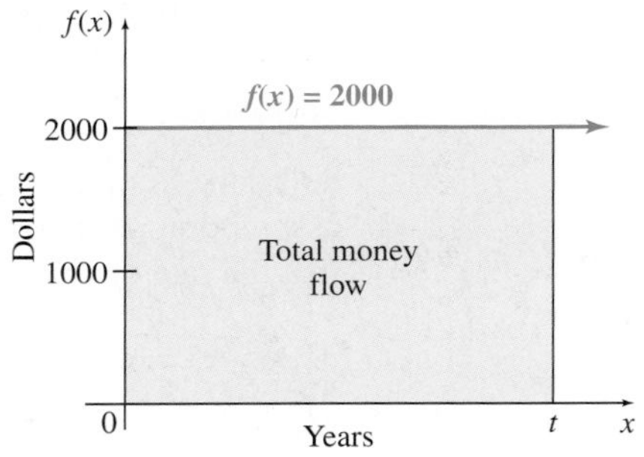

FIGURE 9

The area in the uniform rate example could be found by using an area formula from geometry. For a variable function like the function in Example 1, however, a definite integral is needed to find the total money flow over a specific time interval. For the function $f(x) = 2000e^{.08x}$, for example, the total money flow over a 5-year period would be given by

$$\int_0^5 2000e^{.08x}\,dx \approx 12{,}295.62,$$

or \$12,295.62. See Figure 10.

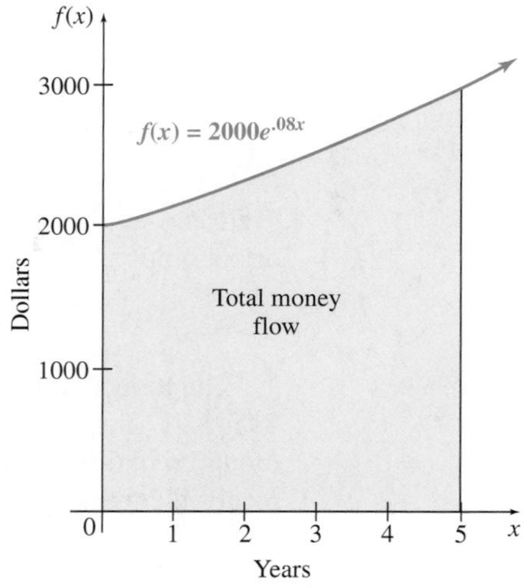

FIGURE 10

TOTAL MONEY FLOW

If $f(x)$ is the rate of money flow, then the **total money flow** over the time interval from $x = 0$ to $x = t$ is given by

$$\int_0^t f(x)\,dx.$$

It should be noted that this "total money flow" does not take into account the interest the money could earn after it is received. It is simply the total income.

Present Value of Money Flow As mentioned earlier, an amount of money that can be deposited today at a specified interest rate to yield a given sum in the future is called the *present value* of this future sum. The future sum may be called the *future value* or *final amount*. To find the **present value of a continuous money flow** with interest compounded continuously, let $f(x)$ represent the rate of the continuous flow. In Figure 11, the time axis from 0 to x is divided into n subintervals, each of width Δx. The amount of money that flows during any interval of time is given by the area between the x-axis and the graph of $f(x)$ over the specified time interval. The area of each subinterval is approximated by the area of a rectangle with height $f(x_i)$, where x_i is the left endpoint of the ith subinterval. The area of each rectangle is $f(x_i)\Delta x$, which (approximately) gives the amount of money flow over that subinterval.

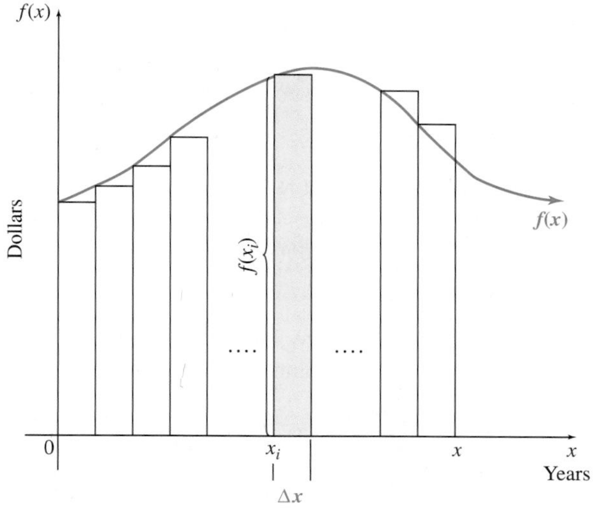

FIGURE 11

Earlier, we saw that the present value P of an amount A compounded continuously for t years at a rate of interest r is $P = Ae^{-rt}$. Letting x_i represent the time (instead of t), and replacing A with $f(x_i)\Delta x$, the present value of the money flow over the ith subinterval is approximately equal to

$$P_i = [f(x_i)\Delta x]e^{-rx_i}.$$

The total present value is approximately equal to the sum

$$\sum_{i=1}^{n}[f(x_i)\Delta x]e^{-rx_i}.$$

This approximation is improved as n increases; taking the limit of the sum as n increases without bound gives the present value

$$P = \lim_{n \to \infty} \sum_{i=1}^{n} [f(x_i)\Delta x]e^{-rx_i}.$$

This limit of a summation is given by the following definite integral.

PRESENT VALUE OF MONEY FLOW

If $f(x)$ is the rate of continuous money flow at an interest rate r for t years, then the present value is

$$P = \int_0^t f(x)e^{-rx}\,dx.$$

To understand present value of money flow, consider an account that earns interest and has a continuous money flow. The present value of the money flow is the amount that would have to be deposited into a second account that has the same interest rate, but does not have a continuous money flow, so the two accounts have the same amount of money after a specified time.

EXAMPLE 2 *Present Value of Income*

A company expects its rate of annual income during the next three years to be given by

$$f(x) = 75{,}000x, \quad 0 \le x \le 3.$$

What is the present value of this income over the 3-year period, assuming an annual interest rate of 8% compounded continuously?

Solution Use the formula for present value given above, with $f(x) = 75{,}000x$, $t = 3$, and $r = .08$.

$$P = \int_0^3 75{,}000xe^{-.08x}\,dx = 75{,}000\int_0^3 xe^{-.08x}\,dx$$

A graphing calculator quickly gives this amount as 288063.8939 or \$288,064. Using integration by parts, verify that

$$\int xe^{-.08x}\,dx = -12.5xe^{-.08x} - 156.25e^{-.08x} + C.$$

Therefore,

$$\begin{aligned}
75{,}000\int_0^3 xe^{-.08x}\,dx &= 75{,}000(-12.5xe^{-.08x} - 156.25e^{-.08x})\Big|_0^3 \\
&= 75{,}000[-12.5(3)e^{-.08(3)} - 156.25e^{-.08(3)} - (0 - 156.25)] \\
&= 75{,}000(-29.498545 - 122.910603 + 156.25) \\
&= 288{,}064,
\end{aligned}$$

or about $288,000. Notice that the actual income over the 3-year period is given by

$$\text{Total money flow} = \int_0^3 75{,}000x\,dx = \left.\frac{75{,}000x^2}{2}\right|_0^3 = 337{,}500,$$

or $337,500. This means that it would take a lump-sum deposit of $288,064 today paying a continuously compounded interest rate of 8% over a 3-year period to equal the total cash flow of $337,500 with interest. This approach is used as a basis for determining insurance claims involving income considerations.

Accumulated Amount of Money Flow at Time t To find the amount of money flow with interest at any time t, start with the formula $A = Pe^{rt}$, and in place of P substitute the expression for present value of money flow. The result is the following formula.

ACCUMULATED AMOUNT OF MONEY FLOW AT TIME t

If $f(x)$ is the rate of money flow at an interest rate r at time x, the amount of flow at time t is

$$A = e^{rt}\int_0^t f(x)e^{-rx}\,dx.$$

Here, the amount of money A represents the accumulated value or final amount of the money flow *including* interest received on the money after it comes in. (Recall, total money flow *does not* take the interest into account.)

It turns out that most money flows can be expressed as (or at least approximated by) exponential or polynomial functions. When these are multiplied by e^{-rx}, the result is a function that can be integrated. The next example illustrates uniform flow, where $f(x)$ is a constant function. (This is a special case of the polynomial function.)

EXAMPLE 3 *Accumulated Amount*

If money is flowing continuously at a constant rate of $2000 per year over 5 years at 12% interest compounded continuously, find the following.

(a) The total amount of the flow over the 5-year period

Solution The total amount is given by $\int_0^t f(x)\,dx$. Here $f(x) = 2000$ and $t = 5$.

$$\int_0^5 2000\,dx = 2000x\Big|_0^5 = 2000(5) = 10{,}000$$

The total money flow over the 5-year period is $10,000.

(b) The accumulated amount, compounded continuously, at time $t = 5$

FOR REVIEW

In this example we use the following two rules for exponents.

1. $a^m \cdot a^n = a^{m+n}$
2. $a^0 = 1$

Solution At $t = 5$ with $r = .12$, the amount is

$$A = e^{rt}\int_0^t f(x)e^{-rx}\,dx = e^{(.12)5}\int_0^5 (2000)e^{-.12x}\,dx$$
$$= (e^{.6})(2000)\int_0^5 e^{-.12x}\,dx = (e^{.6})(2000)\left(\frac{1}{-.12}\right)\left(e^{-.12x}\Big|_0^5\right)$$
$$= \frac{2000e^{.6}}{-.12}(e^{-.6} - 1) = \frac{2000}{-.12}(1 - e^{.6}) \qquad (e^{.6})(e^{-.6}) = 1$$
$$= 13{,}701.98,$$

or \$13,701.98. The answer to part (a), \$10,000, was the amount of money flow over the 5-year period. The \$13,701.98 gives that amount with interest compounded continuously over the 5-year period.

(c) The total interest earned

Solution This is simply the accumulated amount minus the total amount of flow, or

$$\$13{,}701.98 - \$10{,}000.00 = \$3701.98.$$

(d) The present value of the amount with interest

Solution Use $P = \int_0^t f(x)e^{-rx}\,dx$ with $f(x) = 2000$, $r = .12$, and $t = 5$.

$$P = \int_0^5 2000e^{-.12x}\,dx = 2000\left(\frac{e^{-.12x}}{-.12}\right)\Bigg|_0^5$$
$$= \frac{2000}{-.12}(e^{-.6} - 1) = 7519.81$$

The present value of the amount with interest in 5 years is \$7519.81, which can be checked by substituting \$13,701.98 for A in $A = Pe^{rt}$. The present value, P, could have been found by dividing the amount found in (b) by $e^{rt} = e^{.6}$. Check that this would give the same result.

If the rate of money flow is increasing or decreasing exponentially, then $f(x) = Ce^{kx}$, where C is a constant that represents the initial amount and k is the (nominal) continuous rate of change, which may be positive or negative.

EXAMPLE 4 *Accumulated Amount*

A continuous money flow starts at a rate of \$1000 per year and increases exponentially at 2% per year.

(a) Find the accumulated amount at the end of 5 years at 10% interest compounded continuously.

Solution Here $C = 1000$ and $k = .02$, so that $f(x) = 1000e^{.02x}$. Using $r = .10$ and $t = 5$,

$$A = e^{(.10)5}\int_0^5 1000e^{.02x}e^{-.10x}\,dx$$
$$= (e^{.5})(1000)\int_0^5 e^{-.08x}\,dx \qquad e^{.02x}\cdot e^{-.10x} = e^{-.08x}$$

$$= 1000e^{.5}\left(\frac{e^{-.08x}}{-.08}\right)\Bigg|_0^5$$

$$= \frac{1000e^{.5}}{-.08}(e^{-.4} - 1) = \frac{1000}{-.08}(e^{.1} - e^{.5}) = 6794.38,$$

or \$6794.38.

(b) Find the present value at 5% interest compounded continuously.

Solution Using $f(x) = 1000e^{.02x}$ with $r = .05$ and $t = 5$ in the present value expression,

$$P = \int_0^5 1000e^{.02x}e^{-.05x}\,dx$$

$$= 1000\int_0^5 e^{-.03x}\,dx = 1000\left(\frac{e^{-.03x}}{-.03}\Bigg|_0^5\right)$$

$$= \frac{1000}{-.03}(e^{-.15} - 1) = 4643.07,$$

or \$4643.07.

If the rate of change of the continuous money flow is given by the polynomial function $f(x) = a_nx^n + a_{n-1}x^{n-1} + \cdots + a_0$, the expressions for present value and accumulated amount can be integrated term by term using integration by parts.

EXAMPLE 5 *Present Value of Money Flow*

The rate of change of a continuous flow of money is given by

$$f(x) = 1000x^2 + 100x.$$

Find the present value of this money flow at the end of 10 years at 10% compounded continuously.

Solution Evaluate

$$P = \int_0^{10} (1000x^2 + 100x)e^{-.10x}\,dx.$$

Using integration by parts, verify that

$$\int (1000x^2 + 100x)e^{-.10x}\,dx =$$

$$(-10{,}000x^2 - 1000x)e^{-.1x} - (200{,}000x + 10{,}000)e^{-.1x} - 2{,}000{,}000e^{-.1x} + C.$$

Thus,

$$P = (-10{,}000x^2 - 1000x)e^{-.1x} - (200{,}000x + 10{,}000)e^{-.1x} - 2{,}000{,}000e^{-.1x}\Bigg|_0^{10}$$

$$= (-1{,}000{,}000 - 10{,}000)e^{-1} - (2{,}000{,}000 + 10{,}000)e^{-1} - 2{,}000{,}000e^{-1} - (0 - 10{,}000 - 2{,}000{,}000)$$

$$= 163{,}245.21.$$

Alternatively, use a graphing calculator to find the present value of \$163,245.21.

8.3 EXERCISES

Each of the functions in Exercise 1–14 represents the rate of flow of money in dollars per year. Assume a 10-year period at 12% compounded continuously and find the following: **(a)** *the present value;* **(b)** *the accumulated amount at* $t = 10$.

1. $f(x) = 1000$
2. $f(x) = 300$
3. $f(x) = 500$
4. $f(x) = 2000$
5. $f(x) = 400e^{.03x}$
6. $f(x) = 800e^{.05x}$
7. $f(x) = 5000e^{-.01x}$
8. $f(x) = 1000e^{-.02x}$
9. $f(x) = .1x$
10. $f(x) = .5x$
11. $f(x) = .01x + 100$
12. $f(x) = .05x + 500$
13. $f(x) = 1000x - 100x^2$
14. $f(x) = 2000x - 150x^2$

Applications

BUSINESS AND ECONOMICS

15. *Total Money Flow* An investment is expected to yield a uniform continuous rate of money flow of \$20,000 per year for 3 years. Find the final amount at an interest rate of 14% compounded continuously.

16. *Present Value* A real estate investment is expected to produce a uniform continuous rate of money flow of \$8000 per year for 6 years. Find the present value at the following rates, compounded continuously.

a. 12% **b.** 10% **c.** 15%

17. *Money Flow* The rate of a continuous flow of money starts at \$5000 and decreases exponentially at 1% per year for 8 years. Find the present value and final amount at an interest rate of 8% compounded continuously.

18. *Money Flow* The rate of a continuous money flow starts at \$1000 and increases exponentially at 5% per year for 4 years. Find the present value and final amount if interest earned is 11% compounded continuously.

19. *Present Value* A money market fund has a continuous flow of money at a rate of $f(x) = 1500 - 60x^2$, reaching 0 in 5 years. Find the present value of this flow if interest is 10% compounded continuously.

20. *Money Flow* Find the amount of a continuous money flow in 3 years if the rate is given by $f(x) = 1000 - x^2$ and if interest is 10% compounded continuously.

8.4 IMPROPER INTEGRALS

THINK ABOUT IT *If we know the rate at which a pollutant is dumped into a stream, how can we compute the total amount released given that the rate of dumping is decreasing over time?*

In this section we will learn how to answer such questions.

Sometimes it is useful to be able to integrate a function over an infinite period of time. For example, we might want to find the total amount of income generated by an apartment building into the indefinite future, or the total amount of pollution into a bay from a source that is continuing indefinitely. In this section we define integrals with one or more infinite limits of integration that can be used to solve such problems.

The graph in Figure 12(a) shows the area bounded by the curve $f(x) = x^{-3/2}$, the x-axis, and the vertical line $x = 1$. Think of the shaded region

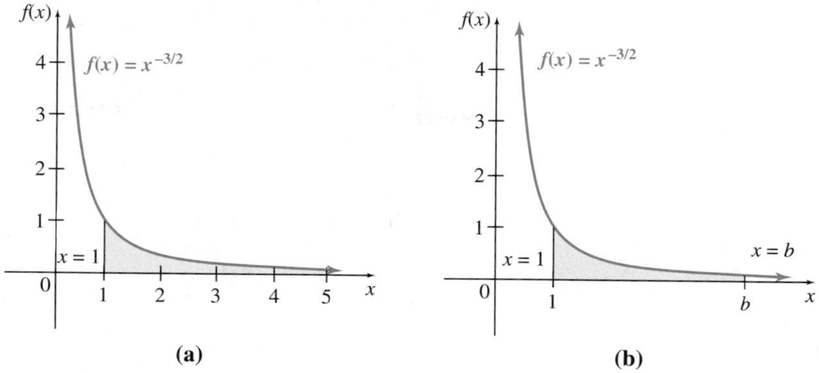

FIGURE 12

below the curve as extending indefinitely to the right. Does this shaded region have an area?

To see if the area of this region can be defined, introduce a vertical line at $x = b$, as shown in Figure 12(b). This vertical line gives a region with both upper and lower limits of integration. The area of this new region is given by the definite integral

$$\int_1^b x^{-3/2}\,dx.$$

By the Fundamental Theorem of Calculus,

$$\begin{aligned}\int_1^b x^{-3/2}\,dx &= \left(-2x^{-1/2}\right)\Big|_1^b \\ &= -2b^{-1/2} - \left(-2 \cdot 1^{-1/2}\right) \\ &= -2b^{-1/2} + 2 = 2 - \frac{2}{b^{1/2}}.\end{aligned}$$

FOR REVIEW

In the section on limits at infinity and curve sketching, we saw that for any positive real number n,

$$\lim_{b\to\infty} \frac{1}{b^n} = 0.$$

Suppose we now let the vertical line $x = b$ in Figure 12(b) move farther to the right. That is, suppose $b \to \infty$. The expression $-2/b^{1/2}$ would then approach 0, and

$$\lim_{b\to\infty}\left(2 - \frac{2}{b^{1/2}}\right) = 2 - 0 = 2.$$

This limit is defined to be the *area* of the region shown in Figure 12(a), so that

$$\int_1^{\infty} x^{-3/2}\,dx = 2.$$

An integral of the form

$$\int_a^{\infty} f(x)\,dx, \qquad \int_{-\infty}^{b} f(x)\,dx, \qquad \text{or} \qquad \int_{-\infty}^{\infty} f(x)\,dx$$

is called an *improper integral*. These **improper integrals** are defined as follows.

IMPROPER INTEGRALS

If f is continuous on the indicated interval and if the indicated limits exist, then

$$\int_a^{\infty} f(x)\,dx = \lim_{b\to\infty}\int_a^b f(x)\,dx,$$

$$\int_{-\infty}^{b} f(x)\,dx = \lim_{a\to-\infty}\int_a^b f(x)\,dx,$$

$$\int_{-\infty}^{\infty} f(x)\,dx = \int_{-\infty}^{c} f(x)\,dx + \int_c^{\infty} f(x)\,dx,$$

for real numbers a, b, and c, where c is arbitrarily chosen.

If the expressions on the right side exist, the integrals are **convergent;** otherwise, they are **divergent.** A convergent integral has a value that is a real number. A divergent integral does not, often because the area under the curve is infinitely large.

EXAMPLE 1 *Improper Integrals*

Find each integral.

(a) $\displaystyle\int_1^{\infty}\frac{dx}{x}$

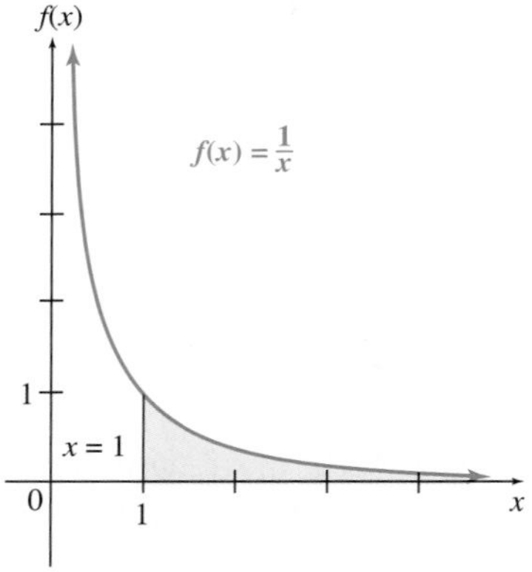

FIGURE 13

Solution A graph of this region is shown in Figure 13. By the definition of an improper integral,

$$\int_1^{\infty}\frac{dx}{x} = \lim_{b\to\infty}\int_1^b\frac{dx}{x}.$$

Find $\displaystyle\int_1^b\frac{dx}{x}$ by the Fundamental Theorem of Calculus.

$$\int_1^b\frac{dx}{x} = \ln|x|\Big|_1^b = \ln|b| - \ln|1| = \ln|b| - 0 = \ln|b|$$

As $b\to\infty$, $\ln|b|\to\infty$, so $\lim\limits_{b\to\infty}\ln|b|$ does not exist. Since the limit does not exist, $\displaystyle\int_1^{\infty}\frac{dx}{x}$ is divergent.

(b) $$\int_{-\infty}^{-2}\frac{1}{x^2}\,dx = \lim_{a\to-\infty}\int_a^{-2}\frac{1}{x^2}\,dx = \lim_{a\to-\infty}\left(\frac{-1}{x}\right)\Bigg|_a^{-2}$$

$$= \lim_{a\to-\infty}\left(\frac{1}{2}+\frac{1}{a}\right) = \frac{1}{2}$$

A graph of this region is shown in Figure 14. Since the limit exists, this integral converges.

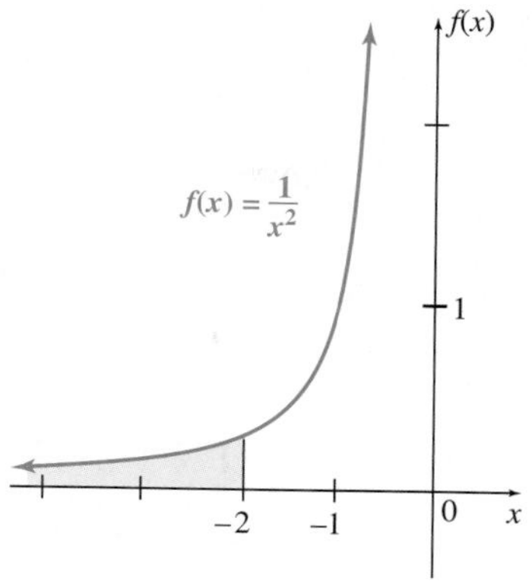

FIGURE 14

It may seem puzzling that the areas under the curves $f(x) = 1/x^{3/2}$ and $f(x) = 1/x^2$ are finite, while $f(x) = 1/x$ has an infinite amount of area. At first glance the graphs of these functions appear similar. The difference is that although all three functions get small as x becomes infinitely large, $f(x) = 1/x$ does not become small enough fast enough.

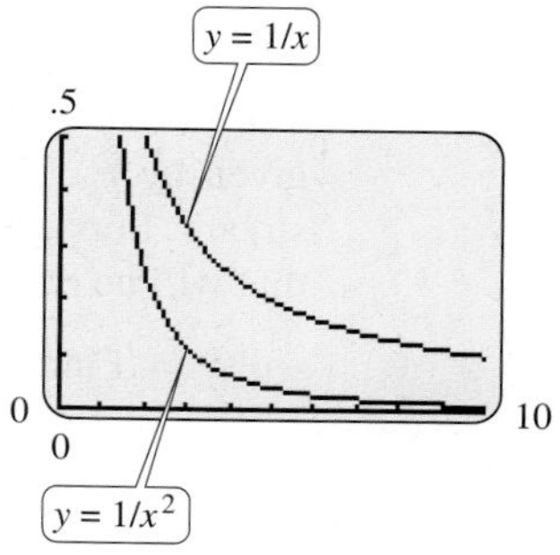

FIGURE 15

In the graphing calculator screen in Figure 15, notice how much faster $1/x^2$ becomes small compared with $1/x$.

CAUTION Since graphing calculators provide only approximations, using them to find improper integrals is tricky and requires skill and care. Although their approximations may be good in some cases, they are wrong in others, and they cannot tell us for certain that an improper integral does not exist. See Exercises 39–41. ■

EXAMPLE 2 *Improper Integral*

Find $\int_{-\infty}^{\infty} 4e^{-3x}\,dx$.

Solution Write the integral as

$$\int_{-\infty}^{\infty} 4e^{-3x}\,dx = \int_{-\infty}^{0} 4e^{-3x}\,dx + \int_{0}^{\infty} 4e^{-3x}\,dx$$

FOR REVIEW

Recall that

$$\lim_{p\to\infty} \frac{1}{e^p} = 0$$

and

$$\lim_{p\to-\infty} \frac{1}{e^p} = \infty.$$

and evaluate each of the two improper integrals on the right. If they both converge, the original integral will equal their sum. To show you all the details while maintaining the suspense, we will evaluate the second integral first.

By definition,

$$\int_0^{\infty} 4e^{-3x}\,dx = \lim_{b\to\infty}\int_0^b 4e^{-3x}\,dx = \lim_{b\to\infty}\left(\frac{-4}{3}e^{-3x}\right)\Bigg|_0^b$$

$$= \lim_{b\to\infty}\left(\frac{-4}{3e^{3b}} + \frac{4}{3}\right) = 0 + \frac{4}{3} = \frac{4}{3}.$$

Similarly, the second integral is evaluated as

$$\int_{-\infty}^{0} 4e^{-3x}\,dx = \lim_{b\to-\infty}\int_b^0 4e^{-3x}\,dx = \lim_{b\to-\infty}\left(\frac{-4}{3}e^{-3x}\right)\Bigg|_b^0$$

$$= \lim_{b\to-\infty}\left(-\frac{4}{3} + \frac{4}{3e^{3b}}\right) = \infty.$$

Since one of the two improper integrals diverges, the original improper integral diverges.

The following examples describe applications of improper integrals.

EXAMPLE 3 *Pollution*

The rate at which a pollutant is being dumped into a small stream at time t is given by P_0e^{-kt}, where P_0 is the amount of pollutant initially released into the stream. Suppose $P_0 = 1000$ and $k = .06$. Find the total amount of the pollutant that will be released into the stream into the indefinite future.

Solution Find

$$\int_0^{\infty} P_0e^{-kt}\,dt = \int_0^{\infty} 1000e^{-.06t}\,dt.$$

Work as above.

$$\int_0^{\infty} 1000e^{-.06t}\,dt = \lim_{b\to\infty}\int_0^b 1000e^{-.06t}\,dt$$

$$= \lim_{b\to\infty}\left(\frac{1000}{-.06}e^{-.06t}\right)\Bigg|_0^b$$

$$= \lim_{b\to\infty}\left(\frac{1000}{-.06e^{.06b}} - \frac{1000}{-.06}e^0\right) = \frac{-1000}{-.06} = 16{,}667$$

A total of 16,667 units of the pollutant eventually will be released.

The *capital value* of an asset is often defined as the present value of all future net earnings of the asset. In other words, suppose an asset provides a continuous money flow that is invested in an account earning a certain rate of interest. A lump sum is invested in a second account earning the same rate of interest, but with no money flow, so that as $t \to \infty$, the amounts in the two accounts approach each other. The lump sum necessary to make this happen is the capital value of the asset. If $R(t)$ gives the annual rate at which earnings are produced by an asset at time t, then the present value formula from Section 3 gives the **capital value** as

$$\int_0^{\infty} R(t)e^{-rt}\,dt,$$

where r is the annual rate of interest, compounded continuously.

EXAMPLE 4 *Capital Value*

Suppose income from a rental property is generated at the annual rate of \$4000 per year. Find the capital value of this property at an interest rate of 10% compounded continuously.

Solution This is a continuous income stream with a rate of flow of \$4000 per year, so $R(t) = 4000$. Also, $r = .10$ or $.1$. The capital value is given by

$$\begin{aligned}\int_0^{\infty} 4000e^{-.1t}\,dt &= \lim_{b\to\infty}\int_0^{b} 4000e^{-.1t}\,dt\\ &= \lim_{b\to\infty}\left(\frac{4000}{-.1}e^{-.1t}\right)\Bigg|_0^{b}\\ &= \lim_{b\to\infty}(-40{,}000e^{-.1b} + 40{,}000) = 40{,}000,\end{aligned}$$

or \$40,000.

8.4 EXERCISES

Determine whether each improper integral converges or diverges, and find the value of each that converges.

1. $\int_2^{\infty} \frac{1}{x^2}\,dx$
2. $\int_5^{\infty} \frac{1}{x^2}\,dx$
3. $\int_1^{\infty} \frac{1}{\sqrt{x}}\,dx$
4. $\int_{16}^{\infty} \frac{-3}{\sqrt{x}}\,dx$
5. $\int_{-\infty}^{-1} \frac{2}{x^3}\,dx$
6. $\int_{-\infty}^{-4} \frac{3}{x^4}\,dx$
7. $\int_1^{\infty} \frac{1}{x^{1.0001}}\,dx$
8. $\int_1^{\infty} \frac{1}{x^{.999}}\,dx$
9. $\int_{-\infty}^{-1} x^{-2}\,dx$
10. $\int_{-\infty}^{-4} x^{-2}\,dx$
11. $\int_{-\infty}^{-1} x^{-8/3}\,dx$
12. $\int_{-\infty}^{-27} x^{-5/3}\,dx$
13. $\int_0^{\infty} 4e^{-4x}\,dx$
14. $\int_0^{\infty} 10e^{-10x}\,dx$
15. $\int_{-\infty}^{0} 4e^{x}\,dx$
16. $\int_{-\infty}^{0} 3e^{4x}\,dx$
17. $\int_{-\infty}^{-1} \ln|x|\,dx$
18. $\int_1^{\infty} \ln|x|\,dx$
19. $\int_0^{\infty} \frac{dx}{(x+1)^2}$
20. $\int_0^{\infty} \frac{dx}{(2x+1)^3}$
21. $\int_{-\infty}^{-1} \frac{2x-1}{x^2-x}\,dx$
22. $\int_1^{\infty} \frac{2x+3}{x^2+3x}\,dx$
23. $\int_2^{\infty} \frac{1}{x\ln x}\,dx$
24. $\int_2^{\infty} \frac{1}{x(\ln x)^2}\,dx$
25. $\int_0^{\infty} xe^{2x}\,dx$
26. $\int_{-\infty}^{0} xe^{3x}\,dx$

27. $\int_{-\infty}^{\infty} x^3 e^{-x^4}\,dx$ (*Hint:* Recall that $\lim_{x\to\infty} x^n e^{-x} = 0$.)

28. $\int_{-\infty}^{\infty} e^{-|x|}\,dx$ (*Hint:* Recall that when $x < 0$, $|x| = -x$.)

29. $\int_{-\infty}^{\infty} \frac{x}{x^2 + 1}\,dx$

30. $\int_{-\infty}^{\infty} \frac{2x + 1}{x^2 + x + 4}\,dx$

Find the area between the graph of the given function and the x-axis over the given interval, if possible.

31. $f(x) = \frac{1}{x - 1}$, for $(-\infty, 0]$

32. $f(x) = e^{-x}$, for $(-\infty, e]$

33. $f(x) = \frac{1}{(x - 1)^2}$, for $(-\infty, 0]$

34. $f(x) = \frac{1}{(x - 1)^3}$, for $(-\infty, 0]$

35. Find $\int_{-\infty}^{\infty} x e^{-x^2}\,dx$.

36. Find $\int_{-\infty}^{\infty} \frac{x}{(1 + x^2)^2}\,dx$.

37. Show that $\int_1^{\infty} 1/x^p\,dx$ converges if $p > 1$ and diverges if $p \le 1$.

38. Example 1(b) leads to a paradox. On the one hand, the unbounded region in that example has an area of 1/2, so theoretically it could be colored with ink. On the other hand, the boundary of that region is infinite, so it cannot be drawn with a finite amount of ink. This seems impossible, because coloring the region automatically colors the boundary. Explain why it is possible to color the region.

39. Consider the functions $f(x) = 1/\sqrt{1 + x^2}$ and $g(x) = 1/\sqrt{1 + x^4}$.

a. Use your calculator to approximate $\int_1^b f(x)\,dx$ for $b = 20, 50, 100, 1000$, and 10,000.

b. Based on your answers from part a, would you guess that $\int_1^{\infty} f(x)\,dx$ is convergent or divergent?

c. Use your calculator to approximate $\int_1^b g(x)\,dx$ for $b = 20, 50, 100, 1000$, and 10,000.

d. Based on your answers from part c, would you guess that $\int_1^{\infty} g(x)\,dx$ is convergent or divergent?

e. Show how the answer to parts b and d might be guessed by comparing the integrals with others whose convergence or divergence is known. (*Hint:* For large x, the difference between $1 + x^2$ and x^2 is relatively small.)

Note: The first integral is indeed divergent, and the second convergent, with an approximate value of .9270.

40. **a.** Use your calculator to approximate $\int_0^b e^{-x^2}\,dx$ for $b = 1, 5, 10$, and 20.

b. Based on your answer to part a, does $\int_0^{\infty} e^{-x^2}\,dx$ appear to be convergent or divergent? If convergent, what seems to be its approximate value?

c. Explain why this integral should be convergent by comparing e^{-x^2} with e^{-x} for $x > 1$.

Note: The integral is convergent, with a value of $\sqrt{\pi}/2$.

41. **a.** Use your calculator to approximate $\int_0^b e^{-.00001x}\,dx$ for $b = 10, 50, 100$, and 1000.

b. Based on your answer to part a, does $\int_0^{\infty} e^{-.00001x}\,dx$ appear to be convergent or divergent?

c. To what value does the integral actually converge?

Applications

BUSINESS AND ECONOMICS.

Capital Value *Find the capital values of the properties in Exercises 42–43.*

42. A castle for which annual rent of \$60,000 will be paid in perpetuity; the interest rate is 8% compounded continuously

43. A fort on a strategic peninsula in the North Sea; the annual rent is \$500,000, paid in perpetuity; the interest rate is 6% compounded continuously

44. Find the capital value of an asset that generates \$6000 yearly income if the interest rate is as follows.

a. 8% compounded continuously

b. 10% compounded continuously

45. An investment produces a perpetual stream of income with a flow rate of

$$R(t) = 1000e^{.02t}.$$

Find the capital value at an interest rate of 7% compounded continuously.

46. Suppose income from an investment starts (at time 0) at \$6000 a year and increases linearly and continuously at a rate of \$200 a year. Find the capital value at an interest rate of 5% compounded continuously.

47. *Scholarship* The Drucker family wants to establish an ongoing scholarship award at a college. Each year in June \$3000 will be awarded, starting 1 year from now. What amount must the Druckers provide the college, assuming funds will be invested at 10% compounded continuously?

LIFE SCIENCES

48. *Drug Reaction* The rate of reaction to a drug is given by

$$r'(x) = 2x^2e^{-x},$$

where x is the number of hours since the drug was administered. Find the total reaction to the drug over all the time since it was administered, assuming this is an infinite time interval. (*Hint:* $\lim_{x\to\infty} x^k e^{-x} = 0$ for all real numbers k.)

49. *Drug Epidemic* In an epidemiological model used to study the spread of drug use, a single drug user is introduced into a population of N non-users. Under certain assumptions, the number of people expected to use drugs as a result of direct influence from each drug user is given by

$$S = N\int_0^\infty \frac{a(1 - e^{-kt})}{k}e^{-bt}\,dt,$$

where a, b, and k are constants.* Find the value of S.

50. *Present Value* When harvesting a population, such as fish, the present value of the resource is given by

$$P = \int_0^\infty e^{-rt}n(t)y(t)dt,$$

where r is a discount factor, $n(t)$ is the net revenue at time t, and $y(t)$ is the harvesting effort.† Suppose $y(t) = K$ and $n(t) = at + b$. Find the present value.

PHYSICAL SCIENCES

Radioactive Waste *Radioactive waste is entering the atmosphere over an area at a decreasing rate. Use the improper integral*

$$\int_0^\infty Pe^{-kt}dt$$

with $P = 50$ *to find the total amount of the waste that will enter the atmosphere for each value of* k.

51. $k = .06$

52. $k = .04$

*Murray, J. D., *Mathematical Biology,* Springer-Verlag, 1989, p. 642, 648.

†Ludwig, Donald, "An Unusual Free Boundary Problem from the Theory of Optimal Harvesting," in *Lectures on Mathematics in the Life Sciences, Vol. 12: Some Mathematical Questions in Biology,* American Mathematical Society, 1979, pp. 173–209.

CHAPTER SUMMARY

This chapter has covered several topics related to integration. In the first section we explored integration by parts as another way to find antiderivatives. The next two sections covered several applications of the integral: volume, average value, and continuous money flow. Finally, we learned to evaluate improper integrals, which have upper or lower limits of ∞ or $-\infty$.

KEY TERMS

8.1 **integration by parts**
column integration
8.2 **solid of revolution**
average value of a function
8.3 **total money flow**
present value of continuous money flow
8.4 **improper integral**
convergent integral
divergent integral

CHAPTER 8 REVIEW EXERCISES

1. Describe the type of integral for which integration by parts is useful.
2. Compare finding the average value of a function with finding the average of n numbers.
3. What is an improper integral? Explain why improper integrals must be treated in a special way.

Find each integral, using techniques from this or the previous chapter.

4. $\int x(8 - x)^{3/2}\, dx$
5. $\int \frac{3x}{\sqrt{x - 2}}\, dx$
6. $\int xe^x\, dx$
7. $\int (x + 2)e^{-3x}\, dx$
8. $\int \ln |2x + 3|\, dx$
9. $\int (x - 1) \ln |x|\, dx$
10. $\int \frac{x}{9 - 4x^2}\, dx$
11. $\int \frac{x}{\sqrt{25 + 9x^2}}\, dx$
12. $\int_1^e x^3 \ln x\, dx$
13. $\int_0^1 x^2 e^{x/2}\, dx$
14. Find the area between $y = (3 + x^2)e^{2x}$ and the x-axis from $x = 0$ to $x = 1$.
15. Find the area between $y = x^3(x^2 - 1)^{1/3}$ and the x-axis from $x = 1$ to $x = 3$.

Find the volume of the solid of revolution formed by rotating each bounded region about the x-axis.

16. $f(x) = 2x - 1, \quad y = 0, \quad x = 3$
17. $f(x) = \sqrt{x - 2}, \quad y = 0, \quad x = 11$
18. $f(x) = e^{-x}, \quad y = 0, \quad x = -2, \quad x = 1$
19. $f(x) = \frac{1}{\sqrt{x - 1}}, \quad y = 0, \quad x = 2, \quad x = 4$
20. $f(x) = 4 - x^2, \quad y = 0, \quad x = -1, \quad x = 1$
21. $f(x) = \frac{x^2}{4}, \quad y = 0, \quad x = 4$
22. A frustum is what remains of a cone when the top is cut off by a plane parallel to the base. Suppose a right circular frustum (that is, one formed from a right circular cone) has a base with radius r, a top with radius $r/2$, and a height h. (See the figure.) Find

the volume of this frustum by rotating about the x-axis the region below the line segment from $(0, r)$ to $(h, r/2)$.

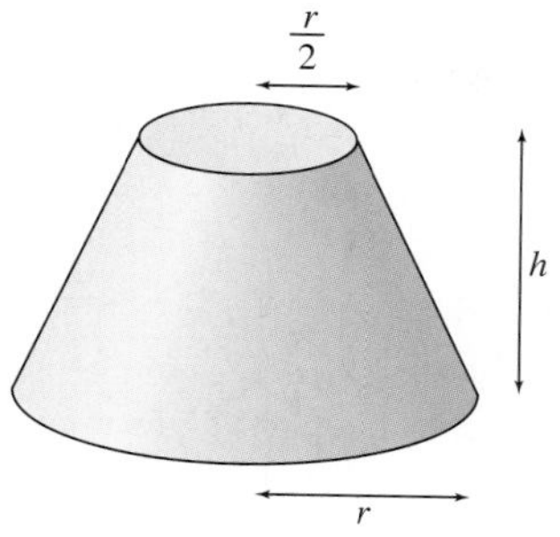

23. How is the average value of a function found?

24. Find the average value of $f(x) = \sqrt{x+1}$ over the interval $[0, 8]$.

25. Find the average value of $f(x) = x^2(x^3+1)^5$ over the interval $[0, 1]$.

Find the value of each integral that converges.

26. $\int_1^{\infty} x^{-1}\,dx$

27. $\int_{-\infty}^{-2} x^{-2}\,dx$

28. $\int_0^{\infty} \frac{dx}{(5x+2)^2}$

29. $\int_1^{\infty} 6e^{-x}\,dx$

30. $\int_{-\infty}^{0} \frac{x}{x^2+3}\,dx$

31. $\int_{10}^{\infty} \ln(2x)\,dx$

Find the area between the graph of each function and the x-axis over the given interval, if possible.

32. $f(x) = \frac{3}{(x-2)^2}$, for $(-\infty, 1]$

33. $f(x) = 3e^{-x}$, for $[0, \infty)$

34. How is the present value of money flow found? The accumulated amount of money flow?

Applications

BUSINESS AND ECONOMICS

35. *Total Revenue* The rate of change of revenue from the sale of x toaster ovens is

$$R'(x) = x(x-50)^{1/2}.$$

Find the total revenue from the sale of the 50th to the 75th ovens.

Present Value of Money Flow *Each function in Exercises 36–39 represents the rate of flow of money (in dollars per year) over the given time period, compounded continuously at the given annual interest rate. Find the present value in each case.*

36. $f(x) = 5000$, 8 years, 9%

37. $f(x) = 25{,}000$, 12 years, 10%

38. $f(x) = 100e^{.02x}$, 5 years, 11%

39. $f(x) = 30x$, 18 months, 5%

Amount of Money at Time t *Assume that each function gives the rate of flow of money in dollars per year over the given period, with continuous compounding at the given rate. Find the accumulated amount at the end of the time period.*

40. $f(x) = 2000$, 5 years, 12% per year

41. $f(x) = 500e^{-.03x}$, 8 years, 10% per year

42. $f(x) = 20x$, 6 years, 12% per year

43. $f(x) = 1000 + 200x$, 10 years, 9% per year

44. *Money Flow* An investment scheme is expected to produce a continuous flow of money, starting at \$1000 and increasing exponentially at 5% a year for 7 years. Find the present value at an interest rate of 11% compounded continuously.

45. *Money Flow* The proceeds from the sale of a building will yield a uniform continuous flow of \$10,000 a year for 10 years. Find the final amount at an interest rate of 10.5% compounded continuously.

46. *Capital Value* Find the capital value of an office building for which annual rent of \$50,000 will be paid in perpetuity, if the interest rate is 9%.

LIFE SCIENCES

47. *Drug Reaction* The reaction rate to a new drug x hours after the drug is administered is

$$r'(x) = .5xe^{-x}.$$

Find the total reaction over the first 5 hours.

48. *Oil Leak Pollution* An oil leak from an uncapped well is polluting a bay at a rate of $f(x) = 100e^{-.05x}$ gallons per year. Use an improper integral to find the total amount of oil that will enter the bay, assuming the well is never capped.

49. *Milk Production* Researchers report that the average amount of milk produced (in kilograms per day) by a 4- to 5-year-old cow weighing 700 kg can be approximated by

$$y = 1.87t^{1.49}e^{-.189(\ln t)^2},$$

where t is the number of days into lactation.*

a. Approximate the total amount of milk produced from $t = 1$ to $t = 321$ using the trapezoidal rule with $n = 8$.

b. Repeat part a using Simpson's rule with $n = 8$.

c. Repeat part a using the integration feature of a graphing calculator, and compare your answer with the answers to parts a and b.

PHYSICAL SCIENCES

50. *Average Temperatures* Suppose the temperature in a river at a point x meters downstream from a factory that is discharging hot water into the river is given by

$$T(x) = 400 - .25x^2.$$

Find the average temperature over each interval.

a. $[0, 10]$ **b.** $[10, 40]$ **c.** $[0, 40]$

*Freeze, Brian S. and Timothy J. Richards, "Lactation Curve Estimation for Use in Economic Optimization Models in the Dairy Industry," *Journal of Dairy Science*, Vol. 75, 1992, pp. 2984–2989.

EXTENDED APPLICATION: Estimating Learning Curves in Manufacturing with Integrals

In the previous chapter you have seen how the trapezoidal rule uses sums of areas of polygons to approximate the area under a smooth curve, that is, a definite integral. In this Extended Application we look at the reverse process, using an integral to estimate a sum, in the context of estimating production costs.

As a manufacturer produces more units of a new product, the individual units generally become cheaper to produce, because with experience, production workers gain skill and speed, and managers spot opportunities for improved efficiency. This decline in unit costs is often called an *experience curve* or *learning curve*. This curve is important when a manufacturer negotiates a contract with a buyer.

Here's an example, based on an actual contract that came before the Armed Services Board of Contract Appeals.* The Navy asked the ITT Defense Communications Division to bid on the manufacture of several different kinds of mobile telephone switchboards, including 280 of the model called the SB 3865. ITT figured that the cost of making a single SB 3865 was around \$300,000. But they couldn't submit a bid of \$300,000 × 280 or \$84 million, because multiple units should have a lower unit price. So ITT used a learning curve to estimate an average unit cost of \$135,296 for all 280 switchboards and submitted a bid of \$135,260 × 280 or \$37.9 million.

The contract gave the Navy an option to purchase 280 SB units over three years, but in fact it bought fewer. Suppose the Navy bought 140 SBs. Should it pay half of the original price of \$37.9 million? No: ITT's bid was based on the efficiencies of a 280-unit run, so 140 units should be repriced to yield *more* than half the full price. A repricing clause in the contract specified that a learning curve would be used to reprice partial orders, and when the Navy ordered less than the full amount, ITT invoked this clause to reprice the switchboards. The question in dispute at the hearing was which learning curve to use.

There are two common learning curve models. The *unit learning curve* model assumes that each time the number of units doubles, say from n to $2n$, the cost of producing the last unit is some constant fraction r of the cost for the nth unit. Usually the fraction r is given as a percent. If $r = 90\%$ (typical for big pieces of hardware), then the contract would refer to a "90% learning curve." The *cumulative learning curve* model assumes that when the number of units doubles, the *average cost* of producing all $2n$ units is some constant fraction of the *average cost* of the first n units. The Navy's contract with ITT didn't specify which model was to be used—it just referred to "a 90% learning curve." The government used the unit model and ITT used the cumulative average model, and ITT calculated a fair price millions of dollars higher than the government's price!

In practice, ITT used a calculator program to make its estimate, and the government used printed tables, but both the program and the tables were derived using calculus. To see how the computation works, we'll derive the government's unit learning curve.

Each unit has a different cost, with the first unit being the most expensive and the 280th unit the least expensive. So the cost of the nth unit, call it $C(n)$, is a function of n. To find a fair price for n units we'll add up all the unit prices. That is, we will compute $C(1) + C(2) + \cdots + C(n-1) + C(n)$. Before we can do that, we need a formula for $C(n)$ in terms of n, but all we know about $C(n)$ is that

$$C(2n) = r \cdot C(n)$$

for *every* n, with $r = .90$. This sort of equation is called a *functional equation:* It relates two different values of the function without giving an explicit formula for the function. In Exercise 3 you'll see how you might discover a solution to this functional equation, but here we'll just give the result:

$$C(n) = C(1) \cdot n^b, \text{ where } b = \frac{\ln r}{\ln 2} \approx -.152.$$

Thus, to find the price for making 280 units, we need to add up all the values of $C(1) \cdot n^b$ as n ranges from 1 to 280. There's just one problem: We don't know $C(1)$! The only numbers that ITT gave were the *average* cost per unit for 280 units, namely \$135,296, and the total price of \$37.9 million. But if we write out the formula for the 280-unit price in terms of $C(1)$, we can figure out $C(1)$ by dividing. Here's how it works.

The cost C is a function of an integer variable n, since the contractor can't deliver fractional units of the hardware. But the function $C(x) = C(1) \cdot x^{-.152}$ is a perfectly good function of the real variable x, and the sum of the first 280 values of $C(n)$ should be close to $\int_1^{280} C(x)\,dx = C(1) \cdot \int_1^{280} x^{-.152}\,dx$. In Exercise 4 you'll see how to derive the following improved estimate:

$$C(1)\left(\int_1^{281} x^{-.152}\,dx + \frac{1 - 281^{-.152}}{2}\right).$$

The integrand is a power function, so you know how to evaluate the integral exactly.

$$\int_1^{281} x^{-.152}\,dx = \frac{1}{.848}x^{.848}\Big|_1^{281} = \frac{1}{.848}(281^{.848} - 1^{.848})$$

Thus, the sum is approximately

$$C(1)\left[\frac{1}{.848}(281^{.848} - 1^{.848}) + \frac{1 - 281^{-.152}}{2}\right] \approx C(1) \cdot 139.75.$$

*For a report on this case, see http://www.starpayn.com/asbca/44791.html; for an introduction to learning curves see Heizer, Jay and Barry Render, *Operations Management*, Prentice-Hall, 2001, or Argote, Linda and Dennis Epple, "Learning Curves in Manufacturing," *Science*, Feb. 23, 1990.

Since ITT's price for 280 units was \$39.7 million,

$$C(1) = \frac{\$39{,}700{,}000}{139.75} \approx \$284{,}000.$$

Now that we know $C(1)$, we can reprice an order of 140 units by adding up the first 140 values of $C(n)$. An estimate exactly like the one above tells us that according to the government's model, a fair price for 140 units is about \$22 million. As we expected, this is more than half the 280-unit price, in fact about \$3 million more.

Exercises

1. According to the formula for $C(n)$, what is the unit price of the 280th unit, to the nearest thousand dollars?
2. Suppose that instead of using natural logarithms to compute b, we use logarithms with a base of 10 and define $b = (\log r)/(\log 2)$. Does this change the value of b?
3. All power functions satisfy an equation similar to our functional equation: If $f(x) = ax^b$, then $f(2x) = a(2x)^b = a2^b \cdot x^b = 2^b \cdot f(x)$. How can you choose a and b to make $C(x) = ax^b$ a solution to the functional equation $C(2n) = r \cdot C(n)$?
4. Figure 16 indicates how you could use the integral

$$\int_1^5 \frac{1}{x}\,dx + \frac{1 - \frac{1}{5}}{2} \text{ as an estimate for the sum } 1 + \frac{1}{2} + \frac{1}{3} + \frac{1}{4}.$$

The graph shows the function $y = \frac{1}{x}$.

a. Write a justification for the integral estimate. (Your argument will also justify the integral estimate for the $C(n)$ sum.) Based on your explanation, does the integral expression overestimate the sum?

b. You know how to integrate the function $1/x$. Compute the integral estimate and the actual value. What is the percentage error in the estimate?

Directions for Group Project

Suppose that you and three other students have an internship with a manufacturing company that is submitting a bid to make several thousand units of some highly technical equipment. The one problem with the bid is that the number of units that will be purchased is only an estimate and that the actual number needed may greatly vary from the estimate. Using the information given above, prepare a presentation for an internal sales meeting that will describe the case listed above and its applications to the bid at hand. Make your presentation realistic in the sense that the product you are manufacturing should have a name, average price, and so on. Then show how integrals can be used to estimate learning curves in this situation and produce a pricing structure for the bid. Presentation software, such as Microsoft Powerpoint, should be used.

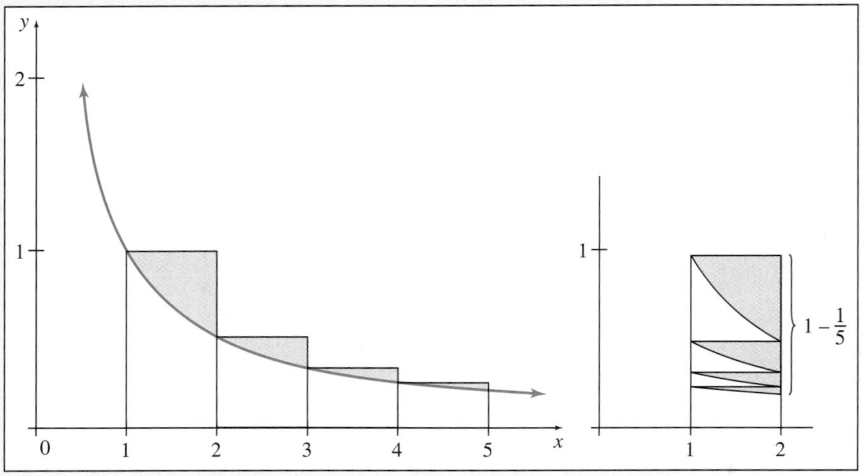

FIGURE 16

CHAPTER

9 Multivariable Calculus

Safe diving requires an understanding of how the increased pressure below the surface affects the body's intake of nitrogen. An exercise in Section 2 of this chapter investigates a formula for nitrogen pressure as a function of two variables, depth and dive time. Partial derivatives tell us how this function behaves when one variable is held constant as the other changes. Dive tables based on the formula help divers to choose a safe time for a given depth, or a safe depth for a given time.

Many of the ideas developed for functions of one variable also apply to functions of more than one variable. In particular, the fundamental idea of derivative generalizes in a very natural way to functions of more than one variable.

9.1 FUNCTIONS OF SEVERAL VARIABLES

THINK ABOUT IT *How are the amounts of labor and capital needed to produce a certain number of items related?*

We will be able to answer this question later in this section using a production function, which depends on the amounts of both labor and capital. That is, production is a function of two independent variables.

If a company produces x items at a cost of \$10 per item, for instance, then the total cost $C(x)$ of producing the items is given by

$$C(x) = 10x.$$

The cost is a function of one independent variable, the number of items produced. If the company produces two products, with x of one product at a cost of \$10 each, and y of another product at a cost of \$15 each, then the total cost to the firm is a function of *two* independent variables, x and y. By generalizing $f(x)$ notation, the total cost can be written as $C(x, y)$, where

$$C(x, y) = 10x + 15y.$$

When $x = 5$ and $y = 12$ the total cost is written $C(5, 12)$, with

$$C(5, 12) = 10 \cdot 5 + 15 \cdot 12 = 230.$$

A general definition follows.

FUNCTION OF TWO OR MORE VARIABLES

The expression $z = f(x, y)$ is a **function of two variables** if a unique value of z is obtained from each ordered pair of real numbers (x, y). The variables x and y are **independent variables,** and z is the **dependent variable.** The set of all ordered pairs of real numbers (x, y) such that $f(x, y)$ exists is the **domain** of f; the set of all values of $f(x, y)$ is the **range.** Similar definitions could be given for functions of three, four, or more independent variables.

EXAMPLE 1 *Evaluating Functions*

Let $f(x, y) = 4x^2 + 2xy + 3/y$ and find the following.

(a) $f(-1, 3)$

Solution Replace x with -1 and y with 3.

$$f(-1, 3) = 4(-1)^2 + 2(-1)(3) + \frac{3}{3} = 4 - 6 + 1 = -1$$

(b) $f(2, 0)$

Solution Because of the quotient $3/y$, it is not possible to replace y with 0, so $f(2, 0)$ is undefined. By inspection, we see that the domain of the function is the set of all (x, y) such that $y \neq 0$.

EXAMPLE 2 *Carbon Dioxide*

Let x represent the number of milliliters (ml) of carbon dioxide released by the lungs in one minute. Let y be the change in the carbon dioxide content of the blood as it leaves the lungs (y is measured in ml of carbon dioxide per 100 ml of blood). The total output of blood from the heart in one minute (measured in ml) is given by C, where C is a function of x and y such that

$$C = C(x, y) = \frac{100x}{y}.$$

Find $C(320, 6)$.

Solution Replace x with 320 and y with 6 to get

$$C(320, 6) = \frac{100(320)}{6}$$
$$\approx 5333 \text{ ml of blood per minute.}$$

FOR REVIEW

Graph the following lines. Refer to Section 1.1 if you need review.

1. $2x + 3y = 6$
2. $x = 4$
3. $y = 2$

EXAMPLE 3 *Evaluating a Function*

Let $f(x, y, z) = 4xz - 3x^2y + 2z^2$. Find $f(2, -3, 1)$.

Solution Replace x with 2, y with -3, and z with 1.

$$f(2, -3, 1) = 4(2)(1) - 3(2)^2(-3) + 2(1)^2 = 8 + 36 + 2 = 46$$

Answers

1.
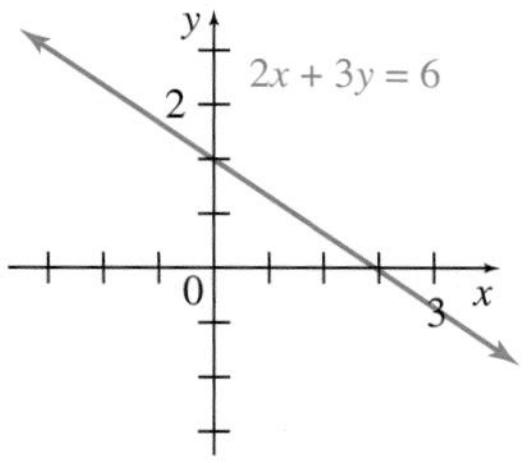

2.
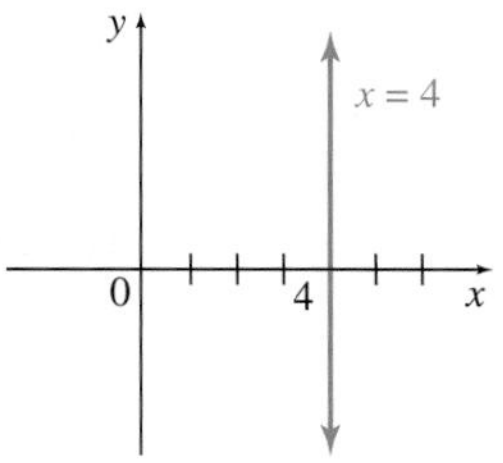

3.
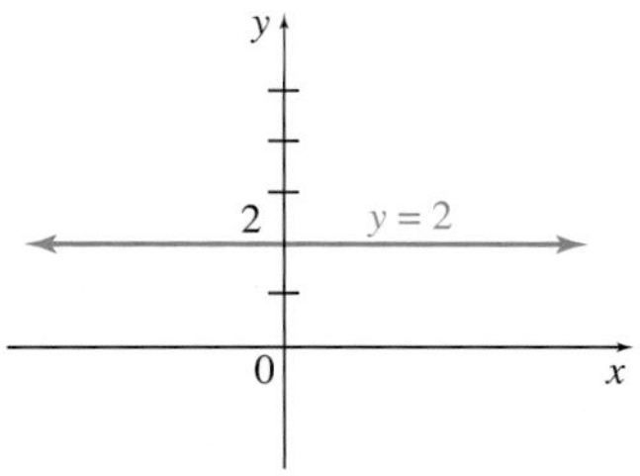

Graphing Functions of Two Independent Variables Functions of one independent variable are graphed by using an x-axis and a y-axis to locate points in a plane. The plane determined by the x- and y-axes is called the *xy-plane*. A third axis is needed to graph functions of two independent variables—the z-axis, which goes through the origin in the xy-plane and is perpendicular to both the x-axis and the y-axis.

Figure 1 shows one possible way to draw the three axes. In Figure 1, the yz-plane is in the plane of the page, with the x-axis perpendicular to the plane of the page.

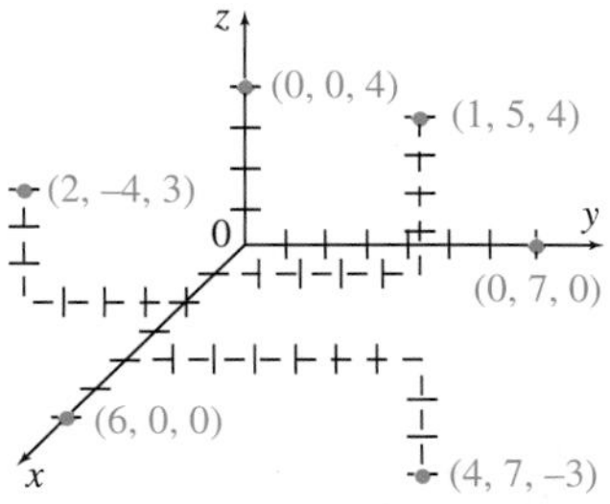

FIGURE 1

Just as we graphed ordered pairs earlier we can now graph **ordered triples** of the form (x, y, z). For example, to locate the point corresponding to the ordered triple $(2, -4, 3)$, start at the origin and go 2 units along the positive x-axis. Then go 4 units in a negative direction (to the left) parallel to the y-axis. Finally, go up 3 units parallel to the z-axis. The point representing $(2, -4, 3)$ is shown in Figure 1, together with several other points. The region of three-dimensional space where all coordinates are positive is called the **first octant.**

In Chapter 1 we saw that the graph of $ax + by = c$ (where a and b are not both 0) is a straight line. This result generalizes to three dimensions.

PLANE

The graph of

$$ax + by + cz = d$$

is a **plane** if a, b, and c are not all 0.

EXAMPLE 4 *Graphing a Plane*

Graph $2x + y + z = 6$.

Solution The graph of this equation is a plane. Earlier, we graphed straight lines by finding x- and y-intercepts. A similar idea helps in graphing a plane. To find the x-intercept, which is the point where the graph crosses the x-axis, let $y = 0$ and $z = 0$.

$$2x + 0 + 0 = 6$$
$$x = 3$$

The point $(3, 0, 0)$ is on the graph. Letting $x = 0$ and $z = 0$ gives the point $(0, 6, 0)$, while $x = 0$ and $y = 0$ lead to $(0, 0, 6)$. The plane through these three points includes the triangular surface shown in Figure 2. This surface is the first-octant part of the plane that is the graph of $2x + y + z = 6$. The surface does not stop at the axes, but extends without bound.

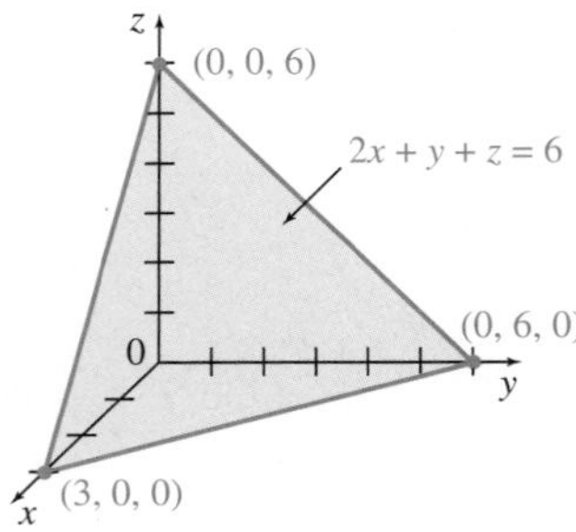

FIGURE 2

EXAMPLE 5 *Graphing a Plane*

Graph $x + z = 6$.

Solution To find the x-intercept, let $z = 0$, giving $(6, 0, 0)$. If $x = 0$, we get the point $(0, 0, 6)$. Because there is no y in the equation $x + z = 6$, there can be no y-intercept. A plane that has no y-intercept is parallel to the y-axis. The first-octant portion of the graph of $x + z = 6$ is shown in Figure 3.

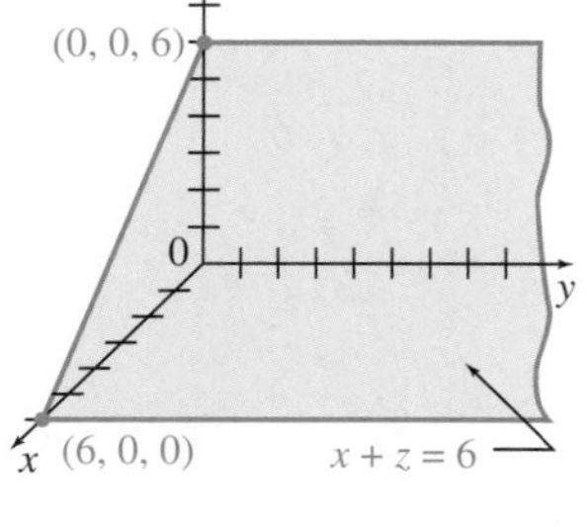

FIGURE 3

EXAMPLE 6 *Graphing Planes*

Graph each function in two variables.

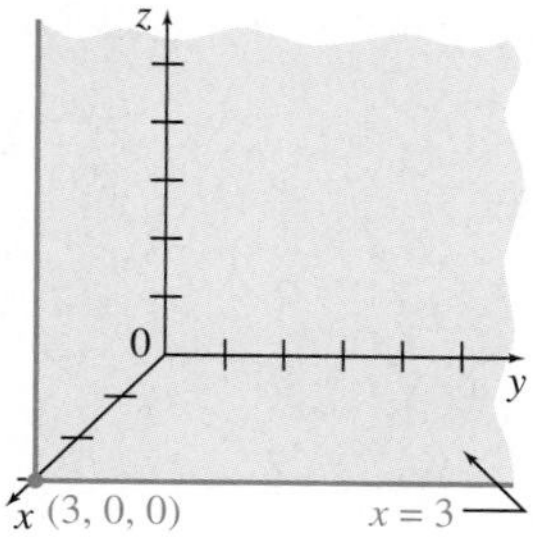

FIGURE 4

(a) $x = 3$

Solution This graph, which goes through $(3, 0, 0)$, can have no y-intercept and no z-intercept. It is therefore a plane parallel to the y-axis and the z-axis and, therefore, to the yz-plane. The first-octant portion of the graph is shown in Figure 4.

(b) $y = 4$

Solution This graph goes through $(0, 4, 0)$ and is parallel to the xz-plane. The first-octant portion of the graph is shown in Figure 5.

(c) $z = 1$

Solution The graph is a plane parallel to the xy-plane, passing through $(0, 0, 1)$. Its first-octant portion is shown in Figure 6.

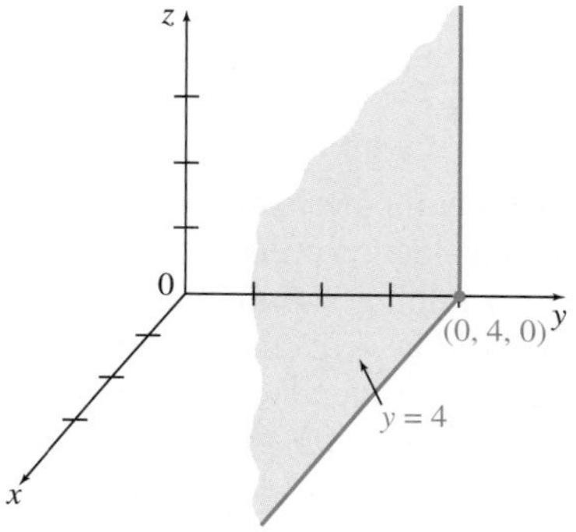

FIGURE 5

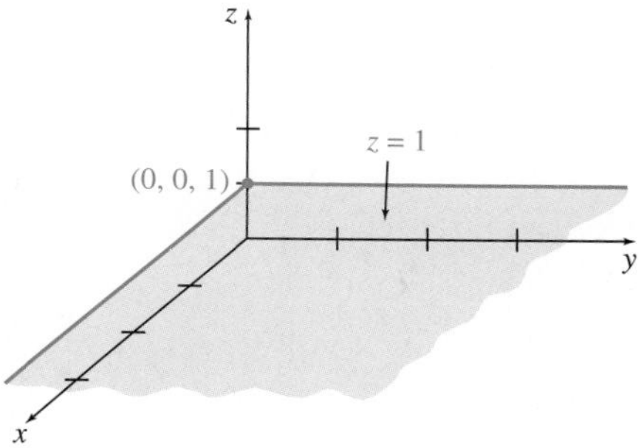

FIGURE 6

The graph of a function of one variable, $y = f(x)$, is a curve in the plane. If x_0 is in the domain of f, the point $(x_0, f(x_0))$ on the graph lies directly above or below the number x_0 on the x-axis, as shown in Figure 7.

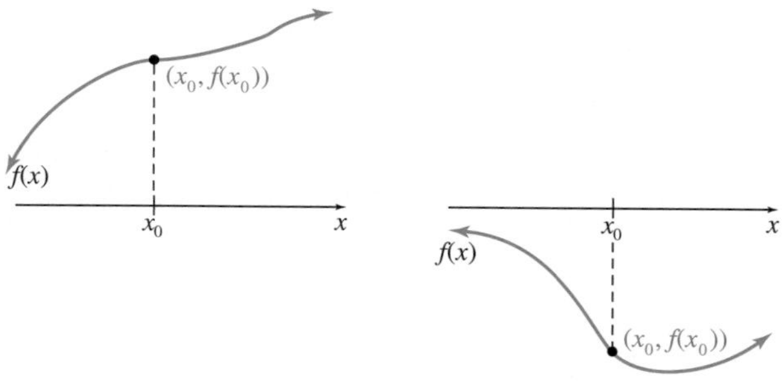

FIGURE 7

The graph of a function of two variables, $z = f(x, y)$, is a **surface** in three-dimensional space. If (x_0, y_0) is in the domain of f, the point $(x_0, y_0, f(x_0, y_0))$ lies directly above or below the point (x_0, y_0) in the xy-plane, as shown in Figure 8.

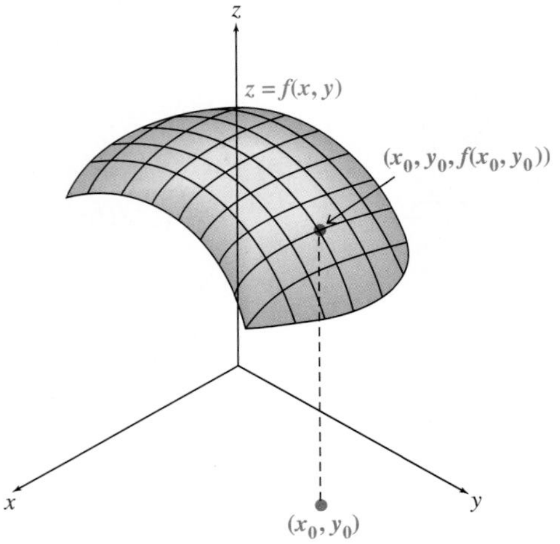

FIGURE 8

Although computer software is available for drawing the graphs of functions of two independent variables, you can often get a good picture of the graph without it by finding various **traces**—the curves that result when a surface is cut by a plane. The ***xy*-trace** is the intersection of the surface with the xy-plane. The ***yz*-trace** and ***xz*-trace** are defined similarly. You can also determine the intersection of the surface with planes parallel to the xy-plane. Such planes are of the form $z = k$, where k is a constant, and the curves that result when they cut the surface are called **level curves.**

EXAMPLE 7 *Graphing a Function*

Graph $z = x^2 + y^2$.

Solution The yz-plane is the plane in which every point has a first coordinate of 0, so its equation is $x = 0$. When $x = 0$, the equation becomes $z = y^2$, which is the equation of a parabola in the yz-plane, as shown in Figure 9(a) on the next page. Similarly, to find the intersection of the surface with the xz-plane (whose equation is $y = 0$), let $y = 0$ in the equation. It then becomes $z = x^2$, which is the equation of a parabola in the xz-plane, as shown in Figure 9(a). The xy-trace (the intersection of the surface with the plane $z = 0$) is the single point $(0, 0, 0)$ because $x^2 + y^2$ is never negative, and is equal to 0 only when $x = 0$ and $y = 0$.

Next, we find the level curves by intersecting the surface with the planes $z = 1$, $z = 2$, $z = 3$, etc. (all of which are parallel to the xy-plane). In each case, the result is a circle:

$$x^2 + y^2 = 1, \qquad x^2 + y^2 = 2, \qquad x^2 + y^2 = 3,$$

and so on, as shown in Figure 9(b). Drawing the traces and level curves on the same set of axes suggests that the graph of $z = x^2 + y^2$ is the bowl-shaped figure, called a **paraboloid,** that is shown in Figure 9(c).

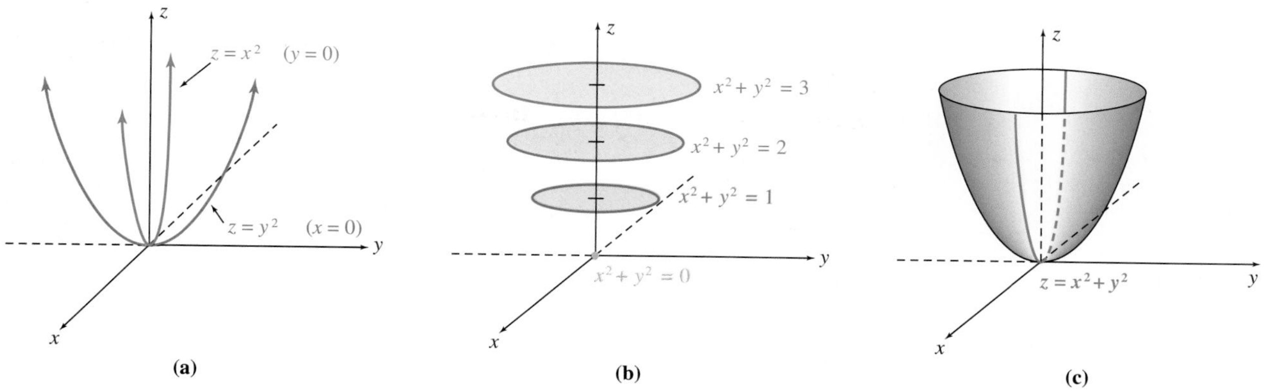

FIGURE 9

Figure 10 shows the level curves from Example 7 plotted in the xy-plane. The picture can be thought of as a topographical map which describes the surface generated by $z = x^2 + y^2$, just as the topographical map in Figure 11 describes the surface of the land in a part of New York state.

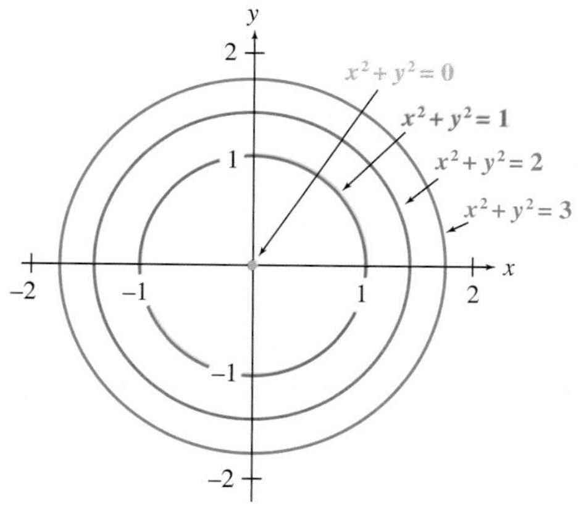

FIGURE 10

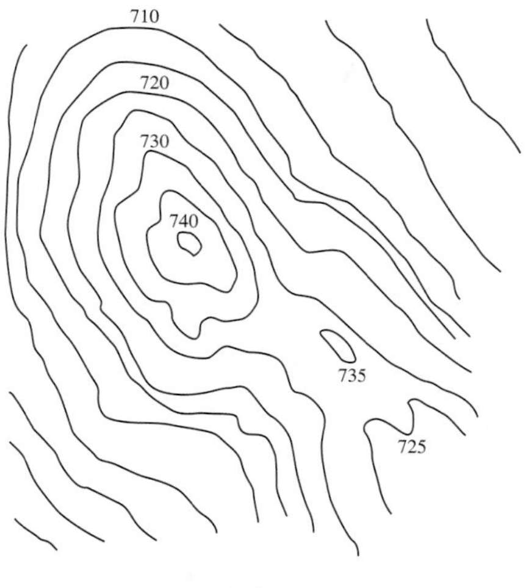

FIGURE 11

One application of level curves occurs in economics with production functions. A **production function** $z = f(x, y)$ is a function that gives the quantity z of an item produced as a function of x and y, where x is the amount of labor and y is the amount of capital (in appropriate units) needed to produce z units. If the production function has the special form $z = P(x, y) = Ax^a y^{1-a}$, where A is a constant and $0 < a < 1$, the function is called a **Cobb-Douglas production function.** For production functions, level curves are used to indicate combinations of the values of x and y that produce the same value of production z.

EXAMPLE 8 *Cobb-Douglas Production Function*

Find the level curve at a production of 100 items for the Cobb-Douglas production function $z = x^{2/3}y^{1/3}$.

Solution Let $z = 100$ to get

$$100 = x^{2/3}y^{1/3}$$

$$\frac{100}{x^{2/3}} = y^{1/3}.$$

Now cube both sides to express y as a function of x.

$$y = \frac{100^3}{x^2} = \frac{1{,}000{,}000}{x^2}$$

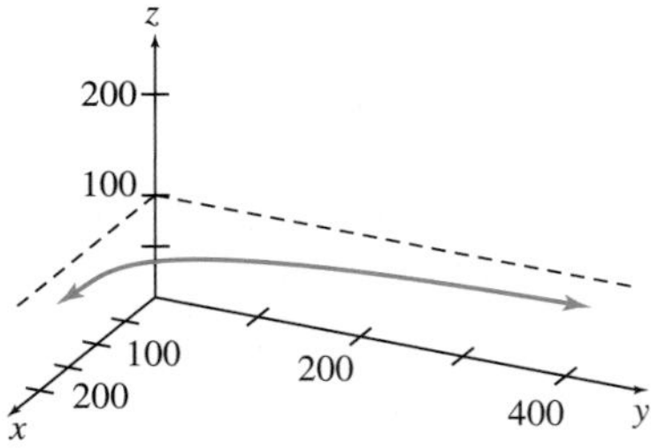

(a)

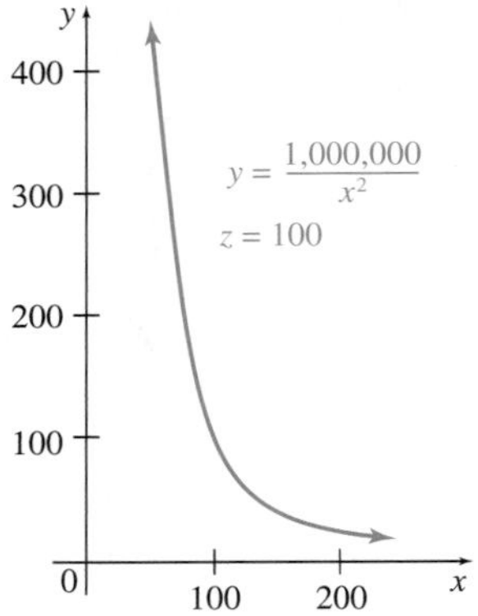

(b)

FIGURE 12

The level curve of height 100 found in Example 8 is shown graphed in three dimensions in Figure 12(a) and on the familiar xy-plane in Figure 12(b). The points of the graph correspond to those values of x and y that lead to production of 100 items.

The curve in Figure 12 is called an *isoquant,* for *iso* (equal) and *quant* (amount). In Example 8, the "amounts" all "equal" 100.

Because of the difficulty of drawing the graphs of more complicated functions, we merely list on the next page some common equations and their graphs. These graphs were drawn by computer, a very useful method of depicting three-dimensional surfaces.

Notice that not all the graphs correspond to functions of two variables. In the ellipsoid, for example, if x and y are both 0, then z can equal c or $-c$, whereas a function can take on only one value. We can, however, interpret the graph as a **level surface** for a function of three variables. Let

$$w(x, y, z) = \frac{x^2}{a^2} + \frac{y^2}{b^2} + \frac{z^2}{c^2}.$$

Then $w = 1$ produces the level surface of the ellipsoid shown, just as $z = c$ gives level curves for the function $z = f(x, y)$.

Another way to draw the graph of a function of two variables is with a graphing calculator. Figure 13 shows the graph of $z = x^2 + y^2$ generated by a TI-89. Figure 14 shows the same graph drawn by the computer program Maple™.

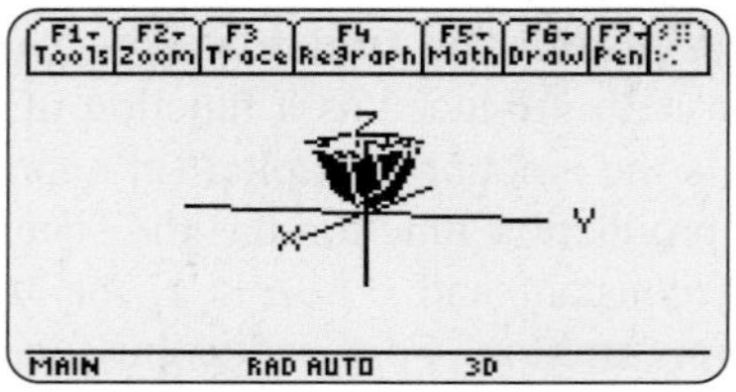

FIGURE 13

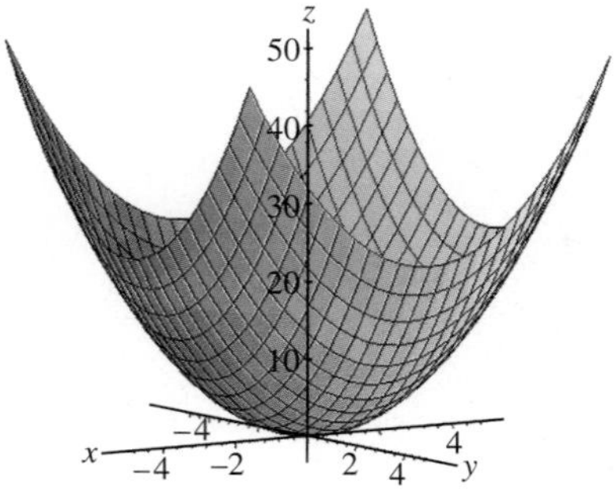

FIGURE 14

Paraboloid, $z = x^2 + y^2$

xy-trace: point
yz-trace: parabola
xz-trace: parabola

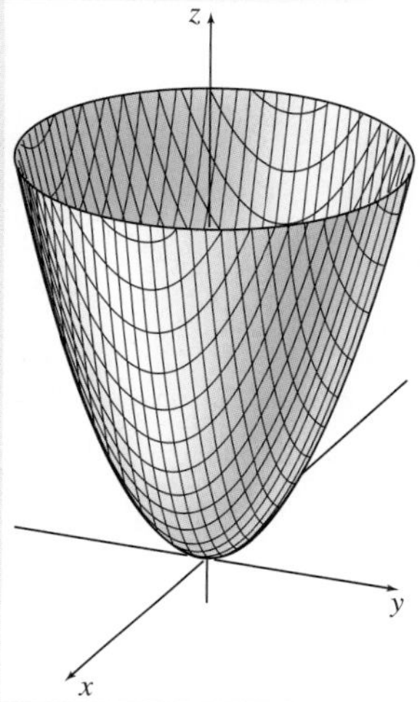

Ellipsoid, $\dfrac{x^2}{a^2} + \dfrac{y^2}{b^2} + \dfrac{z^2}{c^2} = 1$

xy-trace: ellipse
yz-trace: ellipse
xz-trace: ellipse

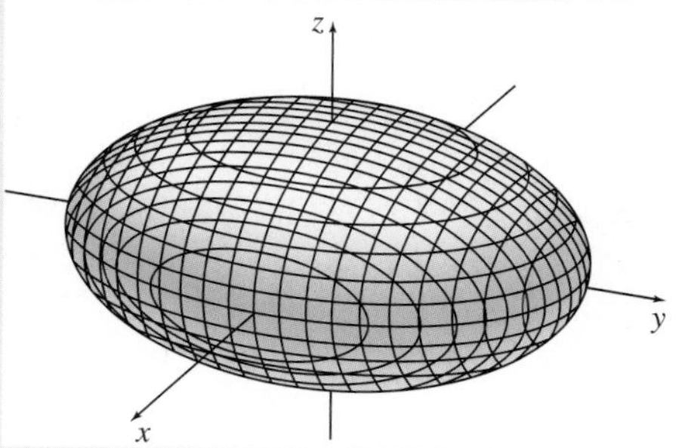

Hyperbolic Paraboloid, $x^2 - y^2 = z$
(sometimes called a *saddle*)

xy-trace: two intersecting lines
yz-trace: parabola
xz-trace: parabola

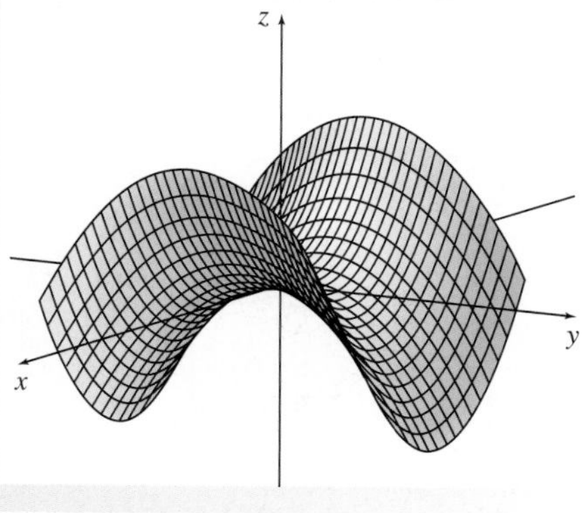

Hyperboloid of Two Sheets, $-x^2 - y^2 + z^2 = 1$

xy-trace: none
yz-trace: hyperbola
xz-trace: hyperbola

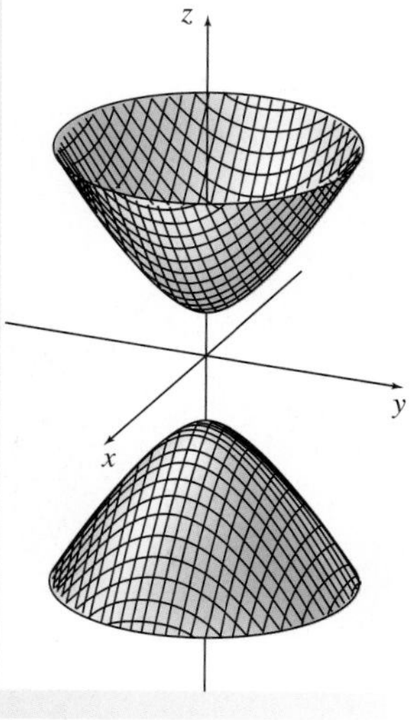

9.1 EXERCISES

1. Let $f(x, y) = 4x + 5y + 3$. Find the following.

a. $f(2, -1)$ **b.** $f(-4, 1)$ **c.** $f(-2, -3)$ **d.** $f(0, 8)$

2. Let $g(x, y) = -x^2 - 4xy + y^3$. Find the following.

a. $g(-2, 4)$ **b.** $g(-1, -2)$ **c.** $g(-2, 3)$ **d.** $g(5, 1)$

3. Let $h(x, y) = \sqrt{x^2 + 2y^2}$. Find the following.

a. $h(5, 3)$ **b.** $h(2, 4)$ **c.** $h(-1, -3)$ **d.** $h(-3, -1)$

4. Let $f(x, y) = \dfrac{\sqrt{9x + 5y}}{\log x}$. Find the following.

a. $f(10, 2)$ **b.** $f(100, 1)$ **c.** $f(1000, 0)$ **d.** $f\left(\frac{1}{10}, 5\right)$

Graph the first-octant portion of each plane.

5. $x + y + z = 6$ **6.** $x + y + z = 12$ **7.** $2x + 3y + 4z = 12$ **8.** $4x + 2y + 3z = 24$

9. $x + y = 4$ **10.** $y + z = 5$ **11.** $x = 2$ **12.** $z = 3$

Graph the level curves in the first octant at heights of $z = 0$, $z = 2$, and $z = 4$ for the following functions.

13. $3x + 2y + z = 18$ **14.** $x + 3y + 2z = 8$

15. $y^2 - x = -z$ **16.** $2y - \dfrac{x^2}{3} = z$

17. Discuss how a function of three variables in the form $w = f(x, y, z)$ might be graphed.

18. Suppose the graph of a plane $ax + by + cz = d$ has a portion in the first octant. What can be said about a, b, c, and d?

19. In the chapter on Nonlinear Functions, the vertical line test was presented, which tells whether a graph is the graph of a function. Does this test apply to functions of two variables? Explain.

20. A graph that was not shown in this section is the *hyperboloid of one sheet,* described by the equation $x^2 + y^2 - z^2 = 1$. Describe it as completely as you can.

Match each equation in Exercises 21–26 with its graph in a–f below and on the next page.

21. $z = x^2 + y^2$ **22.** $z^2 - y^2 - x^2 = 1$ **23.** $x^2 - y^2 = z$

24. $z = y^2 - x^2$ **25.** $\dfrac{x^2}{16} + \dfrac{y^2}{25} + \dfrac{z^2}{4} = 1$ **26.** $z = 5(x^2 + y^2)^{-1/2}$

a.

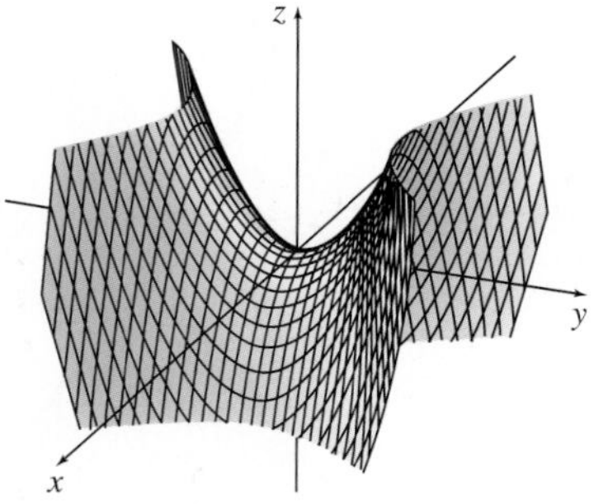

b.

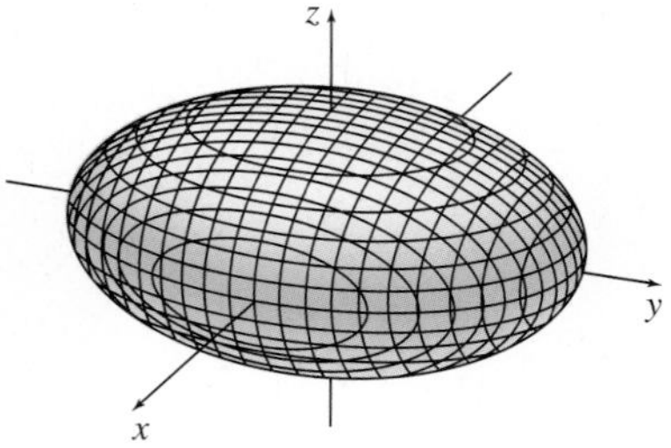

c.

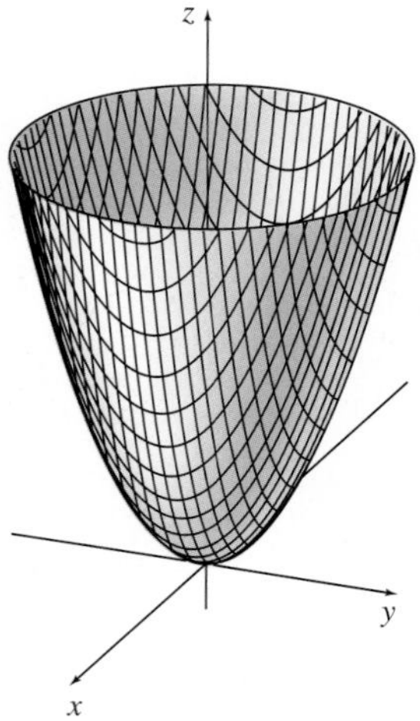

d.

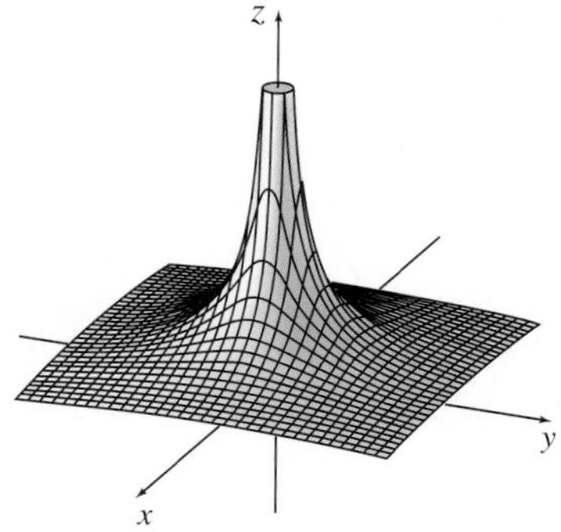

e.

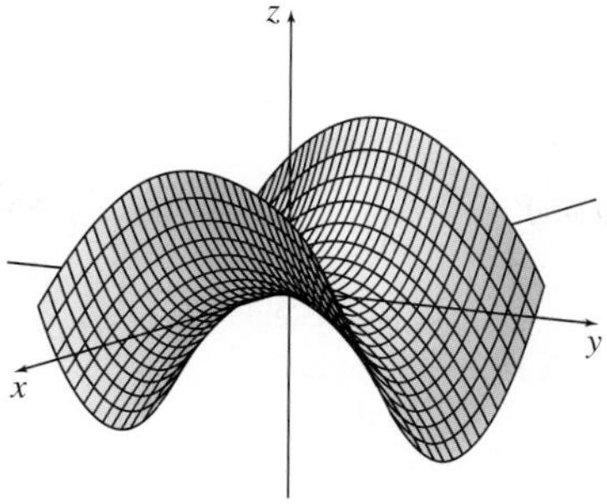

f.

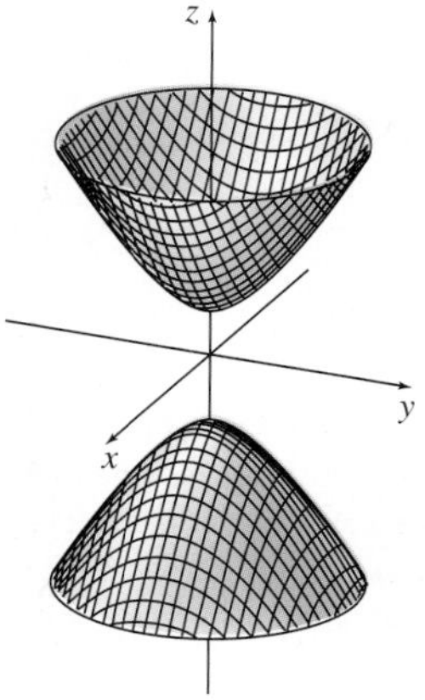

27. Let $f(x, y) = 9x^2 - 3y^2$, and find the following.

a. $\dfrac{f(x + h, y) - f(x, y)}{h}$

b. $\dfrac{f(x, y + h) - f(x, y)}{h}$

c. $\lim\limits_{h \to 0} \dfrac{f(x + h, y) - f(x, y)}{h}$

d. $\lim\limits_{h \to 0} \dfrac{f(x, y + h) - f(x, y)}{h}$

28. Let $f(x, y) = 7x^3 + 8y^2$, and find the following.

a. $\dfrac{f(x + h, y) - f(x, y)}{h}$

b. $\dfrac{f(x, y + h) - f(x, y)}{h}$

c. $\lim\limits_{h \to 0} \dfrac{f(x + h, y) - f(x, y)}{h}$

d. $\lim\limits_{h \to 0} \dfrac{f(x, y + h) - f(x, y)}{h}$

29. Let $f(x, y) = xye^{x^2+y^2}$. Use a graphing calculator or spreadsheet to find each of the following and give a geometric interpretation of the results. (*Hint:* First factor e^2 from the limit and then evaluate the quotient at smaller and smaller values of h.)

a. $\lim\limits_{h \to 0} \dfrac{f(1 + h, 1) - f(1, 1)}{h}$ **b.** $\lim\limits_{h \to 0} \dfrac{f(1, 1 + h) - f(1, 1)}{h}$

30. The following table provides values of the function $f(x, y)$. However, because of potential errors in measurement, the functional values may be slightly inaccurate. Using the statistical package included with a graphing calculator or spreadsheet and critical thinking skills, find the function $f(x, y) = a + bx + cy$ that best estimates the table where a, b, and c are integers. (*Hint:* Do a linear regression on each column with the value of y fixed and then use these four regression equations to determine the coefficient c.)

x \ y	0	1	2	3
0	4.02	7.04	9.98	13.00
1	6.01	9.06	11.98	14.96
2	7.99	10.95	14.02	17.09
3	9.99	13.01	16.01	19.02

Applications

BUSINESS AND ECONOMICS

31. ***Production*** Production of a precision camera is given by

$$P(x, y) = 100\left(\frac{3}{5}x^{-2/5} + \frac{2}{5}y^{-2/5}\right)^{-5},$$

where x is the amount of labor in work-hours and y is the amount of capital. Find the following.

a. What is the production when 32 work-hours and 1 unit of capital are provided?

b. Find the production when 1 work-hour and 32 units of capital are provided.

c. If 32 work-hours and 243 units of capital are used, what is the production output?

Individual Retirement Accounts The multiplier function

$$M = \frac{(1+i)^n(1-t)+t}{[1+(1-t)i]^n}$$

compares the growth of an Individual Retirement Account (IRA) with the growth of the same deposit in a regular savings account. The function M depends on the three variables n, i, and t, where n represents the number of years an amount is left at interest, i represents the interest rate in both types of accounts, and t represents the income tax rate. Values of $M > 1$ *indicate that the IRA grows faster than the savings account. Let* $M = f(n, i, t)$ *and find the following.*

32. Find the multiplier when funds are left for 25 years at 5% interest and the income tax rate is 28%. Which account grows faster?

33. What is the multiplier when money is invested for 25 years at 6% interest and the income tax rate is 33%? Which account grows faster?

Production Find the level curve at a production of 500 for the production functions in Exercises 34 and 35. Graph each function on the xy-plane.

34. The production function z for the United States was once estimated as $z = x^{.7}y^{.3}$, where x stands for the amount of labor and y stands for the amount of capital.

35. A study of the connection between immigration and the fiscal problems associated with the aging of the baby boom generation considered a production function of the form $z = x^{.6}y^{.4}$, where x represents the amount of labor and y the amount of capital.*

36. *Production* For the function in Exercise 34, what is the effect on z of doubling x? Of doubling y? Of doubling both?

37. *Cost* If labor (x) costs \$200 per unit, materials (y) cost \$100 per unit, and capital (z) costs \$50 per unit, write a function for total cost.

LIFE SCIENCES

38. *Oxygen Consumption* The oxygen consumption of a well-insulated mammal that is not sweating is approximated by

$$c = \frac{2.5(T - F)}{m^{.67}},$$

where T is the internal body temperature of the animal (in °C), F is the temperature of the outside of the animal's fur (in °C), and m is the animal's mass (in kilograms).† Find c for the following data.

a. Internal body temperature = 38°C; outside temperature = 6°C; mass = 32 kg

b. Internal body temperature = 40°C; outside temperature = 20°C; mass = 43 kg

39. *Dinosaur Running* An article entitled "How Dinosaurs Ran" explains that the locomotion of different sized animals can be compared when they have the same Froude number, defined as

$$F = \frac{v^2}{gl},$$

where v is the velocity, g is the acceleration of gravity (9.81 m per sec^2), and l is the leg length (in meters).‡

a. One result described in the article is that different animals change from a trot to a gallop at the same Froude number, roughly 2.56. Find the velocity at which this change occurs for a ferret, with a leg length of .09 m, and a rhinoceros, with a leg length of 1.2 m.

b. Ancient footprints in Texas of a sauropod, a large herbivorous dinosaur, are roughly 1 m in diameter, corresponding to a leg length of roughly 4 m. By comparing the stride divided by the leg length with that of various modern creatures, it can be determined that the Froude number for these dinosaurs is roughly .025. How fast were the sauropods traveling?

40. *Body Surface Area* The surface area of a human (in square meters) is approximated by

$$A = .202M^{.425}H^{.725},$$

where M is the mass of the person (in kilograms) and H is the height (in meters).† Find A for the following data.

a. Mass, 72 kg; height, 1.78 m

b. Mass, 65 kg; height, 1.40 m

c. Mass, 70 kg; height, 1.60 m

d. Using your mass and height, find your own surface area.

41. *Deer–Vehicle Accidents* Using data collected by the U.S. Forest Service, the annual number of deer–vehicle acci-

*Storesletten, Kjetil, "Sustaining Fiscal Policy Through Immigration," *Journal of Political Economy*, Vol. 108, No. 2, April 2000, pp. 300–323.

†Exercises 38 and 40 from Clow, Duane J. and N. Scott Urquhart, *Mathematics in Biology*. Copyright © 1974 by W. W. Norton & Company, Inc. Used by permission.

‡Alexander, R. McNeill, "How Dinosaurs Ran," *Scientific American*, Vol. 264, April 1991, p. 4.

dents for any given county in Ohio can be estimated by the function

$$A(L, T, U, C) = 53.02 + .383L + .0015T + .0028U - .0003C,$$

where A is the estimated number of accidents, L is the road length (in kilometers), T is the total county land area (in hundred-acres (Ha)), U is the urban land area (in hundred-acres), and C is the number of hundred-acres of crop land.*

a. Use this formula to estimate the number of deer–vehicle accidents for Mahoning County, where $L = 266$ km, $T = 107{,}484$ Ha, $U = 31{,}697$ Ha, and $C = 24{,}870$ Ha. The actual value was 396.

b. Given the magnitude and nature of the input numbers, which of the variables have the greatest potential to influence the number of deer–vehicle accidents? Explain your answer.

42. *Deer Harvest* Using data collected by the U.S. Forest Service, the annual number of deer that are harvested for any given county in Ohio can be estimated by the function

$$N(R, C) = 329.32 + .0377R - .0171C,$$

where N is the estimated number of harvested deer, R is the rural land area (in hundred-acres), and C is the number of hundred-acres of crop land.*

a. Use this formula to estimate the number of harvested deer for Tuscarawas County, where $R = 141{,}319$ Ha and $C = 37{,}960$ Ha. The actual value in 1995 was 4925 deer harvested.

b. Sketch the graph of this function in the first octant.

43. *Agriculture* Pregnant sows tethered in stalls often show high levels of repetitive behavior, such as bar biting and chain chewing, indicating chronic stress. Researchers from Great Britain have developed a function that estimates the relationship between repetitive behavior, the behavior of sows in adjacent stalls, and food allowances such that

$$\ln(T) = 5.49 - 3.00 \ln(F) + .18 \ln(C),$$

where T is the percent of time spent in repetitive behavior, F is the amount of food given to the sow (in kilograms per day), and C is the percent of time that neighboring sows spent bar biting and chain chewing.†

a. Solve the above expression for T.

b. Find and interpret T when $F = 2$ and $C = 40$.

GENERAL INTEREST

44. *Postage Rates* Extra postage is charged for parcels sent by U.S. mail that are more than 84 in. in length and girth combined. (Girth is the distance around the parcel perpendicular to its length. See the figure.) Express the combined length and girth as a function of L, W, and H.

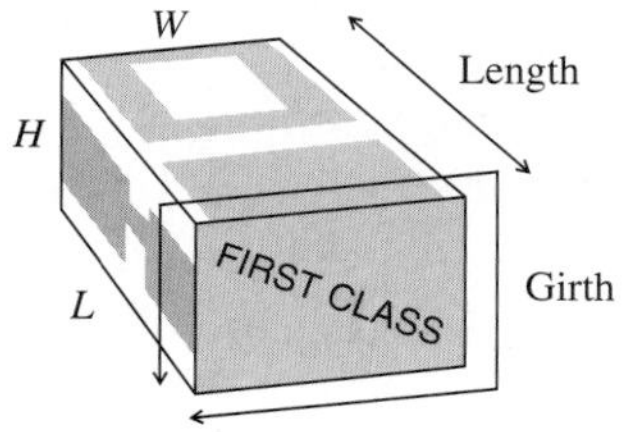

45. *Required Material* Refer to the figure for Exercise 44. Assume L, W, and H are in feet. Write a function in terms of L, W, and H that gives the total material required to build the box.

46. *Elliptical Templates* The holes cut in a roof for vent pipes require elliptical templates. A formula for determining the length of the major axis of the ellipse is given by

$$L = f(H, D) = \sqrt{H^2 + D^2},$$

where D is the (outside) diameter of the pipe and H is the "rise" of the roof per D units of "run"; that is, the slope of the roof is H/D. (See the figure on the next page.) The width of the ellipse (minor axis) equals D. Find the length and width

*Iverson, Aaron and Louis Iverson, "Spatial and Temporal Trends of Deer Harvest and Deer–Vehicle Accidents in Ohio," *Ohio Journal of Science*, 99, 1999, pp. 84–94.
†Appleby, M., A. Lawrence, and A. Illius, "Influence of Neighbours on Stereotypic Behaviour of Tethered Sows," *Applied Animal Behaviour Science*, Vol. 24, 1989, pp. 137–146.

of the ellipse required to produce a hole for a vent pipe with a diameter of 3.75 in. in roofs with the following slopes.

a. 3/4 **b.** 2/5

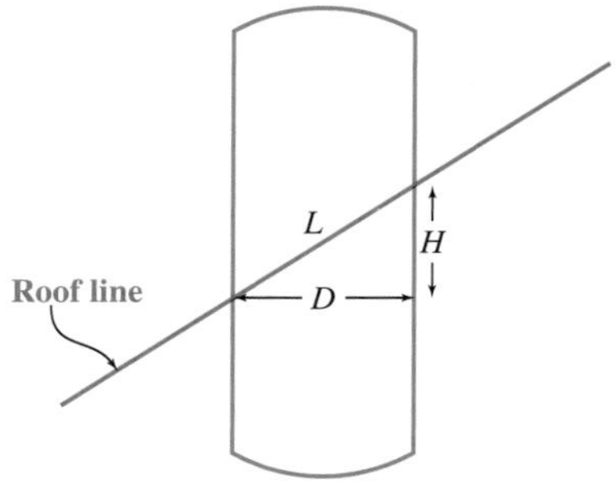

9.2 PARTIAL DERIVATIVES

THINK ABOUT IT *What is the change in productivity if labor is increased by one work-hour? What if capital is increased by one unit?*

FOR REVIEW

You may want to review the chapter on Calculating the Derivative for methods used to find some of the derivatives in this section.

Earlier, we found that the derivative dy/dx gives the rate of change of y with respect to x. In this section, we show how derivatives are found and interpreted for multivariable functions, and we will use that information to answer the questions posed above.

A small firm makes only two products, radios and CD players. The profits of the firm are given by

$$P(x, y) = 40x^2 - 10xy + 5y^2 - 80,$$

where x is the number of radios sold and y is the number of CD players sold. How will a change in x or y affect P?

Suppose that sales of radios have been steady at 10 units; only the sales of CD players vary. The management would like to find the marginal profit with respect to y, the number of CD players sold. Recall that marginal profit is given by the derivative of the profit function. Here, x is fixed at 10. Using this information, we begin by finding a new function, $f(y) = P(10, y)$. Let $x = 10$ to get

$$\begin{aligned} f(y) = P(10, y) &= 40(10)^2 - 10(10)y + 5y^2 - 80 \\ &= 3920 - 100y + 5y^2. \end{aligned}$$

The function $f(y)$ shows the profit from the sale of y CD players, assuming that x is fixed at 10 units. Find the derivative df/dy to get the marginal profit with respect to y.

$$\frac{df}{dy} = -100 + 10y$$

In this example, the derivative of the function $f(y)$ was taken with respect to y only; we assumed that x was fixed. To generalize, let $z = f(x, y)$. An intuitive definition of the *partial derivatives* of f with respect to x and y follows.

PARTIAL DERIVATIVES (INFORMAL DEFINITION)

The **partial derivative of f with respect to x** is the derivative of f obtained by treating x as a variable and y as a constant.

The **partial derivative of f with respect to y** is the derivative of f obtained by treating y as a variable and x as a constant.

The symbols $f_x(x, y)$ (no prime is used), $\partial z/\partial x$, z_x, and $\partial f/\partial x$ are used to represent the partial derivative of $z = f(x, y)$ with respect to x, with similar symbols used for the partial derivative with respect to y.

Generalizing from the definition of the derivative given earlier, partial derivatives of a function $z = f(x, y)$ are formally defined as follows.

PARTIAL DERIVATIVES (FORMAL DEFINITION)

Let $z = f(x, y)$ be a function of two independent variables. Let all indicated limits exist. Then the partial derivative of f with respect to x is

$$f_x(x, y) = \frac{\partial f}{\partial x} = \lim_{h \to 0} \frac{f(x + h, y) - f(x, y)}{h},$$

and the partial derivative of f with respect to y is

$$f_y(x, y) = \frac{\partial f}{\partial y} = \lim_{h \to 0} \frac{f(x, y + h) - f(x, y)}{h}.$$

If the indicated limits do not exist, then the partial derivatives do not exist.

Similar definitions could be given for functions of more than two independent variables.

EXAMPLE 1 *Partial Derivatives*

Let $f(x, y) = 4x^2 - 9xy + 6y^3$. Find $f_x(x, y)$ and $f_y(x, y)$.

Solution To find $f_x(x, y)$, treat y as a constant and x as a variable. The derivative of the first term, $4x^2$, is $8x$. In the second term, $-9xy$, the constant coefficient of x is $-9y$, so the derivative with x as the variable is $-9y$. The derivative of $6y^3$ is zero, since we are treating y as a constant. Thus,

$$f_x(x, y) = 8x - 9y.$$

Now, to find $f_y(x, y)$, treat y as a variable and x as a constant. Since x is a constant, the derivative of $4x^2$ is zero. In the second term, the coefficient of y is $-9x$ and the derivative of $-9xy$ is $-9x$. The derivative of the third term is $18y^2$. Thus,

$$f_y(x, y) = -9x + 18y^2.$$

The next example shows how the chain rule can be used to find partial derivatives.

EXAMPLE 2 *Partial Derivatives*

Let $f(x, y) = \ln |x^2 + y|$. Find $f_x(x, y)$ and $f_y(x, y)$.

Solution Recall the formula for the derivative of a natural logarithm function. If $g(x) = \ln |x|$, then $g'(x) = 1/x$. Using this formula and the chain rule,

$$f_x(x, y) = \frac{1}{x^2 + y} \cdot D_x(x^2 + y) = \frac{1}{x^2 + y} \cdot 2x = \frac{2x}{x^2 + y},$$

and

$$f_y(x, y) = \frac{1}{x^2 + y} \cdot D_y(x^2 + y) = \frac{1}{x^2 + y} \cdot 1 = \frac{1}{x^2 + y}.$$ ▪

The notation

$$f_x(a, b) \qquad \text{or} \qquad \frac{\partial f}{\partial x}(a, b)$$

represents the value of a partial derivative when $x = a$ and $y = b$, as shown in the next example.

EXAMPLE 3 *Evaluating Partial Derivatives*

Let $f(x, y) = 2x^2 + 3xy^3 + 2y + 5$. Find the following.

(a) $f_x(-1, 2)$

Solution First, find $f_x(x, y)$ by holding y constant.

$$f_x(x, y) = 4x + 3y^3$$

Now let $x = -1$ and $y = 2$.

$$f_x(-1, 2) = 4(-1) + 3(2)^3 = -4 + 24 = 20$$

(b) $\dfrac{\partial f}{\partial y}(-4, -3)$

Solution Since $\partial f / \partial y = 9xy^2 + 2$,

$$\frac{\partial f}{\partial y}(-4, -3) = 9(-4)(-3)^2 + 2 = 9(-36) + 2 = -322.$$ ▪

The partial derivative can also be approximated using a small value of h in the difference quotient $[f(x + h, y) - f(x, y)]/h$ or $[f(x, y + h) - f(x, y)]/h$, as we did with functions of single variables. For an application of this, see Exercise 62. To find the derivatives in Example 3 (a) numerically, we would calculate

$$\frac{f(-1 + h, 2) - f(-1, 2)}{h}$$

$$= \frac{[2(-1 + h)^2 + 3(-1 + h)(2^3) + 2(2) + 5] - [2(-1)^2 + 3(-1)(2^3) + 2(2) + 5]}{h}$$

using small values of h. With $h = 10^{-5}$ and 10^{-6}, the results are 20.00002 and 20.000002, respectively. In this example, the difference quotient above can be reduced algebraically to $20 + 2h$, from which the limit is easily seen to be 20. Review parts (c) and (d) of Exercises 27 and 28 of the previous section for examples in which the partial derivative can be computed using the formal definition.

The derivative of a function of one variable can be interpreted as the tangent line to the graph at that point. With some modification, the same is true of partial derivatives of functions of two variables. At a point on the graph of a function of two variables, $z = f(x, y)$, there may be many tangent lines, all of which lie in the same tangent plane, as shown in Figure 15.

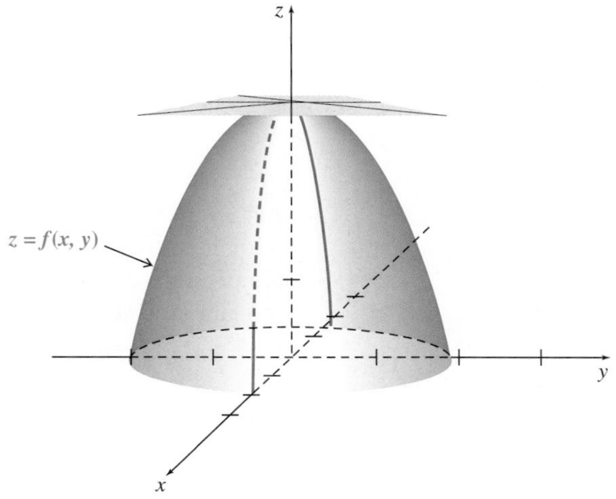

FIGURE 15

In any particular direction, however, there will be only one tangent line. We use partial derivatives to find the slope of the tangent lines in the x- and y-directions as follows.

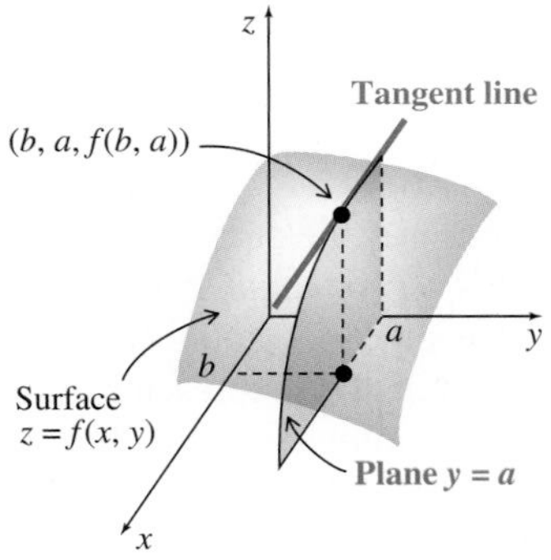

FIGURE 16

Figure 16 shows a surface $z = f(x, y)$ and a plane that is parallel to the xz-plane. The equation of the plane is $y = a$. (This corresponds to holding y fixed.) Since $y = a$ for points on the plane, any point on the curve that represents the intersection of the plane and the surface must have the form $(x, y, z) = (x, a, f(x, a))$. Thus, this curve can be described as $z = f(x, a)$. Since a is constant, $z = f(x, a)$ is a function of one variable. When the derivative of $z = f(x, a)$ is evaluated at $x = b$, it gives the slope of the line tangent to this curve at the point $(b, a, f(b, a))$, as shown in Figure 16. Thus, the partial derivative of f with respect to x, $f_x(b, a)$, gives the rate of change of the surface $z = f(x, y)$ in the x-direction at the point $(b, a, f(b, a))$. In the same way, the partial derivative with respect to y will give the slope of the line tangent to the surface in the y-direction at the point $(b, a, f(b, a))$.

Rate of Change The derivative of $y = f(x)$ gives the rate of change of y with respect to x. In the same way, if $z = f(x, y)$, then $f_x(x, y)$ gives the rate of change of z with respect to x, if y is held constant.

EXAMPLE 4 *Water Temperature*

Suppose that the temperature of the water at the point on a river where a nuclear power plant discharges its hot waste water is approximated by

$$T(x, y) = 2x + 5y + xy - 40,$$

where x represents the temperature of the river water (in degrees Celsius) before it reaches the power plant and y is the number of megawatts (in hundreds) of electricity being produced by the plant.

(a) Find and interpret $T_x(9, 5)$.

Solution First, find the partial derivative $T_x(x, y)$.

$$T_x(x, y) = 2 + y$$

This partial derivative gives the rate of change of T with respect to x. Replacing x with 9 and y with 5 gives

$$T_x(9, 5) = 2 + 5 = 7.$$

Just as marginal cost is the approximate cost of one more item, this result, 7, is the approximate change in temperature of the output water if input water temperature changes by 1 degree, from $x = 9$ to $x = 9 + 1 = 10$, while y remains constant at 5 (500 megawatts of electricity produced).

(b) Find and interpret $T_y(9, 5)$.

Solution The partial derivative $T_y(x, y)$ is

$$T_y(x, y) = 5 + x.$$

This partial derivative gives the rate of change of T with respect to y as

$$T_y(9, 5) = 5 + 9 = 14.$$

This result, 14, is the approximate change in temperature resulting from a 1-unit increase in production of electricity from $y = 5$ to $y = 5 + 1 = 6$ (from 500 to 600 megawatts), while the input water temperature x remains constant at 9°C.

As mentioned in the previous section, if $P(x, y)$ gives the output P produced by x units of labor and y units of capital, $P(x, y)$ is a production function. The partial derivatives of this production function have practical implications. For example, $\partial P/\partial x$ gives the marginal productivity of labor. This represents the rate at which the output is changing with respect to a one-unit change in labor for a fixed capital investment. That is, if the capital investment is held constant and labor is increased by 1 work-hour, $\partial P/\partial x$ will yield the approximate change in the production level. Likewise, $\partial P/\partial y$ gives the marginal productivity of capital, which represents the rate at which the output is changing with respect to a one-unit change in capital for a fixed labor value. So if the labor force is held constant and the capital investment is increased by 1 unit, $\partial P/\partial y$ will approximate the corresponding change in the production level.

EXAMPLE 5 *Production Function*

A company that manufactures computers has determined that its production function is given by

$$P(x, y) = 500x + 800y + 3x^2y + x^3 - \frac{y^4}{4},$$

where x is the size of the labor force (measured in work-hours per week) and y is the amount of capital (measured in units of \$1000) invested. Find the marginal productivity of labor and capital when $x = 50$ and $y = 20$, and interpret the results.

Solution The marginal productivity of labor is found by taking the derivative of P with respect to x.

$$\frac{\partial P}{\partial x} = 500 + 6xy + 3x^2$$

$$\frac{\partial P}{\partial x}(50, 20) = 500 + 6(50)(20) + 3(50)^2 = 14{,}000$$

Thus, if the capital investment is held constant at \$20,000 and labor is increased from 50 to 51 work-hours per week, production will increase by about 14,000 units. In the same way, the marginal productivity of capital is $\partial P/\partial y$.

$$\frac{\partial P}{\partial y} = 800 + 3x^2 - y^3$$

$$\frac{\partial P}{\partial y}(50, 20) = 800 + 3(50)^2 - (20)^3 = 300$$

If work-hours are held constant at 50 hours per week and the capital investment is increased from \$20,000 to \$21,000, production will increase by about 300 units.

Second-Order Partial Derivatives

The second derivative of a function of one variable is very useful in determining relative maxima and minima. **Second-order partial derivatives** (partial derivatives of a partial derivative) are used in a similar way for functions of two or more variables. The situation is somewhat more complicated, however, with more independent variables. For example, $f(x, y) = 4x + x^2y + 2y$ has two first-order partial derivatives,

$$f_x(x, y) = 4 + 2xy \qquad \text{and} \qquad f_y(x, y) = x^2 + 2.$$

Since each of these has two partial derivatives, one with respect to y and one with respect to x, there are *four* second-order partial derivatives of function f. The notations for these four second-order partial derivatives are given below.

SECOND-ORDER PARTIAL DERIVATIVES

For a function $z = f(x, y)$, if the indicated partial derivative exists, then

$$\frac{\partial}{\partial x}\left(\frac{\partial z}{\partial x}\right) = \frac{\partial^2 z}{\partial x^2} = f_{xx}(x, y) = z_{xx} \qquad \frac{\partial}{\partial y}\left(\frac{\partial z}{\partial y}\right) = \frac{\partial^2 z}{\partial y^2} = f_{yy}(x, y) = z_{yy}$$

$$\frac{\partial}{\partial y}\left(\frac{\partial z}{\partial x}\right) = \frac{\partial^2 z}{\partial y \partial x} = f_{xy}(x, y) = z_{xy} \qquad \frac{\partial}{\partial x}\left(\frac{\partial z}{\partial y}\right) = \frac{\partial^2 z}{\partial x \partial y} = f_{yx}(x, y) = z_{yx}$$

NOTE For most functions found in applications and for all of the functions in this book, the second-order partial derivatives $f_{xy}(x, y)$ and $f_{yx}(x, y)$ are equal. This is always true when $f_{xy}(x, y)$ and $f_{yx}(x, y)$ are continuous. Therefore, it is not necessary to be particular about the order in which these derivatives are found. ■

EXAMPLE 6 *Second-Order Partial Derivatives*

Find all second-order partial derivatives for

$$f(x, y) = -4x^3 - 3x^2y^3 + 2y^2.$$

Solution First find $f_x(x, y)$ and $f_y(x, y)$.

$$f_x(x, y) = -12x^2 - 6xy^3 \quad \text{and} \quad f_y(x, y) = -9x^2y^2 + 4y$$

To find $f_{xx}(x, y)$, take the partial derivative of $f_x(x, y)$ with respect to x.

$$f_{xx}(x, y) = -24x - 6y^3$$

Take the partial derivative of $f_y(x, y)$ with respect to y; this gives f_{yy}.

$$f_{yy}(x, y) = -18x^2y + 4$$

Find $f_{xy}(x, y)$ by starting with $f_x(x, y)$, then taking the partial derivative of $f_x(x, y)$ with respect to y.

$$f_{xy}(x, y) = -18xy^2$$

Finally, find $f_{yx}(x, y)$ by starting with $f_y(x, y)$; take its partial derivative with respect to x.

$$f_{yx}(x, y) = -18xy^2$$

EXAMPLE 7 *Second-Order Partial Derivatives*

Let $f(x, y) = 2e^x - 8x^3y^2$. Find all second-order partial derivatives.

Solution Here $f_x(x, y) = 2e^x - 24x^2y^2$ and $f_y(x, y) = -16x^3y$. (Recall: If $g(x) = e^x$, then $g'(x) = e^x$.) Now find the second-order partial derivatives.

$$f_{xx}(x, y) = 2e^x - 48xy^2 \qquad f_{xy}(x, y) = -48x^2y$$
$$f_{yy}(x, y) = -16x^3 \qquad f_{yx}(x, y) = -48x^2y$$

Partial derivatives of functions with more than two independent variables are found in a similar manner. For instance, to find $f_z(x, y, z)$ for $f(x, y, z)$, hold x and y constant and differentiate with respect to z.

EXAMPLE 8 *Second-Order Partial Derivatives*

Let $f(x, y, z) = 2x^2yz^2 + 3xy^2 - 4yz$. Find $f_x(x, y, z)$, $f_y(x, y, z)$, $f_{xz}(x, y, z)$, and $f_{yz}(x, y, z)$.

Solution

$$f_x(x, y, z) = 4xyz^2 + 3y^2$$
$$f_y(x, y, z) = 2x^2z^2 + 6xy - 4z$$

To find $f_{xz}(x, y, z)$, differentiate $f_x(x, y, z)$ with respect to z.

$$f_{xz}(x, y, z) = 8xyz$$

Differentiate $f_y(x, y, z)$ with respect to z to get $f_{yz}(x, y, z)$.

$$f_{yz}(x, y, z) = 4x^2z - 4$$

9.2 EXERCISES

1. Let $z = f(x, y) = 12x^2 - 8xy + 3y^2$. Find the following using the formal definition of the partial derivative.

a. $\frac{\partial z}{\partial x}$ **b.** $\frac{\partial z}{\partial y}$ **c.** $\frac{\partial f}{\partial x}(2, 3)$ **d.** $f_y(1, -2)$

2. Let $z = g(x, y) = 5x + 9x^2y + y^2$. Find the following using the formal definition of the partial derivative.

a. $\frac{\partial g}{\partial x}$ **b.** $\frac{\partial g}{\partial y}$ **c.** $\frac{\partial z}{\partial y}(-3, 0)$ **d.** $g_x(2, 1)$

In Exercises 3–20, find $f_x(x, y)$ and $f_y(x, y)$. Then find $f_x(2, -1)$ and $f_y(-4, 3)$. Leave the answers in terms of e in Exercises 7–10, 15–16, and 19–20.

3. $f(x, y) = -2xy + 6y^3 + 2$

4. $f(x, y) = 4x^2y - 9y^2$

5. $f(x, y) = 3x^3y^2$

6. $f(x, y) = -2x^2y^4$

7. $f(x, y) = e^{x+y}$

8. $f(x, y) = 3e^{2x+y}$

9. $f(x, y) = -5e^{3x-4y}$

10. $f(x, y) = 8e^{7x-y}$

11. $f(x, y) = \frac{x^2 + y^3}{x^3 - y^2}$

12. $f(x, y) = \frac{3x^2y^3}{x^2 + y^2}$

13. $f(x, y) = \ln|1 + 3x^2y^3|$

14. $f(x, y) = \ln|2x^5 - xy^4|$

15. $f(x, y) = xe^{x^2y}$

16. $f(x, y) = y^2e^{x+3y}$

17. $f(x, y) = \sqrt{x^4 + 3xy + y^4 + 10}$

18. $f(x, y) = (7x^2 + 18xy^2 + y^3)^{1/3}$

19. $f(x, y) = \frac{3x^2y}{e^{xy} + 2}$

20. $f(x, y) = (7e^{x+2y} + 4)(e^{x^2} + y^2 + 2)$

Find all second-order partial derivatives for the following.

21. $f(x, y) = 6x^3y - 9y^2 + 2x$

22. $g(x, y) = 5xy^4 + 8x^3 - 3y$

23. $R(x, y) = 4x^2 - 5xy^3 + 12y^2x^2$

24. $h(x, y) = 30y + 5x^2y + 12xy^2$

25. $r(x, y) = \frac{4x}{x + y}$

26. $k(x, y) = \frac{-5y}{x + 2y}$

27. $z = 4xe^y$

28. $z = -3ye^x$

29. $r = \ln|x + y|$

30. $k = \ln|5x - 7y|$

31. $z = x\ln|xy|$

32. $z = (y + 1)\ln|x^3y|$

For the functions defined as follows, find values of x and y such that both $f_x(x, y) = 0$ and $f_y(x, y) = 0$.

33. $f(x, y) = 6x^2 + 6y^2 + 6xy + 36x - 5$

34. $f(x, y) = 50 + 4x - 5y + x^2 + y^2 + xy$

35. $f(x, y) = 9xy - x^3 - y^3 - 6$

36. $f(x, y) = 2200 + 27x^3 + 72xy + 8y^2$

Find $f_x(x, y, z)$, $f_y(x, y, z)$, $f_z(x, y, z)$, and $f_{yz}(x, y, z)$ for the following.

37. $f(x, y, z) = x^2 + yz + z^4$

38. $f(x, y, z) = 3x^5 - x^2 + y^5$

39. $f(x, y, z) = \frac{6x - 5y}{4z + 5}$

40. $f(x, y, z) = \frac{2x^2 + xy}{yz - 2}$

41. $f(x, y, z) = \ln|x^2 - 5xz^2 + y^4|$

42. $f(x, y, z) = \ln|8xy + 5yz - x^3|$

In Exercises 43 and 44, approximate the indicated derivative for each function by using the definition of the derivative with small values of h.

43. $f(x, y) = (x + y/2)^{x+y/2}$

a. $f_x(1, 2)$ **b.** $f_y(1, 2)$

44. $f(x, y) = (x + y^2)^{2x+y}$

a. $f_x(2, 1)$ **b.** $f_y(2, 1)$

Applications

BUSINESS AND ECONOMICS

45. *Manufacturing Cost* Suppose that the manufacturing cost of a precision electronic calculator is approximated by

$$M(x, y) = 40x^2 + 30y^2 - 10xy + 30,$$

where x is the cost of electronic chips and y is the cost of labor. Find the following.

a. $M_y(4, 2)$ **b.** $M_x(3, 6)$ **c.** $(\partial M/\partial x)(2, 5)$

d. $(\partial M/\partial y)(6, 7)$

46. *Revenue* The revenue from the sale of x units of a tranquilizer and y units of an antibiotic is given by

$$R(x, y) = 5x^2 + 9y^2 - 4xy.$$

Suppose 9 units of tranquilizer and 5 units of antibiotic are sold.

a. What is the approximate effect on revenue if 10 units of tranquilizer and 5 units of antibiotic are sold?

b. What is the approximate effect on revenue if the amount of antibiotic sold is increased to 6 units, while tranquilizer sales remain constant?

47. *Sales* A car dealership estimates that the total weekly sales of its most popular model is a function of the car's list price, p, and the interest rate in percent, i, offered by the manufacturer. The approximate weekly sales are given by

$$f(p, i) = 99p - .5pi - .0025p^2.$$

a. Find the weekly sales if the average list price is \$19,400 and the manufacturer is offering an 8% interest rate.

b. Find and interpret $f_p(p, i)$ and $f_i(p, i)$.

c. What would be the effect on weekly sales if the price is \$19,400 and interest rates rise from 8% to 9%?

48. *Marginal Productivity* Suppose the production function of a company is given by

$$P(x, y) = 100\sqrt{x^2 + y^2},$$

where x represents units of labor and y represents units of capital. Find the following when 4 units of labor and 3 units of capital are used.

a. The marginal productivity of labor

b. The marginal productivity of capital

49. *Marginal Productivity* A manufacturer estimates that production (in hundreds of units) is a function of the amounts x and y of labor and capital used, as follows.

$$f(x, y) = \left(\frac{1}{3}x^{-1/3} + \frac{2}{3}y^{-1/3}\right)^{-3}$$

a. Find the number of units produced when 27 units of labor and 64 units of capital are utilized.

b. Find and interpret $f_x(27, 64)$ and $f_y(27, 64)$.

c. What would be the approximate effect on production of increasing labor by 1 unit?

50. *Marginal Productivity* The production function z for the United States was once estimated as

$$z = x^{.7}y^{.3},$$

where x stands for the amount of labor and y the amount of capital. Find the marginal productivity of labor and of capital.

51. *Marginal Productivity* A similar production function for Canada is

$$z = x^{.4}y^{.6},$$

with x, y, and z as in Exercise 50. Find the marginal productivity of labor and of capital.

52. *Marginal Productivity* A manufacturer of automobile batteries estimates that his total production (in thousands of units) is given by

$$f(x, y) = 3x^{1/3}y^{2/3},$$

where x is the number of units of labor and y is the number of units of capital utilized.

a. Find and interpret $f_x(64, 125)$ and $f_y(64, 125)$ if the current level of production uses 64 units of labor and 125 units of capital.

b. What would be the approximate effect on production of increasing labor to 65 units while holding capital at the current level?

c. Suppose that sales have been good and management wants to increase either capital or labor by 1 unit. Which option would result in a larger increase in production?

LIFE SCIENCES

53. *Calorie Expenditure* The average energy expended for an animal to walk or run 1 km can be estimated by the function

$$f(m, v) = 25.92m^{.68} + \frac{3.62m^{.75}}{v},$$

where $f(m, v)$ is the energy used (in kcal per hour), m is the mass (in g), and v is the speed of movement (in km per hour) of the animal.*

a. Find $f(300, 10)$.

b. Find $f_m(300, 10)$ and interpret.

c. If a mouse could run at the same speed that an elephant walks, which animal would expend more energy? How can partial derivatives be used to explore this question?

54. *Oxygen Consumption* The oxygen consumption of a well-insulated mammal that is not sweating is approximated by

$$c = c(T, F, m) = \frac{2.5(T - F)}{m^{.67}} = 2.5(T - F)m^{-.67},$$

where T is the internal body temperature of the animal (in °C), F is the temperature of the outside of the animal's fur (in °C), and m is the animal's mass (in kilograms). Find the approximate change in oxygen consumption under the following conditions.

a. The internal temperature increases from 38°C to 39°C, while the outside temperature remains at 12°C and the mass remains at 30 kg.

b. The internal temperature is constant at 36°C, the outside temperature increases from 14°C to 15°C, and the mass remains at 25 kg.

55. *Body Surface Area* The surface area of a human (in square meters) is approximated by

$$A(M, H) = .202M^{.425}H^{.725},$$

where M is the mass of the person (in kilograms) and H is the height (in meters). Find the approximate change in surface area under the following conditions.

a. The mass changes from 72 kg to 73 kg, while the height remains 1.8 m.

b. The mass remains stable at 70 kg, while the height changes from 1.6 m to 1.7 m.

56. *Blood Flow* In one method of computing the quantity of blood pumped through the lungs in one minute, a researcher first finds each of the following (in milliliters).

b = quantity of oxygen used by the body in one minute

a = quantity of oxygen per liter of blood that has just gone through the lungs

v = quantity of oxygen per liter of blood that is about to enter the lungs

In one minute,

Amount of oxygen used = Amount of oxygen per liter × Liters of blood pumped.

If C is the number of liters of blood pumped through the lungs in one minute, then

$$b = (a - v) \cdot C \quad \text{or} \quad C = \frac{b}{a - v}.$$

a. Find the number of liters of blood pumped through the lungs in one minute if $a = 160$, $b = 200$, and $v = 125$.

b. Find the approximate change in C when a changes from 160 to 161, $b = 200$, and $v = 125$.

c. Find the approximate change in C when $a = 160$, b changes from 200 to 201, and $v = 125$.

d. Find the approximate change in C when $a = 160$, $b = 200$, and v changes from 125 to 126.

e. A change of 1 unit in which the quantity of oxygen produces the greatest change in the liters of blood pumped?

57. *Health* A weight-loss counselor has prepared a program of diet and exercise for a client. If the client sticks to the program, the weight loss that can be expected (in pounds per week) is given by

$$\text{Weight loss} = f(n, c) = \frac{1}{8}n^2 - \frac{1}{5}c + \frac{1937}{8},$$

where c is the average daily calorie intake for the week and n is the number of 40-minute aerobic workouts per week.

a. How many pounds can the client expect to lose by eating an average of 1200 cal per day and participating in four 40-minute workouts in a week?

b. Find and interpret $\partial f / \partial n$.

c. The client currently averages 1100 cal per day and does three 40-minute workouts each week. What would be the approximate impact on weekly weight loss of adding a fourth workout per week?

58. *Health* The body mass index is a number that can be calculated for any individual as follows: Multiply a person's

*Robbins, C., *Wildlife Feeding and Nutrition*, New York: Academic Press, 1983, p. 114.

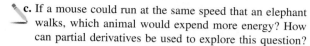

weight by 703 and divide by the person's height squared. That is,

$$B = \frac{703w}{h^2},$$

where w is in pounds and h is in inches.* The National Heart, Lung and Blood Institute uses the body mass index to determine whether a person is "overweight" $(25 \leq B < 30)$ or "obese" $(B \geq 30)$.

a. Calculate the body mass index for a person who weighs 220 lb and is 74 in. tall.

b. Calculate $\frac{\partial B}{\partial w}$ and $\frac{\partial B}{\partial h}$ and interpret.

c. Using the fact that 1 in. = .0254 m and 1 lb = .4555 kg, transform this formula to handle metric units.

59. *Drug Reaction* The reaction to x units of a drug t hours after it was administered is given by

$$R(x, t) = x^2(a - x)t^2e^{-t},$$

for $0 \leq x \leq a$ (where a is a constant). Find the following.

a. $\frac{\partial R}{\partial x}$ **b.** $\frac{\partial R}{\partial t}$ **c.** $\frac{\partial^2 R}{\partial x^2}$ **d.** $\frac{\partial^2 R}{\partial x \partial t}$

e. Interpret your answers to parts a and b.

60. *Scuba Diving* In 1908, J. Haldane constructed diving tables that provide a relationship between the water pressure on body tissues for various water depths and dive times. The tables were successfully used by divers to virtually eliminate decompression sickness. The pressure in atmospheres for a no-stop dive is given by the following formula:†

$$p_l(l, t) = 1 + \frac{l}{33}(1 - 2^{-t/5}),$$

where t is in minutes, l is in feet, and p is in atmospheres (atm).‡

a. Find the pressure at 33 ft for a 10-minute dive.

b. Find $p(33, 10)$ and $p_t(33, 10)$ and interpret. (*Hint:* $D_t(a^t) = \ln(a)a^t$.)

c. Haldane estimated that decompression sickness for no-stop dives could be avoided if the diver's tissue pressure did not exceed 2.15 atm. Find the maximum amount of time that a diver could stay down (time includes going down and coming back up) if he or she wants to dive to a depth of 66 ft.

61. *Wind Chill* In 1941, explorers Paul Siple and Charles Passel discovered that the amount of heat lost when an object is exposed to cold air depends on both the temperature of the air and the velocity of the wind. They developed the *Wind Chill Index* as a way to measure the danger of frostbite while doing outdoor activities. The wind chill can be calculated as follows:

$$W(V, T) = 91.4 - \frac{(10.45 + 6.69\sqrt{V} - .447V)(91.4 - T)}{22},$$

where V is the wind speed in miles per hour and T is the temperature in Fahrenheit for wind speeds between 4 and 45 mph.§

a. Find the wind chill for a wind speed of 20 mph and 10°F.

b. If a weather report indicates that the wind chill is −25°F and the actual outdoor temperature is 5°F, use a graphing calculator to find the corresponding wind speed to the nearest mile per hour.

c. Find $W_V(20, 10)$ and $W_T(20, 10)$ and interpret.

d. Using the table command on a graphing calculator or a spreadsheet, develop a wind chill chart for various wind speeds and temperatures.

62. *Heat Index* The chart on the next page shows the heat index, which combines the effects of temperature with humidity to give a measure of the apparent temperature, or how hot it feels to the body.‖ For example, when the outside temperature is 90°F and the relative humidity is

*The National Institutes of Health

†These estimates are conservative. Please consult modern dive tables before making a dive.

‡Westbrook, David, "The Mathematics of Scuba Diving," *The UMAP Journal*, Vol. 18, No. 2, 1997, pp. 2–19.

§Bosch, William and L. Cobb, "Windchill," *The UMAP Journal*, Vol. 13, No. 3, 1990, pp. 481–489.

‖The Weather Channel: Data from www.weather.com (May 9, 2000).

40%, then the apparent temperature is approximately 93°F. Let $I = f(T, H)$ give the heat index, I, as a function of the temperature T (in degrees Fahrenheit) and the percent humidity H. Estimate the following.

a. $f(90, 30)$ **b.** $f(90, 75)$ **c.** $f(80, 75)$

Heat Index

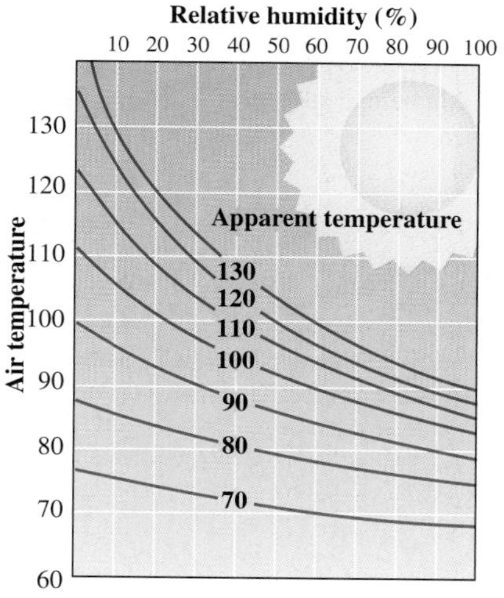

Estimate the following by approximating the partial derivative using a value of $h = 5$ in the difference quotient.

d. $f_T(90, 30)$ **e.** $f_H(90, 30)$ **f.** $f_T(90, 75)$

g. $f_H(90, 75)$

h. Describe in words what your answers in parts d–g mean.

63. *Breath Volume* The following table accompanies the Voldyne® 5000 Volumetric Exerciser. The table gives the typical lung capacity (in milliliters) for women of various ages and heights. Based on the chart, it is possible to conclude that the partial derivative of the lung capacity with respect to age and with respect to height have constant values. What are those values?

	Height (in.)	58″	60″	62″	64″	66″	68″	70″	72″	74″
	20	1900	2100	2300	2500	2700	2900	3100	3300	3500
A	25	1850	2050	2250	2450	2650	2850	3050	3250	3450
G	30	1800	2000	2200	2400	2600	2800	3000	3200	3400
E	35	1750	1950	2150	2350	2550	2750	2950	3150	3350
	40	1700	1900	2100	2300	2500	2700	2900	3100	3300
I	45	1650	1850	2050	2250	2450	2650	2850	3050	3250
N	50	1600	1800	2000	2200	2400	2600	2800	3000	3200
	55	1550	1750	1950	2150	2350	2550	2750	2950	3150
Y	60	1500	1700	1900	2100	2300	2500	2700	2900	3100
E	65	1450	1650	1850	2050	2250	2450	2650	2850	3050
A	70	1400	1600	1800	2000	2200	2400	2600	2800	3000
R	75	1350	1550	1750	1950	2150	2350	2550	2750	2950
S	80	1300	1500	1700	1900	2100	2300	2500	2700	2900

SOCIAL SCIENCES

64. *Education* A developmental mathematics instructor at a large university has determined that a student's probability of success in the university's pass/fail remedial algebra course is a function of s, n, and a, where s is the student's score on the departmental placement exam, n is the number of semesters of mathematics passed in high school, and a is the student's mathematics SAT score. She estimates that p, the probability of passing the course (in percent), will be

$$p = f(s, n, a) = .003a + .1(sn)^{1/2}$$

for $200 \le a \le 800$, $0 \le s \le 10$, and $0 \le n \le 8$. Assuming that the above model has some merit, find the following.

a. If a student scores 8 on the placement exam, has taken 6 semesters of high school math, and has an SAT score of 450, what is the probability of passing the course?

b. Find p for a student with 3 semesters of high school mathematics, a placement score of 3, and an SAT score of 320.

c. Find and interpret $f_n(3, 3, 320)$ and $f_a(3, 3, 320)$.

PHYSICAL SCIENCES

65. *Gravitational Attraction* The gravitational attraction F on a body a distance r from the center of Earth, where r is greater than the radius of Earth, is a function of its mass m and the distance r as follows:

$$F = \frac{mgR^2}{r^2},$$

where R is the radius of Earth and g is the force of gravity—about 32 feet per second per second (ft per sec^2).

a. Find and interpret F_m and F_r.

b. Show that $F_m > 0$ and $F_r < 0$. Why is this reasonable?

66. *Velocity* In 1931, Albert Einstein developed the following formula for adding two velocities, x and y:

$$w(x, y) = \frac{x + y}{1 + \frac{xy}{c^2}},$$

where x and y are in miles per second and c represents the speed of light, 186,282 miles per second.*

a. Suppose that, relative to a stationary observer, a new super space shuttle is capable of traveling at 50,000 miles per second and that, while traveling at this speed, it launches a rocket that travels at 150,000 miles per second. How fast is the rocket traveling relative to the stationary observer?

b. What is the instantaneous rate of change of w with respect to the speed of the space shuttle, x, when the space shuttle is traveling at 50,000 miles per second and the rocket is traveling at 150,000 miles per second?

c. Hypothetically, if a person is driving at the speed of light, c, and she turns on the headlights, what is the velocity of the light coming from the headlights, relative to a stationary observer?

67. *Movement Time* Fitts' law is used to estimate the amount of time it takes for a person, using his or her arm, to pick up a light object, move it, and then place it in a designated target area. Mathematically, Fitts' law for a particular individual is given by

$$T(s, w) = -50 + 105 \log_2\left(\frac{2s}{w}\right),$$

where s is the distance (in feet) the object is moved, w is the width of the area in which the object is being placed, and T is the time (in msec).†

a. Calculate $T(3,0.5)$.

b. Find $T_s(3,0.5)$ and $T_w(3,0.5)$ and interpret these values. (*Hint:* $\log_2 x = \ln x / \ln 2$.)

9.3 MAXIMA AND MINIMA

THINK ABOUT IT *What amounts of sugar and flavoring produce the minimum cost per batch of a soft drink? What is the minimum cost?*

FOR REVIEW
It may be helpful to review the section on relative extrema in the chapter on Graphs and the Derivative at this point. The concepts presented there are basic to what will be done in this section.

One of the most important applications of calculus is finding maxima and minima of functions. Earlier, we studied this idea extensively for functions of a single independent variable; now we will see that extrema can be found for functions of two variables. In particular, an extension of the second derivative test can be defined and used to identify maxima or minima. We begin with the definitions of relative maxima and minima.

RELATIVE MAXIMA AND MINIMA

Let (a, b) be the center of a circular region contained in the xy-plane. Then, for a function $z = f(x, y)$ defined for every (x, y) in the region, $f(a, b)$ is a **relative maximum** if

$$f(a, b) \geq f(x, y)$$

for all points (x, y) in the circular region, and $f(a, b)$ is a **relative minimum** if

$$f(a, b) \leq f(x, y)$$

for all points (x, y) in the circular region.

*Fiore, Greg, "An Out-of-Math Experience: Einstein, Relativity, and the Developmental Mathematics Student," *Mathematics Teacher*, Vol. 93, No. 3, 2000, pp. 194–199.

†Sanders, Mark and Ernest McCormick, *Human Factors in Engineering Design*, 7th ed., New York: McGraw-Hill, 1993, pp. 290–291.

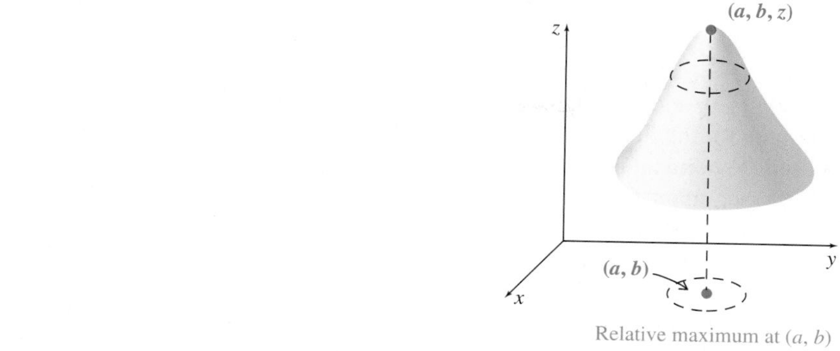

FIGURE 17

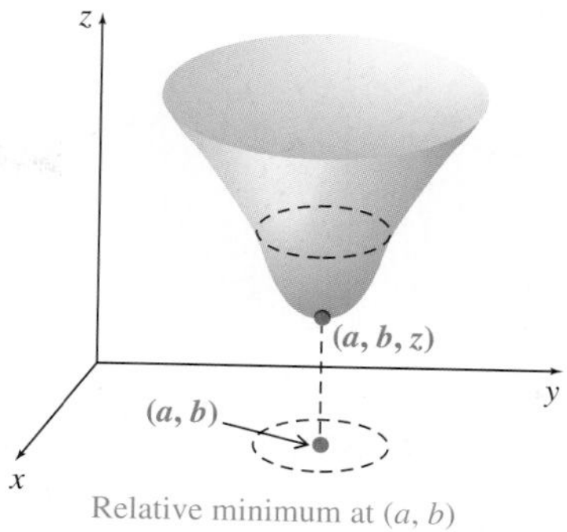

FIGURE 18

As before, the word *extremum* is used for either a relative maximum or a relative minimum. Examples of a relative maximum and a relative minimum are given in Figures 17 and 18.

NOTE When functions of a single variable were discussed, a distinction was made between relative extrema and absolute extrema. The methods for finding absolute extrema are quite involved for functions of two variables, so we will discuss only relative extrema here. In many practical applications the relative extrema coincide with the absolute extrema. Also, in this brief discussion of extrema for multivariable functions, we omit cases where an extremum occurs on a boundary of the domain. ■

As suggested by Figure 19, at a relative maximum the tangent line parallel to the *xz*-plane has a slope of 0, as does the tangent line parallel to the *yz*-plane. (Notice the similarity to functions of one variable.) That is, if the function $z = f(x, y)$ has a relative extremum at (a, b), then $f_x(a, b) = 0$ and $f_y(a, b) = 0$, as stated in the next theorem.

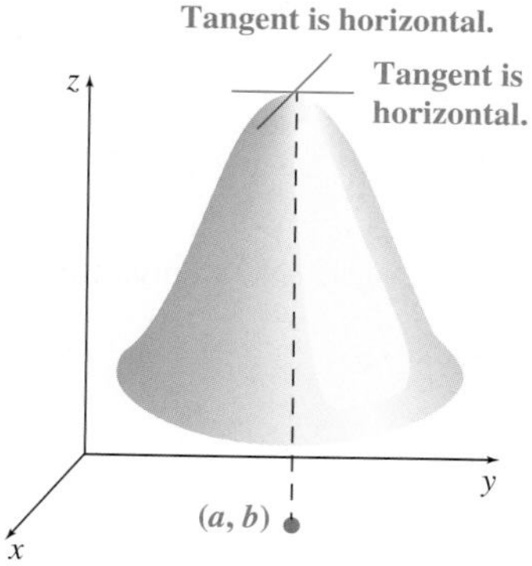

FIGURE 19

LOCATION OF EXTREMA

Let a function $z = f(x, y)$ have a relative maximum or relative minimum at the point (a, b). Let $f_x(a, b)$ and $f_y(a, b)$ both exist. Then

$$f_x(a, b) = 0 \qquad \text{and} \qquad f_y(a, b) = 0.$$

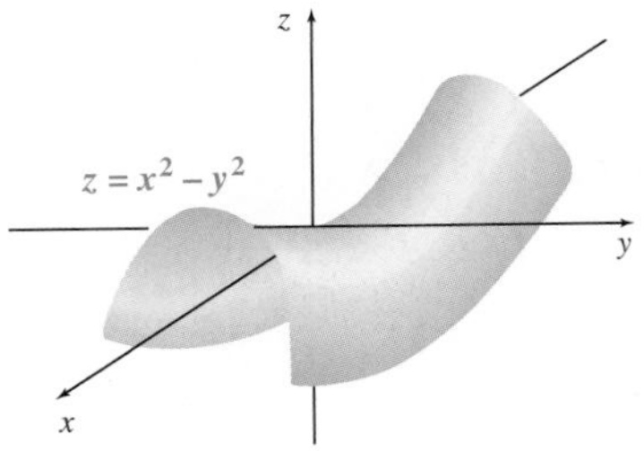

FIGURE 20

Just as with functions of one variable, the fact that the slopes of the tangent lines are 0 is no guarantee that a relative extremum has been located. For example, Figure 20 shows the graph of $z = f(x, y) = x^2 - y^2$. Both $f_x(0, 0) = 0$ and $f_y(0, 0) = 0$, and yet $(0, 0)$ leads to neither a relative maximum nor a relative minimum for the function. The point $(0, 0, 0)$ on the graph of this function is called a **saddle point;** it is a minimum when approached from one direction but a maximum when approached from another direction. A saddle point is neither a maximum nor a minimum.

The theorem on location of extrema suggests a useful strategy for finding extrema. First, locate all points (a, b) where $f_x(a, b) = 0$ and $f_y(a, b) = 0$. Then test each of these points separately, using the test given after the next example. For a function $f(x, y)$, the points (a, b) such that $f_x(a, b) = 0$ and $f_y(a, b) = 0$ are called **critical points.**

NOTE When we discussed functions of a single variable, we allowed critical points to include points from the domain where the derivative does not exist. For functions of more than one variable, to avoid complications, we will only consider cases in which the function is differentiable. ■

EXAMPLE 1 *Critical Points*

Find all critical points for

$$f(x, y) = 6x^2 + 6y^2 + 6xy + 36x - 5.$$

Solution Find all points (a, b) such that $f_x(a, b) = 0$ and $f_y(a, b) = 0$. Here

$$f_x(x, y) = 12x + 6y + 36 \qquad \text{and} \qquad f_y(x, y) = 12y + 6x.$$

Set each of these two partial derivatives equal to 0.

$$12x + 6y + 36 = 0 \qquad \text{and} \qquad 12y + 6x = 0$$

These two equations make up a system of linear equations. We can use the substitution method to solve this system. First, rewrite $12y + 6x = 0$ as follows:

$$\begin{aligned} 12y + 6x &= 0 \\ 6x &= -12y \\ x &= -2y. \end{aligned}$$

Now substitute $-2y$ for x in the other equation and solve for y.

$$\begin{aligned} 12x + 6y + 36 &= 0 \\ 12(-2y) + 6y + 36 &= 0 \\ -24y + 6y + 36 &= 0 \\ -18y + 36 &= 0 \\ -18y &= -36 \\ y &= 2 \end{aligned}$$

From the equation $x = -2y$, $x = -2(2) = -4$. The solution of the system of equations is $(-4, 2)$. Since this is the only solution of the system, $(-4, 2)$ is the only critical point for the given function. By the theorem above, if the function has a relative extremum, it will occur at $(-4, 2)$.

The results of the next theorem can be used to decide whether $(-4, 2)$ in Example 1 leads to a relative maximum, a relative minimum, or neither.

TEST FOR RELATIVE EXTREMA

For a function $z = f(x, y)$, let f_{xx}, f_{yy}, and f_{xy} all exist in a circular region contained in the xy-plane with center (a, b). Further, let

$$f_x(a, b) = 0 \quad \text{and} \quad f_y(a, b) = 0.$$

Define the number D by

$$D = f_{xx}(a, b) \cdot f_{yy}(a, b) - [f_{xy}(a, b)]^2.$$

Then

a. $f(a, b)$ is a relative maximum if $D > 0$ and $f_{xx}(a, b) < 0$;
b. $f(a, b)$ is a relative minimum if $D > 0$ and $f_{xx}(a, b) > 0$;
c. $f(a, b)$ is a saddle point (neither a maximum nor a minimum) if $D < 0$;
d. if $D = 0$, the test gives no information.

This test is comparable to the second derivative test for extrema of functions of one independent variable. The following table summarizes the conclusions of the theorem.

	$f_{xx}(a, b) < 0$	$f_{xx}(a, b) > 0$
$D > 0$	Relative maximum	Relative minimum
$D = 0$	No information	
$D < 0$	Saddle point	

Notice that in parts a and b of the test for relative extrema, it is only necessary to test the second partial $f_{xx}(a, b)$ and not $f_{yy}(a, b)$. This is because if $D > 0$, $f_{xx}(a, b)$ and $f_{yy}(a, b)$ must have the same sign.

EXAMPLE 2 *Relative Extrema*

The previous example showed that the only critical point for the function

$$f(x, y) = 6x^2 + 6y^2 + 6xy + 36x - 5$$

is $(-4, 2)$. Does $(-4, 2)$ lead to a relative maximum, a relative minimum, or neither?

Solution Find out by using the test above. From Example 1,

$$f_x(-4, 2) = 0 \quad \text{and} \quad f_y(-4, 2) = 0.$$

Now find the various second partial derivatives used in finding D. From $f_x(x, y) = 12x + 6y + 36$ and $f_y(x, y) = 12y + 6x$,

$$f_{xx}(x, y) = 12, \quad f_{yy}(x, y) = 12, \quad \text{and} \quad f_{xy}(x, y) = 6.$$

(If these second-order partial derivatives had not all been constants, they would have had to be evaluated at the point $(-4, 2)$.) Now

$$D = f_{xx}(-4, 2) \cdot f_{yy}(-4, 2) - [f_{xy}(-4, 2)]^2 = 12 \cdot 12 - 6^2 = 108.$$

Since $D > 0$ and $f_{xx}(-4, 2) = 12 > 0$, part b of the theorem applies, showing that $f(x, y) = 6x^2 + 6y^2 + 6xy + 36x - 5$ has a relative minimum at $(-4, 2)$. This relative minimum is $f(-4, 2) = -77$.

EXAMPLE 3 *Saddle Point*

Find all points where the function

$$f(x, y) = 9xy - x^3 - y^3 - 6$$

has any relative maxima or relative minima.

Solution First find any critical points. Here

$$f_x(x, y) = 9y - 3x^2 \qquad \text{and} \qquad f_y(x, y) = 9x - 3y^2.$$

Set each of these partial derivatives equal to 0.

$$\begin{aligned} f_x(x, y) &= 0 & \qquad f_y(x, y) &= 0 \\ 9y - 3x^2 &= 0 & 9x - 3y^2 &= 0 \\ 9y &= 3x^2 & 9x &= 3y^2 \\ 3y &= x^2 & 3x &= y^2 \end{aligned}$$

The substitution method can be used again to solve the system of equations

$$\begin{aligned} 3y &= x^2 \\ 3x &= y^2. \end{aligned}$$

The first equation, $3y = x^2$, can be rewritten as $y = x^2/3$. Substitute this into the second equation to get

$$3x = y^2 = \left(\frac{x^2}{3}\right)^2 = \frac{x^4}{9}.$$

Solve this equation as follows.

$$\begin{aligned} 27x &= x^4 \\ x^4 - 27x &= 0 \\ x(x^3 - 27) &= 0 && \text{Factor.} \\ x = 0 \quad \text{or} \quad x^3 - 27 &= 0 && \text{Set each factor equal to 0.} \\ x = 0 \quad \text{or} \quad x^3 &= 27 \\ x = 0 \quad \text{or} \quad x &= 3 && \text{Take the cube root on both sides.} \end{aligned}$$

Use these values of x, along with the equation $3y = x^2$, rewritten as $y = x^2/3$, to find y. If $x = 0$, $y = 0^2/3 = 0$. If $x = 3$, $y = 3^2/3 = 3$. The critical points are $(0, 0)$ and $(3, 3)$. To identify any extrema, use the test. Here

$$f_{xx}(x, y) = -6x, \qquad f_{yy}(x, y) = -6y, \qquad \text{and} \qquad f_{xy}(x, y) = 9.$$

Test each of the possible critical points.

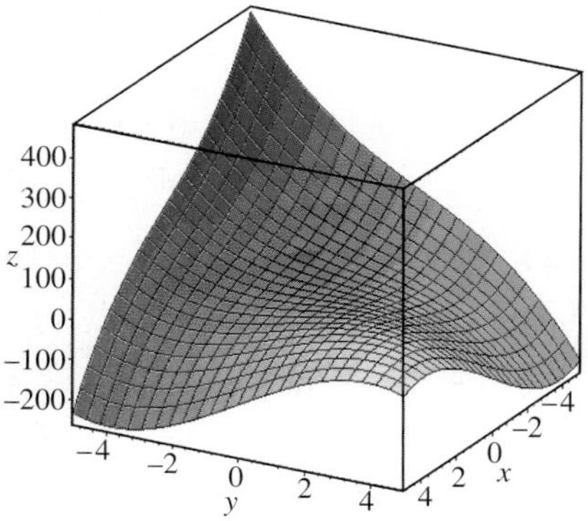

FIGURE 21

For $(0, 0)$:

$f_{xx}(0, 0) = -6(0) = 0$

$f_{yy}(0, 0) = -6(0) = 0$

$f_{xy}(0, 0) = 9$

$D = 0 \cdot 0 - 9^2 = -81.$

Since $D < 0$, there is a saddle point at $(0, 0)$.

For $(3, 3)$:

$f_{xx}(3, 3) = -6(3) = -18$

$f_{yy}(3, 3) = -6(3) = -18$

$f_{xy}(3, 3) = 9$

$D = -18(-18) - 9^2 = 243.$

Here $D > 0$ and $f_{xx}(3, 3) = -18 < 0$; there is a relative maximum at $(3, 3)$.

Notice that these values are in accordance with the graph generated by the computer program Maple™ shown in Figure 21.

EXAMPLE 4 *Production Costs*

A company is developing a new soft drink. The cost in dollars to produce a batch of the drink is approximated by

$$C(x, y) = 2200 + 27x^3 - 72xy + 8y^2,$$

where x is the number of kilograms of sugar per batch and y is the number of grams of flavoring per batch. Find the amounts of sugar and flavoring that result in the minimum cost per batch. What is the minimum cost?

Solution

Method 1: Calculation by Hand

Start with the following partial derivatives.

$$C_x(x, y) = 81x^2 - 72y \quad \text{and} \quad C_y(x, y) = -72x + 16y$$

Set each of these equal to 0 and solve for y.

$$81x^2 - 72y = 0 \qquad -72x + 16y = 0$$
$$-72y = -81x^2 \qquad 16y = 72x$$
$$y = \frac{9}{8}x^2 \qquad y = \frac{9}{2}x$$

Since $(9/8)x^2$ and $(9/2)x$ both equal y, they are equal to each other. Set them equal, and solve the resulting equation for x.

$$\frac{9}{8}x^2 = \frac{9}{2}x$$
$$9x^2 = 36x$$
$$9x^2 - 36x = 0 \qquad \text{Subtract } 36x \text{ from both sides.}$$
$$9x(x - 4) = 0 \qquad \text{Factor.}$$
$$9x = 0 \quad \text{or} \quad x - 4 = 0 \qquad \text{Set each factor equal to 0.}$$

The equation $9x = 0$ leads to $x = 0$ and $y = 0$, which cannot be a minimizer of $C(x, y)$ since, for example, $C(1, 1) < C(0, 0)$. This fact can also be verified by the test for relative extrema. Substitute $x = 4$, the solution of $x - 4 = 0$, into $y = (9/2)x$ to find y.

$$y = \frac{9}{2}x = \frac{9}{2}(4) = 18$$

Now check to see whether the critical point $(4, 18)$ leads to a relative minimum. For $(4, 18)$,

$$C_{xx}(x, y) = 162x = 162(4) = 648, \quad C_{yy}(x, y) = 16, \quad \text{and} \quad C_{xy}(x, y) = -72.$$

Also,

$$D = (648)(16) - (-72)^2 = 5184.$$

Since $D > 0$ and $C_{xx}(4, 18) > 0$, the cost at $(4, 18)$ is a minimum.

To find the minimum cost, go back to the cost function and evaluate $C(4, 18)$.

$$C(x, y) = 2200 + 27x^3 - 72xy + 8y^2$$
$$C(4, 18) = 2200 + 27(4)^3 - 72(4)(18) + 8(18)^2 = 1336$$

The minimum cost for a batch of soft drink is \$1336.00.

Method 2: Spreadsheets

Finding the maximum or minimum of a function of one or more variables can be done using a spreadsheet. The most widely used spreadsheets have built-in solvers that are able to optimize complicated functions. These solvers employ more advanced techniques than we have learned thus far, and they are very efficient and practical to use.

The Solver included with Excel is located in the Tools menu and requires that cells be identified ahead of time for each variable in the problem. It also requires that another cell be identified where the function, in terms of the variable cells, is placed. For example, to solve the above problem, we could identify cells A1 and B1 to represent the variables x and y, respectively. The Solver requires that we place a guess for the answer in these cells. Thus, our initial value or guess will be to place the number 5 in each of these cells. The function must be placed in a cell in terms of cells A1 and B1. If we choose cell A3 to represent the function, in cell A3 we would type "`=2200 + 27*A1^3 - 72*A1*B1 + 8*B1^2`."

We now click on the Tools menu and choose Solver. This solver will attempt to find a solution that either maximizes or minimizes the value of cell A3. Figure 22 illustrates the Solver box and the items placed in it.

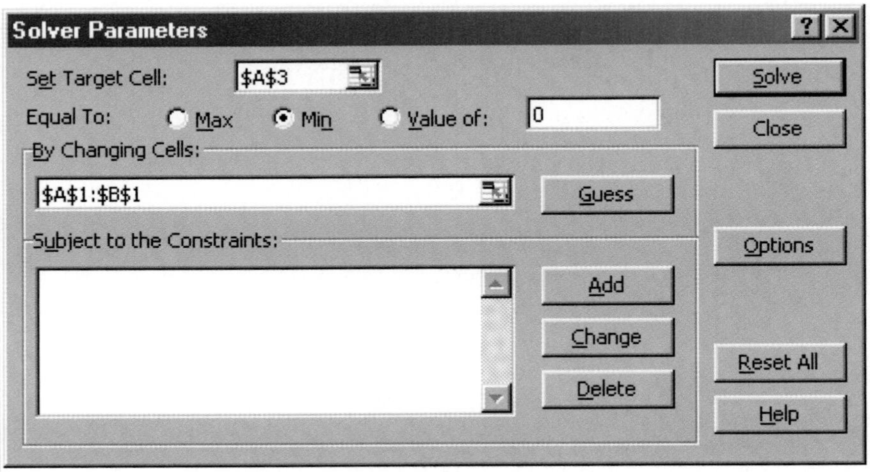

FIGURE 22

To obtain a solution, click on Solve. The rounded solution $x = 4$ and $y = 18$ is located in cells A1 and B1. The minimum cost $C(4, 18) = 1336$ is located in cell A3.

NOTE One must be careful when using Solver because it will not find a maximizer or minimizer of a function if the initial guess is the exact place in which a saddle point occurs. For example, in the problem above, if our initial guess was $(0, 0)$, the Solver would have returned the value of $(0, 0)$ as the place where a minimum occurs. But $(0, 0)$ is a saddle point. Thus, it is always a good idea to run the Solver for two different initial values and compare the solutions. ■

9.3 EXERCISES

Find all points where the functions have any relative extrema. Identify any saddle points.

1. $f(x, y) = xy + x - y$
2. $f(x, y) = 4xy + 8x - 9y$
3. $f(x, y) = x^2 - 2xy + 2y^2 + x - 5$
4. $f(x, y) = x^2 + xy + y^2 - 6x - 3$
5. $f(x, y) = x^2 - xy + y^2 + 2x + 2y + 6$
6. $f(x, y) = x^2 + xy + y^2 + 3x - 3y$
7. $f(x, y) = x^2 + 3xy + 3y^2 - 6x + 3y$
8. $f(x, y) = 5xy - 7x^2 - y^2 + 3x - 6y - 4$
9. $f(x, y) = 4xy - 10x^2 - 4y^2 + 8x + 8y + 9$
10. $f(x, y) = x^2 + xy + 3x + 2y - 6$
11. $f(x, y) = x^2 + xy - 2x - 2y + 2$
12. $f(x, y) = x^2 + xy + y^2 - 3x - 5$
13. $f(x, y) = 2x^3 + 3y^2 - 12xy + 4$
14. $f(x, y) = 5x^3 + 2y^2 - 60xy - 3$
15. $f(x, y) = x^2 + 4y^3 - 6xy - 1$
16. $f(x, y) = 3x^2 + 7y^3 - 42xy + 5$
17. $f(x, y) = e^{xy}$
18. $f(x, y) = x^2 + e^y$
19. Describe the procedure for finding critical points of a function in two independent variables.
20. How are second-order partial derivatives used in finding extrema?

Figures a–f below and on the next page show the graphs of the functions defined in Exercises 21–26. Find all relative extrema for each function, and then match the equation to its graph.

21. $z = -3xy + x^3 - y^3 + \dfrac{1}{8}$
22. $z = \dfrac{3}{2}y - \dfrac{1}{2}y^3 - x^2y + \dfrac{1}{16}$
23. $z = y^4 - 2y^2 + x^2 - \dfrac{17}{16}$
24. $z = -2x^3 - 3y^4 + 6xy^2 + \dfrac{1}{16}$
25. $z = -x^4 + y^4 + 2x^2 - 2y^2 + \dfrac{1}{16}$
26. $z = -y^4 + 4xy - 2x^2 + \dfrac{1}{16}$

a.

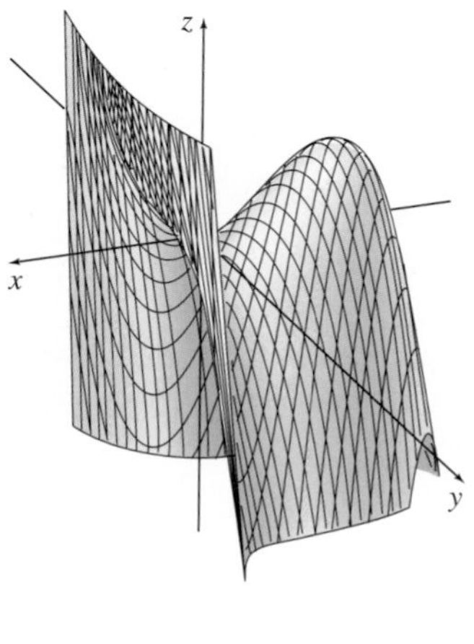

b.

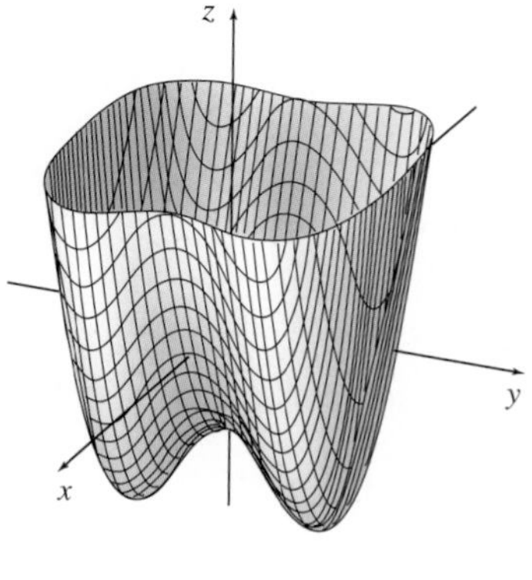

c.

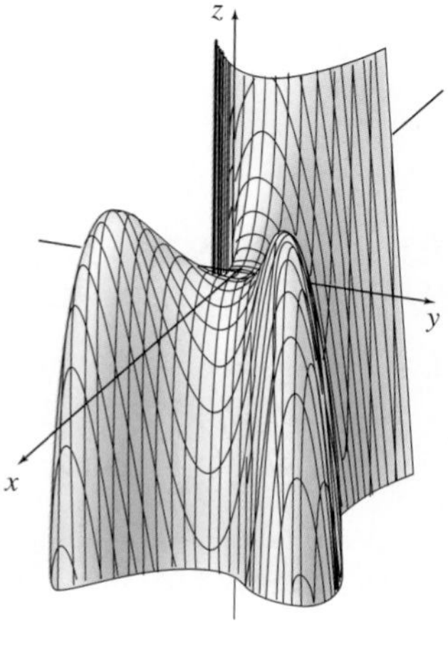

d.

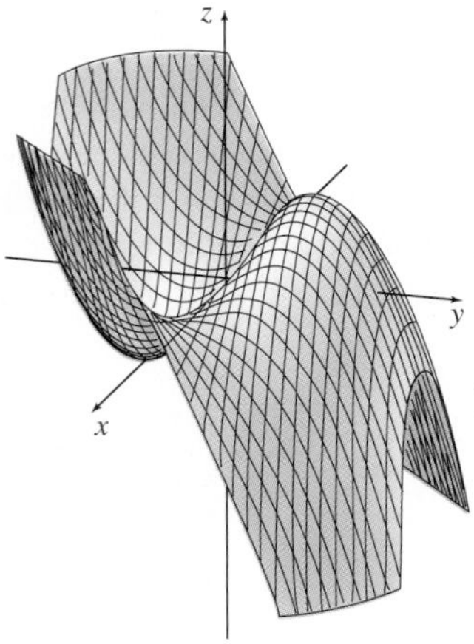

e.

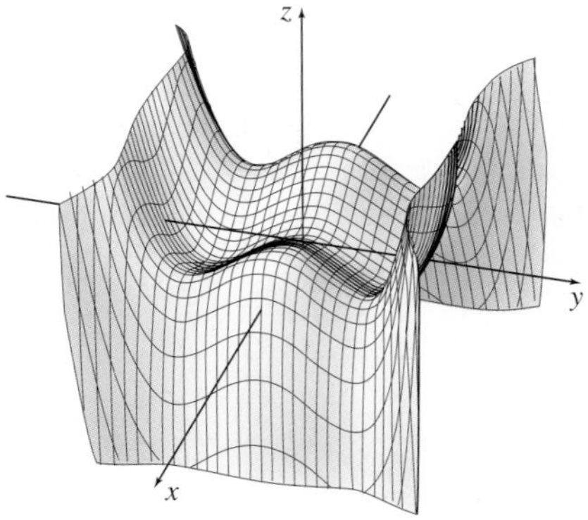

f.

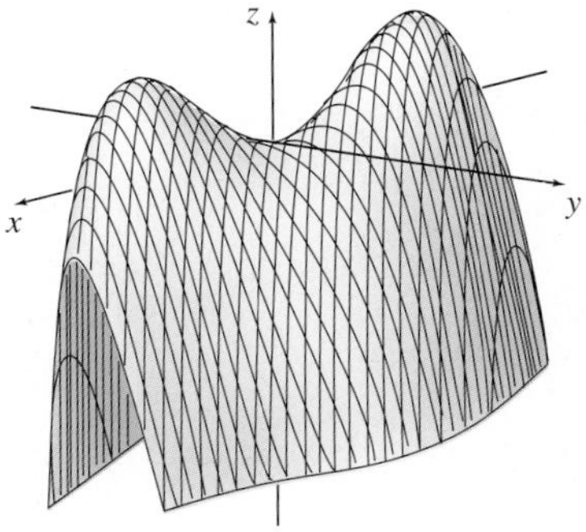

27. Show that $f(x, y) = 1 - x^4 - y^4$ has a relative maximum, even though D in the theorem is 0.

28. Show that $D = 0$ for $f(x, y) = x^3 + (x - y)^2$ and that the function has no relative extrema.

29. A friend taking calculus is puzzled. She remembers that for a function of one variable, if the first derivative is zero at a point and the second derivative is positive, then there must be a relative minimum at the point. She doesn't understand why that isn't true for a function of two variables—that is, why $f_x(x, y) = 0$ and $f_{xx}(x, y) > 0$ doesn't guarantee a relative minimum. Provide an explanation.

30. In Section 1.3, we found the least squares line through a set of n points $(x_1, y_1), (x_2, y_2), \ldots, (x_n, y_n)$ by choosing the slope of the line m and the y-intercept b to minimize the quantity

$$S(m, b) = \sum (mx + b - y)^2,$$

where the summation symbol Σ means that we sum over all the data points. Minimize S by setting $S_m(m, b) = 0$ and $S_b(m, b) = 0$, and then rearrange the results to derive the equations from Section 1.3

$$\left(\sum x\right)b + \left(\sum x^2\right)m = \sum xy$$
$$nb + \left(\sum x\right)m = \sum y.$$

31. Consider the function $f(x, y) = x^2(y + 1)^2 + k(x + 1)^2y^2$.

a. For what values of k is the point $(x, y) = (0, 0)$ a critical point?

b. For what values of k is the point $(x, y) = (0, 0)$ a relative minimum of the function?

Applications
BUSINESS AND ECONOMICS

32. *Profit* Suppose that the profit (in hundreds of dollars) of a certain firm is approximated by

$$P(x, y) = 1000 + 24x - x^2 + 80y - y^2,$$

where x is the cost of a unit of labor and y is the cost of a unit of goods. Find values of x and y that maximize profit. Find the maximum profit.

33. *Labor Costs* Suppose the labor cost (in dollars) for manufacturing a precision camera can be approximated by

$$L(x, y) = \frac{3}{2}x^2 + y^2 - 2x - 2y - 2xy + 68,$$

where x is the number of hours required by a skilled craftsperson and y is the number of hours required by a semiskilled person. Find values of x and y that minimize the labor cost. Find the minimum labor cost.

34. *Cost* The total cost (in dollars) to produce x units of electrical tape and y units of packing tape is given by

$$C(x, y) = 2x^2 + 3y^2 - 2xy + 2x - 126y + 3800.$$

Find the number of units of each kind of tape that should be produced so that the total cost is a minimum. Find the minimum total cost.

35. *Revenue* The total revenue (in hundreds of dollars) from the sale of x spas and y solar heaters is approximated by

$$R(x, y) = 12 + 74x + 85y - 3x^2 - 5y^2 - 5xy.$$

Find the number of each that should be sold to produce maximum revenue. Find the maximum revenue.

36. *Profit* The profit (in thousands of dollars) that Aunt Mildred's Metalworks earns from producing x tons of steel and y tons of aluminum can be approximated by

$$P(x, y) = 36xy - x^3 - 8y^3.$$

Find the amounts of steel and aluminum that maximize the profit, and find the value of the maximum profit.

37. *Time* The time (in hours) that a branch of Amalgamated Entities needs to spend to meet the quota set by the main office can be approximated by

$$T(x, y) = x^4 + 16y^4 - 32xy + 40,$$

where x represents how many thousands of dollars the factory spends on quality control and y represents how many thousands of dollars they spend on consulting. Find the amount of money they should spend on quality control and on consulting to minimize the time spent, and find the minimum number of hours.

SOCIAL SCIENCES

38. *Political Science* The probability that a three-person jury will make a correct decision is given by

$$P(\alpha, r, s) = \alpha[3r^2(1 - r) + r^3] + (1 - \alpha)[3s^2(1 - s) + s^3],$$

where $0 < \alpha < 1$ is the probability that the person is guilty of the crime, r is the probability that a given jury member will vote "guilty" when the defendant is indeed guilty of the crime, and s is the probability that a given jury member will vote "innocent" when the defendant is indeed innocent.*

a. Calculate $P(.9,0.5,0.6)$ and $P(.1,0.8,0.4)$ and interpret your answers.

b. Using common sense and without using calculus, what value of r and s would maximize the jury's probability of making the correct verdict? Do these values depend on α in this problem? Should they? What is the maximum probability?

c. Verify your answer for part b using calculus. (*Hint:* There are two critical points. Argue that the maximum value occurs at one of these points.)

PHYSICAL SCIENCES

39. *Computer Chips* The following table, which illustrates the dramatic increase in the number of transistors in personal computers since 1971, was given in the chapter on Nonlinear Functions, Section 4, Exercise 50.†

Year (since 1971)	Chip	Transistors
0	4004	2250
14	386	275,000
18	486DX	1,180,000
22	Pentium	3,100,000
26	Pentium II	7,500,000
28	Pentium III	24,000,000
29	Pentium 4	42,000,000

a. Since the natural logarithm of both sides of the equation $y = ab^x$ yields $\ln y = \ln a + x \ln b$, we could let $w = \ln y$, $r = \ln a$, and $s = \ln b$ to form $w = r + sx$. Using linear regression, find values for r and s that will fit the data above. (*Hint:* Take the natural logarithm of the values in the transistors column and then use linear regression to find values of r and s that fit the data. Once you know r and s, you can determine the values of a and b by calculating $a = e^r$ and $b = e^s$.)

b. Use the solver capability of a spreadsheet to find a function of the form $y = ab^x$ that fits the data above. (*Hint:* Using the ideas from part a, find values for a and b that minimize the function

$$\begin{aligned} f(a, b) = {} & [\ln(2250) - 0 \ln b - \ln a]^2 \\ & + [\ln(275{,}000) - 14 \ln b - \ln a]^2 \\ & + [\ln(1{,}180{,}000) - 18 \ln b - \ln a]^2 \\ & + [\ln(3{,}100{,}000) - 22 \ln b - \ln a]^2 \\ & + [\ln(7{,}500{,}000) - 26 \ln b - \ln a]^2 \\ & + [\ln(24{,}000{,}000) - 28 \ln b - \ln a]^2 \\ & + [\ln(42{,}000{,}000) - 29 \ln b - \ln a]^2. \end{aligned}$$

c. Compare your answer to this problem with the one found with a graphing calculator in the chapter on Nonlinear Functions, Section 4, Exercise 50.

*Grofman, Bernard, "A Preliminary Model of Jury Decision Making as a Function of Jury Size, Effective Jury Decision Rule, and Mean Juror Judgmental Competence," *Frontiers of Economics*, Vol. 3, 1980, pp. 98–110.

†http://www.intel.com/research/silicon/mooreslaw.htm.

9.4 LAGRANGE MULTIPLIERS

THINK ABOUT IT *What dimensions for a new building will maximize the floor space at a fixed cost?*

In the Applications of Extrema section, it was possible to express problems involving two variables as an equivalent problem requiring only a single variable. This method works well, provided that it is possible to use algebra to express the one variable in terms of the other. It is not always possible to do this, however, and most real applications require more than two variables and one or more additional restrictions, called **constraints.**

An approach that works well when there is a constraint in the problem uses an additional variable, called the **Lagrange multiplier.** For example, in the opening question, suppose a builder wants to maximize the floor space in a new building while keeping the costs fixed at \$500,000. The building will be 40 ft high, with a rectangular floor plan and three stories. The costs, which depend on the dimensions of the rectangular floor plan, are given by

$$\text{Costs} = xy + 20y + 20x + 474{,}000,$$

where x is the width and y the length of the rectangle. Thus, the builder wishes to maximize the area $A(x, y) = xy$ and satisfy the condition

$$xy + 20y + 20x + 474{,}000 = 500{,}000.$$

In addition to maximizing area, then, the builder must keep costs at (or below) \$500,000. We will see how to solve this problem in Example 2 of this section.

A typical problem of this type might require the smallest possible value of the function $z = x^2 + y^2$, subject to the constraint $x + y = 4$. To see how to find this minimum value, we might first graph both the surface $z = x^2 + y^2$ and the plane $x + y = 4$, as in Figure 23 on the next page. The required minimum value is found on the curve formed by the intersection of the two graphs. In this example, it's not too hard to see by symmetry that the minimum value of z occurs when both x and y are 2. In other examples, the solution is not at all obvious, and we need a method that will still give us the solution.

Problems with constraints are often solved by the method of Lagrange multipliers, named for the French mathematician Joseph Louis Lagrange (1736–1813). The proof for the method is complicated and is not given here. The method of Lagrange multipliers is used for problems of the form:

$$\text{Find the relative extrema for } z = f(x, y),$$
$$\text{subject to } g(x, y) = 0.$$

We state the method only for functions of two independent variables, but it is valid for any number of variables.

LAGRANGE MULTIPLIERS

All relative extrema of the function $z = f(x, y)$, subject to a constraint $g(x, y) = 0$, will be found among those points (x, y) for which there exists a value of λ such that

$$F_x(x, y, \lambda) = 0, \qquad F_y(x, y, \lambda) = 0, \qquad F_\lambda(x, y, \lambda) = 0,$$

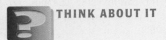

where

$$F(x, y, \lambda) = f(x, y) - \lambda \cdot g(x, y),$$

and all indicated partial derivatives exist.

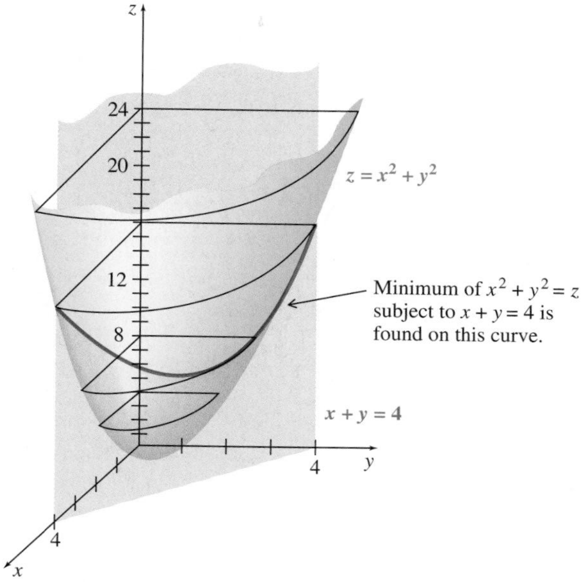

FIGURE 23

In the theorem, the function $F(x, y, \lambda) = f(x, y) - \lambda \cdot g(x, y)$ is called the Lagrange function; λ, the Greek letter *lambda*, is the *Lagrange multiplier*.

CAUTION If the constraint is not of the form $g(x, y) = 0$, it must be put in that form before using the method of Lagrange multipliers. For example, if the constraint is $x^2 + y^3 = 5$, subtract 5 from both sides to get $g(x, y) = x^2 + y^3 - 5 = 0$. ■

EXAMPLE 1 *Lagrange Multipliers*

Find the minimum value of

$$f(x, y) = 5x^2 + 6y^2 - xy,$$

subject to the constraint $x + 2y = 24$.

Solution Go through the following steps.

Step 1 Rewrite the constraint in the form $g(x, y) = 0$.

In this example, the constraint $x + 2y = 24$ becomes

$$x + 2y - 24 = 0,$$

with

$$g(x, y) = x + 2y - 24.$$

Step 2 Form the Lagrange function $F(x, y, \lambda)$, the difference of the function $f(x, y)$ and the product of λ and $g(x, y)$.

Here,

$$\begin{aligned} F(x, y, \lambda) &= f(x, y) - \lambda \cdot g(x, y) \\ &= 5x^2 + 6y^2 - xy - \lambda(x + 2y - 24) \\ &= 5x^2 + 6y^2 - xy - \lambda x - 2\lambda y + 24\lambda. \end{aligned}$$

Step 3 Find $F_x(x, y, \lambda)$, $F_y(x, y, \lambda)$, and $F_\lambda(x, y, \lambda)$.

$$\begin{aligned} F_x(x, y, \lambda) &= 10x - y - \lambda \\ F_y(x, y, \lambda) &= 12y - x - 2\lambda \\ F_\lambda(x, y, \lambda) &= -x - 2y + 24 \end{aligned}$$

Step 4 Form the system of equations $F_x(x, y, \lambda) = 0$, $F_y(x, y, \lambda) = 0$, and $F_\lambda(x, y, \lambda) = 0$.

$$10x - y - \lambda = 0 \qquad \textbf{(1)}$$

$$12y - x - 2\lambda = 0 \qquad \textbf{(2)}$$

$$-x - 2y + 24 = 0 \qquad \textbf{(3)}$$

Step 5 Solve the system of equations from Step 4 for x, y, and λ.

One way to solve this system is to begin by solving each of the first two equations for λ, then set the two results equal and simplify, as follows.

$$10x - y - \lambda = 0 \quad \text{becomes} \quad \lambda = 10x - y$$

$$12y - x - 2\lambda = 0 \quad \text{becomes} \quad \lambda = \frac{-x + 12y}{2}$$

$$10x - y = \frac{-x + 12y}{2} \qquad \text{Set the expressions for } \lambda \text{ equal.}$$

$$\begin{aligned} 20x - 2y &= -x + 12y \\ 21x &= 14y \\ x &= \frac{14y}{21} = \frac{2y}{3} \end{aligned}$$

Now substitute $2y/3$ for x in Equation (3).

$$-x - 2y + 24 = 0$$

$$-\frac{2y}{3} - 2y + 24 = 0 \qquad \text{Let } x = \frac{2y}{3}.$$

$$2y + 6y - 72 = 0 \qquad \text{Multiply by } -3.$$

$$\begin{aligned} 8y &= 72 \\ y &= \frac{72}{8} = 9 \end{aligned}$$

Since $x = 2y/3$ and $y = 9$, $x = 6$. It is not necessary to find the value of λ.

Thus, the minimum value for $f(x, y) = 5x^2 + 6y^2 - xy$, subject to the constraint $x + 2y = 24$, is at the point $(6, 9)$. The minimum value is $f(6, 9) = 612$.

It is desirable to verify that 612 is indeed a minimum for the function. How can we tell that it is not a maximum? The second derivative test from the previous section does not apply to the solutions found by Lagrange multipliers. We could gain some insight by trying a point very close to $(6, 9)$ that also satisfies the constraint $x + 2y = 24$. For example, let $y = 9.1$, so $x = 24 - 2y = 24 - 2(9.1) = 5.8$. Then $f(5.8, 9.1) = 5(5.8)^2 + 6(9.1)^2 - (5.8)(9.1) = 612.28$, which is greater than 612. Because a nearby point has a value larger than 612, the value 612 is probably not a maximum. Another method would be to use a computer to sketch the graph of the function and see that it has a minimum but not a maximum. In practical problems, such as Example 2, it is often obvious whether a function has a minimum or a maximum.

NOTE In Example 1, we solved the system of equations by solving each equation with λ in it for λ. We then set these expressions for λ equal and solved for one of the original variables. This is a good general approach to use in solving these systems of equations, since we are usually not interested in the value of λ.

Before looking at applications of Lagrange multipliers, let us summarize the steps involved in solving a problem by this method.

USING LAGRANGE MULTIPLIERS

1. Write the constraint in the form $g(x, y) = 0$.
2. Form the Lagrange function

$$F(x, y, \lambda) = f(x, y) - \lambda \cdot g(x, y).$$

3. Find $F_x(x, y, \lambda)$, $F_y(x, y, \lambda)$, and $F_\lambda(x, y, \lambda)$.
4. Form the system of equations

$$F_x(x, y, \lambda) = 0, \qquad F_y(x, y, \lambda) = 0, \qquad F_\lambda(x, y, \lambda) = 0.$$

5. Solve the system in Step 4; the relative extrema for f are among the solutions of the system.

To understand why Lagrange multipliers work, consider the curve formed by points in the xy-plane that satisfy $F_\lambda(x, y, \lambda) = -g(x, y) = 0$ (or just $g(x, y) = 0$). Figure 24 on the next page shows how such a curve might look. Crossing this region are curves $f(x, y) = k$ for various values of k. Notice that at the points where the curve $f(x, y) = k$ is tangent to the curve $g(x, y) = 0$, the largest and smallest meaningful values of f occur. It can be shown that this is equivalent to $f_x(x, y) = \lambda g_x(x, y)$ and $f_y(x, y) = \lambda g_y(x, y)$ for some constant λ. In Exercise 17, you are asked to show that this is equivalent to the system of equations found in Step 4 above.

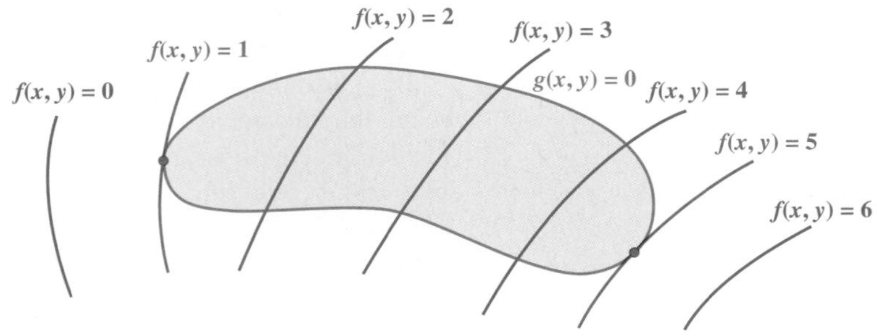

FIGURE 24

Lagrange multipliers are widely used in economics, where a frequent goal is to maximize a utility function, which measures how well consumption satisfies the consumers' desires, subject to constraints on income or time.

EXAMPLE 2 *Lagrange Multipliers*

Complete the solution of the problem given in the introduction to this section. Maximize the area, $A(x, y) = xy$, subject to the cost constraint.

$$xy + 20y + 20x + 474{,}000 = 500{,}000.$$

Solution Go through the five steps presented above.

Step 1 $g(x, y) = xy + 20y + 20x - 26{,}000 = 0$

Step 2 $F(x, y, \lambda) = xy - \lambda(xy + 20y + 20x - 26{,}000)$

Step 3 $F_x(x, y, \lambda) = y - \lambda y - 20\lambda$

$F_y(x, y, \lambda) = x - \lambda x - 20\lambda$

$F_\lambda(x, y, \lambda) = -xy - 20y - 20x + 26{,}000$

Step 4 $y - \lambda y - 20\lambda = 0$ **(4)**

$x - \lambda x - 20\lambda = 0$ **(5)**

$-xy - 20y - 20x + 26{,}000 = 0$ **(6)**

Step 5 Solving Equations (4) and (5) for λ gives

$$\lambda = \frac{y}{y + 20} \quad \text{and} \quad \lambda = \frac{x}{x + 20}$$

$$\frac{y}{y + 20} = \frac{x}{x + 20}$$

$$y(x + 20) = x(y + 20)$$

$$xy + 20y = xy + 20x$$

$$x = y.$$

Now substitute y for x in Equation (6) to get

$$-y^2 - 20y - 20y + 26{,}000 = 0$$

$$-y^2 - 40y + 26{,}000 = 0.$$

Use the quadratic formula to find $y \approx 142.5$. Since $x = y$, $x \approx 142.5$.

The maximum area of $(142.5)^2 \approx 20{,}306$ ft^2 will be achieved if the floor plan is a square with a side of 142.5 ft.

As mentioned earlier, the method of Lagrange multipliers works for more than two independent variables. The next example shows how to find extrema for a function of three independent variables.

EXAMPLE 3 *Volume of a Box*

Find the dimensions of the closed rectangular box of maximum volume that can be produced from 6 ft^2 of material.

Method 1: Lagrange Multipliers

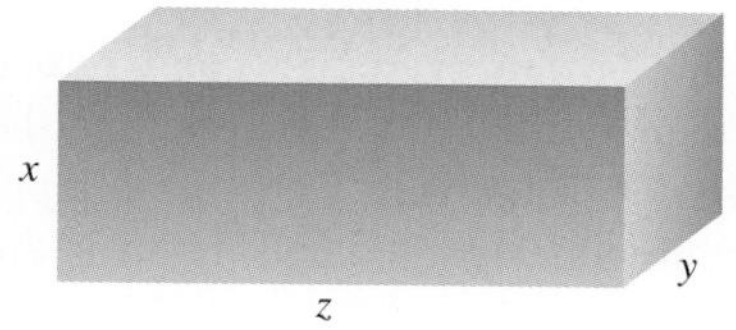

FIGURE 25

Solution Let x, y, and z represent the dimensions of the box, as shown in Figure 25. The volume of the box is given by

$$f(x, y, z) = xyz.$$

As shown in Figure 25, the total amount of material required for the two ends of the box is $2xy$, the total needed for the sides is $2xz$, and the total needed for the top and bottom is $2yz$. Since 6 ft^2 of material is available,

$$2xy + 2xz + 2yz = 6 \quad \text{or} \quad xy + xz + yz = 3.$$

In summary, $f(x, y, z) = xyz$ is to be maximized subject to the constraint $xy + xz + yz = 3$. Go through the steps that were given.

Step 1 $g(x, y, z) = xy + xz + yz - 3 = 0$

Step 2 $F(x, y, z, \lambda) = xyz - \lambda(xy + xz + yz - 3)$

Step 3 $F_x(x, y, z, \lambda) = yz - \lambda y - \lambda z$

$F_y(x, y, z, \lambda) = xz - \lambda x - \lambda z$

$F_z(x, y, z, \lambda) = xy - \lambda x - \lambda y$

$F_\lambda(x, y, z, \lambda) = -xy - xz - yz + 3$

Step 4 $yz - \lambda y - \lambda z = 0$

$xz - \lambda x - \lambda z = 0$

$xy - \lambda x - \lambda y = 0$

$-xy - xz - yz + 3 = 0$

Step 5 Solve each of the first three equations for λ. You should get

$$\lambda = \frac{yz}{y + z}, \qquad \lambda = \frac{xz}{x + z}, \qquad \text{and} \qquad \lambda = \frac{xy}{x + y}.$$

Set these expressions for λ equal, and simplify as follows. Notice in the second and last steps that since none of the dimensions of the box can be 0, we can divide both sides of each equation by x or z.

$$\begin{aligned} \frac{yz}{y + z} &= \frac{xz}{x + z} \\ \frac{y}{y + z} &= \frac{x}{x + z} \\ xy + yz &= xy + xz \\ yz &= xz \\ y &= x \end{aligned} \qquad \text{and} \qquad \begin{aligned} \frac{xz}{x + z} &= \frac{xy}{x + y} \\ \frac{z}{x + z} &= \frac{y}{x + y} \\ zx + zy &= yx + yz \\ zx &= yx \\ z &= y \end{aligned}$$

(Setting the first and third expressions equal gives no additional information.) Thus $x = y = z$. From the fourth equation in Step 4, with $x = y$ and $z = y$,

$$\begin{aligned} -xy - xz - yz + 3 &= 0 \\ -y^2 - y^2 - y^2 + 3 &= 0 \\ -3y^2 &= -3 \\ y^2 &= 1 \\ y &= \pm 1. \end{aligned}$$

The negative solution is not applicable, so the solution of the system of equations is $x = 1$, $y = 1$, $z = 1$. In other words, the box with maximum volume under the constraint is a cube that measures 1 ft on each side.

Method 2: Spreadsheets

Finding extrema of a constrained function of one or more variables can be done using a spreadsheet. In addition to the requirements stated in the last section, the constraint must also be input into the Excel Solver. To do this, we need to input the left-hand or variable part of the constraint into a designated cell. If A5 is the designated cell, then in cell A5 we would type "=A1*B1 + A1*C1 + B1*C1."

We now click on the Tools menu and choose Solver. This solver will attempt to find a solution that either maximizes or minimizes the value of cell A3, depending on which option we choose. Figure 26 illustrates the Solver box and the items placed in it.

To obtain a solution, click on Solve. The solution $x = 1$ and $y = 1$ and $z = 1$ is located in cells A1, B1, and C1, respectively. The maximum volume $f(1, 1, 1) = 1$ is located in cell A3.

NOTE One must be careful when using Solver because the solution may depend on the initial value. Thus, it is always a good idea to run the Solver for two different initial values and compare the solutions. ■

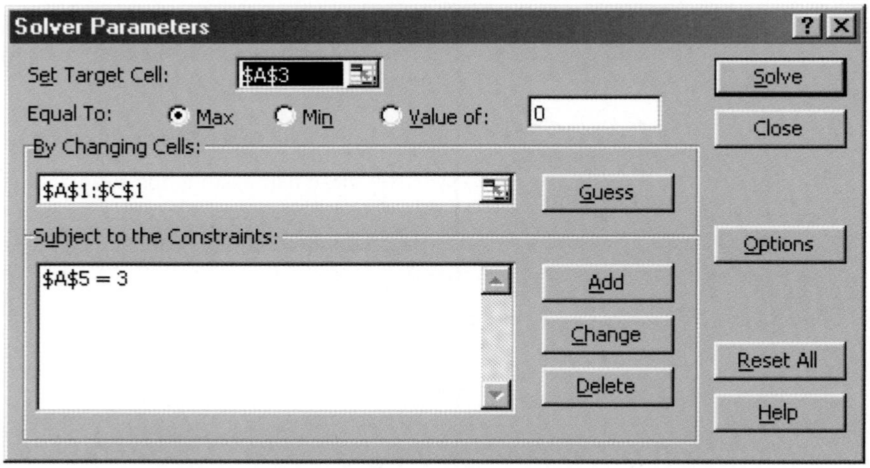

FIGURE 26

9.4 EXERCISES

Find the relative maxima or minima in Exercises 1–10.

1. Maximum of $f(x, y) = 2xy$, subject to $x + y = 12$

2. Maximum of $f(x, y) = 4xy + 2$, subject to $x + y = 24$

3. Maximum of $f(x, y) = x^2y$, subject to $2x + y = 4$

4. Maximum of $f(x, y) = 4xy^2$, subject to $3x - 2y = 5$

5. Minimum of $f(x, y) = x^2 + 2y^2 - xy$, subject to $x + y = 8$

6. Minimum of $f(x, y) = 3x^2 + 4y^2 - xy - 2$, subject to $2x + y = 21$

7. Maximum of $f(x, y) = x^2 - 10y^2$, subject to $x - y = 18$

8. Maximum of $f(x, y) = 12xy - x^2 - 3y^2$, subject to $x + y = 16$

9. Maximum of $f(x, y, z) = xyz^2$, subject to $x + y + z = 6$

10. Maximum of $f(x, y, z) = xy + 2xz + 2yz$, subject to $xyz = 32$

11. Find positive numbers x and y such that $x + y = 18$ and xy^2 is maximized.

12. Find positive numbers x and y such that $x + y = 36$ and x^2y is maximized.

13. Find three positive numbers whose sum is 90 and whose product is a maximum.

14. Find three positive numbers whose sum is 240 and whose product is a maximum.

15. Explain the difference between the two methods we used in Sections 3 and 4 to solve extrema problems.

16. Why is it unnecessary to find the value of λ when using the method explained in this section?

17. Show that the three equations in Step 4 of the box "Using Lagrange Multipliers" are equivalent to the three equations

$$f_x(x, y) = \lambda g_x(x, y),$$
$$f_y(x, y) = \lambda g_y(x, y),$$
$$g(x, y) = 0.$$

Applications

BUSINESS AND ECONOMICS

18. *Maximum Area for Fixed Expenditure* Because of terrain difficulties, two sides of a fence can be built for \$6 per ft, while the other two sides cost \$4 per ft. (See the sketch.) Find the field of maximum area that can be enclosed for \$1200.

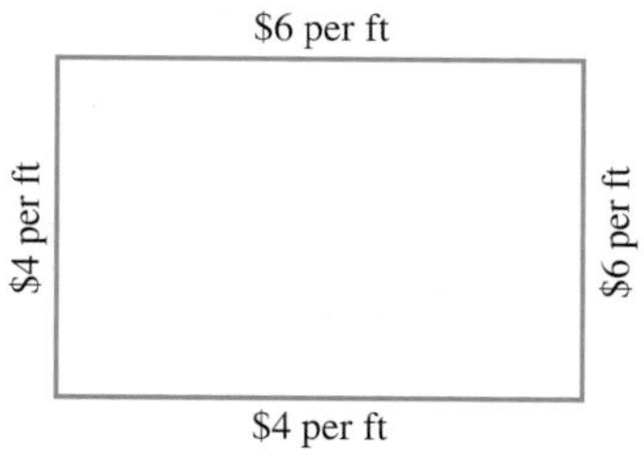

19. *Maximum Area for Fixed Expenditure* To enclose a yard, a fence is built against a large building, so that fencing material is used only on three sides. Material for the ends costs \$8 per ft; material for the side opposite the building costs \$6 per ft. Find the dimensions of the yard of maximum area that can be enclosed for \$1200.

20. *Cost* The total cost to produce x large needlepoint kits and y small ones is given by

$$C(x, y) = 2x^2 + 6y^2 + 4xy + 10.$$

If a total of ten kits must be made, how should production be allocated so that total cost is minimized?

21. *Profit* The profit from the sale of x units of radiators for automobiles and y units of radiators for generators is given by

$$P(x, y) = -x^2 - y^2 + 4x + 8y.$$

Find values of x and y that lead to a maximum profit if the firm must produce a total of 6 units of radiators.

22. *Production* A manufacturing firm estimates that its total production of automobile batteries in thousands of units is

$$f(x, y) = 3x^{1/3}y^{2/3},$$

where x is the number of units of labor and y is the number of units of capital utilized. Labor costs are \$80 per unit, and capital costs are \$150 per unit. How many units each of labor and capital will maximize production, if the firm can spend \$40,000 for these costs?

23. *Production* For another product, the manufacturing firm in Exercise 22 estimates that production is a function of labor x and capital y as follows:

$$f(x, y) = 12x^{3/4}y^{1/4}.$$

If \$25,200 is available for labor and capital, and if the firm's costs are \$100 and \$180 per unit, respectively, how many units of labor and capital will give maximum production?

24. *Area* A farmer has 200 m of fencing. Find the dimensions of the rectangular field of maximum area that can be enclosed by this amount of fencing.

25. *Area* Find the area of the largest rectangular field that can be enclosed with 600 m of fencing. Assume that no fencing is needed along one side of the field.

26. *Surface Area* A cylindrical can is to be made that will hold 250π in^3 of candy. Find the dimensions of the can with minimum surface area.

27. *Surface Area* An ordinary 12-oz beer or soda pop can holds about 25 in^3. Find the dimensions of a can with minimum surface area. Measure a can and see how close its dimensions are to the results you found.

28. *Volume* A rectangular box with no top is to be built from 500 m^2 of material. Find the dimensions of such a box that will enclose the maximum volume.

29. *Surface Area* A 1-lb soda cracker box has a volume of 185 in^3. The end of the box is square. Find the dimensions of such a box that has minimum surface area.

30. *Cost* A rectangular closed box is to be built at minimum cost to hold 27 m^3. Since the cost will depend on the surface area, find the dimensions that will minimize the surface area of the box.

31. *Cost* Find the dimensions that will minimize the surface area (and hence the cost) of a rectangular fish aquarium, open on top, with a volume of 32 ft^3.

32. *Container Construction* A company needs to construct a box with an open top that will be used to transport 400 yd^3 of material, in several trips, from one place to another. Two of the sides and bottom of the box can be made of a free, lightweight material but only 4 yd^2 of the material is available. Because of the nature of the material to be transported, the two ends of the box must be made from a heavyweight material that costs \$20 per yd^2. Each trip costs 10 cents.*

a. Let x, y, and z denote the length, width, and height of the box, respectively. If we want to use all of the free material, show that the total cost in dollars is given by the function

$$f(x, y, z) = \frac{40}{xyz} + 40yz, \text{ subject to the constraint } 2xz + xy = 4.$$

b. Use the solver feature on a spreadsheet to find the dimensions of the box that minimize the transportation cost, subject to the constraint.

SOCIAL SCIENCES

33. *Political Science* The probability that the majority of a three-person jury will convict a guilty person is given by the formula:

$$P(r, s, t) = rs(1 - t) + (1 - r)st + r(1 - s)t + rst$$

subject to the constraint that

$$r + s + t = \alpha,$$

where r, s, and t represent each of the three jury members' probability of reaching a guilty verdict and α is some fixed constant that is generally less than or equal to the number of jurors.†

a. Form the Lagrange function.

b. Find the values of r, s, and t that maximize the probability of convicting a guilty person when $\alpha = .75$.

c. Find the values of r, s, and t that maximize the probability of convicting a guilty person when $\alpha = 3$.

9.5 TOTAL DIFFERENTIALS AND APPROXIMATIONS

THINK ABOUT IT *How do errors in measuring the length and radius of a blood vessel affect the calculation of its volume?*

In the second section of this chapter we used partial derivatives to find the marginal productivity of labor and of capital for a production function. The marginal productivity gives the rate of change of production for a 1-unit change in labor or

*Duffin, R., E. Peterson, and C. Zener, *Geometric Programming-Theory and Application*, New York: Wiley, 1967. Copyright © 1967 John Wiley & Sons, Inc.

†Owen, Guillermo et al., "Proving a Distribution-Free Generalization of the Condorcet Jury Theorem," *Mathematical Social Sciences*, Vol. 17, 1989, pp. 1–16.

capital. To estimate the change in productivity for a small change in both labor and capital, we can extend the concept of differential, introduced in an earlier chapter for functions of one variable, to the concept of *total differential.*

TOTAL DIFFERENTIAL FOR TWO VARIABLES

Let $z = f(x, y)$ be a function of x and y. Let dx and dy be real numbers. Then the **total differential** of z is

$$dz = f_x(x,y) \cdot dx + f_y(x,y) \cdot dy.$$

(Sometimes dz is written df.)

Recall that the differential for a function of one variable $y = f(x)$ is used to approximate the function by its tangent line. This works because a differentiable function appears very much like a line when viewed closely. Similarly, the differential for a function of two variables $z = f(x, y)$ is used to approximate a function by its tangent plane. A differentiable function of two variables looks like a plane when viewed closely, which is why the earth looks flat when you are standing on it.

EXAMPLE 1 *Total Differentials*

Consider the function $z = f(x, y) = 9x^3 - 8x^2y + 4y^3$.

FOR REVIEW

In the chapter on Applications of the Derivative, we introduced the differential. Recall that the differential of a function defined by $y = f(x)$ is

$$dy = f'(x) \cdot dx,$$

where dx, the differential of x, is any real number (usually small). We saw that the differential dy is often a good approximation of Δy, where $\Delta y = f(x + \Delta x) - f(x)$ and $\Delta x = dx$.

(a) Find dz.

Solution First find $f_x(x, y)$ and $f_y(x, y)$.

$$f_x(x, y) = 27x^2 - 16xy \quad \text{and} \quad f_y(x, y) = -8x^2 + 12y^2$$

By the definition,

$$dz = (27x^2 - 16xy)\,dx + (-8x^2 + 12y^2)\,dy.$$

(b) Evaluate dz when $x = 1$, $y = 3$, $dx = .01$, and $dy = -.02$.

Solution Putting these values into the result from part (a) gives

$$\begin{aligned} dz &= [27(1)^2 - 16(1)(3)](.01) + [-8(1)^2 + 12(3)^2](-.02) \\ &= (-21)(.01) + (100)(-.02) \\ &= -2.21. \end{aligned}$$

This result indicates that an increase of .01 in x and a decrease of .02 in y, when $x = 1$ and $y = 3$, will produce an approximate *decrease* of 2.21 in $f(x, y)$.

Approximations Recall that with a function of one variable, $y = f(x)$, the differential dy approximates the change in y, Δy, corresponding to a change in x, Δx or dx. The approximation for a function of two variables is similar.

APPROXIMATIONS

For small values of Δx and Δy,

$$dz \approx \Delta z,$$

where $\Delta z = f(x + \Delta x, y + \Delta y) - f(x, y)$.

EXAMPLE 2 *Approximations*

Approximate $\sqrt{2.98^2 + 4.01^2}$.

Solution Notice that $2.98 \approx 3$ and $4.01 \approx 4$, and we know that $\sqrt{3^2 + 4^2} = \sqrt{25} = 5$. We therefore let $f(x, y) = \sqrt{x^2 + y^2}$, $x = 3$, $dx = -.02$, $y = 4$, and $dy = .01$. We then use dz to approximate $\Delta z = \sqrt{2.98^2 + 4.01^2} - \sqrt{3^2 + 4^2}$.

$$\begin{aligned} dz &= f_x(x, y) \cdot dx + f_y(x, y) \cdot dy \\ &= \left(\frac{1}{2\sqrt{x^2 + y^2}} \cdot 2x\right) dx + \left(\frac{1}{2\sqrt{x^2 + y^2}} \cdot 2y\right) dy \\ &= \left(\frac{x}{\sqrt{x^2 + y^2}}\right) dx + \left(\frac{y}{\sqrt{x^2 + y^2}}\right) dy \\ &= \frac{3}{5}(-.02) + \frac{4}{5}(.01) \\ &= -.004 \end{aligned}$$

Thus, $\sqrt{2.98^2 + 4.01^2} \approx 5 + (-.004) = 4.996$. A calculator gives $\sqrt{2.98^2 + 4.01^2} \approx 4.996048$. The error is approximately .000048.

For small values of dx and dy, the values of Δz and dz are approximately equal. Since $\Delta z = f(x + dx, y + dy) - f(x, y)$,

$$f(x + dx, y + dy) = f(x, y) + \Delta z$$

or

$$f(x + dx, y + dy) \approx f(x, y) + dz.$$

Replacing dz with the expression for the total differential gives the following result.

APPROXIMATIONS BY DIFFERENTIALS

For a function f having all indicated partial derivatives, and for small values of dx and dy,

$$f(x + dx, y + dy) \approx f(x, y) + dz,$$

or

$$f(x + dx, y + dy) \approx f(x, y) + f_x(x, y) \cdot dx + f_y(x, y) \cdot dy.$$

The idea of a total differential can be extended to include functions of three or more independent variables.

TOTAL DIFFERENTIAL FOR THREE VARIABLES

If $w = f(x, y, z)$, then the total differential dw is

$$dw = f_x(x, y, z)\, dx + f_y(x, y, z)\, dy + f_z(x, y, z)\, dz,$$

provided all indicated partial derivatives exist.

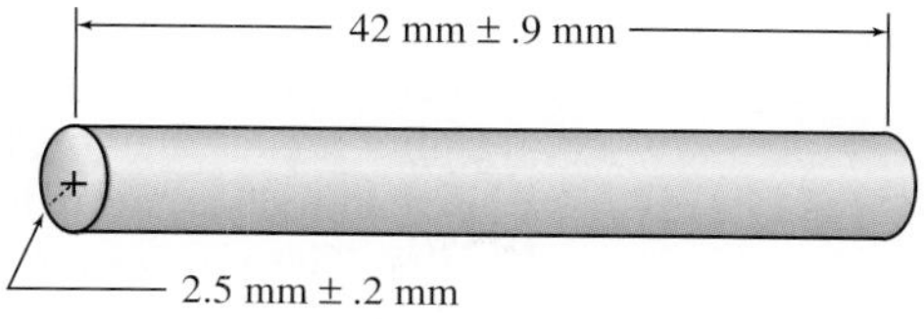

FIGURE 27

EXAMPLE 3 *Blood Vessels*

A short length of blood vessel is in the shape of a right circular cylinder (see Figure 27).

(a) The length of the vessel is measured as 42 mm, and the radius is measured as 2.5 mm. Suppose the maximum error in the measurement of the length is .9 mm, with an error of no more than .2 mm in the measurement of the radius. Find the maximum possible error in calculating the volume of the blood vessel.

Solution The volume of a right circular cylinder is given by $V = \pi r^2 h$. To approximate the error in the volume, find the total differential, dV.

$$dV = (2\pi rh) \cdot dr + (\pi r^2) \cdot dh$$

Here, $r = 2.5$, $h = 42$, $dr = .2$, and $dh = .9$. Substitution gives

$$dV = [(2\pi)(2.5)(42)(.2)] + [\pi(2.5)^2](.9) \approx 149.6.$$

The maximum possible error in calculating the volume is approximately 149.6 mm^3.

(b) Suppose that the errors in measuring the radius and length of the vessel are at most 1% and 3%, respectively. Estimate the maximum percent error in calculating the volume.

Solution To find the percent error, calculate dV/V.

$$\frac{dV}{V} = \frac{(2\pi rh)dr + (\pi r^2)dh}{\pi r^2 h} = 2\frac{dr}{r} + \frac{dh}{h}$$

Because $dr/r = .01$ and $dh/h = .03$,

$$\frac{dV}{V} = 2(.01) + .03 = .05.$$

The maximum percent error in calculating the volume is approximately 5%.

EXAMPLE 4 *Volume of a Can of Beer*

The formula for the volume of a cylinder given in Example 3 also applies to cans of beer, for which $r \approx 1$ in. and $h \approx 5$ in. How sensitive is the volume to changes in the radius compared with changes in the height?

Solution Using the formula for dV from the previous example with $r = 1$ and $h = 5$ gives

$$dV = (2\pi)(1)(5)dr + \pi(1)^2 dh = \pi(10dr + dh).$$

The factor of 10 in front of dr in this equation shows that a small change in the radius has 10 times the effect on the volume as a small change in the height. One author argues that this is the reason that beer cans are so tall and thin.* The brewers can reduce the radius by a tiny amount, and compensate by making the can taller. The resulting can appears larger in volume than the shorter, wider can. (Others have argued that a shorter, wider can does not fit as easily in the hand.)

9.5 EXERCISES

Use the total differential to approximate each quantity. Then use a calculator to approximate the quantity, and give the absolute value of the difference in the two results to 4 decimal places.

1. $\sqrt{3.04^2 + 4.06^2}$

2. $\sqrt{6.07^2 + 7.95^2}$

3. $(1.92^2 + 2.1^2)^{1/3}$

4. $(2.93^2 - .94^2)^{1/3}$

5. $.97e^{.02}$

6. $1.02e^{-.03}$

7. $1.04 \ln .95$

8. $.95 \ln 1.04$

Evaluate dz using the given information.

9. $z = x^2 + 3xy + y^2$; $x = 4, y = -2, dx = .02, dy = -.03$

10. $z = 8x^3 + 2x^2y - y$; $x = 1, y = 3, dx = .01, dy = .02$

11. $z = \dfrac{y^2 + 3x}{y^2 - x}$; $x = 4, y = -4, dx = .01, dy = .03$

12. $z = \ln(x^2 + y^2)$; $x = 2, y = 3, dx = .02, dy = -.03$

Evaluate dw using the given information.

13. $w = \dfrac{5x^2 + y^2}{z + 1}$; $x = -2, y = 1, z = 1, dx = .02, dy = -.03, dz = .02$

14. $w = x\ln(yz) - y\ln\dfrac{x}{z}$; $x = 2, y = 1, z = 4, dx = .03, dy = .02, dz = -.01$

Applications

BUSINESS AND ECONOMICS

15. *Manufacturing* Approximate the amount of aluminum needed for a beverage can of radius 2.5 cm and height 14 cm. Assume the walls of the can are .08 cm thick.

16. *Manufacturing* Approximate the amount of material needed to make a water tumbler of diameter 3 cm and height 9 cm. Assume the walls of the tumbler are .2 cm thick.

17. *Volume of a Coating* An industrial coating .2 in. thick is applied to all sides of a box of dimensions 10 in. by 9 in. by 14 in. Estimate the volume of the coating used.

18. *Manufacturing Cost* The manufacturing cost of a precision electronic calculator is approximated by

$$M(x, y) = 40x^2 + 30y^2 - 10xy + 30,$$

where x is the cost of the chips and y is the cost of labor. Right now, the company spends \$4 on chips and \$7 on

*Colley, Susan Jane, "Calculus in the Brewery," *The College Mathematics Journal*, Vol. 25, No. 3, May 1994, p. 227.

labor. Use differentials to approximate the change in cost if the company spends \$5 on chips and \$6.50 on labor.

19. *Production* The production function for one country is

$$z = x^{.65}y^{.35},$$

where x stands for units of labor and y for units of capital. At present, 50 units of labor and 29 units of capital are available. Use differentials to estimate the change in production if the number of units of labor is increased to 52 and capital is decreased to 27 units.

20. *Production* The production function for another country is

$$z = x^{.8}y^{.2},$$

where x stands for units of labor and y for units of capital. At present, 20 units of labor and 18 units of capital are being provided. Use differentials to estimate the change in production if an additional unit of labor is provided and if capital is decreased to 16 units.

LIFE SCIENCES

21. *Bone Preservative Volume* A piece of bone in the shape of a right circular cylinder is 7 cm long and has a radius of 1.4 cm. It is coated with a layer of preservative .09 cm thick. Estimate the volume of preservative used.

22. *Blood Vessel Volume* A portion of a blood vessel is measured as having length 7.9 cm and radius .8 cm. If each measurement could be off by as much as .15 cm, estimate the maximum possible error in calculating the volume of the vessel.

23. *Blood Volume* In Exercise 56 of Section 2 in this chapter, we found that the number of liters of blood pumped through the lungs in one minute is given by

$$C = \frac{b}{a - v}.$$

Suppose $a = 160$, $b = 200$, $v = 125$. Estimate the change in C if a becomes 145, b becomes 190, and v changes to 130.

24. *Oxygen Consumption* In Exercise 54 of Section 2 of this chapter, we found that the oxygen consumption of a mammal is

$$m = \frac{2.5(T - F)}{w^{.67}}.$$

Suppose T is 38°C, F is 12°C, and w is 30 kg. Approximate the change in m if T changes to 36°C, F changes to 13°C, and w becomes 31 kg.

25. *Dialysis* A model that estimates the concentration of urea in the body for a particular dialysis patient, following a dialysis session, is given by

$$C(t, g) = .6(.96)^{(210t/1500)-1} + \frac{gt}{126t - 900}[1 - (.96)^{(210t/1500)-1}],$$

where t represents the number of minutes of the dialysis session and g represents the rate at which the body generates urea in mg per minute.*

a. Find $C(180, 8)$.

b. Using the total differential, estimate the urea concentration if the dialysis session of part a was cut short by 10 minutes and the urea generation rate was 9 mg per minute. Compare this with the actual concentration. (*Hint:* First, replace the variable g with the number 8, thus reducing the function to one variable. Then use your graphing calculator to calculate the partial derivative $C_t(180, 8)$. A similar procedure can be done for $C_y(180, 8)$.)

26. *Horn Volume* The volume of the horns from bighorn sheep were estimated by researchers using the equation

$$V = \frac{h\pi}{3}(r_1^2 + r_1r_2 + r_2^2),$$

where h is the length of a horn segment (in centimeters) and r_1 and r_2 are the radii of the two ends of the horn segment (in centimeters).†

a. Determine the volume of a segment of horn that is 40 cm long with radii of 5 cm and 3 cm, respectively.

b. Use the total differential to estimate the volume of the segment of horn if the horn segment from part a was actually 42 cm long with radii of 5.1 cm and 2.9 cm, respectively. Compare this with the actual volume.

PHYSICAL SCIENCES

27. *Swimming* The amount of time in seconds it takes for a swimmer to hear a single, hand-held, starting signal is given by the formula

$$t(x, y, p, C) = \frac{\sqrt{x^2 + (y - p)^2}}{331.45 + .6C},$$

*Gotch, Frank, "Clinical Dialysis: Kinetic Modeling in Hemodialysis," *Clinical Dialysis*, 3rd ed., Norwalk: Appleton & Lange, 1995, pp. 156–186.

†Fitzsimmons, N., S. Buskirk, and M. Smith, "Population History, Genetic Variability, and Horn Growth in Bighorn Sheep," *Conservation Biology*, Vol. 9, No. 2, April 1995, pp. 314–323.

where (x, y) is the location of the starter (in meters), $(0, p)$ is the location of the swimmer (in meters), and C is the air temperature (in degrees Celsius).* Assume that the starter is located at the point $(x, y) = (5, -2)$. See the diagram.

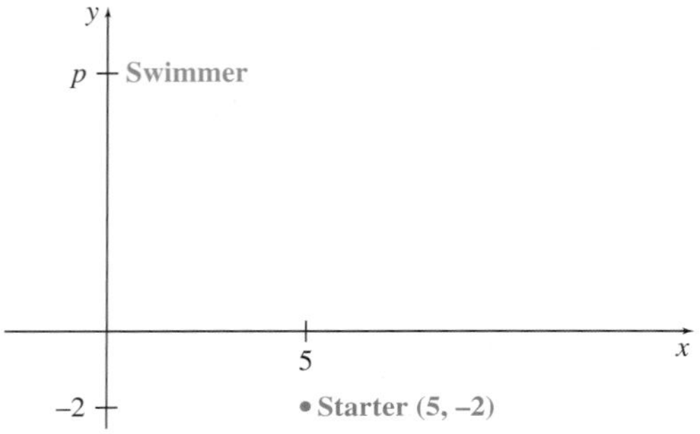

a. Calculate $t(5, -2, 20, 20)$ and $t(5, -2, 10, 20)$. Could the difference in time change the outcome of a race?

b. Calculate the total differential for t if the starter remains stationary, the swimmer moves from 20 m to 20.5 m away from the starter in the y direction, and the temperature decreases from 20°C to 15°C. Interpret your answer.

GENERAL INTEREST

28. *Estimating Area* The height of a triangle is measured as 42.6 cm, with the base measured as 23.4 cm. The measurement of the height can be off by as much as 1.2 cm, and that of the base by no more than .9 cm. Estimate the maximum possible error in calculating the area of the triangle.

29. *Estimating Volume* The height of a cone is measured as 8.4 cm and the radius as 2.9 cm. Each measurement could be off by as much as .1 cm. Estimate the maximum possible error in calculating the volume of the cone.

30. *Estimating Volume* Suppose that in measuring the length, width, and height of a box, there is a maximum 1% error in each measurement. Estimate the maximum error in calculating the volume of the box.

31. *Estimating Volume* Suppose there is a maximum error of $a\%$ in measuring the radius of a cone and a maximum error of $b\%$ in measuring the height. Estimate the maximum percent error in calculating the volume of the cone, and compare this value with the maximum percent error in calculating the volume of a cylinder.

9.6 DOUBLE INTEGRALS

THINK ABOUT IT

How can we find the volume of a bottle with curved sides?

FOR REVIEW

You may wish to review the key ideas of indefinite and definite integrals from the chapter on Integration before continuing with this section. See the review problems at the end of that chapter.

In an earlier chapter, we saw how integrals of functions with one variable may be used to find area. In this section, this idea is extended and used to find volume. We found partial derivatives of functions of two or more variables at the beginning of this chapter by holding constant all variables except one. A similar process is used in this section to find antiderivatives of functions of two or more variables. For example, in

$$\int (5x^3y^4 - 6x^2y + 2)\,dy$$

*Walker, Anita, "Mathematics Makes a Splash: Evaluating Hand Timing Systems," *The HiMAP Pull-Out Section*, Spring 1992, COMAP.

the notation dy indicates integration with respect to y, so we treat y as the variable and x as a constant. Using the rules for antiderivatives gives

$$\int (5x^3y^4 - 6x^2y + 2)\,dy = x^3y^5 - 3x^2y^2 + 2y + C(x).$$

The constant C used earlier must be replaced with $C(x)$ to show that the "constant of integration" here can be any function involving only the variable x. Just as before, check this work by taking the derivative (actually the partial derivative) of the answer:

$$\frac{\partial}{\partial y}[x^3y^5 - 3x^2y^2 + 2y + C(x)] = 5x^3y^4 - 6x^2y + 2 + 0,$$

which shows that the antiderivative is correct.

EXAMPLE 1 *Indefinite Integrals*

Find each indefinite integral.

(a) $\int x(x^2 + y)\,dx$

Solution Multiply x and $x^2 + y$. Then (because of the dx) integrate each term with x as the variable and y as a constant.

$$\int x(x^2 + y)\,dx = \int (x^3 + xy)\,dx$$

$$= \frac{x^4}{4} + \frac{x^2}{2} \cdot y + f(y) = \frac{1}{4}x^4 + \frac{1}{2}x^2y + f(y)$$

(b) $\int x(x^2 + y)\,dy$

Solution Since y is the variable and x is held constant,

$$\int x(x^2 + y)\,dy = \int (x^3 + xy)\,dy = x^3y + \frac{1}{2}xy^2 + g(x).$$

The analogy to integration of functions of one variable can be continued for evaluating definite integrals. We do this by holding one variable constant and using the Fundamental Theorem of Calculus with the other variable.

EXAMPLE 2 *Definite Integrals*

Evaluate each definite integral.

(a) $\int_3^5 (6xy^2 + 12x^2y + 4y)\,dx$

Solution First, find an antiderivative:

$$\int (6xy^2 + 12x^2y + 4y)\,dx = 3x^2y^2 + 4x^3y + 4xy + h(y).$$

Now replace each x with 5, and then with 3, and subtract the results.

$$\begin{aligned}\left[3x^2y^2 + 4x^3y + 4xy + h(y)\right]\Big|_3^5 &= [3 \cdot 5^2 \cdot y^2 + 4 \cdot 5^3 \cdot y + 4 \cdot 5 \cdot y + h(y)] \\ &\quad -[3 \cdot 3^2 \cdot y^2 + 4 \cdot 3^3 \cdot y + 4 \cdot 3 \cdot y + h(y)] \\ &= 75y^2 + 500y + 20y + h(y) \\ &\quad -[27y^2 + 108y + 12y + h(y)] \\ &= 48y^2 + 400y\end{aligned}$$

The *function of integration*, $h(y)$, drops out, just as the constant of integration does with definite integrals of functions of one variable. Thus, the function of integration is not included for definite integrals of more than one variable.

(b) $\int_1^2 (6xy^2 + 12x^2y + 4y)\, dy$

Solution Integrate with respect to y; then substitute 2 and 1 for y and subtract.

$$\begin{aligned}\int_1^2 (6xy^2 + 12x^2y + 4y)\, dy &= (2xy^3 + 6x^2y^2 + 2y^2)\Big|_1^2 \\ &= (2x \cdot 2^3 + 6x^2 \cdot 2^2 + 2 \cdot 2^2) \\ &\quad -(2x \cdot 1^3 + 6x^2 \cdot 1^2 + 2 \cdot 1^2) \\ &= 16x + 24x^2 + 8 - (2x + 6x^2 + 2) \\ &= 14x + 18x^2 + 6\end{aligned}$$

As Example 2 suggests, an integral of the form

$$\int_a^b f(x, y)\, dy$$

produces a result that is a function of x, while

$$\int_a^b f(x, y)\, dx$$

produces a function of y. These resulting functions of one variable can themselves be integrated, as in the next example.

EXAMPLE 3 *Definite Integrals*

Evaluate each integral.

(a) $\int_1^2 \left[\int_3^5 (6xy^2 + 12x^2y + 4y)\, dx\right] dy$

Solution In Example 2(a), we found the quantity in brackets to be $48y^2 + 400y$. Thus,

$$\begin{aligned}\int_1^2 \left[\int_3^5 (6xy^2 + 12x^2y + 4y)\, dx\right] dy &= \int_1^2 (48y^2 + 400y)dy \\ &= (16y^3 + 200y^2)\Big|_1^2 \\ &= 16 \cdot 2^3 + 200 \cdot 2^2 - (16 \cdot 1^3 + 200 \cdot 1^2) \\ &= 128 + 800 - (16 + 200) \\ &= 712.\end{aligned}$$

(b) $\int_3^5 \left[\int_1^2 (6xy^2 + 12x^2y + 4y)\, dy \right] dx$

Solution (This is the same integrand, with the same limits of integration as in part (a), but the order of integration is reversed.)

Use the result from Example 2(b).

$$\int_3^5 \left[\int_1^2 (6xy^2 + 12x^2y + 4y)dy \right] dx = \int_3^5 (14x + 18x^2 + 6)\, dx$$
$$= (7x^2 + 6x^3 + 6x)\Big|_3^5$$
$$= 7 \cdot 5^2 + 6 \cdot 5^3 + 6 \cdot 5 - (7 \cdot 3^2 + 6 \cdot 3^3 + 6 \cdot 3)$$
$$= 175 + 750 + 30 - (63 + 162 + 18) = 712$$

The brackets we have used for the inner integral in Example 3 are not essential because the order of integration is indicated by the order of *dx dy* or *dy dx*. For example, if the integral is written as

$$\int_1^2 \int_3^5 (6xy^2 + 12x^2y + 4y)\, dx\, dy,$$

we first integrate with respect to x, letting x vary from 3 to 5, and then with respect to y, letting y vary from 1 to 2, as in Example 3(a).

The answers in the two parts of Example 3 are equal. It can be proved that for a large class of functions, including most functions that occur in applications, the following equation holds true.

$$\int_a^b \int_c^d f(x, y) dx\, dy = \int_c^d \int_a^b f(x, y) dy\, dx$$

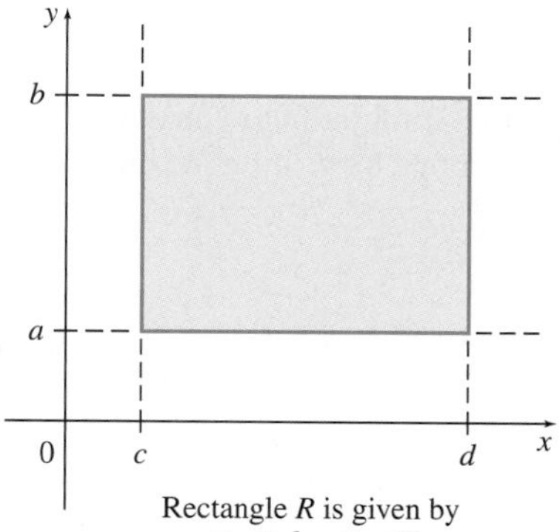

Rectangle R is given by $c \le x \le d$, $a \le y \le b$.

FIGURE 28

Either of these integrals is called an **iterated integral** since it is evaluated by integrating twice, first using one variable and then using the other. The fact that the iterated integrals above are equal makes it possible to define a *double integral*. First, the set of points (x, y), with $c \le x \le d$ and $a \le y \le b$, defines a rectangular region R in the plane, as shown in Figure 28. Then, the *double integral over R* is defined as follows.

DOUBLE INTEGRAL

The **double integral** of $f(x, y)$ over a rectangular region R is written

$$\iint_R f(x, y) dx\, dy \quad \text{or} \quad \iint_R f(x, y) dy\, dx,$$

and equals either

$$\int_a^b \int_c^d f(x, y) dx\, dy \quad \text{or} \quad \int_c^d \int_a^b f(x, y)\, dy\, dx.$$

Extending earlier definitions, $f(x, y)$ is the **integrand** and R is the **region of integration**.

EXAMPLE 4 *Double Integrals*

Find $\iint\limits_R \sqrt{x} \cdot \sqrt{y-2}\, dx\, dy$ over the rectangular region R defined by $0 \le x \le 4, 3 \le y \le 11$.

Solution Integrate first with respect to x; then integrate the result with respect to y.

$$\begin{aligned}
\iint\limits_R \sqrt{x} \cdot \sqrt{y-2}\, dx\, dy &= \int_3^{11} \int_0^4 \sqrt{x} \cdot \sqrt{y-2}\, dx\, dy \\
&= \int_3^{11} \left(\frac{2}{3} x^{3/2} \sqrt{y-2} \right) \Big|_0^4 dy \\
&= \int_3^{11} \left[\frac{2}{3}(4^{3/2})\sqrt{y-2} - \frac{2}{3}(0^{3/2})\sqrt{y-2} \right] dy \\
&= \int_3^{11} \left(\frac{16}{3}\sqrt{y-2} - 0 \right) dy = \int_3^{11} \left(\frac{16}{3}\sqrt{y-2} \right) dy \\
&= \frac{32}{9}(y-2)^{3/2} \Big|_3^{11} = \frac{32}{9}(9)^{3/2} - \frac{32}{9}(1)^{3/2} \\
&= 96 - \frac{32}{9} = \frac{832}{9}
\end{aligned}$$

As a check, integrate with respect to y first. The answer should be the same.

NOTE In the second step of the previous example, it might help you avoid confusion as to whether to put the limits of 0 and 4 into x or y by writing the integral as

$$\int_3^{11} \left(\frac{2}{3} x^{3/2} \sqrt{y-2} \right) \Big|_{x=0}^{x=4} dy.$$

Volume As shown earlier, the definite integral $\int_a^b f(x)\, dx$ can be used to find the area under a curve. In a similar manner, double integrals are used to find the *volume under a surface*. Figure 29 on the next page shows that portion of a surface $f(x, y)$ directly over a rectangle R in the xy-plane. Just as areas were approximated by a large number of small rectangles, volume could be approximated by adding the volumes of a large number of properly drawn small boxes. The height of a typical box would be $f(x, y)$ with the length and width given by dx and dy. The formula for the volume of a box would then suggest the following result.

VOLUME

Let $z = f(x, y)$ be a function that is never negative on the rectangular region R defined by $c \le x \le d, a \le y \le b$. The volume of the solid under the graph of f and over the region R is

$$\iint\limits_R \boldsymbol{f(x, y)\, dx\, dy}.$$

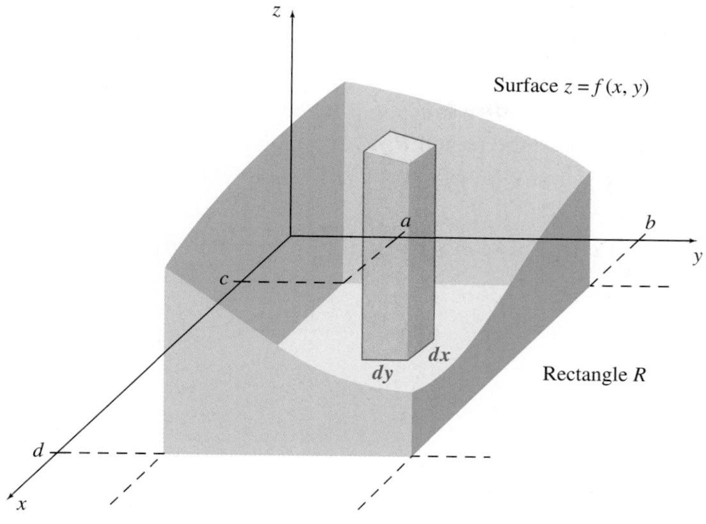

FIGURE 29

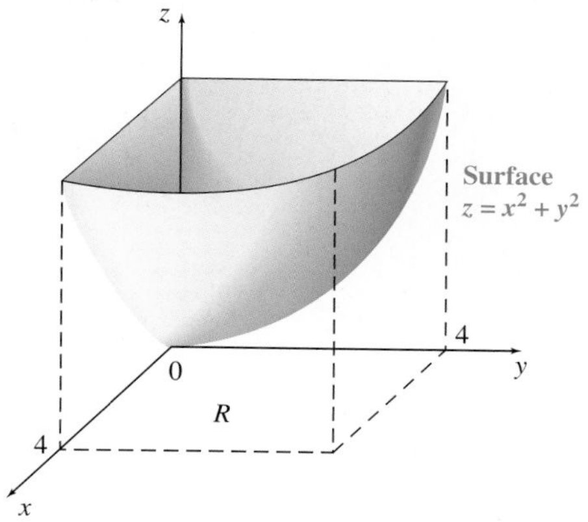

FIGURE 30

EXAMPLE 5 *Volume*

Find the volume under the surface $z = x^2 + y^2$ shown in Figure 30.

Solution By the equation just given, the volume is

$$\iint_R f(x, y)\, dx\, dy,$$

where $f(x, y) = x^2 + y^2$ and R is the region $0 \le x \le 4$, $0 \le y \le 4$. By definition,

$$\begin{aligned}
\iint_R f(x, y)\, dx\, dy &= \int_0^4 \int_0^4 (x^2 + y^2)\, dx\, dy \\
&= \int_0^4 \left(\frac{1}{3}x^3 + xy^2 \right)\Bigg|_0^4 dy \\
&= \int_0^4 \left(\frac{64}{3} + 4y^2 \right) dy = \left(\frac{64}{3}y + \frac{4}{3}y^3 \right)\Bigg|_0^4 \\
&= \frac{64}{3} \cdot 4 + \frac{4}{3} \cdot 4^3 - 0 = \frac{512}{3}.
\end{aligned}$$

EXAMPLE 6 *Perfume Bottle*

A product design consultant for a cosmetics company has been asked to design a bottle for the company's newest perfume. The thickness of the glass is to vary so that the outside of the bottle has straight sides and the inside has curved sides, with flat ends shaped like parabolas on the 4-cm sides, as shown in Figure 31 on the next page. Before presenting the design to management, the consultant needs to make a reasonably accurate estimate of the amount each bottle will hold. If the

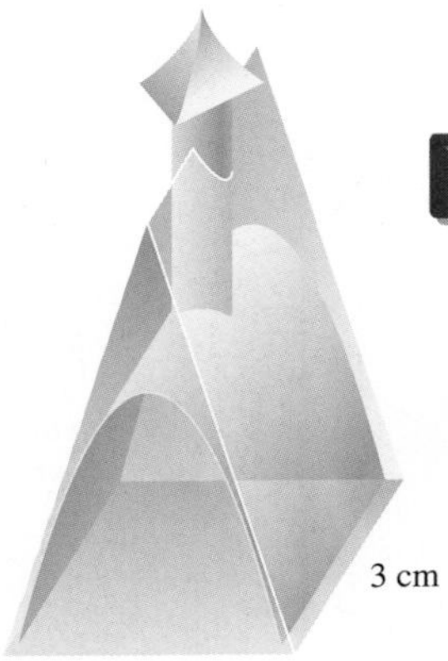

FIGURE 31

base of the bottle is to be 4 cm by 3 cm, and if a cross section of its interior is to be a parabola of the form $z = -y^2 + 4y$, what is its internal volume?

Solution The interior of the bottle can be graphed in three-dimensional space, as shown in Figure 32, where $z = 0$ corresponds to the base of the bottle. Its volume is simply the volume above the region R in the xy-plane and below the graph of $f(x, y) = -y^2 + 4y$. This volume is given by the double integral

$$\int_0^3 \int_0^4 (-y^2 + 4y)\, dy\, dx = \int_0^3 \left(\frac{-y^3}{3} + \frac{4y^2}{2}\right)\Bigg|_0^4 dx$$
$$= \int_0^3 \left(\frac{-64}{3} + 32 - 0\right) dx$$
$$= \frac{32}{3}x\Bigg|_0^3$$
$$= 32 - 0 = 32.$$

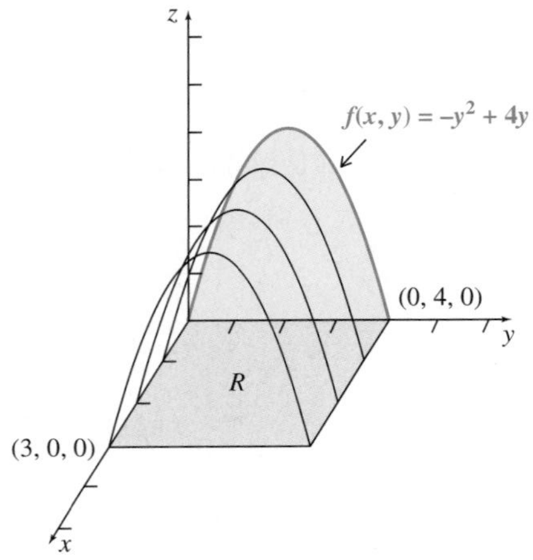

FIGURE 32

The bottle holds 32 cm^3.

Double Integrals Over Other Regions In this section, we found double integrals over rectangular regions by evaluating iterated integrals with constant limits of integration. We can also evaluate iterated integrals with *variable* limits of integration. (Notice in the following examples that the variable limits always go on the *inner* integral sign.)

The use of variable limits of integration permits evaluation of double integrals over the types of regions shown in Figure 33 on the next page. Double integrals over more complicated regions are discussed in more advanced books. Integration over regions such as those in Figure 33 is done with the results of the following theorem.

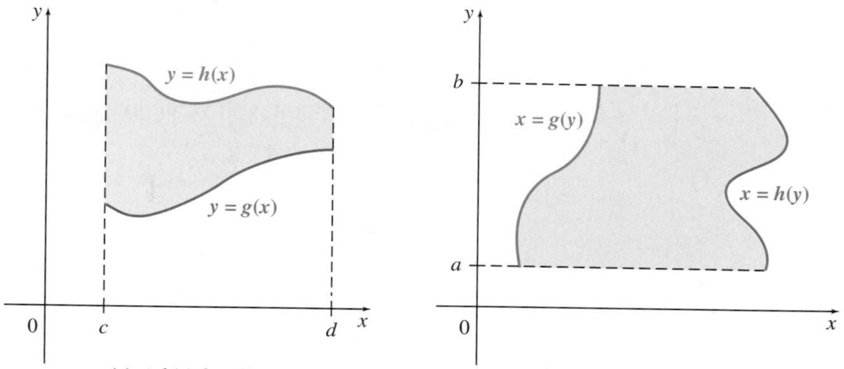

$g(x) \le h(x)$ for all x in $[c, d]$

(a)

$g(y) \le h(y)$ for all y in $[a, b]$

(b)

FIGURE 33

DOUBLE INTEGRALS OVER VARIABLE REGIONS

Let $z = f(x, y)$ be a function of two variables. If R is the region (in Figure 33(a)) defined by $c \le x \le d$ and $g(x) \le y \le h(x)$, then

$$\iint_R f(x, y)\, dy\, dx = \int_c^d \left[\int_{g(x)}^{h(x)} f(x, y)\, dy \right] dx.$$

If R is the region (in Figure 33(b)) defined by $g(y) \le x \le h(y)$ and $a \le y \le b$, then

$$\iint_R f(x, y) dx\, dy = \int_a^b \left[\int_{g(y)}^{h(y)} f(x, y)\, dx \right] dy.$$

EXAMPLE 7 *Double Integrals*

Evaluate $\displaystyle\int_1^2 \int_y^{y^2} xy\, dx\, dy.$

Solution The region of integration is shown in Figure 34. Integrate first with respect to x, then with respect to y.

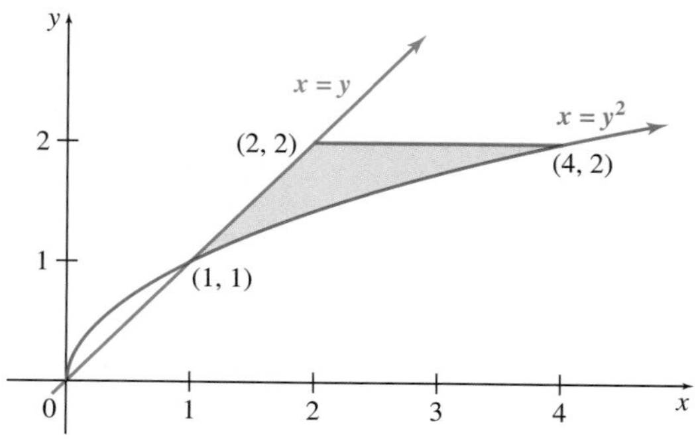

FIGURE 34

$$\int_1^2 \int_y^{y^2} xy\, dx\, dy = \int_1^2 \left(\int_y^{y^2} xy\, dx \right) dy = \int_1^2 \left(\frac{1}{2}x^2 y \right)\Bigg|_y^{y^2} dy$$

Replace x first with y^2 and then with y, and subtract.

$$\begin{aligned}\int_1^2 \int_y^{y^2} xy\, dx\, dy &= \int_1^2 \left[\frac{1}{2}(y^2)^2 y - \frac{1}{2}(y)^2 y \right] dy \\ &= \int_1^2 \left(\frac{1}{2}y^5 - \frac{1}{2}y^3 \right) dy = \left(\frac{1}{12}y^6 - \frac{1}{8}y^4 \right)\Bigg|_1^2 \\ &= \left(\frac{1}{12} \cdot 2^6 - \frac{1}{8} \cdot 2^4 \right) - \left(\frac{1}{12} \cdot 1^6 - \frac{1}{8} \cdot 1^4 \right) \\ &= \frac{64}{12} - \frac{16}{8} - \frac{1}{12} + \frac{1}{8} = \frac{27}{8}\end{aligned}$$

EXAMPLE 8 *Double Integrals*

Let R be the shaded region in Figure 35, and evaluate

$$\iint_R (x + 2y)\, dy\, dx.$$

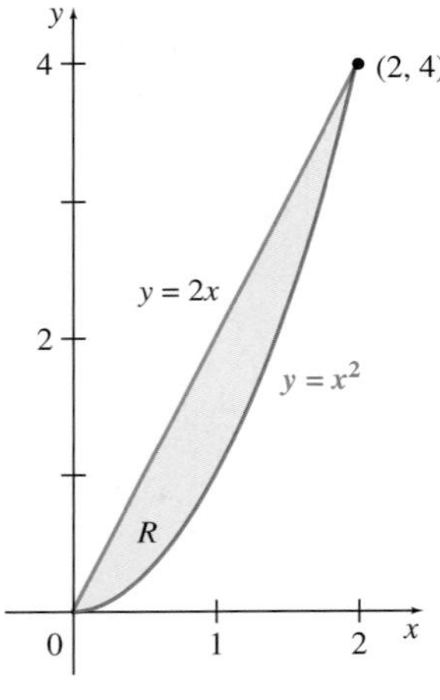

FIGURE 35

Solution Region R is bounded by $h(x) = 2x$ and $g(x) = x^2$, with $0 \le x \le 2$. By the first result in the previous theorem,

$$\begin{aligned}\iint_R (x + 2y)\, dy\, dx &= \int_0^2 \int_{x^2}^{2x} (x + 2y)\, dy\, dx \\ &= \int_0^2 (xy + y^2)\Bigg|_{x^2}^{2x} dx \\ &= \int_0^2 [x(2x) + (2x)^2 - [x \cdot x^2 + (x^2)^2]]\, dx \\ &= \int_0^2 [2x^2 + 4x^2 - (x^3 + x^4)]\, dx \\ &= \int_0^2 (6x^2 - x^3 - x^4)\, dx \\ &= \left(2x^3 - \frac{1}{4}x^4 - \frac{1}{5}x^5 \right)\Bigg|_0^2 \\ &= 2 \cdot 2^3 - \frac{1}{4} \cdot 2^4 - \frac{1}{5} \cdot 2^5 - 0 \\ &= 16 - 4 - \frac{32}{5} = \frac{28}{5}.\end{aligned}$$

In Example 8, the same result would be found if we evaluated the double integral first with respect to x, and then with respect to y. In that case, we would need to define the equations of the boundaries in terms of y rather than x, so R would be defined by $y/2 \le x \le \sqrt{y}$, $0 \le y \le 4$. The resulting integral is

$$\int_0^4 \int_{y/2}^{\sqrt{y}} (x + 2y)\, dx\, dy = \int_0^4 \left(\frac{x^2}{2} + 2xy\right)\Bigg|_{y/2}^{\sqrt{y}} dy$$

$$= \int_0^4 \left[\left(\frac{y}{2} + 2y\sqrt{y}\right) - \left(\frac{y^2}{8} + 2(y/2)y\right)\right] dy$$

$$= \int_0^4 \left(\frac{y}{2} + 2y^{3/2} - \frac{9}{8}y^2\right) dy$$

$$= \left(\frac{y^2}{4} + \frac{4}{5}y^{5/2} - \frac{3}{8}y^3\right)\Bigg|_0^4$$

$$= 4 + \frac{4}{5} \cdot 4^{5/2} - 24$$

$$= \frac{28}{5}.$$

Interchanging Limits of Integration Sometimes it is easier to integrate first with respect to x, and then y, while with other integrals the reverse process is easier. The limits of integration can be reversed whenever the region R is like the region in Figure 35, which has the property that it can be viewed as either type of region shown in Figure 33. In practice, this means that all boundaries can be written in terms of y as a function of x, or by solving for x as a function of y. The next example shows how this process works.

EXAMPLE 9 *Interchanging Limits of Integration*

Evaluate

$$\int_0^{16} \int_{\sqrt{y}}^{4} \sqrt{x^3 + 4}\, dx\, dy.$$

Solution Notice that it is impossible to first integrate this function with respect to x. Thus, we attempt to interchange the limits of integration.

For this integral, region R is given by $\sqrt{y} \le x \le 4$, $0 \le y \le 16$. A graph of R is shown in Figure 36.

The same region R can be written in an alternate way. As Figure 36 shows, one boundary of R is $x = \sqrt{y}$. Solving for y gives $y = x^2$. Also, Figure 36 shows that $0 \le x \le 4$. Since R can be written as $0 \le y \le x^2$, $0 \le x \le 4$, the double integral above can be written

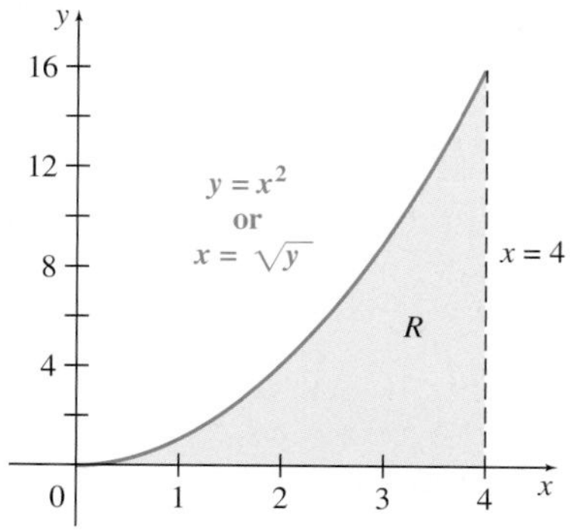

FIGURE 36

$$\int_0^4 \int_0^{x^2} \sqrt{x^3 + 4}\, dy\, dx = \int_0^4 y\sqrt{x^3 + 4}\Bigg|_0^{x^2} dx$$

$$= \int_0^4 x^2\sqrt{x^3 + 4}\, dx$$

$$= \frac{1}{3}\int_0^4 3x^2\sqrt{x^3 + 4}\, dx \qquad \text{Let } u = x^3 + 4.$$

$$= \frac{1}{3}\int_4^{68} u^{1/2}du$$
$$= \frac{2}{9}u^{3/2}\Big|_4^{68}$$
$$= \frac{2}{9}[68^{3/2} - 4^{3/2}]$$
$$\approx 122.83.$$

9.6 EXERCISES

Evaluate each integral.

1. $\int_0^3 (x^3y + y)\, dx$
2. $\int_1^4 (xy^2 - x)\, dy$
3. $\int_4^8 \sqrt{6x + y}\, dx$
4. $\int_3^7 \sqrt{x + 5y}\, dy$
5. $\int_4^5 x\sqrt{x^2 + 3y}\, dy$
6. $\int_3^6 x\sqrt{x^2 + 3y}\, dx$
7. $\int_4^9 \frac{3 + 5y}{\sqrt{x}}\, dx$
8. $\int_2^7 \frac{3 + 5y}{\sqrt{x}}\, dy$
9. $\int_{-1}^1 e^{x+4y}\, dy$
10. $\int_2^6 e^{x+4y}\, dx$
11. $\int_0^5 xe^{x^2+9y}\, dx$
12. $\int_1^6 xe^{x^2+9y}\, dy$

Evaluate each iterated integral. (Many of these use results from Exercises 1–12.)

13. $\int_1^2\int_0^3 (x^3y + y)\, dx\, dy$
14. $\int_0^3\int_1^4 (xy^2 - x)\, dy\, dx$
15. $\int_0^1\int_3^6 x\sqrt{x^2 + 3y}\, dx\, dy$
16. $\int_0^3\int_4^5 x\sqrt{x^2 + 3y}\, dy\, dx$
17. $\int_1^2\int_4^9 \frac{3 + 5y}{\sqrt{x}}\, dx\, dy$
18. $\int_{16}^{25}\int_2^7 \frac{3 + 5y}{\sqrt{x}}\, dy\, dx$
19. $\int_1^2\int_1^2 \frac{dx\, dy}{xy}$
20. $\int_1^4\int_2^5 \frac{dy\, dx}{x}$
21. $\int_2^4\int_3^5 \left(\frac{x}{y} + \frac{y}{3}\right) dx\, dy$
22. $\int_3^4\int_1^2 \left(\frac{6x}{5} + \frac{y}{x}\right) dx\, dy$

Find each double integral over the rectangular region R with the given boundaries.

23. $\iint_R (x + 3y^2)\, dx\, dy;\quad 0 \le x \le 2, 1 \le y \le 5$
24. $\iint_R (4x^3 + y^2)\, dx\, dy;\quad 1 \le x \le 4, 0 \le y \le 2$
25. $\iint_R \sqrt{x + y}\, dy\, dx;\quad 1 \le x \le 3, 0 \le y \le 1$
26. $\iint_R x^2\sqrt{x^3 + 2y}\, dx\, dy;\quad 0 \le x \le 2, 0 \le y \le 3$
27. $\iint_R \frac{2}{(x + y)^2}\, dy\, dx;\quad 2 \le x \le 3, 1 \le y \le 5$
28. $\iint_R \frac{y}{\sqrt{6x + 5y^2}}\, dx\, dy;\quad 0 \le x \le 3, 1 \le y \le 2$
29. $\iint_R ye^{x+y^2}\, dx\, dy;\quad 2 \le x \le 3, 0 \le y \le 2$
30. $\iint_R x^2e^{x^3+2y}\, dx\, dy;\quad 1 \le x \le 2, 1 \le y \le 3$

Find the volume under the given surface $z = f(x, y)$ and above the rectangle with the given boundaries.

31. $z = 6x + 2y + 5;\quad -1 \le x \le 1, 0 \le y \le 3$
32. $z = 9x + 5y + 12;\quad 0 \le x \le 3, -2 \le y \le 1$
33. $z = x^2;\quad 0 \le x \le 1, 0 \le y \le 4$
34. $z = \sqrt{y};\quad 0 \le x \le 4, 0 \le y \le 9$
35. $z = x\sqrt{x^2 + y};\quad 0 \le x \le 1, 0 \le y \le 1$
36. $z = yx\sqrt{x^2 + y^2};\quad 0 \le x \le 4, 0 \le y \le 1$
37. $z = \frac{xy}{(x^2 + y^2)^2};\quad 1 \le x \le 2, 1 \le y \le 4$
38. $z = e^{x+y};\quad 0 \le x \le 1, 0 \le y \le 1$

Although it is often true that a double integral can be evaluated by using either dx or dy first, sometimes one choice over the other makes the work easier. Evaluate the double integrals in Exercises 39 and 40 in the easiest way possible.

39. $\iint_R xe^{xy}\, dx\, dy; \quad 0 \le x \le 2, 0 \le y \le 1$

40. $\iint_R 2x^3e^{x^2y}\, dx\, dy; \quad 0 \le x \le 1, 0 \le y \le 1$

Evaluate each double integral.

41. $\int_2^4 \int_2^{x^2} (x^2 + y^2)\, dy\, dx$

42. $\int_0^5 \int_0^{2y} (x^2 + y)\, dx\, dy$

43. $\int_0^4 \int_0^x \sqrt{xy}\, dy\, dx$

44. $\int_1^4 \int_0^x \sqrt{x + y}\, dy\, dx$

45. $\int_1^2 \int_y^{3y} \frac{1}{x}\, dx\, dy$

46. $\int_1^4 \int_x^{x^2} \frac{1}{y}\, dy\, dx$

47. $\int_0^4 \int_1^{e^x} \frac{x}{y}\, dy\, dx$

48. $\int_0^1 \int_{2x}^{4x} e^{x+y}\, dy\, dx$

Evaluate each double integral. If the function seems too difficult to integrate, try interchanging the limits of integration, as in Exercises 39 and 40.

49. $\int_0^{\ln 2} \int_{e^y}^2 \frac{1}{\ln x}\, dx\, dy$

50. $\int_0^2 \int_{y/2}^1 e^{x^2}\, dx\, dy$

Use the region R with the indicated boundaries to evaluate each double integral.

51. $\iint_R (4x + 7y)\, dy\, dx; \quad 1 \le x \le 3, 0 \le y \le x + 1$

52. $\iint_R (3x + 9y)\, dy\, dx; \quad 2 \le x \le 4, 2 \le y \le 3x$

53. $\iint_R (4 - 4x^2)\, dy\, dx; \quad 0 \le x \le 1, 0 \le y \le 2 - 2x$

54. $\iint_R \frac{dy\, dx}{x}; \quad 1 \le x \le 2, 0 \le y \le x - 1$

55. $\iint_R e^{x/y^2}\, dx\, dy; \quad 1 \le y \le 2, 0 \le x \le y^2$

56. $\iint_R (x^2 - y)\, dy\, dx; \quad -1 \le x \le 1, -x^2 \le y \le x^2$

57. $\iint_R x^3y\, dy\, dx;$ R bounded by $y = x^2, y = 2x$

58. $\iint_R x^2y^2\, dx\, dy;$ R bounded by $y = x, y = 2x, x = 1$

59. $\iint_R \frac{dy\, dx}{y};$ R bounded by $y = x, y = \frac{1}{x}, x = 2$

60. Recall from the Volume and Average Value section in the previous chapter that volume could be found with a single integral. In this section volume is found using a double integral. Explain when volume can be found with a single integral, and when a double integral is needed.

61. Give an example of a region that cannot be expressed by either of the forms shown in Figure 33. (One example is the disk with a hole in the middle between the graphs of $x^2 + y^2 = 1$ and $x^2 + y^2 = 2$ in Figure 10.)

The idea of the average value of a function, discussed earlier for functions of the form $y = f(x)$, can be extended to functions of more than one independent variable. For a function $z = f(x, y)$, the average value *of f over a region R is defined as*

$$\frac{1}{A}\iint_R f(x, y)\, dx\, dy,$$

where A is the area of the region R. Find the average value for each function over the regions R having the given boundaries.

62. $f(x, y) = 5xy + 2y; \quad 1 \le x \le 4, 1 \le y \le 2$

63. $f(x, y) = x^2 + y^2; \quad 0 \le x \le 2, 0 \le y \le 3$

64. $f(x, y) = e^{-5y+3x}; \quad 0 \le x \le 2, 0 \le y \le 2$

65. $f(x, y) = e^{2x+y}; \quad 1 \le x \le 2, 2 \le y \le 3$

Applications

BUSINESS AND ECONOMICS

66. *Packaging* The manufacturer of a fruit juice drink has decided to try innovative packaging in order to revitalize sagging sales. The fruit juice drink is to be packaged in containers in the shape of tetrahedra in which three edges are perpendicular, as shown in the figure. Two of the perpendicular edges will be 3 in. long, and the third edge will be 6 in. long. Find the volume of the container. (*Hint:* The equation of the plane shown in the figure is $z = f(x, y) = 6 - 2x - 2y$.)

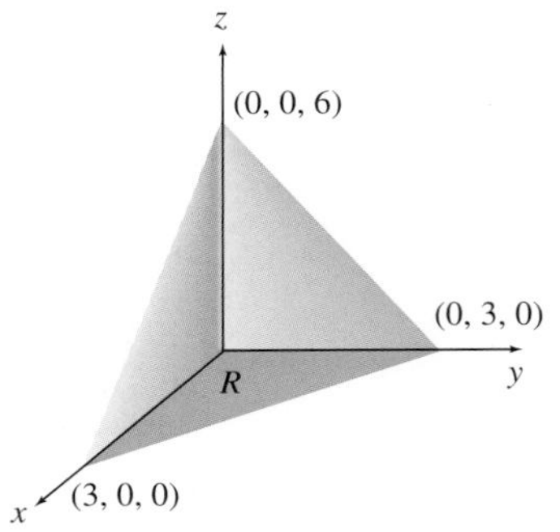

67. *Average Cost* A company's total cost for operating its two warehouses is

$$C(x, y) = \frac{1}{9}x^2 + 2x + y^2 + 5y + 100$$

dollars, where x represents the number of units stored at the first warehouse and y represents the number of units stored at the second. Find the average cost to store a unit if the first warehouse has between 48 and 75 units, and the second has between 20 and 60 units.

68. *Average Production* A production function is given by

$$P(x, y) = 500x^{.2}y^{.8},$$

where x is the number of units of labor and y is the number of units of capital. Find the average production level if x varies from 10 to 50 and y from 20 to 40.

69. *Average Profit* The profit (in dollars) from selling x units of one product and y units of a second product is

$$P = -(x - 100)^2 - (y - 50)^2 + 2000.$$

The weekly sales for the first product vary from 100 units to 150 units, and the weekly sales for the second product vary from 40 units to 80 units. Estimate average weekly profit for these two products.

70. *Average Revenue* A company sells two products. The demand functions of the products are given by

$$q_1 = 300 - 2p_1 \quad \text{and} \quad q_2 = 500 - 1.2p_2,$$

where q_1 units of the first product are demanded at price p_1 and q_2 units of the second product are demanded at price p_2. The total revenue will be given by

$$R = q_1p_1 + q_2p_2.$$

Find the average revenue if the price p_1 varies from \$25 to \$50 and the price p_2 varies from \$50 to \$75.

71. *Time* In an exercise earlier in this chapter, we saw that the time (in hours) that a branch of Amalgamated Entities needs to spend to meet the quota set by the main office can be approximated by

$$T(x, y) = x^4 + 16y^4 - 32xy + 40,$$

where x represents how many thousands of dollars the factory spends on quality control and y represents how many thousands of dollars they spend on consulting. Find the average time if the amount spent on quality control varies from \$0 to \$4000 and the amount spent on consulting varies from \$0 to \$2000.

72. *Profit* In an exercise earlier in this chapter, we saw that the profit (in thousands of dollars) that Aunt Mildred's Metalworks earns from producing x tons of steel and y tons of aluminum can be approximated by

$$P(x, y) = 36xy - x^3 - 8y^3.$$

Find the average profit if the amount of steel produced varies from 0 to 8 tons, and the amount of aluminum produced varies from 0 to 4 tons.

CHAPTER SUMMARY

In this chapter we extended our study of functions to include functions of several variables. The concept of differentiation was generalized to functions of several variables, and partial derivatives were used to determine points where a function of several variables was either maximized or minimized. The chapter introduced the method of Lagrange multipliers to solve constrained problems. The concept of differential, introduced in an earlier chapter for functions of one variable, was generalized to the concept of total differential. Total differentials were then used to approximate the value of a function by its tangent plane. Double integrals were introduced and used to find volume.

KEY TERMS

9.1 function of two variables
ordered triple
first octant
plane
surface
trace
level curves
paraboloid
production function
Cobb-Douglas production function
level surface
ellipsoid
hyperbolic paraboloid
hyperboloid of two sheets

9.2 partial derivative
second-order partial derivative
relative maximum
relative minimum

9.3 saddle point
critical point

9.4 Lagrange multiplier constraints

9.5 total differential

9.6 iterated integral
double integral
integrand
region of integration

CHAPTER 9 REVIEW EXERCISES

1. Describe in words how to take a partial derivative.

2. Describe what a partial derivative means geometrically.

3. Describe what a total differential is and how it is useful.

Find $f(-1, 2)$ and $f(6, -3)$ for the following.

4. $f(x, y) = -4x^2 + 6xy - 3$

5. $f(x, y) = 3x^2y^2 - 5x + 2y$

6. $f(x, y) = \dfrac{x - 3y}{x + 4y}$

7. $f(x, y) = \dfrac{\sqrt{x^2 + y^2}}{x - y}$

Graph the first-octant portion of each plane.

8. $x + y + z = 4$

9. $x + y + 4z = 8$

10. $5x + 2y = 10$

11. $3x + 5z = 15$

12. $x = 3$

13. $y = 2$

14. Let $z = f(x, y) = -5x^2 + 7xy - y^2$. Find the following.

a. $\dfrac{\partial z}{\partial x}$ **b.** $\left(\dfrac{\partial z}{\partial y}\right)(-1, 4)$ **c.** $f_{xy}(2, -1)$

15. Let $z = f(x, y) = \dfrac{x + y^2}{x - y^2}$. Find the following.

a. $\dfrac{\partial z}{\partial y}$ **b.** $\left(\dfrac{\partial z}{\partial x}\right)(0, 2)$ **c.** $f_{xx}(-1, 0)$

Find $f_x(x, y)$ and $f_y(x, y)$.

16. $f(x, y) = 9x^3y^2 - 5x$

17. $f(x, y) = 6x^5y - 8xy^9$

18. $f(x, y) = \sqrt{4x^2 + y^2}$

19. $f(x, y) = \dfrac{2x + 5y^2}{3x^2 + y^2}$

20. $f(x, y) = x^2e^{2y}$

21. $f(x, y) = (y - 2)^2e^{x+2y}$

22. $f(x, y) = \ln|2x^2 + y^2|$

23. $f(x, y) = \ln|2 - x^2y^3|$

Find $f_{xx}(x, y)$ and $f_{xy}(x, y)$.

24. $f(x, y) = 4x^3y^2 - 8xy$

25. $f(x, y) = -6xy^4 + x^2y$

26. $f(x, y) = \dfrac{2x}{x - 2y}$

27. $f(x, y) = \dfrac{3x + y}{x - 1}$

28. $f(x, y) = x^2e^y$

29. $f(x, y) = ye^{x^2}$

30. $f(x, y) = \ln|2 - x^2y|$

31. $f(x, y) = \ln|1 + 3xy^2|$

Find all points where the functions defined below have any relative extrema. Find any saddle points.

32. $z = x^2 + 2y^2 - 4y$

33. $z = x^2 + y^2 + 9x - 8y + 1$

34. $f(x, y) = x^2 + 5xy - 10x + 3y^2 - 12y$

35. $z = x^3 - 8y^2 + 6xy + 4$

36. $z = \frac{1}{2}x^2 + \frac{1}{2}y^2 + 2xy - 5x - 7y + 10$

37. $f(x, y) = 3x^2 + 2xy + 2y^2 - 3x + 2y - 9$

38. $z = x^3 + y^2 + 2xy - 4x - 3y - 2$

39. $f(x, y) = 7x^2 + y^2 - 3x + 6y - 5xy$

40. Describe the three different types of points that might occur when $f_x(x, y) = f_y(x, y) = 0$.

Use Lagrange multipliers to find the extrema of the functions defined in Exercises 41 and 42.

41. $f(x, y) = x^2y; \quad x + y = 4$

42. $f(x, y) = x^2 + y^2; \quad x = y + 2$

43. Find positive numbers x and y, whose sum is 80, such that x^2y is maximized.

44. Find positive numbers x and y, whose sum is 50, such that xy^2 is maximized.

45. Notice in the previous two exercises that we specified that x and y must be positive numbers. Does a maximum exist without this requirement? Explain why or why not.

Evaluate dz using the given information.

46. $z = 2x^2 - 4y^2 + 6xy; \quad x = 2, y = -3, dx = .01, dy = .05$

47. $z = \dfrac{x + 5y}{x - 2y}; \quad x = 1, y = -2, dx = -.04, dy = .02$

Use the total differential to approximate each quantity. Then use a calculator to approximate the quantity, and give the absolute value of the difference in the two results to 4 decimal places.

48. $\sqrt{5.1^2 + 12.05^2}$

49. $\sqrt{4.06}\, e^{.04}$

Evaluate the following.

50. $\displaystyle\int_4^9 \frac{6y - 8}{\sqrt{x}}\, dx$

51. $\displaystyle\int_3^5 e^{2x-7y}\, dx$

52. $\displaystyle\int_0^5 \frac{6x}{\sqrt{4x^2 + 2y^2}}\, dx$

53. $\displaystyle\int_1^3 \frac{y^2}{\sqrt{7x + 11y^3}}\, dy$

Evaluate each iterated integral.

54. $\displaystyle\int_0^2 \int_0^4 (x^2y^2 + 5x)\, dx\, dy$

55. $\displaystyle\int_0^2 \int_0^3 (x + 5y + y^2)\, dy\, dx$

56. $\displaystyle\int_3^4 \int_2^5 \sqrt{6x + 3y}\, dx\, dy$

57. $\displaystyle\int_1^2 \int_3^5 e^{2x-7y}\, dx\, dy$

58. $\displaystyle\int_2^4 \int_2^4 \frac{dx\, dy}{y}$

59. $\displaystyle\int_1^2 \int_1^2 \frac{dx\, dy}{x}$

Find each double integral over the region R with boundaries as indicated.

60. $\displaystyle\iint_R (x^2 + y^2)\, dx\, dy; \quad 0 \le x \le 2, 0 \le y \le 3$

61. $\displaystyle\iint_R \sqrt{2x + y}\, dx\, dy; \quad 1 \le x \le 3, 2 \le y \le 5$

62. $\displaystyle\iint_R \sqrt{y + x}\, dx\, dy; \quad 0 \le x \le 7, 1 \le y \le 9$

63. $\displaystyle\iint_R ye^{y^2 + x}\, dx\, dy; \quad 0 \le x \le 1, 0 \le y \le 1$

Find the volume under the given surface $z = f(x, y)$ and above the given rectangle.

64. $z = x + 9y + 8; \quad 1 \le x \le 6, 0 \le y \le 8$

65. $z = x^2 + y^2; \quad 3 \le x \le 5, 2 \le y \le 4$

Evaluate each double integral. If the function seems too difficult to integrate, try interchanging the limits of integration.

66. $\int_0^1 \int_0^{2x} xy\, dy\, dx$

67. $\int_0^1 \int_0^{x^3} y\, dy\, dx$

68. $\int_0^1 \int_{x^2}^{x} x^3 y\, dy\, dx$

69. $\int_0^1 \int_y^{\sqrt{y}} x\, dx\, dy$

70. $\int_0^2 \int_{x/2}^{1} \frac{1}{y^2+1}\, dy\, dx$

71. $\int_0^8 \int_{x/2}^{4} \sqrt{y^2+4}\, dy\, dx$

Use the region R, with boundaries as indicated, to evaluate the given double integral.

72. $\iint_R (2x+3y)\, dx\, dy;\quad 0 \le y \le 1, y \le x \le 2-y$

73. $\iint_R (2-x^2-y^2)\, dy\, dx;\quad 0 \le x \le 1, x^2 \le y \le x$

Applications

BUSINESS AND ECONOMICS

74. *Charge for Auto Painting* The charge (in dollars) for painting a sports car is given by

$$C(x, y) = 2x^2 + 4y^2 - 3xy + \sqrt{x},$$

where x is the number of hours of labor needed and y is the number of gallons of paint and sealant used. Find the following.

a. The charge for 10 hours and 5 gal of paint and sealant

b. The charge for 15 hours and 10 gal of paint and sealant

c. The charge for 20 hours and 20 gal of paint and sealant

75. *Manufacturing Costs* The manufacturing cost (in dollars) for a medium-sized business computer is given by

$$c(x, y) = 2x + y^2 + 4xy + 25,$$

where x is the memory capacity of the computer in gigabytes (Gb) and y is the number of hours of labor required. For 640 Gb and 6 hours of labor, find the following.

a. The approximate change in cost for an additional 1 Gb of memory

b. The approximate change in cost for an additional hour of labor

76. *Productivity* The production function z for one country is

$$z = x^{.6}y^{.4},$$

where x represents the amount of labor and y the amount of capital. Find the marginal productivity of the following.

a. Labor **b.** Capital

77. *Cost* The cost (in dollars) to manufacture x solar cells and y solar collectors is

$$c(x, y) = x^2 + 5y^2 + 4xy - 70x - 164y + 1800.$$

a. Find values of x and y that produce minimum total cost.

b. Find the minimum total cost.

78. *Cost* The cost (in dollars) to produce x satellite receiving dishes and y transmitters is given by

$$C(x, y) = \ln(x^2 + y) + e^{xy/20}.$$

Production schedules now call for 15 receiving dishes and 9 transmitters. Use differentials to approximate the change in costs if 1 more dish and 1 fewer transmitter are made.

79. *Production Materials* Approximate the amount of material needed to manufacture a cone of radius 2 cm, height 8 cm, and wall thickness .21 cm.

80. *Production Materials* A sphere of radius 2 ft is to receive an insulating coating 1 in. thick. Approximate the volume of the coating needed.

81. *Production Error* The height of a sample cone from a production line is measured as 11.4 cm, while the radius is measured as 2.9 cm. Each of these measurements could be off by .2 cm. Approximate the maximum possible error in the volume of the cone.

82. *Profit* The total profit from 1 acre of a certain crop depends on the amount spent on fertilizer, x, and on hybrid seed, y, according to the model

$$P(x, y) = .01(-x^2 + 3xy + 160x - 5y^2 + 200y + 2600).$$

The budget for fertilizer and seed is limited to \$280.

a. Use the budget constraint to express one variable in terms of the other. Then substitute into the profit function to get a function with one independent variable. Use the method shown in the chapter on Applications of the Derivative to find the amounts spent on fertilizer and seed that will maximize profit. What is the maximum profit per acre? (*Hint:* Throughout this exercise you may ignore the coefficient of .01 until you need to find the maximum profit.)

b. Find the amounts spent on fertilizer and seed that will maximize profit using the first method shown in this chapter. (*Hint:* You will not need to use the budget constraint.)

c. Use the Lagrange multiplier method to solve the original problem.

d. Look for the relationships among these methods.

LIFE SCIENCES

83. *Blood Vessel Volume* A length of blood vessel is measured as 2.7 cm, with the radius measured as .7 cm. If each of these measurements could be off by .1 cm, estimate the maximum possible error in the volume of the vessel.

84. *Total Body Water* Accurate prediction of total body water is critical in determining adequate dialysis doses for patients with renal disease. For African American males, total body water can be estimated by the function

$$T(A, M, S) = -18.37 - .09A + .34M + .25S,$$

where T is the total body water (in liters), A is age (in years), M is mass (in kilograms), and S is height (in centimeters).*

a. Find $T(65, 85, 180)$.

b. Find and interpret $T_A(A, M, S)$, $T_M(A, M, S)$, and $T_S(A, M, S)$.

85. *Brown Trout* Researchers from New Zealand have determined that the length of a brown trout depends on both its mass and age and that the length can be estimated by

$$L(m, t) = (.00082t + .0955)e^{(\ln m + 10.49)/2.842},$$

where $L(m, t)$ is the length of the trout (in centimeters), m is the mass of the trout (in grams), and t is the age of the trout (in years).†

a. Find $L(450, 4)$.

b. Find $L_m(450, 7)$ and $L_t(450, 7)$ and interpret.

86. *Survival Curves* The following figure shows survival curves (percent surviving as a function of age) for females in the United States in 1900 and 1995.‡ Let $f(x, y)$ give the proportion surviving at age x in year y. Use the graph to estimate the following. Interpret each answer in words.

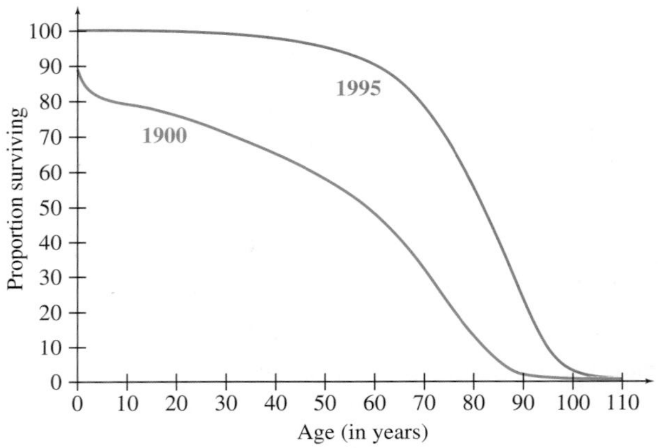

a. $f(60, 1900)$

b. $f(70, 1995)$

c. $f_x(60, 1900)$

d. $f_x(70, 1995)$

GENERAL INTEREST

87. *Area* The bottom of a planter is to be made in the shape of an isosceles triangle, with the two equal sides 3 ft long and the third side 2 ft long. The area of an isosceles trian-

*Chumlea, W., S. Guo, C. Zellar et al., "Total Body Water Reference Values and Prediction Equations for Adults," *Kidney International*, Vol. 59, 2001, pp. 2250–2258.

†Hayes, J., J. Stark, and K. Shearer, "Development and Test of a Whole-Lifetime Foraging and Bioenergetics Growth Model for Drift-Feeding Brown Trout," *Transactions of the American Fisheries Society*, Vol. 129, 2000, pp. 315–332.

‡Data from *Science*, Vol. 291, Feb. 23, 2001, p. 1491.

gle with two equal sides of length a and third side of length b is

$$f(a, b) = \frac{1}{4}b\sqrt{4a^2 - b^2}.$$

a. Find the area of the bottom of the planter.

b. The manufacturer is considering changing the shape so that the third side is 2.5 ft long. What would be the approximate effect on the area?

88. *Surface Area* A closed box with square ends must have a volume of 125 in^3. Find the dimensions of such a box that has minimum surface area.

89. *Area* Find the maximum rectangular area that can be enclosed with 400 ft of fencing, if no fencing is needed along one side.

EXTENDED APPLICATION: Using Multivariable Fitting to Create a Response Surface Design

Suppose you are designing a flavored drink with orange and banana flavors. You want to find the ideal concentrations of orange and banana flavoring agents, but since the concentrations could range from 0% to 100%, you can't try every possibility. A common design technique in the food industry is to make up several test drinks using different combinations of flavorings and have them rated for taste appeal by a panel of tasters. Such ratings are called *hedonic responses* and are often recorded on a 10-point scale from 0 (worst) to 9 (best). One combination will most likely get the highest average score, but since you have only tried a few of the infinite number of flavor combinations, the winning combination on the taste test might be far from the mix that would be the most popular in the market. How can you use the information from your test to locate the best point on the *flavor plane*?

One approach to this problem uses *response surfaces*, three-dimensional surfaces that approximate the data points from your flavor test.* For your test, you might choose mixtures that are spread out over the flavor plane. For example, you could combine low, medium, and high orange with low, medium, and high banana to get 9 different flavors. If you had 15 tasters and used intensities of 20, 50, and 80 for each fruit, the test data might look like the table.

Average Hedonic Scores ($n = 15$)
Banana Intensity (0 to 100)

		20	*50*	*80*
Orange	20	3.2	4.9	2.8
Intensity	50	6.0	7.2	5.1
0 to 100	80	4.5	5.5	4.8

For example, the table shows that the drink with orange intensity 20 and banana intensity 80 got an average flavor rating of 2.8 from the test panel (they didn't like it).

Your test results are points in space, where you can think of the x-axis as the orange axis, the y-axis as banana, and the z-axis as taste score. A three-dimensional bar chart is a common way of displaying data of this kind. Figure 37 is a bar chart of the flavor test results.

Looking at the bar chart, we can guess that the best flavor mix will be somewhere near the middle. We'd like to "drape" a smooth surface over the bars and see where that surface has a maximum. But as with any sample, our tasters are not perfectly representative of the whole population: Our test results give the general shape of the true population response, but each bar includes an error that results from our small sample size. The solution is to fit a *smooth* surface to the data points.

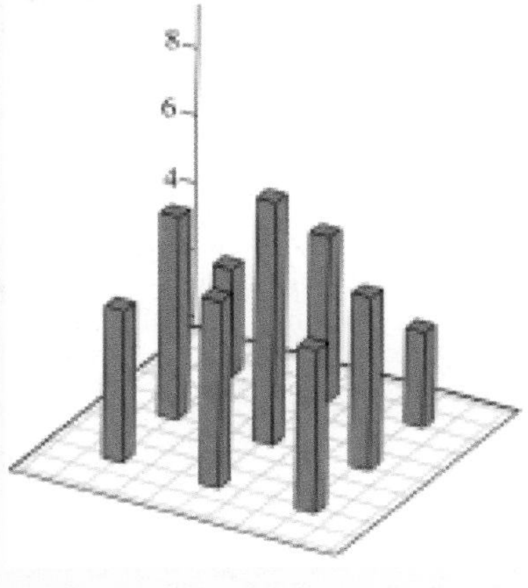

FIGURE 37

In Chapter 1 we used linear functions to model data sets with one independent variable. Here we have two independent variables, and it makes sense to use a *quadratic* function. You've seen many quadratic functions of two variables in the examples and exercises for this chapter, and you know that they can have maxima, minima, and saddle points. We don't know in advance which quadratic shape will give us the best fit, so we'll use the most general quadratic,

$$G(x, y) = Ax^2 + By^2 + Cxy + Dx + Ey + F.$$

Our job is to find the six coefficients, A through F, that give the best fit to our nine data points. As with the least squares line formula you used in Section 1.3, there are formulas for these six coefficients. Most statistical software packages will generate them directly from your data set, and here is the best-fitting quadratic found by one such program:

$$G(x, y) = -.00202x^2 - .00163y^2 + .000194xy + .21380x + .14768y - 2.36204.$$

In this case the response surface shows how the dependent variable, taste rating, *responds* to the two independent variables, orange and banana intensity. Figures 38 and 39 are two views of the surface together with the data: a surface superimposed on the bar chart, and the same surface with the data shown as points in space.

*For a brief introduction to response surfaces, see Devore, Jay L. and Nicholas R. Farnum, *Applied Statistics for Engineers and Scientists*, Duxbury, 1999.

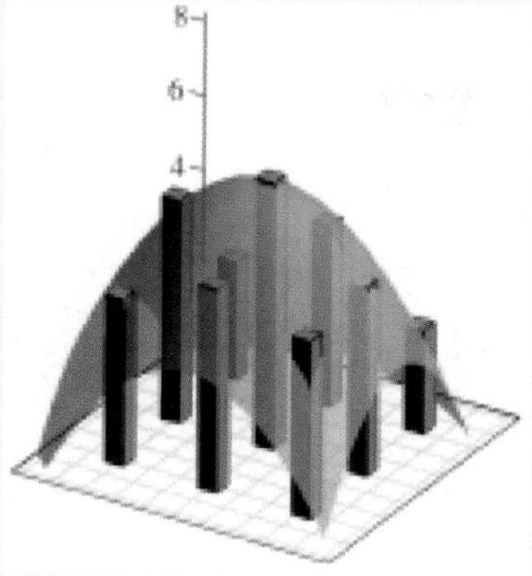

FIGURE 38

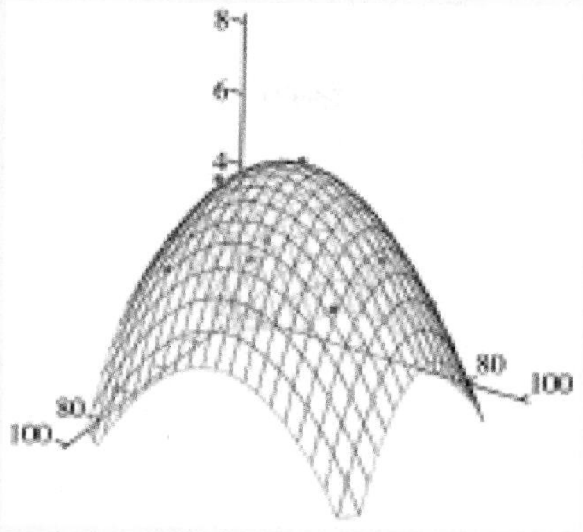

FIGURE 39

In research papers, response surface models are often reported using level curves. A contour map for the surface we have found looks like Figure 40, with orange increasing from left to right and banana from bottom to top.

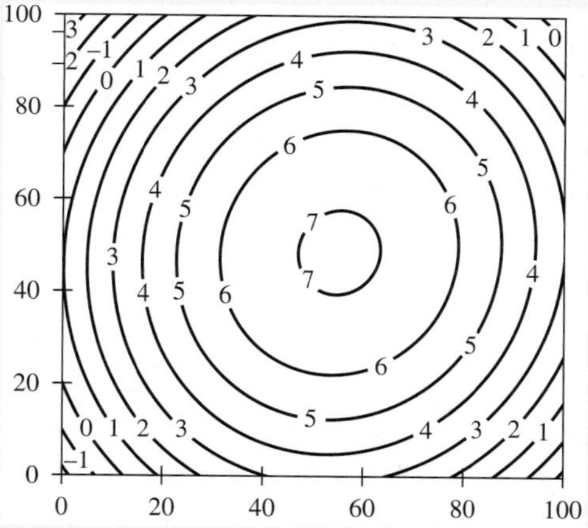

FIGURE 40

It's quite easy to estimate the location of the maximum by marking a point in the middle of the central ellipse and finding its coordinates on the two axes (try it!). You can also use the techniques you learned in the section on Maxima and Minima, computing partial derivatives of G with respect to x and y and solving the resulting linear system. The numbers are awkward, but with some help from a calculator you'll find that the maximum occurs at approximately $(55.3, 48.6)$. So the quadratic model predicts that the most popular drink will have an orange concentration of 55.3 and a banana concentration of 48.6. The model also predicts the public's flavor rating for this drink: we would expect it to be $G(55.3, 48.6)$, which turns out to be about 7.14. When food technologists design a new food, this kind of modeling is often a first step. The next step might be to make a new set of test drinks with concentrations clustered around the point $(55.3, 48.6)$, and use further tests to explore this region of the "flavor plane" in greater detail.

Response surfaces are also helpful for constrained optimization. In the section on Lagrange Multipliers, you saw how Lagrange multipliers could solve problems of the form:

$$\text{Find the relative extrema for } z = f(x, y),$$
$$\text{subject to the constraint } g(x, y) = 0.$$

Sometimes the constraints have a different form: You may have *several* dependent variables that respond to the same inputs, and the design goal is to keep each variable *within a given range*. Here's an example based on the data in U.S. Patent No. 4,276,316, which is titled *Process for Treating Nuts*.* The patent granted to researcher Shri C. Sharma and assigned to CPC International Inc. covers a method for preparing nuts for blanching (that is, having their skins removed). The patent summary reads in part:

> The nuts are heated with a gas at a temperature of 125° to 175°C for 30 to 180 seconds and then immediately cooled to below 35°C within 5 minutes prior to blanching. This provides improved blanching, sorting and other steps in a process for producing products ranging from nuts per se to peanut butters or spreads.

In support of the effectiveness of the method, the patent offers data that describe the effects of nine different combinations of air temperature and treatment time on three variables of interest for blanched peanuts: blanching efficiency, roasted peanut flavor, and overall flavor. Efficiency is given in percent, and the two hedonic variables were rated by tasters on a scale of 0 to 9.

*Patents are online at www.uspto.gov/web/menu/pats.html. You can locate patents by number or carry out a text search of the full patent database.

Air Temperature, °C	Treatment Time, Seconds	Log of Treatment Time	Blanching Efficiency	Roasted Peanut Flavor	Overall Flavor
138	45	3.807	93.18	4.94	5.51
160	120	4.787	94.99	5.24	5.37
149	75	4.317	98.43	5.27	5.10
138	120	4.787	96.42	5.05	5.71
160	45	3.807	96.48	5.17	5.62
127	75	4.317	93.56	4.64	5.04
149	180	5.193	94.99	5.24	5.37
149	30	3.401	87.30	5.43	5.44
171	45	3.807	94.40	4.37	5.18

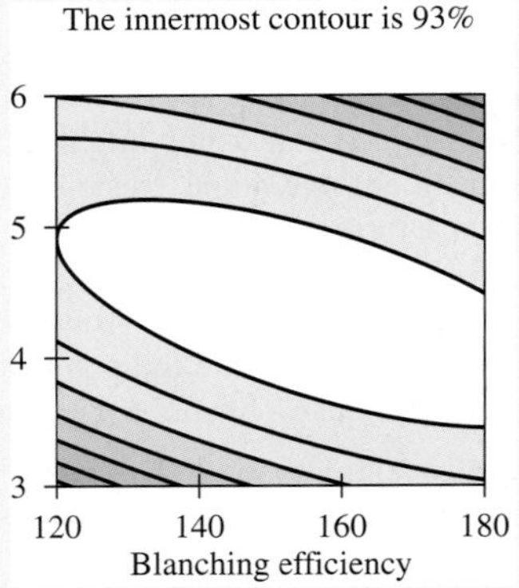

FIGURE 41

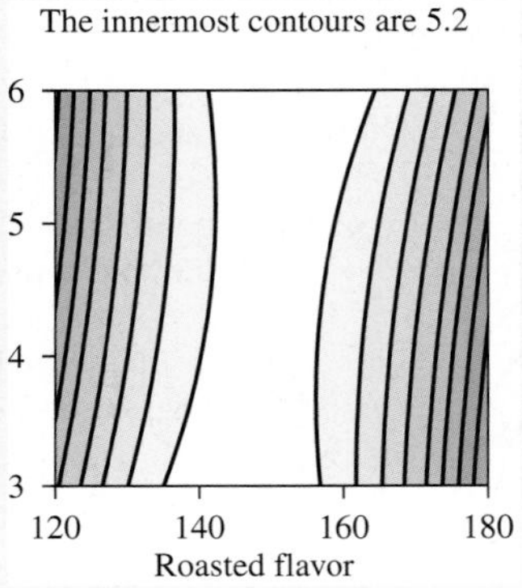

FIGURE 42

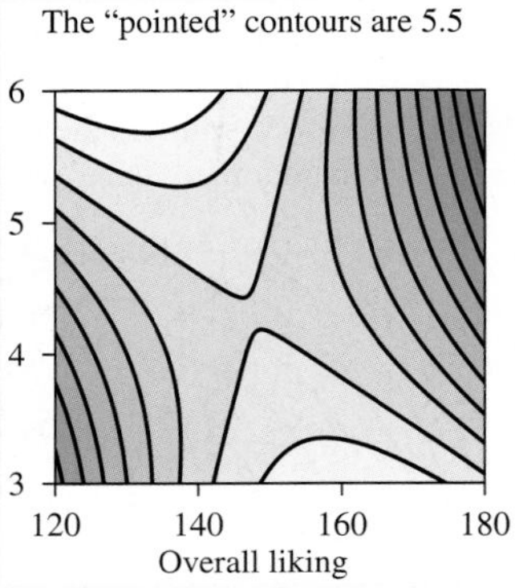

FIGURE 43

The time variable has been converted into a natural logarithm because treatment time effects typically scale with the log of the time. The problem is now to pick a temperature and time range that give the optimum combination of efficiency, roasted flavor, and overall flavor. Each of these three dependent variables responds to the inputs in a different way, and the patent documentation includes quadratic response surfaces for each variable. The lighter shading in Figures 41–43 indicates higher values, which are more desirable. Temperature is plotted across the bottom, and the log of treatment time increases from bottom to top.

Sometimes process designers faced with this kind of problem will combine the dependent variables into a single function by taking a weighted average of their values, and then use a single response surface to optimize this function. Here we look at a different scenario. Suppose we set the following process goals: We want blanching efficiency of at least 93%, a roasted flavor rating of at least 5.2, and an overall flavor rating of at least 5.5. Is there a combination of time and temperature that meets these criteria? If so, what is it?

The first step is to identify the "successful" area on each response surface, which we can do by shading the corresponding region in the contour plot, shown in Figures 44–46.

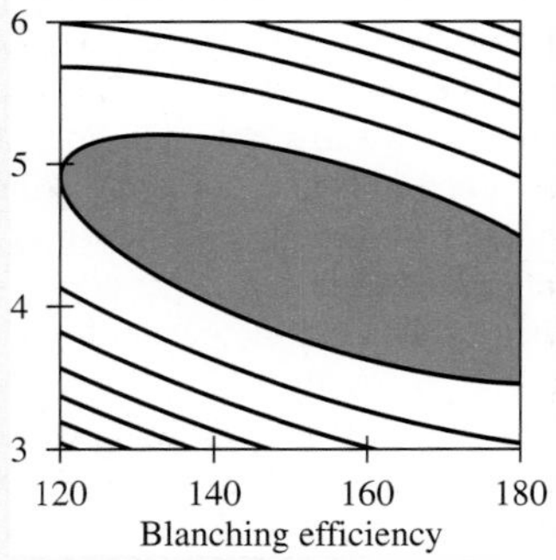

FIGURE 44

Now the strategy is clear: We want to stack the three plots on top of each other and see if the shaded regions overlap. Figure 47 is the result.

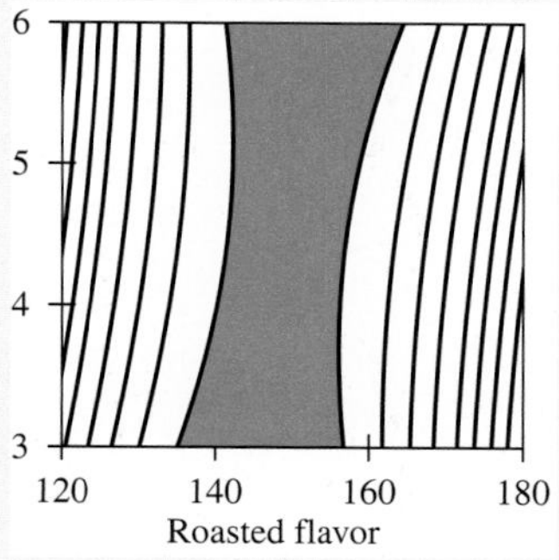

FIGURE 45

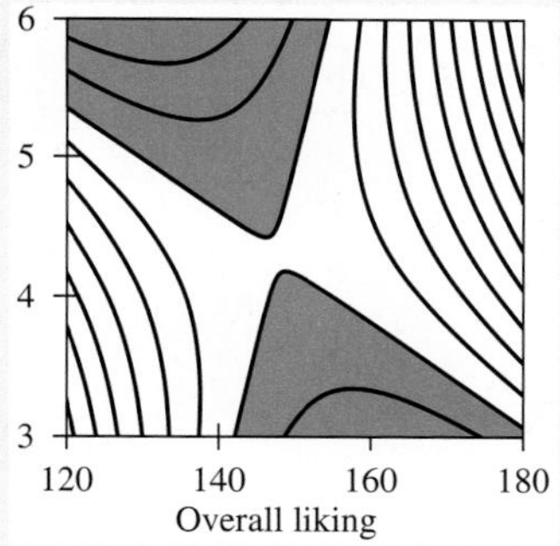

FIGURE 46

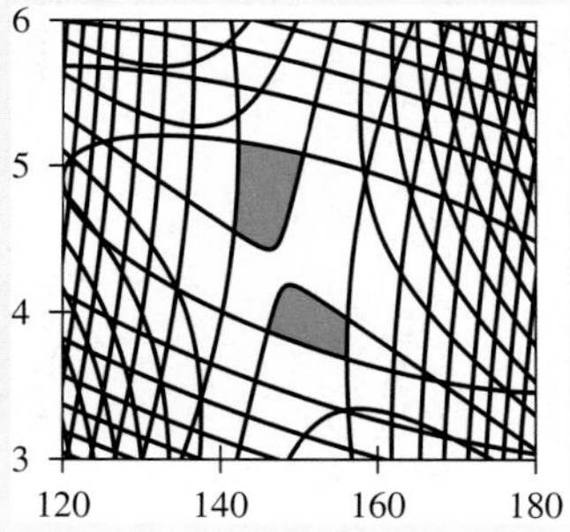

FIGURE 47

So we can see that there are two regions on the temperature–time plane that will work. For example, the upper area of overlap suggests a processing temperature of 140°C to 150°C, with a processing time between 90 and 150 seconds (remember that the numbers on the vertical axis are *natural logarithms* of the time in seconds).

Response surfaces are a standard tool in designing everything from food to machine parts, and we have touched on only a small part of the theory here. Frequently a process depends on more than two independent variables. For example, a soft-drink formula might include three flavorings, an acidifying agent, and a sweetener. The response "surface" now lives in six dimensions and we can no longer draw nice pictures, but the same multivariable mathematics that generated our quadratic response surfaces will lead us to the optimal combination of variables.

Exercises

1. The general quadratic function of two variables has six terms. How many terms are in the general cubic function of two variables?
2. Use the contour plot of orange-banana flavor to estimate the "flavor coordinates" of the best-tasting drink.
3. Find the maximum on the flavor response surface by finding the critical point of the function $G(x, y)$.
4. Without shading or numbers on the contours, how would you know that the point you found in Exercises 2 and 3 represents the best flavor rather than the worst flavor? (*Hint:* Compute the number D as described in the section on Maxima and Minima.)
5. Our best drink has a predicted flavor rating of 7.14, but one of our test drinks got a *higher* rating, 7.2. What's going on?
6. Blanching efficiency has a maximum near the center of the temperature–time plane. What is going on near the center of the plane for the roasted flavor and overall flavor response surfaces? Within the domain plotted, where does overall flavor reach a maximum?
7. In the overall flavor contour plot, if we move one contour toward higher flavor from the "pointed" 5.5 contours, we find curved contours that represent an overall flavor rating of about 5.6. If instead of requiring an overall flavor rating of 5.5 we decided to require a rating of 5.6, what would happen to our process design?
8. Use the last figure to describe the other region in the temperature–time plane that delivers a successful process for preparing nuts for blanching.

Directions for Group Project

Perform an experiment that is similar to the flavored drink example from the text on some other product. For example, you could perform an experiment where you develop hedonic responses for various levels of salt and butter on popcorn. Using technology, to the extent that it is available to you, carry out the analysis of your experiment to determine an optimal mixture of each ingredient.

CHAPTER

10 Differential Equations

When these sky divers open their parachutes, their speed will decrease until air resistance exactly balances the force of gravity. An exercise at the end of this chapter explores solutions to the differential equation that describes free fall with air resistance. The limiting speed with an open parachute is on the order of 10 miles per hour, slow enough for a safe landing.

Suppose that an economist wants to develop an equation that will forecast interest rates. By studying data on previous changes in interest rates, she hopes to find a relationship between the level of interest rates and their rate of change. A function giving the rate of change of interest rates would be the derivative of the function describing the level of interest rates. A **differential equation** is an equation that involves an unknown function $y = f(x)$ and a finite number of its derivatives. Solving the differential equation for y would give the unknown function to be used for forecasting interest rates.

Differential equations have been important in the study of physical science and engineering since the eighteenth century. Among the pioneers in the field of differential equations was the French mathematician Alexis Claude Clairaut (1713–1765). A particular type of equation studied in elementary courses on differential equations bears his name.

More recently, differential equations have become useful in social sciences, life sciences, and economics for solving problems about population growth, ecological balance, and interest rates. In this chapter, we will introduce some methods for solving differential equations and give examples of their applications.

10.1 SOLUTIONS OF ELEMENTARY AND SEPARABLE DIFFERENTIAL EQUATIONS

THINK ABOUT IT *How can we predict the future population of a flock of mountain goats?*

Using differential equations, we will learn to answer such questions.

Usually a solution of an equation is a *number*. A solution of a differential equation, however, is a *function*. For example, the solutions of a differential equation such as

$$\frac{dy}{dx} = 3x^2 - 2x \tag{1}$$

consist of all expressions for y that satisfy the equation. Since the left side of the equation is the derivative of y with respect to x, we can solve the equation for y by finding an antiderivative on each side. On the left, the antiderivative is $y + C_1$. On the right side,

$$\int (3x^2 - 2x)\, dx = x^3 - x^2 + C_2.$$

The solutions of Equation (1) are given by

$$y + C_1 = x^3 - x^2 + C_2$$

or

$$y = x^3 - x^2 + C_2 - C_1.$$

Replacing the constant $C_2 - C_1$ with the single constant C gives

$$y = x^3 - x^2 + C. \tag{2}$$

(From now on we will add just one constant, with the understanding that it represents the difference between the two constants obtained in the two integrations.)

Each different value of C in Equation (2) leads to a different solution of Equation (1), showing that a differential equation can have an infinite number of solutions. Equation (2) is the **general solution** of the differential equation (1). Some of the solutions of Equation (1) are graphed in Figure 1.

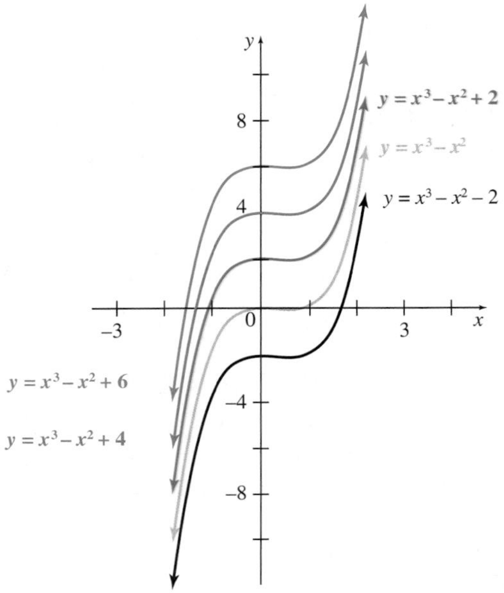

FIGURE 1

The simplest kind of differential equation has the form

$$\frac{dy}{dx} = f(x).$$

Since Equation (1) has this form, the solution of Equation (1) suggests the following generalization.

GENERAL SOLUTION OF $\frac{dy}{dx} = f(x)$

The general solution of the differential equation $dy/dx = f(x)$ is

$$y = \int f(x)\,dx.$$

EXAMPLE 1 *Population*

The population P of a flock of birds is growing exponentially so that

$$\frac{dP}{dx} = 20e^{.05x},$$

where x is time in years. Find P in terms of x if there were 20 birds in the flock initially.

Solution Solve the differential equation:

$$P = \int 20e^{.05x}\,dx = \frac{20}{.05}e^{.05x} + C = 400e^{.05x} + C.$$

Since P is 20 when x is 0,

$$20 = 400e^{0} + C$$
$$-380 = C,$$

and

$$P = 400e^{.05x} - 380.$$

In Example 1, the given information was used to produce a solution with a specific value of C. Such a solution is called a **particular solution** of the given differential equation. The given information, $P = 20$ when $x = 0$, is called an **initial condition.** An **initial value problem** is a differential equation with a value of y given at $x = x_0$, where x_0 is any real number.

Sometimes a differential equation must be rewritten in the form

$$\frac{dy}{dx} = f(x)$$

before it can be solved.

EXAMPLE 2 *Initial Value Problem*

Find the particular solution of

$$\frac{dy}{dx} - 2x = 5,$$

given that $y = 2$ when $x = -1$.

Solution To get this equation into the proper form, add $2x$ to both sides, yielding

$$\frac{dy}{dx} = 2x + 5.$$

The general solution is

$$y = \frac{2x^2}{2} + 5x + C = x^2 + 5x + C.$$

Substituting 2 for y and -1 for x gives

$$2 = (-1)^2 + 5(-1) + C$$
$$C = 6.$$

The particular solution is $y = x^2 + 5x + 6$.

So far in this section, we have used a method that is essentially the same as that used in the section on antiderivatives, when we first started the topic of integration. But not all differential equations can be solved so easily. For example, if interest on an investment is compounded continuously, then the investment grows at a rate proportional to the amount of money present. If A is the amount in an account at time t, then for some constant k, the differential equation

$$\frac{dA}{dt} = kA$$

gives the rate of growth of A with respect to t. This differential equation is different from those discussed previously, which had the form

$$\frac{dy}{dx} = f(x).$$

CAUTION Since the right-hand side of the differential equation for compound interest is a function of A, rather than a function of t, it would be completely invalid to simply integrate both sides as we did before. The previous method only works when the side opposite the derivative is simply a function of the independent variable. ■

The differential equation for compound interest is an example of a more general differential equation we will now learn to solve; namely, those that can be written in the form

$$\frac{dy}{dx} = \frac{f(x)}{g(y)}.$$

Suppose we think of dy/dx as a fraction dy over dx. This is incorrect, of course; the derivative is actually the limit of a small change in y over a small change in x, but the notation is chosen so that this interpretation gives a correct answer, as we shall see. Multiply on both sides by $g(y)\,dx$ to get

$$g(y)\,dy = f(x)\,dx.$$

In this form all terms involving y (including dy) are on one side of the equation and all terms involving x (and dx) are on the other side. A differential equation that can be put into this form is said to be *separable*, since the variables x and y can be separated. After separation, a **separable differential equation** may be solved by integrating each side. This method is known as **separation of variables.**

$$\int g(y)\,dy = \int f(x)\,dx$$

$$G(y) = F(x) + C,$$

where F and G are antiderivatives of f and g. To show that this answer is correct, differentiate implicitly with respect to x.

$$G'(y)\frac{dy}{dx} = F'(x) \qquad \text{Use the chain rule on the left side.}$$

$$g(y)\frac{dy}{dx} = f(x)$$

$$\frac{dy}{dx} = \frac{f(x)}{g(y)}$$

This last equation is the one we set out to solve.

EXAMPLE 3 *Separation of Variables*

Find the general solution of

$$y\frac{dy}{dx} = x^2.$$

Solution Begin by separating the variables to get

$$y\,dy = x^2\,dx.$$

The general solution is found by taking antiderivatives on each side.

$$\int y\,dy = \int x^2\,dx$$

$$\frac{y^2}{2} = \frac{x^3}{3} + C$$

$$y^2 = \frac{2}{3}x^3 + 2C = \frac{2}{3}x^3 + K$$

The constant K was substituted for $2C$ in the last step. The solution is left in implicit form, not solved explicitly for y.

EXAMPLE 4 *Separation of Variables*

Find the general solution of $dy/dx = ky$, where k is a constant.

Solution Separating variables leads to

$$\frac{1}{y}\,dy = k\,dx.$$

To solve this equation, take antiderivatives on each side.

$$\int \frac{1}{y}\,dy = \int k\,dx$$

$$\ln|y| = kx + C$$

Use the definition of logarithm to write the equation in exponential form as

$$|y| = e^{kx+C}.$$

By properties of exponents,

$$|y| = e^{kx}e^{C}.$$

Finally, use the definition of absolute value to get

$$y = e^{kx}e^{C} \qquad \text{or} \qquad y = -e^{kx}e^{C}.$$

Since e^C and $-e^C$ are constants, replace them with the constant M, which may have any nonzero real-number value, to get the single equation

$$y = Me^{kx}.$$

This equation, $y = Me^{kx}$, defines the exponential growth or decay function that was discussed previously (in the chapter on Nonlinear Functions).

CAUTION Notice that $y = 0$ is also a solution to the differential equation in Example 4, but after we divide by y (which is not possible if $y = 0$) and integrate, the resulting equation $|y| = e^{kx+C}$ does not allow y to equal 0. In this example, the lost solution can be recovered in the final answer if we allow M to equal 0, a value that was previously excluded. When dividing by an expression in separation of variables, look for solutions that would make this expression 0 and may be lost. ■

FOR REVIEW

In the chapter on Nonlinear Functions, we saw that the amount of money in an account with interest compounded continuously is given by

$$A = Pe^{rt},$$

where P is the initial amount in the account, r is the annual interest rate, and t is the time in years. Observe that this is the same as the equation for the amount of money in an account derived here, where P and r have been replaced with A_0 and k, respectively.

Recall that equations of the form $y = Me^{kx}$ arise in situations where the rate of change of a quantity is proportional to the amount present at time x; that is, where

$$\frac{dy}{dx} = ky.$$

The constant k is called the **growth rate constant,** while M represents the initial condition, the amount present at time $x = 0$. (A positive value of k indicates growth, while a negative value of k indicates decay.)

Applying the results of Example 4 to the equation discussed earlier,

$$\frac{dA}{dt} = kA,$$

shows that the amount in the account at time t is

$$A = A_0e^{kt},$$

where A_0 is the amount originally invested.

As a model of population growth, the equation $y = Me^{kx}$ is not realistic over the long run for most populations. As shown by graphs of functions of the form $y = Me^{kx}$, with both M and k positive, growth would be unbounded. Additional factors, such as space restrictions or a limited amount of food, tend to inhibit growth of populations as time goes on. In an alternative model that assumes a maximum population of size N, the rate of growth of a population is proportional to how close the population is to that maximum, that is, to the difference between N and x. These assumptions lead to the differential equation

$$\frac{dy}{dx} = k(N - y),$$

the limited growth function mentioned in an earlier chapter. Graphs of limited growth functions look like the graph in Figure 2, where y_0 is the initial population.

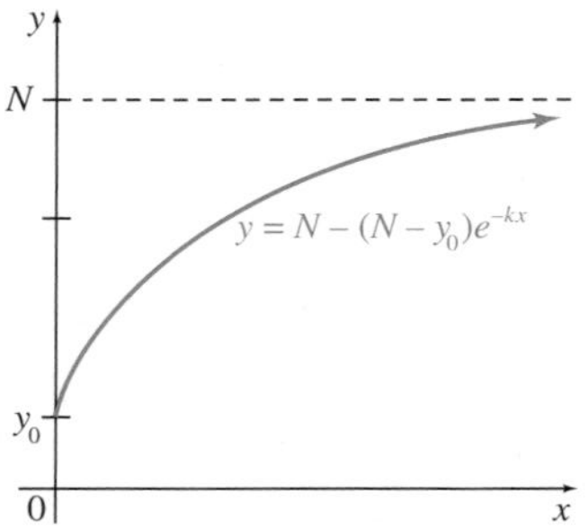

FIGURE 2

EXAMPLE 5 *Population*

A certain nature reserve can support no more than 4000 mountain goats. Assume that the rate of growth is proportional to how close the population is to this maximum, with a growth rate of 20 percent. There are currently 1000 goats in the area.

(a) Write a differential equation for the rate of growth of this population and solve it to obtain a function y describing the population at time x.

Solution Let $N = 4000$ and $k = .20$. The rate of growth of the population is given by

$$\frac{dy}{dx} = .20(4000 - y).$$

To solve for y, first separate the variables.

$$\frac{dy}{4000 - y} = .2\,dx$$

$$\int \frac{dy}{4000 - y} = \int .2\,dx$$

$$-\ln(4000 - y) = .2x + C$$

$$\ln(4000 - y) = -.2x - C$$

$$4000 - y = e^{-.2x - C} = (e^{-.2x})(e^{-C})$$

The absolute value bars are not needed for $\ln(4000 - y)$ since y must be less than 4000 for this population, so that $4000 - y$ is always nonnegative. Let $e^{-C} = B$. Then

$$4000 - y = Be^{-.2x}$$
$$y = 4000 - Be^{-.2x}.$$

Find B by using the fact that $y = 1000$ when $x = 0$.

$$1000 = 4000 - B$$
$$B = 3000$$

Notice that the value of B is the difference between the maximum population and the initial population. Substituting 3000 for B in the equation for y gives

$$y = 4000 - 3000e^{-.2x}.$$

(b) What will the goat population be in 5 years?

Solution In 5 years, the population will be

$$y = 4000 - 3000e^{-(.2)(5)} = 4000 - 3000e^{-1}$$
$$= 4000 - 1103.6 = 2896.4,$$

or about 2900 goats.

Logistic Growth Let y be the size of a certain population at time x. In the standard model for unlimited growth,

$$\frac{dy}{dx} = ky, \tag{3}$$

the rate of growth is proportional to the current population size. The constant k, the growth rate constant, is the difference between the birth and death rates of the population. The unlimited growth model predicts that the population's growth rate is a constant, k.

Growth usually is not unlimited, however, and the population's growth rate is usually not constant because the population is limited by environmental factors to a maximum size N, called the **carrying capacity** of the environment for the species. In the limited growth model already given,

$$\frac{dy}{dx} = k(N - y),$$

the rate of growth is proportional to the remaining room for growth, $N - y$. In the **logistic growth model**

$$\frac{dy}{dx} = k\left(1 - \frac{y}{N}\right)y \tag{4}$$

the rate of growth is proportional to both the current population size y and a factor $(1 - y/N)$ that is equal to the remaining room for growth, $N - y$, divided by N. Equation (4) is called the **logistic equation.** As $y \to 0$, $(1 - y/N) \to 1$, and the differential equation can be approximated as

$$\frac{dy}{dx} = k\left(1 - \frac{y}{N}\right)y \approx k(1)y = ky.$$

In other words, when y is small, the growth of the population behaves as if it were unlimited. On the other hand, as $y \to N$, $(1 - y/N) \to 0$, so

$$\frac{dy}{dx} = k\left(1 - \frac{y}{N}\right)y \approx k(0)y = 0.$$

That is, population growth levels off as y nears the maximum population N. Thus, the logistic Equation (4) is the unlimited growth Equation (3) with a damping factor $(1 - y/N)$ to account for limiting environmental factors when y nears N. Let y_0 denote the initial population size. Under the assumption $0 < y < N$, the general solution of Equation (4) is

$$y = \frac{N}{1 + be^{-kx}}, \tag{5}$$

where $b = (N - y_0)/y_0$ (see Exercise 31). This solution, called a **logistic curve,** is shown in Figure 3.

As expected, the logistic curve begins exponentially and subsequently levels off. Another important feature is the point of inflection $((\ln b)/k, N/2)$, where dy/dx is a maximum (see Exercise 33).

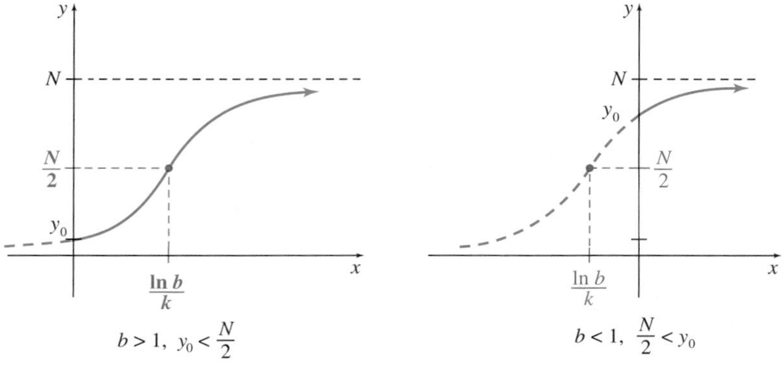

FIGURE 3

Logistic equations arise frequently in the study of populations. In about 1840 the Belgian sociologist P. F. Verhulst fitted a logistic curve to U.S. census figures and made predictions about the population that were subsequently proven to be quite accurate. American biologist Raymond Pearl (circa 1920) found that the growth of a population of fruit flies in a limited space could be modeled by the logistic equation

$$\frac{dy}{dx} = .2y - \frac{.2}{1035}y^2.$$

Some calculators can fit a logistic curve to a set of data points. For example, the TI-83/84 Plus has this capability, listed as `Logistic` in the STAT CALC menu, along with other types of regression. See Exercises 46 and 50.

Logistic growth is an example of how a model is modified over time as new insights occur. The model for population growth changed from the early exponential curve $y = Me^{kx}$ to the logistic curve

$$y = \frac{N}{1 + be^{-kx}}.$$

Many other quantities besides population grow logistically. That is, their initial rate of growth is slow, but as time progresses, their rate of growth increases to a maximum value and subsequently begins to decline and to approach zero.

EXAMPLE 6 *Logistic Curve*

Rapid technological advancements in the last 20 years have made many products obsolete practically overnight. J. C. Fisher and R. H. Pry* successfully described the phenomenon of a technically superior new product replacing another product by the logistic equation

$$\frac{dz}{dx} = k(1 - z)z, \tag{6}$$

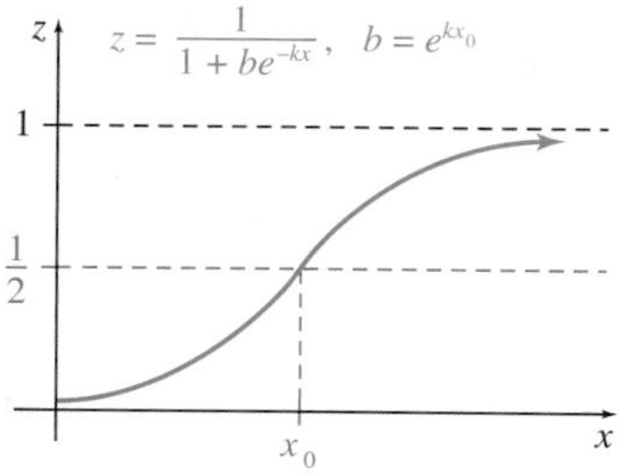

FIGURE 4

where z is the market share of the new product and $1 - z$ is the market share of the other product. The new product will initially have little or no market share; that is, $z_0 \approx 0$. Thus, the constant b in Equation (5) will have to be determined in a different way. Let x_0 be the time at which $z = 1/2$. Under the assumption $0 < z < 1$, the general solution of Equation (6) is

$$z = \frac{1}{1 + be^{-kx}},$$

where $b = e^{kx_0}$ (see Exercise 32). This solution is shown in Figure 4.

The market share of the new product will be growing most rapidly when the new product has captured exactly half the market, and the market share of the older product will be shrinking most rapidly at the same time. Notice that the logistic Equation (4) can be transformed into the simpler logistic Equation (6) by the change of variable $z = y/N$.

10.1 EXERCISES

Find the general solution for each differential equation.

1. $\frac{dy}{dx} = -2x + 3x^2$

2. $\frac{dy}{dx} = 3e^{-2x}$

3. $3x^3 - 2\frac{dy}{dx} = 0$

4. $3x^2 - 3\frac{dy}{dx} = 2$

5. $y\frac{dy}{dx} = x$

6. $y\frac{dy}{dx} = x^2 - 1$

7. $\frac{dy}{dx} = 2xy$

8. $\frac{dy}{dx} = x^2y$

9. $\frac{dy}{dx} = 3x^2y - 2xy$

10. $(y^2 - y)\frac{dy}{dx} = x$

11. $\frac{dy}{dx} = \frac{y}{x}, x > 0$

12. $\frac{dy}{dx} = \frac{y}{x^2}$

13. $\frac{dy}{dx} = y - 5$

14. $\frac{dy}{dx} = 3 - y$

15. $\frac{dy}{dx} = y^2e^x$

16. $\frac{dy}{dx} = \frac{e^x}{e^y}$

Find the particular solution for each initial value problem.

17. $\frac{dy}{dx} + 2x = 3x^2$; $y = 2$ when $x = 0$

18. $x\frac{dy}{dx} = x^2e^{3x}$; $y = \frac{8}{9}$ when $x = 0$

19. $2\frac{dy}{dx} = 4xe^{-x}$; $y = 42$ when $x = 0$

20. $x\frac{dy}{dx} - y\sqrt{x} = 0$; $y = 1$ when $x = 0$

*Fisher, J. C. and R. H. Pry, "A Simple Substitution Model of Technological Change," *Technological Forecasting and Social Change*, Vol. 3, 1971–1972. Copyright © 1972 by Elsevier Science Publishing Co., Inc. Reprinted by permission of the publisher.

21. $\frac{dy}{dx} = \frac{x^2}{y}$; $y = 3$ when $x = 0$

22. $\frac{dy}{dx} = \frac{x^2 + 5}{2y - 1}$; $y = 11$ when $x = 0$

23. $(2x + 3)y = \frac{dy}{dx}$; $y = 1$ when $x = 0$

24. $\frac{dy}{dx} = \frac{2x + 1}{y - 3}$; $y = 4$ when $x = 0$

Find the particular solution for each equation.

25. $\frac{dy}{dx} = 4x^3 - 3x^2 + x$; $y = 0$ when $x = 1$

26. $x^2\frac{dy}{dx} = y$; $y = -1$ when $x = 1$

27. $\frac{dy}{dx} = \frac{y^2}{x}$; $y = 5$ when $x = e$

28. $\frac{dy}{dx} = x^{1/2}y^2$; $y = 12$ when $x = 4$

29. $\frac{dy}{dx} = (y - 1)^2e^{x-1}$; $y = 2$ when $x = 1$

30. $\frac{dy}{dx} = (x + 2)^2e^y$; $y = 0$ when $x = 1$

31. **a.** Solve the logistic Equation (4) in this section by observing that

$$\frac{1}{y} + \frac{1}{N - y} = \frac{N}{(N - y)y}.$$

b. Assume $0 < y < N$. Verify that $b = (N - y_0)/y_0$ in Equation (5), where y_0 is the initial population size.

c. Assume $0 < N < y$ for all y. Verify that $b = (y_0 - N)/y_0$.

32. Suppose that $0 < z < 1$ for all z. Solve the logistic Equation (6) as in Exercise 31. Verify that $b = e^{kx_0}$, where x_0 is the time at which $z = 1/2$.

33. Suppose that $0 < y_0 < N$. Let $b = (N - y_0)/y_0$, and let $y(x) = N/(1 + be^{-kx})$ for all x. Show the following.

a. $0 < y(x) < N$ for all x.

b. The lines $y = 0$ and $y = N$ are horizontal asymptotes of the graph.

c. $y(x)$ is an increasing function.

d. $((\ln b)/k, N/2)$ is a point of inflection of the graph.

e. dy/dx is a maximum at $x_0 = (\ln b)/k$.

34. Suppose that $0 < N < y_0$. Let $b = (y_0 - N)/y_0$ and let

$$y(x) = \frac{N}{1 - be^{-kx}} \quad \text{for all } x \neq \frac{\ln b}{k}.$$

See the figure. Show the following.

a. $0 < b < 1$

b. The lines $y = 0$ and $y = N$ are horizontal asymptotes of the graph.

c. The line $x = (\ln b)/k$ is a vertical asymptote of the graph.

d. $y(x)$ is decreasing on $((\ln b)/k, \infty)$ and on $(-\infty, (\ln b)/k)$.

e. $y(x)$ is concave upward on $((\ln b)/k, \infty)$ and concave downward on $(-\infty, (\ln b)/k)$.

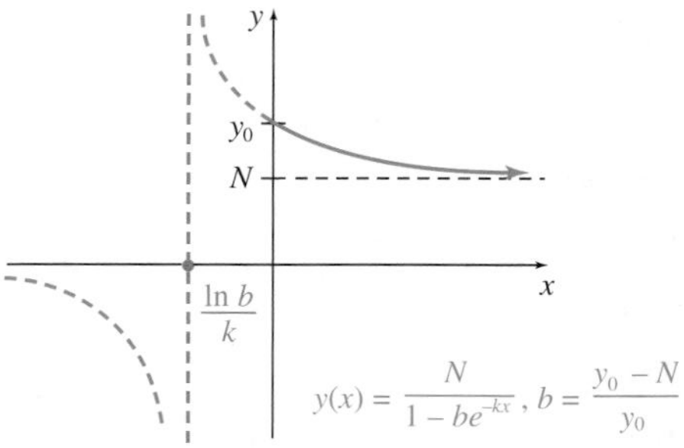

Applications

BUSINESS AND ECONOMICS

35. *Profit* The marginal profit of a certain company is given by

$$\frac{dy}{dx} = \frac{100}{32 - 4x},$$

where x represents the amount of money (in thousands of dollars) that the company spends on advertising. Find the profit for each advertising expenditure if the profit is \$1000 when nothing is spent on advertising.

a. \$3000 **b.** \$5000

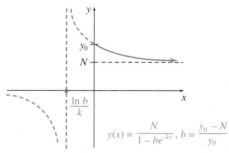

c. Can advertising expenditures ever reach \$8000 according to this model? Explain why or why not.

36. *Sales Decline* Sales (in thousands) of a certain product are declining at a rate proportional to the amount of sales, with a decay constant of 25% per year.

a. Write a differential equation to express the rate of sales decline.

b. Find a general solution to the equation in part a.

c. How much time will pass before sales become 30% of their original value?

37. *Inflation* If inflation grows continuously at a rate of 6% per year, how long will it take for \$1 to lose half its value?

Elasticity of Demand Elasticity of demand was discussed in the chapter on Applications of the Derivative, where it was defined as

$$E = -\frac{p}{q} \cdot \frac{dq}{dp},$$

for demand q and price p. Find the general demand equation $q = f(p)$ for each elasticity function.

(Hint: Set each elasticity function equal to $-\frac{p}{q} \cdot \frac{dq}{dp}$, then solve for q. Write the constant of integration as ln C.)

38. $E = \frac{4p^2}{q^2}$

39. $E = 2$

LIFE SCIENCES

40. *Tracer Dye* The amount of a tracer dye injected into the bloodstream decreases exponentially, with a decay constant of 3% per minute. If 6 cc are present initially, how many cubic centimeters are present after 10 minutes? (Here k will be negative.)

41. *Soil Moisture* The evapotranspiration index I is a measure of soil moisture. An article on 10- to 14-year-old heath vegetation described the rate of change of I with respect to W, the amount of water available, by the equation

$$\frac{dI}{dW} = .088(2.4 - I).*$$

a. According to the article, I has a value of 1 when $W = 0$. Solve the initial value problem.

b. What happens to I as W becomes larger and larger?

42. *Fish Population* An isolated fish population is limited to 5000 by the amount of food available. If there are now 150 fish and the population is growing with a growth constant of 1% a year, find the expected population at the end of 5 years.

Dieting A person's weight depends both on the daily rate of energy intake, say C calories per day, and on the daily rate of energy consumption, typically between 15 and 20 calories per pound per day. Using an average value of 17.5 calories per pound per day, a person weighing w pounds uses 17.5w calories per day. If $C = 17.5w$, then weight remains constant, and weight gain or loss occurs according to whether C is greater or less than 17.5w.†

43. To determine how fast a change in weight will occur, the most plausible assumption is that dw/dt is proportional to the net excess (or deficit) $C - 17.5w$ in the number of calories per day.

a. Assume C is constant and write a differential equation to express this relationship. Use k to represent the constant of proportionality. What does C being constant imply?

b. The units of dw/dt are pounds per day, and the units of $C - 17.5w$ are calories per day. What units must k have?

c. Use the fact that 3500 calories is equivalent to 1 lb to rewrite the differential equation in part a.

d. Solve the differential equation.

e. Let w_0 represent the initial weight and use it to express the coefficient of $e^{-.005t}$ in terms of w_0 and C.

44. (Refer to Exercise 43.) Suppose someone initially weighing 180 lb adopts a diet of 2500 calories per day.

a. Write the weight function for this individual.

b. Graph the weight function on the window $[0, 300]$ by $[120, 200]$. What is the asymptote? This value of w is the equilibrium weight w_{eq}. According to the model, can a person ever achieve this weight?

c. How long will it take a dieter to reach a weight just 2 lb more than w_{eq}?

45. *Bald Eagles* In 1963, the bald eagle was put on the endangered species list. Since then, its numbers have increased, as shown by the table on the next page that gives the estimated number of bald eagle pairs in the world for various years.‡

*Specht, R. L., "Dark Island Heath (Ninety-Mile Plain, South Australia) V: The Water Relationships in Heath Vegetation and Pastures on the Makin Sand," *Australian Journal of Botany*, Vol. 5, No. 2, Sept. 1957, pp. 151–172.

†Segal, Arthur C., "A Linear Diet Model," *The College Mathematics Journal*, Vol. 18, No. 1, Jan. 1987.

‡www.nctm.org/wlme/wlme6/twelve.htm.

Year	Estimated Number of Bald Eagle Pairs
1963	400
1974	800
1981	1200
1984	1800
1986	1900
1988	2500
1989	2700
1990	3000
1991	3400
1992	3700
1993	4000
1994	4500
1996	5000
1999	5800

Use a calculator with logistic regression capability to complete the following.

a. Letting x represent the years since 1900, plot the number of bald eagle pairs on the y-axis against the year on the x-axis. Discuss the appropriateness of fitting a logistic function to these data.

b. Use the logistic regression function on your calculator to determine the logistic equation that best fits the data.

c. Plot the logistic equation from part b on the same graph as the data points. Discuss how well the logistic equation fits the data.

d. Assuming that the logistic equation found in part b continues to be accurate, what seems to be the limiting size of the number of bald eagle pairs in the world?

SOCIAL SCIENCES

46. *Toronto's Jewish Population* The table at the top of the next column gives the population of Toronto's Jewish community at various times.* Use a calculator with logistic regression capability to complete the following.

Year	Population
1901	3103
1911	18,294
1921	34,770
1931	46,751
1941	52,798
1951	66,773
1961	85,000
1971	97,000
1981	128,650
1991	162,605
2001	175,000

a. Letting x represent the years since 1900, plot the population on the y-axis against the year on the x-axis. Discuss the appropriateness of fitting a logistic function to these data.

b. Use the logistic regression function on your calculator to determine the logistic equation that best fits the data.

c. Plot the logistic equation from part a on the same graph as the data points. Discuss how well the logistic equation fits the data.

d. Assuming that the logistic equation found in part b continues to be accurate, what seems to be the limiting size of the Jewish population in Toronto?

e. See Exercise 46 in the section on Exponential Functions (in the chapter on Nonlinear Functions), where we fit a linear and an exponential function to these data. Compare the results of that exercise with the results of this exercise, and discuss which type of function best fits the data.

47. *U.S. Latino Population* A recent report by the U.S. Census Bureau predicts that the U.S. Latino American population will increase from 31.4 million in 2000 to 96.5 million in 2050.† Assuming the unlimited growth model $dy/dt = ky$ fits this population growth, express the population y as a function of the year t. Let 2000 correspond to $t = 0$.

48. *U.S. African American Population* (Refer to Exercise 47.) The report also predicted that the U.S. African American population would increase from 35.5 million in 2000 to 60.6 million in 2050.† Repeat Exercise 47 using this data.

49. *Spread of a Rumor* Suppose the rate at which a rumor spreads—that is, the number of people who have heard the rumor over a period of time—increases with the number of

The Globe and Mail, Feb. 17, 1995 and www.thejewishweek.com.

†*Statistical Abstracts of the United States, 1999*, U.S. Census Bureau, U.S. Dept. of Commerce, Table 12, p. 14.

people who have heard it. If y is the number of people who have heard the rumor, then

$$\frac{dy}{dt} = ky,$$

where t is the time in days.

a. If y is 1 when $t = 0$, and y is 5 when $t = 2$, find k.

Using the value of k from part a, find y for each time.

b. $t = 3$ **c.** $t = 5$ **d.** $t = 10$

50. *World Population* The following table gives the population of the world at various times over the last two centuries, plus projections for the next century.*

Year	Population (billions)
1804	1
1927	2
1960	3
1974	4
1987	5
1999	6
2011	7
2025	8
2041	9
2071	10

Use a calculator with logistic regression capability to complete the following.

a. Use the logistic regression function on your calculator to determine the logistic equation that best fits the data.

b. Plot the logistic function found in part a and the original data in the same window. Does the logistic function seem to fit the data from 1927 on? Before 1927?

c. To get a better fit, subtract .99 from each value of the population in the table. (This makes the population in 1804 small, but not 0 or negative.) Find a logistic function that fits the new data.

d. Plot the logistic function found in part c and the modified data in the same window. Does the logistic function now seem to be a better fit than in part b?

e. Based on the results from parts c and d, predict the limiting value of the world's population as time increases. For comparison, the *New York Times* article predicts a value of 10.73 billion. (*Hint*: After taking the limit, remember to add the .99 that was removed earlier.)

f. Based on the results from parts c and d, predict the limiting value of the world population as you go further and further back in time. Does that seem reasonable? Explain.

51. *Worker Productivity* A company has found that the rate at which a person new to the assembly line produces items is

$$\frac{dy}{dx} = 7.5e^{-.3y},$$

where x is the number of days the person has worked on the line. How many items can a new worker be expected to produce on the eighth day if he produces none when $x = 0$?

PHYSICAL SCIENCES

52. *Radioactive Decay* The amount of a radioactive substance decreases exponentially, with a decay constant of 5% per month.

a. Write a differential equation to express the rate of change.

b. Find a general solution to the differential equation from part a.

c. If there are 90 g at the start of the decay process, find a particular solution for the differential equation from part a.

d. Find the amount left after 10 months.

53. *Snowplow* One morning snow began to fall at a heavy and constant rate. A snowplow started out at 8:00 A.M. At 9:00 A.M. it had traveled 2 miles. By 10:00 A.M. it had traveled 3 miles. Assuming that the snowplow removes a constant volume of snow per hour, determine the time at which it started snowing. (*Hint:* Let t denote the time since the snow started to fall, and let T be the time when the snowplow started out. Let x, the distance the snowplow has traveled, and h, the height of the snow, be functions of t. The assumption that a constant volume of snow per hour is removed implies that the speed of the snowplow times the height of the snow is a constant. Set up and solve differential equations involving dx/dt and dh/dt.)†

The New York Times, Nov. 17, 1996, p. 3.

†This problem first appeared in the *American Mathematical Monthly*, Vol. 44, Dec. 1937.

10.2 LINEAR FIRST-ORDER DIFFERENTIAL EQUATIONS

THINK ABOUT IT *What happens over time to the glucose level in a patient's bloodstream?*

The solution to a linear differential equation gives us an answer.

Recall that $f^{(n)}(x)$ represents the nth derivative of $f(x)$, and that $f^{(n)}(x)$ is called an nth-order derivative. By this definition, the derivative $f'(x)$ is first-order, $f''(x)$ is second-order, and so on. The *order of a differential equation* is that of the highest-order derivative in the equation. In this section only first-order differential equations are discussed.

A **linear first-order differential equation** is an equation of the form

$$\frac{dy}{dx} + P(x)y = Q(x).$$

Many useful models produce such equations. In this section we develop a general method for solving first-order linear differential equations.

For example, to solve the equation

$$x\frac{dy}{dx} + 6y + 2x^4 = 0, \tag{1}$$

we need to get it in the form of a linear first-order differential equation. Thus, dy/dx should have a coefficient of 1. To accomplish this, we divide both sides of the equation by x and rearrange the terms to get the linear differential equation

$$\frac{dy}{dx} + \frac{6}{x}y = -2x^3.$$

This equation is not separable and cannot be solved by the methods discussed so far. (Verify this.) Instead, multiply both sides of the equation by x^6 (the reason will be explained shortly) to get

$$x^6\frac{dy}{dx} + 6x^5y = -2x^9. \tag{2}$$

On the left, $6x^5$, the coefficient of y, is the derivative of x^6, the coefficient of dy/dx. Recall the product rule for derivatives:

$$D_x(uv) = u\frac{dv}{dx} + \frac{du}{dx}v.$$

If $u = x^6$ and $v = y$, the product rule gives

$$D_x(x^6y) = x^6\frac{dy}{dx} + 6x^5y,$$

which is the left side of Equation (2). Substituting $D_x(x^6y)$ for the left side of Equation (2) gives

$$D_x(x^6y) = -2x^9.$$

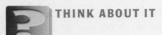

Assuming $y = f(x)$, as usual, both sides of this equation can be integrated with respect to x and the result solved for y to get

$$x^6 y = \int -2x^9\,dx = -2\left(\frac{x^{10}}{10}\right) + C = -\frac{x^{10}}{5} + C$$

$$y = -\frac{x^4}{5} + \frac{C}{x^6}. \tag{3}$$

Equation (3) is the general solution of Equation (2) and therefore of Equation (1).

This procedure has given us a solution, but what motivated our choice of the multiplier x^6? To see where x^6 came from, let $I(x)$ represent the multiplier, and multiply both sides of the general equation

$$\frac{dy}{dx} + P(x)y = Q(x)$$

by $I(x)$:

$$I(x)\frac{dy}{dx} + I(x)P(x)y = I(x)Q(x). \tag{4}$$

The method illustrated above will work only if the left side of the equation is the derivative of the product function $I(x) \cdot y$, which is

$$I(x)\frac{dy}{dx} + I'(x)y. \tag{5}$$

Comparing the coefficients of y in Equations (4) and (5) shows that $I(x)$ must satisfy

$$I'(x) = I(x)P(x),$$

or

$$\frac{I'(x)}{I(x)} = P(x).$$

Integrating both sides of this last equation gives

$$\ln|I(x)| = \int P(x)\,dx + C$$

$$|I(x)| = e^{\int P(x)\,dx + C}$$

or

$$I(x) = \pm\, e^C e^{\int P(x)\,dx}.$$

Only one value of $I(x)$ is needed, so let $C = 0$, so that $e^C = 1$, and use the positive result, giving

$$I(x) = e^{\int P(x)\,dx}.$$

In summary, choosing $I(x)$ as $e^{\int P(x)\,dx}$ and multiplying both sides of a linear first-order differential equation by $I(x)$ puts the equation in a form that can be solved by integration.

INTEGRATING FACTOR

The function $I(x) = e^{\int P(x)\,dx}$ is called an **integrating factor** for the differential equation
$$\frac{dy}{dx} + P(x)y = Q(x).$$

For Equation (1), written as the linear differential equation

$$\frac{dy}{dx} + \frac{6}{x}y = -2x^3,$$

$P(x) = 6/x$, and the integrating factor is

$$I(x) = e^{\int (6/x)\,dx} = e^{6\ln|x|} = e^{\ln|x|^6} = e^{\ln x^6} = x^6.$$

This last step used the fact that, for all positive a, $e^{\ln a} = a$.

In summary, we solve a linear first-order differential equation with the following steps.

SOLVING A LINEAR FIRST-ORDER DIFFERENTIAL EQUATION

1. Put the equation in the linear form $\dfrac{dy}{dx} + P(x)y = Q(x)$.
2. Find the integrating factor $I(x) = e^{\int P(x)\,dx}$.
3. Multiply each term of the equation from Step 1 by $I(x)$.
4. Replace the sum of terms on the left with $D_x[I(x)y]$.
5. Integrate both sides of the equation.
6. Solve for y.

EXAMPLE 1 *Linear Equation*

Give the general solution of $dy/dx + 2xy = x$.

Solution

Step 1 This equation is already in the required form.

Step 2 The integrating factor is

$$I(x) = e^{\int 2x\,dx} = e^{x^2}.$$

Step 3 Multiplying each term by e^{x^2} gives

$$e^{x^2}\frac{dy}{dx} + 2xe^{x^2}y = xe^{x^2}.$$

Step 4 The sum of terms on the left can now be replaced with $D_x(e^{x^2}y)$, to get

$$D_x(e^{x^2}y) = xe^{x^2}.$$

Step 5 Integrating on both sides leads to the general solution

$$e^{x^2}y = \int xe^{x^2}\,dx = \frac{1}{2}e^{x^2} + C$$

Step 6 Divide both sides by e^{x^2}.

$$y = \frac{1}{2} + Ce^{-x^2}.$$

EXAMPLE 2 *Linear Equation*

Solve the initial value problem $2(dy/dx) - 6y - e^x = 0$ if $y = 5$ when $x = 0$.

Solution Write the equation in the required form:

$$\frac{dy}{dx} - 3y = \frac{1}{2}e^x.$$

The integrating factor is

$$I(x) = e^{\int(-3)\,dx} = e^{-3x}.$$

Multiplying each term by $I(x)$ gives

$$e^{-3x}\frac{dy}{dx} - 3e^{-3x}y = \frac{1}{2}e^x e^{-3x}$$

$$e^{-3x}\frac{dy}{dx} - 3e^{-3x}y = \frac{1}{2}e^{-2x}.$$

The left side can now be replaced by $D_x(e^{-3x}y)$ to get

$$D_x(e^{-3x}y) = \frac{1}{2}e^{-2x}.$$

Integrating on both sides gives

$$e^{-3x}y = \int \frac{1}{2}e^{-2x}\,dx = \frac{1}{2}\left(\frac{e^{-2x}}{-2}\right) + C.$$

Now, multiply both sides by e^{3x} to get

$$y = -\frac{e^x}{4} + Ce^{3x},$$

the general solution. Find the particular solution by substituting 0 for x and 5 for y:

$$5 = -\frac{e^0}{4} + Ce^0 = -\frac{1}{4} + C$$

or

$$\frac{21}{4} = C,$$

which leads to the particular solution

$$y = -\frac{e^x}{4} + \frac{21}{4}e^{3x}.$$

NOTE We can write the condition $y = 5$ when $x = 0$ as $y(0) = 5$. We will use this notation from now on. ■

EXAMPLE 3 *Glucose*

Suppose glucose is infused into a patient's bloodstream at a constant rate of a grams per minute. At the same time, glucose is removed from the bloodstream at a rate proportional to the amount of glucose present. Then the amount of glucose, $G(t)$, present at time t satisfies

$$\frac{dG}{dt} = a - KG$$

for some constant K. Solve this equation for G. Does the glucose concentration eventually reach a constant? That is, what happens to G as $t \rightarrow \infty$?*

Solution The equation can be written in the form of the linear first-order differential equation

$$\frac{dG}{dt} + KG = a. \tag{6}$$

The integrating factor is

$$I(t) = e^{\int K dt} = e^{Kt}.$$

Multiply both sides of Equation (6) by $I(t) = e^{Kt}$.

$$e^{Kt}\frac{dG}{dt} + Ke^{Kt}G = ae^{Kt}$$

Write the left side as $D_t(e^{Kt}G)$ and solve for G by integrating on both sides.

$$D_t(e^{Kt}G) = ae^{Kt}$$

$$e^{Kt}G = \int ae^{Kt}dt = \frac{a}{K}e^{Kt} + C$$

Multiply both sides by e^{-Kt} to get

$$G = \frac{a}{K} + Ce^{-Kt}.$$

As $t \rightarrow \infty$,

$$\lim_{t \to \infty} G = \lim_{t \to \infty}\left(\frac{a}{K} + Ce^{-Kt}\right) = \lim_{t \to \infty}\left(\frac{a}{K} + \frac{C}{e^{Kt}}\right) = \frac{a}{K}.$$

Thus, the glucose concentration stabilizes at a/K.

NOTE The equation in Example 3 can also be solved by separation of variables. You are asked to do this in Exercise 22. ■

10.2 EXERCISES

Find the general solution for each differential equation.

1. $\dfrac{dy}{dx} + 2y = 5$

2. $\dfrac{dy}{dx} + 4y = 10$

3. $\dfrac{dy}{dx} + xy = 3x$

4. $\dfrac{dy}{dx} + 2xy = x$

5. $x\dfrac{dy}{dx} - y - x = 0, \quad x > 0$

6. $x\dfrac{dy}{dx} + 2xy - x^2 = 0$

7. $2\dfrac{dy}{dx} - 2xy - x = 0$

8. $3\dfrac{dy}{dx} + 6xy + x = 0$

9. $x\dfrac{dy}{dx} + 2y = x^2 + 3x, \quad x > 0$

10. $x^2\dfrac{dy}{dx} + xy = x^3 - 2x^2, \quad x > 0$

11. $y - x\dfrac{dy}{dx} = x^3, \quad x > 0$

12. $2xy + x^3 = x\dfrac{dy}{dx}$

Solve each differential equation, subject to the given initial condition.

13. $\dfrac{dy}{dx} + y = 2e^x; \quad y(0) = 100$

14. $\dfrac{dy}{dx} + 2y = e^{3x}; \quad y(0) = 50$

*Andrews, Larry C., *Ordinary Differential Equations with Applications*, Scott, Foresman and Company, 1982, p. 79.

15. $\frac{dy}{dx} - xy - x = 0; \quad y(1) = 10$

16. $x\frac{dy}{dx} - 3y + 2 = 0; \quad y(1) = 8$

17. $x\frac{dy}{dx} + 5y = x^2; \quad y(2) = 12$

18. $2\frac{dy}{dx} - 4xy = 5x; \quad y(1) = 10$

19. $x\frac{dy}{dx} + (1 + x)y = 3; \quad y(4) = 50$

20. $\frac{dy}{dx} + 2xy - e^{-x^2} = 0; \quad y(0) = 100$

Applications

LIFE SCIENCES

21. *Population Growth* The logistic equation introduced in Section 1,

$$\frac{dy}{dx} = k\left(1 - \frac{y}{N}\right)y \qquad (7)$$

can be written as

$$\frac{dy}{dx} = cy - py^2, \qquad (8)$$

where c and p are positive constants. Although this is a nonlinear differential equation, it can be reduced to a linear equation by a suitable substitution for the variable y.

a. Letting $y = 1/z$ and $dy/dx = (-1/z^2)dz/dx$, rewrite Equation (8) in terms of z. Solve for z and then for y.

b. Let $z(0) = 1/y_0$ in part a and find a particular solution for y.

c. Find the limit of y as $x \to \infty$. This is the saturation level of the population.

22. *Glucose Level* Solve the glucose level example (Example 3) using separation of variables.

23. *Drug Use* The rate of change in the concentration of a drug with respect to time in a user's blood is given by

$$\frac{dC}{dt} = -kC + D(t),$$

where $D(t)$ is dosage at time t and k is the rate that the drug leaves the bloodstream.*

a. Solve this linear equation to show that, if $C(0) = 0$, then

$$C(t) = e^{-kt}\int_0^1 e^{ky}D(y)\,dy.$$

(*Hint:* To integrate both sides of the equation in Step 5 of "Solving a Linear First-Order Differential Equation," integrate from 0 to t, and change the variable of integration to y.)

b. Show that if $D(y)$ is a constant D, then

$$C(t) = \frac{D(1 - e^{-kt})}{k}.$$

24. *Mouse Infection* A model for the spread of an infectious disease among mice is

$$\frac{dN}{dt} = rN - \frac{\alpha r(\alpha + b + v)}{\beta\left[\alpha - r\left(1 + \frac{v}{b + \gamma}\right)\right]},$$

where N is the size of the population of mice, α is the mortality rate due to infection, b is the mortality rate due to natural causes for infected mice, β is a transmission coefficient for the rate that infected mice infect susceptible mice, v is the rate the mice recover from infection, and γ is the rate that mice lose immunity.† Show that the solution to this equation, with the initial condition $N(0) = (\alpha + b + v)/\beta$, can be written as

$$N(t) = \frac{(\alpha + b + v)}{\beta R}[(R - \alpha)e^{rt} + \alpha],$$

where

$$R = \alpha - r\left(1 + \frac{v}{b + \gamma}\right).$$

SOCIAL SCIENCES

Immigration and Emigration If population is changed either by immigration or emigration, the population model discussed in Section 1 is modified to

$$\frac{dy}{dt} = ky + f(t),$$

where y is the population at time t and $f(t)$ is some (other) function of t that describes the net effect of the emigration/immigration. Assume $k = .02$ and $y(0) = 10{,}000$. Solve this differential equation for y, given the following functions $f(t)$.

25. $f(t) = e^t$

26. $f(t) = e^{-t}$

*Hoppensteadt, F. C. and J. D. Murray, "Threshold Analysis of a Drug Use Epidemic Model," *Mathematical Biosciences*, Vol. 53, No. 1/2, Feb. 1981, pp. 79–87.

†Anderson, Roy M., "The Persistence of Direct Life Cycle Infectious Diseases Within Populations of Hosts," in *Lectures on Mathematics in the Life Sciences, Vol. 12: Some Mathematical Questions in Biology*, American Mathematical Society, 1979, pp. 1–67.

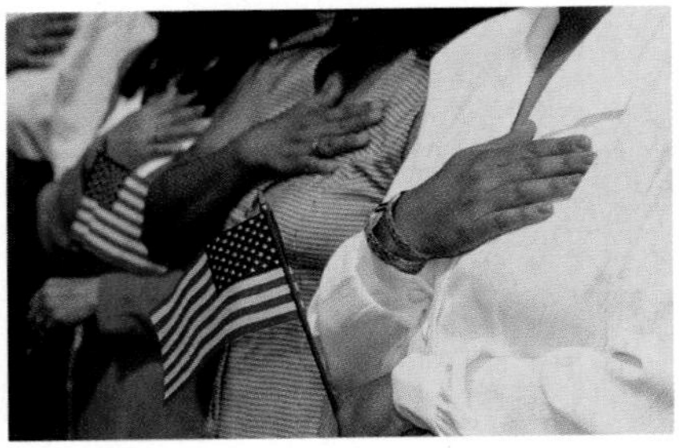

27. $f(t) = -t$

28. $f(t) = t$

PHYSICAL SCIENCES

Newton's Law of Cooling *Newton's law of cooling states that the rate of change of temperature of an object is proportional to the difference in temperature between the object and the surrounding medium. Thus, if T is the temperature of the object after t hours and T_0 is the (constant) temperature of the surrounding medium, then*

$$\frac{dT}{dt} = -k(T - T_0),$$

where k is a constant. Use this equation in Exercises 29–33.

29. Use the method of this section to show that the solution of this differential equation is

$$T = ce^{-kt} + T_0,$$

where c is a constant.

30. Solve the differential equation for Newton's law of cooling using separation of variables.

31. According to the solution of the differential equation for Newton's law of cooling, what happens to the temperature of an object after it has been in a surrounding medium with constant temperature for a long period of time? How well does this agree with reality?

Newton's Law of Cooling *When a dead body is discovered, one of the first steps in the ensuing investigation is for a medical examiner to determine the time of death as closely as possible. Have you ever wondered how this is done? If the temperature of the medium (air, water, or whatever) has been fairly constant and less than 48 hours have passed since the death, Newton's law of cooling can be used. The medical examiner does not actually solve the equation for each case. Instead, a table based on the formula is consulted. Use Newton's law of cooling to work the following exercises.**

32. Assume the temperature of a body at death is 98.6°F, the temperature of the surrounding air is 68°F, and at the end of one hour the body temperature is 90°F.

    a. What is the temperature of the body after 2 hours?

    b. When will the temperature of the body be 75°F?

    c. Approximately when will the temperature of the body be within .01° of the surrounding air?

33. Suppose the air temperature surrounding a body remains at a constant 10°F, $c = 88.6$, and $k = .24$.

    a. Determine a formula for the temperature at any time t.

    b. Use a graphing calculator to graph the temperature T as a function of time t on the window [0, 30] by [0, 100].

    c. When does the temperature of the body decrease more rapidly: just after death, or much later? How do you know?

    d. What will the temperature of the body be after 4 hours?

    e. How long will it take for the body temperature to reach 40°F? Use your calculator graph to verify your answer.

10.3 EULER'S METHOD

THINK ABOUT IT *How many people have heard a rumor 3 hours after it is started?*

This question will be answered in the exercises at the end of this section.

Applications sometimes involve differential equations such as

$$\frac{dy}{dx} = \frac{x + y}{y}$$

that cannot be solved by the methods discussed so far, but approximate solutions to these equations often can be found by numerical methods. For many applica-

*Callas, Dennis and David J. Hildreth, "Snapshots of Applications in Mathematics," *The College Mathematics Journal*, Vol. 26, No. 2, March 1995.

tions, these approximations are quite adequate. In this section we introduce Euler's method, which is only one of numerous mathematical contributions made by Leonhard Euler (1707–1783) of Switzerland. (His name is pronounced "oiler.") He also introduced the $f(x)$ notation used throughout this text. Despite becoming blind during his later years, he was the most prolific mathematician of his era. He published in all mathematical fields. He developed the equation

$$e^{i\theta} = \cos\theta + i\sin\theta,$$

which shows how concepts from three branches of mathematics are interrelated. The number e (in honor of Euler) is the base for the natural logarithm; the number i represents the imaginary unit $\sqrt{-1}$, and the symbols cos and sin are found in the study of trigonometry. (Trigonometric applications are given in Chapter 13 of this book.)

Euler's method of solving differential equations gives approximate solutions to differential equations involving $y = f(x)$ where the initial values of x and y are known; that is, equations of the form

$$\frac{dy}{dx} = g(x, y), \qquad \text{with } y(x_0) = y_0.$$

Geometrically, Euler's method approximates the graph of the solution $y = f(x)$ with a polygonal line whose first segment is tangent to the curve at the point (x_0, y_0), as shown in Figure 5.

FOR REVIEW

In Section 6.6 on Differentials: Linear Approximation, we defined Δy to be the actual change in y as x changed by an amount Δx:

$$\Delta y = f(x + \Delta x) - f(x).$$

The differential dy is an approximation to Δy. We find dy by following the tangent line from the point $(x, f(x))$, rather than by following the actual function. Then dy is found by using the formula $dy = (dy/dx)\,dx$ where $dx = \Delta x$. For example, let $f(x) = x^3$, $x = 1$, and $dx = \Delta x = .2$. Then $dy = f'(x)\,dx = 3x^2\,dx = 3(1^2)(.2) = .6$. The actual change in y as x changes from 1 to 1.2 is

$$\begin{aligned} &f(x + \Delta x) - f(x) \\ &= f(1.2) - f(1) \\ &= 1.2^3 - 1 \\ &= .728. \end{aligned}$$

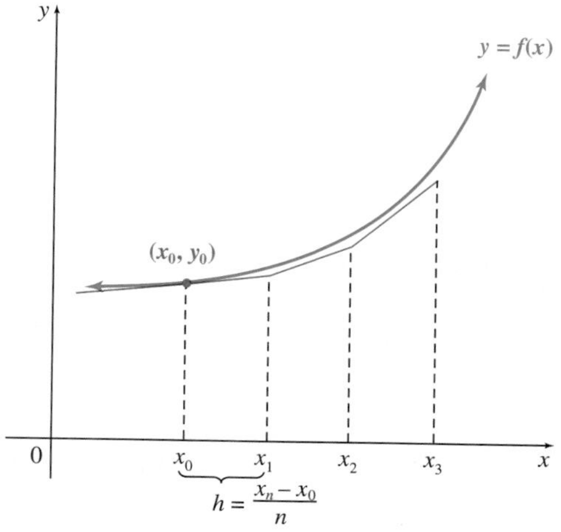

FIGURE 5

To use Euler's method, divide the interval from x_0 to another point x_n into n subintervals of equal width (see Figure 5.) The width of each subinterval is $h = (x_n - x_0)/n$.

Recall from Section 6.6 on Differentials: Linear Approximation that if Δx is a small change in x, then the corresponding small change in y, Δy, is approximated by

$$\Delta y \approx dy = \frac{dy}{dx} \cdot \Delta x.$$

The differential dy is the change in y along the tangent line. On the interval from x_i to x_{i+1}, dy is just $y_{i+1} - y_i$, where y_i is the approximate solution at x_i. We also have $dy/dx = g(x_i, y_i)$ and $\Delta x = h$. Putting these into the previous equation yields

$$y_{i+1} - y_i = g(x_i, y_i)h$$
$$y_{i+1} = y_i + g(x_i, y_i)h.$$

Because y_0 is given, we can use the equation just derived with $i = 0$ to get y_1. We can then use y_1 and the same equation with $i = 1$ to get y_2, and continue in this manner until we get y_n. A summary of Euler's method follows.

EULER'S METHOD

Let $y = f(x)$ be the solution of the differential equation

$$\frac{dy}{dx} = g(x, y), \quad \text{with } y(x_0) = y_0,$$

for $x_0 \le x \le x_n$. Let $x_{i+1} - x_i = h$ and

$$y_{i+1} = y_i + g(x_i, y_i)h,$$

for $0 \le i \le n$. Then

$$f(x_{i+1}) \approx y_{i+1}.$$

As the following examples will show, the accuracy of the approximation varies for different functions. As h gets smaller, however, the approximation improves, although making h too small can make things worse. (See the discussion at the end of this section.) Euler's method is not difficult to program; it then becomes possible to try smaller and smaller values of h to get better and better approximations. Graphing calculator programs for Euler's method are included in *The Graphing Calculator Manual* that is available with this book.

EXAMPLE 1 *Euler's Method*

Use Euler's method to approximate the solution of $dy/dx + 2xy = x$, with $y(0) = 1.5$, for $[0, 1]$. Use $h = .1$.

Method 1: Calculating by Hand

Solution The general solution of this equation was found in Example 1 of the last section, so the results using Euler's method can be compared with the actual solution. Begin by writing the differential equation in the required form as

$$\frac{dy}{dx} = x - 2xy, \qquad \text{so that} \qquad g(x, y) = x - 2xy.$$

Since $x_0 = 0$ and $y_0 = 1.5$, $g(x_0, y_0) = 0 - 2(0)(1.5) = 0$, and

$$y_1 = y_0 + g(x_0, y_0)h = 1.5 + 0(.1) = 1.5.$$

Now $x_1 = .1$, $y_1 = 1.5$, and $g(x_1, y_1) = .1 - 2(.1)(1.5) = -.2$. Then

$$y_2 = 1.5 + (-.2)(.1) = 1.48.$$

The 11 values for x_i and y_i for $0 \le i \le 10$ are shown in the table on the next page, together with the actual values using the result from Example 1 in the last section. (Since the result was only a general solution, replace x with 0 and y with 1.5 to get the particular solution $y = 1/2 + e^{-x^2}$.)

The results in the table look quite good. The graphs in Figure 6 show that the polygonal line follows the actual graph of $f(x)$ quite closely.

	Euler's Method	Actual Solution	Difference
x_i	y_i	$f(x_i)$	$y_i - f(x_i)$
0	1.5	1.5	0
.1	1.5	1.4900498	.0099502
.2	1.48	1.4607894	.0192106
.3	1.4408	1.4139312	.0268688
.4	1.384352	1.3521438	.0322082
.5	1.3136038	1.2788008	.034803
.6	1.232243	1.1976763	.0345667
.7	1.1443742	1.1126264	.0317102
.8	1.0541619	1.0272924	.0268695
.9	.96549595	.94485807	.02063788
1	.88170668	.86787944	.01382724

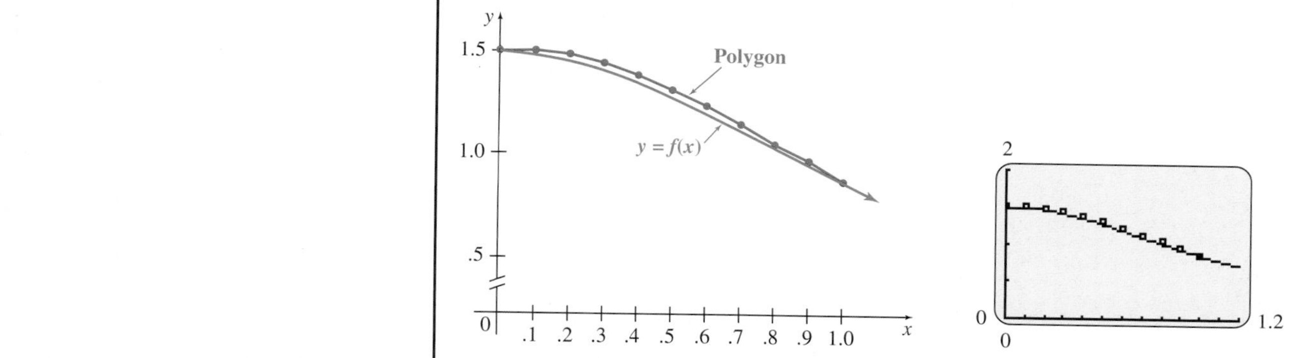

FIGURE 6

Method 2: Graphing Calculator

A graphing calculator can readily implement Euler's method. To do this example on a TI-83/84 Plus, put `X - 2X*Y` into the function Y_1. Start x with the value $x_0 - h$ by storing $-.1$ in X. Store y_0, or 1.5, in Y. Then the command `X + .1 → X:Y + Y1*.1 → Y` gives the next value of y, which is still 1.5. Continue to press the ENTER key to get subsequent values of y. For more details, see *The Graphing Calculator Manual* that is available with this book.

Method 3: Spreadsheet

Euler's method can also be performed on a spreadsheet. In Microsoft Excel, for example, store the values of x in column A and the initial value of y in B1. Then put the command "`=B1+(A1-2*A1*B1)*.1`" into B2 to get the next value of y, using the formula for $g(x, y)$ in this example. Copy this formula into the rest of column B to get subsequent values of y. For more details, see *The Spreadsheet Manual* that is available with this book.

Euler's method produces a very good approximation for this differential equation because the slope of the solution $f(x)$ is not steep in the interval under investigation. The next example shows that such good results cannot always be expected.

EXAMPLE 2 *Euler's Method*

Use Euler's method to solve $dy/dx = 3y + (1/2)e^x$, with $y(0) = 5$, for $[0, 1]$, using 10 subintervals.

Solution This is the differential equation of Example 2 in the last section. The general solution found there, with the initial condition given above, leads to the particular solution

$$y = -\frac{1}{4}e^x + \frac{21}{4}e^{3x}.$$

To solve by Euler's method, start with $g(x, y) = 3y + (1/2)e^x$, $x_0 = 0$, and $y_0 = 5$. For $n = 10$, $h = (1 - 0)/10 = .1$ again, and

$$y_{i+1} = y_i + g(x_i, y_i)h = y_i + \left(3y_i + \frac{1}{2}e^{x_i}\right)h.$$

For y_1, this gives

$$\begin{aligned} y_1 &= y_0 + \left(3y_0 + \frac{1}{2}e^{x_0}\right)h \\ &= 5 + \left[3(5) + \frac{1}{2}(e^0)\right](.1) \\ &= 5 + (15.5)(.1) = 6.55. \end{aligned}$$

Similarly,

$$\begin{aligned} y_2 &= 6.55 + \left[3(6.55) + \frac{1}{2}e^{.1}\right](.1) \\ &= 6.55 + (19.65 + .55258546)(.1) \\ &= 8.57025855. \end{aligned}$$

These and the remaining values for the interval $[0, 1]$ are shown in the table below.

In this example the absolute value of the differences grows very rapidly as x_i gets farther from x_0. See Figure 7 on the next page. These large differences come from the term e^{3x} in the solution; this term grows very quickly as x increases.

Approximate Solution Using $h = .1$

	Euler's Method	Actual Solution	Difference
x_i	y_i	$f(x_i)$	$y_i - f(x_i)$
0	5	5	0
.1	6.55	6.8104660	−.260466
.2	8.570259	9.2607730	−.690514
.3	11.20241	12.575452	−1.373042
.4	14.63062	17.057658	−2.427038
.5	19.09440	23.116687	−4.022287
.6	24.90516	31.305120	−6.399960
.7	32.46781	42.368954	−9.901144
.8	42.30884	57.315291	−15.006451
.9	55.11277	77.503691	−22.390921
1	71.76958	104.76950	−32.999920

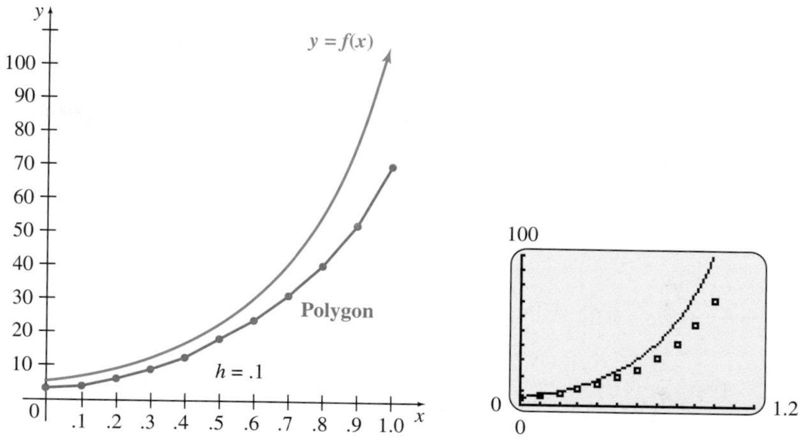

FIGURE 7

As these examples show, numerical methods may produce large errors. The error often can be reduced by using more subintervals of smaller width—letting $n = 100$ or 1000, for example. Approximations for the function in Example 2 with $n = 100$ and $h = (1 - 0)/100 = .01$ are shown in the table below. The approximations are considerably improved.

Approximate Solution Using $h = .01$

	Euler's Method	Actual Solution	Difference
x_i	y_i	$f(x_i)$	$y_i - f(x_i)$
0	5	5	0
.1	6.779419	6.810466	−.031047
.2	9.177102	9.260773	−.083671
.3	12.40634	12.575452	−.169112
.4	16.75386	17.057658	−.303798
.5	22.60505	23.116687	−.511637
.6	30.47795	31.305120	−.827170
.7	41.06884	42.368954	−1.30011
.8	55.31358	57.315291	−2.001711
.9	74.46998	77.503691	−3.033711
1	100.22860	104.769498	−4.540898

We could improve the accuracy of Euler's method by using a smaller h, but there are two difficulties. First, this requires more calculations, and consequently more time. Such calculations are usually done by computer, so the increased time may not matter. But this introduces a second difficulty: The increased number of calculations causes more round-off error, so there is a limit to how small we can make h and still get improvement. The preferred way to get greater accuracy is to use a more sophisticated procedure, such as the Runge-Kutta method. Such methods are beyond the scope of this book but are discussed in numerical analysis and differential equations courses.*

10.3 EXERCISES

Use Euler's method to approximate the indicated function value to 3 decimal places, using $h = .1$.

1. $\dfrac{dy}{dx} = x^2 + y^2$; $y(0) = 1$; find $y(.5)$

2. $\dfrac{dy}{dx} = xy + 2$; $y(0) = 0$; find $y(.5)$

3. $\dfrac{dy}{dx} = 1 + y$; $y(0) = 2$; find $y(.6)$

4. $\dfrac{dy}{dx} = x + y^2$; $y(0) = 0$; find $y(.6)$

5. $\dfrac{dy}{dx} = x + \sqrt{y}$; $y(0) = 1$; find $y(.4)$

6. $\dfrac{dy}{dx} = 1 + \dfrac{y}{x}$; $y(1) = 0$; find $y(1.4)$

7. $\dfrac{dy}{dx} = x\sqrt{1 + y^2}$; $y(1) = 1$; find $y(1.5)$

8. $\dfrac{dy}{dx} = e^{-y} + x$; $y(0) = 0$; find $y(.5)$

Use Euler's method to approximate the indicated function value to 3 decimal places, using $h = .1$. Next, solve the differential equation and find the indicated function value to 3 decimal places. Compare the result with the approximation.

9. $\dfrac{dy}{dx} = -4 + x$; $y(0) = 1$; find $y(.4)$

10. $\dfrac{dy}{dx} = 4x + 3$; $y(1) = 0$; find $y(1.5)$

11. $\dfrac{dy}{dx} = x^2$; $y(0) = 2$; find $y(.5)$

12. $\dfrac{dy}{dx} = \dfrac{1}{x}$; $y(1) = 1$; find $y(1.4)$

13. $\dfrac{dy}{dx} = 2xy$; $y(1) = 1$; find $y(1.6)$

14. $\dfrac{dy}{dx} = x^2y$; $y(0) = 1$; find $y(.6)$

15. $\dfrac{dy}{dx} = ye^x$; $y(0) = 1$; find $y(.3)$

16. $\dfrac{dy}{dx} = \dfrac{x}{y}$; $y(0) = 2$; find $y(.3)$

Use the program for Euler's method in The Graphing Calculator Manual *that is available with this book to construct a table like the ones in the examples for $0 \le x \le 1$, with $h = .2$.*

17. $\dfrac{dy}{dx} = \sqrt[3]{x}$; $y(0) = 0$

18. $\dfrac{dy}{dx} = y$; $y(0) = 1$

19. $\dfrac{dy}{dx} = 1 - y$; $y(0) = 0$

20. $\dfrac{dy}{dx} = x - xy$; $y(0) = .5$

Solve each differential equation and graph the function $y = f(x)$ and the polygonal approximation on the same axes. (The approximations were found in Exercises 17–20.)

21. $\dfrac{dy}{dx} = \sqrt[3]{x}$; $y(0) = 0$

22. $\dfrac{dy}{dx} = y$; $y(0) = 1$

23. $\dfrac{dy}{dx} = 1 - y$; $y(0) = 0$

24. $\dfrac{dy}{dx} = x - xy$; $y(0) = .5$

*For example, see Nagel, R. K., E. B. Saff, and A. D. Snider, *Fundamentals of Differential Equations*, 6th ed., Addison-Wesley, 2004.

25. a. Use Euler's method with $h = .2$ to approximate $f(1)$, where $f(x)$ is the solution to the differential equation

$$\frac{dy}{dx} = y^2; \quad y(0) = 1.$$

b. Solve the differential equation in part a using separation of variables, and discuss what happens to $f(x)$ as x approaches 1.

c. Based on what you learned from parts a and b, discuss what might go wrong when using Euler's method. (More advanced courses on differential equations discuss the question of whether a differential equation has a solution for a given interval in x.)

Applications

Solve Exercises 26–33 using Euler's method.

BUSINESS AND ECONOMICS

26. *Bankruptcy* Suppose 150 small business firms are threatened by bankruptcy. If y is the number bankrupt by time t, then $150 - y$ is the number not yet bankrupt by time t. The rate of change of y is proportional to the product of y and $150 - y$. Let 2000 correspond to $t = 0$. Assume 20 firms are bankrupt at $t = 0$.

a. Write a differential equation using the given information. Use .002 for the constant of proportionality.

b. Approximate the number of firms that are bankrupt in 2004, using $h = 1$.

LIFE SCIENCES

27. *Growth of Algae* The phosphate compounds found in many detergents are highly water soluble and are excellent fertilizers for algae. Assume that there are 3000 algae present at time $t = 0$ and conditions will support at most 100,000 algae. Assume that the rate of growth of algae, in the presence of sufficient phosphates, is proportional both to the number present (in thousands) and to the difference between 100,000 and the number present (in thousands).

a. Write a differential equation using the given information. Use .01 for the constant of proportionality.

b. Approximate the number present when $t = 2$, using $h = .5$.

28. *Immigration* An island is colonized by immigration from the mainland, where there are 100 species. Let the number of species on the island at time t (in years) equal y, where $y = f(t)$. Suppose the rate at which new species immigrate to the island is

$$\frac{dy}{dt} = .02(100 - y^{1/2}).$$

Use Euler's method with $h = .5$ to approximate y when $t = 5$ if there were 10 species initially.

29. *Insect Population* A population of insects, y, living in a circular colony grows at a rate

$$\frac{dy}{dt} = .05y - .1y^{1/2},$$

where t is time in weeks. If there were 60 insects initially, use Euler's method with $h = 1$ to approximate the number of insects after 6 weeks.

30. *Whale Population* Under certain conditions a population may exhibit a polynomial growth rate function. A population of blue whales is growing according to the function

$$\frac{dy}{dt} = -y + .02y^2 + .003y^3.$$

Here y is the population in thousands and t is measured in years. Use Euler's method with $h = 1$ to approximate the population in 4 years if the initial population is 15,000.

31. *Goat Growth* The growth of male Saanen goats can be approximated by the equation

$$\frac{dW}{dt} = -.01189W + .92389W^{.016},$$

where W is the weight (in kilograms) after t weeks.* Find the weight of a goat at 5 weeks, given that the weight at birth is 3.65 kg. Use Euler's method with $h = 1$.

*France, J., J. Kijkstra and M. S. Dhanoa, "Growth Functions and Their Application in Animal Science," *Annales de Zootechnie*, Vol. 45 (Supplement), 1996, pp. 165–174.

SOCIAL SCIENCES

32. *Learning* In an early article describing how people learn, the rate of change of the probability that a person performs a task correctly (p) with respect to time (t) is given by

$$\frac{dp}{dt} = \frac{2k}{\sqrt{m}}(p - p^2)^{3/2},$$

where k and m are constants related to the rate that the person learns the task.* For this exercise, let $m = 4$ and $k = .5$.

a. Letting $p = .1$ when $t = 0$, use the program for Euler's method in *The Graphing Calculator Manual* that is available with this book to construct a table for t_i and p_i like the ones in the examples for $0 \leq x \leq 30$, with $h = 5$.

b. Based on your answer to part a, what does p seem to approach as t increases? Explain why this answer makes sense.

33. *Spread of Rumors* A rumor spreads through a community of 500 people at the rate

$$\frac{dN}{dt} = .2(500 - N)N^{1/2},$$

where N is the number of people who have heard the rumor at time t (in hours). Use Euler's method with $h = .5$ to find the number who have heard the rumor after 3 hours, if only 2 people heard it initially.

10.4 APPLICATIONS OF DIFFERENTIAL EQUATIONS

THINK ABOUT IT *How do the populations of a predator and its prey change over time?*

A Predator-Prey Model The Austrian mathematician A. J. Lotka (1880–1949) and the Italian mathematician Vito Volterra (1860–1940) proposed the following simple model for the way in which the fluctuations of populations of a predator and its prey affect each other.† Let $x = f(t)$ denote the population of the predator and $y = g(t)$ denote the population of the prey at time t. The predator might be a wolf and its prey a moose, or the predator might be a ladybug and the prey an aphid.

Assume that if there were no predators present, the population of prey would increase at a rate py proportional to their number, but that the predators consume the prey at a rate qxy proportional to the product of the number of prey and the number of predators. The net rate of change dy/dt of y is the rate of increase of the prey minus the rate at which the prey are eaten, that is,

$$\frac{dy}{dt} = py - qxy, \tag{1}$$

with positive constants p and q.

Assume that if there were no prey, the predators would starve and their population would *decrease* at a rate rx proportional to their number, but that in the

*Thurstone, L. L., "The Learning Function," *The Journal of General Psychology*, Vol. 3, No. 4, Oct. 1930, pp. 469–493.

†Lotka, A. J., *Elements of Mathematical Biology*, Dover, 1956.

presence of prey the rate of growth of the population of predators is increased by an amount sxy. These assumptions give a second differential equation,

$$\frac{dx}{dt} = -rx + sxy, \tag{2}$$

with additional positive constants r and s.

Equations (1) and (2) form a system of differential equations known as the **Lotka-Volterra equations.** They cannot be solved for x and y as functions of t, but an equation relating the variables x and y can be found. Dividing Equation (1) by Equation (2) gives the separable differential equation

$$\frac{dy}{dx} = \frac{dy/dt}{dx/dt} = \frac{py - qxy}{-rx + sxy}$$

or

$$\frac{dy}{dx} = \frac{y(p - qx)}{x(sy - r)}. \tag{3}$$

Equation (3) is solved for specific values of the constants p, q, r, and s in the next example.

EXAMPLE 1 *Predator-Prey*

Suppose that $x = f(t)$ (hundreds of predators) and $y = g(t)$ (thousands of prey) satisfy the Lotka-Volterra Equations (1) and (2) with $p = 3$, $q = 1$, $r = 4$, and $s = 2$. Suppose that at a time when there are 100 predators $(x = 1)$, there are 1000 prey $(y = 1)$. Find an equation relating x and y.

Solution With the given values of the constants, p, q, r, and s, Equation (3) reads

$$\frac{dy}{dx} = \frac{y(3 - x)}{x(2y - 4)}.$$

Separating the variables yields

$$\frac{2y - 4}{y}\,dy = \frac{3 - x}{x}\,dx,$$

or

$$\int\left(2 - \frac{4}{y}\right)dy = \int\left(\frac{3}{x} - 1\right)dx.$$

Evaluating the integrals gives

$$2y - 4\ln y = 3\ln x - x + C.$$

(It is not necessary to use absolute value for the logarithms since x and y are positive.) Use the initial conditions $x = 1$ and $y = 1$ to find C.

$$2 - 4(0) = 3(0) - 1 + C$$
$$C = 3$$

The desired equation is

$$x + 2y - 3\ln x - 4\ln y = 3. \tag{4}$$

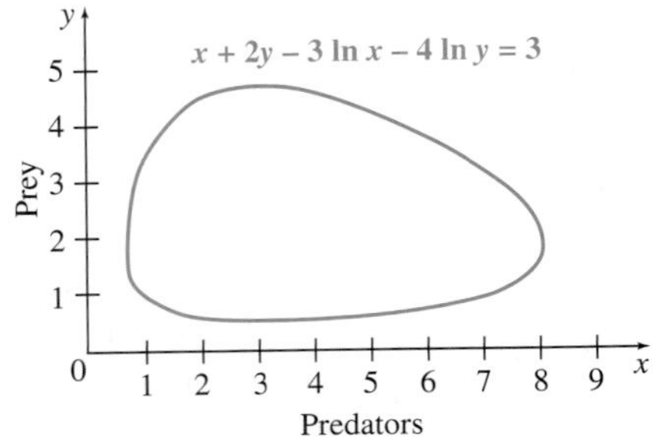

FIGURE 8

A graph of Equation (4), in Figure 8, shows that the variations in the populations of the predator and prey are cyclic. The population of prey tends to rise when the population of predators is low. Then, with the resulting abundance of food, the population of predators increases. As a result, the population of prey decreases again, and this forces the population of predators to decline because of the lack of food. The pattern repeats indefinitely. ■

Arms Race Models   Lewis F. Richardson* proposed a pair of linear first-order differential equations as a model of an arms race between two rival countries X and Y. In order to simplify the situation, the following assumptions are made.

1. X and Y are roughly equal in size.
2. T_0, the annual trade flow from X to Y (in dollars), is a constant.
3. U_0, the annual trade flow from Y to X (in dollars), is a constant.
4. The arms race between X and Y is not affected by the policies of other countries.

Let

t = the time in years,

x = the armaments budget of X in year t (in dollars) minus T_0,

y = the armaments budget of Y in year t (in dollars) minus U_0.

Richardson assumed that x stimulates the growth of y, but the sheer size of x would eventually inhibit its own growth. He made similar assumptions about y. These assumptions led him to the following model:

$$\frac{dx}{dt} = ky - mx + g \qquad \text{and} \qquad \frac{dy}{dt} = kx - my + h, \tag{5}$$

where k, m, g, and h are constants. The constants g and h account for the flow of trade between the two nations and their underlying attitudes toward each other. For example, if the two countries have long-standing grievances and ideological differences that outweigh the value of the constant trade flow, then $g > 0$, $h > 0$,

*Richardson, Lewis F., *Arms and Insecurity*, The Boxwood Press, 1960.

and the armament budgets will grow even when $x = 0 = y$. On the other hand, if the constant trade flow is the dominant factor, then $g < 0$, $h < 0$, and the armament budgets will shrink when $x = 0 = y$. Richardson found that the constant k was roughly proportional to the size of the country, but the constant m was roughly the same for all countries.

To model the mutual stimulation for the growth of armaments, Richardson set $z = x + y$ and added the two equations in (5), obtaining

$$\begin{aligned}\frac{dz}{dt} &= \frac{dx}{dt} + \frac{dy}{dt} \\ &= ky - mx + g + kx - my + h \\ &= k(x + y) - m(x + y) + g + h \\ &= kz - mz + g + h \\ \frac{dz}{dt} &= g + h - (m - k)z. \end{aligned} \tag{6}$$

Solving this differential equation as in Example 3 in Section 10.2 gives the general solution of (6) as

$$z = \frac{g + h}{m - k} + Ce^{-(m-k)t}. \tag{7}$$

Richardson tested his model on the growth of armaments in the years preceding World War I for

X = the bloc consisting of Russia and France,

Y = the bloc consisting of Germany and Austro-Hungary.

He found a remarkably good fit of Equations (6) and (7) to the data.

R. Taagepera et al.* proposed the following model for the arms race (within a finite time span) between Israel and the Arab countries:

$$\frac{dx}{dt} = kx \qquad \text{and} \qquad \frac{dy}{dt} = my, \tag{8}$$

where k and m are constants. By Example 4 in Section 10.1 the general solution of (8) is

$$x = x_0e^{kt} \qquad \text{and} \qquad y = y_0e^{mt}. \tag{9}$$

This model asserts that x stimulates the growth of x, but not of y, and vice versa. Here we are supposing that self-stimulation is the controlling factor, whereas the Richardson model presupposes mutual stimulation. Taagepera et al. found that the self-stimulation model (8)–(9) fit the Israeli-Arab arms race better than the Richardson model (5)–(7). This result is very surprising, for it challenges the common assumption that the driving force in any arms race is mutual stimulation.

Mixing Problems The mixing of two solutions can lead to a first-order differential equation, as the next example shows.

*Taagepera, R., G. M. Shiffler, R. T. Perkins, and D. L. Wagner, "Soviet-American and Israeli-Arab Arms Races and the Richardson Model," *General Systems*, Vol. 20, 1975, pp. 151–158.

EXAMPLE 2 *Salt Concentration*

Suppose a tank contains 100 gal of a solution of dissolved salt and water, which is kept uniform by stirring. If pure water is allowed to flow into the tank at the rate of 4 gal per minute, and the mixture flows out at the rate of 3 gal per minute (see Figure 9), how much salt will remain in the tank after t minutes if 15 lb of salt are in the mixture initially?*

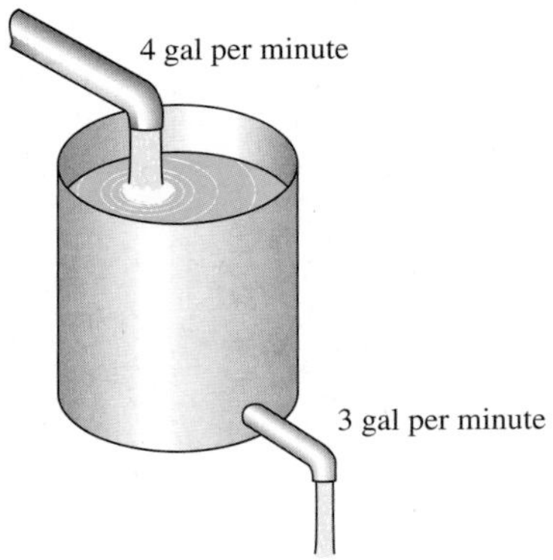

FIGURE 9

Solution Let the amount of salt present in the tank at any specific time be $y = f(t)$. The net rate at which y changes is given by

$$\frac{dy}{dt} = (\text{Rate of salt in}) - (\text{Rate of salt out}).$$

Since pure water is coming in, the rate of salt entering the tank is zero. The rate at which salt is leaving the tank is the product of the amount of salt per gallon (in V gallons) and the number of gallons per minute leaving the tank:

$$\text{Rate of salt out} = \left(\frac{y}{V}\text{ lb per gal}\right)(3\text{ gal per minute}).$$

The differential equation, therefore, can be written as

$$\frac{dy}{dt} = -\frac{3y}{V}; \quad y(0) = 15,$$

where $y(0)$ is the initial amount of salt in the solution. We must take into account the fact that the volume, V, of the mixture is not constant but is determined by

$$\frac{dV}{dt} = (\text{Rate of liquid in}) - (\text{Rate of liquid out}) = 4 - 3 = 1,$$

or

$$\frac{dV}{dt} = 1,$$

from which

$$V(t) = t + C_1.$$

Because the volume is known to be 100 at time $t = 0$, we have $C_1 = 100$, and

$$\frac{dy}{dt} = \frac{-3y}{t + 100}; \quad y(0) = 15,$$

a separable equation with solution

$$\frac{dy}{y} = \frac{-3}{t + 100}dt$$
$$\ln y = -3\ln(t + 100) + C.$$

*Andrews, Larry C., *Ordinary Differential Equations with Boundary Value Problems*, HarperCollins Publishers, Inc., 1991, pp. 85–86. Reprinted by permission of the author.

Since $y = 15$ when $t = 0$,

$$\begin{aligned} \ln 15 &= -3 \ln 100 + C \\ \ln 15 + 3 \ln 10^2 &= C \\ \ln(15 \cdot 10^6) &= C. \end{aligned}$$

Finally,

$$\begin{aligned} \ln y &= \ln(t + 100)^{-3} + \ln(15 \cdot 10^6) \\ &= \ln[(t + 100)^{-3}(15 \cdot 10^6)] \\ y &= \frac{15 \cdot 10^6}{(t + 100)^3}. \end{aligned}$$

Continuous Deposits An amount of money A invested at an annual interest rate r, compounded continuously, grows according to the differential equation

$$\frac{dA}{dt} = rA,$$

where t is time (in years). Suppose regular deposits are made to the account at frequent intervals at a rate of D dollars per year. For simplicity, assume these deposits to be continuous. The differential equation for the growth of the account then becomes

$$\frac{dA}{dt} = rA + D.$$

EXAMPLE 3 *Continuous Deposits*

When Michael was born, his grandfather arranged to deposit \$5000 in an account for him at 8% annual interest compounded continuously. Grandfather plans to add to the account "continuously" at the rate of \$1000 a year. How much will be in the account when Michael is 18?

Solution Here $r = .08$ and $D = 1000$, so the differential equation is

$$\frac{dA}{dt} = .08A + 1000.$$

Separate the variables and integrate on both sides.

$$\begin{aligned} \frac{1}{.08A + 1000} dA &= dt \\ \frac{1}{.08} \ln(.08A + 1000) &= t + C \\ \ln(.08A + 1000) &= .08t + K && K = .08C \\ .08A + 1000 &= e^{.08t + K} \\ .08A &= -1000 + e^{.08t}e^K \\ .08A &= -1000 + Me^{.08t} && M = e^K \\ A &= -12{,}500 + \frac{M}{.08}e^{.08t} \end{aligned}$$

Use the fact that the initial amount deposited was \$5000 to find M.

$$5000 = -12{,}500 + \frac{M}{.08}e^{(.08)0}$$
$$5000 = -12{,}500 + \frac{M}{.08}(1)$$
$$1400 = M$$
$$A = -12{,}500 + \frac{1400}{.08}e^{.08t}$$
$$A = -12{,}500 + 17{,}500e^{.08t}$$

When Michael is 18, $t = 18$, and the amount in the account will be

$$\begin{aligned} A &= -12{,}500 + 17{,}500e^{(.08)18} \\ &= -12{,}500 + 17{,}500e^{1.44} \\ &\approx 61{,}362.18, \end{aligned}$$

or about \$61,400.

Epidemics Under certain conditions, the spread of a contagious disease can be described with the logistic growth model, as in the next example.

EXAMPLE 4 *Spread of an Epidemic*

Consider a population of size N that satisfies the following conditions.

1. Initially there is only one infected individual.
2. All uninfected individuals are susceptible, and infection occurs when an uninfected individual contacts an infected individual.
3. Contact between any two individuals is just as likely as contact between any other two individuals.
4. Infected individuals remain infectious.

Let t = the time (in days) and y = the number of individuals infected at time t. At any moment there are $(N - y)y$ possible contacts between an uninfected individual and an infected individual. Thus, it is reasonable to assume that the rate of spread of the disease satisfies the following logistic equation (discussed in Section 10.1):

$$\frac{dy}{dt} = k\left(1 - \frac{y}{N}\right)y. \qquad \textbf{(10)}$$

As shown in Section 10.1, the general solution of Equation (10) is

$$y = \frac{N}{1 + be^{-kt}}, \qquad \textbf{(11)}$$

where $b = (N - y_0)/y_0$. Since just one individual is infected initially, $y_0 = 1$ here. Substituting these values into Equation (11) gives

$$y = \frac{N}{1 + (N - 1)e^{-kt}} \tag{12}$$

as the general solution of Equation (10).

The infection rate will be a maximum when its derivative is 0, that is, when $d^2y/dt^2 = 0$. Since

$$\frac{dy}{dt} = k\left(1 - \frac{y}{N}\right)y,$$

we have

$$\begin{aligned} \frac{d^2y}{dt^2} &= k\left[\left(1 - \frac{y}{N}\right)(1) + y\left(-\frac{1}{N}\right)\right] \\ &= k\left(1 - \frac{2y}{N}\right). \end{aligned}$$

Set $d^2y/dt^2 = 0$ to get

$$\begin{aligned} 0 &= 1 - \frac{2y}{N} \\ \frac{2y}{N} &= 1 \\ y &= \frac{N}{2}. \end{aligned}$$

The maximum infection rate occurs when exactly half the total population is still uninfected and equals

$$\frac{dy}{dt} = k\left(1 - \frac{N/2}{N}\right)\frac{N}{2} = \frac{kN}{4}.$$

Letting $y = N/2$ in Equation (12) and solving for t shows that the maximum infection rate occurs at time

$$t_m = \frac{\ln(N - 1)}{k}.$$

The graph of dy/dt, shown in Figure 10, is called the *epidemic curve*. It is symmetric about the line $t = t_m$.

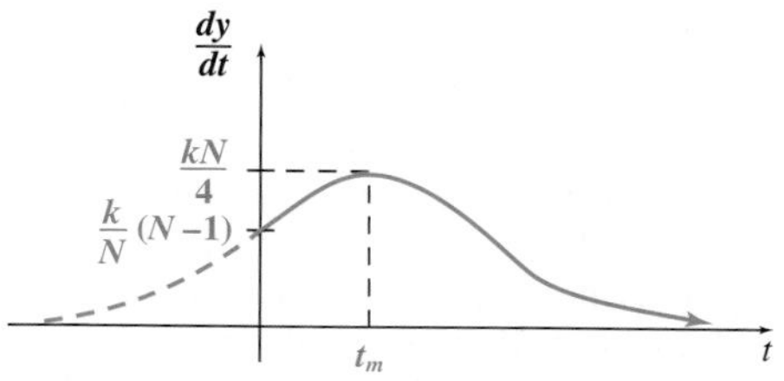

FIGURE 10

10.4 EXERCISES

Applications

BUSINESS AND ECONOMICS

1. *Continuous Deposits* Jessie deposits \$2000 in an IRA at 6% interest compounded continuously for her retirement in 10 years. She intends to make continuous deposits at the rate of \$2000 a year until she retires. How much will she have accumulated at that time?

2. *Continuous Deposits* In Exercise 1, how long will it take Jessie to accumulate \$20,000 in her retirement account?

3. *Continuous Deposits* To provide for a future expansion, a company plans to make continuous deposits to a savings account at the rate of \$10,000 per year, with no initial deposit. The managers want to accumulate \$70,000. How long will it take if the account earns 7% interest compounded continuously?

4. *Continuous Deposits* Suppose the company in Exercise 3 wants to accumulate \$50,000 in 3 years. Find the approximate yearly deposit that will be required.

5. *Continuous Deposits* An investor deposits \$8000 into an account that pays 6% compounded continuously, and then begins to *withdraw* from the account continuously at a rate of \$1200 per year.

   a. Write a differential equation to describe the situation.

   b. How much will be left in the account after 2 years?

   c. When will the account be completely depleted?

LIFE SCIENCES

6. *Competing Species* The system of equations

$$\frac{dy}{dt} = 3y - 2xy$$
$$\frac{dx}{dt} = -2x + 3xy$$

   describes the influence of the populations of two competing species on their growth rates.

   a. Following Example 1, find an equation relating x and y, assuming $y = 2$ when $x = 1$.

   b. Find values of x and y so that both populations are constant. (*Hint:* Set both differential equations equal to 0.)

7. *Symbiotic Species* When two species, such as the rhinoceros and birds pictured in the next column, coexist in a symbiotic (dependent) relationship, they either increase together or decrease together. Typical equations for the growth rates of two such species might be

$$\frac{dx}{dt} = -3x + 4xy$$
$$\frac{dy}{dt} = -3y + xy.$$

   a. Find an equation relating x and y if $x = 3$ when $y = 1$.

   b. Find values of x and y so that both populations are constant. (See Exercise 6.)

8. *Spread of an Epidemic* The native Hawaiians lived for centuries in isolation from other peoples. When foreigners finally came to the islands they brought with them diseases such as measles, whooping cough, and smallpox, which decimated the population. Suppose such an island has a native population of 5000, and a sailor from a visiting ship introduces measles, which has an infection rate of .00005. Also suppose that the model for spread of an epidemic described in Example 4 applies.

   a. Write an equation for the number of natives who remain uninfected. Let t represent time in days.

   b. How many are uninfected after 30 days?

   c. How many are uninfected after 50 days?

   d. When will the maximum infection rate occur?

9. *Spread of an Epidemic* In Example 4, the number of infected individuals is given by Equation (12).

   a. Show that the number of uninfected individuals is given by

$$N - y = \frac{N(N-1)}{N - 1 + e^{kt}}.$$

   b. Graph the equation in part a and Equation (12) on the same axes when $N = 100$ and $k = 1$.

   c. Find the common inflection point of the two graphs.

   d. What is the significance of the common inflection point?

   e. What are the limiting values of y and $N - y$?

10. *Spread of an Epidemic* An influenza epidemic spreads at a rate proportional to the product of the number of people infected and the number not yet infected. Assume that 50 people are infected at the beginning of the epidemic in a community of 10,000 people, and 300 are infected 10 days later.

a. Write an equation for the number of people infected, y, after t days.

b. When will half the community be infected?

11. *Spread of an Epidemic* The Gompertz growth law,

$$\frac{dy}{dt} = kye^{-at},$$

for constants k and a, is another model used to describe the growth of an epidemic. Repeat Exercise 10, using this differential equation with $a = .02$.

12. *Spread of Gonorrhea* Gonorrhea is spread by sexual contact, takes 3 to 7 days to incubate, and can be treated with antibiotics. There is no evidence that a person ever develops immunity. One model proposed for the rate of change in the number of men infected by this disease is

$$\frac{dy}{dt} = -ay + b(f - y)Y,$$

where y is the fraction of men infected, f is the fraction of men who are promiscuous, Y is the fraction of women infected, and a and b are appropriate constants.*

a. Assume $a = 1$, $b = 1$, and $f = .5$. Choose $Y = .01$, and solve for y using $y = .02$ when t is 0 as an initial condition. Round your answer to 3 decimal places.

b. A comparable model for women is

$$\frac{dY}{dt} = -AY + B(F - Y)y,$$

where F is the fraction of women who are promiscuous and A and B are constants. Assume $A = 1$, $B = 1$, and $F = .03$. Choose $y = .1$ and solve for Y, using $Y = .01$ as an initial condition.

SOCIAL SCIENCES

Spread of a Rumor The equation developed in the text for the spread of an epidemic also can be used to describe diffusion of information. In a population of size N, let y be the number who have heard a particular piece of information. Then

$$\frac{dy}{dt} = k\left(1 - \frac{y}{N}\right)y$$

for a positive constant k. Use this model in Exercises 13–15.

13. Suppose a rumor starts among 3 people in a certain office building. That is, $y_0 = 3$. Suppose 500 people work in the building and 50 people have heard the rumor in 2 days. Using Equation (11), write an equation for the number who have heard the rumor in t days. How many people will have heard the rumor in 5 days?

14. A rumor spreads at a rate proportional to the product of the number of people who have heard it and the number who have not heard it. Assume that 5 people in an office with 50 employees heard the rumor initially, and 15 people have heard it 3 days later.

a. Write an equation for the number, y, of people who have heard the rumor in t days.

b. When will 30 employees have heard the rumor?

15. A news item is heard on the late news by 5 of the 100 people in a small community. By the end of the next day 20 people have heard the news. Using Equation (11), write an equation for the number of people who have heard the news in t days. How many have heard the news after 3 days?

16. Repeat Exercise 14 using the Gompertz growth law,

$$\frac{dy}{dt} = kye^{-at},$$

for constants k and a, with $a = .1$.

PHYSICAL SCIENCES

17. *Salt Concentration* A tank holds 100 gal of water that contains 20 lb of dissolved salt. A brine (salt) solution is flowing into the tank at the rate of 2 gal per minute while the solution flows out of the tank at the same rate. The brine solution entering the tank has a salt concentration of 2 lb per gal.

a. Find an expression for the amount of salt in the tank at any time.

b. How much salt is present after 1 hour?

c. As time increases, what happens to the salt concentration?

18. Solve Exercise 17 if the brine solution is introduced at the rate of 3 gal per minute while the rate of outflow remains the same.

19. Solve Exercise 17 if the brine solution is introduced at the rate of 1 gal per minute while the rate of outflow stays the same.

20. Solve Exercise 17 if pure water is added instead of brine.

21. *Chemical in a Solution* Five grams of a chemical is dissolved in 100 L of alcohol. Pure alcohol is added at the rate of 2 L per minute and at the same time the solution is being drained at the rate of 1 L per minute.

a. Find an expression for the amount of the chemical in the mixture at any time.

b. How much of the chemical is present after 30 minutes?

*Bender, Edward A., *An Introduction to Mathematical Modeling*. Copyright © 1978 by John Wiley and Sons, Inc. Reprinted by permission.

22. Solve Exercise 21 if a 10% solution of the same mixture is added instead of pure alcohol.

23. *Soap Concentration* A prankster puts 4 lb of soap in a fountain that contains 200 gal of water. To clean up the mess a city crew runs clear water into the fountain at the rate of 8 gal per minute allowing the excess solution to drain off at the same rate. How long will it be before the amount of soap in the mixture is reduced to 1 lb?

CHAPTER SUMMARY

In this chapter we solved differential equations by finding functions that satisfy a particular equation involving the derivative of that function. We developed the methods of separation of variables and linear first-order equations to solve a wide range of differential equations. We also introduced a numerical method, Euler's method, that employs technology to solve problems in which the other methods cannot be applied. Applications in the chapter included the logistic equation for populations, a predator-prey model, an arms race model, mixing problems, and continuous deposits.

KEY TERMS

differential equation

10.1 general solution
particular solution
initial condition
initial value problem
separable differential equation
separation of variables
growth rate constant
carrying capacity
logistic growth model
logistic equation
logistic curve

10.2 linear first-order differential equation
integrating factor

10.3 Euler's method

10.4 Lotka-Volterra equations

CHAPTER 10 REVIEW EXERCISES

1. What is a differential equation? What is it used for?

2. What is the difference between a particular solution and a general solution to a differential equation?

3. How can you tell that a differential equation is separable? That it is linear?

4. Can a differential equation be both separable and linear? Explain why not, or give an example of an equation that is both.

Classify each equation as separable, linear, both, or neither.

5. $y\dfrac{dy}{dx} = x + y$

6. $\dfrac{dy}{dx} + y^2 = xy^2$

7. $\sqrt{x}\dfrac{dy}{dx} = \dfrac{1 + e^x}{y}$

8. $\dfrac{dy}{dx} = xy + \ln x$

9. $\dfrac{dy}{dx} + x = xy$

10. $\dfrac{x}{y}\dfrac{dy}{dx} = 1 + x^{3/2}$

11. $x\dfrac{dy}{dx} + y = e^x(1 + y)$

12. $\dfrac{dy}{dx} = x^2 + y^2$

Find the general solution for each differential equation.

13. $\dfrac{dy}{dx} = 2x^3 + 6x$

14. $\dfrac{dy}{dx} = x^2 + 5x^4$

15. $\dfrac{dy}{dx} = 4e^x$

16. $\dfrac{dy}{dx} = \dfrac{1}{2x + 3}$

17. $\dfrac{dy}{dx} = \dfrac{3x + 1}{y}$

18. $\dfrac{dy}{dx} = \dfrac{e^x + x}{y - 1}$

19. $\dfrac{dy}{dx} = \dfrac{2y + 1}{x}$

20. $\dfrac{dy}{dx} = \dfrac{3 - y}{e^x}$

21. $\dfrac{dy}{dx} + 5y = 12$

22. $\dfrac{dy}{dx} + xy = 4x$

23. $3\dfrac{dy}{dx} + xy - x = 0$

24. $x\dfrac{dy}{dx} + y - e^x = 0$

Find the particular solution for each initial value problem. (Some solutions may give y implicitly.)

25. $\frac{dy}{dx} = x^2 - 5x; \quad y(0) = 1$

26. $\frac{dy}{dx} = (3x + 2)^2 e^y; \quad y(0) = 0$

27. $\frac{dy}{dx} = 5(e^{-x} - 1); \quad y(0) = 17$

28. $e^x \frac{dy}{dx} - e^x y = x^2 - 1; \quad y(0) = 42$

29. $(5 - 2x)y = \frac{dy}{dx}; \quad y(0) = 2$

30. $\frac{dy}{dx} + x^2 y = x^2; \quad y(0) = 8$

31. $\frac{dy}{dx} = \frac{1 - 2x}{y + 3}; \quad y(0) = 16$

32. $x\frac{dy}{dx} - 2x^2 y + 3x^2 = 0; \quad y(0) = 15$

Find the particular solution for each differential equation.

33. $\frac{dy}{dx} = 4x^3 + 2; \quad y(1) = 3$

34. $\frac{dy}{dx} = \frac{x}{x^2 - 3}; \quad y(2) = 52$

35. $\sqrt{x}\frac{dy}{dx} = xy; \quad y(1) = 4$

36. $x^2\frac{dy}{dx} + 4xy - e^{2x^3} = 0; \quad y(1) = e^2$

37. When is Euler's method useful?

Use Euler's method to approximate the indicated function value for $y = f(x)$ to 3 decimal places, using $h = .2$.

38. $\frac{dy}{dx} = x + y^{-1}; \quad y(0) = 1; \quad$ find $y(1)$

39. $\frac{dy}{dx} = e^x + y; \quad y(0) = 1; \quad$ find $y(.6)$

40. Let $y = f(x)$ and $dy/dx = (x/2) + 4$, with $y(0) = 0$. Use Euler's method with $h = .1$ to approximate $y(.3)$ to 3 decimal places. Then solve the differential equation and find $f(.3)$ to 3 decimal places. Also, find $y_3 - f(x_3)$.

41. Let $y = f(x)$ and $dy/dx = 3 + \sqrt{y}$, with $y(0) = 0$. Construct a table for x_i and y_i like the one in Section 10.3, Example 2, for $[0, 1]$, with $h = .2$. Then graph the polygonal approximation of the graph of $y = f(x)$.

42. What is the logistic equation? Why is it useful?

Applications

BUSINESS AND ECONOMICS

43. *Marginal Sales* The marginal sales (in hundreds of dollars) of a computer software company are given by

$$\frac{dy}{dx} = 5e^{.2x},$$

where x is the number of months the company has been in business. Assume that sales were 0 initially.

a. Find the sales after 6 months.

b. Find the sales after 12 months.

44. *Continuous Withdrawals* A deposit of $10,000 is made to a savings account at 5% interest compounded continuously. Assume that continuous *withdrawals* of $1000 per year are made.

a. Write a differential equation to describe the situation.

b. How much will be left in the account after 1 year?

45. In Exercise 44, approximately how long will it take to use up the account?

46. *Production Rate* The rate at which a new worker in a certain factory produces items is given by

$$\frac{dy}{dx} = .2(125 - y),$$

where y is the number of items produced by the worker per day, x is the number of days worked, and the maximum production per day is 125 items. Assume the worker produced 20 items the first day on the job $(x = 0)$.

a. Find the number of items the new worker will produce in 10 days.

b. According to the function that is the solution of the differential equation, can the worker ever produce 125 items in a day?

LIFE SCIENCES

47. *Effect of Insecticide* After use of an experimental insecticide, the rate of decline of an insect population is

$$\frac{dy}{dt} = \frac{-10}{1 + 5t},$$

where t is the number of hours after the insecticide is applied. Assume that there were 50 insects initially.

a. How many are left after 24 hours?

b. How long will it take for the entire population to die?

48. *Growth of a Mite Population* A population of mites grows at a rate proportional to the number present, y. If the growth constant is 10% and 120 mites are present at time $t = 0$ (in weeks), find the number present after 6 weeks.

49. *Competing Species* Find an equation relating x to y given the following equations, which describe the interaction of two competing species and their growth rates.

$$\frac{dx}{dt} = .2x - .5xy$$
$$\frac{dy}{dt} = -.3y + .4xy$$

Find the values of x and y for which both growth rates are 0.

50. *Smoke Content in a Room* The air in a meeting room of 15,000 ft^3 has a smoke content of 20 parts per million (ppm). An air conditioner is turned on, which brings fresh air (with no smoke) into the room at a rate of 1200 ft^3 per minute and forces the smoky air out at the same rate. How long will it take to reduce the smoke content to 5 ppm?

51. In Exercise 50, how long will it take to reduce the smoke content to 10 ppm if smokers in the room are adding smoke at the rate of 5 ppm per minute?

52. *Spread of Influenza* A small, isolated mountain community with a population of 700 is visited by an outsider who carries influenza. After 6 weeks, 300 people are uninfected.

a. Write an equation for the number of people who remain uninfected at time t (in weeks).

b. Find the number still uninfected after 7 weeks.

c. When will the maximum infection rate occur?

53. *Population Growth* Let

$$y = \frac{N}{1 + be^{-kx}}.$$

If y is y_1, y_2, and y_3 at times x_1, x_2, and $x_3 = 2x_2 - x_1$ (that is, at three equally spaced times), then prove that

$$N = \frac{1/y_1 + 1/y_3 - 2/y_2}{1/(y_1y_3) - 1/y_2^2}.$$

*Population Growth In the following table of U.S. census figures, y is population in millions.**

Year	y	Year	y
1790	3.9	1900	76.0
1800	5.3	1910	92.0
1810	7.2	1920	105.7
1820	9.6	1930	122.8
1830	12.9	1940	131.7
1840	17.1	1950	150.7
1850	23.2	1960	179.3
1860	31.4	1970	203.3
1870	39.8	1980	226.5
1880	50.2	1990	248.7
1890	62.9	2000	281.4

54. Use Exercise 53 and the table to find the following.

a. Find N using the years 1800, 1850, and 1900.

b. Find N using the years 1850, 1900, and 1950.

c. Find N using the years 1870, 1920, and 1970.

d. Explain why different values of N were obtained in parts a–c. What does this suggest about the validity of this model and others?

55. Let $x = 0$ correspond to 1870, and let every decade correspond to an increase in x of 1.

a. Use 1870, 1920, and 1970 to find N, 1870 to find b, and 1920 to find k in the equation

$$y = \frac{N}{1 + be^{-kx}}.$$

b. Estimate the population of the United States in 2000 and compare your estimate to the actual population in 2000.

c. Predict the populations of the United States in 2030 and 2050.

56. Let $x = 0$ correspond to 1790, and let every decade correspond to an increase in x of 1. Use a calculator with logistic regression capability to complete the following.

a. Plot the data points. Do the points suggest that a logistic function is appropriate here?

b. Use the logistic regression function on your calculator to determine the logistic equation that best fits the data.

*U.S. Census Bureau.

c. Plot the logistic equation from part a on the same graph as the data points. How well does the logistic equation seem to fit the data?

d. What seems to be the limiting size of the U.S. population?

SOCIAL SCIENCES

57. *Education* Researchers have proposed that the amount a full-time student is educated (x) changes with respect to the student's age t according to the differential equation

$$\frac{dx}{dt} = 1 - kx,$$

where k is a constant measuring the rate that education depreciates due to forgetting or technological obsolescence.*

a. Solve the equation using the method of separation of variables.

b. Solve the equation using an integrating factor.

c. What does x approach over time?

58. *Spread of a Rumor* A rumor spreads through the offices of a company with 100 employees, starting in a meeting with 4 people. After 3 days, 15 people have heard the rumor.

a. Write an equation for the number of people who have heard the rumor in x days.

b. How many people have heard the rumor in 5 days?

PHYSICAL SCIENCES

59. *Newton's Law of Cooling* A roast at a temperature of 40°F is put in a 300°F oven. After 1 hour the roast has reached a temperature of 150°F. Newton's law of cooling states that

$$\frac{dT}{dt} = k(T - T_F),$$

where T is the temperature of an object, the surrounding medium has temperature T_F at time t, and k is a constant. Use Newton's law to find the temperature of the roast after 2 hours.

60. In Exercise 59, how long does it take for the roast to reach a temperature of 250°F?

61. *Air Resistance* In the section on antiderivatives, we saw that the acceleration of gravity is a constant if air resistance is ignored. But air resistance cannot always be ignored, or parachutes would be of little use. In the presence of air resistance, the equation for acceleration also contains a term roughly proportional to the velocity squared. Since acceleration forces a falling object downward and air resistance pushes it upward, the air resistance term is opposite in sign to the acceleration of gravity. Thus,

$$a(t) = \frac{dv}{dt} = g - kv^2,$$

where g and k are positive constants. Future calculations will be simpler if we replace g and k by the squared constants G^2 and K^2, giving

$$\frac{dv}{dt} = G^2 - K^2v^2.$$

a. Use separation of variables and the fact that

$$\frac{1}{G^2 - K^2v^2} = \frac{1}{2G}\left(\frac{1}{G - Kv} + \frac{1}{G + Kv}\right)$$

to solve the differential equation above. Assume $v < G/K$, which is certainly true when the object starts falling (with $v = 0$). Write your solution in the form of v as a function of t.

b. Find $\lim_{t \to \infty} v(t)$, where $v(t)$ is the solution you found in part a. What does this tell you about a falling object in the presence of air resistance?

c. According to *Harper's Index*, the terminal velocity of a cat falling from a tall building is 60 mph.† Use your answers from part b, plus the fact that 60 mph = 88 ft per second and g, the acceleration of gravity, is 32 ft per second^2, to find a formula for the velocity of a falling cat (in ft per second) as a function of time (in seconds). (*Hint:* Find K in terms of G. Then substitute into the answer from part a.)

*Southwick, Lawrence, Jr. and Stanley Zionts, "An Optimal-Control-Theory Approach to the Education-Investment Decision," *Operations Research*, Vol. 22, 1974, pp. 1156–1174.
†*Harper's*, Oct. 1994, p. 13.

EXTENDED APPLICATION: Pollution of the Great Lakes*

Industrial nations are beginning to face the problems of water pollution. Lakes present a problem, because a polluted lake contains a considerable amount of water that must somehow be cleaned. The main cleanup mechanism is the natural process of gradually replacing the water in the lake. This application deals with pollution in the Great Lakes. The basic idea is to regard the flow in the Great Lakes as a mixing problem.

We make the following assumptions.

1. Rainfall and evaporation balance each other, so the average rates of inflow and outflow are equal.
2. The average rates of inflow and outflow do not vary much seasonally.
3. When water enters the lake, perfect mixing occurs, so that the pollutants are uniformly distributed.
4. Pollutants are not removed from the lake by decay, sedimentation, or in any other way except outflow.
5. Pollutants flow freely out of the lake; they are not retained (as DDT is).

(The first two are valid assumptions; however, the last three are questionable.)

We will use the following variables in the discussion to follow.

V = volume of the lake

P_L = pollution concentration in the lake at time t

P_i = pollution concentration in the inflow to the lake at time t

r = rate of flow

t = time in years

By the assumptions stated above, the net change in total pollutants during the time interval Δt is (approximately)

$$V \cdot \Delta P_L = (P_i - P_L)(r \cdot \Delta t),$$

where ΔP_L is the change in the pollution concentration. Dividing this equation by Δt and by V and taking the limit as $\Delta t \to 0$, we get the differential equation

$$\frac{dP_L}{dt} = \frac{(P_i - P_L)r}{V}.$$

Since we are treating V and r as constants, we replace r/V with k, so the equation can be written as the first-order linear equation

$$\frac{dP_L}{dt} + kP_L = kP_i.$$

The solution is

$$P_L(t) = e^{-kt}\left[P_L(0) + k\int_0^t P_i(x)e^{kx}\,dx\right]. \qquad \textbf{(1)}$$

Figure 11 shows values of $1/k$ for each lake (except Huron) measured in years. *If the model is reasonable*, the numbers in the figure can be used in Equation (1) to determine the effect of various pollution abatement schemes. Lake Ontario is excluded from the discussion because about 84% of its inflow comes from Erie and can be controlled only indirectly.

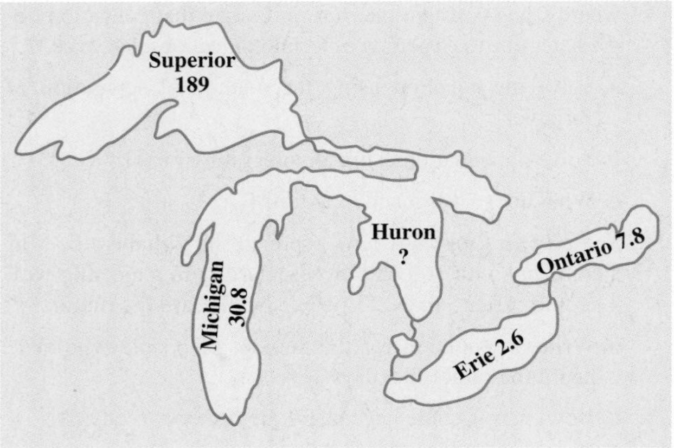

FIGURE 11

The fastest possible cleanup will occur if all pollution inflow ceases. This means that $P_i = 0$. In this case, Equation (1) leads to

$$t = \frac{1}{k}\ln\left(\frac{P_L(0)}{P_L(t)}\right).$$

From this we can tell the length of time necessary to reduce pollution to a given percentage of its present level. For example, from the figure, for Lake Superior $1/k = 189$. Thus, to reduce pollution to 50% of its present level, $P_L(0)$, we want

$$\frac{P_L(t)}{P_L(0)} = .5 \quad \text{or} \quad \frac{P_L(0)}{P_L(t)} = 2,$$

from which

$$t = 189 \ln 2 \approx 131.$$

The following figures, representing years, were found in this way.

*Bender, Edward A., *An Introduction to Mathematical Modeling*. Copyright © 1978 by John Wiley & Sons, Inc. Reprinted by permission from *An Introduction to Mathematical Modeling* by Edward A. Bender (Dover, 2000).

Lake	50%	20%	10%	5%
Erie	2	4	6	8
Michigan	21	50	71	92
Superior	131	304	435	566

Fortunately, the pollution in Lake Superior is quite low at present.

As mentioned before, assumptions 3, 4, and 5 are questionable. For persistent pollutants like DDT, the estimated cleanup times may be too low. For other pollutants, how assumptions 4 and 5 affect cleanup times is unclear. However, the values of $1/k$ given in the figure probably provide rough lower bounds for the cleanup times of persistent pollutants.

Exercises

1. Calculate the number of years to reduce pollution in Lake Erie to each level.
   a. 40%
   b. 30%
2. Repeat Exercise 1 for Lake Michigan.
3. Repeat Exercise 1 for Lake Superior.
4. We claim that Equation (1) is a solution of the differential equation

$$\frac{dP_L}{dt} + kP_L(t) = kP_i(t),$$

where t measures time from the present. The constant $k = r/V$ measures how quickly the water in the lake is replaced through inflow and corresponding outflow. The constant $P_L(0)$ is the current pollution level.

   a. To show that Equation (1) does define a solution of the differential equation, multiply both sides of Equation (1) by e^{kt} and then differentiate both sides with respect to t. Remember from the section on the Fundamental Theorem of Calculus that you can differentiate an integral by using the version of the Fundamental Theorem that says

$$\frac{d}{dt}\int_a^t f(x)\,dx = f(t).$$

   b. When you substitute $t = 0$ into the right-hand side of Equation (1), you should get $P_L(0)$. Do you? What happens to the integral? What happens to the factor of e^{-kt}?

   c. The map indicates a value of 30.8 for Lake Michigan. What value of k does this correspond to? What percent of the water in Lake Michigan is replaced each year by inflow? Which lake has the biggest annual water turnover?

5. Suppose that instead of assuming that all pollution inflow immediately ceases, we model $P_i(t)$ by a decaying exponential of the form $a \cdot e^{-pt}$, where p is a constant that tells us how fast the inflow is being cleaned of pollution. To simplify things, we'll also assume that initially the inflow and the lake have the same pollution concentration, so $a = P_L(0)$. Now substitute $P_L(0)e^{-px}$ for $P_i(x)$ in Equation (1), and evaluate the integral as a function of t.
6. When you simplify the right-hand side of Equation (1) using your new expression for the integral, and then factor out and divide by $P_L(0)$, you'll get the following nice expression for the ratio $P_L(t)/P_L(0)$:

$$\frac{P_L(t)}{P_L(0)} = \frac{1}{k - p}(ke^{-pt} - pe^{-kt}).$$

   a. Suppose that for Lake Michigan the constant p is equal to .02. Use a graph of the ratio $P_L(t)/P_L(0)$ to estimate how long it will take to reduce pollution to 50% of its current value. How does this compare with the time, assuming pollution-free inflow?

   b. If the constant p has the value 0 for Lake Michigan, what does that tell you about the pollution level in the inflow? In this case, what happens to the ratio $P_L(t)/P_L(0)$ over time?

Directions for Group Project

Suppose you and three others are employed by an agency that is concerned about the environmental health of one of the Great Lakes. Choose one of the lakes and collect information about levels of pollution in it. Then, using the information you collected along with the information given in this application, prepare a public presentation for a local community organization that describes the lake and gives possible timelines for reducing pollution in the lake. Use presentation software such as Microsoft Powerpoint.

CHAPTER

11 Probability and Calculus

Though earthquakes may appear to strike at random, the times between quakes can be modeled with an exponential density function. Such *continuous probability models* have many applications in science, engineering, and medicine. In an exercise in Section 1 of this chapter we'll use an exponential density function to describe the times between major earthquakes in southern California, and in Section 3 we will compute the mean and standard deviation for this distribution.

In recent years, probability has become increasingly useful in fields ranging from manufacturing to medicine, as well as in all types of research. The foundations of probability were laid in the seventeenth century by Blaise Pascal (1623–1662) and Pierre de Fermat (1601–1665), who investigated *the problem of the points.* This problem dealt with the fair distribution of winnings in an interrupted game of chance between two equally matched players whose scores were known at the time of the interruption.

Probability has advanced from a study of gambling to a well-developed, deductive mathematical system. In this chapter we give a brief introduction to the use of calculus in probability.

11.1 CONTINUOUS PROBABILITY MODELS

THINK ABOUT IT *What is the probability that there is a bird's nest within .5 kilometers of a given point?*

In this section, we show how calculus is used to find the probability of certain events. Later in the section, we will answer the question posed above. Before discussing probability, however, we need to introduce some new terminology.

Suppose that a bank is studying the transaction times of its tellers. The lengths of time spent on observed transactions, rounded to the nearest minute, are shown in the following table.

Time	1	2	3	4	5	6	7	8	9	10	
Frequency	3	5	9	12	15	11	10	6	3	1	(Total: 75)

The table shows, for example, that 9 of the 75 transactions in the study took 3 minutes, 15 transactions took 5 minutes, and 1 transaction took 10 minutes. Because the time for any particular transaction is a random event, the number of minutes for a transaction is called a **random variable.** The frequencies can be converted to probabilities by dividing each frequency by the total number of transactions (75) to get the results shown in the next table.*

Time	1	2	3	4	5	6	7	8	9	10
Probability	.04	.07	.12	.16	.20	.15	.13	.08	.04	.01

Because each value of the random variable is associated with just one probability, this table defines a function. Such a function is called a **probability function.** The special properties of a probability function are given on the following page.

*One definition of the *probability of an event* is the number of outcomes that favor the event divided by the total number of equally likely outcomes in an experiment.

PROBABILITY FUNCTION OF A RANDOM VARIABLE

If the function f is a probability function with domain $\{x_1, x_2, \ldots, x_n\}$, and $f(x_i)$ is the probability that event x_i occurs, then for $1 \leq i \leq n$,

$$0 \leq f(x_i) \leq 1,$$

and

$$f(x_1) + f(x_2) + \cdots + f(x_n) = 1.$$

Note that $f(x_i) = 0$ implies that event x_i will not occur and $f(x_i) = 1$ implies that event x_i will occur.

The information in the second table can be displayed graphically with a special kind of bar graph called a **histogram.** The bars of a histogram have the same width, and their heights are determined by the probabilities of the various values of the random variable. See Figure 1 on the next page.

The probability function in the second table is a **discrete probability function** because it has a finite domain—the integers from 1 to 10, inclusive. A discrete probability function has a finite domain or an infinite domain that can be listed. For example, if we flip a coin until we get heads, and let the random variables be the number of flips, then the domain is 1, 2, 3, 4, ... On the other hand, the distribution of heights (in inches) of college women includes infinitely many possible measurements, such as 53, 54.2, 66.5, $72.\overline{3}$, and so on, *within some real number interval*. Probability functions with such domains are called *continuous probability distributions*.

CONTINUOUS PROBABILITY DISTRIBUTION

A **continuous random variable** can take on any value in some interval of real numbers. The distribution of this random variable is called a **continuous probability distribution.**

FOR REVIEW

The connection between area and the definite integral is discussed in the chapter on Integration. For example, in that chapter we solved such problems as the following: Find the area between the x-axis and the graph of $f(x) = x^2$ from $x = 1$ to $x = 4$.

Answer: $\int_1^4 x^2 dx = 21$

In the bank example discussed earlier in this section, it would have been possible to time the teller transactions with greater precision—to the nearest tenth of a minute or even to the nearest second (one-sixtieth of a minute), if desired. Theoretically, at least, t could take on any positive real-number value between, say, 0 and 11 minutes. The graph of the probabilities $f(t)$ of these transaction times can be thought of as the continuous curve shown in Figure 1. As indicated in Figure 1, the curve was derived from our table by connecting the points at the tops of the bars in the corresponding histogram and smoothing the resulting polygon into a curve.

For a discrete probability function, the area of each bar (or rectangle) gives the probability of a particular transaction time. Thus, by considering the possible transaction times t as all the real numbers between 0 and 11, the area under the curve of Figure 2 (on the next page) between any two values of t can be interpreted as the probability that a transaction time will be between those two numbers. For example, the shaded region in Figure 2 corresponds to the probability that t is between a and b, written $P(a \leq t \leq b)$.

It was shown earlier that the definite integral of a continuous function f, where $f(x) \geq 0$, gives the area under the graph of $f(x)$ from $x = a$ to $x = b$. If

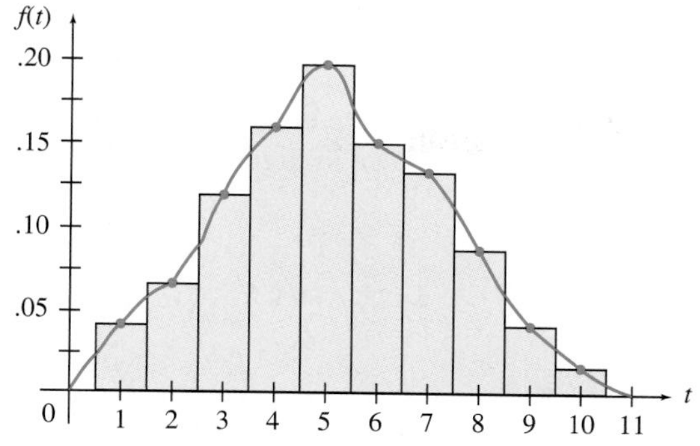

FIGURE 1

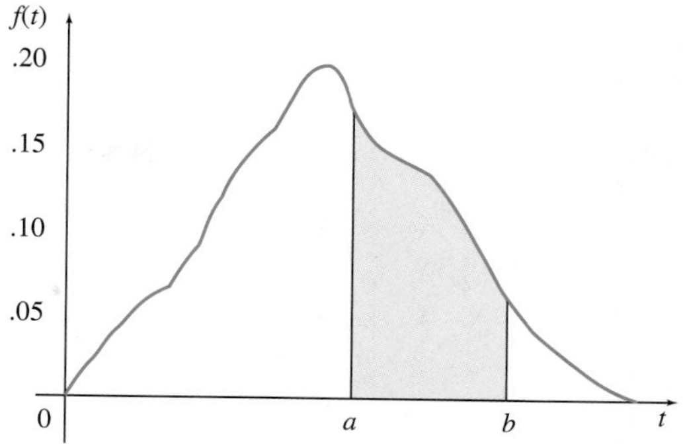

FIGURE 2

a function f can be found to describe a continuous probability distribution, then the definite integral can be used to find the area under the curve from a to b that represents the probability that x will be between a and b.

> If x is a continuous random variable whose distribution is described by the function f on $[a, b]$, then
>
> $$P(a \leq x \leq b) = \int_a^b f(x)\,dx.$$

Probability Density Functions A function f that describes a continuous probability distribution is called a *probability density function.* Such a function must satisfy the following conditions.

> **PROBABILITY DENSITY FUNCTION**
>
> The function f is a **probability density function** of a random variable x in the interval $[a, b]$ if
>
> **1.** $f(x) \geq 0$ for all x in the interval $[a, b]$; and
> **2.** $\int_a^b f(x)\,dx = 1$.

Intuitively, Condition 1 says that the probability of a particular event can never be negative. Condition 2 says that the total probability for the interval must be 1; *something* must happen.

EXAMPLE 1 *Probability Density Function*

(a) Show that the function defined by $f(x) = (3/26)x^2$ is a probability density function for the interval $[1, 3]$.

Solution First, note that Condition 1 holds; that is, $f(x) \geq 0$ for the interval $[1, 3]$. Next show that Condition 2 holds.

$$\int_1^3 \frac{3}{26}x^2 dx = \frac{3}{26}\left(\frac{x^3}{3}\right)\Bigg|_1^3 = \frac{3}{26}\left(9 - \frac{1}{3}\right) = 1$$

Since both conditions hold, $f(x)$ is a probability density function.

(b) Find the probability that x will be between 1 and 2.

Solution The desired probability is given by the area under the graph of $f(x)$ between $x = 1$ and $x = 2$, as shown in Figure 3. The area is found by using a definite integral.

$$P(1 \le x \le 2) = \int_1^2 \frac{3}{26}x^2\,dx = \frac{3}{26}\left(\frac{x^3}{3}\right)\bigg|_1^2 = \frac{7}{26}$$

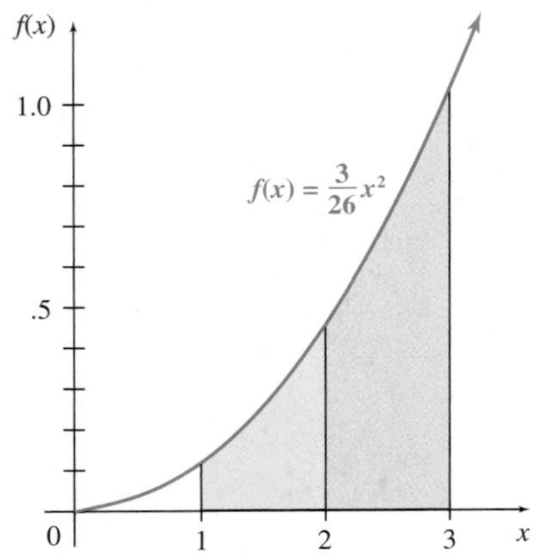

FIGURE 3

Earlier, we noted that determining a suitable function is the most difficult part of applying mathematics to actual situations. Sometimes a function appears to model an application well, but does not satisfy the requirements for a probability density function. In such cases, we may be able to change the function into a probability density function by multiplying it by a suitable constant, as shown in the next example.

EXAMPLE 2 *Probability Density Function*

Is there a constant k such that $f(x) = kx^2$ is a probability density function for the interval $[0, 4]$?

Solution First,

$$\int_0^4 kx^2\,dx = \frac{kx^3}{3}\bigg|_0^4 = \frac{64k}{3}.$$

The integral must be equal to 1 for the function to be a probability density function. To convert it to one, let $k = 3/64$. The function defined by $(3/64)x^2$ for $[0, 4]$ will be a probability density function, since $(3/64)x^2 \ge 0$ for all x in $[0, 4]$ and

$$\int_0^4 \frac{3}{64}x^2 = 1.$$

An important distinction is made between a discrete probability function and a probability density function (which is continuous). In a discrete distribution, the probability that the random variable, x, will assume a specific value is given in the distribution for every possible value of x. In a probability density function, however, the probability that x equals a specific value, say, c, is

$$P(x = c) = \int_c^c f(x)\,dx = 0.$$

For a probability density function, only probabilities of *intervals* can be found. For example, suppose the random variable is the annual rainfall for a given region. The amount of rainfall in one year can take on any value within some continuous interval that depends on the region; however, the probability that the rainfall in a given year will be some specific amount, say 33.25 in., is actually zero.

The definition of a probability density function is extended to intervals such as $(-\infty, b]$, $(-\infty, b)$, $[a, \infty)$, (a, ∞), or $(-\infty, \infty)$ by using improper integrals, as follows.

PROBABILITY DENSITY FUNCTIONS ON $(-\infty, \infty)$

If f is a probability density function for a continuous random variable x on $(-\infty, \infty)$, then

$$P(x \le b) = P(x < b) = \int_{-\infty}^{b} f(x)\,dx,$$

$$P(x \ge a) = P(x > a) = \int_{a}^{\infty} f(x)\,dx,$$

$$P(-\infty < x < \infty) = \int_{-\infty}^{\infty} f(x)\,dx = 1.$$

The total area under the graph of a probability density function of this type must still equal 1.

EXAMPLE 3 *Location of a Bird's Nest*

Suppose the random variable x is the distance (in kilometers) from a given point to the nearest bird's nest, with the probability density function of the distribution given by $f(x) = 2xe^{-x^2}$ for $x \ge 0$.

FOR REVIEW

Improper integrals, those with one or two infinite limits, were discussed in the chapter on Further Techniques and Applications of Integration. The type of improper integral we shall need was defined as

$$\int_{a}^{\infty} f(x)\,dx = \lim_{b\to\infty} \int_{a}^{b} f(x)\,dx.$$

For example,

$$\begin{aligned}\int_{1}^{\infty} x^{-2}\,dx &= \lim_{b\to\infty} \int_{1}^{b} x^{-2}\,dx \\ &= \lim_{b\to\infty} \left(-\frac{1}{x}\bigg|_{1}^{b} \right) \\ &= \lim_{b\to\infty} \left(-\frac{1}{b} + \frac{1}{1} \right) \\ &= 0 + 1 = 1.\end{aligned}$$

(a) Show that $f(x)$ is a probability density function.

Solution Since $e^{-x^2} = 1/e^{x^2}$ is always positive, and $x \ge 0$,

$$f(x) = 2xe^{-x^2} \ge 0.$$

Use substitution to evaluate the definite integral $\int_0^\infty 2xe^{-x^2}\,dx$. Let $u = -x^2$, so that $du = -2x\,dx$, and

$$\begin{aligned}\int 2xe^{-x^2}\,dx &= -\int e^{-x^2}(-2x\,dx) \\ &= -\int e^{u}\,du = -e^{u} = -e^{-x^2}.\end{aligned}$$

Then

$$\begin{aligned}\int_{0}^{\infty} 2xe^{-x^2}\,dx &= \lim_{b\to\infty} \int_{0}^{b} 2xe^{-x^2}\,dx = \lim_{b\to\infty} (-e^{-x^2})\bigg|_{0}^{b} \\ &= \lim_{b\to\infty} \left(-\frac{1}{e^{b^2}} + e^{0} \right) = 0 + 1 = 1.\end{aligned}$$

The function defined by $f(x) = 2xe^{-x^2}$ satisfies the two conditions required of a probability density function.

(b) Find the probability that there is a bird's nest within .5 km of the given point.

Solution Find $P(x \le .5)$ where $x \ge 0$. This probability is given by

$$P(0 \le x \le .5) = \int_{0}^{.5} 2xe^{-x^2}\,dx.$$

Now evaluate the integral. The indefinite integral was found in part (a).

$$P(0 \le x \le .5) = \int_0^{.5} 2xe^{-x^2}\,dx = (-e^{-x^2})\Big|_0^{.5}$$
$$= -e^{-(.5)^2} - (-e^0) = -e^{-.25} + 1$$
$$\approx -.78 + 1 = .22$$

The probability that a bird's nest will be found within .5 km of the given point is about .22.

EXAMPLE 4 *Computing Mortality*

According to the National Center for Health Statistics, if we start with 100,000 people who are 60 years old, we can expect a certain number of them to die within each 5-year time interval, as indicated by the following table.*

Years from Age 60	Midpoint of Interval	Number Dying in Each Interval
0–5	2.5	6457
5–10	7.5	8898
10–15	12.5	12,159
15–20	17.5	15,075
20–25	22.5	18,138
25–30	27.5	18,126
30–35	32.5	13,255
35–40	37.5	6102
40–45	42.5	1790

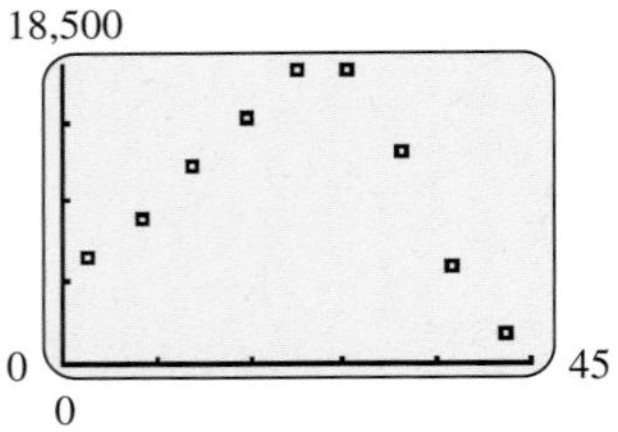

FIGURE 4

(a) Plot the data.

Solution Figure 4 shows that the plot appears to have the shape of a polynomial.

(b) Use the regression feature on a TI-83/84 Plus calculator to find a quartic equation that models the number of years, x, since age 60 and the number of deaths, $N(x)$. Use the midpoints and the number of deaths in each interval from the table above.

Solution The function

$$N(x) = .0681571x^4 - 6.72318x^3 + 184.397x^2 - 1091.40x + 8477.08$$

provided by the calculator models this data quite well, as illustrated by Figure 5.

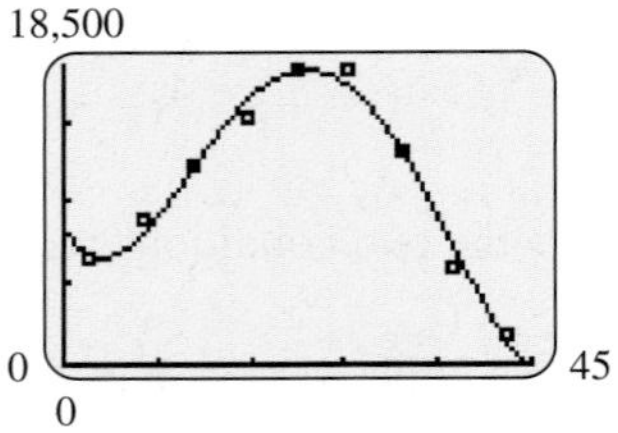

FIGURE 5

(c) Use the answer from part (b) to find a probability density function describing the probability of dying at various ages.

Solution We will construct a density function $S(x) = kN(x)$ by finding a suitable constant k, as we did in Example 2. The function $N(x)$ becomes negative at about $x = 44.29$, so we will restrict the domain of the density func-

**National Vital Statistics Reports,* Vol. 47, No. 19, June 30, 1999, Table 4.

tion to the interval $[0, 44]$. Using the integration feature on our calculator, we find that

$$\int_0^{44} S(x)\,dx = k\int_0^{44} N(x)\,dx = 500{,}669k.$$

We set this equal to 1 to get $k = 1/500{,}669$. The function defined by

$$S(x) = \frac{1}{500{,}669}N(x)$$
$$= \frac{1}{500{,}669}(.0681571x^4 - 6.72318x^3 + 184.397x^2 - 1091.40x + 8477.08)$$

will be a probability density function for $[0, 44]$, because

$$\int_0^{44} S(x)\,dx = 1, \quad \text{and } S(x) \geq 0 \text{ for all } x \text{ in } [0, 44].$$

(d) Find the probability that a randomly chosen 60-year-old person will live at least until age 80.

Solution Again, using the integration feature on our calculator,

$$P(x > 20) = \int_{20}^{44} S(x)\,dx \approx .57.$$

Thus, a 60-year-old person has a 57% chance of living at least until age 80.

Notice that this value could also be estimated from the table by finding the number of people who have not died by age 80 and then dividing this number by 100,000. Thus, according to the table, there are 57,411 people still alive at age 80, representing approximately 57% of the original population. As you can see, our estimate agrees quite well with the actual number.

11.1 EXERCISES

Decide whether the functions defined as follows are probability density functions on the indicated intervals. If not, tell why.

1. $f(x) = \frac{1}{9}x - \frac{1}{18}$; $[2, 5]$

2. $f(x) = \frac{1}{3}x - \frac{1}{6}$; $[3, 4]$

3. $f(x) = \frac{x^2}{21}$; $[1, 4]$

4. $f(x) = \frac{3}{98}x^2$; $[3, 5]$

5. $f(x) = 4x^3$; $[0, 3]$

6. $f(x) = \frac{x^3}{81}$; $[0, 3]$

7. $f(x) = \frac{x^2}{16}$; $[-2, 2]$

8. $f(x) = 2x^2$; $[-1, 1]$

9. $f(x) = \frac{5}{3}x^2 - \frac{5}{90}$; $[-1, 1]$

10. $f(x) = \frac{3}{13}x^2 - \frac{12}{13}x + \frac{45}{52}$; $[0, 4]$

Find a value of k that will make f a probability density function on the indicated interval.

11. $f(x) = kx^{1/2}$; $[1, 4]$

12. $f(x) = kx^{3/2}$; $[4, 9]$

13. $f(x) = kx^2$; $[0, 5]$

14. $f(x) = kx^2$; $[-1, 2]$

15. $f(x) = kx$; $[0, 3]$

16. $f(x) = kx$; $[2, 3]$

17. $f(x) = kx$; $[1, 5]$

18. $f(x) = kx^3$; $[2, 4]$

19. The total area under the graph of a probability density function always equals ________.

20. In your own words, define a random variable.

21. What is the difference between a discrete probability function and a probability density function?

22. Why is $P(x = c) = 0$ for any number c in the domain of a probability density function?

Show that each function defined as follows is a probability density function on the given interval; then find the indicated probabilities.

23. $f(x) = \frac{1}{2}(1 + x)^{-3/2}$; $[0, \infty)$

a. $P(0 \le x \le 2)$ **b.** $P(1 \le x \le 3)$

c. $P(x \ge 5)$

24. $f(x) = e^{-x}$; $[0, \infty)$

a. $P(0 \le x \le 1)$ **b.** $P(1 \le x \le 2)$

c. $P(x \le 2)$

25. $f(x) = (1/2)e^{-x/2}$; $[0, \infty)$

a. $P(0 \le x \le 1)$ **b.** $P(1 \le x \le 3)$

c. $P(x \ge 2)$

26. $f(x) = \frac{20}{(x + 20)^2}$; $[0, \infty)$

a. $P(0 \le x \le 1)$ **b.** $P(1 \le x \le 5)$

c. $P(x \ge 5)$

27. $f(x) = \begin{cases} \dfrac{x^3}{12} & \text{if } 0 \le x \le 2 \\ \dfrac{16}{3x^3} & \text{if } x > 2 \end{cases}$

a. $P(0 \le x \le 2)$ **b.** $P(x \ge 2)$

c. $P(1 \le x \le 3)$

28. $f(x) = \begin{cases} \dfrac{20x^4}{9} & \text{if } 0 \le x \le 1 \\ \dfrac{20}{9x^5} & \text{if } x > 1 \end{cases}$

a. $P(0 \le x \le 1)$ **b.** $P(x \ge 1)$

c. $P(0 \le x \le 2)$

Applications

BUSINESS AND ECONOMICS

29. *Life Span of a Computer Part* The life (in months) of a certain electronic computer part has a probability density function defined by

$$f(x) = \frac{1}{2}e^{-x/2} \quad \text{for } x \text{ in } [0, \infty).$$

Find the probability that a randomly selected component will last the following lengths of time.

a. At most 12 months

b. Between 12 and 20 months

30. *Machine Life* A machine has a useful life of 4 to 9 years, and its life (in years) has a probability density function defined by

$$f(x) = \frac{1}{11}\left(1 + \frac{3}{\sqrt{x}}\right).$$

Find the probabilities that the useful life of such a machine selected at random will be the following.

a. Longer than 6 years **b.** Less than 5 years

c. Between 4 and 7 years

LIFE SCIENCES

31. *Petal Length* The length of a petal, x, on a certain flower varies from 1 cm to 4 cm and has a probability density function defined by

$$f(x) = \frac{1}{2\sqrt{x}}.$$

Find the probabilities that the length of a randomly selected petal will be as follows.

a. Greater than or equal to 3 cm

b. Less than or equal to 2 cm

c. Between 2 cm and 3 cm

32. *Clotting Time of Blood* The clotting time of blood is a random variable x with values from 1 second to 20 seconds and probability density function defined by

$$f(x) = \frac{1}{(\ln 20)x}.$$

Find the following probabilities for a person selected at random.

a. The probability that the clotting time is between 1 and 5 seconds

b. The probability that the clotting time is greater than 10 seconds

33. *Flour Beetles* Researchers who study the abundance of the flour beetle, *Tribolium castaneum*, have developed a probability density function that can be used to estimate the abundance of the beetle in a population. The density function, which is a member of the gamma distribution, is

$$f(x) = 1.185 \cdot 10^{-9} x^{4.5222} e^{-.049846x},$$

where x is the size of the population.*

a. Estimate the probability that a randomly selected flour beetle population is between 0 and 150.

b. Estimate the probability that a randomly selected flour bettle population is between 100 and 200.

34. *Flea Beetles* The mobility of an insect is an important part of its survival. Researchers have determined that the probability that a marked flea beetle, *Phyllotreta cruciferae* and *Phylotreta striolata*, will be recaptured within a certain distance and time after release can be calculated from the probability density function

$$p(x, t) = \frac{e^{-x^2/(4Dt)}}{\int_0^L e^{-u^2/(4Dt)}\,du},$$

where t is the time after release (in hours), x is the distance (in meters) from the release point that recaptures occur, L is the maximum distance from the release point that recaptures can occur, and D is the diffusion coefficient.†

a. If $t = 12$, $L = 6$, and $D = 38.3$, find the probability that a flea beetle will be recaptured within 3 m of the release point.

b. Using the same coefficients, find the probability that a flea beetle will be recaptured between 1 and 5 m of the release point.

35. *Firearms* The number of deaths in the United States caused by firearms for each age group in 1997 is given in the following table.‡

Age interval (years)	Midpoint of interval (year)	Number dying in each interval
0–15	7.5	630
15–25	20	8173
25–35	30	7045
35–45	40	5802
45–55	50	3872
55–65	60	2390
65–75	70	2202
75–85	80	1740
85 +	90 (est)	555
Total		32,409

a. Plot the data. What type of function appears to best match this data?

b. Use the regression feature on your graphing calculator to find a quartic equation that models the number of years, x, since birth and the number of deaths caused by firearms, $N(x)$. Use the midpoint value to estimate the point in each interval when the person died. Graph the function with the plot of the data. Does the function resemble the data?

c. By finding an appropriate constant k, find a function $S(x) = kN(x)$ that is a probability density function describing the probability of death by firearms. (*Hint:*

*Dennis, Brian and Robert F. Costantino, "Analysis of Steady-State Populations with the Gamma Abundance Model: Application to *Tribolium*," *Ecology*, Vol. 69, No. 4, Aug. 1988, pp. 1200–1213.

†Karevia, Peter, "Experimental and Mathematical Analysis of Herbivore Movement: Quantifying the Influence of Plant Spacing and Quality on Foraging Discrimination," *Ecology Monographs*, Vol. 2, No. 3, Sept. 1982, pp. 261–282.

‡*National Vital Statistics Reports*, Vol. 47, No. 19, June 30, 1999, Table 16.

Because the function in part b is negative for values less than 6.8 and greater than 91.5, restrict the domain of the density function to the interval [6.8, 91.5]. That is, integrate the function you found in part b from 6.8 to 91.5.)

d. For a randomly chosen person who was killed by a firearm, find the probabilities that the person killed was less than 25 years old, between 45 and 65 years old, and at least 75 years old, and compare these with the actual probabilities.

SOCIAL SCIENCES

36. *Time to Learn a Task* The time x required for a person to learn a certain task is a random variable with probability density function defined by

$$f(x) = \frac{8}{7(x-2)^2}.$$

The time required to learn the task is between 3 and 10 minutes. Find the probabilities that a randomly selected person will learn the task in the following lengths of time.

a. Less than 4 minutes

b. More than 5 minutes

PHYSICAL SCIENCES

37. *Annual Rainfall* The annual rainfall in a remote Middle Eastern country varies from 0 to 5 in. and is a random variable with probability density function defined by

$$f(x) = \frac{5.5 - x}{15}.$$

Find the following probabilities for the annual rainfall in a randomly selected year.

a. The probability that the annual rainfall is greater than 3 in.

b. The probability that the annual rainfall is less than 2 in.

c. The probability that the annual rainfall is between 1 in. and 4 in.

38. *Earthquakes* The time between major earthquakes in the Southern California region is a random variable with probability density function

$$f(t) = \frac{1}{960}e^{-t/960},$$

where t is measured in days.*

a. Find the probability that the time between a major earthquake and the next one is less than 365 days.

b. Find the probability that the time between a major earthquake and the next one is more than 960 days.

39. *Earthquakes* The time between major earthquakes in the Taiwan region is a random variable with probability density function

$$f(t) = \frac{1}{3650.1}e^{-t/3650.1},$$

where t is measured in days.[†]

a. Find the probability that the time between a major earthquake and the next one is more than 1 year but less than 3 years.

b. Find the probability that the time between a major earthquake and the next one is more than 7300 days.

GENERAL INTEREST

40. *Drunk Drivers* The frequency of alcohol-related traffic fatalities has dropped in recent years, but is still high among young people.[‡] Based on data from the National Highway Traffic Safety Administration, the age of a randomly selected, alcohol-impaired driver in a fatal car crash is a random variable with probability density function given by

$$f(x) = \frac{.1906}{x^{.5012}} \quad \text{for } x \text{ in } [16, 44].$$

Find the following probabilities of the age of such a driver.

a. Less than or equal to 25

b. Greater than or equal to 35

c. Between 21 and 30

41. *Driving Fatalities* We saw in an exercise in the chapter on Calculating the Derivative that driver fatality rates were highest for the youngest and oldest drivers.[§] When adjusted for the number of miles driven by people in each age group, the number of drivers in fatal crashes goes down with age, and the age of a randomly selected driver in a fatal car crash is a random variable with probability density function given by

$$f(x) = .06049e^{-.03211x} \quad \text{for } x \text{ in } [16, 84].$$

Find the following probabilities of the age of such a driver.

*Wang, Jeen-Hwa and Chiao-Hui Kuo, "On the Frequency Distribution of Interoccurrence Times of Earthquakes," *Journal of Seismology,* Vol. 2, 1998, pp. 351–358. Reprinted with kind permission from Kluwer Academic Publishers.

[†]Ibid.

[‡]*Traffic Safety Facts 2001,* National Highway Traffic Safety Administration, p. 36.

[§]www-nrd.nhtsa.dot.gov/pdf/nrd-30/NCSA/RNotes/1998/AgeSex96.pdf.

a. Less than or equal to 25

b. Greater than or equal to 35

c. Between 21 and 30

42. *Length of a Telephone Call* The length of a telephone call (in minutes), x, for a certain town is a continuous random variable with probability density function defined by

$$f(x) = 3x^{-4}, \quad \text{for } x \text{ in } [1, \infty).$$

Find the probabilities for the following situations.

a. The call lasts between 1 and 2 minutes.

b. The call lasts between 3 and 5 minutes.

c. The call lasts longer than 3 minutes.

11.2 EXPECTED VALUE AND VARIANCE OF CONTINUOUS RANDOM VARIABLES

THINK ABOUT IT *What is the average age of a drunk driver in a fatal car crash?*

It often is useful to have a single number, a typical or "average" number, that represents a random variable. The *mean* or *expected value* for a discrete random variable is found by multiplying each value of the random variable by its corresponding probability, as follows.

EXPECTED VALUE

Suppose the random variable x can take on the n values, $x_1, x_2, x_3, \ldots, x_n$. Also, suppose the probabilities that each of these values occurs are, respectively, $p_1, p_2, p_3, \ldots, p_n$. Then the **mean, or expected value,** of the random variable is

$$\mu = x_1p_1 + x_2p_2 + x_3p_3 + \cdots + x_np_n = \sum_{i=1}^{n} x_ip_i.$$

For the banking example in the previous section, the expected value is given by

$$\begin{aligned}\mu &= 1(.04) + 2(.07) + 3(.12) + 4(.16) + 5(.20) + 6(.15) + 7(.13) \\ &\quad + 8(.08) + 9(.04) + 10(.01) \\ &= 5.09.\end{aligned}$$

Thus, the average time a person can expect to spend with the bank teller is 5.09 minutes.

This definition can be extended to continuous random variables by using definite integrals. Suppose a continuous random variable has probability density function f on $[a, b]$. We can divide the interval from a to b into n subintervals of length Δx, where $\Delta x = (b - a)/n$. In the ith subinterval, the probability that the random variable takes a value close to x_i is approximately $f(x_i)\,\Delta x$, and so

$$\mu \approx \sum_{i=1}^{n} x_i f(x_i)\,\Delta x.$$

As $n \to \infty$, the limit of this sum gives the expected value

$$\mu = \int_a^b xf(x)\,dx.$$

The **variance** of a probability distribution is a measure of the *spread* of the values of the distribution. For a discrete distribution, the variance is found by taking the expected value of the squares of the differences of the values of the random variable and the mean. If the random variable x takes the values $x_1, x_2, x_3, \ldots, x_n$, with respective probabilities $p_1, p_2, p_3, \ldots, p_n$ and mean μ, then the variance of x is

$$\text{Var}(x) = \sum_{i=1}^{n} (x_i - \mu)^2 p_i.$$

Think of the variance as the expected value of $(x - \mu)^2$, which measures how far x is from the mean μ. The **standard deviation** of x is defined as

$$\sigma = \sqrt{\text{Var}(x)}.$$

For the banking example in the previous section, the variance and standard deviation are

$$\begin{aligned}\text{Var}(X) &= (1 - 5.09)^2(.04) + (2 - 5.09)^2(.07) + (3 - 5.09)^2(.12)\\ &\quad + (4 - 5.09)^2(.16) + (5 - 5.09)^2(.20) + (6 - 5.09)^2(.15)\\ &\quad + (7 - 5.09)^2(.13) + (8 - 5.09)^2(.08) + (9 - 5.09)^2(.04)\\ &\quad + (10 - 5.09)^2(.01)\\ &= 4.1819\end{aligned}$$

and

$$\sigma = \sqrt{\text{Var}(x)} \approx 2.045.$$

Like the mean or expected value, the variance of a continuous random variable is an integral.

$$\text{Var}(x) = \int_a^b (x - \mu)^2 f(x)\, dx$$

To find the standard deviation of a continuous probability distribution, like that of a discrete distribution, we find the square root of the variance. The formulas for the expected value, variance, and standard deviation of a continuous probability distribution are summarized here.

EXPECTED VALUE, VARIANCE, AND STANDARD DEVIATION

If x is a continuous random variable with probability density function f on $[a, b]$, then the expected value of x is

$$E(x) = \mu = \int_a^b x f(x)\, dx.$$

The variance of x is

$$\text{Var}(x) = \int_a^b (x - \mu)^2 f(x)\, dx,$$

and the standard deviation of x is

$$\sigma = \sqrt{\text{Var}(x)}.$$

Geometrically, the expected value (or mean) of a probability distribution represents the balancing point of the distribution. If a fulcrum were placed at μ on the x-axis, the figure would be in balance. See Figure 6.

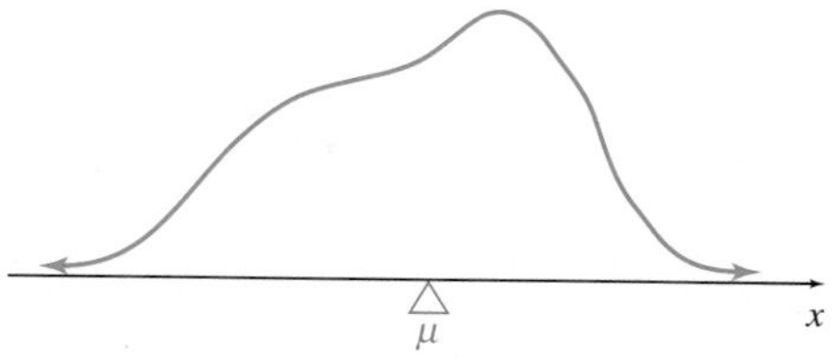

FIGURE 6

The variance and standard deviation of a probability distribution indicate how closely the values of the distribution cluster about the mean. These measures are most useful for comparing different distributions, as in Figure 7.

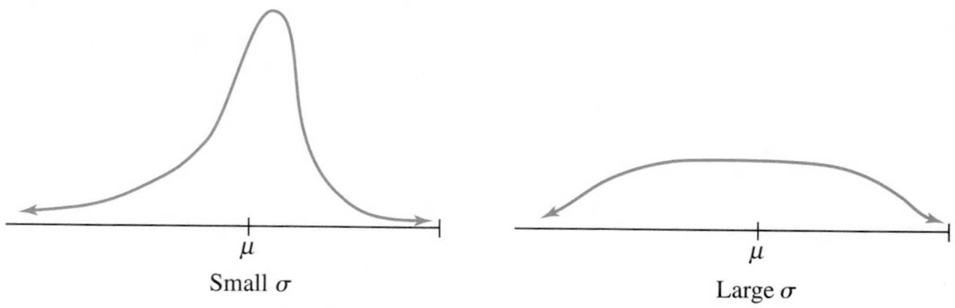

FIGURE 7

EXAMPLE 1 *Expected Value and Variance*

Find the expected value and variance of the random variable x with probability density function defined by $f(x) = (3/26)x^2$ on $[1, 3]$.

Solution By the definition of expected value just given,

$$\begin{aligned}\mu &= \int_1^3 xf(x)dx \\ &= \int_1^3 x\left(\frac{3}{26}x^2\right)dx \\ &= \frac{3}{26}\int_1^3 x^3\,dx \\ &= \frac{3}{26}\left(\frac{x^4}{4}\right)\bigg|_1^3 = \frac{3}{104}(81 - 1) = \frac{30}{13},\end{aligned}$$

or about 2.31.

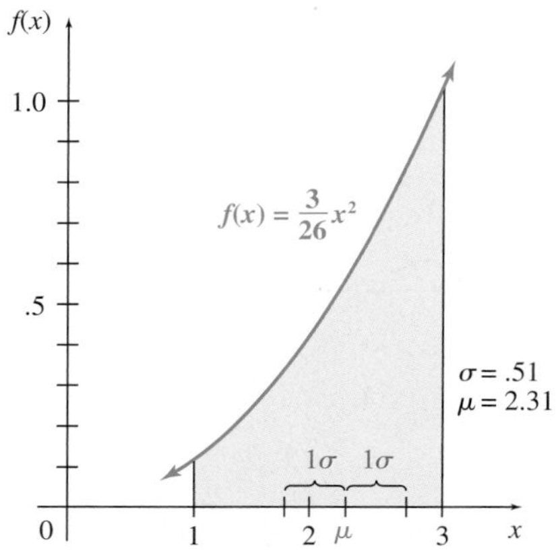

FIGURE 8

The variance is

$$\begin{aligned}
\text{Var}(x) &= \int_1^3 \left(x - \frac{30}{13}\right)^2 \left(\frac{3}{26}x^2\right) dx \\
&= \int_1^3 \left(x^2 - \frac{60}{13}x + \frac{900}{169}\right)\left(\frac{3}{26}x^2\right) dx \qquad \text{Square } \left(x - \tfrac{30}{13}\right). \\
&= \frac{3}{26}\int_1^3 \left(x^4 - \frac{60}{13}x^3 + \frac{900}{169}x^2\right) dx \qquad \text{Multiply.} \\
&= \frac{3}{26}\left(\frac{x^5}{5} - \frac{60}{13}\cdot\frac{x^4}{4} + \frac{900}{169}\cdot\frac{x^3}{3}\right)\Bigg|_1^3 \qquad \text{Integrate.} \\
&= \frac{3}{26}\left[\left(\frac{243}{5} - \frac{60(81)}{52} + \frac{900(27)}{169(3)}\right) - \left(\frac{1}{5} - \frac{60}{52} + \frac{300}{169}\right)\right] \\
&\approx .259.
\end{aligned}$$

From the variance, the standard deviation is $\sigma \approx \sqrt{.259} \approx .51$. The expected value and standard deviation are shown on the graph of the probability density function in Figure 8.

Calculating the variance in the last example was a messy job. An alternative version of the formula for the variance is easier to compute. This alternative formula is derived as follows.

$$\begin{aligned}
\text{Var}(x) &= \int_a^b (x - \mu)^2 f(x)\,dx \\
&= \int_a^b (x^2 - 2\mu x + \mu^2) f(x)\,dx \\
&= \int_a^b x^2 f(x)\,dx - 2\mu \int_a^b x f(x)\,dx + \mu^2 \int_a^b f(x)\,dx \qquad (1)
\end{aligned}$$

By definition,

$$\int_a^b x f(x)\,dx = \mu,$$

and, since $f(x)$ is a probability density function,

$$\int_a^b f(x)\,dx = 1.$$

Substitute back into Equation (1) to get the alternative formula,

$$\text{Var}(x) = \int_a^b x^2 f(x)\,dx - 2\mu^2 + \mu^2 = \int_a^b x^2 f(x)\,dx - \mu^2.$$

ALTERNATIVE FORMULA FOR VARIANCE

If x is a random variable with probability density function f on $[a, b]$, and if $E(x) = \mu$, then

$$\text{Var}(x) = \int_a^b x^2 f(x)\,dx - \mu^2.$$

CAUTION Notice that the term μ^2 comes *after* the dx, and so is *not* integrated. ■

EXAMPLE 2 *Variance*

Use the alternative formula for variance to compute the variance of the random variable x with probability density function defined by $f(x) = 3/x^4$ for $x \geq 1$.

Solution To find the variance, first find the expected value:

$$\mu = \int_1^{\infty} xf(x)dx = \int_1^{\infty} x \cdot \frac{3}{x^4}\, dx = \int_1^{\infty} \frac{3}{x^3}\, dx$$

$$= \lim_{b\to\infty} \int_1^b \frac{3}{x^3}\, dx = \lim_{b\to\infty}\left(\frac{3}{-2x^2}\right)\Bigg|_1^b = \frac{3}{2},$$

or 1.5. Now find the variance by the alternative formula for variance:

$$\mathrm{Var}(x) = \int_1^{\infty} x^2\left(\frac{3}{x^4}\right) dx - \left(\frac{3}{2}\right)^2$$

$$= \int_1^{\infty} \frac{3}{x^2}\, dx - \frac{9}{4}$$

$$= \lim_{b\to\infty} \int_1^b \frac{3}{x^2}\, dx - \frac{9}{4}$$

$$= \lim_{b\to\infty}\left(\frac{-3}{x}\right)\Bigg|_1^b - \frac{9}{4}$$

$$= 3 - \frac{9}{4} = \frac{3}{4}, \quad \text{or } .75.$$

EXAMPLE 3 *Passenger Arrival*

A recent study has shown that airline passengers arrive at the gate with the amount of time (in hours) before the scheduled flight time given by the probability density function $f(x) = (3/4)(2x - x^2)$ for $0 \leq x \leq 2$.

(a) Find and interpret the expected value for this distribution.

Solution The expected value is

$$\mu = \int_0^2 x\left(\frac{3}{4}\right)(2x - x^2)dx = \int_0^2 \left(\frac{3}{4}\right)(2x^2 - x^3)dx$$

$$= \left(\frac{3}{4}\right)\left(\frac{2x^3}{3} - \frac{x^4}{4}\right)\Bigg|_0^2 = \left(\frac{3}{4}\right)\left(\frac{16}{3} - 4\right) = 1.$$

This result indicates that passengers arrive at the gate an average of 1 hour before the scheduled flight time.

(b) Compute the standard deviation.

Solution First compute the variance. We use the alternative formula.

$$\begin{aligned}\text{Var}(x) &= \int_0^2 x^2\left(\frac{3}{4}\right)(2x - x^2)dx - 1^2\\ &= \int_0^2 \left(\frac{3}{4}\right)(2x^3 - x^4)dx - 1\\ &= \left(\frac{3}{4}\right)\left(\frac{x^4}{2} - \frac{x^5}{5}\right)\Bigg|_0^2 - 1\\ &= \left(\frac{3}{4}\right)\left(8 - \frac{32}{5}\right) - 1 = \frac{6}{5} - 1 = \frac{1}{5}\end{aligned}$$

The standard deviation is $\sigma = \sqrt{1/5} \approx .45$.

EXAMPLE 4 *Life Expectancy*

In the previous section of this chapter we used statistics compiled by the National Center for Health Statistics to determine a probability density function that can be used to study the proportion of all 60-year-olds who will be alive in x years. The function is given by

$$S(x) = \frac{1}{500{,}669}(.0681571x^4 - 6.72318x^3 + 184.397x^2 - 1091.40x + 8477.08)$$

for $0 \le x \le 44$.

(a) Find the life expectancy of a 60-year-old person.

Solution Since this is a complicated function that is tedious to integrate analytically, we will employ the integration feature on a TI-83/84 Plus calculator to calculate

$$\mu = \int_0^{44} xS(x)dx \approx 21.32 \text{ years.}$$

According to life tables, the life expectancy of a 60-year-old person is 21.4 years. Our estimate is remarkably accurate given the limited number of data points and the function used in our original analysis. Life expectancy is generally calculated with techniques from life table analysis.*

(b) Find the standard deviation of this probability function.

Solution Using the alternate formula, we first calculate the variance.

$$\text{Var}(x) = \int_0^{44} x^2S(x)dx - \mu^2 = 554.6281 - (21.32)^2 \approx 100.42$$

Thus, $\sigma = \sqrt{\text{Var}(x)} \approx 10.02$ years.

*Chiang, Chin Long, *The Life Table and Its Applications*, Malabar, FL: Krieger Publishers, 1984.

11.2 EXERCISES

In Exercises 1–8, a probability density function of a random variable is defined. Find the expected value, the variance, and the standard deviation. Round answers to the nearest hundredth.

1. $f(x) = \frac{1}{4}$; $[3, 7]$

2. $f(x) = \frac{1}{10}$; $[0, 10]$

3. $f(x) = \frac{x}{8} - \frac{1}{4}$; $[2, 6]$

4. $f(x) = 2(1 - x)$; $[0, 1]$

5. $f(x) = 1 - \frac{1}{\sqrt{x}}$; $[1, 4]$

6. $f(x) = \frac{1}{11}\left(1 + \frac{3}{\sqrt{x}}\right)$; $[4, 9]$

7. $f(x) = 4x^{-5}$; $[1, \infty)$

8. $f(x) = 3x^{-4}$; $[1, \infty)$

9. What information does the mean (expected value) of a continuous random variable give?

10. Suppose two random variables have standard deviations of .10 and .23, respectively. What does this tell you about their distributions?

In Exercises 11–14, the probability density function of a random variable is defined.

a. *Find the expected value to the nearest hundredth.*

b. *Find the variance to the nearest hundredth.*

c. *Find the standard deviation. Round to the nearest hundredth.*

d. *Find the probability that the random variable has a value greater than the mean.*

e. *Find the probability that the value of the random variable is within 1 standard deviation of the mean.*

11. $f(x) = \frac{\sqrt{x}}{18}$; $[0, 9]$

12. $f(x) = \frac{x^{-1/3}}{6}$; $[0, 8]$

13. $f(x) = \frac{1}{2}x$; $[0, 2]$

14. $f(x) = \frac{3}{2}(1 - x^2)$; $[0, 1]$

If x is a random variable with probability density function f on $[a, b]$, then the median *of x is the number m such that*

$$\int_a^m f(x)\,dx = \frac{1}{2}.$$

a. *Find the median of each random variable for the probability density functions in Exercises 15–20.*

b. *In each case, find the probability that the random variable is between the expected value (mean) and the median. The expected value for each of these functions was found in Exercises 1–8.*

15. $f(x) = \frac{1}{4}$; $[3, 7]$

16. $f(x) = \frac{1}{10}$; $[0, 10]$

17. $f(x) = \frac{x}{8} - \frac{1}{4}$; $[2, 6]$

18. $f(x) = 2(1 - x)$; $[0, 1]$

19. $f(x) = 4x^{-5}$; $[1, \infty)$

20. $f(x) = 3x^{-4}$; $[1, \infty)$

Find the expected value, the variance, and the standard deviation, when they exist, for each probability density function.

21. $f(x) = \begin{cases} \frac{x^3}{12} & \text{if } 0 \le x \le 2 \\ \frac{16}{3x^3} & \text{if } x > 2 \end{cases}$

22. $f(x) = \begin{cases} \frac{20x^4}{9} & \text{if } 0 \le x \le 1 \\ \frac{20}{9x^5} & \text{if } x > 1 \end{cases}$

Applications

BUSINESS AND ECONOMICS

23. *Life of a Light Bulb* The life (in hours) of a certain kind of light bulb is a random variable with probability density function defined by

$$f(x) = \frac{1}{58\sqrt{x}} \quad \text{for } x \text{ in } [1,900].$$

a. What is the expected life of such a bulb?

b. Find σ.

c. Find the probability that one of these bulbs lasts longer than 1 standard deviation above the mean.

24. *Machine Life* The life (in years) of a certain machine is a random variable with probability density function defined by

$$f(x) = \frac{1}{11}\left(1 + \frac{3}{\sqrt{x}}\right) \quad \text{for } x \text{ in } [4, 9].$$

a. Find the mean life of this machine.

b. Find the standard deviation of the distribution.

c. Find the probability that a particular machine of this kind will last longer than the mean number of years.

25. *Life of an Automobile Part* The life span of a certain automobile part (in months) is a random variable with probability density function defined by

$$f(x) = \frac{1}{2}e^{-x/2} \quad \text{for } x \text{ in } [0, \infty).$$

a. Find the expected life of this part.

b. Find the standard deviation of the distribution.

c. Find the probability that one of these parts lasts less than the mean number of months.

LIFE SCIENCES

26. *Blood Clotting Time* The clotting time of blood (in seconds) is a random variable with probability density function defined by

$$f(x) = \frac{1}{(\ln 20)x} \quad \text{for } x \text{ in } [1, 20].$$

a. Find the mean clotting time.

b. Find the standard deviation.

c. Find the probability that a person's blood clotting time is within 1 standard deviation of the mean.

27. *Length of a Leaf* The length of a leaf on a tree is a random variable with probability density function defined by

$$f(x) = \frac{3}{32}(4x - x^2) \quad \text{for } x \text{ in } [0, 4].$$

a. What is the expected leaf length?

b. Find σ for this distribution.

c. Find the probability that the length of a given leaf is within 1 standard deviation of the expected value.

28. *Petal Length* The length (in centimeters) of a petal on a certain flower is a random variable with probability density function defined by

$$f(x) = \frac{1}{2\sqrt{x}} \quad \text{for } x \text{ in } [1, 4].$$

a. Find the expected petal length.

b. Find the standard deviation.

c. Find the probability that a petal selected at random has a length more than 2 standard deviations above the mean.

29. *Firearms* In Exercise 35 of the previous section, the probability density function for the number of firearm deaths in the United States in 1997 was found to be

$$S(x) = \frac{1}{343{,}795}(-.00351465x^4 + .792884x^3 - 61.7955x^2 + 1814.54x - 9709.20)$$

where x was the number of years since birth on $[6.8, 91.5]$. Calculate the expected age at which a person will be killed with a firearm, as well as the standard deviation.

30. *Flour Beetles* As we saw in the previous section, a probability density function has been developed to estimate the abundance of the flour beetle, *Tribolium castaneum*. The density function, which is a member of the gamma distribution, is

$$f(x) = 1.185 \cdot 10^{-9}x^{4.5222}e^{-.049846x},$$

where x is the size of the population.* Calculate the expected size of a flour beetle population. (*Hint:* Use 1000 as the upper limit of integration.)

31. *Flea Beetles* As we saw in the previous section, the probability that a marked flea beetle, *Phyllotreta cruciferae* and *Phylotreta striolata*, will be recaptured within a certain distance and time after release can be calculated from the probability density function

$$p(x, t) = \frac{e^{-x^2/(4Dt)}}{\int_0^L e^{-u^2/(4Dt)}\,du},$$

where t is the time (in hours) after release, x is the distance (in meters) from the release point that recaptures occur, L is the maximum distance from the release point that recaptures can occur, and D is the diffusion coefficient.† If $t = 12$, $L = 6$, and $D = 38.3$, find the expected recapture distance.

PHYSICAL SCIENCES

32. *Annual Rainfall* The annual rainfall in a remote Middle Eastern country is a random variable with probability density function defined by

$$f(x) = \frac{5.5 - x}{15}, \quad \text{for } x \text{ in } [0, 5].$$

a. Find the mean annual rainfall.

b. Find the standard deviation.

c. Find the probability of a year with rainfall less than 1 standard deviation below the mean.

33. *Earthquakes* The time between major earthquakes in the Southern California region is a random variable with probability density function defined by

$$f(t) = \frac{1}{960}e^{-t/960},$$

where t is measured in days.‡ Find the expected value and the standard deviation of this probability density function.

GENERAL INTEREST

34. *Drunk Drivers* In the last section, we saw that the age of a randomly selected, alcohol-impaired driver in a fatal car crash is a random variable with probability density function given by

$$f(x) = \frac{.1906}{x^{.5012}} \quad \text{for } x \text{ in } [16, 44].^{\S}$$

a. Find the expected age of a drunk driver in a fatal car crash.

b. Find the standard deviation of the distribution.

c. Find the probability that such a driver will be younger than 1 standard deviation below the mean.

35. *Driving Fatalities* In the last section, we saw that the age of a randomly selected driver in a fatal car crash is a random variable with probability density function given by

$$f(x) = .06049e^{-.03211x} \quad \text{for } x \text{ in } [16, 84].^{\#}$$

a. Find the expected age of a driver in a fatal car crash.

b. Find the standard deviation of the distribution.

c. Find the probability that such a driver will be younger than 1 standard deviation below the mean.

*Dennis, Brian and Robert F. Costantino, "Analysis of Steady-State Population with the Gamma Abundance Model: Application to *Tribolium*," *Ecology*, Vol. 69, No. 4, Aug. 1988, pp. 1200–1213.
†Kareiva, Peter, "Experimental and Mathematical Analyses of Herbivore Movement: Quantifying the Influence of Plant Spacing and Quality on Foraging Discrimination," *Ecology Monographs*, Vol. 2, No. 3, 1982, pp. 261–282.
‡Wang, Jeen-Hwa and Chiao-Hui Kuo, "On the Frequency Distribution of Interoccurrence Times of Earthquakes," *Journal of Seismology*, Vol. 2, 1998, pp. 351–358.
§*Traffic Safety Facts 2001*, National Highway Traffic Safety Administration, p. 36.
#www-nrd.nhtsa.dot.gov/pdf/nrd-30/NCSA/RNotes/1998/AgeSex96.pdf.

11.3 SPECIAL PROBABILITY DENSITY FUNCTIONS

THINK ABOUT IT

What is the probability that the maximum outdoor temperature will be higher than 24°C? What is the probability that a flashlight battery will last longer than 40 hours?

These questions can be answered if the probability density function for the maximum temperature and for the life of the battery are known. In practice, however, it is not feasible to construct a probability density function for each experiment. Instead, a researcher uses one of several probability density functions that are well known, matching the shape of the experimental distribution to one of the known distributions. In this section we discuss some of the most commonly used probability distributions.

Uniform Distribution The simplest probability distribution occurs when the probability density function of a random variable remains constant over the sample space. In this case, the random variable is said to be *uniformly distributed* over the sample space. The probability density function for the **uniform distribution** is defined by

$$f(x) = \frac{1}{b-a} \quad \text{for } x \text{ in } [a, b],$$

where a and b are constant real numbers. The graph of $f(x)$ is shown in Figure 9 below.

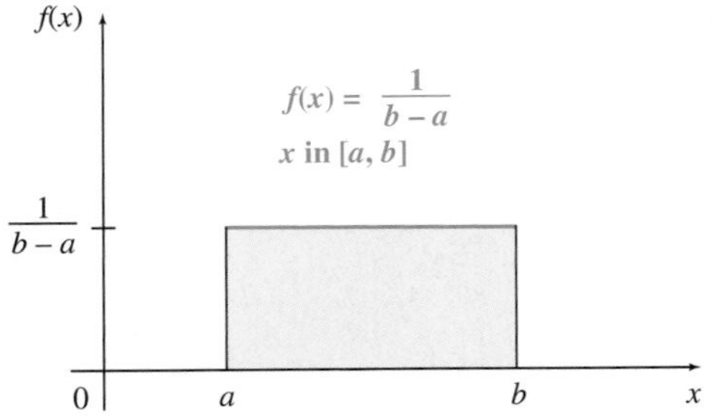

Uniform distribution

FIGURE 9

Since $b - a$ is positive, $f(x) \geq 0$, and

$$\int_a^b \frac{1}{b-a}\,dx = \frac{1}{b-a}x\Big|_a^b = \frac{1}{b-a}(b-a) = 1.$$

Therefore, the function is a probability density function.

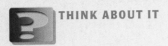

The expected value for the uniform distribution is

$$\mu = \int_a^b \left(\frac{1}{b-a}\right) x\,dx = \left(\frac{1}{b-a}\right)\frac{x^2}{2}\Big|_a^b$$

$$= \frac{1}{2(b-a)}(b^2 - a^2) = \frac{1}{2}(b+a). \qquad b^2 - a^2 = (b-a)(b+a)$$

The variance is given by

$$\text{Var}(x) = \int_a^b \left(\frac{1}{b-a}\right) x^2\,dx - \left(\frac{b+a}{2}\right)^2$$

$$= \left(\frac{1}{b-a}\right)\frac{x^3}{3}\Big|_a^b - \frac{(b+a)^2}{4}$$

$$= \frac{1}{3(b-a)}(b^3 - a^3) - \frac{1}{4}(b+a)^2$$

$$= \frac{b^2 + ab + a^2}{3} - \frac{b^2 + 2ab + a^2}{4} \qquad b^3 - a^3 = (b-a)(b^2 + ab + a^2)$$

$$= \frac{b^2 - 2ab + a^2}{12}.$$

Get a common denominator; subtract.

Thus

$$\text{Var}(x) = \frac{1}{12}(b-a)^2,$$

Factor.

and

$$\sigma = \frac{1}{\sqrt{12}}(b-a).$$

These properties of the uniform distribution are summarized below.

UNIFORM DISTRIBUTION

If x is a random variable with probability density function

$$f(x) = \frac{1}{b-a} \quad \textbf{for } x \textbf{ in } [a, b],$$

then

$$\mu = \frac{1}{2}(b+a) \quad \text{and} \quad \sigma = \frac{1}{\sqrt{12}}(b-a).$$

EXAMPLE 1 *Daily Temperature*

A couple is planning to vacation in San Francisco. They have been told that the maximum daily temperature during the time they plan to be there ranges from 15°C to 27°C. Assume that the probability of any temperature between 15°C and 27°C is equally likely for any given day during the specified time period.

(a) What is the probability that the maximum temperature on the day they arrive will be higher than 24°C?

Solution If the random variable t represents the maximum temperature on a given day, then the uniform probability density function for t is defined by $f(t) = 1/12$ for the interval $[15, 27]$. By definition,

$$P(t > 24) = \int_{24}^{27} \frac{1}{12}\,dt = \frac{1}{12}t\Big|_{24}^{27} = \frac{1}{4}.$$

(b) What average maximum temperature can they expect?

Solution The expected maximum temperature is

$$\mu = \frac{1}{2}(27 + 15) = 21,$$

or 21°C.

(c) What is the probability that the maximum temperature on a given day will be at least 1 standard deviation below the mean?

Solution First find σ.

$$\sigma = \frac{1}{\sqrt{12}}(27 - 15) = \frac{12}{\sqrt{12}} = \sqrt{12} = 2\sqrt{3} \approx 3.5.$$

One standard deviation below the mean indicates a temperature of $21 - 3.5 = 17.5$°C.

$$P(T \le 17.5) = \int_{15}^{17.5} \frac{1}{12}\,dt = \frac{1}{12}t\Big|_{15}^{17.5} \approx .21$$

The probability is .21 that the temperature will not exceed 17.5°C.

Exponential Distribution The next distribution is very important in reliability and survival analysis. When manufactured items and living things have a constant failure rate over a period of time, the exponential distribution is used to describe their probability of failure. In this case, the random variable is said to be *exponentially distributed* over the sample space. The probability density function for the **exponential distribution** is defined by

$$f(x) = ae^{-ax} \qquad \text{for } x \text{ in } [0, \infty),$$

where a is a positive constant. The graph of $f(x)$ is shown in Figure 10.

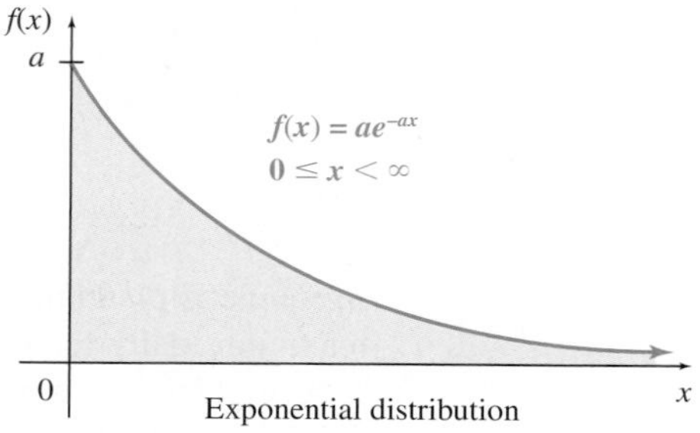

FIGURE 10

Here $f(x) \geq 0$, since e^{-ax} and a are both positive for all values of x. Also,

$$\int_0^\infty ae^{-ax}\,dx = \lim_{b\to\infty}\int_0^b ae^{-ax}\,dx$$

$$= \lim_{b\to\infty}(-e^{-ax})\Big|_0^b = \lim_{b\to\infty}\left(\frac{-1}{e^{ab}} + \frac{1}{e^0}\right) = 1,$$

so the function is a probability density function.

The expected value and the standard deviation of the exponential distribution can be found using integration by parts. The results are given below.

EXPONENTIAL DISTRIBUTION

If x is a random variable with probability density function

$$f(x) = ae^{-ax} \quad \text{for } x \text{ in } [0, \infty),$$

then

$$\mu = \frac{1}{a} \quad \text{and} \quad \sigma = \frac{1}{a}.$$

EXAMPLE 2 *Flashlight Battery*

Suppose the useful life (in hours) of a flashlight battery is the random variable t, with probability density function given by the exponential distribution

$$f(t) = \frac{1}{20}e^{-t/20} \quad \text{for } t \geq 0.$$

(a) Find the probability that a particular battery, selected at random, has a useful life of less than 100 hours.

Solution The probability is given by

$$P(t \leq 100) = \int_0^{100} \frac{1}{20}e^{-t/20}\,dt = \frac{1}{20}(-20e^{-t/20})\Big|_0^{100}$$

$$= -(e^{-100/20} - e^0) = -(e^{-5} - 1)$$

$$\approx 1 - .0067 = .9933.$$

(b) Find the expected value and standard deviation of the distribution.

Solution Use the formulas given above. Both μ and σ equal $1/a$, and since $a = 1/20$ here,

$$\mu = 20 \quad \text{and} \quad \sigma = 20.$$

This means that the average life of a battery is 20 hours, and no battery lasts less than 1 standard deviation below the mean.

(c) What is the probability that a battery will last longer than 40 hours?

Solution The probability is given by

$$P(t > 40) = \int_{40}^\infty \frac{1}{20}e^{-t/20}\,dt = \lim_{b\to\infty}(-e^{-t/20})\Big|_{40}^b = \frac{1}{e^2} \approx .1353,$$

or about 14%.

Normal Distribution The **normal distribution**, with its well-known bell-shaped graph, is undoubtedly the most important probability density function. It is widely used in various applications of statistics. The random variables associated with these applications are said to be normally distributed. The probability density function for the normal distribution has the following characteristics.

NORMAL DISTRIBUTION

If μ and σ are real numbers, $\sigma > 0$, and if x is a random variable with probability density function defined by

$$f(x) = \frac{1}{\sigma\sqrt{2\pi}} e^{-(x-\mu)^2/(2\sigma^2)} \quad \text{for } x \text{ in } (-\infty, \infty),$$

then

$$E(x) = \mu \quad \text{and} \quad \text{Var}(x) = \sigma^2, \quad \text{with } \textbf{standard deviation } \sigma.$$

Notice that the definition of the probability density function includes σ, which is the standard deviation of the distribution.

Advanced techniques can be used to show that

$$\int_{-\infty}^{\infty} \frac{1}{\sigma\sqrt{2\pi}} e^{-(x-\mu)^2/(2\sigma^2)}\, dx = 1.$$

Deriving the expected value and standard deviation for the normal distribution also requires techniques beyond the scope of this text.

Each normal probability distribution has associated with it a bell-shaped curve, called a **normal curve,** such as the one in Figure 11. Each normal curve is symmetric about a vertical line through the mean, μ. Vertical lines at points $+1\sigma$ and -1σ from the mean show the inflection points of the graph. A normal curve never touches the x-axis; it extends indefinitely in both directions.

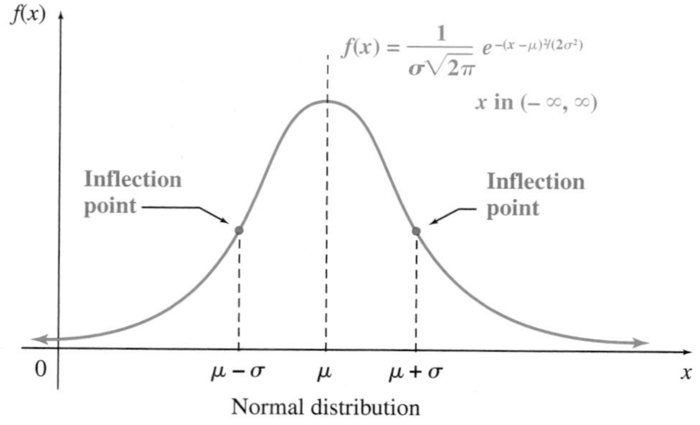

FIGURE 11

The development of the normal curve is credited to the Frenchman Abraham De Moivre (1667–1754). Three of his publications dealt with probability and associated topics: *Annuities upon Lives* (which contributed to the development of actuarial studies), *Doctrine of Chances*, and *Miscellanea Analytica*.

Many different normal curves have the same mean. In such cases, a larger value of σ produces a "flatter" normal curve, while smaller values of σ produce more values near the mean, resulting in a "taller" normal curve. See Figure 12.

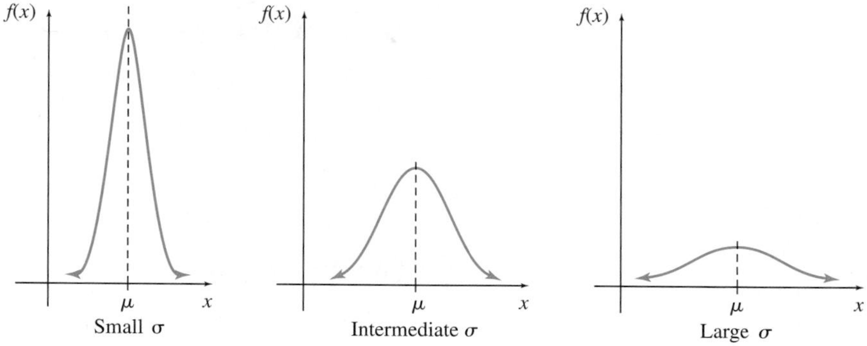

FIGURE 12

It would be far too much work to calculate values for the normal probability distribution for various values of μ and σ. Instead, values are calculated for the **standard normal distribution,** which has $\mu = 0$ and $\sigma = 1$. The graph of the standard normal distribution is shown in Figure 13.

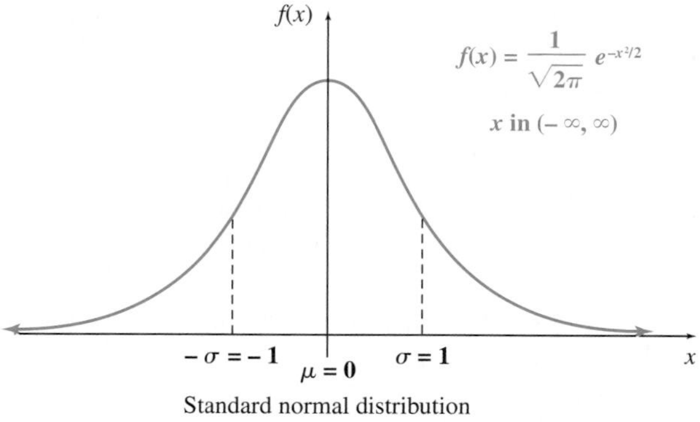

FIGURE 13

Probabilities for the standard normal distribution come from the definite integral

$$\int_a^b \frac{1}{\sqrt{2\pi}} e^{-x^2/2}\,dx.$$

Since $f(x) = e^{-x^2/2}$ does not have an antiderivative that can be expressed in terms of functions used in this course, numerical methods are used to find values of this definite integral. A table in the appendix of this book gives areas under the standard normal curve, along with a sketch of the curve. Each value in this table is the total area under the standard normal curve to the left of the number z.

If a normal distribution does not have $\mu = 0$ and $\sigma = 1$, we use the following theorem, which is proven in Exercise 21.

z-SCORES THEOREM

Suppose a normal distribution has mean μ and standard deviation σ. The area under the associated normal curve that is to the left of the value x is exactly the same as the area to the left of

$$z = \frac{x - \mu}{\sigma}$$

for the standard normal curve.

Using this result, the table can be used for *any* normal distribution, regardless of the values of μ and σ. The number z in the theorem is called a **z-score.**

EXAMPLE 3 *Normal Distribution*

A normal distribution has mean 35 and standard deviation 5.9. By computing the areas under the associated normal curve, find the following probabilities.

(a) $P(x < 40)$

Solution Find the appropriate z-score using $x = 40$, $\mu = 35$, and $\sigma = 5.9$. Round to the nearest hundredth.

$$z = \frac{40 - 35}{5.9} = \frac{5}{5.9} \approx .85$$

Look up .85 in the normal curve table in the Appendix. The corresponding area is .8023. Thus, the shaded area shown in Figure 14 is .8023. This area represents 80.23% of the total area under the normal curve and the proportion of observations that are less than 40.

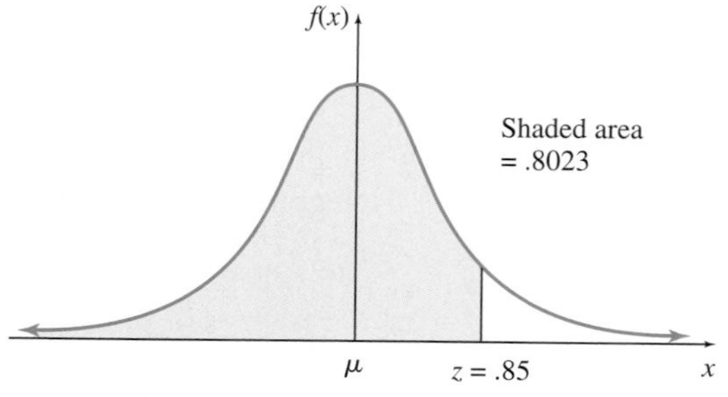

FIGURE 14

(b) $P(x > 32)$

Solution

$$z = \frac{32 - 35}{5.9} = \frac{-3}{5.9} \approx -.51$$

The area to the *left* of $z = -.51$ is .3050, so the area to the *right is* $1 - .3050 = .6950$. See Figure 15. Thus, $P(x > 32) = .6950$.

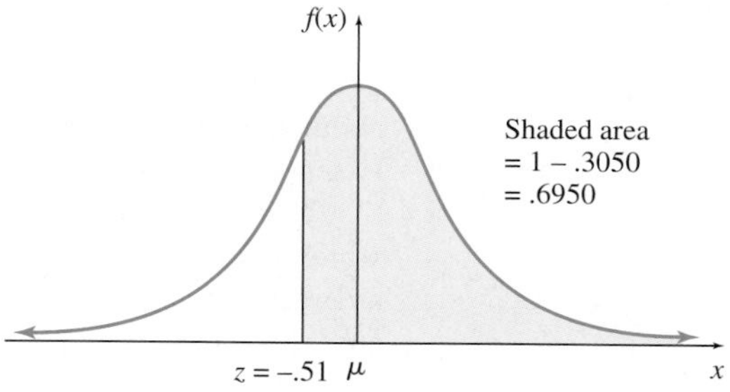

FIGURE 15

(c) $P(30 \le x \le 33)$

Solution Find z-scores for both values.

$$z = \frac{30 - 35}{5.9} = \frac{-5}{5.9} \approx -.85 \quad \text{and} \quad z = \frac{33 - 35}{5.9} = \frac{-2}{5.9} \approx -.34$$

Start with the area to the left of $z = -.34$ and subtract the area to the left of $z = -.85$. Thus,

$$P(30 \le x \le 33) = .3669 - .1977 = .1692.$$

The required area is shaded in Figure 16.

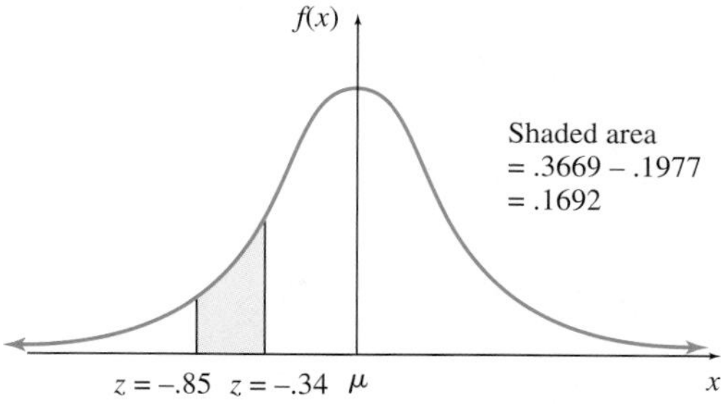

FIGURE 16

As an alternative to using the normal curve table, we can use a graphing calculator. Enter the formula for the normal distribution into the calculator, using $\mu = 35$ and $\sigma = 5.9$. Plot the function in a window that contains at least 4 standard deviations to the left and right of μ; for this example, we will let $0 \le x \le 60$. Then use the integration feature (under CALC on a TI-83/84 Plus) to find the area under the curve.

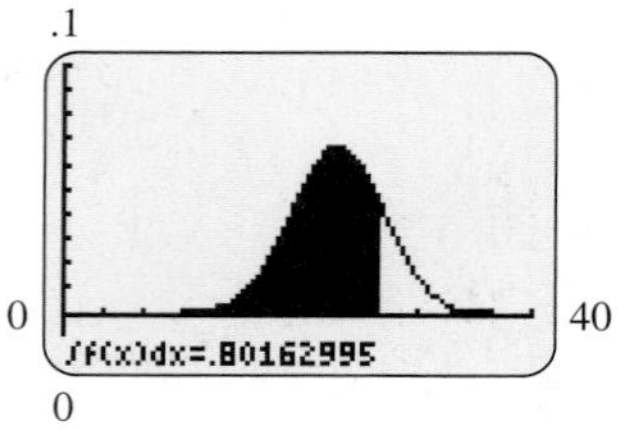

FIGURE 17

The result is shown in Figure 17. In place of $-\infty$, we have used $x = 0$ as the left endpoint. This is far enough to the left of $\mu = 35$ that it can be considered as $-\infty$ for all practical purposes. You can verify that choosing a lower limit even further to the left makes little or no difference in the answer. In fact, the answer of .80162995 is more accurate than the answer of .8023 that we found in Example 3, where we needed to round $5/5.9 \approx .8474576$ to .85 in order to use the table.

We could get the answer on a TI-83/84 Plus without generating a graph using the command `fnInt(Y₁, X, 0, 40)`, where Y_1 is the formula for the normal distribution with $\mu = 35$ and $\sigma = 5.9$.

The numerical integration method works with any probability density function. In addition, many graphing calculators are programmed with information about specific density functions, such as the normal. We can solve the first part of Example 3 on the TI-83/84 Plus by entering `normalcdf(−1E99, 40, 35, 5.9)`. The calculator responds with .8016300043. (`-1E99` stands for $-1 \cdot 10^{99}$, which the calculator uses for $-\infty$.) If you use this method in the exercises, your answers will differ slightly from those in the back of the book, which were generated using the normal curve table in the Appendix.

The z-scores are actually standard deviation multiples; that is, a z-score of 2.5 corresponds to a value 2.5 standard deviations above the mean. For example, looking up $z = 1.00$ and $z = -1.00$ in the table shows that

$$.8413 - .1587 = .6826,$$

so that 68.26% of the area under a normal curve is within 1 standard deviation of the mean. Also, using $z = 2.00$ and $z = -2.00$,

$$.9772 - .0228 = .9544,$$

meaning 95.44% of the area is within 2 standard deviations of the mean. These results, summarized in Figure 18, can be used to get a quick estimate of results when working with normal curves.

Manufacturers make use of the fact that a normal random variable is almost always within 3 standard deviations of the mean to design control charts. When a sample of items produced by a machine has a mean farther than 3 standard devia-

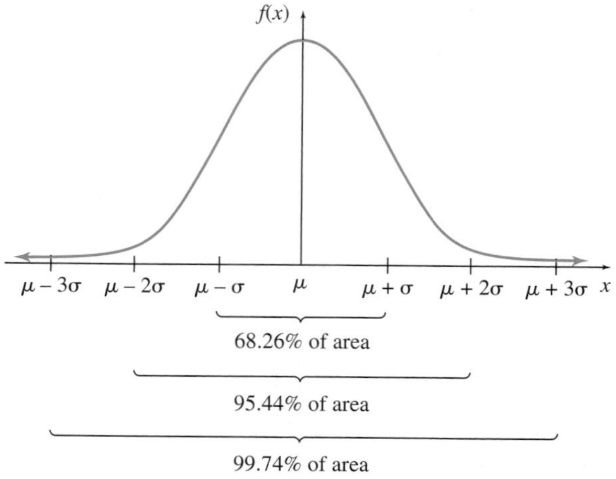

FIGURE 18

tions from the desired specification, the machine is assumed to be out of control, and adjustments are made to ensure that the items produced meet the tolerance required.

EXAMPLE 4 *Lead Poisoning*

Historians and biographers have collected evidence suggesting that President Andrew Jackson suffered from lead poisoning. Recently, researchers measured the amount of lead in samples of Jackson's hair from 1815. The results of this experiment showed that Jackson had a mean lead level of 130.5 ppm.*

(a) If levels of lead in hair samples from that time period follow a normal distribution with mean 93 and standard deviation 16,† find the probability that a randomly selected person from this time period would have a lead level of 130.5 ppm or higher. Does this provide evidence that Jackson suffered from lead poisoning during this time period?

Solution $P(x > 130.5) = P\left(z > \dfrac{130.5 - 93}{16}\right) = P(z > 2.34) \approx .01$

Since this probability is so low, it is likely that Jackson suffered from lead poisoning during this time period.‡

(b) Today's normal lead levels follow a normal distribution with approximate mean of 10 ppm and standard deviation of 5 ppm.§. By today's standards, calculate the probability that a randomly selected person from today would have a lead level of 130.5 ppm or higher. From this, can we conclude that Andrew Jackson had lead poisoning?

Solution $P(x > 130.5) = P\left(z > \dfrac{130.5 - 10}{5}\right) = P(z > 24.1) \approx 0$

By today's standards, which may not be valid for this experiment, Jackson certainly suffered from lead poisoning.

11.3 EXERCISES

Find the mean of the distribution, the standard deviation of the distribution, and the probability that the random variable is between the mean and 1 *standard deviation above the mean.*

1. The length (in centimeters) of the leaf of a certain plant is a continuous random variable with probability density function defined by

$$f(x) = \frac{5}{4} \quad \text{for } x \text{ in } [4, 4.8].$$

*Deppisch, Lidwig, Jose Centeno, David Gemmel, and Norca Torres, "Andrew Jackson's Exposure to Mercury and Lead," *JAMA,* Vol. 282, No. 6, Aug. 11, 1999, pp. 569–571.

†Weiss, D., B. Whitten, and D. Leddy, "Lead Content of Human Hair (1871–1971)," *Science*, Vol. 178, 1972, pp. 69–70.

‡Although this provides evidence that Andrew Jackson had elevated lead levels, the authors of the paper concluded that Andrew Jackson did not die from lead poisoning.

§Iyengar, V. and J. Woittiez, "Trace Elements in Human Clinical Specimens," *Clinical Chemistry,* Vol. 34, 1988, pp. 474–481.

2. The price of an item (in dollars) is a continuous random variable with probability density function defined by

$$f(x) = 2 \quad \text{for } x \text{ in } [1.25, 1.75].$$

3. The length of time (in years) until a particular radioactive particle decays is a random variable t with probability density function defined by

$$f(t) = .03e^{-.03t} \quad \text{for } t \text{ in } [0, \infty).$$

4. The length of time (in years) that a seedling tree survives is a random variable t with probability density function defined by

$$f(t) = .05e^{-.05t} \quad \text{for } t \text{ in } [0, \infty).$$

5. The length of time (in days) required to learn a certain task is a random variable t with probability density function defined by

$$f(t) = e^{-t} \quad \text{for } t \text{ in } [0, \infty).$$

6. The distance (in meters) that seeds are dispersed from a certain kind of plant is a random variable x with probability density function defined by

$$f(x) = .1e^{-.1x} \quad \text{for } x \text{ in } [0, \infty).$$

Find the proportion of observations of a standard normal distribution that are between the mean and the given number of standard deviations above the mean.

7. 3.50

8. 1.68

Find the proportion of observations of a standard normal distribution that are between the given z-scores.

9. 1.28 and 2.05

10. -2.13 and $-.04$

Find a z-score satisfying the conditions given in Exercises 11–14. (Hint: Use the table backwards.)

11. 10% of the total area is to the left of z.

12. 2% of the total area is to the left of z.

13. 18% of the total area is to the right of z.

14. 22% of the total area is to the right of z.

15. Describe the standard normal distribution. What are its characteristics?

16. What is a z-score? How is it used?

17. Describe the shape of the graph of each probability distribution.

a. Uniform **b.** Exponential **c.** Normal

In the exercises for the second section of this chapter, we defined the median of a probability distribution as an integral. The median also can be defined as the number m such that $P(x \le m) = P(x \ge m)$.

18. Find an expression for the median of the uniform distribution.

19. Find an expression for the median of the exponential distribution.

20. Verify the expected value and standard deviation of the exponential distribution given in the text.

21. Prove the z-scores theorem. (*Hint:* Write the formula for the normal distribution with mean μ and standard deviation σ, using t instead of x as the variable. Then write the integral representing the area to the left of the value x, and make the substitution $u = (t - \mu)/\sigma$.)

22. Show that a normal random variable has inflection points at $x = \mu - \sigma$ and $x = \mu + \sigma$.

23. Use Simpson's rule with $n = 100$, or use the integration feature on a graphing calculator, to approximate the following integrals.

a. $\int_0^{50} .5e^{-.5x}\,dx$ **b.** $\int_0^{50} .5xe^{-.5x}\,dx$ **c.** $\int_0^{50} .5x^2e^{-.5x}\,dx$

24. Use your results from Exercise 23 to verify that, for the exponential distribution with $a = 5$, the total probability is 1, and both the mean and the standard deviation are equal to $1/a$.

25. Use Simpson's rule with $n = 100$, or the integration feature on a graphing calculator, to approximate the following for the standard normal probability distribution. Use limits of -4 and 4 in place of $-\infty$ and ∞.

a. The mean **b.** The standard deviation

26. A very important distribution for analyzing the reliability of manufactured goods is the Weibull distribution, whose probability density function is defined by

$$f(x) = abx^{b-1}e^{-ax^b} \quad \text{for } x \text{ in } [0, \infty),$$

where a and b are constants. Notice that when $b = 1$, this reduces to the exponential distribution. The Weibull distribution is more general than the exponential, because it applies even when the failure rate is not constant. Use Simpson's rule with $n = 100$, or the integration feature on a graphing calculator, to approximate the following for the Weibull distribution with $a = 4$ and $b = 1.5$. Use a limit of 3 in place of ∞.

a. The mean **b.** The standard deviation

Applications

BUSINESS AND ECONOMICS

27. *Insurance Sales* The amount of insurance (in thousands of dollars) sold in a day by a particular agent is uniformly distributed over the interval [10, 85].

a. What amount of insurance does the agent sell on an average day?

b. Find the probability that the agent sells more than \$50,000 of insurance on a particular day.

28. *Fast-Food Outlets* The number of new fast-food outlets opening during June in a certain city is exponentially distributed, with a mean of 5.

a. Give the probability density function for this distribution.

b. What is the probability that the number of outlets opening is between 2 and 6?

29. *Sales Expense* A salesperson's monthly expenses (in thousands of dollars) are exponentially distributed, with an average of 4.25 (thousand dollars).

a. Give the probability density function for the expenses.

b. Find the probability that the expenses are more than \$10,000.

In Exercises 30–32, assume a normal distribution.

30. *Machine Accuracy* A machine that fills quart bottles with apple juice averages 32.8 oz per bottle, with a standard deviation of 1.1 oz. What are the probabilities that the amount of juice in a bottle is as follows?

a. Less than 1 qt

b. At least 1 oz more than 1 qt

31. *Machine Accuracy* A machine produces screws with a mean length of 2.5 cm and a standard deviation of .2 cm. Find the probabilities that a screw produced by this machine has lengths as follows.

a. Greater than 2.7 cm

b. Within 1.2 standard deviations of the mean

32. *Customer Expenditures* Customers at a certain pharmacy spend an average of \$54.40, with a standard deviation of \$13.50. What are the largest and smallest amounts spent by the middle 50% of these customers?

LIFE SCIENCES

33. *Insect Life Span* The life span of a certain insect (in days) is uniformly distributed over the interval [20, 36].

a. What is the expected life of this insect?

b. Find the probability that one of these insects, randomly selected, lives longer than 30 days.

34. *Location of a Bee Swarm* A swarm of bees is released from a certain point. The proportion of the swarm located at least 2 m from the point of release after 1 hour is a random variable that is exponentially distributed, with $a = 2$.

a. Find the expected proportion under the given conditions.

b. Find the probability that fewer than $1/3$ of the bees are located at least 2 m from the release point after 1 hour.

35. *Digestion Time* The digestion time (in hours) of a fixed amount of food is exponentially distributed, with $a = 1$.

a. Find the mean digestion time.

b. Find the probability that the digestion time is less than 30 minutes.

36. *Pygmy Heights* The average height of a member of a certain tribe of pygmies is 3.2 ft, with a standard deviation of .2 ft. If the heights are normally distributed, what are the largest and smallest heights of the middle 50% of this population?

37. *Finding Prey* H. R. Pulliam found that the time (in minutes) required by a predator to find a prey is a random variable that is exponentially distributed, with $\mu = 25$.*

a. According to this distribution, what is the longest time within which the predator will be 90% certain of finding a prey?

b. What is the probability that the predator will have to spend more than 1 hour looking for a prey?

38. *Life Expectancy* According to the National Center for Health Statistics, the life expectancy for a 55-year-old African American female is 24.8 years.† Assuming that from age 55, the survival of African American females follows an exponential distribution, determine the following probabilities.

a. The probability that a randomly selected 55-year-old African American female will live beyond 80 years of age (at least 25 more years)

b. The probability that a randomly selected 55-year-old African American female will live less than 20 more years

39. *Life Expectancy* According to the National Center for Health Statistics, life expectancy for a 70-year-old African American male is 11.5 years.‡ Assuming that from age 70, the survival of African American males follows an exponential distribution, determine the following probabilities.

a. The probability that a randomly selected 70-year-old African American male will live beyond 90 years of age

b. The probability that a randomly selected 70-year-old African American male will live between 10 and 20 more years

40. *Mercury Poisoning* Historians and biographers have collected evidence that suggests that President Andrew Jackson suffered from mercury poisoning. Recently, researchers measured the amount of mercury in samples of Jackson's hair from 1815. The results of this experiment showed that Jackson had a mean mercury level of 6.0 ppm.§

a. If levels of mercury in hair samples from that time period followed a normal distribution with mean 6.9 and standard deviation 4.6,‖ find the probability that a randomly selected person from that time period would have a mercury level of 6.0 ppm or higher. Discuss whether this provides evidence that Jackson suffered from mercury poisoning during this time period.

b. Today's accepted normal mercury levels follow a normal distribution with approximate mean .6 ppm and standard deviation .3 ppm.# By today's standards, how likely is it that a randomly selected person from today would have a mercury level of 6.0 ppm or higher? Discuss whether we can conclude from this that Andrew Jackson suffered from mercury poisoning.

*Pulliam, H. R., "On the Theory of Optimal Diets," *American Naturalist,* Vol. 108, 1974, pp. 59–74.
†*National Vital Statistics Reports,* Vol. 47, No. 19, June 30, 1999, Table 5.
‡Ibid.
§Deppisch, Lidwig, Jose Centeno, David Gemmel, and Norca Torres, "Andrew Jackson's Exposure to Mercury and Lead," *JAMA,* Vol. 282, No. 6, August 11, 1999, pp. 569–571.
‖Suzuki, T., T. Hongo, M. Morita, and R. Yamamoto, "Elemental Contamination of Japanese Women's Hair from Historical Samples," *Sci. Total Environ.,* Vol. 39, 1984, pp. 81–91.
#Iyengar, V. and J. Woittiez, "Trace Elements in Human Clinical Specimens," *Clinical Chemistry,* Vol. 34, 1988, pp. 474–481.

SOCIAL SCIENCES

41. *Dating a Language* Over time, the number of original basic words in a language tends to decrease as words become obsolete or are replaced with new words. In 1950, C. Feng and M. Swadesh established that of the original 210 basic ancient Chinese words from 950 A.D., 167 were still being used.* The proportion of words that remain after t millennia is a random variable that is exponentially distributed with $a = .229$.

a. Find the life expectancy and standard deviation of a Chinese word.

b. Calculate the probability that a randomly chosen Chinese word will remain after 2000 years.

PHYSICAL SCIENCES

42. *Rainfall* The rainfall (in inches) in a certain region is uniformly distributed over the interval $[32, 44]$.

a. What is the expected number of inches of rainfall?

b. What is the probability that the rainfall will be between 38 and 40 in.?

43. *Dry Length Days* Researchers have shown that the number of successive dry days that occur after a rainstorm for particular regions of Catalonia, Spain, is a random variable that is distributed exponentially with a mean of 8 days.†

a. Find the probability that 10 or more successive dry days occur after a rainstorm.

b. Find the probability that fewer than 2 dry days occur after a rainstorm.

44. *Earthquakes* The proportion of the times (in days) between major earthquakes in the north-south seismic belt of China is a random variable that is exponentially distributed, with $a = 1/609.5$.‡

a. Find the expected number of days and the standard deviation between major earthquakes for this region.

b. Find the probability that the time between a major earthquake and the next one is more than 1 year.

CHAPTER SUMMARY

In this chapter, we gave a brief introduction to the use of calculus in the study of probability. In particular, the conceptual framework of a random variable and its connection to a probability function was given. Integration techniques were used to determine probabilities, expected value, and variance of continuous random variables. Three probability density functions that have a wide range of applications—uniform, exponential, and normal—were studied in detail.

KEY TERMS

11.1 **random variable**
probability function
histogram
discrete probability function
continuous random variable
continuous probability distribution
probability density function

11.2 **mean**
expected value
variance
standard deviation

11.3 **uniform distribution**
exponential distribution
normal distribution
normal curve
standard normal distribution
***z*-score**

CHAPTER 18 REVIEW EXERCISES

1. In a probability function, the y-values (or function values) represent ____________.

2. Define a continuous random variable.

*Lo Bello, Anthony and Maurice Weir, "Glottochronology: An Application of Calculus to Linguistics," *The UMAP Journal,* Vol. 3., No. 1, Spring 1982, pp. 85–99.
†Lana, X. and A. Burgueno, "Daily Dry-Wet Behaviour in Catalonia (NE Spain) from the Viewpoint of Markov Chains," *International Journal of Climatology,* Vol. 18, 1998, pp. 793–815.
‡Wang, Jeen-Hwa and Chiao-Hui Kuo, "On the Frequency Distribution of Interoccurrence Times of Earthquakes," *Journal of Seismology,* Vol. 2, 1998, pp. 351–358.

3. Give the two conditions that a probability density function for $[a, b]$ must satisfy.

Decide whether each function defined as follows is a probability density function for the given interval.

4. $f(x) = \frac{1}{27}(2x + 4);\ [1, 4]$

5. $f(x) = \sqrt{x};\ [4, 9]$

6. $f(x) = .1;\ [0, 10]$

7. $f(x) = e^{-x};\ [0, \infty)$

In Exercises 8 and 9, find a value of k that will make $f(x)$ define a probability density function for the indicated interval.

8. $f(x) = k\sqrt{x};\ [1, 4]$

9. $f(x) = kx^2;\ [0, 3]$

10. The probability density function of a random variable x is defined by

$$f(x) = 1 - \frac{1}{\sqrt{x - 1}} \quad \text{for } x \text{ in } [2, 5].$$

Find the following probabilities.

a. $P(x \geq 3)$ **b.** $P(x \leq 4)$ **c.** $P(3 \leq x \leq 4)$

11. The probability density function of a random variable x is defined by

$$f(x) = \frac{1}{10} \quad \text{for } x \text{ in } [10, 20].$$

Find the following probabilities.

a. $P(x \leq 12)$ **b.** $P(x \geq 31/2)$ **c.** $P(10.8 \leq x \leq 16.2)$

12. Describe what the expected value or mean of a probability distribution represents geometrically.

13. The probability density functions shown in the graphs have the same mean. Which has the smallest standard deviation?

a. **b.** **c.**

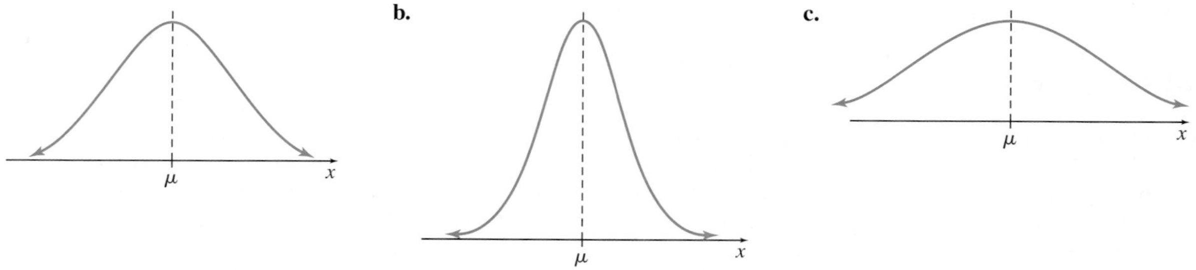

For the probability density functions defined in Exercises 14–17, find the expected value, the variance, and the standard deviation.

14. $f(x) = \frac{1}{5};\ [4, 9]$

15. $f(x) = \frac{2}{9}(x - 2);\ [2, 5]$

16. $f(x) = \frac{1}{7}\left(1 + \frac{2}{\sqrt{x}}\right);\ [1, 4]$

17. $f(x) = 5x^{-6};\ [1, \infty)$

18. The probability density function of a random variable is defined by $f(x) = 4x - 3x^2$ for x in $[0, 1]$. Find the following for the distribution.

a. The mean **b.** The standard deviation

c. The probability that the value of the random variable will be less than the mean

d. The probability that the value of the random variable will be within 1 standard deviation of the mean

19. Find the median of the random variable of Exercise 18. (See Exercises 15–20 in the section on Expected Value and Variance of Continuous Random Variables.) Then find the probability that the value of the random variable will lie between the median and the mean of the distribution.

For Exercises 20 and 21, find **(a)** *the mean of the distribution,* **(b)** *the standard deviation of the distribution, and* **(c)** *the probability that the value of the random variable is within 1 standard deviation of the mean.*

20. $f(x) = \frac{5}{112}(1 - x^{-3/2})$ for x in $[1, 25]$

21. $f(x) = .01e^{-.01x}$ for x in $[0, \infty)$

In Exercises 22–27, find the proportion of observations of a standard normal distribution for each region.

22. The region to the right of $z = 1.53$

23. The region to the left of $z = -.49$

24. The region between $z = -1.47$ and $z = 1.03$

25. The region between $z = -.98$ and $z = -.15$

26. The region that is up to 2.5 standard deviations above the mean

27. The region that is up to 1.2 standard deviations below the mean

28. Find a z-score so that 21% of the area under the normal curve is to the left of z.

29. Find a z-score so that 52% of the area under the normal curve is to the right of z.

The topics in this short chapter involved much of the material studied earlier in this book, including functions, domain and range, exponential functions, area and integration, improper integrals, integration by parts, and numerical integration. For the following special probability density functions, give

a. *the type of distribution;*

b. *the domain and range;*

c. *the graph;*

d. *the mean and standard deviation;*

e. $P(\mu - \sigma \le x \le \mu + \sigma)$.

30. $f(x) = .05$ for x in $[10, 30]$

31. $f(x) = e^{-x}$ for x in $[0,\infty)$

32. $f(x) = \frac{e^{-x^2}}{\sqrt{\pi}}$ for x in $(-\infty, \infty)$ (*Hint:* $\sigma = 1/\sqrt{2}$.)

Applications

BUSINESS AND ECONOMICS

33. *Mutual Funds* The price per share (in dollars) of a particular mutual fund is a random variable x with probability density function defined by

$$f(x) = \frac{3}{4}(x^2 - 16x + 65) \quad \text{for } x \text{ in } [8, 9].$$

Find the probability that the price will be less than \$8.50.

34. *Machine Repairs* The time (in years) until a certain machine requires repairs is a random variable t with probability density function defined by

$$f(x) = \frac{5}{112}(1 - t^{-3/2}) \quad \text{for } t \text{ in } [1, 25].$$

Find the probability that no repairs are required in the first three years by finding the probability that a repair will be needed in years 4 through 25.

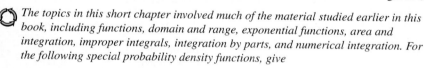

35. *Product Repairs* The number of repairs required by a new product each month is exponentially distributed, with an average of 8.

a. What is the probability density function for this distribution?

b. Find the expected number of repairs per month.

c. Find the standard deviation.

d. What is the probability that the number of repairs per month will be between 5 and 10?

36. *Retail Outlets* The number of new outlets for a clothing manufacturer is an exponential distribution with probability density function defined by

$$f(x) = \frac{1}{6}e^{-x/6} \quad \text{for } x \text{ in } [0, \infty).$$

Find the following for this distribution.

a. The mean

b. The standard deviation

c. The probability that the number of new outlets will be greater than the mean

37. *Useful Life of an Appliance Part* The useful life of a certain appliance part (in hundreds of hours) is 46.2, with a standard deviation of 15.8. Find the probability that one such part would last for at least 6000 (60 hundred) hours. Assume a normal distribution.

LIFE SCIENCES

38. *Movement of a Released Animal* The distance (in meters) that a certain animal moves away from a release point is exponentially distributed, with a mean of 100 m. Find the probability that the animal will move no farther than 100 m away.

39. *Weight Gain of Rats* The weight gain (in grams) of rats fed a certain vitamin supplement is a continuous random variable with probability density function defined by

$$f(x) = \frac{8}{7}x^{-2} \quad \text{for } x \text{ in } [1, 8].$$

a. Find the mean of the distribution.

b. Find the standard deviation of the distribution.

c. Find the probability that the value of the random variable is within 1 standard deviation of the mean.

40. *Body Temperature of a Bird* The body temperature (in degrees Celsius) of a particular species of bird is a continuous random variable with probability density function defined by

$$f(x) = \frac{6}{15{,}925}(x^2 + x) \quad \text{for } x \text{ in } [20, 25].$$

a. What is the expected body temperature of this species?

b. Find the probability of a body temperature below the mean.

41. *Snowfall* The snowfall (in inches) in a certain area is uniformly distributed over the interval [2, 40].

a. What is the expected snowfall?

b. What is the probability of getting more than 20 inches of snow?

42. *Heart Muscle Tension* In a pilot study on tension of the heart muscle in dogs, the mean tension was 2.4 g, with a standard deviation of .4 g. Find the probability of a tension of less than 1.9 g. Assume a normal distribution.

43. *Average Birth Weight* The average birth weight of infants in the United States is 7.8 lb, with a standard deviation of 1.1 lb. Assuming a normal distribution, what is the probability that a newborn will weigh more than 9 lb?

44. *Suicides* The number of suicides in the United States caused by firearms for each age group in 1997 is given in the table below.*

Age Interval (years)	Midpoint of Interval (year)	Number Dying in Each Interval
5–15	10	127
15–25	20	2587
25–35	30	3010
35–45	40	3321
45–55	50	2647
55–65	60	1859
65–75	70	1906
75–85	80	1608
85 +	90 (est)	494
Total		17,559

a. Plot the data. What type of function appears to best match this data?

b. Use the regression feature on your graphing calculator to find a quartic equation that models the number of years, x, since birth and the number of suicides caused by firearms, $N(x)$. Use the midpoint value to estimate the point in each interval when the person died. Graph the function with the plot of the data. Does the function resemble the data?

*National Vital Statistics Reports, Vol. 47, No. 19, June 30, 1999, Table 16.

c. By finding an appropriate constant k, find a function $S(x) = kN(x)$ that is a probability density function describing the probability of suicide by firearm. (*Hint:* Integrate the function you found in part b from 9.7 to 93.2 years.)

d. For a randomly chosen person who committed suicide with a firearm, find the probabilities that the person was between 25 and 35 years old, between 45 and 65 years old, and at least 55 years old. Compare these with the actual probabilities.

e. Estimate the expected age of suicide.

f. Estimate the median of this distribution. (*Hint:* See Exercises 15–20 in the section on Expected Value and Variance of Continuous Random Variables.)

PHYSICAL SCIENCES

45. *Earthquakes* The time between major earthquakes in the Taiwan region is a random variable with probability density function defined by

$$f(t) = \frac{1}{3650.1} e^{-t/3650.1},$$

where t is measured in days.* Find the expected value and standard deviation of this probability density function.

GENERAL INTEREST

46. *State-Run Lotteries* The average state "take" on lotteries is 40%, with a standard deviation of 13%. Assuming a normal distribution, what is the probability that a state-run lottery will have a "take" of more than 50%?

*Wang, Jeen-Hwa and Chiao-Hui Kuo, "On the Frequency Distribution of Interoccurrence Times of Earthquakes," *Journal of Seismology,* Vol. 2, 1998, pp. 351–358.

EXTENDED APPLICATION: Exponential Waiting Times

We have seen in this chapter how probabilities that are spread out over continuous time intervals can be modeled by continuous probability density functions. The exponential distribution you met in the last section of this chapter is often used to model *waiting times,* the gaps between events that are randomly distributed in time, such as decays of a radioactive nucleus or arrivals of customers in the waiting line at a bank. In this application we investigate some properties of the exponential family of distributions.

Suppose that in a badly run subway system, the times between arrivals of subway trains at your station are exponentially distributed with a mean of 10 minutes. Sometimes trains arrive very close together, sometimes far apart, but if you keep track over many days, you'll find that the *average* time between trains is 10 minutes. According to the last section of this chapter, the exponential distribution with density function $f(t) = ae^{-at}$ has mean $1/a$, so the probability density function for our interarrival times is

$$f(t) = \frac{1}{10}e^{-t/10}.$$

First let's see what these waiting times look like. We have used a random-number generator from a statistical software package to draw 25 waiting times from this distribution. Figure 19 shows cumulative arrival times, which is what you would observe if you recorded the arrival time of each train measured in minutes from an arbitrary 0 point.

You can see that 25 trains arrive in a span of about 260 minutes, so the average interarrival time was indeed close to 10 minutes. You may also notice that there are some large gaps and some cases where trains arrived very close together.

To get a better feeling for the distribution of long and short interarrival times, look at the following list, which gives the 25 interarrival times in minutes, sorted from smallest to largest.

.016	4.398	15.659
.226	4.573	15.954
.457	5.415	16.403
.989	9.570	18.978
1.576	10.413	20.736
1.988	10.916	33.013
2.738	13.109	39.073
3.133	13.317	
3.895	14.622	

You can see that there were some very short waits. (In fact, the shortest time between trains is only 1 second, which means our model needs to be adjusted somehow to allow for the time trains spend stopped in the station.) The longest time between trains was 39 minutes, almost four times as long as the average! Although the exponential model exaggerates the irregularities of typical subway service, the problem of pile-ups and long gaps is very real for public transportation, especially for bus routes which are subject to unpredictable traffic delays. Anyone who works at a customer service job is also familiar with this behavior: The waiting line at a bank may be empty for minutes at a stretch, and then several customers walk in at nearly the same time. In this case, the customer interarrival times are exponentially distributed.

Planners who are involved with scheduling need to understand this "clumping" behavior. One way to explore it is to find probabilities for ranges of interarrival times. Here integrals are the natural tool. For example, if we want to estimate the fraction of interarrival times that will be less than 2 minutes, we compute

$$\frac{1}{10}\int_0^2 e^{-t/10}\,dt = 1 - e^{-1/5} \approx .181.$$

So on average, 18% of the interarrival times will be less than 2 minutes, which indicates that clustering of trains will be a problem in our system. (If you have ridden a system like the one in New York City, you may have boarded a train that was ordered to "stand by" for several minutes to spread out a cluster of trains.) We can also compute the probability of a gap of 30 minutes or longer. It will be

$$\frac{1}{10}\int_{30}^{\infty} e^{-t/10}\,dt = e^{-3} \approx .05.$$

So in a random sample of 25 interarrival times we might expect one or two long waits, and our simulation, which includes times of 33 and 39 minutes, is not a fluke. Of course, the rider's experience depends on when she arrives at the station, which is another random input to our model. If she arrives in the middle of a cluster, she'll get a train right away, but if she arrives at the beginning of a long gap she may have a half-hour wait. So we would also like to model the rider's *waiting time,* the time between the rider's arrival at the station and the arrival of the next train.

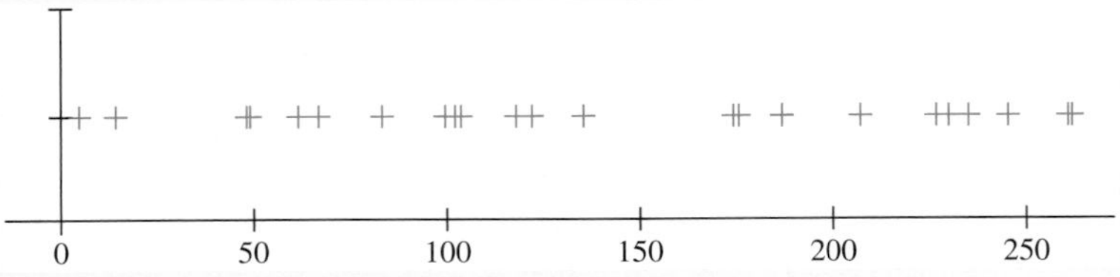

FIGURE 19

A remarkable fact about the exponential distribution is that if our passenger arrives at the station at a random time, the distribution of the rider's waiting times is *the same* as the distribution of interarrival times (that is, exponential with mean 10 minutes). At first this seems paradoxical; since she usually arrives between trains, she should wait less, on average, than the average time between trains. But remember that she's more likely to arrive at the station in one of those long gaps. In our simulation, 72 out of 160 minutes is taken up with long gaps, and even if the rider arrives at the middle of such a gap she'll still wait longer than 15 minutes. Because of this feature the exponential distribution is often called *memoryless:* If you dip into the process at random, it is as if you were starting all over. If you arrive at the station just as a train leaves, your waiting time for the next one still has an exponential distribution with mean 10 minutes. The next train doesn't "know" anything about the one that just left.*

Because the riders' waiting times are exponential, the calculations we have already made tell us what riders will experience: A wait of less than 2 minutes has probability .18. The average wait is 10 minutes, but long waits of more than 30 minutes are not all that rare (probability .05).

Customers waiting for service care about the average wait, but they may care even more about the *predictability* of the wait. In this chapter we stated that the standard deviation for an exponential distribution is the same as the mean, so in our model the standard deviation of riders' waiting times will be 10 minutes. This indicates that a wait of twice the average length is not a rare event. (See Exercise 3.)

Let's compare the experience of riders on our exponential subway with the experience of riders of a perfectly regulated service in which trains arrive *exactly* 10 minutes apart. We'll still assume that the passenger arrives at random. But now the waiting time is uniformly distributed on the time interval [0 minutes, 10 minutes]. This uniform distribution has density function

$$f(t) = \begin{cases} \dfrac{1}{10} & \text{for } 0 \le t \le 10 \\ 0 & \text{otherwise} \end{cases}$$

The mean waiting time is

$$\int_0^{10} \frac{1}{10} \cdot t\,dt = 5 \text{ minutes}$$

and the standard deviation of the waiting times is

$$\sqrt{\int_0^{10} \frac{(t-5)^2}{10}\,dt} = \sqrt{\frac{25}{3}} \approx 2.89 \text{ minutes.}$$

Clearly the rider has a better experience on this system. Even though the same average number of trains is running per hour as in the exponential subway, the average wait for the uniform subway is only 5 minutes with a standard deviation of 2.89 minutes, and no one ever waits longer than 10 minutes!

Any subway run is subject to unpredictable accidents and variations, and this random input is always pushing the riders' waiting times toward the exponential model. Indeed, even with uniform scheduling of trains, there will be service bottlenecks because the exponential distribution is also a reasonable model (over a short time period) for interarrival times of *passengers* entering the station. The goal of schedulers is to move passengers efficiently in spite of random train delays and random input of passengers. One proposed solution, the PRT or personal rapid transit system, uses small vehicles holding just a few passengers that can be scheduled to match a fluctuating demand.

The subway scheduling problem is part of a branch of statistics called *queueing theory*, the study of any process in which inputs arrive at a service point and wait in a line or queue to be served. Examples include telephone calls arriving at a customer service center, our passengers entering the subway station, packets of information traveling through the Internet, and even pieces of code waiting for a processor in a multiprocessor computer. The following Web sites provide a small sampling of work in this very active research area.

- *http://www2.uwindsor.ca/~hlynka/queue.html* (A collection of information on queueing theory)
- *http://faculty.washington.edu/jbs/itrans/ingsim.htm* (an article on scheduling a PRT)
- *http://byte.com/art/9506/sec8/art9.htm* (an article on queueing theory in computer network design)

Exercises

1. If x is a continuous random variable, $P(a \le x \le b)$ is the same as $P(a < x < b)$. Since these are different events, how can they have the same probability?
2. Someone who rides the subway back and forth to work each weekday makes about 40 trips a month. On the exponential subway, how many times a month can this commuter expect a wait longer than half an hour?
3. Find the probability that a rider of the exponential subway waits more than 20 minutes for a train; that is, find the probability of a wait more than twice as long as the average.
4. On the exponential subway, what is the probability that a randomly arriving passenger has a wait of between 9 and 10 minutes? What is the corresponding probability on the uniform subway?
5. If our system is aiming for an average interarrival time of 10 minutes, we might set a tolerance of plus or minus 2 minutes and try to keep the interarrival times between 8 and 12 minutes. Under the exponential model, what fraction of interarrival times fall in this range? How about under the uniform model?

*See Chapter 1 in Volume 2 of Feller, William, *An Introduction to Probability Theory and Its Applications,* 2nd ed., New York: Wiley, 1971.

6. Most mathematical software includes routines for generating "pseudo-random" numbers (that is, numbers that behave randomly even though they are generated by arithmetic). That's what we used to simulate the exponential waiting times for our subway system. But a source on the Internet (http://www.fourmilab.ch/hotbits/) delivers random numbers based on the times between decay events in a sample of Krypton-85. As noted above, the waiting times between decay events have an exponential distribution, so we can see what nature's random numbers look like. Here's a short sample:

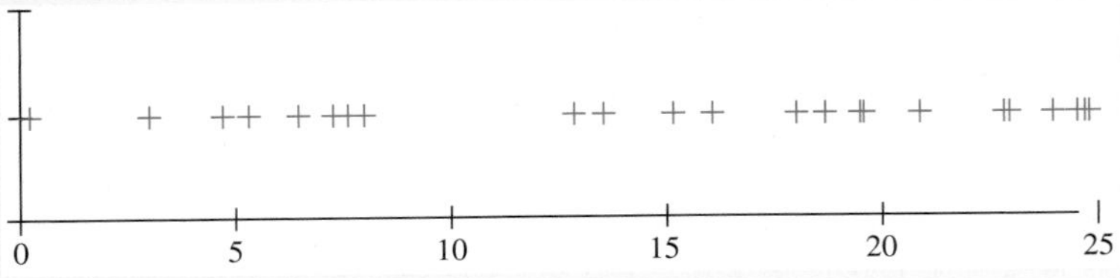

Actually, this source builds its random numbers from random bits, that is, 0's and 1's that occur with equal probability. See if you can think of a way of turning a sequence of exponential waiting times into a random sequence of 0's and 1's.

Directions for Group Project

Find a situation in which you and your group can gather actual wait times, such as a bus stop, doctor's office, teller line at a bank, or check-out line at a grocery store. Collect data on interarrival/service times and determine the mean service time. Using this average, determine whether the data appears to follow an exponential distribution. Develop a table that lists the percentage of the time that particular waiting times occur using both the data and the exponential function. Construct a poster that could be placed near the location where people wait that estimates the waiting time for service.

CHAPTER

12 Sequences and Series

In sports that place high stress on the body, such as basketball, professional athletes often have relatively short playing careers. In Section 2 of this chapter we look at annuities, a kind of investment that allows a player to use her current high earnings to purchase a sequence of guaranteed annual payments that will begin when she retires.

A function whose domain is the set of natural numbers, such as

$$a(n) = 2n, \quad \text{for } n = 1, 2, 3, 4, \ldots$$

is a **sequence.** The sequence $a(n) = 2n$ can be written by listing its *terms,* $2, 4, 6, 8, \ldots, 2n, \ldots$. The letter n is used instead of x as a variable to emphasize the fact that the domain includes only natural numbers. For the same reason, a is used instead of f to name the function.

Sequences have many different applications; one example is the sequence of payments used to pay off a car loan or a home mortgage. (For most practical problems, the domain is a *subset* of the set of natural numbers.) This use of sequences is discussed in Section 12.2. The remaining sections of this chapter cover topics related to sequences.

12.1 GEOMETRIC SEQUENCES

THINK ABOUT IT *If a person saved 1¢ on June 1, 2¢ on June 2, 4¢ on June 3, and so forth, continuing the pattern of saving twice as much each day as the previous day, how much would she have saved by the last day of June?*

We will answer this question in Example 4 of this section.

In our definition of sequence we used the example $a(n) = 2n$. The range values of this sequence function,

$$a(1) = 2, \quad a(2) = 4, \quad a(3) = 6, \ldots,$$

are called the **elements** or **terms** of the sequence. Instead of writing $a(5)$ for the fifth term of a sequence, it is customary to write a_5; for the sequence above

$$a_5 = 10.$$

In the same way, for the sequence above, $a_1 = 2$, $a_2 = 4$, $a_8 = 16$, $a_{20} = 40$, and $a_{51} = 102$.

The symbol a_n is used for the **general** or ***n*th term** of a sequence. For example, for the sequence $4, 7, 10, 13, 16, \ldots$ the general term might be given by $a_n = 1 + 3n$. This formula for a_n can be used to find any term of the sequence that might be needed. For example, the first three terms of the sequence are

$$a_1 = 1 + 3(1) = 4, \qquad a_2 = 1 + 3(2) = 7, \qquad \text{and} \qquad a_3 = 1 + 3(3) = 10.$$

Also, $a_8 = 25$ and $a_{12} = 37$.

EXAMPLE 1 *Sequence*

Find the first four terms of the sequence having general term $a_n = -4n + 2$.

Solution Replace n, in turn, with 1, 2, 3, and 4.

$$\text{If } n = 1, \quad a_1 = -4(1) + 2 = -4 + 2 = -2.$$
$$\text{If } n = 2, \quad a_2 = -4(2) + 2 = -8 + 2 = -6.$$
$$\text{If } n = 3, \quad a_3 = -4(3) + 2 = -12 + 2 = -10.$$
$$\text{If } n = 4, \quad a_4 = -4(4) + 2 = -16 + 2 = -14.$$

The first four terms of this sequence are $-2, -6, -10$, and -14.

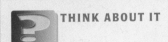

A sequence in which each term after the first is found by *multiplying* the preceding term by the same number is called a **geometric sequence.** The ratio of any two consecutive terms is a constant r,

$$r = \frac{a_{n+1}}{a_n}, \quad \text{where } n \geq 1,$$

called the **common ratio.** For example, to find r in the following sequence:

$$3, -6, 12, -24, 48, -96, \ldots$$

take $-6/3$ or $12/(-6)$ or $-24/12$, etc. and get $r = -2$. Thus, it is a geometric sequence in which each term after the first is found by multiplying the preceding term by the number -2, the common ratio.

If a is the first term of a geometric sequence and r is the common ratio, then the second term is given by $a_2 = ar$ and the third term by $a_3 = a_2r = ar^2$. Also, $a_4 = ar^3$ and $a_5 = ar^4$. These results are generalized below.

GENERAL TERM OF A GEOMETRIC SEQUENCE

If a geometric sequence has first term a and common ratio r, then

$$a_n = ar^{n-1}.$$

EXAMPLE 2 *Geometric Sequences*

Find the indicated term for each geometric sequence.

(a) Find a_7 for 6, 24, 96, 384, ...

Solution Here $a = a_1 = 6$. To verify that the sequence is geometric, divide each term except the first by the preceding term.

$$\frac{24}{6} = \frac{96}{24} = \frac{384}{96} = 4$$

Since the ratio is constant, the sequence is geometric with $r = 4$. To find a_7, use the formula for a_n with $n = 7$, $a = 6$, and $r = 4$.

$$a_7 = 6(4)^{7-1} = 6(4)^6 = 6(4096) = 24{,}576$$

(b) Find a_6 for 8, -16, 32, -64, 128, ...

Solution As before, verify that $r = -16/8 = 32/(-16) = -64/32 = 128/(-64) = -2$. Here $a = a_1 = 8$, so

$$a_6 = 8(2)^{6-1} = 8(-2)^5 = 8(-32) = -256.$$

In the next section we will need to know how to find the sum of the first n terms of a geometric sequence. To get a general rule for finding such a sum, begin by writing the sum S_n of the first n terms of a geometric sequence with first term $a = a_1$ and common ratio r as

$$S_n = a_1 + a_2 + a_3 + \cdots + a_n.$$

Since $a_n = ar^{n-1}$, this sum can be written as

$$S_n = a + ar + ar^2 + \cdots + ar^{n-1}. \tag{1}$$

Using summation notation, we can write S_n as

$$S_n = \sum_{i=0}^{n-1} ar^i.$$

If $r = 1$, all the terms are equal to a, and $S_n = n \cdot a$, the correct result for this case. If $r \neq 1$, multiply both sides of Equation (1) by r, obtaining

$$rS_n = ar + ar^2 + ar^3 + \cdots + ar^n. \tag{2}$$

Now subtract corresponding sides of Equation (1) from Equation (2):

$$rS_n - S_n = ar^n - a.$$

Factoring yields

$$S_n(r - 1) = a(r^n - 1),$$

and dividing by $r - 1$ on both sides gives

$$S_n = \frac{a(r^n - 1)}{r - 1}.$$

This result is summarized below.

SUM OF THE FIRST n TERMS OF A GEOMETRIC SEQUENCE

If a geometric sequence has first term a and common ratio r, then the sum of the first n terms, S_n, is given by

$$S_n = \frac{a(r^n - 1)}{r - 1}, \quad \textbf{where } r \neq 1.$$

EXAMPLE 3 *Summing Terms of a Geometric Sequence*

Find the sum of the first six terms of the geometric sequence 3, 12, 48, ...

Method 1: Using the Formula

Solution Here $a = a_1 = 3$ and $r = 4$. Find S_6, the sum of the first six terms, by the formula above.

$$S_6 = \frac{3(4^6 - 1)}{4 - 1} \qquad \text{Let } n = 6, a = 3, r = 4.$$

$$= \frac{3(4096 - 1)}{3}$$

$$= 4095$$

Method 2: Using a Calculator

The sum of a sequence can be conveniently calculated on a TI-83/84 Plus calculator using the command `sum(seq(3*4^N, N, 0, 5))` as shown in Figure 1. Notice that to find the sum of the first six terms of the sequence we begin with `N = 0` and end with `N = 5` for a total of 4095.

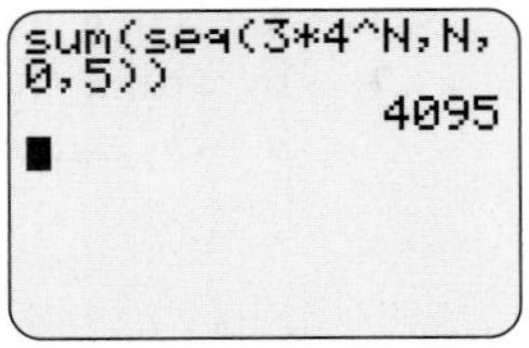

FIGURE 1

EXAMPLE 4 *Savings*

A person saved 1¢ on June 1, 2¢ on June 2, 4¢ on June 3, and so forth, continuing the pattern of saving twice as much each day as the previous day. How much would she have saved by the end of June?

Solution This is a geometric sequence with $a = 1$ and $r = 2$. We want to find S_{30}.

$$S_{30} = \frac{1(2^{30} - 1)}{2 - 1} = 1{,}073{,}741{,}823$$

By the end of June, the person will have saved 1,073,741,823 cents, or \$10,737,418.23!

12.1 EXERCISES

List the first n terms of the geometric sequence satisfying the following conditions.

1. $a_1 = 2, r = 3, n = 4$

2. $a_1 = 4, r = 2, n = 5$

3. $a_1 = 1/2, r = 4, n = 4$

4. $a_1 = 2/3, r = 6, n = 3$

5. $a_3 = 6, a_4 = 12, n = 5$

6. $a_2 = 9, a_3 = 3, n = 4$

Find a_5 and a_n for the following geometric sequences.

7. $a_1 = 4, r = 3$

8. $a_1 = 8, r = 4$

9. $a_1 = -3, r = -5$

10. $a_1 = -4, r = -2$

11. $a_2 = 3, r = 2$

12. $a_3 = 6, r = 3$

13. $a_4 = 64, r = -4$

14. $a_4 = 81, r = -3$

For each sequence that is geometric, find r and a_n.

15. 6, 12, 24, 48, ...

16. 4, 16, 64, 256, ...

17. 3/4, 3/2, 3, 6, 12, ...

18. −7, −5, −3, −1, 1, 3, ...

19. 4, 8, −16, 32, 64, −128, ...

20. 6, 8, 10, 12, 14, ...

21. −5/8, 5/12, −5/18, 5/27, ...

22. 7/4, −7/12, 7/36, −7/108, ...

Find the sum of the first five terms of each geometric sequence.

23. 3, 6, 12, 24, ...

24. 5, 20, 80, 320, ...

25. 12, −6, 3, −3/2, ...

26. 18, −3, 1/2, −1/12, ...

27. $a_1 = 3, r = -2$

28. $a_1 = -4, r = 3$

29. $a_1 = 8.423, r = 2.859$

30. $a_1 = -3.772, r = -1.553$

Use the formula for the sum of the first n terms of a geometric sequence to evaluate the following sums.

31. $\sum_{i=0}^{4} 3(2^i)$

32. $\sum_{i=0}^{3} 2(3^i)$

33. $\sum_{i=0}^{4} \frac{1}{2}(4^i)$

34. $\sum_{i=0}^{4} \frac{3}{2}(2^i)$

35. $\sum_{i=0}^{4} \frac{4}{3}(3^i)$

36. $\sum_{i=0}^{4} \frac{5}{3}(3^i)$

37. $\sum_{i=0}^{8} 64\left(\frac{1}{2}\right)^i$

38. $\sum_{i=0}^{6} 81\left(\frac{2}{3}\right)^i$

Applications

BUSINESS AND ECONOMICS

39. *Depreciation* A certain machine annually loses 30% of the value it had at the beginning of that year. If its initial value is \$10,000, find its value at the following times.

a. The end of the fifth year

b. The end of the eighth year

40. *Income* An oil well produced \$4,000,000 of income its first year. Each year thereafter, the well produced half as much income as the previous year. What is the total amount of income produced by the well in 6 years?

41. *Savings* Suppose you could save \$1 on January 1, \$2 on January 2, \$4 on January 3, and so on. What amount would you save on January 31? What would be the total amount of your savings during January?

42. *Depreciation* Each year a machine loses 20% of the value it had at the beginning of the year. Find the value of the machine at the end of 6 years if it cost \$100,000 new.

LIFE SCIENCES

43. *Population* The population of a certain colony of bacteria increases by 10% each hour. After 5 hours, what is the percent increase in the population over the initial population?

PHYSICAL SCIENCES

44. *Radioactive Decay* The half-life of a radioactive substance is the time it takes for half the substance to decay. Suppose the half-life of a substance is 3 years and that 10^{15} molecules of the substance are present initially. How many molecules will be unchanged after 15 years?

45. *Rotation of a Wheel* A bicycle wheel rotates 400 times per minute. If the rider removes his or her feet from the pedals, the wheel will start to slow down. Each minute, it will rotate only 3/4 as many times as in the preceding minute. How many times will the wheel rotate in the fifth minute after the rider's feet are removed from the pedals?

GENERAL INTEREST

46. *Thickness of a Paper Stack* A piece of paper is .008 in. thick.

    a. Suppose the paper is folded in half, so that its thickness doubles, for 12 times in a row. How thick is the final stack of paper?

    b. Suppose it were physically possible to fold the paper 50 times in a row. How thick would the final stack of paper be?

47. *Sports* In the NCAA Men's Basketball Tournament, 64 teams are initially paired off. By playing a series of single-elimination games, a champion is crowned.*

    a. Write a geometric sequence whose sum determines the number of games that must be played to determine the champion team.

    b. How many games must be played to produce the champion?

    c. Generalize parts a and b to a tournament where 2^n teams are initially present.

    d. Discuss a quick way to determine the answers to parts b and c, based on the fact that each game produces one loser, and all teams except the champion lose one game.

48. *Game Shows* Some game shows sponsor tournaments where in each game, three individuals play against each other, yielding one winner and two losers. The winners of three such games then play each other, until the final game of three players produces a tournament winner.† Suppose 81 people begin such a tournament.

    a. Write a geometric sequence whose sum determines the number of games that must be played to determine the tournament champion.

    b. How many games must be played to produce the champion?

    c. Generalize parts a and b to a tournament where 3^n players are initially present.

    d. Further generalize parts a and b to a tournament where t^n players are initially present.

12.2 ANNUITIES: AN APPLICATION OF SEQUENCES

THINK ABOUT IT

Suppose \$1500 is deposited at the end of each year for the next 6 years in an account paying 8% per year, compounded annually. How much will be in the account after 6 years?

Such a sequence of equal payments made at equal periods of time is called an **annuity.** If each payment is made at the end of a period, and if the frequency of payments is the same as the frequency of compounding, the annuity is called an

*Schielack, Vincent. "Tournaments and Geometric Sequences," *Mathematics Teacher*, Vol. 86, No. 2, Feb. 1993, pp. 127–129.

†Ibid.

ordinary annuity. The time between payments is the **payment period,** and the time from the beginning of the first period to the end of the last period is called the **term** of the annuity. The **amount** of the annuity, the final sum on deposit in the account, is defined as the sum of the compound amounts of all the payments, compounded to the end of the term.

Figure 2 shows the annuity described above. To find the amount of this annuity, look at each of the \$1500 payments separately. The first of these payments will produce a compound amount of

$$1500(1 + .08)^5 = 1500(1.08)^5$$

at the end of 6 years. Use 5 as the exponent instead of 6 because the money is deposited at the *end* of the first year and thus earns interest for only 5 years.

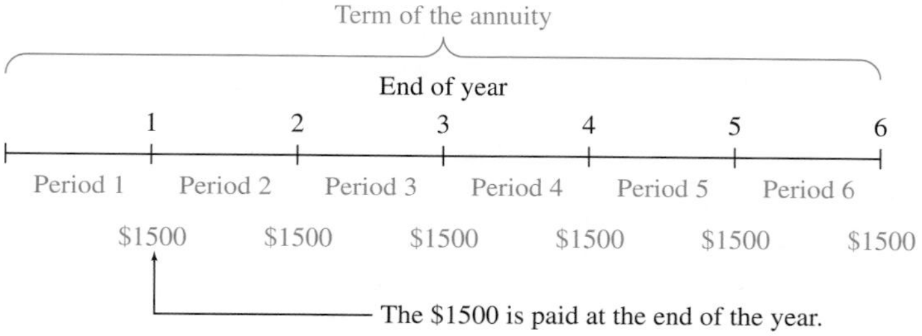

FIGURE 2

The second payment of \$1500 will produce a compound amount (at the end of 5 years) of $1500(1.08)^4$. As shown in Figure 3, the total amount of the annuity is

$$1500(1.08)^5 + 1500(1.08)^4 + 1500(1.08)^3 + 1500(1.08)^2 \\ + 1500(1.08)^1 + 1500. \tag{1}$$

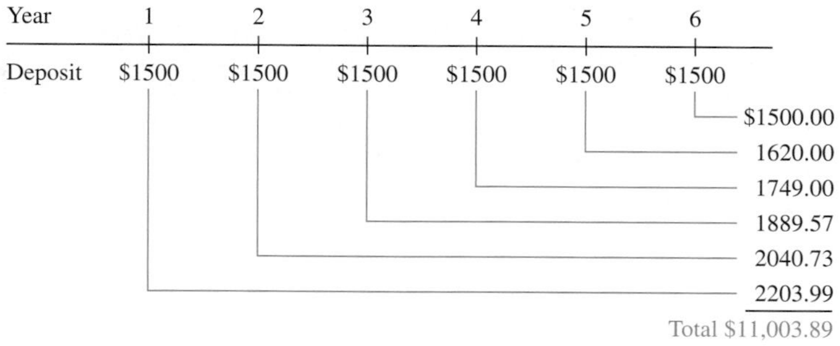

FIGURE 3

(The last payment earns no interest.) Reversing the order of the terms, so the last term is first, shows that Equation (1) is the sum of the terms of a geometric sequence with $a = 1500$, $r = 1.08$, and $n = 6$. Using the formula for the sum of the first n terms of a geometric sequence gives

$$\begin{aligned}1500 + 1500(1.08)^1 + 1500(1.08)^2 + 1500(1.08)^3 \\ + 1500(1.08)^4 + 1500(1.08)^5 &= \frac{1500(1.08^6 - 1)}{1.08 - 1} \\ &\approx \$11{,}003.89.\end{aligned}$$

To generalize this result, suppose that R dollars are paid into an account at the end of each period for n periods, at a rate of interest i per period. The first payment of R dollars will produce a compound amount of $R(1 + i)^{n-1}$ dollars, the second payment will produce $R(1 + i)^{n-2}$ dollars, and so on; the final payment earns no interest and contributes just R dollars to the total. If S represents the future value (or sum) of the annuity, then

$$S = R(1 + i)^{n-1} + R(1 + i)^{n-2} + R(1 + i)^{n-3} + \cdots + R(1 + i) + R,$$

or, written in reverse order,

$$S = R + R(1 + i)^1 + R(1 + i)^2 + \cdots + R(1 + i)^{n-1}.$$

This is the sum of the first n terms of the geometric sequence having first term R and common ratio $1 + i$. Using the formula for the sum of the first n terms of a geometric sequence gives

$$S = \frac{R[(1 + i)^n - 1]}{(1 + i) - 1} = \frac{R[(1 + i)^n - 1]}{i} = R\left[\frac{(1 + i)^n - 1}{i}\right].$$

The quantity in brackets is commonly written $s_{\overline{n}|i}$ (read "s-angle-n at i"), so that

$$S = R \cdot s_{\overline{n}|i}.$$

Values of $s_{\overline{n}|i}$ can be found by a calculator. The TI-83/84 Plus has a special `Finance` menu with this formula built in. For more details, see *The Graphing Calculator Manual* that is available with this book.

Our work with annuities can be summarized as follows.

AMOUNT OF AN ANNUITY

The amount S of an annuity of payments of R dollars each, made at the end of each period for n consecutive interest periods at a rate of interest i per period, is given by

$$S = R\left[\frac{(1 + i)^n - 1}{i}\right] \quad \text{or} \quad S = R \cdot s_{\overline{n}|i}.$$

EXAMPLE 1 *Annuity*

Susan Gallier is an athlete who feels that her playing career will last 7 more years. To prepare for her future, she deposits \$22,000 at the end of each year for 7 years in an account paying 8% compounded annually. How much will she have on deposit after 7 years?

Solution Her payments form an ordinary annuity with $R = 22{,}000$, $n = 7$, and $i = .08$. The amount of this annuity is (by the formula on the previous page)

$$S = 22{,}000\left[\frac{(1.08)^7 - 1}{.08}\right].$$

The number in brackets, $s_{\overline{7}|.08}$, is 8.9228033, so that

$$S = 22{,}000(8.9228033) = 196{,}301.67,$$

or \$196,301.67.

EXAMPLE 2 *Annuity*

Suppose \$1000 is deposited at the end of each 6-month period for 5 years in an account paying 6% per year compounded semiannually. Find the amount of the annuity.

Solution Interest of $6\%/2 = 3\%$ is earned semiannually. In 5 years there are $5 \times 2 = 10$ semiannual periods. Since $s_{\overline{10}|.03} = [(1.03)^{10} - 1]/.03 = 11.46388$, the \$1000 deposits will produce a total of

$$S = 1000(11.46388) = 11{,}463.88,$$

or \$11,463.88.

The formula for S involves the variables R, i, and n. The next example shows how to solve for one of these other variables.

EXAMPLE 3 *Annuity*

Nancy Hart wants to buy an expensive video camera three years from now. She plans to deposit an equal amount at the end of each quarter for three years in order to accumulate enough money to pay for the camera. Nancy expects the camera to cost \$2400 at that time. The bank pays 6% interest per year compounded quarterly. Find the amount of each of the 12 equal deposits she must make.

Solution This example describes an ordinary annuity with $S = 2400$, $i = .015$ $(6\%/4 = 1.5\%)$, and $n = 3 \cdot 4 = 12$ periods. The unknown here is the amount of each payment, R. By the formula for the amount of an annuity given above,

$$2400 = R \cdot s_{\overline{12}|.015}.$$
$$2400 = R(13.04121)$$
$$R = 184.03, \qquad \text{Divide both sides by 13.04121.}$$

or \$184.03.

Sinking Fund A **sinking fund** is a fund set up to receive periodic payments; these periodic payments plus the interest on them are designed to produce a given total at some time in the future. As an example, a corporation might set up a sinking fund to receive money that will be needed to pay off a loan in the future. The deposits in Examples 1 and 3 form sinking funds.

EXAMPLE 4 *Sinking Fund*

The Moores are close to retirement. They agree to sell an antique urn to a local museum for \$17,000. Their tax adviser suggests that they defer receipt of this money until they retire, 5 years in the future. (At that time, they might well be in a lower tax bracket.) The museum agrees to pay them the \$17,000 in a lump sum in 5 years. Find the amount of each payment the museum must make into a sinking fund so that it will have the necessary \$17,000 in 5 years. Assume that the museum can earn 8% compounded annually on its money and that the payments are made annually.

Solution These payments make up an ordinary annuity. The annuity will amount to \$17,000 in 5 years at 8% compounded annually, so

$$17{,}000 = R \cdot s_{\overline{5}|.08}$$

$$R = \frac{17{,}000}{s_{\overline{5}|.08}} = \frac{17{,}000}{5.86660} \approx 2897.76,$$

or \$2897.76. If the museum deposits \$2897.76 at the end of each year for 5 years in an account paying 8% compounded annually, it will have the needed \$17,000. This result is shown in the following table. In other cases, the last payment might differ slightly from the others due to rounding R to the nearest penny.

Payment Number	Amount of Deposit	Interest Earned	Total in Account
1	\$2897.76	\$0	\$2897.76
2	\$2897.76	\$231.82	\$6027.34
3	\$2897.76	\$482.19	\$9407.29
4	\$2897.76	\$752.58	\$13,057.63
5	\$2897.76	\$1044.61	\$17,000.00

Present Value of an Annuity As shown above, if a deposit of R dollars is made at the end of each period for n periods, at a rate of interest i per period, then the account will contain

$$S = R \cdot s_{\overline{n}|i} = R\left[\frac{(1 + i)^n - 1}{i}\right]$$

dollars after n periods. Now suppose we want to find the *lump sum* P that must be deposited today at a rate of interest i per period in order to produce the same amount S after n periods.

First recall that P dollars deposited today will amount to $P(1 + i)^n$ dollars after n periods at a rate of interest i per period. This amount, $P(1 + i)^n$, should be the same as S, the amount of the annuity. Substituting $P(1 + i)^n$ for S in the formula above gives

$$P(1 + i)^n = R\left[\frac{(1 + i)^n - 1}{i}\right].$$

To solve this equation for P, multiply both sides of the equation by $(1 + i)^{-n}$.

$$P = R(1 + i)^{-n}\left[\frac{(1 + i)^n - 1}{i}\right]$$

Use the distributive property and the fact that $(1 + i)^{-n}(1 + i)^n = 1$.

$$P = R\left[\frac{(1 + i)^{-n}(1 + i)^n - (1 + i)^{-n}}{i}\right] = R\left[\frac{1 - (1 + i)^{-n}}{i}\right]$$

The amount P is called the **present value of the annuity.** The quantity in brackets is abbreviated as $a_{\overline{n}|i}$.

PRESENT VALUE OF AN ANNUITY

The present value P of an annuity of payments of R dollars each, made at the end of each period for n consecutive interest periods at a rate of interest i per period is given by

$$P = R\left[\frac{1 - (1 + i)^{-n}}{i}\right] \quad \text{or} \quad P = R \cdot a_{\overline{n}|i}.$$

EXAMPLE 5 *Present Value*

What lump sum deposited today at 6% interest compounded annually will yield the same total amount as payments of \$1500 at the end of each year for 12 years, also at 6% compounded annually?

Solution Find the present value of an annuity of \$1500 per year for 12 years at 6% compounded annually. From the present value formula, $a_{\overline{12}|.06} = [1 - (1.06)^{-12}]/.06 = 8.383844$, so

$$P = 1500(8.383844) = 12{,}575.77,$$

or \$12,575.77. A lump sum deposit of \$12,575.77 today at 6% compounded annually will yield the same total after 12 years as deposits of \$1500 at the end of each year for 12 years at 6% compounded annually.

Check this result as follows. The compound amount in 12 years of a deposit of \$12,575.77 today at 6% compounded annually can be found by the formula $A = P(1 + i)^n$:

$$12{,}575.77(1.06)^{12} = (12{,}575.77)(2.012196) \approx 25{,}304.91,$$

or \$25,304.91. On the other hand, a deposit of \$1500 into an annuity at the end of each year for 12 years, at 6% compounded annually, gives an amount of

$$1500[(1.06)^{12} - 1]/.06 = 1500(16.86994) = 25{,}304.91,$$

or \$25,304.91.

In summary, there are two ways to have \$25,304.91 in 12 years at 6% compounded annually—a single deposit of \$12,575.77 today, or payments of \$1500 at the end of each year for 12 years.

The formula above can be used if the lump sum is known and the periodic payment of the annuity must be found. The next example shows how to do this.

EXAMPLE 6 *Payments*

A used car costs \$6000. After a down payment of \$1000, the balance will be paid off in 36 monthly payments with interest of 12% per year, compounded monthly. Find the amount of each payment.

Solution A single lump sum payment of \$5000 today would pay off the loan, so \$5000 is the present value of an annuity of 36 monthly payments with interest of 12%/12 = 1% per month. We can find R, the amount of each payment, by using the formula

$$P = R \cdot a_{\overline{n}|i}$$

and replacing P with 5000, n with 36, and i with .01. From the present value formula, $a_{\overline{36}|.01} = 30.10751$, so

$$5000 = R(30.10751)$$
$$R \approx 166.07$$

or \$166.07. Monthly payments of \$166.07 each will be needed.

Amortization A loan is **amortized** if both the principal and interest are paid by a sequence of equal periodic payments. In Example 6 above, a loan of \$5000 at 12% interest compounded monthly could be amortized by paying \$166.07 per month for 36 months, or (it turns out) \$131.67 per month for 48 months.

EXAMPLE 7 *Amortization*

A speculator agrees to pay \$15,000 for a parcel of land. Payments will be made twice each year for 4 years at an interest rate of 12% compounded semiannually.

(a) Find the amount of each payment.

Solution If the speculator immediately paid \$15,000, there would be no need for any payments at all. Thus, \$15,000 is the present value of an annuity of R dollars, with $2 \cdot 4 = 8$ periods, and $i = 12\%/2 = 6\% = .06$ per period. If P is the present value of an annuity,

$$P = R \cdot a_{\overline{n}|i}.$$

In this example, $P = 15{,}000$, with

$$15{,}000 = R \cdot a_{\overline{8}|.06}$$

or

$$R = \frac{15{,}000}{a_{\overline{8}|.06}}.$$
$$= \frac{15{,}000}{6.20979} \approx 2415.54,$$

or \$2415.54. Each payment is \$2415.54.

(b) Find the portion of the first payment that is applied to the reduction of the debt.

Solution Interest is 12% per year, compounded semiannually. During the first period, the entire \$15,000 is owed. Interest on this amount for 6 months (1/2 year) is found by the formula for simple interest, $I = Prt$, so that

$$I = 15{,}000(.12)\left(\frac{1}{2}\right) = 900,$$

or \$900. At the end of 6 months, the speculator makes a payment of \$2415.54; since \$900 of this represents interest, a total of

$$\$2415.54 - \$900 = \$1515.54$$

is applied to the reduction of the original debt.

(c) Find the balance due after 6 months.

Solution The original balance due is \$15,000. After 6 months, \$1515.54 is applied to reduction of the debt. The debt owed after 6 months is

$$\$15{,}000 - \$1515.54 = \$13{,}484.46.$$

(d) How much interest is owed for the second 6-month period? How much will be applied to the debt?

Solution A total of \$13,484.46 is owed for the second 6 months. Interest on this amount is

$$I = 13{,}484.46(.12)\left(\frac{1}{2}\right) \approx 809.07,$$

or \$809.07. A payment of \$2415.54 is made at the end of this period; a total of

$$\$2415.54 - \$809.07 = \$1606.47$$

is applied to the reduction of the debt.

Continuing this process gives the *amortization schedule* shown below. As the schedule shows, each payment is the same, except perhaps for a small adjustment in the final payment. Payment 0 represents the original amount of the loan.

Payment Number	Amount of Payment	Interest for Period	Portion to Principal	Principal at End of Period
0	—	—	—	\$15,000.00
1	\$2415.54	\$900	\$1515.54	\$13,484.46
2	\$2415.54	\$809.07	\$1606.47	\$11,877.99
3	\$2415.54	\$712.68	\$1702.86	\$10,175.13
4	\$2415.54	\$610.51	\$1805.03	\$8370.10
5	\$2415.54	\$502.21	\$1913.33	\$6456.77
6	\$2415.54	\$387.41	\$2028.13	\$4428.64
7	\$2415.54	\$265.72	\$2149.82	\$2278.82
8	\$2415.54	\$136.73	\$2278.82	\$0

The unpaid balance of a loan after x payments is equivalent to the present value of an annuity after $n - x$ consecutive payments, and is given by the function

$$y = R\left[\frac{1 + (1 + i)^{-(n-x)}}{i}\right]$$

For Example 7, the unpaid balance after two payments is

$$y = 2415.54\left[\frac{1 - (1 + .12/2)^{-(8-2)}}{.12/2}\right] \approx 11{,}877.99,$$

or $11,877.99.

This formula can also be used to produce a graph of the unpaid balance. For Example 7, the graph of

$$y = 2415.54\left[\frac{1 - (1 + .12/2)^{-(8-x)}}{.12/2}\right]$$

is shown in Figure 4.

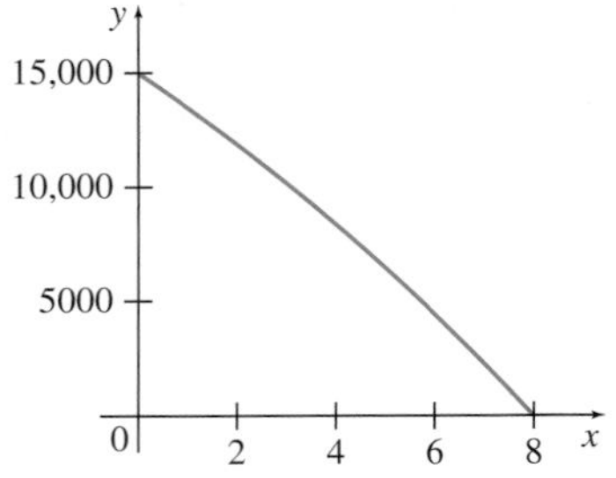

FIGURE 4

EXAMPLE 8 *Amortization*

The Millers buy a house for $74,000, with a down payment of $16,000. Interest is charged at 10.25% per year for 30 years. Find the amount of each monthly payment to amortize the loan.

Method 1: Calculation by Hand

Solution Here, the present value, P, is 58,000 (or 74,000 − 16,000). Also, $i = .1025/12 \approx .0085416667$, and $n = 12 \cdot 30 = 360$. The monthly payment R must be found. From the formula for the present value of an annuity,

$$\begin{aligned} 58{,}000 &= R \cdot a_{\overline{360}|.0085416667} \\ &= R\left[\frac{1 - (1 + .0085416667)^{-360}}{.0085416667}\right] \\ &= R\left(\frac{1 - .0467967507}{.0085416667}\right) \\ &= R\left(\frac{.9532032493}{.0085416667}\right), \end{aligned}$$

or

$$R \approx 519.74.$$

Monthly payments of $519.74 will be required to amortize the loan.

Method 2: Graphing Calculator

We can find the monthly payments to amortize this loan using the Finance Application of a TI-83/84 Plus calculator. To solve this problem, press the APPS button on the calculator and then select the `Finance` option. To input the particular information into the application, select the `TVM Solver` as shown in Figure 5 on the next page and then press ENTER. Then input the relevant values needed for the Solver, as shown in Figure 6. Note that the value of `PMT` is zero in `TVM Solver`. At this point, the particular value of `PMT` does not matter since we are going to calculate that value. Once the information is input into the solver, you must press 2nd QUIT to leave `TVM Solver`. To find the payment, press APPS, then `Finance`, and then select the `tvm_Pmt` button and press ENTER. The result shown in Figure 7 agrees with our work above.

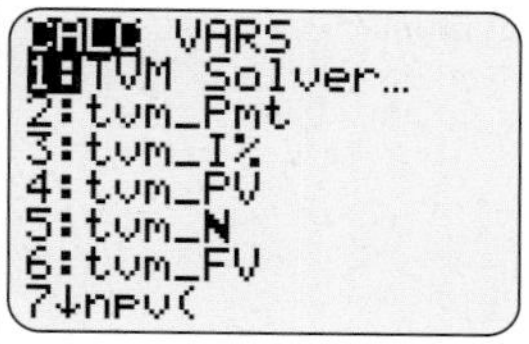

FIGURE 5

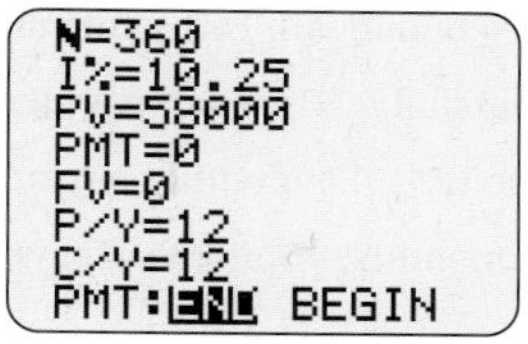

FIGURE 6

FIGURE 7

12.2 EXERCISES

Find the value of each ordinary annuity. (Interest is compounded annually.)

1. $R = 100, i = .06, n = 10$

2. $R = 1000, i = .06, n = 12$

3. $R = 10{,}000, i = .05, n = 19$

4. $R = 100{,}000, i = .08, n = 23$

5. $R = 8500, i = .06, n = 30$

6. $R = 11{,}200, i = .08, n = 25$

Find the value of each ordinary annuity based on the information given.

7. $R = 9200$, 16% interest compounded semiannually for 7 years

8. $R = 3700$, 12% interest compounded semiannually for 11 years

9. $R = 800$, 12% interest compounded semiannually for 12 years

10. $R = 4600$, 16% interest compounded quarterly for 9 years

Find the periodic payments that will amount to the given sums under the given conditions.

11. $S = \$10{,}000$; interest is 8% compounded annually; payments are made at the end of each year for 12 years.

12. $S = \$100{,}000$; interest is 16% compounded semiannually; payments are made at the end of each semiannual period for 9 years.

13. $S = \$50{,}000$; interest is 12% compounded quarterly; payments are made at the end of each quarter for 8 years.

Find the present value of each ordinary annuity.

14. Payments of \$1000 are made annually for 9 years at 8% compounded annually.

15. Payments of \$5000 are made annually for 11 years at 6% compounded annually.

16. Payments of \$890 are made annually for 16 years at 8% compounded annually.

17. Payments of \$1400 are made annually for 8 years at 8% compounded annually.

18. Payments of \$10,000 are made semiannually for 15 years at 10% compounded semiannually.

19. Payments of \$50,000 are made quarterly for 10 years at 8% compounded quarterly.

Find the lump sum deposited today that will yield the same total amount as payments of \$10,000 *at the end of each year for* 15 *years, at the following interest rates. Interest is compounded annually.*

20. 4%

21. 5%

22. 6%

23. 8%

Find the payments necessary to amortize each loan.

24. \$1000, 8% compounded annually, 9 annual payments

25. \$2500, 16% compounded quarterly, 6 quarterly payments

26. \$41,000, 12% compounded semiannually, 10 semiannual payments

27. \$90,000, 8% compounded annually, 12 annual payments

28. \$5500, 18% compounded monthly, 24 monthly payments

29. \$45,000, 18% compounded monthly, 36 monthly payments

Applications

BUSINESS AND ECONOMICS

30. *Amount of an Annuity* Pat Pearson deposits \$12,000 at the end of each year for 9 years in an account paying 8% interest compounded annually. Find the final amount she will have on deposit.

31. *Amount of an Annuity* Pat's brother-in-law works in a bank that pays 6% compounded annually. If she deposits her money in this bank instead of the one in Exercise 30, how much will she have in her account?

32. *Amount of an Annuity* Referring to Exercises 30 and 31, how much would Pat lose over 9 years by using her brother-in-law's bank instead of the bank in Exercise 30?

33. *Sinking Fund* Jack Wells needs \$10,000 in 8 years. What amount should he deposit at the end of each quarter at 16% compounded quarterly to accumulate the \$10,000?

34. *Sinking Fund* Find Jack Wells's quarterly deposit (see Exercise 33) if the money is deposited at 12% compounded quarterly.

35. *Sinking Fund* Tonya McCarley wants to buy a car that she estimates will cost \$18,000 in 6 years. How much money must she deposit at the end of each quarter at 12% interest compounded quarterly in order to have enough in 6 years to pay for her car?

36. *Sinking Fund* Harv's Meats will need to buy a new deboner machine in 4 years. At that time Harv expects the machine to cost \$12,000. To accumulate enough money to pay for the machine, Harv decides to deposit a sum of money at the end of each 6-month period in an account paying 16% compounded semiannually. How much should each payment be?

Individual Retirement Accounts With Individual Retirement Accounts (IRAs), a worker whose income does not exceed certain limits can deposit \$2000 *annually, with taxes deferred on the principal and interest. To attract depositors, banks have been advertising the amount that would accumulate by retirement. Suppose a* 40*-year-old person deposits* \$2000 *per year until age* 65*. Find the total in the account with the interest rates stated in Exercises 37–40. Assume semiannual compounding with payments of \$1000 made at the end of each semiannual period.*

37. 6% **38.** 8% **39.** 10% **40.** 12%

41. *Sinking Fund* A firm must pay off \$40,000 worth of bonds in 7 years. Find the amount of each annual payment to be made into a sinking fund, if the money earns 6% compounded annually.

42. *Sinking Fund* What payment should be made in Exercise 41 if the firm can get interest of 8% compounded annually?

43. *Sinking Fund* Pam Snow sells some land in Nevada. She will be paid a lump sum of \$60,000 in 7 years. Until then, the buyer pays 8% interest, compounded quarterly.

a. Find the amount of each quarterly interest payment.

b. The buyer sets up a sinking fund so that enough money will be present to pay off the \$60,000. The buyer wants to make semiannual payments into the sinking fund; the account pays 6% compounded semiannually. Find the amount of each payment into the fund.

c. Prepare a table showing the amount in the sinking fund after each deposit.

44. *Sinking Fund* John Remington bought a rare stamp for his collection. He agreed to pay a lump sum of \$4000 after 5 years. Until then, he pays 6% interest, compounded semiannually.

a. Find the amount of each semiannual interest payment.

b. John sets up a sinking fund so that enough money will be present to pay off the \$4000. He wants to make annual payments into the fund. The account pays 8% compounded annually. Find the amount of each payment into the fund.

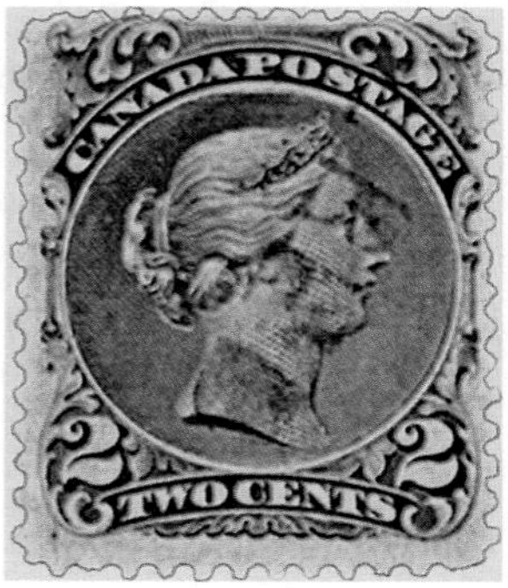

c. Prepare a table showing the amount in the sinking fund after each deposit.

45. *Present Value of an Annuity* In his will the late Mr. Hudspeth said that each child in his family could have an annuity of \$2000 at the end of each year for 9 years, or the equivalent present value. If money can be deposited at 8% compounded annually, what is the present value?

46. *Lottery Winnings* In the "Million Dollar Lottery," a winner is paid a million dollars at the rate of \$50,000 per year for 20 years. Assume that these payments form an ordinary annuity and that the lottery managers can invest money at 6% compounded annually. Find the lump sum that the management must put away to pay off the "million dollar" winner.

47. *Car Payments* Leslie Ulman buys a new car costing \$16,000. She agrees to make payments at the end of each month for 4 years. If she pays 12% interest, compounded monthly, what is the amount of each payment? Find the total amount of interest Leslie will pay.

48. *Present Value of an Annuity* What lump sum deposited today at 5% compounded annually for 8 years will provide the same amount as \$1000 deposited at the end of each year for 8 years at 6% compounded annually?

49. *Investment* In 1995, Oseola McCarty donated \$150,000 to the University of Southern Mississippi to establish a scholarship fund.* What is unusual about her is that the entire amount came from what she was able to save each month from her work as a washer woman, a job she began in 1916 at the age of 8, when she dropped out of school.

a. How much would Ms. McCarty have to put into her savings account at the end of every 3 months to accumulate \$150,000 over 79 years? Assume she received an interest rate of 5.25% compounded quarterly.

b. Answer part a using a 2% and a 7% interest rate.

50. *Lottery Winnings* In most states, the winnings of million-dollar lottery jackpots are divided into equal payments given annually for 20 years. (In Colorado, the results are distributed over 25 years.)† This means that the present value of the jackpot is worth less than the stated prize, with the actual value determined by the interest rate at which the money could be invested.

a. Find the present value of a \$1 million lottery jackpot distributed in equal annual payments over 20 years, using an interest rate of 5%.

b. Find the present value of a \$1 million lottery jackpot distributed in equal annual payments over 20 years, using an interest rate of 9%.

c. Calculate the answer for part a using the 25-year distribution time in Colorado.

d. Calculate the answer for part b using the 25-year distribution time in Colorado.

House Payments Find the monthly house payment necessary to amortize each of the loans in Exercises 51–54. Then find the unpaid balance after 5 years for each loan.

51. \$249,560 at 7.75% for 25 years

52. \$170,892 at 8.11% for 30 years

53. \$153,762 at 8.45% for 30 years

54. \$196,511 at 7.57% for 25 years

55. *Annuity* When Ms. Thompson died, she left \$25,000 to her husband, which he deposited at 6% compounded annually. He wants to make annual withdrawals from the account so that the money (principal and interest) is gone in exactly 8 years.

a. Find the amount of each withdrawal.

b. Find the amount of each withdrawal if the money must last 12 years.

**The New York Times*, Nov. 12, 1996, pp. A1, A22.
†Gould, Louis, "Ticket to Trouble," *The New York Times Magazine*, April 23, 1995, p. 39.

56. *Annuity* The trustees of a college have accepted a gift of \$150,000. The donor has directed the trustees to deposit the money in an account paying 6% per year, compounded semiannually. The trustees may withdraw an equal amount of money at the end of each 6-month period; the money must last 5 years.

a. Find the amount of each withdrawal.

b. Find the amount of each withdrawal if the money must last 7 years.

57. *Amortization* An insurance firm pays \$4000 for a new printer for its computer. It amortizes the loan for the printer in 4 annual payments at 8% compounded annually. Prepare an amortization schedule for this machine.

58. *Amortization* Certain large semitrailer trucks cost \$72,000 each. Ace Trucking buys such a truck and agrees to pay for it with a loan that will be amortized with 9 semiannual payments at 9.5% compounded semiannually. Prepare an amortization schedule for this truck.

59. *Amortization* A printer manufacturer charges \$1048 for a high-speed printer. A firm of tax accountants buys 8 of these machines. They make a down payment of \$1200 and agree to amortize the balance with monthly payments at 10.5% compounded monthly for 4 years. Prepare an amortization schedule showing the first six payments.

60. *Amortization* When Marissa Raffaele opened her law office, she bought \$14,000 worth of law books and \$7200 worth of office furniture. She paid \$1200 down and agreed to amortize the balance with semiannual payments for 5 years at 8% compounded semiannually. Prepare an amortization schedule for this purchase.

12.3 TAYLOR POLYNOMIALS AT 0

THINK ABOUT IT *How can we determine the length of time before a machine part must be replaced?*

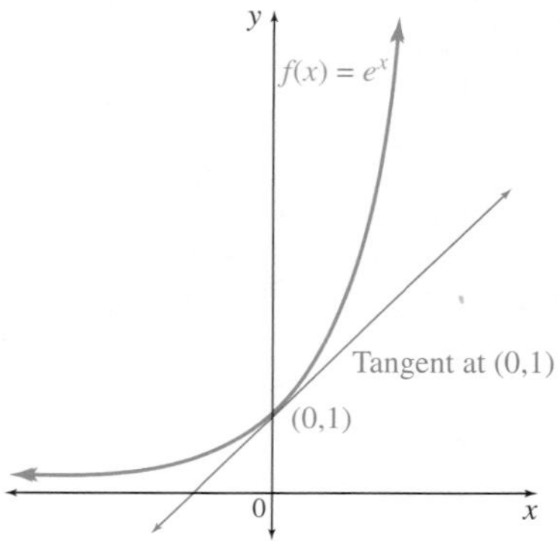

FIGURE 8

We shall see in the exercises that a Taylor polynomial can be used to approximate the answer to this question.

Although exponential and logarithmic functions are quite different from polynomials, they can be closely approximated by polynomials. These approximating polynomials are called **Taylor polynomials** after British mathematician Brook Taylor (1685–1731), who published his work on them in 1715.

One of the most useful nonpolynomial functions is the exponential function $f(x) = e^x$. Let us begin our discussion of Taylor polynomials by finding polynomials that approximate e^x for values of x close to 0. As a first approximation to e^x, choose the straight line that is tangent to the graph of $f(x) = e^x$ at the point $(0, 1)$. (See Figure 8.) Since the slope of a tangent line is given by the derivative of the function, and the derivative of $f(x) = e^x$ is $f'(x) = e^x$, the slope of the tangent line at $x = 0$ is $f'(0) = e^0 = 1$.

The tangent line goes through $(0, f(0)) = (0, 1)$ and has slope $f'(0) = 1$. By the point-slope form of the equation of a line, the equation of the tangent line is

$$\begin{aligned} y - y_1 &= m(x - x_1) \\ y - f(0) &= f'(0)(x - 0) \\ y &= f(0) + f'(0) \cdot x, \end{aligned}$$

or, after substituting 1 for $f(0)$ and 1 for $f'(0)$,

$$y = 1 + x.$$

If $P_1(x)$ is used to represent $1 + x$, then

$$P_1(x) = 1 + x$$

is called the *Taylor polynomial of degree 1* for $f(x) = e^x$ at $x = 0$.

To be useful, $P_1(x)$ should approximate e^x for values of x close to 0. To check the accuracy of this approximation, compare values of $P_1(x)$ and values of e^x, for x close to 0, in the following table.

x	$P_1(x) = 1 + x$	$f(x) = e^x$
−1	0	.3678794412
−.1	.9	.904837418
−.01	.99	.9900498337
−.001	.999	.9990004998
0	1	1
.001	1.001	1.0010005
.01	1.01	1.010050167
.1	1.1	1.105170918
1	2	2.718281828

This table agrees with the graph in Figure 8: the polynomial $P_1(x)$ is a good approximation for $f(x) = e^x$ only when x is close to 0.

For the polynomial $P_1(x)$,

$$P_1(0) = f(0) \qquad \text{and} \qquad P_1'(0) = f'(0);$$

that is, $P_1(x)$ and $f(x)$ are equal at 0 and their derivatives are equal at 0. A better approximation could be found with a curve. Since $P_1(x)$ is first-degree, we use a second-degree polynomial and require the second derivatives to be equal at 0. If $P_1(x)$ is written as

$$P_1(x) = a_0 + a_1 x,$$

with $a_0 = f(0)$ and $a_1 = f'(0)$, then a second-degree polynomial can be written as

$$P_2(x) = a_0 + a_1 x + a_2 x^2,$$

where $a_0 = f(0)$, $a_1 = f'(0)$, and $a_2 = f''(0)/2$.

Just as above, we want

$$P_2(0) = f(0) \qquad \text{and} \qquad P_2'(0) = f'(0),$$

but we also want

$$P_2''(0) = f''(0).$$

Since $P_2(0) = a_0$, and $f(0) = 1$, $a_0 = 1$. Also, $P_2'(x) = a_1 + 2a_2 x$, so $P_2'(0) = a_1$. Since $f'(0) = e^0 = 1$, we also must have $a_1 = 1$. Finally, $P_2''(x) = 2a_2$. Since $f''(x) = e^x$, $P_2''(0) = 2a_2$ and $f''(0) = 1$, so that

$$2a_2 = 1$$
$$a_2 = \frac{1}{2}.$$

Our second approximation, the *Taylor polynomial of degree 2* for $f(x) = e^x$ at $x = 0$, is thus

$$P_2(x) = 1 + x + \frac{1}{2}x^2.$$

A graph of this polynomial, along with the graph of $f(x) = e^x$, is shown in Figure 9.

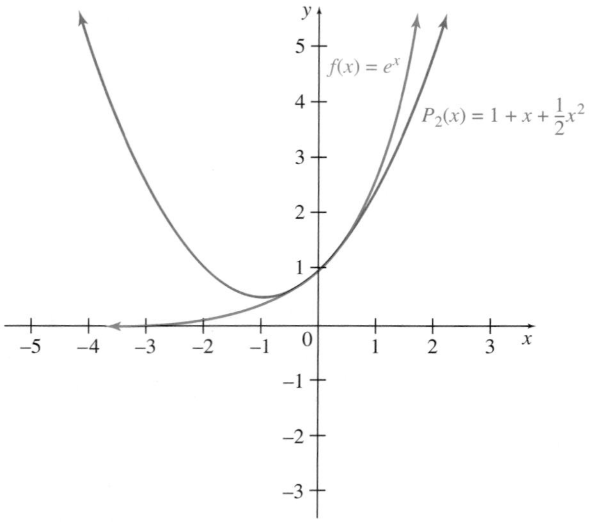

FIGURE 9

As above, the accuracy of this approximation can be checked with a table comparing values of $P_2(x)$ with those of $P_1(x)$ and $f(x)$.

x	$P_1(x) = 1 + x$	$P_2(x) = 1 + x + \frac{1}{2}x^2$	$f(x) = e^x$
−1	0	.5	.3678794412
−.1	.9	.905	.904837418
−.01	.99	.99005	.9900498337
−.001	.999	.9990005	.9990004998
0	1	1	1
.001	1.001	1.0010005	1.0010005
.01	1.01	1.01005	1.010050167
.1	1.1	1.105	1.105170918
1	2	2.5	2.718281828

Although the approximations provided by $P_2(x)$ are better than those provided by $P_1(x)$, they are still accurate only for values of x close to 0. A better approximation would be given by a polynomial $P_3(x)$ that equals $f(x)$ when

$x = 0$, and also has the first, second, and *third* derivatives of $P_3(x)$ and $f(x) = e^x$ equal when $x = 0$. If we let

$$P_3(x) = a_0 + a_1x + a_2x^2 + a_3x^3,$$

FOR REVIEW

Recall that $P_3^{(n)}(x)$ represents the nth derivative of $P_3(x)$.

we can find the first three derivatives:

$$P_3^{(1)}(x) = a_1 + 2a_2x + 3a_3x^2$$
$$P_3^{(2)}(x) = 2a_2 + 6a_3x$$
$$P_3^{(3)}(x) = 6a_3.$$

Letting $x = 0$ in $P_3(x)$ and in each derivative, in turn, yields

$$P_3(0) = a_0, \qquad P_3^{(1)}(0) = a_1, \qquad P_3^{(2)}(0) = 2a_2, \qquad P_3^{(3)}(0) = 6a_3.$$

Since $f(0) = 1$, $f^{(1)}(0) = 1$, $f^{(2)}(0) = 1$, and $f^{(3)}(0) = 1$ for $f(x) = e^x$,

$$a_0 = 1, \qquad a_1 = 1, \qquad a_2 = \frac{1}{2}, \qquad \text{and} \qquad a_3 = \frac{1}{6},$$

with

$$P_3(x) = 1 + x + \frac{1}{2}x^2 + \frac{1}{6}x^3.$$

A graph of $f(x) = e^x$ and $P_3(x)$ is shown in Figure 10.

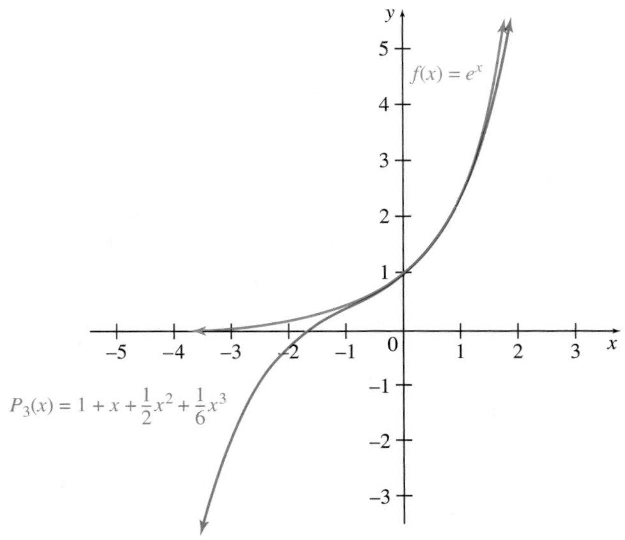

FIGURE 10

To generalize the work above, let $f(x) = e^x$ be approximated by

$$P_n(x) = a_0 + a_1x + a_2x^2 + \cdots + a_nx^n,$$

where

$$
\begin{aligned}
P_n(0) &= f(0), \\
P_n^{(1)}(0) &= f^{(1)}(0) \\
&\vdots \\
P_n^{(n)}(0) &= f^{(n)}(0).
\end{aligned}
$$

Taking derivatives of $P_n(x)$ gives

$$
\begin{aligned}
P_n^{(1)}(x) &= a_1 + 2a_2x + 3a_3x^2 + \cdots + n \cdot a_nx^{n-1} \\
P_n^{(2)}(x) &= 2a_2 + 2 \cdot 3a_3x + \cdots + n(n-1)a_nx^{n-2} \\
P_n^{(3)}(x) &= 2 \cdot 3a_3 + \cdots + n(n-1)(n-2)a_nx^{n-3} \\
&\vdots \\
P_n^{(n)}(x) &= n(n-1)(n-2)(n-3)\cdots 3 \cdot 2 \cdot 1 \cdot a_n = n!a_n.
\end{aligned}
$$

The symbol $n!$ (read "n-factorial") is used for the product

$$n(n-1)(n-2)(n-3)\cdots 3 \cdot 2 \cdot 1.$$

For example, $3! = 3 \cdot 2 \cdot 1 = 6$, while $5! = 120$. By convention, $0! = 1$.* If we use factorials and replace x with 0, the various derivatives of $P_n(x)$ become

$$
\begin{aligned}
P_n^{(1)}(0) &= 1!a_1 \\
P_n^{(2)}(0) &= 2!a_2 \\
P_n^{(3)}(0) &= 3!a_3 \\
&\vdots \\
P_n^{(n)}(0) &= n!a_n.
\end{aligned}
$$

For every value of n, $f^{(n)}(0) = 1$. Setting corresponding derivatives equal gives

$$
\begin{aligned}
1!a_1 &= 1 \\
2!a_2 &= 1 \\
3!a_3 &= 1 \\
&\vdots \\
n!a_n &= 1,
\end{aligned}
$$

from which

$$a_1 = \frac{1}{1!}, \quad a_2 = \frac{1}{2!}, \quad a_3 = \frac{1}{3!}, \quad \ldots, \quad a_n = \frac{1}{n!}.$$

*The symbol $n!$ for the product

$$n(n-1)(n-2)(n-3)\cdots 3 \cdot 2 \cdot 1$$

came into use during the late 19th century, although it was by no means the only symbol for n-factorial. Another popular symbol was $\underline{|n}$. The exclamation point notation has won out, probably because it is more convenient for printers of textbooks.

Finally, the *Taylor polynomial of degree n* for $f(x) = e^x$ at $x = 0$ is

$$P_n(x) = 1 + \frac{1}{1!}x + \frac{1}{2!}x^2 + \frac{1}{3!}x^3 + \cdots + \frac{1}{n!}x^n.$$

Using the convention that the *zeroth derivative* of $y = f(x)$ is just f itself, and using Σ to represent a sum, we can write this result in the following way.

TAYLOR POLYNOMIAL FOR $f(x) = e^x$

The **Taylor polynomial of degree *n*** for $f(x) = e^x$ at $x = 0$ is

$$P_n(x) = \sum_{i=0}^{n} \frac{f^{(i)}(0)}{i!}x^i.$$

Taylor polynomials of degree up to 10 for $f(x) = e^x$ are shown in Figure 11.

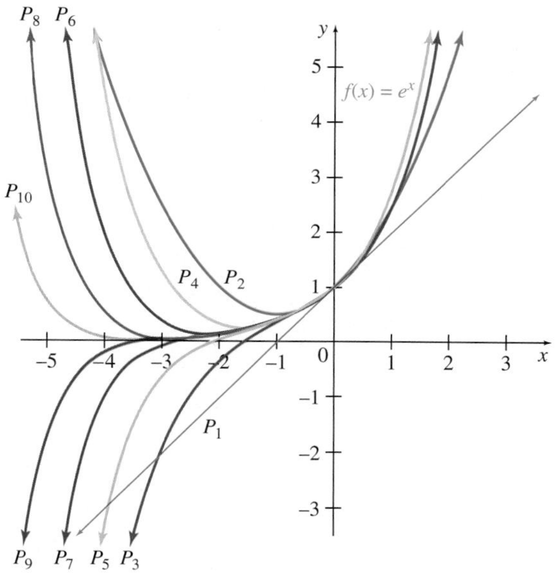

FIGURE 11

Graphing calculators simplify the creation of a sequence of Taylor polynomials. For example, to create Taylor polynomials of degree 1, 2, and 3 for e^x on a TI-83/84 Plus, let `Y1 = 1 + X, Y2 = Y1 + X^2/2`, and `Y3 = Y2 + X^3/6`.

EXAMPLE 1 *Taylor Polynomial*

Use a Taylor polynomial of degree 5 to approximate $e^{-.2}$.

Solution In the work above, we found Taylor polynomials for e^x at $x = 0$. As the graphs in Figure 11 suggest, these polynomials can be used to find approximate values of e^x for values of x near 0. The Taylor polynomial of degree 5 for $f(x) = e^x$ is

$$P_5(x) = 1 + \frac{1}{1!}x + \frac{1}{2!}x^2 + \frac{1}{3!}x^3 + \frac{1}{4!}x^4 + \frac{1}{5!}x^5.$$

Replacing x with $-.2$ gives

$$1 + \frac{1}{1!}(-.2) + \frac{1}{2!}(-.2)^2 + \frac{1}{3!}(-.2)^3 + \frac{1}{4!}(-.2)^4 + \frac{1}{5!}(-.2)^5 \approx .8187307.$$

Using a calculator to evaluate $e^{-.2}$ directly gives $.8187308$, which agrees with our approximation to 6 decimal places.

Generalizing our work in finding the Taylor polynomials for $f(x) = e^x$ leads to the following definition of Taylor polynomials for any appropriate function f.

TAYLOR POLYNOMIAL OF DEGREE n

Let f be a function that can be differentiated n times at 0. The Taylor polynomial of degree n for f at 0 is

$$P_n(x) = f(0) + \frac{f^{(1)}(0)}{1!}x + \frac{f^{(2)}(0)}{2!}x^2 + \frac{f^{(3)}(0)}{3!}x^3 + \cdots + \frac{f^{(n)}(0)}{n!}x^n.$$

NOTE Because of the $f(0)$ term, a Taylor polynomial of degree n has $n + 1$ terms. ■

EXAMPLE 2 *Taylor Polynomial*

Let $f(x) = \sqrt{x + 1}$. Find the Taylor polynomial of degree 4 at $x = 0$.

Solution To find the Taylor polynomial of degree 4, use the first four derivatives of f, evaluated at 0. Arrange the work as follows.

Derivative	Value at 0
$f(x) = \sqrt{x + 1} = (x + 1)^{1/2}$	$f(0) = 1$
$f^{(1)}(x) = \frac{1}{2}(x + 1)^{-1/2} = \frac{1}{2(x + 1)^{1/2}}$	$f^{(1)}(0) = \frac{1}{2}$
$f^{(2)}(x) = -\frac{1}{4}(x + 1)^{-3/2} = \frac{-1}{4(x + 1)^{3/2}}$	$f^{(2)}(0) = -\frac{1}{4}$
$f^{(3)}(x) = \frac{3}{8}(x + 1)^{-5/2} = \frac{3}{8(x + 1)^{5/2}}$	$f^{(3)}(0) = \frac{3}{8}$
$f^{(4)}(x) = -\frac{15}{16}(x + 1)^{-7/2} = \frac{-15}{16(x + 1)^{7/2}}$	$f^{(4)}(0) = -\frac{15}{16}$

Now use the definition of a Taylor polynomial.

$$P_4(x) = f(0) + \frac{f^{(1)}(0)}{1!}x + \frac{f^{(2)}(0)}{2!}x^2 + \frac{f^{(3)}(0)}{3!}x^3 + \frac{f^{(4)}(0)}{4!}x^4$$

$$= 1 + \frac{1/2}{1!}x + \frac{-1/4}{2!}x^2 + \frac{3/8}{3!}x^3 + \frac{-15/16}{4!}x^4$$

$$= 1 + \frac{1}{2}x - \frac{1}{8}x^2 + \frac{1}{16}x^3 - \frac{5}{128}x^4$$

EXAMPLE 3 *Approximation*

Use the result of Example 2 to approximate $\sqrt{.9}$.

Solution To approximate $\sqrt{.9}$, we must evaluate $f(-.1) = \sqrt{-.1 + 1} = \sqrt{.9}$. Using $P_4(x)$ from Example 2, with $x = -.1$, gives

$$\begin{aligned} P_4(-.1) &= 1 + \frac{1}{2}(-.1) - \frac{1}{8}(-.1)^2 + \frac{1}{16}(-.1)^3 - \frac{5}{128}(-.1)^4 \\ &= 1 - .05 - .00125 - .0000625 - .000003906 = .948683594. \end{aligned}$$

Thus, $\sqrt{.9} \approx .948683594$. A calculator gives a value of .9486832981 for the square root of .9.

EXAMPLE 4 *Taylor Polynomial*

Find the Taylor polynomial of degree n at $x = 0$ for

$$f(x) = \frac{1}{1 - x}.$$

Solution As above, find the first n derivatives, and evaluate each at 0.

Derivative	Value at 0
$f(x) = \frac{1}{1-x} = (1-x)^{-1}$	$f(0) = 1$
$f^{(1)}(x) = -1(1-x)^{-2}(-1) = (1-x)^{-2}$	$f^{(1)}(0) = 1 = 1!$
$f^{(2)}(x) = 2(1-x)^{-3}$	$f^{(2)}(0) = 2 = 2!$
$f^{(3)}(x) = 3!(1-x)^{-4}$	$f^{(3)}(0) = 3!$
$f^{(4)}(x) = 4!(1-x)^{-5}$	$f^{(4)}(0) = 4!$

Continuing this process,

$$f^{(n)}(x) = n!(1-x)^{-1-n} \quad \text{and} \quad f^{(n)}(0) = n!.$$

By the definition of Taylor polynomials,

$$\begin{aligned} P_n(x) &= 1 + \frac{1!}{1!}x + \frac{2!}{2!}x^2 + \frac{3!}{3!}x^3 + \frac{4!}{4!}x^4 + \cdots + \frac{n!}{n!}x^n \\ &= 1 + x + x^2 + x^3 + x^4 + \cdots + x^n. \end{aligned}$$

EXAMPLE 5 *Approximation*

Use a Taylor polynomial of degree 4 to approximate $1/.98$.

Solution Use the function f from Example 4, with $x = .02$, to get

$$f(.02) = \frac{1}{1 - .02} = \frac{1}{.98}.$$

Based on the result obtained in Example 4,

$$P_4(x) = 1 + x + x^2 + x^3 + x^4,$$

with $$P_4(.02) = 1 + (.02) + (.02)^2 + (.02)^3 + (.02)^4$$
$$= 1 + .02 + .0004 + .000008 + .00000016$$
$$= 1.02040816.$$

A calculator gives $1/.98 = 1.020408163$.

12.3 EXERCISES

For the functions defined as follows, find the Taylor polynomials of degree 4 at 0.

1. $f(x) = e^{-2x}$

2. $f(x) = e^{3x}$

3. $f(x) = e^{x+1}$

4. $f(x) = e^{-x}$

5. $f(x) = \sqrt{x+9}$

6. $f(x) = \sqrt{x+16}$

7. $f(x) = \sqrt[3]{x-1}$

8. $f(x) = \sqrt[3]{x+8}$

9. $f(x) = \sqrt[4]{x+1}$

10. $f(x) = \sqrt[4]{x+16}$

11. $f(x) = \ln(1-x)$

12. $f(x) = \ln(1+2x)$

13. $f(x) = \ln(1+2x^2)$

14. $f(x) = \ln(1-x^3)$

15. $f(x) = xe^{-x}$

16. $f(x) = x^2e^x$

17. $f(x) = (9-x)^{3/2}$

18. $f(x) = (1+x)^{3/2}$

19. $f(x) = \dfrac{1}{1+x}$

20. $f(x) = \dfrac{1}{x-1}$

Use Taylor polynomials of degree 4 at $x = 0$, found in Exercise 1–14 above, to approximate the quantities in Exercises 21–34. Round answers to 4 decimal places.

21. $e^{-.02}$

22. $e^{.03}$

23. $e^{1.01}$

24. $e^{-.04}$

25. $\sqrt{8.9}$

26. $\sqrt{16.01}$

27. $\sqrt[3]{-1.02}$

28. $\sqrt[3]{7.98}$

29. $\sqrt[4]{1.03}$

30. $\sqrt[4]{15.97}$

31. ln .99

32. ln 1.04

33. ln 1.0002

34. ln .999

35. **a.** Generalize the result of Example 2 to show that if x is small compared with a,

$$(a+x)^{1/n} \approx a^{1/n} + \frac{xa^{1/n}}{na}.$$

b. Use the result of part a to approximate $\sqrt[3]{66}$, and compare with the actual value.

36. Find a polynomial of degree 3 such that $f(0) = 3$, $f'(0) = 6$, $f''(0) = 12$, and $f'''(0) = 24$.

37. Find a polynomial of degree 4 such that $f(0) = 1$, $f'(0) = 1$, $f''(0) = 2$, $f'''(0) = 6$, and $f^{(4)}(0) = 24$. Generalize this result to a polynomial of degree n, assuming that $f^{(n)}(0) = n!$

Applications

BUSINESS AND ECONOMICS

38. *Duration* Let D represent *duration*, a term in finance that measures the length of time an investor must wait to receive half of the value of a cash flow stream totaling S dollars. Let r be the rate of interest and V the value of the investment. The value of S can be calculated by two formulas that are approximately equal:

$$S \approx V(1 + rD)$$

and $$S \approx V(1 + r)^D.$$

a. Show that the first approximation follows from the second by using the Taylor polynomial of degree 1 for the function $f(r) = (1 + r)^D$.*

*This exercise was contributed by Robert D. Campbell of the Frank G. Zarb School of Business at Hofstra University.

b. For $V = \$1000$, $r = .1$, and $D = 3.2$, calculate and compare the two expressions for S.

39. *Replacement Time for a Part* A book on management science gives the equation

$$\frac{e^{\lambda N}}{\lambda} - N = \frac{1}{\lambda} + k$$

to determine N, the time until a particular part can be expected to need replacing. (λ and k are constants for a particular machine.) To find a useful approximate value for N when λN is near 0, go through the following steps.

a. Find a Taylor polynomial of degree 2 at $N = 0$ for $e^{\lambda N}$.

b. Substitute this polynomial into the given equation and solve for N.

In Exercises 40–44, use a Taylor polynomial of degree 2 at $x = 0$ to approximate the desired value. Compare your answers with the results obtained by direct substitution.

40. *Profit* The profit (in thousands of dollars) when x thousand tons of apples are sold is

$$P(x) = \frac{20 + x^2}{50 + x}.$$

Find $P(.3)$.

41. *Profit* The profit (in thousands of dollars) from the sale of x thousand packages of note paper is

$$P(x) = \ln(100 + 3x).$$

Find $P(.6)$ if ln 100 is given as 4.605.

42. *Cost* For a certain electronic part, the cost to make the part declines as more parts are made. Suppose that the cost (in dollars) to manufacture the xth part is

$$C(x) = e^{-x/50}.$$

Find $C(5)$.

43. *Revenue* Revenue from selling agricultural products often increases at a slower and slower rate as more of the products are sold. Suppose the revenue from the xth unit of a product is

$$R(x) = 500 \ln\left(4 + \frac{x}{50}\right).$$

Find $R(10)$ if ln 4 is given as 1.386.

LIFE SCIENCES

44. *Amount of a Drug in the Bloodstream* The amount (in milliliters) of a certain drug in the bloodstream x minutes after being administered is

$$A(x) = \frac{6x}{1 + 10x}.$$

Find $A(.05)$.

45. *Species Survival* According to a text on species survival, the probability P that a certain species survives is given by

$$P = 1 - e^{-2k},$$

where k is a constant. Use a Taylor polynomial to show that if k is small, P is approximately $2k$.

12.4 INFINITE SERIES

THINK ABOUT IT

If some fraction of a particular gene in a population experiences a mutation each generation, can we expect that the entire population will have this mutation over time? The answer to this question is found by considering the sum of an infinite series.

A repeating decimal such as .66666... is really the sum of an infinite number of terms:

$$\begin{aligned} .66666\ldots &= .6 + .06 + .006 + .0006 + \cdots \\ &= \frac{6}{10} + \frac{6}{10^2} + \frac{6}{10^3} + \frac{6}{10^4} + \cdots. \end{aligned}$$

In this section we will show how an infinite number of terms can sometimes be added to get a finite sum by a limit process. To do this, we need the following definition.

INFINITE SERIES

An **infinite series** is an expression of the form

$$\boldsymbol{a_1 + a_2 + a_3 + a_4 + \cdots + a_n + \cdots = \sum_{i=1}^{\infty} a_i.}$$

To find the sum $a_1 + a_2 + a_3 + a_4 + \cdots + a_n + \cdots$, first find the sum S_n of the first n terms, called the ***n*th partial sum.** For example,

$$\begin{aligned}
S_1 &= a_1 \\
S_2 &= a_1 + a_2 \\
S_3 &= a_1 + a_2 + a_3 \\
&\ \ \vdots \\
S_n &= a_1 + a_2 + a_3 + \cdots + a_n = \sum_{i=1}^{n} a_i.
\end{aligned}$$

EXAMPLE 1 *Partial Sums*

Find the first five partial sums for the sequence

$$1, \frac{1}{2}, \frac{1}{4}, \frac{1}{8}, \frac{1}{16}, \ldots.$$

Solution By the definition of partial sum,

$$\begin{aligned}
S_1 &= 1 \\
S_2 &= 1 + \frac{1}{2} = \frac{3}{2} \\
S_3 &= 1 + \frac{1}{2} + \frac{1}{4} = \frac{7}{4} \\
S_4 &= 1 + \frac{1}{2} + \frac{1}{4} + \frac{1}{8} = \frac{15}{8} \\
S_5 &= 1 + \frac{1}{2} + \frac{1}{4} + \frac{1}{8} + \frac{1}{16} = \frac{31}{16}.
\end{aligned}$$

As n gets larger, the partial sum $S_n = a_1 + a_2 + \cdots + a_n$ includes more and more terms from the infinite series. It is thus reasonable to define the *sum of the infinite series* as $\lim_{n\to\infty} S_n$, if it exists.

SUM OF THE INFINITE SERIES

Let $S_n = a_1 + a_2 + a_3 + \cdots + a_n$ be the nth partial sum for the series $a_1 + a_2 + a_3 + \cdots + a_n + \cdots$. Suppose

$$\lim_{n\to\infty} S_n = L$$

for some real number L. Then L is called the **sum of the infinite series** $a_1 + a_2 + a_3 + \cdots + a_n + \cdots$, and the infinite series **converges.** If no such limit exists, then the infinite series has no sum and **diverges.**

Infinite Geometric Series Some good examples of convergent and divergent series come from the study of infinite geometric series, which are the sums of the terms of geometric sequences, discussed in this chapter's first section. For example,

$$1, \frac{1}{2}, \frac{1}{4}, \frac{1}{8}, \frac{1}{16}, \ldots, \frac{1}{2^n}, \ldots$$

is a geometric sequence with first term $a_1 = 1$ and common ratio $r = 1/2$. The first five partial sums for this sequence were found in Example 1. To find S_n, the nth partial sum, use the formula given in the first section: The sum of the first n terms of a geometric sequence having first term $a = a_1$ and common ratio r is

$$S_n = \frac{a(r^n - 1)}{r - 1}.$$

For any value of n, S_n can be found for the geometric sequence above by using the formula with $a = 1$ and $r = 1/2$.

$$S_n = \frac{a(r^n - 1)}{r - 1} = \frac{1\left[\left(\frac{1}{2}\right)^n - 1\right]}{\frac{1}{2} - 1}$$

$$= \frac{\left(\frac{1}{2}\right)^n - 1}{-\frac{1}{2}} = -2\left[\left(\frac{1}{2}\right)^n - 1\right] = 2\left[1 - \left(\frac{1}{2}\right)^n\right]$$

As n gets larger and larger, that is, as $n \to \infty$, the value of $(1/2)^n$ gets closer and closer to 0, so that

$$\lim_{n\to\infty}\left(\frac{1}{2}\right)^n = 0.$$

Using properties of limits from Chapter 3,

$$\lim_{n\to\infty} S_n = \lim_{n\to\infty} 2\left[1 - \left(\frac{1}{2}\right)^n\right] = 2(1 - 0) = 2.$$

By the definition of the sum of an infinite series,

$$1 + \frac{1}{2} + \frac{1}{4} + \frac{1}{8} + \cdots = 2,$$

and the series converges.

To generalize from this example, start with the formula for the sum of the first n terms of a geometric sequence.

$$S_n = \frac{a(r^n - 1)}{r - 1}.$$

If r is in the interval $(-1, 1)$, then $\lim_{n \to \infty} r^n = 0$. (Consider what happens to a small number as you raise it to a larger and larger power.) In that case,

$$\lim_{n \to \infty} S_n = \lim_{n \to \infty} \frac{a(r^n - 1)}{r - 1} = \frac{a(0 - 1)}{r - 1}$$
$$= \frac{a}{r - 1}.$$

On the other hand, if $r > 1$, then $\lim_{n \to \infty} r^n = \infty$. (Consider what happens to a large number as you raise it to a larger and larger power.) In that case,

$$\lim_{n \to \infty} S_n = \infty,$$

and the series diverges because the terms of the series are getting larger and larger. If $r < -1$, then $\lim_{n \to \infty} r^n$ does not exist, because r^n becomes larger and larger in magnitude while alternating in sign, and the same thing happens to the partial sums, so the series diverges. If $r = 1$, all the terms of the series equal a, so the series diverges (except in the trivial case when $a = 0$.) Finally, if $r = -1$, the terms of the series alternate between a and $-a$, and the partial sums alternate between a and 0, so the series diverges.

SUM OF A GEOMETRIC SERIES

The infinite geometric series

$$a + ar + ar^2 + ar^3 + \cdots$$

converges, if r is in $(-1, 1)$, to the sum

$$\frac{a}{1 - r}.$$

The series diverges if r is not in $(-1, 1)$.

EXAMPLE 2 *Geometric Series*

Show that each series converges. Find each sum.

(a) $3 + \frac{3}{8} + \frac{3}{64} + \frac{3}{512} + \cdots$

Solution This is a geometric series, with $a = a_1 = 3$ and $r = 1/8$. Since r is in $(-1, 1)$, the series converges and has sum

$$\frac{a}{1-r} = \frac{3}{1-1/8} = \frac{3}{7/8} = 3 \cdot \frac{8}{7} = \frac{24}{7}.$$

(b) $\frac{3}{4} - \frac{9}{16} + \frac{27}{64} - \frac{81}{256} + \cdots$

Solution This geometric series has $a = a_1 = 3/4$ and $r = -3/4$. Since r is in $(-1, 1)$, the series converges. The sum of the series is

$$\frac{3/4}{1-(-3/4)} = \frac{3/4}{1+3/4} = \frac{3/4}{7/4} = \frac{3}{7}.$$

EXAMPLE 3 *Geometric Series*

The series

$$1 + 1.1 + (1.1)^2 + (1.1)^3 + (1.1)^4 + \cdots + (1.1)^{n-1} + \cdots$$

is a geometric series with common ratio $r = 1.1$. Since $r > 1$, the series diverges. (The partial sums S_n will eventually exceed any preassigned number, no matter how large.)

EXAMPLE 4 *Multiplier Effect*

Suppose a company spends \$1,000,000 on payroll in a certain city. Suppose also that the employees of the company reside in the city. Assume that on the average the inhabitants of this city spend 80% of their income in the same city and save the remaining 20% or spend it elsewhere. Then 80% of the original \$1,000,000, or $(.80)(\$1{,}000{,}000) = \$800{,}000$ also will be spent in that city. An additional 80% of this \$800,000, or \$640,000, will in turn be spent in the city, as will 80% of the \$640,000, and so on. These amounts, \$1,000,000, \$800,000, \$640,000, \$512,000, and so on, form an infinite series with $a = a_1 = \$1{,}000{,}000$ and $r = .80$. The sum of these amounts is

$$\frac{a}{1-r} = \frac{\$1{,}000{,}000}{1-.80} = \$5{,}000{,}000.$$

The original \$1,000,000 payroll leads to a total expenditure of \$5,000,000 in the city. In economics, the quotient of these numbers, $\$5{,}000{,}000/\$1{,}000{,}000 = 5$, is called the *multiplier*.

EXAMPLE 5 *Mutation*

Retinoblastoma is a kind of cancer of the eye in children. Medical researchers believe that the disease depends on a single dominant gene, say A. Let a be the normal gene. It is believed that a fraction m of the population, $m = 2 \times 10^{-5}$, per generation will experience *mutation*, a sudden unaccountable change, of a into A. (We exclude the possibility of back mutations of A into a.) With medical care, approximately 70% of those affected with the disease survive. According to past data, the survivors reproduce at half the normal rate. The net fraction of affected persons who produce offspring is thus $r = 35\% = .35$. Since gene A is extremely rare, practically all the affected persons are of genotype *Aa*, so that we may neglect the few individuals of genotype *AA*.

m = fraction of population with disease due to mutation in this generation

mr = fraction of population with disease due to mutation in the previous generation

mr^2 = fraction of population with disease due to mutation two generations ago

mr^n = fraction of population with disease due to mutation n generations ago

The total fraction p of the population having the disease in this generation is thus

$$p = m + mr + mr^2 + \cdots + mr^n + \cdots$$

Use the formula for the sum of an infinite geometric series to find

$$p = \frac{m}{1 - r} = \frac{2 \times 10^{-5}}{1 - (.35)} \approx 3.1 \times 10^{-5}.$$

The fraction of the population having retinoblastoma is about 3×10^{-5}, or about 50% more than the fraction of each generation that experiences mutation. ■

12.4 EXERCISES

Identify which geometric series converge. Give the sum of each convergent series.

1. $12 + 6 + 3 + \dfrac{3}{2} + \cdots$

2. $1 + .9 + .81 + .729 + \cdots$

3. $9 + 18 + 36 + 72 + \cdots$

4. $3 + 9 + 27 + 81 + \cdots$

5. $16 + 4 + 1 + \cdots$

6. $81 + 27 + 9 + 3 + 1 + \cdots$

7. $100 + 10 + 1 + \cdots$

8. $128 + 64 + 32 + \cdots$

9. $\dfrac{3}{4} + \dfrac{3}{8} + \dfrac{3}{16} + \cdots$

10. $\dfrac{4}{5} + \dfrac{2}{5} + \dfrac{1}{5} + \cdots$

11. $\dfrac{1}{3} - \dfrac{2}{9} + \dfrac{4}{27} - \dfrac{8}{81} + \cdots$

12. $1 + \dfrac{1}{1.01} + \dfrac{1}{(1.01)^2} + \cdots$

13. $e - 1 + \dfrac{1}{e} - \dfrac{1}{e^2} + \cdots$

14. $e + e^2 + e^3 + e^4 + \cdots$

The nth term of a sequence is given. Calculate the first five partial sums.

15. $a_n = \dfrac{1}{n}$

16. $a_n = \dfrac{1}{n + 1}$

17. $a_n = \dfrac{1}{2n + 5}$

18. $a_n = \dfrac{1}{3n - 1}$

19. $a_n = \dfrac{1}{(n + 1)(n + 2)}$

20. $a_n = \dfrac{1}{(n + 3)(2n + 1)}$

21. The following classical formulas for computing the value of π were developed by François Viète (1540–1603) and Gottfried von Leibniz (1646–1716), respectively:*

$$\frac{2}{\pi} = \frac{\sqrt{2}}{2} \cdot \frac{\sqrt{2 + \sqrt{2}}}{2} \cdot \frac{\sqrt{2 + \sqrt{2 + \sqrt{2}}}}{2} \cdots$$

and

$$\frac{\pi}{4} = 1 - \frac{1}{3} + \frac{1}{5} - \frac{1}{7} + \cdots.$$

*Dence, Joseph and Thomas Dence, "A Rapidly Converging Recursive Approach to Pi," *Mathematics Teacher,* Vol. 86, No. 2, Feb. 1993, pp. 121–124. Boyer, Carl, *A History of Mathematics,* Princeton University Press, 1985.

a. Multiply the first three terms of Viète's formula together and compare this with the sum of the first four terms of Leibniz's formula. Which formula is more accurate?

b. Use the table function on a graphing calculator or a spreadsheet to determine how many terms of the second formula must be added together to produce the same accuracy as the product of the first three terms of the first formula. [*Hint:* On a TI-83/84 Plus, use the command `Y1=4*sum(seq((−1)^(N−1)/(2N−1), N,1,X))`.]

Applications

BUSINESS AND ECONOMICS

22. *Present Value* In Section 8.3, we computed the present value of a continuous flow of money. Suppose that instead of a continuous flow, an amount C is deposited each year, and the annual interest rate is r. Then the present value of the cash flow over n years is

$$P = C(1+r)^{-1} + C(1+r)^{-2} + C(1+r)^{-3} + \cdots + C(1+r)^{-n}.$$

a. Show that the present value can be simplified to

$$P = C\frac{(1+r)^n - 1}{r(1+r)^n}.$$

b. Show that the present value, taken over an infinite amount of time, is given by $P = C/r$.

23. *Production Orders* A sugar factory receives an order for 1000 units of sugar. The production manager thus orders production of 1000 units of sugar. He forgets, however, that the production of sugar requires some sugar (to prime the machines, for example), and so he ends up with only 900 units of sugar. He then orders an additional 100 units, and receives only 90 units. A further order for 10 units produces 9 units. Finally seeing he is wrong, the manager decides to try mathematics. He views the production process as an infinite geometric series with $a_1 = 1000$ and $r = .1$.

a. Using this, find the number of units of sugar that he should have ordered originally.

b. Afterwards, the manager realizes a much simpler solution to his problem. If x is the amount of sugar he orders, and he only gets 90% of what he orders, he should solve $.9x = 1000$. What is the solution?

c. Explain why the answers to parts a and b are the same.

PHYSICAL SCIENCES

24. *Distance* Mitzi drops a ball from a height of 10 m and notices that on each bounce the ball returns to about 3/4 of its previous height. About how far will the ball travel before it comes to rest?

25. *Rotation of a Wheel* After a person pedaling a bicycle removes his or her feet from the pedals, the wheel rotates 400 times the first minute. As it continues to slow down, in each minute it rotates only 3/4 as many times as in the previous minute. How many times will the wheel rotate before coming to a complete stop?

26. *Pendulum Arc Length* A pendulum bob swings through an arc 40 cm long on its first swing. Each swing thereafter, it swings only 80% as far as on the previous swing. How far will it swing altogether before coming to a complete stop?

GENERAL INTEREST

27. *Perimeter* A sequence of equilateral triangles is constructed as follows: The first triangle has sides 2 m in length. To get the next triangle, midpoints of the sides of the previous triangle are connected. If this process could be continued indefinitely, what would be the total perimeter of all the triangles?

28. *Area* What would be the total area of all the triangles of Exercise 27, disregarding the overlaps?

29. *Zeno's Paradox* In the fifth century B.C., the Greek philosopher Zeno posed a paradox involving a race between Achilles (the fastest runner at the time) and a tortoise. The tortoise was given a head start, but once the race began, Achilles quickly reached the point where the tortoise had started. By then the tortoise had moved on to a new point. Achilles quickly reached that second point, but the tortoise had now moved to another point. Zeno concluded that Achilles could never reach the tortoise because every time he reached the point where the tortoise had been, the tortoise had moved on to a new point. This conclusion was absurd, yet people had trouble finding an error in Zeno's logic.

a. Suppose Achilles runs 10 m per second, the tortoise runs 1 m per second and the tortoise has a 10 m head start. How much time does it take for Achilles to reach the tortoise's starting point, and how far has the tortoise run during this time?

b. After Achilles reaches the tortoise's starting point, the tortoise has moved to a new point. How much time does it take for Achilles to reach this new point, and how far does the tortoise run during this time?

c. Continue the reasoning in parts a and b, and sum the infinite intervals of time to find how long it takes Achilles to catch up with the tortoise.

d. Determine the time it takes for Achilles to reach the tortoise using a simple algebraic method. (*Hint:* Distance = rate × time, and Achilles runs 10 m farther than the tortoise.)

e. Explain the error in Zeno's reasoning.

12.5 TAYLOR SERIES

THINK ABOUT IT *How many years will it take to double an amount invested at 9% annual interest?*

Using Taylor series, we derive in this section the rule of 70 and the rule of 72 to answer such questions.

As we saw in the previous section, the sum of the infinite geometric series having first term a and common ratio r is

$$\frac{a}{1-r} \quad \text{for } r \text{ in } (-1, 1).$$

If the first term of an infinite geometric series is $a = 1$ and the common ratio is x, then the series is written

$$1 + x + x^2 + x^3 + x^4 + \cdots + x^{n-1} + \cdots.$$

If x is in $(-1, 1)$, then by the formula for the sum of an infinite geometric series, the sum of this series is

$$\frac{1}{1-x}.$$

That is,

$$\frac{1}{1-x} = 1 + x + x^2 + x^3 + x^4 + \cdots \quad \text{for } x \text{ in } (-1, 1).$$

The interval $(-1, 1)$ is called the **interval of convergence** for the series. This series is not an approximation for $1/(1-x)$; the sum of the series is actually *equal to* $1/(1-x)$ for any x in $(-1, 1)$.

Earlier in this chapter, we found that the Taylor polynomial of degree n at $x = 0$ for $1/(1-x)$ is

$$P_n(x) = 1 + x + x^2 + x^3 + x^4 + \cdots + x^n.$$

Since the series given above for $1/(1-x)$ is just an extension of this Taylor polynomial, it seems natural to call the series a *Taylor series.*

TAYLOR SERIES

If all derivatives of a function f exist at 0, then the **Taylor series** for f at 0 is defined to be

$$f(0)+f^{(1)}(0)x+\frac{f^{(2)}(0)}{2!}x^2+\frac{f^{(3)}(0)}{3!}x^3+\cdots.$$

The particular Taylor series at 0 is also called a **Maclaurin series.** Scotsman Colin Maclaurin (1698–1746) used this series in his work *Treatise of Fluxions*, published in 1742. In this text, we will only consider Taylor series at 0. Taylor series at other points, as well as methods for finding the interval of convergence, are beyond the scope of this text. For more information, see *Thomas' Calculus, Updated*, 10th ed., by George B. Thomas, Ross L. Finney, Maurice D. Weir, and Frank Giordano, Addison-Wesley, 2003.

EXAMPLE 1 *Taylor Series*

Find the Taylor series for $f(x)=e^x$ at 0.

Solution Work as in Section 12.3.

Derivative	Value at 0
$f(x)=e^x$	$f(0)=1$
$f^{(1)}(x)=e^x$	$f^{(1)}(0)=1$
$f^{(2)}(x)=e^x$	$f^{(2)}(0)=1$
$f^{(3)}(x)=e^x$	$f^{(3)}(0)=1$
$\cdot$	$\cdot$
$\cdot$	$\cdot$
$\cdot$	$\cdot$
$f^{(n)}(x)=e^x$	$f^{(n)}(0)=1$

Using the definition given in this section, the Taylor series for $f(x)=e^x$ is

$$1+x+\frac{1}{2!}x^2+\frac{1}{3!}x^3+\frac{1}{4!}x^4+\cdots+\frac{1}{n!}x^n+\cdots.$$

While we cannot prove it here, the interval of convergence is $(-\infty,\infty)$.

The process for finding the interval of convergence for a given Taylor series is discussed in advanced calculus courses. Three of the most common Taylor series are listed below, along with their intervals of convergence. (As is customary, these series are written so that the initial term is a *zeroth* term.)

COMMON TAYLOR SERIES

$f(x)$	*Taylor Series*	*Interval of Convergence*
e^x	$1 + x + \frac{1}{2!}x^2 + \frac{1}{3!}x^3 + \cdots + \frac{1}{n!}x^n + \cdots$	$(-\infty, \infty)$
$\ln(1 + x)$	$x - \frac{x^2}{2} + \frac{x^3}{3} - \frac{x^4}{4} + \cdots + \frac{(-1)^n x^{n+1}}{n+1} + \cdots$	$(-1, 1]$
$\frac{1}{1 - x}$	$1 + x + x^2 + x^3 + \cdots + x^n + \cdots$	$(-1, 1)$

Operations on Taylor Series The first n terms of a Taylor series form a polynomial. Because of this, we would expect many of the operations on polynomials to generalize to Taylor series; some properties of series concerning these operations are given in the following theorems.

OPERATIONS ON TAYLOR SERIES

Let f and g be functions having Taylor series with

$$f(x) = a_0 + a_1x + a_2x^2 + a_3x^3 + \cdots + a_nx^n + \cdots$$

and

$$g(x) = b_0 + b_1x + b_2x^2 + b_3x^3 + \cdots + b_nx^n + \cdots.$$

1. The Taylor series for $f + g$ is
$$(a_0 + b_0) + (a_1 + b_1)x + (a_2 + b_2)x^2 + \cdots + (a_n + b_n)x^n + \cdots,$$
for all x in the interval of convergence of both f and g. (Convergent series may be added term by term.)
2. For a real number c, the Taylor series for $c \cdot f(x)$ is
$$c \cdot a_0 + c \cdot a_1x + c \cdot a_2x^2 + \cdots + c \cdot a_nx^n + \cdots,$$
for all x in the interval of convergence of f.
3. For any positive integer k, the Taylor series for $x^k \cdot f(x)$ is
$$a_0x^k + a_1x^k \cdot x + a_2x^k \cdot x^2 + \cdots + a_nx^k \cdot x^n + \cdots = a_0x^k + a_1x^{k+1} + a_2x^{k+2} + \cdots + a_nx^{k+n} + \cdots,$$
for all x in the interval of convergence of f.

These properties follow from the properties of derivatives and from the definition of a Taylor series.

EXAMPLE 2 *Taylor Series*

Find Taylor series for the following functions.

(a) $f(x) = 5e^x$

Solution Use property 2 above, with $c = 5$, along with the Taylor series for $f(x) = e^x$ given earlier. The Taylor series for $5e^x$ is

$$5 \cdot 1 + 5 \cdot x + 5 \cdot \frac{1}{2!}x^2 + 5 \cdot \frac{1}{3!}x^3 + \cdots + 5 \cdot \frac{1}{n!}x^n + \cdots$$
$$= 5 + 5x + \frac{5}{2!}x^2 + \frac{5}{3!}x^3 + \cdots + \frac{5}{n!}x^n + \cdots$$

for all x in $(-\infty, \infty)$.

(b) $f(x) = x^3 \ln(1 + x)$

Solution Use the Taylor series for $\ln(1 + x)$ with Property 3. With $k = 3$, this gives the Taylor series for $x^3 \ln(1 + x)$.

$$x^3 \cdot x - x^3 \cdot \frac{1}{2}x^2 + x^3 \cdot \frac{1}{3}x^3 - x^3 \cdot \frac{1}{4}x^4 + \cdots + \frac{x^3(-1)^n \cdot x^{n+1}}{n+1} + \cdots$$
$$= x^4 - \frac{1}{2}x^5 + \frac{1}{3}x^6 - \frac{1}{4}x^7 + \cdots + \frac{(-1)^n x^{4+n}}{n+1} + \cdots$$

To see why the properties are so useful, try writing the Taylor series for $x^3 \ln(1 + x)$ directly from the definition of a Taylor series.

The final property of Taylor series is perhaps the most useful of all.

COMPOSITION WITH TAYLOR SERIES

Let a function f have a Taylor series such that

$$f(x) = a_0 + a_1x + a_2x^2 + a_3x^3 + \cdots + a_nx^n + \cdots.$$

Then replacing each x with $g(x) = cx^k$ for some constant c and positive integer k gives the Taylor series for $f[g(x)]$:

$$a_0 + a_1g(x) + a_2[g(x)]^2 + a_3[g(x)]^3 + \cdots + a_n[g(x)]^n + \cdots.$$

The interval of convergence of this new series may be different from that of the first series.

EXAMPLE 3 *Composition with Taylor Series*

Find the Taylor series for each function.

(a) $f(x) = e^{-x^2/2}$

Solution We know that the Taylor series for e^x is

$$1 + x + \frac{1}{2!}x^2 + \frac{1}{3!}x^3 + \cdots + \frac{1}{n!}x^n + \cdots$$

for all x in $(-\infty, \infty)$. Use the composition property above and replace each x with $-x^2/2$ to get the Taylor series for $e^{-x^2/2}$.

$$1 + \left(-\frac{x^2}{2}\right) + \frac{1}{2!}\left(-\frac{x^2}{2}\right)^2 + \frac{1}{3!}\left(-\frac{x^2}{2}\right)^3 + \cdots + \frac{1}{n!}\left(-\frac{x^2}{2}\right)^n + \cdots$$
$$= 1 - \frac{1}{2}x^2 + \frac{1}{2!2^2}x^4 - \frac{1}{3!2^3}x^6 + \cdots + \frac{(-1)^n}{n!2^n}x^{2n} + \cdots$$

The Taylor series for $e^{-x^2/2}$ has the same interval of convergence, $(-\infty, \infty)$, as the Taylor series for e^x.

(b) $f(x) = \dfrac{1}{1 + 4x}$

Solution Write $1/(1 + 4x)$ as

$$\frac{1}{1+4x} = \frac{1}{1-(-4x)},$$

which is $1/(1 - x)$ with x replaced with $-4x$. Start with

$$\frac{1}{1-x} = 1 + x + x^2 + x^3 + \cdots + x^n + \cdots,$$

which converges for x in $(-1, 1)$, and replace each x with $-4x$ to get

$$\begin{aligned}\frac{1}{1+4x} &= \frac{1}{1-(-4x)} \\ &= 1 + (-4x) + (-4x)^2 + (-4x)^3 + \cdots + (-4x)^n + \cdots \\ &= 1 - 4x + 16x^2 - 64x^3 + \cdots + (-1)^n 4^n x^n + \cdots.\end{aligned}$$

The interval of convergence of the original series is $(-1, 1)$, or $-1 < x < 1$. Replacing x with $-4x$ gives

$$-1 < -4x < 1 \qquad \text{or} \qquad \frac{1}{4} > x > -\frac{1}{4},$$

so that the interval of convergence of the new series is $(-1/4, 1/4)$.

(c) $f(x) = \dfrac{1}{2 - x^2}$

Solution This function most nearly matches $1/(1 - x)$. To get 1 in the denominator, instead of 2, divide the numerator and denominator by 2.

$$\frac{1}{2-x^2} = \frac{1/2}{1-x^2/2}$$

Thus, we can find the Taylor series for $1/(2 - x^2)$ by starting with the Taylor series for $1/(1 - x)$, multiplying each term by $1/2$, and replacing each x with $x^2/2$.

$$\begin{aligned}\frac{1}{2-x^2} &= \frac{1/2}{1-x^2/2} \\ &= \frac{1}{2}\cdot 1 + \frac{1}{2}\cdot\left(\frac{x^2}{2}\right) + \frac{1}{2}\left(\frac{x^2}{2}\right)^2 + \frac{1}{2}\left(\frac{x^2}{2}\right)^3 + \cdots + \frac{1}{2}\left(\frac{x^2}{2}\right)^n + \cdots \\ &= \frac{1}{2} + \frac{x^2}{4} + \frac{x^4}{8} + \frac{x^6}{16} + \cdots + \frac{x^{2n}}{2^{n+1}} + \cdots\end{aligned}$$

The Taylor series for $1/(1 - x)$ is valid when $-1 < x < 1$. Replacing x with $x^2/2$ gives

$$-1 < \frac{x^2}{2} < 1 \quad \text{or} \quad -2 < x^2 < 2.$$

This inequality is satisfied by any x in the interval $(-\sqrt{2}, \sqrt{2})$.

Although we do not go into detail in this book, the Taylor series we discuss may be differentiated and integrated term by term. This result is used in the next example.

EXAMPLE 4 *Integrating a Taylor Series*

The standard normal curve of statistics is given by

$$f(x) = \frac{1}{\sqrt{2\pi}}e^{-x^2/2}.$$

Find the area bounded by this curve and the lines $x = 0$, $x = 1$, and the x-axis.

Solution The desired area is shown in Figure 12 below. By earlier methods, this area is given by the definite integral

$$\int_0^1 \frac{1}{\sqrt{2\pi}}e^{-x^2/2}dx = \frac{1}{\sqrt{2\pi}}\int_0^1 e^{-x^2/2}dx.$$

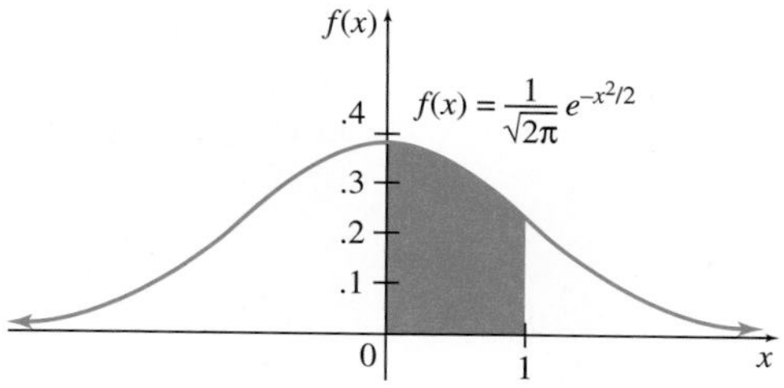

FIGURE 12

This integral cannot be evaluated by any method we have used, but recall that Example 3 gave the Taylor series for $f(x) = e^{-x^2/2}$:

$$e^{-x^2/2} = 1 - \frac{1}{2}x^2 + \frac{1}{2!2^2}x^4 - \frac{1}{3!2^3}x^6 + \cdots + \frac{(-1)^n}{n!2^n}x^{2n} + \cdots.$$

An approximation to $\int_0^1 e^{-x^2/2}dx$ can be found by integrating this series term by term. Using, say, the first six terms of this series gives

$$\begin{aligned}
\frac{1}{\sqrt{2\pi}}\int_0^1 e^{-x^2/2}\,dx &\approx \frac{1}{\sqrt{2\pi}}\int_0^1\left(1 - \frac{1}{2}x^2 + \frac{1}{2!2^2}x^4 - \frac{1}{3!2^3}x^6 + \frac{1}{4!2^4}x^8 - \frac{1}{5!2^5}x^{10}\right)dx\\
&= \frac{1}{\sqrt{2\pi}}\int_0^1\left(1 - \frac{1}{2}x^2 + \frac{1}{8}x^4 - \frac{1}{48}x^6 + \frac{1}{384}x^8 - \frac{1}{3840}x^{10}\right)dx\\
&= \frac{1}{\sqrt{2\pi}}\left(x - \frac{1}{6}x^3 + \frac{1}{40}x^5 - \frac{1}{336}x^7 + \frac{1}{3456}x^9 - \frac{1}{42{,}240}x^{11}\right)\Bigg|_0^1\\
&= \frac{1}{\sqrt{2\pi}}\left(1 - \frac{1}{6} + \frac{1}{40} - \frac{1}{336} + \frac{1}{3456} - \frac{1}{42{,}240} - 0\right)\\
&\approx \frac{1}{\sqrt{2(3.1416)}}(.855623)\\
&\approx .3413.
\end{aligned}$$

This result agrees with the value .3413 obtained from the normal curve table in the Appendix. We could have obtained a more accurate result by using more terms of the Taylor series.

In the chapter titled Nonlinear Functions, we saw that the **doubling time** (in years) for a quantity that increases at an annual rate r is given by

$$n = \frac{\ln 2}{\ln(1 + r)},$$

and we approximated n using the rule of 70 and the rule of 72. Now we can derive these rules by using a Taylor series. As shown in the list of common Taylor series,

$$\ln(1 + x) = x - \frac{x^2}{2} + \frac{x^3}{3} - \frac{x^4}{4} + \cdots$$

for x in $(-1, 1]$. Further,

$$\ln(1 + x) = x - \left(\frac{x^2}{2} - \frac{x^3}{3}\right) - \left(\frac{x^4}{4} - \frac{x^5}{5}\right) - \cdots < x$$

because each term in parentheses is positive for x in $(-1, 1]$. Therefore, for $0 < r < 1$, the doubling time,

$$n = \frac{\ln 2}{\ln(1 + r)} = \frac{\ln 2}{r - \frac{r^2}{2} + \frac{r^3}{3} - \frac{r^4}{4} + \cdots},$$

is just slightly larger than the quotient

$$\frac{\ln 2}{r} \approx \frac{.693}{r} = \frac{69.3}{100r}.$$

Since the actual value of

$$\ln(1 + r) = r - \frac{r^2}{2} + \frac{r^3}{3} - \frac{r^4}{4} + \cdots$$

is slightly smaller than r for $0 < r < 1$, the quotients

$$\frac{70}{100r} \quad \text{and} \quad \frac{72}{100r}$$

give good approximations for the doubling time, the *rule of 70* and the *rule of 72*.

RULE OF 70 AND RULE OF 72

Rule of 70 If a quantity is increasing at a constant rate r compounded annually, where $.001 \le r \le .05$,

$$\textbf{Doubling time} \approx \frac{\mathbf{70}}{\mathbf{100}r} \textbf{ years.}$$

Rule of 72 If a quantity is increasing at a constant rate r compounded annually, where $.05 < r \le .12$, then

$$\textbf{Doubling time} \approx \frac{\mathbf{72}}{\mathbf{100}r} \textbf{ years.}$$

The rule of 70 is used by demographers because populations usually grow at rates of less than 5 percent. The rule of 72 is preferred by economists and investors, since money frequently grows at a rate of between 5 percent and 12 percent. Because the difference between compounding continuously and compounding several times a year is small, both the rule of 70 and the rule of 72 can be used to approximate the doubling time in any interval.

EXAMPLE 5 *Doubling Time*

Find the doubling time for an investment at each interest rate.

(a) 9%

Solution By the formula for doubling time, at an interest rate of 9%, money will double in

$$\frac{\ln 2}{\ln(1 + .09)} \approx 8 \text{ years.}$$

(b) 1%

Solution At an interest rate of 1%, money will double in

$$\frac{\ln 2}{\ln(1 + .01)} \approx 70 \text{ years.}$$

(c) The rule of 70 predicts that at a growth rate of 1%, a population will double in 70 years, in agreement with part b. The rule of 72 predicts that at an interest rate of 9%, money will double in 8 years, in agreement with part a. ■

The following table gives the actual doubling time n in years for various growth rates r, together with the approximate doubling times given by the rules of 70 and 72.

r	.001	.005	.01	.02	.03	.04	.05	.06	.07	.08	.09	.10	.11	.12
n	693	139	69.7	35.0	23.4	17.7	14.2	11.9	10.2	9.0	8.0	7.3	6.6	6.1
$\dfrac{70}{100r}$	700	140	70	35	23.3	17.5	14	11.7	10	8.8	7.8	7.0	6.4	5.8
$\dfrac{72}{100r}$	720	144	72	36	24	18	14.4	12	10.3	9	8	7.2	6.5	6

The last row in the table is particularly easy to compute because 72 has so many integral divisors. Therefore, the rule of 72 is frequently used by economists and investors for any interest rate r.

It can be shown that the rule of 70 will give the doubling time with an error of 2% or less if $.001 \leq n \leq .05$, and the rule of 72 will give the doubling time with a 2% error or less if $.05 < r \leq .12$. The above table shows the accuracy of the approximations, and the graph in Figure 13 on the next page shows that the graphs of

$$\frac{\ln 2}{\ln(1 + r)}, \qquad \frac{70}{100r}, \qquad \text{and} \qquad \frac{72}{100r}$$

are virtually indistinguishable over the domains just indicated.

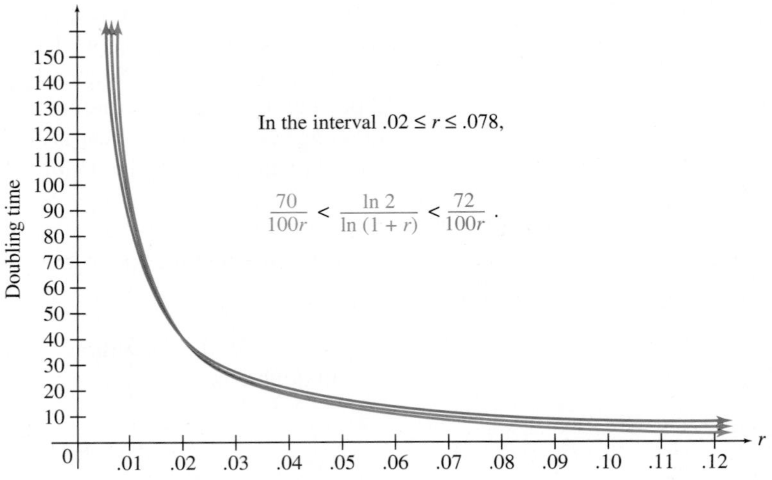

FIGURE 13

12.5 EXERCISES

Find the Taylor series for the functions defined as follows. Give the interval of convergence for each series.

1. $f(x) = \dfrac{5}{2-x}$

2. $f(x) = \dfrac{-3}{4-x}$

3. $f(x) = \dfrac{8x}{1+3x}$

4. $f(x) = \dfrac{7x}{1+2x}$

5. $f(x) = \dfrac{x^2}{4-x}$

6. $f(x) = \dfrac{9x^4}{1-x}$

7. $f(x) = \ln(1+4x)$

8. $f(x) = \ln\left(1 - \dfrac{x}{2}\right)$

9. $f(x) = e^{4x^2}$

10. $f(x) = e^{-3x^2}$

11. $f(x) = x^3e^{-x}$

12. $f(x) = x^5e^x$

13. $f(x) = \dfrac{2}{1+x^2}$

14. $f(x) = \dfrac{6}{3+x^2}$

15. $f(x) = \dfrac{e^x + e^{-x}}{2}$

16. $f(x) = \dfrac{e^x - e^{-x}}{2}$

17. $f(x) = \ln(1+2x^4)$

18. $f(x) = \ln(1-5x^2)$

19. Use the fact that

$$\frac{1+x}{1-x} = \frac{1}{1-x} + \frac{x}{1-x}$$

to find a Taylor series for $(1+x)/(1-x)$.

20. By properties of logarithms,

$$\ln\left(\frac{1+x}{1-x}\right) = \ln(1+x) - \ln(1-x).$$

Use this to find a Taylor series for $\ln[(1+x)/(1-x)]$.

21. Use the Taylor series for e^x to suggest that

$$e^x \approx 1 + x + \frac{x^2}{2}$$

for all x close to zero.

22. Use the Taylor series for e^{-x} to suggest that

$$e^{-x} \approx 1 - x + \frac{x^2}{2}$$

for all x close to zero.

23. Use the Taylor series for e^x to show that

$$e^x \geq 1 + x$$

for all x.

Use the method in Example 4 (with five terms of the appropriate Taylor series) to approximate the areas of the following regions.

24. The region bounded by $f(x) = e^{x^2}$, $x = 0$, $x = 1/2$, and the x-axis

25. The region bounded by $f(x) = 1/(1 - x^2)$, $x = 0$, $x = 1/2$, and the x-axis

26. The region bounded by $f(x) = 1/(1 - \sqrt{x})$, $x = 1/4$, $x = 1/2$, and the x-axis

27. The region bounded by $f(x) = e^{\sqrt{x}}$, $x = 0$, $x = 1$, and the x-axis

As mentioned in Example 4, the equation of the standard normal curve is

$$f(x) = \frac{1}{\sqrt{2\pi}} e^{-x^2/2}.$$

Use the method in Example 4 (with five terms of the Taylor series) to approximate the area of the region bounded by the normal curve, the x-axis, $x = 0$, and the values of x in Exercise 28 and 29.

28. $x = .5$

29. $x = .75$

Applications

BUSINESS AND ECONOMICS

30. *Investment* William Hoffman has invested \$10,000 in a certificate of deposit that has a 4.25% annual interest rate. Determine the doubling time for this investment using the doubling-time formula. How does this compare with the estimate given by the rule of 70?

31. *Investment* It is anticipated that a bank stock that Mary Reynolds has invested \$10,000 in will achieve an annual interest rate of 15%. Determine the doubling time for this investment using the doubling-time formula. How does this compare with the estimate given by the rule of 72?

LIFE SCIENCES

32. *Infant Mortality* Infant mortality is an example of a relatively rare event that can be described by the *Poisson distribution*, for which the probability of x occurrences is given by

$$f(x) = \frac{\lambda^x e^{-\lambda}}{x!}, \qquad x = 0, 1, 2, \ldots$$

a. Verify that f describes a probability distribution by showing that

$$\sum_{x=0}^{\infty} f(x) = 1.$$

b. Calculate the expected value for f, given by

$$\sum_{x=0}^{\infty} x f(x).$$

c. Between 2000 and 2005, the U.S. infant mortality rate was 7 per 1000 live births.* Assuming that this is the expected value for a Poisson distribution, find the probability that in a random sample of 1000 live births, there were fewer than 4 cases of infant mortality.

GENERAL INTEREST

33. *Baseball* In the year 2000, the chance that a U.S. major league baseball player was foreign born was 1 in 5.† Suppose we begin to randomly select major league players

**The New York Times 2003 Almanac*, p. 481.
†"Harper's Index," *Harper's*, July 2000, p. 13.

until we find one who is foreign born. Such an experiment can be described by the *geometric distribution*, for which the probability of success after x tries is given by

$$f(x) = (1 - p)^{x-1}p, \quad x = 1, 2, 3, \ldots$$

where p is the probability of success on a given try. (*Note:* This formula is only accurate if the number of baseball players is very large, compared with the number that we select before meeting one who is foreign born.)

a. Verify that f describes a probability distribution by showing that

$$\sum_{x=1}^{\infty} f(x) = 1.$$

b. Calculate the expected value for f, given by

$$\sum_{x=1}^{\infty} xf(x).$$

(*Hint:* Let $g(z) = p\sum_{x=1}^{\infty} z^x$, and evaluate $g'(1 - p)$.)

c. On average, how many major league baseball players would you have to meet before meeting one who is foreign born?

d. What is the probability that you meet a foreign-born player within the first three major league players that you meet?

34. *Trouble* In the Milton Bradley game Trouble™, each player takes turns pressing a "popper" that contains a single die. To begin moving a game piece around the board a player must first pop a 6 on the die. The number of tries required to get a 6 can be described by the geometric distribution. (See Exercise 33.)

a. Using the result of Exercise 33b, what is the expected number of times a popper must be pressed before a success occurs?

b. What is the probability that you will have to press the popper four or more times before a 6 pops up?

12.6 NEWTON'S METHOD

THINK ABOUT IT *How can the true interest rate be found, given the amount loaned, the number of payments, and the amount of each payment?*

We will answer this question in an exercise using the technique developed in this section.

Given a function f, a number r such that $f(r) = 0$ is called a zero of f. For example, if $f(x) = x^2 - 4x + 3$, then $f(3) = 0$ and $f(1) = 0$, so that both 3 and 1 are zeros of f. The zeros of linear and quadratic functions can be found with the methods of algebra. More complicated methods exist for finding zeros of third-degree or fourth-degree polynomial functions, but there is no general method for finding zeros of higher-degree polynomials.

In practical applications of mathematics, it is seldom necessary to find *exact* zeros of a function; usually a decimal approximation is all that is needed. We have seen earlier how a graphing calculator may be used to find approximate values of zeros. In this section, we will explore a calculus-based method to do the same. The method provides a sequence of values, $c_1, c_2, \ldots,$ whose limit is the true value in a wide variety of applications. Of course, you may simply prefer to use the `zero` feature on your graphing calculator, but Newton's method is the basis for some of the techniques used by mathematicians to solve more complex problems.

The zeros of a differentiable function f can be approximated as follows. Find a closed interval $[a, b]$ so that $f(a)$ and $f(b)$ are of opposite sign, one positive and one negative. As suggested by Figure 14 on the next page, this means there must exist at least one value c in the interval (a, b) such that $f(c) = 0$. This number c is a zero of the function f.

To find an approximate value for c, first make a guess for c. Let c_1 be the initial guess. (See Figure 15.) Then locate the point $(c_1, f(c_1))$ on the graph of $y = f(x)$ and identify the tangent line at this point. This tangent line will cut

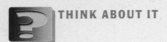

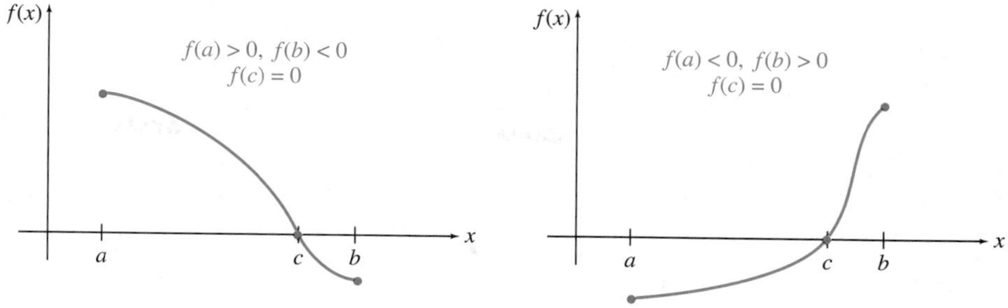

FIGURE 14

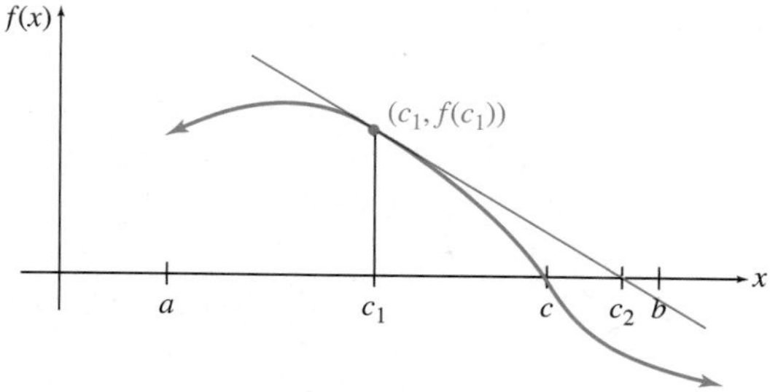

FIGURE 15

the x-axis at a point c_2. The number c_2 is often a better approximation of c than was c_1.

To locate c_2, first find the equation of the tangent line through $(c_1, f(c_1))$. The slope of this tangent line is $f'(c_1)$. The point-slope form of the equation of the tangent line is

$$y - f(c_1) = f'(c_1)(x - c_1).$$

At the point $x = c_2$, $y = 0$. Substituting into the equation of the tangent line gives

$$0 - f(c_1) = f'(c_1)(c_2 - c_1)$$

or

$$-\frac{f(c_1)}{f'(c_1)} = c_2 - c_1,$$

from which

$$c_2 = c_1 - \frac{f(c_1)}{f'(c_1)}.$$

If $f'(c_1)$ should be 0, the tangent line would be horizontal and not cut the x-axis. For this reason, assume $f'(c_1) \neq 0$. This new value, c_2, is usually a better approximation to c than was c_1. To improve the approximation further, locate the tangent line to the curve at $(c_2, f(c_2))$. Let this tangent cut the x-axis at c_3. (See Figure 16 on the next page). Find c_3 by a process similar to that used above: if $f'(c_2) \neq 0$,

$$c_3 = c_2 - \frac{f(c_2)}{f'(c_2)}.$$

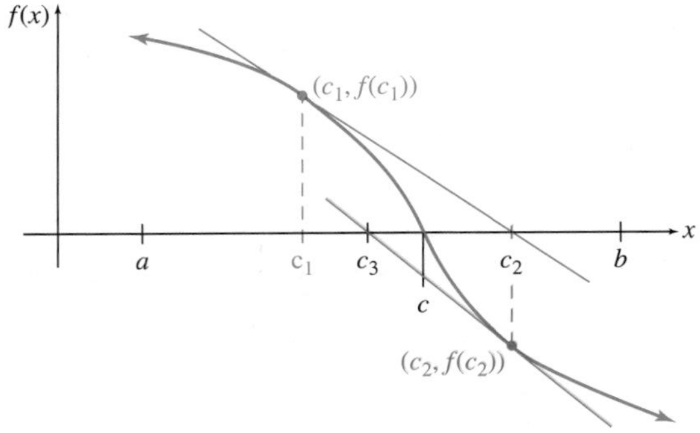

FIGURE 16

The approximation to c often can be improved by repeating this process as many times as desired. In general, if c_n is an approximation to c, a better approximation, c_{n+1}, frequently can be found by the following formula.

NEWTON'S METHOD

If $f'(c_n) \neq 0$, then

$$c_{n+1} = c_n - \frac{f(c_n)}{f'(c_n)}.$$

This process of first obtaining a rough approximation for c, then replacing it successively by approximations that are often better, is called **Newton's method,** named after Sir Isaac Newton, the codiscoverer of calculus. An early version of this method appeared in his work *Method of Fluxions*, published in 1736.

EXAMPLE 1 *Newton's Method*

Approximate a solution for the equation

$$3x^3 - x^2 + 5x - 12 = 0$$

in the interval $[1, 2]$.

Solution Let $f(x) = 3x^3 - x^2 + 5x - 12$, so that $f'(x) = 9x^2 - 2x + 5$. Check that $f(1) < 0$ with $f(2) > 0$. Since $f(1)$ and $f(2)$ have opposite signs, there is a solution for the equation in the interval $(1, 2)$. As an initial guess, let $c_1 = 1$. A better guess, c_2, can be found as follows.

$$c_2 = c_1 - \frac{f(c_1)}{f'(c_1)} = 1 - \frac{-5}{12} = 1.4167$$

A third approximation, c_3, can now be found.

$$c_3 = c_2 - \frac{f(c_2)}{f'(c_2)} = 1.4167 - \frac{1.6066}{20.230} = 1.3373$$

In the same way,

$$c_4 = 1.3373 - \frac{.072895}{18.421} = 1.3333 \quad \text{and} \quad c_5 = 1.3333 - \frac{-6.111 \cdot 10^{-4}}{18.333} = 1.3333.$$

Subsequent approximations yield no further accuracy, either to the 4 decimal places to which we have rounded or to the digits displayed in a TI-83/84 Plus calculator. Thus $x = 1.3333$ is a reasonably accurate solution of $3x^3 - x^2 + 5x - 12 = 0$. (The exact solution is $4/3$.)

Newton's method is easily implemented on a graphing calculator. For the previous example on a TI-83/84 Plus, start by storing 1 in X, the function $f(x)$ in Y_1, and the function $f'(x)$ in Y_2. The command X − Y_1/Y_2 → X gives the next value of x. Continue to press the ENTER key for subsequent calculations.

In Example 1 we had to go through five steps to get the degree of accuracy that we wanted. The solutions of similar polynomial equations usually can be found in about as many steps, although other types of equations might require more steps, particularly if the initial guess is far from the true solution.

In any case, if a solution can be found by Newton's method, it usually can be found by a computer in a small fraction of a second. But in some cases, the method will not find the solution, or will only do so for a good initial guess. Figure 17 shows an example in which Newton's method does not give a solution. Because of the symmetry of the graph in Figure 17, all the odd steps (c_3, c_5, and so forth) give c_1, while all the even steps (c_4, c_6, and so forth) give c_2, so the approximations never approach the true solution. Such cases are rare in practice. If you find that Newton's method is not producing a solution, verify that there is a solution, and then try a better initial guess.

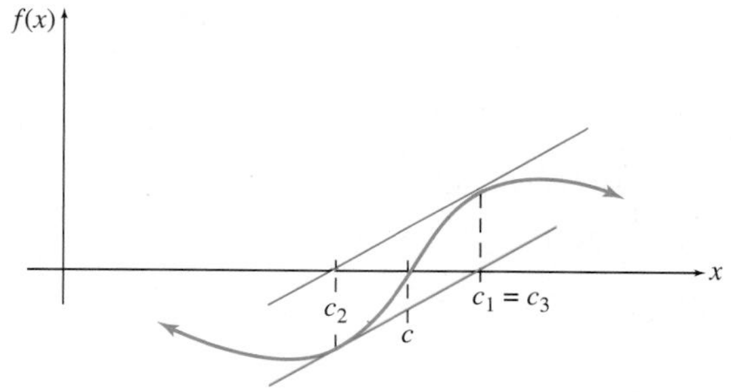

FIGURE 17

Newton's method also can be used to approximate the values of radicals, as shown by the next example.

EXAMPLE 2 *Approximation*

Approximate $\sqrt{12}$ to the nearest thousandth.

Solution First, note that $\sqrt{12}$ is a solution of the equation $x^2 - 12 = 0$. Therefore, let $f(x) = x^2 - 12$, so that $f'(x) = 2x$. Since $3 < \sqrt{12} < 4$, use $c_1 = 3$ as the first approximation to $\sqrt{12}$. A better approximation is given by c_2:

$$c_2 = 3 - \frac{-3}{6} = 3.5.$$

Now find c_3 and c_4:

$$c_3 = 3.5 - \frac{.25}{7} = 3.464,$$

$$c_4 = 3.464 - \frac{-.0007}{6.928} = 3.464.$$

Since $c_3 = c_4 = 3.464$, to the nearest thousandth, $\sqrt{12} = 3.464$.

12.6 EXERCISES

Use Newton's method to find a solution for each equation in the given intervals. Find all solutions to the nearest hundredth.

1. $x^2 - 2x - 2 = 0;\quad [2, 3]$

2. $x^2 - 6x + 4 = 0;\quad [5, 6]$

3. $3x^3 - 5x^2 + 8x - 4 = 0;\quad [0, 1]$

4. $5x^3 - 2x^2 - 2x - 7 = 0;\quad [1, 2]$

5. $3x^3 - 4x^2 - 4x - 7 = 0;\quad [2, 3]$

6. $3x^3 - 14x^2 + 17x - 22 = 0;\quad [3, 4]$

7. $3x^4 - 5x^3 - 16x^2 - 19x - 11 = 0;\quad [-2, -1], [3, 4]$

8. $2x^4 + 7x^3 + 6x^2 + 7x - 6 = 0;\quad [0, 1], [-3, -2]$

9. $x^{1/3} - x^2 + 3 = 0;\quad [-3, 0]$

10. $x^{1/3} - x^2 + 3 = 0;\quad [0, 3]$

11. $e^x + x - 2 = 0;\quad [0, 3]$

12. $e^{2x} + 3x - 4 = 0;\quad [0, 3]$

13. $x^2e^{-x} + x^2 - 2 = 0;\quad [0, 3]$

14. $x^2e^{-x} + x^2 - 2 = 0;\quad [-3, 0]$

15. $\ln x + x - 2 = 0;\quad [1, 4]$

16. $2 \ln x + x - 3 = 0;\quad [1, 4]$

Use Newton's method to find each root to the nearest thousandth.

17. $\sqrt{2}$ **18.** $\sqrt{3}$ **19.** $\sqrt{11}$ **20.** $\sqrt{15}$ **21.** $\sqrt{250}$

22. $\sqrt{300}$ **23.** $\sqrt[3]{9}$ **24.** $\sqrt[3]{15}$ **25.** $\sqrt[3]{100}$ **26.** $\sqrt[3]{121}$

Use Newton's method to find the critical points for the functions defined as follows. Approximate them to the nearest hundredth. Decide whether each critical point leads to a relative maximum or a relative minimum.

27. $f(x) = x^3 - 3x^2 - 18x + 4$

28. $f(x) = x^3 + 9x^2 - 6x + 4$

29. $f(x) = x^4 - 3x^3 + 6x - 1$

30. $f(x) = x^4 + 2x^3 - 5x + 2$

Applications

BUSINESS AND ECONOMICS

31. *Break-Even Point* For a particular product, the revenue and cost functions are

$$R(x) = \sqrt{64 - x^2} \quad \text{and} \quad C(x) = x - 1.$$

Approximate the break-even point to the nearest hundredth.

32. *Manufacturing* A new manufacturing process produces savings of

$$S(x) = x^3 + 5x^2 + 9$$

dollars after x years, with increased costs of

$$C(x) = x^2 + 40x + 20$$

dollars. For how many years, to the nearest hundredth, should the process be used?

33. *True Annual Interest Rate* Federal government regulations require that people loaning money to consumers disclose the true annual interest rate of the loan. The formulas for calculating this interest rate are very complex. For example, suppose P dollars is loaned, with the money to be repaid in n monthly payments of M dollars each. Then the true annual interest rate is found by solving the equation

$$\frac{1 - (1 + i)^{-n}}{i} - \frac{P}{M} = 0$$

for i, the monthly interest rate, and then multiplying i by 12 to get the true annual rate. This equation can best be solved by Newton's method, as explained in the next few exercises. (This is how the financial function IRR (Internal Rate of Return) is computed in Microsoft Excel.)

a. Let $f(i) = \dfrac{1-(1+i)^{-n}}{i} - \dfrac{P}{M}$. Find $f'(i)$.

b. Form the quotient $f(i)/f'(i)$.

c. Suppose that $P = \$4000$, $n = 24$, and $M = \$197$. Let the initial guess for i be $i_1 = .01$. Use Newton's method and find i_2.

d. Find i_3. (*Note:* For the accuracy required by federal law, it is usually sufficient to stop after two successive values of i differ by no more than 10^{-7}.)

Find i_2 and i_3.

34. $P = \$600$, $M = \$57$, $n = 12$, $i_1 = .02$

35. $P = \$15{,}000$, $M = \$337$, $n = 60$, $i_1 = .01$

12.7 L'HOSPITAL'S RULE

We began our study of calculus with a discussion of *limits*. For example,

$$\lim_{x\to 1} \frac{x^2+1}{x+4} = \frac{2}{5},$$

or, in words, as x gets closer and closer to 1, the quotient $(x^2+1)/(x+4)$ gets closer and closer to $2/5$. In this section we will use derivatives to find limits of quotients of functions that could not easily be found before.

If we try to find

$$\lim_{x\to 1} \frac{x^2-1}{x-1}$$

by evaluating the numerator and denominator at $x = 1$, we get

$$\frac{1^2-1}{1-1} = \frac{0}{0},$$

an **indeterminate form.** Any attempt to assign a value to $0/0$ leads to a meaningless result. The limit exists, however; as shown earlier, it is found by factoring.

$$\lim_{x\to 1} \frac{x^2-1}{x-1} = \lim_{x\to 1} \frac{(x+1)(x-1)}{x-1} = \lim_{x\to 1}(x+1) = 2$$

As a second example,

$$\lim_{x\to 1} \frac{\ln x}{(x-1)^2}$$

also leads to the indeterminate form $0/0$. Selecting values of x close to 1 and using a calculator gives the following table.

x	.99	.999	.9999	1.0001	1.001	1.01
$\dfrac{\ln x}{(x-1)^2}$	−100.5	−1000.5	−10,000.5	9999.5	999.5	99.5

As this table suggests, $\lim_{x\to 1}(\ln x)/(x-1)^2$ does not exist.

In the first example, trying to find $\lim_{x\to 1}(x^2-1)/(x-1)$ by evaluating the expression at $x = 1$ led to the indeterminate form $0/0$, but factoring the expression led to the actual limit, 2. Evaluating $\lim_{x\to 1}(\ln x)/(x-1)^2$ in the second example led to the indeterminate form 0/0, but using a table of values showed that this limit did not exist. **L'Hospital's rule** gives a quicker way to decide whether a quotient with the indeterminate form 0/0 has a limit.

L'HOSPITAL'S RULE

Let f and g be functions and let a be a real number such that

$$\lim_{x\to a} f(x) = 0 \qquad \text{and} \qquad \lim_{x\to a} g(x) = 0.$$

Let f and g have derivatives that exist at each point in some open interval containing a.

If $\displaystyle\lim_{x\to a}\frac{f'(x)}{g'(x)} = L$, then $\displaystyle\lim_{x\to a}\frac{f(x)}{g(x)} = L$.

If $\displaystyle\lim_{x\to a}\frac{f'(x)}{g'(x)}$ does not exist because $\displaystyle\left|\frac{f'(x)}{g'(x)}\right|$ becomes large without bound for values of x near a, then $\displaystyle\lim_{x\to a}\frac{f(x)}{g(x)}$ also does not exist.

A partial proof of this rule is given at the end of this section. L'Hospital's rule is another example of a mathematical misnomer. Although named after the Marquis de l'Hospital (1661–1704), it was actually developed by Johann Bernoulli (1667–1748) in a textbook published in 1696. (Johann Bernoulli was the brother of Jakob Bernoulli, mentioned in the section on Antiderivatives.) L'Hospital was a student of Bernoulli and published, with a financial arrangement, the works of his teacher under his own name.

EXAMPLE 1 *L'Hospital's Rule*

Find $\displaystyle\lim_{x\to 1}\frac{x^2-1}{x-1}$.

Solution It is very important to first make sure that the conditions of l'Hospital's rule are satisfied. Here

$$\lim_{x\to 1}(x^2-1) = 0 \qquad \text{and} \qquad \lim_{x\to 1}(x-1) = 0.$$

Since the limits of both numerator and denominator are 0, the rule does apply. Now take the derivatives of both numerator and denominator separately. (Do *not* use the quotient rule for derivatives.)

$$\text{If } f(x) = x^2 - 1, \quad \text{then } f'(x) = 2x;$$
$$\text{if } g(x) = x - 1, \quad \text{then } g'(x) = 1.$$

Find the limit of the quotient of the derivatives.

$$\lim_{x\to 1}\frac{f'(x)}{g'(x)} = \lim_{x\to 1}\frac{2x}{1} = \lim_{x\to 1} 2x = 2$$

By l'Hospital's rule, this result is the desired limit:

$$\lim_{x \to 1} \frac{x^2 - 1}{x - 1} = 2.$$

EXAMPLE 2 *L'Hospital's Rule*

Find $\lim_{x \to 1} \frac{\ln x}{(x - 1)^2}$.

Solution Make sure that l'Hospital's rule applies.

$$\lim_{x \to 1} \ln x = \ln 1 = 0 \qquad \text{and} \qquad \lim_{x \to 1} (x - 1)^2 = 0$$

Since the conditions of l'Hospital's rule are satisfied, we can now take the derivatives of the numerator and denominator separately.

$$D_x(\ln x) = \frac{1}{x} \qquad \text{and} \qquad D_x[(x - 1)^2] = 2(x - 1)$$

Next, we find the limit of the quotient of these derivatives:

$$\lim_{x \to 1} \frac{1/x}{2(x - 1)} = \lim_{x \to 1} \frac{1}{2x(x - 1)} \text{ does not exist.}$$

By l'Hospital's rule, this means that

$$\lim_{x \to 1} \frac{\ln x}{(x - 1)^2} \text{ does not exist.}$$

Before looking at more examples of l'Hospital's rule, consider the following summary.

USING L'HOSPITAL'S RULE

1. Be sure that $\lim_{x \to a} f(x)/g(x)$ leads to the indeterminate form $0/0$.
2. Take the derivatives of f and g separately.
3. Find the limit of $f'(x)/g'(x)$; this limit, if it exists, equals the limit of $f(x)/g(x)$.
4. If necessary, apply l'Hospital's rule more than once.

EXAMPLE 3 *L'Hospital's Rule*

Find $\lim_{x \to 0} \frac{x^3}{e^x - 1}$.

Solution The limit in the numerator is 0, as is the limit in the denominator, so that l'Hospital's rule applies. Taking derivatives separately in the numerator and denominator gives

$$\lim_{x \to 0} \frac{3x^2}{e^x} = \frac{0}{e^0} = \frac{0}{1} = 0.$$

By l'Hospital's rule,

$$\lim_{x\to 0} \frac{x^3}{e^x - 1} = 0.$$

EXAMPLE 4 *L'Hospital's Rule*

Find $\lim\limits_{x\to 0} \dfrac{e^x - x - 1}{x^2}$.

Solution Find the limit in both the numerator and denominator to verify that l'Hospital's rule applies. Then take derivatives of both the numerator and denominator separately.

$$\lim_{x\to 0} \frac{e^x - 1}{2x} = \frac{e^0 - 1}{2 \cdot 0} = \frac{1 - 1}{0} = \frac{0}{0}$$

The result is still the indeterminate form $0/0$; use l'Hospital's rule a second time. Taking derivatives of $e^x - 1$ and $2x$ gives

$$\lim_{x\to 0} \frac{e^x}{2} = \frac{e^0}{2} = \frac{1}{2}.$$

Finally, by l'Hospital's rule,

$$\lim_{x\to 0} \frac{e^x - x - 1}{x^2} = \frac{1}{2}.$$

EXAMPLE 5 *L'Hospital's Rule*

Find $\lim\limits_{x\to 1} \dfrac{x^2 - 1}{\sqrt{x}}$.

Solution Taking derivatives of the numerator and denominator separately gives

$$\lim_{x\to 1} \frac{2x}{(1/2)x^{-1/2}} = \lim_{x\to 1} 4x^{3/2} = 4 \cdot 1^{3/2} = 4 \cdot 1 = 4. \qquad \text{Incorrect}$$

Unfortunately, 4 is the wrong answer. What happened? We did not verify that the conditions of l'Hospital's rule were satisfied. In fact,

$$\lim_{x\to 1}(x^2 - 1) = 0, \qquad \text{but} \qquad \lim_{x\to 1} \sqrt{x} = 1 \neq 0.$$

Since l'Hospital's rule does not apply, we must use another method to find the limit. By substitution,

$$\lim_{x\to 1} \frac{x^2 - 1}{\sqrt{x}} = \frac{1^2 - 1}{\sqrt{1}} = \frac{0}{1} = 0.$$

Proof of l'Hospital's Rule Because the proof of l'Hospital's rule is too advanced for this text we will not prove it here. We will, however, prove the theorem for the special case where f, g, f', and g' are continuous on some open interval containing a, and $g'(a) \neq 0$. With these assumptions, the fact that

$$\lim_{x\to a} f(x) = 0 \qquad \text{and} \qquad \lim_{x\to a} g(x) = 0$$

means that both $f(a) = 0$ and $g(a) = 0$. Thus,

$$\lim_{x \to a} \frac{f(x)}{g(x)} = \lim_{x \to a} \frac{f(x) - f(a)}{g(x) - g(a)}, \qquad f(a) = 0 \text{ and } g(a) = 0$$

where we subtracted 0 in both the numerator and denominator. Multiplying the numerator and denominator by $1/(x - a)$ gives

$$\lim_{x \to a} \frac{f(x)}{g(x)} = \lim_{x \to a} \frac{\dfrac{f(x) - f(a)}{x - a}}{\dfrac{g(x) - g(a)}{x - a}}.$$

By a property of limits, this becomes

$$\lim_{x \to a} \frac{f(x)}{g(x)} = \frac{\lim\limits_{x \to a} \dfrac{f(x) - f(a)}{x - a}}{\lim\limits_{x \to a} \dfrac{g(x) - g(a)}{x - a}}.$$

By the definition of the derivative, the limit in the numerator is $f'(a)$, and the limit in the denominator is $g'(a)$. From our assumption that both f' and g' are continuous, and if $g'(a) \neq 0$, the quotient on the right above becomes

$$\frac{f'(a)}{g'(a)} = \frac{\lim\limits_{x \to a} f'(x)}{\lim\limits_{x \to a} g'(x)} = \lim_{x \to a} \frac{f'(x)}{g'(x)}.$$

Thus,

$$\lim_{x \to a} \frac{f(x)}{g(x)} = \lim_{x \to a} \frac{f'(x)}{g'(x)},$$

which is what we wanted to show.

12.7 EXERCISES

Use l'Hospital's rule where applicable to find each limit.

1. $\lim\limits_{x \to 1} \dfrac{x^3 + x^2 - x - 1}{x^2 - x}$

2. $\lim\limits_{x \to 3} \dfrac{x^3 + x^2 - 11x - 3}{x^2 - 3x}$

3. $\lim\limits_{x \to 0} \dfrac{x^5 - 2x^3 + 4x^2}{8x^5 - 2x^2 + 5x}$

4. $\lim\limits_{x \to 0} \dfrac{8x^6 + 3x^4 - 9x}{9x^7 - 2x^4 + x^3}$

5. $\lim\limits_{x \to 2} \dfrac{\ln(x - 1)}{x - 2}$

6. $\lim\limits_{x \to 0} \dfrac{\ln(x + 1)}{x}$

7. $\lim\limits_{x \to 0} \dfrac{e^x - 1}{x^4}$

8. $\lim\limits_{x \to 0} \dfrac{e^{2x} - 1}{5x^2 - x}$

9. $\lim\limits_{x \to 0} \dfrac{xe^x}{e^x - 1}$

10. $\lim\limits_{x \to 0} \dfrac{xe^{-x}}{2e^{2x} - 2}$

11. $\lim\limits_{x \to 0} \dfrac{e^x}{2x^3 + 9x^2 - 11x}$

12. $\lim\limits_{x \to 0} \dfrac{e^x}{8x^5 - 3x^4}$

13. $\lim\limits_{x \to 0} \dfrac{\sqrt{2 + x} - \sqrt{2}}{x}$

14. $\lim\limits_{x \to 0} \dfrac{\sqrt{9 + x} - 3}{x}$

15. $\lim\limits_{x \to 4} \dfrac{\sqrt{x} - 2}{x - 4}$

16. $\lim\limits_{x \to 9} \dfrac{\sqrt{x} - 3}{x - 9}$

17. $\lim\limits_{x \to 8} \dfrac{\sqrt[3]{x} - 2}{x - 8}$

18. $\lim\limits_{x \to 27} \dfrac{\sqrt[3]{x} - 3}{x - 27}$

19. $\lim\limits_{x \to 1} \dfrac{x^9 - 1}{x - 1}$

20. $\lim\limits_{x \to 2} \dfrac{x^7 - 128}{x - 2}$

21. $\lim\limits_{x \to 0} \dfrac{e^x + e^{-x} - 2}{x}$

22. $\lim\limits_{x \to 0} \dfrac{e^x - 1 + x}{e^{-x} - 1 - x}$

23. $\lim\limits_{x \to 3} \dfrac{\sqrt{x^2 + 7} - 4}{x^2 - 9}$

24. $\lim\limits_{x \to 5} \dfrac{\sqrt{x^2 + 11} - 6}{x^2 - 25}$

25. $\lim\limits_{x \to 0} \dfrac{1 + \frac{1}{3}x - (1 + x)^{1/3}}{x^2}$

26. $\lim\limits_{x \to 2} \dfrac{2 - \sqrt{x + 2}}{2x^2 - 10x + 8}$

27. $\lim\limits_{x \to 0} \dfrac{\sqrt{1 + x} - \sqrt{1 - x}}{x}$

28. $\lim\limits_{x \to 0} \dfrac{\sqrt{3 - x} - \sqrt{3 + x}}{x}$

29. $\lim\limits_{x \to 0} \dfrac{\sqrt{x^2 - 5x + 4}}{x}$

30. $\lim\limits_{x \to 1} \dfrac{\sqrt{x^2 + 5x + 9}}{x - 1}$

31. $\lim\limits_{x \to 0} \dfrac{(5 + x)\ln(x + 1)}{e^x - 1}$

32. $\lim\limits_{x \to 0} \dfrac{(7 - x)\ln(1 - x)}{e^{-x} - 1}$

In Exercises 33–36, first get a common denominator; then find the limits that exist.

33. $\lim\limits_{x \to 0} \left(\dfrac{1}{x} - \dfrac{1}{x^2}\right)$

34. $\lim\limits_{x \to 0} \left(\dfrac{1}{x^2} + \dfrac{1}{x^3}\right)$

35. $\lim\limits_{x \to 0} \left(\dfrac{1}{x} + \dfrac{1}{\sqrt[3]{x}}\right)$

36. $\lim\limits_{x \to 0} \left(\dfrac{1}{\sqrt[5]{x}} - \dfrac{1}{x}\right)$

37. Explain what is wrong with the following calculation using l'Hospital's rule.

$$\lim_{x \to 0} \frac{x^2}{x^2 + 3} = \lim_{x \to 0} \frac{2x}{2x} = 1$$

CHAPTER SUMMARY

We have provided a brief introduction to the topics of sequences, series, and l'Hospital's rule. Geometric sequences are comparatively simple to analyze and arise in various applications, including annuities. We next investigated infinite series, as well as a particular form known as Taylor series. Because Taylor series have an infinite number of terms, it is often more practical to take a small number of terms, creating Taylor polynomials. We then discussed Newton's method, which produces a sequence that approaches a zero of a function. Finally, l'Hospital's rule provides a method for evaluating certain limits.

KEY TERMS

12.1 sequence
element
term
general term
*n*th term
geometric sequence
common ratio

12.2 annuity
ordinary annuity
payment period
term of an annuity
amount of an annuity
sinking fund
present value of an annuity
amortization

12.3 Taylor polynomial
Taylor polynomial of degree n

12.4 infinite series
*n*th partial sum
sum of an infinite series
convergence
divergence

12.5 interval of convergence
Taylor series
Maclaurin series
doubling time
rule of 70
rule of 72

12.6 Newton's method

12.7 indeterminate form
l'Hospital's rule

CHAPTER 12 REVIEW EXERCISES

Find Taylor polynomials of degree 4 at 0 for the functions defined as follows.

1. $f(x) = e^{2-x}$

2. $f(x) = 5e^{2x}$

3. $f(x) = \sqrt{x + 1}$

4. $f(x) = \sqrt[3]{x + 27}$

5. $f(x) = \ln(2 - x)$

6. $f(x) = \ln(3 + 2x)$

7. $f(x) = (1 + x)^{2/3}$

Use Taylor polynomials of degree 4 at $x = 0$, found in Exercises 1–7 above, to approximate the quantities in Exercises 8–14. Round to 4 decimal places.

8. $e^{1.99}$

9. $5e^{.02}$

10. $\sqrt{.98}$

11. $\sqrt[3]{26.98}$

12. $\ln 2.03$ ($\ln 2 = .69315$)

13. $\ln 2.96$ ($\ln 3 = 1.09861$)

14. $(1.01)^{2/3}$

Identify the geometric series that converge. Give the sum of each convergent series.

15. $9 - 6 + 4 - 8/3 + \cdots$

16. $2 + 1.4 + .98 + .686 + \cdots$

17. $3 + 9 + 27 + 81 + \cdots$

18. $4 + 4.8 + 5.76 + 6.912 + \cdots$

19. $\frac{2}{5} - \frac{2}{25} + \frac{2}{125} - \frac{2}{625} + \cdots$

20. $1 + \frac{1}{.99} + \frac{1}{(.99)^2} + \frac{1}{(.99)^3} + \cdots$

In Exercises 21–22, the nth term of a sequence is given. Calculate the first five partial sums.

21. $a_n = \frac{1}{2n - 1}$

22. $a_n = \frac{1}{(n + 2)(n + 3)}$

Use the Taylor series given in the text to find the Taylor series for the functions defined as follows. Give the interval of convergence of each series.

23. $f(x) = \frac{4}{3 - x}$

24. $f(x) = \frac{2x}{1 + 3x}$

25. $f(x) = \frac{x^2}{x + 1}$

26. $f(x) = \frac{3x^3}{2 - x}$

27. $f(x) = \ln(1 - 2x)$

28. $f(x) = \ln\left(1 + \frac{1}{3}x\right)$

29. $f(x) = e^{-2x^2}$

30. $f(x) = e^{-5x}$

31. $f(x) = 2x^3e^{-3x}$

32. $f(x) = x^6e^{-x}$

Use l'Hospital's rule, where applicable, to find each limit.

33. $\lim_{x \to 2} \frac{x^3 - x^2 - x - 2}{x^2 - 4}$

34. $\lim_{x \to 0} \frac{x^3 - 4x^2 + 6x}{3x}$

35. $\lim_{x \to -5} \frac{x^3 - 3x^2 + 4x - 1}{x^2 - 25}$

36. $\lim_{x \to 0} \frac{\ln(3x + 1)}{x}$

37. $\lim_{x \to 0} \frac{5e^x - 5}{x^3 - 8x^2 + 7x}$

38. $\lim_{x \to 0} \frac{\sqrt{5 + x} - \sqrt{5}}{x}$

39. $\lim_{x \to 0} \frac{-xe^{2x}}{e^{2x} - 1}$

40. $\lim_{x \to 16} \frac{\sqrt{x} - 4}{x - 16}$

41. $\lim_{x \to 0} \frac{1 + 2x - (1 + x)^{1/2}}{x^3}$

42. $\lim_{x \to 0} \frac{\sqrt{5 + x} - \sqrt{5 - x}}{2x}$

In Exercises 43–46, first get a common denominator; then find the limits that exist.

43. $\lim_{x \to 0}\left(\frac{1}{2x} - \frac{1}{3x}\right)$

44. $\lim_{x \to 0}\left(\frac{1}{x} + \frac{1}{x^2}\right)$

45. $\lim_{x \to 0}\left(\frac{1}{x} - \frac{1}{\sqrt{x}}\right)$

46. $\lim_{x \to 0}\left(\frac{1}{\sqrt[3]{x}} + \frac{1}{\sqrt{x}}\right)$

Use Newton's method to find a solution to the nearest hundredth for each equation in the given interval.

47. $x^3 - 8x^2 + 18x - 12 = 0$; $[4, 5]$

48. $3x^3 - 4x^2 - 4x - 7 = 0$; $[2, 3]$

49. $x^4 + 3x^3 - 4x^2 - 21x - 21 = 0$; $[2, 3]$

50. $x^4 + x^3 - 14x^2 - 15x - 15 = 0$; $[3, 4]$

Use Newton's method to approximate each radical to the nearest thousandth.

51. $\sqrt{46.8}$ **52.** $\sqrt{39.5}$ **53.** $\sqrt[3]{-123.8}$ **54.** $\sqrt[4]{70.9}$

Applications

BUSINESS AND ECONOMICS

55. *Total Income* A mine produced $600,000 of income during its first year. Each year thereafter, income increased by 20%. Find the total income produced in the first 5 years of the mine's life.

56. *Sinking Fund* In 4 years, Byron Hopkins must pay a pledge of $5000 to his church's building fund. He wants to set up a sinking fund to accumulate that amount. What should each semiannual payment into the fund be at 8% compounded semiannually?

57. *Annuity* Michelle Cook deposits $491 at the end of each quarter for 9 years. If the account pays 9.4% compounded quarterly, find the final amount in the account.

58. *Annuity* J. Euclid deposits $1526.38 at the end of each 6-month period in an account paying 7.6% compounded semiannually. How much will be in the account after 5 years?

59. *Amortization* Amy Schwartz borrows $20,000 from the bank to help her expand her business. She agrees to repay the money in equal payments at the end of each year for 9 years. Interest is at 8.9% compounded annually. Find the amount of each payment.

60. *Amortization* Robert Koch wants to expand his pharmacy. To do this, he takes out a bank loan of $49,275 and agrees to repay it at 12.2% compounded monthly over 48 months. Find the amount of each payment necessary to amortize this loan.

House Payments Find the monthly house payments for the following mortgages.

61. $156,890 at 7.74% for 25 years

62. $177,110 at 8.45% for 30 years

63. *Investment* John Hoffner has invested $15,000 in a certificate of deposit that has a 3.5% annual interest rate. Determine the doubling time for this investment using the doubling-time formula. How does this compare with the estimate given by the rule of 70?

64. *Investment* It is anticipated that a bank stock that David Chadwick has invested $15,000 in will achieve an annual interest rate of 10%. Determine the doubling time for this investment using the doubling-time formula. How does this compare with the estimate given by the rule of 72?

CHAPTER

13 The Trigonometric Functions

The time when the sun sets depends on both your location on the globe and on the time of year. Because Earth's motion around the Sun is periodic, the sunset time for a particular location is a periodic function of time measured in days. We explore this trigonometric model in the exercises for Section 1 in this chapter.

Throughout this book we have discussed many different types of functions, including linear, quadratic, exponential, and logarithmic functions. In this chapter we introduce the *trigonometric functions,* which differ in a fundamental way from those previously studied: the trigonometric functions describe periodic or repetitive relationships.

An example of a periodic relationship is given by an electrocardiogram (EKG), a graph of a human heartbeat. The EKG in Figure 1 shows electrical impulses from a heart.* Each small square represents .04 second, and each large square represents .2 second. How often does this heart beat?

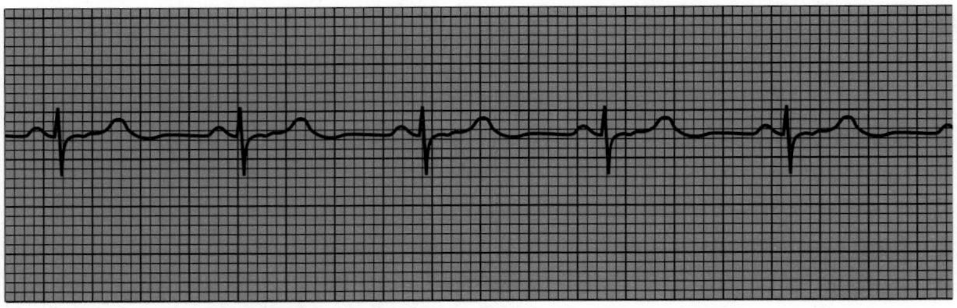

FIGURE 1

Trigonometric functions describe many natural phenomena and are important in the study of optics, heat, electronics, acoustics, and seismology. Also, many algebraic functions have integrals involving trigonometric functions

13.1 DEFINITIONS OF THE TRIGONOMETRIC FUNCTIONS

THINK ABOUT IT *How far from a camera should an object be to put it in focus?*

In Exercise 81 in this section, we will use trigonometry to answer this question.

The angle is one of the basic concepts of trigonometry. The definition of an angle depends on that of a ray: a **ray** is the portion of a line that starts at a given point and continues indefinitely in one direction. Figure 2 shows a line through the two points A and B. The portion of the line AB that starts at A and continues through and past B is called ray AB. Point A is the **endpoint** of the ray.

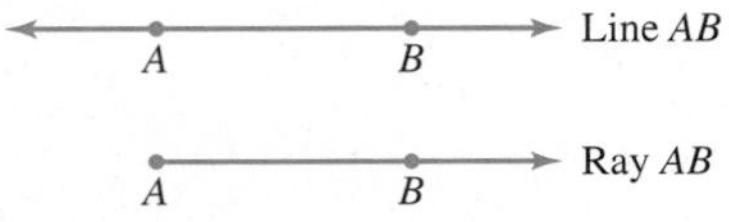

FIGURE 2

An **angle** is formed by rotating a ray about its endpoint. The initial position of the ray is called the **initial side** of the angle, and the endpoint of the ray is called the **vertex** of the angle. The location of the ray at the end of its rotation is called the **terminal side** of the angle. See Figure 3.

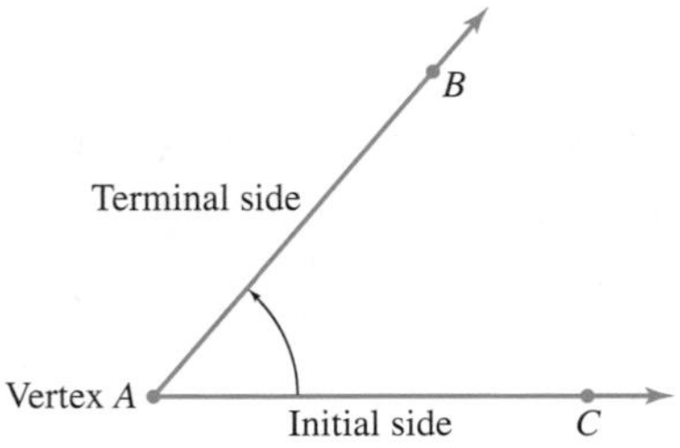

FIGURE 3

An angle can be named by its vertex. For example, the angle in Figure 3 can be called angle A. An angle also can be named by using three letters, with the vertex letter in the middle. For example, the angle in Figure 3 could be named angle BAC or angle CAB.

*EKG courtesy of Nancy Schiller, Phoenix, Arizona.

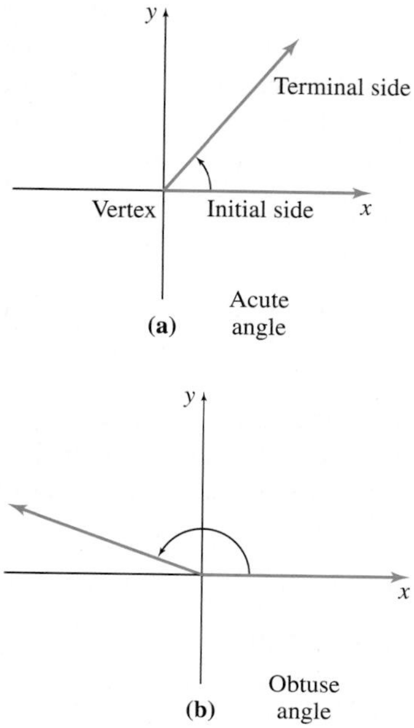

FIGURE 4

An angle is in **standard position** if its vertex is at the origin of a coordinate system and if its initial side is along the positive x-axis. The angles in Figures 4 and 5 are in standard position. An angle in standard position is said to be in the quadrant of its terminal side. For example, the angle in Figure 4(a) is in quadrant I, while the angle in Figure 4(b) is in quadrant II.

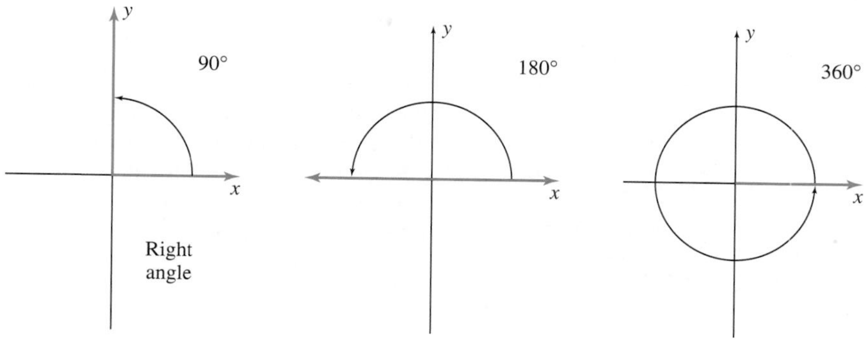

FIGURE 5

Notice that the angles in Figures 3 and 4 are formed with a counterclockwise rotation from the positive x-axis. This is true for any positive angle. A negative angle is measured clockwise from the positive x-axis, as we shall see in Example 5.

Degree Measure

The sizes of angles are often indicated in *degrees.* **Degree measure** has remained unchanged since the Babylonians developed it 4000 years ago. In degree measure, 360 degrees represents a complete rotation of a ray. *One degree,* written 1°, is 1/360 of a rotation. Also, 90° is 90/360 or 1/4 of a rotation, and 180° is 180/360 or 1/2 of a rotation. See Figure 5.

An angle having a degree measure between 0° and 90° is called an **acute angle.** An angle of 90° is a **right angle.** An angle having measure more than 90° but less than 180° is an **obtuse angle,** while an angle of 180° is a **straight angle.** See Figures 4 and 5.

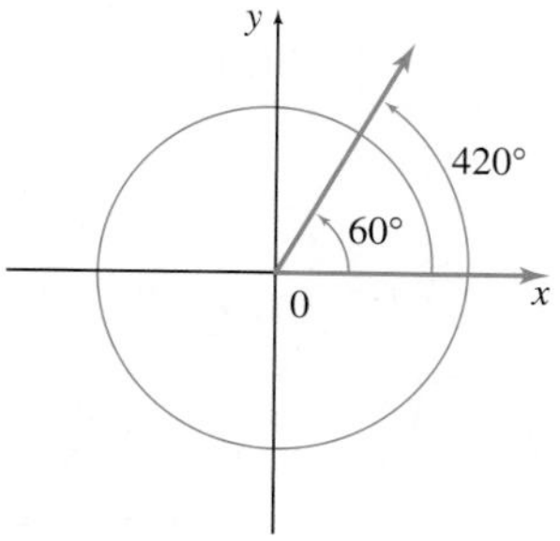

FIGURE 6

A complete rotation of a ray results in an angle of measure 360°. But there is no reason why the rotation need stop at 360°. By continuing the rotation, angles of measure larger than 360° can be produced. The angles in Figure 6 have measures 60° and 420°. These two angles have the same initial side and the same terminal side, but different amounts of rotation.

Radian Measure

While degree measure is best for some applications, this system of angle measurement is not the best for calculus. To keep the formulas for derivatives as simple as possible, it is better to measure angles with **radian measure.** To see how this alternative system for measuring angles is obtained, look at angle θ (the Greek letter *theta*) in Figure 7(a) on the next page. The angle θ is in standard position; Figure 7(a) also shows a circle of radius r, centered at the origin.

The vertex of θ is at the center of the circle in Figure 7(a). Angle θ cuts a piece of the circle called an **arc.** *One radian* is the measure of an angle that has its vertex at the center of a circle and that cuts an arc on the circle equal in length to the radius of the circle. Thus, the measure of angle θ is defined to be 1 radian.

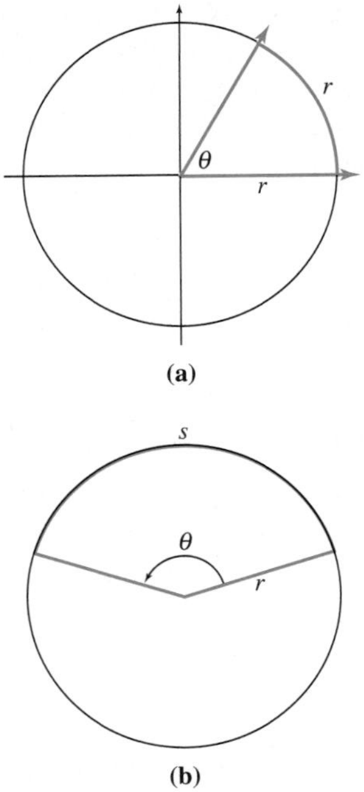

FIGURE 7

The term *radian* comes from the phrase *radial angle.* Two 19th century scientists, mathematician Thomas Muir and physicist James Thomson, are credited with the development of the radian as a unit of angular measure.

Generalizing, the radian measure of a central angle θ (see Figure 7(b)) cutting off an arc of length s in a circle of radius r is defined as follows:

$$\text{Radian measure of } \theta = \frac{\text{Length of arc}}{\text{Radius}} = \frac{s}{r}.$$

In this definition, the units of measure of the length of the arc and the radius cancel, leaving a quotient without units. For this reason, a real number can be thought of as the radian measure of some angle.

Since the circumference of a circle is 2π times the radius of the circle, the radius could be marked off 2π times around the circle. Therefore, an angle of 360°—that is, a complete circle—cuts off an arc equal in length to 2π times the radius of the circle, or

$$360° = 2\pi \text{ radians}.$$

This result gives a basis for comparing degree and radian measure.

Since an angle of 180° is half the size of an angle of 360°, an angle of 180° would have half the radian measure of an angle of 360°, or

$$180° = \frac{1}{2}(2\pi) \text{ radians} = \pi \text{ radians}.$$

$$\mathbf{180° = \pi \textbf{ radians}}$$

Since π radians = 180°, divide both sides by π to find the degree measure of 1 radian.

1 RADIAN

$$\mathbf{1 \textbf{ radian} = \left(\frac{180°}{\pi}\right)}$$

This quotient is approximately 57.29578°. Since 180° = π radians, we can find the radian measure of 1° by dividing by 180° on both sides.

1 DEGREE

$$\mathbf{1° = \frac{\pi}{180} \textbf{ radians}}$$

One degree is approximately equal to .0174533 radians.

Graphing calculators and many scientific calculators have the capability of changing from degree to radian measure or from radian to degree measure. If your calculator has this capability, you can practice using it with the angle measures in Example 1. *The most important thing to remember when using a calcula-*

tor to work with angle measures is to be sure the calculator mode is set for degrees or radians, as appropriate.

EXAMPLE 1 *Equivalent Angles*

Convert degree measures to radians and radian measure to degrees.

(a) 45°

Solution Since $1^\circ = \pi/180$ radians,

$$45^\circ = 45\left(\frac{\pi}{180}\right) \text{ radians} = \frac{45\pi}{180} \text{ radians} = \frac{\pi}{4} \text{ radians.}$$

The word *radian* is often omitted, so the answer could be written as just $45^\circ = \pi/4$.

(b) $\dfrac{9\pi}{4}$

Solution Since 1 radian $= 180^\circ/\pi$,

$$\frac{9\pi}{4} \text{ radians} = \frac{9\pi}{4}\left(\frac{180^\circ}{\pi}\right) = 405^\circ.$$

The following table shows the equivalent radian and degree measure for several angles that we will encounter frequently.

Degrees	0°	30°	45°	60°	90°	180°	270°	360°
Radians	0	$\pi/6$	$\pi/4$	$\pi/3$	$\pi/2$	π	$3\pi/2$	2π

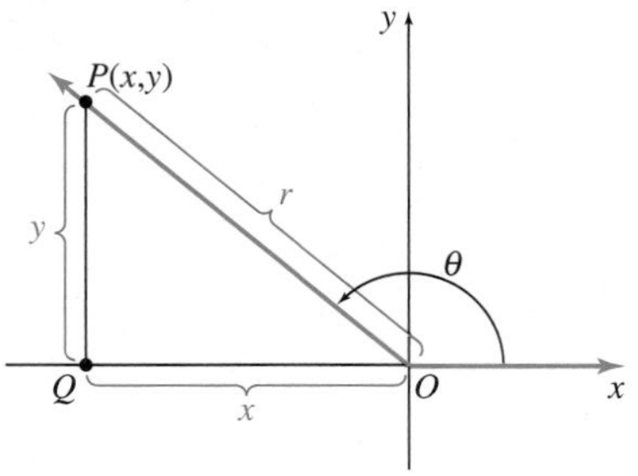

FIGURE 8

The Trigonometric Functions To define the six basic trigonometric functions, we start with an angle θ in standard position, as shown in Figure 8. Next, we choose an arbitrary point P having coordinates (x, y), located on the terminal side of angle θ. (The point P must not be the vertex of θ.)

Drawing a line segment perpendicular to the x-axis from P to point Q sets up a right triangle having vertices at O (the origin), P, and Q. The distance from P to O is r. Since the distance from P to O can never be negative, $r > 0$. The six **trigonometric functions** of angle θ are defined as follows.

TRIGONOMETRIC FUNCTIONS

Let (x, y) be a point other than the origin on the terminal side of an angle θ in standard position. Let r be the distance from the origin to (x, y). Then

$$\textbf{sine } \theta = \sin\theta = \frac{y}{r} \qquad \textbf{cosecant } \theta = \csc\theta = \frac{r}{y}$$

$$\textbf{cosine } \theta = \cos\theta = \frac{x}{r} \qquad \textbf{secant } \theta = \sec\theta = \frac{r}{x}$$

$$\textbf{tangent } \theta = \tan\theta = \frac{y}{x} \qquad \textbf{cotangent } \theta = \cot\theta = \frac{x}{y}.$$

From these definitions, it is easy to prove the following elementary trigonometric identities.

ELEMENTARY TRIGONOMETRIC IDENTITIES

$$\csc\theta = \frac{1}{\sin\theta} \qquad \sec\theta = \frac{1}{\cos\theta} \qquad \cot\theta = \frac{1}{\tan\theta} \qquad \tan\theta = \frac{\sin\theta}{\cos\theta}$$

$$\cot\theta = \frac{\cos\theta}{\sin\theta} \qquad \sin^2\theta + \cos^2\theta = 1$$

EXAMPLE 2 *Values of Trigonometric Functions*

The terminal side of an angle α (the Greek letter *alpha*) goes through the point $(8, 15)$. Find the values of the six trigonometric functions of angle α.

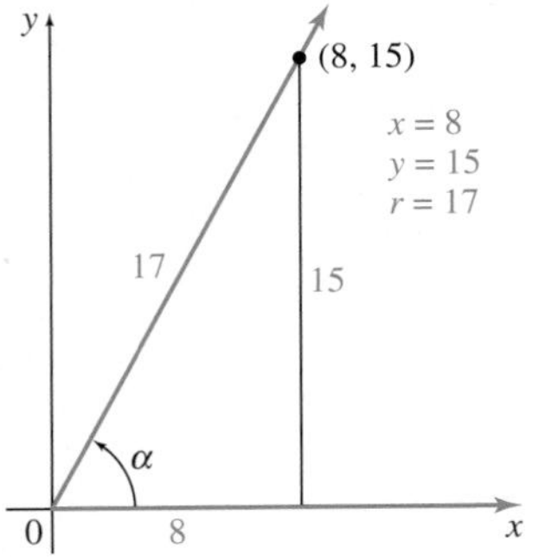

FIGURE 9

Solution Figure 9 shows angle α and the triangle formed by dropping a perpendicular from the point $(8, 15)$. To reach the point $(8, 15)$, begin at the origin and go 8 units to the right and 15 units up, so that $x = 8$ and $y = 15$. To find the radius r, use the Pythagorean theorem:* In a triangle with a right angle, if the longest side of the triangle is r and the shorter sides are x and y, then

$$r^2 = x^2 + y^2,$$

or

$$r = \sqrt{x^2 + y^2}.$$

(Recall that $\sqrt{a}$ represents the *positive* square root of a.)

Substituting the known values $x = 8$ and $y = 15$ in the equation gives

$$r = \sqrt{8^2 + 15^2} = \sqrt{64 + 225} = \sqrt{289} = 17.$$

We have $x = 8$, $y = 15$, and $r = 17$. The values of the six trigonometric functions of angle α are found by using the definitions given on the previous page.

$$\sin\alpha = \frac{y}{r} = \frac{15}{17} \qquad \tan\alpha = \frac{y}{x} = \frac{15}{8} \qquad \sec\alpha = \frac{r}{x} = \frac{17}{8}$$

$$\cos\alpha = \frac{x}{r} = \frac{8}{17} \qquad \cot\alpha = \frac{x}{y} = \frac{8}{15} \qquad \csc\alpha = \frac{r}{y} = \frac{17}{15}$$

EXAMPLE 3 *Values of Trigonometric Functions*

Find the values of the six trigonometric functions for an angle of $\pi/2$.

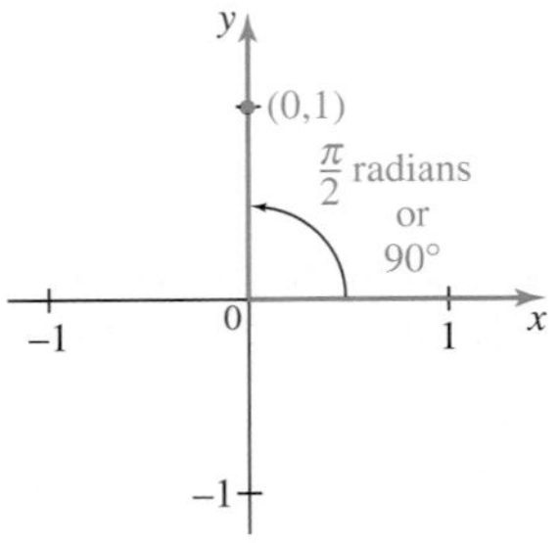

FIGURE 10

Solution Select any point on the terminal side of an angle of measure $\pi/2$ radians (or 90°). See Figure 10. Selecting the point $(0, 1)$ gives $x = 0$ and $y = 1$. Check that $r = 1$ also. Then

$$\sin\frac{\pi}{2} = \frac{1}{1} = 1 \qquad \tan\frac{\pi}{2} = \frac{1}{0}\ (\text{undefined}) \qquad \sec\frac{\pi}{2} = \frac{1}{0}\ (\text{undefined})$$

$$\cos\frac{\pi}{2} = \frac{0}{1} = 0 \qquad \cot\frac{\pi}{2} = \frac{0}{1} = 0 \qquad \csc\frac{\pi}{2} = \frac{1}{1} = 1.$$

*Although one of the most famous theorems in mathematics is named after the Greek mathematician Pythagoras, there is much evidence that the relationship between the sides of a right triangle was known long before his time. The Babylonian mathematical tablet identified as *Plimpton 322* has been determined to be essentially a list of *Pythagorean triples*—sets of three numbers a, b, and c that satisfy the equation $a^2 + b^2 = c^2$.

Methods similar to the procedure in Example 3 can be used to find the values of the six trigonometric functions for the angles with measure 0, π, and $3\pi/2$. These results are summarized in the following table. The table shows that the results for 2π are the same as those for 0.

θ (in radians)	θ (in degrees)	$\sin\theta$	$\cos\theta$	$\tan\theta$	$\cot\theta$	$\sec\theta$	$\csc\theta$
0	0°	0	1	0	Undefined	1	Undefined
$\pi/2$	90°	1	0	Undefined	0	Undefined	1
π	180°	0	-1	0	Undefined	-1	Undefined
$3\pi/2$	270°	-1	0	Undefined	0	Undefined	-1
2π	360°	0	1	0	Undefined	1	Undefined

NOTE When considering the trigonometric functions, it is customary to use x (rather than θ) for the domain elements, as we did with earlier functions, and to write $y = \sin x$ instead of $y = \sin\theta$. ■

Special Angles The values of the trigonometric functions for most angles must be found by using a calculator with trigonometric keys. For a few angles called **special angles,** however, the function values can be found exactly. These values are found with the aid of two kinds of right triangles that will be described in this section.

30°–60°–90° TRIANGLE

In a right triangle having angles of 30°, 60°, and 90°, the hypotenuse is always twice as long as the shortest side, and the middle side has a length that is $\sqrt{3}$ times as long as that of the shortest side. Also, the shortest side is opposite the 30° angle.

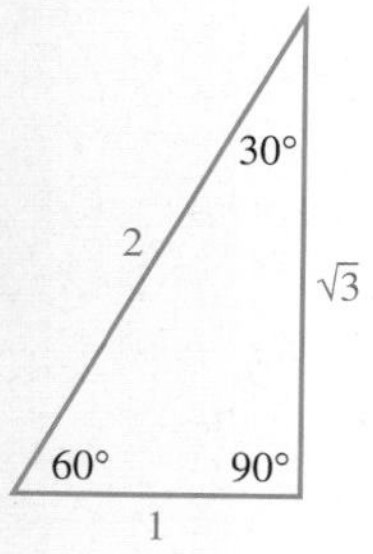

EXAMPLE 4 *Values of Trigonometric Functions*

Find the values of the trigonometric functions for an angle of $\pi/6$ radians.

Solution Since $\pi/6$ radians $= 30°$, find the necessary values by placing a 30° angle in standard position, as in Figure 11. Choose a point P on the terminal side of the angle so that $r = 2$. From the description of 30°–60°–90° triangles, P will have coordinates $(\sqrt{3}, 1)$, with $x = \sqrt{3}$, $y = 1$, and $r = 2$. Using the definitions of the trigonometric functions gives the following results.

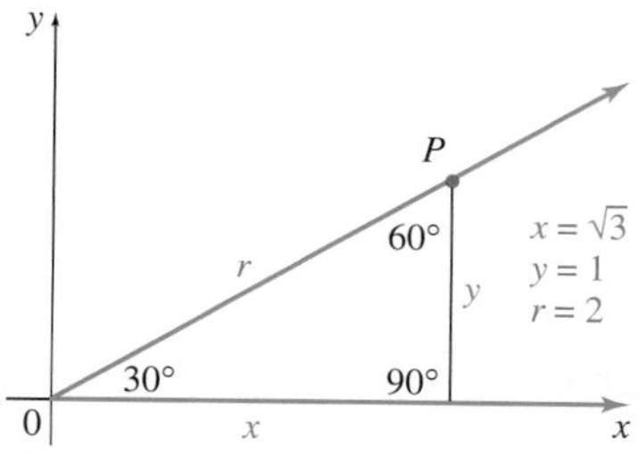

FIGURE 11

$$\sin\frac{\pi}{6} = \frac{1}{2} \qquad \tan\frac{\pi}{6} = \frac{1}{\sqrt{3}} = \frac{\sqrt{3}}{3} \qquad \sec\frac{\pi}{6} = \frac{2}{\sqrt{3}} = \frac{2\sqrt{3}}{3}$$

$$\cos\frac{\pi}{6} = \frac{\sqrt{3}}{2} \qquad \cot\frac{\pi}{6} = \sqrt{3} \qquad \csc\frac{\pi}{6} = 2$$

We can find the trigonometric function values for 45° angles by using the properties of a right triangle having two sides of equal length.

45°–45°–90° TRIANGLE

In a 45°–45° right triangle, the hypotenuse has a length that is $\sqrt{2}$ times as long as the length of either of the shorter (equal) sides.

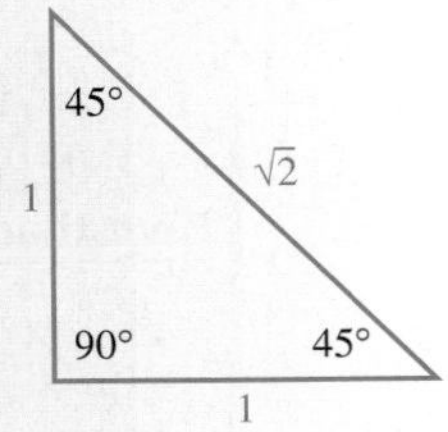

EXAMPLE 5 *Values of Trigonometric Functions*
Find the trigonometric function values for an angle of $-\pi/4$.

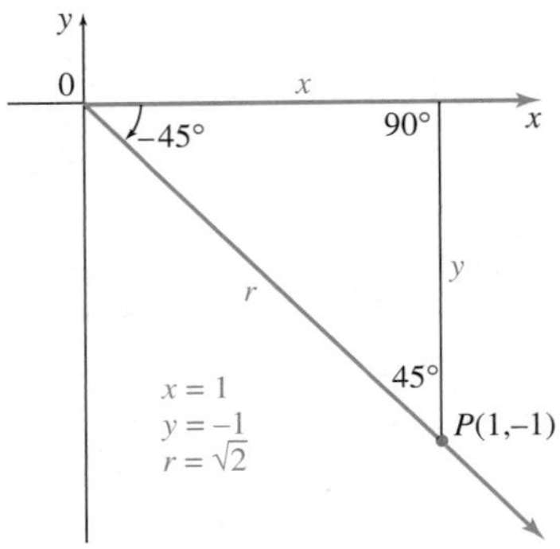

Solution Place an angle of $-\pi/4$ radians, or $-45°$, in standard position, as in Figure 12. Choose point P on the terminal side so that $r = \sqrt{2}$. By the description of $45°-45°-90°$ triangles, P has coordinates $(1, -1)$, with $x = 1$, $y = -1$, and $r = \sqrt{2}$.

$$\sin\left(-\frac{\pi}{4}\right) = -\frac{1}{\sqrt{2}} = -\frac{\sqrt{2}}{2} \qquad \tan\left(-\frac{\pi}{4}\right) = -1 \qquad \sec\left(-\frac{\pi}{4}\right) = \sqrt{2}$$

$$\cos\left(-\frac{\pi}{4}\right) = \frac{1}{\sqrt{2}} = \frac{\sqrt{2}}{2} \qquad \cot\left(-\frac{\pi}{4}\right) = -1 \qquad \csc\left(-\frac{\pi}{4}\right) = -\sqrt{2}$$

FIGURE 12

For angles other than the special angles of 30°, 45°, 60°, and their multiples, a calculator should be used. Many calculators have keys labeled sin, cos, and tan. To get the other trigonometric functions, use the fact that $\sec x = 1/\cos x$, $\csc x = 1/\sin x$, and $\cot x = 1/\tan x$. (The x^{-1} key is also useful here.)

CAUTION Whenever you use a calculator to compute trigonometric functions, check whether the calculator is set on radians or degrees. If you want one and your calculator is set on the other, you will get erroneous answers. Most calculators have a way of switching back and forth; check the calculator manual for details. On the TI-83/84 Plus, press the MODE button and then select Radian or Degree. ■

EXAMPLE 6 *Values of Trigonometric Functions*
Use a calculator to verify the following results.

(a) $\sin 10° \approx .1736$

(b) $\cos 48° \approx .6691$

(c) $\tan 82° \approx 7.1154$

(d) $\sin .2618 \approx .2588$

(e) $\cot 1.2043 = 1/\tan 1.2043 \approx 1/2.6053 \approx .3838$

(f) $\sec r.7679 = 1/\cos r.7679 \approx 1/.71937 \approx 1.3901$

Graphs of the Trigonometric Functions Because of the way the trigonometric functions are defined (using a circle), the same function values will be obtained for any two angles that differ by 2π radians (or 360°). For example,

$$\sin(x + 2\pi) = \sin x \qquad \text{and} \qquad \cos(x + 2\pi) = \cos x$$

for any value of x. Because of this property, the trigonometric functions are *periodic functions.*

PERIODIC FUNCTION

A function $y = f(x)$ is **periodic** if there exists a positive real number a such that

$$f(x) = f(x + a)$$

for all values of x in the domain of the function. The smallest positive value of a is called the **period** of the function.*

Intuitively, a function with period a repeats itself over intervals of length a. Once we know what the graph looks like over one period of length a, we know what the entire graph looks like by simply repeating. Because sine is periodic with period 2π, the graph is found by first finding the graph on the interval between 0 and 2π and then repeating as many times as necessary.

To find values of $y = \sin x$ for values of x between 0 and 2π, think of a point moving counterclockwise around a circle, tracing out an arc for angle x. The value of $\sin x$ gradually increases from 0 to 1 as x increases from 0 to $\pi/2$. The value of $\sin x$ then decreases back to 0 as x goes from $\pi/2$ to π. For $\pi < x < 2\pi$, $\sin x$ is negative. A few typical values from these intervals are given in the following table, where decimals have been rounded to the nearest tenth.

x	0	$\pi/4$	$\pi/2$	$3\pi/4$	π	$5\pi/4$	$3\pi/2$	$7\pi/4$	2π
$\sin x$	0	.7	1	.7	0	−.7	−1	−.7	0

Plotting the points from the table of values and connecting them with a smooth curve gives the solid portion of the graph in Figure 13. Since $y = \sin x$ is periodic, the graph continues in both directions indefinitely, as suggested by the dashed lines. The solid portion of the graph in Figure 13 gives the graph over one period.

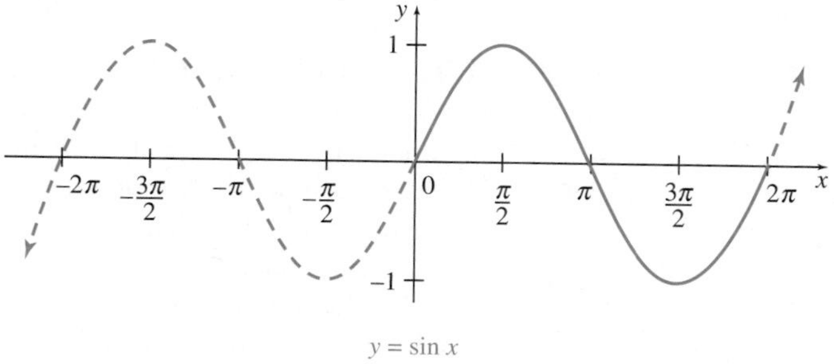

FIGURE 13

*Some authors define the period of the function as any value of a that satisfies $f(x) = f(x + a)$.

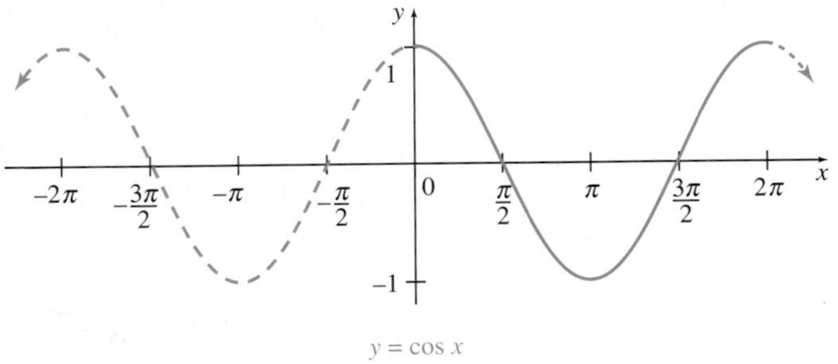

FIGURE 14

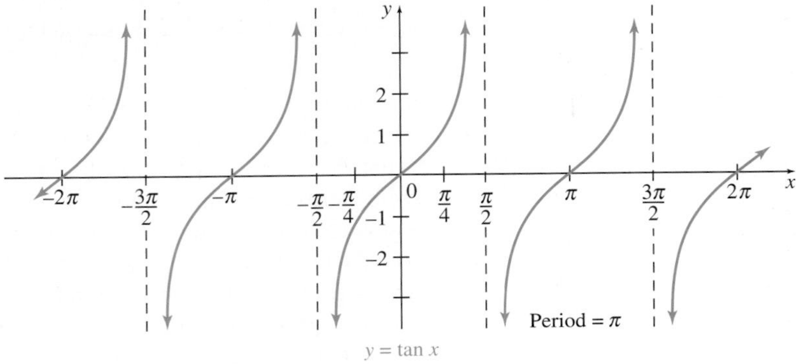

FIGURE 15

The graph of $y = \cos x$ in Figure 14 above can be found in much the same way. Again, the period is 2π. (These graphs could also be drawn using a graphing calculator or a computer.)

Finally, Figure 15 shows the graph of $y = \tan x$. Since $\tan x$ is undefined (because of zero denominators) for $x = \pi/2, 3\pi/2, -\pi/2$, and so on, the graph has vertical asymptotes at these values. As the graph suggests, the tangent function is periodic, with a period of π.

The graphs of the secant, cosecant, and cotangent functions are not used as often as these three, so they are not given here.

Translating Graphs of Sine and Cosine Functions In an earlier section we saw that the graph of the function $y = f(x + c) + d$ was simply the graph of $y = f(x)$ translated c units horizontally and d units vertically. The same facts hold true with trigonometric functions. The constants a, b, c, and d affect the graphs of the functions $y = a\sin(bx + c) + d$ and $y = a\cos(bx + c) + d$ in a similar manner.

In addition, the constants a, b, and c have particular properties. Since the sine and cosine functions range between -1 and 1, the value of a, whose absolute value is called the **amplitude,** can be interpreted as half the difference between the maximum and minimum values of the function.

The period of the function is determined by the constant b, which we will assume to be greater than 0. Recall that the period of both $\sin x$ and $\cos x$ is 2π. The value of $b > 0$ will increase or decrease the period, depending on its value.

A similar phenomenon occurs when $b < 0$, but it is not covered in this textbook. Thus, the graph of $y = \sin(bx)$ will look like that of $y = \sin x$, but with a period of $T = 2\pi/b$. The results are similar for $y = \cos(bx)$.

The quantity c/b is called the **phase shift** and corresponds to the number of units that the graph of $\sin bx$ or $\cos bx$ is shifted horizontally. The constant d determines the vertical shift of $\sin x$ or $\cos x$.

EXAMPLE 7 *Graphing Trigonometric Functions*

Graph each function.

(a) $y = \sin 3x$

Solution The graph of this function has amplitude $a = 1$ and no vertical or horizontal shifts. The period of this function is $T = 2\pi/b = 2\pi/3$. Hence, the graph of $y = \sin 3x$ is the same as $y = \sin x$ except that the period is different. See Figure 16.

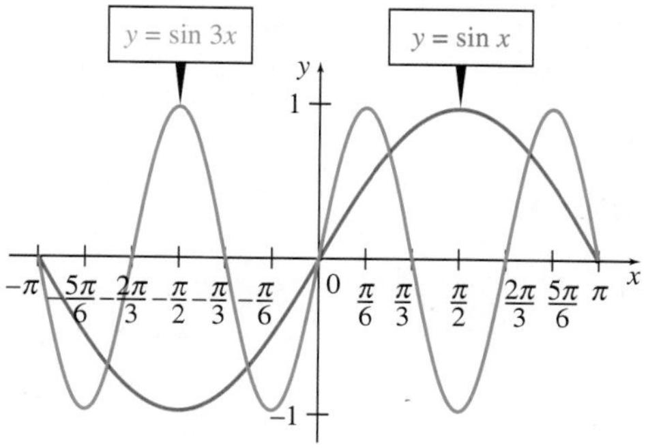

FIGURE 16

(b) $f(x) = 4\cos\left(\frac{1}{2}x + \frac{\pi}{4}\right) - 1$

Solution The amplitude is $a = 4$. The graph of $f(x)$ is shifted down 1 unit vertically. The phase shift is $c/b = (\pi/4)/(1/2) = \pi/2$. This shifts the graph $\pi/2$ units to the left, relative to the graph $g(x) = \cos((1/2)x)$. The period of $f(x)$ is $2\pi/(1/2) = 4\pi$. Making these translations on $y = \cos x$ leads to Figure 17.

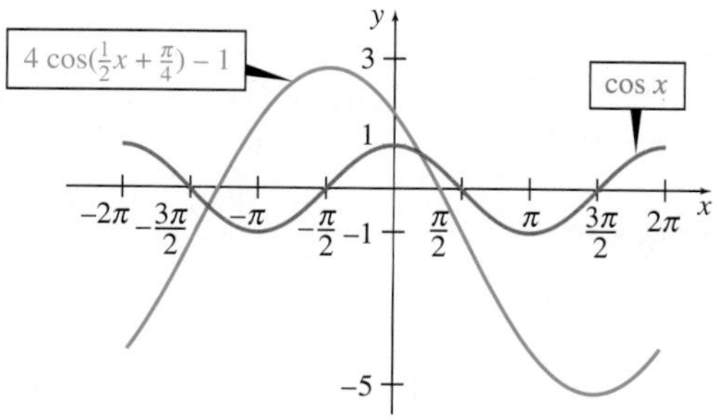

FIGURE 17

EXAMPLE 8 *Music*

A change in pressure on the eardrum occurs when a pure musical tone is played. For some tones, the pressure on the eardrum follows the sine curve

$$P(t) = .004 \sin\left(2\pi f t + \frac{\pi}{7}\right),$$

where P is the pressure in pounds per square foot at time t seconds and f is the frequency on the sound wave in cycles per second.* When $P(t)$ is positive there is an increase in pressure and the eardrum is pushed inward; when $P(t)$ is negative there is a decrease in pressure and the eardrum is pushed outward.

(a) Graph the pressure on the eardrum for Middle C, which has a frequency of $f = 261.63$ cycles per second, on $[0,r.005]$.

Solution A graphing calculator graph of this function is given in Figure 18.

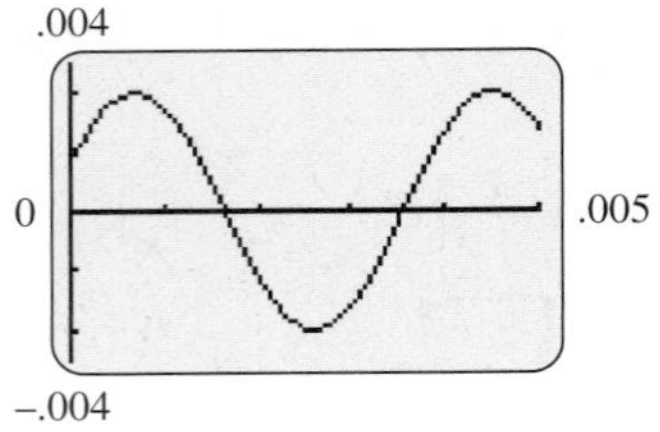

FIGURE 18

(b) Determine analytically the values of t for which $P = 0$ on $[0,r.005]$.

Solution Since the sine function is zero for multiples of π, we can determine the value(s) of t where $P = 0$ by setting $523.26\pi t + \pi/7 = n\pi$, where n is an integer, and solving for t. After some algebraic manipulations,

$$t = \frac{n - \frac{1}{7}}{523.26}$$

and $P = 0$ when $n = 0, \pm 1, \pm 2, \ldots$. However, only values of $n = 1$ or $n = 2$ produce values of t that lie in the interval $[0,r.005]$. Thus, $P = 0$ when $t \approx .0016$ and $.0035$, corresponding to $n = 1$ and $n = 2$, respectively.

(c) Determine the period T of $P(t)$. What is the relationship between the period and frequency of the tone?

Solution The period is $T = 2\pi/b = 2\pi/(523.26\pi) = 1/261.63 \approx .004$. This implies that the period of the pressure equation is the reciprocal of the frequency. That is, $T = 2\pi/b = 2\pi/(2\pi f) = 1/f$.

EXAMPLE 9 *Sunrise*

The table on the next page lists the approximate number of minutes after midnight, Eastern Standard Time, that the sun rises in Boston for specific days of the year.†

(a) Plot the data. Is it reasonable to assume that the times of sunrise are periodic?

Solution Figure 19 shows a graphing calculator plot of the data. Because of the cyclical nature of the days of the year, it is reasonable to assume that the data are periodic.

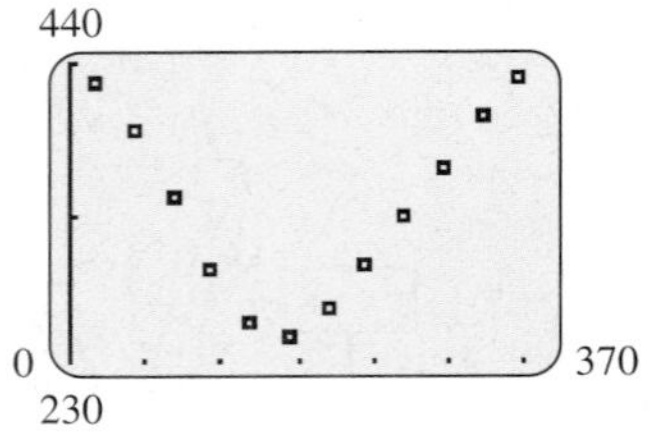

FIGURE 19

(b) Find a trigonometric function of the form $s(x) = a \sin(bx + c) + d$ that models this data when x is the day of the year and $s(x)$ is the number of minutes past midnight, Eastern Standard Time, that the sun rises. Use the data from the table on the next page.

*Roederer, Juan, *The Physics and Psychophysics of Music: An Introduction,* New York: Springer-Verlag, 1995.

†Thomas, Robert, *The Old Farmer's Almanac,* 2000.

Day of the Year	Sunrise (minutes after midnight)
21	428
52	393
81	345
112	293
142	257
173	247
203	266
234	298
265	331
295	365
326	403
356	431

Solution The function $s(x)$, derived by a TI-83/84 Plus using the sine regression function under the STAT-CALC menu, is given by

$$s(x) = 92.1414 \sin(.016297x + 1.80979) + 342.934.$$

Figure 20 shows that this function fits the data well.

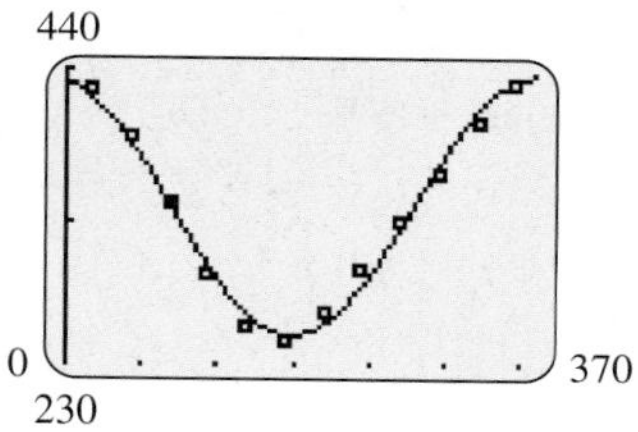

FIGURE 20

(c) Estimate the time of sunrise for days 30, 90, and 240. Round answers to the nearest minute.

Solution

$$\begin{aligned} s(30) &= 92.1414 \sin(.016297(30) + 1.80979) + 342.934 \\ &= 412 \text{ minutes} = 412/60 \text{ hours} = 6.867 \text{ hours} \\ &= 6 \text{ hours} + .867(60) \text{ minutes} = 6{:}52 \text{ A.M.} \end{aligned}$$

Similarly,

$$\begin{aligned} s(90) &= 333 \text{ minutes} = 5{:}33 \text{ A.M.} \\ s(240) &= 288 \text{ minutes} + 60 \text{ minutes (daylight savings)} \\ &= 5{:}48 \text{ A.M.} \end{aligned}$$

(d) Estimate the days of the year that the sun rises at 5:45 A.M.

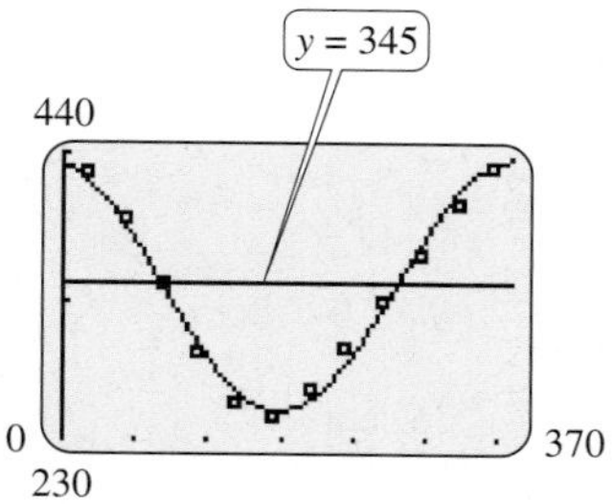

FIGURE 21

Solution Figure 21 shows the graphs of $s(x)$ and $y = 345$ (corresponding to a sunrise of 5:45 A.M.). These graphs first intersect on day 80. However, because of daylight savings time, to find the second value we find where the graphs of $s(x)$ and $y = 345 - 60 = 285$ intersect. These graphs intersect on day 233. Thus, the sun rises at approximately 5:45 A.M. on the 80th and 233rd days of the year.

13.1 EXERCISES

Convert the following degree measures to radians. Leave answers as multiples of π.

1. $60°$ **2.** $90°$ **3.** $150°$ **4.** $135°$

5. $210°$ **6.** $300°$ **7.** $390°$ **8.** $480°$

Convert the following radian measures to degrees.

9. $\frac{7\pi}{4}$ **10.** $\frac{2\pi}{3}$ **11.** $\frac{11\pi}{6}$ **12.** $-\frac{\pi}{4}$

13. $\frac{8\pi}{5}$ **14.** $\frac{7\pi}{10}$ **15.** $\frac{4\pi}{15}$ **16.** 5π

Find the values of the six trigonometric functions for the angles in standard position having the points in Exercises 17–20 on their terminal sides.

17. $(-3, 4)$ **18.** $(-12, -5)$ **19.** $(6, 8)$ **20.** $(-7, 24)$

In quadrant I, x, y, and r are all positive, so that all six trigonometric functions have positive values. In quadrant II, x is negative and y is positive (r is always positive). Thus, in quadrant II, sine is positive, cosine is negative, and so on. For Exercises 21–24, complete the following table of values for the signs of the trigonometric functions.

	Quadrant of θ	$\sin\theta$	$\cos\theta$	$\tan\theta$	$\cot\theta$	$\sec\theta$	$\csc\theta$
21.	I	+					
22.	II						
23.	III						
24.	IV						

For Exercises 25–32, complete the following table. Use the 30°–60°–90° and 45°–45°–90° triangles. Do not use a calculator.

	θ	$\sin\theta$	$\cos\theta$	$\tan\theta$	$\cot\theta$	$\sec\theta$	$\csc\theta$
25.	30°	$1/2$	$\sqrt{3}/2$			$2\sqrt{3}/3$	
26.	45°			1	1		
27.	60°		$1/2$	$\sqrt{3}$		2	
28.	120°	$\sqrt{3}/2$		$-\sqrt{3}$			$2\sqrt{3}/3$
29.	135°	$\sqrt{2}/2$	$-\sqrt{2}/2$			$-\sqrt{2}$	$\sqrt{2}$
30.	150°		$-\sqrt{3}/2$	$-\sqrt{3}/3$			2
31.	210°	$-1/2$		$\sqrt{3}/3$	$\sqrt{3}$		-2
32.	240°	$-\sqrt{3}/2$	$-1/2$			-2	$-2\sqrt{3}/3$

Find the following function values without using a calculator.

33. $\sin\frac{\pi}{3}$ **34.** $\cos\frac{\pi}{6}$ **35.** $\tan\frac{\pi}{4}$ **36.** $\cot\frac{\pi}{3}$

37. $\sec\frac{\pi}{6}$ **38.** $\sin\frac{\pi}{2}$ **39.** $\cos 3\pi$ **40.** $\sec\pi$

41. $\sin\frac{4\pi}{3}$ **42.** $\tan\frac{3\pi}{4}$ **43.** $\csc\frac{5\pi}{4}$ **44.** $\cos 5\pi$

45. $\tan -\frac{\pi}{3}$ **46.** $\cot -\frac{2\pi}{3}$ **47.** $\sin -\frac{7\pi}{6}$ **48.** $\cos -\frac{\pi}{6}$

Use a calculator to find the following function values.

49. sin 39° **50.** cos 58° **51.** tan 82° **52.** tan 54°

53. sin .4014 **54.** tan 1.0123 **55.** cos 1.4137 **56.** sin 1.5359

Find the amplitude (a) and period (T) of each function.

57. $f(x) = \cos(3x)$

58. $g(t) = 5\sin\left(\frac{\pi}{6}t - 2\right)$

59. $s(x) = 3\sin(880\pi t - 7)$

Graph each function defined as follows over a two-period interval.

60. $y = 2\sin x$

61. $y = 2\cos x$

62. $y = -\sin x$

63. $y = -\frac{1}{2}\cos x$

64. $y = 3\cos\left(2x - \frac{\pi}{2}\right) + 1$

65. $y = 4\sin\left(\frac{1}{2}x + \pi\right) + 2$

66. $y = \frac{1}{2}\tan x$

67. $y = -3\tan x$

Applications

BUSINESS AND ECONOMICS

68. *Sales* Sales of snowblowers are seasonal. Suppose the sales of snowblowers in one region of the country are approximated by

$$S(t) = 500 + 500\cos\left(\frac{\pi}{6}t\right),$$

where t is time in months, with $t = 0$ corresponding to November. Find the sales for a–e.

a. November **b.** January **c.** February

d. May **e.** August **f.** Graph $y = S(t)$.

LIFE SCIENCES

69. *Monkey Eyes* In a study of how monkeys' eyes pursue a moving object, an image was moved sinusoidally through a monkey's field of vision with an amplitude of 2° and a period of .350 seconds.*

a. Find an equation giving the position of the image in degrees as a function of time in seconds.

b. After how many seconds does the image reach its maximum amplitude?

c. What is the position of the object after 2 seconds?

70. *Transylvania Hypothesis* The "Transylvania hypothesis" claims that the full moon has an effect on health-related behavior. A study investigating this effect found a significant relationship between the phase of the moon and the number of general practice consultations nationwide, given by

$$y = 100 + 1.8\cos\left[\frac{(x - 6)\pi}{14.77}\right],$$

where y is the number of consultations as a percentage of the daily mean and x is the days since the last full moon.†

a. What is the period of this function? What is the significance of this period?

b. There was a full moon on December 8, 2003. On what day in December 2003 does this formula predict the maximum number of consultations? What percent increase would be predicted for that day?

c. What does the formula predict for December 21, 2003?

71. *Air Pollution* The amount of pollution in the air fluctuates with the seasons. It is lower after heavy spring rains and higher after periods of little rain. In addition to this seasonal fluctuation, the long-term trend in many areas is upward. An idealized graph of this situation is shown in the figure. Trigonometric functions can be used to describe the fluctuating part of the pollution levels. Powers of the number e can be used to show the long-term growth. In fact, the pollution level in a certain area might be given by

$$P(t) = 7(1 - \cos 2\pi t)(t + 10) + 100e^{.2t},$$

where t is time in years, with $t = 0$ representing January 1 of the base year. Thus, July 1 of the same year would be represented by $t = .5$, while October 1 of the following year would be represented by $t = 1.75$. Find the pollution levels on the following dates.

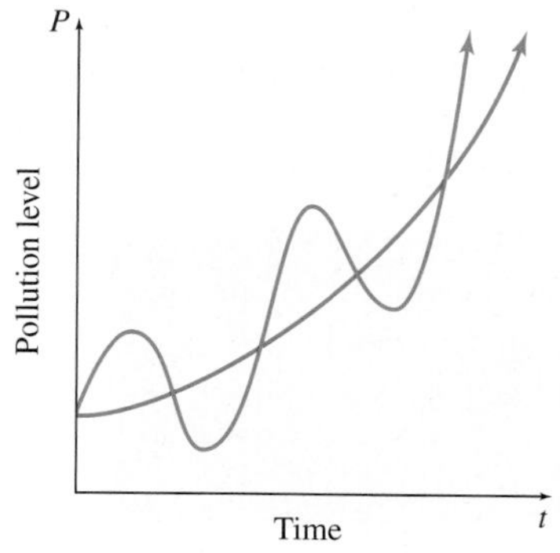

*Churchland, M. M. and S. G. J. Lisberger, "Experimental and Computational Analysis of Monkey Smooth Pursuit Eye Movements," *Journal of Neurophysiology,* Vol. 86, No. 2, Aug. 2001, pp. 741–759.

†Neal, R. D. and M. Colledge, "The Effect of the Full Moon on General Practice Consultation Rates," *Family Practice,* Vol. 17, No. 6, Dec. 2000, pp. 472–474.

a. January 1, base year

b. July 1, base year

c. January 1, following year

d. July 1, following year

72. *Air Pollution* Using a computer or a graphing calculator, sketch the function for air pollution given in Exercise 71 over the interval [0, 6].

PHYSICAL SCIENCES

Light Rays *When a light ray travels from one medium, such as air, to another medium, such as water or glass, the speed of the light changes, and the direction that the ray is traveling changes. (This is why a fish under water is in a different position from the place at which it appears to be.) These changes are given by Snell's law,*

$$\frac{c_1}{c_2} = \frac{\sin \theta_1}{\sin \theta_2},$$

where c_1 is the speed in the first medium, c_2 is the speed in the second medium, and θ_1 and θ_2 are the angles shown in the figure.

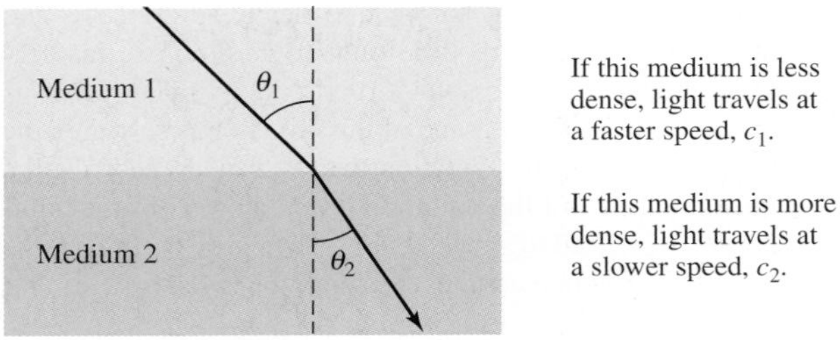
Medium 1
θ1
Medium 2
θ2
If this medium is less dense, light travels at a faster speed, c1.
If this medium is more dense, light travels at a slower speed, c2.

In Exercises 73 and 74, assume that $c_1 = 3 \times 10^8$ m per second, and find the speed of light in the second medium.

73. $\theta_1 = 39°, \theta_2 = 28°$

74. $\theta_1 = 46°, \theta_2 = 31°$

Sound *Pure sounds produce single sine waves on an oscilloscope. Find the period of each sine wave in the photographs in Exercises 75 and 76. On the vertical scale each square represents .5, and on the horizontal scale each square represents 30°.*

75.

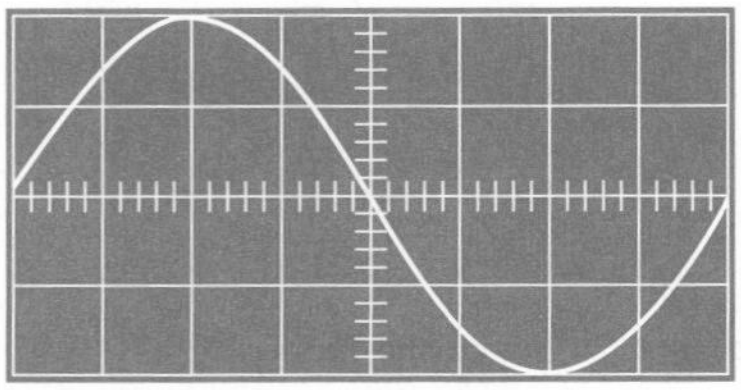

76.

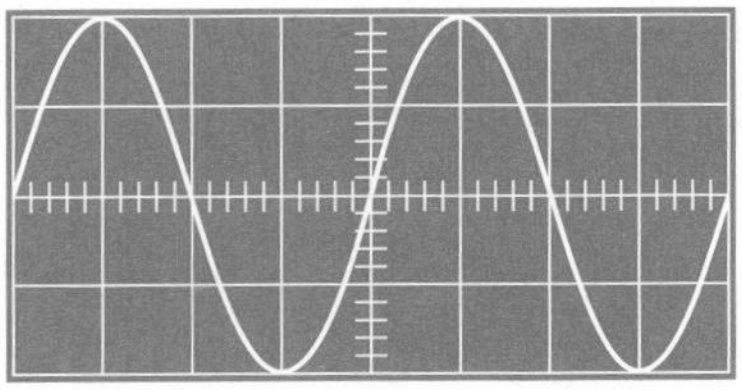

77. *Sound* Suppose the A key above Middle C is played as a pure tone. For this tone,

$$P(t) = .002 \sin(880\pi t),$$

where $P(t)$ is the change of pressure (in pounds per square foot) on a person's eardrum at time t (in seconds).*

a. Graph this function on [0,r.003].

b. Determine analytically the values of t for which $P = 0$ on [0,r.003] and check graphically.

c. Determine the period T of $P(t)$ and the frequency of the A note.

78. *Temperature* The maximum afternoon temperature (in degrees Fahrenheit) in a given city is approximated by

$$T(x) = 60 - 30 \cos(x/2),$$

where x represents the month, with $x = 0$ representing January, $x = 1$ representing February, and so on. Use a calculator to find the maximum afternoon temperature for the following months.

a. January **b.** March **c.** October

d. June **e.** August

79. *Temperature* A mathematical model for the temperature in Fairbanks is

*Roederer, Juan, *The Physics and Psychophysics of Music: An Introduction,* New York: Springer-Verlag, 1995.

$$T(x) = 37 \sin\left[\frac{2\pi}{365}(x - 101)\right] + 25,$$

where $T(x)$ is the temperature (in degrees Celsius) on day x, with $x = 0$ corresponding to January 1 and $x = 365$ corresponding to December 31.*

Use a calculator to estimate the temperature for a–d.

a. March 1 (Day 60) **b.** April 1 (Day 91)

c. Day 101 **d.** Day 150

e. Find maximum and minimum values of T.

f. Find the period, T.

80. *Sunset* The number of minutes after noon, Eastern Standard Time, that the sun sets in Boston for specific days of the year is approximated in the following table.†

Day of the Year	Sunset (minutes after noon)
21	283
52	323
81	358
112	393
142	425
173	445
203	434
234	396
265	343
295	292
326	257
356	255

a. Plot the data. Is it reasonable to assume that the times of sunset are periodic?

b. Use a calculator with trigonometric regression to find a trigonometric function of the form $s(x) = a \sin(bx + c) + d$ that models this data when x is the day of the year and $s(x)$ is the number of minutes past noon, Eastern Standard Time, that the sun sets.

c. Estimate the time of sunset for days 60, 120, 240. Round answers to the nearest minute. (*Hint:* Don't forget about daylight savings time.)

d. Use part b to estimate the days of the year that the sun sets at 6:00 P.M. In reality, the days are close to 82 and 290.

81. *Cameras* In the Kodak Customer Service Pamphlet AA-26, *Optical Formulas and Their Applications,* the near and far limits of the depth of field (how close or how far away an object can be placed and still be in focus) are given by

$$w_1 = \frac{u^2(\tan\theta)}{L + u(\tan\theta)} \quad \text{and} \quad w_2 = \frac{u^2(\tan\theta)}{L - u(\tan\theta)}.$$

In these equations, θ represents the angle between the lens and the "circle of confusion," which is the circular image on the film of a point that is not exactly in focus. (The pamphlet suggests letting $\theta = 1/30°$.) L is the diameter of the lens opening, which is found by dividing the focal length by the f-stop. (This is camera jargon you need not worry about here.) For this problem, let the focal length be 50 mm, or .05 m; if the lens is set at f/8, then $L = .05/8 = .00625$ m. Finally, u is the distance to the object being photographed. Find the near and far limits of the depth of field when the object being photographed is 6 m from the camera.

82. *Measurement* A surveyor standing 48 m from the base of a building measures the angle to the top of the building and finds it to be 37.4°. (See the figure.) Use trigonometry to find the height of the building.

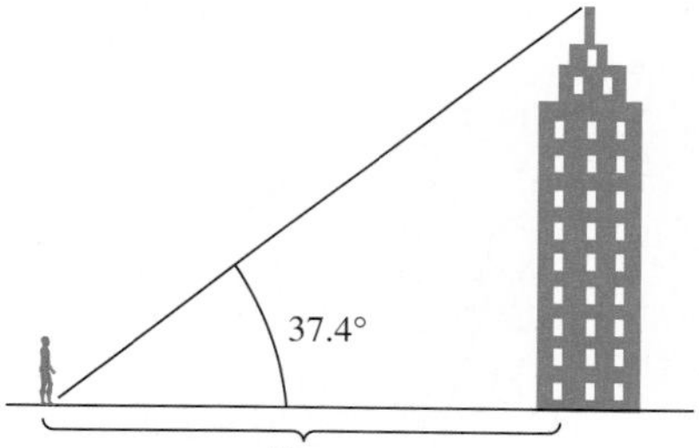

83. *Measurement* Phyllis Crittenden stands on a cliff at the edge of a canyon. On the opposite side of the canyon is another cliff equal in height to the one she is on. (See the figure at the top of the next page.) By dropping a rock and timing its fall, she determines that it is 80 ft to the bottom of the canyon. She also determines that the angle to the base of the opposite cliff is 24°. How far is it to the opposite side of the canyon?

84. *Whitewater Rafting* A mathematics textbook author rafting down the Colorado River was told by a guide that the river dropped an average of 26 ft per mile as it ran through Cataract Canyon. Find the average angle of the river with the horizontal in degrees. (*Hint:* Find the tangent of the

*Lando, Barbara and Clifton Lando, "Is the Graph of Temperature Variation a Sine Curve?" *Mathematics Teacher,* Vol. 70, Sept. 1977, pp. 534–537.
†Thomas, Robert, *The Old Farmer's Almanac,* 2000.

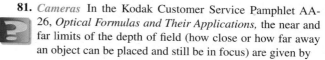

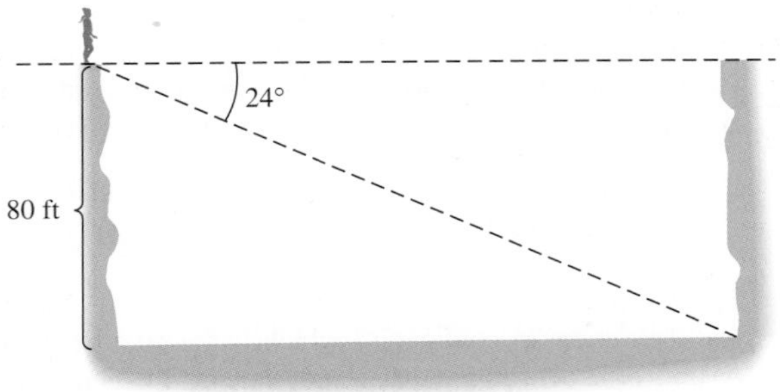

angle, and then use a calculator to find the angle where the tangent has that value. There are 5280 ft in a mile. Be sure your calculator is set on degrees.)

85. *Computer Drawing* A mathematics professor wanted to use a computer drawing program to draw a picture of a regular pentagon (a five-sided figure with sides of equal length and with equal angles). He first made a 1-in. base by drawing a line from $(0, 0)$ to $(1, 0)$. (See the figure.) He then needed to find the coordinates of the other three vertex points. Use trigonometry to find them. (*Hint:* The sum of the exterior angles of any polygon is 360°.)

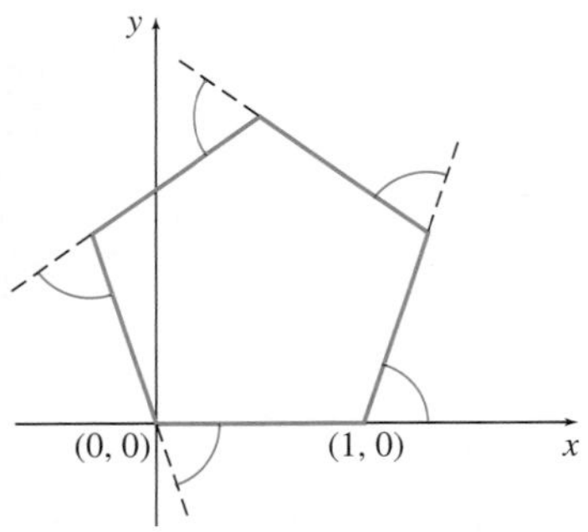

GENERAL INTEREST

86. *Amusement Rides* A proud father is attempting to take a picture of his daughters while they are riding on a merry-go-round. Horses on this particular ride move up and down as the ride progresses according to the function

$$h(t) = \sin\left(\frac{t}{\pi} - 2\right) + 4,$$

where $h(t)$ represents the height of the horse's nose at time t, relative to the merry-go-round platform. However, because of safety fencing surrounding the ride, it is only possible to get a good picture when the height of the horse's nose is between 3.5 and 4 ft off the merry-go-round platform. Find the first time interval that the father has to take the picture.

13.2 DERIVATIVES OF TRIGONOMETRIC FUNCTIONS

THINK ABOUT IT *How long must a ladder be to reach over a 9-foot-high fence and lean against a nearby building?*

In Exercise 50 in this section, we will use trigonometry to answer this question. First, we derive formulas for the derivatives of some of the trigonometric functions. All these derivatives can be found from the formula for the derivative of $y = \sin x$.

We will need to use the following identities, which are listed without proof, to find the derivatives of the trigonometric functions.

BASIC IDENTITIES

$$\sin^2 x + \cos^2 x = 1$$

$$\tan x = \frac{\sin x}{\cos x}$$

$$\sin(x + y) = \sin x \cos y + \cos x \sin y$$
$$\sin(x - y) = \sin x \cos y - \cos x \sin y$$
$$\cos(x + y) = \cos x \cos y - \sin x \sin y$$
$$\cos(x - y) = \cos x \cos y + \sin x \sin y$$

The derivative of $y = \sin x$ also depends on the value of

$$\lim_{x \to 0} \frac{\sin x}{x}.$$

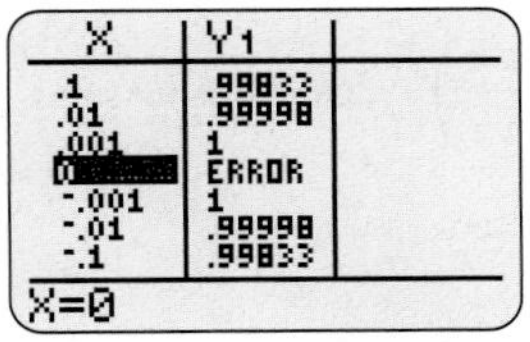

FIGURE 22

To estimate this limit, find the quotient $(\sin x)/x$ for various values of x close to 0. (Be sure that your calculator is set for radian measure.) For example, we used the TABLE feature of the TI-83/84 Plus calculator to get the values of this quotient shown in Figure 22 as x approaches 0 from either side. Note that, although the calculator shows the quotient equal to 1 for $x = \pm.001$, it is an approximation—the value is not exactly 1. Why does the calculator show ERROR when $x = 0$?

These results suggest, and it can be proven, that

$$\lim_{x \to 0} \frac{\sin x}{x} = 1.$$

In Example 1, this limit is used to obtain another limit. Then the derivative of $y = \sin x$ can be found.

EXAMPLE 1 *Trigonometric Limit*

Find $\displaystyle\lim_{h \to 0} \frac{\cos h - 1}{h}$.

Solution Use the limit above and some trigonometric identities.

$$\lim_{h \to 0} \frac{\cos h - 1}{h} = \lim_{h \to 0} \frac{(\cos h - 1)}{h} \cdot \frac{(\cos h + 1)}{(\cos h + 1)}$$
$$= \lim_{h \to 0} \frac{\cos^2 h - 1}{h(\cos h + 1)}$$
$$= \lim_{h \to 0} \frac{-\sin^2 h}{h(\cos h + 1)} \qquad \cos^2 h = 1 - \sin^2 h$$
$$= \lim_{h \to 0} (-\sin h)\left(\frac{\sin h}{h}\right)\left(\frac{1}{\cos h + 1}\right)$$
$$= (0)(1)\left(\frac{1}{1 + 1}\right)$$

$$\lim_{h \to 0} \frac{\cos h - 1}{h} = 0$$

FOR REVIEW

Recall from the section on Limits: When taking the limit of a product, if the limit of each factor exists, the limit of the product is simply the product of the limits.

We can now find the derivative of $y = \sin x$ by using the general definition for the derivative of a function f given in Chapter 3:

$$f'(x) = \lim_{h \to 0} \frac{f(x+h) - f(x)}{h},$$

provided this limit exists. By this definition, the derivative of $f(x) = \sin x$ is

$$
\begin{aligned}
f'(x) &= \lim_{h \to 0} \frac{\sin(x+h) - \sin x}{h} \\
&= \lim_{h \to 0} \frac{\sin x \cdot \cos h + \cos x \cdot \sin h - \sin x}{h} && \text{Identity for } \sin(x+h) \\
&= \lim_{h \to 0} \frac{(\sin x \cdot \cos h - \sin x) + \cos x \cdot \sin h}{h} && \text{Rearrange terms.} \\
&= \lim_{h \to 0} \frac{\sin x(\cos h - 1) + \cos x \cdot \sin h}{h} && \text{Factor.} \\
f'(x) &= \lim_{h \to 0}\left(\sin x \frac{\cos h - 1}{h}\right) + \lim_{h \to 0}\left(\cos x \frac{\sin h}{h}\right) && \text{Limit rule for sums} \\
&= (\sin x)(0) + (\cos x)(1) \\
&= \cos x.
\end{aligned}
$$

This result is summarized below.

FOR REVIEW

Recall that the symbol $D_x[f(x)]$ means the derivative of $f(x)$ with respect to x.

DERIVATIVE OF $\sin x$

$$D_x(\sin x) = \cos x$$

We can use the chain rule to find derivatives of other sine functions, as shown in the following examples.

EXAMPLE 2 *Derivatives of* $\sin x$

Find the derivative of each function.

(a) $y = \sin 6x$

Solution By the chain rule,

$$
\begin{aligned}
\frac{dy}{dx} &= (\cos 6x) \cdot D_x(6x) \\
&= (\cos 6x) \cdot 6 \\
\frac{dy}{dx} &= 6 \cos 6x.
\end{aligned}
$$

(b) $y = 5 \sin(9x^2 + 2) + \cos\left(\frac{\pi}{7}\right)$

Solution

$$
\begin{aligned}
\frac{dy}{dx} &= [5 \cos(9x^2 + 2)] \cdot D_x(9x^2 + 2) + 0 \\
&= [5 \cos(9x^2 + 2)]18x
\end{aligned}
$$

$$\frac{dy}{dx} = 90x\cos(9x^2 + 2)$$

EXAMPLE 3 *Chain Rule*

Find $D_x(\sin^4 x)$.

Solution The expression $\sin^4 x$ means $(\sin x)^4$. By the chain rule,

$$D_x(\sin^4 x) = 4 \cdot \sin^3 x \cdot D_x(\sin x)$$
$$= 4\sin^3 x \cos x.$$

The derivative of $y = \cos x$ is found from trigonometric identities and from the fact that $D_x(\sin x) = \cos x$. First, use the identity for $\sin(x - y)$ to get

$$\sin\left(\frac{\pi}{2} - x\right) = \sin\frac{\pi}{2} \cdot \cos x - \cos\frac{\pi}{2} \cdot \sin x$$
$$= 1 \cdot \cos x - 0 \cdot \sin x$$
$$= \cos x.$$

In the same way, $\cos\left(\frac{\pi}{2} - x\right) = \sin x$. Therefore,

$$D_x(\cos x) = D_x\left[\sin\left(\frac{\pi}{2} - x\right)\right].$$

By the chain rule,

$$D_x\left[\sin\left(\frac{\pi}{2} - x\right)\right] = \cos\left(\frac{\pi}{2} - x\right) \cdot D_x\left(\frac{\pi}{2} - x\right)$$
$$= \cos\left(\frac{\pi}{2} - x\right) \cdot (-1) \qquad \pi/2 \text{ is constant.}$$
$$= -\cos\left(\frac{\pi}{2} - x\right)$$
$$= -\sin x.$$

DERIVATIVE OF $\cos x$

$$D_x(\cos x) = -\sin x$$

EXAMPLE 4 *Derivatives of $\cos x$*

Find each derivative.

(a) $D_x[\cos(3x)] = -\sin(3x) \cdot D_x(3x) = -3\sin 3x$

(b) $D_x(\cos^4 x) = 4\cos^3 x \cdot D_x(\cos x) = 4\cos^3 x(-\sin x)$
$= -4\sin x\cos^3 x$

(c) $D_x(3x \cdot \cos x)$

Solution Use the product rule.

$$\begin{aligned} D_x(3x \cdot \cos x) &= 3x(-\sin x) + (\cos x)(3) \\ &= -3x \sin x + 3 \cos x \end{aligned}$$

As mentioned in the list of basic identities at the beginning of this section, $\tan x = (\sin x)/\cos x$. The derivative of $y = \tan x$ can be found by using the quotient rule to find the derivative of $y = (\sin x)/\cos x$.

$$\begin{aligned} D_x(\tan x) = D_x\left(\frac{\sin x}{\cos x}\right) &= \frac{\cos x \cdot D_x(\sin x) - \sin x \cdot D_x(\cos x)}{\cos^2 x} \\ &= \frac{\cos x(\cos x) - \sin x(-\sin x)}{\cos^2 x} \\ &= \frac{\cos^2 x + \sin^2 x}{\cos^2 x} \\ &= \frac{1}{\cos^2 x} = \sec^2 x \end{aligned}$$

The last step follows from the definitions of the trigonometric functions, which could be used to show that $1/\cos x = \sec x$. A similar calculation leads to the derivative of $\cot x$.

DERIVATIVES OF $\tan x$ AND $\cot x$

$$D_x(\tan x) = \sec^2 x$$
$$D_x(\cot x) = -\csc^2 x$$

EXAMPLE 5 *Derivatives of $\tan x$ and $\cot x$*

Find each derivative.

(a) $D_x(\tan 9x) = \sec^2 9x \cdot D_x(9x) = 9 \sec^2 9x$

(b) $D_x(\cot^6 x) = 6 \cot^5 x \cdot D_x(\cot x) = -6 \cot^5 x \csc^2 x$

(c) $D_x(\ln |6 \tan x|) = \dfrac{D_x(6 \tan x)}{6 \tan x} = \dfrac{6 \sec^2 x}{6 \tan x} = \dfrac{\sec^2 x}{\tan x}$

Using the facts that $\sec x = 1/\cos x$ and $\csc x = 1/\sin x$, it is possible to use the quotient rule to find the derivative of each of these functions. In Exercises 34 and 35 at the end of this section, you will be asked to verify the following.

DERIVATIVES OF $\sec x$ AND $\csc x$

$$D_x \sec x = \sec x \tan x$$
$$D_x \csc x = -\csc x \cot x$$

EXAMPLE 6 *Derivatives of $\sec x$ and $\csc x$*

Find each derivative.

(a) $D_x(x^2 \sec x) = x^2 \sec x \tan x + 2x \sec x$

(b) $D_x(\csc e^{2x}) = -\csc e^{2x} \cot e^{2x} \cdot D_x(e^{2x}) = -\csc e^{2x}\cot e^{2x} \cdot (2e^{2x})$
$= -2e^{2x} \csc e^{2x} \cot e^{2x}$

EXAMPLE 7 *Derivatives of Trigonometric Functions*

Find the derivative of each function at the specified value of x.

(a) $f(x) = \sin(\pi e^x)$, when $x = 0$

Solution Using the chain rule, the derivative of $f(x)$ is

$$f'(x) = \cos(\pi e^x) \cdot \pi e^x.$$

Thus,

$$f'(0) = \cos(\pi e^0) \cdot \pi e^0 = (-1)\pi(1) = -\pi.$$

(b) $g(x) = e^x \sin(\pi x)$, when $x = 0$

Solution Using the product rule, the derivative of $g(x)$ is

$$g'(x) = e^x \cos(\pi x)\pi + \sin(\pi x)e^x.$$

Thus,

$$g'(0) = e^0\cos(\pi 0)\pi + \sin(\pi 0)e^0 = \pi.$$

(c) $h(x) = \tan(\cot x)$, when $x = \frac{\pi}{4}$

Solution Using the chain rule, the derivative of $h(x)$ is

$$h'(x) = \sec^2[\cot(x)] \cdot [-\csc^2(x)].$$

Thus,

$$h'\left(\frac{\pi}{4}\right) = \sec^2\left[\cot\left(\frac{\pi}{4}\right)\right] \cdot \left[-\csc^2\left(\frac{\pi}{4}\right)\right] = \sec^2(1) \cdot \left(-\frac{1}{2}\right) \approx -1.71.$$

(d) $k(x) = \tan x \cot x$, when $x = \frac{\pi}{4}$.

Solution Since $k(x) = \tan x \cot x = \frac{\sin x}{\cos x} \cdot \frac{\cos x}{\sin x} = 1$, $k'(x) = 0$. In particular, $k'\left(\frac{\pi}{4}\right) = 0$.

EXAMPLE 8 *Carbon Dioxide Levels*

At Mauna Loa, Hawaii, atmospheric carbon dioxide levels in parts per million (ppm) have been measured regularly since 1958. The function defined by

$$L(x) = .022x^2 + .55x + 316 + 3.5\sin(2\pi x),$$

can be used to model these levels, where x is in years and $x = 0$ corresponds to 1960.*

(a) Find $L(25)$, $L(35.5)$, and $L(50.2)$.

*Nilsson, A., *Greenhouse Earth,* New York: Wiley, 1992.

Solution

$$L(25) = .022(25)^2 + .55(25) + 316 + 3.5\sin(50\pi) = 343.5 \text{ ppm},$$
$$L(35.5) = .022(35.5)^2 + .55(35.5) + 316 + 3.5\sin(71\pi) = 363.25 \text{ ppm},$$
$$L(50.2) = .022(50.2)^2 + .55(50.2) + 316 + 3.5\sin(100.4\pi) = 402.38 \text{ ppm}$$

(b) Find $L'(50.2)$.

Solution Since $L'(x) = .044x + .55 + 7\pi\cos(2\pi x)$, $L'(50.2)$ is given by

$$L'(50.2) = .044(50.2) + .55 + 7\pi\cos(100.4\pi) = 9.55 \text{ ppm per year.}$$

EXAMPLE 9 *Volume*

The owners of a boarding stable wish to construct a watering trough from which the horses can drink. They have 9 ft by 9 ft pieces of metal, which they can bend into three parts to make the bottom and sides of the trough, as shown in Figure 23(a). They can then weld pieces of scrap metal to the ends to form a trough. At what angle θ should they bend the metal to create the largest possible volume? What is the largest possible volume?

Solution The volume of the trough is its length, 9 ft, times the cross-sectional area. (This is true for any shape with parallel ends and straight sides. For example, the volume of a cylinder is the height times the area of the circular ends, or $h\pi r^2$.) Notice from Figure 23(b) that the cross-sectional area can be broken up into a rectangle with base 3 and height y, and two triangles, each with base x, height y, and angle θ. Since $x/3 = \cos\theta$, we have $x = 3\cos\theta$. Similarly, since $y/3 = \sin\theta$, we have $y = 3\sin\theta$. Therefore,

$$\begin{aligned} A &= 3y + 2\cdot\frac{1}{2}xy \\ &= 3(3\sin\theta) + (3\cos\theta)(3\sin\theta) \\ &= 9(\sin\theta + \cos\theta\sin\theta). \end{aligned}$$

Because the volume is the length times the area,

$$\begin{aligned} V &= 9A = 9\cdot 9(\sin\theta + \cos\theta\sin\theta) \\ &= 81(\sin\theta + \cos\theta\sin\theta), \end{aligned}$$

where $0 \le \theta \le \pi/2$. To find the maximum volume, set the derivative equal to 0.

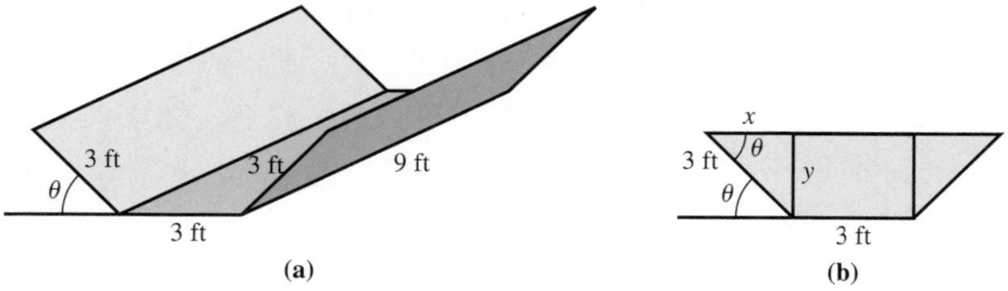

FIGURE 23

$$\frac{dV}{d\theta} = 81[\cos\theta + \cos\theta\cos\theta + \sin\theta(-\sin\theta)] \quad \text{Product rule}$$
$$= 81(\cos\theta + \cos^2\theta - \sin^2\theta)$$
$$= 81[\cos\theta + \cos^2\theta - (1 - \cos^2\theta)] \quad \text{Use } \sin^2 x + \cos^2 x = 1.$$
$$= 81(2\cos^2\theta + \cos\theta - 1) \quad \text{Rearrange terms.}$$
$$= 81(2\cos\theta - 1)(\cos\theta + 1) \quad \text{Factor.}$$

Notice in the third line of the above derivation that we used a trigonometric identity to put the expression entirely in terms of $\cos\theta$. To make $dV/d\theta = 0$, set either factor equal to 0.

$$2\cos\theta - 1 = 0 \qquad \cos\theta + 1 = 0$$
$$\cos\theta = \frac{1}{2} \qquad \cos\theta = -1$$
$$\theta = \frac{\pi}{3} \qquad \text{No solution on } [0, \pi/2]$$

The only value of x for which $dV/d\theta = 0$ is $\theta = \pi/3$, where $\cos\theta = 1/2$, $\sin\theta = \sqrt{3}/2$, and

$$V = 81\left(\frac{\sqrt{3}}{2} + \frac{1}{2}\cdot\frac{\sqrt{3}}{2}\right)$$
$$= \frac{243\sqrt{3}}{4} \approx 105.2 \text{ ft}^3$$

We must also check the endpoints. At $\theta = 0$, we have $V = 0$, while at $\theta = \pi/2$, we have $V = 81$. Thus the maximum volume of about 105.2 ft^3 is achieved with $\theta = \pi/3$.

13.2 EXERCISES

Find the derivatives of the functions defined as follows.

1. $y = 2\sin 6x$

2. $y = -\cos 4x + \cos\frac{\pi}{4}$

3. $y = 12\tan(9x + 1)$

4. $y = -3\cos(8x^2 + 2)$

5. $y = \cos^4 x$

6. $y = -9\sin^5 x$

7. $y = \tan^5 x$

8. $y = 2\cot^4 x$

9. $y = -5x \cdot \sin 4x$

10. $y = 6x \cdot \sec 3x$

11. $y = \frac{\csc x}{x}$

12. $y = \frac{\tan x}{2x + 4}$

13. $y = \sin e^{5x}$

14. $y = \cos(4e^{2x})$

15. $y = e^{\sin x}$

16. $y = -8e^{\tan x}$

17. $y = \sin(\ln 4x^2)$

18. $y = \cos(\ln|2x^3|)$

19. $y = \ln|\sin x^2|$

20. $y = \ln|\tan^2 x|$

21. $y = \frac{2\sin x}{3 - 2\sin x}$

22. $y = \frac{4\cos x}{2 - \cos x}$

23. $y = \sqrt{\frac{\sin x}{\sin 3x}}$

24. $y = \sqrt{\frac{\cos 4x}{\cos x}}$

25. $y = 2\csc x - 3\tan 4x - 7\cos\left(\frac{1}{8}x\right) + e^{3x}$

26. $y = [\sin 3x + \cot(x^3)]^8$

In Exercises 27–32, recall that the slope of the tangent line to a graph is given by the derivative of the function. Find the slope of the tangent line to the graph of each equation at the given point. You may wish to use a graphing calculator to support your answers.

27. $y = \sin x; \quad x = 0$

28. $y = \sin x; \quad x = \pi/4$

29. $y = \cos x; \quad x = \pi/2$

30. $y = \cos x; \quad x = -\pi/4$

31. $y = \tan x; \quad x = 0$

32. $y = \cot x; \quad x = \pi/2$

33. Find the derivative of $\cot x$ by using the quotient rule and the fact that $\cot x = \cos x/\sin x$.

34. Verify that the derivative of $\sec x$ is $\sec x \tan x$. (*Hint:* Use the fact that $\sec x = 1/\cos x$.)

35. Verify that the derivative of $\csc x$ is $-\csc x \cot x$. (*Hint:* Use the fact that $\csc x = 1/\sin x$.)

36. In the discussion of the limit of the quotient $(\sin x)/x$, explain why the calculator gave ERROR for the value of $(\sin x)/x$ when $x = 0$.

Applications

BUSINESS AND ECONOMICS

37. *Revenue from Seasonal Merchandise* The revenue received from the sale of electric fans is seasonal, with maximum revenue in the summer. Let the revenue received from the sale of fans be approximated by

$$R(x) = 100 \cos 2\pi x + 120,$$

where x is time in years, measured from July 1.

a. Find $R'(x)$.

b. Find $R'(x)$ for August 1. (*Hint:* August 1 is 1/12 of a year from July 1.)

c. Find $R'(x)$ for January 1.

d. Find $R'(x)$ for June 1.

e. Discuss whether the answers in parts b–d are reasonable for this model.

LIFE SCIENCES

38. *Swing of a Runner's Arm* A runner's arm swings rhythmically according to the equation

$$y = \frac{\pi}{8} \cos 3\pi\left(t - \frac{1}{3}\right),$$

where y denotes the angle between the actual position of the upper arm and the downward vertical position (as shown in the figure*) and where t denotes time (in seconds).

a. Graph y as a function of t.

b. Calculate the velocity and the acceleration of the arm.

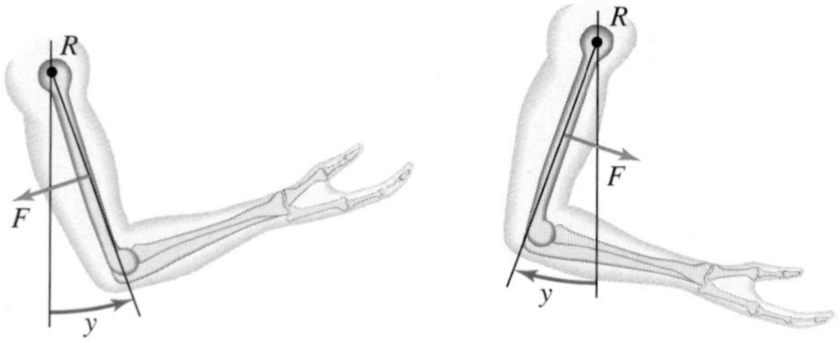

c. Verify that the angle y and the acceleration d^2y/dt^2 are related by the differential equation

$$\frac{d^2y}{dt^2} + 9\pi^2 y = 0.$$

d. Apply the fact that the force exerted by the muscle as the arm swings is proportional to the acceleration of y, with a positive constant of proportionality, to find the direction of the force (counterclockwise or clockwise) at $t = 1$ second, $t = 4/3$ seconds, and $t = 5/3$ seconds. What is the position of the arm at each of these times?

39. *Swing of a Jogger's Arm* A jogger's arm swings according to the equation

$$y = \frac{1}{5} \sin \pi(t - 1).$$

Proceed as directed in parts a–d of the preceding exercise, with the following exceptions: in part c, replace the differential equation with

$$\frac{d^2y}{dt^2} + \pi^2 y = 0,$$

*Art for Exercises 38 and 39 from De Sapio, Rodolfo, *Calculus for the Life Sciences,* Copyright © 1976, 1978 by W. H. Freeman and Company. Reprinted by permission.

and in part d, consider the times $t = 1.5$ seconds, $t = 2.5$ seconds, and $t = 3.5$ seconds.

40. *Carbon Dioxide Levels* At Barrow, Alaska, atmospheric carbon dioxide levels (in parts per million) can be modeled using the function defined by

$$C(x) = .04x^2 + .6x + 330 + 7.5\sin(2\pi x),$$

where x is in years and $x = 0$ corresponds to 1960.*

a. Graph C on $[0, 25]$.

b. Find $C(25)$, $C(35.5)$, and $C(50.2)$.

c. Find $C'(50.2)$ and interpret.

d. L is the sum of a quadratic function and a sine function. What is the significance of each of these functions? Discuss what physical phenomena may be responsible for each function.

41. *Population Growth* Many biological populations, both plant and animal, experience seasonal growth. For example, an animal population might flourish during the spring and summer and die back in the fall. The population, $f(t)$, at time t, is often modeled by

$$f(t) = f(0)e^{c\sin(t)},$$

where $f(0)$ is the size of the population when $t = 0$. Suppose that $f(0) = 1000$ and $c = 2$. Find the functional values in parts a–d.

a. $f(.2)$ **b.** $f(1)$ **c.** $f'(0)$ **d.** $f'(.2)$

e. Graph $f(t)$.

f. Find the maximum and minimum values of $f(t)$ and the values of t where they occur.

42. *Piston Velocity* The distance s of a piston from the center of the crankshaft as it rotates in a 1937 John Deere B engine with respect to the angle θ of the connecting rod, as indicated by the figure, is given by the formula

$$s(\theta) = 2.625\cos\theta + 2.625(15 + \cos^2\theta)^{1/2},$$

where s is measured in inches and θ in radians.†

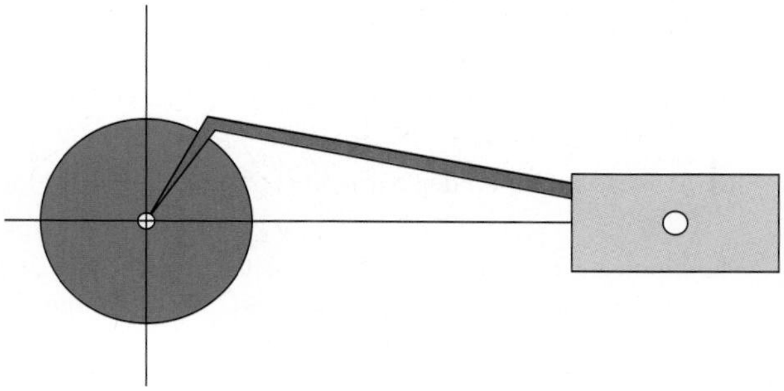

a. Find $s\left(\frac{\pi}{2}\right)$.

b. Find $\frac{ds}{d\theta}$.

c. Find the value(s) of θ where the maximum velocity of the piston occurs.

43. *Sound* If a string with a fundamental frequency of 110 hertz is plucked in the middle, it will vibrate at the odd harmonics of 110, 330, 550, ... hertz but not at the even harmonics of 220, 440, 660, ... hertz. The resulting pressure P on the eardrum caused by the string can be approximated using the equation

$$P(t) = .003\sin(220\pi t) + \frac{.003}{3}\sin(660\pi t) + \frac{.003}{5}\sin(1100\pi t) + \frac{.003}{7}\sin(1540\pi t),$$

where P is in pounds per square foot at a time of t seconds after the string is plucked.‡

a. Graph $P(t)$ on $[0, .01]$.

b. Find $P'(.002)$ and interpret.

*Zeilik, M., S. Gregory, and E. Smith, *Introductory Astronomy and Astrophysics,* Saunders College Publishing, 1992.

†Drost, John and Robert Kunferman, "Related Rates Challenge Problem: Calculate the Velocity of a Piston," *The AMATYC Review,* Vol. 21, No. 1, 1999, pp. 17–21.

‡Benade, Arthur, *Fundamentals of Musical Acoustics,* New York: Oxford University Press, 1976, and Juan Roederer, *The Physics and Psychophysics of Music: An Introduction,* New York: Springer-Verlag, 1995.

44. *Ground Temperature* Mathematical models of ground temperature variation usually involve Fourier series or other sophisticated methods. However, the elementary model

$$u(x,t) = T_0 + A_0 e^{-ax} \cos\left(\frac{\pi}{6}t - ax\right)$$

has been developed for temperature $u(x,t)$ at a given location at a variable time t (in months) and a variable depth x (in centimeters) beneath Earth's surface, T_0 is the annual average surface temperature, and A_0 is the amplitude of the seasonal surface temperature variation.*

Assume that $T_0 = 16°C$ and $A_0 = 11°C$ at a certain location. Also assume that $a = .00706$ in cgs (centimeter-gram-second) units.

a. At what minimum depth x is the amplitude of $u(x,t)$ at most 1°C?

b. Suppose we wish to construct a cellar to keep wine at a temperature between 14°C and 18°C. What minimum depth will accomplish this?

c. At what minimum depth x does the ground temperature model predict that it will be winter when it is summer at the surface, and vice versa? That is, when will the phase shift correspond to $1/2$ year?

d. Show that the ground temperature model satisfies the *heat equation*

$$\frac{\partial u}{\partial t} = k\frac{\partial^2 u}{\partial x^2},$$

where k is a constant.

45. *Flying Gravel* The grooves or tread in a tire occasionally pick up small pieces of gravel, which then are often thrown into the air as they work loose from the tire. When following behind a vehicle on a highway with loose gravel, it is possible to determine a safe distance to travel behind the vehicle so that your automobile is not hit with flying debris by analyzing the function

$$y = x\tan\alpha - \frac{16x^2}{V^2}\sec^2\alpha,$$

where y is the height (in feet) of a piece of gravel that leaves the bottom of a tire at an angle α relative to the roadway, x is the horizontal distance (in feet) of the gravel, and V is the velocity of the automobile (in feet per second).†

a. If a car is traveling 30 mph (44 ft per second), find the height of a piece of gravel thrown from a car tire at an angle of $\pi/4$, when the stone is 40 ft from the car.

b. Putting $y = 0$ and solving for x gives the distance that the gravel will fly. Show that the function that gives a relationship between x and the angle α for $y = 0$ is given by

$$x = \frac{V^2}{32}\sin(2\alpha).$$

(*Hint:* $2\sin\alpha\cos\alpha = \sin(2\alpha)$.)

c. Using part b, if the gravel is thrown from the car at an angle of $\pi/3$ and initial velocity of 44 ft per second, determine how far the gravel will travel.

d. Find $dx/d\alpha$ and use it to determine the value of α that gives the maximum distance that a stone could fly.

e. Find the maximum distance that a stone can fly from a car that is traveling 60 mph.

46. *Engine Velocity* As shown in Exercise 42, a formula that can be used to determine the distance of a piston with respect to the crankshaft for a 1937 John Deere B engine is

$$s(\theta) = 2.625\cos(\theta) + 2.625(15 + \cos^2\theta)^{1/2},$$

where s is measured in inches and θ in radians.‡

a. Given that the angle θ is changing with respect to time, that is, it is a function of t, use the chain rule to find the derivative of s with respect to t, ds/dt.

b. Use part a, with $\theta = 4.944$ and $d\theta/dt = 1340$ rev per minute, to find the maximum velocity of the engine. Express your answer in miles per hour. (*Hint:* 1340 rev per minute = 505,168.1 rad per hour. Use this value and then convert your answer from inches to miles, where 1 mile = 5280 ft.)

47. *Motion of a Particle* A particle moves along a straight line. The distance of the particle from the origin at time t is given by

$$s(t) = \sin t + 2\cos t.$$

Find the velocity at the following times.

a. $t = 0$ **b.** $t = \pi/2$ **c.** $t = \pi$

Find the acceleration at the following times.

d. $t = 0$ **e.** $t = \pi/2$ **f.** $t = \pi$

*Corbitt, Mary Kay and C. Edwards, "Mathematical Modeling and Cool Buttermilk in the Summer," *Applications in School Mathematics 1979 Yearbook,* ed. by Sidney Sharron and Robert Hays, Reston, VA, NCTM, 1979, p. 221.

†Bolt, Brian, "Tennis, Golf, and Loose Gravel: Insight from Easy Math Models," *UMAP Journal,* Vol. 4, No. 1, Spring 1983, pp. 6–18.

‡Drost, John and Robert Kunferman, "Related Rates Challenge Problem: Calculate the Velocity of a Piston," *The AMATYC Review,* Vol. 21, No. 1, all 1999, pp. 17–21.

GENERAL INTEREST

48. *Rotating Lighthouse* The beacon on a lighthouse 40 m from a straight shoreline rotates twice per minute. (See the figure.)

a. How fast is the beam moving along the shoreline at the moment when the light beam and the shoreline are at right angles? (*Hint:* This is a related rate exercise. Find an equation relating θ, the angle between the beam of light and the line from the lighthouse to the shoreline, and x, the distance along the shoreline from the point on the shoreline closest to the lighthouse and the point where the beam hits the shoreline. You need to express $d\theta/dt$ in radians per minute.)

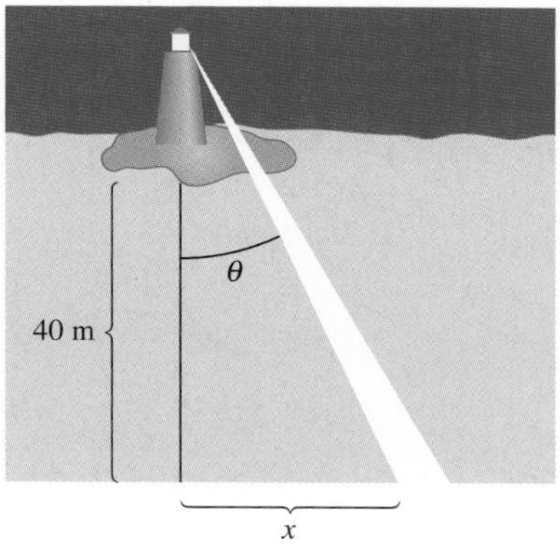

b. In part a, how fast is the beam moving along the shoreline when the beam hits the shoreline 40 m from the point on the shoreline closest to the lighthouse?

49. *Rotating Camera* A television camera on a tripod 60 ft from a road is filming a car carrying the president of the United States. (See the figure.) The car is moving along the road at 600 ft per minute.

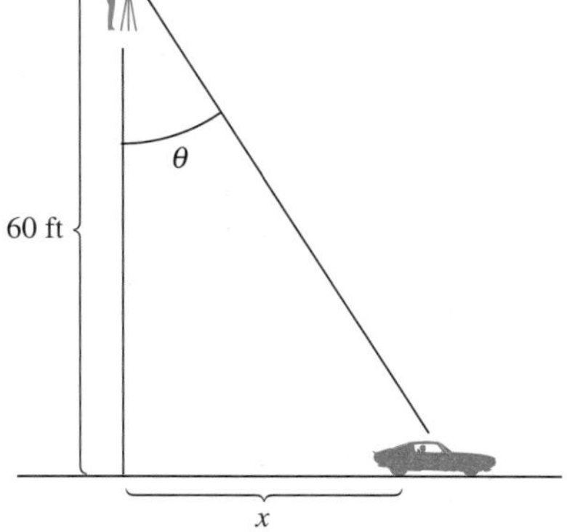

a. How fast is the camera rotating (in revolutions per minute) when the car is at the point on the road closest to the camera? (See the hint for Exercise 48.)

b. How fast is the camera rotating 6 seconds after the moment in part a?

50. *Ladder* A thief tries to enter a building by placing a ladder over a 9-ft-high fence so it rests against the building, which is 2 ft back from the fence. (See the figure below.) What is the length of the shortest ladder that can be used? (*Hint:* Let θ be the angle between the ladder and the ground. Express the length of the ladder in terms of θ, and then find the value of θ that minimizes the length of the ladder.)

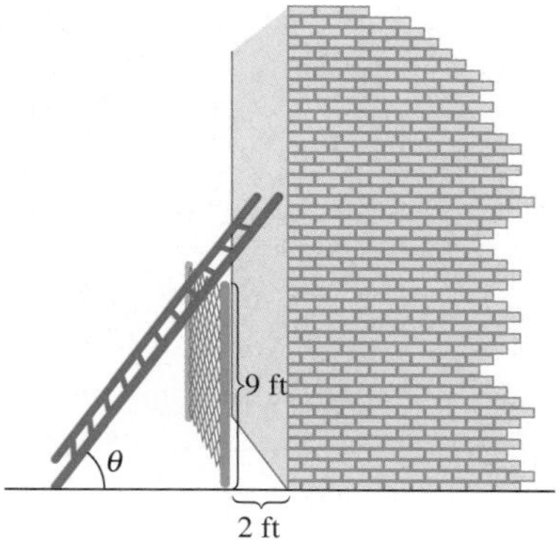

51. *Ladder* A janitor in a hospital needs to carry a ladder around a corner connecting a 10-ft-wide corridor and a 5-ft-wide corridor. (See the figure below.) What is the longest such ladder that can make it around the corner? (*Hint:* Find the narrowest point in the corridor by minimizing the length of the ladder as a function of θ, the angle the ladder makes with the 5-ft-wide corridor.)

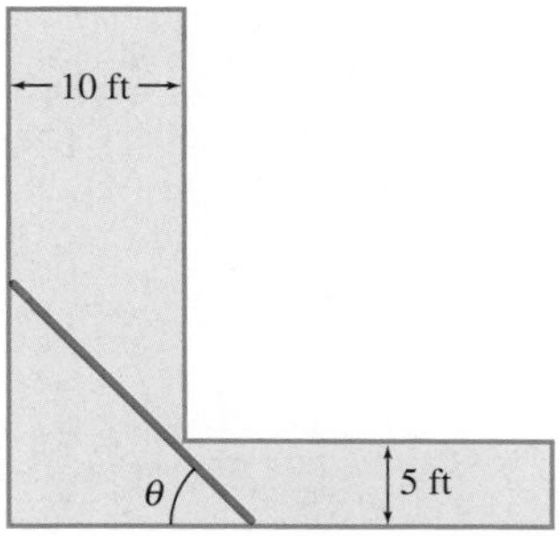

13.3 INTEGRALS OF TRIGONOMETRIC FUNCTIONS

THINK ABOUT IT *Given a sales equation, how many snowblowers are sold in a year?*

In Exercise 39 in this section, we will use trigonometry to answer this question.

Any differentiation formula leads to a corresponding formula for integration. In particular, the formulas of the last section lead to the following indefinite integrals.

BASIC TRIGONOMETRIC INTEGRALS

$$\int \sin x\, dx = -\cos x + C \qquad \int \cos x\, dx = \sin x + C$$

$$\int \sec^2 x\, dx = \tan x + C \qquad \int \csc^2 x\, dx = -\cot x + C$$

$$\int \sec x \tan x\, dx = \sec x + C \qquad \int \csc x \cot x\, dx = -\csc x + C$$

EXAMPLE 1 *Integrals of* $\sin x$ *and* $\cos x$

Find each integral.

(a) $\int \sin 7x\, dx$

Solution Use substitution. Let $u = 7x$, so that $du = 7\, dx$. Then

$$\begin{aligned}\int \sin 7x\, dx &= \frac{1}{7}\int \sin 7x(7\, dx)\\ &= \frac{1}{7}\int \sin u\, du\\ &= -\frac{1}{7}\cos u + C\\ &= -\frac{1}{7}\cos 7x + C.\end{aligned}$$

(b) $\int \cos \frac{2}{3}x\, dx = \frac{3}{2}\int \cos \frac{2}{3}x\left(\frac{2}{3}\, dx\right) = \frac{3}{2}\sin \frac{2}{3}x + C$

(c) $\int \sin^2 x \cos x\, dx$

Solution Let $u = \sin x$, with $du = \cos x\, dx$. This gives

$$\int \sin^2 x \cos x\, dx = \int u^2\, du = \frac{1}{3}u^3 + C.$$

Replacing u with $\sin x$ gives

$$\int \sin^2 x \cos x\, dx = \frac{1}{3}\sin^3 x + C.$$

(d) $\displaystyle\int \frac{\sin x}{\sqrt{\cos x}}\,dx$

Solution Rewrite the integrand as

$$\int (\cos x)^{-1/2} \sin x\, dx.$$

If $u = \cos x$, then $du = -\sin x\, dx$, with

$$\begin{aligned}\int (\cos x)^{-1/2} \sin x\, dx &= -\int (\cos x)^{-1/2}(-\sin x\, dx)\\ &= -\int u^{-1/2}\, du\\ &= -2u^{1/2} + C\\ &= -2\cos^{1/2} x + C.\end{aligned}$$

(e) $\displaystyle\int \sec^2 12x\, dx = \frac{1}{12}\int \sec^2 12x\,(12\, dx) = \frac{1}{12}\tan 12x + C$

(f) $\displaystyle\int e^{3x} \sec e^{3x} \tan e^{3x}\, dx$

Solution Use substitution. Let $u = e^{3x}$, so that $du = 3e^{3x}\, dx$. Then,

$$\begin{aligned}\int e^{3x} \sec e^{3x} \tan e^{3x}\, dx &= \frac{1}{3}\int \sec e^{3x} \tan e^{3x}(3e^{3x}\, dx)\\ &= \frac{1}{3}\int \sec u \tan u\, du\\ &= \frac{1}{3}\sec u + C\\ &= \frac{1}{3}\sec e^{3x} + C.\end{aligned}$$

As in Chapter 7, we can find the area under a curve by setting up an appropriate definite integral.

EXAMPLE 2 *Area Under the Curve*

Find the shaded area in Figure 24.

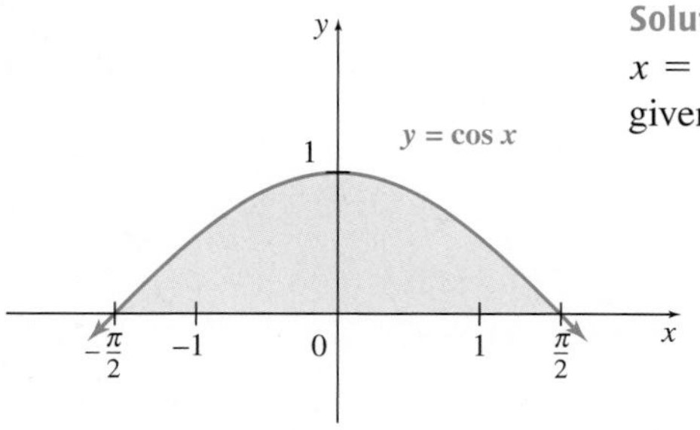

FIGURE 24

Solution The shaded area in Figure 24 is bounded by $y = \cos x$, $y = 0$, $x = -\pi/2$, and $x = \pi/2$. By the Fundamental Theorem of Calculus, this area is given by

$$\begin{aligned}\int_{-\pi/2}^{\pi/2} \cos x\, dx &= \sin x\Big|_{-\pi/2}^{\pi/2}\\ &= \sin\frac{\pi}{2} - \sin\left(-\frac{\pi}{2}\right)\\ &= 1 - (-1)\\ &= 2.\end{aligned}$$

By symmetry, the same area could be found by evaluating

$$2\int_0^{\pi/2} \cos x\, dx.$$

The area in Example 2 could also be found using the definite integral feature of a graphing calculator, entering the expression $\cos x$, the variable x, and the limits of integration.

The method of integration by parts discussed in Chapter 8 is often useful for finding certain integrals involving trigonometric functions.

EXAMPLE 3 *Integration by Parts*

Find $\int 2x \sin x\, dx$.

Solution Let $u = 2x$ and $dv = \sin x\, dx$. Then $du = 2\, dx$ and $v = -\cos x$. Use the formula for integration by parts,

$$\int u\, dv = uv - \int v\, du,$$

to get

$$\begin{aligned}\int 2x \sin x\, dx &= -2x \cos x - \int (-\cos x)(2\, dx)\\ &= -2x \cos x + 2\int \cos x\, dx\\ &= -2x \cos x + 2 \sin x + C.\end{aligned}$$

Check the result by differentiating. (This integral could also have been found by using column integration.)

As mentioned earlier, $\tan x = (\sin x)/\cos x$, so that

$$\int \tan x\, dx = \int \frac{\sin x}{\cos x}\, dx.$$

To find $\int \tan x\, dx$, let $u = \cos x$, with $du = -\sin x\, dx$. Then

$$\int \tan x\, dx = \int \frac{\sin x}{\cos x}\, dx = -\int \frac{du}{u} = -\ln |u| + C.$$

Replacing u with $\cos x$ gives the formula for integrating $\tan x$. The integral for $\cot x$ is found in a similar way.

INTEGRALS OF $\tan x$ AND $\cot x$

$$\int \tan x\, dx = -\ln |\cos x| + C$$

$$\int \cot x\, dx = \ln |\sin x| + C$$

EXAMPLE 4 *Integrals of* $\tan x$ *and* $\cot x$

(a) $\int \tan 6x\,dx = \frac{1}{6}\int \tan 6x(6\,dx) = -\frac{1}{6}\ln|\cos 6x| + C$

(b) $\int x\cot x^2\,dx = \frac{1}{2}\int(\cot x^2)(2x\,dx) = \frac{1}{2}\ln|\sin x^2| + C$

EXAMPLE 5 *Natural Gas Consumption*

The monthly residential consumption of natural gas in the United States in 2002 is found in the following table.*

Index	Month	Consumption (billion cubic feet)
1	January	820
2	February	718
3	March	665
4	April	417
5	May	256
6	June	161
7	July	127
8	August	117
9	September	125
10	October	252
11	November	484
12	December	773

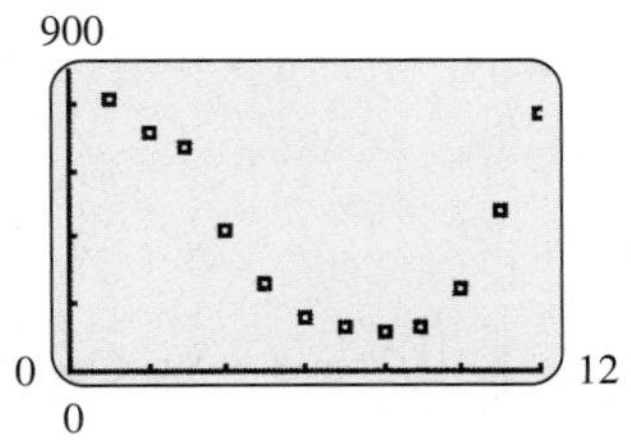

FIGURE 25

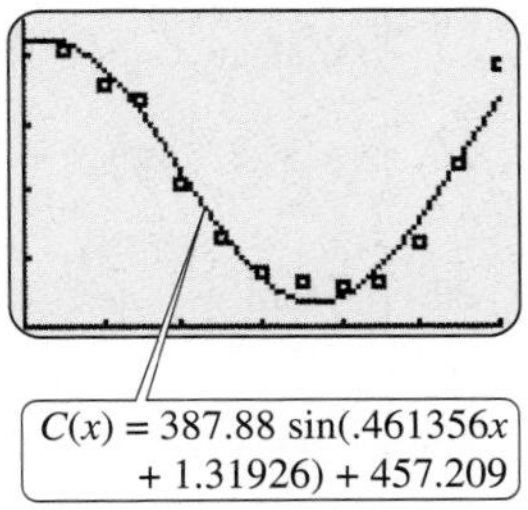

FIGURE 26

(a) Plot the data. Is it reasonable to assume that the monthly consumption of energy is periodic?

Solution Figure 25 shows a graphing calculator plot of the data. Because of the cyclical nature of the four seasons, it is reasonable to assume that the data are periodic.

(b) Find a trigonometric function of the form $C(x) = a\sin(bx + c) + d$ that models this data when x is the month of the year and $C(x)$ is the natural gas consumption. Use the table.

Solution The function $C(x)$, derived by a TI-83/84 Plus calculator, is given by $C(x) = 387.88\sin(.461356x + 1.31926) + 457.209$. Figure 26 shows that this function fits the data well.

(c) Estimate the consumption for the month of September and compare it to the actual value.

Solution $C(9) = 175.82$. The actual value is 125.

*Energy Information Administration, *Natural Gas Monthly,* October 2003.

(d) Estimate the rate at which the consumption is changing in September.

Solution $C'(x) = (.461356)387.88 \cos(.461356x + 1.31926)$

$= 178.95 \cos(.461356x + 1.31926)$, and

$C'(9) = 123.16$ billion ft^3 per month

(e) Estimate the total natural gas consumption for the year for residential customers and compare it to the actual value.

Solution To estimate the total consumption of residential customers for 2002, we use integration as follows.

$$\int_0^{12} C(x)\, dx = \int_0^{12} 387.88 \sin(.461356x + 1.31926) + 457.209\, dx$$

$$= -\frac{387.88}{.461356} \cos(.461356x + 1.31926) + 457.209x \Big|_0^{12} \approx 4989$$

The actual value is 4915.

(f) What would you expect the period of a function that models annual natural gas consumption to be? What is the period of the function found in part b?

Solution If we assume that the annual natural gas consumption is periodic, we would expect the period to be 12 so that it repeats itself every 12 months. The period for the function given above is $T = 2\pi/.461356 \approx 13.62$. Although this value is different from 12, we must remember that the derived function was based on finding the trigonometric function that best describes the given data of 12 points. This suggests that although energy usage is seasonal, other factors that affect energy usage are not accounted for in the model. For example, a very cold winter might be followed by a mild one, so that the natural gas consumption 12 months later is significantly lower.　■

13.3 EXERCISES

Find each integral.

1. $\displaystyle\int \cos 5x\, dx$

2. $\displaystyle\int \sin 8x\, dx$

3. $\displaystyle\int (5 \cos x + 2 \sin x)\, dx$

4. $\displaystyle\int (7 \sin x - 8 \cos x)\, dx$

5. $\displaystyle\int x \sin x^2\, dx$

6. $\displaystyle\int 2x \cos x^2\, dx$

7. $\displaystyle -\int 6 \sec^2 2x\, dx$

8. $\displaystyle -\int 2 \csc^2 8x\, dx$

9. $\displaystyle\int \sin^7 x \cos x\, dx$

10. $\displaystyle\int \sin^6 x \cos x\, dx$

11. $\displaystyle\int \sqrt{\sin x}\, (\cos x)\, dx$

12. $\displaystyle\int \frac{\cos x}{\sqrt{\sin x}}\, dx$

13. $\displaystyle\int \frac{\sin x}{1 + \cos x}\, dx$

14. $\displaystyle\int \frac{\cos x}{1 - \sin x}\, dx$

15. $\displaystyle\int x^5 \cos x^6\, dx$

16. $\displaystyle\int (x + 2)^4 \sin(x + 2)^5\, dx$

17. $\displaystyle\int \tan \frac{1}{4}x\, dx$

18. $\displaystyle\int \cot\left(-\frac{3}{8}x\right) dx$

19. $\displaystyle\int x^2 \cot x^3\, dx$

20. $\displaystyle\int \frac{x}{4} \tan\left(\frac{x}{4}\right)^2 dx$

21. $\displaystyle\int e^x \sin e^x\, dx$

22. $\displaystyle\int e^{-x} \tan e^{-x}\, dx$

23. $\displaystyle\int e^x \csc e^x \cot e^x\, dx$

24. $\displaystyle\int x^4 \sec x^5 \tan x^5\, dx$

25. $\displaystyle\int -6x \cos 5x\, dx$

26. $\displaystyle\int 9x \sin 2x\, dx$

27. $\displaystyle\int 8x \sin x\, dx$

28. $\displaystyle\int -11x \cos x\, dx$

29. $\int -6x^2 \cos 8x \, dx$ **30.** $\int 10x^2 \sin \frac{x}{2} \, dx$

Evaluate each definite integral. Use the integration feature of a graphing calculator, if you wish, to support your answers.

31. $\int_0^{\pi/4} \sin x \, dx$ **32.** $\int_{-\pi/2}^{0} \cos x \, dx$ **33.** $\int_0^{\pi/3} \tan x \, dx$

34. $\int_{\pi/4}^{\pi/2} \cot x \, dx$ **35.** $\int_{\pi/2}^{2\pi/3} \cos x \, dx$ **36.** $\int_{\pi/4}^{3\pi/4} \sin x \, dx$

For Exercises 37 and 38, use the integration feature on a graphing calculator and successively larger values of b to estimate $\int_0^\infty f(x)dx$.

37. $\int_0^b e^{-x} \sin x \, dx$ **38.** $\int_0^b e^{-x} \cos x \, dx$

Applications

BUSINESS AND ECONOMICS

39. *Sales* Sales of snowblowers are seasonal. Suppose the sales of snowblowers in one region of the country are approximated by

$$S(t) = 500 + 500 \cos\left(\frac{\pi}{6}t\right),$$

where t is time (in months), with $t = 0$ corresponding to November. The figure below shows a graph of S. Use a definite integral to find total sales over a year.

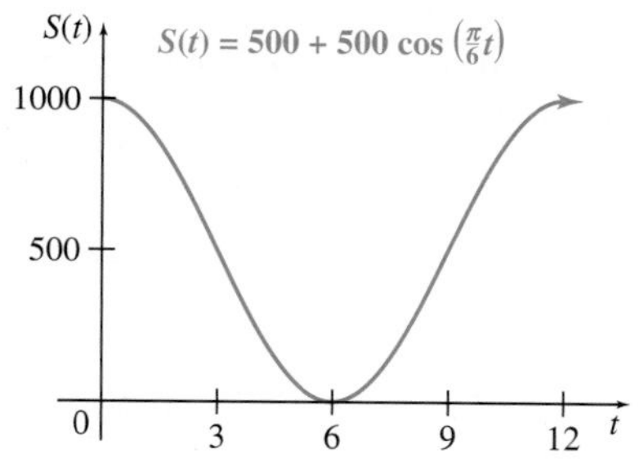

40. *Energy Consumption* The monthly residential consumption of natural gas in Pennsylvania for 2002 is found in the following table.*

Index	Month	Consumption (million cubic feet)
1	January	39,679
2	February	33,339
3	March	31,731
4	April	22,201
5	May	12,211
6	June	7274
7	July	5177
8	August	4465
9	September	5155
10	October	11,310
11	November	24,408
12	December	40,690

a. Plot the data. Is it reasonable to assume that the monthly consumption of energy is periodic?

b. Find a trigonometric function of the form

$$C(x) = a \sin(bx + c) + d$$

that models this data when x is the month of the year and $C(x)$ is the natural gas consumption.

c. Estimate the total natural gas consumption for the year for residential customers in Pennsylvania and compare it to the actual value.

LIFE SCIENCES

41. *Migratory Animals* The number of migratory animals (in hundreds) counted at a certain checkpoint is given by

$$T(t) = 50 + 50 \cos\left(\frac{\pi}{6}t\right),$$

*Energy Information Administration, *Natural Gas Monthly,* October 2003.

where t is time in months, with $t = 0$ corresponding to July. The figure below shows a graph of T. Use a definite integral to find the number of animals passing the checkpoint in a year.

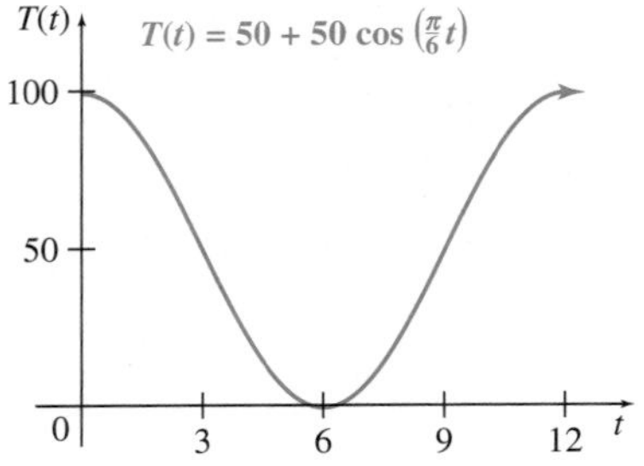

PHYSICAL SCIENCES

42. *Voltage* The electrical voltage from a standard wall outlet is given as a function of time t by

$$V(t) = 170 \sin(120\,\pi t).$$

This is an example of alternating current, which is electricity that reverses direction at regular intervals. The common method for measuring the level of voltage from an alternating current is the *root mean square,* which is given by

$$\text{Root mean square} = \sqrt{\frac{\int_0^T V^2(t)\,dt}{T}},$$

where T is one period of the current.

a. Verify that $T = 1/60$ second for $V(t)$ given above.

b. You may have seen that the voltage from a standard wall outlet is 120 volts. Verify that this is the root mean square value for $V(t)$ given above. (*Hint:* Use the trigonometric identity $\sin^2 x = (1 - \cos 2x)/2$. This identity can be derived by letting $y = x$ in the basic identity for $\cos(x + y)$, and then eliminating $\cos^2 x$ by using the identity $\cos^2 x = 1 - \sin^2 x$.)

43. *Length of Day* The following function can be used to estimate the number of minutes of daylight in Boston for any given day of the year.

$$N(t) = 183.549 \sin(.0172t - 1.329) + 728.124,$$

where t is the day of the year.* Use this function to estimate the total amount of daylight in a year and compare it to the total amount of daylight reported to be 4467.57 hours.

CHAPTER SUMMARY

In this chapter we introduced the trigonometric functions and we studied some of their properties, including their periodic or repetitive nature. We saw that trigonometric functions describe many natural phenomena and have many applications in business, economics, and science. We used the techniques of earlier chapters to find derivatives and integrals of trigonometric functions. Finally, differentiation and integration of trigonometric functions were used to analyze a variety of applications.

KEY TERMS

13.1
ray
endpoint
angle
vertex
initial side
terminal side
standard position
degree measure
acute angle
right angle
obtuse angle
straight angle
arc
radian measure
trigonometric functions
sine
cosine
tangent
cotangent
secant
cosecant
special angles
periodic functions
period
amplitude
phase shift

*Thomas, Robert, *The Old Farmer's Almanac,* 2000.

CHAPTER 13 REVIEW EXERCISES

1. What is the relationship between the degree measure and the radian measure of an angle?

2. Under what circumstances should radian measure be used instead of degree measure? Degree measure instead of radian measure?

3. Describe in words how each of the six trigonometric functions is defined.

4. At what angles (given as rational multiples of π) can you determine the exact values for the trigonometric functions?

Convert the following degree measures to radians. Leave answers as multiples of π.

5. $90°$ **6.** $120°$ **7.** $210°$ **8.** $270°$ **9.** $360°$ **10.** $420°$

Convert the following radian measures to degrees.

11. 7π **12.** $\frac{3\pi}{4}$ **13.** $\frac{9\pi}{20}$ **14.** $\frac{7\pi}{15}$ **15.** $\frac{13\pi}{20}$ **16.** $\frac{11\pi}{15}$

Find each function value without *using a calculator.*

17. $\sin 60°$ **18.** $\tan 120°$ **19.** $\cos(-30°)$ **20.** $\sec 45°$ **21.** $\csc 120°$

22. $\cot 300°$ **23.** $\sin \frac{\pi}{6}$ **24.** $\cos \frac{2\pi}{3}$ **25.** $\sec \frac{5\pi}{4}$ **26.** $\csc \frac{7\pi}{3}$

Find each function value.

27. $\sin 47°$ **28.** $\cos 59°$ **29.** $\tan 81°$ **30.** $\sin(-32°)$

31. $\sin 1.4661$ **32.** $\cos .3142$ **33.** $\cos .5934$ **34.** $\tan 1.2915$

35. Because the derivative of $y = \sin x$ is $dy/dx = \cos x$, the slope of $y = \sin x$ varies from ____________ to ____________.

Graph one period of each function.

36. $y = 3 \cos x$ **37.** $y = \frac{1}{2} \tan x$ **38.** $y = -\tan x$ **39.** $y = -2 \sin x$

Find the derivative of each function.

40. $y = -4 \sin 7x$ **41.** $y = 6 \tan 3x$ **42.** $y = \tan(4x^2 + 3)$ **43.** $y = \cot(9 - x^2)$

44. $y = 3 \cos^6 x$ **45.** $y = 2 \sin^4(4x^2)$ **46.** $y = \cot(4x^5)$ **47.** $y = \cos(1 + x^2)$

48. $y = x^2 \csc x$ **49.** $y = e^{-x} \sin x$ **50.** $y = \frac{\sin x - 1}{\sin x + 1}$ **51.** $y = \frac{\cos^2 x}{1 - \cos x}$

52. $y = \frac{x - 2}{\sec x}$ **53.** $y = \frac{\tan x}{1 + x}$ **54.** $y = \ln |\cos x|$ **55.** $y = \ln |5 \sin x|$

Find each integral.

56. $\int \sin 2x \, dx$ **57.** $\int \cos 3x \, dx$ **58.** $\int \tan 9x \, dx$

59. $\int \sec^2 5x \, dx$ **60.** $\int 5 \sec^2 x \, dx$ **61.** $\int 4 \csc^2 x \, dx$

62. $\int x \sin 3x^2 \, dx$ **63.** $\int 5x \sec 2x^2 \tan 2x^2 \, dx$ **64.** $\int \sqrt{\cos x} \sin x \, dx$

65. $\int \sin^4 x \cos x \, dx$ **66.** $\int x \tan 11x^2 \, dx$ **67.** $\int x^2 \cot 8x^3 \, dx$

68. $\int (\sin x)^{5/2} \cos x \, dx$

69. $\int (\cos x)^{-4/3} \sin x \, dx$

70. $\int \sec^2 3x \tan 3x \, dx$

Find each definite integral.

71. $\int_0^{\pi/2} \cos x \, dx$

72. $\int_{\pi/2}^{\pi} \sin x \, dx$

73. $\int_0^{2\pi} (10 + 10 \cos x) \, dx$

74. $\int_0^{2\pi} (5 + 5 \sin x) \, dx$

Applications

LIFE SCIENCES

75. *Blood Pressure* A person's blood pressure at time t (in seconds) is given by

$$P(t) = 90 + 15 \sin 144\pi t.$$

Find the maximum and minimum values of P on the interval $[0, 1/72]$. Graph one period of $y = P(t)$.

Blood Vessel System *The body's system of blood vessels is made up of arteries, arterioles, capillaries, and veins. The transport of blood from the heart through all organs of the body and back to the heart should be as efficient as possible. One way this can be done is by having large enough blood vessels to avoid turbulence, with blood cells small enough to minimize viscosity.*

In Exercises 76–89, we will find the value of angle θ (see the figure) such that total resistance to the flow of blood is minimized. Assume that a main vessel of radius r_1 runs along the horizontal line from A to B. A side artery, of radius r_2, heads for a point C. Choose point B so that CB is perpendicular to AB. Let CB = s and let D be the point where the axis of the branching vessel cuts the axis of the main vessel.

According to Poiseuille's law, the resistance R in the system is proportional to the length L of the vessel and inversely proportional to the fourth power of the radius r. That is,

$$R = k \cdot \frac{L}{r^4}, \qquad (2)$$

*where k is a constant determined by the viscosity of the blood. Let AB = L_0, AD = L_1, and DC = L_2.**

76. Use right triangle BDC to find $\sin \theta$.

77. Solve the result of Exercise 76 for L_2.

78. Find $\cot \theta$ in terms of s and $L_0 - L_1$.

79. Solve the result of Exercise 78 for L_1.

80. Write an expression similar to Equation (2) for the resistance R_1 along AD.

81. Write a formula for the resistance along DC.

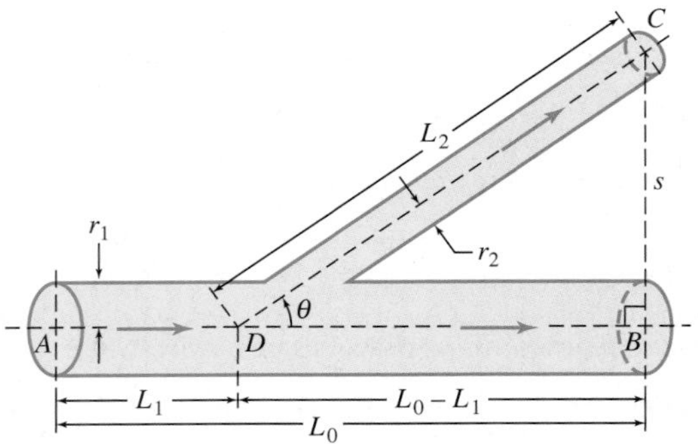

82. The total resistance R is given by the sum of the resistances along AD and DC. Use your answers to Exercises 80 and 81 to write an expression for R.

83. In your formula for R, replace L_1 with the result of Exercise 79 and L_2 with the result of Exercise 77. Simplify your answer.

84. Find $dR/d\theta$. Simplify your answer. (Remember that k, L_1, L_0, s, r_1, and r_2 are constants.)

85. Set $dR/d\theta$ equal to 0.

86. Multiply through by $(\sin^2 \theta)/s$.

87. Solve for $\cos \theta$.

88. Suppose $r_1 = 1$ cm and $r_2 = 1/4$ cm. Find $\cos \theta$ and then find θ.

89. Find θ if $r_1 = 1.4$ cm and $r_2 = .8$ cm.

PHYSICAL SCIENCES

90. *Simple Harmonic Motion* The differential equation $s''(t) = -B^2 s(t)$ approximately describes the motion of a pendulum, known as *simple harmonic motion*. Verify that

$$s(t) = A \cos(Bt + C)$$

satisfies this differential equation.

*Art from Batschelet, Edward, *Introduction to Mathematics for Life Scientists*. Copyright © 1971 by Springer-Verlag New York, Inc. Reprinted by permission.

91. *Temperature* The table lists the average monthly temperatures in Vancouver, Canada.*

Month	Jan	Feb	Mar	Apr	May	June
Temperature	36	39	43	48	55	59

Month	July	Aug	Sep	Oct	Nov	Dec
Temperature	64	63	57	50	43	39

These average temperatures cycle yearly and change only slightly over many years. Because of the repetitive nature of temperatures from year to year, they can be modeled with a sine function. Some graphing calculators have a sine regression feature. If the table is entered into a calculator, the points can be plotted automatically, as shown in the early chapters of this book with other types of functions.

a. Use a graphing calculator to plot the ordered pairs (month, temperature) in the interval $[0, 12]$ by $[30, 70]$.

b. Use a graphing calculator with a sine regression feature to find an equation of the sine function that models this data.

c. Graph the equation from part b.

92. *Tennis* It is possible to model the flight of a tennis ball that has just been served down the center of the court by the equation

$$y = x \tan \alpha - \frac{16x^2}{V^2} \sec^2 \alpha + h,$$

where y is the height (in feet) of a tennis ball that is being served at an angle α relative to the horizontal axis, x is the horizontal distance (in feet) that the ball has traveled, h is the height of the ball when it leaves the server's racket and V is the velocity of the tennis ball when it leaves the server's racket.†

a. If a tennis ball is served from a height of 9 ft and the net is 3 ft high and 39 ft away from the server, does the tennis ball that is hit with a velocity of 50 mph (approximately 73 ft per second) make it over the net if it is served at an angle of $\pi/24$?

b. When $y = 0$, the corresponding value of x gives the total distance that the tennis ball has traveled while in flight (provided that it cleared the net). For a serving height of 9 ft, the equation for calculating the distance traveled is given by

$$x = \frac{V^2 \sin \alpha \cos \alpha \pm V^2 \cos^2 \alpha \sqrt{\tan^2 \alpha + \frac{576}{V^2} \sec^2 \alpha}}{32}.$$

Use the TABLE function on a graphing calculator or a spreadsheet to determine a range of angles for which the tennis ball will clear the net and travel between 39 and 60 ft when it is hit with an initial velocity of 44 ft per second.

c. Because calculating $dx/d\alpha$ is so complicated analytically, use a graphing calculator to estimate this derivative when the initial velocity is 44 ft per second and $\alpha = \pi/8$. Interpret your answer.

93. *Energy Usage* A mathematics textbook author has determined that her monthly gas usage y approximately follows the sine curve

$$y = 12.5 \sin\left(\frac{\pi}{6}(x + 1.2)\right) + 14.7,$$

where y is measured in thousands of cubic feet (MCF) and x is the month of the year ranging from 1 to 12.

a. Graph this function.

b. Find the approximate gas usage for the months of February and July.

c. Find dy/dx, when $x = 7$. Interpret your answer.

GENERAL INTEREST

94. *Area* A 6-ft board is placed against a wall as shown in the figure on the next page, forming a triangular-shaped area beneath it. At what angle θ should the board be placed to make the triangular area as large as possible?

*Miller, A. and J. Thompson, *Elements of Meteorology,* Columbus: Charles E. Merrill Publishing Company, 1975.

†Bolt, Brian, "Tennis, Golf, and Loose Gravel: Insight from Easy Math Models," *UMAP Journal,* Vol. 4, No. 1, Spring 1983, pp. 6–18.

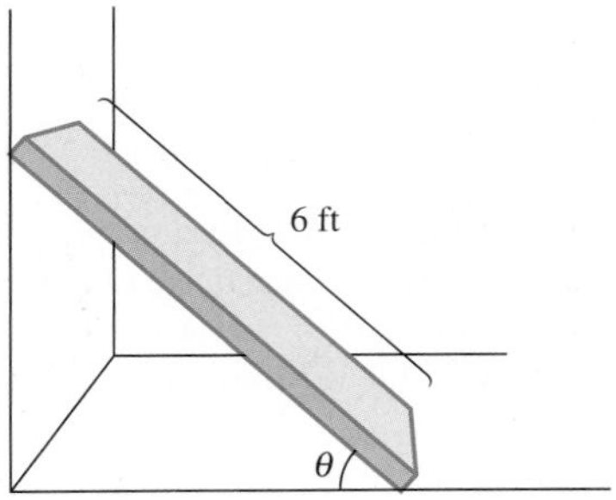

95. *Mercator's World Map* Before Gerardus Mercator designed his map of the world in 1569, sailors who traveled in a fixed compass direction could follow a straight line on a map only over short distances. Over long distances, such a course would be a curve on existing maps, which tried to make area on the map proportional to the actual area. Mercator's map greatly simplified navigation: even over long distances, straight lines on the map corresponded to fixed compass bearings. This was accomplished by distorting distances. On Mercator's map, the distance of an object from the equator to a parallel at latitude θ is given by

$$D(\theta) = k\int_0^{\theta} \sec x \, dx,$$

where k is a constant of proportionality. Calculus had not yet been discovered when Mercator designed his map; he approximated the distance between parallels of latitude by hand.*

a. Verify that

$$\frac{d}{dx}\ln|\sec x + \tan x| = \sec x.$$

b. Verify that

$$\frac{d}{dx}(-\ln|\sec x - \tan x|) = \sec x.$$

c. Using parts a and b, give two different formulas for $\int \sec x \, dx$. Explain how they can both be correct.

d. Los Angeles has a latitude of 34°03′N. (The 03′ represents 3 minutes of latitude. Each minute of latitude is 1/60 of a degree.) If Los Angeles is to be 7 in. from the equator on a Mercator map, how far from the equator should we place New York City, which has a latitude of 40°45′N?

e. Repeat part d for Miami, which has a latitude of 25°46′N.

f. If you do not live in Los Angeles, New York City, or Miami, repeat part d for your town or city.

*Rickey, V. Frederick and Philip M. Tuchinsky, "An Application of Geography to Mathematics: History of the Integral of the Secant," *Mathematics Magazine,* Vol. 53, May 1980, pp. 162–166.

EXTENDED APPLICATION: The Shortest Time and the Cheapest Path

In an application at the end of Section 13.1 we stated Snell's law relating the angle of refraction when light passes from one medium to another to the speed of light in each medium. Figure 27 represents the relationship graphically.

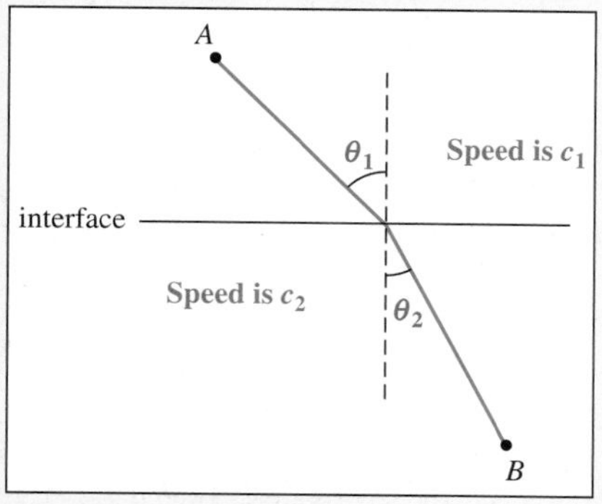

FIGURE 27

If the speed of light in the upper medium is c_1, and in the lower medium the speed is c_2, then the speeds are related to the angles (called *angles of refraction*) by the equation

$$\frac{c_1}{c_2} = \frac{\sin \theta_1}{\sin \theta_2}.$$

You might think this law is of interest only to physicists, but the same minimization problem shows up in other contexts, such as planning the path of a pipeline or road. First we'll use some calculus to derive Snell's law, and then look at some applications.

Let's see why the law, or something like it, should be true. Snell's law is based on the fact that in traveling from A to B, a light ray will follow the path that takes the *minimum time*. If the speeds c_1 and c_2 are equal, the shortest path will also be the fastest, so the best route is a straight line from A to B. In this case the angles θ_1 and θ_2 are equal, since they are vertical angles. But if $c_1 > c_2$, the light ray will "do better" by spending more time in the upper medium, where it travels faster, and less time in the lower medium. Therefore the point where it crosses the interface will move to the right, which will make θ_1 greater than θ_2, as shown in Figure 28.

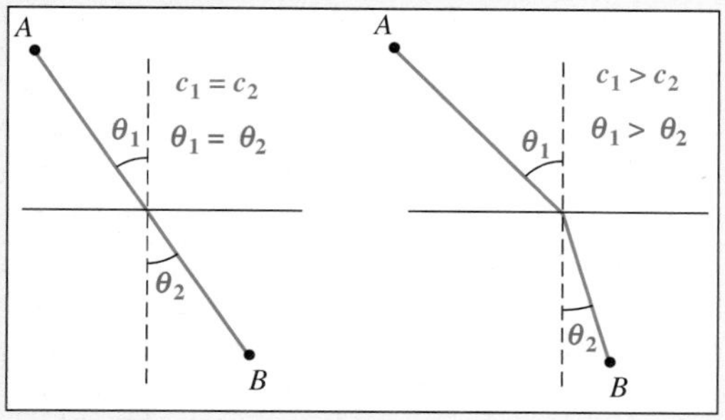

FIGURE 28

The sine is increasing on the interval $(0, \pi/2)$, so $\theta_1 > \theta_2$ implies $\sin \theta_1 > \sin \theta_2$, and Snell's law at least agrees with our intuitive reasoning about how the angles and speeds should be related. Let's get a more complete picture of the geometric setup. Figure 29 illustrates the relationship.

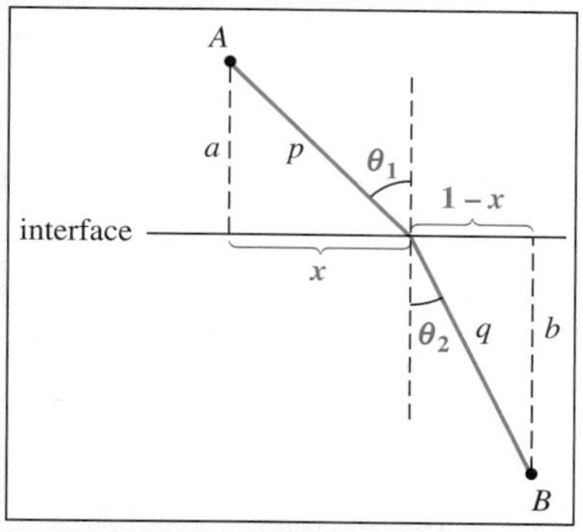

FIGURE 29

Since time is distance divided by speed, the total transit time from A to B is

$$\frac{p}{c_1} + \frac{q}{c_2}.$$

This is the expression we want to minimize, but before we can use the techniques we learned in Chapter 6 we need to choose a variable. We might try expressing both distances in terms of x and come up with an expression for the total time T as a function of x:

$$T(x) = \frac{\sqrt{a^2 + x^2}}{c_1} + \frac{\sqrt{b^2 + (1 - x)^2}}{c_2}$$

The prospect of differentiating this expression with respect to x—setting the derivative equal to 0—and solving for x is not attractive. We'll zoom in on the point where the light ray crosses the interface, and look at what happens if we move this crossing point a little bit to the right. Using the delta notation we introduced back in Chapter 1, the picture looks like Figure 30.

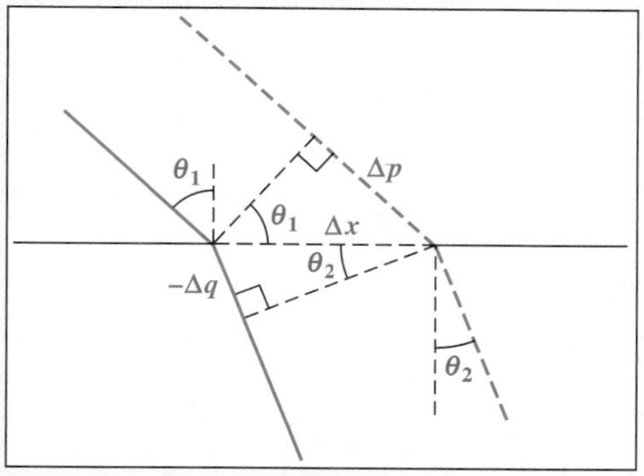

FIGURE 30

If we increase x by a small amount, Δx, the upper path gets *longer* by Δp and the lower path gets *shorter* by Δq. (Since Δq is negative, we label the triangle side with $-\Delta q$ so that it will be a positive length.) When we have found the best crossing point, the amount we add to the time by making the upper path longer must be exactly balanced by the amount of time we save by making the lower path shorter, which means that

$$\frac{\Delta p}{c_1} = \frac{-\Delta q}{c_2} \quad \text{or} \quad \frac{c_1}{c_2} = \frac{\Delta p}{-\Delta q}.$$

Using the two small right triangles, we find that $\Delta p = \Delta x \sin\theta_1$ and $-\Delta q = \Delta x \sin\theta_2$, so

$$\frac{c_1}{c_2} = \frac{\Delta p}{-\Delta q} = \frac{\Delta x \sin\theta_1}{\Delta x \sin\theta_2} = \frac{\sin\theta_1}{\sin\theta_2},$$

which is Snell's law.

Some of the equations in this derivation are only approximate, and we will look at these approximations more closely in the exercises. But engineers use this kind of reasoning with small increments all the time, and we could make it precise and rigorous using the language of differentials introduced in Section 6.6. The argument with increments shows why the sines of the angles appear in Snell's law: They measure the rate at which moving along the interface changes the lengths of the upper and lower path segments.

Let's use Snell's law to solve an optimization problem that involves *cost* rather than *time*. A road is to be built from town A to town B. Part of the region between the towns is swampy land, over which the road will have to be elevated on a causeway. The rest of the territory is dry land on which a conventional road can be built. The territory looks like Figure 31.

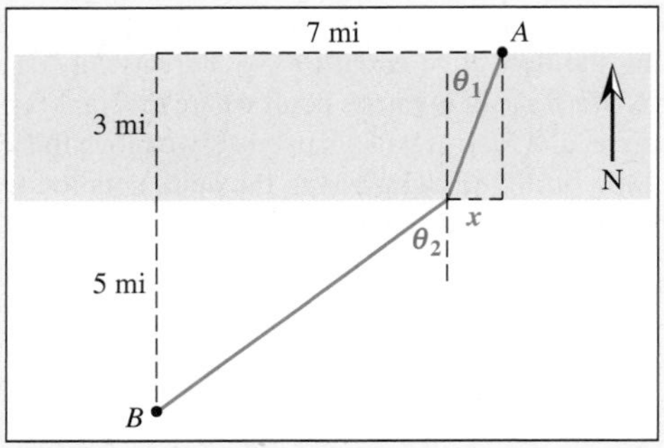

FIGURE 31

Suppose building a road over a swamp is three times more expensive per mile than building it over land. We can reduce the cost by making the road a bit longer than a straight connection and shortening the portion built over the swamp, but what is the right tradeoff? Where should the road emerge from the swamp? How long is the distance x?

This is a classic calculus problem, often solved as an exercise in simplifying complicated derivative expressions, but now that we have Snell's law, the minimization is already done. The preferred medium is the cheaper one, so our equation will look like this:

$$\frac{1}{3} = \frac{\sin\theta_1}{\sin\theta_2} \quad \text{or} \quad \sin\theta_2 = 3\sin\theta_1.$$

Of course we don't know θ_1 and θ_2, but we have enough information to figure out x. Using the basic identities and writing everything in terms of x, we get

$$\frac{7-x}{\sqrt{(7-x)^2+5^2}} = 3\frac{x}{\sqrt{x^2+3^2}}$$

The solver on your calculator will find the root easily; it's $x \approx .806$ miles. As we expected, the cheapest route goes almost perpendicularly across the swamp. The angle θ_1 has tangent equal to $.806/3$, which means that θ_1 is about 15°.

Exercises

1. In Figure 30, we drew the two rays coming from point A as if they were nearly parallel. Why can't they be parallel? If they *were* parallel, how would you prove that the two angles labeled θ_1 are equal? How could you make the rays more nearly parallel?
2. In Figure 30, we claim that the segment labeled Δp is the change in the length of the ray from point A. Is it? How could you improve the approximation?
3. If you wear glasses, you've probably been offered the choice between glass and "high-index plastic" for your lenses. The typical high-index plastic lens has an *index of refraction* of 1.6, which means that the ratio (speed of light in air)/(speed of light in plastic) is equal to 1.6. What percent of the speed of light in air is the speed of light in 1.6-index plastic?
4. Without doing the calculation, describe the cheapest route for the road between A and B in the case where swamp construction is 100 times as expensive as construction over land.
5. In the road construction example, what would change in the equation for x if construction over land was actually *more expensive* than construction over the swamp? Find the best x for the case where dry-land construction is twice as expensive as swamp construction, and show that the corresponding route lies to the west of the straight-line route.
6. A light ray that reaches you when you look at a sunset is bent by the same process of refraction that Snell's law describes. The higher density and higher water content of the air close to Earth cause light to travel more slowly

closer to Earth, so as it moves through the atmosphere the light ray is bent down toward Earth. Rather than happening all at once at a sharp interface between one medium and another, this *atmospheric refraction* happens gradually, so the light follows a curved path. Light rays coming at an angle of 90° to the vertical (that is, directly from the horizon) are bent by an angle of about .57°. The diameter of the sun's disk as we see it is about .53°. When you see the sun begin to set, where is it actually located?*

Directions for Group Project

Prepare a demonstration of Snell's law that illustrates the phenomenon of light refraction. Your demonstration should include an explanation of what Snell's law is, why it is true, and some of its uses. Assume that the audience for this presentation has a conceptual understanding of angles but no formal studies in either trigonometry or calculus. Be sure to use Exercises 1–6 in making your presentation. Presentation software such as Microsoft Powerpoint should be used.

*See the U.S. Naval Observatory's Web site at http://aa.usno.navy.mil/AA/faq/docs/RST_defs.html.

Appendix

TABLE 1 FORMULAS FROM GEOMETRY

PYTHAGOREAN THEOREM
For a right triangle with legs of lengths a and b and hypotenuse of length c, $a^2 + b^2 = c^2$.

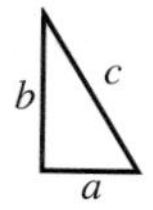

CIRCLE
Area: $A = \pi r^2$
Circumference: $C = 2\pi r$

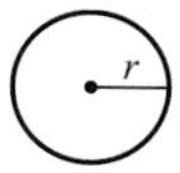

RECTANGLE
Area: $A = lw$
Perimeter: $P = 2l + 2w$

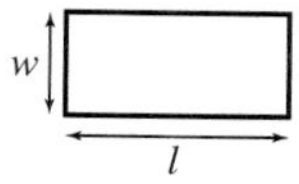

TRIANGLE
Area: $A = \dfrac{1}{2}bh$

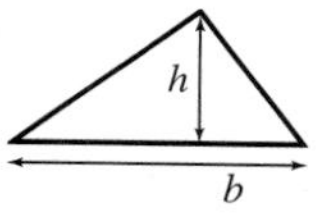

SPHERE
Volume: $V = \dfrac{4}{3}\pi r^3$
Surface area: $A = 4\pi r^2$

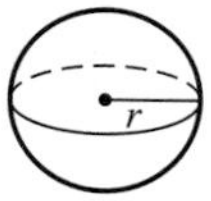

CONE
Volume: $V = \dfrac{1}{3}\pi r^2 h$

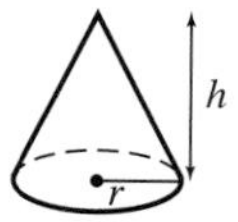

RECTANGULAR BOX
Volume: $V = lwh$
Surface area: $A = 2lh + 2wh + 2lw$

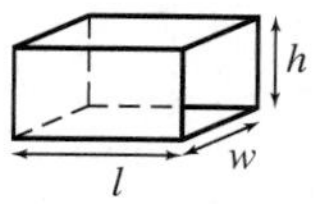

CIRCULAR CYLINDER
Volume: $V = \pi r^2 h$
Surface area: $A = 2\pi r^2 + 2\pi rh$

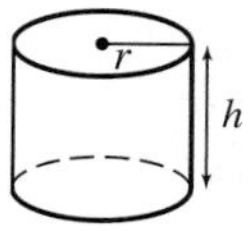

TRIANGULAR CYLINDER
Volume: $V = \dfrac{1}{2}bhl$

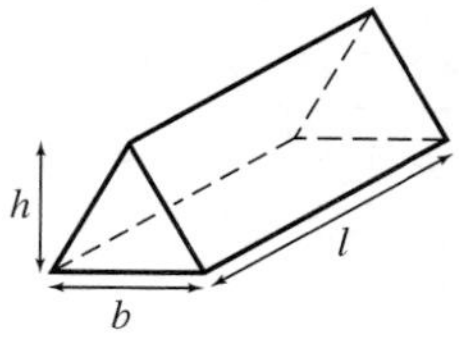

GENERAL INFORMATION ON SURFACE AREA
To find the surface area of a figure, break down the total surface area into the individual components and add up the areas of the components. For example, a rectangular box has six sides, each of which is a rectangle. A circular cylinder has two ends, each of which is a circle, plus the side, which forms a rectangle when opened up.

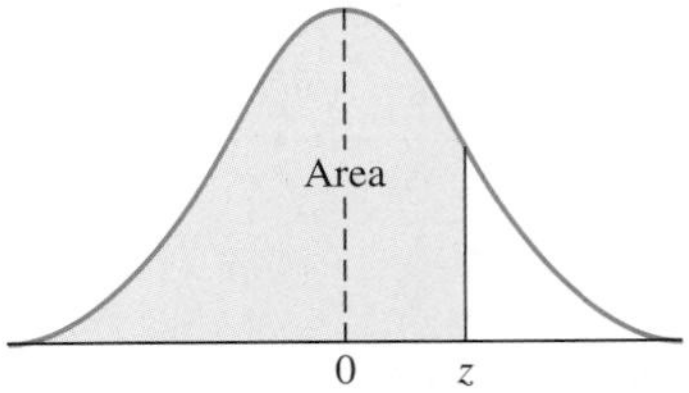

TABLE 2 AREA UNDER A NORMAL CURVE TO THE LEFT OF z, WHERE $z = \dfrac{x - \mu}{\sigma}$

z	.00	.01	.02	.03	.04	.05	.06	.07	.08	.09
−3.4	.0003	.0003	.0003	.0003	.0003	.0003	.0003	.0003	.0003	.0002
−3.3	.0005	.0005	.0005	.0004	.0004	.0004	.0004	.0004	.0004	.0003
−3.2	.0007	.0007	.0006	.0006	.0006	.0006	.0006	.0005	.0005	.0005
−3.1	.0010	.0009	.0009	.0009	.0008	.0008	.0008	.0008	.0007	.0007
−3.0	.0013	.0013	.0013	.0012	.0012	.0011	.0011	.0011	.0010	.0010
−2.9	.0019	.0018	.0017	.0017	.0016	.0016	.0015	.0015	.0014	.0014
−2.8	.0026	.0025	.0024	.0023	.0023	.0022	.0021	.0021	.0020	.0019
−2.7	.0035	.0034	.0033	.0032	.0031	.0030	.0029	.0028	.0027	.0026
−2.6	.0047	.0045	.0044	.0043	.0041	.0040	.0039	.0038	.0037	.0036
−2.5	.0062	.0060	.0059	.0057	.0055	.0054	.0052	.0051	.0049	.0048
−2.4	.0082	.0080	.0078	.0075	.0073	.0071	.0069	.0068	.0066	.0064
−2.3	.0107	.0104	.0102	.0099	.0096	.0094	.0091	.0089	.0087	.0084
−2.2	.0139	.0136	.0132	.0129	.0125	.0122	.0119	.0116	.0113	.0110
−2.1	.0179	.0174	.0170	.0166	.0162	.0158	.0154	.0150	.0146	.0143
−2.0	.0228	.0222	.0217	.0212	.0207	.0202	.0197	.0192	.0188	.0183
−1.9	.0287	.0281	.0274	.0268	.0262	.0256	.0250	.0244	.0239	.0233
−1.8	.0359	.0352	.0344	.0336	.0329	.0322	.0314	.0307	.0301	.0294
−1.7	.0446	.0436	.0427	.0418	.0409	.0401	.0392	.0384	.0375	.0367
−1.6	.0548	.0537	.0526	.0516	.0505	.0495	.0485	.0475	.0465	.0455
−1.5	.0668	.0655	.0643	.0630	.0618	.0606	.0594	.0582	.0571	.0559
−1.4	.0808	.0793	.0778	.0764	.0749	.0735	.0722	.0708	.0694	.0681
−1.3	.0968	.0951	.0934	.0918	.0901	.0885	.0869	.0853	.0838	.0823
−1.2	.1151	.1131	.1112	.1093	.1075	.1056	.1038	.1020	.1003	.0985
−1.1	.1357	.1335	.1314	.1292	.1271	.1251	.1230	.1210	.1190	.1170
−1.0	.1587	.1562	.1539	.1515	.1492	.1469	.1446	.1423	.1401	.1379
−.9	.1841	.1814	.1788	.1762	.1736	.1711	.1685	.1660	.1635	.1611
−.8	.2119	.2090	.2061	.2033	.2005	.1977	.1949	.1922	.1894	.1867
−.7	.2420	.2389	.2358	.2327	.2296	.2266	.2236	.2206	.2177	.2148
−.6	.2743	.2709	.2676	.2643	.2611	.2578	.2546	.2514	.2483	.2451
−.5	.3085	.3050	.3015	.2981	.2946	.2912	.2877	.2843	.2810	.2776

TABLE 2 AREA UNDER A NORMAL CURVE (continued)

z	.00	.01	.02	.03	.04	.05	.06	.07	.08	.09
−.4	.3446	.3409	.3372	.3336	.3300	.3264	.3228	.3192	.3156	.3121
−.3	.3821	.3783	.3745	.3707	.3669	.3632	.3594	.3557	.3520	.3483
−.2	.4207	.4168	.4129	.4090	.4052	.4013	.3974	.3936	.3897	.3859
−.1	.4602	.4562	.4522	.4483	.4443	.4404	.4364	.4325	.4286	.4247
−.0	.5000	.4960	.4920	.4880	.4840	.4801	.4761	.4721	.4681	.4641
.0	.5000	.5040	.5080	.5120	.5160	.5199	.5239	.5279	.5319	.5359
.1	.5398	.5438	.5478	.5517	.5557	.5596	.5636	.5675	.5714	.5753
.2	.5793	.5832	.5871	.5910	.5948	.5987	.6026	.6064	.6103	.6141
.3	.6179	.6217	.6255	.6293	.6331	.6368	.6406	.6443	.6480	.6517
.4	.6554	.6591	.6628	.6664	.6700	.6736	.6772	.6808	.6844	.6879
.5	.6915	.6950	.6985	.7019	.7054	.7088	.7123	.7157	.7190	.7224
.6	.7257	.7291	.7324	.7357	.7389	.7422	.7454	.7486	.7517	.7549
.7	.7580	.7611	.7642	.7673	.7704	.7734	.7764	.7794	.7823	.7852
.8	.7881	.7910	.7939	.7967	.7995	.8023	.8051	.8078	.8106	.8133
.9	.8159	.8186	.8212	.8238	.8264	.8289	.8315	.8340	.8365	.8389
1.0	.8413	.8438	.8461	.8485	.8508	.8531	.8554	.8577	.8599	.8621
1.1	.8643	.8665	.8686	.8708	.8729	.8749	.8770	.8790	.8810	.8830
1.2	.8849	.8869	.8888	.8907	.8925	.8944	.8962	.8980	.8997	.9015
1.3	.9032	.9049	.9066	.9082	.9099	.9115	.9131	.9147	.9162	.9177
1.4	.9192	.9207	.9222	.9236	.9251	.9265	.9278	.9292	.9306	.9319
1.5	.9332	.9345	.9357	.9370	.9382	.9394	.9406	.9418	.9429	.9441
1.6	.9452	.9463	.9474	.9484	.9495	.9505	.9515	.9525	.9535	.9545
1.7	.9554	.9564	.9573	.9582	.9591	.9599	.9608	.9616	.9625	.9633
1.8	.9641	.9649	.9656	.9664	.9671	.9678	.9686	.9693	.9699	.9706
1.9	.9713	.9719	.9726	.9732	.9738	.9744	.9750	.9756	.9761	.9767
2.0	.9772	.9778	.9783	.9788	.9793	.9798	.9803	.9808	.9812	.9817
2.1	.9821	.9826	.9830	.9834	.9838	.9842	.9846	.9850	.9854	.9857
2.2	.9861	.9864	.9868	.9871	.9875	.9878	.9881	.9884	.9887	.9890
2.3	.9893	.9896	.9898	.9901	.9904	.9906	.9909	.9911	.9913	.9916
2.4	.9918	.9920	.9922	.9925	.9927	.9929	.9931	.9932	.9934	.9936
2.5	.9938	.9940	.9941	.9943	.9945	.9946	.9948	.9949	.9951	.9952
2.6	.9953	.9955	.9956	.9957	.9959	.9960	.9961	.9962	.9963	.9964
2.7	.9965	.9966	.9967	.9968	.9969	.9970	.9971	.9972	.9973	.9974
2.8	.9974	.9975	.9976	.9977	.9977	.9978	.9979	.9979	.9980	.9981
2.9	.9981	.9982	.9982	.9983	.9984	.9984	.9985	.9985	.9986	.9986
3.0	.9987	.9987	.9987	.9988	.9988	.9989	.9989	.9989	.9990	.9990
3.1	.9990	.9991	.9991	.9991	.9992	.9992	.9992	.9992	.9993	.9993
3.2	.9993	.9993	.9994	.9994	.9994	.9994	.9994	.9995	.9995	.9995
3.3	.9995	.9995	.9995	.9996	.9996	.9996	.9996	.9996	.9996	.9997
3.4	.9997	.9997	.9997	.9997	.9997	.9997	.9997	.9997	.9997	.9998

TABLE 3 INTEGRALS

(C is an arbitrary constant.)

1. $\int x^n\,dx = \frac{x^{n+1}}{n+1} + C \quad (\text{if } n \neq -1)$

2. $\int e^{kx}\,dx = \frac{e^{kx}}{k} + C$

3. $\int \frac{a}{x}\,dx = a \ln|x| + C$

4. $\int \ln|ax|\,dx = x(\ln|ax| - 1) + C$

5. $\int \frac{1}{\sqrt{x^2+a^2}}\,dx = \ln\left|x + \sqrt{x^2+a^2}\right| + C$

6. $\int \frac{1}{\sqrt{x^2-a^2}}\,dx = \ln\left|x + \sqrt{x^2-a^2}\right| + C$

7. $\int \frac{1}{a^2-x^2}\,dx = \frac{1}{2a} \cdot \ln\left|\frac{a+x}{a-x}\right| + C \quad (a \neq 0)$

8. $\int \frac{1}{x^2-a^2}\,dx = \frac{1}{2a} \cdot \ln\left|\frac{x-a}{x+a}\right| + C \quad (a \neq 0)$

9. $\int \frac{1}{x\sqrt{a^2-x^2}}\,dx = -\frac{1}{a} \cdot \ln\left|\frac{a+\sqrt{a^2-x^2}}{x}\right| + C \quad (a \neq 0)$

10. $\int \frac{1}{x\sqrt{a^2+x^2}}\,dx = -\frac{1}{a} \cdot \ln\left|\frac{a+\sqrt{a^2+x^2}}{x}\right| + C \quad (a \neq 0)$

11. $\int \frac{x}{ax+b}\,dx = \frac{x}{a} - \frac{b}{a^2} \cdot \ln|ax+b| + C \quad (a \neq 0)$

12. $\int \frac{x}{(ax+b)^2}\,dx = \frac{b}{a^2(ax+b)} + \frac{1}{a^2} \cdot \ln|ax+b| + C \quad (a \neq 0)$

13. $\int \frac{1}{x(ax+b)}\,dx = \frac{1}{b} \cdot \ln\left|\frac{x}{ax+b}\right| + C \quad (b \neq 0)$

14. $\int \frac{1}{x(ax+b)^2}\,dx = \frac{1}{b(ax+b)} + \frac{1}{b^2} \cdot \ln\left|\frac{x}{ax+b}\right| + C \quad (b \neq 0)$

15. $\int \sqrt{x^2+a^2}\,dx = \frac{x}{2}\sqrt{x^2+a^2} + \frac{a^2}{2} \cdot \ln\left|x + \sqrt{x^2+a^2}\right| + C$

16. $\int x^n \cdot \ln|x|\,dx = x^{n+1}\left[\frac{\ln|x|}{n+1} - \frac{1}{(n+1)^2}\right] + C \quad (n \neq -1)$

17. $\int x^n e^{ax}\,dx = \frac{x^n e^{ax}}{a} - \frac{n}{a} \cdot \int x^{n-1} e^{ax}\,dx + C \quad (a \neq 0)$

TABLE 4 INTEGRALS INVOLVING TRIGONOMETRIC FUNCTIONS

18. $\int \sin u\, du = -\cos u + C$

19. $\int \cos u\, du = \sin u + C$

20. $\int \sec^2 u\, du = \tan u + C$

21. $\int \csc^2 u\, du = -\cot u + C$

22. $\int \sec u \tan u\, du = \sec u + C$

23. $\int \csc u \cot u\, du = -\csc u + C$

24. $\int \tan u\, du = \ln |\sec u| + C$

25. $\int \cot u\, du = \ln |\sin u| + C$

26. $\int \sec u\, du = \ln |\sec u + \tan u| + C$

27. $\int \csc u\, du = \ln |\csc u - \cot u| + C$

28. $\int \sin^n u\, du = -\frac{1}{n} \sin^{n-1} u \cos u + \frac{n-1}{n} \int \sin^{n-2} u\, du \quad (n \neq 0)$

29. $\int \cos^n u\, du = \frac{1}{n} \cos^{n-1} u \sin u + \frac{n-1}{n} \int \cos^{n-2} u\, du \quad (n \neq 0)$

30. $\int \tan^n u\, du = \frac{1}{n-1} \tan^{n-1} u - \int \tan^{n-2} u\, du \quad (n \neq 1)$

31. $\int \sec^n u\, du = \frac{1}{n-1} \tan u \sec^{n-2} u + \frac{n-2}{n-1} \int \sec^{n-2} u\, du \quad (n \neq 1)$

32. $\int \sin au \sin bu\, du = \frac{\sin(a-b)u}{2(a-b)} - \frac{\sin(a+b)u}{2(a+b)} + C, \quad |a| \neq |b|$

33. $\int \cos au \cos bu\, du = \frac{\sin(a-b)u}{2(a-b)} + \frac{\sin(a+b)u}{2(a+b)} + C, \quad |a| \neq |b|$

34. $\int \sin au \cos bu\, du = -\frac{\cos(a-b)u}{2(a-b)} - \frac{\cos(a+b)u}{2(a+b)} + C, \quad |a| \neq |b|$

35. $\int u \sin u\, du = \sin u - u \cos u + C$

36. $\int u^n \sin u\, du = -u^n \cos u + n \int u^{n-1} \cos u\, du$

37. $\int e^{au} \sin bu\, du = \frac{e^{au}}{a^2 + b^2}(a \sin bu - b \cos bu) + C$

38. $\int e^{au} \cos bu\, du = \frac{e^{au}}{a^2 + b^2}(a \cos bu + b \sin bu) + C$

Answers to Selected Exercises

Answers to selected writing exercises are provided.

CHAPTER R ALGEBRA REFERENCE

Exercises R.1 (page xxii)

1. $-x^2 + x + 9$ **2.** $-6y^2 + 3y + 10$ **3.** $-14q^2 + 11q - 14$ **4.** $9r^2 - 4r + 19$ **5.** $-.327x^2 - 2.805x - 1.458$ **6.** $-2.97r^2 - 8.083r + 7.81$ **7.** $-18m^3 - 27m^2 + 9m$ **8.** $12k^2 - 20k + 3$ **9.** $25r^2 + 5rs - 12s^2$ **10.** $18k^2 - 7kq - q^2$ **11.** $(6/25)y^2 + (11/40)yz + (1/16)z^2$ **12.** $(15/16)r^2 - (7/12)rs - (2/9)s^2$ **13.** $144x^2 - 1$ **14.** $36m^2 - 25$ **15.** $27p^3 - 1$ **16.** $6p^3 - 11p^2 + 14p - 5$ **17.** $8m^3 + 1$ **18.** $12k^4 + 21k^3 - 5k^2 + 3k + 2$ **19.** $m^2 + mn - 2n^2 - 2km + 5kn - 3k^2$ **20.** $2r^2 - 7rs + 3s^2 + 3rt - 4st + t^2$ **21.** $x^3 + 6x^2 + 11x + 6$ **22.** $x^3 - 2x^2 - 5x + 6$ **23.** $9a^2 + 6ab + b^2$ **24.** $x^3 - 6x^2y + 12xy^2 - 8y^3$

Exercise R.2 (page xxv)

1. $8a(a^2 - 2a + 3)$ **2.** $3y(y^2 + 8y + 3)$ **3.** $5p^2(5p^2 - 4pq + 20q^2)$ **4.** $10m^2(6m^2 - 12mn + 5n^2)$ **5.** $(m + 7)(m + 2)$ **6.** $(x + 5)(x - 1)$ **7.** $(z + 4)(z + 5)$ **8.** $(b - 7)(b - 1)$ **9.** $(a - 5b)(a - b)$ **10.** $(s - 5t)(s + 7t)$ **11.** $(y - 7z)(y + 3z)$ **12.** $6(a - 10)(a + 2)$ **13.** $3m(m + 3)(m + 1)$ **14.** $(2x + 1)(x - 3)$ **15.** $(3a + 7)(a + 1)$ **16.** $(2a - 5)(a - 6)$ **17.** $(5y + 2)(3y - 1)$ **18.** $(7m + 2n)(3m + n)$ **19.** $2a^2(4a - b)(3a + 2b)$ **20.** $4z^3(8z + 3a)(z - a)$ **21.** $(x + 8)(x - 8)$ **22.** $(3m + 5)(3m - 5)$ **23.** $(11a + 10)(11a - 10)$ **24.** Prime **25.** $(z + 7y)^2$ **26.** $(m - 3n)^2$ **27.** $(3p - 4)^2$ **28.** $(a - 6)(a^2 + 6a + 36)$ **29.** $(2r - 3s)(4r^2 + 6rs + 9s^2)$ **30.** $(4m + 5)(16m^2 - 20m + 25)$ **31.** $(x - y)(x + y)(x^2 + y^2)$ **32.** $(2a - 3b)(2a + 3b)(4a^2 + 9b^2)$

Exercise R.3 (page xxviii)

1. $z/2$ **2.** $5p/2$ **3.** $8/9$ **4.** $3/(t - 3)$ **5.** $2(x + 2)/x$ **6.** $4(y + 2)$ **7.** $(m - 2)/(m + 3)$ **8.** $(r + 2)/(r + 4)$ **9.** $(x + 4)/(x + 1)$ **10.** $(z - 3)/(z + 2)$ **11.** $(2m + 3)/(4m + 3)$ **12.** $(2y + 1)/(y + 1)$ **13.** $(3k)/5$ **14.** $(25p^2)/9$ **15.** $6/(5p)$ **16.** 2 **17.** $2/9$ **18.** $3/10$ **19.** $2(a + 4)/(a - 3)$ **20.** $2/(r + 2)$ **21.** $(k + 2)/(k + 3)$ **22.** $(m + 6)/(m + 3)$ **23.** $(m - 3)/(2m - 3)$ **24.** $(2n - 3)/(2n + 3)$ **25.** 1 **26.** $(6 + p)/(2p)$ **27.** $(8 - y)/(4y)$ **28.** $137/(30m)$ **29.** $(3m - 2)/[m(m - 1)]$ **30.** $(r - 12)/[r(r - 2)]$ **31.** $14/[3(a - 1)]$ **32.** $23/[20(k - 2)]$ **33.** $(7x + 9)/[(x - 3)(x + 1)(x + 2)]$ **34.** $y^2/[(y + 4)(y + 3)(y + 2)]$ **35.** $k(k - 13)/[(2k - 1)(k + 2)(k - 3)]$ **36.** $m(3m - 19)/[(3m - 2)(m + 3)(m - 4)]$ **37.** $(4a + 1)/[a(a + 2)]$ **38.** $(5x^2 + 4x - 4)/[x(x - 1)(x + 1)]$

Exercise R.4 (page xxxiv)

1. 12 **2.** $-2/7$ **3.** -12 **4.** $3/4$ **5.** $-7/8$ **6.** $-6/11$ **7.** -1 **8.** $-10/19$ **9.** $-3, -2$ **10.** $-1, 3$ **11.** 4 **12.** $-2, 5/2$ **13.** $-1/2, 4/3$ **14.** $2, 5$ **15.** $-4/3, 4/3$ **16.** $-4, 1/2$ **17.** $0, 4$ **18.** $(5 + \sqrt{13})/6 \approx 1.434, (5 - \sqrt{13})/6 \approx .232$ **19.** $(1 + \sqrt{33})/4 \approx 1.686, (1 - \sqrt{33})/4 \approx -1.186,$ **20.** $(-1 + \sqrt{5})/2 \approx .618, (-1 - \sqrt{5})/2 \approx -1.618$ **21.** $5 + \sqrt{5} \approx 7.236, 5 - \sqrt{5} \approx 2.764$ **22.** $(-6 + \sqrt{26})/2 \approx -.450, (-6 - \sqrt{26})/2 \approx -5.550$ **23.** $1, 5/2$ **24.** No real number solutions **25.** $(-1 + \sqrt{73})/6 \approx 1.257, (-1 - \sqrt{73})/6 \approx -1.591$ **26.** $-1, 0$ **27.** 3 **28.** 12 **29.** $-59/6$ **30.** $-11/5$ **31.** No real number solutions **32.** $-5/2$ **33.** $2/3$ **34.** 1 **35.** $(-13 - \sqrt{185})/4 \approx -6.650, (-13 + \sqrt{185})/4 \approx .150$ **36.** No solution **37.** $-15/4$

Exercise R.5 (page xl)

1. $(-\infty, 0)$ 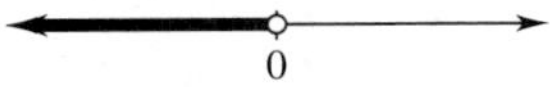 **2.** $[-3, \infty)$  **3.** $[1, 2)$ 

4. $(-5, -4]$  **5.** $(-\infty, -9)$  **6.** $[6, \infty)$ 

7. $-4 < x < 3$ **8.** $2 \le x < 7$ **9.** $x \le -1$ **10.** $x > 3$ **11.** $-2 \le x < 6$ **12.** $0 < x < 8$ **13.** $x \le -4$ or $x \ge 4$
14. $x < 0$ or $x \ge 3$ **15.** $(-\infty, -1]$  **16.** $(-\infty, 1)$ 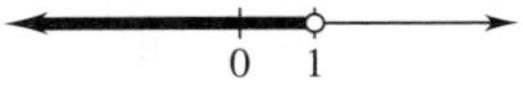

17. $(-1, \infty)$ 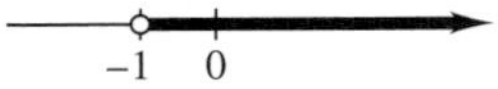 **18.** $(-\infty, 1]$ 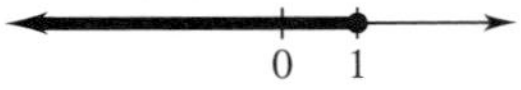 **19.** $(1/5, \infty)$ 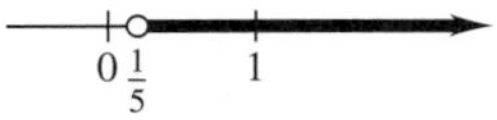

20. $(1/3, \infty)$ 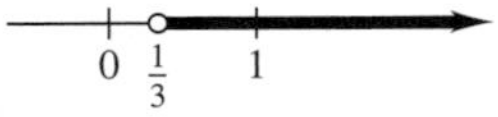 **21.** $(-5, 6)$  **22.** $[7/3, 4]$ 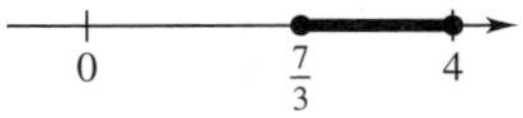

23. $[-11/2, 7/2]$ 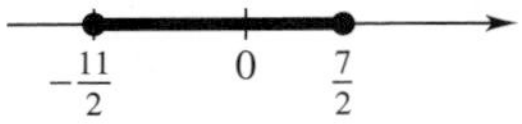 **24.** $[-1, 2]$ 

25. $[-17/7, \infty)$  **26.** $(-\infty, 50/9]$ 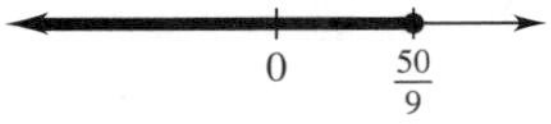

27. $(-2, 4)$  **28.** $(-\infty, -6] \cup [1, \infty)$ 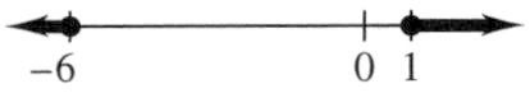

29. $(1, 2)$ 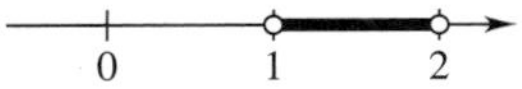 **30.** $(-\infty, -4) \cup (1/2, \infty)$ 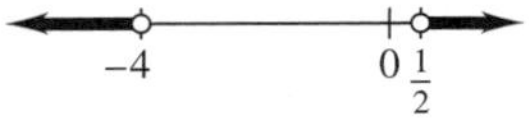

31. $[1, 6]$  **32.** $[-3/2, 5]$ 

33. $(-\infty, -1/2) \cup (1/3, \infty)$ 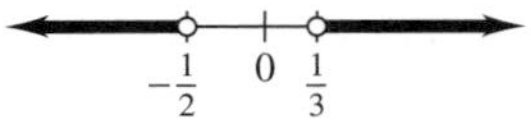 **34.** $[-1/2, 2/5]$ 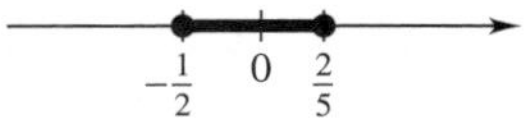

35. $[-3, 1/2]$ 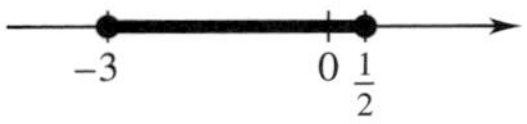 **36.** $(-\infty, -2) \cup (5/3, \infty)$ 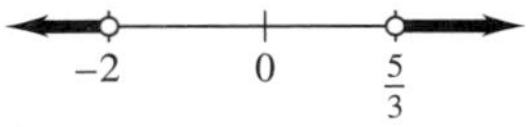

37. $[-5, 5]$  **38.** $(-\infty, 0) \cup (16, \infty)$  **39.** $(-5, 3]$

40. $(-\infty, -1) \cup (1, \infty)$ **41.** $(-\infty, -2)$ **42.** $(-2, 3/2)$ **43.** $[-8, 5)$ **44.** $(-\infty, -3/2) \cup [-13/9, \infty)$ **45.** $(-2, \infty)$
46. $(-\infty, -1)$ **47.** $(-\infty, -1) \cup (-1/2, 1) \cup (2, \infty)$ **48.** $(-4, -2) \cup (0, 2)$ **49.** $(1, 3/2]$
50. $(-\infty, -2) \cup (-2, 2) \cup [4, \infty)$

Exercises R.6 (page xliv)

1. $1/64$ **2.** $1/81$ **3.** 1 **4.** 1 **5.** $-1/9$ **6.** $1/9$ **7.** $49/4$ **8.** $27/64$ **9.** $1/3^6$ **10.** 8^5 **11.** $1/10^8$ **12.** 5 **13.** x^2 **14.** y^3
15. 2^3k^3 **16.** $1/(3z^7)$ **17.** $x^2/(2y)$ **18.** $m^3/5^4$ **19.** a^3b^6 **20.** $d^6/(2^2c^4)$ **21.** x^4/y^4 **22.** b/a^3 **23.** $(a + b)/(ab)$
24. $(1 - ab^2)/b^2$ **25.** $2(m - n)/[mn(m + n^2)]$ **26.** $(3n^2 + 4m)/(mn^2)$ **27.** $xy/(y - x)$ **28.** $x^4y^4/(x^2 + y^2)^2$ **29.** 9
30. 3 **31.** 4 **32.** -25 **33.** $2/3$ **34.** $4/3$ **35.** $1/32$ **36.** $1/5$ **37.** $4/3$ **38.** $1000/1331$ **39.** 2^2 **40.** $27^{1/3}$ **41.** 4^2 **42.** 1
43. r **44.** $12^3/y^8$ **45.** $1/(2^2 \cdot 3k^{5/2})$ or $1/(12k^{5/2})$ **46.** $1/(2p^2)$ **47.** $a^{2/3}b^2$ **48.** $y/(x^{4/3}z^{1/2})$ **49.** $h^{1/3}t^{1/5}/k^{2/5}$ **50.** m^3p/n
51. $4x(x^2 + 2)(-x^3 + 6x - 1)$ **52.** $6x(x^3 + 7)(-2x^3 - 5x + 7)$ **53.** $3x^3(x^2 - 1)^{-1/2}$ **54.** $3(6x + 2)^{-1/2}(27x + 5)$
55. $(2x + 5)(x^2 - 4)^{-1/2}(4x^2 + 5x - 8)$ **56.** $(4x^2 + 1)(2x - 1)^{-1/2}(36x^2 - 16x + 1)$

Exercises R.7 (page xlix)

1. 5 **2.** 6 **3.** -5 **4.** $5\sqrt{2}$ **5.** $20\sqrt{5}$ **6.** $4y^2\sqrt{2y}$ **7.** $7\sqrt{2}$ **8.** $9\sqrt{3}$ **9.** $2\sqrt{5}$ **10.** $-2\sqrt{7}$ **11.** $5\sqrt[3]{2}$ **12.** $7\sqrt[3]{3}$
13. $3\sqrt[3]{4}$ **14.** $xyz^2\sqrt{2x}$ **15.** $7rs^2t^5\sqrt{2r}$ **16.** $2x^2yz\sqrt[3]{2x^2yz^2}$ **17.** $x^2yz^2\sqrt[4]{y^3z^3}$ **18.** $ab\sqrt{ab}(b - 2a^2 + b^3)$

19. $p^2\sqrt{pq}(pq - q^4 + p^2)$ **20.** $5\sqrt{7}/7$ **21.** $-2\sqrt{3}/3$ **22.** $-\sqrt{3}/2$ **23.** $\sqrt{2}$ **24.** $-3(1 + \sqrt{5})/4$ **25.** $-5(2 + \sqrt{6})/2$ **26.** $-2(\sqrt{3} + \sqrt{2})$ **27.** $(\sqrt{10} - \sqrt{3})/7$ **28.** $(\sqrt{r} + \sqrt{3})/(r - 3)$ **29.** $5(\sqrt{m} + \sqrt{5})/(m - 5)$ **30.** $\sqrt{y} + \sqrt{5}$ **31.** $\sqrt{z} + \sqrt{11}$ **32.** $-2x - 2\sqrt{x(x + 1)} - 1$ **33.** $[p^2 + p + 2\sqrt{p(p^2 - 1)} - 1]/(-p^2 + p + 1)$ **34.** $-1/[2(1 - \sqrt{2})]$ **35.** $-2/[3(1 + \sqrt{3})]$ **36.** $-1/[2x - 2\sqrt{x(x + 1)} + 1]$ **37.** $(-p^2 + p + 1)/[p^2 + p - 2\sqrt{p(p^2 - 1)} - 1]$ **38.** $|4 - x|$ **39.** $|2y + 1|$ **40.** Cannot be simplified **41.** Cannot be simplified

CHAPTER 1 LINEAR FUNCTIONS

Exercises 1.1 (page 14)

1. $3/5$ **3.** Not defined **5.** 2 **7.** $5/9$ **9.** Not defined **11.** 0 **13.** 2 **15.** $y = -2x + 5$ **17.** $y = 1$ **19.** $y = -(1/3)x + 10/3$ **21.** $y = -(3/5)x + 59/30$ **23.** $x = -8$ **25.** $y = (2/3)x - 2$ **27.** $x = -6$ **29.** $y = -(1/3)x + 11/3$ **31.** $y = x - 7$ **33.** $y = (2/3)x + 2$ **35.** No **39.** a **41.** -4

45.

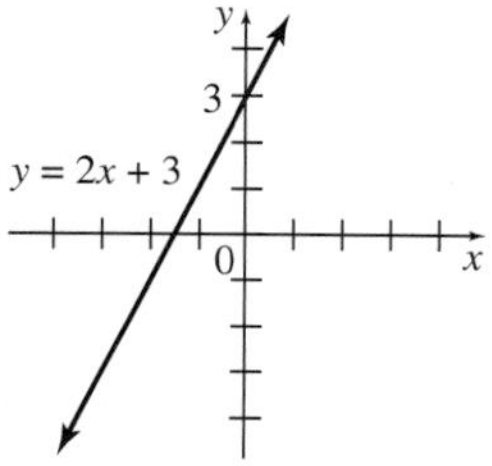

47.

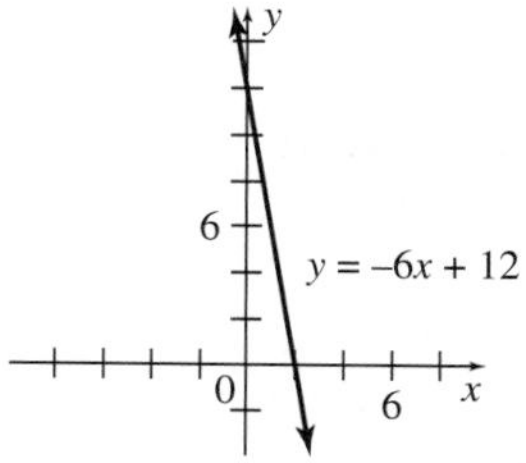

49.

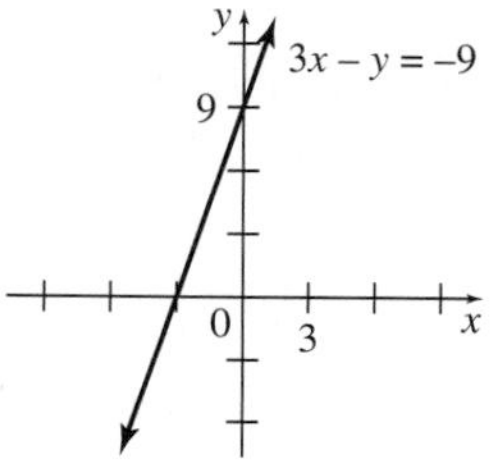

51.

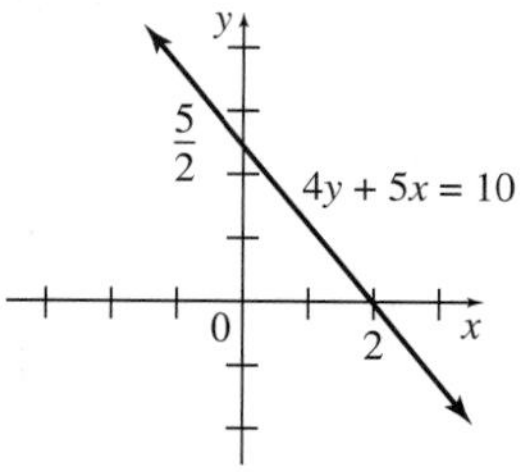

53.

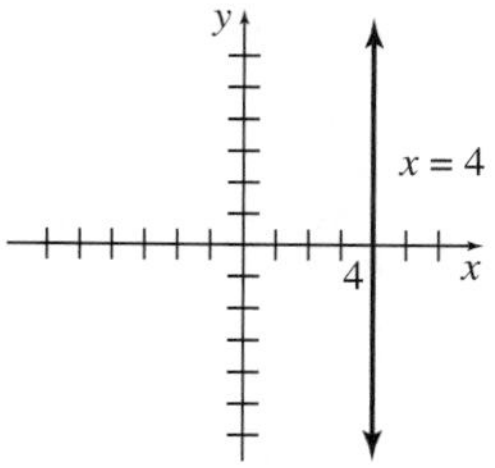

55.

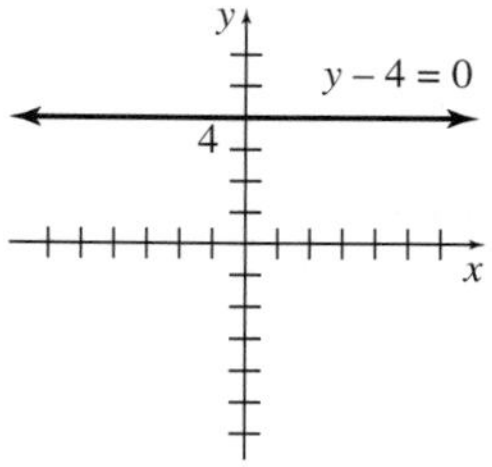

57.

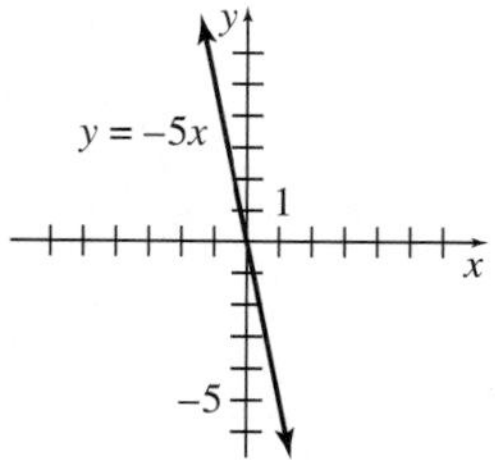

59.

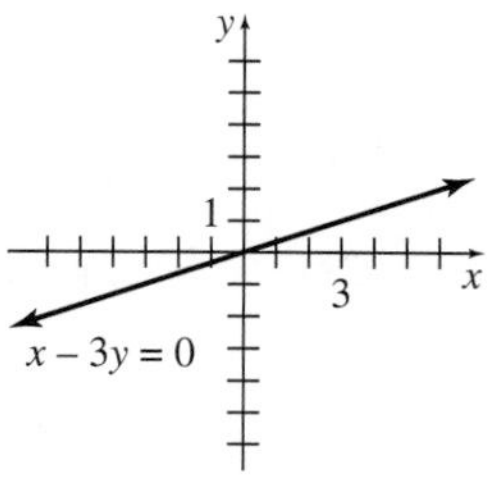

61. a.

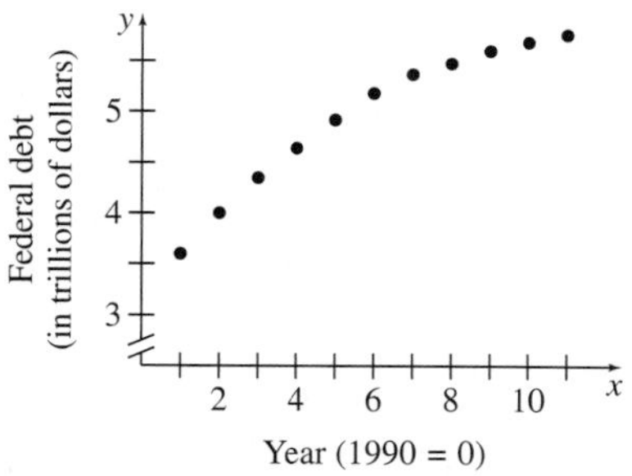

The debt is increasing and the data appear to be nearly linear.

b.

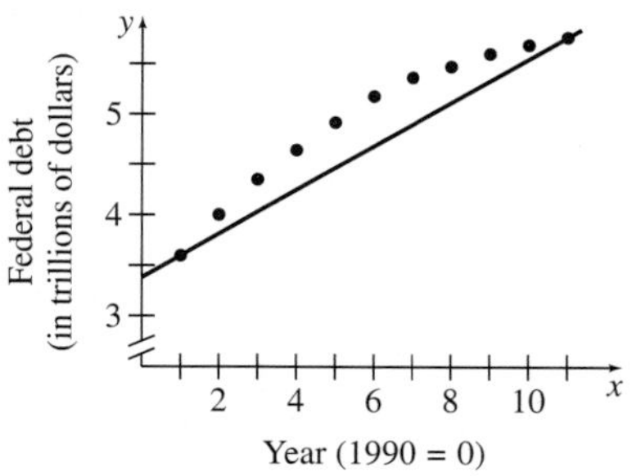

$y = .2171x + 3.3819$. The slope .2171 indicates that the federal debt was increasing at a rate of .2171 trillion dollars per year.

c. 5.9871 trillion dollars. The result is less than the actual debt.

63. a. $y \approx .053x - .043$ **b.** About 10.2 yr **65.** Approximately 4.3 m/sec **67.** 23 **69. a.** $y = .19x + 1.1$ **b.** About 7.4 million **71. a.** $y = 6.62x + 100$ **b.** 219.16; it is less than the actual CPI. **c.** It increases 6.62 per year. **73. a.** There appears to be a linear relationship. **b.** $y = 76.9x$ **c.** About 780 megaparsecs (about 1.5×10^{22} mi) **d.** About 12.4 billion yr

75. a.

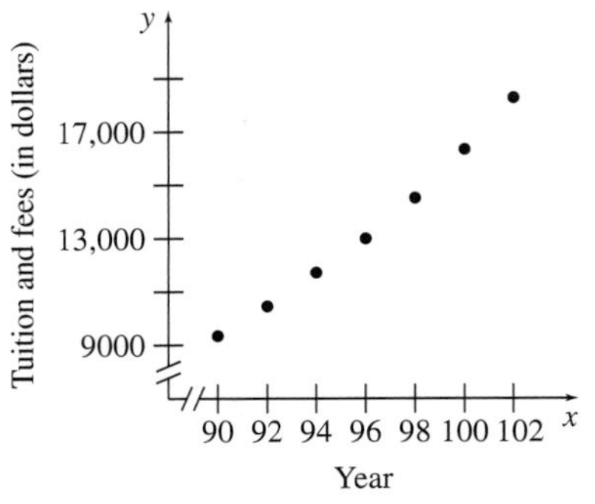

Yes

b. $y = 699.2x - 53{,}588$; the slope of 699.2 indicates that tuition and fees have increased approximately $699 per yr. **c.** The year 2020 is too far in the future to make accurate predictions based on the given data. Many factors could affect college costs by then.

Exercises 1.2 (page 27)

1. True **3.** True **9.** If $C(x)$ is the cost of hauling a trailer for x mi, then $C(x) = 45 + 2x$. **11.** If $R(x)$ is the cost of renting a car for x mi, then $R(x) = 44 + .28x$. **13.** $C(x) = 25x + 400$ **15.** $C(x) = 120x + 12{,}500$

19. a.

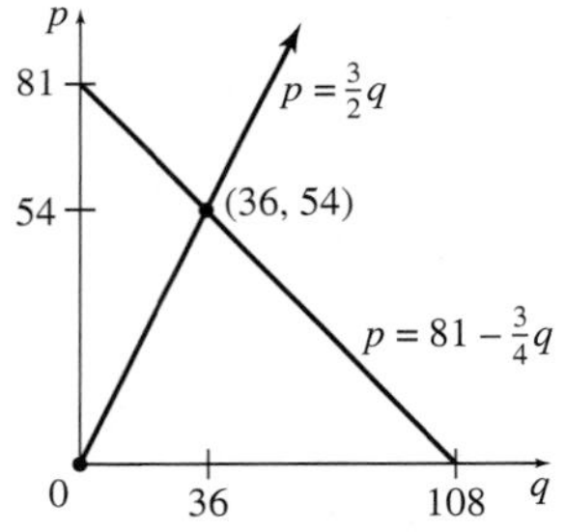

b. 36 batches; $54 **21. a.**

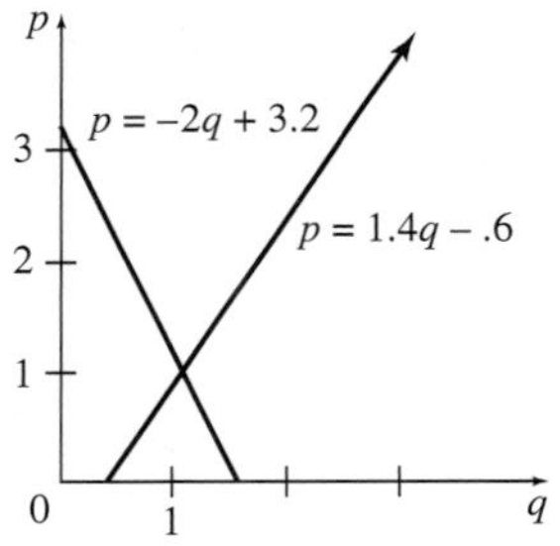

b. About 1120 lb; about $.96

23. a. $C(x) = 2.15x + 525$ **b.** 188 **c.** 545 **25. a.** $C(x) = 500{,}000 + 4.75x$ **b.** $500,000 **c.** $975,000 **d.** $4.75; each additional items costs $4.75 to produce **27. a.** 2 units **b.** $980 **c.** 52 units **29.** Break-even quantity is 45 units; don't produce; $P(x) = 20x - 900$ **31.** Break-even quantity is −50 units; impossible to make a profit here since $C(x) > R(x)$ for all positive x; $P(x) = -10x - 500$ (always a loss!) **33. a.** 14.4°C **b.** −28.9°C **c.** 122°F **35.** −40°

Exercises 1.3 (page 37)

3. c **5. a.** $Y = 32.3x - 2672.6$ **b.** 7.189 billion **c.** 2006 **d.** $r = .999$, which indicates that the line fits the data points very well.

7. a.

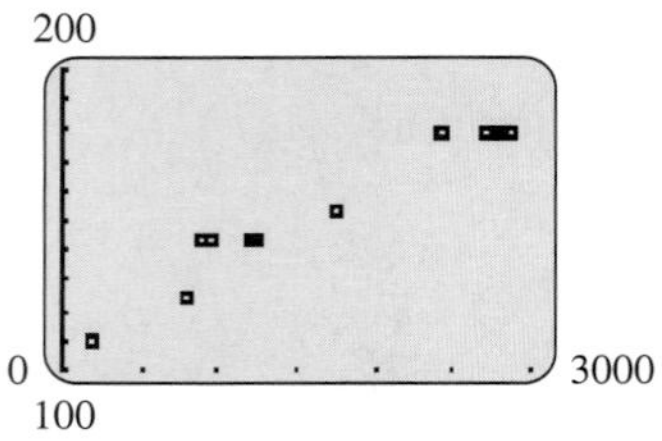

Yes **b.** .972; yes **c.** $Y = 113 + .0243x$; 2.43¢ per mi **9. a.** $Y = .129x + 2.15$; **b.** $Y = .106x + 4.12$; $r = .436$

11. a.

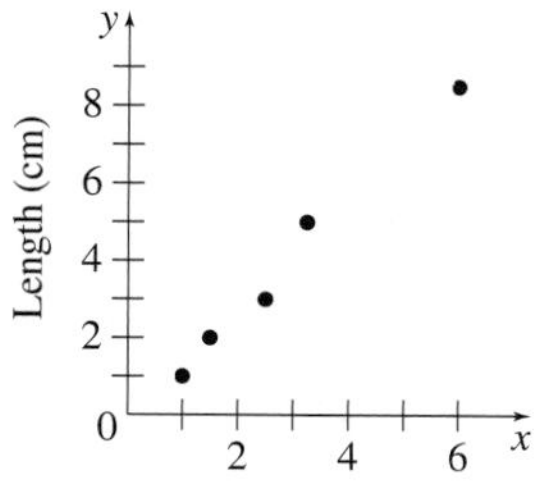

b. $Y = 1.585x - .487$

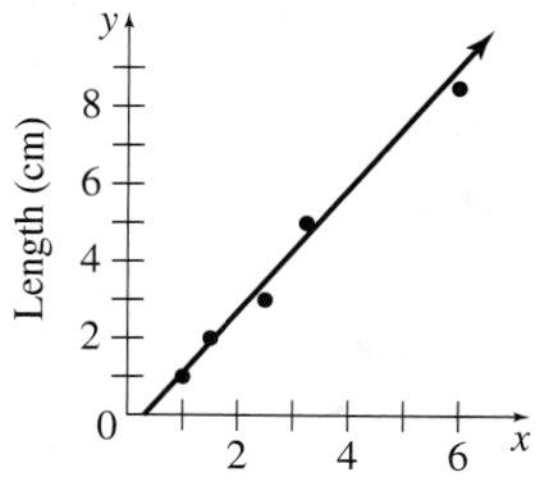

c. No; it gives negative values for small widths.

d. .999 **13.** .159; no **15. a.** $Y = -.0067x + 14.75$ **b.** 12 **c.** 11 **d.** $-.13$ **e.** There is no linear relationship.

17. a.

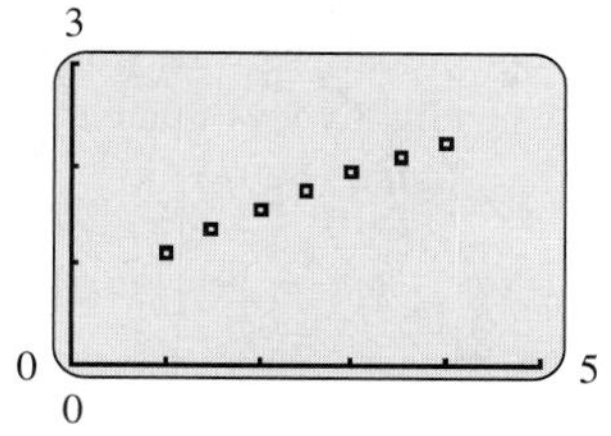

b. $Y = .366x + .803$; the line seems to fit the data.

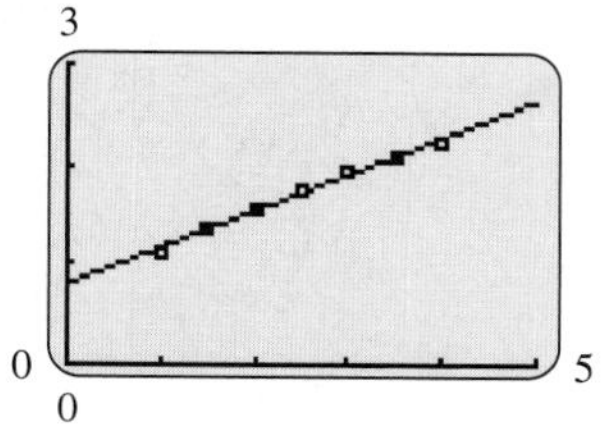

c. $r = .995$ indicates a good fit, which confirms the conclusion in part b. **19. a.** $-.995$; yes **b.** $Y = -.0769x + 5.91$ **c.** 2.07 points

21. a. 3.44 mph **b.**

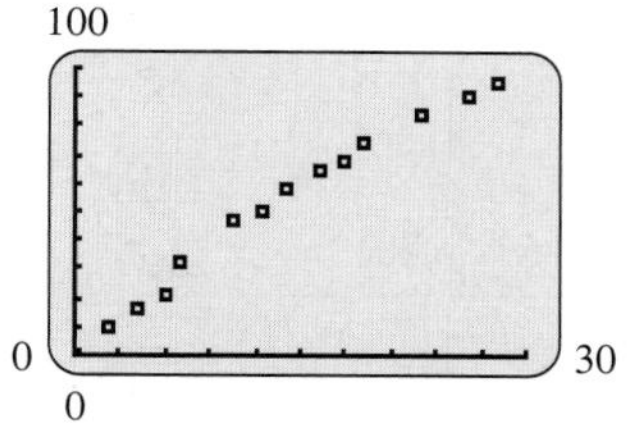

Yes **c.** $Y = 3.39x + 4.32$ **d.** .994; yes **e.** 3.39 miles per hour

Chapter 1 Review Exercises (page 44)

3. 1/3 **5.** $-2/11$ **7.** $-2/3$ **9.** 0 **11.** -3 **13.** $y = (2/3)x - 13/3$ **15.** $y = -(5/4)x + 17/4$ **17.** $x = -1$

19. $y = 3x - 7$ **21.** $y = -10$ **23.** $x = -7$ **25.**

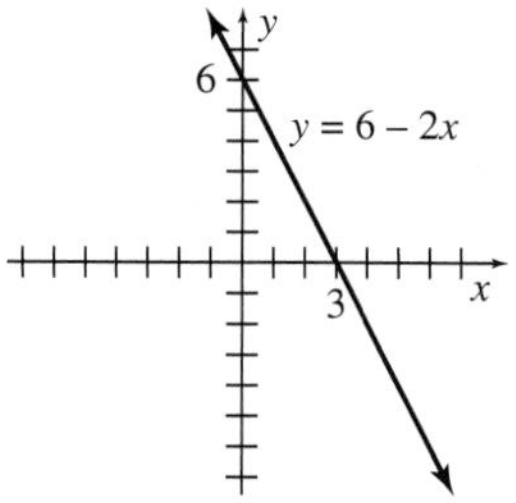

27.

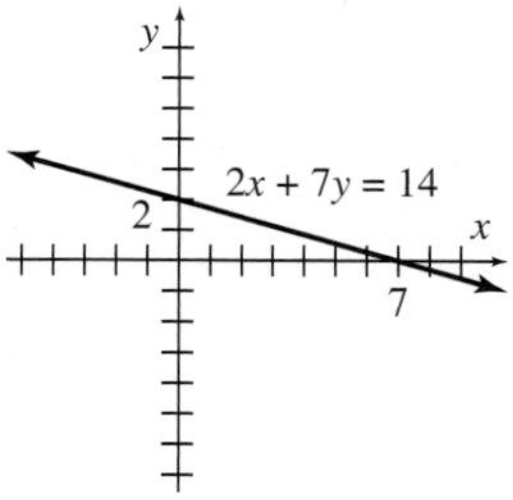

29.

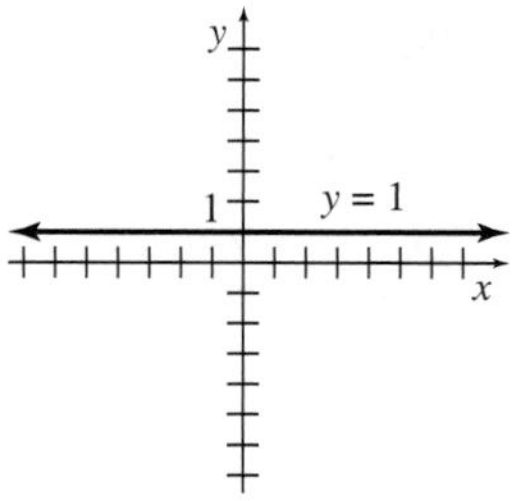

31.

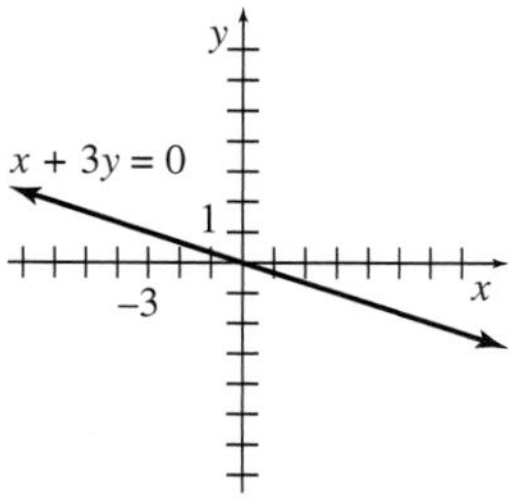

33. a. 7/6; 9/2 **b.** 2; 2 **c.** 5/2; 1/2 **d.**

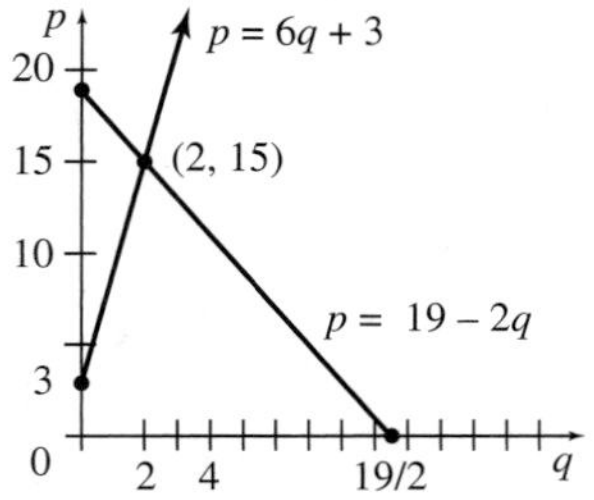

e. \$15 **f.** 2

35. $D(q) = -.5q + 72.50$ **37.** $C(x) = 30x + 60$ **39.** $C(x) = 30x + 85$ **41. a.** 5 cartons **b.** \$2000 **43.** $y = 8.3x - 736.3$ **45.** $I(x) = 771.7x - 35{,}022$ **47. a.** .929; yes **b.**

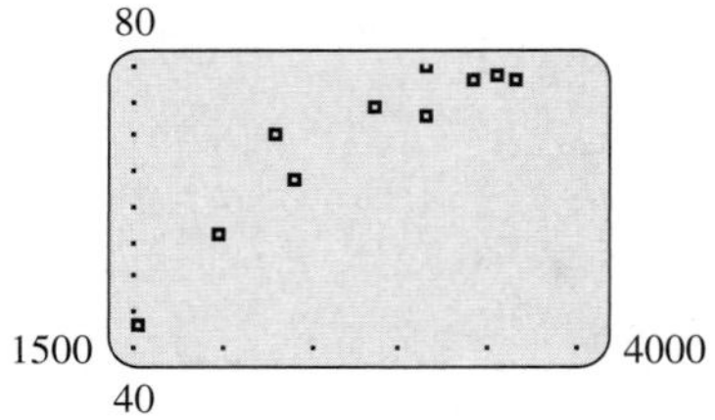

Yes **c.** $Y = .0156x + 25.9$ **d.** About 76 years, which is slightly lower than the actual figure

49. $y = -.41x + 157.79$ **51. a.** .835; yes, but the fit is not very good **b.**

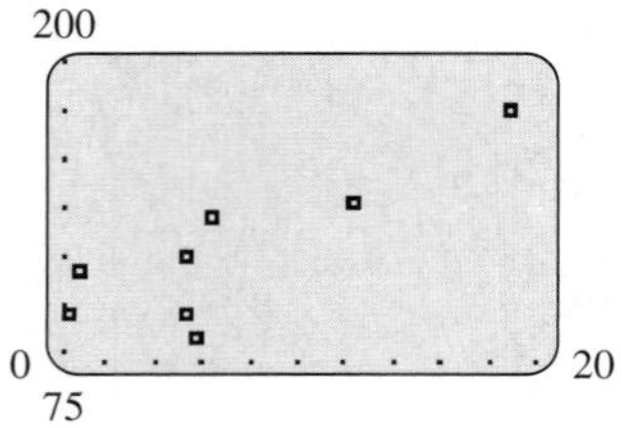

c. $Y = 4.23x + 91.32$ **d.** About \$4230

CHAPTER 2 NONLINEAR FUNCTIONS

Exercises 2.1 (page 59)

1. Not a function **3.** Function **5.** Function **7.** Not a function **9.** $(-2, -1), (-1, 1), (0, 3), (1, 5), (2, 7), (3, 9)$; range: $\{-1, 1, 3, 5, 7, 9\}$

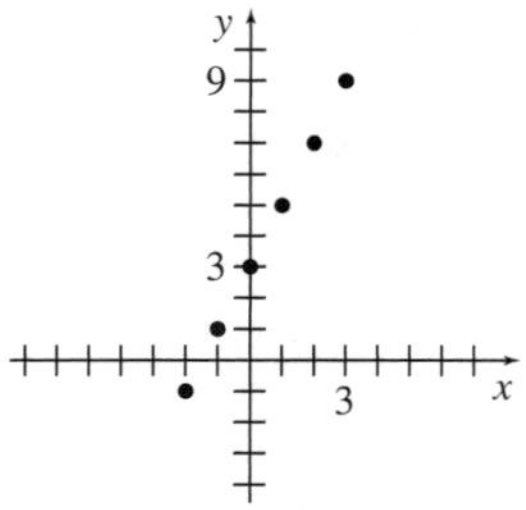

11. $(-2, 3/2), (-1, 2), (0, 5/2), (1, 3), (2, 7/2), (3, 4)$; range: $\{3/2, 2, 5/2, 3, 7/2, 4\}$

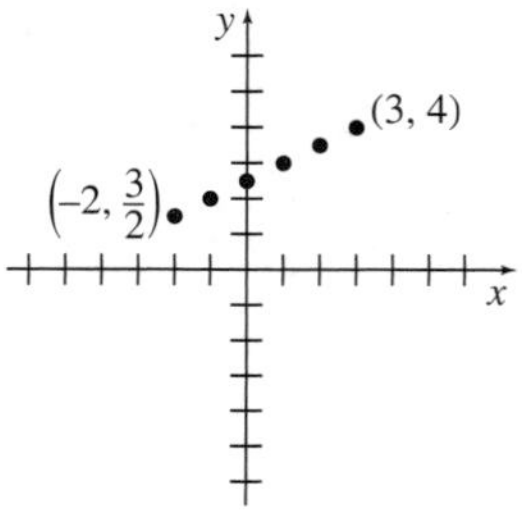

13. $(-2, 2), (-1, 0), (0, 0), (1, 2), (2, 6), (3, 12)$; range: $\{0, 2, 6, 12\}$

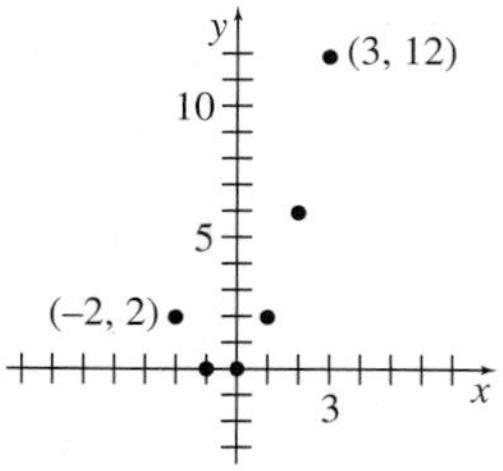

15. $(-2, 4), (-1, 1), (0, 0), (1, 1), (2, 4), (3, 9)$; range: $\{0, 1, 4, 9\}$

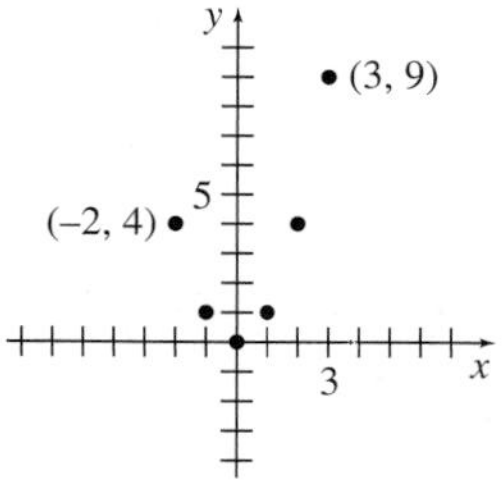

17. $(-2, 1), (-1, 1/2), (0, 1/3), (1, 1/4), (2, 1/5), (3, 1/6)$; range: $\{1, 1/2, 1/3, 1/4, 1/5, 1/6\}$

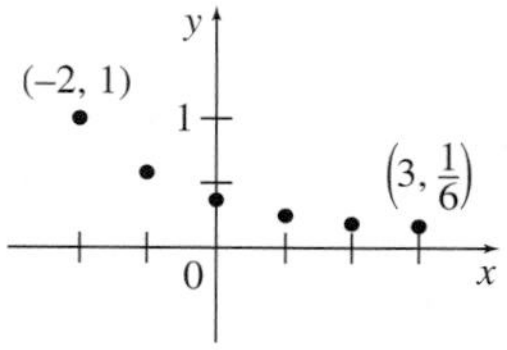

19. $(-2, -3), (-1, -3/2), (0, -3/5), (1, 0), (2, 3/7), (3, 3/4)$; range: $\{-3, -3/2, -3/5, 0, 3/7, 3/4\}$

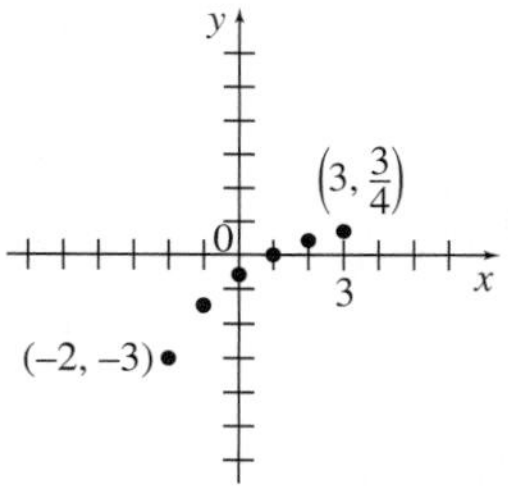

21. $(-\infty, \infty)$ **23.** $(-\infty, \infty)$ **25.** $[-4, 4]$ **27.** $[3, \infty)$ **29.** $(-\infty, -2) \cup (-2, 2) \cup (2, \infty)$ **31.** $(-\infty, \infty)$ **33.** $(-\infty, -1] \cup [5, \infty)$ **35.** $(-\infty, 2) \cup (4, \infty)$ **37.** Domain: $[-5, 4)$; range: $[-2, 6]$ **39.** Domain: $(-\infty, \infty)$; range: $(-\infty, 12]$ **41. a.** 5 **b.** $-7/4$ **c.** $-a^2 + 5a + 1$ **d.** $-4/m^2 + 10/m + 1$ or $(-4 + 10m + m^2)/m^2$ **e.** 0, 5 **43. a.** $9/2$ **b.** 0 **c.** $(2a + 1)/(a - 2)$ **d.** $(4 + m)/(2 - 2m)$ **e.** -3 **45.** Domain: $[-2, 4]$; range: $[0, 4]$ **a.** 0 **b.** 4 **c.** 3 **d.** $-1.5, 1.5, 2.5$ **47.** Domain: $[-2, 4]$; range: $[-3, 2]$ **a.** -3 **b.** -2 **c.** -1 **d.** 2.5 **49.** $6m^2 - 36m + 52$ **51.** $r^2 + 2rh + h^2 - 2r - 2h + 5$ **53.** $9/q^2 - 6/q + 5$ or $(9 - 6q + 5q^2)/q^2$ **55.** Function **57.** Not a function **59.** Function **61. a.** $x^2 + 2xh + h^2 - 4$ **b.** $2xh + h^2$ **c.** $2x + h$ **63. a.** $2x^2 + 4xh + 2h^2 - 4x - 4h - 5$ **b.** $4xh + 2h^2 - 4h$ **c.** $4x + 2h - 4$ **65. a.** $1/(x + h)$ **b.** $-h/[x(x + h)]$ **c.** $-1/[x(x + h)]$ **67. a.** The years **b.** The number of Internet users **c.** 101 million **d.** Doman: $1995 \le x \le 2001$; range: $16{,}000{,}000 \le y \le 553{,}000{,}000$ **69. a.** \$52 **b.** \$52 **c.** \$52 **d.** \$79 **e.** \$106 **f.**

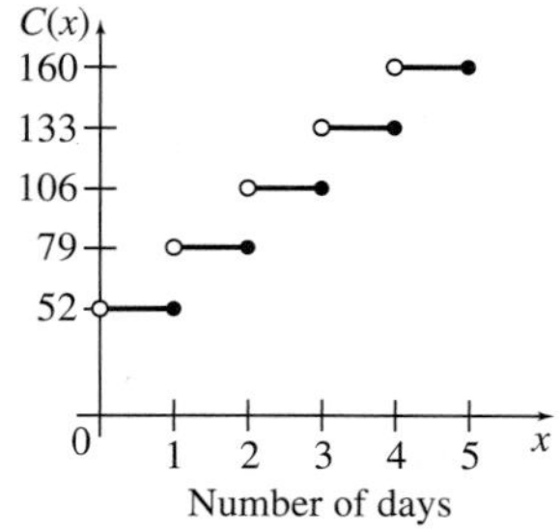

g. Yes **h.** No

71. a. i. 66 kcal/day **ii.** 222 kcal/day **b.** $g(z) = .454z$ **73. a.** 1880; 50% **b.** 1965; 35% **75. a.** $A = (3000 - w)w$ **b.** $0 \le w \le 3000$ **c.**

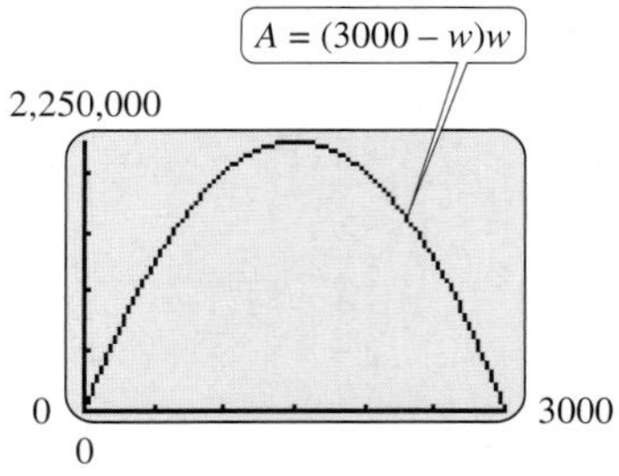

Exercises 2.2 (page 72)

3. D **5.** B **7.** E **9.** Vertex is $(5, -4)$; axis is $x = 5$; x-intercepts are 3 and 7; y-intercept is 21

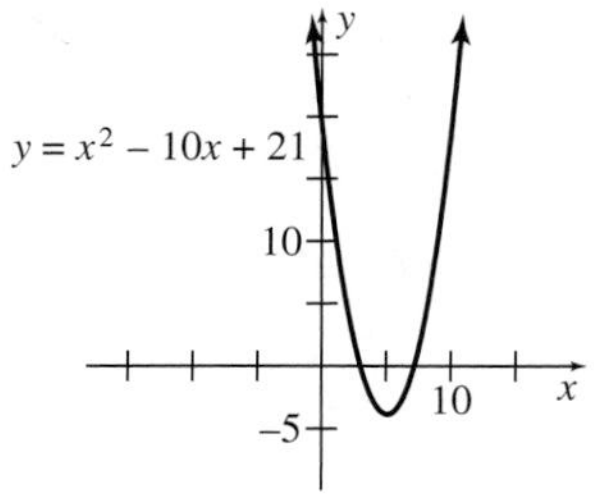

11. Vertex is $(2, 1)$; axis is $x = 2$; x-intercepts are $2 \pm \sqrt{3}/3 \approx 2.58$ or 1.42; y-intercept is -11

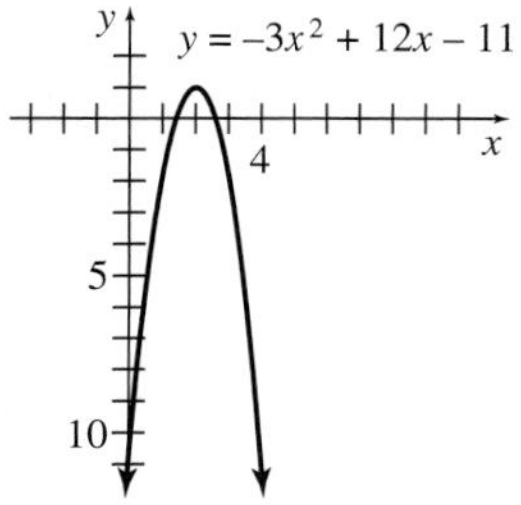

13. Vertex is $(3, 3)$; axis is $x = 3$; x-intercepts are $3 \pm \sqrt{3} \approx 4.73$ or 1.27; y-intercept is -6

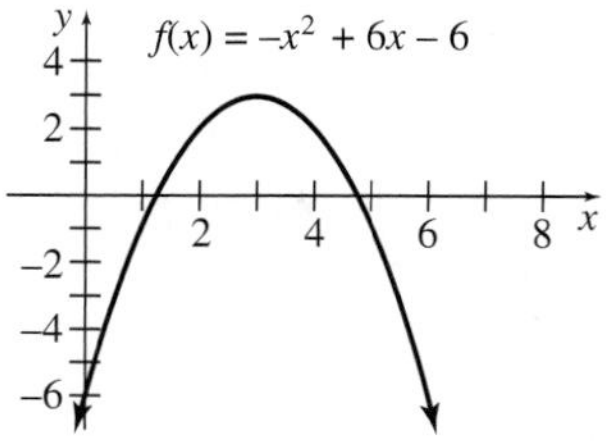

15. Vertex is $(4, 12)$; axis is $x = 4$; x-intercepts are 6 and 2; y-intercept is -36

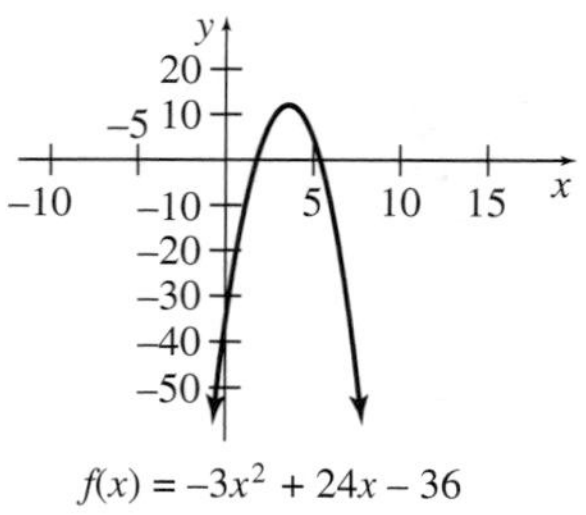

17. Vertex is $(-2, -2)$; axis is $x = -2$; x-intercepts are $-2 \pm 2\sqrt{5}/5 \approx -1.11$ or -2.89; y-intercept is 8

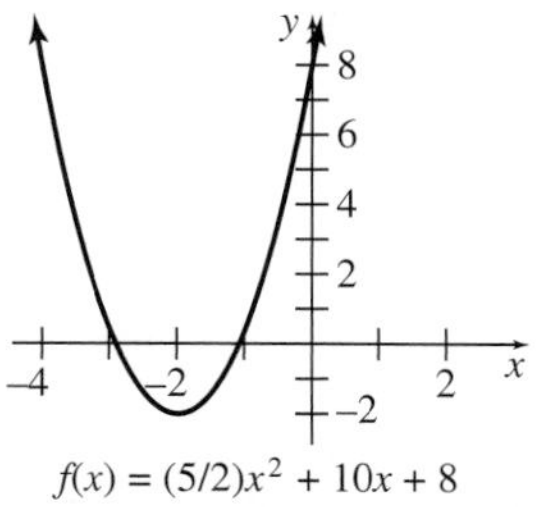

19. Vertex is $(-1, -3)$; axis is $x = -1$; no x-intercepts; y-intercept is $-7/2$

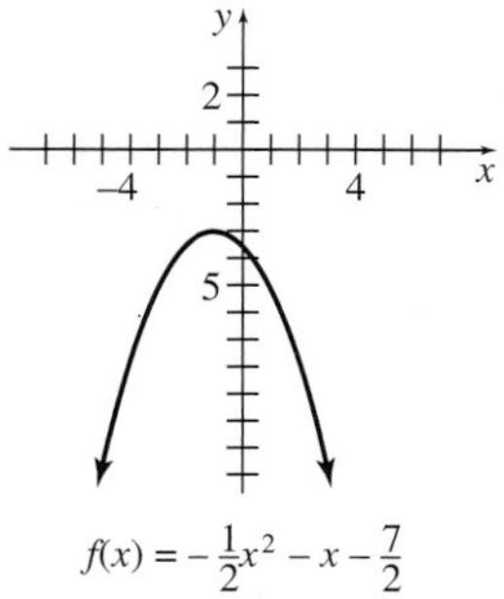

21. A **23.** B **25.**

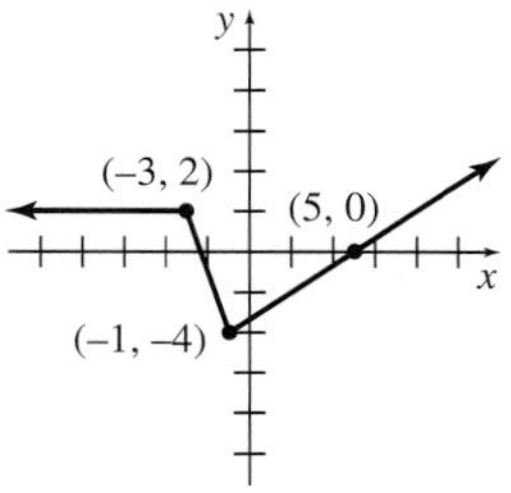

27.

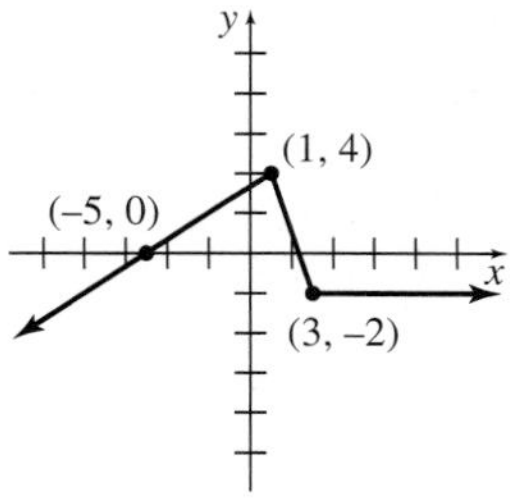

29.

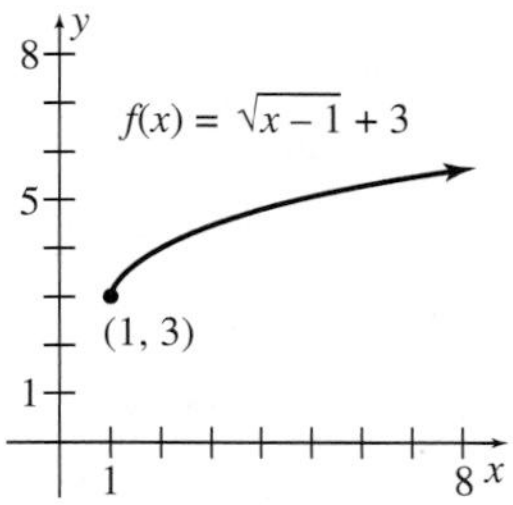

31.

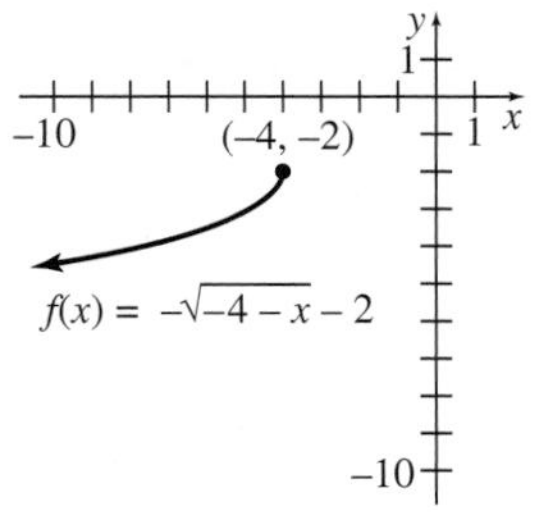

33.

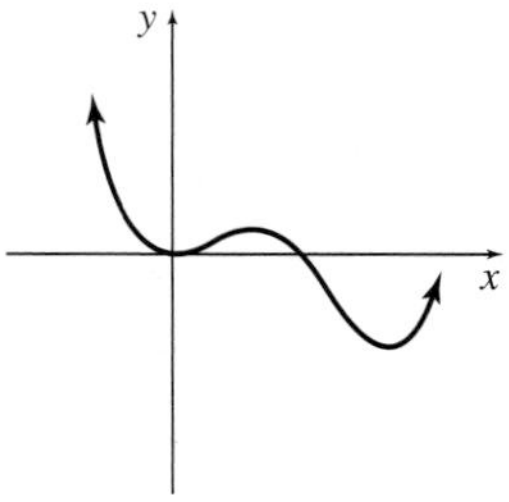

35.

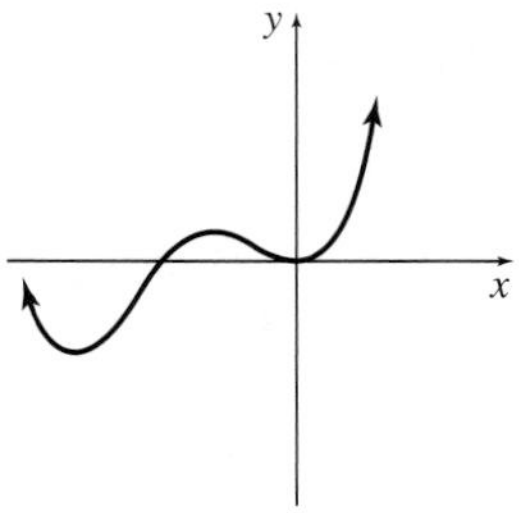

37.

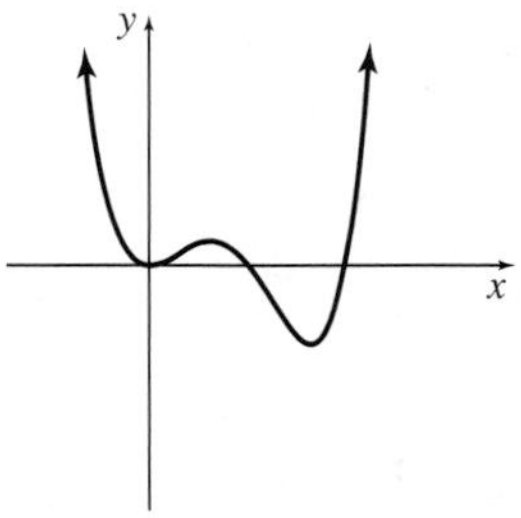

39.

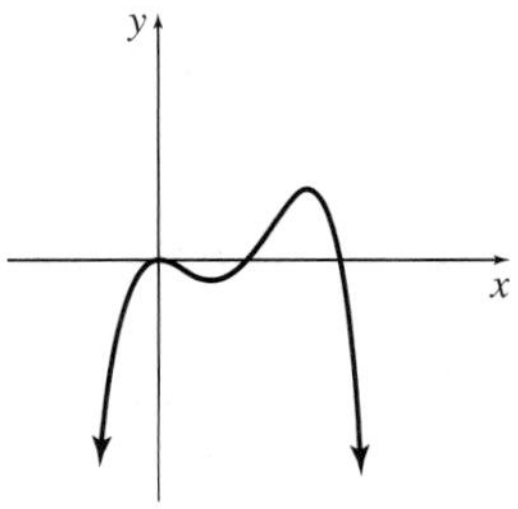

41. a. r **b.** $-r$ **c.** $-r$ **43. a.**

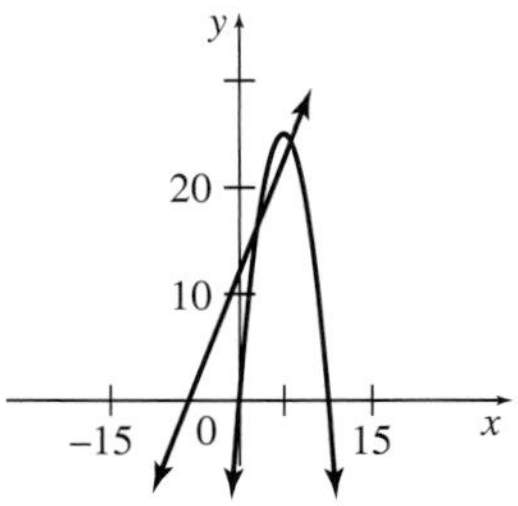

b. 2 **c.** 25 **d.** 4

45. a.

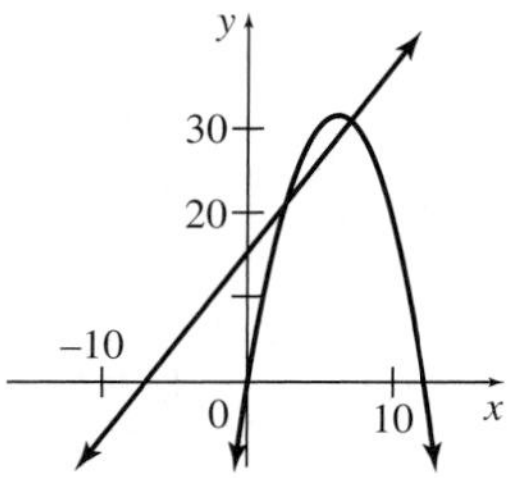

b. 2.5 **c.** 31.25 **d.** 5 **47.** Maximum revenue is \$5625; 25 seats are unsold.

49. a. $R(x) = x(500 - x) = 500x - x^2$ **b.**

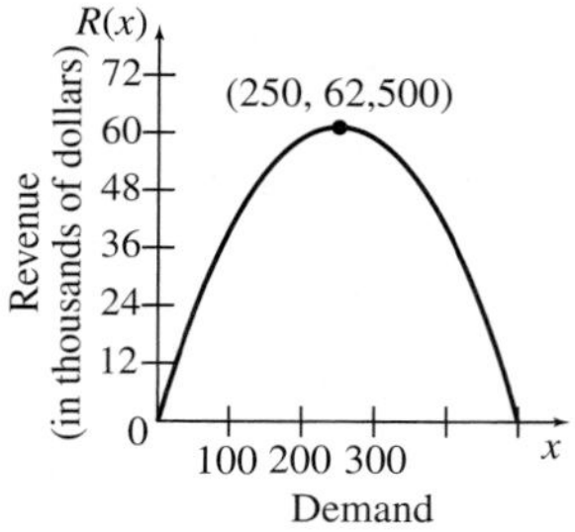

c. \$250 **d.** \$62,500

51. a. $600 + 60x$ **b.** $80 - x$ **c.** $R(x) = 48{,}000 + 4200x - 60x^2$ **d.** 35 **e.** \$121,500 **53. a.** 87 yr **b.** 98 yr
55. a. 2.024 mm; 6.104 mm **b.** 8.48 mm; 39.4 weeks after conception
57. a.

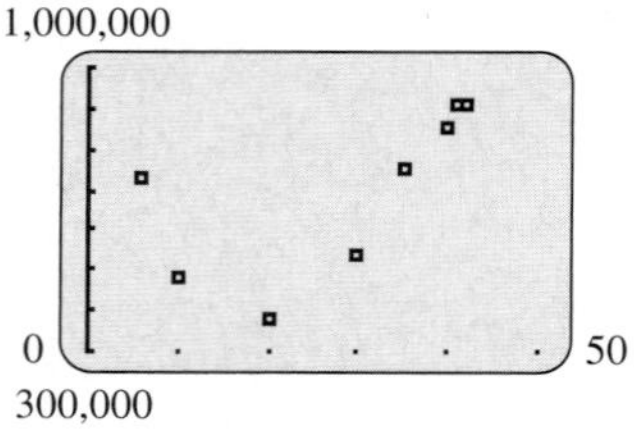

b. A quadratic models it best.
c. $y = 1174.69x^2 - 48{,}242x + 913{,}439$

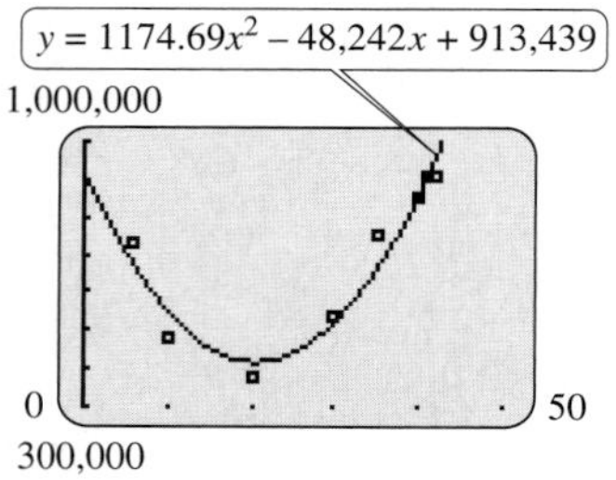

d. $y = 1107.53(x - 20)^2 + 376{,}300$ **e.**

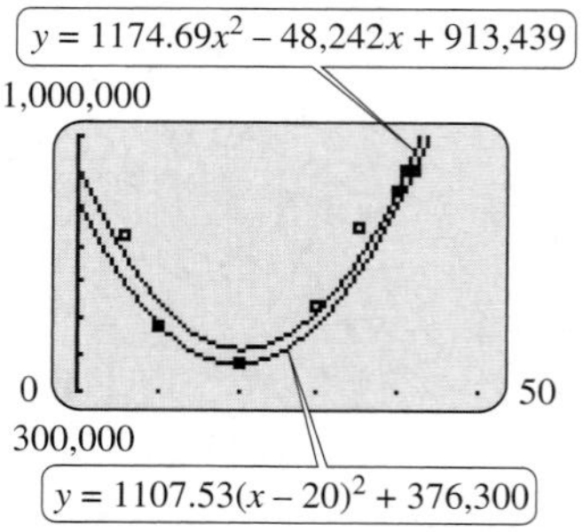

59. a. 61.7 ft **b.** 43 mph **61.** 49 yr; 3.98 **63.** 6400 ft^2 **65.** $y = (4/27)x^2$; $6\sqrt{3}$ ft $\approx$ 10.39 ft

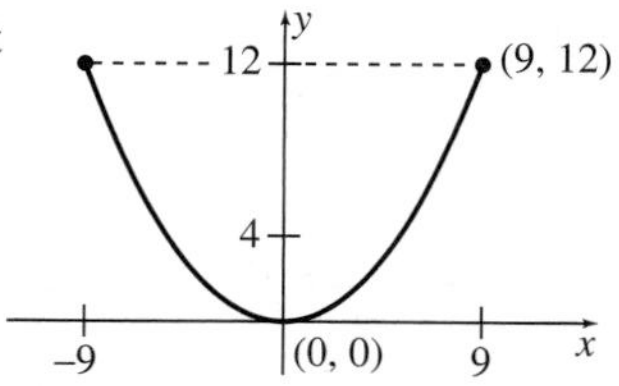

$y = (4/27)x^2$; $6\sqrt{3}$ ft $\approx$ 10.39 ft

Exercises 2.3 (page 83)

3.

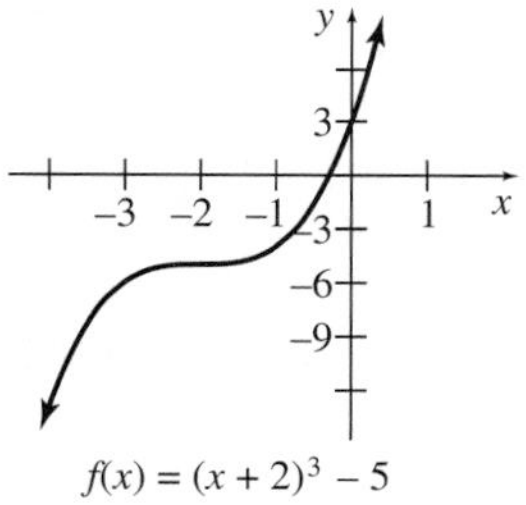

5.

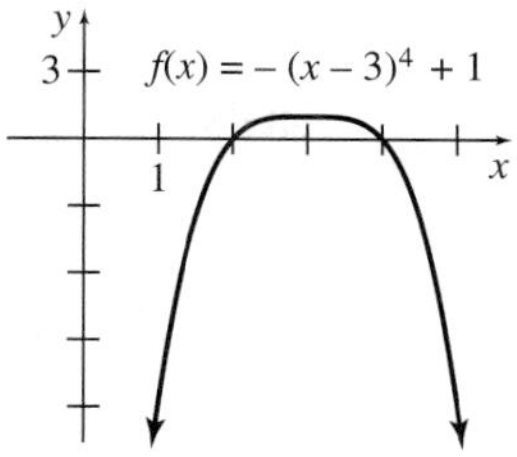

7. D **9.** E **11.** I **13.** G **15.** A **17.** D **19.** E

21. 4, 6, etc. (true degree $= 4$); $+$ **23.** 5, 7, etc. (true degree $= 5$); $+$ **25.** 7, 9, etc. (true degree $= 7$); $-$

27. Horizontal asymptote: $y = 0$; vertical asymptote: $x = 3$; no x-intercept; y-intercept $= 4/3$

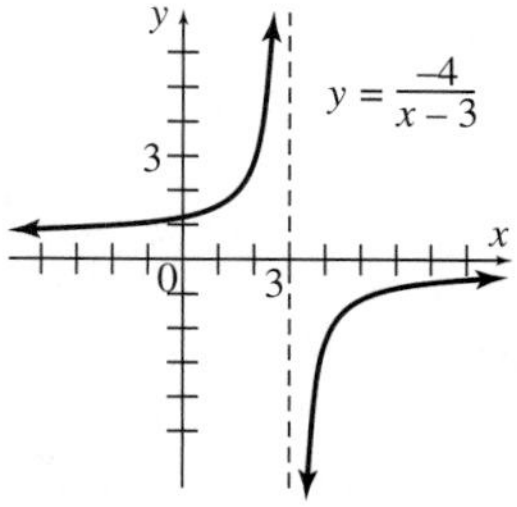

29. Horizontal asymptote: $y = 0$; vertical asymptote: $x = -3/2$; no x-intercept; y-intercept $= 2/3$

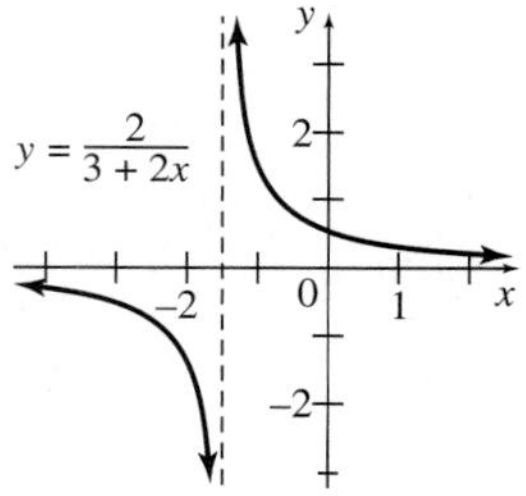

31. Horizontal asymptote: $y = 3$; vertical asymptote: $x = 1$; x-intercept $= 0$; y-intercept $= 0$

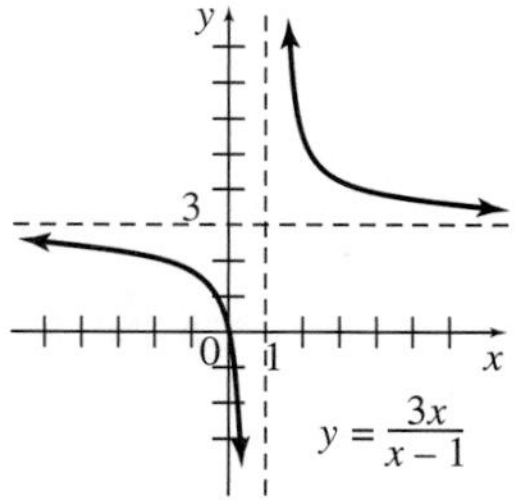

33. Horizontal asymptote: $y = 1$; vertical asymptote: $x = 4$; x-intercept $= -1$; y-intercept $= -1/4$

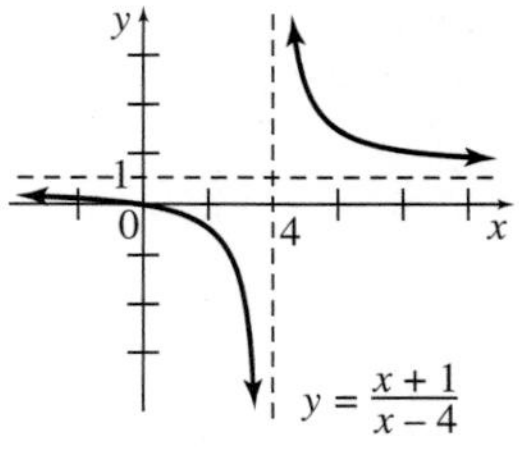

35. Horizontal asymptote: $y = -2/5$; vertical asymptote: $x = -4$; x-intercept $= 1/2$; y-intercept $= 1/20$

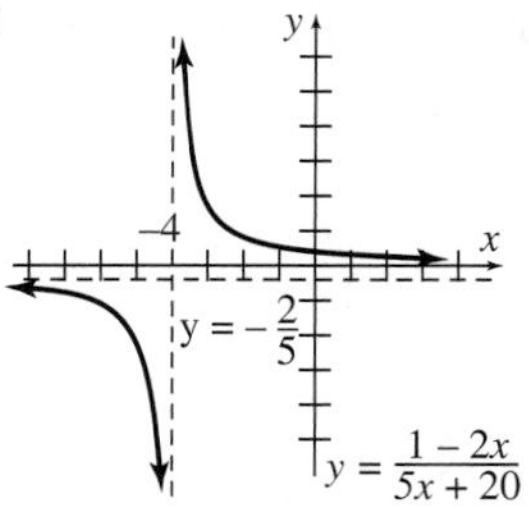

37. Horizontal asymptote: $y = -1/3$; vertical asymptote: $x = -2$; x-intercept $= -4$; y-intercept $= -2/3$

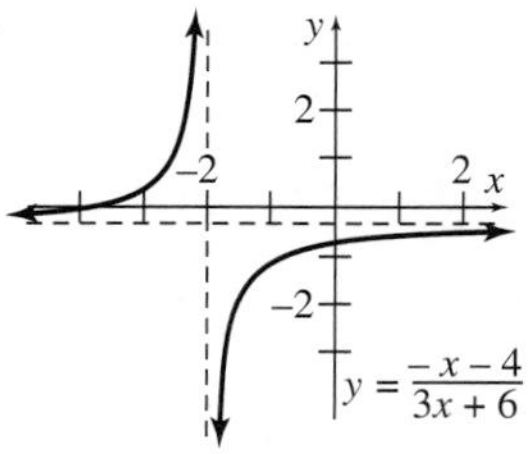

39. One possible answer is $y = 2x/(x - 1)$. **41. a.** 0 **b.** 2, −3 **d.** $(x + 1)(x - 1)(x + 2)$ **e.** $3(x + 1)(x - 1)(x + 2)$ **f.** $(x - a)$ **43. a.** Two; one at $x = -1.4$ and one at $x = 1.4$ **b.** Three; one at $x = -1.414$, one at $x = 1.414$ and one at $x = 1.442$ **45. a.** 440, 400, 338, 259, 210, 176 **b.** $x = -225$, $y = 0$ **c.** $y = 488.\overline{8}$ **d.**

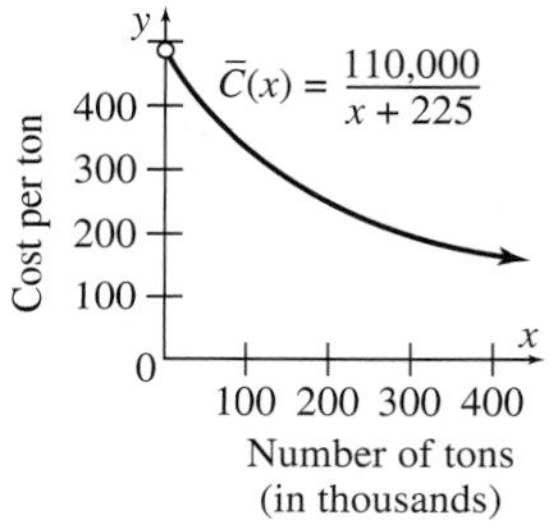

47. $f_1(x) = x(100 - x)/25$,
$f_2(x) = x(100 - x)/10$,
$f(x) = x^2(100 - x)^2/250$

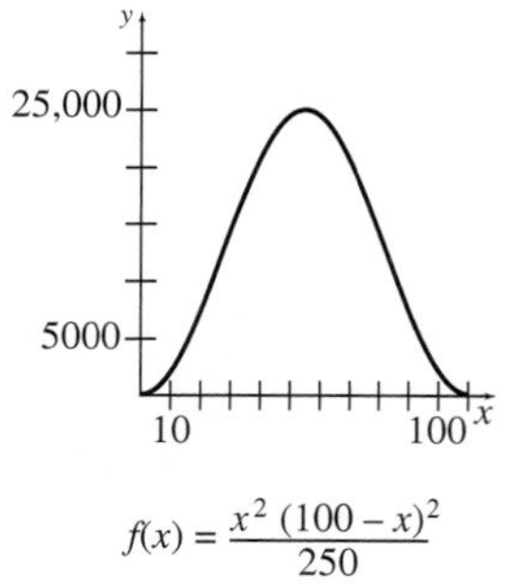

49. a. \$6700; \$15,600; \$26,800; \$60,300; \$127,300; \$328,300; \$663,300 **b.** No **c.**

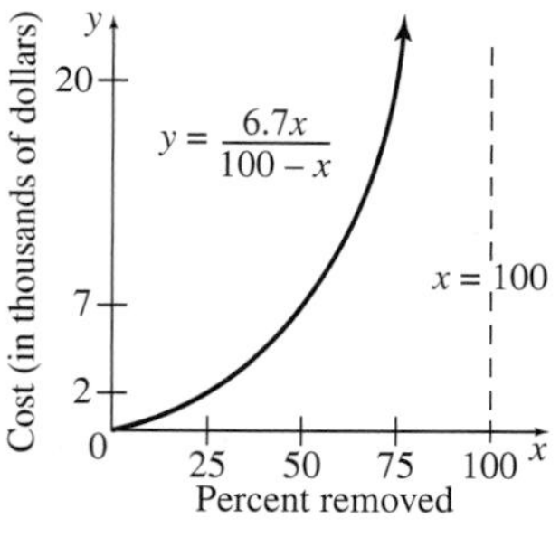

51. a. $a = 337/d$ **b.** 8.32 (using $k = 337$) **53. a.**

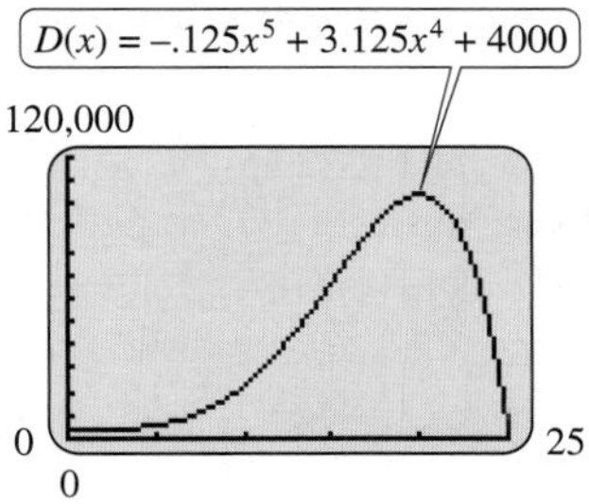

b. 1905 to 1925; 1905 to 1910; 1925 to 1930

55. a.

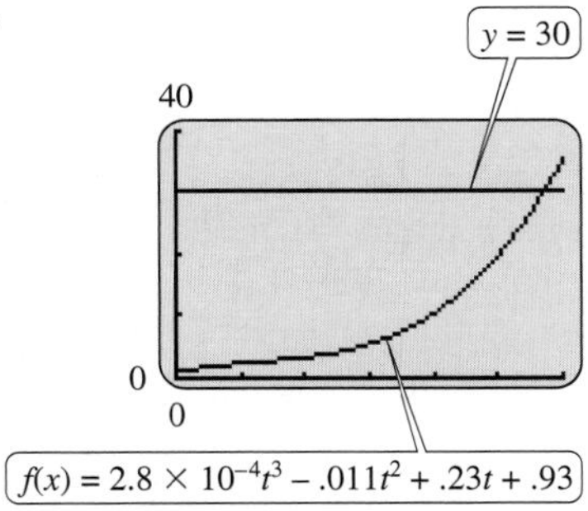

b. 1986 **57. a.** $[0, \infty)$ **b.**

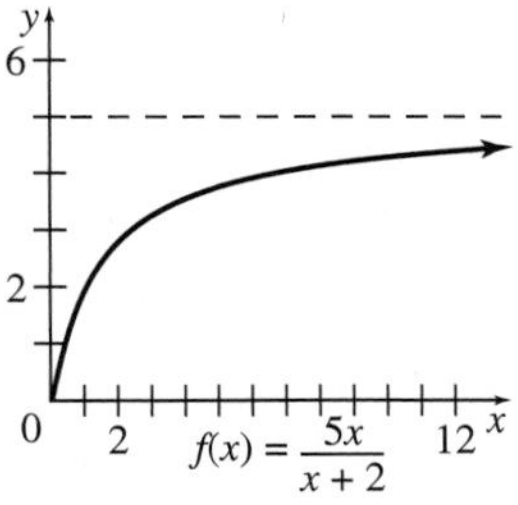

d. Maximum growth rate

59. a. 1.80, .812, .366 **b.**

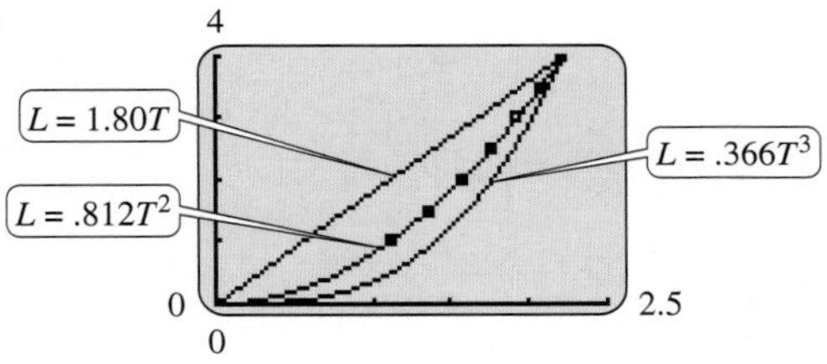

c. 2.48 sec **d.** The period increases by a factor of $\sqrt{2}$. **e.** $L = .822T^2$, which is very close to the function found in part b.

61. a.

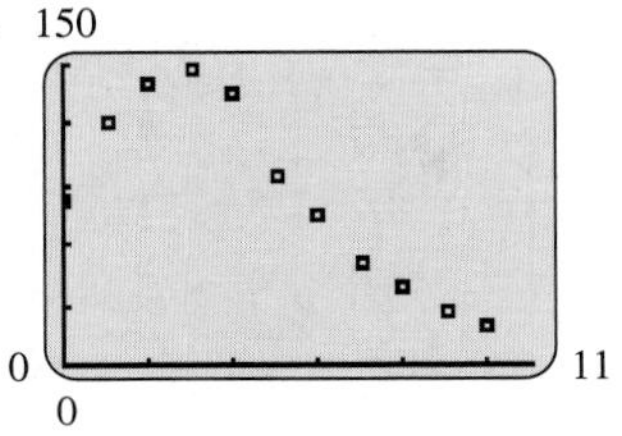

b. $y = .779526x^3 - 13.7908x^2 + 54.2770x + 81.5594$
$y = -.037005x^4 + 1.51962x^3 - 18.4164x^2 + 63.5282x + 78.8951$

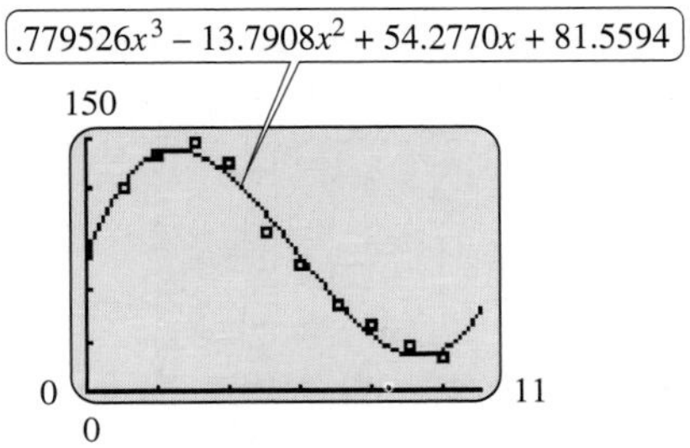

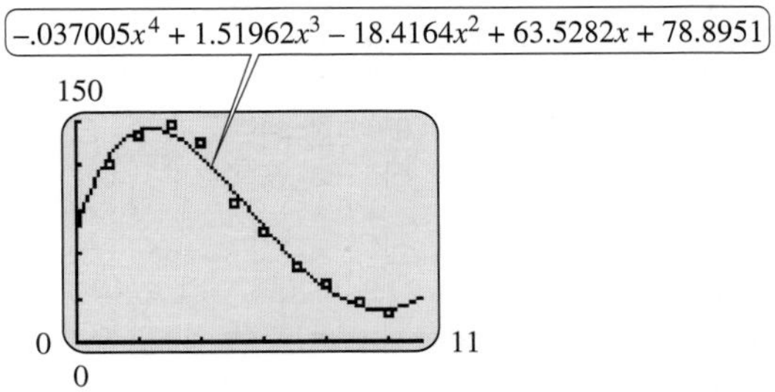

Exercises 2.4 (page 99)

1. 2, 4, 8, 16, 32, ..., 1024; $1.125899907 \times 10^{15}$ **3.** E **5.** C **7.** F **9.** A **11.** C **13.** -3 **15.** -2 **17.** 6 **19.** $7/4$ **21.** -2 **23.** $2, -2$ **25.** $0, -1$ **27.** $0, -1/3$ **33. a.** \$3382.26 **b.** \$3439.16 **c.** \$3468.55 **d.** \$3488.50 **35.** Choose the 8% investment, which would yield \$111.30 additional interest. **37. a.** \$10.94 **b.** \$11.27 **c.** \$11.62 **39.** 6.30%
41. a. 1, .92, .85, .78, .72, .66, .61, .56, .51, .47, .43 **b.**

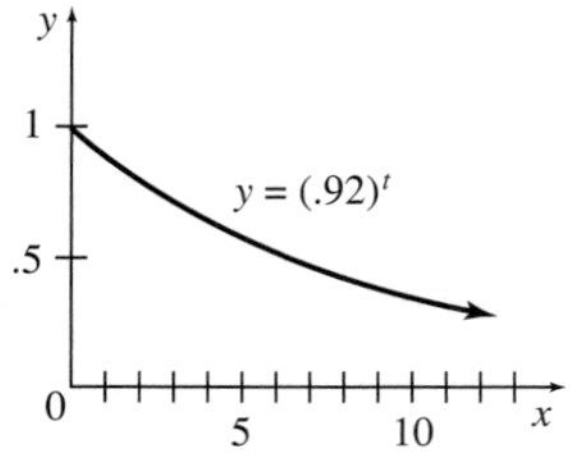

c. About \$384,000 **d.** About \$98

43. 1.25 **45. a.** 4000 bacteria **b.** 500 bacteria **c.** Every 1/3 hr or 20 min **d.** In 2 hr **47. a.** 2.6, 2.7; yes **b.** $C(x) = .05(1.040)^{x-1950}$ **c.**

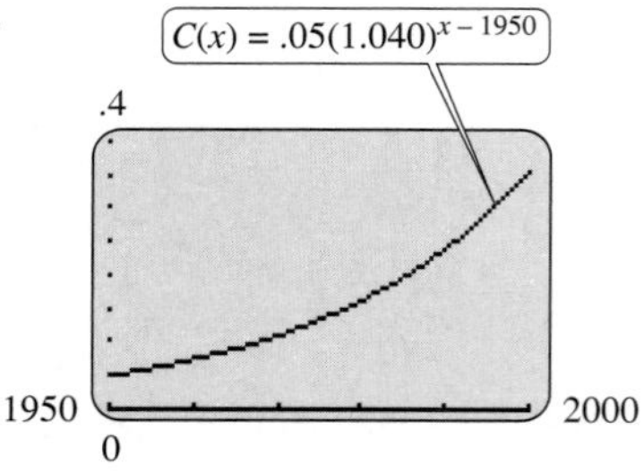

$C(1975) = .133$, which appears identical to the value on the graph. **d.** 4.0% **e.** 2008

49. a. $P = 1013e^{-1.34 \cdot 10^{-4}x}$; $P = -.0748x + 1013$; $P = 1/(2.79 \cdot 10^{-7}x + 9.87 \cdot 10^{-4})$
b. $P = 1013e^{-1.34 \cdot 10^{-4}x}$ is the best fit.

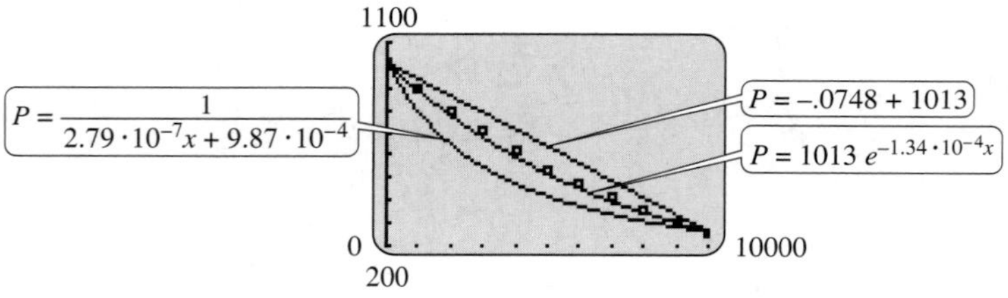

c. 829 millibars, 232 millibars
d. $P = 1038(.99998661)^x$. This is slightly different from the function found in part b, which can be rewritten as $P = 1013(.99998660)^x$.

Exercises 2.5 (page 113)

1. $\log_2 8 = 3$ **3.** $\log_3 81 = 4$ **5.** $\log_3(1/9) = -2$ **7.** $2^7 = 128$ **9.** $25^{-1} = 1/25$ **11.** $10^4 = 10{,}000$ **13.** 2 **15.** 3 **17.** -2 **19.** $-2/3$ **21.** 1 **23.** $5/3$ **25.** $\log_3 4$ **27.** $\log_9 7 + \log_9 m$ **29.** $1 + \log_3 p - \log_3 5 - \log_3 k$ **31.** $\ln 5 + (1/2)\ln 2 - (1/4)\ln 7$ **33.** $3a$ **35.** $2c + 3a + 1$ **37.** 1.86 **39.** -3.28 **41.** $x = 1/5$ **43.** $z = 2/3$ **45.** $r = 49$ **47.** $x = 1$ **49.** No solution **51.** 3 **53.** 5 **55.** $x = 1.79$ **57.** $y = 1.24$ **59.** $z = 2.10$ **61.** $x = 7.41$ **63.** $x < -2$ or $x > 2$ **67. a.** 23.4 yr **b.** 11.9 yr **c.** 9.0 yr **d.** 23.3 yr; 12 yr; 9 yr **69.** 6.25% **71.** 2029 **73. a.** About .693 **b.** ln 2 **c.** Yes **75. a.** About 1.099 **b.** About 1.386 **77.** About every 7 hr, $T = 3 \ln 5/\ln 2$ **79.** 2011 **81.** $s/n = 2^{C/B} - 1$ **83.** No; 1/10 **85. a.** 1000 times greater **b.** 1,000,000 times greater

Exercises 2.6 (page 124)

7. 15.87% **9.** 11.63% **11.** \$4537.71 **13.** \$5248.14 **15.** 9.20% **17.** 6.17% **19. a.** \$257,107.67 **b.** \$49,892.33 **c.** \$68,189.54 **21. a.** The 10% investment compounded quarterly **b.** \$622.56 **c.** 10.38% and 10.24% **d.** 2.95 yr **e.** 3 yr **23.** No; 17.2% **25. a.** 200 **b.** About 1/2 year **c.** No **d.** Yes; 1000 **27. a.** $P(t) = .002784e^{.007295t}$ **b.** 2804 **c.** No; it is too small. Exponential growth does not accurately describe population growth for the world over a long period of time.

29. No; 6.6% **31. a.** $y = 50{,}000e^{-.102t}$ **b.** About 6.8 hours **35.** About 4100 years old **37.** About 1600 years **39. a.** 1.9 g **b.** Approximately 7000 years **41. a.** $y = 25.0e^{-.00497t}$ **b.** 139 days **43.** .5% **45. a.** $y = 10e^{.0095t}$ **b.** 42.7°C **47.** About 30 minutes

Chapter 2 Review Exercises (page 129)

5. $(-3, 20), (-2, 9), (-1, 2), (0, -1), (1, 0), (2, 5), (3\ 14)$; range: $\{-1, 0, 2, 5, 9, 14, 20\}$

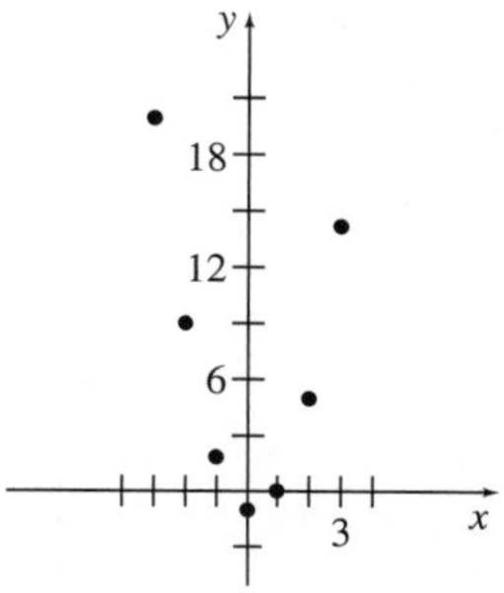

7. a. -28 **b.** -12 **c.** -28 **d.** $-r^2 - 3$ **9. a.** 17 **b.** 4 **c.** $5k^2 - 3$ **d.** $-9m^2 + 12m + 1$ **e.** $5x^2 + 10xh + 5h^2 - 3$ **f.** $-x^2 - 2xh - h^2 + 4x + 4h + 1$ **g.** $10x + 5h$ **h.** $-2x - h + 4$ **11.** $(-\infty, -3) \cup (3, \infty)$ **13.**

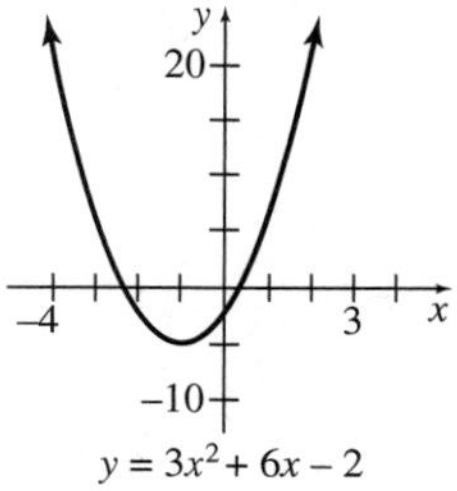

15.

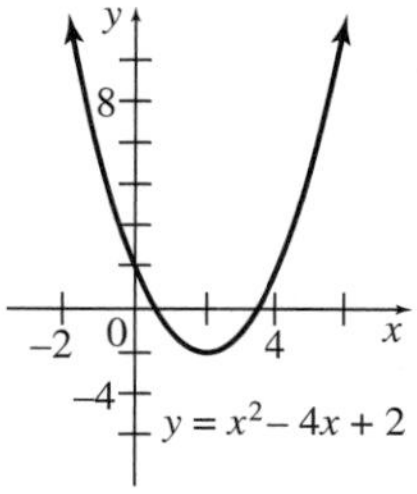

17.

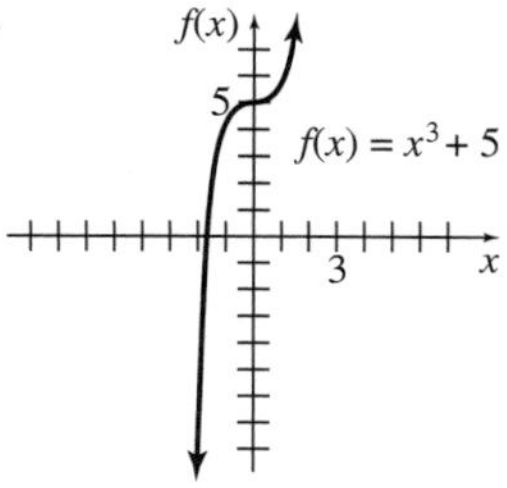

19.

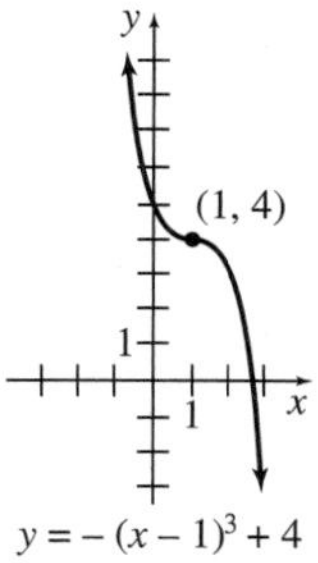

21.

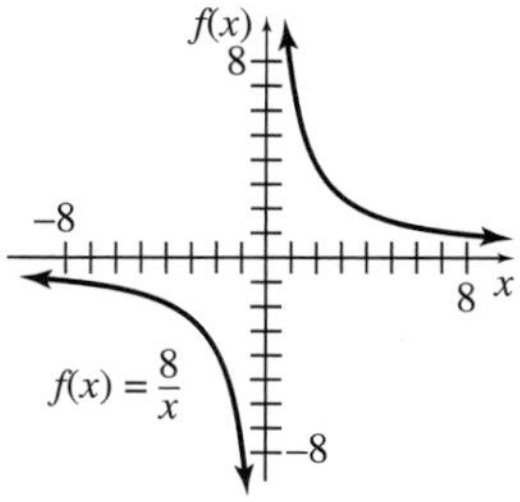

23.

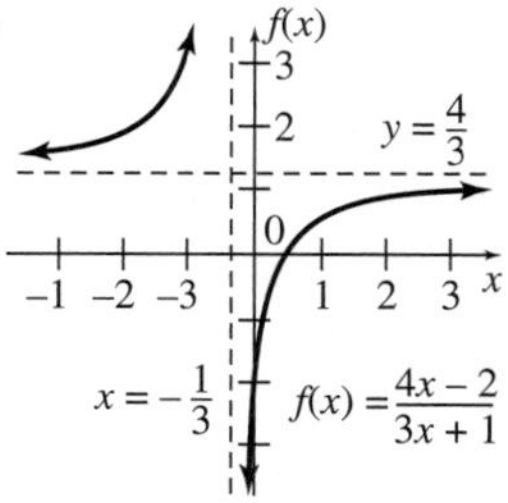

25.

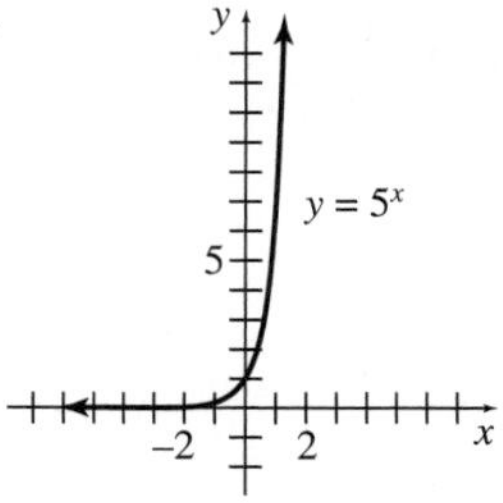

27.

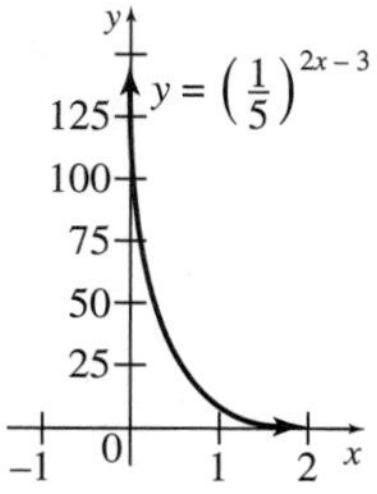

29.

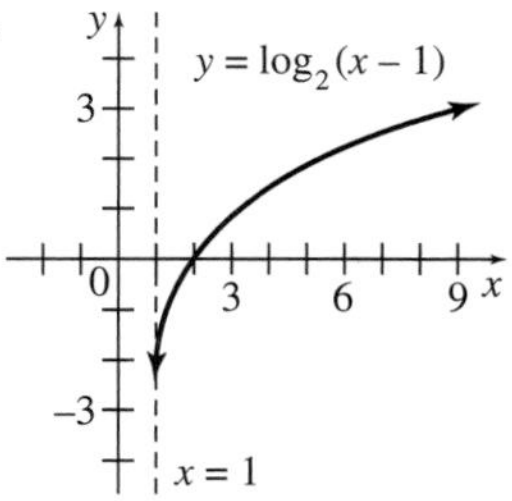

31.

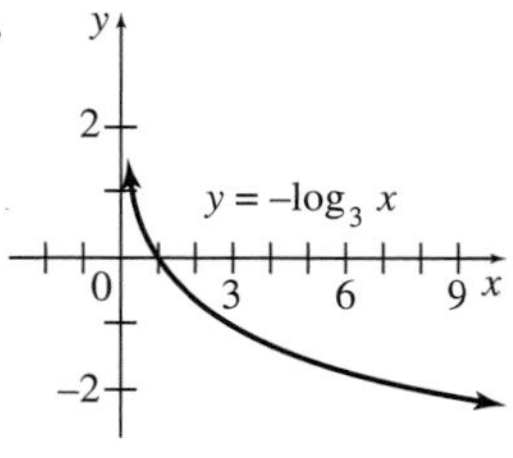

33. -1 **35.** 2 **37.** $\log_2 64 = 6$ **39.** $\ln 1.09417 = .09$

41. $2^5 = 32$ **43.** $e^{4.41763} = 82.9$ **45.** 4 **47.** $1/2$ **49.** $\log_5(21k^4)$ **51.** $\log_2(x^2/m^3)$ **53.** $p = 1.416$ **55.** $m = -1.807$ **57.** $x = -3.305$ **59.** $m = 1.7547$ **61.** $k = 2$ **63.** $p = 3/4$ **65. a.** $(-\infty, \infty)$ **b.** $(0, \infty)$ **c.** 1 **d.** None **e.** $y = 0$ **f.** Greater than 1 **g.** Between 0 and 1 **69. a.** \$28,000 **b.** \$7000 **c.** \$63,000 **d.**

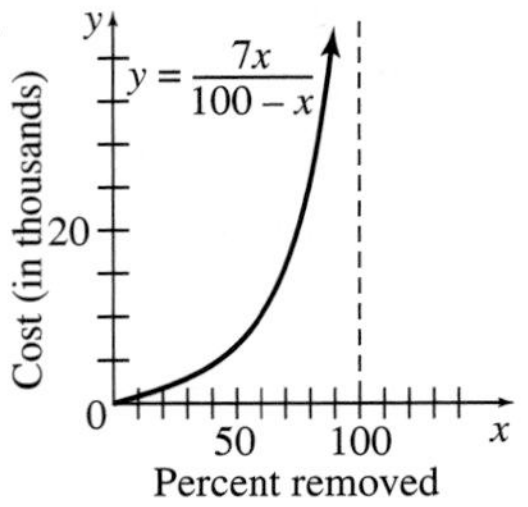

e. No.

71. \$1692.28 **73.** 70 quarters or 17.5 years; 111 quarters or 27.75 years **75.** \$16,668.75 **77.** \$17,901.90 **79.** 9.38% **81.** \$1494.52 **83.** \$21,828.75 **85.** About 9.59% **87. a.** $n = 1200 - 10p$ **b.** $R = p(1200 - 10p)$ **c.** $20 \leq p \leq 120$ **d.** $R = (1200n - n^2)/10$ **e.** $0 \leq n \leq 1000$ **f.** \$60 **g.** 600 **h.** \$36,000 **i.**

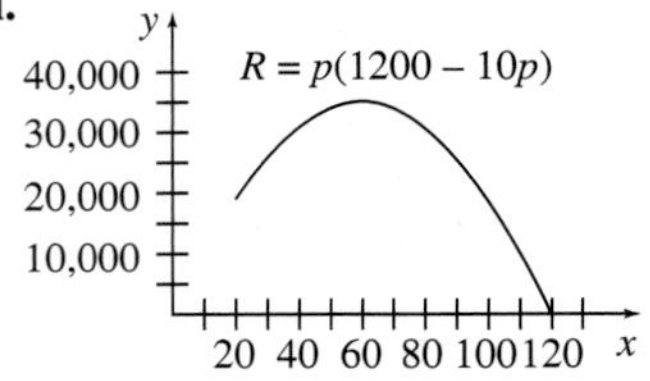

j. The revenue starts at \$20,0000 when the price is \$20, rises to a maximum of \$36,000 when the price is \$60, and falls to 0 when the price is \$120.

89. a.

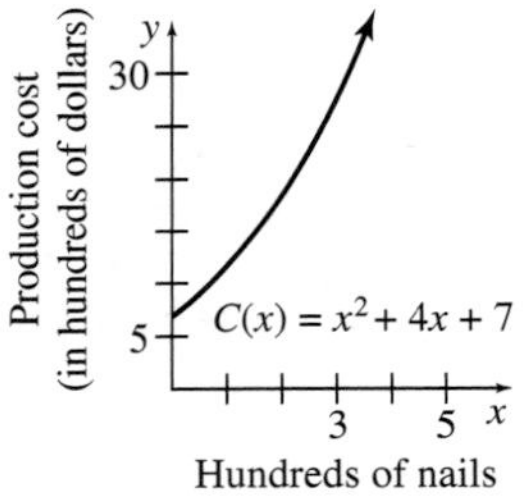

b. $2x + 5$ **c.** $A(x) = x + 4 + 7/x$ **d.** $1 - 7/[x(x + 1)]$

91. a. $1/8$. The fraction of radiation let in is 1 over the SPF rating. **b.**

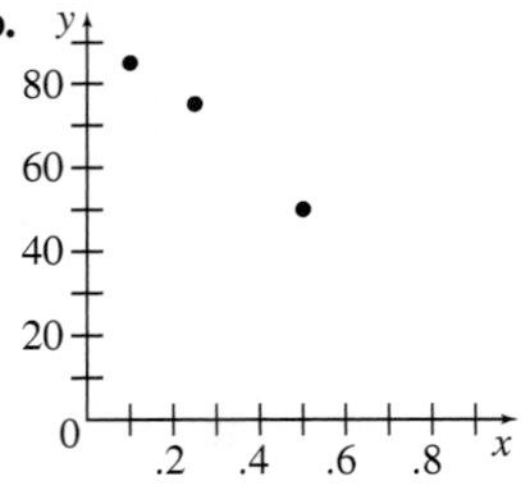

c. UVB $= 1 - 1/\text{SPF}$ **d.** 12.5% **e.** 3.3% **f.** It decreases to 0.

93. a. $[0, 36]$ **b.** Decreasing **c.**

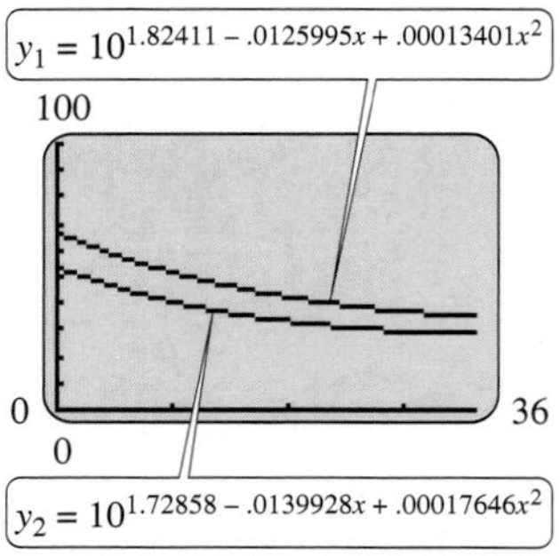

d. 10.7 breaths per minute

95. a. $y = 15{,}000e^{.0313t}$ **b.** About 35 years **97.** About 7.7 m **99. a.** When it is first injected; .08 g **b.** Never **c.** It approaches $c/a = .0769$ g. **101. a.** 0 yr **b.** 1.85×10^9 yr **c.** As r increases, t increases, but at a slower and slower rate. As r decreases, t decreases at a faster and faster rate. **103. a.** $x = .9$ means the speed is 10% slower on the return trip. $x = 1.1$ means the speed is 10% faster on the return trip. **c.** The formula for v_{aver} is a rational function with a horizontal asymptote at $v_{aver} = 2v$. This means that as the return velocity becomes greater, the average velocity approaches twice the velocity on the first part of the trip, and can never exceed twice that velocity. **105. a.** $P = 5.48D$; $P = 1.00D^{1.5}$; $P = .182D^2$ **b.** $P = 1.00D^{1.5}$ is the best fit. **c.** 248.3 yr **d.** $P = 1.00D^{1.5}$, the same as the function found in part b.

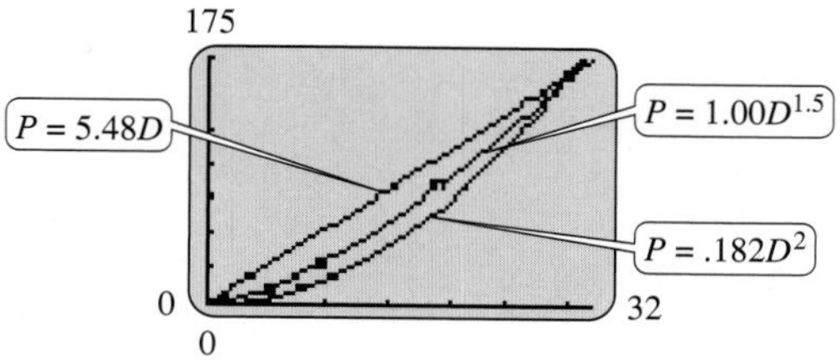

107. a. $y = 29.6(1.0439)^t$ **b.** $y = 29.02(1.04695)^t$ **c.**

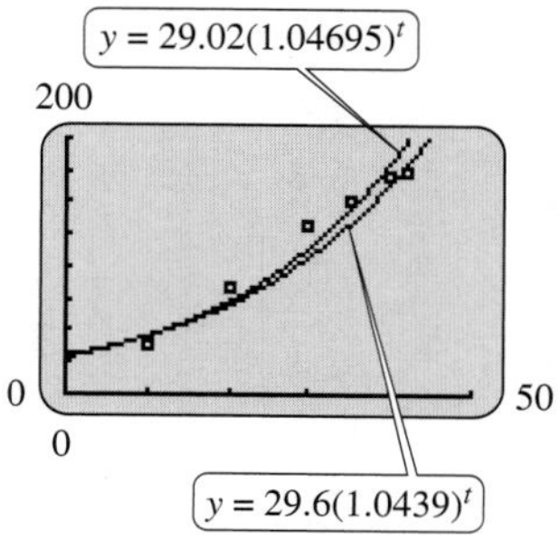

d. $y = .04087t^2 + 2.1025t + 24.153$; $y = -.003020t^3 + .23273t^2 - .88343t + 29.119$

e.

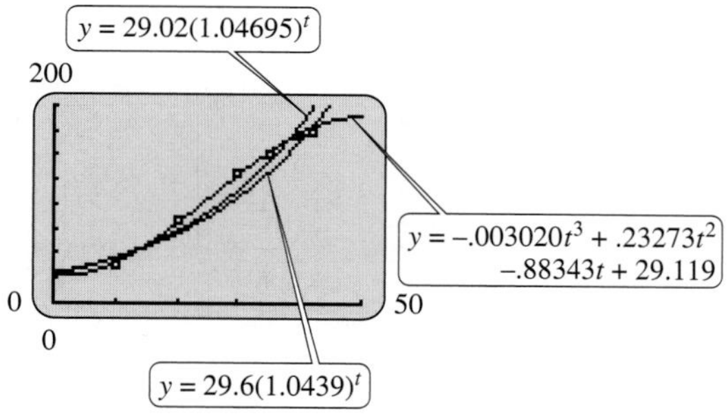

CHAPTER 3 THE DERIVATIVE

Exercises 3.1 (page 152)

1. c **3.** b **5.** 3 **7.** 0 **9. a.** -1; $-1/2$; does not exist; does not exist **b.** $-1/2$; $-1/2$; $-1/2$; $-1/2$ **11.** 3 **15.** 2 **17.** 10 **19.** Does not exist **21.** 8 **23.** 2 **25.** 4 **27.** 256 **29.** 3/2 **31.** 6 **33.** 3/2 **35.** -5 **37.** $-1/9$ **39.** 1/10 **41.** $2x$ **43.** 3/5 **45.** 1/2 **47.** 0 **49.** ∞ **51.** $-\infty$ **53. a.** Does not exist **b.** $x = -2$ **c.** If $x = a$ is an asymptote for the graph of $f(x)$, then $\lim_{x \to a} f(x)$ does not exist. **57. a.** 0 **b.** $y = 0$ **59. a.** $-\infty$ **b.** $x = 0$ **63.** 5 **65.** .3333 or 1/3 **67. a.** 1.5 **69. a.** -2 **71. a.** 8 **75. a.** 7.25 cents **b.** 7 cents **c.** 7.25 cents **d.** Does not exist **e.** 7.25 cents **77.** 75 items; the number of items a new employee produces gets closer and closer to 75 as the number of days of training increases. **79.** $R/(i - g)$ **81. a.** 36.2 cm; the depth of the contaminated sediment layer deposited below the bottom of the lake in 1970 is 36.2 cm. **b.** 155 cm; all of the sediment in the lake is within 155 cm of the bottom of the lake. **83. a.** .572 **b.** .526 **c.** .503 **d.** .5; the numbers in a, b, and c give the probability that the legislator will vote yes on the second, fourth, and eight votes. In d, as the number of roll calls increases, the probability of a yes vote approaches .5 but is never less than .5.

Exercises 3.2 (page 164)

1. $a = -1$; **a.** $1/2$ **b.** $1/2$ **c.** $1/2$ **d.** $f(a)$ does not exist. **3.** $a = 1$: **a.** -2 **b.** -2 **c.** -2 **d.** 2 **5.** $a = -5$: **a.** ∞ **b.** $-\infty$ **c.** Limit does not exist. **d.** $f(a)$ does not exist; $a = 0$: **a.** 0 **b.** 0 **c.** 0 **d.** $f(a)$ does not exist. **7.** $a = 0$, limit does not exist; $a = 2$, limit does not exist. **9.** $a = 2, 4$ **11.** Nowhere **13.** $a = -2$; limit does not exist. **15.** $a < 1$; limit does not exist. **17.** $0 \le a \le 1$; limit does not exist. **19. a.**

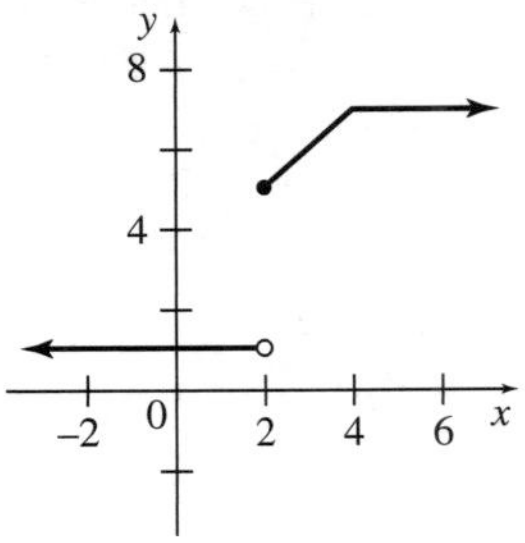

b. 2 **c.** 1, 5

21. a.

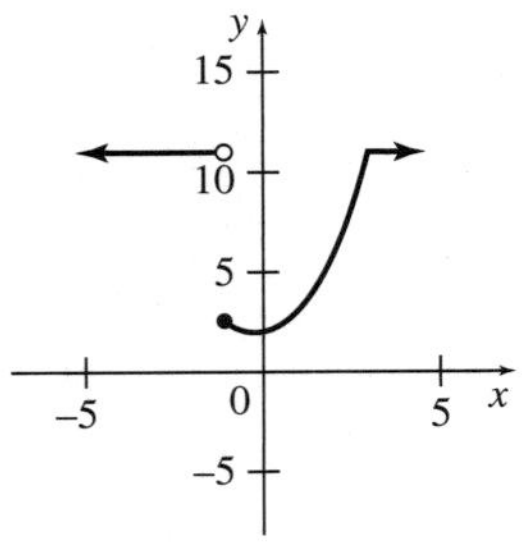

b. -1 **c.** 11, 3 **23. a.**

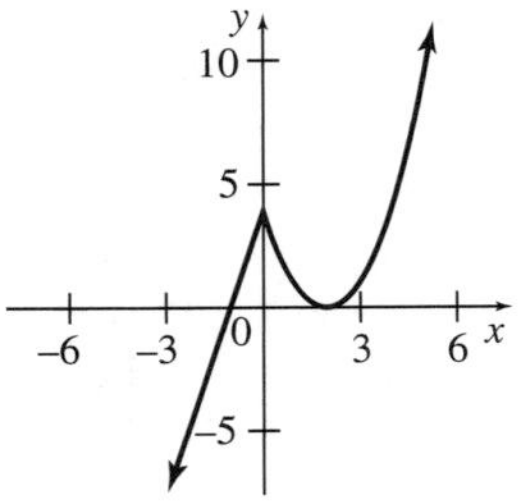

b. None **25.** $2/3$ **27.** 4

31. Discontinuous at $x = 1.2$ **33. a.** \$500 **b.** \$1500 **c.** \$1000 **d.** Does not exist **e.** Discontinuous at $x = 10$; a change in shifts **f.** 15 **35. a.** \$96 **b.** \$150 **c.** \$120 **d.** At $x = 100$ **37. a.** \$1.70 **b.** \$2.60 **c.** Limit does not exist. **d.** \$2.60 **e.** 1, 2, 3, 4, 5, 6, 7 **f.**

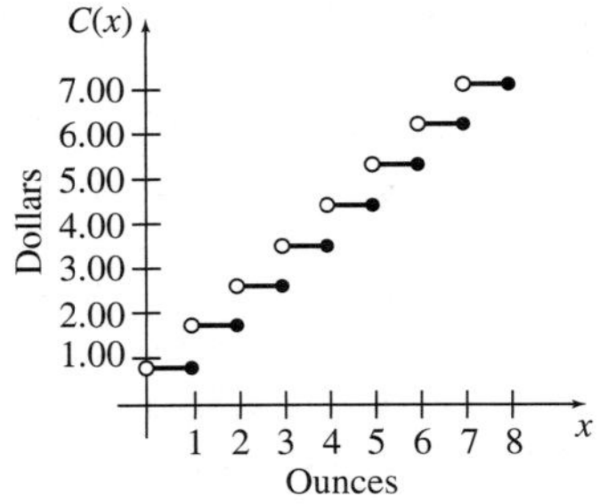

39. a. About 687 g **b.** No **c.** 3000

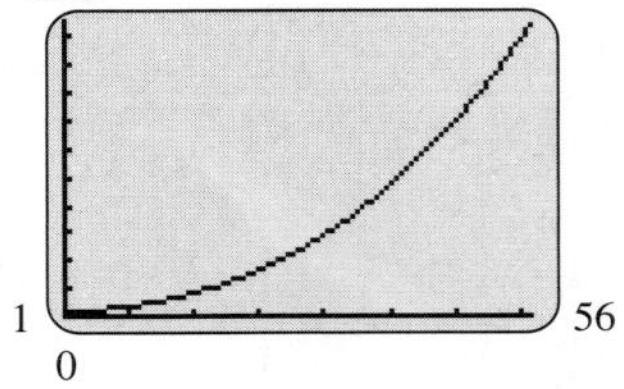

Exercises 3.3 (page 173)

1. 5 **3.** 8 **5.** $1/3$ **7.** $-1/3$ **9.** 17 **11.** 5 **13.** 2 **15.** 2 **17.** 6.7726 **19.** 1.1207 **23. a.** Approximately $-$\$6.75 billion **b.** Approximately \$14.58 billion **c.** Approximately \$0 billion **25. a.** \$3000 **b.** 0 **c.** $-$\$1800 **d.** Sales increase in years 0–4, then stay constant until year 7, then decrease. **e.** Many answers are possible; one example might be CDs. **27. a.** \$700 per item **b.** \$500 per item **c.** \$300 per item **d.** \$1100 per item **29. a.** -25 boxes per dollar **b.** -20 boxes per dollar **c.** -30 boxes per dollar **d.** Demand is decreasing. Yes, a higher price usually reduces demand. **31.** All numbers are approximations. **a.** Replacement level by 2050: 75 million people per year; replacement level by 2030: 65 million people per year; replacement level by 2010: 55 million people per year. The projection for replacement-level fertility by 2010 predicts the smallest growth in world population. **b.** Replacement level by 2050: 10 million people per year; replacement level by 2030: 7.5 million people per year; replacement level by 2010: 7.5 million people per year. By 2050 the three projections show almost the same rate of change in world population.

33. a. .288 mm per wk **b.** .348 mm per wk **c.**

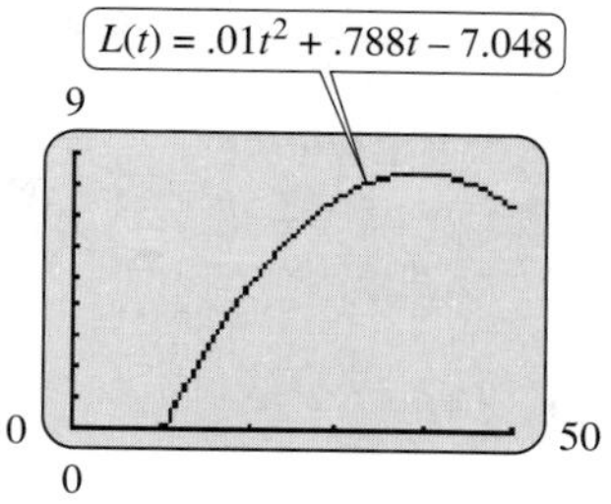

35. a. .08 kg per day **b.** .09 kg per day

c.

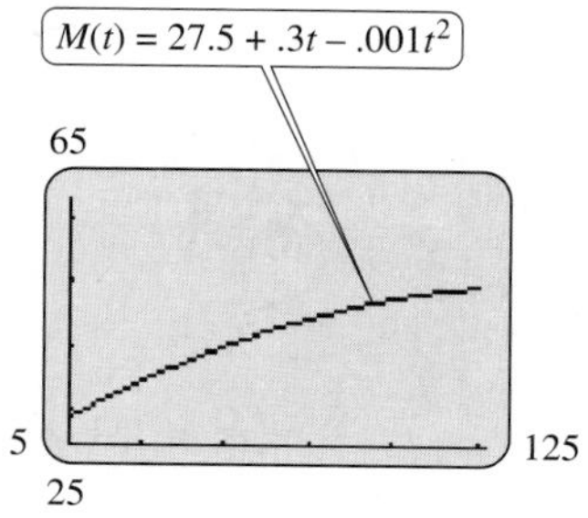

37. a. -2% per year; $.85\%$ per year **b.** $-.575\%$ per year **39. a.** 5 ft per sec **b.** 2 ft per sec **c.** 3 ft per sec **d.** 5 ft per sec **e.** **(i)** 2.5 ft per sec; **(ii)** 2.5 ft per sec **f.** **(i)** 4 ft per sec; **(ii)** 4 ft per sec **41. a.** 15 ft per sec **b.** 14 ft per sec **c.** 13 ft per sec

Exercises 3.4 (page 191)

1. a. 0 **b.** 1 **c.** -1 **d.** Does not exist **e.** m **3.** At $x = -2$ **5.** 2 **7.** $1/4$ **9.** 0 **11.** $-8x + 11$; 27; 11; -13 **13.** $2/x^2$; $1/2$; does not exist; $2/9$ **15.** $1/(2\sqrt{x})$; does not exist; does not exist; $1/(2\sqrt{3})$ **17.** $y = 8x - 9$ **19.** $y = -(5/4)x + 5$ **21.** $y = (2/3)x + 6$ **23.** -5; -117; 35 **25.** 8; 8; 8 **27.** $1/2$; $1/128$; $2/9$ **29.** $1/(2\sqrt{2})$; $1/8$; does not exist **31.** 0 **33.** -3; -1; 0; 2; 3; 5 **35. a.** $(a, 0)$ and (b, c) **b.** $(0, b)$ **c.** $x = 0$ and $x = b$ **37. a.** Distance **b.** Velocity **39.** 56.6625 **41.** $-.0158$ **45. a.** $-4p - 4$ **b.** -44; demand is decreasing at a rate of about 44 items for each increase in price of \$1. **47. a.** \$16 per table **b.** \$16 **c.** \$15.998 (or \$16) **d.** The marginal revenue gives a good approximation of the actual revenue from the sale of the 1001st table. **49.** Answers are in billions of dollars. **a.** 961, 2526, 3806 **b.** 106, 189, -8, -318 **51.** 1000; the population is increasing at a rate of 1000 shellfish per time unit. 570; the population is increasing more slowly at 570 shellfish per time unit. 250; the population is increasing at a much slower rate of 250 shellfish per time unit. **53. a.** 1690 m per sec **b.** 4.84 days per m per sec; an increase in velocity from 1700 m per sec to 1701 m per sec indicates an approximate increase in the age of the cheese of 4.84 days. **55. a.** .75, 3 **b.** 1033; the oven temperature is increasing at 1033° per hour. **c.** 0; the oven temperature is not changing. **d.** -1033; the oven temperature is decreasing at 1033° per hour. **57.** About 0 mph per second for the hands and 640 mph per second for the bat. This represents the acceleration of the hands and the bat at the moment when the velocities are equal.

Exercises 3.5 (page 201)

3. $f{:}Y_2$; $f'{:}Y_1$ **5.** $f{:}Y_1$; $f'{:}Y_2$ **7.**

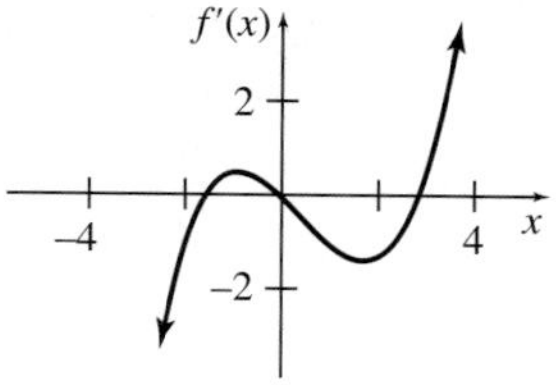

9.

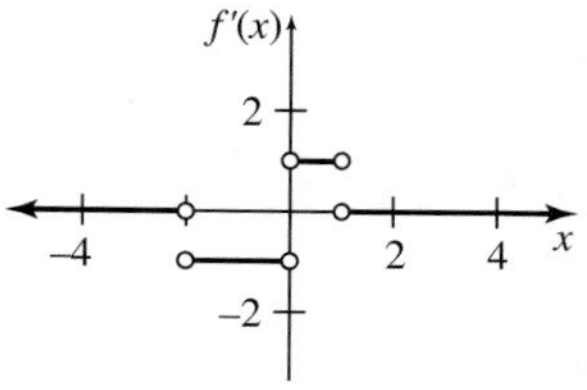

11.

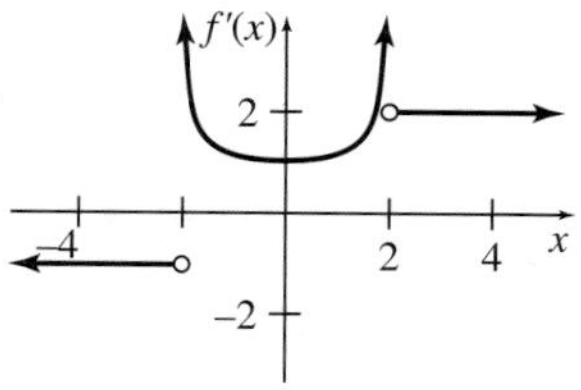

13.

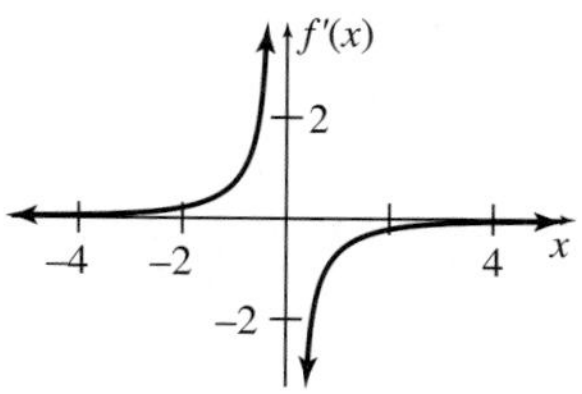

15.

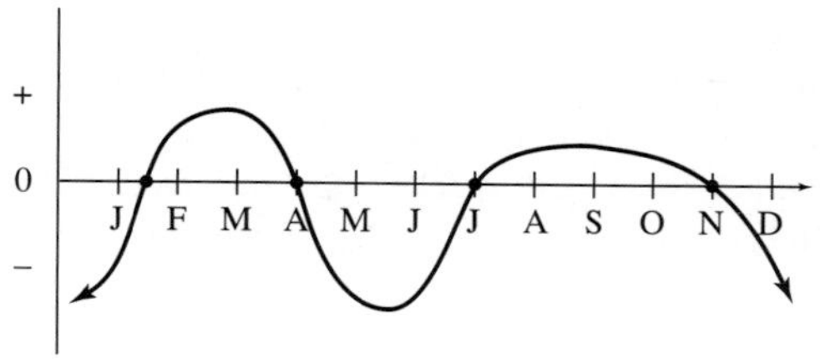

17.

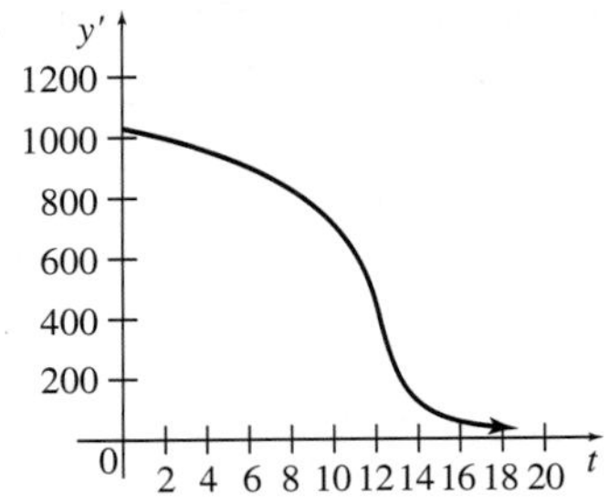

19.

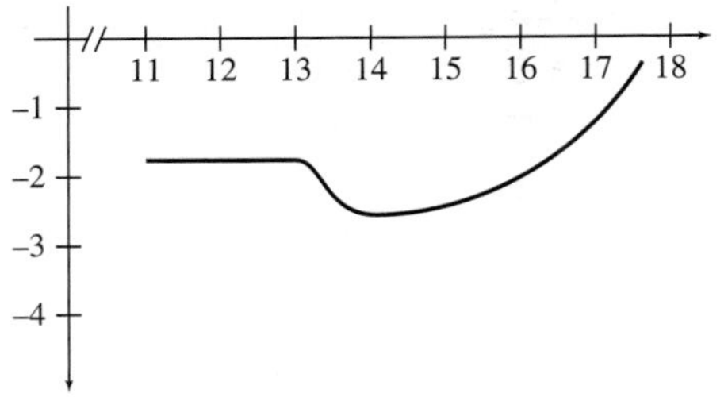

About 9 cm; about 2.6 cm less per year

Chapter 3 Review Exercises (page 205)

5. a. 4 **b.** 4 **c.** 4 **d.** 4 **7. a.** ∞ **b.** $-\infty$ **c.** Does not exist **d.** Does not exist **9.** ∞ **11.** $17/3$ **13.** 8 **15.** -13 **17.** $1/6$ **19.** $1/5$ **21.** $3/4$ **23.** Discontinuous at x_2 and x_4 **25.** 0, does not exist, does not exist; $1/2$, does not exist, does not exist **27.** -5, does not exist, does not exist **29.** Continuous everywhere **31. a.**

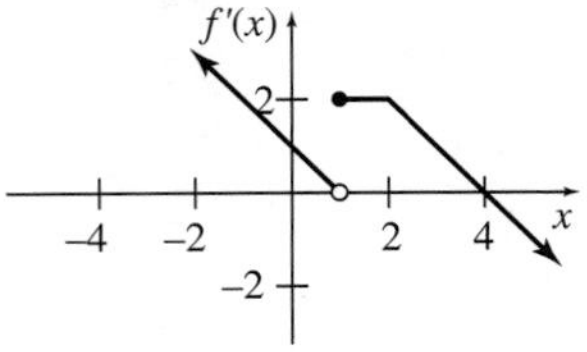

b. 1 **c.** 0, 2

33. 2 **35.** 30; 12 **37.** $9/77$; $18/49$ **39.** 4 **41.** .5171 **43.**

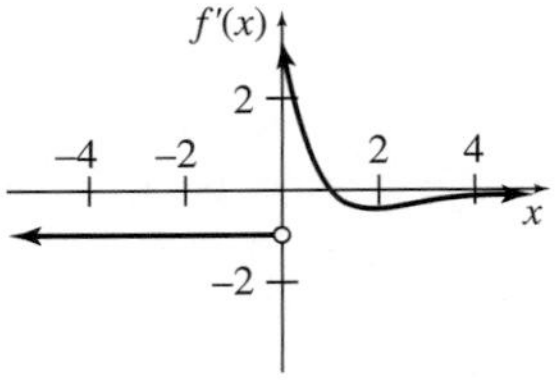

45.

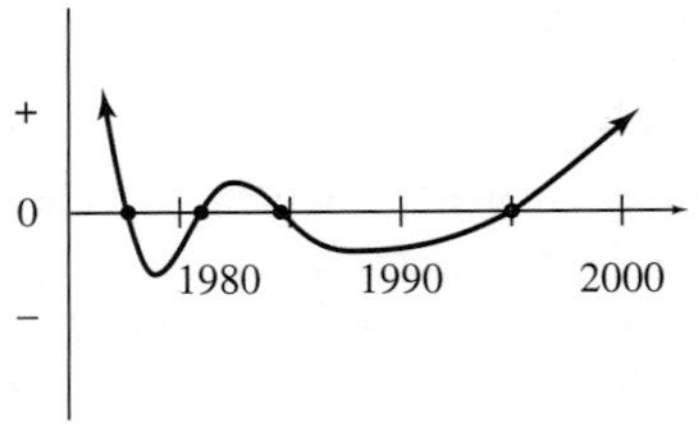

1997 earnings: approximately \$10 per hour; 1997 rate of change of earnings: \$2 per hour per year

47. a. \$150 **b.** \$187.50 **c.** \$189 **d.**

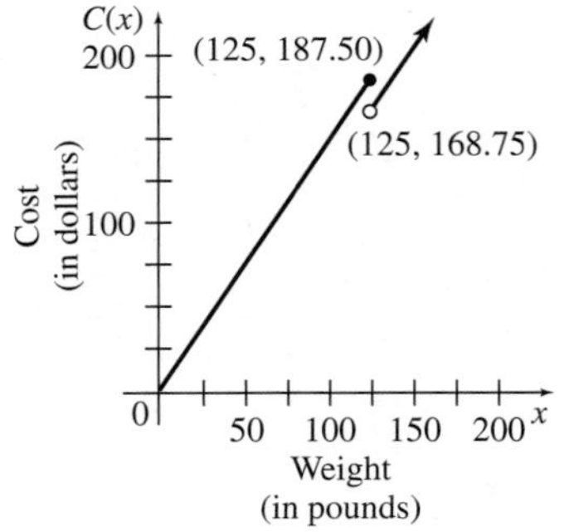

e. Discontinuous at x = \$125 **f.** \$1.50 **g.** \$1.50 **h.** \$1.35 **i.** 1.5; when 100 lb are purchased, an additional pound will cost \$1.50 more. **j.** 1.35; when 140 lb are purchased, an additional pound will cost \$1.35 more.

49. b. $x = 7.5$ **c.** The marginal cost equals the average cost at the point where the average cost is smallest. **51. a.** .02; the risk of heart attack is going up at a rate of .02 per 1000 people for each increase in the blood cholesterol of 1 mg/dL **b.** .15; the risk of heart attack is going up at a rate of .15 per 1000 people for each increase in the blood cholesterol of 1 mg/dL **c.** .065 per 1000 people per mg/dL of cholesterol **53. a. i.** About 70 m per minute **ii.** About 80 m per minute

b.

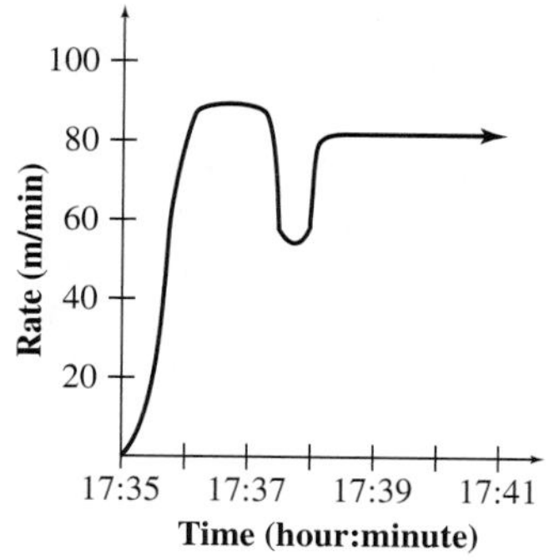

55. a. 1; the ball is rising 1 ft for each foot it travels horizontally. **b.** -2.7; the ball is dropping 2.7 ft for each foot it travels horizontally.

CHAPTER 4 CALCULATING THE DERIVATIVE

Exercises 4.1 (page 223)

1. $dy/dx = 30x^2 - 18x + 6$ **3.** $dy/dx = 4x^3 - 15x^2 + (2/9)x$ **5.** $f'(x) = 9x^{.5} - 2x^{-.5}$ or $9x^{.5} - 2/x^{.5}$ **7.** $dy/dx = 4x^{-1/2} + (9/2)x^{-1/4}$ or $4/x^{1/2} + 9/(2x^{1/4})$ **9.** $g'(x) = -30x^{-6} + x^{-2}$ or $-30/x^6 + 1/x^2$ **11.** $dy/dx = -5x^{-6} + 2x^{-3} - 5x^{-2}$ or $-5/x^6 + 2/x^3 - 5/x^2$ **13.** $f'(t) = -4t^{-2} - 6t^{-4}$ or $-4/t^2 - 6/t^4$ **15.** $dy/dx = -18x^{-7} - 5x^{-6} + 14x^{-3}$ or $-18/x^7 - 5/x^6 + 14/x^3$ **17.** $h'(x) = -x^{-3/2}/2 + 21x^{-5/2}$ or $-1/(2x^{3/2}) + 21/x^{5/2}$ **19.** $dy/dx = 2x^{-4/3}/3$ or $2/(3x^{4/3})$ **21.** $g'(x) = (3/2)\sqrt{x} - 1/\sqrt{x}$ **23.** $h'(x) = 6x^5 - 12x^3 + 6x$ **27.** $-(9/2)x^{-3/2} - 3x^{-5/2}$ or $-9/(2x^{3/2}) - 3/x^{5/2}$ **29.** $-14/3$ **31.** -28; $y = -28x + 34$ **33.** $5/2$ **35.** $(4/9, 20/9)$ **37.** $-3, -7$ **39.** $(5 \pm \sqrt{7})/3$ **41.** $(-1, 6), (4, -59)$ **43.** 38 **45. a.** 2 **b.** $1/2$ **c.** $[-1, \infty)$ **d.** $[0, \infty)$ **49. a.** 57.18 **b.** $-.88, .88$ **51. a.** 30 **b.** 4.8 **c.** -10 **53.** \$990 **55. a.** $C'(x) = 2$ **b.** $R'(x) = 6 - x/500$ **c.** $P'(x) = 4 - x/500$ **d.** $x = 2000$ **e.** \$4000 **57.** $M'(x) = 9.132x^2 - 759.2x + 14{,}274.5$ **a.** 2743 **b.** 1598 **c.** 11,983 **d.** 29,675 **59. a.** 450 **b.** 325 **c.** The blood sugar level is decreasing at a rate of 4 points per unit of insulin. **d.** The blood sugar level is decreasing at a rate of 10 points per unit of insulin. **61. a.** 220 g **b.** $28\frac{2}{3}$ g per cm; when the circumference of the brain is 30 cm, its mass is increasing by $28\frac{2}{3}$ g with every 1 cm increase in circumference. **63. a.** [18, 44] **b.** $l'(x) = .2356 - .005348x$ **c.** .1019 cm per wk **65.** $R = (2/3)R_0$ **67. a.** $v(t) = 22t + 4$ **b.** 4; 114; 224 **69. a.** $v(t) = 12t^2 + 16t$ **b.** 0; 380; 1360 **71. a.** -32 ft per sec; -64 ft per sec **b.** ln 3 sec **c.** -96 ft per sec **73. a.** 1.3275 g/cm^3 **b.** $-.43$ g per cm^3; when the level of the Dead Sea decreases to 50% of its current level, the density of the brine is decreasing at the rate of .43 g per cm^3.

Exercises 4.2 (page 234)

1. $dy/dx = 18x^2 - 6x + 4$ **3.** $dy/dx = 8x - 20$ **5.** $k'(t) = 4t^3 - 4t$ **7.** $dy/dx = (3/2)x^{1/2} + (1/2)x^{-1/2} + 2$ or $3x^{1/2}/2 + 1/(2x^{1/2}) + 2$ **9.** $p'(y) = -8y^{-5} + 15y^{-6} + 30y^{-7}$ **11.** $f'(x) = 53/(3x + 8)^2$ **13.** $dy/dx = -17/(4 + t)^2$ **15.** $dy/dx = (x^2 - 2x - 1)/(x - 1)^2$ **17.** $f'(t) = (-4t^2 - 22t - 12)/(t^2 - 3)^2$ **19.** $g'(x) = (x^2 + 6x - 14)/(x + 3)^2$ **21.** $p'(t) = [-\sqrt{t}/2 - 1/(2\sqrt{t})]/(t - 1)^2$ or $(-t - 1)/[2\sqrt{t}(t - 1)^2]$ **23.** $dy/dx = (5\sqrt{x}/2 - 3/\sqrt{x})/x$ or $(5x - 6)/(2x\sqrt{x})$ **25.** $g'(y) = (-1.1y^{2.9} - 2.5y^{1.5} + 2.8y^{.4})/(y^{2.5} + 2)^2$ **27.** 59 **29.** In the first step, the numerator should be $(x^2 - 1)2 - (2x + 5)(2x)$. **31.** $y = -2x + 9$ **35.** $x = -.828, 4.828$ **37. a.** \$2.24 per book **b.** \$1.39 per book **c.** $(5x - 6)/(2x^2 + 3x)$ per book **d.** $\overline{P}'(x) = (-10x^2 - 24x + 18)/(2x^2 + 3x)^2$ **41. a.** $s'(x) = m/(m + nx)^2$ **b.** $1/2560 \approx .000391$ mm per ml **43. a.** $N'(t) = 6t^2 - 80t + 200$ **b.** -56 million per hr **c.** 46 million per hr **d.** The population first declines, and then increases. **45. a.** $dW/dH = (H^2 - 1.86H - 17.7351)/(H - .93)^2$ **b.** 5.24 m **c.** Crows apply optimal foraging techniques. **47. a.** .1173 **b.** 2.625

Exercises 4.3 (page 243)

1. 1122 **3.** 97 **5.** $256k^2 + 48k + 2$ **7.** $(3x + 95)/8$; $(3x + 280)/8$ **9.** $1/x^2$; $1/x^2$ **11.** $\sqrt{8x^2 - 4}$; $8x + 10$ **13.** $\sqrt{(x - 1)/x}$; $-1/\sqrt{x + 1}$ **17.** If $f(x) = x^{2/5}$ and $g(x) = 5 - x$, then $y = f[g(x)]$. **19.** If $f(x) = -\sqrt{x}$ and $g(x) = 13 + 7x$, then $y = f[g(x)]$. **21.** If $f(x) = x^{1/3} - 2x^{2/3} + 7$ and $g(x) = x^2 + 5x$, then $y = f[g(x)]$.

23. $dy/dx = 3(8x^4 - 3x^2)^2(32x^3 - 6x)$ **25.** $k'(x) = -288x(12x^2 + 5)^5$ **27.** $s'(t) = (1215/2)t^2(3t^3 - 8)^{1/2}$ **29.** $g'(t) = -63t^2/(2\sqrt{7t^3 - 1})$ **31.** $m'(t) = -6(5t^4 - 1)(45t^4 - 1)$ **33.** $dy/dx = 3x^2(3x^4 + 1)(11x^4 + 32x + 1)$ **35.** $q'(y) = 2y(y^2 + 1)^{1/4}(9y^2 + 4)$ **37.** $dy/dx = 60x^2/(2x^3 + 1)^3$ **39.** $r'(t) = 2(5t - 6)^3(15t^2 + 18t + 40)/(3t^2 + 4)^2$ **41.** $dy/dx = (-18x^2 + 2x + 1)/(2x - 1)^6$ **43. a.** -2 **b.** $-24/7$ **45.** $y = (3/5)x + 16/5$ **47.** $y = x$ **49.** 1, 3 **53.** $D(c) = (-c^2 + 10c + 12{,}475)/25$ **55. a.** \$101.22 **b.** \$111.86 **c.** \$117.59 **57. a.** $-$\$1050 **b.** $-$\$457.06 **59.** $P[f(a)] = 18a^2 + 24a + 9$ **61. a.** $A[r(t)] = A(t) = 4\pi t^2$; this function gives the area of the pollution in terms of the time since the pollutants were first emitted. **b.** 32π; at 12 P.M. the area of pollution is changing at the rate of 32π mi^2 per hour **63. a.** $-.5$ **b.** $-1/54 \approx -.02$ **c.** $-1/128 \approx -.008$ **d.** Always decreasing; the derivative is negative for all $t \geq 0$. **65. a.** 34 minute **b.** $-(108/17)\pi$ mm^3 per minute, $-(72/17)\pi$ mm^2 per minute

Exercises 4.4 (page 251)

1. $dy/dx = 4e^{4x}$ **3.** $dy/dx = -16e^{2x}$ **5.** $dy/dx = -16e^{x+1}$ **7.** $dy/dx = 2xe^{x^2}$ **9.** $dy/dx = 12xe^{2x^2}$ **11.** $dy/dx = 16xe^{2x^2-4}$ **13.** $dy/dx = xe^x + e^x = e^x(x + 1)$ **15.** $dy/dx = 2(x - 3)(x - 2)e^{2x}$ **17.** $dy/dx = (2xe^x - x^2e^x)/e^{2x} = x(2 - x)/e^x$ **19.** $dy/dx = [x(e^x - e^{-x}) - (e^x + e^{-x})]/x^2$ **21.** $dp/dx = 8000e^{-.2t}/(9 + 4e^{-.2t})^2$ **23.** $f'(z) = 4(2z + e^{-z^2})(1 - ze^{-z^2})$ **25.** $dy/dx = -(\ln 2)\,2^{-x}$ **27.** $dy/dx = -6x(10^{3x^2-4}) \ln 10$ **29.** $ds/dt = (5 \ln 7)7^{\sqrt{t-2}}/(2\sqrt{t - 2})$ **31.**

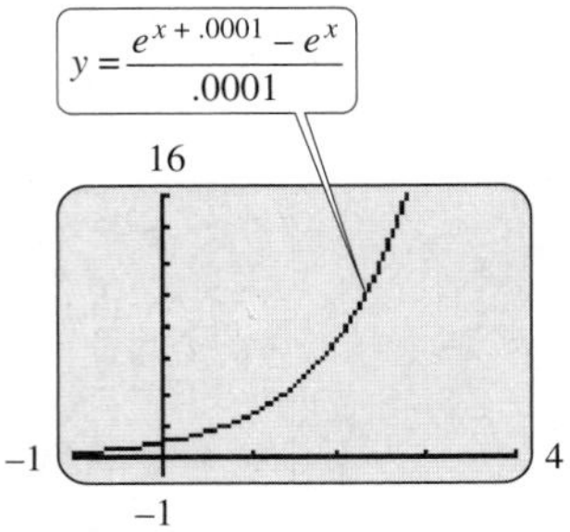

33. a. 20 **b.** 6 **c.** The rate of change of sales is decreasing. **d.** No, but it gets closer and closer to 0 as t increases. **35. a.** 3.07 **b.** -1.93 **c.** Public awareness increased at first and then decreased. **37. a.** 100% **b.** 94% **c.** 89% **d.** 83% **e.** -3.045 **f.** -2.865 **g.** The percent of these cars on the road is decreasing, but at a slower rate as they age. **39. a.** $G(t) = 10{,}000/(1 + 49e^{-.1t})$ **b.** 359; 34.6 **c.** 4276; 245 **d.** 9891; 10.8 **e.** It increases for a while and then gradually decreases to 0. **41. a.** .005 **b.** .0007 **c.** .000013 **d.** $-.022$ **e.** $-.0029$ **f.** $-.000054$ **43. a.** .026%; .286%; 3.130% **b.** .0025% per year; .0274% per year; .300% per year **c.** The percentage of people in each of the age groups that die in a given year is increasing as indicated by the answers in parts a and b. A person who is 75 years old has a 3% chance of dying during the year and the rate is increasing by almost .3%. The formula implies that everyone will be dead by age 112. **45. a.** .027 **b.** 2024 **c.** The marginal increase in the proportion per year in 2002 is approximately .008. **47. a.** Approaches 589 mm **b.** About 15 years **c.** About 32.2 mm per year; the rate of growth of a cutlassfish increases by approximately 32.2 mm during its fourth year of life. **d.**

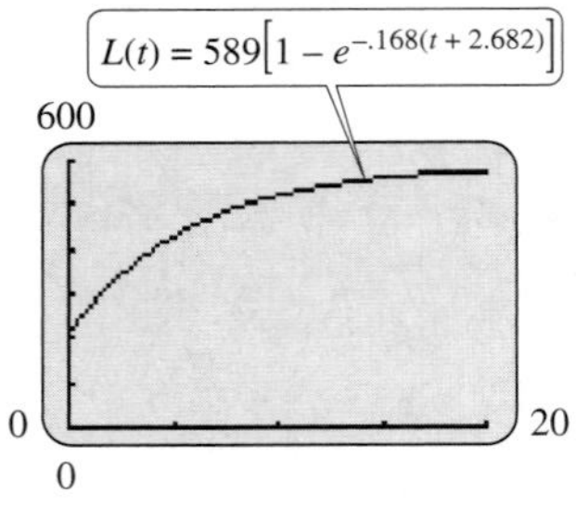

49. a. 36.8 **b.** .00454 **c.** Approximately 0 **d.** $100e^{-.1N}$ is always positive, since powers of e are never negative. This means that repetition always makes a habit stronger. **51. a.** -44.9 gm per yr **b.** -24.1 gm per year **c.** -6.98 gm per year **d.** It is approaching 0. **e.** No. **53. a.** 218.7 seconds **b.** The record is decreasing by .047 seconds per year at the end of 2005. **c.** 218 seconds. If the estimate is correct, then this is the least amount of time that it will ever take for a human to run a mile.

Exercises 4.5 (page 260)

1. $dy/dx = 1/x$ **3.** $dy/dx = -1/(3 - x)$ or $1/(x - 3)$ **5.** $dy/dx = (4x - 7)/(2x^2 - 7x)$ **7.** $dy/dx = 1/[2(x + 5)]$ **9.** $dy/dx = 3(2x^2 + 5)/[x(x^2 + 5)]$ **11.** $dy/dx = -3x/(x + 2) - 3\ln(x + 2)$ **13.** $ds/dt = t + 2t \ln|t|$ **15.** $dy/dx = [2x - 4(x + 3)\ln(x + 3)]/[x^3(x + 3)]$ **17.** $dy/dx = (4x + 7 - 4x \ln x)/[x(4x + 7)^2]$ **19.** $dy/dx = (6x \ln x - 3x)/(\ln x)^2$ **21.** $dy/dx = 4(\ln|x + 1|)^3/(x + 1)$ **23.** $dy/dx = 1/(x \ln x)$ **25.** $dy/dx = e^{x^2}/x + 2xe^{x^2} \ln x$

27. $dy/dx = (xe^x \ln x - e^x)/[x(\ln x)^2]$ **29.** $g'(z) = 3(e^{2z} + \ln z)^2(2ze^{2z} + 1)/z$ **31.** $dy/dx = 2/[(\ln 10)(2x - 3)]$
33. $dy/dx = 1/(x \ln 10)$ **35.** $dy/dx = 1/[(\ln 7)(2x - 3)]$ **37.** $dy/dx = 5(4x - 1)/[(2 \ln 2)(2x^2 - x)]$
39. $dz/dy = 10^y/[(\ln 10)y] + (\log y)(\ln 10)10^y$ **43.**

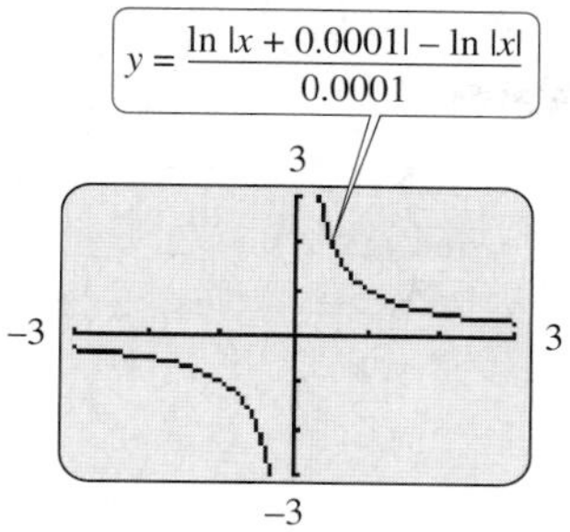

49. a. $dR/dx = 100 + 50(\ln x - 1)/(\ln x)^2$ **b.** \$112.48 **c.** To decide whether it is reasonable to sell additional items
51. a. $-.19396$ **b.** $-.06099$ **53. a.** $N(t) = 1000e^{9.8901e^{-e^{2.54197-.2167t}}}$ **b.** 1,307,416 bacteria per hour; the number of bacteria is increasing at a rate of 1,307,416 per hour, 20 hours after the experiment began. **c.**

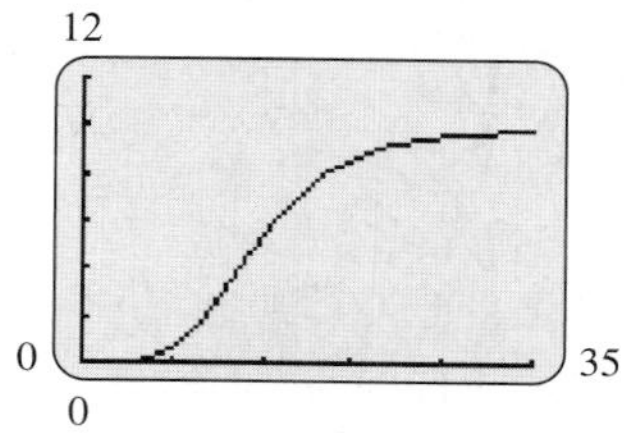

d.

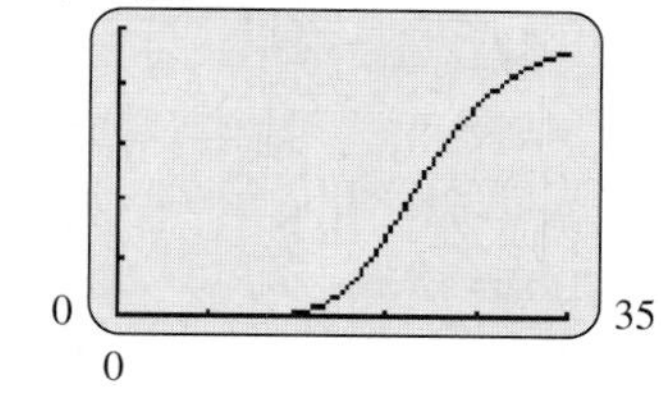

e. 9.8901; $1000e^{9.8901} \approx 19{,}734{,}033$ **55. b. i.** 3.343 **ii.** 1.466 **c. i.** $-.172$ **ii.** $-.0511$
57. 26.9; 13.1 **59. a.** 817 vehicles per hr, -41.2 vehicles per hr per ft
b. 522 vehicles per hr, -20.9 vehicles per hr

Chapter 4 Review Exercises (page 265)

1. $dy/dx = 10x - 7$ **3.** $dy/dx = 14x^{4/3}$ **5.** $f'(x) = -3x^{-4} + (1/2)x^{-1/2}$ or $-3/x^4 + 1/(2x^{1/2})$ **7.** $k'(x) = 15/(x + 5)^2$
9. $dy/dx = (x^2 - 2x)/(x - 1)^2$ **11.** $f'(x) = 12(3x - 2)^3$ **13.** $dy/dx = 1/(2t - 5)^{1/2}$ **15.** $dy/dx = 3(2x + 1)^2(8x + 1)$
17. $r'(t) = (-15t^2 + 52t - 7)/(3t + 1)^4$ **19.** $p'(t) = t(t^2 + 1)^{3/2}(7t^2 + 2)$ **21.** $dy/dx = -12e^{2x}$ **23.** $dy/dx = -6x^2e^{-2x^3}$
25. $dy/dx = 10xe^{2x} + 5e^{2x} = 5e^{2x}(2x + 1)$ **27.** $dy/dx = 2x/(2 + x^2)$ **29.** $dy/dx = (x - 3 - x \ln |3x|)/[x(x - 3)^2]$
31. $dy/dx = [e^x(x + 1)(x^2 - 1) \ln(x^2 - 1) - 2x^2e^x]/[(x^2 - 1)[\ln(x^2 - 1)]^2]$ **33.** $ds/dt = 2(t^2 + e^t)(2t + e^t)$
35. $dy/dx = -6x(\ln 10) \cdot 10^{-x^2}$ **37.** $q'(z) = (3z^2 + 1)/[(\ln 2)(z^3 + z + 1)]$ **41.** -2; $y = -2x + 9$
43. $-5/9$; $y = -(5/9)x + 16/9$ **45.** $-4/5$; $y = -(4/5)x - 13/5$ **47.** $2e$; $y = 2ex - e$ **49.** 2; $y = 2x - e$
53. $\overline{C}'(x) = (-x - 2)/[2x^2(x + 1)^{1/2}]$ **55.** $\overline{C}'(x) = (x^2 + 3)^2(5x^2 - 3)/x^2$ **57.** $\overline{C}'(x) = [e^{-x}(x + 1) - 10]/x^2$
59. a. 55/3; sales will increase by \$55 million when \$3000 more is spent on research. **b.** 65/4; sales will increase by \$65 million when \$4000 more is spent on research. **c.** 15; sales will increase by \$15 million when \$1000 more is spent on research. **d.** As more is spent on research, the increase in sales is decreasing. **61. a.** -9.5; costs will decrease by \$9500 for the next \$100 spent on training. **b.** -2.375; costs will decrease by \$2375 for the next \$100 spent on training. **c.** Decreasing **63.** \$218.65. The balance increases by roughly \$218.65 for every 1% increase in the interest rate when the rate is 5%. **65.** 50,000 per year **67. a.** 28.1 cm **b.** 4.34 cm per year **c.** 205 g **d.** 21.2 g per cm **e.** 92.0 g per year **69.** .0129. The concentration of CFC-11 is increasing at a rate of .0129 ppb per year in 1998. **71.** .247; the production of corn is increasing at a rate of .247 billion bushels a year in 2000.
73. a. $-.4677$ fatalities per 1000 licensed drivers per 100 million miles per year; at the age of 20, each extra year results in a decrease of .4677 fatalities per 1000 licensed drivers per 100 million miles. **b.** .003672 fatalities per 1000 licensed drivers per 100 million miles per year; at the age of 60, each extra year results in an increase of .003672 fatalities per 1000 licensed drivers per 100 million miles.

CHAPTER 5 GRAPHS AND THE DERIVATIVE

Exercises 5.1 (page 278)

1. a. $(1, \infty)$ **b.** $(-\infty, 1)$ **3. a.** $(-\infty, -2)$ **b.** $(-2, \infty)$ **5. a.** $(-\infty, -4), (-2, \infty)$ **b.** $(-4, -2)$ **7. a.** $(-7, -4), (-2, \infty)$ **b.** $(-\infty, -7), (-4, -2)$ **9. a.** $3/2$ **b.** $(-\infty, 3/2)$ **c.** $(3/2, \infty)$ **11. a.** $-3, 4$ **b.** $(-\infty, -3), (4, \infty)$ **c.** $(-3, 4)$ **13. a.** $-3/2, 4$ **b.** $(-\infty, -3/2), (4, \infty)$ **c.** $(-3/2, 4)$ **15. a.** $-2, -1, 0$ **b.** $(-2, -1), (0, \infty)$ **c.** $(-\infty, -2), (-1, 0)$ **17. a.** None **b.** None **c.** $(-\infty, \infty)$ **19. a.** None **b.** None **c.** $(-\infty, -1), (-1, \infty)$ **21. a.** 0 **b.** $(0, \infty)$ **c.** $(-\infty, 0)$ **23. a.** 0 **b.** $(0, \infty)$ **c.** $(-\infty, 0)$ **25. a.** 7 **b.** $(7, \infty)$ **c.** $(3, 7)$ **27. a.** $0, 2/5$ **b.** $(0, 2/5)$ **c.** $(-\infty, 0), (2/5, \infty)$ **31.** Vertex: $(-b/(2a), (4ac - b^2)/(4a))$; increasing on $(-\infty, -b/(2a))$, decreasing on $(-b/(2a), \infty)$ **33.** On $(0, \infty)$; nowhere; nowhere **35. a.** Nowhere **b.** Everywhere **37. a.** $(0, 3)$ **b.** $(3, 3.9)$ **39.** After 9 days **41. a.** $(0, 3)$ **b.** $(3, \infty)$ (*Remember:* x must be at least 0.) **43.** Increasing $(55, 130)$; decreasing nowhere **45.** $(0, \infty)$ **47. a.** About $(1945, 1966), (1970, 1974)$ **b.** About $(1986, 2002)$ **49. a.** Negative, because f is decreasing **b.** mpg/lb, or miles per gallon per pound

Exercises 5.2 (page 292)

1. Relative minimum of -4 at 1 **3.** Relative maximum of 3 at -2 **5.** Relative maximum of 3 at -4; relative minimum of 1 at -2 **7.** Relative maximum of 3 at -4; relative minimum of -2 at -7 and -2 **9.** Relative minimum of -44 at -6 **11.** Relative maximum of -8 at -3; relative minimum of -12 at -1 **13.** Relative maximum of $827/96$ at $-1/4$; relative minimum of $-377/6$ at -5 **15.** Relative maximum of -4 at 0; relative minimum of -85 at 3 and -3 **17.** Relative maximum of 0 at $8/5$ **19.** Relative maximum of 1 at -1; relative minimum of 0 at 0 **21.** No relative extrema **23.** Relative maximum of 0 at 1; relative minimum of 8 at 5 **25.** Relative maximum of -2.46 at -2; relative minimum of -3 at 0 **27.** No relative extrema **29.** $(2, 7)$ **31.** Relative maximum of 6.211 at .085; relative minimum of -57.607 at 2.161 **33.** Relative minimum at $x = 5$

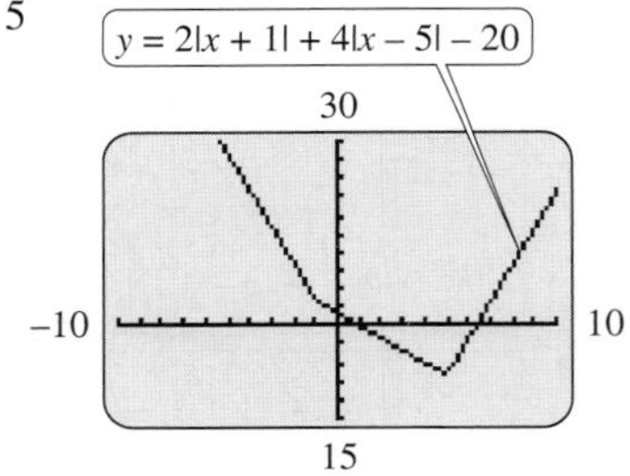

35. a. 15 **b.** \$40 **c.** \$375 **37. a.** 100 **b.** \$14.72 **c.** \$635.76 **39.** $q = 10$; $p \approx \$73.58$ **41.** 100 units **43. a.** 9.68 weeks; 8.54 kg **b.** b/c; $a(b/c)^b e^{-b}$ **45.** 1.3 hr **47.** 10 **49. a.** 67 ft **b.** 4.05 sec

Exercises 5.3 (page 306)

1. $f''(x) = 18x$; 0; 36 **3.** $f''(x) = 36x^2 - 30x + 4$; 4; 88 **5.** $f''(x) = 6$; 6; 6 **7.** $f''(x) = 2/(1 + x)^3$; 2; 2/27 **9.** $f''(x) = -1/[4(x + 4)^{3/2})]$; $-1/32$; $-1/[4(6^{3/2}] \approx -.0170$ **11.** $f''(x) = -(6/5)x^{-7/5}$ or $-6/(5x^{7/5})$; $f''(0)$ does not exist; $-6/[5(2^{7/5})] \approx -.4547$ **13.** $f''(x) = 20x^2e^{-x^2} - 10e^{-x^2}$; -10; $70e^{-4} \approx 1.282$ **15.** $f''(x) = (-3 + 2 \ln x)/(4x^3)$; does not exist; $-.050$ **17.** $f'''(x) = -24x$; $f^{(4)}(x) = -24$ **19.** $f'''(x) = 240x^2 + 144x$; $f^{(4)}(x) = 480x + 144$ **21.** $f'''(x) = 18(x + 2)^{-4}$ or $18/(x + 2)^4$; $f^{(4)}(x) = -72(x + 2)^{-5}$ or $-72/(x + 2)^5$ **23.** $f'''(x) = -36(x - 2)^{-4}$ or $-36/(x - 2)^4$; $f^{(4)}(x) = 144(x - 2)^{-5}$ or $144/(x - 2)^5$ **25. a.** $f'(x) = 1/x$; $f''(x) = -1/x^2$; $f'''(x) = 2/x^3$; $f^{(4)}(x) = -6/x^4$; $f^{(5)}(x) = 24/x^5$ **b.** $f^{(n)}(x) = (-1)^{n-1}[1 \cdot 2 \cdot 3 \cdots (n - 1)]/x^n$ or, using factorial notation, $f^{(n)}(x) = (-1)^{n-1}(n - 1)!/x^n$ **27.** Concave upward on $(2, \infty)$; concave downward on $(-\infty, 2)$; point of inflection at $(2, 3)$ **29.** Concave upward on $(-\infty, -1)$ and $(8, \infty)$; concave downward on $(-1, 8)$; points of inflection at $(-1, 7)$ and $(8, 6)$ **31.** Concave upward on $(2, \infty)$; concave downward on $(-\infty, 2)$; no points of inflection **33.** Always concave upward; no points of inflection **35.** Concave upward on $(-\infty, 3/2)$; concave downward on $(3/2, \infty)$; point of inflection at $(3/2, 525/2)$ **37.** Concave upward on $(5, \infty)$; concave downward on $(-\infty, 5)$; no points of inflection **39.** Concave upward on $(-10/3, \infty)$; concave downward on $(-\infty, -10/3)$; point of inflection at $(-10/3, -250/27)$ **41.** Never concave upward; always concave downward; no inflection points **43.** Concave upward on $(-\infty, 0)$ and $(1, \infty)$; concave downward on $(0, 1)$; points of inflection at $(0, 0)$ and $(1, -3)$ **45. a.**

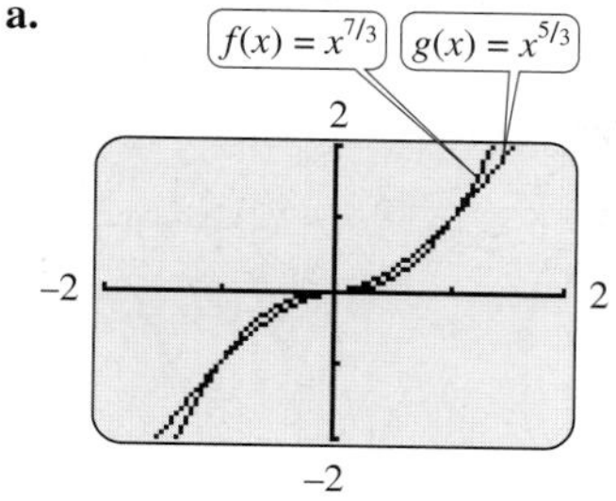

c. $f''(0) = 0$, while $g''(0)$ is undefined. **d.** No.

47. Approaches 0; approaches ∞ **49.** Relative minimum at 6 **51.** Relative maximum at 0; relative minimum at $4/3$ **53.** Critical number at 0, but neither a maximum nor minimum there **55. a.** Minimum at about $-.4$ and 4.0; maximum at about 2.4 **b.** Increasing on about $(-.4, 2.4)$ and about $(4.0, \infty)$; decreasing on about $(-\infty, -.4)$ and $(2.4, 4.0)$ **c.** About .7 and 3.3 **d.** Concave upward on about $(-\infty, 4.7)$ and $(3.3, \infty)$; concave downward on about $(.7, 3.3)$ **57. a.** Maximum at 1; minimum at -1 **b.** Increasing on $(-1, 1)$; decreasing on $(-\infty, -1)$ and $(1, \infty)$ **c.** About 1.7 and -1.7 and 0 **d.** Concave upward on about $(-1.7, 0)$ and $(1.7, \infty)$; concave downward on about $(-\infty, -1.7), (0, 1.7)$ **59.** $(14, 26{,}688)$ **61.** $(1.11, 13.5)$ **63.** $1/(2M), 1/(3M); U(M) = \sqrt{M}$ indicates a greater aversion to risk. **65. a.** Initial population **b.** Inflection point **c.** Maximum carrying capacity **67.** $c(t)$ is increasing and concave downward, $c'(t) > 0, c''(t) < 0.$ **69. a.** After 3 hr **b.** 2/9% **71.** (4.96981, 2600) **73.** 301 days; the time when the rate of growth begins to slow down. **75.** (32.01, 26.41) **77.** 50 **79. a.** 343.25 ft **b.** About 9 sec, about -148 ft per sec **81. a.** 190 ft **b.** 19 ft per sec; 34 ft per sec **c.** The car stops if $v(t) = 0$, but here $v(t) > 0$ for all nonnegative t. **d.** 3 ft per sec^2; 3 ft per sec^2 **e.** Velocity is increasing; acceleration is constant.

Exercises 5.4 (page 318)

1. 0 **3.**

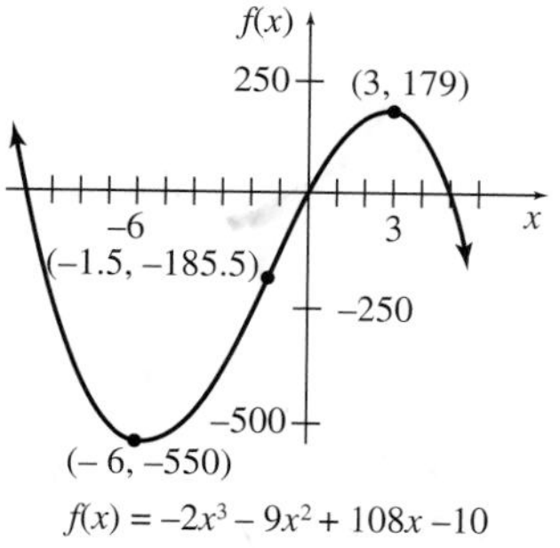

5.

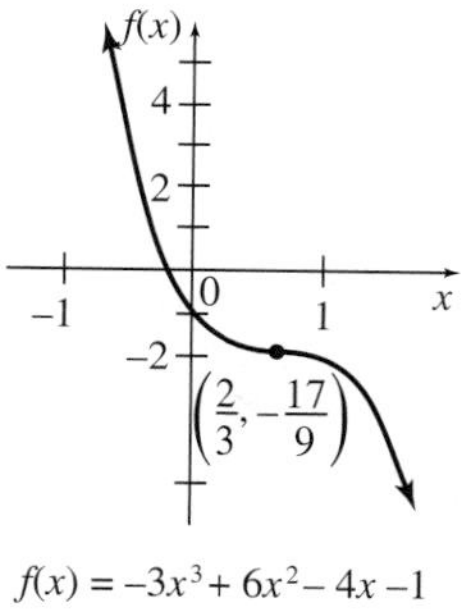

7.

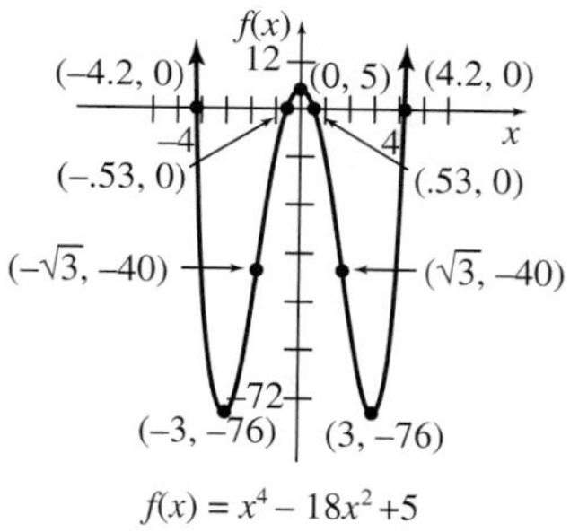

9.

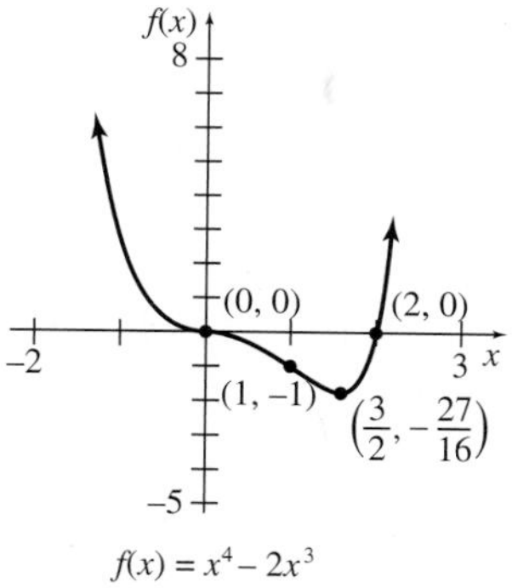

11.

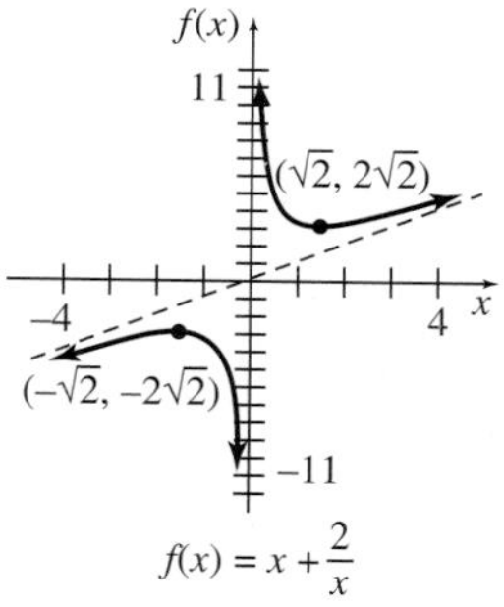

13.

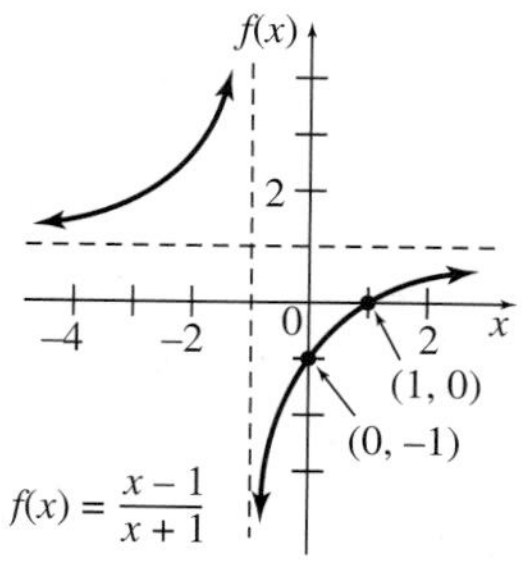

15.

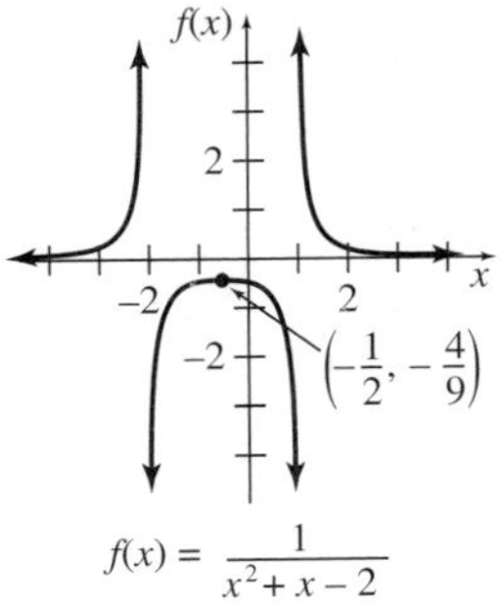

17.

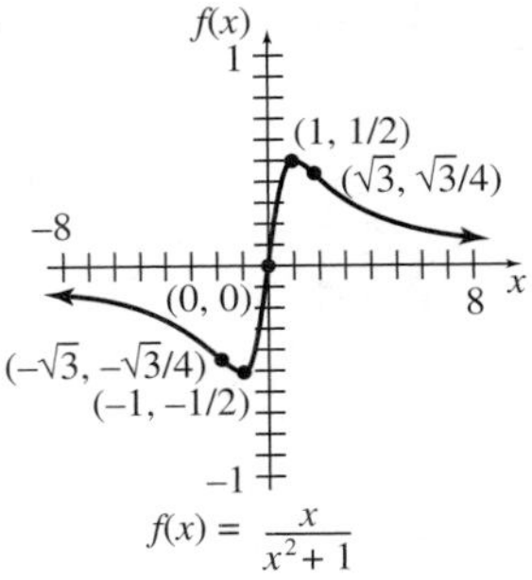

19.

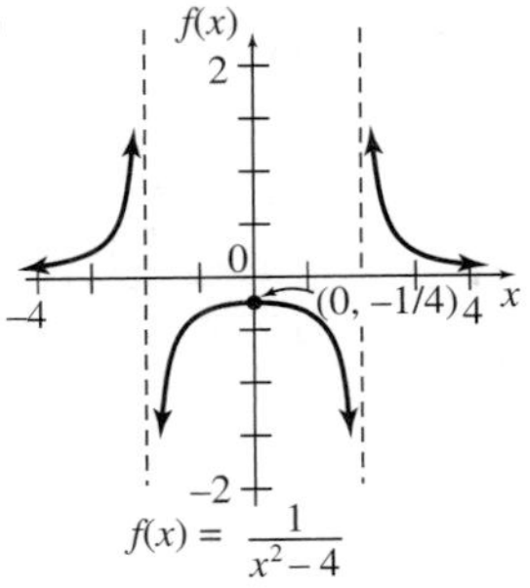

21.

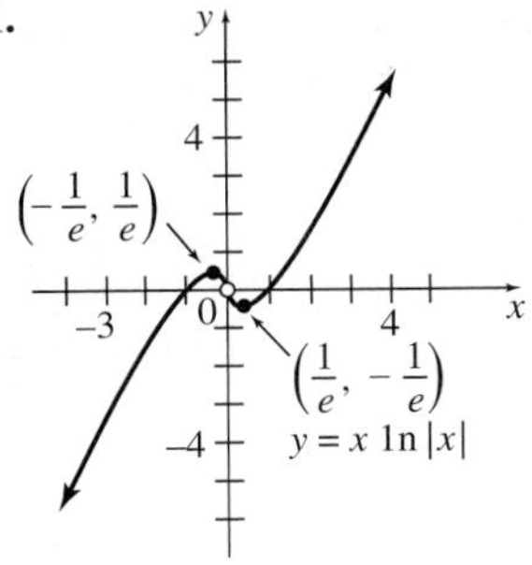

23.

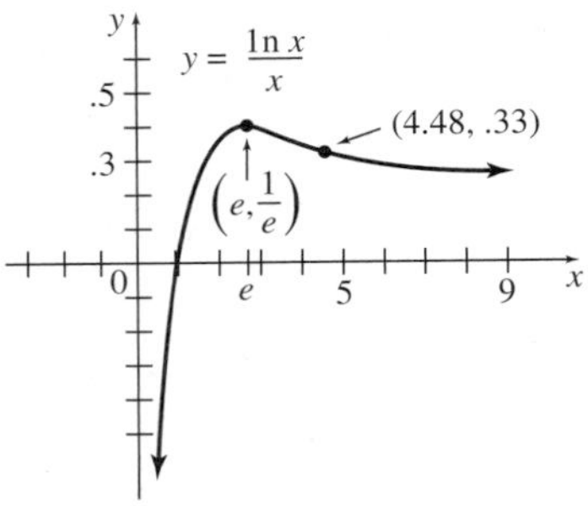

25.

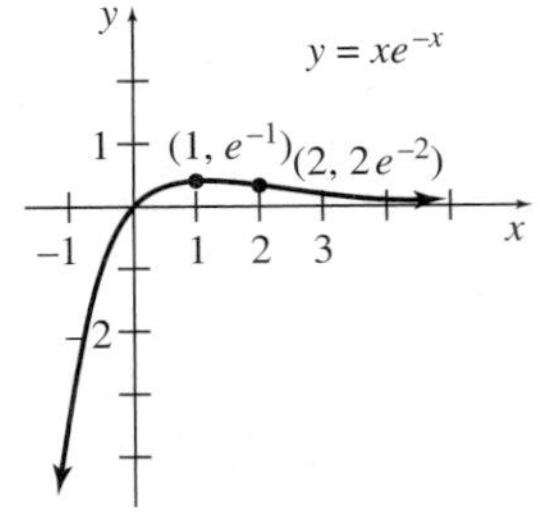

27.

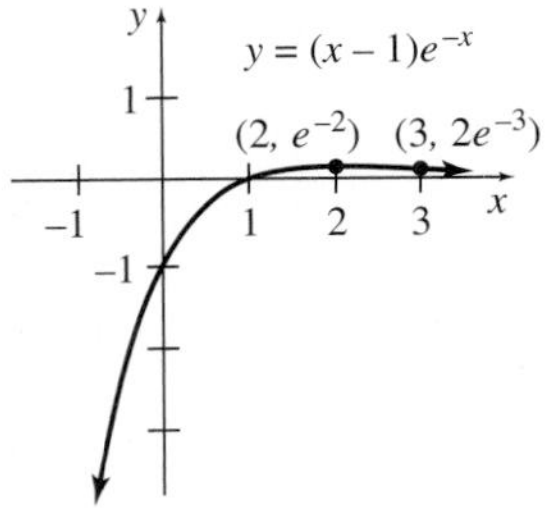

29.

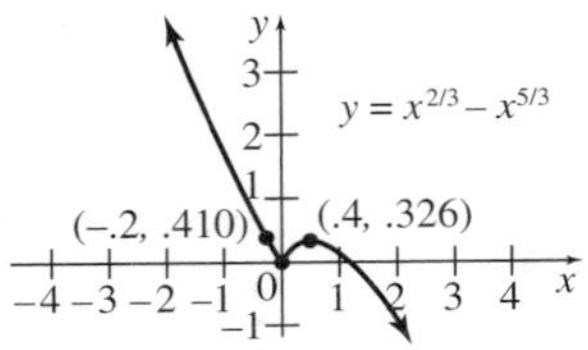

31. 3, 7, 15 **33.** 17, 19, 23, 25, 27 **35.** In Exercises 35–39, other answers are possible.

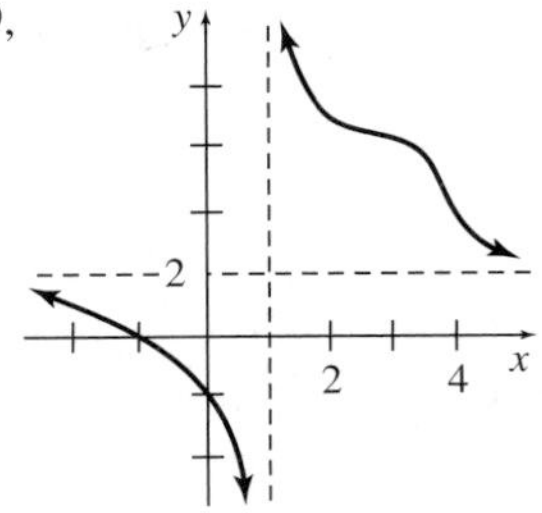

37.

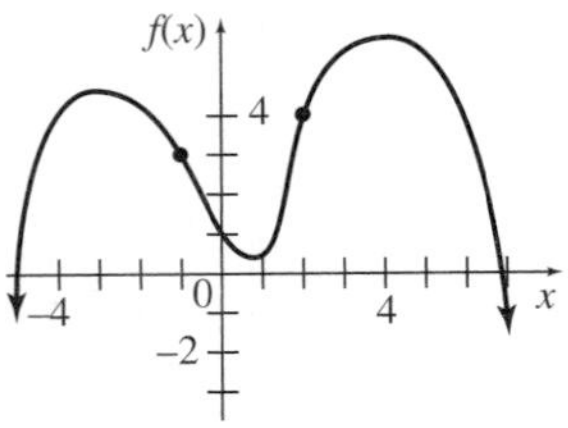

39.

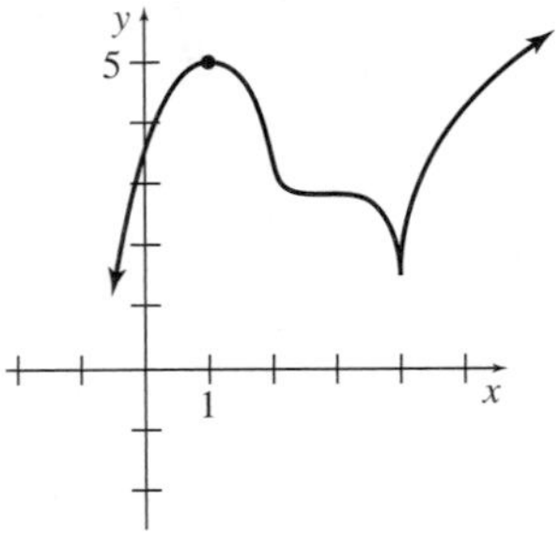

Chapter 5 Review Exercises page (320)

5. Increasing on $(5/2, \infty)$; decreasing on $(-\infty, 5/2)$ **7.** Increasing on $(-4, 2/3)$; decreasing on $(-\infty, -4)$ and $(2/3, \infty)$ **9.** Never increasing; decreasing on $(-\infty, 4)$ and $(4, \infty)$ **11.** Decreasing on $(-\infty, -1)$; increasing on $(1, \infty)$ **13.** Relative maximum of -4 at 2 **15.** Relative minimum of -7 at 2 **17.** Relative maximum of 101 at -3; relative minimum of -24 at 2 **19.** Relative maximum at $(-.618, .206)$; relative minimum at $(1.618, 13.203)$ **21.** $f''(x) = 36x^2 - 10$; 26; 314 **23.** $f''(x) = -68(2x + 3)^{-3}$ or $-68/(2x + 3)^3$; $-68/125$; $68/27$

25. $f''(x) = (t^2 + 1)^{-3/2}$ or $1/(t^2 + 1)^{3/2}$; $1/2^{3/2} \approx .354$; $1/10^{3/2} \approx .032$

27.

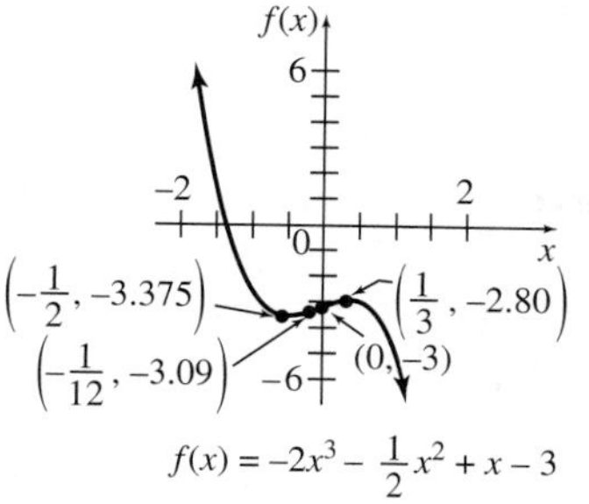

29.

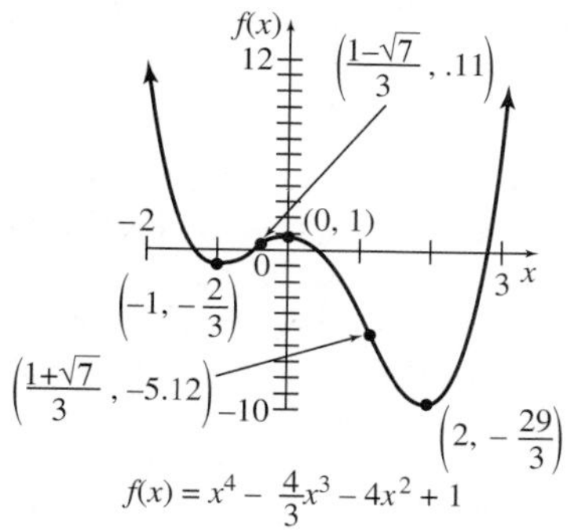

31.

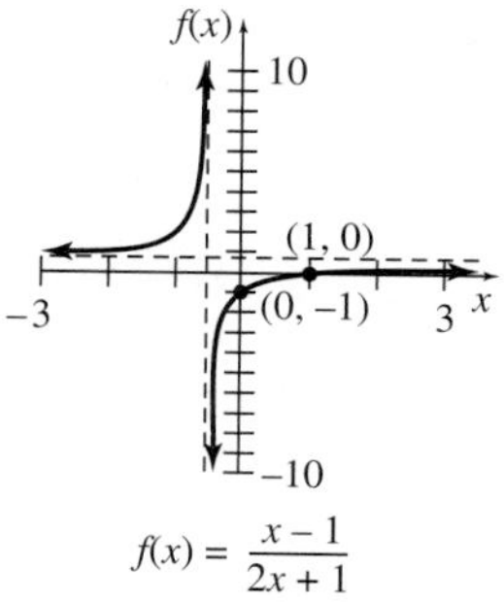

33.

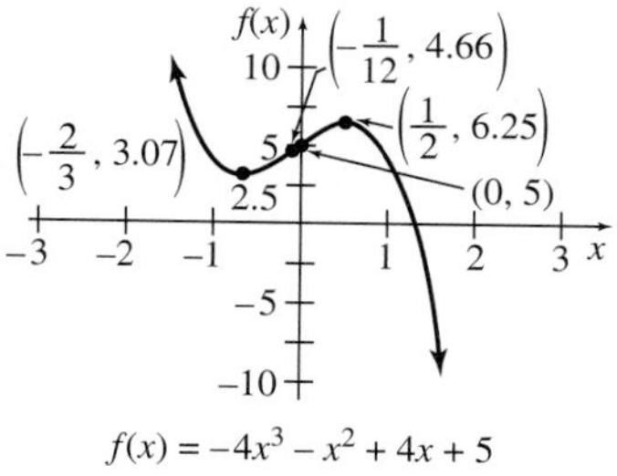

35.

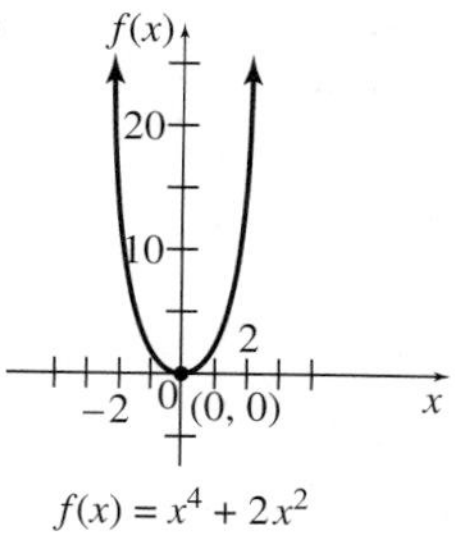

37.

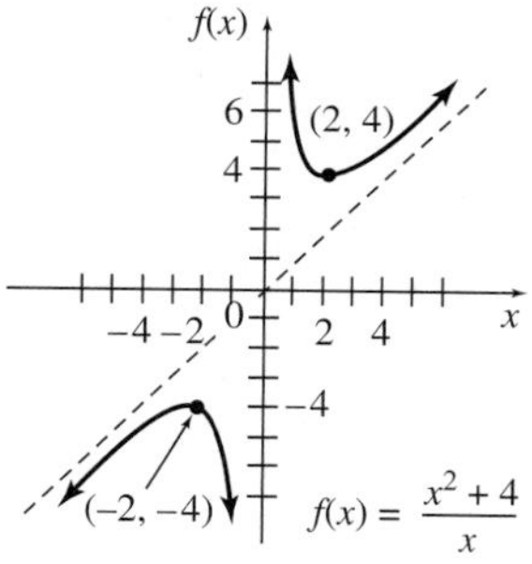

39.

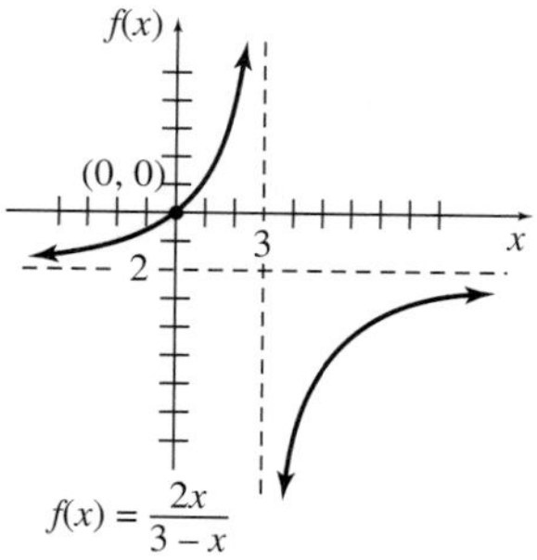

In Exercise 41, other answers are possible.

41.

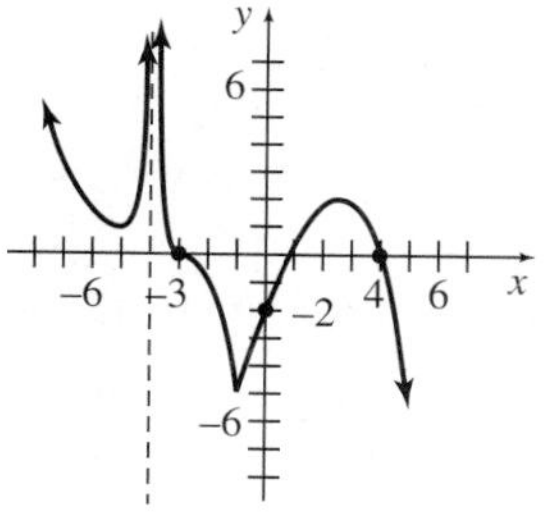

43. a. Both are negative. **45. a.** $P(x) = -x^3 + 7x^2 + 49x$ **b.** 7 brushes **c.** \$229 **d.** \$343 **e.** $x = 7/3$; between 2 and 3 brushes **47. a.** Metabolic rate and life span are increasing and concave downward. Heartbeat is decreasing and concave upward.

49.

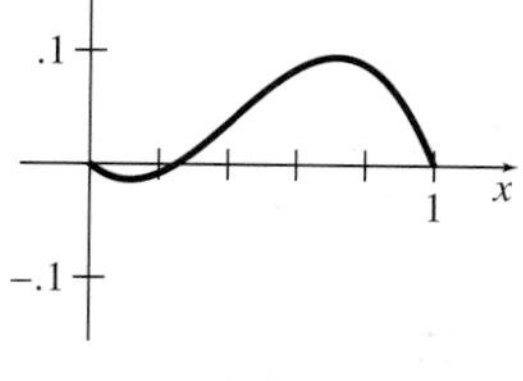

51. a. 1486 ml per square meter; for males with 1.88 m^2 of surface area, the red cell volume increases approximately 1486 ml for each additional square meter of surface area. **b.** 1.57 m^2; 2593 ml (Hurley); 2484 ml (Pearson et al.) **c.** 1578 ml per m^2; for males with 1.57 m^2 of surface area, the red cell volume increases approximately 1578 ml for each additional square meter of surface area. **53.** 7.405 yr; the age at which the rate of learning to pass the test begins to slow down. **55. a.** $v(t) = 512 - 32t$; $a(t) = -32$ **b.** 4096 ft **c.** After 32 sec; -512 ft per sec

CHAPTER 6 APPLICATIONS OF THE DERIVATIVE

Exercises 6.1 (page 331)

1. Absolute maximum at x_3; no absolute minimum **3.** No absolute extrema **5.** Absolute minimum at x_1; no absolute maximum **7.** Absolute maximum at x_1; absolute minimum at x_2 **11.** Absolute maximum at $x = 5$; absolute minimum at $x = 0$ and $x = 3$ **13.** Absolute maximum at $x = -4$; absolute minimum at $x = 1$ **15.** Absolute maximum at $x = 0$; absolute minimum at $x = -3$ and $x = 3$ **17.** Absolute maximum at $x = 0$; absolute minimum at $x = 3$ **19.** Absolute maximum at $x = 1 + \sqrt{2} \approx 2.4$; absolute minimum at $x = 1$ **21.** Absolute maximum at $x = -2$ and $x = 2$; absolute minimum at $x = 0$ **23.** Absolute maximum at $x = -2$; absolute minimum at $x = 5/2$ **25.** Absolute maximum at $x = .6085$; absolute minimum at $x = -1$ **27.** Absolute minimum at $x = 2$; no absolute maximum **29.** Absolute maximum at $x = 3$; no absolute minimum **31.** Absolute maximum at $x = 4$; absolute minimum at $x = -2$ **33. a.** Relative maxima of 9540 in 1992, 8362 in 1996, 8516 in 2001; relative minima of 8000 in 1990, 6986 in 1995, 6813 in 1999 **b.** Annual bank robberies reached an absolute maximum of 9540 in 1992 and an absolute minimum of 6813 in 1999. **35.** Maximum profit of \$20,000 occurs when 100 units are made per week. **37. a.** 341 **b.** 859.4 **39.** 20 units **41.** 300 units **43.** 6 mo; 6% **45.** About 7.2 mm **47.** 25; 16.1 **49.** The piece formed into a circle should have length $12\pi/(4 + \pi)$ ft, or about 5.28 ft. **51. b.** $1/2$ **c.** To decide how to phrase a message to get maximum information content

Exercises 6.2 (page 341)

1. a. $y = 100 - x$ **b.** $P = x(100 - x)$ **c.** $[0, 100]$ **d.** $dP/dx = 100 - 2x$; $x = 50$ **e.** $P(0) = 0$; $P(100) = 0$; $P(50) = 2500$ **f.** 2500; 50 and 50 **3. a.** $y = 150 - x$ **b.** $P = x^2(150 - x)$ **c.** $[0, 150]$ **d.** $dP/dx = 300x - 3x^2$; $x = 0, x = 100$ **e.** $P(0) = 0, P(100) = 500{,}000, P(150) = 0$ **f.** 100; 50; 500,000 **5.** $A(x) = x^2/2 + 2x - 3 + 35/x$; $x = 2.722$ **7. a.** $R(x) = 100{,}000x - 100x^2$ **b.** 500 **c.** 25,000,000 cents **9. a.** $1200 - 2x$ **b.** $A(x) = 1200x - 2x^2$ **c.** 300 m **d.** 180,000 m^2 **11.** 405,000 m^2 **13.** \$1000 **15.** In 5 days; \$490 **17. a.** 80 **b.** \$32,000 **19.** 20 cm by 20 cm by 40 cm; \$7200 **23.** 2/3 ft (or 8 in.) **25.** $3\sqrt{6} + 3$ by $2\sqrt{6} + 2$ **27.** Point A **29.** Radius = 5.206 cm, height = 11.75 cm **31.** Radius = 5.242 cm; height = 11.58 cm **33. a.** 12 days **b.** 50 per ml **c.** 1 day **d.** 81.365 per ml **35.** 12.98 thousand **39.** 237.10 **41.** Point P is at Point L. **45.** $(56 - 2\sqrt{21})/7 \approx 6.7$ mi

Exercises 6.3 (page 353)

3. c **5.** 280 **7.** 60 **9.** 95 **11.** 913 **13.** 10 runs **15. a.** $E = p/(200 - p)$ **b.** 25 **17. a.** $E = 2p^2/(7500 - p^2)$ **b.** 25,000 **19. a.** $E = 5/q$ **b.** 5 **21. a.** $E = .5$; inelastic; total revenue increases as price increases. **b.** $E = 8$; elastic; total revenue decreases as price increases. **23. a.** $-.287$ **b.** 1.4% **27. a.** k

Exercises 6.4 (page 359)

1. $dy/dx = -4x/(3y)$ **3.** $dy/dx = (-6x - 4y)/(4x + y)$ **5.** $dy/dx = 3x^2/(2y)$ **7.** $dy/dx = -3x(2 + y)^2/2$ **9.** $dy/dx = -y^{1/2}/x^{1/2}$ **11.** $dy/dx = (4x^3y^3 + 6x^{1/2})/(9y^{1/2} - 3x^4y^2)$ **13.** $dy/dx = (5 - 2xye^{x^2y})/(x^2e^{x^2y} - 4)$ **15.** $dy/dx = y(2xy^3 - 1)/(1 - 3x^2y^3)$ **17.** $y = (3/4)x + 25/4$ **19.** $y = x + 2$ **21.** $y = x/64 + 7/4$ **23.** $y = (5/2)x - 1/2$ **25.** $y = 1$ **27.** $y = -2x + 7$ **29. a.** $y = -(3/4)x + 25/2$; $y = (3/4)x - 25/2$ **b.**

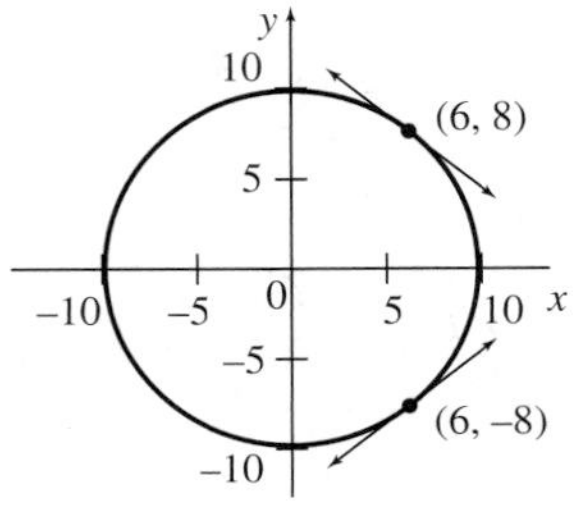

31. $y = -(2/11)x + 15/11$ **33.** $y = (11/12)x - 5/6$ **35.** $du/dv = -2u^{1/2}/(2v + 1)^{1/2}$ **37. a.** $dq/dp = -2p/q$; the rate of change of demand with respect to price **b.** $dp/dq = -q/(2p)$; the rate of change of price with respect to demand **39.** $R'(w) = -29.0716w^{-1.43}$ **41.** $-y/(ax)$ **43.** $ds/dt = (-s + 6\sqrt{st})/(8s\sqrt{st} + t)$

Exercises 6.5 (page 366)

1. $-15/2$ **3.** $-5/7$ **5.** $1/5$ **7.** $-3/2$ **9.** \$200 per month **11. a.** Revenue is increasing at a rate of \$180 per day. **b.** Cost is increasing at a rate of \$50 per day. **c.** Profit is increasing at a rate of \$130 per day. **13.** Demand is decreasing at a rate of approximately 98 units per unit time. **15.** .067 mm per min **17.** About 1.9849 g per day **19. a.** $105.15\ m^{-.25}\ dm/dt$ **b.** About 52.89 kcal per day^2 **21.** 25.6 crimes per month **23.** .008 **25. a.** 50 mph **b.** About 47.15 mph **27.** -16π in^3 per hr **29.** 62.5 ft per min **31.** $\sqrt{2} \approx 1.41$ ft per sec

Exercises 6.6 (page 374)

1. -2.6 **3.** .1 **5.** .130 **7.** $-.023$ **9.** 12.0417; 12.0416; .0001 **11.** .995; .9950; 0 **13.** 1.01; 1.0101; .0001 **15.** .05; .0488; .0012 **17. a.** -34 thousand lb **b.** -169.2 thousand lb **19.** \$.23 **21.** About 9600 in^3 **23. a.** Alcohol concentration increases by about .33 tenths of a percent. **b.** Alcohol concentration decreases by about .13 tenths of a percent. **25. a.** .347 million **b.** $-.022$ million **27.** 1568π mm^3 **29.** 80π mm^2 **31. a.** About 9.3 kg **b.** About 9.5 kg **33.** -7.2π cm^3 **35.** $\pm$ 1.224 in^2 **37.** $\pm$ .116 in^3

Chapter 6 Review Exercises (page 376)

1. Absolute maximum of 29/4 at 5/2; absolute minimum of 5 at 1 and 4 **3.** Absolute maximum of 39 at -3; absolute minimum of $-319/27$ at 5/3 **7. a.** Maximum = .37; minimum = 0 **b.** Maximum = .35; minimum = .13 **11.** $dy/dx = (y - 3y^2)/(4y^2 + x)$ **13.** $dy/dx = 16(y-1)^{1/2}/(3x^{1/3})$ **15.** $dy/dx = -(30 + 50x)/3$ **19.** 272 **21.** -2 **25.** .00204 **29.** 2 m by 4 m by 4 m **31.** 3 in. **33.** 8000 **35.** 80 **37.** 56π ft^2 per min **39. a.**

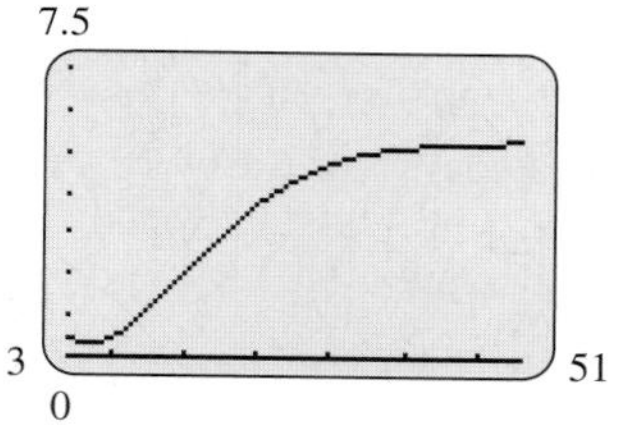

b. About the 15th day

41. 8/3 ft per min **43.** $21/16 = 1.3125$ ft per min **45.** $\pm$ .736 in^2 **47.** $1.25 + 2 \ln 1.5$

CHAPTER 7 INTEGRATION

Exercises 7.1 (page 391)

1. They differ only by a constant. **5.** $6k + C$ **7.** $z^2 + 3z + C$ **9.** $t^3/3 - 2t^2 + 5t + C$ **11.** $z^4 + z^3 + z^2 - 6z + C$ **13.** $10z^{3/2}/3 + C$ **15.** $x^4/4 - 3x^2/2 + C$ **17.** $8v^{3/2}/3 - 6v^{5/2}/5 + C$ **19.** $4u^{5/2} - 4u^{7/2} + C$ **21.** $-1/z + C$ **23.** $-1/(2y^2) - 2y^{1/2} + C$ **25.** $9/t - 2 \ln |t| + C$ **27.** $-1/(3x) + C$ **29.** $-15e^{-.2x} + C$ **31.** $3 \ln |x| - 8e^{-.5x} + C$ **33.** $\ln |t| + 2t^3/3 + C$ **35.** $e^{2u}/2 + 2u^2 + C$ **37.** $x^3/3 + x^2 + x + C$ **39.** $6x^{7/6}/7 + 3x^{2/3}/2 + C$ **41.** $10^x/(\ln 10) + C$ **43.** $f(x) = 3x^{5/3}/5$ **45.** $C(x) = 2x^2 - 5x + 8$ **47.** $C(x) = 3e^{.01x} + 5$ **49.** $C(x) = 3x^{5/3}/5 + 2x + 114/5$ **51.** $C(x) = 5x^2/2 - \ln |x| - 153.50$ **53.** $p = 175 - .01x - .01x^2$ **55.** $p = 500 - .1\sqrt{x}$ **57. a.** $f(x) = .8295x^2 + .743x + 4.4$ **b.** Approximately 154 million subscribers. **59. a.** $P(x) = -40 + 4x - 3x^2 + x^3$ **b.** \$312 **61.** $a \ln x - bx + c$ **63. a.** $f(x) = .024x^3 - .836x^2 + 11.5x + 26$ **b.** 91% **65.** $v(t) = t^3/3 + t + 6$ **67.** $s(t) = -16t^2 + 6400$; 20 sec **69.** $s(t) = 2t^{5/2} + 3e^{-t} + 1$

Exercises 7.2 page (401)

3. $2(2x+3)^5/5 + C$ **5.** $-(2m+1)^{-2}/2 + C$ **7.** $-(x^2 + 2x - 4)^{-3}/3 + C$ **9.** $(z^2 - 5)^{3/2}/3 + C$ **11.** $-2e^{2P} + C$ **13.** $e^{2x^3}/2 + C$ **15.** $e^{2t-t^2}/2 + C$ **17.** $-e^{1/z} + C$ **19.** $(x^4 + 4x^2 + 7)^9/36 + C$ **21.** $-1/[2(x^2 + x)^2] + C$ **23.** $(p+1)^7/7 - (p+1)^6/6 + C$ **25.** $2(u-1)^{3/2}/3 + 2(u-1)^{1/2} + C$ **27.** $(x^2 + 12x)^{3/2}/3 + C$ **29.** $[\ln(t^2 + 2)]/2 + C$ **31.** $(1 + \ln x)^3/3 + C$ **33.** $(1/2) \ln(e^{2x} + 5) + C$ **37a.** $R(x) = (x^2 + 50)^3/3 + 137{,}919.33$ **b.** 7 **39. a.** $C(x) = 6 \ln(5x^2 + e) + 4$ **b.** No **41. a.** $N(t) = 155.3337e^{.3218873t} + 144.666$ **b.** 7537 **43. a.** $S(t) = 127.57e^{.07t} + 2.07$ **b.** About 10 years

Exercises 7.3 (page 411)

3. a. 56 **b.** $\int_0^8 (2x+1)dx$ **7. a.** 15.5 **b.** 16.5 **c.** 16 **d.** 16 **9. a.** 10 **b.** 10 **c.** 10 **d.** 11 **11. a.** 8.22 **b.** 15.48 **c.** 11.85 **d.** 10.96 **13. a.** 6.70 **b.** 3.15 **c.** 4.93 **d.** 4.17 **15.** 12.5 **17.** $9\pi/2$ **19.** 6 **21.** **b.** .385 **c.** .33835 **d.** .334334 **e.** .333333 **23. a.** About 33 billion barrels **b.** About 32 billion barrels **c.** About 66 billion barrels **25. a.** About \$154,000 **b.** About \$127,000 **27.** About 35.8 L **29.** About 1900 ft **31. a.** About 1230 BTUs **b.** About 230 BTUs **33. a.** 9 ft **b.** 2 sec **c.** 4.6 ft **d.** Between 3 and 3.5 sec **35.** 22.5 and 18 ft

Exercises 7.4 (page 424)

1. -6 **3.** $3/2$ **5.** $28/3$ **7.** 13 **9.** $1/3$ **11.** 76 **13.** $4/3$ **15.** $112/25$ **17.** $20e^{-.2} - 20e^{-.3} + 3\ln 3 - 3\ln 2 \approx 2.775$ **19.** $e^{10}/5 - e^5/5 - 1/2 \approx 4375.1$ **21.** $91/3$ **23.** $447/7 \approx 63.857$ **25.** $(\ln 2)^2/2 \approx .24023$ **27.** 49 **29.** $1/4 - 1/(3+e) \approx .075122$ **31.** 42 **33.** 76 **35.** $41/2$ **37.** $e^2 - 3 + 1/e \approx 4.021$ **39.** 1 **41.** $23/3$ **43.** $e^2 - 2e + 1 \approx 2.9525$ **45. a.** Yes **b.** Yes **51. a.** 2.92530 **b.** 14.98998 **53. a.** $(9000/8)(17^{4/3} - 2^{4/3}) \approx \$46.341$ **b.** $(9000/8)(26^{4/3} - 17^{4/3}) \approx \$37,477$ **c.** It is slowly increasing without bound. **55.** No **57. a.** 1.37 ft **b.** .32 ft **59. a.** 14.26 **b.** 3.55 **61. b.** $\int_0^{60} n(x)\,dx$ **c.** $2(51^{3/2} - 26^{3/2})/15 \approx 30.885$ million **63. a.** $Q(R) = \pi k R^4/2$ **b.** $.04k$ mm per min **65. b.** About 505,000 kJ/W$^{.67}$ **67. a.** About 263 million; the total population aged 0 to 90 **b.** About 68 million **69. a.** $c'(t) = 1.2e^{.04t}$ **b.** $\int_0^{10} 1.2e^{.04t}\,dt$ **c.** $30e^{.4} - 30 \approx 14.75$ billion **d.** About 12.8 yr **e.** About 14.4 yr

Exercises 7.5 (page 436)

1. 15 **3.** 4 **5.** $23/3$ **7.** 366.1667 **9.** $4/3$ **11.** $5 + \ln 6 \approx 6.792$ **13.** $6\ln(3/2) - 6 + 2e^{-1} + 2e \approx 2.6051$ **15.** $e^2 - e - \ln 2 \approx 3.978$ **17.** $1/2$ **19.** $1/20$ **21.** $3(2^{4/3})/2 - 3(2^{7/3})/7 \approx 1.6199$ **23.** $-1.9241, -.4164, .6650$ **25. a.** 8 yr **b.** About \$148 **c.** About \$771 **27. a.** 39 days **b.** \$3369.18 **c.** \$484.02 **d.** \$2885.16 **29.** 12,931.66 **31.** 27 **33. a.**

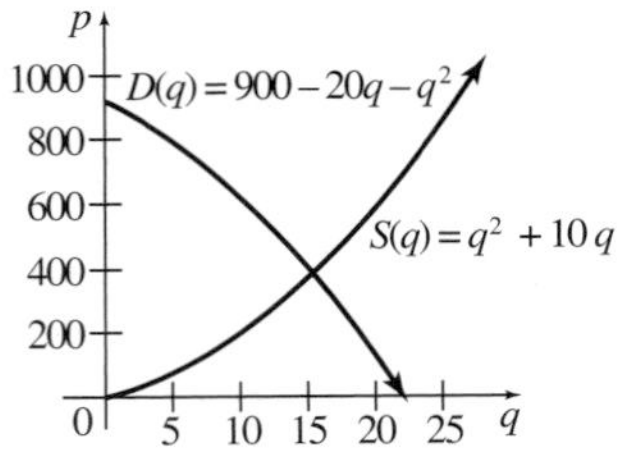

b. (15, 375) **c.** \$4500 **d.** \$3375 **35. a.** 12 **b.** \$5616, \$1116 **c.** \$1872, \$1503 **d.** \$387

37. a. About 71.25 gal **b.** About 25 hr **c.** About 105 gal **d.** About 47.91 hr **39. a.** .019; the lower 10% of the income producers earn 1.9% of the total income of the population. **b.** .184; the lower 40% of the income producers earn 18.4% of the total income of the population. **c.**

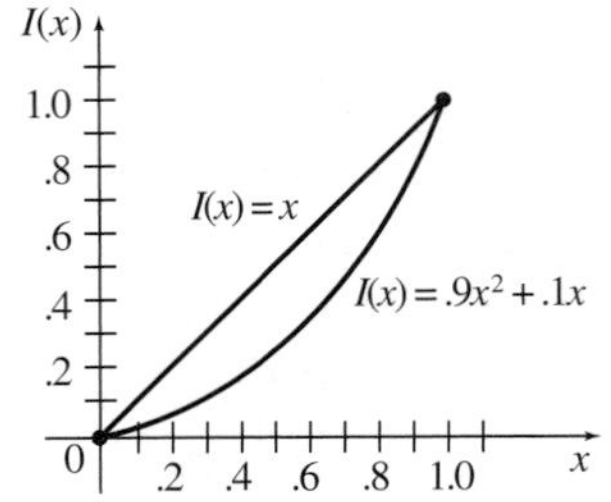

d. .15 **e.** Income is distributed less equally in 2001 than in 1968.

Exercises 7.6 (page 445)

1. a. 2.7500 **b.** 2.6667 **c.** $8/3 \approx 2.6667$ **3. a.** 1.6833 **b.** 1.6222 **c.** $\ln 5 \approx 1.6094$ **5. a.** 16 **b.** 14.6667 **c.** $44/3 \approx 14.6667$ **7. a.** .9436 **b.** .8374 **c.** $4/5 = .8$ **9. a.** .895892 **b.** .860997 **c.** $e/2 - 1/2 \approx .859141$ **11. a.** 5.9914 **b.** 6.1672 **c.** 6.2832; Simpson's rule **13.** b is true. **15. a.** .2 **b.** .220703, .205200, .201302, .200325, .020703, .005200, .001302, .000325 **c.** $p = 2$ **17. a.** .2 **b.** .2005208, .2000326, .2000020, .2000001, .0005208, .0000326, .0000020, .0000001

c. $p = 4$ **19.** $M = .7355$; $S = .8048$ **21. a.**

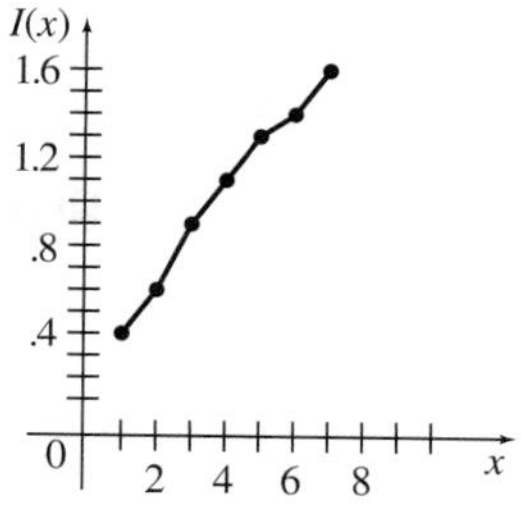

b. 6.3 **c.** 6.27 **23. a.** 2.4759 **b.** 2.3572
25. About 30 mcg(h)/ml; this represents the total amount of drug available to the patient for each ml of blood.
27. About 9 mcg(h)/ml; this represents the total effective amount of the drug available to the patient for each ml of blood. **29. a.** $y = b_0(t/7)^{b_1}e^{-b_2 t/7}$ **b.** About 1212 kg; about 1231 kg **c.** About 1224 kg; about 1250 kg

31. a.

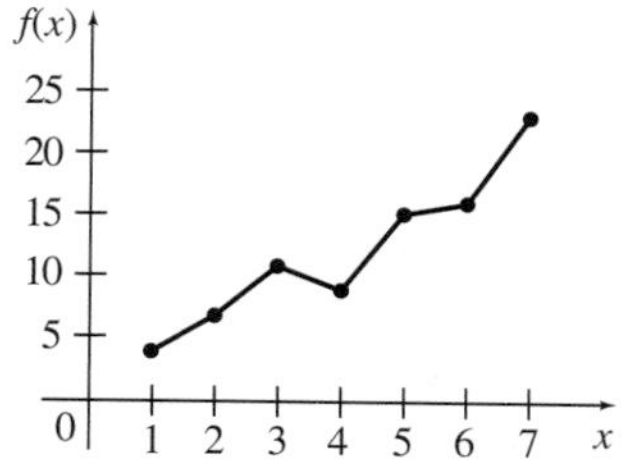

b. 71.5 **c.** 69.0 **33.** 3979.24 **35. a.** .682689 **b.** .954500 **c.** .997300

Chapter 7 Review Exercises (page 450)

5. $x^2 + 3x + C$ **7.** $x^3/3 - 3x^2/2 + 2x + C$ **9.** $2x^{3/2} + C$ **11.** $2x^{3/2}/3 + 9x^{1/3} + C$ **13.** $2x^{-2} + C$ **15.** $-3e^{2x}/2 + C$ **17.** $e^{3x^2}/6 + C$ **19.** $(3 \ln |x^2 - 1|)/2 + C$ **21.** $-(x^3 + 5)^{-3}/9 + C$ **23.** $-e^{-3x^4}/12 + C$ **25.** 20 **27.** 24 **29. a.** $s(T) - s(0)$ **b.** $\int_0^T v(t)\,dt = s(T) - s(0)$ is equivalent to the Fundamental Theorem with $a = 0$ and $b = T$ because $s(t)$ is an antiderivative of $v(t)$. **31.** 12 **33.** $72/25 \approx 2.88$ **35.** $2 \ln 3$ or $\ln 9 \approx 2.1972$ **37.** $2e^4 - 2 \approx 107.1963$ **39.** $\pi/32$ **41.** $128/7$ **43.** $5 - e^{-4} \approx 4.982$ **45.** $1/6$ **47.** 32 **49.** 10.463; 10.197 **51.** .60111 **53.** 2.8946 **55. a.** 0 **b.** 0 **57.** $C(x) = \ln |x + 1| + 18$ **59.** About 26.3 yr **61. a.** $f(x) = .14x^2 + .54x + 98.36$ **b.** 125, which differs from the actual value by .3 **63.** 2.5 yr; about \$99,000 **65.** $50 \ln 17 \approx 141.66$; about 142 people **67. a.** About 4600 pM **b.** About 2800 pM **c.** The area under the curve is about 64% more for the fasting sheep. **69.** About 8120 **71.** Approximately 4800 degree-days (the actual value according to the National Weather Service is 4868 degree-days.)

CHAPTER 8 FURTHER TECHNIQUES AND APPLICATIONS OF INTEGRATION

Exercises 8.1 (page 466)

1. $xe^x - e^x + C$ **3.** $-5xe^{-3x}/3 - 5e^{-3x}/9 + 3e^{-3x} + C$ or $-5xe^{-3x}/3 + 22e^{-3x}/9 + C$ **5.** $-5e^{-1} + 3 \approx 1.1606$ **7.** $11 \ln 2 - 3 \approx 4.6246$ **9.** $(x^2 \ln x)/2 - x^2/4 + C$ **11.** $e^4 + e^2 \approx 61.9872$ **13.** $x^2e^{2x}/2 - xe^{2x}/2 + e^{2x}/4 + C$ **15.** $243/8 - 3\sqrt[3]{2}/4 \approx 29.4301$ **17.** $4x^2 \ln(5x) + 7x \ln(5x) - 2x^2 - 7x + C$ **19.** $2x^2(x + 2)^{3/2}/3 - 8x(x + 2)^{5/2}/15 + 16(x + 2)^{7/2}/105 + C$ **21.** .13077 **23.** $9 \ln |x + \sqrt{x^2 + 9}| + C$ **25.** $-(3/11) \ln |(11 + \sqrt{121 - x^2})/x| + C$ **27.** $-1/(4x + 3) - (1/3) \ln |x/(4x + 3)| + C$ **33. a.** $(2/3)x(x + 1)^{3/2} - (4/15)(x + 1)^{5/2} + C$ **b.** $(2/5)(x + 1)^{5/2} - (2/3)(x + 1)^{3/2} + C$ **35.** $(169/2) \ln 13 - 42 \approx \$174.74$ **37.** $15(5e^6 + 1)/2 \approx 15{,}136$ **39.** About 219 kJ

Exercises 8.2 (page 474)

1. $8\pi/3$ **3.** $364\pi/3$ **5.** $386\pi/27$ **7.** $3\pi/2$ **9.** 18π **11.** $\pi(e^4 - 1)/2 \approx 84.19$ **13.** $\pi \ln 4 \approx 4.36$ **15.** $3124\pi/5$ **17.** $16\pi/15$ **19.** $4\pi/3$ **21.** $4\pi r^3/3$ **23.** $\pi r^2 h$ **25.** $19/3$ **27.** $38/15$ **29.** $e - 1 \approx 1.718$ **31.** $(5e^4 - 1)/8 \approx 33.999$ **33.** 3.9372 **35.** \$16.88 **37.** 200 cases **39. a.** $110e^{-.1} - 120e^{-.2} \approx 1.2844$ **b.** $210e^{-1.1} - 220e^{-1.2} \approx 3.6402$ **c.** $330e^{-2.3} - 340e^{-2.4} \approx 2.2413$ **41. a.** $7(6 \ln 6 - 5) \approx 40.254$ **b.** $7(11 \ln 11 - 10)/2 \approx 57.319$ **c.** $7(16 \ln 16 - 15)/3 \approx 68.510$

Exercises 8.3 (page 484)

1. a. \$5823.38 **b.** \$19,334.31 **3. a.** \$2911.69 **b.** \$9667.15 **5. a.** \$2637.47 **b.** \$8756.70 **7. a.** \$27,979.55 **b.** \$92,895.37 **9. a.** \$2.34 **b.** \$7.78 **11. a.** \$582.57 **b.** \$1934.21 **13. a.** \$9480.41 **b.** \$31,476.07 **15.** \$74,565.94 **17.** \$28,513.76; \$54,075.81 **19.** \$4175.52

Exercises 8.4 (page 489)

1. $1/2$ **3.** Divergent **5.** -1 **7.** 10,000 **9.** 1 **11.** $3/5$ **13.** 1 **15.** 4 **17.** Divergent **19.** 1 **21.** Divergent **23.** Divergent **25.** Divergent **27.** 0 **29.** Divergent **31.** Divergent **33.** 1 **35.** 0 **39. a.** 2.8081, 3.7239, 4.4170, 6.7195, 9.0221 **b.** Divergent **c.** .8770, .9070, .9170, .9260, .9269 **d.** Convergent **41. a.** 9.9995, 49.9875, 99.9500, 995.0166 **b.** Divergent **c.** 100,000 **43.** \$8,333,333.33 **45.** \$20,000 **47.** \$30,000 **49.** $Na/[b(b+k)]$ **51.** 833.33

Chapter 8 Review Exercises (page 492)

5. $6x(x-2)^{1/2} - 4(x-2)^{3/2} + C$ **7.** $-(3x+7)e^{-3x}/9 + C$ **9.** $(x^2/2 - x)\ln|x| - x^2/4 + x + C$ **11.** $(1/9)\sqrt{25+9x^2} + C$ **13.** $10e^{1/2} - 16 \approx .48721$ **15.** $234/7 \approx 33.43$ **17.** $81\pi/2 \approx 127.23$ **19.** $\pi \ln 3 \approx 3.451$ **21.** $64\pi/5 \approx 40.21$ **25.** $7/2$ **27.** $1/2$ **29.** $6/e \approx 2.207$ **31.** Divergent **33.** 3 **35.** $16{,}250/3 \approx 5416.67$ **37.** \$174,701.45 **39.** \$32.11 **41.** \$5534.28 **43.** \$30,035.17 **45.** \$176,919.15 **47.** .480 **49. a.** About 8208 kg **b.** About 8430 kg **c.** About 8558 kg

CHAPTER 9 MULTIVARIABLE CALCULUS

Exercises 9.1 (page 506)

1. a. 6 **b.** -8 **c.** -20 **d.** 43 **3. a.** $\sqrt{43}$ **b.** 6 **c.** $\sqrt{19}$ **d.** $\sqrt{11}$

5.

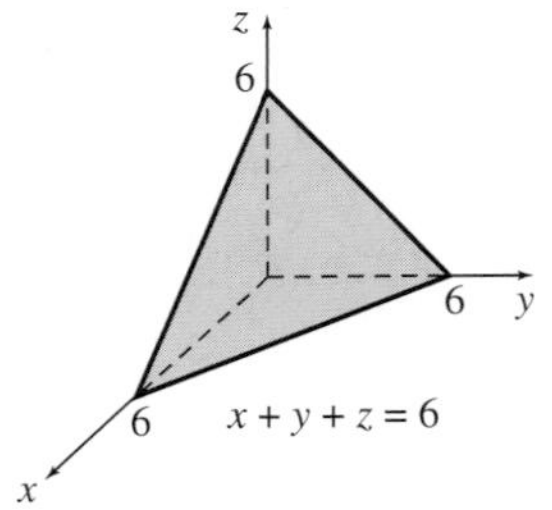

7.

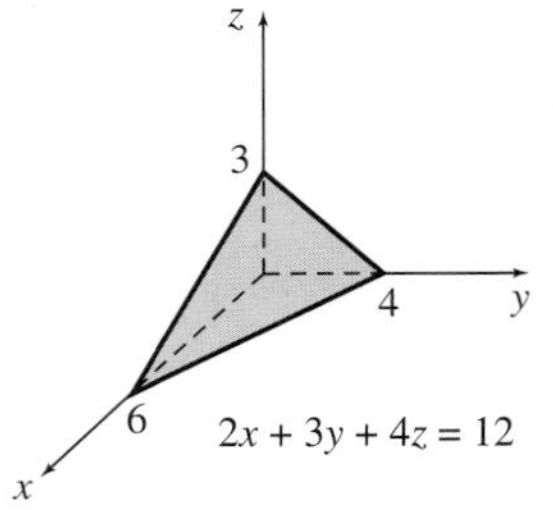

9.

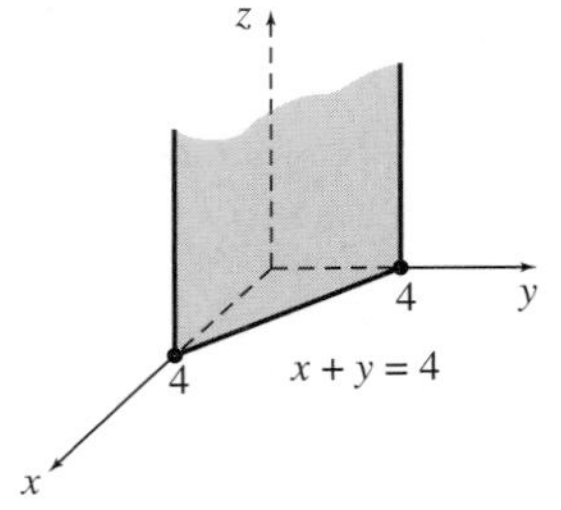

11.

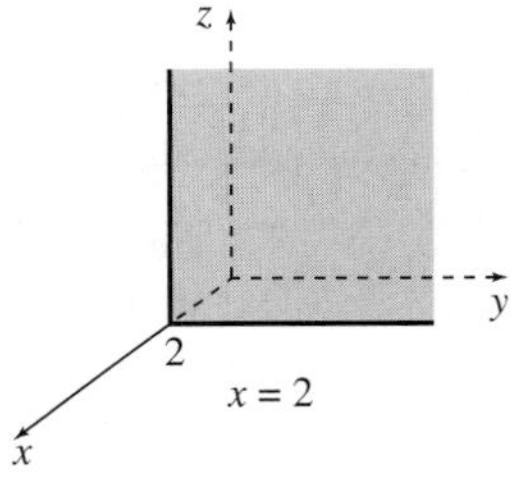

13.

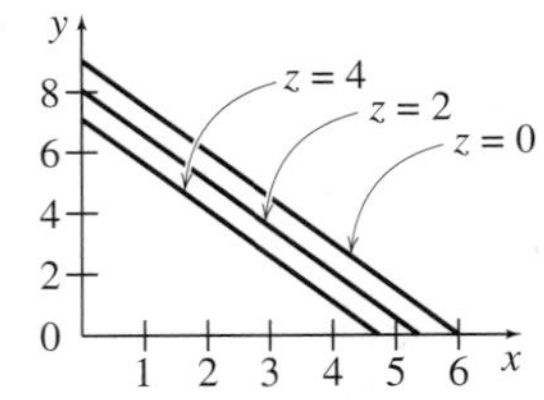

15.

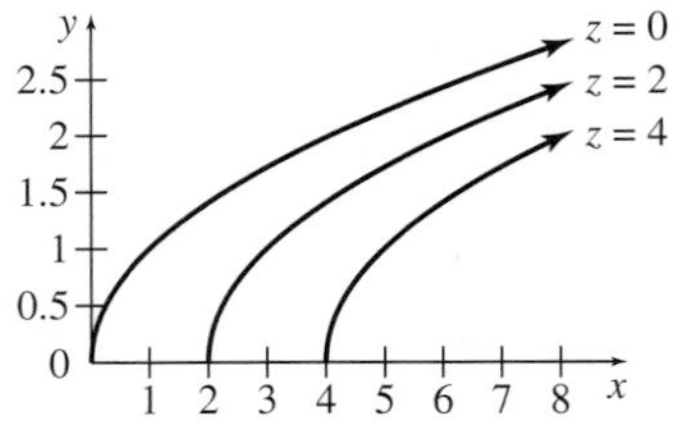

21. c **23.** e **25.** b **27. a.** $18x + 9h$ **b.** $-6y - 3h$ **c.** $18x$ **d.** $-6y$ **29. a.** $3e^2$; slope of tangent line in the direction of x at $(1, 1)$ **b.** $3e^2$; slope of tangent line in the direction of y at $(1, 1)$ **31. a.** 1987 (rounded) **b.** 595 (rounded) **c.** 359,768 (rounded) **33.** 1.197 (rounded); the IRA account grows faster.

35. $y = 500^{5/2}/x^{3/2} \approx 5{,}590{,}170/x^{3/2}$

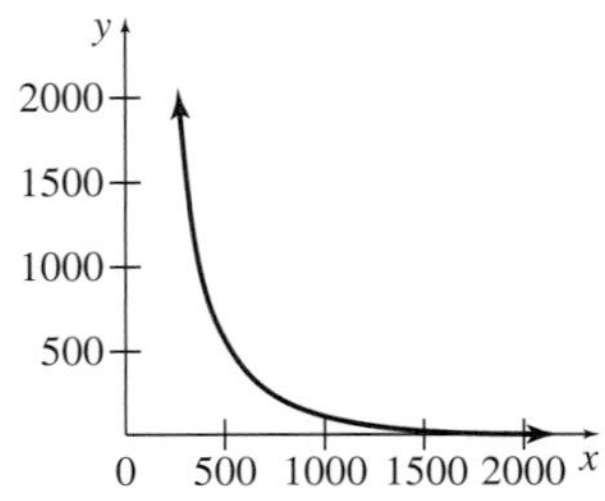

37. $C(x, y, z) = 200x + 100y + 50z$ **39. a.** 1.5 m per sec, 5.5 m per sec **b.** 1 m per sec **41. a.** 397 accidents **43. a.** $T = 242.257\, C^{.18}/F^3$ **b.** 58.82; a tethered sow spends nearly 59% of the time doing repetitive behavior when she is fed 2 kg of food a day and neighboring sows spend 40% of the time doing repetitive behavior. **45.** $g(L, M, H) = 2LM + 2MH + 2LH$ ft^2

Exercises 9.2 (page 517)

1. a. $24x - 8y$ **b.** $-8x + 6y$ **c.** 24 **d.** -20 **3.** $f_x(x, y) = -2y$; $f_y(x, y) = -2x + 18y^2$; 2; 170 **5.** $f_x(x, y) = 9x^2y^2$; $f_y(x, y) = 6x^3y$; 36; -1152 **7.** $f_x(x, y) = e^{x+y}$; $f_y(x, y) = e^{x+y}$; e^1 or e; e^{-1} or $1/e$ **9.** $f_x(x, y) = -15e^{3x-4y}$; $f_y(x, y) = 20e^{3x-4y}$; $-15e^{10}$; $20e^{-24}$ **11.** $f_x(x, y) = (-x^4 - 2xy^2 - 3x^2y^3)/(x^3 - y^2)^2$; $f_y(x, y) = (3x^3y^2 - y^4 + 2x^2y)/(x^3 - y^2)^2$; $-8/49$; $-1713/5329$ **13.** $f_x(x, y) = 6xy^3/(1 + 3x^2y^3)$; $f_y(x, y) = 9x^2y^2/(1 + 3x^2y^3)$; 12/11; 1296/1297 **15.** $f_x(x, y) = e^{x^2y}(2x^2y + 1)$; $f_y(x, y) = x^3e^{x^2y}$; $-7e^{-4}$; $-64e^{48}$ **17.** $f_x(x, y) = (1/2)(x^4 + 3xy + y^4 + 10)^{-1/2}(4x^3 + 3y)$; $f_y(x, y) = (1/2)(x^4 + 3xy + y^4 + 10)^{-1/2}(3x + 4y^3)$; $29/(2\sqrt{21})$; $48/\sqrt{311}$ **19.** $f_x(x, y) = [6xy(e^{xy} + 2) - 3x^2y^2e^{xy}]/(e^{xy} + 2)^2$; $f_y(x, y) = [3x^2(e^{xy} + 2) - 3x^3ye^{xy}]/(e^{xy} + 2)^2$; $-24(e^{-2} + 1)/(e^{-2} + 2)^2$; $(624e^{-12} + 96)/(e^{-12} + 2)^2$ **21.** $f_{xx}(x, y) = 36xy$; $f_{yy}(x, y) = -18$; $f_{xy}(x, y) = f_{yx}(x, y) = 18x^2$ **23.** $R_{xx}(x, y) = 8 + 24y^2$; $R_{yy}(x, y) = -30xy + 24x^2$; $R_{xy}(x, y) = R_{yx}(x, y) = -15y^2 + 48xy$ **25.** $r_{xx}(x, y) = -8y/(x + y)^3$; $r_{yy}(x, y) = 8x/(x + y)^3$; $r_{xy}(x, y) = r_{yx}(x, y) = (4x - 4y)/(x + y)^3$ **27.** $z_{xx} = 0$; $z_{yy} = 4xe^y$; $z_{xy} = z_{yx} = 4e^y$ **29.** $r_{xx} = -1/(x + y)^2$; $r_{yy} = -1/(x + y)^2$; $r_{xy} = r_{yx} = -1/(x + y)^2$ **31.** $z_{xx} = 1/x$; $z_{yy} = -x/y^2$; $z_{xy} = z_{yx} = 1/y$ **33.** $x = -4, y = 2$ **35.** $x = 0, y = 0$; or $x = 3, y = 3$ **37.** $f_x(x, y, z) = 2x$; $f_y(x, y, z) = z$; $f_z(x, y, z) = y + 4z^3$; $f_{yz}(x, y, z) = 1$ **39.** $f_x(x, y, z) = 6/(4z + 5)$; $f_y(x, y, z) = -5/(4z + 5)$; $f_z(x, y, z) = -4(6x - 5y)/(4z + 5)^2$; $f_{yz}(x, y, z) = 20/(4z + 5)^2$ **41.** $f_x(x, y, z) = (2x - 5z^2)/(x^2 - 5xz^2 + y^4)$; $f_y(x, y, z) = 4y^3/(x^2 - 5xz^2 + y^4)$; $f_z(x, y, z) = -10xz/(x^2 - 5xz^2 + y^4)$; $f_{yz}(x, y, z) = 40xy^3z/(x^2 - 5xz^2 + y^4)^2$ **43. a.** 6.773 **b.** 3.386 **45. a.** 80 **b.** 180 **c.** 110 **d.** 360 **47. a.** \$902,100 **b.** $f_p(p, i) = 99 - .5i - .005p$; $f_i(p, i) = -.5p$; the rate at which weekly sales are changing per unit of change in price when the interest rate remains constant $(f_p(p, i))$ or interest rate when the price remains constant $(f_i(p, i))$ **c.** A weekly sales decrease of \$9700 **49. a.** 46.656 hundred units **b.** $f_x(27, 64) = .6912$ hundred units and is the rate at which production is changing when labor changes by 1 unit (from 27 to 28) and capital remains constant; $f_y(27, 64) = .4374$ hundred units and is the rate at which production is changing when capital changes by 1 unit (from 64 to 65) and labor remains constant. **c.** Production would increase at a rate of $f_x(x, y) = (1/3)x^{-4/3}[(1/3)x^{-1/3} + (2/3)y^{-1/3})^{-4}]$. **51.** $.4x^{-.6}y^{.6}$; $.6x^{.4}y^{-.4}$ **53. a.** 1279 kcal per hr **b.** 2.91 kcal per hr per g; the instantaneous rate of change of energy usage for a 300-kg animal traveling at 10 km per hr is about 2.9 kcal per hr per g. **55. a.** .0112 **b.** .783 **57. a.** 4.125 lb **b.** $\partial f/\partial n = n/4$; the rate of change of weight loss per unit change in workouts **c.** An additional loss of 3/4 lb **59. a.** $(2ax - 3x^2)t^2e^{-t}$ **b.** $x^2(a - x)(2t - t^2)e^{-t}$ **c.** $(2a - 6x)t^2e^{-t}$ **d.** $(2ax - 3x^2)(2t - t^2)e^{-t}$ **e.** $\partial R/\partial x$ gives the rate of change of the reaction per unit of change in the amount of drug administered. $\partial R/\partial t$ gives the rate of change of the reaction for a 1-hour change in the time after the drug is administered. **61. a.** -24.9°F **b.** 15 mph **c.** $W_V(20, 10) = -1.11$; while holding the temperature fixed at 10°F, the wind chill decreases approximately 1.1°F when the wind velocity increases by 1 mph; $W_T(20, 10) = 1.43$; while holding the wind velocity fixed at 20 mph, the wind chill increases approximately 1.43°F if the actual temperature increases from 10°F to 11°F.
d. Sample table

T/V	5	10	15	20
30	27	16	9	4
20	16	3	-5	-11
10	6	-9	-18	-25
0	-5	-21	-32	-39

63. -10 ml per year, 100 ml per in. **65. a.** $F_m = gR^2/r^2$; the rate of change in force per unit change in mass while the distance is held constant; $F_r = -2mgR^2/r^3$; the rate of change in force per unit change in distance while the mass is held constant **67. a.** 326 **b.** $T_s(3, .5) = 50.49$ msec per ft. If the distance to move an object increases from 3 ft to 4 ft, while keeping w fixed at .5, the approximate increase in movement time is 50.49 msec. $T_w(3, .5) = -303$ msec per ft^2. If the target area increases by 1 ft^2, while keeping s fixed at 3 ft, the approximate decrease in movement time is 303 msec.

Exercises 9.3 (page 529)

1. Saddle point at $(1, -1)$ **3.** Relative minimum at $(-1, -1/2)$ **5.** Relative minimum at $(-2, -2)$ **7.** Relative minimum at $(15, -8)$ **9.** Relative maximum at $(2/3, 4/3)$ **11.** Saddle point at $(2, -2)$ **13.** Saddle point at $(0, 0)$; relative minimum at $(4, 8)$ **15.** Saddle point at $(0, 0)$; relative minimum at $(9/2, 3/2)$ **17.** Saddle point at $(0, 0)$ **21.** Relative maximum of 9/8 at $(-1, 1)$; saddle point at $(0, 0)$; a **23.** Relative minima of $-33/16$ at $(0, 1)$ and at $(0, -1)$; saddle point at $(0, 0)$; b

25. Relative maxima of 17/16 at $(1, 0)$ and $(-1, 0)$; relative minima of $-15/16$ at $(0, 1)$ and $(0, -1)$; saddle points at $(0, 0)$, $(-1, 1)$, $(1, -1)$, $(1, 1)$, and $(-1, -1)$; e **31. a.** all values of k **b.** $k \geq 0$ **33.** Minimum cost of \$59 when $x = 4$, $y = 5$ **35.** Sell 9 spas and 4 solar heaters for a maximum revenue of \$51,500 **37.** \$2000 on quality control and \$1000 on consulting, for a minimum time of 8 hours **39. a.** $y = 2491(1.3869)^x$ **b.** Same as a **c.** Same as a

Exercises 9.4 (page 539)

1. $f(6, 6) = 72$ **3.** $f(4/3, 4/3) = 64/27 \approx 2.4$ **5.** $f(5, 3) = 28$ **7.** $f(20, 2) = 360$ **9.** $f(3/2, 3/2, 3) = 81/4 = 20.25$ **11.** $x = 6$, $y = 12$ **13.** 30, 30, 30 **19.** 37.5 ft by 100 ft **21.** $x = 2$, $y = 4$ **23.** 189 units of labor and 35 units of capital **25.** 45,000 m^2 **27.** Radius $\approx$ 1.58 in.; height $\approx$ 3.17 in. **29.** 5.70 in. by 5.70 in. by 5.70 in. **31.** 4 ft by 4 ft for the base; 2 ft for the height **33. a.** $F(r, s, t, \lambda) = rs(1 - t) + (1 - r)st + r(1 - s)t + rst - \lambda(r + s + t - \alpha)$ **b.** $r = s = t = .25$ **c.** $r = s = t = 1.0$

Exercises 9.5 (page 544)

1. 5.072; 5.0720; 0 **3.** 2.0067; 2.0080; .0013 **5.** .99; .9896; .0004 **7.** $-.05$; $-.0533$; .0033 **9.** $-.2$ **11.** .0311 **13.** $-.335$ **15.** 20.73 cm^3 **17.** 142.4 in^3 **19.** .0769 units **21.** 6.65 cm^3 **23.** 2.98 L **25a.** .265 **b.** actual .282; approximation .282 **27. a.** .066 sec .038 sec; In a close race, this could certainly affect the outcome. **b.** .00199 sec; This is the approximate change in time when the temperature decreases 5 degrees and the swimmer stands .5 m farther away from the starter in the y direction, while the starter's position is held fixed. **29.** 5.98 cm^3 **31.** $2a + b$, the same as for a cylinder

Exercises 9.6 (page 556)

1. $93y/4$ **3.** $(1/9)[(48 + y)^{3/2} - (24 + y)^{3/2}]$ **5.** $(2x/9)[(x^2 + 15)^{3/2} - (x^2 + 12)^{3/2}]$ **7.** $6 + 10y$ **9.** $(1/4)e^{x+4} - (1/4)e^{x-4}$ **11.** $(1/2)e^{25+9y} - (1/2)e^{9y}$ **13.** 279/8 **15.** $(2/45)(39^{5/2} - 12^{5/2} - 7533)$ **17.** 21 **19.** $(\ln 2)^2$ **21.** $8 \ln 2 + 4$ **23.** 256 **25.** $(4/15)(33 - 2^{5/2} - 3^{5/2})$ **27.** $-2 \ln(6/7)$ or $2 \ln(7/6)$ **29.** $(1/2)(e^7 - e^6 - e^3 + e^2)$ **31.** 48 **33.** 4/3 **35.** $(2/15)(2^{5/2} - 2)$ **37.** $(1/4) \ln(17/8)$ **39.** $e^2 - 3$ **41.** $97{,}632/105 \approx 929.83$ **43.** 128/9 **45.** ln 3 **47.** 64/3 **49.** 1 **51.** 116 **53.** 10/3 **55.** $7(e - 1)/3$ **57.** 16/3 **59.** $4 \ln 2 - 2$ **63.** 13/3 **65.** $(e^7 - e^6 - e^5 + e^4)/2$ **67.** \$2583 **69.** \$933.33 **71.** 78.4 hr

Chapter 9 Review Exercises (page 559)

5. 21; 936 **7.** $-\sqrt{5}/3$; $\sqrt{5}/3$ **9.**

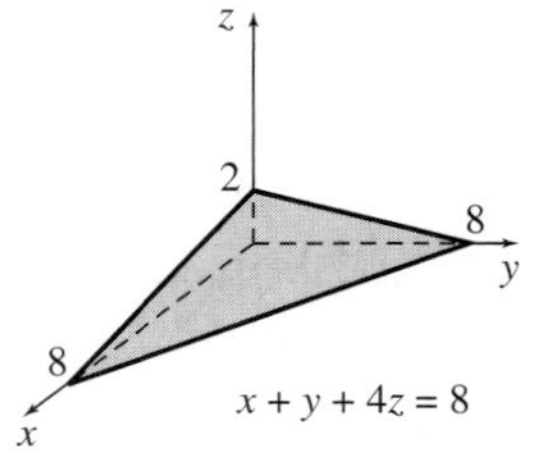

11.

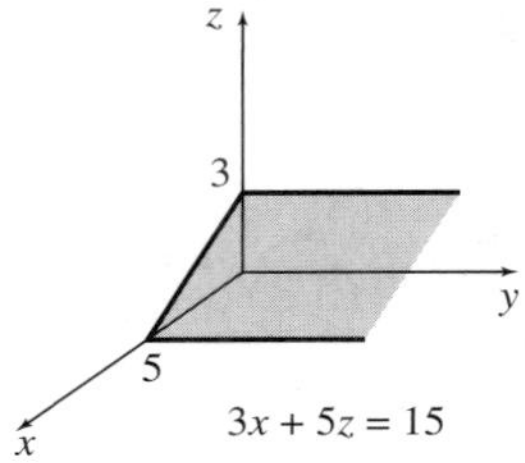

13.

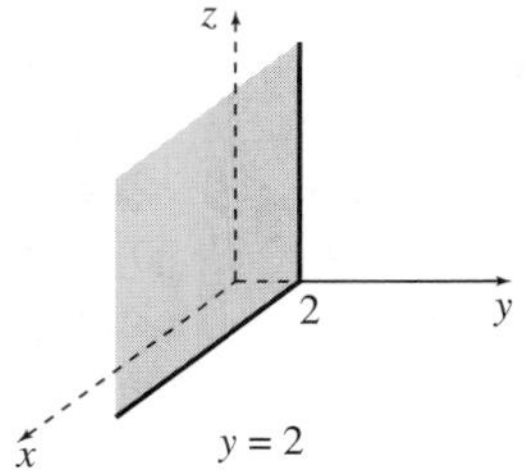

15. a. $4xy/(x - y^2)^2$ **b.** $-1/2$ **c.** 0 **17.** $f_x(x, y) = 30x^4y - 8y^9$; $f_y(x, y) = 6x^5 - 72xy^8$ **19.** $f_x(x, y) = (-6x^2 + 2y^2 - 30xy^2)/(3x^2 + y^2)^2$; $f_y(x, y) = (30x^2y - 4xy)/(3x^2 + y^2)^2$ **21.** $f_x(x, y) = (y - 2)^2e^{x+2y}$; $f_y(x, y) = 2(y - 2)(y - 1)e^{x+2y}$ **23.** $f_x(x, y) = -2xy^3/(2 - x^2y^3)$; $f_y(x, y) = -3x^2y^2/(2 - x^2y^3)$ **25.** $f_{xx}(x, y) = 2y$; $f_{xy}(x, y) = -24y^3 + 2x$ **27.** $f_{xx}(x, y) = 2(3 + y)/(x - 1)^3$; $f_{xy}(x, y) = -1/(x - 1)^2$ **29.** $f_{xx}(x, y) = 2ye^{x^2}(2x^2 + 1)$; $f_{xy}(x, y) = 2xe^{x^2}$ **31.** $f_{xx}(x, y) = -9y^4/(1 + 3xy^2)^2$; $f_{xy}(x, y) = 6y/(1 + 3xy^2)^2$ **33.** Relative minimum at $(-9/2, 4)$ **35.** Relative maximum at $(-3/4, -9/32)$; saddle point at $(0, 0)$ **37.** Relative minimum at $(4/5, -9/10)$ **39.** Relative minimum at $(-8, -23)$ **41.** Minimum of 0 at $(0, 4)$; maximum of 256/27 at $(8/3, 4/3)$ **43.** $x = 160/3$, $y = 80/3$ **45.** No **47.** $-.0168$ **49.** 2.095; 2.0972; .0022 **51.** $(e^{10-7y} - e^{6-7y})/2$ **53.** $(2/33)[(7x + 297)^{1/2} - (7x + 11)^{1/2}]$ **55.** 69 **57.** $(e^3 + e^{-8} - e^{-4} - e^{-1})/14$ **59.** ln 2 **61.** $(2/15)(11^{5/2} - 8^{5/2} - 7^{5/2} + 32)$ **63.** $(e^2 - 2e + 1)/2$ **65.** 308/3 **67.** 1/14 **69.** 1/12 **71.** $16(5\sqrt{5} - 1)/3$ **73.** 26/105 **75. a.** \$26 **b.** \$2572 **77. a.** Relative minimum at $(11, 12)$ **b.** \$431 **79.** 7.92 cm^3 **81.** 15.6 cm^3 **83.** 1.3 cm^3 **85. a.** About 33.98 cm **b.** About .027 cm per g; about .28 cm/year; the approximate change in the length of a trout if its mass increases from 450 to 451 g while age is held constant at 7 years is .027 cm; the approximate change in the length of a trout if its age increases from 7 to 8 years while mass is held constant at 450 g is .28 cm. **87. a.** 2.83 ft^2 **b.** An increase of .6187 ft^2 **89.** 20,000 ft^2 with dimensions 100 ft by 200 ft

CHAPTER 10 DIFFERENTIAL EQUATIONS

Exercises 10.1 (page 577)

1. $y = -x^2 + x^3 + C$ **3.** $y = 3x^4/8 + C$ **5.** $y^2 = x^2 + C$ **7.** $y = ke^{x^2}$ **9.** $y = ke^{x^3 - x^2}$ **11.** $y = Mx$ **13.** $y = Me^x + 5$ **15.** $y = -1/(e^x + C)$ **17.** $y = x^3 - x^2 + 2$ **19.** $y = -2xe^{-x} - 2e^{-x} + 44$ **21.** $y^2 = 2x^3/3 + 9$ **23.** $y = e^{x^2+3x}$ **25.** $y = x^4 - x^3 + x^2/2 - 1/2$ **27.** $y = -5/(5 \ln |x| - 6)$ **29.** $y = (e^{x-1} - 3)/(e^{x-1} - 2)$ **35. a.** \$1011.75 **b.** \$1024.52 **c.** No **37.** About 11.6 yr **39.** $q = C/p^2$ **41. a.** $I = 2.4 - 1.4e^{-.088W}$ **b.** I approaches 2.4. **43. a.** $dw/dt = k(C - 17.5w)$ The calorie intake per day is constant. **b.** lb/calorie **c.** $dw/dt = (C - 17.5w)/3500$ **d.** $w = C/17.5 - e^{-.005M}e^{-.005t}/17.5$ **e.** $w = C/17.5 + (w_0 - C/17.5)e^{-.005t}$

45. a.

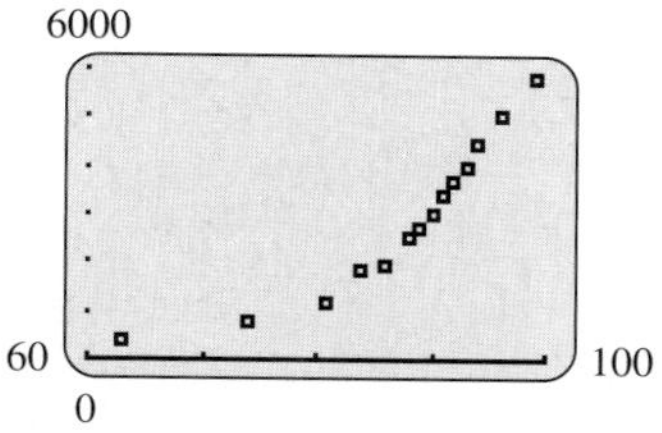

The data seems appropriate for a logistic function.

b. $y = 11{,}074/(1 + 151{,}378e^{-.1219x})$ **c.**

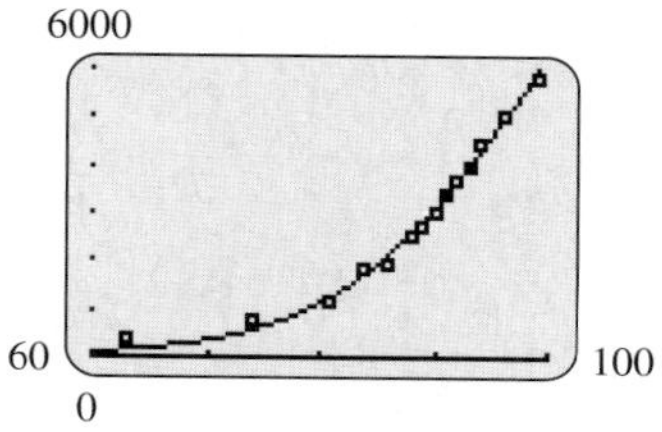

The logistic equation fits the data well.

d. 11,074 **47.** $y = 31.4e^{.02245t}$ **49. a.** $k \approx .8$ **b.** 11 **c.** 55 **d.** about 3000 **51.** About 10 **53.** 7:22:55 A.M.

Exercises 10.2 (page 586)

1. $y = 5/2 + Ce^{-2x}$ **3.** $y = 3 + Ce^{-x^2/2}$ **5.** $y = x \ln x + Cx$ **7.** $y = -1/2 + Ce^{x^2/2}$ **9.** $y = x^2/4 + x + Cx^{-2}$ **11.** $y = -x^3/2 + Cx$ **13.** $y = e^x + 99e^{-x}$ **15.** $y = -1 + 11e^{(x^2-1)/2}$ **17.** $y = x^2/7 + 2560/(7x^5)$ **19.** $y = (3 + 197e^{4-x})/x$ **21. a.** $y = c/(p + Kce^{-cx})$ **b.** $y = cy_0/[py_0 + (c - py_0)e^{-cx}]$ **c.** c/p **25.** $y = 1.02e^t + 9999e^{.02t}$ (rounded) **27.** $y = 50t + 2500 + 7500e^{.02t}$ **31.** The temperature approaches T_0, the temperature of the surrounding medium. **33. a.** $T = 88.6e^{-.24t} + 10$ **b.**

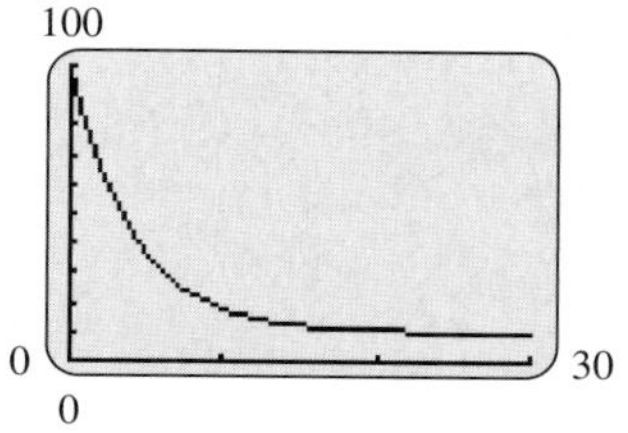

c. Just after death—the graph shows that the most rapid decrease occurs in the first few hours. **d.** About 43.9°F **e.** About 4.5 hours

Exercises 10.3 (page 594)

1. 1.837 **3.** 4.315 **5.** 1.491 **7.** 2.022 **9.** −.540; −.520 **11.** 2.030; 2.042 **13.** 3.806; 4.759 **15.** 1.371; 1.419

17.

x_i	y_i	$y(x_i)$	$y_i - y(x_i)$
0	0	0	0
.2	0	.08772053	−.08772053
.4	.11696071	.22104189	−.10408118
.6	.26432197	.37954470	−.11522273
.8	.43300850	.55699066	−.12398216
1.0	.61867206	.75000000	−.13132794

19.

x_i	y_i	$y(x_i)$	$y_i - y(x_i)$
0	0	0	0
.2	.2	.1812692	.0187308
.4	.36	.32967995	.03032005
.6	.488	.45118836	.03681164
.8	.5904	.55067104	.03972896
1.0	.67232	.63212056	.04019944

21.

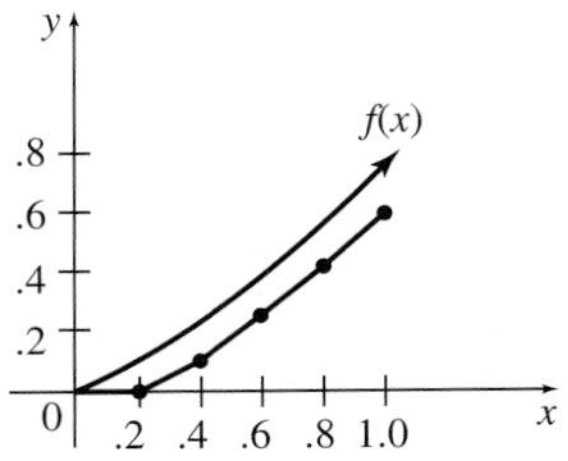

23.

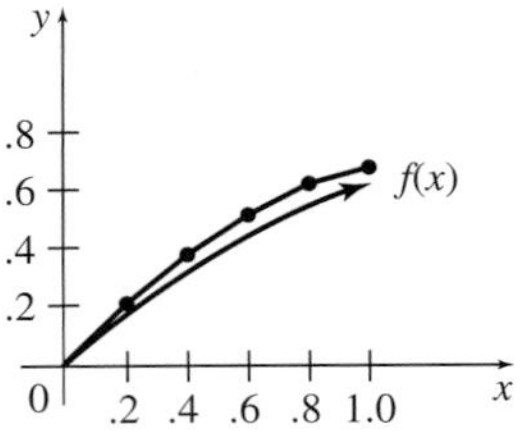

25. a. 4.109 **b.** $y = 1/(1 - x)$; y approaches ∞

27. a. $dy/dt = .01y(100 - y) = y - .01y^2$ **b.** About 14,000 **29.** About 75 **31.** About 8.07 kg **33.** About 360 people

Exercises 10.4 (page 604)

1. \$31,048.20 **3.** About 5.7 years **5. a.** $dA/dt = .06A - 1200$ **b.** \$6470.04 **c.** 8.51 years
7. a. $3 \ln x - 3 \ln y - x + 4y = 1 + \ln 27$ **b.** $x = 3, y = 3/4$ **9. b.**

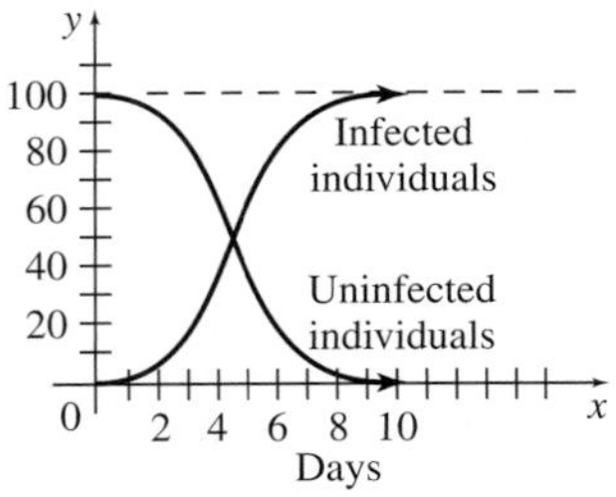

c. (4.6 days, 50 people) **d.** This is the point where the number of infected people equals the number of uninfected people. **e.** For y it is 100 people. For $N - y$ it is 0 people. **11. a.** $y \approx 982{,}000e^{-9.89e^{-.02t}}$ **b.** In about 31 days **13.** $y = 1500/(497e^{-1.456t} + 3)$; about 449 **15.** $y = 100/(1 + 19e^{-1.558t})$; about 85 **17. a.** $y = 200 - 180e^{-.02t}$ **b.** About 146 lb **c.** It increases, approaching 200 lb. **19. a.** $y = 2(100 - t) - .018(100 - t)^2$ **b.** About 51 lb **c.** It increases at first but then decreases. **21. a.** $y = 500/(t + 100)$ **b.** About 3.8 g **23.** 34.7 minutes

Chapter 10 Review Exercises (page 606)

5. Neither **7.** Separable **9.** Both **11.** Linear **13.** $y = x^4/2 + 3x^2 + C$ **15.** $y = 4e^x + C$ **17.** $y^2 = 3x^2 + 2x + C$ **19.** $y = (Cx^2 - 1)/2$ **21.** $y = 12/5 + Ce^{-5x}$ **23.** $y = 1 + Ce^{-x^2/6}$ **25.** $y = x^3/3 - 5x^2/2 + 1$ **27.** $y = -5e^{-x} - 5x + 22$ **29.** $y = 2e^{5x-x^2}$ **31.** $y^2 + 6y = 2x - 2x^2 + 352$ **33.** $y = x^4 + 2x$ **35.** $y = 2.054e^{(2/3)x^{3/2}}$ **39.** 2.608

41.

x_i	y_i
0	0
.2	.6
.4	1.355
.6	2.188
.8	3.084
1.0	4.035

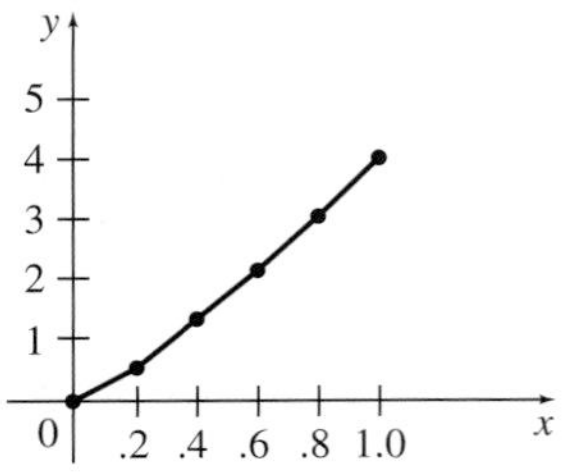

43. a. \$5800 **b.** \$25,100 **45.** About 13.9 years **47. a.** About 40 **b.** About 1.44×10^{10} hours **49.** $.2 \ln y - .5y = -.3 \ln x + .4x + C$; $x = 3/4$ units, $y = 2/5$ units **51.** It is not possible (t is negative). **55. a.** $N = 329$, $b = 7.23$; $k = .247$ **b.** $y \approx 255$ million, which is less than the table value of 281.4 million **c.** About 289 million for 2030, about 303 million for 2050 **57. a. and b.** $x = 1/k + Ce^{-kt}$ **c.** $1/k$ **59.** 213°
61. a. $v = (G/K)(e^{2GKt} - 1)/(e^{2GKt} + 1)$ **b.** G/K **c.** $v = 88(e^{.727t} - 1)/(e^{.727t} + 1)$

CHAPTER 11 PROBABILITY AND CALCULUS

Exercises 11.1 (page 619)

1. Yes **3.** Yes **5.** No; $\int_0^3 4x^3\,dx \neq 1$ **7.** No; $\int_{-2}^{2} x^2/16\,dx \neq 1$ **9.** No; $f(x) < 0$ for some x values in $[-1, 1]$. **11.** $k = 3/14$ **13.** $k = 3/125$ **15.** $k = 2/9$ **17.** $k = 1/12$ **19.** 1 **23. a.** .4226 **b.** .2071 **c.** .4082 **25. a.** .3935 **b.** .3834 **c.** .3679 **27. a.** 1/3 **b.** 2/3 **c.** 295/432 **29. a.** .9975 **b.** .0024 **31. a.** .2679 **b.** .4142 **c.** .3178 **33. a.** About .81 **b.** About .49 **35. a.**

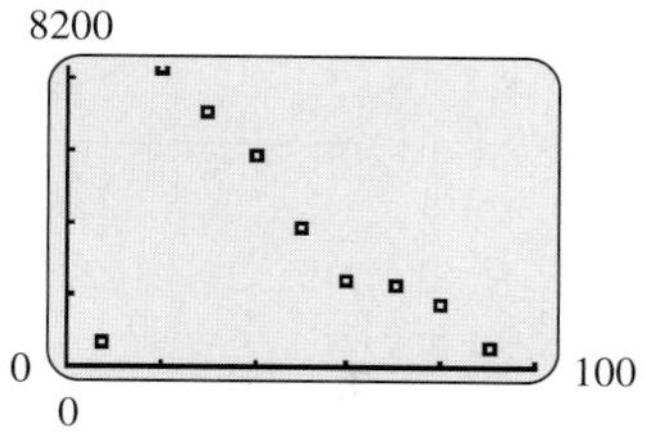

polynomial function

b. $N(x) = -.00351465x^4 + .792884x^3 - 61.7955x^2 + 1814.54x - 9709.20$

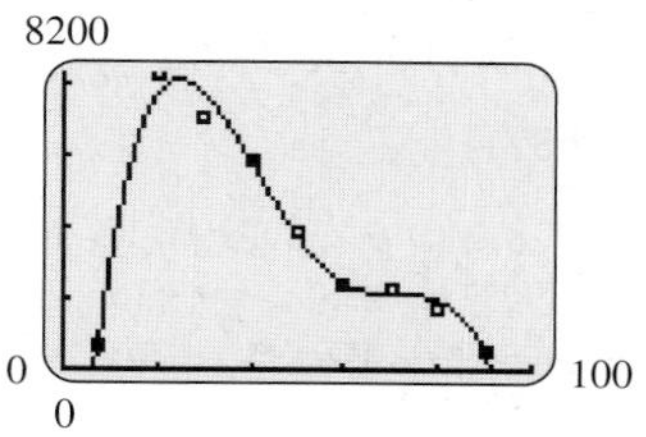

The function models the data well.

c. $S(x) = (1/343{,}795)(-.00351465x^4 + .792884x^3 - 61.7955x^2 + 1814.54x - 9709.20)$
d. Estimates .300; .179; .072; actual .272; .193; .071. The probabilities are closer where the curve fits the data better.
37. a. .2 **b.** .6 **c.** .6 **39. a.** About .16 **b.** About .14 **41. a.** .2829 **b.** .4853 **c.** .2409

Exercises 11.2 (page 629)

1. $\mu = 5$; $\text{Var}(x) \approx 1.33$; $\sigma \approx 1.15$ **3.** $\mu = 14/3 \approx 4.67$; $\text{Var}(x) \approx .89$; $\sigma \approx .94$ **5.** $\mu = 2.83$; $\text{Var}(x) \approx .57$; $\sigma \approx .76$ **7.** $\mu = 4/3 \approx 1.33$; $\text{Var}(x) \approx 2/9 \approx .22$; $\sigma \approx .47$ **11. a.** 5.40 **b.** 5.55 **c.** 2.36 **d.** .54 **e.** .60 **13. a.** $4/3 \approx 1.33$ **b.** .22 **c.** .47 **d.** .56 **e.** .63 **15. a.** 5 **b.** 0 **17. a.** 4.83 **b.** .055 **19. a.** $\sqrt[4]{2} \approx 1.19$ **b.** .18 **21.** 16/5; does not exist; does not exist **23. a.** 310.3 hours **b.** 267 hours **c.** .206 **25. a.** 2 months **b.** 2 months **c.** .632 **27. a.** 2 **b.** .89 **c.** .62 **29.** About 38.35 years; about 19.9 years **31.** About 2.99 m **33.** 960 days; 960 days **35. a.** 38.5 years **b.** 17.6 years **c.** .1641

Exercises 11.3 (page 641)

1. a. 4.4 cm **b.** .23 cm **c.** .29 **3. a.** 33.33 years **b.** 33.33 years **c.** .23 **5. a.** 1 day **b.** 1 day **c.** .23 **7.** 49.98% **9.** 8.01% **11.** −1.28 **13.** .92 **19.** $m = -\ln f.5/a$ or $\ln 2/a$ **23. a.** 1.00000 **b.** 1.99999 **c.** 8.00003 **25. a.** $\mu \approx 0$ **b.** $\sigma = .999433 \approx 1$ **27. a.** \$47,500 **b.** .47 **29. a.** $f(x) = .235e^{-.235x}$ on $[0, \infty)$ **b.** .095 **31. a.** .1587 **b.** .7698 **33. a.** 28 days **b.** .375 **35. a.** 1 hour **b.** .39 **37. a.** About 58 minutes **b.** .09 **39. a.** About .18 **b.** About .24 **41. a.** 4.36 millennia; 4.36 millennia **b.** About .63 **43. a.** About .29 **b.** About .22

Chapter 11 Review Exercises (page 645)

1. Probabilities **3.** 1. $f(x) \geq 0$ for all x in $[a, b]$; 2. $\int_a^b f(x)dx = 1$ **5.** Not a probability density function **7.** Probability density function **9.** $k = 1/9$ **11. a.** $1/5 = .2$ **b.** $9/20 = .45$ **c.** .54 **13.** b **15.** $\mu = 4$; $\text{Var}(x) \approx .5$; $\sigma \approx .71$

17. $\mu = 5/4$; $\text{Var}(x) = 5/48 \approx .10$; $\sigma \approx .32$ **19.** $m = .60$; .02 **21. a.** 100 **b.** 100 **c.** .86 **23.** 31.21% **25.** 27.69%
27. 11.51% **29.** $-.05$ **31. a.** Exponential **b.** Domain: $[0, \infty]$, range: $(0, 1]$ **c.**

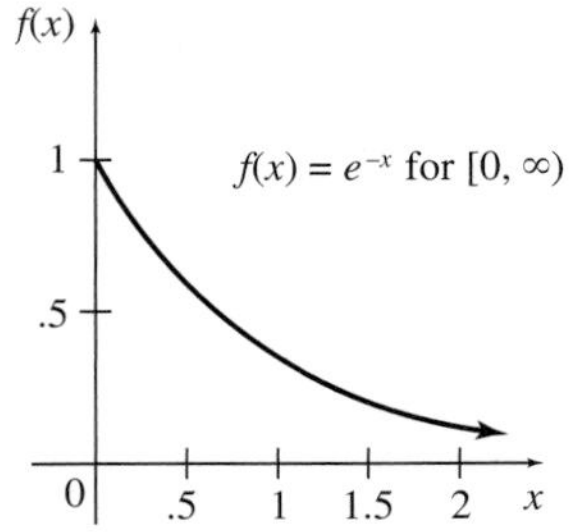

d. $\mu = 1$; $\sigma = 1$ **e.** .86 **33.** .406 **35. a.** $f(x) = e^{-x/8}/8$; $[0, \infty)$ **b.** 8 **c.** 8 **d.** .249 **37.** .1922 **39. a.** 2.377 g **b.** 1.533 g **c.** .851 **41. a.** 21 in. **b.** .526 **43.** .1379 **45.** 3650.1 days; 3650.1 days

CHAPTER 12 SEQUENCES AND SERIES

Exercises 12.1 (page 657)

1. 2, 6, 18, 54 **3.** 1/2, 2, 8, 32 **5.** 3/2, 3, 6, 12, 24 **7.** $a_5 = 324$; $a_n = 4 \cdot 3^{n-1}$ **9.** $a_5 = -1875$; $a_n = -3(-5)^{n-1}$ **11.** $a_5 = 24$; $a_n = (3/2) \cdot 2^{n-1}$ **13.** $a_5 = -256$; $a_n = -(-4)^{n-1}$ **15.** $r = 2$; $a_n = 6 \cdot 2^{n-1}$ **17.** $r = 2$; $a_n = (3/4) \cdot 2^{n-1}$ **19.** Not geometric **21.** $r = -2/3$; $a_n = (-5/8)(-2/3)^{n-1}$ **23.** 93 **25.** 33/4 **27.** 33 **29.** 860.95 **31.** 93 **33.** 341/2 **35.** 484/3 **37.** 511/4 **39. a.** \$1681 **b.** \$576 **41.** $\$2^{30}$ or \$1,073,741,824; $\$2^{31} - \$1$ **43.** 61% **45.** About 95 times **47. a.** $1 + 2 + 2^2 + 2^3 + 2^4 + 2^5$ **b.** 63 **c.** $1 + 2 + 2^2 + \cdots + 2^{n-1} = 2^n - 1$

Exercises 12.2 (page 667)

(*Note:* Answers in this section may differ by a few cents, depending on how calculators are used.) **1.** \$1318.08 **3.** \$305,390.04 **5.** \$671,994.58 **7.** \$222,777.27 **9.** \$40,652.46 **11.** \$526.95 **13.** \$952.33 **15.** \$39,434.37 **17.** \$8045.29 **19.** \$1,367,773.96 **21.** \$103,796.58 **23.** \$85,594.79 **25.** \$476.90 **27.** \$11,942.55 **29.** \$1626.86 **31.** \$137,895.79 **33.** \$159.49 **35.** \$522.85 **37.** \$112,796.87 **39.** \$209,348.00 **41.** \$4765.40 **43. a.** \$1200 **b.** \$3511.58 **c.**

Payment Number	Amount of Deposit	Interest Earned	Total
1	\$3511.58	\$0	\$3511.58
2	\$3511.58	\$105.35	\$7128.51
3	\$3511.58	\$213.86	\$10,853.95
4	\$3511.58	\$325.62	\$14,691.15
5	\$3511.58	\$440.73	\$18,643.46
6	\$3511.58	\$559.30	\$22,714.34
7	\$3511.58	\$681.43	\$26,907.35
8	\$3511.58	\$807.22	\$31,226.15
9	\$3511.58	\$936.78	\$35,674.51
10	\$3511.58	\$1070.24	\$40,256.33
11	\$3511.58	\$1207.69	\$44,975.60
12	\$3511.58	\$1349.27	\$49,836.45
13	\$3511.58	\$1495.09	\$54,843.12
14	\$3511.59	\$1645.29	\$60,000.00

45. \$12,493.78 **47.** \$421.34; \$4224.32

49. a. \$32.49 **b.** \$195.52; \$10.97 **51.** \$1885.00; \$229,612.44 **53.** \$1176.85; \$146,764.95 **55. a.** \$4025.90 **b.** \$2981.93

57.

Payment Number	Amount of Payment	Interest for Period	Portion to Principal	Principal at End of Period
0	—	—	—	\$4000
1	\$1207.68	\$320.00	\$887.68	\$3112.32
2	\$1207.68	\$248.99	\$958.69	\$2153.63
3	\$1207.68	\$172.29	\$1035.39	\$1118.24
4	\$1207.70	\$89.46	\$1118.24	\$0

59.

Payment Number	Amount of Payment	Interest for Period	Portion to Principal	Principal at End of Period
0	—	—	—	—
1	183.93	62.86	121.07	7062.93
2	183.93	61.80	122.13	6940.80
3	183.93	60.73	123.20	6817.60
4	183.93	59.65	124.28	6693.32
5	183.93	58.57	125.36	6567.96
6	183.93	57.47	126.46	6441.50

Exercises 12.3 (page 678)

1. $1 - 2x + 2^2x^2/2! - 2^3x^3/3! + 2^4x^4/4!$ or $1 - 2x + 2x^2 - (4/3)x^3 + (2/3)x^4$
3. $e + ex + ex^2/2! + ex^3/3! + ex^4/4!$ or $e + ex + ex^2/2 + ex^3/6 + ex^4/24$ **5.** $3 + x/6 - x^2/216 + x^3/3888 - 5x^4/279{,}936$
7. $-1 + x/3 + x^2/9 + 5x^3/81 + 10x^4/243$ **9.** $1 + x/4 - 3x^2/32 + 7x^3/128 - 77x^4/2048$ **11.** $-x - x^2/2 - x^3/3 - x^4/4$
13. $2x^2 - 2x^4$ **15.** $x - x^2 + x^3/2 - x^4/6$ **17.** $27 - 9x/2 + x^2/8 + x^3/432 + x^4/10{,}368$ **19.** $1 - x + x^2 - x^3 + x^4$
21. .9802 **23.** 2.7456 **25.** 2.9833 **27.** −1.0066 **29.** 1.0074 **31.** −.0101 **33.** .0002 **35. b.** 4.04167; actual value is 4.04124. **37.** $f(x) = 1 + x + x^2 + x^3 + x^4$; $f(x) = 1 + x + \cdots + x^{n-1} + x^n$ **39. a.** $1 + \lambda N + \lambda^2N^2/2$ **b.** $N = \sqrt{2k/\lambda}$
41. 4.623 thousand dollars, or \$4623 **43.** About 718

Exercises 12.4 (page 684)

1. Converges to 24 **3.** $r = 2$; diverges **5.** Converges to 64/3 **7.** Converges to 1000/9 **9.** Converges to 3/2
11. Converges to 1/5 **13.** Converges to $e^2/(e + 1)$ **15.** $S_1 = 1$; $S_2 = 3/2$; $S_3 = 11/6$; $S_4 = 25/12$; $S_5 = 137/60$
17. $S_1 = 1/7$; $S_2 = 16/63$; $S_3 = 239/693$; $S_4 = 3800/9009$; $S_5 = 22{,}003/45{,}045$
19. $S_1 = 1/6$; $S_2 = 1/4$; $S_3 = 3/10$; $S_4 = 1/3$; $S_5 = 5/14$ **21. a.** First 3.12; second 2.90 **b.** 38 **23. a.** $10{,}000/9 \approx 1111$ units **b.** $10{,}000/9 \approx 1111$ units **25.** 1600 rotations **27.** 12 m **29. a.** 1 sec, 1 m **b.** .1 sec, .1 m **c.** 10/9 sec **d.** 10/9 sec

Exercises 12.5 (page 694)

1. $5/2 + 5x/2^2 + 5x^2/2^3 + 5x^3/2^4 + \cdots + 5x^n/2^{n+1} + \cdots$; $(-2, 2)$
3. $8x - 8 \cdot 3x^2 + 8 \cdot 3^2x^3 - 8 \cdot 3^3x^4 + \cdots + (-1)^n \cdot 8 \cdot 3^nx^{n+1} + \cdots$; $(-1/3, 1/3)$
5. $x^2/4 + x^3/4^2 + x^4/4^3 + x^5/4^4 + \cdots + x^{n+2}/4^{n+1} + \cdots$; $(-4, 4)$
7. $4x - 4^2x^2/2 + 4^3x^3/3 - 4^4x^4/4 + \cdots + (-1)^n4^{n+1}x^{n+1}/(n + 1) + \cdots$; $(-1/4, 1/4]$
9. $1 + 4x^2 + 4^2x^4/2! + 4^3x^6/3! + \cdots + 4^nx^{2n}/n! + \cdots$; $(-\infty, \infty)$
11. $x^3 - x^4 + x^5/2! - x^6/3! + \cdots + (-1)^nx^{n+3}/n! + \cdots$; $(-\infty, \infty)$
13. $2 - 2x^2 + 2x^4 - 2x^6 + \cdots + (-1)^n2x^{2n} + \cdots$; $(-1, 1)$
15. $1 + x^2/2! + x^4/4! + x^6/6! + \cdots + x^{2n}/(2n)! + \cdots$; $(-\infty, \infty)$

17. $2x^4 - 2^2x^8/2 + 2^3x^{12}/3 - 2^4x^{16}/4 + \cdots + (-1)^n \cdot 2^{n+1} \cdot x^{4n+4}/(n+1) + \cdots; (-1/\sqrt[4]{2}, 1/\sqrt[4]{2}]$
19. $1 + 2x + 2x^2 + 2x^3 + \cdots + 2x^n + \cdots$ **25.** .5492 **27.** 1.9972 **29.** .2734 **31.** About 5 years; about 4 years 10 months; differ by 2 months **33. b.** $1/p$ **c.** 5 **d.** .488

Exercises 12.6 (page 700)

1. 2.73 **3.** .67 **5.** 2.33 **7.** −1; 3.67 **9.** −1.37 **11.** .44 **13.** 1.25 **15.** 1.56 **17.** 1.414 **19.** 3.317 **21.** 15.811 **23.** 2.080 **25.** 4.642 **27.** Relative maximum at −1.65; relative minimum at 3.65 **29.** Relative minima at −.71 and 1.77; relative maximum at 1.19 **31.** 6.13 **33. a.** $f'(i) = [-1 + ni(1+i)^{-n-1} + (1+i)^{-n}]/i^2$
b. $[Mi - Mi(1+i)^{-n} - Pi^2]/\{M[in(1+i)^{-n-1} - 1 + (1+i)^{-n}]\}$ **c.** .013712295 **d.** .01383273
35. $i_2 = .01036283$; $i_3 = .01036541$

Exercises 12.7 (page 705)

1. 4 **3.** 0 **5.** 1 **7.** Does not exist **9.** 1 **11.** Does not exist **13.** $1/(2\sqrt{2})$ **15.** 1/4 **17.** 1/12 **19.** 9 **21.** 0 **23.** 1/8 **25.** 1/9 **27.** 1 **29.** Does not exist **31.** 5 **33.** Does not exist **35.** Does not exist **37.** $\lim_{x\to 0}(x^2+3) \neq 0$, so l'Hospital's rule does not apply.

Chapter 12 Review Exercises (page 706)

1. $e^2 - e^2x + e^2x^2/2! - e^2x^3/3! + e^2x^4/4!$ **3.** $1 + x/2 - x^2/8 + x^3/16 - 5x^4/128$ **5.** $\ln 2 - x/2 - x^2/8 - x^3/24 - x^4/64$
7. $1 + 2x/3 - x^2/9 + 4x^3/81 - 7x^4/243$ **9.** 5.1010 **11.** 2.9993 **13.** 1.0852 **15.** Converges to 27/5 **17.** $r = 3$; diverges
19. Converges to 1/3 **21.** $S_1 = 1$; $S_2 = 4/3$; $S_3 = 23/15$; $S_4 = 176/105$; $S_5 = 563/315$
23. $4/3 + 4x/3^2 + 4x^2/3^3 + 4x^3/3^4 + \cdots + 4x^n/3^{n+1} + \cdots; (-3, 3)$
25. $x^2 - x^3 + x^4 - x^5 + \cdots + (-1)^n x^{n+2} + \cdots; (-1, 1)$
27. $-2x - 2^2x^2/2 - 2^3x^3/3 - 2^4x^4/4 - \cdots - 2^{n+1}x^{n+1}/(n+1) - \cdots; [-1/2, 1/2)$
29. $1 - 2x^2 + 2^2x^4/2! - 2^3x^6/3! + \cdots + (-1)^n 2^n x^{2n}/n! + \cdots; (-\infty, \infty)$
31. $2x^3 - 2\cdot 3x^4 + 2\cdot 3^2x^5/2! - 2\cdot 3^3x^6/3! + \cdots + (-1)^n \cdot 2 \cdot 3^n x^{n+3}/n! + \cdots; (-\infty, \infty)$ **33.** 7/4 **35.** Does not exist
37. 5/7 **39.** −1/2 **41.** Does not exist **43.** Does not exist **45.** Does not exist **47.** 4.73 **49.** 2.65 **51.** 6.841 **53.** −4.984
55. \$4,464,960 **57.** \$27,320.71 **59.** \$3322.43 **61.** \$1184.01 **63.** About 20 years 2 months; about 20 years; differ by 2 months

CHAPTER 13 THE TRIGONOMETRIC FUNCTIONS

Exercises 13.1 (page 721)

1. $\pi/3$ **3.** $5\pi/6$ **5.** $7\pi/6$ **7.** $13\pi/6$ **9.** 315° **11.** 330° **13.** 288° **15.** 48°
Note: In Exercises 17–23 we give the answers in the following order: sine, cosine, tangent, cotangent, secant, and cosecant.
17. 4/5; −3/5; −4/3; −3/4; −5/3; 5/4 **19.** 4/5; 3/5; 4/3; 3/4; 5/3; 5/4 **21.** + + + + + + **23.** − − + + − −
25. $\sqrt{3}/3$; $\sqrt{3}$; 2 **27.** $\sqrt{3}/2$; $\sqrt{3}/3$; $2\sqrt{3}/3$ **29.** −1; −1 **31.** $-\sqrt{3}/2$; $-2\sqrt{3}/3$ **33.** $\sqrt{3}/2$ **35.** 1 **37.** $2\sqrt{3}/3$ **39.** −1
41. $-\sqrt{3}/2$ **43.** $-\sqrt{2}$ **45.** $-\sqrt{3}$ **47.** 1/2 **49.** .6293 **51.** 7.1154 **53.** .3907 **55.** .1564 **57.** $a = 1$; $T = 2\pi/3$
59. $a = 3$; $T = 1/440$ **61.**

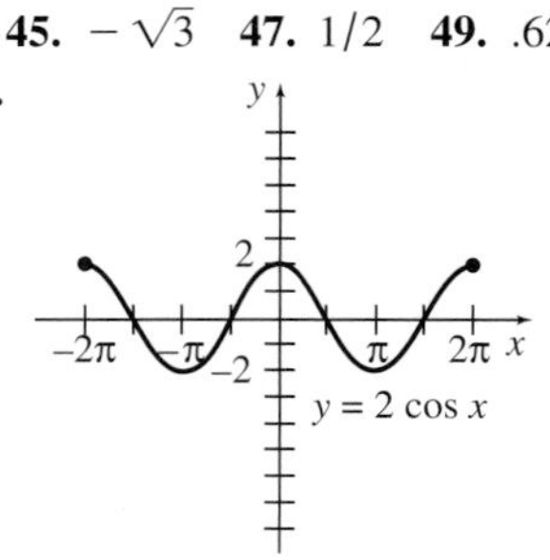

63.

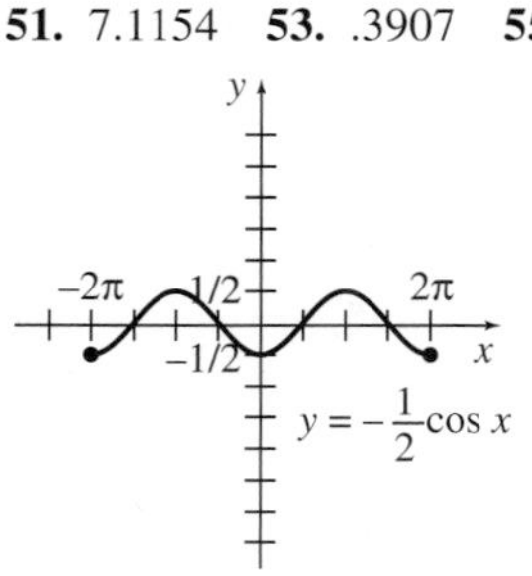

65.

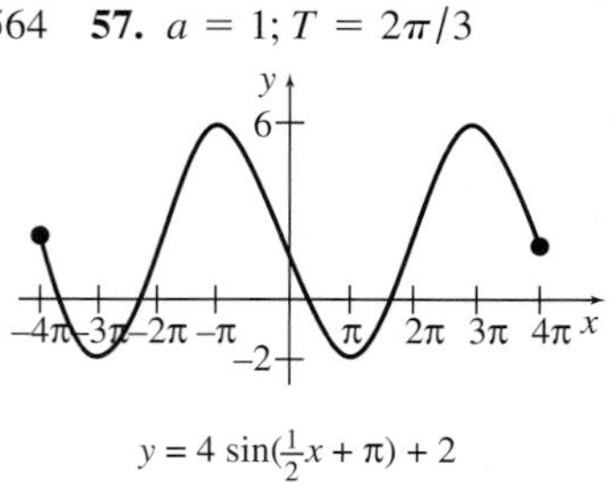

67.

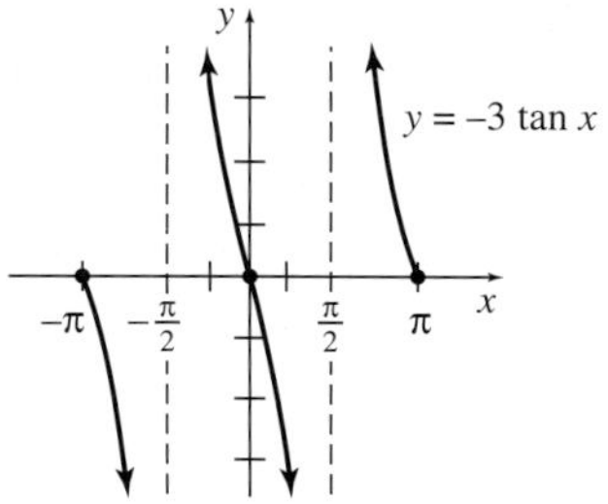

69. a. $y = 2 \sin(2\pi x/.350)$ **b.** .0875 second **c.** $-1.95°$ **71. a.** 100 **b.** 258 **c.** 122 **d.** 296 **73.** 2.2×10^8 m per second **75.** 240°

77. a.

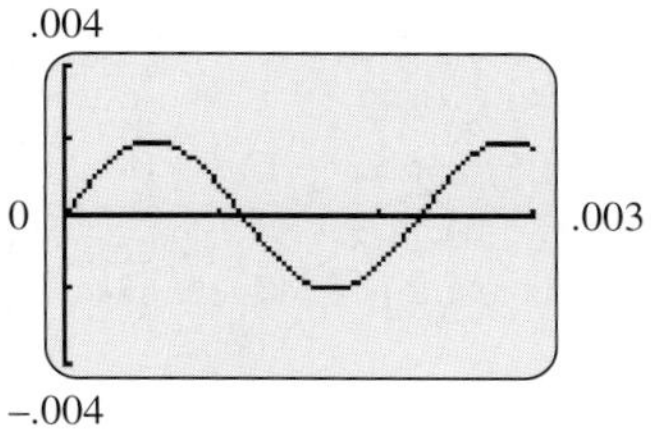

b. 0, 1/880, 1/440 **c.** $T = 1/440$; frequency: 440 cycles per second **79. a.** About 1°C **b.** About 19°C **c.** 25°C **d.** About 53°C **e.** 62°C and −12°C **f.** 365 **81.** 2.2 m; 7.6 m **83.** 180 ft **85.** (1.309,ƒ.951), (.5, 1.539), (−.309,ƒ.951)

Exercises 13.2 (page 733)

1. $dy/dx = 12 \cos 6x$ **3.** $dy/dx = 108 \sec^2(9x + 1)$ **5.** $dy/dx = -4 \cos^3 x \sin x$ **7.** $dy/dx = 5 \tan^4 x \sec^2 x$ **9.** $dy/dx = -5(4x \cos 4x + \sin 4x)$ **11.** $dy/dx = -(x \csc x \cot x + \csc x)/x^2$ **13.** $dy/dx = 5e^{5x} \cos e^{5x}$ **15.** $dy/dx = (\cos x)e^{\sin x}$ **17.** $dy/dx = (2/x) \cos(\ln 4x^2)$ **19.** $dy/dx = (2x \cos x^2)/\sin x^2$ or $2x \cot x^2$ **21.** $dy/dx = (6 \cos x)/(3 - 2 \sin x)^2$ **23.** $dy/dx = \sqrt{\sin 3x}\,(\sin 3x \cos x - 3 \sin x \cos 3x)/(2\sqrt{\sin x}\,(\sin^2 3x))$ **25.** $dy/dx = -2 \csc x \cot x - 12 \sec^2 4x + (7/8) \sin(x/8) + 3e^{3x}$ **27.** 1 **29.** −1 **31.** 1 **33.** $-\csc^2 x$ **35.** $-\csc x \cot x$ **37. a.** $R'(x) = -200\pi \sin 2\pi x$ **b.** -100π **c.** 0 **d.** 100π **39. a.**

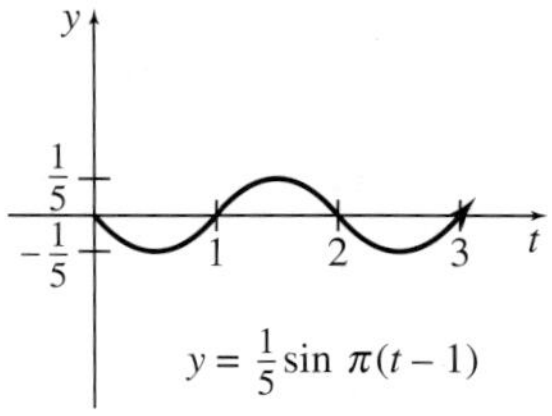

b. $v = dy/dt = (\pi/5) \cos \pi(t - 1)$; $a = d^2y/dt^2 = (-\pi^2/5) \sin \pi(t - 1)$ **d.** At $t = 1.5$, acceleration is negative, arm is moving clockwise and is at an angle of 1/5 radian from vertical; at $t = 2.5$, acceleration is positive, arm is moving counterclockwise and is at an angle of −1/5 radian from vertical; at $t = 3.5$, acceleration is negative, arm is moving clockwise and is at an angle of 1/5 radian from vertical. **41. a.** About 1490 **b.** About 5381 **c.** 2000 **d.** About 2916 **e.**

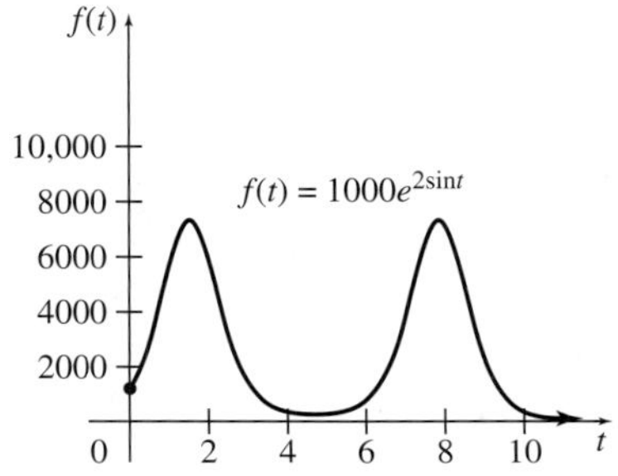

f. Maximum is 7390 when $t = \pi/2 + 2\pi n$, where n is any integer; minimum is 135 when $t = 3\pi/2 + 2\pi n$.

43. a.

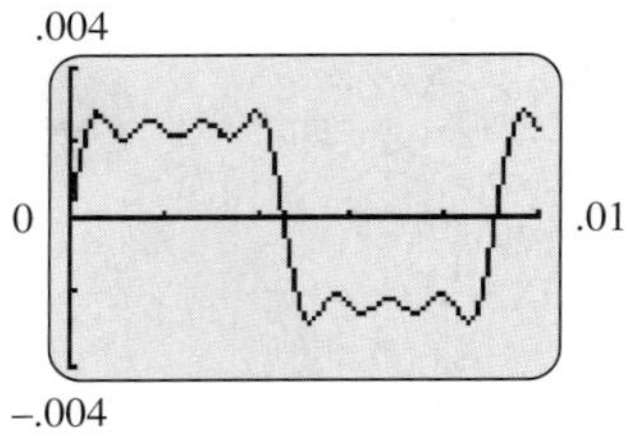

b. The pressure is decreasing at a rate of 1.05 lb per ft^2 per sec when $t = .002$. **45. a.** 13.55 ft **c.** 52.39 ft **d.** $dx/d\alpha = (V^2/16)\cos(2\alpha)$ and x is maximized when $\alpha = \pi/4$. **e.** 242 ft **47. a.** 1 **b.** -2 **c.** -1 **d.** -2 **e.** -1 **f.** 2 **49. a.** $5/\pi$ rev per minute **b.** $5/(2\pi)$ rev per minute **51.** 20.81 ft

Exercises 13.3 (page 742)

1. $(1/5)\sin 5x + C$ **3.** $5\sin x - 2\cos x + C$ **5.** $(-\cos x^2)/2 + C$ **7.** $-3\tan 2x + C$ **9.** $(1/8)\sin^8 x + C$ **11.** $(2/3)\sin^{3/2} x + C$ **13.** $-\ln|1 + \cos x| + C$ **15.** $(1/6)\sin x^6 + C$ **17.** $-4\ln|\cos(x/4)| + C$ **19.** $(1/3)\ln|\sin x^3| + C$ **21.** $-\cos e^x + C$ **23.** $-\csc e^x + C$ **25.** $(-6/5)x\sin 5x - (6/25)\cos 5x + C$ **27.** $-8x\cos x + 8\sin x + C$ **29.** $(-3/4)x^2\sin 8x - (3/16)x\cos 8x + (3/128)\sin 8x + C$ **31.** $1 - \sqrt{2}/2$ **33.** $-\ln(1/2)$ or $\ln 2$ **35.** $\sqrt{3}/2 - 1$ **37.** $1/2$ **39.** 6000 **41.** 60,000 **43.** 4430 hours; this result is relatively close to the actual value.

Chapter 13 Review Exercises (page 745)

5. $\pi/2$ **7.** $7\pi/6$ **9.** 2π **11.** 1260° **13.** 81° **15.** 117° **17.** $\sqrt{3}/2$ **19.** $\sqrt{3}/2$ **21.** $2\sqrt{3}/3$ **23.** $1/2$ **25.** $-\sqrt{2}$ **27.** .7314 **29.** 6.314 **31.** .9945 **33.** .8290 **35.** $-1, 1$ **37.**

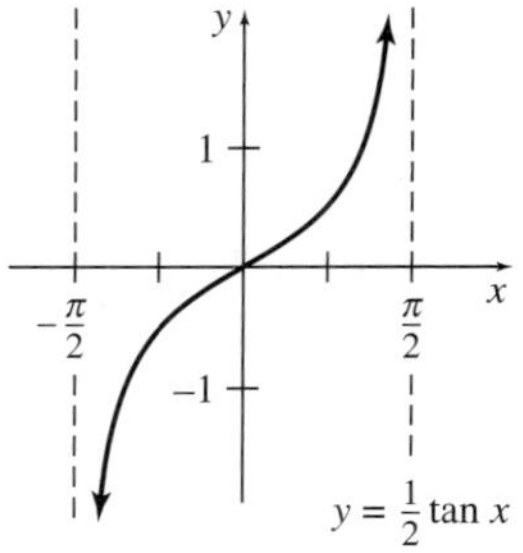

39.

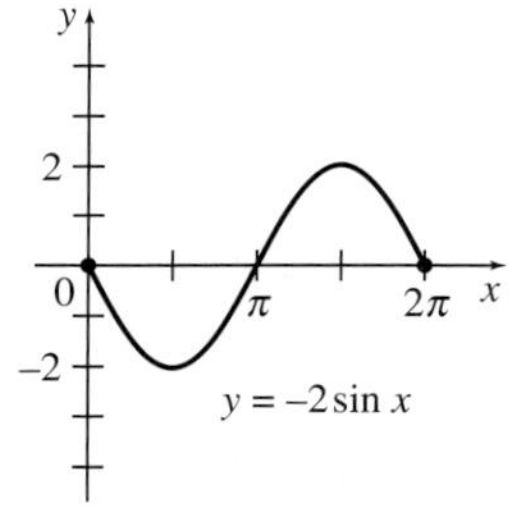

41. $dy/dx = 18\sec^2 3x$ **43.** $dy/dx = 2x\csc^2(9 - x^2)$ **45.** $dy/dx = 64x\sin^3(4x^2)\cos(4x^2)$ **47.** $dy/dx = -2x\sin(1 + x^2)$ **49.** $dy/dx = e^{-x}(\cos x - \sin x)$ **51.** $dy/dx = (-2\cos x\sin x + \cos^2 x\sin x)/(1 - \cos x)^2$ **53.** $dy/dx = (\sec^2 x + x\sec^2 x - \tan x)/(1 + x)^2$ **55.** $dy/dx = (\cos x)/(\sin x)$ or $\cot x$ **57.** $(1/3)\sin 3x + C$ **59.** $(1/5)\tan 5x + C$ **61.** $-4\cot x + C$ **63.** $(5/4)\sec 2x^2 + C$ **65.** $(1/5)\sin^5 x + C$ **67.** $(1/24)\ln|\sin 8x^3| + C$ **69.** $3(\cos x)^{-1/3} + C$ **71.** 1 **73.** 20π **75.** 105; 75

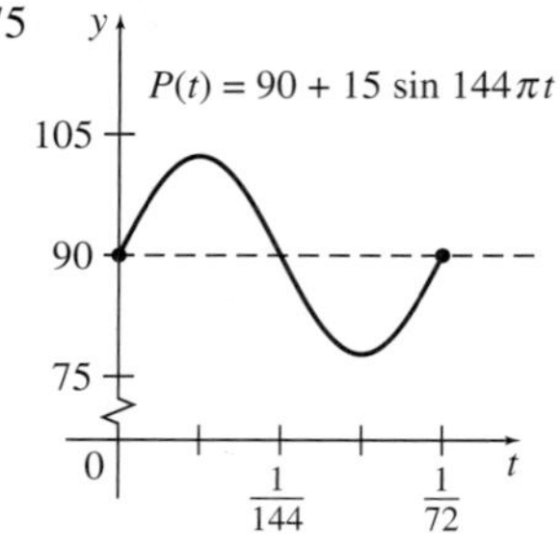

77. $L_2 = s/\sin\theta$ **79.** $L_1 = L_0 - s\cot\theta$

81. $R_2 = k \cdot L_2/r_2^4$ **83.** $R = k(L_0 - s \cot \theta)/r_1^4 + ks/(r_2^4 \sin \theta)$ **85.** $0 = ks \csc^2\theta/r_1^4 - (ks \cos \theta)/(r_2^4 \sin^2\theta)$
87. $\cos \theta = r_2^4/r_1^4$ **89.** 84° to the nearest degree **91. a.**

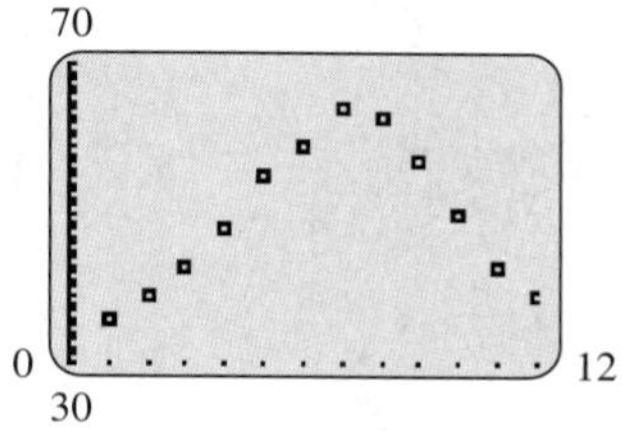

b. $y = 13.1 \sin(.534x - 2.24) + 49.9$

c.

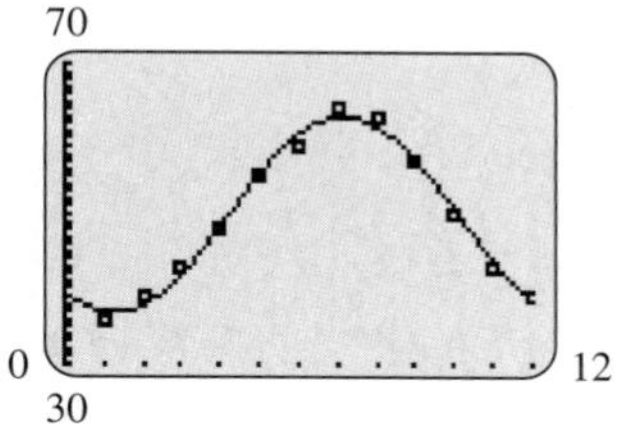

93. a.

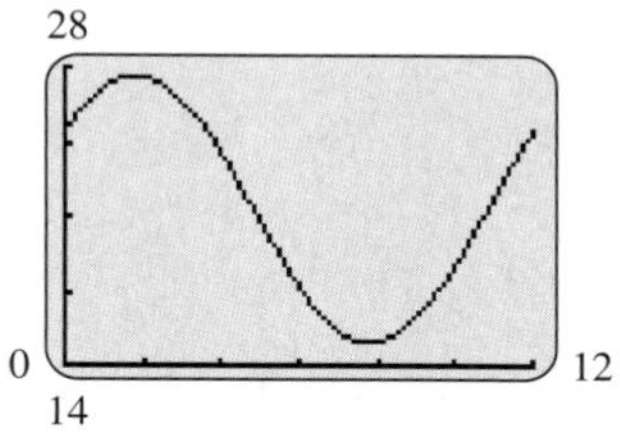

b. 27.13 MCF; 3.28 MCF **c.** −2.66 MCF/month; in July, the monthly gas usage is decreasing by 2.66 MCF per month. **95. d.** 8.63 in. **e.** 5.15 in.

Photo Acknowledgments

page xix ©Digital Vision **page 1** ©Digital Vision **page 17** ©PhotoDisc RF **page 21** ©PhotoDisc **page 25** ©PhotoDisc **page 39** ©PhotoDisc **page 42** ©PhotoDisc **page 44** ©PhotoDisc **page 49** ©Robert Holmes/Corbis **page 62** ©Corbis RF **page 88** ©PhotoDisc RF **page 115** ©Brandon D. Cole/Corbis **page 135** Courtesy of NASA **page 139** ©Digital Vision **page 172** Courtesy Karla Harby **page 175** ©Digital Vision RF **page 187** Courtesy Karla Harby **page 195** ©Digital Vision RF **page 202** ©PhotoDisc RF **page 210** ©PhotoDisc RF **page 211** ©PhotoDisc RF **page 226** ©PhotoDisc RF **page 236** ©PhotoDisc RF **page 246** ©PhotoDisc RF **page 262** Courtesy Beckett's Music, Ltd. **page 269** ©PhotoDisc Red RF **page 280** ©PhotoDisc RF **page 290** ©PhotoDisc Red RF **page 295** ©PhotoDisc RF **page 304** ©Corbis RF **page 310** Courtesy Sara Anderson **page 323** Courtesy Joe Nonneman **page 325** ©PhotoDisc Red RF **page 334** Courtesy Environmental Protection Association **page 340** ©PhotoDisc RF **page 354** ©PhotoDisc RF **page 374** Courtesy www.tjdesigns.com **page 378** ©Beth Anderson **page 379** ©Eyewire RF **page 381** ©David Butow/Corbis SABA **page 390** ©Corbis RF **page 402** ©PhotoDisc RF **page 416** Courtesy Karla Harby **page 417** ©Mike King/Corbis **page 428** ©PhotoDisc Blue RF **page 453** ©PhotoDisc Blue RF **page 459** ©Corbis RF **page 468** ©PhotoDisc RF **page 473** ©Digital Vision RF **page 476** ©PhotoDisc RF **page 491** ©PhotoDisc RF **page 494** ©PhotoDisc RF **page 497** ©PhotoDisc RF **page 509** ©PhotoDisc RF **page 520** ©Digital Vision RF **page 546** ©PhotoDisc RF **page 561** ©PhotoDisc RF **page 568** ©Digital Vision RF **page 571** ©PhotoDisc RF **page 575** ©PhotoDisc RF **page 581** ©Digital Vision RF **page 588** ©PhotoDisc RF **page 604** ©PhotoDisc RF **page 609** ©DV/BrX RF **page 612** ©PhotoDisc RF **page 621** ©PhotoDisc RF **page 630** ©PhotoDisc RF **page 644** ©Digital Vision RF **page 649** ©Beth Anderson **page 653** ©Greg Fiume/New Sport/Corbis TL **page 668** ©Saskatoon Stamp Center **page 669** (left) ©AP/Wideworld/Steve Coleman TL **page 669** (right) ©Corbis RF **page 679** ©PhotoDisc Blue RF **page 686** ©PhotoDisc RF **page 708** ©PhotoDisc RF **page 709** ©Beth Anderson **page 724** ©PhotoDisc Red RF **page 735** ©PhotoDisc RF **page 747** ©PhotoDisc RF

Index

4.1 *Sum or Difference Rule* If $f(x) = u(x) \pm v(x)$, then

$$f'(x) = u'(x) \pm v'(x).$$

4.2 *Product Rule* If $f(x) = u(x) \cdot v(x)$, then

$$f'(x) = u(x) \cdot v'(x) + v(x) \cdot u'(x).$$

4.2 *Quotient Rule* If $f(x) = \dfrac{u(x)}{v(x)}$, and $v(x) \neq 0$, then

$$f'(x) = \frac{v(x) \cdot u'(x) - u(x) \cdot v'(x)}{[v(x)]^2}.$$

4.3 *Chain Rule* If y is a function of u, say $y = f(u)$, and if u is a function of x, say $u = g(x)$, then $y = f(u) = f[g(x)]$, and

$$\frac{dy}{dx} = \frac{dy}{du} \cdot \frac{du}{dx}.$$

4.3 *Chain Rule (Alternate Form)* If $y = f[g(x)]$, then $dy/dx = f'g(x)] \cdot g'(x)$.

4.4 *Exponential Function*

$$D_x[a^{g(x)}] = (\ln a)a^{g(x)}g'(x)$$

$$D_x[e^{g(x)}] = e^{g(x)}g'(x)$$

4.5 *Logarithmic Function*

$$D_x[\log_a|g(x)|] = \frac{1}{\ln a} \cdot \frac{g'(x)}{g(x)}$$

$$D_x[\ln |g(x)|] = \frac{g'(x)}{g(x)}$$

5.2 First Derivative Test Let c be a critical number for a function f. Suppose that f is continuous on (a, b) and differentiable on (a, b) except possibly at c, and that c is the only critical number for f in (a, b).

1. $f(c)$ is a relative maximum of f if the derivative $f'(x)$ is positive in the interval (a, c) and negative in the interval (c, b).

2. $f(c)$ is a relative minimum of f if the derivative $f'(x)$ is negative in the interval (a, c) and positive in the interval (c, b).

5.3 Second Derivative Test Let f'' exist on some open interval containing c, and let $f'(c) = 0$.

1. If $f''(c) > 0$, then $f(c)$ is a relative minimum.

2. If $f''(c) < 0$, then $f(c)$ is a relative maximum.

3. If $f''(c) = 0$, then the test gives no information about extrema.

7.2 General Power Rule for Integrals For $u = f(x)$ and $du = f'(x)\,dx$,

$$\int u^n\,du = \frac{u^{n+1}}{n+1} + C.$$

7.2 Indefinite Integral of e^u If $u = f(x)$, then $du = f'(x)\,dx$ and

$$\int e^u\,du = e^u + C.$$

7.2 Indefinite Integral of u^{-1} If $u = f(x)$, then $du = f'(x)\,dx$ and

$$\int u^{-1}\,du = \int \frac{du}{u} = \ln|u| + C.$$

7.4 Fundamental Theorem of Calculus Let f be continuous on the interval $[a, b]$, and let F be *any* antiderivative of f. Then

$$\int_a^b f(x)\,dx = F(b) - F(a) = F(x)\Big|_a^b.$$

7.6 Trapezoidal Rule Let f be a continuous function on $[a, b]$ and let $[a, b]$ be divided into n equal subintervals by the points $a = x_0, x_1, x_2, \ldots, x_n = b$. Then, by the trapezoidal rule,

$$\int_a^b f(x)\,dx \approx \left(\frac{b-a}{n}\right)\left[\frac{1}{2}f(x_0) + f(x_1) + \cdots + f(x_{n-1}) + \frac{1}{2}f(x_n)\right].$$

7.6 Simpson's Rule Let f be a continuous function on $[a, b]$ and let $[a, b]$ be divided into an even number n of equal subintervals by the points $a = x_0, x_1, x_2, \ldots, x_n = b$. Then, by Simpson's rule,

$$\int_a^b f(x)\,dx \approx \frac{b-a}{3n}[f(x_0) + 4f(x_1) + 2f(x_2) + 4f(x_3) + \cdots + 2f(x_{n-2}) + 4f(x_{n-1}) + f(x_n)].$$

8.1 Integration by Parts If u and v are differentiable functions, then

$$\int u\,dv = uv - \int v\,du.$$

8.4 Improper Integrals If f is continuous on the indicated interval and if the indicated limits exist, then

$$\int_a^\infty f(x)\,dx = \lim_{b\to\infty}\int_a^b f(x)\,dx,$$

$$\int_{-\infty}^b f(x)\,dx = \lim_{a\to-\infty}\int_a^b f(x)\,dx,$$

$$\int_{-\infty}^\infty f(x)\,dx = \int_{-\infty}^c f(x)\,dx + \int_c^\infty f(x)\,dx,$$

for real numbers a, b, and c, where c is arbitrarily chosen.

9.3 Test for Relative Extrema For a function $z = f(x, y)$, let f_{xx}, f_{yy}, and f_{xy} all exist in a circular region contained in the xy-plane with center (a, b). Further, let

$$f_x(a, b) = 0 \qquad \text{and} \qquad f_y(a, b) = 0.$$

Define the number D by

$$D = f_{xx}(a, b)\cdot f_{yy}(a, b) - [f_{xy}(a, b)]^2.$$